**10.** Estimating $\tilde{\mu}$:

      lower limit = the $j$th smallest value in the sample

      upper limit = the $j$th largest value in the sample

The confidence coefficient is approximately:

$$[1 - 2 \cdot P(X \le j - 1)] \cdot 100\%$$

for $X$ a binomial random variable for $n$ trials and $p = 0.5$.

**11.** Estimating $\tilde{\mu}_1 - \tilde{\mu}_2$:

      lower limit = the $j$th smallest $x_i - y_i$

      upper limit = the $j$th largest $x_i - y_i$

The confidence coefficient is approximately:

$$\left[ 1 - 2 \cdot P\left( W \le j - 1 + \frac{n_1(n_1 + 1)}{2} \right) \right] \cdot 100\%$$

where the probability is calculated supposing the $x$ and $y$ distributions are the same.

## Test Statistics

**1.** Large sample tests on $\mu$: $\quad Z = \dfrac{\bar{X} - \mu_0}{\dfrac{S}{\sqrt{n}}}$

**2.** Small sample tests on $\mu$: $\quad T = \dfrac{\bar{X} - \mu_0}{\dfrac{S}{\sqrt{n}}}$

**3.** Large sample tests on $P$: $\quad Z = \dfrac{\hat{P} - P_0}{\sqrt{\dfrac{P_0(1 - P_0)}{n}}}$

**4.** Tests on $\sigma^2$: $\quad X^2 = \dfrac{(n - 1)S^2}{\sigma_0^2}$

**5.** Large sample tests on $\mu_1 - \mu_2$: $\quad Z = \dfrac{\bar{X} - \bar{Y}}{\sqrt{\dfrac{S_1^2}{n_1} + \dfrac{S_2^2}{n_2}}}$

**6.** Large sample tests on $\mu_D$: $\quad Z = \dfrac{\bar{D}\sqrt{n}}{S_d}$

**7.** Small sample tests on $\mu_D$: $\quad T = \dfrac{\bar{D}\sqrt{n}}{S_d}$

**8.** Large sample tests on $P_1 - P_2$: $\quad Z = \dfrac{\hat{P}_1 - \hat{P}_2}{\sqrt{\hat{P}(1 - \hat{P})}\sqrt{\dfrac{1}{n_1} + \dfrac{1}{n_2}}}$ where $\hat{P}$ is the "pooled" sample proportion

**9.** Tests of equality of $r$ distributions: $\quad X^2 = \Sigma \dfrac{O_{ij}^2}{E_{ij}} - n$

**10.** Tests of equality of $r$ means: $\quad F = \dfrac{S_b^2}{S_p^2}$

**11.** Tests of independence: $\quad X^2 = \Sigma \dfrac{O_{ij}^2}{E_{ij}} - n$

**12.** Tests on $\beta_1$: $\quad T = \dfrac{B_1}{\sqrt{\dfrac{S_e^2}{\Sigma(x_i - \bar{x})^2}}}$

**13.** Tests on $\mu_{\hat{y}|x^*}$: $\quad T = \dfrac{\hat{Y} - \mu_0}{\sqrt{S_e^2}\sqrt{\dfrac{1}{n} + \dfrac{(x^* - \bar{x})^2}{\Sigma(x_i - \bar{x})^2}}}$

**14.** For tests of $H_0$: $\quad \tilde{\mu} = \tilde{\mu}_0$

    **a.** Reject if $b^* \ge b_c$, and $H_a$: $\quad \tilde{\mu} > \tilde{\mu}_0$

    **b.** Reject if $b^* \le b_c$, and $H_a$: $\quad \tilde{\mu} < \tilde{\mu}_0$

    **c.** Reject if $b^* \le$ lower $b_c$ or $b^* \ge$ upper $b_c$, and $H_a$: $\quad \tilde{\mu} \ne \tilde{\mu}_0$

**15.** For tests of $H_0$: $\quad \tilde{\mu}_1 - \tilde{\mu}_2 = 0$

    **a.** Use lower percentage points in Appendix Table 8, if $H_a$: $\quad \tilde{\mu}_1 - \tilde{\mu}_2 < 0$

    **b.** Use both upper and lower percentage points in Appendix Table 8, if $H_a$: $\quad \tilde{\mu}_1 - \tilde{\mu}_2 \ne 0$

    **c.** Use upper percentage points in Appendix Table 8, if $H_a$: $\quad \tilde{\mu}_1 - \tilde{\mu}_2 > 0$

# A First Course in
# STATISTICS

# A First Course in
# STATISTICS

**GENE R. SELLERS**
*Sacramento City College*

**STEPHEN B. VARDEMAN**
*Iowa State University*

**ADELBERT F. HACKERT**
*Sacramento City College*

HarperCollins*Publishers*

**Sponsoring Editor:** George Duda
**Developmental Editor:** Margaret Prullage, Linda Youngman
**Project Editor:** Ann-Marie Buesing
**Assistant Art Director:** Julie Anderson
**Text and Cover Design:** Tessing Design Inc.
**Photo Researcher:** Rosemary Hunter
**Production:** Jeanie Berke, Linda Murray, and Helen Driller
**Compositor:** Syntax International
**Printer and Binder:** R. R. Donnelley & Sons Company
**Cover Printer:** Lehigh Press Lithographers

If you want further help with this course, you may want to obtain a copy of the *Study Guide and Student's Solutions Manual* and the *Statistical Utility Disk*. The first item contains solutions to half the odd-numbered exercises plus more aids, and the second item contains programs for solving the many problems in the text. Your college bookstore either has the manual or can order it for you, and you may obtain a copy of the disk through your instructor.

A First Course in Statistics 3e

**Library of Congress Cataloging-in-Publication Data**

Sellers, Gene R.
    A first course in statistics / Gene R. Sellers, Stephen B.
Vardeman, Adelbert F. Hackert. -- 3rd ed.
        p.    cm.
    Includes index.
    ISBN 0-673-38124-2
    1. Statistics.    I. Vardeman, Stephen B.    II. Hackert, Adelbert F.
III. Title.
QA276. 12.S45   1992
519.5--dc20

94  9 8 7 6 5 4 3

# Preface

The number of students enrolled in introductory statistics courses has increased dramatically in recent years. One reason for the increase is that a growing number of academic areas either require, or strongly recommend, that students take at least one course in statistics. As a consequence, we find students in our statistics classes who are pursuing degrees in nursing, human resources management, social sciences, physical and life sciences, medicine, law, and many other areas.

We are certain that statistics is receiving more attention in all academic areas because it is one course that devotes considerable attention to decision making when faced with uncertainty. In a statistics course we learn that decisions often must be made about vital issues on the basis of incomplete information. This important element in a statistics course, called *inferential*, is a body of knowledge that can significantly affect the way individuals attempt to solve problems in virtually any area of employment.

We have tried to produce a conceptual text that is capable of reaching students in a dignified way. Most students can comprehend the meaning of descriptive measures, the implications of probability, the purpose of confidence intervals, the power of hypothesis testing, and other topics when these concepts are presented in language they understand. For example, the exercises primarily center on activities to which students can relate. Because statistical concepts abound in the world around us, we have selected newspaper clippings that discuss serious topics for mature audiences. Further, the data sets in the back of the text contain data on lung cancer and the New York Stock Exchange. The exercises linked to these data sets provide students with hands-on experience in using sample data to make appropriate statistical inferences, in ways that help students appreciate the seriousness of such inferences.

## Features of the Third Edition

### Accurate and class-tested materials

We have carefully monitored every aspect of the text that has any impact on the correctness of the presentation. That is, we have attempted to maintain the readability of the text without sacrificing its statistical accuracy. We have endeavored to use symbols that are widely accepted for every statistical measure or variable, from mean to random variable.

A draft of the manuscript was class tested by Carla Thompson at Tulsa Junior College, as well as by one of the authors. The students' responses were used to reinforce the pedagogy in some places, to add a few more illustrative examples, and to carefully refine the exercise sets.

## New and expanded topics

Since the last edition was published, a few new topics have become standard in a first course in statistics. Included in the list of new topics are stem-and-leaf diagrams and box-plot diagrams, found in Chapter 2. In Chapter 6, the use of $p$-values in hypothesis tests and tests of variances is new. Determining sample sizes needed for certain confidence coefficients is new to Chapter 5.

Some topics in the last edition have been expanded to improve certain areas of statistics. For example, in Section 2-4, the descriptive measures of position now include deciles and quartiles, since quartiles are needed for box-plot diagrams. Another example is the use of bell-shaped populations in Section 2-3 and the percentages of data points within one and two standard deviations of the mean to give a better understanding of the value of knowing a standard deviation.

## Chapter openers

Every chapter opens with one or two clippings from newspapers, magazines, and journals. These are followed by a few paragraphs that explain how the clippings illustrate topics studied in that chapter. These openers can help give students an awareness of the fact that statistics and statistical concepts are part of everyday life.

## Key topics

Each section begins with a list of key topics identifying the skills to be learned within the section. The key topics are then used to separate the section into subsections.

## Statistics in action

The key topics are followed by a narrative called "Statistics in Action" that provides a practical overview of the statistical concepts to be studied in that section. For example, in the section on large $n$ confidence intervals for a mean, we give a story about Susan Thompson, field worker for the California Department of Fish and Game. It includes data she collected to construct a confidence interval estimate of the mean length of trout that had been planted in a lake two years previously. Further, the details of the story are referred to throughout the section as each concept is presented.

## Career profiles

Another important new feature is the inclusion of the "Career Profiles." These biographical statements from people who use statistical concepts and procedures on the job include information on how these people were directed to their current careers, benefits they have enjoyed as a result of that choice, and some advice to students. These biographical sketches are designed to help students realize the relevance of statistics to a variety of occupations. They also may suggest some occupational choices that students have not yet considered.

## Improved design and format

There have been several changes in the design and format of the text to improve its effectiveness as a teaching device. First, all equations, definitions, theorems, and other items of importance are enclosed inside boxes with headings. These displays make it easy for students to find the most important items in each section.

Second, the examples have been improved with *discussion* statements and *side comments*. The labeled discussion component included in many examples is a verbal statement on how a problem is going to be solved. The side comments are brief statements or equations that explain how a given step in the solution followed the previous one.

Finally, an expanded chapter summary includes not only review exercises but also a restatement of the definitions, equations, procedures, and symbols used in the chapter. In addition, the summary provides a comparison of two or more major topics developed within the chapter. The comparison includes a grid of the components integrated within each topic.

# Exercises

## Changes in emphasis

In recent years there has been a major change from a computationally based statistics course to a more interpretive one. With this shift in mind, the exercises in this edition have been constructed so that students are required to write word statements that describe the meaning of numerical calculations. For example, exercises may ask for a meaning of a confidence interval in the context of a given problem, or the meaning of the rejection of an $H_0$, or the meaning of a computed probability. The exercises continually probe the students for a statement of their understanding of what they have just calculated.

## Two sets of exercises

Each section contains two exercise sets. Set A, which comprises practice exercises, is similar to the illustrative examples in the text. These exercises provide students with the kind of activities necessary to learn the principles studied in a given section.

Set B, which comprises extension exercises, is supplementary to the practice exercises. Some of these sets include exercises that require taking random samples from Data Sets I and II in Appendix B. The sample data are then applied to the concepts studied in that section. These exercises provide in-class statistics projects by bringing the populations into the classroom and then requiring students to sample from the data.

These sets also include exercises based on newspaper clippings and projects that students can carry out on campus. Finally, a few additional topics such as Chebychev's theorem and the Poisson distributions are included in these sets for those instructors who want to cover them.

## Two populations of data

Appendix B contains two sets of actual data. Data Set I is a record of some of the data about lung cancer collected by a team of researchers in southern Louisiana. Data Set II contains stock prices recently reported on the New York Stock Exchange. Throughout the text there are numerous exercises in which students use random numbers to sample from these sets. The students then use the statistics from their samples to practice using the statistical concepts studied in that section. These exercises provide concrete illustrations of many concepts, such as the variableness of sample means and proportions, and the mechanics involved in constructing confidence intervals and conducting hypothesis tests.

# Emphasis on Technology

## Minitab lab

No introduction to statistics can be complete without some mention of the power of the computer in statistical analysis. To this end, another end-of-chapter element is a short discussion of how MINITAB, a popular software program, can be used to investigate some of the concepts and exercises introduced in the chapter.* These sections consist of a discussion of a few MINITAB commands and illustrations of collateral MINITAB output. The exercises in these sections are built around sample output and are designed to help the student interpret such output. In the **Study Guide and Student's Solutions Manual**, we discuss the newest version of MINITAB (Release 8).

## Calculator activities: scientific and graphing

The calculator has had a significant impact on the teaching of statistics. The statistics programs on scientific and graphing calculators reduce the calculations of many descriptive measures to button-pushing exercises. Many students, however,

---

* MINITAB is a registered trademark of MINITAB, Inc.

do not know how to use their calculators and are unable to get the necessary information from the calculator manual. The calculator activities at the end of each chapter contain keystroke instructions with corresponding calculator displays that lead students through the process. These instructions are appropriate for most algebraic calculators. The graphing calculator activity provides similar instructions for the Casio® and Texas Instruments® models. Additional instructions are given for graphic displays that these calculators provide.

## Supplements

The **Instructor's Resource and Test Manual** includes answers to even-numbered exercises; four forms of chapter tests, ready for duplication; essays on using the data sets and the student projects in the classroom; and a description of three levels of course instruction—basic, moderate, and advanced—plus assignments for those levels.

A set of nearly fifty **overhead transparencies** is provided. These transparencies include graphs, tables, formulas, and other items from the text.

The **Statistical Utility Disk** by Hackert is available for instructors and students in IBM and Macintosh versions. Instructors will be given a copy of the disk upon adoption of the textbook. With this disk students may generate random files or enter their own data sets. The disk contains programs for constructing histograms and for conducting hypothesis tests on one and two samples for means, variances, and proportions. The tests can be carried out on both one and two tail tests. There are also programs for $\chi^2$ tests of independence, linear correlation and regression, and ANOVA. Sample data can also be used to construct confidence intervals for one and two sample data. Instructions for student use of the disk to solve specific exercises in the text are included in the **Study Guide and Student's Solutions Manual**.

The **Statistical Toolkit** by Tony Patricelli, Northeastern Illinois University, has been updated in both the IBM and Apple versions. This program, on a two-sided disk, has been class-tested and praised for its simplicity and ease of use for students in elementary statistics. Students may wish to use this program to help them with tedious calculations or to show step-by-step solutions that facilitate understanding of the concepts in question, thus allowing for practice and immediate reinforcement. Graphics portions of the program can be used by instructors for classroom demonstrations to illustrate such concepts as probability defined as areas under curves or the central limit theorem. A site license is available.

The **HarperCollins Test Generator** is one of the top testing programs on the market for IBM and Macintosh computers. It enables instructors to select questions for any section in the text or to use a ready-made test for each chapter. Instructors may generate tests in multiple-choice or open-response formats, scramble the order of questions while printing, and produce twenty-five versions of each test. The system features printed graphics and accurate mathematical symbols. The program also allows instructors to choose problems randomly from a section or problem type or to choose questions manually while viewing them on the screen, with the option to regenerate variables if desired. The editing feature allows instructors to customize the chapter data disks by adding their own problems.

An **Instructor's Solutions Manual**, written by Sellers, contains complete solutions to every exercise in the text.

A **Study Guide and Student's Solutions Manual**, written by Hackert and Sellers, is available for the text. It includes a chapter summary that identifies all the key elements in the chapter, complete solutions for every other odd-numbered exercise (according to the pattern 1, 5, 9, . . . ,), a set of test readiness exercises with solutions to be used as practice tests, instruction on the use of the statistical utility disk produced by Hackert, and a set of ongoing dialogues from a fictitious college statistics class.

We would like to express our appreciation to many individuals who were instrumental in bringing this project to a successful completion. First and foremost we want to thank George Duda, acquisitions editor at HarperCollins Publishers. His efforts to produce a text that is complete in meeting the needs of instructors who teach a first course in statistics and students who take the course are clearly seen in the final product. We also would like to thank others at HarperCollins who participated in the production of this text: developmental team Linda Youngman and Marge Prullage, project editor Ann Buesing, art director Julie Anderson, and Ellen Keith and Tammy McClenning for commenting on the career essays from a student's perspective.

We are also grateful for the many reviewers whose comments were helpful in improving the quality and the pedagogy in the text:

Chris Burditt, Napa Valley College
Carolyn Case, Vincennes University
Deann Christianson, University of the Pacific
Joan Garfield, University of Minnesota, Twin Cities
Stuart Goff, Keene State College
Alice Hemingway, Clark College
Shu-ping Hodgson, Central Michigan University
Francine Johnson, Coppin State College
Marlene J. Kovaly, Florida Community College at Jacksonville
Ronald LaBorde, Marian College
Jack Leddon, Dundalk Community College
Lou Levy, Northland Pioneer College
Albert Liberi, Westchester Community College
Norman Martin, Northern Arizona University
Carolyn Meitler, Concorida University, Wisconsin

Donald E. Miller, Saint Mary's College
Luis F. Moreno, Broome Community College
Thomas A. Mowry, Diablo Valley College
Mike Perry, Appalachian State University
Ronald Persky, Christopher Newport College
Shirley Stanley, Schenectady County Community College
Carla Thompson, Tulsa Junior College
James E. Walker, American River College
Ray William, Northern Arizona University
August Zarcone, College of DuPage

We want to thank Carla Thompson for class testing the manuscript, and Ronald Hatton for writing the calculator activity sections. We especially want to thank the following people for their contributions to the career profiles sections: Judith C. Schafer, Christy Chuang-Stein, Craig Barbarino, Charles B. Sampson, Myla Hunt, Roger M. Sauter, James Inglis, Kirk Wolter, and James A. Adams.

Last, but certainly not least, we want to thank Linda Sellers for typing the original manuscript and the several revisions. For her many hours of work and tireless efforts, we want to dedicate this edition of the text to her.

## Acknowledgments

The reviewers of the manuscript offered many suggestions that improved the quality of this text. However, we know there are many instructors who have found a technique or procedure for teaching some topics in statistics that are more effective than the ones we have presented in this text. As a consequence, we would appreciate any comments that you have regarding the material. You may write Steve Vardeman at Iowa State University, Ames, Iowa, 50010, or Del Hackert and Gene Sellers at Sacramento City College, Sacramento, California, 95822.

Gene R. Sellers
Stephen B. Vardeman
Adelbert F. Hackert

# Contents

# Note to the Student

We believe that statistics can be one of the most rewarding courses that you will take in your entire college career. This statement is not intended to be trite. Instead, it is meant to challenge you to approach this course with an expectation of learning valuable and exciting information that may change your educational goals.

Throughout the text we have included data references from newspapers, magazines, and journals to show you the pervasiveness of statistics in the world around us. Frequently the data are merely summary statements of activities that involve numbers, such as business and sports reports. Other data, however, are collected to support a statistical inference such as an estimation of some unknown quantity, or rejection of some previously claimed value. In any case, data are used to give credibility to some investigation that was carried out. By the time you complete this course, you will have the necessary critical and analytical tools to apply statistical concepts and procedures to similar sets of data.

Each chapter begins with some of these references, and we attempt to make the material interesting with sections called "Statistics in Action." As appropriate, we have highlighted procedures in a step-by-step fashion within each section and then summarized the topics in detail at the end of each chapter. These summaries can be used as you prepare for examinations.

Some of our students have shared with us their frustrations at not being able to develop an effective strategy for studying statistics. For this reason we suggest the following steps that should maximize the value of your study time and bring you success in the course.

**Step 1** Before attempting any homework problems, *first read the section of the text currently under study*. Many important items of information given by the instructor in class will be reinforced by this simple task.

**Step 2** Study the information that is given inside boxes. *Examine every word*, since these words were carefully selected to support some fact. Many boxes include formulas, and these formulas should be studied until every symbol is recognized and the purpose of the formula is understood. We suggest that you write these formulas on 3″ × 5″ cards.

**Step 3** Study the illustrative examples in each section. They contain side comments that explain how each step in the solution leads to the next one.

**Step 4** Next, work the assigned problems. If an odd-numbered problem is assigned, you can check the answer in the back of the text. If an even-numbered problem is assigned, you can work the previous odd-numbered one and check the answer to see if you are on the right track. Should you have difficulty, consider again the illustrative example that is representative of the problem you are attempting to solve.

**Step 5** Whether or not they are assigned, consider some of the activities suggested in the Set B exercises. Some of these are projects that can be carried out on your campus. Others use the data sets in the Appendix of the text. Both activities provide personal experiences that enhance the understanding of the topics covered in that section.

Taking these steps and listening to your instructor will help alleviate any "stat anxiety" that some of you may feel. You might even be tempted to take additional courses in statistics and consider a career in this lucrative and exciting field. At the end of each chapter, we have included a career profile to further illustrate that statistics is not just a body of facts to be learned and memorized, but something that is used by real people in many different occupations. If you want more information on a particular career in one of the profiles, feel free to write to the author of the article. Furthermore, you can receive the pamphlet *Careers in Statistics* by writing to the American Statistical Association, 1429 Duke Street, Alexandria, VA, 22314-3402. Finally, if you have any comments regarding the text, please write to any of the authors.

If you want further help in studying statistics, we have prepared two instructional aides. The first is a **Study Guide and Student's Solutions Manual** that was written by Hackert and Sellers. Included are complete solutions to half of the odd-numbered exercises. These solutions will help you maximize your study time, and possibly reduce frustration. The second is a disk that contains programs for solving the many problems in the text. Free for you to use on school computers or your own personal computer, this disk will provide you with endless practice doing problems similar to those in the text.

# Applications Index

The following is a representative sample of the exercises in the text based on the content categories identified in this index. A more complete applications index of all the exercises is given in the **Instructor's Resource and Test Manual**. The following list contains application problems followed by the page numbers in bold and the exercise numbers in parentheses.

*Business and Economics*
Ages of new savings account members **122** (5)
Amounts of federal income taxes paid **132** (14)
Average salary of employees **75** (4)
Checking vehicles for automobile insurance **310** (5)
Closing prices of stocks in Data Set II **60** (4)
Defective saws in a production run **239** (21)
Dollar amounts of stocks for sale **76** (13)
Gasoline prices at service stations **350** (5)
Homes for sale **106** (12)
Homes for sale **24** (6)
Items at a department store **24** (5)
Manufacturers and mpg's of cars **194** (4)
Mean tips per day **469** (13)
Median years of service of state workers **687** (9)
Mileage ratings for tires **285** (3 & 4)
Noting the manufacturer of a car **154** (26)
Part-time job incomes for college students **392** (3)
Pay rates of two trade unions **491** (6)
Price and change data in Data Set II **184** (2)
Prime locations in supermarkets **293** (12)
Probabilities associated with job stability **183** (28)

Rankings of quality-price of electric knives **717** (11)
Renting one-bedroom apartments **103** (2)
Selecting coins from a bank **165** (5)
Selling a condominium **183** (27)
Shelf prices at two supermarkets **74** (17)
Standard and Poor's Averages **12** (2)
Stock prices on American Stock Exchange **21** (16)
Stock prices for 60 corporations **105** (10)
The per account profit on investments **418** (21)
Time clock to reduce tardiness **29** (3)
Trading stamps to increase sales **30** (18)

*Chemistry, Physics, and Engineering*
Accelerations of two model cars **492** (10)
Ampere rating of a breaker **154** (23)
Braking distances for four brake designs **590** (4)
Breaking forces of nylon rope **359** (18)
Capacitances of mylar capacitors **58** (11)
Compound SS707 data **632** (15)
Cotton versus nylon thread **30** (16)
Cresote on fence posts **30** (22)
Defectives in lots of K47T960 transistors **115** (16)
Diameters of A749C rivets **58** (12)
Difference in braking distances **513** (30)
Dry cell batteries **263** (28)
Electrical device to improve mileage **30** (13)
Engine size and mpg data **649** (5)
Fluoride to reduce cavities **30** (15)
Hardness test and diameters of pieces **194** (4)

# A First Course in
# STATISTICS

# 1

# Introduction to Statistics

Figures 1-1 and 1-2 are examples of articles and data summaries that are as commonplace in today's newspapers and magazines as the comics, sports results, and reports on local and national politics. Frequently readers scan such reports with little thought as to how such information was gathered. Furthermore, such reports are frequently read with little understanding of the terms used in them. An elementary course in statistics will (in addition to other benefits) help remedy such deficiencies. In Chapter 1, we will study many of the basic terms used in statistics and commonly seen in news articles like those in Figures 1-1 and 1-2.

## 1-1 Some Basic Terms

**Key Topics**

1. *The definition of statistics*

2. *Projects in descriptive statistics*

3. *Projects in inferential statistics*

4. *The definition of a population*

5. *The definition of a sample*

6. *The definition of a simple random sample*

7. *Using a table of random digits*

S T A T I S T I C S   I N   A C T I O N

Clyde Daniels is principal of Ezra Pound Elementary School. He recently received a directive from the District Office to survey all students in his school and report on the number of brothers and sisters living with each student. He was instructed to tell the students to include all brothers and sisters, whether they had the same mother and father (full brothers and sisters) or only the same mother or same father (half brothers and sisters). The information was needed for a funding request being prepared for the state legislature. With the able assistance of the homeroom teachers, Clyde was able to compile the report shown in Table 1-1.

**TABLE 1-1   Information on Brother and Sister Relationships of Current Students at Ezra Pound Elementary School**

| Number of Students | Number of Brothers and Sisters |
|---|---|
| 137 | 0 |
| 203 | 1 |
| 129 | 2 |
| 75 | 3 |
| 28 | 4 |
| 8 | 5 or more |

B   Gilda Johnson is a reporter for the Daily Digest newspaper. She was instructed to get public opinion on a proposed sports complex. A joint committee composed of city council members and county supervisors has suggested that the complex be funded by selling municipal bonds. The bonds would be paid for by a 1 percent sales tax levied within the county. Gilda visited several locations around the city in which there is always heavy pedestrian traffic. She interviewed more than 400 individuals and compiled the results in Table 1-2.

TABLE 1-2   Results of Public Opinion Poll on Sports Complex

| Opinions on Sports Complex | Number of Responses in this Category |
|---|---|
| Strongly in favor of complex | 49 |
| Moderately in favor of complex | 119 |
| No opinion on complex | 182 |
| Moderately against complex | 55 |
| Strongly against complex | 31 |

## The definition of statistics

The functions of any society depend on information. Businesses, governments, educators, sports organizations, and researchers are only a few of the many segments of a society that depend on timely and accurate information. In statistics such information is usually called *data*.

### DEFINITION 1-1 Statistics

**Statistics** is the study of ways to collect, describe, draw conclusions, and make projections (inferences) from data.

In this text, the activities of data collection and data description will be referred to as *descriptive statistics*. The activities of drawing conclusions and making projections from data using the mathematical theory of chance will be referred to as *inferential statistics*.

## Projects in descriptive statistics

Any planned activity that results in data being collected and described and/or summarized can be called a project in descriptive statistics. The activity carried out at Ezra Pound Elementary School and reported in Table 1-1 is an example of a project in descriptive statistics.

## Five tasks usually included in a descriptive statistics project

Task 1 Define the problem related to some topic of interest.

Task 2 Devise a means to collect/observe data.

Task 3 Collect data related to the topic of interest.

Task 4 Calculate some numerical measures that describe the data.

Task 5 Present the data in some display.

Figure 1-3 contains data on the 50 states of the United States of America and the District of Columbia. The data include the capitals of each state, the land areas in square miles and square kilometers, and the estimated populations of each state based on the 1989 census.

Figure 1-4 is a pie chart of expenses and capital requirements for the Northern Section of Kaiser Foundation Health Plan, Inc. The percentage of the total expenses and capital requirements for each category has been calculated, such as Hospital Services: 31.4 percent. The area of each slice of the pie chart represents that category's percentage of the total expenses and capital requirements.

Figure 1-5 is an example of summaries of professional sports that are frequently seen in daily newspapers. The teams are listed in order based on their winning percentage (Pct.), that is, the number of games won divided by the number of games played. To illustrate, for Boston:

$$\text{Pct} = \frac{29}{35} = 0.829$$

Some other descriptive measures shown in the display include:

Won (W)          Number of games won

Lost (L)           Number of games lost

Games Behind (GB)   Number of games behind the division leaders

## Projects in inferential statistics

Any planned activity resulting in a set of data being used to predict or draw a conclusion about an unknown quantity can be called a project in inferential statistics. The activity carried out by Gilda Johnson and reported in Table 1-2 is an example of a project in inferential statistics. Based on the data Gilda recorded in the table, she inferred that 39 percent of the residents favored the proposed sports complex.

### EXPENSES & CAPITAL REQUIREMENTS

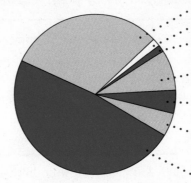

. **Hospital services:** 31.4% = $790,799,000

. **Health Plan administration:** 1.6% = $39,605,000

. **Community services (research, medical education, charitable care):** 0.9% = $23,103,000

. **Outpatient pharmacies and optical sevices:** 8.0% = $201,269,000

. **Other benefits (out-of-plan, ambulance, etc.):** 4.7% = $117,732,000

. **Capital requirements (construction & renovation of facilities, new major equipment, etc.):** 4.9% = $124,861,000

**Medical services (inpatient/outpatient):** 48.5% = $1,224,406,000

**Total expenses & capital requirements:** $2,521,775,000

**Figure 1-4.** "Expenses & Capital Requirements" from *Planning for Health*, Summer 1990. Reprinted by permission of Kaiser Permanente.

**Figure 1-3.** "U.S. Area and Population," *1990 Britannica World Data*, p. 740. Copyright © 1990 by Encyclopaedia Britannica, Inc. Reprinted by permission.

### Area and population

| States | Capitals | area[1] sq mi | sq km | population 1989 estimate[2] |
|---|---|---|---|---|
| Alabama | Montgomery | 51,705 | 133,915 | 4,150,000 |
| Alaska | Juneau | 591,004 | 1,530,693 | 565,000 |
| Arizona | Phoenix | 114,000 | 295,259 | 3,649,000 |
| Arkansas | Little Rock | 53,187 | 137,754 | 2,414,000 |
| California | Sacramento | 158,706 | 411,047 | 28,607,000 |
| Colorado | Denver | 104,091 | 269,594 | 3,393,000 |
| Connecticut | Hartford | 5,018 | 12,997 | 3,257,000 |
| Delaware | Dover | 2,044 | 5,294 | 658,000 |
| Florida | Tallahassee | 58,664 | 151,939 | 12,535,000 |
| Georgia | Atlanta | 58,910 | 152,576 | 6,524,000 |
| Hawaii | Honolulu | 6,471 | 16,760 | 1,121,000 |
| Idaho | Boise | 83,564 | 216,430 | 1,013,000 |
| Illinois | Springfield | 57,871 | 149,885 | 11,599,000 |
| Indiana | Indianapolis | 36,413 | 94,309 | 5,542,000 |
| Iowa | Des Moines | 56,275 | 145,752 | 2,780,000 |
| Kansas | Topeka | 82,277 | 213,096 | 2,485,000 |
| Kentucky | Frankfort | 40,409 | 104,659 | 3,742,000 |
| Louisiana | Baton Rouge | 47,752 | 123,677 | 4,510,000 |
| Maine | Augusta | 33,265 | 86,156 | 1,203,000 |
| Maryland | Annapolis | 10,460 | 27,091 | 4,665,000 |
| Massachusetts | Boston | 8,284 | 21,455 | 5,863,000 |
| Michigan | Lansing | 97,102 | 251,493 | 9,266,000 |
| Minnesota | St. Paul | 86,614 | 224,329 | 4,298,000 |
| Mississippi | Jackson | 47,689 | 123,514 | 2,680,000 |
| Missouri | Jefferson City | 69,697 | 180,514 | 5,163,000 |
| Montana | Helena | 147,046 | 380,847 | 808,000 |
| Nebraska | Lincoln | 77,355 | 200,349 | 1,590,000 |
| Nevada | Carson City | 110,561 | 286,352 | 1,049,000 |
| New Hampshire | Concord | 9,279 | 24,032 | 1,116,000 |
| New Jersey | Trenton | 7,787 | 20,168 | 7,827,000 |
| New Mexico | Santa Fe | 121,593 | 314,924 | 1,595,000 |
| New York | Albany | 52,735 | 136,583 | 17,761,000 |
| North Carolina | Raleigh | 52,669 | 136,412 | 6,602,000 |
| North Dakota | Bismarck | 70,702 | 183,117 | 664,000 |
| Ohio | Columbus | 44,787 | 115,998 | 10,787,000 |
| Oklahoma | Oklahoma City | 69,956 | 181,185 | 3,285,000 |
| Oregon | Salem | 97,073 | 251,418 | 2,750,000 |
| Pennsylvania | Harrisburg | 46,043 | 119,251 | 11,844,000 |
| Rhode Island | Providence | 1,212 | 3,139 | 996,000 |
| South Carolina | Columbia | 31,113 | 80,582 | 3,507,000 |
| South Dakota | Pierre | 77,116 | 199,730 | 708,000 |
| Tennessee | Nashville | 42,144 | 109,152 | 4,933,000 |
| Texas | Austin | 266,807 | 691,027 | 17,451,000 |
| Utah | Salt Lake City | 84,899 | 219,887 | 1,750,000 |
| Vermont | Montpelier | 9,614 | 24,900 | 557,000 |
| Virginia | Richmond | 40,767 | 105,586 | 6,068,000 |
| Washington | Olympia | 68,139 | 176,479 | 4,612,000 |
| West Virginia | Charleston | 24,231 | 62,758 | 1,871,000 |
| Wisconsin | Madison | 66,215 | 171,496 | 4,803,000 |
| Wyoming | Cheyenne | 97,809 | 253,324 | 503,000 |
| **District** | | | | |
| Dist. of Columbia | — | 69 | 179 | 615,000 |
| TOTAL | | 3,679,192[3] | 9,529,063 | 247,732,000[3] |

## NBA STANDINGS

### EASTERN CONFERENCE

#### Atlantic Division

| | W | L | Pct. | GB |
|---|---|---|---|---|
| Boston | 29 | 6 | .829 | — |
| Philadelphia | 22 | 14 | .611 | 7½ |
| Washington | 15 | 18 | .455 | 13 |
| New York | 15 | 19 | .441 | 13½ |
| New Jersey | 10 | 24 | .294 | 18½ |
| Miami | 9 | 26 | .257 | 20 |

#### Central Division

| | W | L | Pct. | GB |
|---|---|---|---|---|
| Chicago | 26 | 10 | .722 | — |
| Detroit | 26 | 11 | .703 | ½ |
| Milwaukee | 25 | 12 | .676 | 1½ |
| Atlanta | 20 | 15 | .571 | 5½ |
| Indiana | 14 | 21 | .400 | 11½ |
| Charlotte | 11 | 22 | .333 | 13½ |
| Cleveland | 11 | 25 | .306 | 15 |

### WESTERN CONFERENCE

#### Midwest Division

| | W | L | Pct. | GB |
|---|---|---|---|---|
| San Antonio | 24 | 8 | .750 | — |
| Utah | 24 | 12 | .667 | 2 |
| Houston | 19 | 17 | .528 | 7 |
| Dallas | 12 | 22 | .353 | 13 |
| Minnesota | 11 | 22 | .333 | 13½ |
| Orlando | 10 | 26 | .278 | 16 |
| Denver | 7 | 28 | .200 | 18½ |

#### Pacific Division

| | W | L | Pct. | GB |
|---|---|---|---|---|
| Portland | 31 | 7 | .816 | — |
| Phoenix | 23 | 11 | .676 | 6 |
| L.A. Lakers | 22 | 11 | .667 | 6½ |
| Golden State | 18 | 16 | .529 | 11 |
| Seattle | 15 | 18 | .455 | 13½ |
| L.A. Clippers | 13 | 24 | .351 | 17½ |
| Sacramento | 8 | 25 | .242 | 20½ |

**Figure 1-5.**  Team standings in the National Basketball Association.

## Five tasks usually included in an inferential statistics project

**Task 1** Define the problem related to some topic of interest.

**Task 2** Devise a means to collect/observe data.

**Task 3** Collect data related to the topic of interest.

**Task 4** Calculate some numerical measures that describe the data.

**Task 5** Make a prediction or draw a conclusion based on the measures obtained in the data.

Tasks 1 through 4 are identical with those involved in a project in descriptive statistics. The major difference between the two types of projects occurs in Task 5. In an inferential statistics project, the data are used to *predict* or *draw a conclusion* about an unknown quantity.

To illustrate, Gilda can use the results of her interviews to make some conclusion about the opinions of all adults in the county regarding the proposed sports complex. To understand the implications of making a prediction or drawing a conclusion, it is necessary to know something about *probability*—the mathematical theory of chance. This topic will be studied in Chapter 3.

Figure 1-6 is an example of inferences made daily by the National Weather Service (NWS). The NWS collects data related to wind speeds and directions, local high and low temperatures, locations of low and high pressure systems, and data from weather satellites. An analysis of such data then leads to a forecast of weather conditions for subsequent days.

Figure 1-7 contains a prediction regarding the likelihood of a major earthquake somewhere in the eastern two-thirds of the United States. We may assume that the prediction is based on data routinely collected and studied by the national earthquake engineering center. Paragraph 4 is a statement regarding the probability of the earthquake. Such a reference to what is expected to happen *in the future* is the indicator that identifies this report as inferential and not descriptive.

## The definition of a population

The word *population* is often used in both descriptive and inferential statistics.

### DEFINITION 1-2 Population

A **population** is a group of objects about which information is to be gained. An upper case $N$ will be used for the number of objects in a population.

The data reported by Clyde Daniels in Table 1-1 are based on an interview with every child in the school. Thus the source of the data is a population, and for this population: $N = 137 + 203 + 129 + 75 + 28 + 8 = 580$.

**Sierra:** Partly cloudy in the north and fair in the south Sierra today, tonight and Wednesday. Yosemite Valley high 56, low 32.

**Bay Area:** Partial clearing today after low clouds and fog today redeveloping tonight. Highs to mid-60s, lows to 40s. Clearing Wednesday after low clouds and fog, highs in the 60s.

**Coastal Regions:** Mostly sunny after morning low clouds and patchy fog today redeveloping tonight and Wednesday. Highs to 60s, lows to 40s. Wednesday's high to 65.

**Figure 1-6.** An illustration of a weather forecast. Courtesy of National Weather Service.

# Big quake due in East, expert says

WASHINGTON — A major earthquake is nearly certain to strike the eastern two-thirds of the nation in the next 20 years, threatening havoc in a region unprepared for it, the head of the national earthquake engineering center said Tuesday.

Robert L. Ketter declined to pin down a specific location for the temblor but said likely sites include the areas of Memphis, Tenn., Charleston, S.C., Boston and New York City.

The probability of a destructive quake occurring at any particular spot is low, he said.

"However, the probability of one occurring somewhere in the eastern United States before the year 2000 can be considered better than 75 percent to 95 percent. Before the year 2010, nearly 100 percent," he told a symposium on the federal response to earthquakes.

"Terra firma is an illusion," added Russell Christesen, chairman of the board of Ebasco Services Inc., an engineering company that sponsored the gathering.

**Associated Press**

**Figure 1-7.** A prediction of a major earthquake in the East. "Big Quake due in East, expert says," *The Sacramento Bee,* October 5, 1988, p. A3. Reprinted by permission of Associated Press.

Sometimes populations are only "conceptual." For example, in Figure 1-6, the data are related to conditions that affect weather. We cannot "see" individual objects in this "population." However, when a population consists of concrete objects, it is important that the population be well defined. By well defined, we mean that general agreement exists as to which objects belong to the population and which do not.

▶ **Example 1**  Determine whether the following populations are well defined.

**a.** The population of stocks listed on the New York Stock Exchange on a given day.
**b.** The population of good books.

**Solution**  **a.** Many daily newspapers list the stocks on the New York Stock Exchange. Thus the objects in the population can be identified and the population is well defined.
**b.** There is no standard or widely accepted definition of a good book. Therefore the objects of such a population cannot be readily identified and the population is not well defined.  ◀

There are instances when each object in a population is examined in order to gain data. An investigation that attempts to include all objects in a population is called a *census*. The following identify populations that regularly undergo a census.

| Population | When a Census Might Be Taken |
|---|---|
| Current student body at Florida State University | At the start and close of each term |
| The registered vehicles in Ohio | Each year when registrations are renewed |
| Workers in New York State with a taxable income | Each year to assess a state income tax |

## The definition of a sample

A census is rarely a practical way to obtain data from a large population. Most projects in statistics involve studying only some of the objects in a population. Any subset of a population selected to study is called a *sample*.

### DEFINITION 1-3 Sample

A **sample** is a subset of a population used to gain information about the whole population. A lower case $n$ is used for the number of objects in a sample.

The data reported by Gilda Johnson in Table 1-2 are based on the sample of individuals she interviewed. That is, only a part of the population within the county was polled about their opinions on the sports complex. Thus the source of the data is a sample, and for this sample:

$$n = 49 + 119 + 182 + 55 + 31 = 436$$

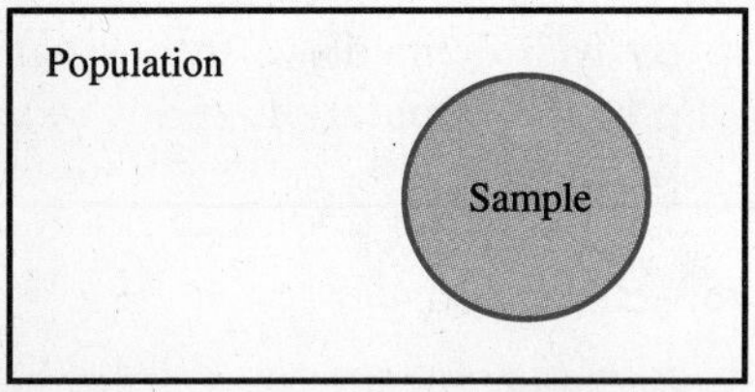

The relationship between a population and a sample is illustrated in Figure 1-8. In such a display called a *Venn diagram*, the population is represented by a rectangle. The sample selected from the population is displayed as a circle drawn within the rectangle.

Population

Sample

**Figure 1-8.** A diagram of a population and a sample.

## The definition of a simple random sample

In a typical statistics project involving a population of concrete objects, a sample of size $n$ is selected from the population of size $N$ under investigation. The observed values of any characteristic of each object in the sample are called *data*. A specific value in a set of data is called a *data point*.

To use the data obtained from a sample to gain information about the population from which it was selected, the sample data points should in some sense be representative of the population data points. Selecting a *simple random sample* is a way to try to achieve this goal.

### DEFINITION 1-4 Simple Random Sample

A **simple random sample** of size $n$ is a collection of $n$ objects selected from a concrete population in such a way that each set of $n$ objects in the population has the same chance of being selected.

To illustrate the concept of a simple random sample, consider a group of six people:

Ann, Beth, Carl, Dan, Ed, and Fred

Suppose two people need to be selected as delegates to attend a conference. There are 15 different pairs that can be formed using these six people:

| | | |
|---|---|---|
| (Ann, Beth) | (Beth, Carl) | (Carl, Ed) |
| (Ann, Carl) | (Beth, Dan) | (Carl, Fred) |
| (Ann, Dan) | (Beth, Ed) | (Dan, Ed) |
| (Ann, Ed) | (Beth, Fred) | (Dan, Fred) |
| (Ann, Fred) | (Carl, Dan) | (Ed, Fred) |

Based on Definition 1-4, the pair selected would be a simple random sample if each of these 15 pairs had the same chance of being selected.

### Four facts about a simple random sample

**Fact 1** Whether a physical process used to produce a sample satisfies Definition 1-4 is to some degree a subjective judgment.

This statement means that not all statisticians might agree that a given process qualifies as "random." As a consequence, we may hear or read of disagreements among statisticians regarding the randomness of samples selected by some processes. However, two physical processes will be discussed later that are commonly accepted as satisfying the condition of "same chance."

**Fact 2** A random sample is not always representative of a population.

It is possible that a given random sample has objects with characteristics unlike most objects in the population from which it is selected. However, our intuition should be that random samples are representative in some average or long-run sense.

**Fact 3** A simple random sample is obtained objectively.

This statement asserts that the process used to select the sample determines which objects of the population are included in the sample. The personal preferences of the person taking the sample (conscious or unconscious) do not enter into sample selection. As a consequence, random sampling guards against an element called *investigator bias*.

**Fact 4** The element of chance is introduced (in a manageable fashion) into an investigation when one selects a simple random sample.

As we will see in later chapters, this element of chance permits one to draw reasonable conclusions based on data from a sample by applying ideas from probability.

Two types of procedures that can be used to obtain a simple random sample are:

Procedure 1    Mechanical methods

Procedure 2    Use of a table of random digits.

▶ **Example 2**    Discuss a mechanical method of randomly selecting two people from: Ann, Beth, Carl, Dan, Ed, and Fred.

Solution    List the names of these six people on slips of paper of the same size and shape. Fold the papers in exactly the same way, put them in a box and mix thoroughly. Without looking, select two slips from the box. The two people whose names are on the slips would make up the sample.    ◀

▶ **Example 3**    Debbie Williams is a quality control inspector in the stereo receiver section of a major electronics manufacturing firm. She wants to select five receivers from a production run of 100 receivers. Identify a method to select a random sample.

Solution    One method that Debbie could use would be to write the serial number of each receiver on a 3 × 5 in. card. The 100 cards would then be thoroughly mixed and five cards selected one after another. The receivers whose serial numbers appeared on the cards would make up the required sample.    ◀

### Using a table of random digits

A table of random digits can also be used to randomly select a sample. Appendix Table 1 has several pages of random digits. A portion of this table is in Figure 1-9.

```
91 30 70 69 91   19 07 22 42 10   36 69 95 37 28   28 82 53 57 93   28 97 66 62 52
68 43 49 46 88   84 47 31 36 22   62 12 69 84 08   12 84 38 25 90   09 81 59 31 46
48 90 81 58 77   54 74 52 45 91   35 70 00 47 54   83 82 45 26 92   54 13 05 51 60
06 91 34 51 97   42 67 27 86 01   11 88 30 95 28   63 01 19 89 01   14 97 44 03 44
10 45 51 60 19   14 21 03 37 12   91 34 23 78 21   88 32 58 08 51   43 66 77 08 83

12 88 39 73 43   65 02 76 11 84   04 28 50 13 92   17 97 41 50 77   90 71 22 67 69
21 77 83 09 76   38 80 73 69 61   31 64 94 20 96   63 28 10 20 23   08 81 64 74 49
19 52 35 95 15   65 12 25 96 59   86 28 36 82 58   69 57 21 37 98   16 43 59 15 29
67 24 55 26 70   35 58 31 65 63   79 24 68 66 86   76 46 33 42 22   26 65 59 08 02
60 58 44 73 77   07 50 03 79 92   45 13 42 65 29   26 76 08 36 37   41 32 64 43 44

53 85 34 13 77   36 06 69 48 50   58 83 87 38 59   49 36 47 33 31   96 24 04 36 42
24 63 73 87 36   74 38 48 93 42   52 62 30 79 92   12 36 91 86 01   03 74 28 38 73
83 08 01 24 51   38 99 22 28 15   07 75 95 17 77   97 37 72 75 85   51 97 23 78 67
16 44 42 43 34   36 15 19 90 73   27 49 37 09 39   85 13 03 25 52   54 84 65 47 59
60 79 01 81 57   57 17 86 57 62   11 16 17 85 76   45 81 95 29 79   65 13 00 48 60
```

**Figure 1-9.**   Part of a table of random digits.

The computer disk that accompanies this text has a program for generating random numbers, and the calculator activity section in this chapter provides instructions for using a hand calculator to get random numbers.

In a table of random digits, each of the digits 0 through 9 has the same chance of appearing at any particular location in the table. Furthermore, if a given digit in the table is known, then such knowledge does not help in predicting which digit occurs at other locations in the table.

## Three steps for working a table of random digits

**Step 1** Use a block of $D$ digits to label each object in the population.

For example, if there are from 1 through 9999 objects in the population, then a block of $D = 4$ digits would be assigned to each object.

**Step 2** Assign to each object a different label of $D$ digits.

For example, if there are 4560 objects to be labeled, then the objects would be labeled consecutively as follows:

$$0001, 0002, 0003, \ldots, 4558, 4559, 4560$$

**Step 3** Begin at the point where you left off when the table was last used. Move through the table along a row from left to right and from top to bottom on each page. Take $D$ digits at a time until the desired number of labels have been selected.

For example, if ten objects are to be selected from the 4560 objects mentioned in step 2, then the objects selected by the table would make up the sample. Obviously, if a given block of four digits did not identify an object in this population (such as 8420), then that block of digits would be ignored. Also, blocks that are repeats of others already selected would be ignored.

▶ **Example 4**   Sergeant Bruno is a drill instructor at Grass Hills High School Military Academy. He wants to randomly select five students from his platoon of 60 to serve on mess duty in the kitchen. Discuss how Sgt. Bruno could use a table of random digits.

**Solution**   **Discussion.** The random digits in Figure 1-9 will be used to solve this problem.

**Step 1**  There are 60 students in the platoon. A two-digit number will be assigned to each student, that is:

$$D = 2$$

**Step 2**  To each student, assign one of the following numbers:

$$01, 02, 03, 04, \ldots, 58, 59, 60$$

**Step 3**   Suppose that Sgt. Bruno has already used the first five rows of digits. Therefore he starts at the beginning of the sixth row in the table:

12   88   39   73   43   65   02   76   11   Stop

The numbers 12, 39, 43, 02, and 11 name the five students in the sample. The numbers 88, 73, 65, and 76 are rejected, since these numbers are not paired with any students.   ◄

# Exercises 1-1   Set A

*In Exercises 1–6, identify whether each news item is an example of descriptive statistics or inferential statistics. Make specific references to the news item to justify your answer.*

**1.**

> **Tuesday's Results**
> Atlanta 117, Indiana 106
> Golden State 112, New Jersey 111, OT
> Miami 104, Orlando 102
> Portland 132, Minnesota 117
> Phoenix 127, Washington 97
> Utah 124, San Antonio 102
> Seattle 146, Denver 99
> LA Lakers 128, Charlotte 103

Figure 1-10.

**2.**

**Standard & Poor's averages**

|          | High   | Low    | Close  | Chg   |
|----------|--------|--------|--------|-------|
| Indust   | 369.56 | 362.88 | 366.64 | -2.92 |
| Transpt  | 227.73 | 223.45 | 226.22 | -1.51 |
| Utilities| 137.85 | 134.04 | 136.07 | -1.78 |
| Financl  | 22.24  | 21.74  | 22.01  | -0.23 |
| 500 Stocks | 315.23 | 309.35 | 312.49 | -2.74 |

Figure 1-11.

**3.**

THE SPEEDY MICROWAVE oven is no longer a kitchen rarity. More than 12 million of them were sold last year, and if your home hasn't got one yet, most surveys say, it will soon.

Besides being a timesaver, the microwave oven also saves nutrients. Research has shown, for example, that fruits and vegetables retain more of their vitamin C, color and texture when microwaved than when boiled or baked.

Figure 1-12.

**4.**

A 1986 survey found that more than 60% of U.S. women wear size 12 or larger, and 31% wear size 16 or larger. And government statistics show the average U.S. adult weighs 6 pounds more now than in the '60s.

**Figure 1-13.**

**5.**

The cold, sweet facts about ice cream and us: We ate 923 million gallons of it last year — more than 15 quarts per person — most of it (31%) vanilla, only 8% of it chocolate. Our top toppings: chocolate fudge and hot fudge. Our top ice cream-eating region: frosty New England. Our top eaters: adults.

**Figure 1-14.**

**6.**

### CSAA spot checks show motorists using safety belts

A high percentage of California drivers and passengers continues to use safety belts according to CSAA spot checks. In the most recent checks of commuter-hour traffic, 71 percent of drivers and 62 percent of front seat passengers were observed wearing safety belts. During non-commute hours, 67.1 percent of drivers and 63.2 percent of front seat passengers wore safety belts.

**Figure 1-15.**

*In Exercises 7–20, determine whether each population is well defined.*

**7.** The public schools in Houston, Texas.

**8.** The days last year during which it rained in New York City.

**9.** Expensive homes in Los Angeles County.

**10.** Nutritious breakfast cereals.

**11.** Major league teams that have won the World Series.

**12.** Tennis players who have won titles at Wimbledon.

**13.** Major universities in the United States.

**14.** Economy cars manufactured in the United States.

**15.** Blue-collar workers in Ohio.

**16.** Graduating students last spring at the University of Texas.

**17.** Movies that won the Academy Award for Best Picture.

**18.** Movies that are classified as "good western movies."

**19.** Professional athletes who have won the Associated Press award for best male and best female athlete of the year.

**20.** The ten best male and female tennis players in the past 20 years.

*In Exercises 21–26, state a circumstance in which a census of each population might be taken.*

**21.** The population of licensed drivers in the state of Pennsylvania.

**22.** The population of homes in Denver, Colorado.

**23.** The population of students in grades kindergarten through 12 in the Chicago city school system.

**24.** The population of hourly employees for Granite Construction Company.

**25.** The population of inmates at San Quentin Prison.

**26.** The population of patients at Sutter Memorial Hospital.

*In Exercises 27–32, identify a method whereby a simple random sample can be selected from each population. Specifically state the manner in which you would identify each member of the population and the process you would use to select the members of the sample. Answers in text are only suggestions.*

**27.** The population is the faculty at Kent State University. The president wants to sample the opinions of 30 faculty members regarding a proposed change in the regulations governing graduate assistant stipends.

**28.** The manager of a supermarket in St. Louis wants to sample 50 customers for their opinions about a change in the store hours.

**29.** Harry Daniels is the foreman of the warehouse for Giant Discount Stores, Inc. A shipment of 10,000 toasters just arrived at the warehouse and Harry wants to randomly sample 20 toasters to inspect for defects.

**30.** Sharon Gorman is a state inspector in charge of checking for faulty emission controls on vehicles in a large city in the state. To investigate the frequency of violations, she wants to sample 50 vehicles within the city.

**31.** Steve Rienowski needs to sample 25 water meters of private homes in a large metropolitan area to study water consumption. Each meter in the area has a six digit number that is registered with the Water Resources Department.

**32.** The manager of 25,000 acres of cotton currently being grown by the Delta Growers Cooperation needs to randomly select 30 plots measuring $10 \times 10$ feet to estimate the yield of cotton from the total acreage.

*In Exercises 33–36, identify at least one way in which the method used to select the sample violates the definition of simple random sampling.*

**33.** Angela Rose is a newsperson for a local television station. To obtain a sample of public opinion about a proposed change in a local ordinance, she telephoned 150 people in the city.

**34.** Sylvia Nunnes is an editor for a major women's magazine. To obtain data about the current attitudes and behavior patterns of women on sex, she places a questionnaire in this month's issue of the magazine for readers to complete and mail to the publisher.

**35.** Gary Yoder is a personnel manager at Central Motors Incorporated. To obtain the opinions of employees about a recent change in the company's policy regarding paid vacations, outside his office door he places a box into which employees can place their opinions.

**36.** Curly Cox is a postal inspector who checks trays of outgoing mail from various post offices. He needs to determine the classification of each item of mail that is inspected. He selects every tenth item of mail in each tray.

# Exercises 1-1   Set B

*Appendix B contains two large data sets. Exercises 1 and 2 refer to these sets.*

**1.** Use a table of random digits to select a sample of 40 individuals from Data Set I.
   a. Identify the numbers of women and men in the sample.
   b. Identify the ages of the youngest and oldest individuals in the sample.
   c. Identify the number of individuals in the sample that have some form of cancer.

2. Use a table of random digits to select a sample of 40 stock prices from Data Set II.
   a. Identify the number of stock prices in the sample that are less than $10 per share.
   b. Identify the number of stock prices in the sample that declined in price on this day of trading.

*Exercises 3 and 4 identify two projects that can be carried out using the students attending your college.*

3. Survey 40 students. Ask each student his/her current class standing, that is, freshman, sophomore, junior, senior, or graduate student.
   a. Do you think your sampling technique satisfies the definition of simple random sampling?
   b. How many students in your sample are juniors?

   c. How many students in your sample are graduate students?
   d. Compare/contrast your results with others in your class. What do you observe?

4. Survey 40 students. Ask the students how many textbooks they had to buy for the current term.
   a. Do you think your sampling technique satisfies the definition of simple random sampling?
   b. How many students in your sample had to buy fewer than five texts?
   c. How many students in your sample had to buy between eight and 12 textbooks, inclusive?
   d. Compare/contrast your results with others in your class. What do you observe?

*In Exercises 5 and 6, identify three questions you might ask those people making the inferences about the data on which their claims are based.*

5. A supermarket chain claims "our prices are the lowest of all supermarkets in town."

6. A pharmaceutical corporation claims "most doctors prescribe our brand of pain reliever."

# 1-2  Types of Data

**Key Topics**

*1. Quantitative versus qualitative data*

*2. Discrete versus continuous quantitative data*

*3. Associating numbers with qualitative data*

## S T A T I S T I C S   I N   A C T I O N

Robin Gordon is the administrative assistant for the Director of the Highway Patrol in a large eastern state. Recently she was asked to compile data on several details present in vehicular accidents involving one or more fatalities. Included were the following:

1. The number of vehicles involved in the accident.
2. The age of the driver responsible for the accident.
3. The sex of the driver responsible for the accident.

To obtain data for the report, Robin programmed the Management Information System (MIS) for the Highway Patrol to randomly select 150 reports from the files. The only stipulation required in the selection was that the report was a vehicular accident involving at least one fatality. From these reports, Robin compiled the data in Table 1-3.

**TABLE 1-3   Analysis of Data on 150 Vehicular Accidents Involving Fatalities**

| Number of Vehicles Involved | Total | Age of Responsible Driver | Total | Sex of Responsible Driver | Total |
|---|---|---|---|---|---|
| 1 | 30 | Under 21 | 42 | Female | 23 |
| 2 | 63 | 21–30 | 54 | Male | 127 |
| 3 | 33 | 31–50 | 27 | | |
| 4 | 18 | 51–65 | 15 | | |
| More than 4 | 6 | Over 65 | 12 | | |

## Quantitative versus qualitative data

Definition 1-5 identifies the basic difference between quantitative and qualitative data.

### DEFINITION 1-5 Quantitative Versus Qualitative Data

**Quantitative data** are obtained when some aspect of the objects in a population or sample is measured or counted. The information gathered is *numerical* and the data points are *numbers*.

**Qualitative data** are obtained when the objects in a population or sample are separated into categories based on clearly defined characteristics, attributes, attitudes, or opinions. The information gathered is *nonnumerical*.

In Table 1-3, the "Number of Vehicles Involved" and "Age of Responsible Driver" are quantitative. The "Sex of Responsible Driver" is qualitative.

▶ **Example 1**   The following questions are included on application forms for a student financial assistance program at Grand Islands Community College.

**a.** What is your current class standing?
**b.** How many relatives (mother, father, sisters, and brothers) are there in your family?
**c.** What is the combined annual income of your mother and father?

Identify the data from these questions as quantitative or qualitative.

**Solution**   **a.** Class standing separates students into categories such as freshman and sophomore. Thus the data from this question are qualitative.

**b.** The number of relatives in one's family is counted. Thus the data from this question are quantitative.

**c.** Annual income is expressed in numbers of dollars. Therefore the data from this question are quantitative.   ◄

▶ **Example 2**    The residents in an unincorporated section of a county in Illinois are going to vote in November on whether to incorporate and become a city. Christine Cross, a TV newscaster, wants to poll the residents in this area on their attitudes toward incorporation. Identify possible categories into which Christine can put the responses.

Solution    **Discussion.** The data that Christine will collect are nonnumerical because she wants to learn about the attitudes of the people she will poll.

Christine could restrict the responses to three categories:

> Category 1    Favor incorporation
>
> Category 2    Oppose incorporation
>
> Category 3    No opinion on incorporation

Or Christine could refine the responses by dividing the Favor and Oppose categories:

> Category 1    Strongly favor incorporation
>
> Category 2    Moderately favor incorporation
>
> Category 3    No opinion on incorporation
>
> Category 4    Moderately oppose incorporation
>
> Category 5    Strongly oppose incorporation   ◄

▶ **Example 3**    The trucks that haul logs harvested from national forests must stop at inspection stations. At these stations, measurements of the logs are taken to provide a record of the trees removed. Name three aspects of logs on a particular truck that would yield quantitative data.

Solution    **Discussion.** To answer this question, we must consider characteristics of logs that can be measured or counted. The following are only representative of the kinds of answers that might be given.

**a.** By counting the number of rings on a log, the age of the tree from which the log was cut could be approximated.

**b.** The length and circumference of a log might be measured to estimate the number of board-feet of lumber in the log.

**c.** The weight of the logs on the truck might also be of interest. This measurement could be obtained by subtracting the truck's empty weight from the weight with the logs on it.   ◄

## Discrete versus continuous quantitative data

All quantitative data are numerical. However, it is useful to separate quantitative data into *discrete* and *continuous types*, based on the kinds of numbers that are possible as data points.

---

### DEFINITION 1-6 Discrete and Continuous Data

**a.** Quantitative data are discrete when only certain separated numbers are possible as data points.

Discrete data are usually obtained when some aspect of the objects of a population or sample is *counted*.

**b.** Quantitative data are continuous when any number on an interval of numbers is possible.

Continuous data are usually obtained when some aspect of the objects of a population or sample is *measured*.

▶ **Example 4**　A simple random sample of ten shoppers at Sunrise Shopping Mall were each asked the following question: "How many bank or department store credit cards do you have?" Are the data discrete or continuous?

**Solution**　Since the shoppers were asked to count the number of credit cards they had, the possible data points were 0, 1, 2, 3, . . . . Thus the data were discrete.　◀

When continuous data are obtained in an investigation, the possible values of the data points will occur on an *interval of numbers*. To specify an interval, the variable $x$ can be used to represent each of the possible values in the interval. If $a$ and $b$ are used to represent the smallest and largest possible data points in a given set, then the interval can be written using inequalities as follows:

$$a \leq x \leq b \quad \text{read "}x\text{ is any number between } a \text{ and } b\text{, inclusive."}$$

The inequalities indicate there is no "gap" or "space" between the numbers in the set. When continuous data are collected, any apparent gaps between possible data points are the result of rounding the numbers.

To illustrate, suppose the weights of students at St. Matthew's Elementary School are being measured. The weight of each child would be rounded to some number before it was recorded.

| Possible Data Points in Set | Data Points Are Rounded |
|---|---|
| 72 and 108 pounds | to the nearest pound |
| 72.5 and 108.3 pounds | to the nearest tenth of a pound |
| 72.58 and 108.35 pounds | to the nearest hundredth of a pound |

When quantitative data are examined, the observed values of the data points may be listed in a set. To make such a list for the general case, the variable $x$ can be used to represent all the data points. Subscripts on $x$ are then used to distinguish between the different numbers. To illustrate, suppose a sample yields $n$ data points:

$x_1$   read "$x$ sub-one," is the first data point.

$x_2$   read "$x$ sub-two," is the second data point.

$\vdots$

$x_n$   read "$x$ sub-$n$," is the last data point.

All the data points in a given set can be listed as follows:

$$\{x_1, x_2, x_3, x_4, \ldots, x_n\}$$

A subscript $i$ is used to symbolize the "$i$-th data point" in the set. Thus $x_i$ represents all the data points in the set.

To illustrate the use of subscripts, suppose the number of credit cards carried by the ten shoppers in Example 4 are in the following set:

$$\{2, 1, 1, 0, 3, 1, 0, 4, 2, 1\}$$

We can identify each data point with an $x_i$ and array them as follows:

$$x_1 = 2 \quad x_2 = 1 \quad x_3 = 1 \quad x_4 = 0 \quad x_5 = 3$$
$$x_6 = 1 \quad x_7 = 0 \quad x_8 = 4 \quad x_9 = 2 \quad x_{10} = 1$$

▶ **Example 5**   The population under investigation is the vehicle inventory of a major Ford dealer in a large metropolitan area in the south.

**a.** Name two features of a vehicle that would yield discrete quantitative data.

**b.** Name two features of a vehicle that would yield continuous quantitative data.

**Solution**   **a. Discussion.** To generate discrete quantitative data, look for features of a vehicle that could be *counted*.

1. The number of cylinders in the engine.   Numbers such as 4, 6, and 8 are possible data points.

2. The number of doors on the vehicle.   Numbers such as 2 and 4

**b. Discussion.** To generate continuous quantitative data, look for features of a vehicle that could be *measured*.

1. The length of the wheelbase in inches or centimeters.   Such as 65 in. or 180 cm.

2. The weight in pounds or kilograms.   Such as 4526 lb or 2880 kg.   ◀

## Associating numbers with qualitative data

Sometimes an artificial scale of numbers is used to express data that are basically qualitative but nevertheless involve a natural order. The numbers are called *ordinal data*.

One reason qualitative data are so quantified is that comparisons can then be made between data points. For example, consider the intrinsic quality of intelligence in individuals. To compare intelligence, we may give an "intelligence test." The test scores are examples of ordinal data because they are used in an attempt to quantify something that is basically qualitative. There is no meaningful way, for example, to judge whether the difference between scores of 110 and 120 is "the same as" the difference between scores of 120 and 130.

▶ **Example 6**     Barbara Davis teaches a lecture class of 250 students in Psychology 209. On the last test, she gave 30 questions. She wants to rate each test question on a scale of 1 through 4, where 1 is easy and 4 is difficult. Discuss a method whereby she might use the results of the test to give an ordinal data point to each question.

**Solution**     One method that can be used to rate each question is based on the percentage of the students that got the answer right. The following list is one possible set of ratings.

**a.** A question is rated 4 if 25% or fewer got the answer correct.
**b.** A question is rated 3 if 26% to 50% got the answer correct.
**c.** A question is rated 2 if 51% to 75% got the answer correct.
**d.** A question is rated 1 if 76% or more got the answer correct.     ◀

## Exercises 1-2   Set A

*In Exercises 1–10, identify possible categories into which the members of a population or sample under investigation could be placed.*

1. The political party affiliations of the students in a classical humanities class are being examined. Into what categories might the affiliations be placed?

2. The religious preferences of the students in a human sexuality class are being investigated. Into what categories could the preferences be placed?

3. The national parks in the United States are being studied by a sixth grade class. Into what categories can these parks be placed?

4. The types of automobiles produced by the major car manufacturers in the world are being studied in a twelfth grade auto mechanics class. What are possible categories into which automobiles might be placed?

5. The flowers in the Botanical Garden Section of Central Park are being discussed at this week's meeting of the Women's Auxiliary Club. Into what categories might the flowers be placed?

6. The animals in the city zoo need to receive their annual checkups and inoculations. The zoo's veterinarian wants to separate the population into categories to organize the program. Into what categories might the animals be separated?

7. Donald Redman is currently studying the population of patients at a major hospital. He wants to organize the investigation by separating the patients into categories. What are possible categories that Donald might specify?

8. Barry Tucker is dean of students at a large community college. He is surveying the current student population and wants to organize the survey by separating the students into categories. What are possible categories that Barry might use?

9. Judy Novack is studying the current members of the U.S. Congress. To organize the study, she wants to put the members of Congress into categories. What are possible categories that Judy might use?

10. Natalie Gross is investigating the crimes committed last year in the county in which she lives. To organize the investigation, she wants to put the crimes into categories. What are possible categories that Natalie might use?

*In Exercises 11–20, for the objects identified:*
*a. Identify a feature of the object that can yield continuous quantitative data.*
*b. Identify a feature of the object that can yield discrete quantitative data.*

11. The population of hollywood junipers being grown by Matsuii Nursery.

12. The population of employee cars at the Burbank plant of Hi-Tech Incorporated.

13. The population of books at the Mill Valley Public Library.

14. The population of tires produced last night by the third shift at the Akron plant of Better Tires Incorporated.

15. The population of television sets in the warehouse of Consumer Outlet Stores in Kansas City.

16. The population of stock prices on the American Stock Exchange.

17. The population of heads of iceberg lettuce recently shipped from a ranch in the Imperial Valley.

18. The population of inland lakes identified by name in the 50 states of the United States.

19. The population of students at Roger Bacon Elementary School.

20. The population of residents in a condominium park in a suburb of Boston.

*In Exercises 21–28 either a report is cited or an activity described. For each exercise, answer parts (a) and (b).*
*a. Are the data in this exercise quantitative or qualitative? If the data are quantitative, state whether they are discrete or continuous.*
*b. Write a brief statement justifying your answer to part (a).*

21. Five coins are simultaneously tossed many times. The number of heads obtained in each toss is recorded.

22. An article reports that the average American worker steals 3 hours and 45 minutes per week from the employer.

23. An article reports that the typical American consumes 182.5 gallons of liquids a year, including 55.6 gallons of water, 32.8 gallons of coffee, 31.6 gallons of soft drinks, and 24.2 gallons of milk.

24. According to the U.S. Office of Education, one of every five students who has taken out federally guaranteed loans to attend school is defaulting on the loan.

**25.** A record of the teams in the four divisions of the National Hockey League contains data on games won (W), lost (L), and tied (T).

**26.** A pair of dice is rolled and the sum of the spots on the tops of the cubes are added.

**27.** A table contains quantitative data reported by the National Weather Service. The data includes the high and low temperatures and precipitation on a given day.

**28.** A report from The Kline Brokerage House contains data on several stocks listed on the New York Stock Exchange. The data includes the current prices of the stocks and the number of shares traded yesterday.

*In Exercises 29–32, identify each data point with an $x_i$.*

**29.** Ten sample lots of Harmer E49C extension springs yielded the following number of defective springs based on a quality control inspection test:

$$\{3, 0, 1, 0, 5, 2, 1, 2, 3, 1\}$$

**30.** Twelve sample lots of fingerling trout at a state fish hatchery yielded the following number of fish in the samples that were less than the minimum length allowable for use in stocking lakes in the state:

$$\{5, 8, 2, 10, 5, 4, 4, 3, 7, 6, 5, 2\}$$

**31.** A survey of 15 patients' records at a major hospital was taken. The number of days the patients were in the hospital for the most recent admission was noted. The following list contains the data obtained from the records:

$$\{2, 5, 11, 3, 4, 2, 5, 1, 12, 8, 2, 9, 3, 6, 2\}$$

**32.** Sixteen members of the Elk Grove Chamber of Commerce had their cholesterol levels tested at a recent meeting of the council. The results of the tests are as follows:

$$\{235, 210, 248, 186, 157, 286, 193, 242, 176, 302, 240, 216, 195, 274, 166, 233\}$$

*Exercises 33–38 contain situations in which qualitative aspects of populations need to be evaluated.*
*a. Name a specific feature of each population that can be used to make the evaluation.*
*b. Identify categories that might be used to generate ordinal data for the evaluation.*

**33.** A judge at a sheep dog trial needs to evaluate the control the handlers exhibit over their dogs. She wants to quantify "control."

**34.** A sportscaster is studying the difficulty of major golf courses in the United States. She wants to quantify "difficulty."

**35.** The editor of a car magazine wants to compare the riding comfort of this year's car models. The editor wants to quantify "comfort."

**36.** The judges of a talent show are comparing several dance acts. Each act consists of a man and a woman performing an original dance routine. The judges want to quantify "performance."

**37.** The coach of a girls' softball team is studying the players for signs of a commitment to improving themselves and the team. The coach wants to quantify "commitment."

**38.** The editor of a newspaper column entitled "The Gourmet Diner" is visiting local restaurants to rate them on the quality of the overall eating experience. She wants to quantify "quality."

## Exercises 1-2    Set B

1. Figure 1-16 contains temperature and rainfall data on 83 U.S. cities.
   a. Are the temperature data discrete or continuous?
   b. Are the rainfall data discrete or continuous?

| NATION | HI | Lo | Prc |
|---|---|---|---|
| Albany | 43 | 29 | — |
| Albuquerque | 47 | 27 | — |
| Amarillo | 48 | 28 | .04 |
| Anchorage | -03 | -14 | .02 |
| Atlanta | 54 | 34 | .21 |
| Atlantic City | 54 | 24 | — |
| Austin | 66 | 47 | — |
| Baltimore | 55 | 27 | — |
| Billings | 44 | 35 | — |
| Birmingham | 57 | 32 | .18 |
| Bismarck | 38 | 14 | — |
| Boise | 43 | 36 | .22 |
| Boston | 43 | 29 | — |
| Brownsville | 76 | 55 | — |
| Buffalo | 44 | 36 | — |
| Burlington | 38 | 29 | .01 |
| Casper | 42 | 23 | — |
| Charlotte | 59 | 28 | — |
| Cheyenne | 39 | 16 | — |
| Chicago | 37 | 22 | — |
| Cincinnati | 50 | 34 | .04 |
| Cleveland | 52 | 30 | — |
| Columbia | 64 | 29 | — |
| Columbus | 51 | 32 | — |
| Dallas | 58 | 46 | .17 |
| Dayton | 50 | 36 | — |
| Denver | 43 | 22 | — |
| Des Moines | 29 | 26 | — |
| Detroit | 41 | 29 | — |
| Duluth | 23 | 10 | — |
| El Paso | 57 | 25 | — |
| Evansville | 45 | 30 | .09 |
| Fairbanks | -37 | -45 | — |
| Fargo | 31 | 22 | — |
| Flagstaff | 48 | 23 | — |
| Grand Rapids | 35 | 33 | — |
| Great Falls | 45 | 33 | — |
| Hartford | 42 | 18 | — |
| Helena | 45 | 31 | — |
| Honolulu | 83 | 69 | — |
| Houston | 59 | 47 | 1.34 |
| Indianapolis | 47 | 33 | .10 |
| Jacksonville | 75 | 45 | — |
| Juneau | 33 | 25 | — |
| Kansas City | 34 | 32 | .20 |
| Las Vegas | 71 | 40 | — |
| Little Rock | 42 | 39 | .39 |
| Louisville | 51 | 34 | .16 |
| Lubbock | 55 | 31 | .02 |
| Memphis | 52 | 39 | .13 |
| Miami | 75 | 71 | .07 |
| Milwaukee | 36 | 30 | — |
| Minneapolis | 24 | 18 | — |
| Nashville | 50 | 32 | — |
| New Orleans | 70 | 53 | .59 |
| New York | 50 | 34 | — |
| Norfolk | 62 | 37 | — |
| North Platte | 50 | 18 | — |
| Oklahoma City | 53 | 38 | .15 |
| Omaha | 26 | 14 | — |
| Orlando | 70 | 56 | 1.31 |
| Philadelphia | 55 | 29 | — |
| Phoenix | 69 | 46 | — |
| Pittsburgh | 55 | 33 | — |
| Portland, Ore. | 50 | 43 | .57 |
| Providence | 46 | 24 | — |
| Raleigh | 61 | 28 | — |
| Reno | 53 | 30 | — |
| Richmond | 60 | 31 | — |
| St. Louis | 42 | 37 | .08 |
| Salt Lake City | 38 | 33 | .49 |
| San Antonio | 68 | 48 | — |
| Santa Fe | 39 | 22 | — |
| Savannah | 73 | 41 | — |
| Seattle | 48 | 44 | .02 |
| Shreveport | 52 | 44 | 2.25 |
| Sioux Falls | 29 | 27 | — |
| Spokane | 43 | 23 | .07 |
| Syracuse | 46 | 34 | — |
| Tulsa | 45 | 44 | .37 |
| Washington | 59 | 34 | — |
| Wichita | 47 | 33 | — |
| Wilkes-Barre | 48 | 22 | — |

**Figure 1-16.**

2. Figure 1-17 contains data on stock prices listed on the New York Stock Exchange. Identify the following data as discrete or continuous:
   a. 52-week High—The highest price of the stock for the past 52 weeks
   b. Div—The most recent dividend earned by a stock
   c. Sales 100s—The number of stocks (in 100s) that were traded on the previous day of trading
   d. Chg—The change in the price of the stock (in $\frac{1}{8}$ths of a dollar) as a result of the previous day's trading

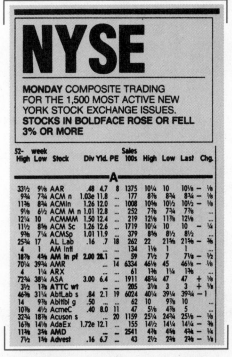

**Figure 1-17.**

## WATER TABLES

As of 7:33 a.m. Tuesday

**Sacramento River**
SHASTA — Capacity 4,552,100 acre-feet; elevation 931.99 feet; inflow 3,749 cfs; outflow 4,198 cfs; storage 1,595,518 acre-feet; storage year ago this date 2,209,164 acre-feet.

**Trinity River**
TRINITY DAM — Capacity 2,500,000 acre-feet; elevation 2249.95 feet; inflow 687 cfs; outflow 337 cfs; storage 954,710 acre-feet; storage year ago this date 1,303,774 acre-feet.

**Feather River**
OROVILLE DAM — Capacity 3,538,000 acre-feet; elevation 655.68 feet; inflow 729 cfs; outflow 2,202 cfs; storage 947,175 acre-feet; storage year ago this date 1,864,972 acre-feet.

**American River**
FOLSOM DAM — Capacity 1,000,000 acre-feet; elevation 352.79 feet; inflow 706 cfs; outflow 933 cfs; storage 156,868 acre-feet; storage year ago this date 326,857 acre-feet. NIMBUS DAM — Capacity 8,800 acre-feet; elevation 123.11 feet; inflow 747 cfs; outflow 519 cfs; storage 7,842 acre-feet.

**Figure 1-18.**

3. Figure 1-18 contains water flow data on four rivers and water storage data on five dams. Identify the following data as discrete or continuous:
   a. Dam capacity in acre-feet.
   b. Dam elevation in feet.
   c. Water inflow to the dam in cfs (cubic feet per second).

4. Figure 1-19 contains demographic data on the United States as of 1989.
   a. Are the population data discrete or continuous?
   b. Are the density data discrete or continuous?
   c. Are the urban-rural data quantitative or qualitative?
   d. Are the sex distribution data quantitative or qualitative?

**Demography**
*Population* (1989): 248,777,000.
*Density* (1989): persons per sq mi 67.6, persons per sq km 26.1.
*Urban–rural* (1987): urban 76.7%; rural 23.3%.
*Sex distribution* (1988): male 48.71%; female 51.29%.
*Age breakdown* (1988): under 15, 21.5%; 15–29, 24.2%; 30–44, 23.2%; 45–59, 14.2%; 60–74, 11.7%; 75 and over, 5.1%.
*Population projection:* (2000) 268,834,000; (2010) 283,174,000.
*Doubling time:* 95 years.
*Composition* by race (1988): white 84.3%; black 12.5%; other races 3.2%.
*Religious affiliation* (1987): Christian 87.1%, of which Protestant 49.1%, Roman Catholic 29.6%, other Christian 8.4%; Jewish 2.7%; Muslim 1.9%; Hindu 0.2%; nonreligious 6.6%; atheist 0.2%; other 1.3%.

**Figure 1-19.** "Demography," *1990 Britannica World Data*, p. 740. Copyright © 1990 by Encyclopaedia Britannica, Inc. Reprinted by permission.

*Exercises 5 and 6 identify two projects that can be carried out in the community in which you live.*

5. Select 30 items that could be purchased at a major department store.
   a. Describe your technique for sampling the items.
   b. Identify possible categories into which the items could be placed based on some qualitative features.
   c. Identify two items of discrete quantitative data that can be obtained from each item.
   d. Identify two items of continuous quantitative data that can be obtained from each item.
   e. Compare your results with others in the class.

6. Obtain a copy of the current real estate section of your local newspaper. Select 25 homes listed for sale in your community.
   a. Describe your technique for sampling the homes.
   b. Identify two different categories into which these homes could be placed based on qualitative features.
   c. Identify two aspects of a home that yield discrete quantitative data.
   d. Identify two aspects of a home that yield continuous quantitative data.
   e. Compare your results with others in the class.

# 1-3 Experimentation

**Key Topics**

1. *A statistical study involving experimentation*

2. *Two kinds of experimental designs*

3. *The element of confounding in experimentation*

4. *Two principles of experimentation to reduce confounding*

5. *Possible problems in experimentation*

S T A T I S T I C S   I N   A C T I O N

Fred Tyler is marketing manager for a supermarket chain with 245 retail outlets in three midwestern states. He was recently contacted by representatives of a trading stamp firm. These representatives presented a plan whereby trading stamps might increase the sales volume in the stores. The plan sounded promising, but Fred wanted some data before asking the executive board to implement the plan at all outlets.

With the help of a statistical consultant, Fred devised the following experiment. He first separated the 245 stores into possible test sites of 5 stores each. He made sure that the total sales volume of each site was approximately the same. To each of the forty-nine possible sites he assigned a two-digit number. Using a random number table, he selected six test sites. Thus a total of six sites containing 30 stores were used in the experiment. Fred monitored the advertising campaigns in the six sites to ensure they were the same.

Then, in three of the sites chosen at random from the six, Fred had the stores introduce the trading stamps. He had the store managers support the stamps with publicity in newspapers, radio commercials, and billboard ads. In the other three sites, Fred did not introduce trading stamps, but increased the advertising budgets to the amounts spent in the test sites with stamps. He further monitored the items featured in all advertisements to ensure equity in sales promotions at all six sites.

The experiment lasted 4 weeks. At the end of that time, he collected the total sales figures for all test sites for the 4 weeks before the test and the 4 weeks of the test. The compiled data are shown in Table 1-4.

**TABLE 1-4   Total Sales of Test Sites Involved in the Trading Stamps Experiment**

|  | Total Sales for 4 Weeks Before the Test | Total Sales for 4 Weeks During the Test |
|---|---|---|
| *The Three Sites that Used Trading Stamps* | $414,450 | $498,270 |
| *The Three Sites that Did Not Use Trading Stamps* | $442,200 | $478,425 |

## A statistical study involving experimentation

Much information is gained about the world in which we live by observing how various things react to forces applied to them. The process whereby such information is obtained is called *experimentation*.

### DEFINITION 1-7 A Statistical Study Involving Experimentation

A statistical study involving experimentation is different from one involving only observation and recording in that the investigator manipulates or applies external forces to the objects of a sample drawn from a population of interest.

The terms *manipulates* and *applies external forces* imply that the investigator either directly or indirectly does something to the objects being investigated. This terminology will be used for both possibilities.

For Fred Tyler, the population of interest is the 49 sites in the supermarket chain. The external force in the experiment is the trading stamps and the possible effect of the stamps on the total sales in the stores.

## Two kinds of experimental designs

The two experimental designs that we will consider involve only one or two steps. We will call these designs (A) and (B), and they are diagramed in Figure 1-20. The arrows ($\longrightarrow$) in the diagrams are read "followed by."

(A)   **Treatment** $\longrightarrow$ **Observation**
        "Treatment followed by observation."

(B)   **Observation** $\longrightarrow$ **Treatment** $\longrightarrow$ **Observation**
        "Observation followed by treatment followed by observation."

**Figure 1-20.**   Experimental designs (A) and (B).

Consider the experiment that Fred Tyler conducted on the three sites where the trading stamps were introduced. He collected the total sales figures for the 4-week period before the introduction of stamps. Then he collected the total sales figures for the 4-week period during which stamps were used. Thus Fred was using design (B).

▶ **Example 1**      Ms. Jean Vrechek is a mathematics instructor at Glass Valley Community College. She wants to know whether electronic calculators can improve students' abilities to learn logarithms in her college algebra classes. Use design (A) and describe an experiment to determine whether calculators are effective.

Solution    **Discussion.** Design (A) requires a treatment phase before any observations are taken. For this experiment, the treatment would be teaching logarithms by using calculators.

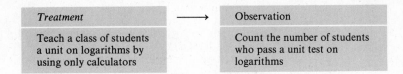

▶ **Example 2**    Use design (B) and describe an experiment to determine whether calculators are effective in teaching logarithms.

Solution    **Discussion.** Design (B) uses an observation phase before the treatment phase. The treatment phase is then followed by a second observation phase. For this experiment, the treatment would be teaching logarithms by using calculators.

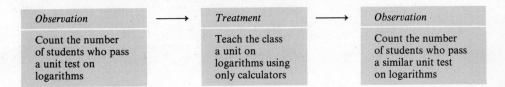

## The element of confounding in experimentation

In many experiments, there are other factors not accounted for in the basic designs (A) and (B). These factors may contribute more or less to the results of the experiment, but they are not identified in the experiment. When such factors are present, the possibility of *confounding* exists.

### DEFINITION 1-8 Confounding

**Confounding** occurs when the effects of a factor of interest in an experiment get confused with those of factors not of interest.

▶ **Example 3**    Name two factors that might produce confounding in Ms. Vrechek's experiment.

Solution    **Discussion.** A confounding factor would be anything that might increase (or decrease) the chance of students learning logarithms.

**a.** The general mathematical abilities of the students in the class

   If success is achieved, then maybe it is because the class contains mostly students with above average abilities in mathematics. If such is the case, then other classes may not experience the same degree of success.

**b.** The general effectiveness of Ms. Vrechek to teach mathematics

If success is achieved, then it may be that Ms. Vrechek is just a good mathematics instructor. Perhaps she would have had similar success using logarithm tables.   ◄

## Two principles of experimentation to reduce confounding

There are two principles of good experimentation that are intended to reduce confounding. The first principle is *control*. Control gives an investigator a way of comparing a treated group of objects with a group that is untreated. The second principle is *randomization*. Randomization is used to balance the effects of extraneous factors between the treatment and control groups. When these principles are applied to experimental designs (A) and (B), we get the modified designs (A*) and (B*) in Figure 1-21.

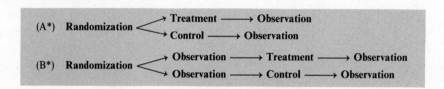

**Figure 1-21.**   Modified experimental designs (A*) and (B*).

In the trading stamp experiment, the treatment group was the three sites that used stamps. The control group was the three sites that did not use stamps. Notice other factors that Fred Tyler used to reduce confounding:

1. He used equitable advertising budgets for treatment and control groups.
2. He had both groups feature the same items in their advertising.
3. He used three different test sites for treatment and control groups.
4. He used random numbers to select the test sites and also to determine which ones were given trading stamps.

► **Example 4**   For Ms. Vrechek's experiment on the use of calculators versus nonmechanical means of teaching logarithms:

**a.** Discuss a method of introducing control into the experiment using design (A*).
**b.** Discuss a method of introducing randomization into the experiment.

**Solution**   **a.** To introduce control into the experiment, Ms. Vrechek could teach the same course to two different groups of students. She would then teach one group using calculators (the treatment group) and the other group without using calculators (the control group). Then at the observation stage of the experiment, she would give both groups the same unit test on logarithms.

**b.** Once Ms. Vrechek was given the list of students in her college algebra class, she could then use a table of random numbers to divide the students into two groups. One group would be the treatment group, and the other group would be the control group. She could then work out the details of how to meet with the two groups at different times, so that the treatment and control groups would be instructed solely by the method prescribed for that group. ◀

## Possible problems in experimentation

Many experiments use living creatures as subjects. Such experiments can create moral, ethical, and legal problems. Animals rights activists protest the use of animals for experiments that subject animals to pain and possible death. The government's use of soldiers in the 1950s to test the effect of excessive radiation on people has been severely criticized. Other examples could also be cited.

Finally, the psychological nature of human beings must be taken into consideration when planning an experiment. It is widely believed that if a technician recording measurements knows which experimental units have received the treatment, she or he will tend to perceive them more favorably than the control units. Also, many have observed a *placebo effect* when experimenting with humans. That is, when told things will get better, humans often perceive that they have improved, although there is no real change in physical conditions.

## Exercises 1-3 Set A

*In Exercises 1–14:*
*a. Use design (A) or (B) and describe an experiment for each situation.*
*b. For each experiment name two factors that may cause confounding.*

1. A manufacturer needs to determine whether an oxide coating will prevent the bleaching effect of the sun on a painted surface.

2. A physical education instructor wonders whether exercise alone, without a change in diet, can cause weight reduction in an individual.

3. A personnel manager would like to know whether a time clock will reduce the tardiness of workers.

4. A statistics instructor wants to find out whether the class will perform better on tests if he regularly collects and grades homework assignments.

5. An avid golfer is interested in knowing whether using clubs with graphite shafts can improve her score.

6. A manufacturer needs to determine whether a battery developed by its research division can provide 500 miles of continuous operation for its electric cars before it needs to be recharged.

7. An advertisement claims that an elastic belt can help someone lose 24 pounds or more. A government agency is responsible for checking such advertisement claims and needs to check the effectiveness of the belt.

8. The management of a regional transit system wonders whether the gas mileages of its buses can be increased by inflating the tires by 10 percent more than the present pressure.

9. A research laboratory has developed a "hair restoration gel" that the researchers hope can prevent balding caused by testosterone accumulation. Laboratory personnel would like to test whether morning and evening messages with the gel can prevent such balding.

10. The trainer of a professional football team has developed an elastic knee support that he feels can reduce damage to knee ligaments. How can the effectiveness of the device be tested?

11. An engineer for Farm Implements Corporation is currently working on a revised model of the mounted plows that the corporation manufactures. He wishes to find out whether a particular seam on the plow can be riveted or whether it requires a weld.

12. A nutritionist at a large medical facility surmises that massive doses of vitamin C taken regularly can reduce the incidence of the common cold among the members of the staff.

13. Connie Tucker bought an electrical device that reportedly will increase the gasoline mileage of her car. She wants to test whether the device is actually effective in increasing the mileage of her car.

14. Barry Tucker supervises a maintenance staff at a large southern university. He has developed a training program that reportedly will increase the productivity of his staff. How should he determine whether the program is really effective in increasing productivity?

*In Exercises 15–22, use design (A\*) or (B\*) and describe an experiment for each situation.*

15. A chemist thinks that a fluoride compound might reduce tooth cavities when added to a toothpaste.

16. A clothing manufacturer wonders whether the new line of casual pants can be stitched with cotton thread, or whether a more expensive nylon thread must be used.

17. The police department in a large city needs to investigate whether the operating cost of its vehicles can be reduced by using more expensive radial tires instead of bias-belted tires.

18. The owners of a supermarket chain wonder whether profits can be increased by introducing trading stamps into the stores.

19. The mathematics department at a large university has been requested to find out whether a new "self-paced" program in basic mathematics is effective in teaching remedial mathematics to college students.

20. The manufacturer of a field corn fertilizer has added some new chemicals to the basic formula. It needs to measure whether corn yields are increased as a result of the new additives.

21. A project is proposed at a medical school to be carried on for several years to determine whether a low cholesterol diet can effectively reduce the incidence of heart attacks in a population of individuals defined as "at high risk." (These people show symptoms that are frequently observed in heart attack patients.)

22. A rancher wonders whether it is worth the extra time and expense required to put the wood preservative creosote on a fence post before using it on a fence line. She decides to base her decision on the difference in time it takes for treated and untreated posts to become unusable because of rotting.

## Exercises 1-3    Set B

1. Suppose you want to conduct an experiment to show the following:

   "Students who study in groups do better in college than students who study alone."

   Furthermore, suppose you get 70 students at your school to help with the experiment.
   a. Identify a method of randomly dividing the students into two groups of 35 each.
   b. What would the treatment group do as their part of the experiment?
   c. What would the control group do as their part of the experiment?
   d. What would you do at the observation stage of the experiment?
   e. Identify several confounding factors that may be present in the experiment.

2. Suppose you want to conduct an experiment to show the following:

   "Swimming is more effective than jogging in maintaining proper weight and a strong heart."

   Furthermore, suppose you get 80 students at your school to help with the experiment.
   a. Identify a method of randomly dividing the students into two groups of 35 each.
   b. What would the treatment group do for their part of the experiment?
   c. What would the control group do for their part of the experiment?
   d. What would you do at the observation stage of the experiment?
   e. Identify several confounding factors that may be present in the experiment.

# Chapter 1 Summary

## Definitions

1. **Statistics** is the study of ways to **collect**, **describe**, and **draw conclusions** from data.
2. A **population** is a group of objects about which information is to be gained.
3. A **sample** is part of a population used to gain information about the whole population.
4. A **simple random sample** of size $n$ is selected from a population in such a way that each group of size $n$ has an equal chance of being selected.
5. **Quantitative data** are numerical. **Qualitative data** are fundamentally nonnumerical.
6. Quantitative data are **discrete** when only certain numbers are possible as data points. Quantitative data are **continuous** when any number on an interval of numbers is possible.
7. A **statistical study involving experimentation** is fundamentally different from one involving only observation and recording because the investigator manipulates or applies external forces to the objects of a sample drawn from a population of interest.
8. **Confounding** occurs when the effects of a factor of interest in an experiment get confused with the effects of factors not of interest.

## Symbols

$N$        the number of objects in a population

$n$        the number of objects in a sample

$x_i$       the $i$-th data point in a set of data

$\longrightarrow$   in an experimental design means "followed by"

# Comparing Quantitative Data Versus Qualitative Data

A Population of Objects Can
Usually Be Examined to Yield
Both Quantitative and Qualitative Data

Quantitative data are numerical and result from activities that *measure* some aspect of the objects, or *count* something related to the objects.

Quantitative data are divided into two types based on the numbers possible in the data sets.

Discrete Data

Only certain specified, separated numbers are possible data points.

Discrete data are always (but not exclusively) the result of a *counting activity*.

Continuous Data

Any number on an interval of numbers is possible data point.

Continuous data are always (but not exclusively) the result of a *measuring activity*.

Qualitative data are nonnumerical and result from activities that examine some opinion, characteristic, attribute, or other nonquantitative aspect of the objects.

Once the *categories* of characteristics, or attributes, and so on are identified, then the objects of the population can be separated into the specified categories. Membership of each object in a particular category is based solely on the object having the *quality* defining the category.

Qualitative data are sometimes assigned numbers, producing ordinal data. For example, objects in a population are sometimes assigned numbers on the basis of their desirability.

# Chapter 1 Review Exercises

*In Exercises 1 and 2, identify whether the articles are examples of descriptive or inferential statistics.*

**1.**

**By JOHN SPITLER**
SACRAMENTO UNION STAFF WRITER

A decline in the 1990 income for several key Yolo County crops was offset by increased plantings of other commodities, including cannery tomatoes, enabling the region to register a slight gain over 1989 gross agricultural income levels, reported the Yolo County Agricultural Department.

With water scarce and tomato prices high, Yolo County growers opted to increase plantings from 50,000 acres in 1989 to more than 59,00 acres in 1990, the county agricultural department reported.

**Figure 1-22.** A copy of part of an annual report by the Yolo County Agricultural Department. *The Sacramento Union,* May 17, 1991, p. B1.

*In Exercises 3 and 4, determine whether each population is well defined.*

**3.** The population of days last year that New Orleans had good weather.

**4.** The population of universities in the United States with annual enrollment of more than 10,000 students.

*In Exercises 5 and 6, identify a method whereby a simple random sample can be selected from each population.*

**5.** Gary Kersting manages a flock of 80,000 chickens on an egg ranch in Oklahoma. He wants to collect data on the number of eggs collected in one week for a sample of 50 chickens. Each hen is housed in a cage that is identified by a five-digit number on the cage.

**2.**

# So ... What's the Price of a New You?

While cosmetic surgery used to be a luxury only the rich and famous could afford, times have changed. Today, a new nose or chin won't cost you an arm and a leg.

Because costs vary from doctor to doctor and hospital to hospital. The following list is intended only as a general guide to cosmetic surgery price ranges.

* Facelift: from $2,000-$3,500 depending on the extent of surgery.
* Nose reshaping: from $2,000-$3,000.
* Smoothing sags and wrinkles around the eyes: from $1,000-$2,000.
* Sanding of facial scars/wrinkles: from $1,000-$2,000.
* Breast enlargement: from $2,000-$3,000.
* Breast uplift: from $2,000-$3,000.

**Figure 1-23.** Some costs of cosmetic surgery. *"So ... What's the Price of a New You?"* On Call, *Methodist Hospital of Sacramento, Summer, 1987.* Reprinted by permission.

**6.** Asian students comprise a large subset of the students enrolled in a large western community college. May Fong, a counselor at the school, wants to study the unit load and types of programs of these students. To perform the study, she wants to randomly sample 50 programs of only Asian students.

*In Exercises 7 and 8, identify possible categories into which the members of a population or sample under investigation could be placed.*

7. Alexis Turner is studying the air traffic that uses Chicago's O'Hare International Airport. The population being studied is the aircraft that arrive and leave the terminal. What are possible categories into which Alexis might put the aircraft to help in the study?

8. Hector Juarez is studying the sports program sponsored by the city recreation department in a large midwestern town. The population being studied is the activities such as softball, swimming, and soccer conducted by this department. What are possible categories into which Hector might put such activities to help in the study?

*In Exercises 9 and 10, for the objects identified:*
*a. Identify a feature of the object that can yield continuous quantitative data.*
*b. Identify a feature of the object that can yield discrete quantitative data.*

9. The population of 4-year colleges and universities in the United States.

10. The population of animals housed in the San Diego Zoo.

*Exercises 11 and 12 describe situations in which qualitative data of populations need to be evaluated.*
*a. Name a specific feature of each population that can be used to make the evaluation.*
*b. Identify categories that might be used to generate ordinal data for the evaluation.*

11. Valdimir Zukow is a judge at gymnastic events. His specialty is the balance beam event. Valdimir wants to quantify a contestant's "performance."

12. Heather Barnes is a movie critic for a local newspaper. Heather wants to quantify a movie's "presentation."

13. Use design (A) or (B) to diagram an experiment that will attempt to show that regular applications of an iron supplement solution will improve the condition of gardenia plants. Identify two possible confounding factors.

14. Use design (A*) or (B*) to diagram an experiment that will attempt to show that students in remedial reading and writing classes will learn better when the classes are taught as a combined class than when they are taught as two separate classes. Identify two possible confounding factors.

## Using Computers

**Key Topics**

1. *The worksheet*

2. *Communicating with Minitab*

3. *Placing data on the worksheet*

4. *Computer output*

Many excellent computer software programs have been written to help the statistician quickly and easily process raw data. Even if you do not have access to one of these, it is important that you know what a computer can do as well as how to read and interpret computer output.

Although computer programs differ, all have some features in common. The software called Minitab that has been chosen for inclusion in this text has features that are typical of most of the better programs.

### The worksheet

```
MTB> PRINT C1-C3

Row     C1    C2    C3
  1     15    20    13
  2     17    22    12
  3     21    19    14
  4     18    19    15
  5     30    21
  6     19
```

**Figure 1-24.**

Data are stored in Minitab in a "worksheet." Think of this worksheet as a sheet of paper ruled into rows and columns. The rows are designated by the natural numbers 1, 2, 3, . . . , and the columns by C1, C2, C3, . . . . Figure 1-24 shows a worksheet containing three columns of data.

### Communicating with Minitab

All good statistical computer programs involve a dialogue between the user (the human) and the computer. This dialogue consists of:

| | |
|---|---|
| Prompts | signals from the computer to the user requesting instructions or input of data |
| Commands | instructions from the user to the computer |
| Error messages | when a user error is made, most programs identify the error and provide an opportunity for it to be corrected |

Figure 1-25 is an example of a dialogue between the user and Minitab.

```
MTB> SET C1
DATA> 12 13 12 14 13
DATA> END
MTB>
```

**Figure 1-25.**

First Minitab gave the ready prompt MTB>. The user responded with SET C1 and pressed the ENTER key. This command tells Minitab that data is to be put into column 1. The Enter key was pressed to turn control of the computer back to Minitab. Minitab then gave the DATA> prompt after which the user typed in the data. When all the data was in, the user pressed the ENTER key again. Minitab responded with the DATA> prompt. If there had been more data to go in column 1, the user would have typed it in. In this case there was no more data, so the user typed END. The last ready prompt shows that Minitab is awaiting further instructions.

## Placing data on the worksheet

Data may be placed on the worksheet in three ways:

1. From the computer keyboard
2. Generated randomly by Minitab
3. Retrieved from previously stored files

Following is an example of the instructions that could be used to put 150 random integers between 0 and 100 in column 2 of the worksheet.

```
MTB> RANDOM 150 C2;
SUBC> INTEGER 0 100.
```

The command RANDOM causes Minitab to create a very long list of numbers that appear to be random. Then starting at an arbitrary point in this list, Minitab selects the specified number of numbers. The subcommand INTEGER tells Minitab to use an algebraic formula to convert each of the selected numbers into an integer between 0 and 100.

## Computer output

Minitab permits two ways of outputting data or the results of manipulating data. One of these is be means of the computer screen and other is by means of a printer connected to the computer.

To have the contents of the worksheet shown on the screen, the user gives the PRINT command followed by the numbers of the columns that he or she wishes to see. This command was used to obtain the display of Figure 1-26.

To obtain a printed copy of any Minitab output, the user gives the PAPER command before the command requesting the output. The following would cause the contents of columns 1 through 5 to be shown on the screen and at the same time printed by the printer:

```
MTB> PAPER
MTB> PRINT C1-C5
MTB> NOPAPER
```

The command NOPAPER must be given to disengage the printer after a printout.

```
MTB> PRINT C1-C3
ROW    C1    C2    C3
  1    15    20    13
  2    17    22    12
  3    21    19    14
  4    18    19    15
  5    30    21
  6    19
  7    17
```

**Figure 1-26.** A sample Minitab worksheet.

## Exercises

1. What values are in column 1 of the Minitab worksheet after the following?

```
MTB> SET C1
DATA> 320 415 216 127
DATA> 210 115
DATA> END
```

2. Only quantitative data can be put on the Minitab worksheet. Following is part of the data collected from a sample of students at a 4-year college. Suggest a way that the data for sex and class could be entered in a worksheet.

| Student | Sex | Height in Inches | Weight in Pounds | Class |
|---------|-----|------------------|------------------|-------|
| 1 | M | 72 | 165 | Jr. |
| 2 | F | 63 | 105 | Soph. |
| 3 | M | 70 | 159 | Sr. |

3. Tell how Minitab could be used to select the random sample required by Exercise 1 in Exercises 1-1 Set B.

4. The following Minitab printout shows a sample of data from a study of tooth retention. The variables are:

```
      TEETH - the number of teeth retained
      SEX - 1 = male 2 = female
      AGE - age to nearest year
MTB> PRINT C1-C3
ROW   TEETH   SEX   AGE
 1      26     1    30
 2      31     1    31
 3      30     2    32
 4      31     1    35
 5      18     1    29
 6      27     1    31
 7      30     2    45
 8      29     1    24
 9      31     1    40
10      24     2    43
```

a. Identify each of the variables as quantitative or qualitative.
b. Identify each of the variables as discrete or continuous.
c. How many males are included in the study?

## SCIENTIFIC

Throughout this text we have exercises that require sampling from Data Sets I and II in Appendix B. A scientific calculator with a random number key (labeled [RAN #]) can be used to obtain a list of numbers that identifies which objects to select for a given sample. On most calculators only two keys are needed to get a random number.

Before pressing the [RAN #] key, we need to press a functions key such as [2nd], [SHIFT], or [INVERSE]. On many calculators the random number appears as a three-digit decimal, such as 0.093, 0.852, and 0.690. On a Hewlett Packard calculator you may be able to fix more than three digits for a random number. For example, fixing six digits on the display and then using the random number key will yield numbers such as 0.401135, 0.537996, and 0.919182.

Since we need a four-digit random number to sample from both Data Sets I and II, the program in the following example can be used to get such a number from a calculator programmed to give only three-digit random numbers.

▶ **Example**    List five random numbers with four digits each

**Solution**

| | Press | Comments |
|---|---|---|
| Step 1 | [MODE] 7 4 | This fixes a four-decimal digit display. |
| Step 2 | [SHIFT] [RAN #] | Display shows a three-digit number as a decimal. |
| Step 3 | [+] | Display shows the random number with a 0 in the fourth place. |
| Step 4 | [SHIFT] [RAN #] | Display shows another three-digit random number. |
| Step 5 | [÷] 10 | This moves the decimal point in the second random number. |
| Step 6 | [=] | Display shows a random number as a four-place decimal. Use the four digits for the random number. |

Using these six steps yielded the following random numbers:

5009   2828   3471   9668   2391   ◀

**Note 1**   The above steps may require a rounding to take place that yields a number such as 1.0325. You may reject the number, or use 0325 as the random number.

**Note 2**   The divisor in step 5 can be modified to increase the number of digits in the random number to five or six. Divide by 100 to get five digits and by 1000 to get six digits.

# GRAPHING

The following steps can be used to program the Casio Graphics calculator to generate a list of four-digit random numbers:

| | Key Sequence | Comments |
|---|---|---|
| Step 1 | MODE 7 4 EXE | Fixes a four-decimal place number. |
| Step 2 | SHIFT RAN # | The display shows a three-digit random number. |
| | + SHIFT RAN # | The display shows another three-digit number. |
| | ÷ 10 EXE | The display shows a four-digit random number. |
| Step 3 | EXE | The display shows another four-digit random number. |

Continued pressing of EXE key will display new random numbers. The routine may yield a number such as 1.0035, which can be rejected, or the 0035 may be used as the random number.

The following steps can be used to program the TI 81 calculator to generate a list of four-digit random numbers:

| | Key Sequence | Comments |
|---|---|---|
| Step 1 | MODE | Displays MODE options. |
| Step 2 | ▼ | The word FLOAT is highlighted. |
| Step 3 | ► ► ► ► ► | Highlights 4 to show four-decimal digits. |
| Step 4 | ENTER | Fixes the display for four-decimal digits. |
| Step 5 | MATH | Displays math functions menu. |
| Step 6 | ► ► ► | Highlights PRB. |
| Step 7 | 1 | Selects the random number option. |
| Step 8 | ENTER | Displays a four-digit random number. |

Continued pressing of ENTER key will display a new random number.

Name: Judith C. Schafer
Occupation: Manager, Exploration and Production Statistics
Employer: American Petroleum Institute, Washington, DC

I'd like to be able to say that I became interested in statistics because math was always my favorite subject in school, math was something that I've always felt comfortable with, and because I've always known how useful math can be.

The truth is, I developed a classic case of "math anxiety" early on. As a result, math was my least favorite subject, and dealing with it made me very uncomfortable (to the point that I usually felt "clinically" stupid). I never really believed that I was going to need Algebra II/Trig or Analytical Geometry in the "real world."

It wasn't until I took an introductory statistics course in college that I realized how useful this stuff can be. Statistics, it seemed to me, was math with a purpose. It could be used to solve all kinds of practical problems across a wide variety of disciplines, ranging from psychology to engineering. In fact, I became interested enough in statistics that I forgot I was afraid of math, and I discovered that I actually had an aptitude for the subject.

At the American Petroleum Institute, where I manage a small group of analysts, we rely on statistics a lot. My staff and I provide oil and gas well drilling and environmental information on the U.S. petroleum industry to various groups including Congress, the news media, and consumers. This involves designing and conducting surveys to collect data, and developing methodologies using statistical analysis to estimate data.

Many of the techniques that we employ are relatively simple. For example, we typically begin a project by looking at scatter plots of the data, measures of central tendency (mean, range), and measures of variability (standard deviation). This gives us a great deal of specific information about the data, which we attempt to describe in a linear regression mode. Then, using $R$-squares and $p$-values, along with plots of the residuals, we examine and fine-tune the model to ensure that our assumptions about the data are being met and that the proper variables are being used.

This is how we estimate the number of U.S. wells being drilled, how deep they are, and how much the petroleum industry spent to drill them. These data are used by other organizations and petroleum industry analysts to conduct economic and financial assessments, develop forecasts, and examine trends.

We are also increasing our use of statistics to measure the industry's environmental performance and to keep track of its environmental progress. Some of the areas we study include air toxins and oil spills and how to prevent them.

If this sounds intimidating, take the word of a former math anxiety victim, it really isn't. Statistics are a powerful real-world tool *that anyone can use*. They help you do your job better and faster. And the subject isn't difficult to learn if you're willing to spend some time studying it.

My advice would be to try statistics. You just might like it a lot.

## Canada

*Official name:* Canada.
*Form of government:* federal multiparty parliamentary state with two legislative houses (Senate [104]; House of Commons [295]).
*Chief of state:* British Monarch represented by governor-general.
*Head of government:* Prime Minister.
*Capital:* Ottawa.
*Official languages:* English; French.
*Official religion:* none.
*Monetary unit:* 1 Canadian dollar (Can$) = 100 cents; valuation (Oct. 2, 1989) 1 U.S.$ = Can$1.18; 1 £ = Can$1.91.

### Area and population

| | | area | | population |
| --- | --- | --- | --- | --- |
| | | sq mi | sq km | 1989 estimate[1] |
| **Provinces** | **Capitals** | | | |
| Alberta | Edmonton | 248,800 | 644,390 | 2,423,200 |
| British Columbia | Victoria | 358,971 | 929,730 | 3,044,200 |
| Manitoba | Winnipeg | 211,723 | 548,360 | 1,083,300 |
| New Brunswick | Fredericton | 27,834 | 72,090 | 717,600 |
| Newfoundland | Saint John's | 143,510 | 371,690 | 569,200 |
| Nova Scotia | Halifax | 20,402 | 52,840 | 885,700 |
| Ontario | Toronto | 344,090 | 891,190 | 9,546,200 |
| Prince Edward Island | Charlottetown | 2,185 | 5,660 | 130,000 |
| Quebec | Quebec | 523,859 | 1,356,790 | 6,679,000 |
| Saskatchewan | Regina | 220,348 | 570,700 | 1,007,100 |
| | | | | 53,100 |
| | | | | 25,700 |
| **Territories** | | 1,271,442 | 3,293,020 | 26,164,200[2] |
| Northwest Territories | Yellowknife | 184,931 | 478,970 | |
| Yukon Territory | Whitehorse | 3,558,096 | 9,215,430 | |
| TOTAL LAND AREA | | 291,579 | 755,180 | |
| INLAND WATER | | 3,849,675 | 9,970,610 | |
| TOTAL AREA | | | | |

### Demography

*Population* (1989): 26,189,000.
*Density*[3] (1989): persons per sq mi 7.4, persons per sq km 2.8.
*Urban-rural* (1985): urban 75.9%; rural 24.1%.
*Sex distribution* (1988): male 49.32%; female 50.68%.
*Age breakdown*[4] (1986): under 15, 21.4%; 15–29, 25.8%; 30–44, 22.9%; 45–59, 14.9%; 60–74, 10.9%; 75 and over, 4.1%.
*Population projection:* (2000) 29,110,000; (2010) 32,047,000.
*Doubling time:* 99 years.
*Ethnic origin* (1986): British 34.4%; French 25.7%; German 3.6%; Italian 2.8%; Ukrainian 1.7%; Amerindian and Inuktitut (Eskimo) 1.5%; Chinese 1.4%; Dutch 1.4%; multiple origin and other 27.5%[5].
*Religious affiliation* (1981): Roman Catholic 46.5%; Protestant 41.2%; Eastern Orthodox 1.5%; Jewish 1.2%; Muslim 0.4%; Hindu 0.3%; Sikh 0.3%; nonreligious 7.4%; other 1.2%.
*Major metropolitan areas* (1986): Toronto 3,427,168; Montreal 2,921,357; Vancouver 1,380,729; Ottawa–Hull 819,263; Edmonton 785,465; Calgary 671,326; Winnipeg 625,304; Quebec 603,267; Hamilton 557,029; Saint Catharines–Niagara 343,258.

### Vital statistics

*Birth rate* per 1,000 population (1987): 14.4 (world avg. 27.1); (1985) legitimate 83.8%; illegitimate 16.2%.
*Death rate* per 1,000 population (1987): 7.2 (world avg. 9.9).
*Natural increase rate* per 1,000 population (1987): 7.2 (world avg. 17.2).
*Total fertility rate* (avg. births per childbearing woman; 1985): 1.7.
*Marriage rate* per 1,000 population (1986): 7.4.
*Divorce rate* per 1,000 population (1985): 2.4.
*Life expectancy at birth* (1983–85): male 72.9 years; female 79.8 years.
*Major causes of death* per 100,000 population (1986): diseases of the circulatory system 313.2; malignant neoplasms (cancers) 187.5; diseases of respiratory system 59.0; accidents and violence 54.3.

**Figure 2-1.** "Canada," *1990 Britannica World Data*, p. 583. Copyright © 1990 by Encyclopaedia Britannica, Inc. Reprinted by permission.

## How the Odds Vary On State Bar Exams

LAW GRADS can't go for the gold in salaries until they first pass the dreaded bar. And their odds vary enormously depending on the state where they've chosen to practice.

Last summer, the test was a cinch in Montana, which led the nation with a pass rate of 96.7%, according to the National Conference of Bar Examiners. In contrast, California failed 49.7% of those who took its test.

New York passed 65.4% of its 6,710 test takers. But the District of Columbia flunked more than half of those who sat for the grueling two-day exam.

Why the disparities? Each state writes the part of the exam that covers state law and sets its own pass rate for the part that's uniform nationally. So each state controls its own pass rate.

States with low pass rates maintain that they are just trying to keep standards high, not place artificial limits on the number of lawyers in their state.

California is already spilling over with 114,200 attorneys, close to 15% of all the lawyers in the U.S. Nonetheless, says James Tippin, executive director of the state's committee of bar examiners, "We're not in the business of controlling the number who pass or fail."

—JILL ABRAMSON, STEPHEN J. ADLER and LAURIE P. COHEN

**Figure 2-2.** "How the Odds Vary On State Bar Exams" by Jill Abramson, Stephen J. Adler, and Laurie P. Cohen, *The Wall Street Journal*, 7/29/88, p. 15. Copyright © 1988 Dow Jones & Company, Inc. Reprinted by permission of *The Wall Street Journal*. All Rights Reserved Worldwide.

# 2

# Data
# Summarization

Statistical reports are frequently made based on large quantities of data.
Figure 2-1 contains some of the information reported in the 1990
Britannica Book of the Year on Canada. Figure 2-2 is a report on the
success of lawyer candidates who took a bar examination in several
states. Both articles contain numbers that describe in some way the
original data. For example, in the "Vital statistics" part of Figure 2-1, we
read that the Canadian birth rate per 1000 population (1987) was 14.4.
(A rate is a descriptive measure.) In Figure 2-2, we read that 96.7% was
the pass rate for the lawyer candidates in Montana. (Pass rate is a
descriptive measure.) In this chapter, we will study some of the measures
that are used to describe sets of data.

## 2-1  Tables and Graphs Displaying Quantitative Data

**Key Topics**

*1. Frequency tables*

*2. Histograms*

*3. Frequency polygons*

*4. Stem-and-leaf diagrams*

S T A T I S T I C S   I N   A C T I O N

Maria Rivera is a member of the quality control staff at a large corporation that manufactures electric appliances and accessories. She recently participated in the development of a unit designed to measure the burn-time of incan-

| | | | | | | | | | |
|---|---|---|---|---|---|---|---|---|---|
| 15.3 | 6.9 | 16.1 | 13.5 | 18.2 | 15.0 | 9.8 | 12.4 | 10.9 | 15.6 |
| 10.3 | 15.3 | 12.7 | 16.8 | 8.5 | 14.1 | 6.8 | 16.3 | 14.2 | 19.7 |
| 13.4 | 17.8 | 11.9 | 18.3 | 12.0 | 16.6 | 11.7 | 15.2 | 15.8 | 10.4 |
| 15.6 | 14.9 | 7.7 | 15.9 | 14.6 | 10.0 | 16.9 | 14.2 | 6.2 | 17.3 |

**Figure 2-3.**  Burn-times in seconds for the light bulb test.

descent light bulbs. (The burn-time of a bulb is the number of hours the bulb remains lighted under ordinary operating conditions.) The test unit subjected a bulb to a surge of power that accelerated the life of the bulb. As a consequence, the useful life of a bulb in hours could be predicted by the test lasting only a few seconds.

Yesterday Maria sampled 40 bulbs from a large production run of 100 watt bulbs. She subjected each bulb to the test and recorded the test times for each bulb. The times, measured in seconds, for the 40 bulbs are listed in Figure 2-3.

To organize the 40 data points, Maria put them in Table 2-1. For a geometric representation of the data, she constructed the histogram in Figure 2-4 and the frequency polygon in Figure 2-5.

**TABLE 2-1   A Frequency Table for the Burn-Times Data**

| Class Number | Class Limits | Class Boundaries | Class Mark | Tally Marks | Frequency | Relative Frequency |
|---|---|---|---|---|---|---|
| 1 | 6.0–8.9 | 5.95–8.95 | 7.45 | ⊞ | 5 | $\frac{5}{40}$ or 12.5% |
| 2 | 9.0–11.9 | 8.95–11.95 | 10.45 | ⊞ \|\| | 7 | $\frac{7}{40}$ or 17.5% |
| 3 | 12.0–14.9 | 11.95–14.95 | 13.45 | ⊞ ⊞ | 10 | $\frac{10}{40}$ or 25% |
| 4 | 15.0–17.9 | 14.95–17.95 | 16.45 | ⊞ ⊞ ⊞ | 15 | $\frac{15}{40}$ or 37.5% |
| 5 | 18.0–20.9 | 17.95–20.95 | 19.45 | \|\|\| | 3 | $\frac{3}{40}$ or 7.5% |

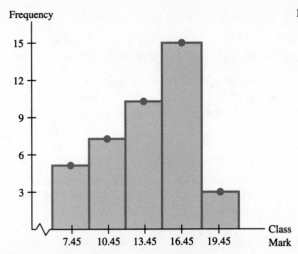

**Figure 2-4.** A histogram of the burn-time data.

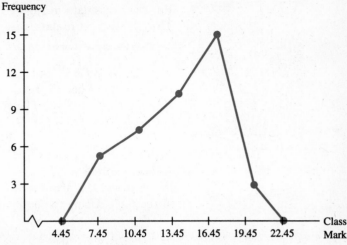

**Figure 2-5.** A frequency polygon of the tabled data.

## Frequency tables

The first step in organizing a set of data in a frequency table is deciding on a set of classes. Although the choice of classes is somewhat arbitrary, we recommend the following guidelines.

### Determining a set of classes for a frequency table

Guideline 1   The range of possible values in each class should be the same.
This recommendation gives each class the *same width*. If a uniform width is used for the classes in a frequency table, then the corresponding widths of the bars on a corresponding histogram are also the same.

Guideline 2   The number of classes appropriate is somewhat dependent on the number of data points.
This guideline states that the number of classes may increase as the number of data points increases. However, it is rare that fewer than five or more than 12 classes are used.

### DEFINITION 2-1 The Range of a Set of Data

Range = largest value − smallest value

The symbol $R$ is frequently used for the range.

For the times data in Figure 2-3, the largest value is 19.7 and the smallest is 6.2. Thus for this set of data:

$$R = 19.7 - 6.2 = 13.5 \qquad R = \text{largest value} - \text{smallest value}$$

Because 40 is a relatively small number of data points, Maria used five classes in Table 2-1. When the range is divided by the number of classes, an approximate class width is obtained. (The class width is formally defined in Definition 2-2.)

$$\text{Class width} \approx \frac{13.5}{5} = 2.7$$

A class width of at least 2.7 would be appropriate for these data. In Definition 2-2, the class width is identified as the difference between the boundaries of each class in the table. Using the boundaries of class 1:

$$\text{Class width} = 8.95 - 5.95 = 3.0$$

Thus Maria used a class width that is slightly larger than the minimum required. As a consequence, she was able to use 6.0 as the lower limit of class 1, a value that is 0.2 less than the smallest data point in the set. The product of the class width and the number of classes must be greater than the range of the data. For the burn-times data:

$$(3.0)(5) = 15.0 \quad \text{and} \quad 15.0 > 13.5$$

Items (a) through (g) identify the seven headings in the frequency table.

a. Class number — The numerical label assigned to a given class

b. Class limits — The smallest and largest values that would be put in a given class

  The limits are possible values for data points. For this reason, no two limits can have the same value.

c. Class boundary — A value midway between an upper limit of one class and a lower limit of the next class

  A boundary can be found by adding the limits that determine the boundary and dividing the sum by two. Thus the boundary between classes 1 and 2 is:

$$\text{Boundary} = \frac{8.9 + 9.0}{2} = 8.95$$

d. Class mark — The value exactly in the middle of a class

  A class mark can be found by adding the limits or boundaries of a class and dividing the sum by 2. Using the limits of class 1 to illustrate:

$$\text{Class mark} = \frac{6.0 + 8.9}{2} = 7.45$$

e. Tally mark      A mark for each data point in a class

       By using tally marks, it is possible to go through a set of data once and put each data point in its appropriate class.

f. Frequency      The number of data points in a class

       The symbol $f_i$ is used to represent the frequency of the $i$th class. Using class 2 to illustrate:

$$f_2 = 7$$

g. Relative frequency      The proportion of the data points in a class

       Using class 2 to illustrate:

$$\text{Relative frequency} = \frac{f_2}{n} = \frac{7}{40} \quad \frac{\text{The number of data points in class 2}}{\text{The total number of data points}}$$

$$= 0.175 \quad \text{Written as a decimal}$$

$$= 17.5\% \quad \text{Written as a percentage}$$

## DEFINITION 2-2 Class Width

Class width = upper boundary of a class − lower boundary of the same class

The symbol $w$ is frequently used for the class width.

Using Table 2-1:

| Class Number | Upper Boundary | Lower Boundary | Class Width | |
|---|---|---|---|---|
| 1 | $w = 8.95 -$ | $5.95 =$ | 3.0 | The width of each class is 3.0; thus there is uniform width. |
| 2 | $w = 11.95 -$ | $8.95 =$ | 3.0 | |
| 3 | $w = 14.95 -$ | $11.95 =$ | 3.0 | |
| 4 | $w = 17.95 -$ | $14.95 =$ | 3.0 | |
| 5 | $w = 20.95 -$ | $17.95 =$ | 3.0 | |

A set of data can usually be grouped in frequency tables in a variety of ways. In this text, a recommended number of classes, class width, and lower limit of class 1 will be specified for every example and exercise. The corresponding answers to these examples and exercises will be based on the given recommendations.

▶ **Example 1**    Susan Hendsley teaches Economics 2B at Skyline Community College. She currently has $N = 60$ students in the fall semester class. In the last class session, Susan gave a ten-point quiz on the previous homework assignment. The results are shown below.

$$
\begin{array}{cccccccccccc}
7 & 5 & 4 & 8 & 7 & 4 & 6 & 9 & 1 & 7 & 8 & 6 \\
6 & 8 & 7 & 9 & 2 & 6 & 7 & 4 & 5 & 9 & 4 & 8 \\
9 & 4 & 8 & 5 & 7 & 10 & 5 & 2 & 7 & 7 & 1 & 5 \\
4 & 10 & 6 & 4 & 8 & 8 & 7 & 1 & 6 & 2 & 9 & 4 \\
8 & 6 & 4 & 9 & 5 & 8 & 9 & 7 & 5 & 7 & 6 & 7
\end{array}
$$

Put these data in a frequency table using five classes, a class width of 2, and a lower limit of 1 for class 1.

Solution    **Discussion.** The data are whole numbers. Therefore a class width of 2 means that 2 different possible data points can go in each class. For example:

> Data points 1 and 2 go in class 1
>
> Data points 3 and 4 go in class 2

and so on.

The boundary between classes 1 and 2 is $\dfrac{2 + 3}{2} = \dfrac{5}{2} = 2.5$. The other class boundaries can be similarly calculated. The class mark of class 1 is $\dfrac{1 + 2}{2} = \dfrac{3}{2} = 1.5$. The other class marks can be similarly calculated. The results are shown in Table 2-2.    ◀

**TABLE 2-2    A Frequency Table for A Quiz Data**

| Class Number | Class Limits | Class Boundaries | Class Mark | Tally Marks | Frequency ($f_i$) | Relative Frequency ($f_i/n$) |
|---|---|---|---|---|---|---|
| 1 | 1–2 | 0.5–2.5 | 1.5 | ﾱﾱ I | 6 | $\frac{6}{60}$ or 10% |
| 2 | 3–4 | 2.5–4.5 | 3.5 | ﾱﾱ IIII | 9 | $\frac{9}{60}$ or 15% |
| 3 | 5–6 | 4.5–6.5 | 5.5 | ﾱﾱ ﾱﾱ ﾱﾱ | 15 | $\frac{15}{60}$ or 25% |
| 4 | 7–8 | 6.5–8.5 | 7.5 | ﾱﾱ ﾱﾱ ﾱﾱ ﾱﾱ I | 21 | $\frac{21}{60}$ or 35% |
| 5 | 9–10 | 8.5–10.5 | 9.5 | ﾱﾱ IIII | 9 | $\frac{9}{60}$ or 15% |

▶ **Example 2**    The supervisor of a state fish hatchery recently sampled 45 trout from a holding tank and measured the length of each fish to the nearest tenth of a centimeter. The data are listed on the next page.

| 40.8 | 40.4 | 40.5 | 40.9 | 40.7 | 40.0 | 40.4 | 41.2 | 41.0 |
| 40.6 | 40.9 | 40.6 | 40.7 | 41.2 | 40.6 | 40.7 | 40.4 | 40.9 |
| 40.8 | 40.4 | 40.7 | 41.2 | 40.5 | 41.3 | 40.0 | 40.0 | 40.7 |
| 40.7 | 40.6 | 40.2 | 41.0 | 41.0 | 40.5 | 40.8 | 41.3 | 40.8 |
| 41.0 | 40.5 | 41.0 | 40.5 | 41.2 | 40.9 | 41.0 | 40.8 | 40.6 |

Put these data in a frequency table using five classes, a class width of 0.3, and a lower limit of 40.0 for class 1.

**Solution**   **Discussion.** The data are numbers to one decimal place. Therefore a class width of 0.3 means that three different data points can go in each class. For example, data points 40.0, 40.1, and 40.2 go in class 1. The boundary between classes 1 and 2 is:

$$\frac{40.2 + 40.3}{2} = 40.25$$

The class mark of class 1 is:

$$\frac{40.0 + 40.2}{2} = \frac{80.2}{2} = 40.1$$

The result is Table 2-3. Notice that the relative frequencies were rounded to the nearest percent.   ◄

TABLE 2-3   A Frequency Table for the Fish Data

| Class Number | Class Limits | Class Boundaries | Class Mark | Tally Marks | Frequency $(f_i)$ | Relative Frequency $(f_i/n)$ |
|---|---|---|---|---|---|---|
| 1 | 40.0–40.2 | 39.95–40.25 | 40.1 | \|\|\|\| | 4 | 9% |
| 2 | 40.3–40.5 | 40.25–40.55 | 40.4 | ⊞⊞ \|\|\|\| | 9 | 20% |
| 3 | 40.6–40.8 | 40.55–40.85 | 40.7 | ⊞⊞ ⊞⊞ ⊞⊞ \| | 16 | 36% |
| 4 | 40.9–41.1 | 40.85–41.15 | 41.0 | ⊞⊞ ⊞⊞ | 10 | 22% |
| 5 | 41.2–41.4 | 41.15–41.45 | 41.3 | ⊞⊞ \| | 6 | 13% |

## Histograms

A next step in summarizing quantitative data might be to make a graphic representation of the data. There are several types of graphs that can be used such as bar graphs, line graphs, and circle graphs.

Two types of graphs that will be studied at this time are *histograms* and *frequency polygons*. These provide a sense of the "shape" of a distribution of data points. The shape of a distribution reflects the way data points in a set may be grouped more densely around certain values in the set than around other values. For example, in some distributions, the data points appear to bunch toward either the low end or high end of the set. In other distributions, the data points appear to cluster around a central value. Such tendencies can be seen in histograms and frequency polygons of the distributions.

---

### Steps for constructing a histogram

**Step 1** On a horizontal axis draw a tick mark for each class in the frequency table.

    To show the uniform width of each class, make the distance between the marks the same and label them with the class marks.

**Step 2** Draw a vertical axis at the left end of the horizontal one.

    Mark off an appropriate scale on the vertical axis by using the greatest frequency in the table as a guideline.

**Step 3** Construct a bar centered on each class mark to a height that shows the frequency of data points in that class.

    The edges of each bar meet the horizontal axis at the boundary values of that class.

**Step 4** Label the horizontal axis "Class Marks" and the vertical axis "Frequency."

---

To illustrate these four steps, a histogram of the fish data in Table 2-3 is shown in Figure 2-6.

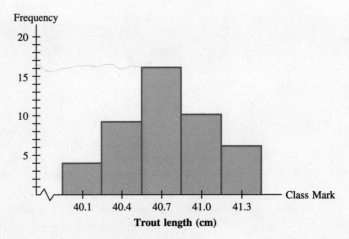

**Figure 2-6.**   A histogram of the fish data.

**Steps for constructing a histogram of fish data**

**Step 1**  The five tick marks on the horizontal axis are evenly spaced and labeled with the class marks. The broken line at the left end shows that a part of the axis beginning at 0 has been deleted.

**Step 2**  The vertical axis has 20 tick marks, more than the greatest frequency of 16 in class 3.

**Step 3**  Five bars are centered on the five class marks, and the height of each bar shows the frequency of data points in that class.

**Step 4**  The horizontal axis is labeled "Class Marks," and the vertical axis is labeled "Frequency."

Notice how the histogram reflects the shape of the distribution of fish data. The middle bar is the tallest one, which shows that the fish have lengths clustering around some central value. The heights of the bars get shorter on either side of the middle one. We say that such a distribution is "bell-shaped." Other commonly occurring shapes of histograms are identified and labeled in Figure 2-7.

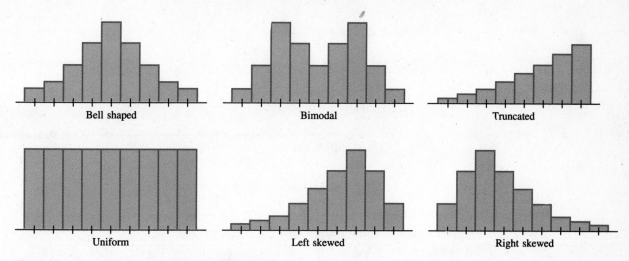

**Figure 2-7.**  Six commonly occurring shapes of histograms.

## Frequency polygons

The shape of a distribution of data points can also be seen in a frequency polygon. Most of the steps used to construct a histogram can also be used to construct a frequency polygon.

## Steps for constructing a frequency polygon

**Step 1**   Draw horizontal and vertical axes similar to the ones used for a histogram.

**Step 2**   Label both axes as if a histogram was being made. Include class marks for classes immediately to the left of the first class and to the right of the last class in the frequency table.

   The additional class marks belong to classes with zero frequencies, but their class marks are needed to construct the polygon.

**Step 3**   Plot a point directly above each class mark.

   The height of the point above the horizontal axis corresponds to the frequency of that class.

**Step 4**   Connect from left to right the points plotted in step 3 with line segments.

Using these four steps, a frequency polygon for the fish data in Table 2-3 is shown in Figure 2-8. Notice that the class marks of the classes with zero frequencies are 39.8 and 41.6.

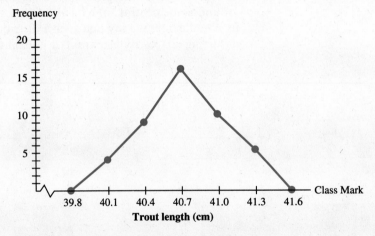

**Figure 2-8.**   A frequency polygon of the fish data.

▶ **Example 3**   The weights at birth of 75 babies at a large metropolitan hospital in the East were measured to the nearest one-tenth of a pound.

| 7.6 | 8.5 | 5.7 | 8.0 | 10.2 | 6.5 | 7.0 | 7.7 | 5.0 | 6.8 |
|-----|-----|-----|-----|------|-----|-----|-----|-----|-----|
| 6.2 | 6.9 | 8.7 | 5.9 | 7.5 | 10.8 | 6.0 | 9.1 | 7.3 | 9.8 |
| 8.0 | 6.6 | 10.0 | 4.5 | 7.3 | 5.9 | 8.5 | 6.9 | 9.0 | 6.1 |
| 5.6 | 8.2 | 6.3 | 7.2 | 9.2 | 6.3 | 7.5 | 7.9 | 5.5 | 5.9 |
| 9.7 | 6.5 | 5.8 | 11.4 | 6.9 | 9.5 | 6.8 | 7.4 | 9.8 | 7.8 |
| 7.6 | 4.8 | 8.8 | 6.3 | 8.3 | 7.3 | 10.6 | 5.4 | 7.2 | 6.7 |
| 6.6 | 11.0 | 7.5 | 9.2 | 5.9 | 9.8 | 7.9 | 8.6 | 6.8 | 7.3 |
| 9.0 | 7.2 | 10.2 | 7.3 | 8.0 | | | | | |

**a.** Make a frequency table for the data. Use a class width of 1.2 pounds, a lower limit of 4.5 pounds for class 1, and six classes.

**b.** Construct a histogram.

**c.** Construct a frequency polygon.

**d.** What general shape best describes the data?

**Solution**   **a.** For class 1:

Lower limit is 4.5 (given)

Lower boundary is 4.45 (= 4.50 − 0.05)

Upper boundary is 5.65 (= 4.45 + 1.2)

Upper limit is 5.6 (= 5.65 − 0.05)

A completed table is given in Table 2-4.

**TABLE 2-4   A Frequency Table for the Weights at Birth of 75 Babies**

| Class Number | Class Limits | Class Boundaries | Class Mark | Tally Marks | Frequency ($f_i$) | Relative Frequency ($f_i/n$) |
|---|---|---|---|---|---|---|
| 1 | 4.5–5.6 | 4.45–5.65 | 5.05 | ⊬⊬⊬ I | 6 | 8% |
| 2 | 5.7–6.8 | 5.65–6.85 | 6.25 | ⊬⊬⊬ ⊬⊬⊬ ⊬⊬⊬ ⊬⊬⊬ | 20 | 27% |
| 3 | 6.9–8.0 | 6.85–8.05 | 7.45 | ⊬⊬⊬ ⊬⊬⊬ ⊬⊬⊬ ⊬⊬⊬ ⊬⊬⊬ | 25 | 33% |
| 4 | 8.1–9.2 | 8.05–9.25 | 8.6 | ⊬⊬⊬ ⊬⊬⊬ II | 12 | 16% |
| 5 | 9.3–10.4 | 9.25–10.45 | 9.85 | ⊬⊬⊬ III | 8 | 11% |
| 6 | 10.5–11.6 | 10.45–11.65 | 11.05 | IIII | 4 | 5% |

**b.** A histogram for the grouped data is shown in Figure 2-9.

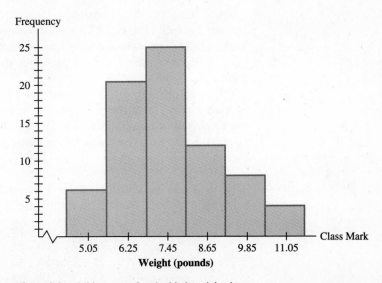

**Figure 2-9.**   A histogram for the birth weight data.

**c.** A frequency polygon is given in Figure 2-10.

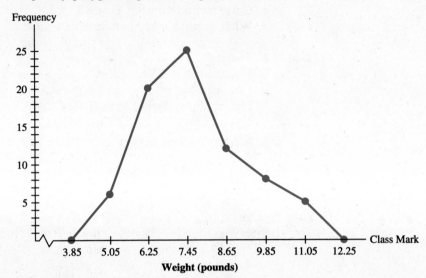

**Figure 2-10.**    A frequency polygon for the birth weight data.

**d.** The data are bunched toward the left end of the distribution and tail off to the right. Thus the data are said to be *skewed right*.    ◄

## Stem-and-leaf diagrams

A set of quantitative data can also be summarized using a stem-and-leaf diagram. In such a display, the "stem" is usually drawn vertically and the "leaves" are drawn horizontally. The stem is formed by drawing a line and writing only the leading digit or digits of the data to its left. Then data points with the same leading digits form part of the same leaf.

If three or more digits are used, then more than just the first digit of the data may be used to form the stem. In a stem-and-leaf diagram, individual data points are displayed. As such, the exact data points can be recovered from the diagram. In such a diagram, the shape of the data set can be seen from the lengths of the leaves.

► **Example 4**   A sixth grade class at Lancaster Elementary School was given a standardized arithmetic test last week. The results are listed below:

| | | | | | | | | |
|---|---|---|---|---|---|---|---|---|
| 30 | 40 | 27 | 52 | 59 | 35 | 45 | 63 | 46 |
| 53 | 41 | 22 | 60 | 58 | 32 | 67 | 64 | 48 |
| 15 | 56 | 29 | 58 | 40 | 38 | 56 | 47 | 43 |
| 67 | 52 | 38 | 19 | 59 | 36 | 61 | 55 | 49 |

Construct a stem-and-leaf diagram of these data.

**Solution**    **Discussion.** The smallest and largest data points are 15 and 67, respectively. Each data point must therefore have a 1, 2, 3, 4, 5, or 6 as the tens digit. These digits will be used to form the stem of the diagram. On a horizontal line with the 1 will be the units digits of all data points with 1 as the tens digit. Similarly, on a horizontal line with the 2 will be the units digits of all data points with 2 as the tens digit. The process continues for 3, 4, 5, and 6. The result is the diagram in Figure 2-11.

The leaf next to the 5 is the longest, and the leaves adjacent to it are longer than the leaves at the top of the stem. The effect gives a left skewed shape to the data set.  ◄

| The stem is the tens digit in the data points. | 1 | 5 9 | | | | | | | | The leaves are only the units digits in the data points. |
|---|---|---|---|---|---|---|---|---|---|---|
| | 2 | 7 2 9 | | | | | | | |
| | 3 | 0 5 2 8 8 6 | | | | | | | |
| | 4 | 0 5 6 1 8 0 7 3 9 | | | | | | | |
| | 5 | 2 9 3 8 6 8 6 2 9 5 | | | | | | | |
| | 6 | 3 0 7 4 7 1 | | | | | | | |

**Figure 2-11.**  A stem-and-leaf diagram of the 36 test scores.

► **Example 5**    The records of the families that live in Valley High condominiums contain an estimated annual income of each family. The data are listed below in thousands of dollars.

| | | | | | | | |
|---|---|---|---|---|---|---|---|
| $30.9 | $23.2 | $21.8 | $17.8 | $28.5 | $35.5 | $42.5 | $29.3 |
| 35.8 | 32.9 | 25.0 | 31.5 | 34.4 | 45.0 | 24.8 | 32.0 |
| 23.9 | 28.4 | 36.5 | 30.6 | 19.6 | 41.9 | 33.8 | 37.7 |
| 39.8 | 15.9 | 26.5 | 38.2 | 48.5 | 38.0 | 29.7 | 44.2 |
| 27.6 | 39.4 | 31.8 | 49.2 | 33.5 | 43.0 | 22.5 | 34.7 |

Construct a stem-and-leaf diagram for these data.

**Solution**    **Discussion.** The smallest and largest data points are $15.9 and $49.2, respectively. Each data point must therefore have 1, 2, 3, or 4 as the tens digit. Using these digits to label the stem would yield only four leaves in the display. However, if both the tens and units digits were used to label the stem, too many leaves would be required. By using the tens digits and a range of values for the units digits, a meaningful display can be obtained.

Using the possible data points from $20.0 through $29.9 to illustrate:

For $20.0–24.9, label the stem (20–24) 2       The leaves will be listed using
For $25.0–29.9, label the stem (25–29) 2       the units and tenths digits.

The result of this procedure is shown in Figure 2-12.

Notice that the data set is bell shaped.  ◄

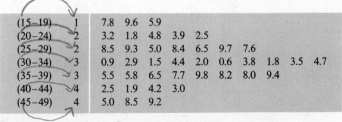

| (15–19) | 1 | 7.8 9.6 5.9 |
|---|---|---|
| (20–24) | 2 | 3.2 1.8 4.8 3.9 2.5 |
| (25–29) | 2 | 8.5 9.3 5.0 8.4 6.5 9.7 7.6 |
| (30–34) | 3 | 0.9 2.9 1.5 4.4 2.0 0.6 3.8 1.8 3.5 4.7 |
| (35–39) | 3 | 5.5 5.8 6.5 7.7 9.8 8.2 8.0 9.4 |
| (40–44) | 4 | 2.5 1.9 4.2 3.0 |
| (45–49) | 4 | 5.0 8.5 9.2 |

**Figure 2-12.**  A stem-and-leaf diagram of the annual incomes.

# Exercises 2-1   Set A

*In Exercises 1–8:*

a. *Put the data in a frequency table. Use the recommended number of classes, class width, and lower limit for class 1.*

b. *Construct a frequency histogram.*

c. *Construct a frequency polygon.*

d. *Name the general shape that best describes the data.*

1. The following data are the scores of a quiz in Accounting IB.

   | 10 | 7 | 8 | 4 | 5 | 6 | 6 | 9 | 1 | 7 | 9 | 5 | 6 | 3 | 8 |
   |----|---|---|---|---|---|---|---|---|---|---|---|---|---|---|
   | 4  | 6 | 10| 5 | 9 | 7 | 6 | 2 | 6 | 5 | 4 | 8 | 7 | 5 | 6 |

   Use five classes, a class width of 2, and a lower limit of 1 for class 1.

2. Thirty-six cans of dog food were randomly sampled from a large production run. The contents of each can were weighed to the nearest gram.

   | 410 | 409 | 400 | 403 | 412 | 410 | 413 | 406 | 409 |
   |-----|-----|-----|-----|-----|-----|-----|-----|-----|
   | 412 | 406 | 413 | 404 | 409 | 408 | 404 | 407 | 412 |
   | 413 | 401 | 400 | 413 | 406 | 402 | 411 | 410 | 410 |
   | 413 | 410 | 410 | 410 | 411 | 412 | 409 | 410 | 413 |

   Use five classes, a class width of 3 grams, and a lower limit of 400 for class 1.

3. Elizabeth Turner is a waitress at the Old Ironsides Club. She recently polled 40 patrons about their ages for a project she was doing in her statistics class at college. She obtained the following data, where the ages were rounded to the nearest one-tenth of a year:

   | 23.5 | 28.5 | 30.6 | 24.5 | 19.3 | 25.1 | 31.8 | 22.6 |
   |------|------|------|------|------|------|------|------|
   | 32.4 | 19.8 | 34.5 | 25.7 | 32.6 | 27.2 | 24.4 | 29.4 |
   | 26.7 | 38.5 | 23.8 | 30.3 | 21.0 | 37.0 | 27.5 | 30.8 |
   | 36.3 | 24.3 | 33.6 | 27.3 | 26.4 | 20.2 | 34.7 | 26.0 |
   | 21.5 | 38.2 | 28.8 | 24.9 | 34.2 | 26.9 | 29.9 | 23.2 |

   Use five classes, a class width of 4.0 years, and a lower limit of 19.0 for class 1.

4. Michelle Jasper is an inspector for the Timken Corporation. She is the final inspector of roller bearings. She recently sampled 40 bearings from a large production run and weighed each bearing to the nearest one-tenth of a gram.

   | 40.9 | 40.6 | 41.3 | 41.0 | 41.2 | 40.3 | 40.0 | 40.9 | 41.0 | 40.4 |
   |------|------|------|------|------|------|------|------|------|------|
   | 41.2 | 40.7 | 40.4 | 40.8 | 40.9 | 40.4 | 41.3 | 40.6 | 41.2 | 40.7 |
   | 41.0 | 41.0 | 41.1 | 40.2 | 40.6 | 41.3 | 40.0 | 40.1 | 41.3 | 40.5 |
   | 41.3 | 41.0 | 40.9 | 41.2 | 41.1 | 41.0 | 41.0 | 41.0 | 41.3 | 40.8 |

   Use five classes, a class width of 0.3 grams, and a lower limit of 40.0 for class 1.

6.20-8.89   11.60-14.29
8.90-11.59   14.30-16.99

5. Aaron Ward needed a set of brake shoes for his car. He therefore contacted 36 automotive stores in the area and got the following prices (in dollars and cents) for the same brand of brake shoes:

| | | | | | | | |
|---|---|---|---|---|---|---|---|
| 6.39 | 12.50 | 17.98 | 11.50 | 18.11 | 9.95 | 18.11 | 13.90 |
| 16.00 | 11.95 | 19.46 | 12.57 | 15.50 | 13.27 | 17.60 | 12.50 |
| 18.19 | 15.90 | 14.76 | 16.50 | 14.44 | 15.00 | 14.50 | 18.00 |
| 14.30 | 13.00 | 13.17 | 11.50 | 12.60 | 17.00 | 13.60 | 8.50 |
| 9.48 | 16.50 | 15.48 | 15.95 | | | | |

Use five classes, a class width of $2.70, and a lower limit of $6.20 for class 1.

6. Shirley Chan works in a retail chain store that specializes in athletic shoes. On a given Saturday, she recorded the prices (in dollars and cents) of 100 pairs of shoes that were sold.

| | | | | | | | | | |
|---|---|---|---|---|---|---|---|---|---|
| 39.99 | 20.00 | 27.99 | 36.99 | 14.99 | 44.99 | 21.99 | 39.99 | 20.00 | 29.99 |
| 36.99 | 47.99 | 19.99 | 29.99 | 57.99 | 32.99 | 47.99 | 57.99 | 16.99 | 34.99 |
| 16.99 | 29.99 | 39.99 | 29.99 | 41.99 | 17.99 | 27.99 | 49.99 | 29.99 | 42.99 |
| 29.99 | 30.00 | 34.99 | 12.99 | 39.99 | 21.99 | 27.99 | 17.99 | 49.99 | 33.99 |
| 47.99 | 19.99 | 39.99 | 27.99 | 20.00 | 47.99 | 33.99 | 20.00 | 29.99 | 46.99 |
| 44.99 | 30.00 | 32.99 | 57.99 | 39.99 | 23.99 | 14.99 | 27.99 | 39.99 | 20.00 |
| 24.99 | 39.99 | 21.99 | 41.99 | 15.99 | 34.99 | 49.99 | 21.99 | 49.99 | 34.99 |
| 39.99 | 12.99 | 34.99 | 47.99 | 32.99 | 21.99 | 29.99 | 29.99 | 41.99 | 59.99 |
| 27.99 | 36.99 | 42.99 | 16.99 | 29.99 | 59.99 | 37.99 | 44.99 | 29.99 | 30.00 |
| 19.99 | 32.99 | 41.99 | 27.99 | 49.99 | 34.99 | 17.99 | 39.99 | 15.99 | 21.99 |

6.20
2.70
8.90

Use seven classes, a class width of $7.00, and a lower limit of $12.00 for class 1.

7. A flock of chickens raised for sale through supermarkets was sampled, and the weights of 100 chickens were measured to the nearest one-tenth of a pound.

| | | | | | | | | | |
|---|---|---|---|---|---|---|---|---|---|
| 4.2 | 5.6 | 3.7 | 5.2 | 5.6 | 4.4 | 4.7 | 3.8 | 5.0 | 4.8 |
| 3.4 | 4.3 | 4.1 | 4.8 | 4.8 | 4.0 | 4.7 | 4.2 | 3.0 | 3.8 |
| 4.4 | 5.1 | 4.5 | 3.9 | 4.9 | 4.3 | 4.8 | 4.5 | 3.6 | 3.5 |
| 4.7 | 4.9 | 4.5 | 4.2 | 4.2 | 4.8 | 5.7 | 4.1 | 4.6 | 4.9 |
| 4.7 | 4.4 | 4.7 | 4.8 | 3.2 | 5.0 | 4.6 | 4.7 | 4.0 | 3.8 |
| 4.3 | 3.7 | 4.5 | 4.0 | 3.5 | 5.0 | 4.9 | 3.9 | 5.2 | 4.3 |
| 5.2 | 4.7 | 4.1 | 4.8 | 5.2 | 4.6 | 4.3 | 3.0 | 5.6 | 3.7 |
| 4.6 | 4.1 | 5.0 | 3.9 | 4.4 | 4.8 | 4.4 | 4.9 | 3.9 | 4.8 |
| 4.5 | 3.2 | 5.6 | 4.5 | 4.3 | 4.7 | 3.7 | 4.2 | 4.6 | 4.2 |
| 4.8 | 3.6 | 4.4 | 4.0 | 3.4 | 4.6 | 3.8 | 5.1 | 4.9 | 5.7 |

Use seven classes, a class width of 0.4 pounds, and a lower limit of 3.0 pounds for class 1.

8. A resident of an eastern state intends to enroll in a university in the West. To prepare for her move, she sampled the newspapers for the cost of renting an unfurnished studio apartment in that area. The data she gathered on the dollar cost of 1 month's rent are listed on the following page.

$$
\begin{array}{cccccccccc}
475 & 325 & 575 & 500 & 350 & 495 & 445 & 540 & 375 & 450 \\
525 & 450 & 424 & 650 & 515 & 550 & 325 & 650 & 500 & 425 \\
425 & 395 & 475 & 275 & 425 & 600 & 565 & 450 & 525 & 375 \\
325 & 550 & 470 & 525 & 415 & 530 & 495 & 395 & 725 & 475 \\
500 & 675 & 375 & 435 & 550 & 350 & 750 & 525 & 600 & 540 \\
425 & 525 & 550 & 475 & 450 & 565 & 500 & 600 & 410 & 435 \\
600 & 285 & 515 & 425 & 675 & 495 & 420 & 550 & 470 & 500 \\
400 & 535 & 450 & 375 & 650 & 445 & 525 & 300 & 550 & 425 \\
\end{array}
$$

Use six classes, a class width of $80, and a lower limit of $275 for class 1.

*In Exercises 9–14:*

*a. Construct a stem-and-leaf diagram of the data. Use the hints given for the stem of the diagram.*
*b. What general shape best describes the data set?*

9. Fifty community college students who were enrolled in 12 or more units and were also working a regular job were recently surveyed. The following data are the numbers of hours per week each student was scheduled to work:

$$
\begin{array}{cccccccccc}
10 & 12 & 25 & 10 & 8 & 16 & 6 & 30 & 20 & 16 \\
20 & 20 & 25 & 22 & 25 & 25 & 15 & 12 & 18 & 15 \\
8 & 35 & 20 & 8 & 20 & 20 & 20 & 28 & 30 & 34 \\
24 & 15 & 25 & 6 & 12 & 10 & 32 & 40 & 20 & 15 \\
30 & 24 & 40 & 25 & 35 & 30 & 18 & 32 & 14 & 21 \\
\end{array}
$$

Use the tens digits of the data for the stem. [Hint: Write 6 as 06.]

10. A technician at Crowl Fort Sutter Rehabilitation Center sampled the records at the center and got the following ages (in years) of current patients:

$$
\begin{array}{ccccccccc}
49 & 76 & 70 & 83 & 75 & 69 & 81 & 37 & 60 \\
37 & 30 & 48 & 60 & 30 & 52 & 75 & 54 & 43 \\
59 & 74 & 49 & 42 & 36 & 40 & 56 & 66 & 52 \\
39 & 59 & 38 & 83 & 46 & 55 & 48 & 78 & 58 \\
\end{array}
$$

Use the tens digits of the data for the stem.

11. Janet Kwan works in a government defense laboratory. She sampled 30 mylar capacitors from a recent shipment and measured the capacitances (in microfarads) to three decimal places. The following data were recorded:

$$
\begin{array}{cccccccc}
0.631 & 0.672 & 0.643 & 0.633 & 0.618 & 0.630 & 0.622 & 0.626 \\
0.640 & 0.624 & 0.659 & 0.646 & 0.627 & 0.654 & 0.633 & 0.668 \\
0.661 & 0.645 & 0.630 & 0.640 & 0.657 & 0.615 & 0.626 & 0.639 \\
0.622 & 0.630 & 0.643 & 0.652 & 0.679 & 0.612 & & \\
\end{array}
$$

Use the tenths and hundredths digits of the data for the stem.

12. Deborah Pollack is a quality control inspector at Sanford Rivet Manufacturers. From a recent production run of type A749C rivets, she measured the diameters (in mm) of 36 rivets. The following measurements were listed:

| | | | | | | | | |
|---|---|---|---|---|---|---|---|---|
| 6.64 | 6.76 | 6.81 | 6.76 | 6.78 | 6.72 | 6.79 | 6.72 | 6.68 |
| 6.70 | 6.74 | 6.70 | 6.66 | 6.82 | 6.76 | 6.70 | 6.74 | 6.60 |
| 6.81 | 6.70 | 6.66 | 6.78 | 6.64 | 6.79 | 6.75 | 6.66 | 6.76 |
| 6.72 | 6.62 | 6.74 | 6.67 | 6.70 | 6.72 | 6.77 | 6.72 | 6.75 |

For the stem use (6.60–6.64) 6.6, (6.65–6.69) 6.6, (6.70–6.74) 6.7, and so on.

**13.** A local chapter of the Chinese Outdoorsmen Association, in cooperation with the State Fish and Game Department, sampled the bluegill fish population in a lake near the city. On a Sunday morning, members of the association caught, measured, and released 50 fish. The data below are the lengths of the fish measured in inches.

| | | | | | | | | | |
|---|---|---|---|---|---|---|---|---|---|
| 3.24 | 3.72 | 3.63 | 3.23 | 3.07 | 3.43 | 3.64 | 2.25 | 3.02 | 3.34 |
| 2.56 | 4.35 | 4.21 | 2.39 | 4.49 | 3.50 | 3.22 | 3.38 | 3.88 | 2.51 |
| 3.68 | 2.88 | 3.35 | 3.58 | 3.98 | 2.92 | 3.57 | 3.00 | 2.96 | 2.71 |
| 3.65 | 2.45 | 3.26 | 2.77 | 2.15 | 4.26 | 2.63 | 3.81 | 3.27 | 4.38 |
| 4.07 | 3.47 | 3.51 | 3.58 | 2.65 | 3.52 | 3.39 | 2.63 | 4.31 | 4.33 |

Construct a stem-and-leaf diagram of these data in which the stem is written (2.0–2.4) 2, (2.5–2.9) 2, (3.0–3.4) 3, and so on.

**14.** The weights (in ounces) of bottles of cola being filled at a plant in Detroit, Michigan, are periodically checked. The weights of the randomly selected bottles are measured to the nearest one-tenth of an ounce. The weights of the last 40 bottles sampled are listed as follows:

| | | | | | | | | | |
|---|---|---|---|---|---|---|---|---|---|
| 16.7 | 16.3 | 15.9 | 16.3 | 15.7 | 16.6 | 16.4 | 15.9 | 16.2 | 16.1 |
| 16.3 | 16.4 | 16.1 | 16.5 | 16.2 | 16.1 | 16.3 | 16.4 | 16.1 | 16.2 |
| 15.8 | 16.7 | 16.2 | 15.9 | 15.6 | 16.7 | 15.6 | 16.6 | 16.2 | 15.9 |
| 16.2 | 16.0 | 16.4 | 15.7 | 16.2 | 16.4 | 15.9 | 16.0 | 16.5 | 16.0 |

Construct a stem-and-leaf diagram of these data in which the stem is written (15.6–15.7) 15, (15.8–15.9) 15, (16.0–16.1) 16, and so on.

*In Exercises 15–22, identify the general shape that you feel would best describe a distribution of data collected from each population. Justify your answer.*

**15.** The annual income of residents in New York City last year.

**16.** The selling prices last year of residential properties in Dade County, Florida.

**17.** The yields per acre of corn grown last year in Iowa.

**18.** The butterfat contents in quarts of homogenized milk in a large production run at a major dairy in Chicago.

**19.** The rates at which sales taxes are levied by the states in the United States.

**20.** The distances (in feet) that all sixth graders in the Kansas City public schools can throw a softball.

**21.** The ages of drivers who had moving traffic violations last year in Los Angeles County, California.

**22.** The weights of a large flock of chickens prepared for sale at a supermarket chain in Alabama.

# Exercises 2-1    Set B

*In Exercises 1 and 2, use the data in Figure 2-13.*

1. For the games won (W) columns:
   a. Put the 27 data points in a frequency table of five classes.
   b. Construct a frequency histogram.
   c. Construct a frequency polygon.
   d. What general shape best describes this data?
   e. Which conference has the greatest percentage of teams in classes 1 and 2?
   f. Which conference has the greatest percentage of teams in classes 4 and 5?

2. For the games lost (L) columns:
   a. Put the 27 data points in a frequency table of five classes.
   b. Construct a frequency histogram.
   c. Construct a frequency polygon.
   d. What general shape best describes this data?
   e. Which conference has the greatest percentage of teams in classes 1 and 2?
   f. Which conference has the greatest percentage of teams in classes 4 and 5?

3. Use Data Set I in Appendix B and:
   a. Use random numbers to sample $n = 50$ individuals. Let

   $$x_i = \text{the age of the } i\text{th person selected.}$$

   b. Compute $R$, the range of the ages in the sample.
   c. Construct a frequency table of six classes for the data.
   d. Construct a histogram for the data.

4. Use Data Set II in Appendix B and:
   a. Use random numbers to sample $n = 50$ stocks. Let $x_i = $ the closing price of the $i$th stock selected.
   b. Compute $R$, the range of the prices in the sample.
   c. Construct a frequency table of six classes for the data.
   d. Construct a histogram for the data.

## Basketball

### Western Conference

| PACIFIC DIVISION | W | L | Pct. | GB | Hm | Aw | Strk |
|---|---|---|---|---|---|---|---|
| Portland | 31 | 7 | .816 | — | 17-2 | 14-5 | W1 |
| Phoenix | 23 | 11 | .676 | 6 | 13-4 | 10-7 | W4 |
| LA Lakers | 22 | 11 | .667 | 6½ | 14-4 | 8-7 | W3 |
| Golden State | 18 | 16 | .529 | 11 | 12-4 | 6-12 | L1 |
| Seattle | 15 | 18 | .455 | 13½ | 10-6 | 5-12 | L2 |
| LA Clippers | 13 | 24 | .351 | 17½ | 10-7 | 3-17 | W1 |
| KINGS | 8 | 25 | .242 | 20½ | 7-9 | 1-16 | W1 |
| MIDWEST DIVISION | | | | | | | |
| San Antonio | 24 | 8 | .750 | — | 12-2 | 12-6 | W2 |
| Utah | 24 | 12 | .667 | 2 | 15-3 | 9-9 | L1 |
| Houston | 19 | 17 | .528 | 7 | 12-5 | 7-12 | L3 |
| Dallas | 12 | 22 | .353 | 13 | 7-9 | 5-13 | L3 |
| Minnesota | 11 | 22 | .333 | 13½ | 7-9 | 4-13 | W2 |
| Orlando | 10 | 26 | .278 | 16 | 8-9 | 2-17 | W1 |
| Denver | 7 | 28 | .200 | 18½ | 5-11 | 2-17 | W1 |

### Eastern Conference

| ATLANTIC DIVISION | W | L | Pct. | GB | Hm | Aw | Strk |
|---|---|---|---|---|---|---|---|
| Boston | 29 | 6 | .829 | — | 19-1 | 10-5 | L1 |
| Philadelphia | 22 | 14 | .611 | 7½ | 14-4 | 8-10 | W2 |
| New York | 15 | 19 | .441 | 13½ | 8-11 | 7-8 | L1 |
| Washington | 15 | 18 | .455 | 13 | 10-5 | 5-13 | W2 |
| New Jersey | 10 | 24 | .294 | 18½ | 8-11 | 2-13 | L10 |
| Miami | 9 | 26 | .257 | 20 | 6-11 | 3-15 | L2 |
| CENTRAL DIVISION | | | | | | | |
| Chicago | 26 | 10 | .722 | — | 17-3 | 9-7 | W6 |
| Detroit | 26 | 11 | .703 | ½ | 16-1 | 10-10 | W10 |
| Milwaukee | 25 | 12 | .676 | 1½ | 18-1 | 7-11 | L4 |
| Atlanta | 20 | 15 | .571 | 5½ | 13-6 | 7-9 | W1 |
| Indiana | 14 | 21 | .400 | 11 | 12-5 | 2-16 | W2 |
| Charlotte | 11 | 22 | .333 | 13 | 8-11 | 3-11 | L1 |
| Cleveland | 11 | 25 | .306 | 14½ | 7-11 | 4-14 | L11 |

**Figure 2-13.** National Basketball Association standings on January 15, 1991.

*Exercises 5 and 6 are projects that students can carry out to use the concepts presented in this section.*

**5.** Obtain the ages of 40 students on your campus.
  a. Put the data in a frequency table of five classes.
  b. Construct a histogram using the grouped data.

c. What shape best describes the data?
d. Compare your results with those of other students in the class.

**6.** Obtain the heights of 40 students on your campus. Repeat (a) through (d) in Exercise 5 for these data.

Another column that is sometimes included in a frequency table is entitled *cumulative frequency*. The numbers in this column indicate the number of data points in the set that are less than the upper boundary of a given class. A graph of the cumulative frequency is called an *ogive* (pronounced "ojīv"). To construct an ogive, the horizontal axis is labeled with the boundaries in the table. The cumulative frequencies of the upper boundaries are plotted, and the dots are connected with line segments. We will use the burn-times data collected by Maria Rivera in Table 2-1 to illustrate. Figure 2-14 is an ogive for these data.

Data from Table 2-1

| Class Number | Class Boundaries | Frequency | Cumulative Frequency | |
|---|---|---|---|---|
| 1 | 5.95–8.95 | 5 | less than 8.95 | 5 |
| 2 | 8.95–11.95 | 7 | less than 11.95 | 12 (= 7 + 5) |
| 3 | 11.95–14.95 | 10 | less than 14.95 | 22 (= 10 + 7 + 5) |
| 4 | 14.95–17.95 | 15 | less than 17.95 | 37 (= 15 + 10 + 7 + 5) |
| 5 | 17.95–20.95 | 3 | less than 20.95 | 40 (= 3 + 15 + 10 + 7 + 5) |

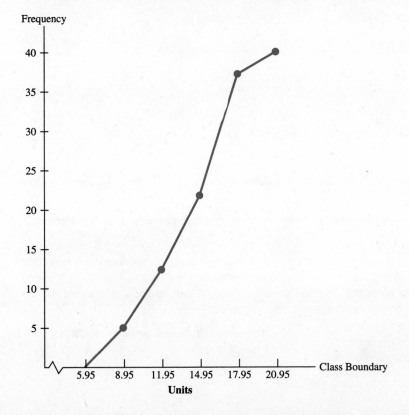

**Figure 2-14.** An ogive for the burn-times data.

*In Exercises 7–10, use the data in Exercises 1–4 in the Set A section:*

*a. Complete a cumulative frequency column for the data.*

*b. Construct an ogive for the data.*

  **7.** Use the quiz scores in Accounting IB.

  **8.** Use the weights of the 36 cans of dog food.

  **9.** Use the ages of the 40 patrons at the Old Ironside Club.

**10.** Use the weights of the 40 roller bearings.

## 2-2 Measures of Central Tendency

**Key Topics**

1. *Definitions of parameters and statistics*

2. *The mean as a measure of central tendency*

3. *A physical interpretation of the mean*

4. *The median as a measure of central tendency*

5. *A physical interpretation of the median*

6. *The mode as a measure of central tendency*

## STATISTICS IN ACTION

**A**    Glenn Fleming is the adult group leader of Troop 2301 of the Boy Scouts of America. Three truckloads of trees were delivered to the lot where the troop holds its annual Christmas tree sale. Glenn wanted to estimate the average height of the trees in the current shipment. He therefore marked the bottoms of 40 trees before they were unloaded. As the trees were unloaded, he had the marked ones separated and their heights measured to the nearest inch. The measurements are listed in Figure 2-15. Based on the sampled trees, Glenn had a sign painted for the sales lot that declared:

> "Christmas trees for sale. Average height is
> 5 feet, 7 inches."

| 70 | 65 | 68 | 64 | 71 | 69 | 74 | 65 | 64 | 66 | 77 | 67 | 65 | 73 | 65 |
| 63 | 65 | 72 | 66 | 61 | 70 | 63 | 68 | 70 | 62 | 64 | 67 | 62 | 75 | 65 |
| 70 | 59 | 74 | 66 | 65 | 71 | 60 | 70 | 65 | 72 | | | | | |

**Figure 2-15.**   Heights of 40 trees (in inches) selected by Glenn Fleming.

B   Gladys Fong is the director of a local office of a national real estate corporation. To update an advertising brochure on the cost of homes in the region served by her office, she instructed a staff member to randomly sample the prices of 30 houses sold in this region over the last 12 months. She specified that only the selling prices of houses with three bedrooms and two baths should be included in the sample. The activity yielded the data points in Figure 2-16. After Gladys analyzed the data, she changed the brochure to read as follows:

"The average cost of a three-bedroom and two-bath house in the area is $94,500."

| 93,500 | 137,200 | 67,900 | 102,750 | 88,600 | 310,400 | 98,500 | 85,900 |
| 206,480 | 91,300 | 87,300 | 95,500 | 142,700 | 189,300 | 84,100 | 93,175 |
| 72,990 | 195,200 | 79,950 | 419,300 | 105,000 | 83,600 | 88,800 | |
| 149,900 | 74,500 | 194,250 | 91,950 | 274,800 | 86,750 | 117,350 | |

**Figure 2-16.**   Selling prices (in dollars) of 30 houses in the past 12 months.

C   Ruby Thompson manages the Womens Casual Wear department of a major department store in St. Louis. She has access to a computer that records every sales and credit transaction in her department. Consequently she maintains an ongoing inventory of the merchandise in the department. Ruby accessed the records to get a printout of the sales of a particular dress featured in the store's fall catalog. The sizes of the dresses in the printout are listed in Figure 2-17. Based on these data, Ruby concluded the "average size" of women who bought this dress was 10.

| 8 | 10 | 6 | 12 | 10 | 8 | 8 | 14 | 10 | 6 | 8 | 14 | 4 | 10 |
| 6 | 4 | 8 | 10 | 4 | 12 | 10 | 10 | 6 | 8 | 8 | 6 | 12 | 8 |
| 10 | 10 | 6 | 8 | 12 | 10 | 6 | 4 | 10 | 12 | 12 | 8 | 14 | 10 |
| 6 | 10 | 4 | 8 | 10 | 6 | 10 | 8 | 8 | 12 | 14 | | | |

**Figure 2-17.**   Sizes of dresses on the computer printout.

## Definitions of parameters and statistics

Measures of central tendency are examples of summary measures used to describe data. If the data compose a population, then the measure is called a *parameter*. However, if the data compose a sample, then the measure is called *a statistic*.

---

### DEFINITION 2-3 Parameter and Statistic

---

A **parameter** is a quantitative measure that describes a **population**.
A **statistic** is a quantitative measure that describes a **sample**.

The terms parameter and statistic have similar meanings; however, a parameter is a characteristic of a population and a statistic is a characteristic of a sample. Furthermore, a parameter is some constant value for a given population, but for random samples drawn from that population, the corresponding statistics can vary from sample to sample.

## The mean as a measure of central tendency

One measure of center that is often used to draw conclusions from data is the mean. A definition of the mean and formulas for computing the mean of a population and the mean of a sample are given in Definition 2-4. However, before defining the mean, we need to discuss briefly the *summation symbol* used in the formula to compute the mean. Consider the following six data points:

$$x_1 = 9 \quad x_2 = 13 \quad x_3 = 7 \quad x_4 = 12 \quad x_5 = 9 \quad x_6 = 16$$

The summation symbol, written $\sum$, can be used to indicate the sum of these data points.

$$\begin{aligned}
\sum_{i=1}^{6} x_i &= x_1 + x_2 + x_3 + x_4 + x_5 + x_6 \\
&= 9 + 13 + 7 + 12 + 9 + 16 \\
&= 66
\end{aligned}$$

Read $\sum_{i=1}^{6} x_i$ as "the sum of the $x$ sub $i$'s, for $i$ from 1 through 6." For a general set of $n$ data points, the symbol is most properly written as $\sum_{i=1}^{n} x_i$. However, for simplicity, the $i = 1$ and $n$ will be deleted, and we will simply write $\sum x_i$.

## DEFINITION 2-4 The Mean of a Set of Quantitative Data

The mean of a set of data is the sum of the data points divided by the number of data points.

(a) If the set of data is a finite population of $N$ data points, then the mean is denoted by $\mu$ (read "mu"), and

$$\mu = \frac{\sum x_i}{N} \begin{cases} \sum x_i \text{ stands for the sum of the data points.} \\ N \text{ stands for the number of data points.} \end{cases}$$

(b) If the set of data is a sample of $n$ data points, then the mean is denoted by $\bar{x}$ (read "x-bar"), and

$$\bar{x} = \frac{\sum x_i}{n} \begin{cases} \text{The sum of the data points} \\ \text{The number of data points} \end{cases}$$

Notice that the formulas for $\mu$ and $\bar{x}$ are essentially the same. The capital letter $N$ signifies the size of the population of data and the lower case $n$, the size of the sample of data.

For the 40 trees that Glenn Fleming sampled from the three truckloads:

$$\bar{x} = \frac{70 + 65 + 68 + \cdots + 72}{40} \qquad \frac{\text{Add the heights of the trees}}{\text{The number of trees in the sample}}$$

$$= \frac{2688}{40} \qquad \text{The sum is 2688.}$$

$$= 67.2 \text{ inches} \qquad \text{The mean height of the 40 trees}$$

Thus Glenn used the mean as the measure of central tendency for the trees.

▶ **Example 1**   Angela Rodriques just completed a course in college algebra. Her semester grade will be based on the mean of the following nine test scores:

$$94 \quad 97 \quad 86 \quad 75 \quad 89 \quad 72 \quad 83 \quad 75 \quad 82$$

Find Angela's mean test score.

**Solution**   **Discussion.** The nine data points can be taken to be a population.

$$\mu = \frac{94 + 97 + \cdots + 82}{9} \qquad \frac{\text{The sum of the data points}}{\text{The number of data points}}$$

$$= \frac{753}{9} \qquad \text{The sum is 753.}$$

$$= 83.7 \qquad \text{Round to one decimal place.}$$

Thus Angela had a test average of about 83.7.   ◀

▶ **Example 2**     Norris Wilcox works for a stock broker in Dallas, Texas. He recommends to his clients common stocks that reportedly have high dividend rates. The following are a sample of dividend rates for some of the stocks he recommended last month. Find the mean of these data.

$$
\begin{array}{ccccccccc}
3.9 & 5.8 & 6.8 & 7.0 & 2.6 & 4.4 & 3.4 & 3.6 & 4.8 \\
4.8 & 4.8 & 7.0 & 5.1 & 4.0 & 3.7 & 5.0 & 5.6 & \\
2.7 & 5.6 & 5.6 & 8.4 & 5.3 & 6.0 & 5.6 & 6.0 &
\end{array}
$$

**Solution**

$$
\begin{aligned}
\bar{x} &= \frac{3.9 + 5.8 + \cdots + 6.0}{25} \\
&= \frac{127.5}{25} \\
&= 5.1
\end{aligned}
$$

Thus the mean dividend rate for these stocks is 5.1.     ◀

## A physical interpretation of the mean

The mean is a "balance point" for a set of data. To illustrate this concept, consider the following six data points:

$$
3, \quad 4, \quad 5, \quad 7, \quad 8, \quad 9
$$

$$
\mu = \frac{3 + 4 + 5 + 7 + 8 + 9}{6} = \frac{36}{6} = 6 \qquad \text{The mean is 6.}
$$

Suppose that each data point represents the location of a unit mass on a rod as shown in Figure 2-18. Assume the rod has no mass. When a fulcrum is placed at 6 (the mean of the six locations), the system of masses will balance.

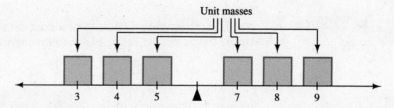

**Figure 2-18.**   Six unit masses on a rod.

If a given set of data has some data points that are quite different from the majority of the data points (often called "outliers"), then the mean will be shifted away from the majority toward the extreme values in the set. To illustrate, consider the following population of seven data points.

$$2, \quad 2, \quad 2, \quad 3, \quad 3, \quad 3, \quad 13$$

$$\mu = \frac{2 + 2 + 2 + 3 + 3 + 3 + 13}{7} = \frac{28}{7} = 4 \qquad \text{The mean is 4.}$$

In Figure 2-19, the seven data points are displayed as unit masses on a rod with no mass. The fulcrum is placed at the location of the mean of the data. Notice that the single mass at 13 balances the six masses to the left of the mean. The reason is that this single mass is nine units to the right of the mean, whereas the sum of the distances the other six masses are from the mean is also 9 ($= 2 + 2 + 2 + 1 + 1 + 1$).

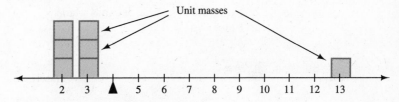

**Figure 2-19.**   Seven unit masses on a rod.

## The median as a measure of central tendency

The median is another commonly used measure of center.

### DEFINITION 2-5 The Median of a Set of Quantitative Data

The **median** of a set of data is the data point in the middle of the **ordered array** of an odd number of data points, or the mean of the two in the middle of the **ordered array** of an even number of data points.

(a) If the set of data is a finite population of $N$ data points, then the median is denoted by $\tilde{\mu}$ (read "mu-tilde").
(b) If the set of data is a sample of $n$ data points, then the median is denoted by $\tilde{x}$ (read "x-tilde").

The selling prices of the 30 houses in Figure 2-19 include two outliers. The $419,300 house is located on 5 acres with a horse barn and riding arena. The $310,400 house is near a country club and golf course in an exclusive area. As a consequence, Gladys used the median as a representative price of the houses in the list.

$$67,900, \quad 72,990, \quad \ldots, \quad 93,500, \quad 95,500, \quad \ldots, \quad 310,400, \quad 419,300$$

15th data point ⎯⎯⎯⎯⎯⎯⎯|     |⎯⎯⎯ 16th data point

$$\tilde{x} = \frac{93,500 + 95,500}{2} = \$94,500$$

The corresponding mean selling price is:

$$\bar{x} = \frac{67{,}900 + 72{,}990 + \cdots + 419{,}300}{30} \approx \$133{,}630$$

▶ **Example 3**   Twenty students at Central Valley Community College were surveyed regarding their typical commute times between home and school. The times, rounded to the nearest whole minute, are listed below.

$$
\begin{array}{ccccccccc}
10 & 6 & 14 & 14 & 18 & 15 & 6 & 15 & 3 & 12 \\
15 & 12 & 16 & 8 & 14 & 13 & 16 & 12 & 10 & 15
\end{array}
$$

**a.** Compute $\bar{x}$ to one decimal place.
**b.** Find $\tilde{x}$.

Solution   **a.**  $\bar{x} = \dfrac{10 + 6 + \cdots + 15}{20}$      $\bar{x} = \dfrac{\text{The sum of the data points}}{\text{The number of data points}}$

$\phantom{\bar{x}}\ = \dfrac{244}{20}$      The sum of the data points is 244.

$\phantom{\bar{x}}\ = 12.2$      The mean is 12.2 minutes.

**b.**  $\begin{array}{cccccccccc} 3 & 6 & 6 & 8 & 10 & 10 & 12 & 12 & 12 & 13 \\ 14 & 14 & 14 & 15 & 15 & 15 & 15 & 16 & 16 & 18 \end{array}$

The tenth and eleventh items in the array are 13 and 14.

$$\tilde{x} = \frac{13 + 14}{2} = 13.5 \qquad \text{The median is 13.5 minutes.} \qquad ◀$$

## A physical interpretation of the median

The median of a set of data is physically interpreted as a number such that approximately one-half the area contained by a well-made histogram or frequency polygon is located on either side of the value. To illustrate, the approximate locations of the medians for three distributions are shown in Figure 2-20.

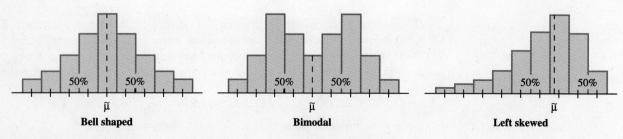

**Figure 2-20.**   Approximate locations of the medians of three different distributions of data.

The median, unlike the mean, is not sensitive to extreme values, as illustrated by the following sets of data:

A = {2, 3, 3, 4, 4, 4, 5, 5, 6}

B = {2, 3, 3, 4, 4, 4, 5, 15, 26}

C = {2, 3, 3, 4, 4, 4, 5, 25, 36}

Set B is set A with 5 and 6 replaced by 15 and 26, respectively. Set C is set A with 5 and 6 replaced by 25 and 36, respectively.

| For A | For B | For C |
|---|---|---|
| $\bar{x} = \dfrac{36}{9} = 4$ | $\bar{x} = \dfrac{66}{9} \approx 7.3$ | $\bar{x} = \dfrac{86}{9} \approx 9.6$ |
| $\tilde{x}$ is 4. | $\tilde{x}$ is still 4. | $\tilde{x}$ is still 4. |

The means of B and C are greater than the mean of A because of the larger values in the data sets. However, these larger values do not affect the medians because 4 remains the middle number in the ordered arrays.

## The mode as a measure of central tendency

The mode is a third commonly used measure of center. The mode is useful when it is important to know which value in a set of data has the greatest frequency.

### DEFINITION 2-6 The Mode of a Set of Quantitative Data

The **mode** of a set of quantitative data is the data point that has the greatest frequency.

For the dress size data in Figure 2-17, we have the following frequencies:

| | | | | | |
|---|---|---|---|---|---|
| Size 4 | 5 sold | Size 6 | 9 sold | Size 8 | 13 sold |
| Size 10 | 15 sold | Size 12 | 7 sold | Size 14 | 4 sold |

For these data:

$$\bar{x} = \frac{4 + 4 + \cdots + 14}{53} \approx 8.8$$

$\tilde{x}$ is the value of the 27th data point, namely 8. The mode is 10, the dress size with the greatest frequency.

Recall that Ruby stated the "average" for the data was 10. Thus she was identifying the data point with the greatest frequency as being the one that most accurately describes the central tendency of the data.

Some data sets may have two different values that have the greatest frequency in the set. That is, the frequencies of these two data points are the same, but the frequencies are also greater than those of any other data points in the set. Both values are called modes, and the data set is called *bimodal*.

If more than two different values have the greatest frequency, then we will say that the set of data has no mode. As a consequence, exercise instructions will be stated as follows:

"If one exists, identify the mode."

▶ **Example 4**   If one exists, identify the mode for each of the following data sets.

**a.** 2,  5,  5,  7,  7,  7,  8,  10
**b.** 3,  4,  4,  4,  4,  5,  6,  7,  8,  8,  8,  8,  9

Solution   **a.** The number 7 has a frequency of three, more than any other value in the set. Thus the mode is 7.
**b.** The numbers 4 and 8 both have a frequency of four, more than any other values in the set. Thus 4 and 8 are both modes and the data set is bimodal.   ◀

▶ **Example 5**   For the first 25 days in July, the highest temperature readings (in degrees Fahrenheit) in Springfield were as follows:

$$
\begin{array}{cccccccc}
80 & 70 & 88 & 94 & 96 & 88 & 90 & 85 & 101 \\
102 & 98 & 78 & 86 & 88 & 91 & 89 & 90 \\
90 & 92 & 98 & 90 & 90 & 95 & 97 & 102
\end{array}
$$

**a.** Find the mean high temperature to one decimal place.
**b.** Find the median high temperature.
**c.** If one exists, identify the modal high temperature.

Solution   **a.** $\bar{x} = \dfrac{80 + 70 + \cdots + 102}{25}$

$= \dfrac{2268}{25}$

$= 90.72$     The mean high temperature is about 90.7°.

**b.** Arraying the data from smallest to largest:

$$
\begin{array}{cccccccccccc}
70 & 78 & 80 & 85 & 86 & 88 & 88 & 88 & 89 & 90 & 90 & 90 & 90 \\
90 & 91 & 92 & 94 & 95 & 96 & 97 & 98 & 98 & 101 & 102 & 102
\end{array}
$$

The median is the value of the thirteenth item in the array.

$\tilde{x} = 90$     The median high temperature is 90°.

**c.** The data point 90 has a frequency of five. Because this value has the greatest frequency, the modal high temperature is 90°.   ◀

# Exercises 2-2 Set A

*In Exercises 1–16:*
*a. Compute the mean of the data.*
*b. Find the median of the data.*
*c. If one exists, identify the mode (or modes if the data set is bimodal).*

1. Thirteen members of the Adonis Health Club are participating in a weight loss program. For the first 5 weeks of the program the following weight losses (in pounds) were recorded:

$$9, \quad 4, \quad 5, \quad 15, \quad 4, \quad 1, \quad 8, \quad 2, \quad 11, \quad 0, \quad 8, \quad 7, \quad 4$$

2. During the months of October and November, the 12 members of Kappa Phi sorority kept a record of the number of dates had by each member. The following is a list of the data:

$$9, \quad 5, \quad 10, \quad 4, \quad 9, \quad 0, \quad 13, \quad 8, \quad 9, \quad 4, \quad 11, \quad 2$$

3. The produce manager of one of the stores in a nationwide supermarket chain sampled 15 oranges from a recent shipment and weighed each one to the nearest gram. The following data are the weights (in grams) recorded:

$$80 \quad 73 \quad 68 \quad 77 \quad 73 \quad 85 \quad 68 \quad 79 \quad 73$$
$$76 \quad 83 \quad 76 \quad 73 \quad 74 \quad 82$$

4. The lengths of ten trout to the nearest one-tenth of 1 centimeter are as follows:

$$32.3, \quad 21.6, \quad 22.9, \quad 27.4, \quad 24.4, \quad 29.2, \quad 33.3, \quad 24.9, \quad 22.4, \quad 26.7$$

*Mean 26.5 no mode*
*med 25.8*

5. In the last golf tournament at Pebble Beach, the scores of the top 24 golfers at the end of the third day of the tournament were as follows:

$$209 \quad 211 \quad 214 \quad 205 \quad 209 \quad 216 \quad 215 \quad 215$$
$$209 \quad 216 \quad 212 \quad 211 \quad 208 \quad 212 \quad 213 \quad 216$$
$$211 \quad 209 \quad 214 \quad 212 \quad 211 \quad 210 \quad 209 \quad 216$$

6. On a recent Sunday, the 26 teams of the National Football League that played on that day scored the following numbers of points:

$$27 \quad 26 \quad 27 \quad 7 \quad 17 \quad 17 \quad 34 \quad 30 \quad 23$$
$$10 \quad 27 \quad 17 \quad 34 \quad 10 \quad 20 \quad 3 \quad 21 \quad 20$$
$$16 \quad 14 \quad 21 \quad 16 \quad 28 \quad 24 \quad 43 \quad 14$$

7. For the week that ended July 24, the high selling prices (in dollars) of the most active stocks in New York Stock Exchange composite trading for the week were the following:

$$5\frac{1}{4} \quad 27\frac{1}{2} \quad 31\frac{5}{8} \quad 24\frac{7}{8} \quad 34\frac{1}{4} \quad 47\frac{1}{2}$$
$$34\frac{5}{8} \quad 165\frac{7}{8} \quad 23\frac{7}{8} \quad 39\frac{3}{4} \quad 57\frac{1}{4} \quad 52\frac{1}{2}$$

8. A stock and bond portfolio includes shares of stocks of 15 companies listed on the New York Stock Exchange. Last week, the selling prices (in dollars) of each of the stocks were as follows:

$$37\tfrac{1}{2} \quad 51\tfrac{1}{4} \quad 32\tfrac{1}{8} \quad 28\tfrac{1}{4} \quad 9\tfrac{7}{8} \quad 8\tfrac{1}{8} \quad 49\tfrac{1}{4} \quad 69$$
$$3\tfrac{1}{4} \quad 22 \quad 16\tfrac{1}{4} \quad 13 \quad 108\tfrac{1}{8} \quad 16\tfrac{1}{4} \quad 20\tfrac{5}{8}$$

9. Last Friday, the 30 reporting stations for the National Weather Service in the West and Northwest Regions reported the following low temperatures (in degrees Fahrenheit):

| | | | | | | | | | |
|---|---|---|---|---|---|---|---|---|---|
| 33 | 52 | 59 | 35 | 62 | 57 | 53 | 47 | 58 | 49 |
| 50 | 52 | 32 | 44 | 55 | 62 | 52 | 48 | 55 | 69 |
| 52 | 62 | 51 | 54 | 42 | 53 | 45 | 75 | 70 | 53 |

10. Tess Caldwell is a swimming coach at a health club. She wanted to evaluate the effect of a training session on the heart rate of adult swimmers. To obtain data, she selected 30 adults and had them swim one of the "strenuous" exercise sets. At the end of each set, she recorded the heart rates of the volunteers. The results (beats/min) are recorded below.

| | | | | | | | | | |
|---|---|---|---|---|---|---|---|---|---|
| 132 | 140 | 180 | 178 | 164 | 156 | 180 | 160 | 140 | 144 |
| 186 | 168 | 140 | 182 | 168 | 140 | 186 | 180 | 160 | 190 |
| 156 | 168 | 182 | 168 | 180 | 172 | 168 | 140 | 148 | 168 |

11. Desiree Baker is a quality control inspector for a company that makes round plastic disks. She recently sampled 30 GTO108 disks and weighed them (in grams) to the nearest one-tenth of a gram. The weights are listed below.

| | | | | | | |
|---|---|---|---|---|---|---|
| 2.3 | 2.6 | 2.5 | 2.7 | 2.6 | 2.5 | 3.0 |
| 2.1 | 2.5 | 2.4 | 2.0 | 2.5 | 2.9 | 2.5 |
| 2.5 | 3.0 | 3.1 | 2.4 | 2.5 | 3.0 | 2.9 |
| 2.5 | 2.6 | 3.0 | 2.4 | 3.1 | 2.5 | 3.0 | 2.8 | 2.5 |

12. Janice Polanski is an engineer for Abernathy Valve Manufacturers. She recently tested the relief pressure (in pounds per square inch, psi) of 35 3T9OG relief valves. The measurements were rounded to the nearest whole number and are listed below.

| | | | | | |
|---|---|---|---|---|---|
| 72 | 83 | 82 | 92 | 70 | 72 |
| 91 | 71 | 87 | 95 | 69 | 80 |
| 77 | 92 | 78 | 89 | 67 | 82 |
| 87 | 80 | 90 | 73 | 79 | 83 |
| 74 | 75 | 86 | 66 | 85 | 96 |
| 88 | 64 | 96 | 67 | 81 | |

13. Benjamin Harris is executive vice president of Wholesome Groceries Inc, a small chain of supermarkets in the Midwest. To get a sense of the average age of employees in these

stores, he randomly sampled employee records of cashiers, meat cutters, produce workers, and stock personnel. He did not include the ages of any personnel in management. The results of the survey are as follows:

```
17  22  21  28  22  26  21  24
34  23  20  23  42  27  19  33
21  27  18  20  19  29  19  31
36  27  19  35  19  31  49  45
```

14. Lorna Phillips just completed a class in English composition. One of the activities in the course required estimating the average number of letters in English words. She used *The American Heritage Dictionary* to randomly select 60 words and then counted the number of letters in each word. The results of the activity are as follows:

```
 6   5  12   6   8   8   7  14   5   6
 5   6   5  10   8   5   5   4   8  10
 9   6   7   9  10   6   3   6   6   5
10  11   6   9  11  10   4   8   5   6
 4   5  10   3   7   7   9   9  11   4
 9   6   5   4  10   5  14   9  10   5
```

15. Chandar Jackson is an analyst for the Health Facilities Commission of a western state. Her job requires analyzing the disclosure reports of long-term care facilities. She recently collected the following data (in dollars and cents) on the revenue in dollars per person per day of health care facilities with 1- to 59-bed capacities.

```
39.02  38.08  38.89  40.81  39.45  38.64
27.27  39.70  39.72  39.85  26.87  38.53
39.06  28.62  38.99  39.94  35.35  32.15
37.59  39.45  39.03  38.90  39.55  39.09
38.72  38.17  38.81  33.28  38.99  40.82
36.59  37.05  35.50  27.53  39.53  38.56
```

16. Theresa Evans is a waitress at the Old Spaghetti Factory. She wanted to estimate the average amount spent per customer for a meal at this restaurant. With the cooperation of seven other waiters and waitresses, she collected the following data over a period of several nights. The data are the average amounts (in dollars and cents) per customer for 36 receipts.

```
3.59  4.71  4.95  6.74  5.33  6.76  4.80  5.64  5.95
4.47  5.32  4.97  7.72  5.70  6.82  5.03  6.39  3.96
6.12  5.91  5.21  5.93  4.85  7.12  5.69  4.33  5.35
7.18  5.35  7.18  6.44  4.83  4.61  4.93  3.98  5.02
```

In Chapter 7, a statistic $\bar{d}$ (read "*d*-bar") is computed, where $\bar{d}$ is the mean of a sample of *differences*.

$$\bar{d} = \frac{\sum d_i}{n} \qquad \frac{\text{The sum of the differences}}{\text{The number of differences}}$$

*In Exercises 17–20, compute $\bar{d}$.*

17. Michael Sanford wanted to compare the shelf prices of items at supermarkets A and B. He sampled 25 items sold by both stores. He then computed *d* for each item, where:

$$d = (\text{price of item at A}) - (\text{price of same item at B})$$

A list of the 25 differences (in cents) is given below.

| | | | | | | | | |
|---|---|---|---|---|---|---|---|---|
| 0.10 | 0.50 | 0.16 | 0.20 | 0.03 | 0.22 | 0.33 | −0.20 | 0.16 |
| 0.04 | 0.02 | 0.16 | 0.02 | 0.19 | 0.10 | 0.04 | 0.05 | |
| −0.15 | 0.00 | 0.56 | −0.26 | 0.20 | 0.24 | 0.42 | 0.12 | |

Compute $\bar{d}$ to the nearest cent and state at which market you would prefer to shop.

18. Twenty members of a health clinic are currently participating in a weight-loss program. The weight (in pounds) of each participant was taken at the beginning of the program. The weight of each participant was measured again this week and the differences in weights (in pounds) are recorded below, where:

$$d = (\text{weight at start of program}) - (\text{current weight})$$

| | | | | | | | | | |
|---|---|---|---|---|---|---|---|---|---|
| 15 | 8 | 11 | 21 | 16 | 20 | 16 | 18 | 38 | 9 |
| 43 | 14 | 30 | 39 | 50 | 23 | 2 | 23 | 32 | 5 |

Compute $\bar{d}$ to the nearest one-tenth of a pound.

19. A class on speed-reading was offered last semester through a local community college. At the start of the class, the words-per-minute reading rate for each student was measured. The reading rate was again measured at the end of the class. A difference was computed for each student, where:

$$d = (\text{words/minute rate after class}) - (\text{words/minute before class})$$

A list of the 30 differences is given below.

| | | | | | | | | | |
|---|---|---|---|---|---|---|---|---|---|
| 17 | 5 | 40 | 15 | 5 | 111 | 26 | 21 | 22 | 33 |
| 18 | 14 | 13 | 30 | 60 | 20 | 8 | 24 | 11 | 100 |
| 17 | 6 | 8 | 32 | 45 | 2 | 40 | 30 | 23 | 40 |

Compute $\bar{d}$ to the nearest whole number.

20. A smog-control device was developed by Automotive Electronics Corporation. To check the effect of the device on the miles-per-gallon rating of a car, 30 different cars were selected. The difference in the numbers of miles per gallon each car got without versus with the device was computed. The data are listed on the following page in miles per gallon to the nearest one-tenth.

| 1.1 | 0.8 | 0.5 | 0.5 | 1.4 | 0.2 | 0.7 | 2.2 | 1.2 | 0.6 |
| 0.4 | 0.4 | 2.6 | 1.2 | 1.5 | 2.6 | 0.8 | 1.5 | 0.6 | 0.3 |
| 0.5 | 1.5 | 1.2 | 1.9 | 0.2 | 1.2 | 1.3 | 2.3 | 0.3 | 1.6 |

Compute $\bar{d}$ to one decimal place.

21. In a collective bargaining session in which a new salary contract is being negotiated, the labor representative claims that the "average" salary is $5.35 per hour. The management representative claims that the "average" salary is $8.35 per hour. Under what conditions could the two parties both be correct?

22. A potential manufacturer of jigsaw puzzles has sufficient capital to invest in just one stamping machine to cut the pieces of a puzzle. Previous experience concerning puzzle purchases indicates that people purchase puzzles with a mode of 750 pieces, a median of 1000 pieces, and a mean of 1250 pieces. Which number should the manufacturer use for the initial production and why?

23. To please the employees in an office, they are polled about the ideal temperature for the office. The distribution of choices has the following: mode = 73° F, median = 72° F, and mean = 74° F.
    a. To make roughly the same number of people feel it's too cool as the number who feel it's too warm, the thermostat should be set at what temperature?
    b. To make the greatest number of people comfortable, regardless of anyone else, the thermostat should be set at what temperature?

## Exercises 2-2   Set B

*In Exercises 1–10, indicate whether each of the indicated "averages" is more likely to be a mean, a median, or a mode.*

1. She graduated from college with a 3.25 grade-point average.

2. He finished the season with a 172 bowling average.

3. The average age in the United States is now more than 30 years.

4. The average salary of employees in this corporation is $18,500.

5. The average shoe size for adult men is 10 medium.

6. The average grade level at which entering freshmen read is grade 11.

7. The average high temperature in Phoenix last month was 96.3° F.

8. The average number of children per family in Buffalo, New York, is 2.4.

9. The average weight of a 5 ft 10 in. adult man with a medium-sized frame is 163 lb.

10. The average commute time to work for the 3572 workers at Ad Tech Electronics is 18.8 minutes.

11. Use a table of random numbers to select two different samples of 30 data points each from Data Set I, Appendix B. Call the first one sample A and the second one sample B. Record the ages of the 30 individuals selected for both samples in two different lists.
   a. Compute $\bar{x}_A$ and $\bar{x}_B$.
   b. Are the two sample means the same?
   c. Compute $\tilde{x}_A$ and $\tilde{x}_B$.
   d. Are the two sample medians the same?

12. Use a table of random numbers to select two different samples of 25 data points each from Data Set II, Appendix B. Call the first one sample A and the second one sample B. Record the stock prices in the Close column for both samples in two different lists.
   a. Compute $\bar{x}_A$ and $\bar{x}_B$.
   b. Are the two sample means the same?
   c. Compute $\tilde{x}_A$ and $\tilde{x}_B$.
   d. Are the two sample medians the same?

*In Exercises 13 and 14, use the data in Figure 2-21. These data are for N = 23 companies that offered stock for sale to the public in August 1988.*

13. In the Size ($ mil.) column are the dollar amounts of the stocks offered for sale. For example, ACM Government Spectrum offered $270,000,000 in stock.
   a. Find the mean dollar amount.
   b. Find the median dollar amount.

### INITIAL PUBLIC OFFERINGS: MAY 1988

| O-T-C sym. | Company | Business/product | Offering Size ($mil.) | Offering Price | First aftermarket bid | Underwriter |
|---|---|---|---|---|---|---|
| SI[a] | ACM Government Spectrum | Closed-end fund | 270.00 | 10.00 | 11⅜ | Paine Webber |
| BFSI | BFS Bankorp | S&L | 2.32 | 7.00 | 7 | Merrill Lynch |
| BMRG | BMR Financial Group | Bank holding co. | 8.55 | 9.50 | 9½ | Alex. Brown |
| LTG[b] | Catalina Lighting | Mfrs. light fixtures | 3.22 | 3.50 | 3⅞ | Werbel-Roth |
| CBOG | Community Bancshares | Bank holding co. | 7.25 | 7.25 | 7¼ | Robinson-Humphrey |
| CMAT | Compumat | Mkts. computers & software | 3.80 | 4.75 | 4⅞ | Oberweis |
| CPF[a] | Comstock Partners Strategy | Closed-end fund | 1,100.00 | 10.00 | 10 | Merrill Lynch |
| CYTBU | Cytrx Biopool Ltd. | R&D of vascular-syst. prods. | 6.00 | 6.00[c] | 6¼ | D. H. Blair |
| FRMG | Firstmiss Gold | Dvlps. gold mine | 13.00 | 8.00 | 7¾ | Shearson Lehman |
| FCX[a] | Freeport McMoRan | Holding co. | 65.63 | 17.50 | 17¾ | Kidder, Peabody |
| HBUF | Homestyle Buffet | Restaurant chain | 3.80 | 6.50 | 7 | Oberweis |
| ICAR | Intercargo | Ships cargo | 4.64 | 6.25 | 6½ | Blunt Ellis |
| MYOTU | Myo-Tech | Mfr. & mkts. medical testing device | 3.75 | 5.00[c] | 5½ | D. H. Blair |
| PIF[a] | Pru Intermed Income | Closed-end fund | 470.00 | 10.00 | 10 | Prudential-Bache |
| RELY | Relational Technology | Mkts. software | 28.00 | 14.00 | 14⅝ | Goldman, Sachs |
| RESP | Resperonics | Designs & mfrs. respirator prods. | 6.00 | 12.00 | 12 | Parker/Hunter |
| RCHFA | Richfood Holding | Food distr. | 19.38 | 7.75 | 8 | Paine Webber |
| SGHI | Silk Greenhouse | Operates synthetic-flower stores | 12.08 | 11.00 | 13½ | Robertson Coleman |
| BID[b] | Sotheby's Holdings | Auctions art | 52.26 | 18.00 | 18 | Salomon Bros. |
| TSQ[b] | T2 Medical | Operates home-care infusion therapy units | 12.00 | 8.00 | 8 | Drexel Burnham |
| TNN[a] | Teleconnect Co. | Diversified telephone syst. | 57.60 | 18.00 | 18 | Merrill Lynch |
| TJCK | Timberjack | Mfrs. logging equipment | 13.80 | 12.00 | 12 | Kidder, Peabody |
| USM[b] | United States Cellular | Owns & operates cellular phone service | 42.00 | 15.00 | 15 | Salomon Bros. |

Initial public offerings sold at $1 or less not included; issues without underwriters not included. First aftermarket bid is approximate. [a] trading on NYSE; [b] trading on AMEX; [c] units of 3 common + 3 warrants. Offering data source: *GOING PUBLIC: THE IPO Reporter*, a publication of The Dealer's Digest Inc., 150 Broadway, New York, NY 10038.

**Figure 2-21.** Investing in growth companies. "Initial Public Offerings," May, 1988, *INC.* Magazine, August, 1988. Copyright © 1988 by Goldhirst Group, Inc., 38 Commercial Wharf, Boston, MA 02110. Reprinted by permission.

   c. Are there any extreme data points in the set that would cause the mean and median to be considerably different? If so, identify these data points.

**14.** In the Price column are the prices at which these stock were initially offered.
 a. Find the mean stock price.
 b. Find the median stock price.

 c. Are there any extreme data points in the set that would cause the mean and median to be considerably different? If so, identify these data points.

## 2-3 Measures of Variability

**Key Topics**

*1. The range as a measure of variability*

*2. The variance of a population*

*3. The standard deviation of a population*

*4. The variance and standard deviation of a sample*

*5. The standard deviation and bell-shaped populations*

S T A T I S T I C S   I N   A C T I O N

Captain Brenda Leply is a crew chief at a large military base. Her crew is responsible for keeping some mechanical systems operative at all times. Equipment failure at the wrong time could result in large losses of personnel and equipment. As a consequence, her crew follows a PMP (Preventive Maintenance Program) that periodically replaces key mechanical components to guard against unwanted breakdowns.

   One such component is the K420GB bearing. Brenda has two possible suppliers for this bearing. To help her decide which supplier to use, Brenda has her crew subject 20 bearings, 10 from each supplier, to normal operating conditions. They are instructed to record the number of hours the bearings operate properly before failing. The data from the tests are listed in Figure 2-22.

   The means ($\bar{x}_a$ and $\bar{x}_b$) and standard deviations ($s_a$ and $s_b$) were computed. As will be seen in this section, the

standard deviation measures the amount of variability in a set of data. Comparing the means, $\bar{x}_b$ is 24 hours more than $\bar{x}_a$ ($= 516 - 492$ hours). However, the standard deviation $s_b$ is more than four times as large as $s_a$ (41 hours compared with 10 hours). Because there was less variability in the data for supplier A, Brenda decided to have them supply the facility with the K420GB bearings. Brenda knew that the smaller variability in supplier A bearings made their service lives more predictable.

| Supplier A | | | | | Supplier B | | | | |
|---|---|---|---|---|---|---|---|---|---|
| 490 | 482 | 496 | 505 | 498 | 475 | 494 | 552 | 487 | 583 |
| 488 | 510 | 485 | 476 | 490 | 446 | 513 | 549 | 528 | 533 |
| $\bar{x}_a = 492$ hours and | | | | | $\bar{x}_b = 516$ hours and | | | | |
| $s_a = 10$ hours | | | | | $s_b = 41$ hours | | | | |

**Figure 2-22.** The results of the test on the K420GB bearing.

## The range as a measure of variability

The range is a simple way to measure variability in a set of quantitative data. This measure was introduced in Section 2-1 and its definition is restated here.

### DEFINITION 2-1 The Range of a Set of Data

The **range** of a set of quantitative data is the difference between the largest and smallest values in the set. A capital letter $R$ is used to represent the range.

$$R = \text{largest value} - \text{smallest value}$$

For the bearing data in Figure 2-22:

$$R_a = 510 - 476 = 34 \text{ hours} \qquad \text{Range in data for supplier A}$$
$$R_b = 583 - 446 = 137 \text{ hours} \qquad \text{Range in data for supplier B}$$

▶ **Example 1**   Figure 2-23 contains data on the stock prices that had the greatest percentage change in price for a particular day of trading last year.

**a.** Find the range in the percentage changes for the **Ups** data.
**b.** Find the range in the percentage changes for the **Downs** data.

### Percentage Leaders

| # | Name (UPS) | Last | Chg | Pct. | # | Name (DOWNS) | Last | Chg | Pct. |
|---|---|---|---|---|---|---|---|---|---|
| 1 | CyclopsInd | 19¾ + | 7½ | Up 61.2 | 1 | HarmanInt | 5¼ − | 1¼ | Off 19.2 |
| 2 | NEstFdl pf | 6⅜ + | 1⅜ | Up 27.5 | 2 | Hibernia | 4¼ − | 1 | Off 19.0 |
| 3 | Tonka | 3¾ + | ¾ | Up 25.0 | 3 | viLTV pfB | 1⅛ − | ¼ | Off 18.2 |
| 4 | NestFdl | 2⅛ + | ⅜ | Up 21.4 | 4 | Glenfed | 4⅝ − | ⅞ | Off 15.9 |
| 5 | SPS Tech | 28¾ + | 3¾ | Up 15.0 | 5 | HotelInv | 1⅜ − | ¼ | Off 15.4 |
| 6 | CtytrstBcp | 1 + | ⅛ | Up 14.3 | 6 | FtFidBcp pfD | 48½ − | 8½ | Off 14.9 |
| 7 | DimeSvNY | 2 + | ¼ | Up 14.3 | 7 | Nat Stand | 1¾ − | ¼ | Off 12.5 |
| 8 | DigitalEq | 59⅜ + | 6⅝ | Up 12.6 | 8 | NewsCorp | 5⅝ − | ⅝ | Off 10.0 |
| 9 | ChmBank B | 1⅛ + | ⅛ | Up 12.5 | 9 | CarriageInd | 4 − | ⅜ | Off 8.6 |
| 10 | CrossldSv pfB | 1⅛ + | ⅛ | Up 12.5 | 10 | AmShipB | 1⅜ − | ⅛ | Off 8.3 |

**Figure 2-23.** Percentage leaders in stock prices on the New York Stock Exchange for 1 day's trading.

**Solution**   **a.** For the Ups data, the largest and smallest values are 61.2% and 12.5%, respectively. For this set:

$$R = 61.2\% - 12.5\% \qquad R = \text{largest value} - \text{smallest value}$$
$$= 48.7\%$$

**b.** For the Downs data, the largest and smallest values are 19.2% and 8.3%, respectively. For this set:

$$R = 19.2\% - 8.3\%$$
$$= 10.9\% \quad \blacktriangleleft$$

Because only the smallest and largest values in a set of data are used to calculate the range, it is quite sensitive to extreme values. Two measures of variability that are less sensitive to extreme values, though by no means completely insensitive, are the variance and standard deviation. These measures are based on the average of the squares of the deviations of the data from the mean.

## The variance of a population

Consider set A, consisting of five data points:

$$A = \{1, 5, 6, 8, 10\}$$

For these data:

$$\mu = \frac{1 + 5 + 6 + 8 + 10}{5} = \frac{30}{5} = 6$$

To measure the amount by which a data point *deviates* from $\mu$, we would subtract $\mu$ from it.

### The Deviation of $X_i$ from $\mu$

If $x_i$ is the $i$th data point in a set with mean $\mu$, then

$$x_i - \mu$$

measures the **deviation** of $x_i$ from $\mu$.

TABLE 2-5   The Deviations of Data Points in A from 6

| $x_i$ | $x_i - \mu$ |
|-------|-------------|
| 1 | $1 - 6 = -5$ |
| 5 | $5 - 6 = -1$ |
| 6 | $6 - 6 = 0$ |
| 8 | $8 - 6 = 2$ |
| 10 | $10 - 6 = 4$ |

The deviations of the data points in A from $\mu$ are shown in Table 2-5.

For $x_1 = 1$, the deviation is $-5$. Thus 1 is 5 units less than the mean. For $x_3 = 6$, the deviation is 0. Thus 6 is 0 units from the mean. For $x_5 = 10$, the deviation is 4. Thus 10 is 4 units greater than the mean.

The summation symbol can be used to indicate the sum of these deviations.

$$\sum_{i=1}^{5} (x_i - 6) = (1 - 6) + (5 - 6) + \cdots + (10 - 6) = 0$$

## Using the Summation Symbol to Indicate a Sum of Deviations

For a set of $N$ data points wth mean $\mu$:

$$\sum_{i=1}^{N} (x_i - \mu) \text{ indicates the sum of the deviations of the } x_i\text{'s from } \mu.$$

For simplicity:

$$\sum_{i=1}^{N} (x_i - \mu) \text{ is written } \sum(x_i - \mu)$$

For the data in A, the sum of the deviations from $\mu$ is zero. Such a zero sum is not unique to the data in A. When the sum of the deviations from the mean is calculated for any set of data, the total will always equal zero. For this reason initially although the concept of "average deviation" as a measure of variability might appear to have merit, we need to modify it to get a useful number. To obtain a nonzero value, two operations could be performed on each $(x_i - \mu)$ before computing the sum. One is to *square each deviation*, and the other is to take *the absolute value*.

For any $x_i$ and $\mu$:

$(x_i - \mu)^2 \geq 0$     The square of the deviations is nonnegative.

$|x_i - \mu| \geq 0$     The absolute value of the deviations is nonnegative.

The statistical measure of variability called the *variance* is based on the sum of the squares of the deviations. The measure of variability called the *average absolute deviation* is based on the sum of the absolute values. We will focus our study on the variance here, but the average absolute deviation is considered briefly in the B exercises in this section.

Table 2-6   The Squares of the Deviations of the Data in A from 6

| $x_i$ | $x_i - 6$ | $(x_i - 6)^2$ |
|-------|-----------|---------------|
| 1 | $1 - 6 = -5$ | $(-5)^2 = 25$ |
| 5 | $5 - 6 = -1$ | $(-1)^2 = 1$ |
| 6 | $6 - 6 = 0$ | $0^2 = 0$ |
| 8 | $8 - 6 = 2$ | $2^2 = 4$ |
| 10 | $10 - 6 = 4$ | $4^2 = 16$ |

Consider the squares of the deviations for the data in A, shown in Table 2-6. The summation symbol can be used to indicate the sum of the squares of these deviations:

$$\sum_{i=1}^{5} (x_i - 6)^2 = (1 - 6)^2 + (5 - 6)^2 + \cdots + (10 - 6)^2$$
$$= 25 + 1 + 0 + 4 + 16$$
$$= 46$$

### Using Summation Symbol to Show Sum of Squared Deviations from Mean

For a set of $N$ data points with mean $\mu$:

$\displaystyle\sum_{i=1}^{N} (x_i - \mu)^2$ indicates the **sum of the squares of the deviations** of the $x_i$'s from $\mu$.

For simplicity:

$$\sum_{i=1}^{N} (x_i - \mu)^2 \text{ is written } \sum(x_i - \mu)^2.$$

If $\sum(x_i - \mu)^2$ is divided by the number of data points, then the mean of the squares of the deviations is obtained. This number is called the *variance* for the data.

### DEFINITION 2-7 The Variance of a Population of Quantitative Data

The **variance** of a finite population of quantitative data is the sum of the squares of the differences of each data point from the mean divided by the number of data points. The symbol $\sigma^2$, read "sigma squared," is used for a population variance.

$$\sigma^2 = \frac{\sum(x_i - \mu)^2}{N}$$

Consider sets A and B:

$$A = \{1, 5, 6, 8, 10\} \quad \text{and} \quad B = \{-1, 1, 6, 9, 15\}$$

The means of the data in both A and B are 6. Intuitively we might say that the data in B are more variable (or dispersed) than the data in A. This greater variability will show up in the calculations of the variances.

| | Set A | | | Set B | |
|---|---|---|---|---|---|
| $x_i$ | $x_i - 6$ | $(x_i - 6)^2$ | $x_i$ | $x_i - 6$ | $(x_i - 6)^2$ |
| 1 | $1 - 6 = -5$ | 25 | $-1$ | $-1 - 6 = -7$ | 49 |
| 5 | $5 - 6 = -1$ | 1 | 1 | $1 - 6 = -5$ | 25 |
| 6 | $6 - 6 = 0$ | 0 | 6 | $6 - 6 = \ 0$ | 0 |
| 8 | $8 - 6 = 2$ | 4 | 9 | $9 - 6 = \ 3$ | 9 |
| 10 | $10 - 6 = 4$ | 16 | 15 | $15 - 6 = \ 9$ | 81 |
| | $\sum(x - 6)^2 = 46$ | | | $\sum(x - 6)^2 = 164$ | |
| | $\sigma^2 = \frac{46}{5} = 9.2$ | | | $\sigma^2 = \frac{164}{5} = 32.8$ | |

Thus the greater variability of the data in B is reflected in the greater variance. That is, $32.8 > 9.2$.

▶ **Example 2**    Alice Lavin and Bob Clouse are students in Dr. Vandenberg's Biology 210 class.
They have both taken the same nine quizzes and have gotten the following scores:

$$\text{Alice} \quad 1 \quad 8 \quad 9 \quad 9 \quad 9 \quad 9 \quad 9 \quad 9 \quad 9$$
$$\text{Bob} \quad \; 1 \quad 2 \quad 3 \quad 3 \quad 5 \quad 7 \quad 7 \quad 8 \quad 9$$

**a.** Compute the range for both sets of scores.
**b.** Compute the variance for Alice's scores to two decimal places.
**c.** Compute the variance for Bob's scores to two decimal places.

**Solution**    **Discussion.**  It appears that there is more variability in Bob's scores than in Alice's.
We will see that the range does not detect the greater variability, but the variance
does.

**a.** For both sets of scores the largest and smallest data points are 9 and 1, respectively. Thus:

$$R = 9 - 1 = 8$$

**b.** For Alice's scores:

$$\mu = \frac{1 + 8 + 9 + \cdots + 9}{9}$$

$$= \frac{72}{9}$$

$$= 8$$

| $x_i$ | $x_i - 8$ | $(x_i - 8)^2$ |
|-------|-----------|---------------|
| 1 | $1 - 8 = -7$ | $(-7)^2 = 49$ |
| 8 | $8 - 8 = 0$ | $0^2 = 0$ |
| 9 | $9 - 8 = 1$ | $1^2 = 1$ |
| 9 | $9 - 8 = 1$ | $1^2 = 1$ |
| 9 | $9 - 8 = 1$ | $1^2 = 1$ |
| 9 | $9 - 8 = 1$ | $1^2 = 1$ |
| 9 | $9 - 8 = 1$ | $1^2 = 1$ |
| 9 | $9 - 8 = 1$ | $1^2 = 1$ |
| 9 | $9 - 8 = 1$ | $1^2 = 1$ |
| | | $\sum(x_i - 8)^2 = 56$ |
| | $\sigma^2 = \frac{56}{9} = 6.22$ | |

**c.** For Bob's scores:

$$\mu = \frac{1 + 2 + 3 + \cdots + 9}{9}$$

$$= \frac{45}{9}$$

$$= 5$$

| $x_i$ | $x_i - 5$ | $(x_i - 5)^2$ |
|-------|-----------|---------------|
| 1 | $1 - 5 = -4$ | $(-4)^2 = 16$ |
| 2 | $2 - 5 = -3$ | $(-3)^2 = 9$ |
| 3 | $3 - 5 = -2$ | $(-2)^2 = 4$ |
| 3 | $3 - 5 = -2$ | $(-2)^2 = 4$ |
| 5 | $5 - 5 = 0$ | $0^2 = 0$ |
| 7 | $7 - 5 = 2$ | $2^2 = 4$ |
| 7 | $7 - 5 = 2$ | $2^2 = 4$ |
| 8 | $8 - 5 = 3$ | $3^2 = 9$ |
| 9 | $9 - 5 = 4$ | $4^2 = 16$ |
| | | $\sum(x_i - 5)^2 = 66$ |
| | $\sigma^2 = \frac{66}{9} = 7.33$ | |

The variance for Bob's scores (7.33) is greater than the variance for Alice's
scores (6.22).

There are alternate formulas for computing $\sigma^2$ that do not require that $\mu$ be computed first.

### Alternate Formulas for Computing $\sigma^2$

$$\sigma^2 = \frac{N\sum x_i^2 - (\sum x_i)^2}{N^2} \quad \text{and} \quad \sigma^2 = \frac{\sum x_i^2 - \frac{1}{N}(\sum x_i)^2}{N}$$

$$\sum x_i^2 = x_1^2 + x_2^2 + x_3^2 + \cdots + x_N^2 \quad \longrightarrow \quad \text{Square each data point, then sum the squares.}$$

$$(\sum x_i)^2 = (x_1 + x_2 + x_3 + \cdots + x_N)^2 \quad \longrightarrow \quad \text{Add the data points, then square the sum.}$$

▶ **Example 3** Compute the variance for Bob's scores using the alternate formula:

$$1, \quad 2, \quad 3, \quad 3, \quad 5, \quad 7, \quad 7, \quad 8, \quad 9$$

Solution **Discussion.** A table displaying $x$ and $x^2$ for each data point is a good way to organize the calculations.

| $x_i$ | $x_i^2$ |
|-------|---------|
| 1 | $1^2 = 1$ |
| 2 | $2^2 = 4$ |
| 3 | $3^2 = 9$ |
| 3 | $3^2 = 9$ |
| 5 | $5^2 = 25$ |
| 7 | $7^2 = 4$ |
| 7 | $7^2 = 49$ |
| 8 | $8^2 = 64$ |
| 9 | $9^2 = 81$ |

$$\sum x_i^2 = 1 + 4 + 9 + \cdots + 81$$
$$= 291$$
$$\sum x_i = 1 + 2 + 3 + \cdots + 9$$
$$= 45$$
$$(\sum x_i)^2 = 45^2$$
$$= 2025$$
$$N = 9$$
$$N^2 = 81$$

$$\sigma^2 = \frac{9(291) - 2025}{81} \qquad \sigma^2 = \frac{N\sum x_i^2 - (\sum x_i)^2}{N^2}$$

$$= \frac{2619 - 2025}{81} \qquad \text{Do the multiplication first.}$$

$$= \frac{594}{81} \qquad \text{Now do the subtraction.}$$

$$\approx 7.33 \qquad \text{Divide to two decimal places.}$$

As in Example 2, the variance to two decimal places is 7.33. ◀

### The standard deviation of a population

The units of a variance are the squares of the original units. For that reason, it is common practice to take the positive square root of the variance and use it as a measure of variability. This measure is called the **standard deviation**.

---

### DEFINITION 2-8 The Population Standard Deviation

The **standard deviation** of a finite population of quantitative data is the positive square root of the variance.

$$\sigma = \sqrt{\sigma^2} \qquad \qquad \sigma^2 \text{ is the variance of the data set.}$$

$$\sigma = \sqrt{\frac{\sum(x_i - \mu)^2}{N}} \qquad \text{Using the formula in Definition 2-7}$$

$$\sigma = \sqrt{\frac{N\sum x_i^2 - (\sum x_i)^2}{N^2}} \qquad \text{Using one of the alternate formulas for } \sigma^2$$

▶ **Example 4**   An assistant manager at a fast-food restaurant recently used a stopwatch to measure the time between the start of a customer placing an order and getting the food. She carried out this activity for $N = 25$ customers. The data, given in seconds, are listed below.

$$
\begin{array}{ccccccccc}
38 & 46 & 15 & 32 & 37 & 56 & 28 & 73 & 29 \\
36 & 51 & 21 & 28 & 19 & 46 & 21 & 26 & \\
21 & 45 & 18 & 14 & 26 & 64 & 55 & 41 &
\end{array}
$$

Find the standard deviation to one decimal place.

**Solution**

| Row 1 of Data | | Row 2 of Data | | Row 3 of Data | |
|---|---|---|---|---|---|
| $x_i$ | $x_i^2$ | $x_i$ | $x_i^2$ | $x_i$ | $x_i^2$ |
| 38 | 1444 | 36 | 1296 | 21 | 441 |
| 46 | 2116 | 51 | 2601 | 45 | 2025 |
| 15 | 225 | 21 | 441 | 18 | 324 |
| 32 | 1024 | 28 | 784 | 14 | 196 |
| 37 | 1369 | 19 | 361 | 26 | 676 |
| 56 | 3136 | 46 | 2116 | 64 | 4096 |
| 28 | 784 | 21 | 441 | 55 | 3025 |
| 73 | 5329 | 26 | 676 | 41 | 1681 |
| 29 | 841 | | | | |

For the given set of data:

$$\sum x_i = 886 \qquad \sum x_i^2 = 37{,}448 \qquad N = 25$$

$$(\sum x_i)^2 = 784{,}996 \qquad N^2 = 625$$

$$\sigma = \sqrt{\frac{25(37{,}448) - 784{,}996}{625}} \qquad \sigma = \sqrt{\frac{N\sum x_i^2 - (\sum x_i)^2}{N^2}}$$

$$= \sqrt{\frac{151{,}204}{625}} \qquad \qquad \text{In the numerator, multiply first, then subtract.}$$

$$= \sqrt{241.92\ldots} \qquad \qquad \text{Divide.}$$

$$\approx 15.6 \text{ seconds} \qquad \qquad \text{Take the square root and round to one decimal place.} \qquad ◀$$

## The variance and standard deviation of a sample

The variance and standard deviation of a sample of data can be used to describe the variability in the sample. These measures are computed by using the differences between each $x_i$ and $\bar{x}$.

### DEFINITION 2-9 The Variance and Standard Deviation of a Sample

The symbols $s^2$ and $s$ are used respectively for the variance and standard deviation of a sample. If $\bar{x}$ is the mean and $n$ is the number of data points, then:

**Alternate Formulas**

$$s^2 = \frac{\sum(x_i - \bar{x})^2}{n - 1} \quad \text{or} \quad s^2 = \frac{n\sum x_i^2 - (\sum x_i)^2}{n(n - 1)}$$

$$s = \sqrt{\frac{\sum(x_i - \bar{x})^2}{n - 1}} \quad \text{or} \quad s = \sqrt{\frac{n\sum x_i^2 - (\sum x_i)^2}{n(n - 1)}}$$

There are two facts regarding the sample variance requiring discussion.

### Facts concerning the sample variance

**Fact 1** The denominator of the expression used to compute $s^2$ is $n - 1$.

The reason $n - 1$ (and not $n$) is used is beyond the scope of this discussion. However, we should point out that when $n$ is *large*, the numerical difference between $s^2$ based on $n - 1$ and $s^2$ based on $n$ is small. That is, $\frac{\sum(x_i - \bar{x})^2}{n - 1} \approx \frac{\sum(x_i - \bar{x})^2}{n}$, when $n$ is large. Some calculators with statistics functions have a $\sigma_n$ key for a population standard deviation and a $\sigma_{n-1}$ key for a sample standard deviation.

**Fact 2** The variance of a random sample can be used to estimate a population variance.

We frequently must work on populations for which the variance is not known. If a population has a variance of $\sigma^2$ and a simple random sample is taken from that population, then $s^2$ can be used to estimate $\sigma^2$.

▶ **Example 5**     Stacy Phillips enlisted the help of 30 students at her school who promised to record the amount of time they spent studying on a Saturday and Sunday. The individual times were rounded to the nearest whole hour and are recorded below.

$$
\begin{array}{cccccccccc}
5 & 4 & 7 & 10 & 9 & 6 & 4 & 6 & 6 & 8 \\
3 & 8 & 8 & 9 & 7 & 5 & 7 & 8 & 4 & 10 \\
6 & 7 & 4 & 5 & 3 & 5 & 8 & 6 & 5 & 4
\end{array}
$$

Compute to one decimal place:

**a.** $\bar{x}$     **b.** $s$

**Solution**   **a.** $\bar{x} = \dfrac{5 + 4 + \cdots + 4}{30}$     $\bar{x} = \dfrac{\sum x_i}{n}$

$= \dfrac{187}{30}$

$= 6.2$ hours          Rounded to one decimal

**b.**

| Row 1 | | Row 2 | | Row 3 | |
|---|---|---|---|---|---|
| $x_i$ | $x_i^2$ | $x_i$ | $x_i^2$ | $x_i$ | $x_i^2$ |
| 5 | 25 | 3 | 9 | 6 | 36 |
| 4 | 16 | 8 | 64 | 7 | 49 |
| 7 | 49 | 8 | 64 | 4 | 16 |
| 10 | 100 | 9 | 81 | 5 | 25 |
| 9 | 81 | 7 | 49 | 3 | 9 |
| 6 | 36 | 5 | 25 | 5 | 25 |
| 4 | 16 | 7 | 49 | 8 | 64 |
| 6 | 36 | 8 | 64 | 6 | 36 |
| 6 | 36 | 4 | 16 | 5 | 25 |
| 8 | 64 | 10 | 100 | 4 | 16 |

$s^2 = \dfrac{30(1281) - 187^2}{30(29)}$          $\sum x_i = 187; \sum x_i^2 = 1281; n = 30; n - 1 = 29$

$= \dfrac{38,430 - 34,969}{870}$          Do the multiplications and squaring first.

$= \dfrac{3,461}{870}$          Now subtract in the numerator.

$= 3.9782\ldots$          Do not round yet.

$s = \sqrt{3.9782\ldots}$          Take the positive square root.

$s = 2.0$ hours          Round to one decimal place.

Thus the mean number of hours the 30 students studied this weekend is 6.2 hours and the sample standard deviation is 2.0 hours.     ◀

## The standard deviation and bell-shaped populations

Many populations of quantitative data are reasonably bell shaped. For such populations, the percentage of data points that are within specified numbers of standard deviations of the mean can be approximated using a tabled *theoretical distribution*.

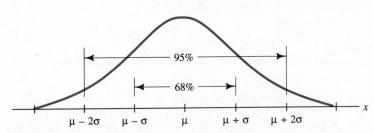

To illustrate, in Figure 2-24, an idealized histogram for a bell-shaped distribution is shown. As indicated, the mean is $\mu$ and the standard deviation is $\sigma$. For this idealized population, approximately 68 percent of the data are within one standard deviation of the mean, and approximately 95 percent are within two standard deviations.

**Figure 2-24.** A theoretical bell-shaped distribution with mean $\mu$ and standard deviation $\sigma$.

▶ **Example 6**  The distribution of heights of adult men in the United States is approximately bell-shaped with a mean of 70 inches (5 feet, 10 inches) and a standard deviation of 4 inches.

**a.** Sketch and label a graph of this distribution. On the horizontal axis, label the five points with coordinates $\mu$, $\mu + \sigma$, $\mu - \sigma$, $\mu + 2\sigma$, and $\mu - 2\sigma$.
**b.** Approximately 68 percent of adult men have heights between what values?
**c.** Approximately 95 percent of adult men have heights between what values?

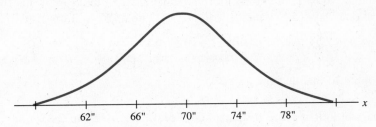

**Figure 2-25.** The distribution of heights of adult men in the United States is bell shaped with a mean of 70 inches and a standard deviation of 4 inches.

**Solution**  **a.** In Figure 2-25, a bell-shaped curve is drawn. The $x$ on the right end of the axis represents height. The point on the axis under the peak of the curve locates the mean and is labeled 70″. The standard deviation is approximately 4″.

$$\mu + \sigma \quad \text{becomes} \quad 70'' + 4'' = 74'' \text{ (6 feet, 2 inches)}$$
$$\mu + 2\sigma \quad \text{becomes} \quad 70'' + 8'' = 78'' \text{ (6 feet, 6 inches)}$$
$$\mu - \sigma \quad \text{becomes} \quad 70'' - 4'' = 66'' \text{ (5 feet, 6 inches)}$$
$$\mu - 2\sigma \quad \text{becomes} \quad 70'' - 8'' = 62'' \text{ (5 feet, 2 inches)}$$

The approximate locations of these points are labeled in Figure 2-25.
- b. Based on the graph, approximately 68 percent of adult men in the United States have heights between 5 feet, 6 inches and 6 feet, 2 inches.
- c. Based on the graph, approximately 95 percent of adult men in the United States have heights between 5 feet, 2 inches and 6 feet, 6 inches.   ◄

## Exercises 2-3    Set A

1. Monica Wong is currently taking college algebra. The instructor frequently gives quizzes on the homework. On the past seven quizzes, Monica got the following scores:

$$2 \quad 4 \quad 5 \quad 7 \quad 8 \quad 10 \quad 13$$

Compute to one decimal place:
a. $R$      b. $\sigma^2$      c. $\sigma$

2. Norman Atherton is currently taking Physics 10B. On the 11 laboratory assignments for the semester, he got the following scores:

$$0 \quad 3 \quad 4 \quad 4 \quad 6 \quad 9 \quad 12 \quad 13 \quad 15 \quad 21 \quad 23$$

Compute to one decimal place:
a. $R$      b. $\sigma^2$      c. $\sigma$

3. The owner of a small retail clothing outlet employs 20 people. As part of their personnel file, he asked each one to record to the nearest one-tenth of a mile the distance they travel one way from home to work. The 20 items of data are listed below.

$$2.5 \quad 15.0 \quad 7.2 \quad 10.8 \quad 6.1 \quad 7.5 \quad 7.4 \quad 9.6 \quad 7.1 \quad 5.4$$
$$6.5 \quad 6.8 \quad 4.4 \quad 4.7 \quad 1.2 \quad 5.0 \quad 4.5 \quad 20.5 \quad 1.6 \quad 14.0$$

Compute to one decimal place:
a. $R$      b. $\sigma^2$      c. $\sigma$

4. A class of fourth grade students kept accurate records on the amount of time each student spent watching television for 1 week. The times (in hours) are listed below.

$$18.2 \quad 27.3 \quad 23.8 \quad 21.0 \quad 10.5 \quad 23.8 \quad 11.2 \quad 25.2 \quad 19.6$$
$$18.9 \quad 25.9 \quad 25.2 \quad 13.3 \quad 23.1 \quad 9.8 \quad 25.9 \quad 13.3$$
$$28.1 \quad 22.4 \quad 28.0 \quad 32.9 \quad 25.2 \quad 14.7 \quad 16.1 \quad 21.0$$

Compute to one decimal place:
a. $R$      b. $\sigma^2$      c. $\sigma$

5. An accountant for a major horse race track records the amounts returned for a $2 bet on each of the winning horses. Last Saturday and Sunday nights, winning tickets were paid the following amounts (in dollars and cents) in the 18 races (nine races each night):

| 14.00 | 9.00 | 11.20 | 17.60 | 105.00 | 4.80 | 12.80 | 4.20 | 23.40 |
|-------|------|-------|-------|--------|------|-------|------|-------|
| 20.20 | 47.60 | 18.20 | 7.00 | 10.00 | 6.80 | 17.60 | 4.20 | 8.40 |

Compute to the nearest cent:

a. $R$    b. $\sigma^2$    c. $\sigma$

6. To get the best deal on an oil change (plus oil filter) and lubrication, the owner of a Honda Accord contacted 25 service stations and repair facilities for the cost of such service. The data (in dollars and cents) she obtained are listed below.

| 20.00 | 21.00 | 23.00 | 17.95 | 19.95 | 24.00 | 15.95 | 17.95 | 14.50 |
|-------|-------|-------|-------|-------|-------|-------|-------|-------|
| 19.00 | 20.00 | 20.50 | 17.00 | 23.00 | 18.50 | 22.00 | 18.00 | |
| 17.60 | 18.50 | 20.00 | 16.00 | 18.50 | 21.50 | 24.00 | 15.00 | |

Compute to the nearest cent:

a. $R$    b. $\sigma^2$    c. $\sigma$

7. As part of a study of shad that spawn in the American River, a biology major caught, measured, and released 25 of these fish last season. The lengths, measured to the nearest one-tenth of an inch, are recorded below.

| 12.5 | 21.3 | 13.1 | 17.8 | 14.6 | 23.4 | 20.6 | 19.7 | 18.5 |
|------|------|------|------|------|------|------|------|------|
| 17.4 | 19.4 | 16.8 | 19.7 | 20.5 | 21.3 | 18.2 | 13.6 | |
| 15.6 | 17.7 | 20.2 | 17.4 | 18.6 | 12.9 | 22.1 | 14.3 | |

Compute to the nearest one-tenth of an inch:

a. $R$    b. $\sigma^2$    c. $\sigma$

8. The 30 employees of the Sheldon Real Estate Firm recently had a party to celebrate the holidays. The topic of age at the time of the most recent marriage came up. (Some of the people had been married before their current marriage.) A survey yielded the following ages in years at which the members of the group were most recently married:

| 24 | 34 | 50 | 27 | 23 | 28 | 26 | 46 | 24 | 27 |
|----|----|----|----|----|----|----|----|----|----|
| 20 | 32 | 23 | 31 | 28 | 26 | 36 | 30 | 28 | 24 |
| 31 | 25 | 27 | 21 | 39 | 21 | 30 | 22 | 33 | 29 |

Compute to one decimal place:

a. $R$    b. $\sigma^2$    c. $\sigma$

9. The students in an engineering statistics class were instructed how to measure the thickness of the paper in their texts. The experiment yielded the following data (in millimeters):

| 0.0848 | 0.0817 | 0.0972 | 0.0830 | 0.0818 |
|--------|--------|--------|--------|--------|
| 0.0804 | 0.0910 | 0.0814 | 0.0807 | 0.0895 |
| 0.0815 | 0.0791 | 0.0789 | 0.0857 | 0.0817 |
| 0.0833 | 0.0807 | 0.0829 | 0.0914 | 0.0815 |

Compute to four decimal places:

a. $\bar{x}$    b. $\tilde{x}$    c. $s$

10. A warehouse of electrical equipment contains a large supply of 50-foot rolls of wire. A random sample of 20 rolls of this wire was taken and resistances were measured. The data collected are listed below (in ohms).

| | | | | |
|---|---|---|---|---|
| 0.73 | 0.55 | 0.89 | 0.62 | 0.94 |
| 0.66 | 1.03 | 0.80 | 0.78 | 0.58 |
| 1.08 | 0.70 | 0.60 | 0.54 | 0.82 |
| 0.48 | 0.75 | 0.67 | 0.65 | 0.56 |

Compute to two decimal places:
a. $\bar{x}$     b. $\tilde{x}$     c. $s$

11. Kemal Sirhan recently moved to a large city in the South. He needed to store part of his belongings while he looked for a house to buy. He contacted 20 storage facilities and got the following quotations (in dollars) for the cost of renting a 10 by 20-ft storage space for 1 month:

| | | | | | | | | | |
|---|---|---|---|---|---|---|---|---|---|
| 55 | 45 | 42 | 82 | 65 | 67 | 65 | 60 | 60 | 80 |
| 59 | 53 | 58 | 65 | 69 | 75 | 72 | 48 | 87 | 63 |

Compute to one decimal place:
a. $\bar{x}$     b. $\tilde{x}$     c. $s$

12. Hely Carter is a reporter on a large city newspaper. She has been writing a series of articles on the average cost of some services in the area. To estimate the cost of hair-care services, she contacted 30 hair centers to get the cost of a "haircut" and "blowdry." The results of the survey (in dollars) are listed below.

| | | | | | | | | | |
|---|---|---|---|---|---|---|---|---|---|
| 19 | 20 | 14 | 25 | 15 | 10 | 14.5 | 20 | 15 | 20 |
| 9 | 19 | 12.5 | 17.5 | 25 | 18 | 20 | 25 | 17 | 21 |
| 20 | 20 | 17.5 | 15 | 25 | 25 | 15 | 30 | 13 | 12 |

Compute to one decimal place:
a. $\bar{x}$     b. $\tilde{x}$     c. $s$

13. Karen LaBonte is the president of a large community college. She asked a librarian to obtain data and estimate the average number of hours per week that students study in the school library. The librarian surveyed 40 students across the campus and obtained the following data on the numbers of hours each student studied in the library last week:

| | | | | | | | | | | | | |
|---|---|---|---|---|---|---|---|---|---|---|---|---|
| 2.5 | 2 | 1 | 3.5 | 17 | 3 | 2 | 1.5 | 4 | 0 | 9 | 0 | 2 | 1 |
| 1.5 | 3 | 5 | 5.5 | 2 | 3 | 2 | 6.5 | 2 | 12 | 4.5 | 3 | 16 |
| 0 | 15 | 10 | 2.5 | 0 | 0 | 5 | 0 | 4 | 2.5 | 3 | 0 | 5 |

Compute to one decimal place:
a. $\bar{x}$     b. $\tilde{x}$     c. $s$

14. Thu Nguyen and three of her friends were stopped by campus security on the way to their dormitory and asked whether they were high school students because, as the officer said, "they looked so young." Thu wondered whether Asian girls were preceived as

"younger" because of their height. She consequently sampled 25 Asian girls on campus and obtained the following heights to the nearest inch:

$$
\begin{array}{ccccccc}
5'0'' & 5'5'' & 4'10'' & 5'1'' & 4'10'' & 4'11'' & 5'4'' \\
5'4'' & 5'2'' & 4'11'' & 5'5'' & 5'1'' & 5'0'' & 5'3'' \\
5'0'' & 5'5'' & 5'3'' & 5'2'' & 5'3'' & 5'4'' & \\
5'5'' & 4'10'' & 5'3'' & 5'1'' & 5'1'' & &
\end{array}
$$

Compute to the nearest inch:

a. $\bar{x}$    b. $\tilde{x}$    c. $s$

15. Diane Chopko, student in the evening program at River Valley Community College, sensed that students in the evening college were older than students in the day college. She therefore sampled 40 students across the campus and obtained the following ages to the nearest year of those surveyed:

$$
\begin{array}{cccccccccc}
22 & 27 & 19 & 30 & 33 & 31 & 28 & 27 & 51 & 49 \\
32 & 37 & 25 & 22 & 46 & 39 & 26 & 35 & 41 & 28 \\
31 & 29 & 40 & 54 & 37 & 32 & 20 & 25 & 45 & 29 \\
39 & 28 & 43 & 50 & 50 & 37 & 26 & 23 & 40 & 28
\end{array}
$$

Compute to one decimal place:

a. $\bar{x}$    b. $\tilde{x}$    c. $s$

16. A produce manager of a supermarket recently got a large shipment of strawberries that were packaged in small baskets. The regulation weight of a filled basket is 12.5 ounces. The manager sampled 30 of the baskets and weighed each one to the nearest one-tenth of an ounce. The results are listed below.

$$
\begin{array}{cccccccccc}
13.0 & 12.8 & 13.2 & 13.1 & 12.7 & 12.5 & 13.5 & 14.0 & 12.5 & 13.2 \\
12.9 & 13.1 & 13.2 & 13.0 & 13.8 & 12.2 & 12.9 & 13.0 & 12.7 & 12.5 \\
12.8 & 12.4 & 12.9 & 12.7 & 12.6 & 13.1 & 13.4 & 13.3 & 12.8 & 13.8
\end{array}
$$

Compute to one decimal place:

a. $\bar{x}$    b. $\tilde{x}$    c. $s$

In Exercises 17–20:

a. Sketch and label a graph of each bell-shaped distribution. On the horizontal axis, label the five points with coordinates $\mu$, $\mu + \sigma$, $\mu - \sigma$, $\mu + 2\sigma$, and $\mu - 2\sigma$.

b. Based on the graph, approximately 68 percent of the data points lie between what values?

c. Based on the graph, approximately 95 percent of the data points lie between what values?

17. The distribution of heights of adult women in the United States is bell-shaped with a mean of 5'5.0'' and a standard deviation of 3.5''.

18. The distribution of test scores on an Adult Intelligence Scale is bell-shaped with a mean of 60.8 and a standard deviation of 14.6.

19. A large shipment of Magic Transparent Tape dispensers arrived at the warehouse of a discount drugstore chain. The relative frequency distribution of tape lengths in the dispensers is approximately bell-shaped with a mean of 20.50 meters and a standard deviation of 0.20 meters.

20. A large production run of Hickory Smoked Barbecue Sauce has a relative frequency distribution of weights that is approximately bell-shaped with a mean of 510.0 grams and a standard deviation of 4.5 grams.

## Exercises 2-3    Set B

*For Exercises 1 and 2, use random numbers to select individuals from Data Set I, Appendix B.*

1. Select the ages of 20 females.
   a. Compute the mean age to one decimal place.
   b. Compute the standard deviation to one decimal place.

2. Select the ages of 20 males.
   a. Compute the mean age to one decimal place.
   b. Compute the standard deviation to one decimal place.

*For Exercises 3 and 4, use random numbers to select n = 10 stocks from Data Set II, Appendix B.*

3. List the High stock prices (52 weeks) for each of the stocks in the sample. Compute to the nearest one-tenth of a dollar:
   a. The mean stock price.
   b. The standard deviation.

4. List the Low stock prices (52 weeks) for each of the stocks in the sample. Compute to the nearest one-tenth of a dollar:
   a. The mean stock price.
   b. The standard deviation.

*In Exercises 5 and 6, use the data in Figure 2-26.*

5. The columns headed G report the number of games in which each player has appeared.
   a. For the Atlanta Hawks, compute the mean and standard deviation for the games played data.
   b. For the Boston Celtics, compute the mean and standard deviation for the games played data.

c. Of these two teams, which one averages more games per player?

d. Of these two teams, which one has greater variability in games per player?

6. The columns headed Fg report the number of field goals scored by each player.
   a. For the Atlanta Hawks, compute the mean and standard deviation for the field goals data.
   b. For the Boston Celtics, compute the mean and standard deviation for the field goals data.
   c. Which team has the higher average per player?
   d. Which team shows greater variability in field goals among its players?

### ATLANTA HAWKS

| Player | G | Min | Fg | Fga | Fg% | Ft | Fta | Ft% | 3Fg | 3Fga | Orb | Reb | Ast | Stl | To | Blk | Pf | Pts | Ppg |
|--------|---|-----|----|----|-----|----|----|-----|-----|------|-----|-----|-----|-----|----|----|----|-----|-----|
| Wilkins | 33 | 1239 | 303 | 641 | .473 | 197 | 238 | .828 | 35 | 93 | 91 | 275 | 112 | 68 | 79 | 27 | 64 | 838 | 25.4 |
| Rivers | 31 | 1017 | 187 | 411 | .455 | 93 | 110 | .845 | 32 | 112 | 16 | 99 | 154 | 59 | 50 | 20 | 90 | 499 | 16.1 |
| Battle | 34 | 826 | 173 | 370 | .468 | 131 | 153 | .856 | 7 | 24 | 18 | 67 | 105 | 23 | 44 | 4 | 63 | 484 | 14.2 |
| Willis | 34 | 1065 | 202 | 409 | .494 | 64 | 92 | .696 | 2 | 5 | 109 | 313 | 50 | 23 | 61 | 19 | 108 | 470 | 13.8 |
| Webb | 29 | 714 | 99 | 229 | .432 | 55 | 64 | .859 | 14 | 43 | 15 | 58 | 147 | 38 | 54 | 3 | 70 | 267 | 9.2 |
| Malone | 34 | 702 | 104 | 224 | .464 | 104 | 126 | .825 | 0 | 1 | 89 | 240 | 39 | 10 | 60 | 26 | 49 | 312 | 9.2 |
| Robinson | 27 | 466 | 73 | 167 | .437 | 35 | 60 | .583 | 2 | 10 | 15 | 52 | 89 | 17 | 58 | 8 | 41 | 183 | 6.8 |
| Ferrell | 30 | 402 | 71 | 135 | .526 | 39 | 56 | .696 | 0 | 1 | 35 | 57 | 26 | 11 | 30 | 7 | 51 | 181 | 6.0 |
| Koncak | 29 | 831 | 74 | 151 | .490 | 13 | 20 | .650 | 1 | 3 | 39 | 163 | 67 | 34 | 18 | 37 | 111 | 162 | 5.6 |
| McCormick | 27 | 380 | 51 | 101 | .505 | 34 | 45 | .756 | 0 | 1 | 28 | 99 | 16 | 4 | 20 | 10 | 46 | 136 | 5.0 |
| Moncrief | 30 | 435 | 48 | 102 | .471 | 39 | 45 | .867 | 10 | 30 | 14 | 59 | 30 | 24 | 26 | 5 | 39 | 145 | 4.8 |
| Wilson | 15 | 88 | 10 | 35 | .286 | 3 | 10 | .300 | 0 | 2 | 8 | 23 | 8 | 4 | 9 | 1 | 9 | 23 | 1.5 |

### BOSTON CELTICS

| Player | G | Min | Fg | Fga | Fg% | Ft | Fta | Ft% | 3Fg | 3Fga | Orb | Reb | Ast | Stl | To | Blk | Pf | Pts | Ppg |
|--------|---|-----|----|----|-----|----|----|-----|-----|------|-----|-----|-----|-----|----|----|----|-----|-----|
| McHale | 35 | 1049 | 275 | 469 | .586 | 126 | 156 | .808 | 7 | 22 | 72 | 255 | 64 | 14 | 77 | 69 | 100 | 683 | 19.5 |
| Bird | 31 | 1157 | 241 | 507 | .475 | 87 | 100 | .870 | 27 | 68 | 24 | 270 | 239 | 60 | 87 | 30 | 54 | 596 | 19.2 |
| Lewis | 35 | 1277 | 263 | 533 | .493 | 126 | 148 | .851 | 0 | 6 | 45 | 187 | 98 | 47 | 72 | 37 | 105 | 652 | 18.6 |
| Parish | 35 | 1029 | 204 | 343 | .595 | 102 | 129 | .791 | 0 | 1 | 103 | 354 | 19 | 32 | 65 | 52 | 98 | 510 | 14.6 |
| Gamble | 35 | 1062 | 209 | 341 | .613 | 88 | 110 | .800 | 0 | 2 | 37 | 109 | 110 | 43 | 57 | 15 | 87 | 506 | 14.5 |
| Shaw | 35 | 1230 | 196 | 384 | .510 | 103 | 129 | .798 | 1 | 11 | 56 | 177 | 251 | 52 | 100 | 11 | 78 | 496 | 14.2 |
| Brown | 35 | 711 | 109 | 214 | .509 | 36 | 42 | .857 | 2 | 11 | 14 | 70 | 155 | 29 | 51 | 9 | 66 | 256 | 7.3 |
| Pinckney | 23 | 258 | 41 | 74 | .554 | 19 | 21 | .905 | 0 | 1 | 32 | 66 | 20 | 17 | 11 | 11 | 33 | 101 | 4.4 |
| Smith | 21 | 146 | 37 | 80 | .463 | 11 | 15 | .733 | 3 | 13 | 6 | 14 | 15 | 3 | 14 | 2 | 8 | 88 | 4.2 |
| Kleine | 32 | 357 | 40 | 91 | .440 | 18 | 26 | .692 | 0 | 0 | 27 | 90 | 12 | 8 | 22 | 8 | 44 | 98 | 3.1 |
| Popson | 14 | 48 | 11 | 24 | .458 | 9 | 10 | .900 | 0 | 4 | 9 | 1 | 5 | 1 | 5 | 1 | 10 | 31 | 2.2 |
| Vrankovic | 15 | 76 | 10 | 23 | .435 | 6 | 9 | .667 | 0 | 0 | 6 | 21 | 1 | 1 | 16 | 14 | 24 | 26 | 1.7 |

**Figure 2-26.**    Statistics for the Atlanta Hawks and Boston Celtics basketball teams.

The Russian mathematician P. L. Chebyshev (1821–1894) proved a theorem regarding the fraction of data points that must be within a specified number of standard deviations of the mean. The theorem is a statement of a minimum fractional part. For many sets of data, the actual fractional part of the data points within a certain number of standard deviations of the mean is greater than the minimum guaranteed by Chebyshev's theorem.

## Chebyshev's Theorem

For any set of data, the fraction of the data points falling within $k$ standard deviations of the mean is at least $1 - \dfrac{1}{k^2}$, for $k > 1$.

*In Exercises 7–10, use Chebyshev's theorem to determine the minimum percentage of the data points that will be within the indicated number of standard deviations of the mean. Write answers correct to the nearest percent.*

**7.** 1.5 standard deviations

**8.** 2.0 standard deviations

**9.** 2.25 standard deviations

**10.** 2.75 standard deviations

## DEFINITION 2-10 The Average Absolute Deviation

The average absolute deviation of a finite population of quantitative data is the mean of the absolute values of the deviations of the $x_i$'s and $\mu$. The symbol for average absolute deviation is AD.

$$AD = \frac{\sum |x_i - \mu|}{N}$$

*In Exercises 11 and 12, compute to one decimal place the average absolute deviation.*

**11.** 0   3   4   4   6   9   12   13   15   21   23

**12.** 2   4   5   7   8   10   13   13   16   20   25

**13.** Compute the average absolute deviation for the mileage data of Exercise 3 in Exercises 2-3 Set A and compare it with $\sigma$.

**14.** Compute to the nearest cent the average absolute deviation for the winning tickets data of Exercise 5 in Exercises 2-3 Set A and compare it with $\sigma$.

*Exercises 15 and 16 are possible projects that can be carried out at your school.*

15. Obtain the number of units being carried this term by 30 students.
    a. Compute the mean of the data.
    b. Find the median of the data.
    c. Which measure of center is the better one for these data?
    d. Compute the standard deviation for the data.
    e. How many of the data points are within one standard deviation of the mean?
    f. How many of the data points are within two standard deviations of the mean?
    g. Compare your results with those of other students in the class.

16. Obtain the selling prices of 30 casebound texts in your college store. For these data, do parts (a) through (g) given in Exercise 15.

## 2-4  Measures of Position

**Key Topics**

1. *The z score as a measure of position*

2. *The percentile as a measure of position*

3. *Deciles and quartiles as measures of position*

4. *The box-whisker plot as a display of data*

5. *Using z scores for bell-shaped distributions*

S T A T I S T I C S   I N   A C T I O N

Paul Gould is the counselor for international students at Adams Community College. Luyen Luu, a new student from Hong Kong, came to his office yesterday to get advice on placement in a mathematics course. Her transcripts showed that she had scored a 36.5 on the mathematics portion of a test she had taken in her country. Paul was not familiar with the test, and therefore, the score had no meaning for him as to what course Luyen was qualified to take.

Paul called Jessica Saunders, his acquaintance at the regional office of a company that provides educational testing services across the United States. Jessica said she was familiar with the test Luyen had taken. She further had access to conversions of raw scores on this test to descriptive measures called percentiles.

Later that day, a phone call from Jessica provided Paul with the information he needed. The test was used as a calculus placement examination, and Luyen's score of 36.5 placed her at the 89th percentile. Based on this information, Paul felt confident that Luyen should enroll in the first of a three-term calculus sequence offered at Adams.

## The z score as a measure of position

Data points from different distributions can be compared by means of their $z$ scores.

### DEFINITION 2-11 The z Score (or Standard Score) of x

If $x$ is a data point in a population of data with mean $\mu$ and standard deviation $\sigma$, then the $z$ **score** of $x$ is:

$$z = \frac{x - \mu}{\sigma}$$

In the equation in Definition 2-11, the quantity $x - \mu$ measures the amount by which $x$ differs (or deviates) from $\mu$.

If $x - \mu < 0$, then $x$ is *less than* the mean.

If $x - \mu = 0$, then $x$ is *equal to* the mean.

If $x - \mu > 0$, then $x$ is *greater than* the mean.

The quotient $\frac{x - \mu}{\sigma}$ measures *the number of standard deviations* in the value $x - \mu$. For example, if $x$ is one and one-half standard deviations greater than the mean, then the quotient will be 1.5. For most sets of data, $z$ scores that are less than $-3.0$ or greater than $+3.0$ will rarely be obtained.

In this text, a $z$ score will always be written to two decimal places. The reason for this is that $z$ scores will generally be used in conjunction with Appendix Table 4. In this table, all values of $z$ are written to two decimal places.

▶ **Example 1**    Figure 2-27 contains data on the estimated life expectancies for males and females in $N = 60$ selected countries. For these sets of data, the following means and standard deviations have been computed:

| *Male* | *Female* |
|---|---|
| $\mu = 65.7$ years $\sigma = 7.2$ years | $\mu = 71.0$ years $\sigma = 8.4$ years |

**a.** Compute the $z$ score for the life expectancy of males in India.

**b.** Compute the $z$ score for the life expectancy of females in Mexico.

**c.** Relative to life expectanceis for their own sexes, which of the two groups considered in (a) and (b) has the greater life expectancy?

Table IV. Life Expectancy at Birth
in Years, for Selected Countries

| Country | Period | Male | Female |
|---|---|---|---|
| **Africa** | | | |
| Burundi | 1980–85[1] | 45.3 | 48.6 |
| Egypt | 1980–85[1] | 55.9 | 58.4 |
| Ivory Coast | 1980–85[1] | 46.9 | 50.2 |
| Kenya | 1980–85[1] | 56.3 | 60.0 |
| Nigeria | 1980–85[1] | 48.3 | 51.7 |
| Swaziland | 1980–85[1] | 46.8 | 50.0 |
| **Asia** | | | |
| China | 1980 | 66.0 | 69.0 |
| Hong Kong | 1982 | 71.9 | 77.6 |
| India | 1981 | 53.9 | 52.9 |
| Indonesia | 1980–85[1] | 51.2 | 53.9 |
| Israel | 1980 | 72.1 | 75.7 |
| Japan | 1982 | 74.2 | 79.7 |
| Kuwait | 1980–85[1] | 68.1 | 72.9 |
| Pakistan | 1980–85[1] | 54.4 | 54.2 |
| Taiwan | 1981 | 69.7 | 74.6 |
| Thailand | 1980–85[1] | 59.5 | 65.1 |
| **Europe** | | | |
| Albania | 1978–79 | 66.8 | 71.4 |
| Austria | 1981 | 69.3 | 76.4 |
| Belgium | 1980 | 69.0 | 75.0 |
| Bulgaria | 1980 | 69.0 | 75.0 |
| Czechoslovakia | 1980 | 66.8 | 74.0 |
| Denmark | 1980–81 | 71.1 | 77.2 |
| Finland | 1980 | 69.2 | 77.6 |
| France | 1981 | 70.2 | 78.5 |
| Germany, East | 1980 | 68.7 | 74.6 |
| Germany, West | 1979–81 | 69.9 | 76.6 |
| Greece | 1980 | 71.0 | 75.0 |
| Hungary | 1981 | 66.0 | 73.4 |
| Iceland | 1979–80 | 73.7 | 79.7 |
| Ireland | 1980 | 70.0 | 75.0 |
| Italy | 1980 | 70.0 | 76.0 |
| Netherlands, The | 1980 | 72.4 | 79.2 |
| Norway | 1980–81 | 72.5 | 79.2 |
| Poland | 1981 | 67.1 | 75.2 |
| Portugal | 1980 | 66.0 | 74.0 |
| Romania | 1980 | 68.0 | 73.0 |
| Spain | 1980 | 70.0 | 76.0 |
| Sweden | 1981 | 73.1 | 79.1 |
| Switzerland | 1980 | 72.0 | 78.0 |
| United Kingdom | 1978–80 | 70.2 | 76.2 |
| Yugoslavia | 1979 | 67.5 | 72.9 |
| **North America** | | | |
| Canada | 1980 | 70.0 | 77.0 |
| Costa Rica | 1980 | 68.0 | 72.0 |
| Cuba | 1980 | 71.0 | 74.0 |
| Martinique | 1980–85[1] | 67.8 | 73.0 |
| Mexico | 1980 | 62.0 | 67.0 |
| Panama | 1980 | 68.0 | 72.0 |
| Puerto Rico | 1979 | 69.6 | 76.1 |
| Trinidad and Tobago | 1980 | 66.0 | 72.0 |
| United States | 1982 | 70.8 | 78.2 |
| **Oceania** | | | |
| Australia | 1980 | 70.0 | 76.0 |
| New Zealand | 1980 | 70.0 | 76.0 |
| **South America** | | | |
| Argentina | 1980–85[1] | 66.8 | 73.2 |
| Brazil | 1980 | 60.0 | 64.0 |
| Chile | 1980–85[1] | 63.8 | 70.4 |
| Peru | 1980–85[1] | 56.7 | 59.7 |
| Suriname | 1980–85[1] | 66.3 | 71.5 |
| Uruguay | 1980–85[1] | 67.1 | 73.7 |
| Venezuela | 1980 | 64.0 | 69.0 |
| **U.S.S.R.** | 1981 | 65.0 | 74.0 |

[1]Projection.
Sources: United Nations, *World Statistics in Brief*, statistical pocketbook, 6th edition (1981), *World Population Trends and Prospects by Country, 1950–2000*, summary report of the 1978 assessment; official country sources.

**Figure 2-27.** "Life Expectancy at Birth In Years, for Selected Countries," *1984 Britannica Book of the Year*, p. 289. Copyright © 1984 Encyclopaedia Britannica, Inc. Reprinted by permission.

**Solution**   **a.** For males in India, $x = 53.9$ years.

$$z = \frac{53.9 - 65.7}{7.2} = -1.64, \text{ to two places.}$$

Thus males in India have a life expectancy about 1.64 standard deviations below the world per country average.

**b.** For females in Mexico, $x = 67.0$ years

$$z = \frac{67.0 - 71.0}{8.4} = -0.48, \text{ to two places.}$$

Thus females in Mexico have a life expectancy about 0.48 standard deviations below the world per country average.

**c.** Because females in Mexico have a $z$ score that is greater than males in India, they have relatively speaking a greater life expectancy.   ◄

## The percentile as a measure of position

Another measure of position is the *percentile*.

---

### DEFINITION 2-12 The $m$th Percentile of a Distribution of Data

If $x$ is a value such that approximately $m$ percent of a set of quantitative data is less than $x$, then $x$ is the value of the **$m$th percentile** of the distribution.

---

Recall that the median divides a distribution so that approximately 50 percent of the data points are less. The median is therefore the 50th percentile of a distribution.

► **Example 2**   An independent research agency conducted mileage tests on 50 new model cars, small vans, and trucks. The following data were rounded to the nearest whole number of miles per gallon, and are ranked in order from lowest to highest:

| | | | | | | | | | |
|---|---|---|---|---|---|---|---|---|---|
| 9 | 10 | 12 | 12 | 13 | 13 | 14 | 15 | 15 | 16 |
| 17 | 18 | 18 | 19 | 19 | 19 | 20 | 20 | 20 | 20 |
| 21 | 22 | 22 | 22 | 23 | 24 | 24 | 25 | 25 | 25 |
| 26 | 27 | 27 | 28 | 28 | 28 | 29 | 29 | 30 | 30 |
| 31 | 32 | 33 | 35 | 38 | 40 | 41 | 42 | 43 | 45 |

**a.** Find an approximate 20th percentile.
**b.** Find an approximate 60th percentile.

**Solution**   **a.** To locate the 20th percentile, compute 20 percent of 50. 20% of 50 = $(0.20)(50) = 10$. There are 10 data points ranked below the 20th percentile. The 11th item in the ranked data is 17. Thus 17 is the approximate 20th percentile for the data.

**b.** To locate the 60th percentile, compute 60 percent of 50. 60% of 50 = (0.60)(50) = 30. There are 30 data points ranked below the 60th percentile. The 31st item in the ranked data is 26. Thus 26 is the approximate 60th percentile for the data.    ◄

Percentiles put about 1 percent of the data points in classes extending from percentile to percentile. If a population has fewer than 100 data points (such as in Example 2), then some of the classes will be empty. For such sets of data, the percentiles separate the data into "too many" classes. There are similar groupings of data sets in which larger percentages of data are put into each class. Two such groupings are determined by the deciles and quartiles.

## Deciles and quartiles as measures of position

When a set of data is divided by deciles, approximately 10 percent of the data points are put into each of ten classes. When a set of data is divided by quartiles, then approximately 25 percent of the data points are put into each of four classes. In Figure 2-28, (a) and (b) histograms are drawn for a set of data that has been divided by deciles and quartiles.

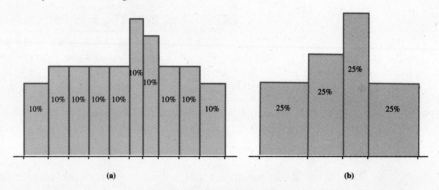

**Figure 2-28.**   (a) A set of data divided by deciles. (b) A set of data divided by quartiles.

► **Example 3**   The mileage data from Example 2 are listed below.

$$
\begin{array}{cccccccccc}
9 & 10 & 12 & 12 & 13 & 13 & 14 & 15 & 15 & 16 \\
17 & 18 & 18 & 19 & 19 & 19 & 20 & 20 & 20 & 20 \\
21 & 22 & 22 & 22 & 23 & 24 & 24 & 25 & 25 & 25 \\
26 & 27 & 27 & 28 & 28 & 28 & 29 & 29 & 30 & 30 \\
31 & 32 & 33 & 35 & 38 & 40 & 41 & 42 & 43 & 45
\end{array}
$$

**a.** Find an approximate 7th decile.
**b.** Find an approximate 1st quartile.

**Solution**    **a. Discussion.** The 7th decile is the 70th percentile. To locate the 7th decile, compute 70 percent of 50,

$$(0.70)(50) = 35$$

There are 35 data points ranked below the 7th decile. The 36th item in the ranked data is 28.

    The median divides a data set into two groups with approximately 50 percent of the data in each group. Therefore $\tilde{\mu}$ will be the 2nd quartile. The first quartile is the 25th percentile and the third quartile is the 75th percentile. However, for these special percentiles we will use a convention for calculation that is slightly different from the one illustrated in Example 2. That is, to determine the first quartile (written $Q_1$) and the third quartile (written $Q_3$), we can use the following procedure.

---

### Determining $Q_1$, $Q_2$, and $Q_3$

**Step 1**  Determine $\tilde{\mu}$, the middle value in the ordered array of an odd number of data points, or the mean of the two middle values in the ordered array of an even number of data points. This is $Q_2$.

**Step 2**  Determine the median of the smallest $n/2$ (or $N/2$) data points if $n$ (or $N$) is even, or the smallest $(n - 1)/2$ (or $(N - 1)/2$) if $n$ (or $N$) is odd. This value is $Q_1$ and can be determined by the same procedure used to determine $\tilde{\mu}$.

**Step 3**  Determine the median of the largest $n/2$ (or $N/2$) data points if $n$ (or $N$) is even, or the largest $(n - 1)/2$ (or $(N - 1)/2$) data points if $n$ (or $N$) is odd. This value is $Q_3$ and can be determined by the same procedure used to determine $\tilde{\mu}$.

---

**b.** Using the procedure described above:

**Step 1**  There are 50 data points. The 25th and 26th values in the ordered array are 23 and 24.

$$\tilde{\mu} = \frac{23 + 24}{2} = 23.5$$

**Step 2**  For the 25 data points less than 23.5, $Q_1$ is the middle value in the ordered array; that is, the 13th value. Thus, $Q_1$ is 18.   ◄

## The box-whisker plot as a display of data

The quartiles of a set of quantitative data can be used to construct a box-whisker plot, a visual display of the distribution about the median.

## DEFINITION 2-13 A Box-Whisker Plot (or Box-Plot)

A box-whisker plot is a visual display made using $Q_1$, $\tilde{\mu}$, $Q_3$, and the largest and smallest data values.

To illustrate a box-plot, we will use the data on the heights of the 40 trees measured by Glenn Fleming in Section 2-2. These data are arrayed from smallest to largest in Figure 2-29.

| 59 | 60 | 61 | 62 | 62 | 63 | 63 | 64 | 64 | 64 |
|----|----|----|----|----|----|----|----|----|----|
| 65 | 65 | 65 | 65 | 65 | 65 | 65 | 65 | 66 | 66 |
| 66 | 67 | 67 | 68 | 68 | 69 | 70 | 70 | 70 | 70 |
| 70 | 71 | 71 | 72 | 72 | 73 | 74 | 74 | 75 | 77 |

**Figure 2-29.**  The data (in inches) on heights arrayed from smallest to largest.

**Step 1**  The 20th and 21st values are 66 and 66.

$$\tilde{\mu} = \frac{66 + 66}{2} = 66$$

**Step 2**  For the 20 smallest values, 64 and 65 are the middle values.

$$Q_1 = \frac{64 + 65}{2} = 64.5$$

**Step 3**  For the 20 largest values, 70 and 70 are the middle values.

$$Q_3 = \frac{70 + 70}{2} = 70$$

In Figure 2-30, a scale is drawn with evenly spaced tick marks. The largest and smallest values are used to label the endpoints of the scale. The values of $Q_1$, $\tilde{\mu}$, and $Q_3$ are also shown.

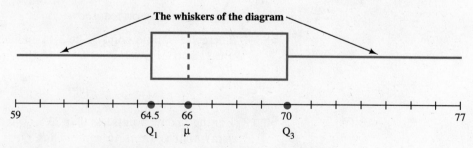

**Figure 2-30.**  A box-whisker plot of the data (in inches) on heights of trees.

A rectangle is drawn above the scale between 64.5 and 70, and divided at 66. Because these are the values of the first and third quartiles respectively, 50 percent of the data points are between 64.5 and 70. We can therefore imagine that one-half of the data points are "inside the box." Two segments are attached to the box and extend left and right to the smallest and largest values of the data. These segments appear to be "whiskers" attached to the box. Hence the terminology, "box-whisker plot."

▶ **Example 4** Dr. Joseph Grim had 48 students who took the final exam in his Chemistry 1-A class. Their scores are listed below.

84  76  65  93  90  84  92  52  85  89  86  58
93  66  73  82  88  75  83  71  67  76  79  95
69  62  55  84  61  77  56  80  85  73  89  90
87  76  73  82  60  58  77  94  62  70  78  64

**a.** Array the test scores using a stem-and-leaf diagram.
**b.** For the set of test scores, compute $Q_1$, $\tilde{\mu}$, and $Q_3$.
**c.** Construct a box-plot of the test scores.

**Solution** **a.** The stem-and-leaf diagram is shown in Figure 2-31.

| 5 | 2 5 6 8 8 |
|---|---|
| 6 | 0 1 2 2 4 5 6 7 9 |
| 7 | 0 1 3 3 3 5 6 6 6 7 7 8 9 |
| 8 | 0 2 2 3 4 4 4 5 5 6 7 8 9 9 |
| 9 | 0 0 2 3 3 4 5 |

**Figure 2-31.** A stem-and-leaf diagram of the test scores data.

**b.** **Step 1** The 24th and 25th values are 77 and 77.

$$\tilde{\mu} = \frac{77 + 77}{2} = 77$$

**Step 2** For the 24 smallest values, the middle values are 66 and 67, respectively.

$$Q_1 = \frac{66 + 67}{2} = 66.5$$

**Step 3** For the 24 largest values, the middle values are 85 and 86.

$$Q_3 = \frac{85 + 86}{2} = 85.5$$

**c.** The box-plot is shown in Figure 2-32.   ◀

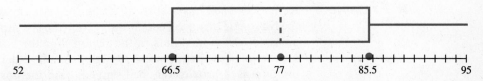

52          66.5          77          85.5          95

**Figure 2-32.** A box-plot of the test scores.

## Using z scores for bell-shaped distributions

If $x$ is an item of data from a bell-shaped distribution, then its $z$ score can be used to approximate the fraction of the distribution below $x$. Appendix Table 4 has the values of $z$ for fractions from 0.0002 through 0.9998. When $x$ is converted to a $z$ score, then the fraction associated with $z$ by Appendix Table 4 is the corresponding fraction for $x$.

| z | 0.00 | 0.01 | 0.02 | 0.03 | 0.04 | 0.05 | 0.06 | 0.07 | 0.08 | 0.09 |
|---|---|---|---|---|---|---|---|---|---|---|
| 0.0 | 0.5000 | 0.5040 | 0.5080 | 0.5120 | 0.5160 | 0.5199 | 0.5239 | 0.5279 | 0.5319 | 0.5359 |
| 0.1 | 0.5398 | 0.5438 | 0.5478 | 0.5517 | 0.5557 | 0.5596 | 0.5636 | 0.5675 | 0.5714 | 0.5753 |
| 0.2 | 0.5793 | 0.5832 | 0.5871 | 0.5910 | 0.5948 | 0.5987 | 0.6026 | 0.6064 | 0.6103 | 0.6141 |
| 0.3 | 0.6179 | 0.6217 | 0.6255 | 0.6293 | 0.6331 | 0.6368 | 0.6406 | 0.6443 | 0.6480 | 0.6517 |
| 0.4 | 0.6554 | 0.6591 | 0.6628 | 0.6664 | 0.6700 | 0.6736 | 0.6772 | 0.6808 | 0.6844 | 0.6879 |
| 0.5 | 0.6915 | 0.6950 | 0.6985 | 0.7019 | 0.7054 | 0.7088 | 0.7123 | 0.7157 | 0.7190 | 0.7224 |
| 0.6 | 0.7257 | 0.7291 | 0.7324 | 0.7357 | 0.7389 | 0.7422 | 0.7454 | 0.7486 | 0.7517 | 0.7549 |
| 0.7 | 0.7580 | 0.7611 | 0.7642 | 0.7673 | 0.7704 | 0.7734 | 0.7764 | 0.7794 | 0.7823 | 0.7852 |
| 0.8 | 0.7881 | 0.7910 | 0.7939 | 0.7967 | 0.7995 | 0.8023 | 0.8051 | 0.8078 | 0.8106 | 0.8133 |
| 0.9 | 0.8159 | 0.8186 | 0.8212 | 0.8238 | 0.8264 | 0.8289 | 0.8315 | 0.8340 | 0.8365 | 0.8389 |

**Figure 2-33.**   A portion of Appendix Table 4.

A portion of Appendix Table 4 is shown in Figure 2-33. To use this table:

**Step 1**  Locate the first two digits under the column headed $z$.

**Step 2**  Locate the third digit along the top row of the table.

**Step 3**  Move to the right in the row determined in step 1, to the column determined in step 2. The entry in the table is the fraction of the distribution below $x$.

Items (a) and (b) show how the table values are obtained for two values of $z$.

| Values of z | Row Number | Column Number | Table Value |
|---|---|---|---|
| a.  0.44 | 0.4 | 0.04 | $0.6700 = 67\%$ |
| b.  0.92 | 0.9 | 0.02 | $0.8212 \approx 82\%$ |

Based on examples (a) and (b), if $x$ is the value of a data point from a bell-shaped distribution and the $z$ score for $x$ is 0.44 or 0.92, then $x$ is the 67th or 82nd percentile, respectively, of the distribution.

▶ **Example 5**   Kelly O'Connell has a horse that she shows in dressage in the western region of the National Dressage Federation. During the last show season, her horse Colombo amassed 682 show points in the region. For the horses in this region, the mean number of show points was 578 with a standard deviation of 140, and the distribution was approximately bell-shaped. Find the position for Colombo's score in the distribution.

**Solution**   **Convert the value of $x$ to a $z$ score.**
With $x = 682$, $\mu = 578$, and $\sigma = 140$:

$$z = \frac{682 - 578}{140} \qquad z = \frac{x - \mu}{\sigma}$$

$$= 0.74 \qquad \text{Round to two decimal places.}$$

**Find the value for $z$ in Appendix Table 4.**
Using Figure 2-33,      To the right of 0.7
the table value is 0.7704.      Under the column headed 0.04

**Round the table value to two decimal places and write it as a percent.**

$$0.7704 \approx 0.77$$

$$= 77\%$$

In the distribution of show points, Colombo's score is the 77th percentile.   ◀

## Exercises 2-4   Set A

1. Deraugh Manion scored 82 on an English literature test in a class in which the mean was 74.0 and the standard deviation was 6.8. Clayton Turner scored a 76 on a similar English literature test in a class in which the mean was 63.7 and the standard deviation was 8.4.
   a. Change Deraugh's score to a $z$ score.
   b. Change Clayton's score to a $z$ score.
   c. Relatively speaking, which student scored higher on this test?

2. Araceli Silva lives in a city in which the average cost of rent for a one-bedroom apartment is $379 with a standard deviation of $65. Brenda Kantor lives in a city in which the average cost of rent for a similar apartment is $469 with a standard deviation of $87. Araceli and Brenda live in one-bedroom apartments and pay $325 and $410, respectively.
   a. Change Araceli's rent to a $z$ score.
   b. Change Brenda's rent to a $z$ score.
   c. Relatively speaking, which woman has the cheaper rent?

3. Wilma and Jesse run the 100-meter dash for their college track teams. In the last track meet, Wilma ran her dash in 12.2 seconds and Jesse ran his dash in 10.5 seconds. In the distributions of times for the 100-meter dash at this meet:

| For the Women Runners | For the Men Runners |
| --- | --- |
| $\mu = 13.0$ seconds<br>$\sigma = 0.8$ seconds | $\mu = 10.8$ seconds<br>$\sigma = 0.5$ seconds |

   a. Change Wilma's time to a $z$ score.
   b. Change Jesse's time to a $z$ score.
   c. Relatively speaking, which runner had the faster time?

4. Angela and Peter run, respectively, the 3-mile and 4-mile event for their college track teams. In the last track meet, which included runners from 25 schools, Angela ran her event in 19.6 minutes and Peter ran his event in 20.2 minutes. In the distributions of times for these events at this meet:

| For the Women Runners | For the Men Runners |
|---|---|
| $\mu = 20.5$ minutes<br>$\sigma = 2.3$ minutes | $\mu = 21.7$ minutes<br>$\sigma = 1.9$ minutes |

   a. Change Angela's time to a z score.
   b. Change Peter's time to a z score.
   c. Relatively speaking, which runner had the faster time?

5. The high temperature in city P yesterday was 98 degrees. For this month of the year, the average high temperature in city P is 95 degrees with a standard deviation of 8.5 degrees. The high temperature in city C yesterday was 82 degrees. For this month of the year, the average high temperature in city C is 80 with a standard deviation of 5.6 degrees.
   a. Change city P's temperature to a z score.
   b. Change city C's temerature to a z score.
   c. Relatively speaking, which city had the higher temperature?

6. In country A, the life expectancy for men is 70.1 years, and the standard deviation for the distribution of ages at death is 12.8 years. In country B, the life expectancy for men is 53.6 years, and the standard deviation for the distribution of ages at death is 16.8 years.
   a. Change to a z score the age at death of a man from country A who died at age 78 years.
   b. Change to a z score the age at death of a man from country B who died at age 64 years.
   c. Relatively speaking, which man was older at the time of his death?

7. At Riverdale Community College, the mean grade point average (gpa) of the current student body is 2.87 with a standard deviation of 0.52. To two decimal places, compute the following gpa's:
   a. A gpa with a z score of 1.35.
   b. A gpa with a z score of 0.70.
   c. A gpa with a z score of $-0.42$.
   d. A gpa with a z score of $-1.10$.

8. Nicron Corporation has approximately 8,500 employees. The average length of service for these employees is 12.9 years with a standard deviation of 6.4 years. To the nearest one-tenth of a year, compute the following lengths of service:
   a. A time in service with a z score of 1.82.
   b. A time in service with a z score of 0.55.
   c. A time in service with a z score of $-1.48$.
   d. A time in service with a z score of $-0.26$.

**9.** The following scores on the final examination in Logic 101 were recorded:

$$
\begin{array}{cccccccccc}
84 & 76 & 65 & 93 & 90 & 84 & 92 & 52 & 85 & 79 \\
76 & 75 & 82 & 56 & 93 & 90 & 80 & 75 & 70 & 66 \\
75 & 80 & 66 & 46 & 71 & 78 & 75 & 72 & 59 & 85 \\
96 & 69 & 76 & 57 & 61 & 83 & 90 & 84 & 76 & 73 \\
90 & 86 & 50 & 58 & 73 & 84 & 53 & 73 & 78 & 64
\end{array}
$$

a. Rank the data from lowest to highest score.
b. Find the 40th percentile for the data.
c. Find the 66th percentile for the data.
d. Find the 3rd decile.
e. Find the 9th decile.
f. Find the 1st quartile, $Q_1$.
g. Find the 3rd quartile, $Q_3$.

**10.** The Advantage Financial Planning Service Corporation maintains a portfolio for its clients that includes stocks from 60 corporations listed on the New York Stock Exchange. The closing prices (in dollars) of each stock after yesterday's trading are listed below.

$$
\begin{array}{ccccccccc}
16 & 10\frac{1}{2} & 6\frac{1}{8} & 12\frac{1}{2} & 24\frac{3}{4} & 30\frac{1}{4} & 18\frac{1}{4} & 27\frac{1}{4} & 27\frac{5}{8} \\
9\frac{3}{4} & 37\frac{3}{4} & 19\frac{1}{4} & 14 & 9 & 13\frac{3}{8} & 17 & 8\frac{5}{8} & 88\frac{3}{4} \\
13\frac{1}{4} & 14\frac{1}{2} & 7\frac{7}{8} & 32\frac{1}{2} & 5\frac{1}{4} & 28\frac{1}{2} & 11\frac{1}{8} & 15 & 41\frac{1}{4} \\
26\frac{1}{2} & 10 & 5\frac{3}{8} & 29\frac{1}{4} & 3\frac{1}{2} & 7\frac{1}{8} & 9 & 25\frac{5}{8} & 8\frac{3}{4} \\
76\frac{7}{8} & 17 & 14\frac{1}{2} & 13\frac{3}{4} & 30\frac{1}{2} & 16\frac{1}{2} & 19 & 36\frac{3}{4} & 27 \\
10\frac{1}{4} & 50\frac{1}{2} & 35 & 27\frac{5}{8} & 119\frac{1}{2} & 26\frac{1}{4} & 9\frac{1}{2} & 23 & 41 \\
98 & 24\frac{7}{8} & 7\frac{3}{4} & 27\frac{1}{2} & 12\frac{1}{8} & 115\frac{1}{2} & & &
\end{array}
$$

a. Rank the data from lowest to highest price.
b. Find the 30th percentile for the data.
c. Find the 65th percentile for the data.
d. Find the 2nd decile.
e. Find the 8th decile.
f. Find the 1st quartile, $Q_1$.
g. Find the 3rd quartile, $Q_3$.

**11.** A new formula for paint was tested by applying it to 25 different surfaces. The times, measured in minutes, for the paint to dry on these surfaces were recorded. The measurements, rounded to the nearest minute, are listed below.

$$
\begin{array}{ccccccccccccc}
106 & 97 & 93 & 119 & 96 & 89 & 110 & 98 & 87 & 103 & 99 & 96 & 79 \\
98 & 104 & 88 & 103 & 78 & 112 & 96 & 108 & 92 & 95 & 106 & 87 & 
\end{array}
$$

a. Array the scores using a stem-and-leaf diagram.
b. For the set of scores, compute $Q_1$, $\tilde{\mu}$, and $Q_3$.
c. Construct a box-plot of the scores.

12. The prices of 30 homes for sale in the Greenhaven Area were selected from the home buyers section of a local newspaper. The prices, rounded to the nearest thousand dollars, are listed below.

$$
\begin{array}{cccccccccc}
123 & 105 & 98 & 109 & 115 & 126 & 135 & 88 & 110 & 126 \\
102 & 119 & 132 & 124 & 118 & 97 & 115 & 138 & 105 & 95 \\
129 & 89 & 107 & 106 & 110 & 103 & 123 & 92 & 115 & 128
\end{array}
$$

a. Array the prices using a stem-and-leaf diagram.
b. For the set of prices, compute $Q_1$, $\tilde{\mu}$, and $Q_3$.
c. Construct a box-plot of the prices.

13. A group of fishermen visited a lake that is known for its trout. The following data are the lengths of the fish caught by the group. The lengths have been rounded to the nearest one-tenth of an inch.

$$
\begin{array}{cccccccccc}
15.4 & 13.9 & 12.8 & 13.5 & 16.7 & 17.3 & 14.3 & 13.8 & 12.7 & 13.0 \\
12.8 & 14.5 & 16.7 & 14.8 & 13.7 & 15.8 & 13.4 & 15.7 & 14.1 & 16.9 \\
16.5 & 15.0 & 13.5 & 15.7 & 14.3 & 13.2 & 14.6 & 13.8 & 15.2 & 15.3 \\
13.2 & 17.8 & 14.9 & 16.4 & 13.5 & 15.4 & 12.9 & 14.7 & 13.9 & 14.2
\end{array}
$$

a. Array the lengths using a stem-and-leaf diagram.
b. For the set of lengths, compute $Q_1$, $\tilde{\mu}$, and $Q_3$.
c. Construct a box-plot of the lengths.

14. The weights (in pounds) of the players on a National Football League team are listed below.

$$
\begin{array}{ccccccccccc}
282 & 255 & 250 & 245 & 235 & 270 & 225 & 293 & 205 & 227 & 209 \\
255 & 208 & 254 & 210 & 263 & 275 & 265 & 220 & 195 & 230 & 195 \\
196 & 215 & 193 & 202 & 196 & 198 & 215 & 231 & 260 & 234 & 262 \\
195 & 270 & 215 & 267 & 198 & 251 & 246 & 219 & 265 & 245 & 210
\end{array}
$$

a. Array the weights using a stem-and-leaf diagram.
b. For the set of weights, compute $Q_1$, $\tilde{\mu}$, and $Q_3$.
c. Construct a box-plot of the weights.

15. A standardized test produces scores with a mean of 500 and a standard deviation of 100. The distribution of scores on this test is known to be bell shaped.
a. Find the fraction of scores below 570.
b. Find the fraction of scores below 495.

16. As presently adjusted, a robotic welder at Cumberland Tractor Manufacturers is making spot welds with a mean shear strength of 12,750 pounds and a standard deviation of shear strength of 200 pounds. The distribution of shear strengths is bell shaped.
a. Find the fraction of spot welds with shear strength of less than 12,500 pounds.
b. Find the fraction of spot welds with shear strength of less than 12,900 pounds.

17. Lisa and Kathy are sales representatives for a large medical supply firm. Last year, Lisa earned $32,980 and Kathy earned $22,460 on sales. The distribution of earnings based on sales for the staff

of this firm is bell shaped with a mean of $25,600 and a standard deviation of $6,275.

a. Find the fraction of earnings less than Lisa's.

b. Find the fraction of earnings less than Kathy's.

18. Tom and Harry are graduating seniors at a major university in the South. For the graduating class,

the mean grade point average (gpa) is 2.94 with a standard deviation of 0.37. Tom's final gpa is 2.78 and Harry's final gpa is 3.32.

a. Find the fraction of gpa's below Tom's.

b. Find the fraction of gpa's below Harry's.

## Exercises 2-4    Set B

*In Exercises 1 and 2, use the data on life expectancies in Figure 2-27.*

1. The life expectancies for men run from 45.3 years (Burundi) through 74.2 years (Japan).
   a. Array the 60 data points from smallest to largest.
   b. Determine $Q_1$.
   c. Determine $\tilde{\mu}$.
   d. Determine $Q_3$.
   e. Construct a box-plot for the data.
   f. Let $x_1 = 45.3$, $x_2 = Q_1$, $x_3 = \tilde{\mu}$, $x_4 = Q_3$, and $x_5 = 74.2$. Compute $x_2 - x_1$, $x_3 - x_2$, $x_4 - x_3$, and $x_5 - x_4$.
   g. Which of the four intervals defined by the quartiles has the smallest length?

2. The life expectancies for women run from 48.6 years (Burundi) through 79.7 years (Iceland and Japan).
   a. Array the 60 data points from smallest to largest.
   b. Determine $Q_1$.
   c. Determine $\tilde{\mu}$.
   d. Determine $Q_3$.
   e. Construct a box-plot for the data.
   f. Let $x_1 = 48.6$, $x_2 = Q_1$, $x_3 = \tilde{\mu}$, $x_4 = Q_3$, and $x_5 = 79.7$. Compute $x_2 - x_1$, $x_3 - x_2$, $x_4 - x_3$, and $x_5 - x_4$.
   g. Which of the four intervals defined by the quartiles has the smallest length?

*In Exercises 3 and 4, use the area (sq mi) information. In Figure 2-34, for the 50 states in the United States:*

$$\mu = 73,582 \text{ sq mi} \quad and \quad \sigma = 87,488 \text{ sq mi}$$

### Area and population

| States | Capitals | area[1] sq mi | area[1] sq km | population 1989 estimate[2] |
|---|---|---|---|---|
| Alabama | Montgomery | 51,705 | 133,915 | 4,150,000 |
| Alaska | Juneau | 591,004 | 1,530,693 | 565,000 |
| Arizona | Phoenix | 114,000 | 295,259 | 3,649,000 |
| Arkansas | Little Rock | 53,187 | 137,754 | 2,414,000 |
| California | Sacramento | 158,706 | 411,047 | 28,607,000 |
| Colorado | Denver | 104,091 | 269,594 | 3,393,000 |
| Connecticut | Hartford | 5,018 | 12,997 | 3,257,000 |
| Delaware | Dover | 2,044 | 5,294 | 658,000 |
| Florida | Tallahassee | 58,664 | 151,939 | 12,535,000 |
| Georgia | Atlanta | 58,910 | 152,576 | 6,524,000 |
| Hawaii | Honolulu | 6,471 | 16,760 | 1,121,000 |
| Idaho | Boise | 83,564 | 216,430 | 1,013,000 |
| Illinois | Springfield | 57,871 | 149,885 | 11,599,000 |
| Indiana | Indianapolis | 36,413 | 94,309 | 5,542,000 |
| Iowa | Des Moines | 56,275 | 145,752 | 2,780,000 |
| Kansas | Topeka | 82,277 | 213,096 | 2,485,000 |
| Kentucky | Frankfort | 40,409 | 104,659 | 3,742,000 |
| Louisiana | Baton Rouge | 47,752 | 123,677 | 4,510,000 |
| Maine | Augusta | 33,265 | 86,156 | 1,203,000 |
| Maryland | Annapolis | 10,460 | 27,091 | 4,665,000 |
| Massachusetts | Boston | 8,284 | 21,455 | 5,863,000 |
| Michigan | Lansing | 97,102 | 251,493 | 9,266,000 |
| Minnesota | St. Paul | 86,614 | 224,329 | 4,298,000 |
| Mississippi | Jackson | 47,689 | 123,514 | 2,680,000 |
| Missouri | Jefferson City | 69,697 | 180,514 | 5,163,000 |
| Montana | Helena | 147,046 | 380,847 | 808,000 |
| Nebraska | Lincoln | 77,355 | 200,349 | 1,590,000 |
| Nevada | Carson City | 110,561 | 286,352 | 1,049,000 |
| New Hampshire | Concord | 9,279 | 24,032 | 1,116,000 |
| New Jersey | Trenton | 7,787 | 20,168 | 7,827,000 |
| New Mexico | Santa Fe | 121,593 | 314,924 | 1,595,000 |
| New York | Albany | 52,735 | 136,583 | 17,761,000 |
| North Carolina | Raleigh | 52,669 | 136,412 | 6,602,000 |
| North Dakota | Bismarck | 70,702 | 183,117 | 664,000 |
| Ohio | Columbus | 44,787 | 115,998 | 10,787,000 |
| Oklahoma | Oklahoma City | 69,956 | 181,185 | 3,285,000 |
| Oregon | Salem | 97,073 | 251,418 | 2,750,000 |
| Pennsylvania | Harrisburg | 46,043 | 119,251 | 11,844,000 |
| Rhode Island | Providence | 1,212 | 3,139 | 996,000 |
| South Carolina | Columbia | 31,113 | 80,582 | 3,507,000 |
| South Dakota | Pierre | 77,116 | 199,730 | 708,000 |
| Tennessee | Nashville | 42,144 | 109,152 | 4,933,000 |
| Texas | Austin | 266,807 | 691,027 | 17,451,000 |
| Utah | Salt Lake City | 84,899 | 219,887 | 1,750,000 |
| Vermont | Montpelier | 9,614 | 24,900 | 557,000 |
| Virginia | Richmond | 40,767 | 105,586 | 6,068,000 |
| Washington | Olympia | 68,139 | 176,479 | 4,612,000 |
| West Virginia | Charleston | 24,231 | 62,758 | 1,871,000 |
| Wisconsin | Madison | 66,215 | 171,496 | 4,803,000 |
| Wyoming | Cheyenne | 97,809 | 253,324 | 503,000 |

| District | | | | |
|---|---|---|---|---|
| Dist. of Columbia | — | 69 | 179 | 615,000 |
| TOTAL | | 3,679,192[3] | 9,529,063 | 247,732,000[3] |

**Figure 2-34.** "U.S. Area and Population," *1990 Britannica World Data*, p. 740. Copyright © 1990 by Encyclopaedia Britannica, Inc. Reprinted by permission.

*In Figure 2-35, for the 49 provinces in Poland:*

$$\mu = 2,464 \text{ sq mi} \quad and \quad \sigma = 856 \text{ sq mi}$$

3. a. Compute the z score for Ohio in the United States.
   b. Compute the z score for Krosno in Poland.
   c. Relatively speaking, which region has the greater area?

4. a. Compute the z score for Nevada in the United States.
   b. Compute the z score for Opole in Poland.
   c. Relatively speaking, which region has the greater area?

*In Exercises 5 and 6, use the population (1988 and 1989 estimates) information. In Figure 2-34, for the 50 states in the United States:*

$$\mu \approx 4,942,000 \quad and \quad \sigma \approx 5,296,000$$

*In Figure 2-35, for the 49 provinces in Poland:*

$$\mu \approx 762,050 \quad and \quad \sigma \approx 588,140$$

5. a. Compute the z score for Virginia in the United States.
   b. Compute the z score for Bielsko in Poland.
   c. Relatively speaking, which region has the greater population?

6. a. Compute the z score for Oklahoma in the United States.
   b. Compute the z score for Konin in Poland.
   c. Relatively speaking, which region has the greater population?

| Area and population | | area | | population |
|---|---|---|---|---|
| **Provinces** | **Capitals** | sq mi | sq km | 1988[1] estimate |
| Biała Podlaska | Biała Podlaska | 2,065 | 5,348 | 301,100 |
| Białystok | Białystok | 3,882 | 10,055 | 679,900 |
| Bielsko | Bielsko Biala | 1,430 | 3,704 | 884,700 |
| Bydgoszcz | Bydgoszcz | 3,996 | 10,349 | 1,096,600 |
| Chełm | Chełm | 1,493 | 3,866 | 243,500 |
| Ciechanów | Ciechanów | 2,456 | 6,362 | 422,200 |
| Częstochowa | Częstochowa | 2,387 | 6,182 | 771,300 |
| Elbląg | Elbląg | 2,356 | 6,103 | 472,200 |
| Gdańsk | Gdańsk | 2,855 | 7,394 | 1,419,800 |
| Gorzów | Gorzów Wielkopolski | 3,276 | 8,484 | 490,700 |
| Jelenia Góra | Jelenia Góra | 1,690 | 4,378 | 513,800 |
| Kalisz | Kalisz | 2,514 | 6,512 | 703,700 |
| Katowice | Katowice | 2,568 | 6,650 | 3,970,800 |
| Kielce | Kielce | 3,556 | 9,211 | 1,115,600 |
| Konin | Konin | 1,984 | 5,139 | 463,400 |
| Koszalin | Koszalin | 3,270 | 8,470 | 498,400 |
| Kraków | Kraków | 1,256 | 3,254 | 1,216,600 |
| Krosno | Krosno | 2,202 | 5,702 | 483,700 |
| Legnica | Legnica | 1,559 | 4,037 | 502,700 |
| Leszno | Leszno | 1,604 | 4,154 | 380,400 |
| Łódź | Łódź | 588 | 1,523 | 1,148,400 |
| Łomża | Łomża | 2,581 | 6,684 | 342,200 |
| Lublin | Lublin | 2,622 | 6,792 | 997,300 |
| Nowy Sącz | Nowy Sącz | 2,153 | 5,576 | 679,400 |
| Olsztyn | Olsztyn | 4,759 | 12,327 | 738,600 |
| Opole | Opole | 3,295 | 8,535 | 1,022,700 |
| Ostrołęka | Ostrołęka | 2,509 | 6,498 | 389,500 |
| Piła | Piła | 3,168 | 8,205 | 472,600 |
| Piotrków | Piotrków Trybunalski | 2,419 | 6,266 | 639,200 |
| Płock | Płock | 1,976 | 5,117 | 512,800 |
| Poznań | Poznań | 3,147 | 8,151 | 1,316,100 |
| Przemyśl | Przemyśl | 1,713 | 4,437 | 400,500 |
| Radom | Radom | 2,816 | 7,294 | 736,500 |
| Rzeszów | Rzeszów | 1,698 | 4,397 | 703,700 |
| Siedlce | Siedlce | 3,281 | 8,499 | 642,000 |
| Sieradz | Sieradz | 1,880 | 4,869 | 403,700 |
| Skierniewice | Skierniewice | 1,529 | 3,960 | 413,200 |
| Słupsk | Słupsk | 2,878 | 7,453 | 403,900 |
| Suwałki | Suwałki | 4,050 | 10,490 | 458,900 |
| Szczecin | Szczecin | 3,854 | 9,981 | 958,700 |
| Tarnobrzeg | Tarnobrzeg | 2,426 | 6,283 | 586,900 |
| Tarnów | Tarnów | 1,603 | 4,151 | 651,800 |
| Toruń | Toruń | 2,065 | 5,348 | 649,600 |
| Wałbrzych | Wałbrzych | 1,609 | 4,168 | 739,300 |
| Warszawa | Warszawa | 1,463 | 3,788 | 2,431,800 |
| Włocławek | Włocławek | 1,700 | 4,402 | 428,000 |
| Wrocław | Wrocław | 2,427 | 6,287 | 1,121,100 |
| Zamość | Zamość | 2,695 | 6,980 | 489,900 |
| Zielona Góra | Zielona Góra | 3,424 | 8,868 | 654,900 |
| TOTAL | | 120,727 | 312,683 | 37,764,300 |

**Figure 2-35.** "Poland Area and Population," *1990 Britannica World Data*, p. 696. Copyright © 1990 by Encyclopaedia Britannica, Inc. Reprinted by permission.

**Key Topics**

# 2-5 Proportion of a Population of Qualitative Data

1. *P as a measure of composition of a population*

2. *$\hat{p}$ as a measure of composition of a sample*

STATISTICS IN ACTION

David Garrison is a newsperson on station WHOZ-TV. In the upcoming June primary election, the following are the five candidates for the Democratic nomination for county supervisor from the third district:

George Aufmann — a retired attorney

Frieda Baker — the incumbent supervisor

Ernestine Carroll — the chancellor of the city school system

David Donnelly — a college political science instructor

Kim Eppenson — president of Eppenson Land Developers Inc.

David wanted to report on the preferences of voters on the candidates. He therefore instructed three of his staff members to get a list of Democratic voters in the third district and to poll by telephone a random sample of the individuals on the list. At the end of 2 days, the staff had compiled the preferences of 286 voters. The results of the poll are shown in Table 2-7.

To report these data in a meaningful way, David changed the totals to relative frequencies. The ratios were then changed to percents.

George Aufmann — Percent in favor $= \frac{23}{286} = 8\%$

Frieda Baker — Percent in favor $= \frac{91}{286} \approx 32\%$

Ernestine Carroll — Percent in favor $= \frac{63}{286} \approx 22\%$

David Donnelly — Percent in favor $= \frac{28}{286} \approx 10\%$

Kim Eppenson — Percent in favor $= \frac{38}{286} \approx 13\%$

Undecided — Percent undecided $= \frac{43}{286} = 15\%$

At the subsequent news broadcast, David reported to the viewing audience, "Incumbent Frieda Baker is ahead of her challengers for the Democratic nomination for supervisor of the third district with 32% of the prospective votes. However, 15% of the voters are still undecided."

**TABLE 2-7** The Results of the Telephone Poll

| Candidate | Tally Marks | Totals |
|-----------|-------------|--------|
| George Aufmann | 卌 卌 卌 卌 ||| | 23 |
| Frieda Baker | 卌 卌 卌 卌 卌 卌 卌 卌 \| <br> 卌 卌 卌 卌 卌 卌 卌 卌 卌 | 91 |
| Ernestine Carroll | 卌 卌 卌 卌 卌 卌 卌 卌 卌 卌 卌 卌 ||| | 63 |
| David Donnelly | 卌 卌 卌 卌 卌 ||| | 28 |
| Kim Eppenson | 卌 卌 卌 卌 卌 卌 卌 ||| | 38 |
| Undecided | 卌 卌 卌 卌 卌 卌 卌 卌 ||| | 43 |

## *P* as a measure of composition of a population

The measure frequently used to describe a population of qualitative data is $P$, the proportion of a population in a particular category of interest.

---

### DEFINITION 2-14 The Proportion of a Population, *P*

If A is the subset of a finite population consisting of those elements that have some characteristic, attitude, or attribute of interest, then $P_A$ is the **proportion of the population** with the characteristic, and

$$P_A = \frac{n(A)}{N}$$

$n(A)$ ⟶ The number of objects in A.

$N$ ⟶ The number of objects in the population.

There are three ways in which $P_A$ can be written.

### Three forms for writing $P_A$

**a.** A common fraction in reduced form.
**b.** A decimal fraction, rounded off, if necessary.
**c.** A percentage.

The percentage form for $P_A$ is probably the one that is most commonly used. For example, as illustrated above in the comments by David Garrison, the results of polls are usually given in percentages.

▶ **Example 1**    Last Tuesday, 2225 issues were traded on the stock exchange. Of these issues:

    950 had stock prices that advanced (A)

  1240 had stock prices that declined (D)

     35 had stock prices that were unchanged (U)

Compute to the nearest one-tenth of 1 percent:

**a.** $P_A$    **b.** $P_D$    **c.** $P_U$

Solution    **a.** $P_A = \dfrac{950}{2225}$    $\dfrac{950 \text{ issues advanced}}{2225 \text{ issues traded}}$

$\quad\quad\quad = 0.427$    Rounded to three places

$\quad\quad\quad = 42.7\%$    Written as a percentage

**b.** $P_D = \dfrac{1240}{2225}$    $\dfrac{1240 \text{ issues declined}}{2225 \text{ issues traded}}$

$\quad\quad\quad = 0.557$    Rounded to three places

$\quad\quad\quad = 55.7\%$    Written as a percentage

**c.** $P_U = \dfrac{35}{2225}$    $\dfrac{35 \text{ issues were unchanged}}{2225 \text{ issues traded}}$

$\quad\quad\quad = 0.016$    Rounded to three places

$\quad\quad\quad = 1.6\%$    Written as a percentage    ◀

2-5 Proportion of a Population of Qualitative Data

▶ **Example 2**   Figure 2-36 contains data regarding religious affiliation in the United States. Calculate the following proportions to the nearest one-tenth of 1 percent:

**a.** $P_C$, the proportion of individuals affiliated with the Roman Catholic religion.
**b.** $P_J$, the proportion of individuals affiliated with the Jewish religion.

**Solution**   **Discussion.**  For the population in Figure 2-36, $N = 248,780,000$, the approximate number of residents in the United States at the time the data were collected.

**a.** With $n(C) = 73,740,000$ and $N = 248,780,000$:

| United States | |
|---|---|
| Christian[1] | 216,640,000 |
| Protestant | 122,170,000 |
| Roman Catholic | 73,740,000 |
| Anglican | 5,900,000 |
| Eastern Orthodox | 5,400,000 |
| Jewish | 6,770,000 |
| Muslim | 4,730,000 |
| Hindu | 550,000 |
| atheist and | |
| nonreligious | 16,840,000 |
| other | 3,250,000 |

$$P_C = \frac{73,740,000}{248,780,000}$$

$$= 0.29640 \ldots$$

$$\approx 29.6\%$$

About 29.6 percent of U.S. residents are affiliated with the Roman Catholic religion.

**b.** With $n(J) = 6,770,000$ and $N = 248,780,000$:

$$P_J = \frac{6,770,000}{248,780,000}$$

$$= 0.0272 \ldots$$

$$\approx 2.7\%$$

**Figure 2-36.**  "United States," *1990 Britannica World Data,* p. 779. Copyright © 1990 by Encyclopaedia Britannica, Inc. Reprinted by permission.

About 2.7 percent of U.S. residents are affiliated with the Jewish religion.   ◀

## $\hat{p}$ as a measure of composition of a sample

Recall that $\bar{x}$ is a statistic that is the mean of a sample of quantitative data. In a similar way $\hat{p}$ (read "$p$-hat") is a statistic that describes the proportion of a sample of qualitative data with a characteristic of interest.

---

### DEFINITION 2-15 The Proportion of a Sample, $\hat{p}$

---

If A is the subset of a sample of size $n$ consisting of those elements that have some characteristic, attitude, or attribute of interest, then $\hat{p}_A$ is the **proportion of the sample** with the characteristic, and

$$\hat{p}_A = \frac{n(A)}{n}$$

To illustrate the use of this symbol, consider again the data in Table 2-7. Let $\hat{p}_B$ be the proportion of voters in the sample that are in favor of Frieda Baker. With $n(B) = 91$ and $n = 286$:

$$\hat{p}_B = \frac{91}{286} = 0.3181 \ldots \approx 32\%$$

The statistic $\hat{p}$, like $\bar{x}$, is a quantity that can change from sample to sample. For example, if David Garrison authorized another sample of voter preferences, the sample proportions obtained would probably differ from the ones shown in Table 2-7.

▶ **Example 3**    A report from the National Park Service, recently, stated that smog from Los Angeles and the San Joaquin Valley had affected more than one-half the trees in national parks in California. Suppose that an ecologist for the National Park Service in the state of Washington examines 740 trees in a national park in her district and finds evidence of smog damage to 163 of the trees. Use these data to estimate to the nearest percent the proportion of trees in this area with smog damage.

**Solution**    If A is the set of smog damaged trees, then $n(A) = 163$. With $n = 740$:

$$\hat{p}_A = \frac{163}{740} \qquad \hat{p}_A = \frac{n(A)}{n}$$

$$= 0.22 \qquad \text{Rounded to two places}$$

$$= 22\% \qquad \text{Written as a percentage}$$

Based on these data, she estimated that 22 percent of the trees in her area show evidence of smog damage.    ◀

## Exercises 2-5    Set A

*In Exercises 1–4, use the results of a survey of the 75 students in Dr. Warren's classical humanities course for the current fall semester. Write each value of P to the nearest percent.*

**1.** The class standings of the students were as follows:

12 were freshmen (f)

33 were sophomores (s)

25 were juniors (j)

5 were seniors (e)

Find: a. $P_f$    b. $P_s$    c. $P_j$    d. $P_e$

**2.** The ethnic backgrounds of the students were noted as follows:

24 were Caucasian (c)

21 were Asian (a)

15 were black (b)

11 were hispanic (h)

4 were other (t)

Find: a. $P_c$    b. $P_a$    c. $P_b$    d. $P_h$    e. $P_t$

3. The employment status of each student was as follows:

   10 were working full time (f)

   28 were working part-time (p)

   37 were not working (n)

   Find: a. $P_f$ b. $P_p$ c. $P_n$

4. The students' opinions about whether the local power company should continue to operate its nuclear power plant to generate electricity were as follows:

   10 students were strongly in favor of continued use (A)

   28 students were moderately in favor of continued use (B)

   15 students had no opinion on the question (C)

   19 students were moderately opposed to continued use (D)

   3 students were strongly opposed to continued use (E)

   Find: a. $P_A$ b. $P_B$ c. $P_C$ d. $P_D$ e. $P_E$

5. Betty Pappas has 120 books in her personal library. The books include 60 novels (n), 20 history books (h), 28 educational books (e), and 12 sports books (s). Compute the following proportions to the nearest percent:
   a. $P_n$   b. $P_h$   c. $P_e$   d. $P_s$

6. Luong Ho is a nutritionist studying the eating habits of teenagers in the United States. He is currently working with the 1375 students in grades 10 through 12 at Clayton High School. A survey of these students yields the following data:

   440 eat breakfast every morning (e)

   495 frequently eat breakfast (f)

   275 occasionally eat breakfast (o)

   165 never eat breakfast (n)

   Compute the following proportions to the nearest percent:
   a. $P_e$   b. $P_f$   c. $P_o$   d. $P_n$

| United States | | | |
|---|---|---|---|
| American Indian or Alaska Native languages | 430,000 | Korean | 340,000 |
|  |  | Lithuanian | 80,000 |
|  |  | Norwegian | 130,000 |
| Arabic | 270,000 | Persian | 130,000 |
| Armenian | 120,000 | Philippine languages | 600,000 |
| Asian Indian languages | 310,000 | Polish | 940,000 |
| Chinese | 760,000 | Portuguese | 420,000 |
| Czech | 140,000 | Russian | 200,000 |
| Dutch | 180,000 | Serbo-Croatian | 180,000 |
| ● English | 220,620,000 | Slovak | 100,000 |
| Finnish | 80,000 | Spanish | 13,910,000 |
| French | 1,870,000 | Swedish | 120,000 |
| German | 1,880,000 | Thai | 110,000 |
| Greek | 480,000 | Ukrainian | 140,000 |
| Hungarian | 200,000 | Vietnamese | 240,000 |
| Italian | 1,870,000 | Yiddish | 360,000 |
| Japanese | 400,000 | Other | 1,190,000 |

**Figure 2-37.** "Major languages spoken in the United States." *1990 Britannica World Data*, pp. 777–778. Copyright © 1990 by Encyclopaedia Britannica, Inc. Reprinted by permission.

7. Use the major language data in Figure 2-37. Use $N = 248,780,000$ for the size of the population. Compute to the nearest one-tenth of 1 percent:
   a. $P_S$, the proportion of the population that speaks Spanish
   b. $P_E$, the proportion of the population identified as speaking French or German.

8. Use the religious affiliation data in Figure 2-38. Use $N = 1,104,275,000$ for the size of the population. Compute to the nearest one-tenth of 1 percent:
   a. $P_f$, the proportion of the population that are Chinese folk-religionist.
   b. $P_c$, the proportion of the population that are Buddhist or Muslim.

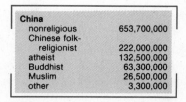

| China | |
|---|---|
| nonreligious | 653,700,000 |
| Chinese folk-religionist | 222,000,000 |
| atheist | 132,500,000 |
| Buddhist | 63,300,000 |
| Muslim | 26,500,000 |
| other | 3,300,000 |

**Figure 2-38.** "Religious affiliations in China." *1990 Britannica World Data*, pp. 777–778. Copyright © 1990 by Encyclopaedia Britannica, Inc. Reprinted by permission.

9. The molecular formula for sugar is $C_{12}H_{22}O_{11}$, meaning that a sugar molecule contains 45 atoms. Of these:

> 12 are carbon atoms (C)
>
> 22 are hydrogen atoms (H)
>
> 11 are oxygen atoms (O)

To the nearest one-tenth of 1 percent, find:
a. $P_C$     b. $P_H$     c. $P_O$

10. The compound trinitrotoluene (trī nī′trō tol′yüēn′) (TNT) has the molecular formula $C_7H_5O_6N_3$, meaning that a TNT molecule has 21 atoms. Of these:

> 7 are carbon atoms (C)
>
> 5 are hydrogen atoms (H)
>
> 6 are oxygen atoms (O)
>
> 3 are nitrogen atoms (N)

To the nearest one-tenth of 1 percent, find:
a. $P_C$     b. $P_H$     c. $P_O$     d. $P_N$

11. The major responsibility of a courtesy clerk at a supermarket is to bag and carry out the groceries of customers. He has noticed that customers frequently comment, "I don't remember where I parked my car." To estimate the proportion of customers who have "lost their cars," the clerk kept a record for several days. During this time, he noted:

> 396 customers went directly to their cars from the store (d)
>
> 69 customers lost their cars and wandered around to find them (l)

For this sample of customers find:
a. $\hat{p}_d$     b. $\hat{p}_l$

12. Last week, a large shipment of colored Christmas tree bulbs was delivered to a discount outlet store. The manager sampled 50 of the bulbs to check for color. The results were 20 red bulbs (r), 15 blue bulbs (b), 9 green bulbs (g), and 6 white bulbs (w). For this sample, compute:
a. $\hat{p}_r$     b. $\hat{p}_b$     c. $\hat{p}_g$     d. $\hat{p}_w$

13. Craig Stiles was doing a study of prescription drugs. He enlisted the assistance of several pharmacists who recorded the prescriptions they filled over several days. A total of 540 prescriptions were examined. The results of the study showed the following.

> 96 prescriptions were for Valium (v)
>
> 96 prescriptions were for Dyazide (d)
>
> 62 prescriptions were for Tylenol with codeine (t)
>
> 48 prescriptions were for penicillin (p)
>
> 41 prescriptions were for tetracycline (c)
>
> 197 prescriptions were for other drugs

For this sample of prescriptions, compute to the nearest one-tenth of 1 percent:
a. $\hat{p}_v$     b. $\hat{p}_d$     c. $\hat{p}_t$     d. $\hat{p}_p$     e. $\hat{p}_c$

14. A discussion took place in Lacy Van Spake's sociology class as to whether single women should be permitted to adopt children. Maria Torres, who is single and has been considering adopting a child, decided to check public opinion on the matter. She surveyed 200 adults at a large shopping mall and obtained the following results.

> 158 said "yes," single women should be permitted to adopt children (y)
>
> 37 said "no," only married couples should adopt children (n)
>
> 5 were unsure of how they felt about the issue (u)

For this sample of opinions, compute to the nearest one-tenth of 1 percent:
a. $\hat{p}_y$     b. $\hat{p}_n$     c. $\hat{p}_u$

15. Over a period of several days, the following sales in the soft-drink section of a large supermarket were recorded.

    192 cartons of Coca-Cola products (c)

    106 cartons of Pepsi-Cola products (p)

    78 cartons of 7-Up products (u)

    56 cartons of Dr. Pepper products (d)

    43 cartons of other soft-drink products

Compute to the nearest one-tenth of 1 percent:
a. $\hat{p}_c$    b. $\hat{p}_p$    c. $\hat{p}_u$    d. $\hat{p}_d$

16. Jordan Electronics Manufacturing produces a model K47T960 transistor. Periodic sampling of these transistors is done and the number of defectives in each sample is recorded. For the last 130 lots checked, the following data were recorded.

    38 lots had 0 defective transistors

    54 lots had 1 defective transistor

    22 lots had 2 defective transistors

    13 lots had 3 defective transistors

    3 lots had 4 defective transistors

To the nearest one-tenth of 1 percent, compute:
a. $\hat{p}_0$    b. $\hat{p}_2$

17. Tim and Tom were discussing how severely their respective statistics instructors graded. Tom wanted to compare the number of students who were passing the course from each of the two instructors, using this as a measure of severity. Tim insisted that a comparison of the proportion of the students passing the course in each section was a better measure of the instructors' severity. Who was correct, Tim or Tom? Justify your answer.

18. Discuss the offer of an instructor of statistics to award course grades to each student on the basis of the letter grade that the student had received most often during the term. For example, if a student collected 30% A grades, 30% B grades, and 40% C grades, the semester grade would be a C.

## Exercises 2-5    Set B

*In Exercises 1 and 2, use the Cancer Research data in Data Set I, Appendix B. Use random numbers to select three different samples of 50 individuals each (sample A, sample B, and sample C).*

1. Let $\hat{p}_f$ represent the proportion of females in a sample.
   a. Compute $\hat{p}_f$ for sample A.
   b. Compute $\hat{p}_f$ for sample B.
   c. Compute $\hat{p}_f$ for sample C.
   d. Are the three values of $\hat{p}_f$ for samples A, B, and C the same?

2. Let $\hat{p}_c$ represent the proportion of individuals with cancer in a sample.
   a. Compute $\hat{p}_c$ for sample A.
   b. Compute $\hat{p}_c$ for sample B.
   c. Compute $\hat{p}_c$ for sample C.
   d. Are the three values of $\hat{p}_c$ for samples A, B, and C the same?

*In Exercises 3 and 4, use the stock prices in Data Set II, Appendix B. Use random numbers to select three different samples of 40 stocks each (sample A, sample B, and sample C).*

3. Let $\hat{p}_i$ represent the proportion of stock prices in a sample that increased (+).
   a. Compute $\hat{p}_i$ for sample A.
   b. Compute $\hat{p}_i$ for sample B.
   c. Compute $\hat{p}_i$ for sample C.
   d. Are the three values of $\hat{p}_i$ for samples A, B, and C the same?

**4.** Let $\hat{p}_u$ represent the proportion of stock prices in a sample that were unchanged (0).
   a. Compute $\hat{p}_u$ for sample A.
   b. Compute $\hat{p}_u$ for sample B.
   c. Compute $\hat{p}_u$ for sample C.
   d. Are the three values of $\hat{p}_u$ for samples A, B, and C the same?

**5.** Select $n = 40$ students on your campus. To each student, say the following:

> "Hi, I'm a student in a statistics class, and I'm taking a survey. Would you please answer two questions?" If you get a yes response, ask the following questions and record the responses:

   1. "In what academic area is your proposed major?"
   2. "Are you an out-of-state student?"

Use the results of your survey to compute the following:
   a. $\hat{p}_e$, the proportion of students that are in engineering.
   b. $\hat{p}_o$, the proportion of students that are from out-of-state.

**6.** Position yourself in a strategic spot near a major road or freeway during the peak commute time. Select $n = 50$ cars. Do not use trucks, buses, or other such vehicles in the survey. Record the number of passengers riding in each car. Use the results of your survey to compute the following:
   a. $\hat{p}_o$, the proportion of cars carrying one passenger.
   b. $\hat{p}_t$, the proportion of cars carrying two passengers.

# 2-6  Approximating Means and Variances from a Frequency Table

**Key Topics**

*1. Approximating $\sum x$ and $\sum x^2$*

*2. Approximating means and variances for grouped data*

## STATISTICS IN ACTION

Melissa Graves-Robin owns and operates the Wilton Feed and Supply Store. Many of her customers live on ranchettes of from 2 to 5 acres. These "city farmers" frequently buy some livestock such as horses, some sheep, or a couple of cattle. As a consequence, Melissa often sells hay to these customers a few bales at a time, since they do not want to store large quantities of hay on their property.

Last week Glenn Wilson delivered a truck and trailer load of alfalfa hay. Frequently a customer will ask how much a bale of hay weighs, so Melissa asked Glenn for an accounting of the number of bales and total weight of the hay. Glenn said that computers had recently been installed on the equipment that picks up the bales from the fields and transports them to the storage barns. The computers automatically weigh and record the weight of each bale as it is loaded, and the foreman compiles the data from the several loads that comprise a truck and trailer load. The data for the current load are in Table 2-8.

To approximate the mean weight of a bale on this load, Melissa used the frequencies and class marks of the tabled data to obtain Table 2-9.

For this set of data:

$$\sum f_i \cdot x_i = 2140 + 7056 + \cdots + 8509 = 50{,}230, \; n = 420,$$

and the approximated mean is $\dfrac{50{,}230}{420} = 119.5 \ldots \approx 120.$

Melissa would therefore tell her customers that the weight of a bale of hay is about 120 pounds.

**TABLE 2-8   Frequency Table for Hay Delivered**

| Class | Limits of Class | Frequency of Class |
|-------|-----------------|--------------------|
| 1 | 105–109 | 20 |
| 2 | 110–114 | 63 |
| 3 | 115–119 | 83 |
| 4 | 120–124 | 187 |
| 5 | 125–129 | 67 |

**TABLE 2-9   Calculations Used to Approximate the Mean Weight**

| Class | Class Marks, $x_i$ | Frequencies, $f_i$ | $x_i \cdot f_i$ |
|-------|--------------------|--------------------|-----------------|
| 1 | 107 | 20 | 2,140 |
| 2 | 112 | 63 | 7,056 |
| 3 | 117 | 83 | 9,711 |
| 4 | 122 | 187 | 22,814 |
| 5 | 127 | 67 | 8,509 |

# Approximating $\sum x$ and $\sum x^2$

When a set of data is grouped in a frequency table, each data point is assumed to have approximately the value of the class mark of its class. To illustrate, consider the hay data in Table 2-9. There are 20 bales in the population, with weights between 105 and 109 pounds. When approximating the mean of the 420 bales in the population, the product of the frequency and the class mark of class 1 is used to estimate the sum of the weights of these bales. Let $x_1, x_2, \ldots, x_{20}$ represent the weights of the 20 bales in class 1.

$$\underbrace{x_1 + x_2 + x_3 + \cdots + x_{20}}_{\substack{\text{The actual sum} \\ \text{of the 20 weights}}} \approx \underbrace{(20)(107)}_{\substack{\text{The product of the frequency} \\ \text{and class mark}}}$$

Similarly, the total weight of the 420 bales is approximated by the sum of the products of the class frequencies and corresponding class marks.

$$\text{420 data points} \quad \sum_{i=1}^{420} x_i \approx \sum_{i=1}^{5} f_i \cdot x_i \quad \text{5 classes in table}$$

For the hay data, 50,230 pounds is an approximation of the weights of the 420 bales. Thus the 120-pound mean is an approximation of the mean of the population.

To calculate the standard deviation, we need the sum of the squares of the data points. If the data are grouped in a frequency table, then the sum of squares can be approximated by using the squares of the class marks and the frequencies of the classes. Using the 20 bales in class 1 of Table 2-9 to illustrate:

$$\underbrace{x_1^2 + x_2^2 + x_3^2 + \cdots + x_{20}^2}_{\substack{\text{The actual sum} \\ \text{of the squares}}} \approx \underbrace{(20)(107)^2}_{\substack{\text{The product of the frequency} \\ \text{and square of class mark}}}$$

Similarly, the sum of the squares of the 420 weights is approximated by the sum of the products of the class frequencies and corresponding squares of the class marks.

420 data points          $$\sum_{i=1}^{420} x_i^2 \approx \sum_{i=1}^{5} f_i \cdot x_i^2$$          5 classes in table

## Approximating $\sum x_i$ and $\sum x_i^2$ for Grouped Data

For $N$ data points in a frequency table with $m$ classes:

$$\sum_{i=1}^{N} x_i \approx \sum_{i=1}^{m} f_i \cdot x_i,$$     where $x_i$ is the class mark of the $i$th class.

$$\sum_{i=1}^{N} x_i^2 \approx \sum_{i=1}^{m} f_i \cdot x_i^2,$$     where $x_i^2$ is the square of the class mark of the $i$th class.

## Approximating means and variances for grouped data

For data in a frequency table, approximations for means and variances can usually be obtained with equations (1) through (4) below. In these equations:

$f_i$ stands for the frequency of the $i$th class.

$x_i$ stands for the corresponding class mark.

$x_i^2$ stands for the square of the corresponding class mark.

$\sum f_i$ stands for $N$ or $n$, the total number of data points.

## Approximating Means and Variances for Grouped Data

| Approximate Values | Exact Values |
|---|---|
| 1. $\mu \approx \dfrac{\sum f_i x_i}{\sum f_i}$ | $\mu = \dfrac{\sum x_i}{N}$ |
| 2. $\sigma^2 \approx \dfrac{(\sum f_i)(\sum f_i x_i^2) - (\sum f_i x_i)^2}{(\sum f_i)^2}$ | $\sigma^2 = \dfrac{N\sum x_i^2 - (\sum x_i)^2}{N^2}$ |
| 3. $\bar{x} \approx \dfrac{\sum f_i x_i}{\sum f_i}$ | $\bar{x} = \dfrac{\sum x_i}{n}$ |
| 4. $s^2 \approx \dfrac{(\sum f_i)(\sum f_i x_i^2) - (\sum f_i x_i)^2}{(\sum f_i)(\sum f_i - 1)}$ | $s^2 = \dfrac{n\sum x_i^2 - (\sum x_i)^2}{n(n-1)}$ |

▶ **Example 1**    Greg Odjakjian works for Lipscomb Pharmaceutical Company. He used an electronic analyzer to determine the number of milligrams (mg) of compound K in 36 capsules of a sedative drug produced by the company. The measurements were rounded to two decimal places and the data were recorded in Table 2-10.

**TABLE 2-10    A Record of Milligrams of Compound K in Capsules**

| Class Number | Class Limits | Class Mark, $x_i$ | Tally Marks | Frequency, $f_i$ |
|---|---|---|---|---|
| 1 | 2.00–2.39 | 2.195 | ⅢⅡ ∣∣∣∣ | 9 |
| 2 | 2.40–2.79 | 2.595 | ⅢⅡ ∣∣∣∣ | 9 |
| 3 | 2.80–3.19 | 2.995 | ⅢⅡ ∣∣∣∣ | 9 |
| 4 | 3.20–3.59 | 3.395 | ⅢⅡ ∣∣ | 7 |
| 5 | 3.60–3.99 | 3.795 | ∣∣ | 2 |

**a.** Use the table to approximate the mean amount of compound K.

**b.** Approximate $s^2$ using the table.

**Solution**    **Discussion.** The work will be organized using Table 2-11:

**TABLE 2-11    Calculations Needed to Approximate $\bar{x}$ and $s^2$**

| Class Number | Class Limits | Class Mark, $x_i$ | Square of Class Mark, $x_i^2$ | Frequency, $f_i$ | $f_i x_i$ | $f_i x_i^2$ |
|---|---|---|---|---|---|---|
| 1 | 2.00–2.39 | 2.195 | 4.818025 | 9 | 19.755 | 43.362225 |
| 2 | 2.40–2.79 | 2.595 | 6.734025 | 9 | 23.355 | 60.606225 |
| 3 | 2.80–3.19 | 2.995 | 8.970025 | 9 | 26.955 | 80.730225 |
| 4 | 3.20–3.59 | 3.395 | 11.526025 | 7 | 23.765 | 80.682175 |
| 5 | 3.60–3.99 | 3.795 | 14.402025 | 2 | 7.590 | 28.804050 |

**a.** $\bar{x} \approx \dfrac{19.755 + \cdots + 7.590}{36}$

$\approx \dfrac{101.42}{36}$

$\approx 2.82$, to two decimal places

Thus there was an average of about 2.82 mg of compound K in the capsules.

**b.** $\sum f_i = 36,$   $\sum f_i x_i = 101.42,$   and   $\sum f_i x_i^2 = 294.1849$:

$$s^2 \approx \frac{36(294.1849) - 101.42^2}{(36)(35)}$$

$\approx 0.24$, to two decimal places

Thus the variance for the distribution of compound K was approximately 0.24 mg². ◄

▶ **Example 2**   Cynthia Kishimoto bowls in a Friday night mixed doubles league with her husband, Fred. Cynthia grouped the scores of the first 48 games she bowled this season in Table 2-12:

**TABLE 2-12**   Calculations Needed to Approximate $\mu$ and $\sigma^2$

| Class Number | Class Limits | Class Mark, $x_i$ | Tally Marks | Frequency, $f_i$ | $f_i x_i$ | $f_i x_i^2$ |
|---|---|---|---|---|---|---|
| 1 | 125–140 | 132.5 | ||||  | 5 | 662.5 | 87,781.25 |
| 2 | 141–156 | 148.5 | ||||  |||| ||| | 13 | 1,930.5 | 286,679.25 |
| 3 | 157–172 | 164.5 | ||||  |||| | | 11 | 1,809.5 | 297,662.75 |
| 4 | 173–188 | 180.5 | ||||  |||| | 9 | 1,624.5 | 293,222.25 |
| 5 | 189–204 | 196.5 | ||||  |||| | 10 | 1,965.0 | 386,122.50 |

     **a.** Approximate $\mu$ to one decimal place.
     **b.** Approximate $\sigma^2$ to two decimal places.
     **c.** Approximate $\sigma$ to one decimal place.

**Solution**   **a.** $\mu \approx \dfrac{662.5 + 1930.5 + \cdots + 1965.0}{48}$

$$\approx \frac{7992.0}{48} \qquad\qquad \mu \approx \frac{\sum f_i x_i}{\sum f_i}$$

$$\approx 166.5 \qquad\qquad \text{To one decimal place}$$

Thus for these games, Cynthia averaged 166.5.

**b.** $\sigma^2 \approx \dfrac{(48)(1{,}351{,}468) - 7992^2}{48^2}$

$$\approx \frac{998{,}400}{2304} \qquad\qquad \sigma^2 \approx \frac{(\sum f_i)(\sum f_i x_i^2) - (\sum f_i x_i)^2}{(\sum f_i)^2}$$

$$\approx 433.33 \qquad\qquad \text{To two decimal places}$$

**c.** $\sigma \approx \sqrt{433.33} \qquad\qquad \sigma = \sqrt{\sigma^2}$

$$\approx 20.8 \qquad\qquad \text{To one decimal place}$$

Thus the standard deviation for the distribution of scores is about 20.8.    ◀

# Exercises 2-6   Set A

**1.** Table 2-13 contains the test scores of 40 students in a Philosophy 2A class.
   a. Complete the table.
   b. Approximate $\mu$ to one decimal place.
   c. Approximate $\sigma^2$ to two decimal places.
   d. Approximate $\sigma$ to one decimal place.

**TABLE 2-13**   Test Scores in Philosophy 2A

| Class Number | Class Limits | Class Mark, $x_i$ | Frequency, $f_i$ | $f_i x_i$ | $x_i^2$ | $f_i x_i^2$ |
|---|---|---|---|---|---|---|
| 1 | 50–59 | 54.5 | 4 | | | |
| 2 | 60–69 | 64.5 | 5 | | | |
| 3 | 70–79 | 74.5 | 18 | | | |
| 4 | 80–89 | 84.5 | 7 | | | |
| 5 | 90–99 | 94.5 | 6 | | | |

TABLE 2-14   The Distances that 40 Students at St. Patrick's Elementary School Live from School

| Class Number | Class Limits | Class Mark, $x_i$ | Frequency, $f_i$ | $f_i x_i$ | $x_i^2$ | $f_i x_i^2$ |
|---|---|---|---|---|---|---|
| 1 | 1.0–1.9 | 1.45 | 3 | | | |
| 2 | 2.0–2.9 | 2.45 | 10 | | | |
| 3 | 3.0–3.9 | 3.45 | 15 | | | |
| 4 | 4.0–4.9 | 4.45 | 7 | | | |
| 5 | 5.0–5.9 | 5.45 | 5 | | | |

2. Table 2-14 contains the distances (in miles) that 40 students at St. Patrick's Elementary school must travel between school and home.
   a. Complete the table.
   b. Approximate $\bar{x}$ to the nearest one-tenth of 1 mile.
   c. Approximate $s^2$ to the nearest one-hundredth of 1 mile$^2$. (Remember: variance is measured in *squared units* of the data.)
   d. Approximate $s$ to the nearest one-tenth of 1 mile.

TABLE 2-15   The Weights of 75 Jars of Applesauce

| Class Number | Class Limits | Class Mark, $x_i$ | Frequency, $f_i$ | $f_i x_i$ | $x_i^2$ | $f_i x_i^2$ |
|---|---|---|---|---|---|---|
| 1 | 700–704 | | 4 | | | |
| 2 | 705–709 | | 12 | | | |
| 3 | 710–714 | | 22 | | | |
| 4 | 715–719 | | 18 | | | |
| 5 | 720–724 | | 16 | | | |
| 6 | 725–729 | | 3 | | | |

3. Table 2-15 contains the weights (in grams) of 75 Bonnie Hubbard No. 305 jars of applesauce.
   a. Complete the table.
   b. Approximate $\bar{x}$ to the nearest gram.
   c. Approximate $s^2$ to the nearest gram$^2$.
   d. Approximate $s$ to the nearest gram.

TABLE 2-16   The Commute Times Between Work and Home

| Class Number | Class Limits | Class Mark, $x_i$ | Frequency, $f_i$ | $f_i x_i$ | $x_i^2$ | $f_i x_i^2$ |
|---|---|---|---|---|---|---|
| 1 | 25–26 | | 6 | | | |
| 2 | 27–28 | | 22 | | | |
| 3 | 29–30 | | 18 | | | |
| 4 | 31–32 | | 9 | | | |
| 5 | 33–34 | | 8 | | | |
| 6 | 35–36 | | 2 | | | |

4. Table 2-16 contains the commute time (in minutes) between work and home over the last 3 months for a state worker in Indiana.
   a. Complete the table.
   b. Approximate $\mu$ to the nearest one-tenth of 1 minute.
   c. Approximate $\sigma^2$ to the nearest one-tenth of 1 minute$^2$.
   d. Approximate $\sigma$ to the nearest one-tenth of 1 minute.

5. Mary Ellen Murnin is a clerk at a large bank in the North Area. Last month she opened savings accounts for 64 new members. The ages of the members (in years) are listed below.

| | | | | | | | | | | | | | | | |
|---|---|---|---|---|---|---|---|---|---|---|---|---|---|---|---|
| 30 | 47 | 64 | 31 | 28 | 60 | 33 | 29 | 27 | 69 | 41 | 60 | 30 | 63 | 33 | 30 |
| 23 | 58 | 38 | 28 | 29 | 51 | 29 | 59 | 29 | 47 | 53 | 43 | 46 | 48 | 28 | 29 |
| 64 | 20 | 26 | 69 | 42 | 41 | 36 | 38 | 55 | 20 | 34 | 40 | 36 | 61 | 37 | 24 |
| 31 | 52 | 43 | 27 | 35 | 25 | 69 | 28 | 61 | 51 | 31 | 38 | 64 | 43 | 38 | 33 |

a. Put the data in a frequency table of five classes, a class width of 10, and a lower limit of 20 for class 1.
b. Approximate $\mu$ to one decimal place.
c. Approximate $\sigma^2$ to two decimal places.
d. Approximate $\sigma$ to one decimal place.
e. Use a calculator to calculate $\mu$ and $\sigma$ and compare with the approximate values.

6. Earl Karn works in the bakery department of a supermarket. During the month of April, he recorded the following costs (in dollars and cents) of special-order cakes that are usually decorated sheet cakes for parties:

| | | | | | | | | |
|---|---|---|---|---|---|---|---|---|
| 19.50 | 21.50 | 20.00 | 11.00 | 15.90 | 14.90 | 11.90 | 11.90 | 21.75 |
| 11.90 | 14.90 | 10.90 | 25.75 | 29.00 | 21.50 | 15.90 | 29.50 | 14.90 |
| 15.90 | 15.90 | 23.25 | 13.90 | 11.90 | 15.90 | 21.50 | 15.90 | 21.50 |
| 21.50 | 19.50 | 11.90 | 15.90 | 18.90 | 15.90 | 27.50 | 13.90 | 11.90 |
| 22.50 | 18.90 | 23.50 | 11.90 | 15.90 | 20.00 | 27.50 | 14.90 | 19.50 |

a. Put the data in a frequency table of five classes, a class width of $4.00, and a lower limit of $10.00 for class 1.
b. Approximate $\mu$ to the nearest 1 cent.
c. Approximate $\sigma^2$ to the nearest 1 cent$^2$.
d. Approximate $\sigma$ to the nearest 1 cent.
e. Use a calculator to calculate $\mu$ and $\sigma$ and compare with the approximate values.

7. Bill Bass works in a retail chain store that specializes in athletic shoes. On Saturday, May 17, he recorded the following prices (in dollars and cents) of shoes sold on that day:

| | | | | | | | | | |
|---|---|---|---|---|---|---|---|---|---|
| 19.99 | 29.99 | 47.99 | 9.99 | 49.99 | 36.99 | 36.99 | 16.99 | 39.99 | 27.99 |
| 29.99 | 39.99 | 9.99 | 34.99 | 17.99 | 20.99 | 9.99 | 47.99 | 39.99 | 21.99 |
| 32.99 | 29.99 | 42.99 | 33.99 | 29.99 | 33.99 | 14.99 | 29.99 | 41.99 | 49.99 |
| 39.99 | 17.99 | 36.99 | 29.99 | 47.99 | 49.99 | 39.99 | 19.99 | 19.99 | 23.99 |
| 9.99 | 49.99 | 34.99 | 41.99 | 37.99 | 9.99 | 47.99 | 17.99 | 21.99 | 32.99 |
| 19.99 | 41.99 | 12.99 | 27.99 | 16.99 | 29.99 | 15.99 | 9.99 | 9.99 | 44.99 |

a. Put the data in a frequency table of six classes, a class width of $7.00, and a lower limit of $9.00 for class 1.
b. Approximate $\mu$ to the nearest 1 cent.
c. Approximate $\sigma^2$ to the nearest 1 cent$^2$.
d. Approximate $\sigma$ to the nearest 1 cent.
e. Use a calculator to calculate $\mu$ and $\sigma$ and compare with the approximate values.

8. Doug Gilbert is an intern who works in the Emergency Room of a metropolitan hospital. On Sunday, February 10, he recorded the following ages in years of the patients that received medical help in the ER of this hospital:

$$
\begin{array}{cccccccccc}
4 & 31 & 25 & 66 & 90 & 20 & 41 & 60 & 48 & 15 \\
20 & 12 & 16 & 82 & 27 & 12 & 39 & 16 & 66 & 22 \\
23 & 23 & 64 & 17 & 18 & 21 & 23 & 21 & 1 & 8 \\
21 & 52 & 60 & 17 & 2 & 13 & 52 & 16 & 42 & 75 \\
6 & 10 & 23 & 31 & 48 & 8 & 18 & 19 & 5 & 68 \\
\end{array}
$$

a. Put the data in a frequency table of six classes, a class width of 15, and a lower limit of 1 for class 1.
b. Approximate $\mu$ to the nearest year.
c. Approximate $\sigma^2$ to the nearest year$^2$.
d. Approximate $\sigma$ to the nearest year.
e. Use a calculator to calculate $\mu$ and $\sigma$ and compare with the approximate values.

9. The following data are the grade point averages of 36 graduating seniors at Manitoba Polytechnic School.

$$
\begin{array}{ccccccccc}
2.41 & 2.33 & 3.50 & 3.00 & 2.46 & 3.43 & 2.66 & 3.61 & 3.75 \\
3.05 & 3.53 & 2.06 & 2.61 & 2.00 & 3.09 & 2.08 & 2.70 & 3.56 \\
2.46 & 3.25 & 2.26 & 2.22 & 3.04 & 3.36 & 2.78 & 2.83 & 2.08 \\
2.76 & 3.06 & 3.50 & 2.95 & 2.21 & 2.73 & 2.89 & 2.15 & 2.81 \\
\end{array}
$$

a. Put the data in a frequency table of five classes, a class width of 0.40, and a lower limit of 2.00 for class 1.
b. Approximate $\bar{x}$ to two decimal places.
c. Approximate $s^2$ to two decimal places.
d. Approximate $s$ to two decimal places.
e. Use a calculator to calculate and compare with the approximate values.

10. A survey was taken of the 32 counties in a western state to determine the number of sworn peace officers each county had per 1000 population. The following ratios were rounded to the nearest one-tenth:

$$
\begin{array}{cccccccc}
1.5 & 4.1 & 2.1 & 1.5 & 1.8 & 1.5 & 2.4 & 2.7 \\
4.6 & 1.6 & 5.4 & 4.7 & 1.5 & 3.0 & 1.3 & 1.7 \\
1.9 & 1.6 & 1.0 & 1.4 & 3.9 & 1.1 & 1.8 & 1.6 \\
1.5 & 1.2 & 1.6 & 1.5 & 5.0 & 1.8 & 1.6 & 3.3 \\
\end{array}
$$

a. Put the data in a frequency table of five classes, a class width of 0.9, and a lower limit of 1.0 for class 1.
b. Approximate $\mu$ to one decimal place.
c. Approximate $\sigma^2$ to two decimal places.
d. Approximate $\sigma$ to one decimal place.
e. Use a calculator to calculate $\mu$ and $\sigma$ and compare with the approximate values.

# Exercises 2-6   Set B

1. Use a table of random numbers to sample 50 individuals from the population in Data Set I, Appendix B. Record the ages of these individuals.
   a. Put the 50 data points in a frequency table of five classes. Use uniform class widths.
   b. Approximate the mean for the grouped data.
   c. Use a calculator to find the exact value of the mean of the data.
   d. Compute the percent difference between the approximate and actual values of the mean based on the actual value:

   $$\frac{\text{approximate value} - \text{actual value}}{\text{actual value}} \quad \text{(write as a percent)}$$

   e. Approximate $s$ for the grouped data to one decimal place.
   f. With a calculator, calculate $s$ to one decimal place.
   g. Compute the percent difference between the approximate and actual values of $s$ based on the actual value.

2. Use a table of random numbers to sample 50 stocks from Data Set II, Appendix B. Record the closing prices of the selected stocks.
   a. Put the 50 data points in a frequency table of five classes. Use uniform class widths.
   b. Approximate the mean for the grouped data.
   c. Use a calculator to find the exact value of the mean of the data.
   d. Compute the percent difference between the approximate and actual values of the mean based on the actual value:

   $$\frac{\text{approximate value} - \text{actual value}}{\text{actual value}} \quad \text{(write as a percent)}$$

   e. Approximate $s$ for the grouped data to one decimal place.
   f. Calculate $s$ with a calculator to one decimal place.
   g. Compute the percent difference between the approximate and actual values of $s$ based on the actual value.

*In Exercises 3 and 4, use the data in Figure 2-39.*

3. a. Put the Yesterday H (high) temperatures in a frequency table of six classes. Use uniform class widths.
   b. Approximate $\mu$ to the nearest whole degree.
   c. Approximate $\sigma$ to the nearest whole degree.
   d. What percent of the temperatures fall within one standard deviation of $\mu$?
   e. What percent of the temperatures fall within two standard deviations of $\mu$?

**4.** a. Put the Yesterday L (low) temperatures in a frequency table of six classes. Use uniform class widths.

b. Approximate $\mu$ to the nearest whole degree.

c. Approximate $\sigma$ to the nearest whole degree.

d. What percent of the temperatures fall within one standard deviation of $\mu$?

e. What percent of the temperatures fall within two standard deviations of $\mu$?

| City | Yesterday H/L | Prcp. | Today H/L | Sky | Tomorrow H/L | Sky |
|---|---|---|---|---|---|---|
| Albany NY | 25/4 | | 37/21 | CY | 37/24 | RN |
| Albuquerque | 43/32 | | 46/24 | CL | 46/23 | PC |
| Amarillo | 44/27 | .05 | 48/24 | CL | 47/24 | PC |
| Anchorage | 7/3 | | 5/-5 | PC | 0/-7 | PC |
| Atlanta | 54/24 | | 54/35 | RN | 49/42 | CY |
| Atlantic City | 40/14 | | 46/32 | PC | 43/41 | RN |
| Austin | 70/41 | | 57/43 | WN | 60/35 | PC |
| Baltimore | 45/19 | | 51/29 | CL | 46/41 | RN |
| Baton Rouge | 62/30 | | 53/48 | RN | 54/40 | PC |
| Billings | 48/36 | | 46/36 | WN | 40/30 | WN |
| Birmingham | 53/23 | | 50/35 | RN | 46/41 | CY |
| Bismarck | 34/18 | | 36/18 | PC | 31/17 | PC |
| Boise | 41/28 | .04 | 39/32 | RN | 38/26 | PC |
| Boston | 33/18 | | 41/29 | PC | 41/34 | RN |
| Brownsville | 78/53 | | 72/60 | WN | 73/45 | PC |
| Buffalo | 37/28 | .01 | 40/31 | CY | 43/32 | RN |
| Burlington | 29/7 | | 34/22 | CY | 36/22 | SN |
| Casper | 43/24 | | 40/24 | CY | 34/19 | SN |
| Charleston SC | 57/41 | | 64/40 | PC | 63/53 | RN |
| Charleston WV | 55/27 | | 53/29 | CY | 48/41 | RN |
| Charlotte | 52/26 | | 54/32 | CY | 50/44 | RN |
| Cheyenne | 39/28 | | 42/23 | CY | 32/22 | SN |
| Chicago | 38/27 | | 38/30 | CY | 37/34 | SN |
| Cincinnati | 47/31 | | 48/33 | CY | 41/37 | RN |
| Cleveland | 38/30 | | 46/30 | CY | 41/33 | RN |
| Colorado Springs | 42/30 | | 39/18 | PC | 36/19 | SN |
| Columbia SC | 59/25 | | 61/32 | PC | 57/46 | RN |
| Columbus OH | 45/32 | | 48/32 | CY | 42/36 | RN |
| Concord NH | 26/-8 | | 36/16 | CY | 34/20 | RN |
| Dallas | 61/44 | | 48/41 | CY | 54/33 | PC |
| Dayton | 46/30 | | 47/32 | CY | 39/34 | RN |
| Denver | 46/26 | | 43/19 | PC | 34/21 | SN |
| Des Moines | 43/26 | | 39/21 | CY | 38/25 | CY |
| Detroit | 38/32 | | 40/29 | CY | 37/29 | RN |
| Duluth | 22/17 | .05 | 24/15 | CY | 29/19 | CY |
| El Paso | 60/34 | | 55/30 | CL | 57/28 | CL |
| Eugene | 54/50 | .40 | 55/45 | CY | 54/39 | CY |
| Fairbanks | -33/-42 | | -35/-45 | PC | -25/-38 | PC |
| Fargo | 21/16 | | 28/12 | PC | 26/18 | PC |
| Flagstaff | 45/31 | | 45/20 | PC | 40/22 | PC |
| Grand Junction | 29/22 | .11 | 28/20 | SN | 28/17 | SN |
| Hartford | 32/2 | | 40/19 | PC | 39/30 | RN |
| Helena | 43/35 | | 44/33 | SN | 40/28 | PC |
| Honolulu | 84/68 | .01 | 82/65 | CL | 82/65 | CL |
| Houston | 65/44 | .12 | 55/51 | CY | 60/36 | PC |
| Indianapolis | 44/32 | | 41/32 | CY | 37/34 | RN |
| Jackson MS | 57/30 | | 57/42 | RN | 47/41 | CY |
| Juneau | 33/30 | | 38/27 | RN | 40/35 | RN |
| Kansas City | 44/28 | | 40/33 | CY | 38/26 | CY |
| Las Vegas | 67/44 | | 65/38 | CL | 63/37 | CL |
| Little Rock | 53/37 | | 45/41 | RN | 50/33 | PC |
| Louisville | 52/33 | | 45/34 | RN | 40/37 | RN |
| Madison | 38/21 | | 35/24 | CY | 35/24 | SN |
| Medford | 52/45 | .05 | 50/42 | FG | 46/38 | FG |
| Memphis | 58/35 | | 47/42 | RN | 44/40 | CY |
| Miami Beach | 72/58 | | 78/64 | PC | 80/67 | PC |
| Milwaukee | 37/27 | | 36/26 | CY | 36/30 | SN |
| Mpls-St Paul | 32/26 | | 29/24 | FG | 30/24 | CY |
| Nashville | 55/29 | | 47/35 | RN | 44/40 | CY |
| New Orleans | 60/33 | | 61/48 | RN | 55/45 | PC |
| New York City | 38/22 | | 47/32 | PC | 47/38 | RN |
| Newark | 37/19 | | 48/29 | PC | 47/38 | RN |
| Norfolk | 51/26 | | 55/35 | CY | 55/45 | RN |
| North Platte | 54/24 | | 46/22 | PC | 41/22 | PC |
| Oklahoma City | 55/34 | | 45/38 | CY | 44/31 | CY |
| Omaha | 44/28 | | 39/24 | PC | 38/25 | CY |
| Orlando | 70/45 | | 73/52 | PC | 75/62 | RN |
| Philadelphia | 40/22 | | 46/29 | PC | 47/39 | RN |
| Phoenix | 73/46 | | 67/43 | CL | 64/45 | CL |
| Pittsburgh | 43/28 | | 46/31 | PC | 44/37 | RN |
| Portland ME | 29/1 | | 35/19 | CY | 34/22 | RN |
| Portland OR | 58/42 | .20 | 55/45 | CY | 54/42 | CY |
| Providence | 35/13 | | 43/24 | PC | 42/34 | RN |
| Raleigh | 53/24 | | 57/30 | PC | 52/46 | RN |
| Rapid City | 48/29 | | 47/28 | CY | 42/27 | CY |
| Reno | 58/28 | | 53/26 | CL | 51/24 | CL |
| Richmond | 51/23 | | 53/30 | CL | 50/43 | RN |
| St Louis | 49/36 | | 43/37 | RN | 40/30 | SN |
| Salt Lake City | 43/28 | .01 | 36/27 | SN | 34/23 | CY |
| San Antonio | 73/47 | | 60/44 | WN | 61/36 | PC |
| Santa Fe | 41/26 | | 39/18 | CL | 43/18 | PC |
| Sault Ste. Marie | 30/27 | .33 | 29/19 | CY | 28/18 | CY |
| Seattle | 51/46 | .26 | 49/46 | RN | 47/44 | PC |
| Sioux Falls | 36/21 | | 40/24 | CY | 35/22 | CY |
| Spokane | 41/33 | .03 | 41/34 | RN | 39/31 | PC |
| Syracuse | 34/14 | .02 | 38/27 | CY | 42/28 | RN |
| Tampa-St. Pete | 68/45 | | 74/53 | PC | 71/65 | PC |
| Tucson | 65/38 | | 65/34 | CL | 62/36 | CL |
| Tulsa | 55/35 | | 46/41 | RN | 45/31 | CY |
| Washington | 46/29 | | 50/33 | CL | 47/41 | RN |
| Wichita | 45/25 | | 43/31 | CY | 45/23 | CY |
| Wilmington | 42/20 | | 47/28 | CL | 46/38 | RN |

**Figure 2-39.** Temperatures in 92 major cities in the United States.

# Chapter 2 Summary

## Definitions

1. The **range** of a set of data is      (the largest value) − (the smallest value)
2. The **class width** is

   (the upper boundary of a class) − (the lower boundary of the same class)
3. A **parameter** is a quantitative measure that describes some aspect of a **population**. A **statistic** is a quantitative measure that describes some aspect of a **sample**.
4. The **mean** of a set of quantitative data is the sum of the data points divided by the number of data points.
5. The **median** of a set of quantitative data is the value of the item of data in the middle of the ordered array of data points.
6. The **mode** of a set of quantitative data is the data point that has the greatest frequency.
7. The **variance** $\sigma^2$ of a finite population of quantitative data is the sum of the squares of the differences of each data point from the mean divided by the number of data points.
8. The **standard deviation** $\sigma$ of a finite population of quantitative data is the positive square root of the variance.
9. The symbols $s^2$ and $s$ are used respectively for the **variance** and **standard deviation of a sample** of quantitative data.
10. The **average deviation** of a finite population of quantitative data is the mean of the absolute values of the deviations of the $x_i$'s from $\mu$.
11. If $x$ is a data point in a population of data with mean $\mu$ and standard deviation $\sigma$, then the **z score** of $x$ is the number of standard deviations that $x$ deviates from $\mu$.
12. If $x$ is a data point in a distribution of quantitative data such that $m$ percent of the data is less than $x$, then $x$ is the value of the **mth percentile** of the distribution.
13. A **box-whisker plot** is a visual display made using $Q_1$, $\tilde{\mu}$, $Q_3$, and the smallest and largest points in a data set.
14. The **parameter** $P$ is the proportion of a population of objects that are in a specified category based on some characteristic, attitude, or attribute.
15. The **statistic** $\hat{p}$ is the proportion of a sample of objects that are in a specified category based on some characteristic, attitude, or attribute.

## Formulas and Procedures

$$\mu = \frac{\sum x_i}{N} \quad \frac{\text{The sum of the data points}}{\text{The number of data points}}$$

$$\bar{x} = \frac{\sum x_i}{n} \quad \frac{\text{The sum of the data points}}{\text{The number of data points}}$$

$\tilde{\mu}$ is the middle data point of an odd number of data points, or the mean of the two in the middle of an even number of data points, after the data have been arrayed from smallest to largest.

$\tilde{x}$ is the middle data point (or the mean of the two in the middle) of a sample of data that have been ordered from largest to smallest.

$$\sigma^2 = \frac{\sum(x_i - \mu)^2}{N} \quad \text{or} \quad \sigma^2 = \frac{N\sum x_i^2 - (\sum x_i)^2}{N^2}$$

$$\sigma = \sqrt{\sigma^2} = \sqrt{\frac{N\sum x_i^2 - (\sum x_i)^2}{N^2}}$$

$$s^2 = \frac{\sum(x_i - \bar{x})^2}{n - 1} \quad \text{or} \quad s^2 = \frac{n\sum x_i^2 - (\sum x_i)^2}{n(n - 1)}$$

$$s = \sqrt{s^2} = \sqrt{\frac{n\sum x_i^2 - (\sum x_i)^2}{n(n - 1)}}$$

$$z = \frac{x - \mu}{\sigma}$$

To determine the fraction of values below $x$ in a bell-shaped distribution, change $x$ to a $z$ score, then read the fraction from Appendix Table 4.

$$P = \frac{n(A)}{N}$$

$$\hat{p} = \frac{n(A)}{n}$$

For data grouped in a frequency table:

$$\mu \approx \frac{\sum f_i x_i}{\sum f_i} \quad \text{and} \quad \sigma^2 \approx \frac{(\sum f_i)(\sum f_i x_i^2) - (\sum f_i x_i)^2}{(\sum f_i)^2} \quad \text{and}$$

$$\bar{x} \approx \frac{\sum f_i x_i}{\sum f_i} \quad \text{and} \quad s^2 = \frac{(\sum f_i)(\sum f_i x_i^2) - (\sum f_i x_i)^2}{(\sum f_i)(\sum f_i - 1)}$$

## Symbols

$\mu$    the mean of a population

$\bar{x}$    the mean of a sample

$\tilde{\mu}$    the median of a population

$\tilde{x}$    the median of a sample

$R$    the range of a set of quantitative data

$\sigma^2$    the variance of a population

$\sigma$    the standard deviation of a population

$s^2$    the variance of a sample

$s$    the standard deviation of a sample

$z$    the $z$ score of a data point $x$ in a population with mean $\mu$ and standard deviation $\sigma$

$Q_i$    the $i$th quartile, $i = 1, 2, 3$

$\approx$    approximately equal to

$\bar{d}$    the mean of a sample of differences

$P$    the proportion of a population with a specified characteristic

$\hat{p}$    the proportion of a sample with a specified characteristic

## Comparing Descriptive Measures for Quantitative and Qualitative Data

For Quantitative Data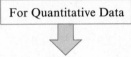

For Qualitative Data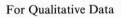

*Measures of Central Tendency*

**mean ($\bar{\mu}$ or $\bar{x}$)**   the balance point

**median ($\tilde{\mu}$ or $\tilde{x}$)**   the geometric middle

**mode (or bimode)**   most frequently occurring data point (or points)

*Measure of Composition*

**proportion ($P$ or $\hat{p}$)**   the proportion of data in each specified category into which data are separated

*Measures of Variation*

**range (R)**   largest data point minus smallest data point

**variance ($\sigma^2$ or $s^2$)**   the average of the squares of the deviations of the data points about the mean

**standard deviation ($\sigma$ or $s$)**   the square root of the variance

*Measures of Position*

**z score**   measures the number of standard deviations $\sigma$ a data point $x$ is from $\mu$

**percentiles**   divide the population into 100 parts with equal frequencies

**deciles**   divide the population into 10 parts with equal frequencies

**quartiles**   divide the population into 4 parts with equal frequencies

*Methods for Displaying*

**frequency table**   using a suitable number of classes and uniform class widths

**histogram**   horizontal axis labeled with class marks of frequency table and vertical axis the frequencies of each class

**stem-and-leaf diagram**   using suitable values from data set as the stem

**box-plot**   using the median of the data, the quartiles, and the smallest and largest data points

# Chapter 2 Review Exercises

1. Carie Petrakis was a volunteer recently on a television show that raised funds for a new children's hospital. Her job was to record the names and amounts of money that were pledged in phone calls to the studio. The following list includes the amounts (in dollars) pledged during her time on the program:

| | | | | | | | | | |
|---|---|---|---|---|---|---|---|---|---|
| 14 | 10 | 42 | 20 | 100 | 55 | 15 | 75 | 25 | 20 |
| 50 | 20 | 75 | 10 | 60 | 20 | 35 | 8 | 10 | 45 |
| 20 | 68 | 100 | 10 | 7 | 5 | 15 | 20 | 25 | 10 |
| 15 | 25 | 20 | 75 | 25 | 20 | 40 | 7 | 10 | 75 |
| 40 | 75 | 8 | 35 | 5 | 15 | 28 | 50 | 75 | 30 |

   a. Put the data in a frequency table of five classes. Use a class width of 20 and a lower limit of 5 for class 1.
   b. Construct a histogram.
   c. Construct a frequency polygon.
   d. What general shape best describes these data?

2. David Haddad is a physical therapist at a Veterans Administration Hospital. He developed an exercise program to increase the cardiac efficiency of physically handicapped patients. As part of the program, he administered a pretest to the 60 members of the program. The scores on this test are listed below.

| | | | | | | | | | |
|---|---|---|---|---|---|---|---|---|---|
| 84 | 80 | 68 | 76 | 88 | 80 | 84 | 78 | 92 | 84 |
| 96 | 80 | 84 | 88 | 70 | 64 | 92 | 80 | 68 | 80 |
| 80 | 56 | 80 | 94 | 76 | 80 | 70 | 56 | 84 | 79 |
| 56 | 80 | 90 | 100 | 88 | 74 | 88 | 84 | 80 | 96 |
| 84 | 96 | 68 | 76 | 58 | 80 | 84 | 72 | 92 | 102 |
| 88 | 70 | 100 | 80 | 84 | 78 | 84 | 92 | 64 | 98 |

   a. Put the data in a frequency table of six classes. Use a class width of 8 and a lower limit of 56 for class 1.
   b. Construct a histogram.
   c. Construct a frequency polygon.
   d. What general shape best describes the data?

3. Gayle Conlin is teaching a typing class. She gives 3-minute tests regularly to determine her students' typing speeds. The average number of words typed in a 3-minute test is computed, and then the number of errors on the paper are subtracted to give a net-words-per-minute score for the test. The 36 results of the last test are listed below.

| | | | | | | | | | | | |
|---|---|---|---|---|---|---|---|---|---|---|---|
| 43 | 51 | 40 | 42 | 47 | 38 | 35 | 44 | 36 | 52 | 38 | 29 |
| 34 | 46 | 42 | 34 | 43 | 29 | 50 | 47 | 53 | 28 | 43 | 26 |
| 43 | 37 | 33 | 51 | 48 | 40 | 38 | 29 | 36 | 52 | 28 | 40 |

Construct a stem-and-leaf diagram in which the tens digits of the data are used for the stem.

4. Carol Matsuii and her friends have been trying to reduce their living expenses. One area that they have been studying is the amount they spend each day on lunch. The following data are the amounts (in dollars and cents) spent on lunch one day in April:

    2.90  3.00  3.16  2.91  3.90  2.30  2.87  2.96  2.36  3.00  3.87
    3.20  3.42  3.00  1.90  2.30  3.12  2.20  3.00  2.16  2.55  3.40

For this set of data:
a. Compute the mean amount to the nearest 1 cent.
b. Find the median amount to the nearest 1 cent.
c. If one exists, give the mode.

5. A quality control inspector for a major food processing corporation recently sampled twenty-five 10-lb bags of flour and measured the contents of each bag to the nearest one-hundredth of a pound. The data are listed below.

    10.38  10.13  10.19  10.25  10.25  10.19  10.31  10.12  10.08
    10.18  10.00  10.15  10.12  10.25  10.08  10.06  10.25
    10.35  10.10  10.35  10.19  10.22  10.14  10.19  10.23

For this set of data:
a. Compute $\bar{x}$.     b. Find $\tilde{x}$.     c. If one exists, give the mode.

6. Jennifer Luong teaches sixth grade at Lewiston Elementary School. She recently asked her students to identify the three items they most wanted this year for Christmas. She also had them estimate (in dollars) what they thought the three items would cost. The dollar amounts of the estimates are listed below.

     50   75   60    40    50  100   85    75  125    80
    120   90   85    75    55  125   45   150   75   110
     95   75   65   150   100   75   90    65  130    85

For these dollar amounts:
a. Compute $\mu$ to the nearest dollar.
b. Compute $R$.
c. Compute $\sigma^2$ to the nearest one-tenth of 1 dollar$^2$.
d. Compute $\sigma$ to the nearest tenth of 1 dollar.

7. Grayson's Construction Supply Company stocks boxes of welding electrodes. Recently they got a large shipment of these electrodes. A warehouse foreman weighed 24 of the boxes to the nearest one-tenth of a pound. A list of the weights is given below.

    46.8  50.3  49.8  54.2  47.5  52.6  49.5  52.8
    48.0  50.5  53.9  48.6  52.8  54.0  49.0  52.2
    49.7  51.3  52.5  50.0  48.5  51.4  54.0  47.1

For this set of weights, compute to one decimal place:
a. $\bar{x}$     b. $s^2$     c. $s$

**8.** The community of Fahrnsworth recently had a drive to raise money for a youth center. The members of the Elks Lodge set up a "Golf-ball-driving contest" in which a person could, for a dollar contribution, hit a golf ball with a driver. The longest drives in categories A, B, and C, based on golfing ability, would win a major prize. The proceeds would go to the youth center fund. For this activity the following data (in yards) were recorded:

|  | Category A | Category B | Category C |
|---|---|---|---|
| Length of winning drive | 310 | 274 | 203 |
| Average length of drives in this category | 282 | 241 | 142 |
| Standard deviation of distribution of lengths of drives in this category | 14.3 | 18.7 | 26.5 |

a. Change the length of the winning drive in category A to a $z$ score.
b. Change the length of the winning drive in category B to a $z$ score.
c. Change the length of the winning drive in category C to a $z$ score.
d. Relatively speaking, which category winner hit the longest drive?

**9.** The local newspaper prints the Hi/Lo temperatures reported by national weather stations in 70 major cities in the United States. The lowest temperatures recorded yesterday at these stations are listed below.

```
28  33  43  16  31  41  49  60  45  26  47  31  36  36
70  42  33  23  56  45  39  33  48  50  45  36  48  22
64  33  40  42  27  40   4  21  32  27  42  27  20  73
65  47  51  51  37  48  57  53  44  60  76  44  46  35
57  63  48  50  37  51  37  62  42  56  40  25  41  29
```

a. Rank the data from lowest to highest temperature.
b. Find the 33rd percentile for the data.
c. Find the 85th percentile for the data.
d. Find the 2nd decile.
e. Find the 7th decile.
f. Find $Q_1$.
g. Find $Q_3$.

**10.** The numbers of grams of carbohydrates in a 1-ounce serving of 32 breakfast cereals are listed below.

```
12  23  50  37  32  28  43  19  27  32  46
57  31  29  18  26  35  26  38  53  15  24
33  42  33  27  46  17  45  20  41  22
```

a. Array the data using a stem-and-leaf diagram.
b. For the set of weights, compute $Q_1$, $\tilde{\mu}$, and $Q_3$.
c. Construct a box-plot of the weights.

**11.** Georgene and Carla play right field and second base, respectively, on the college softball team. In the athletic association of which their team is a member, the mean batting average of all the players is 0.285 with a standard deviation 0.075, and the distribution of averages is bell shaped. Georgene finished the season with a 0.385 batting average, and Carla finished with a 0.260 average.

     a. Find the fraction of batters with a batting average worse than Georgene's.

     b. Find the fraction of batters with a batting average worse than Carla's.

**12.** The 175 employees of Granite Construction Company were polled as to which professional sports they most enjoyed watching, whether in person or on television.

     56 named football (f)

     42 named baseball (b)

     35 named basketball (t)

     21 named golf (g)

     21 named some other sport (o)

To the nearest percent, find:

a. $P_f$     b. $P_b$     c. $P_t$     d. $P_g$     e. $P_o$

**13.** Heather Nordstrom sampled 80 candy-coated chocolate candies from a large display in a supermarket. She recorded the following counts of the numbers of candies with the indicated colored shells:

     25 candies were yellow (y)

     19 candies were brown (b)

     14 candies were red (r)

     12 candies were green (g)

     10 candies were orange (o)

To the nearest one-tenth of 1 percent, find:

a. $\hat{p}_y$     b. $\hat{p}_b$     c. $\hat{p}_r$     d. $\hat{p}_g$     e. $\hat{p}_o$

**14.** The 30 residents in American River Parkway Condominiums paid the following amounts (in dollars) last year in federal income taxes:

| | | | | | | | | | |
|---|---|---|---|---|---|---|---|---|---|
| 2780 | 3790 | 560 | 2320 | 5480 | 3610 | 3270 | 1810 | 3700 | 2940 |
| 1690 | 4320 | 3150 | 3910 | 2530 | 4090 | 2310 | 4260 | 1920 | 2610 |
| 5230 | 2960 | 1280 | 4460 | 5070 | 4390 | 3060 | 4350 | 3330 | 4180 |

     a. Put the data in a frequency table of five classes, a class width of $1000, and a lower limit of $500 for class 1.

     b. Approximate $\mu$ to the nearest dollar.

     c. Approximate $\sigma^2$ to the nearest dollar.

     d. Approximate $\sigma$ to the nearest dollar.

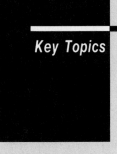

**Key Topics**

## Descriptive Statistics

1. *Histograms*

2. *Stem-and-leaf plots*

3. *Box-plots*

4. *Descriptive statistics*

The data on birth weights introduced in Example 3 of Section 2-1 will be used to illustrate how Minitab produces histograms, stem-and-leaf plots, box-plots, and related statistics from a sample of data. Assume that the data are in column 1 of the Minitab worksheet.

```
MTB> HIST C1
Histogram of C1  N = 75
Midpoint    Count
      5        4   ****
      6       14   **************
      7       21   *********************
      8       14   **************
      9       10   **********
     10        8   ********
     11        4   ****
```

**Figure 2-40.** Histogram showing weights of newborn babies.

## Histograms

Figure 2-40 is a computer printout of a histogram of the data and of the Minitab command used to create it.

Minitab selects a number of intervals appropriate for the data and plots with an asterisk (*) data points that fall within an interval or on its lower boundary. It also lists the midpoint of each interval and a count of the data points in the interval.

## Stem-and-leaf plots

Figure 2-41 shows a stem-and-leaf plot of the birth weights data.

There are two basic differences between this display and the one you would produce by hand for the same data. Minitab ignores decimal points and makes plots as though each data point were an integer. To determine the actual value of each number in the plot, look at the information, "Leaf Unit = 0.10," at the top of the plot. This leaf unit indicates that each number in the plot is really one-tenth as large as it appears to be. For instance, the smallest weight in the plot appears to be 45, but it is really 4.5.

```
MTB> STEM C1

Stem-and-leaf of C1    N = 75
Leaf Unit = 0.10

   2     4 58
   4     5 04
  12     5 56789999
  18     6 012333
  29     6 55667888999
 (10)    7 0222333334
  36     7 555667899
  27     8 00023
  22     8 55678
  17     9 00122
  12     9 57888
   7    10 022
   4    10 68
   2    11 04
```

**Figure 2-41.** Stem-and-leaf plot of weights of newbron babies.

A second difference is a "depth" column printed at the left of the plot. The number in this column before each stem tells how many data points are on that stem or beyond it. For instance, the 12 on the third line from the top tells us that there are 12 data points on or above that stem, and the 7 on the third line from the bottom tells us that there are seven data points on that stem or below it. The 10 in parentheses tells us that there are ten data points in that stem and that the median data point is to be found on that stem.

## Box-plots

A box-plot of the birth weights is shown in Figure 2-42. Because Minitab selects a "nice" scale, it is not usually easy to determine exactly from such a plot either the quartile values or the range.

## Descriptive statistics

Minitab provides a complete printout of relevant statistics from a sample. The one for the weights of newborn babies is shown in Figure 2-43.

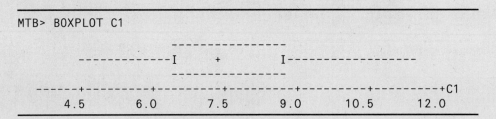

**Figure 2-42.** Box-plot of weights of newborn babies.

```
MTB> DESCRIBE C1
              N      MEAN    MEDIAN    TRMEAN     STDEV    SEMEAN
C1           75     7.607     7.300     7.567     1.572     0.182

             MIN      MAX        Q1        Q3
C1         4.500   11.400     6.500     8.700
```

**Figure 2-43.** Statistics from the sample of weights.

TRMEAN is the 5% trimmed mean found by taking the mean after the lowest 5% and the highest 5% of the data points are eliminated from the sample. SEMEAN is an estimate of the standard deviation of the means of all samples of the same size that could be made from the population of weights of newborn babies. This statistic will be discussed extensively in a later chapter.

## Exercises

1. Refer to the information given in Figures 2-40 through 2-43.
   a. Would you say that the given data has a distribution that is approximately bell shaped?
   b. How many baby's weights are there in this sample?
   c. What is the range of the weights?
   d. What is the significance of $Q1 = 6.5$?

2. Figure 2-44 shows a stem-and-leaf plot, a box-plot, and statistics of the 1970 populations (in 100,000s) of several large metropolitan areas in the United States.
   a. What do you think might be the significance of the asterisk in the box-plot?
   b. Which of the three displays shows most clearly the shape of the distribution?
   c. Is there anything in the statistics, by themselves, that suggests the presence of outliers?

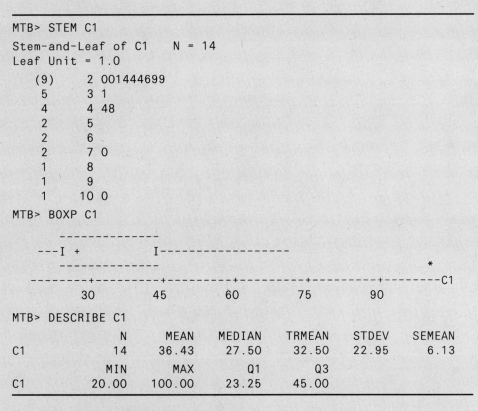

Figure 2-44.  Population (in 100,000's) of certain metropolitan areas in U.S.

## SCIENTIFIC

### Using a calculator to find the mean and standard deviation

Most scientific and business calculators have a statistics program that can be used to find the mean and standard deviation of a set of data. For most calculators, once the calculator has been set in the statistics mode, the data points are entered, one at a time, by using a *summation key* ($\sum$), a *memory addition key* (M+), or a *data key* (DATA). As the data are entered, the calculator stores in one cell the sum of the data ($\sum x_i$), in another cell the sum of the squares of the data ($\sum x_i^2$), and in another cell the number of data points (N or n). Depending on the calculator, the display will either show how many data points have been entered (1, 2, 3, 4, ...) or the value of the last data point entered. You need to check the manual for your calculator to get complete details.

| Steps | Display Shows |
|---|---:|
| Enter 17 | 17. |
| Press M+ | 17. |
| Enter 22 | 22. |
| Press M+ | 22. |
| Enter 31 | 31. |
| Press M+ | 31. |
| Enter 19 | 19. |
| Press M+ | 19. |
| Enter 20 | 20. |
| Press M+ | 20. |
| Enter 38 | 38. |
| Press M+ | 38. |
| Enter 19 | 19. |
| Press M+ | 19. |
| Press SHIFT, then $\bar{x}$ | 23.71428571 |
| Press SHIFT, then $x\sigma_n$ | 7.205440122 |

► **Example**  For the data below use a calculator to find:

**a.** the mean, to one decimal place.
**b.** the standard deviation, to one decimal place:

$$17 \quad 22 \quad 31 \quad 19 \quad 20 \quad 38 \quad 19$$

**Solution**  **Discussion.** The following instructions apply to Casio fx calculators. Check the manual for your calculator for similar commands.

To set the calculator in the statistics program, press MODE and then the $\cdot$ key. To clear statistics memories, press SHIFT KAC.

Display shows:

**DEG**        **SD**
**0.**

To one decimal place, the mean is 23.7.
To one decimal place, the standard deviation is 7.2.
Notice on this calculator there is a $x\sigma_{n-1}$ key.

| Press SHIFT, then $x\sigma_{n-1}$ | 7.782764842 |
|---|---:|

The $x\sigma_n$ key calculates a *population standard deviation*, and the $x\sigma_{n-1}$ key calculates a *sample standard deviation*.

# GRAPHING

The Casio Graphics calculator can find a mean ($\bar{x}$), a standard deviation (both $\sigma$ and $s$), and graphics such as bar graphs, line graphs, and a bell-shaped curve. To illustrate the procedure, we will use the following example.

▶ **Example 1**   For the data points:

$$2.1 \quad 3.4 \quad 3.7 \quad 4.1 \quad 4.2 \quad 4.9 \quad 5.2 \quad 5.5$$

**a.** Compute the mean $\bar{x}$.
**b.** Compute the standard deviation $s$.
**c.** Construct a histogram.

**Solution**

| Calculator Key Strokes | Comments |
|---|---|
| [SHIFT] [MODE] [×] | Sets SD2 mode |
| [RANGE] 0 [EXE] 6 [EXE] 1 [EXE] | The x-axis of graph |
| 0 [EXE] 5 [EXE] 1 [EXE] | The y-axis of graph |
| [MODE] [ · ] 6 [EXE] | Sets graph for six bars |
| [SHIFT] [Scl] [EXE] | Clears memories |
| 2.1 [DT] 3.4 [DT] 3.7 [DT] . . . | This enters data points (Enter number, then press [DT]) |
| [SHIFT] [Cls] [EXE] | Clears graphics display |
| **a.** [SHIFT] [$\bar{x}$] [EXE] | 4.1375 is the mean |
| **b.** [SHIFT] [$x\sigma_{n-1}$] [EXE] | 1.09 . . . is the standard deviation ($\sigma_{n-1}$ is used for $s$) |
| [ALPHA] [$n$] [EXE] | The number of data points is 8. |
| **c.** [GRAPH] [EXE] | The histogram is displayed. |

Notes:
   1. The key strokes [GRAPH] [SHIFT] [LINE] [EXE] will display a line graph.
   2. The key strokes [GRAPH] [SHIFT] [LINE] 1 will display a bell-shaped curve.
   3. If the same data point needs to be entered several times, then two methods are available. To illustrate, suppose 3.2, 3.2, 3.2, and 3.2 need to be entered.

*Method a*   Press 3.2, then [DT] [DT] [DT] and [DT]

*Method b*   Press 3.2; 4 [DT]

4. The following calculations are automatically stored in the calculator as data points are entered, and can be retrieved with an appropriate sequence of keys: $\boxed{\sum x^2}$ and $\boxed{\sum x}$.

5. If the number of bars needs to be changed, then the memories must be cleared and the data reentered.

6. To regain the use of all the keys for normal computation, press $\boxed{\text{MODE}}$ $\boxed{+}$

◀

▶ **Example 2**    For the data:

$$5.9 \quad 15.7 \quad 15.2 \quad 17.8 \quad 11.4 \quad 17.8 \quad 19.7 \quad 2.6 \quad 16.6$$

**a.** Compute $\bar{x}$.
**b.** Compute $s$.
**c.** Construct a histogram.

| Calculator Key Strokes | Comments |
|---|---|
| $\boxed{\text{SHIFT}}$ $\boxed{\text{MODE}}$ $\boxed{\times}$ | Sets SD2 mode |
| $\boxed{\text{RANGE}}$ $0$ $\boxed{\text{EXE}}$ $20$ $\boxed{\text{EXE}}$ $2$ $\boxed{\text{EXE}}$ | The $x$-axis of graph |
| $0$ $\boxed{\text{EXE}}$ $10$ $\boxed{\text{EXE}}$ $1$ $\boxed{\text{EXE}}$ | The $y$-axis of graph |
| $\boxed{\text{MODE}}$ $\boxed{\cdot}$ $6$ $\boxed{\text{EXE}}$ | Sets graph for six bars |
| $\boxed{\text{SHIFT}}$ $\boxed{\text{Scl}}$ $\boxed{\text{EXE}}$ | Clears memories. |
| $\boxed{\text{SHIFT}}$ $\boxed{\text{Cls}}$ $\boxed{\text{EXE}}$ | Clears graphics display |
| $5.9$ $\boxed{\text{DT}}$ $15.7$ $\boxed{\text{DT}}$ $\ldots$ | Enter data points |
| **a.** $\boxed{\text{SHIFT}}$ $\boxed{\bar{x}}$ $\boxed{\text{EXE}}$ | $13.63\ldots$ is the mean |
| **b.** $\boxed{\text{SHIFT}}$ $\boxed{x\sigma_{n-1}}$ $\boxed{\text{EXE}}$ | $5.84\ldots$ is the standard deviation. |
| $\boxed{\text{ALPHA}}$ $\boxed{n}$ $\boxed{\text{EXE}}$ | 9 is the number of data points. |
| **c.** $\boxed{\text{GRAPH}}$ $\boxed{\text{EXE}}$ | The histogram is displayed. |

◀

## TI-81 single variable statistics

The following example illustrates the key strokes for the Texas Instrument TI-81.

▶ **Example**    Find the mean and standard deviation for the following data points:

$$13, \quad 17, \quad 19, \quad 19, \quad 15, \quad 16, \quad 11, \quad 14, \quad 14, \quad 14$$

| Key Strokes | | Comments |
|---|---|---|
| [2nd] [STAT] | | Stat menu |
| [▶] [▶] | | Highlight **DATA** |
| 2 [ENTER] | | ClrStat    clears memories |
| [2nd] [STAT] [▶] [▶] 1 | | Calculator set to accept data |
| 13 [ENTER] [ENTER] | | Enters the value 13 |
| 17 [ENTER] [ENTER] | | Enters the value 17 |
| 19 [ENTER] 2 [ENTER] | | Enters the value 19 twice |
| 15 [ENTER] [ENTER] | | Enters the value 15 |
| 16 [ENTER] [ENTER] | | Enters the value 16 |
| 11 [ENTER] [ENTER] | | Enters the value 11 |
| 14 [ENTER] 3 [ENTER] | | Enters the value 14 three times |
| [2nd] [STAT] 1 [ENTER] | | Display shows 6 items of interest |
| | | $\bar{x} = 15.2$ |
| | | $s_x = 2.573 \ldots$ |

To graph the data points

| Key Strokes | | Comments |
|---|---|---|
| [RANGE] 0 [ENTER] 20 [ENTER] 1 [ENTER] | | $x$-axis **RANGE** |
| 0 [ENTER] 5 [ENTER] 1 [ENTER] | | $y$-axis **RANGE** |
| 1 [ENTER] | | Best resolution |
| [Y=] | | Del or turn off OLD functions. |
| [2nd] [DRAW] 1 [ENTER] | | Clears screen |
| [2nd] [STAR] [▶] | | Draws menu |
| 1 [ENTER] | | Displays histogram |

Notes:

1. Change the parameters of the range to give different presentations of the histogram. The data does *not* need to be re-entered.

2. New data points can be added and the calculator will recalculate and regraph using the old and the new data.   ◀

**Name:** Christy Chuang-Stein, Ph.D.

**Occupation:** Senior Biostatistics Scientist

**Employer:** The Upjohn Company, Research Support Biostatistics Kalamazoo, Michigan

**B**oth my parents are educators and enjoy great respect from their children. At a very young age, I decided to grow up to be just like them and become a teacher too. Thanks to my mother who spent numerous hours with me during my summer vacations in grade school, I became not only comfortable with numbers but also very fond of them. My interest in mathematics led to a major in the subject. Toward the latter part of my college years, I grew more interested in the applications of math, especially the inferential power of mathematical thinking. This, plus my fondness for numbers, influenced my graduate study in statistics. As the years went by, my aspiration to become a teacher narrowed to a career as a university professor. I patted myself on the back the day I began teaching statistics at the University of Rochester (Rochester, New York) for the fulfillment of my childhood career dream.

My career path took a turn when I joined The Upjohn Company in 1985. The Upjohn Company is a drug company whose primary mission is to develop safe and effective drugs. Before a drug can be marketed for public consumption, it must undergo various tests on animals and human subjects. Testing on humans in a clinic setting is called a clinical trial. My job at The Upjohn Company consists of helping physicians design clinical trials, overseeing the flow of data, and analyzing the data when the trials are completed. Clinical trials have different objectives. Some are to study the safety of the drug and others to find the optimal dose and schedule. Large clinical trials usually include a standard therapy for the underlying disease and randomize patients to receive either the test drug or the standard therapy. The goal is to evaluate the efficacy of the test drug in a controlled environment. Depending on the study designs, statistical techniques ranging from the paired $t$ test to analysis of variance to more sophisticated response surface modeling are used.

My current job has a research component. I find it gratifying to be able to develop new statistical methods that can facilitate the drug development process. The methods can be efficient study designs that require fewer study participants or new analytical tools that result in more powerful test results. I also find it exciting to work on a drug candidate that shows very promising potential use because of its potential impact on human beings.

I have taught elementary statistics to students like you. Some of you might react to the course the same way that many people do when they find out what I do for a living. Unfortunately, *number scare* has deprived many students of the joy of learning what statistics is about. In addition, *computer scare* has diminished the chance for many to realize that numbers can be quite manageable and fun. For those of you who have spent hours practicing your swing on a golf course or your turn on a downhill ski slope, you know very well that only practice can bring familiarity, which in turn will bring confidence, fun, and pride. There is no myth about numbers and there is no myth about computers. They are tools to help us do our jobs better. Enroll in a statistics course with an open mind, attend the lectures regularly, be diligent in your practice, and find ways to apply your newly learned knowledge in the real world. You will find that statistics is not as formidable as many people think. In the end, you are the one who will benefit from your own effort.

# 1 in 17 trillion?
# Hey, pretty good odds

Coincidences, those surprising and often eerie events that add spice to everyday life, may not be so unusual after all, researchers say.

After spending 10 years collecting thousands of stories of coincidences and analyzing them, two Harvard statisticians report that virtually all coincidences can be explained by some simple rules.

Some of the analyses performed by them or other statisticians showed that events that appeared to be extremely unlikely were almost to be expected.

When a woman won the New Jersay Lottery twice in four months, the event was widely reported as an amazing coincidence that beat odds of 1 in 17 trillion.

But when carefully analyzed, it turned out that the chance that such an event could happen to someone somewhere in the United States was more like 1 in 30.

It was an example of what the authors, Persi Diaconis, a professor of mathematics at Harvard University, and Frederick Mosteller, an emeritus mathematics professor at Harvard, call "the law of very large numbers."

That long-understood law of statistics states, in their formulation:

"With a large enough sample, any outrageous thing is apt to happen."

Some of the findings are in the December issue of The Journal of the American Statistical Association; others are now appearing in other professional journals.

**Figure 3-1.** From "1-in-a-Trillion Coincidence You Say? Not Really Experts Find" by Gina Kolata, *The New York Times*, February 27, 1990. Copyright © 1990 by The New York Times Company. Reprinted by permission.

# Probability

Figure 3-1 is part of an article on some work recently reported by Persi Diaconis, a professor of mathematics at Harvard University, and Frederick Mosteller, an emeritus mathematics professor at Harvard. Like many of us, these mathematicians were intrigued by "stories of coincidences." As stated in the article, "some of the analyses performed by them or other statisticians showed that events that appeared to be extremely unlikely were almost to be expected."

Also found in the article are the words "chance" and "odds." These terms are used in *probability*, which can be described as the mathematical description of chance. We need to understand something about this branch of mathematics because much of the work of inferential statistics is based on probability. Probability theory allows us to quantify the reliability of inferences we make based on random samples.

# 3-1  Some Basic Terms in Probability

**Key Topics**

1. *An experiment*

2. *An outcome of an experiment*

3. *A sample space for an experiment*

4. *An event*

5. *Mutually exclusive events*

6. *The complement of an event*

## S T A T I S T I C S   I N   A C T I O N

Pam Fullerton is running for the state senate of a large midwestern state. The current polls in the local newspaper suggest that she is trailing her opponent, Ben Whitney. However, Pam is a veteran politician and knows that there are many important limitations on polls as useful predictors of election outcomes. When she takes into account such issues as typical underrepresentation of certain classes of voters in such polls and the tendencies for races to tighten during the final weeks of a campaign, she is not discouraged. She feels she can realistically claim at this time to have "an even chance" of winning the election.

## An experiment

The statement in Definition 3-1 identifies what mathematicians mean by an experiment in the context of probability theory.

### DEFINITION 3-1 Experiment

An **experiment** is any activity whose result is not predetermined but instead involves chance.

Examples (A) through (D) describe four experiments.

A. Select a stock from those listed on the New York Stock Exchange and note yesterday's closing price.
B. Select a student from the current student body at Penn State University and determine the student's weight to the nearest pound.
C. Select ten telephones from a large production run and determine how many of the telephones have defects.
D. Select five cars from a particular faculty parking lot at The Ohio State University and determine how many were manufactured in the United States.

## An outcome of an experiment

Each time an experiment is performed, some particular result is observed. This result is called an *outcome*.

### DEFINITION 3-2 Outcome

An **outcome** is any result obtained when performing an experiment.

Examples (A*) through (D*) identify possible outcomes for the experiments given in examples (A) through (D).

A*. The closing price of the stock was $12\frac{5}{8}$.
B*. The weight of the student is 115 pounds.
C*. Two of the ten telephones have defects.
D*. Three of the five cars were manufactured in the United States.

## A sample space for an experiment

Each outcome in examples (A*) through (D*) is only one of many possible outcomes that could occur in that experiment. The set of all possible outcomes for any experiment is called the *sample space*.

### DEFINITION 3-3 Sample Space

A **sample space** is a specification of all possible outcomes of an experiment. The symbol $S$ will be used to represent a sample space.

## Three methods of representing a sample space

| List or roster method | Write out descriptions of all the possible outcomes. Sometimes these outcomes are listed using set notation. |
|---|---|
| Tree diagram | Use a network of *branches* and *nodes* that show the possible outcomes as complete paths through the tree. |
| Venn diagram | Represent the sample space with a rectangle. |

The list and tree methods are practical only if the sample space contains relatively few outcomes.

To illustrate and compare the three methods, recall the activity discussed in Section 1-2 in which two people were randomly sampled from a set of six people. This activity qualifies as a chance experiment. The names, and abbreviations we will use for the names, are listed below.

Ann—A   Beth—B   Carl—C   Dan—D   Ed—E   Fred—F

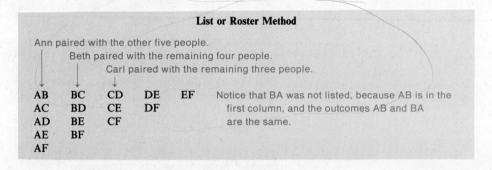

**List or Roster Method**

Ann paired with the other five people.
    Beth paired with the remaining four people.
       Carl paired with the remaining three people.

| | | | | | |
|---|---|---|---|---|---|
| AB | BC | CD | DE | EF | Notice that BA was not listed, because AB is in the |
| AC | BD | CE | DF | | first column, and the outcomes AB and BA |
| AD | BE | CF | | | are the same. |
| AE | BF | | | | |
| AF | | | | | |

In the tree diagram in Figure 3-2, the line segments are the *branches*, and the points where the branches are connected are the *nodes*.

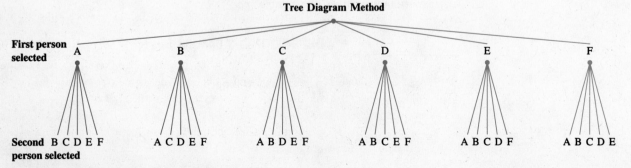

**Tree Diagram Method**

First person selected: A   B   C   D   E   F

Second person selected: B C D E F   A C D E F   A B D E F   A B C E F   A B C D F   A B C D E

**Figure 3-2.** A tree diagram of the sample space.

*Venn Diagram Method*

| | | | |
|---|---|---|---|
| AB | AF | BF | DE |
| AC | BC | CD | DF |
| AD | BD | CE | EF |
| AE | BE | CF | |

*S*   The rectangle represents the sample space and the outcomes are listed inside.

▶ **Example 1**   A boat on Folsom Lake carried four women, three children and one man. A park ranger stopped the boat and told the passengers that the boat was overloaded. Two people would have to be put on shore to make the boat legal. The eight passengers each put their names on slips of paper and two were selected. The people whose names were on the slips were put on shore.

**a.** Use the list method to identify the 28 possible outcomes.
**b.** How many of the outcomes included two children?

**Solution**   **Discussion.** To distinguish among the four women, use the symbols $w_1$, $w_2$, $w_3$, and $w_4$. Similarly use $c_1$, $c_2$, and $c_3$ to distinguish among the three children. Use $m$ to represent the one man. We will understand the symbols $w_1 w_2$ and $w_2 w_1$ to both represent the outcome that woman 1 and woman 2 were selected. Because both symbols represent the same outcome, only one is listed in the sample space.

**a.** The 28 possible outcomes are listed in the following way to help keep track of the different combinations:

$$w_1 w_2 \quad w_2 w_3 \quad w_3 w_4 \quad w_4 c_1 \quad c_1 c_2 \quad c_2 c_3 \quad c_3 m$$
$$w_1 w_3 \quad w_2 w_4 \quad w_3 c_1 \quad w_4 c_2 \quad c_1 c_3 \quad c_2 m$$
$$w_1 w_4 \quad w_2 c_1 \quad w_3 c_2 \quad w_4 c_3 \quad c_1 m$$
$$w_1 c_1 \quad w_2 c_2 \quad w_3 c_3 \quad w_4 m$$
$$w_1 c_2 \quad w_2 c_3 \quad w_3 m$$
$$w_1 c_3 \quad w_2 m$$
$$w_1 m$$

**b.** The three outcomes inside the triangle above list only children. Thus three of the 28 outcomes include two children.   ◀

If a legal coin is flipped, only the following two outcomes are possible:

Head (H)     A head is showing on the upturned face.

Tail (T)     A tail is showing on the upturned face.

When two or more such coins are flipped, an outcome consists of combinations of heads and tails on the coins.

▶ **Example 2**   An experiment consists of flipping three coins and noting the combination of heads and tails on the coins.

**a.** Show a sample space with a tree diagram.
**b.** How many of the outcomes include two heads and one tail?

**Solution**   **Discussion.** Two segments are drawn from a common point to show the two possible outcomes (H or T) for the first coin. From the ends of each of these segments, two more segments are drawn to show the two possible outcomes for the second coin. The process is continued for the third coin. An outcome for the experiment is the sequence of Hs and Ts along a complete "branch" of the tree. The result is the diagram in Figure 3-3.

**a.**

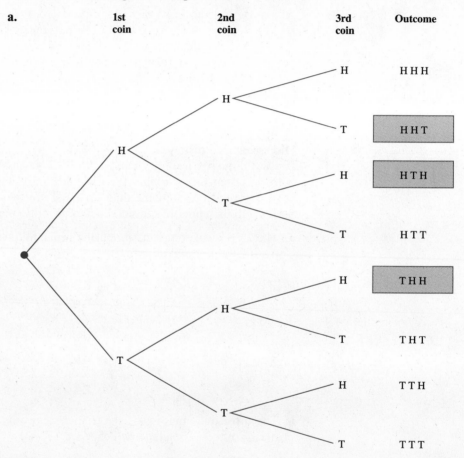

**Figure 3-3.**   A tree diagram for the coin-toss experiment.

**b.** The three boxed outcomes in Figure 3-3 consist of two heads and one tail.   ◀

## An event

Frequently some of the possible outcomes of an experiment are of more interest than others. A group of such outcomes can be called an event.

## DEFINITION 3-4 Event

An **event** is any subset of the outcomes in a sample space. Capital letters $A, B, C, \ldots$ are used to represent events.

For the experiment in Example 1, two people from the boat were selected to remain on shore. The following are examples of possible events:

*A*: Two women were selected.

*B*: One woman and one child were selected.

*C*: One child and the man were selected.

As previously mentioned, a sample space can sometimes be conveniently represented by a Venn diagram as shown in Figure 3-4. If such a representation is used, then an event (or events) can be shown as a closed region (or regions), such as event *A* in Figure 3-4. Any outcomes of *S* that satisfy the conditions specifying *A* are inside the circle. Any outcomes that do not satisfy *A* are outside the circle but inside the rectangle.

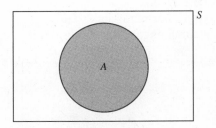

**Figure 3-4.** A sample space *S* with an event *A*.

▶ **Example 3**  The digits 1, 2, 3, 4, and 5 are each painted on disks of the same size and shape. The disks are put into a box and two are selected, one after the other. The number on the first disk is used as the tens digit in a two-digit number. The number on the second disk is used as the units digit. *A* contains the outcomes that are even numbers. Use a Venn diagram to display the sample space and *A*.

**Solution**  The outcomes of the experiment are listed in the following display:

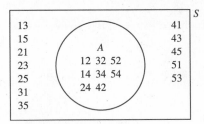

**Figure 3-5.** Event *A* consists of even-numbered outcomes.

| 12 | 21 | 31 | 41 | 51 |
|----|----|----|----|----|
| 13 | 23 | 32 | 42 | 52 |
| 14 | 24 | 34 | 43 | 53 |
| 15 | 25 | 35 | 45 | 54 |

These outcomes are also displayed in the Venn diagram in Figure 3-5.  ◀

## Mutually exclusive events

Suppose events *A* and *B* are defined for some experiment. Furthermore, suppose these events have the characteristic that when it is known that one of them has occurred, then it is possible to conclude that the other one did not occur. Such events are called *mutually exclusive* or *disjoint*.

## DEFINITION 3-5 Mutually Exclusive Events

If $A$ and $B$ are events defined for the same experiment, then $A$ and $B$ are **mutually exclusive** or **disjoint** if no outcome can simultaneously satisfy the conditions specifying both $A$ and $B$.

Figure 3-6 is a Venn diagram of mutually exclusive events. Notice that no outcome of $S$ can simultaneously be found inside both circles.

Figure 3-7 is a Venn diagram of events that are not mutually exclusive. The outcomes of $S$ in the overlapping region belong simultaneously to both events.

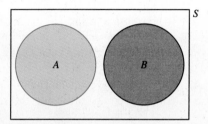

**Figure 3-6.**   $A$ and $B$ are mutually exclusive.

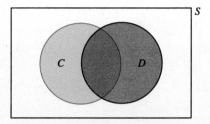

**Figure 3-7.**   $C$ and $D$ are not mutually exclusive.

To determine whether events $A$ and $B$ are mutually exclusive, ask the following question:

"If an outcome in $A$ (or $B$) is obtained, is it possible that the outcome is also in $B$ (or $A$)?"

If the answer is yes, then the events are not mutually exclusive. If the answer is no, then the events are mutually exclusive.

▶ **Example 4**   Carey is studying the records of patients in a large county hospital. She has identified the following events:

$A$:   The patient is female.

$B$:   The patient is male.

$C$:   The patient has blond hair.

**a.** Are $A$ and $B$ mutually exclusive events?
**b.** Are $A$ and $C$ mutually exclusive events?

**Solution**   **a.** For $A$ and $B$ we ask the following question: "If a patient is female, can the patient also be male?"

The answer is no; therefore $A$ and $B$ are mutually exclusive.

**b.** For $A$ and $C$ we ask the following question: "If a patient is female, can the patient also have blond hair?"

It is reasonable to assume that a female patient might have blond hair. Therefore the answer is yes, and $A$ and $C$ are not mutually exclusive.  ◀

## The complement of an event

Suppose event $A$ is defined for some experiment. Every time the experiment is performed, an outcome will be obtained that is either in $A$ or not in $A$. The outcomes of $S$ that are not in $A$ are in the *complement of A*.

---

### DEFINITION 3-6 The Complement of *A*

The outcomes in $S$ that do not satisfy the conditions specifying an event $A$ are contained in an event called **the complement of** $A$. The symbol $A'$ (read "$A$ prime") represents the complement of $A$.

---

The shaded region in Figure 3-8 consists of the outcomes in the complement of $A$. The complement of $A$ is sometimes called "not-$A$," because the outcomes in $A'$ are "not-in-$A$."

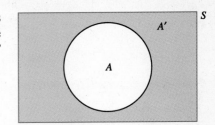

**Figure 3-8.** The shaded region represents $A'$.

▶ **Example 5**   Each of the letters a, b, c, d, e, f, and g are painted on a disk. The disks are put in a box, one is randomly selected, and the letter on the disk is noted. The following events are defined for this experiment:

$A$:   the letter is in "probability."
$B$:   the letter is in "statistics."
$C$:   the letter is in "frequency."

**a.** List the outcomes in $A$.
**b.** List the outcomes in $B$.
**c.** List the outcomes in $C$.
**d.** List the outcomes in $B'$.
**e.** Draw a Venn diagram showing $A$, $B$, and $C$.
**f.** Which two of the three events are mutually exclusive?

**Solution**    **Discussion.** For the given experiment:

$$S = \{a, b, c, d, e, f, g\}.$$

**a.** The letters in $S$ that are in "probability" are a and b.

$$A = \{a, b\}$$

**b.** The letters in $S$ that are in "statistics" are a and c.

$$B = \{a, c\}$$

**c.** The letters in $S$ that are in "frequency" are c, e, and f.

$$C = \{c, e, f\}$$

**d.** The letters in $S$ that are not in $B$ are b, d, e, f, and g.

$$B' = \{b, d, e, f, g\}$$

**e.** A diagram of $S$, $A$, $B$, and $C$ is shown in Figure 3-9.

**f.** $A$ and $C$ have no letters in common. Therefore $A$ and $C$ are mutually exclusive. ◄

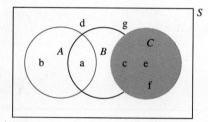

**Figure 3-9.** A Venn diagram of $S$, $A$, $B$, and $C$.

# Exercises 3-1    Set A

*In Exercises 1–6, use the list method to specify a sample space for each experiment.*

**1.** Each of the letters v, w, x, y, and z are painted on disks of similar size and the disks placed in a box. Two disks are simultaneously selected and an outcome consists of noting the letters on the disks.
  a. List the ten outcomes in a sample space.
  b. How many outcomes have an x or y?
  c. How many outcomes have an x and y?

**2.** A box contains four red $(r_1, r_2, r_3, r_4)$ and two blue $(b_1, b_2)$ marbles. Two marbles are simultaneously removed from the box and an outcome consists of noting which two marbles are chosen.
  a. List the 15 outcomes in a sample space.
  b. How many of the outcomes have at least one red marble?
  c. How many of the outcomes have marbles of the same color?

**3.** At a luncheon, seven women were polled as to the number of children each one had. The poll yielded four women each with no children $(x_1, x_2, x_3, x_4)$, two women each with one child $(y_1, y_2)$, and one woman with two children $(z)$. Subsequently two of these women were selected at random.
  a. List the 21 outcomes in a sample space.
  b. How many outcomes have women with no children?
  c. How many outcomes have women with at least one child?

4. Linda McGreavy has eight tulip bulbs. She knows that four of the bulbs grow red flowers ($r_1, r_2, r_3, r_4$), three grow yellow flowers ($y_1, y_2, y_3$), and one grows a purple flower ($p$). Two bulbs are randomly selected.
   a. List the 28 outcomes in a sample space.
   b. How many of the outcomes have two red flowers?
   c. How many of the outcomes have two flowers of the same color?

5. A box contains three nickels ($n_1, n_2, n_3$) and three dimes ($d_1, d_2, d_3$). Three coins are simultaneously removed from the box.
   a. List the 20 different outcomes in a sample space.
   b. How many outcomes have a value of 25 cents?
   c. How many outcomes have a value of at least 25 cents?

6. Seven babies were born yesterday at St. Luke's Hospital. Three of the babies were black ($b_1, b_2, b_3$), two were white ($c_1, c_2$), and two were Hispanic ($h_1, h_2$). Three of the babies will be randomly selected to have a picture taken for the local newspaper.
   a. List the 35 outcomes in a sample space.
   b. How many outcomes include one baby of each race?
   c. How many outcomes include exactly one black baby?

*In Exercises 7–12, use a tree diagram to specify the sample space for each experiment.*

7. Each of the digits 1, 2, 3, and 4 is painted on a disk and the disks placed in a box. Two disks are selected, one after the other, to form a two-digit number.
   a. Use a tree diagram to show the 12 outcomes.
   b. How many of the outcomes are even numbers?
   c. How many of the outcomes are numbers greater than 20?

8. Each of the letters a, b, c, d, and e are painted on a disk and placed in a box. Two disks are randomly selected, one after the other, and the letters on the disks are noted.
   a. Use a tree diagram to show the 20 outcomes.
   b. How many outcomes contain the letter b?

c. How many outcomes contain at least one vowel?

9. Joyce O'Brien sells insurance for a national insurance corporation. On Tuesday she has four interviews scheduled. At each interview she will sell a policy ($s$) or will not sell a policy ($n$).
   a. Use a tree diagram to show 16 possible outcomes.
   b. How many of the outcomes include exactly two sales?
   c. How many of the outcomes include at least one sale?

10. Sammy Laporte plays forward on the school basketball squad. Suppose that in an upcoming game he has the opportunity to shoot four foul shots. On each attempt, he makes the basket ($b$) or does not make the basket ($n$).
   a. Use a tree diagram to show the 16 possible outcomes.
   b. How many of the outcomes include at least two baskets?
   c. How many of the outcomes include at most one basket?

11. Coach Chuck Rago has nine players able to play the three backfield positions on the football team. Four of the players are seniors ($s_1, s_2, s_3, s_4$), three are juniors ($j_1, j_2, j_3$), and two are sophomores ($m_1, m_2$). Coach Rago plans to use one player from each of the three classes on the starting team.
   a. Use a tree diagram to show 24 possible outcomes.
   b. How many of the outcomes include player $s_1$?
   c. How many of the outcomes include players $s_2$ and $j_2$?

12. Renee Ortega has a new fall wardrobe that consists of four blouses ($b_1, b_2, b_3, b_4$), three skirts ($s_1, s_2, s_3$), and two pairs of boots ($t_1, t_2$). The wardrobe is coordinated so that any combination of blouse, skirt, and boots is stylish.
   a. Use a tree diagram to show 24 possible outfit combinations.
   b. How many combinations include boots $t_1$?
   c. How many combinations include $b_1$ paired with $s_2$?

*In Exercises 13–20, describe three events for each experiment. (Answers may vary.)*

**13.** A student at a large university is selected and the student's academic major is noted.

**14.** An employee at Watts Electronic Corporation is selected and the number of years of employment is noted.

**15.** A U.S. senator is randomly selected and the state from which the senator was elected is noted.

**16.** A voter in New York City is randomly selected and the political party affiliation of the voter is noted.

**17.** A house for sale in Buffalo, New York, is randomly selected and the appraised value of the house is noted.

**18.** A textbook from the bookstore at the University of Texas is selected and the course for which the text is assigned is noted.

**19.** A stock from the New York Stock Exchange is randomly selected at the close of a day of trading and the effect of the day's trading on the stock's price is noted.

**20.** A car from the records of the Department of Motor Vehicles in the state of Indiana is selected and the manufacturer of the car is noted.

*In Exercises 21–34, determine whether or not the pairs of events are mutually exclusive.*

**21.** A student is randomly selected at Penn State University:

        *A*:   The student is an electrical engineering major.

        *B*:   The student is female.

**22.** A student at City Community College is randomly selected:

        *A*:   The student is registered in a statistics class.

        *B*:   The student is registered in a sociology class.

**23.** A circuit breaker is randomly selected from the stock of an electronics supply house and the ampere rating is noted:

        *A*:   The breaker is rated 15 amp.

        *B*:   The breaker is rated 20 amp.

**24.** A bulb from a large bin of assorted Christmas tree bulbs is randomly selected:

        *A*:   The bulb is red.

        *B*:   The bulb is green.

**25.** A horse is randomly selected from the registration list of thoroughbreds at the Jockey Club:

        *A*:   The horse is a mare.

        *B*:   The horse's color is gray.

**26.** A car is randomly selected from the employee parking lot at Universal Steel Corporation:

        *A*:   The car was imported from Germany.

        *B*:   The car has a six-cylinder engine.

**27.** The weather conditions at O'Hare International Airport are checked for a given day:

> $A$: The lowest termperature for the day is 50° Fahrenheit.
>
> $B$: It snows that day.

**28.** A pollster at a shopping mall asked a female shopper how many children she has:

> $A$: The shopper has two children.
>
> $B$: The shopper has no children.

**29.** Two cars are stopped in a large metropolitan area to check whether the passengers are wearing seat belts:

> $A$: The passengers in the first car are not wearing seat belts.
>
> $B$: The passengers in the second car are not wearing seat belts.

**30.** Two roller bearings are randomly selected from a large production run at the Timken Roller Bearing Company:

> $A$: The first roller bearing has a defect.
>
> $B$: The second roller bearing has a defect.

*In Exercises 31–34, do the following for each experiment:*
*a. List the outcomes in S.*
*b. List the outcomes in A.*
*c. List the outcomes in B.*
*d. Draw a Venn diagram showing S, A, and B.*

**31.** The digits 6, 7, and 8 are each painted on a disk. The disks are placed in a box and two are selected, one after the other, to form a two-digit number.

> $A$: The number is even.
>
> $B$: The number is divisible by 3.

**32.** The letters a, b, c, d, e, f, and g are each painted on a disk. The disks are placed in a box and one disk is selected.

> $A$: The letter is in the word "graduate."
>
> $B$: The letter is in the word "federal."

**33.** A solid with eight faces has one of the digits 1, 2, 3, 4, 5, 6, 7, and 8 painted on each of the faces. The solid is rolled and the outcome is the number on the upturned face.

> $A$: The digit is an odd number.
>
> $B$: The digit is in the zip code 56980.

**34.** A box contains six fuses for a car's electrical system. Four of the fuses are good $(g_1, g_2, g_3, g_4)$ and two are defective $(d_1, d_2)$. Two fuses are randomly selected.

> $A$: Both fuses are good.
>
> $B$: Both fuses are defective.

## Exercises 3-1   Set B

Many experiments in statistics involve *sampling without replacement* from a set of objects. The term *without replacement* means that elements removed from the set are not returned to the set before additional elements are selected. The total number of different samples that are possible is the number of outcomes in the sample space. (Two samples are different if they have at least one element not in common and order is not important.) The number of outcomes in the sample space can be calculated with a formula that uses *factorial notation.*

### DEFINITION *n* Factorial

If *n* is a positive integer (1, 2, 3, . . .), then *n*! (*n* factorial) means:

$$n! = n(n-1)(n-2)\ldots 3 \cdot 2 \cdot 1$$

Thus *n*! stands for the product of *n* and all the positive integers less than *n*. Also, by convention

$$0! = 1 \quad \text{and} \quad 1! = 1.$$

Examples (a) and (b) illustrate factorial notation.

    **a.**  $4! = 4 \cdot 3 \cdot 2 \cdot 1 = 24$

    **b.**  $10! = 10 \cdot 9 \cdot 8 \cdot 7 \cdot 6 \cdot 5 \cdot 4 \cdot 3 \cdot 2 \cdot 1 = 3,628,800$

**Note**   Most scientific calculators have an *n*! or *x*! key. These calculators will display the products in scientific notation for any factorial larger than 14! Values returned for factorials larger than 14! are only approximations.

*In Exercises 1–12, evaluate each expression:*

**1.** 5!

**2.** 7!

**3.** 12!

**4.** 15!

**5.** $\dfrac{8!}{3!}$

**6.** $\dfrac{10!}{7!}$

**7.** $\dfrac{13!}{9!}$

**8.** $\dfrac{20!}{19!}$

**9.** $\dfrac{6!}{4!2!}$

**10.** $\dfrac{9!}{3!6!}$

**11.** $\dfrac{14!}{7!7!}$

**12.** $\dfrac{19!}{17!2!}$

---

## The (Number of) Combinations of *n* Things Taken *r* at a Time

---

Suppose an experiment consists of sampling without replacement *r* elements from a set of *n* elements (where $r \leq n$). The number of outcomes in the sample space is:

$$\binom{n}{r} = \frac{n!}{r!(n-r)!}$$

which can be read "the combination of *n* elements taken *r* at a time."

To illustrate the formula, recall the experiment of Example 1:

"Select two people from the boat containing eight people."

With $n = 8$ and $r = 2$, the number of different possible outcomes is:

$$\binom{8}{2} = \frac{8!}{2!(8-2)!} \qquad \binom{n}{r} = \frac{n!}{r!(n-r)!}$$

$$= \frac{8 \cdot 7 \cdot 6!}{2 \cdot 1 \cdot 6!} \qquad \frac{8! = 8 \cdot 7 \cdot 6!}{(8-2)! = 6!}$$

$$= 4 \cdot 7 \qquad \frac{8}{2} = 4 \quad \text{and} \quad \frac{6!}{6!} = 1$$

$$= 28 \qquad \text{The answer is 28.}$$

*In Exercises 13–20, evaluate each expression.*

13. $\binom{10}{2}$      14. $\binom{12}{2}$      15. $\binom{7}{3}$      16. $\binom{9}{3}$

17. $\binom{8}{0}$      18. $\binom{13}{0}$      19. $\binom{25}{1}$      20. $\binom{42}{1}$

21. In a game called keno, a machine selects 20 numbers from 1 through 80. For this game a player selects ten numbers from 1 through 80.
    a. Compute the number of different ways in which ten numbers can be selected from the 80.
    b. Compute the number of different ways in which 20 numbers can be selected from the 80.

22. One state in the United States has a lottery called 6/49. In this game a machine selects six balls from a basket containing 49 balls numbered 1 through 49.

The first five numbers are the numbers for that game, and the last is a bonus number. A player also selects six numbers. If any five of a player's numbers match the numbers for that game, she/he wins the big prize. If any five of a player's numbers match four of the numbers for that game plus the bonus number, the player gets a reduced prize.

a. Calculate $\binom{49}{5}$.      b. Calculate $\binom{49}{6}$.

# 3-2 Assessing Probabilities

**Key Topics**

1. *The classical notion of chance*

2. *The long-run relative frequency notation of chance*

3. *Assigning probabilities by subjective judgment*

4. *Some basic rules of probability*

---

## S T A T I S T I C S   I N   A C T I O N

---

**A**    For rest and relaxation from the pressures of being a corporate lawyer, Susan Siegrist spends an occasional weekend in Las Vegas, Nevada. She enjoys playing a game of chance called blackjack. Last night she had a hand of cards that would win if she got a face card from the dealer. Because she regularly keeps track of the previously played cards, Susan knew that 12 of the 30 cards still held by the dealer were face cards. She therefore felt she had a reasonable chance of winning the game and bet accordingly.

---

**B**    Juanita Ramirez is a quality control inspector at a plant that makes electronic components for telecommunication equipment. She recently sampled components for telephones from a large production run to check for defectives. Records from past inspections suggest that 8 percent of these components have some flaw that causes them to be defective. Juanita therefore knows that the chance a given component is flawed is rather low.

---

**C**    Charles Dawkins is employed by the National Weather Service in Des Moines, Iowa. He just completed a study of the current weather conditions in the area. Based on these and the expert knowledge he has of the typical effects of similar conditions on future weather, Charles feels there is an excellent chance of severe thunderstorms in the next 24 to 48 hours. These conditions also suggest a slight chance of tornadoes during this period.

---

Examples A, B, and C illustrate three commonly used methods for assigning numbers to the likelihood that a particular event might occur. These three methods will be studied in this section.

## The classical notion of chance

So-called games of chance, such as those played with dice, playing cards, and so on, provided much of the incentive for developing a theory of probability. Some of these games consist of experiments in which it is commonly assumed that each outcome is *equally likely*.

### DEFINITION 3-7 Equally Likely Outcomes

The term **equally likely** means that each possible outcome of an experiment is assigned the same probability of occurring.

Examples (a) through (c) are associated with games of chance. The outcomes of such experiments are typically assumed to be equally likely.

  a. **Draw a card from a deck of 52 cards**
     Each card in the deck has an equal chance of being selected.
  b. **Roll a pair of dice**
     An outcome is the pair of counts of the spots on the upturned faces. Each of the 36 possible outcomes is equally likely.
  c. **Spin an American roulette wheel with 38 similar slots**
     When the wheel is turned in one direction and the ball in the opposite direction, the ball has an equal chance of coming to rest in any one of the 38 slots.

### The Probability of Event A Based on Equally Likely Outcomes

Let $S$ be the sample space for a random experiment in which the outcomes are assumed to be equally likely and let $A$ be some event defined for the experiment. The probability of $A$, written $P(A)$, is given by the following equation:

$$P(A) = \frac{\text{the number of outcomes in } A}{\text{the number of outcomes in } S} = \frac{n(A)}{n(S)}$$

In the game of chance played by Susan Siegrist in Example A, each of the remaining 30 cards has an equal chance of being selected. Because 12 of these cards are face cards,

$$P(\text{face card}) = \frac{12}{30} = 40\%$$

Thus, Susan had a 40 percent chance of being dealt a winning card.

▶ **Example 1**   A pair of legal dice is rolled. An outcome consists of the pair of counts of spots showing on the up-turned faces.

**a.** Display the 36 outcomes in a table.
**b.** Determine $P$(three total spots showing).
**c.** Determine $P$(seven total spots showing).
**d.** Determine $P$(13 total spots showing).

| | | Number of Spots on First Die | | | | | |
|---|---|---|---|---|---|---|---|
| | | *1* | *2* | *3* | *4* | *5* | *6* |
| Number of Spots on Second Die | *1* | 2 | 3 | 4 | 5 | 6 | 7 |
| | *2* | 3 | 4 | 5 | 6 | 7 | 8 |
| | *3* | 4 | 5 | 6 | 7 | 8 | 9 |
| | *4* | 5 | 6 | 7 | 8 | 9 | 10 |
| | *5* | 6 | 7 | 8 | 9 | 10 | 11 |
| | *6* | 7 | 8 | 9 | 10 | 11 | 12 |

Total Number of Spots on Both Dice

**Figure 3-10.**   The 36 possible outcomes when a pair of dice is rolled.

Solution   **a.** The 36 possible outcomes are shown in Figure 3-10.
**b.** There are two outcomes that total three spots:

| First die | Second die | Total | |
|---|---|---|---|
| 2 spots + | 1 spot | = 3 spots | Two outcomes |
| 1 spot + | 2 spots | = 3 spots | |

$$P(3) = \frac{2}{36} = \frac{1}{18} \approx 6\%$$

**c.** In Figure 3-10, we can count six outcomes that total 7. Thus:

$$P(7) = \frac{6}{36} = \frac{1}{6} \approx 17\%$$

**d.** There are no outcomes in $S$ that total 13 spots.

$$P(13) = \frac{0}{36} = 0 \quad \blacktriangleleft$$

Notice that the probabilities in Example 1 can be computed without actually rolling a pair of dice. Such probabilities are called *theoretical* because they exist in theory, whether or not the experiment is ever actually carried out.

▶ **Example 2**   A bag contains eight marbles of the same size. Four are red ($r_1$, $r_2$, $r_3$, $r_4$), two are blue ($b_1$, $b_2$), and two are white ($w_1$, $w_2$). Two marbles are simultaneously selected from the bag.

a. Make a table of the 28 different possible outcomes.
b. Determine $P$(two red marbles).
c. Determine $P$(both marbles are of the same color).

**Solution**

**Discussion.** Because the marbles are of the same size, it is reasonable to assume that each pair of marbles has an equal chance of being selected.

$$
\begin{array}{llllllll}
r_1r_2 & r_2r_3 & r_3r_4 & r_4b_1 & b_1b_2 & b_2w_1 & w_1w_2 \\
r_1r_3 & r_2r_4 & r_3b_1 & r_4b_2 & b_1w_1 & b_2w_2 \\
r_1r_4 & r_2b_1 & r_3b_2 & r_4w_1 & b_1w_2 \\
r_1b_1 & r_2b_2 & r_3w_1 & r_4w_2 \\
r_1b_2 & r_2w_1 & r_3w_2 \\
r_1w_1 & r_2w_2 \\
r_1w_2
\end{array}
$$

**Figure 3-11.** The 28 possible outcomes when two marbles are selected.

a. The 28 outcomes are listed in Figure 3-11.
b. Six of the outcomes have two red marbles.

$$P(\text{two red marbles}) = \tfrac{6}{28} \approx 21\%$$

c. Six outcomes involve two red marbles. One outcome involves two blue marbles. One outcome involves two white marbles. Therefore eight outcomes involve marbles of the same color.

$$P(\text{both marbles are of the same color}) = \tfrac{8}{28} \approx 29\% \quad \blacktriangleleft$$

## The long-run relative frequency notion of chance

Figure 3-12 is a drawing of a coin. One side of this coin is painted red and the other side blue. An experiment consists of flipping the coin and an outcome is the color of the upturned face. The following single outcome events are defined for this experiment:

$A$: The red side is showing.

$B$: The blue side is showing.

**Figure 3-12.** A coin with a red side and a blue side.

This coin has been tampered with so that event $A$ is more likely to occur than $B$. That is, the two possible outcomes of this experiment are not equally likely.

Although the exact probabilities of $A$ and $B$ may never be determined, estimates of the probabilities can be found by actually repeating the experiment a number of times. To illustrate, suppose the coin was flipped 25 times, and 16 times the red side was showing and 9 times the blue side was showing. Based on these numbers, the following values might be given for $P(A)$ and $P(B)$:

$$P(A) = \frac{16}{25} = 0.64 \quad \text{and} \quad P(B) = \frac{9}{25} = 0.36$$

Therefore an estimate of the likelihood that the next flip of the coin will show the red side is 0.64. To impove on the 0.64 and 0.36 estimates, we could carry out the experiment a greater number of times. For example, assume the coin was flipped 1000 times and yielded 680 outcomes for $A$ and 320 for $B$. Revised estimates for $A$ and $B$ would then be:

$$P(A) = \frac{680}{1000} = 0.68 \quad \text{and} \quad P(B) = \frac{320}{1000} = 0.32$$

These estimates are probably closer to the "real" probabilities of $A$ and $B$ because they are based on more repetitions of the experiment. The law of large numbers is a formal statement of this idea.

## The Law of Large Numbers

As the number of times that an experiment is repeated increases, the relative frequency with which an event occurs will tend to approach the theoretical probability for the event.

Some people prefer to use special symbols to distinguish a probability derived from a relative frequency from a theoretical probability. For simplicity, in this text, the symbol $P(A)$ will be used for the probability of $A$, regardless of how the numerical value is derived.

In Example B, Juanita Ramirez would use 8 percent for the probability that a given component had a defect. This probability is based on the relative frequency of defects in previous samples of similar components. As more samples are taken under fixed planned conditions and defects are counted, the records can be updated and an improved probability estimate can be determined.

▶ **Example 3**   The employees of Gigantic Corporation can be divided into four categories based on job descriptions.

$A$:   Assembly workers

$B$:   Clerical workers

$C$:   Maintenance workers

$D$:   Management workers

A sample of 80 employee records yielded 45 in category $A$, 20 in category $B$, 11 in category $C$, and 4 in category $D$. A random experiment consists of selecting an employee of Gigantic Corporation. Find the following probabilities:

**a.** $P(A)$.

**b.** $P(D)$.

**Solution**   **Discussion.** The probabilities will be estimated using the relative frequencies for the four categories obtained in the sample of 80 employees.

**a.** $P(A) = \dfrac{45}{80} \approx 56\%$     45 assembly workers in sample

**b.** $P(D) = \dfrac{4}{80} \approx 5\%$     4 management workers in sample     ◀

## Assigning probabilities by subjective judgment

Frequently the probability of some event needs to be assessed, but there is no long-run record of repetitions of the experiment on which the event is defined, and no sample space where an equally likely outcomes assignment of probabilities is obvious. Thus, neither the long-run relative frequency method for estimating probabilities nor the classical method for assigning probabilities can be used. In such cases, evaluation of probabilities is frequently made on a subjective basis.

In Example C, Charles Dawkins needed to assess the probability of severe thunderstorms for the Des Moines area during the next 24 to 48 hours. Although he could base his probability partially on data gathered by the National Weather Service, he would also rely heavily on his accumulated knowledge of the effects of current weather conditions on future weather in this area. People who are very knowledgeable in their jobs often develop the ability to subjectively assess probabilities of certain work-related events occurring.

▶ **Example 4** Karin Sandberg is a heart specialist in a large metropolitan hospital. She has just examined a heart attack patient in the emergency ward. The concerned wife of the patient has just asked Karin what the chance is that her husband will recover. What factors might Karin use to arrive at an estimate of the probability of recovery for this patient?

**Solution** Some of the more obvious factors that Karin might consider are the following:

1. The patient's age, weight, blood pressure, and so on.
2. The patient's medical record of previous heart attacks.
3. The patient's history of smoking and drinking.
4. Results of tests of patient's condition that reveal factors such as severity of the attack and cholesterol levels in blood.  ◀

The experienced doctor could undoubtedly add several other factors to this list and use them to make subjective assignments of probabilities in a case like Example 4.

## Some basic rules of probability

In this section, the classical, relative frequency, and subjective judgment methods have been used to assign basic probabilities. Regardless of which method is used, certain mathematical rules of probability must be followed in assigning probabilities to events. Four of these rules are identified at this time.

## Four mathematical rules of probability

1. If $A$ contains no outcomes, then $P(A) = 0$.
   A probability of 0 means the event cannot occur.

2. If $A$ contains all the outcomes in $S$, then $P(A) = 1$.
   A probability of 1 means the event is certain to occur.

3. For any event $A$, $0 \le P(A) \le 1$.
   The minimum and maximum probabilities possible are 0 and 1, respectively.

4. For complementary events $A$ and $A'$, $P(A) + P(A') = 1$
   Each time the experiment is carried out, either $A$ or $A'$ will occur.

▶ **Example 5**   The letters in "statistics" are painted on disks of similar size and placed in a box. One disk is randomly selected. Compute the following probabilities.

   **a.** $P$(letter s)
   **b.** $P$(vowel)
   **c.** $P$(not a letter t)
   **d.** $P$(letter z)

**Solution**   **a.** $P(\text{letter s}) = \dfrac{3}{10}$    $\dfrac{3 \text{ disks labeled s}}{10 \text{ disks in the box}}$

   **b.** $P(\text{vowel}) = \dfrac{3}{10}$    a and i are vowels.

   **c.** $P(\text{not a letter t}) = \dfrac{7}{10}$    $\dfrac{7 \text{ disks not labeled t}}{10 \text{ disks in the box}}$

   **d.** $P(\text{letter z}) = 0$    There are no zs in statistics.   ◀

Frequently the probabilities assigned to events are displayed in a Venn diagram.

▶ **Example 6**   Use the probabilities shown in Figure 3-13 to compute the following:

   **a.** $P(A)$.
   **b.** $P(A')$.
   **c.** $P(B)$.
   **d.** $P(B')$.

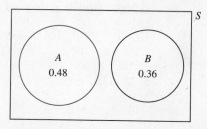

**Figure 3-13.**   A Venn diagram for Example 6.

**Solution**    The numbers in the diagram specify the probabilities of obtaining outcomes in the respective regions.

  **a.** The number inside $A$ is 0.48.

$$P(A) = 0.48$$

  **b.** $P(A') = 1 - 0.48$     $P(A') = 1 - P(A)$
  $= 0.52$

  **c.** The number inside $B$ is 0.36.

$$P(B) = 0.36$$

  **d.** $P(B') = 1 - 0.36$     $P(B') = 1 - P(B)$
  $= 0.64$   ◄

## Exercises 3-2    Set A

*In Exercises 1–8, find the probabilities of the events for the following experiments with equally likely outcomes.*

1.  The digits 0, 1, 2, . . . , 9 are painted on disks of similar size and shape. The disks are placed in a container and one is randomly selected. Assign probabilities to the following events:
    a.  *A*:   The disk has the digit 5.
    b.  *B*:   The disk has an odd number.
    c.  *C*:   The disk has a whole number.
    d.  *D*:   The disk has a two-digit number.
    e.  *E*:   The disk has a digit appearing in the current year.

2.  The digits 1, 2, 3, 4, 5, and 6 are painted on disks of similar size and shape. The disks are placed in a container and two are randomly selected, one after the other, to form a two-digit number. The first digit is used for the tens place and the second digit is used for the ones place.
    a.  List the 30 outcomes in the sample space.
    b.  Find $P(A)$, where $A$ is the event that the number is even.
    c.  Find $P(B)$, where $B$ is the event that the number is less than 25.
    d.  Find $P(C)$, where $C$ is the event that the ones digit is greater than the tens digit.
    e.  Find $P(D)$, where $D$ is the event that the number is greater than 70.

3.  The letters a, b, c, d, e, f, and g are painted on disks of similar size and shape. One disk is randomly selected. Find the probabilities of the following events:
    a.  *A*:   The letter is a vowel.
    b.  *B*:   The letter is a consonant.
    c.  *C*:   The letter is in the word "mathematics."
    d.  *D*:   The letter is in the word "statistics."
    e.  *E*:   The letter is in the word "mississippi."

4.  The letters in "abominable" are painted on disks of similar size and shape. One disk is randomly selected. Find the probabilities of the following events:
    a.  *A*:   The letter is b.
    b.  *B*:   The letter is a vowel.
    c.  *C*:   The letter is in the word "event."
    d.  *D*:   The letter is contained in "mutually exclusive."

5.  A bank contains two quarters $(q_1, q_2)$, three dimes $(d_1, d_2, d_3)$, and four nickels $(n_1, n_2, n_3, n_4)$. Two coins are randomly selected.
    a.  List the 36 different possible outcomes.
    b.  Find $P(15 \text{ cents})$.
    c.  Find $P(\text{more than 25 cents})$.
    d.  Find $P(\text{less than 20 cents})$.

6.  Eight people recently went on a picnic. Two of the people were women $(w_1, w_2)$, two were men

$(m_1, m_2)$, and four were children $(c_1, c_2, c_3, c_4)$. Two volunteers were needed to return to the parking lot to bring back the ice chest. To make the selection fair, a random process was used to obtain the volunteers.

a. List the 28 outcomes in the sample space.

b. Find $P$(two children).

c. Find $P$(one woman and one man).

d. Find $P$(not two women).

7. Two solids each have eight faces of equal area. The digits 1 through 8 are each painted on one face of each solid. The solids are rolled and an outcome is the sum of the numbers on the up-turned faces.

a. List the 64 outcomes in the sample space.

b. Find $P$(a sum of ten).

c. Find $P$(a sum of 7).

d. Find $P$(the sum is an even number).

e. Find $P$(the sum is less than 6).

8. Five coins are tossed and whether heads (H) or tails (T) occur is noted.

a. List the 32 outcomes in a sample space.

b. Find $P$(four heads and one tail).

c. Find $P$(two heads and three tails).

d. Find $P$(at least one head).

*In Exercises 9–16, use the given data to estimate the indicated probabilities.*

9. Thirty workers in service occupations were surveyed and put into the following categories:

   $A$:  Workers employed in private homes

   $B$:  Workers employed by food services

   $C$:  Workers in all other service areas

The survey yielded the following record:

   5 workers in category $A$

   8 workers in category $B$

   17 workers in category $C$

An additional worker in a service occupation is to be randomly selected. Estimate the following probabilities based on the given data.

a. $P(A)$    b. $P(B')$

10. At a convention, a group of 40 psychologists were surveyed as to the number of times they had been married and they were put into the following categories:

   $A$:  Never been married

   $B$:  Married exactly one time

   $C$:  Married exactly two times

   $D$:  Married exactly three times

The survey yielded the following record:

   5 psychologists in category $A$

   20 psychologists in category $B$

   12 psychologists in category $C$

   3 psychologists in category $D$

An additional psychologist at the conference is to be randomly selected. Estimate the following probabilities based on the given data.

a. $P(B)$.

b. $P(C)$.

c. $P(A')$.

11. At a large community college in the East, 2500 students last year completed Business Mathematics 101. The records of 70 of these students were surveyed and the data in Table 3-1 were obtained:

TABLE 3-1    Grades in Business Mathematics 101

| Final Grade | Frequency |
|:---:|:---:|
| A | 10 |
| B | 21 |
| C | 28 |
| D | 8 |
| F | 3 |

A record of one of the 2500 students is to be randomly selected. Estimate the following probabilities based on the given data:

a. $P$(the student got a final grade of C)

b. $P$(the student got an A or B)

c. $P$(the student did not get a D)

**12.** At Sweets Candy Company, a thin, colored, candy shell is used to coat chocolate candies. A sample of 80 such candies was randomly selected from a large production run, yielding the data in Table 3-2.

TABLE 3-2   Colors of Candies in Sample

| Color | Frequency |
|-------|-----------|
| Yellow | 15 |
| Orange | 25 |
| Green | 18 |
| Red | 8 |
| Brown | 14 |

One additional candy from this production run is to be randomly selected. Estimate the following probabilities based on the given data:
a. $P$(the candy is coated orange)
b. $P$(the candy is coated yellow or green)
c. $P$(the candy is not coated brown)

**13.** Glow Light Corporation manufactures 75-watt bulbs. From a large production run of these bulbs, a sample of 40 is selected and the length of time each bulb remained lighted before burning out was recorded.

    5 bulbs burned less than 100 hours

    12 bulbs burned between 100 and 125 hours

    16 bulbs burned between 125 and 150 hours

    7 bulbs burned more than 150 hours

One additional bulb from this production run is to be randomly selected. Estimate the following probabilities based on the given data:
a. $P$(the bulb will burn less than 100 hours)
b. $P$(the bulb will burn between 100 and 150 hours)
c. $P$(the bulb will burn less than 150 hours)

**14.** A fixed quantity of ore from a copper mine was refined and the quantity of copper obtained was weighed to the nearest gram. The data from 45 repetitions of the procedure yielded the following data:

8 specimens yielded less than 100 grams of copper.

14 specimens yielded between 100 and 150 grams of copper.

20 specimens yielded between 150 and 200 grams of copper.

3 specimens yielded more than 200 grams of copper.

Another specimen of ore is to be obtained and refined. Estimate the following probabilities based on the given data:
a. $P$(the amount of copper will be between 150 and 200 grams)
b. $P$(the amount of copper will be less than 200 grams)
c. $P$(the amount of copper will be more than 100 grams)

**15.** The question of whether or not the legal age for getting a driver's license should be raised to 18 years was asked of 350 residents in St. Louis. The results are given in Table 3-3.

TABLE 3-3   Results of Survey on Legal Age for Driver's License

| Opinion | Frequency |
|---------|-----------|
| Strongly favored | 85 |
| Moderately favored | 90 |
| Favored | 125 |
| Moderately opposed | 40 |
| Strongly opposed | 10 |

An additional resident of St. Louis is to be selected and asked whether the legal age for getting a driver's license should be raised. Estimate the following probabilities based on the given data:
a. $P$(the person favors raising the age limit)
b. $P$(the person opposes raising the age limit)
c. $P$(the person is in favor of raising the age limit)

**16.** At an Air Force base, the number of aircraft landing or taking off between 0800 and 2000 hours was recorded. The data in Table 3-4 were obtained from the study.

TABLE 3-4　Results of Aircraft Landings Study

| Number of Activities | Frequency |
|---|---|
| Less than 50 | 12 |
| From 50 through 100 | 28 |
| From 101 through 150 | 40 |
| From 151 through 200 | 32 |
| More than 200 | 8 |

On an upcoming day at this Air Force base, the number of activities will be recorded. Estimate the following probabilities based on the given data:

a. $P$ (between 151 and 200 activities will be recorded)
b. $P$ (less than 101 activities will be recorded)
c. $P$ (more than 150 activities will be recorded)

*In Exercises 17–22, list at least three factors that might be considered in subjectively assessing the probability of the indicated event.*

**17.** Elizabeth Murray is an entering freshman at Grand Central University. She would like to assess the probability of event $A$.

　　$A$:　Elizabeth graduates from Grand Central U.

**18.** Brad Rogers is a lawyer who was court appointed to defend an accused burglar. The district attorney has offered to plea bargain, but Brad wants to assess the probability of event $A$ before entering into negotiations.

　　$A$:　Brad can successfully defend the accused.

Patricia Quinlin is the coach for the Redwood College women's softball team. The opposing team the winning run on first base and their best is coming to bat. The team's best pitcher is mound, but a relief pitcher is prepared to the game. Patricia needs to assess the of event $A$.

rting pitcher can get the next

**20.** Ed Roberts has an opportunity to invest some money in a partnership opening a laundromat in a developing section of the city. Ed needs to assess the probability of event $A$.

　　$A$:　The business venture will be profitable.

**21.** Dr. Nan Williams is an orthopedic surgeon. Her latest patient is a high school senior with a knee injured in a football game. She is planning on performing surgery on the knee. Nan needs to assess the probability of event $A$.

　　$A$:　The knee will recover sufficiently to permit the player to accept a football scholarship.

**22.** Peter Clovin is a county supervisor. A proposed zoning change is scheduled for the next meeting of the board. Peter needs to assess the probability of event $A$.

　　$A$:　A majority of next election voters supports the zoning change.

*In Exercises 23–26, use the values stated in the Venn diagrams and compute the indicated probabilities.*

**23.** a. $P(A)$
　　b. $P(B)$
　　c. $P(A')$
　　d. $P(B')$

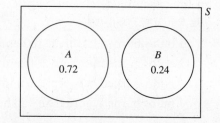

Figure 3-14.

**24.** a. $P(A)$
　　b. $P(B)$
　　c. $P(A')$
　　d. $P(B')$

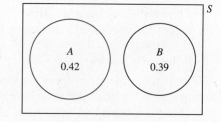

Figure 3-15.

**25.** a. $P(A)$
b. $P(A')$
c. $P(B)$
d. $P(B')$
e. $P(C)$
f. $P(C')$

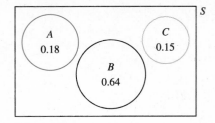

**Figure 3-16.**

**26.** a. $P(A)$
b. $P(A')$
c. $P(B)$
d. $P(B')$
e. $P(C)$
f. $P(C')$

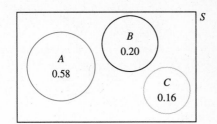

**Figure 3-17.**

## Exercises 3-2    Set B

*In Exercises 1–4, identify which of the rules of probability is violated by the indicated choice of numerical probabilities.*

**1.** A coin has been biased so that a head is more likely to occur than a tail each time the coin is tossed. Based on a large number of tosses, the following estimates are given for the coin:

$$P(\text{head}) = 0.68 \quad P(\text{tail}) = 0.38$$

**2.** A candidate for county supervisor authorizes a survey of the voters in the county to estimate the chance she will get elected. The staff members that conducted the survey gave her the following estimated probabilities:

$$P(\text{candidate will win}) = 0.57$$
$$P(\text{candidate will lose}) = 0.48$$

**3.** A small bank contains only dimes and quarters. The owner of the bank claims that when a coin is removed from the bank, then the following probabilities apply:

$$P(\text{dime}) = 0.62 \quad \text{and} \quad P(\text{quarter}) = 0.35$$

**4.** An instructor of college algebra told the class that everyone in the class was getting a B or C grade. Furthermore, if a student in the class was randomly selected and the grade of that student was determined, then the following probabilities would apply:

$$P(B \text{ grade}) = 0.33 \quad \text{and} \quad P(C \text{ grade}) = 0.66$$

*Exercise 5 refers to Figure 3-18 and Exercise 6 refers to Figure 3-19. State why the Venn diagrams cannot accurately depict the probabilities of the indicated events. You may make use of the fact that the sum of the probabilities of a set of mutually exclusive events making up a sample space must be 1.*

**5.**

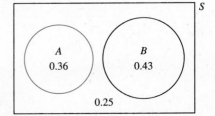

**Figure 3-18.**

**6.**

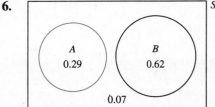

**Figure 3-19.**

*In Exercises 7–10, use the values indicated on the Venn diagrams in Figures 3-20 through 3-23 to compute the indicated probabilities. You may make use of the fact that the probability of an event made up of several mutually exclusive parts can be calculated as the sum of the probabilities of those parts.*

**7.** a. $P(A)$
    b. $P(B)$
    c. $P(A')$
    d. $P(B')$

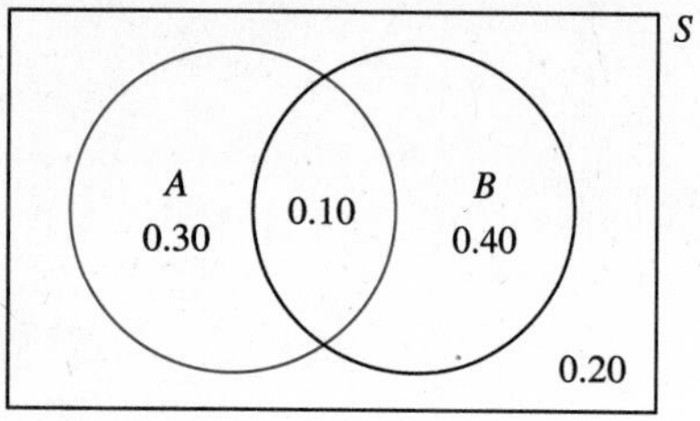

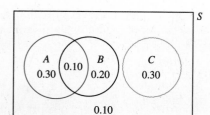

**Figure 3-20.**

**8.** a. $P(A)$
    b. $P(B)$
    c. $P(A')$
    d. $P(B')$

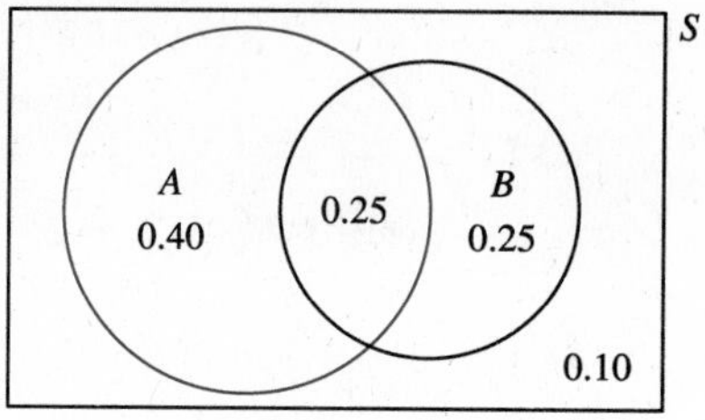

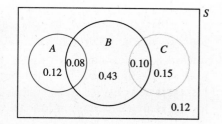

**Figure 3-21.**

**9.** a. $P(A)$
    b. $P(A')$
    c. $P(B)$
    d. $P(B')$
    e. $P(C)$
    f. $P(C')$

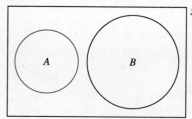

**Figure 3-22.**

**10.** a. $P(A)$
    b. $P(A')$
    c. $P(B)$
    d. $P(B')$
    e. $P(C)$
    f. $P(C')$

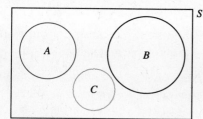

**Figure 3-23.**

*In Exercises 11 and 12, use the given probabilities to label the Venn diagrams. Again, you may make use of the fact that the sum of the probabilities of several mutually exclusive events making up S must be 1.*

**11.** $P(A) = 0.22$
    $P(B) = 0.54$

**Figure 3-24.**

**12.** $P(A) = 0.31$
    $P(B) = 0.40$
    $P(C) = 0.22$

**Figure 3-25.**

*In Exercises 13 and 14, use the stated probabilities and those displayed in the figures to completely label the Venn diagrams. You may make use of these facts:*

  *1. The probability of an event made up of several mutually exclusive parts can be calculated as the sum of the probabilities of those parts.*

  *2. The sum of the probabilities of several mutually exclusive events making up S must be 1.*

**13.** $P(A) = 0.64$
$P(B) = 0.55$

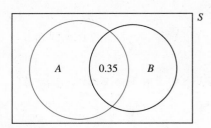

Figure 3-26.

**14.** $P(A) = 0.48$
$P(B) = 0.63$
$P(C) = 0.25$

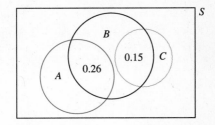

Figure 3-27.

## 3-3 The Addition Rule

**Key Topics**

1. *The event (A or B)*

2. *The event (A and B)*

3. *The addition rule for computing P(A or B)*

4. *P(A or B) when A and B are mutually exclusive*

5. *Using the addition rule to find P(A), P(B), or P(A and B)*

## STATISTICS IN ACTION

Caralee Woods is a sales representative for a pharmaceutical corporation that markets veterinary supplies. There are two days remaining in the current business cycle and to earn the maximum bonus rate offered by the company, Caralee must sell an additional $37,480 in products. She can reach this goal by successfully completing one or the other of two pending sales. She has identified these two potential sales as events *A* and *B*.

  Event *A*:  A sale to Monolith Veterinary Supplies

  Event *B*:  A sale to Stinson's Animal Clinic

Caralee is an experienced sales representative and feels the following values are reasonable estimates for the probabilities of these events:

$P(A) = 0.50$    A 50% chance of making the Monolith sale

$P(B) = 0.40$    A 40% chance of making the Stinson sale

Based on these estimates, would it be appropriate to say that Caralee has a 0.90 probability of earning the maximum bonus rate for this business cycle? (Recall that she needs to make only one of the two sales to earn the maximum rate.) In symbols we might be tempted to write:

$$P(A \text{ or } B) = 0.50 + 0.40$$
$$= 0.90$$

Although such a conclusion might initially seem plausible, we will learn there are other factors that we need to know about *A* and *B* before we can compute $P(A \text{ or } B)$. That is, for some events *A* and *B*, the actual probability of (*A* or *B*) will be less than the sum of the probabilities of *A* and *B*. The details of the so-called *addition rule for probability* will be studied in this section.

## The event (*A* or *B*)

Definition 3-8 identifies the outcomes in the event (*A* or *B*).

### DEFINITION 3-8 The Event (*A* or *B*)

An outcome is in the event (**A or B**) if it is in *A* only, or in *B* only, or in both *A* and *B*.

The Venn diagrams in Figure 3-28, (a), (b), and (c) are geometric displays of the outcomes in (*A* or *B*). As shown, any outcome of *S* in the shaded portion of the diagram is in (*A* or *B*).

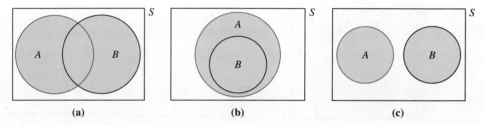

|     |     |     |
| --- | --- | --- |
| (a) | (b) | (c) |

**Figure 3-28.**    The shaded portions represent (*A* or *B*).

## The event (*A* and *B*)

Definition 3-9 identifies the outcomes in the event (*A* and *B*).

### DEFINITION 3-9 The Event (*A* and *B*)

An outcome is in the event (**A and B**), if it is simultaneously in *A* and *B*.

The Venn diagrams in Figure 3-29(a), (b), and (c) are geometric displays of the outcomes in (*A* and *B*). As shown, only outcomes of *S* in the shaded portion of the diagram are in (*A* and *B*). In (c), there are no outcomes in (*A* and *B*), because *A* and *B* are mutually exclusive.

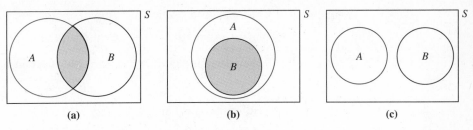

**Figure 3-29.**   The shaded portions represent (*A* and *B*).

▶ **Example 1**   A bag contains four similar disks numbered 1, 2, 3, and 4. One disk is selected and the number is recorded as the tens digit of a two-digit number. The first disk is replaced and a second disk is selected, and the number is used as the ones digit. The following events are defined:

*A*:   The number is even.

*B*:   The number is less than 30.

**a.** List the outcomes in *A*.        **c.** List the outcomes in (*A* or *B*).
**b.** List the outcomes in *B*.        **d.** List the outcomes in (*A* and *B*).

**Solution**   **Discussion.** Notice that because the disks are selected with replacement, the numbers 11, 22, 33, and 44 are possible outcomes.

**a.** Numbers with 2 and 4 in the units position are even.

$$A = \{12, 14, 22, 24, 32, 34, 42, 44\}$$

**b.** Numbers less than 30 have 1 or 2 in the tens position.

$$B = \{11, 12, 13, 14, 21, 22, 23, 24\}$$

**c.** (*A* or *B*) consists of those outcomes in *A*, or *B*, or both *A* and *B*.

$$(A \text{ or } B) = \{11, 12, 13, 14, 21, 22, 23, 24, 32, 34, 42, 44\}.$$

**d.** (*A* and *B*) consists of those outcomes in both *A* and *B*.

$$(A \text{ and } B) = \{12, 14, 22, 24\}. \quad ◀$$

▶ **Example 2**   An experiment consists of randomly selecting a letter from the set {a, b, c, d, e, f, g}. The following events are defined:

*A*:   The letter is in the word "delicious."

*B*:   The letter is in the word "ghastly."

**a.** List the outcomes in *A*.        **c.** List the outcomes in (*A* or *B*).
**b.** List the outcomes in *B*.        **d.** List the outcomes in (*A* and *B*).

Solution   **a.** Of the letters in "delicious," only d, e, and c are in $S$.

$$A = \{c, d, e\}.$$

**b.** Of the letters in "ghastly," only g and a are in $S$.

$$B = \{a, g\}.$$

**c.** Based on parts (a) and (b):

$$(A \text{ or } B) = \{a, c, d, e, g\}.$$

**d.** There are no outcomes of $S$ that are simultaneously in $(A \text{ and } B)$.

$(A \text{ and } B) = \varnothing$, the null, or empty set.   ◄

## The addition rule for computing *P(A or B)*

The experiment in Example 1 involves selecting, with replacement, two digits from the set $\{1, 2, 3, 4\}$. When $A$ is the event the number is even. When $B$ is the event the number is less than 20. A listing of the outcomes in $S$, $A$, $B$, $(A \text{ or } B)$, and $(A \text{ and } B)$ is given below.

| Outcomes in S | | | | Outcomes in A | | | | Outcomes in B | |
|---|---|---|---|---|---|---|---|---|---|
| 11 | 21 | 31 | 41 | | | | | 11 | 21 |
| 12 | 22 | 32 | 42 | 12 | 22 | 32 | 42 | 12 | 22 |
| 13 | 23 | 33 | 43 | | | | | 13 | 23 |
| 14 | 24 | 34 | 44 | 14 | 24 | 34 | 44 | 14 | 24 |

| Outcomes in (A or B) | | | | Outcomes in (A and B) | |
|---|---|---|---|---|---|
| 11 | 21 | | | | |
| 12 | 22 | 32 | 42 | 12 | 22 |
| 13 | 23 | | | | |
| 14 | 24 | 34 | 44 | 14 | 24 |

Using the classical notion of chance:

$$P(A) = \frac{8}{16} \qquad \frac{8 \text{ outcomes in } A}{16 \text{ outcomes in } S}$$

$$P(B) = \frac{8}{16} \qquad \frac{8 \text{ outcomes in } B}{16 \text{ outcomes in } S}$$

$$P(A \text{ or } B) = \frac{12}{16} \qquad \frac{12 \text{ outcomes in } (A \text{ or } B)}{16 \text{ outcomes in } S}$$

$$P(A \text{ and } B) = \frac{4}{16} \qquad \frac{4 \text{ outcomes in } (A \text{ and } B)}{16 \text{ outcomes in } S}$$

These four probabilities can be used to write the following equation:

$$\frac{12}{16} = \frac{8}{16} + \frac{8}{16} - \frac{4}{16}$$

**$P(A \text{ or } B) = P(A) + P(B) - P(A \text{ and } B)$**

Notice that $P(A \text{ and } B)$ is subtracted from the sum $P(A) + P(B)$. The subtraction is necessary because the outcomes in $(A \text{ and } B)$ were counted twice, namely in $A$, and then again in $B$. The subtraction "uncounts" these outcomes once, and therefore reduces $P(A) + P(B)$ to the correct value for $P(A \text{ or } B)$. The equation above for $P(A \text{ or } B)$ is true in general and is called *the addition rule of probability*.

### The Addition Rule of Probability

If $A$ and $B$ are two events defined in a sample space $S$, then:

$$P(A \text{ or } B) = P(A) + P(B) - P(A \text{ and } B)$$

Consider again Caralee Woods in the Statistics in Action portion of this section. To estimate $P(A \text{ or } B)$ for the events she defined, she needs to estimate the probability that both of the pending sales are successfully completed. This probability should then be subtracted from 0.90 to obtain the probability that she earns the maximum bonus.

For example, suppose Caralee feels there is a 20 percent chance that both the Monolith and Stinson sales are made. That is,

$$P(A \text{ and } B) = 0.20$$

With this probability known:

$$P(A \text{ or } B) = 0.50 + 0.40 - 0.20 = 0.70$$

The probability that Caralee earns the maximum bonus rate this business cycle would be approximately 70 percent.

▶ **Example 3**    A midwest automobile dealer has a large inventory of used cars and trucks. Records indicate that 50 percent of these vehicles are more than 3 years old and 40 percent are imports. Furthermore 30 percent are imports that are more than 3 years old. One of these vehicles is randomly selected. Find the probability that the vehicle is more than 3 years old or is an import.

**Solution**    **Discussion.** Represent the first two events described above as $A$ and $B$, and then identify the given probabilities. Finally use the addition rule to compute $P(A \text{ or } B)$.

$A$:   The vehicle is more than 3 years old.

$B$:   The vehicle is an import.

$$P(A) = 0.50$$                    50% of the vehicles are more than
                                    3 years old.

$$P(B) = 0.40$$                    40% of the vehicles are imports.

$$P(A \text{ and } B) = 0.30$$      30% are imports that are more than
                                    3 years old.

$$P(A \text{ or } B) = 0.50 + 0.40 - 0.30$$   $P(A \text{ or } B) = P(A) + P(B) - P(A \text{ and } B)$

$$= 0.60$$

Thus the probability is about 60 percent that the randomly selected vehicle will be more than 3 years old or an import.   ◀

## $P(A$ or $B)$ when $A$ and $B$ are mutually exclusive

Mutually exclusive events have no outcomes in common. Therefore if $A$ and $B$ are mutually exclusive:

$$P(A \text{ and } B) = 0$$    The probability is 0 that an outcome will be found
                               simultaneously in both of two mutually exclusive events.

This fact can be combined with the general addition rule to produce a special formula for mutually exclusive events.

### Addition Rule of Probability for Two Mutually Exclusive Events

If $A$ and $B$ are mutually exclusive events, then:

$$P(A \text{ or } B) = P(A) + P(B)$$

▶ **Example 4**   A card is randomly selected from a deck of 52 playing cards. Compute the probability that the card is a heart or a spade.

Solution   **Discussion.** For this experiment, the following events are defined:

$A$:   The card is a heart.

$B$:   The card is a spade.

Because no card can be simultaneously a heart and a spade, the events are mutually exclusive. The modified addition rule can therefore be used to compute $P(A$ or $B)$.

$$P(\text{heart or spade}) = \frac{13}{52} + \frac{13}{52}$$    13 cards are hearts and 13 cards are spades.

$$= \frac{26}{52} \quad \text{or} \quad \frac{1}{2} \quad ◀$$

The modified addition rule of probability can be extended to more than two mutually exclusive events.

---

### Addition Rule of Probability for $m$ Mutually Exclusive Events

If $A_1, A_2, \ldots, A_m$ are $m$ mutually exclusive events, that is, no outcome is in more than one of these events, then:

$$P(A_1 \text{ or } A_2 \text{ or } \ldots \text{ or } A_m) = P(A_1) + P(A_2) + \cdots + P(A_m)$$

---

▶ **Example 5**   Capital Nursery received a large shipment of tulip bulbs from Holland, Michigan. According to the invoice, 15 percent of the bulbs will produce red flowers, 20 percent will produce yellow flowers, and 10 percent will produce purple flowers. One bulb is randomly selected from this shipment and planted. Compute the probability that the flower produced will be red, or yellow, or purple.

**Solution**   For this experiment, the following events with the stated probabilities can be defined:

$A$:   The flower is red.      $P(A) = 0.15$

$B$:   The flower is yellow.   $P(B) = 0.20$

$C$:   The flower is purple.   $P(C) = 0.10$

Because these events are mutually exclusive, the special addition rule can be used.

$$P(A \text{ or } B \text{ or } C) = 0.15 + 0.20 + 0.10$$
$$= 0.45$$

Thus the probability is about 45 percent that the flower will be red, or yellow, or purple.   ◀

▶ **Example 6**   Use Figure 3-30 and compute the following probabilities:

**a.** $P(A)$
**b.** $P(B)$
**c.** $P(A \text{ and } B)$
**d.** $P(A \text{ or } B)$

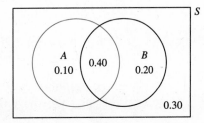

**Figure 3-30.**   A Venn diagram of $A$ and $B$.

Solution     **Discussion.** There are two mutually exclusive regions that make up event $A$. Therefore $P(A)$ is determined by adding the numbers in these two regions. The same is true for event $B$.

**a.** $P(A) = 0.1 + 0.4$
    $= 0.5$     50% of the outcomes are in $A$.

**b.** $P(B) = 0.2 + 0.4$
    $= 0.6$     60% of the outcomes are in $B$.

**c.** $P(A \text{ and } B) = 0.4$     40% of the outcomes are in $(A \text{ and } B)$.

**d.** $P(A \text{ or } B) = 0.5 + 0.6 - 0.4$     $P(A \text{ or } B) = P(A) + P(B) - P(A \text{ and } B)$
    $= 0.7$          70% of the outcomes are in $(A \text{ or } B)$.

Notice that in part (d) we could have simply added the numbers in the three mutually exclusive regions that comprise $A$ or $B$.     ◄

## Using the addition rule to find $P(A)$, $P(B)$, or $P(A$ and $B)$

**The addition rule of probability involves the following four terms:**

> **Term 1** $P(A \text{ or } B)$
> **Term 2** $P(A)$
> **Term 3** $P(B)$
> **Term 4** $P(A \text{ and } B)$
>
> When any three of these probabilities are known, the fourth one can be computed.

► **Example 7**     Linda is going shopping at the Sunrise Shopping Mall. She has defined the following events:

$A$:   She buys a pair of jeans.

$B$:   She buys a pair of boots.

By her own estimates:

$$P(A) = 0.7 \quad \text{and} \quad P(A \text{ or } B) = 0.9$$

If the probability is 0.4 that she buys both jeans and boots, i.e. $P(A \text{ and } B) = 0.4$, find the probability that she buys a pair of boots.

Solution     **Discussion.** From the given information, $P(A \text{ or } B)$, $P(A)$, and $P(A \text{ and } B)$ are known. The unknown probability is $P(B)$.

$$0.9 = 0.7 + P(B) - 0.4 \qquad P(A \text{ or } B) = P(A) + P(B) - P(A \text{ and } B)$$
$$0.9 = 0.3 + P(B) \qquad\qquad 0.7 - 0.4 = 0.3$$
$$0.6 = P(B) \qquad\qquad\qquad \text{Subtract } 0.3 \text{ from both sides.}$$

Based on the given probabilities, the probability is 0.6 or 60 percent that Linda buys a pair of boots.  ◄

## Exercises 3-3   Set A

*In Exercises 1–4, for each random experiment and events A and B:*
*a. List the outcomes in A.*
*b. List the outcomes in B.*
*c. List the outcomes in (A or B).*
*d. List the outcomes in (A and B).*

1. A two-digit number is formed by randomly selecting, with replacement, digits from the set $\{6, 7, 8, 9\}$. The following events are defined:

    $A$:  The two digits are the same.

    $B$:  The number is odd.

2. A two-letter "word" is formed by randomly selecting, with replacement, letters from the set $\{a, b, c, d\}$. The following events are defined:

    $A$:  The first letter in the word is a vowel.

    $B$:  The second letter in the word is a vowel.

3. One letter is selected from the set $\{a, b, c, d, e\}$. The following events are defined:

    $A$:  The letter is in the word "slab."

    $B$:  The letter is in the word "pad."

4. One digit is selected from the set $\{0, 1, 2, 3, 4\}$. The following events are defined:

    $A$:  The digit is in the zip code 94137-3861.

    $B$:  The digit is in the phone number (936)555-8629

*In Exercises 5–8, list the outcomes in the indicated events.*

5. Robin Bullock owns a taxi company in Winnemucca, Nevada. For cabs, she has three General Motors products $(x_1, x_2, x_3)$, two Ford products $(y_1, y_2)$, and one import $(z)$. Two cabs are randomly selected to pick up passengers at a local cassino. There are 15 possible outcomes in a sample space of interest. The following events are defined:

    $A$:  Both are General Motors products.

    $B$:  One is a Ford product and one is the import.

    $C$:  The cars are made by different automakers.

    List the outcomes in the following events:
    a. $A$            b. $B$
    c. $C$            d. $(A$ or $B)$
    e. $(A$ or $C)$   f. $(B$ or $C)$
    g. $(A$ and $B)$  h. $(B$ and $C)$

**6.** Of the eight most active stocks on the New York Stock Exchange on a given day of trading, four of the stock prices advanced ($a_1, a_2, a_3, a_4$), three of the stock prices declined ($d_1, d_2, d_3$), and one remained unchanged ($u$). Two of the stocks in this list are selected, without replacement. There are 28 possible outcomes in a sample space. The following events are defined:

> $A$:   The prices of both stocks advanced.
>
> $B$:   One stock price declined and the other remained unchanged.
>
> $C$:   The prices of both stocks declined.

List the outcomes in the following events:

a.  $A$         b.  $B$
c.  $C$         d.  ($A$ or $B$)
e.  ($A$ or $C$)    f.  ($B$ or $C$)
g.  ($A$ and $B$)    h.  ($B$ and $C$)

**7.** A box contains two quarters ($q_1, q_2$), three dimes ($d_1, d_2, d_3$), and four nickels ($n_1, n_2, n_3, n_4$). An experiment consists of selecting two coins from the box without replacement. There are 36 possible outcomes for the experiment. The following events are defined:

> $A$:   The sum of the values of the selected coins is 35 cents.
>
> $B$:   The sum of the values of the selected coins is 15 cents.
>
> $C$:   The sum of the values of the selected coins is less than 20 cents.

List the outcomes in the following events:

a.  $A$         b.  $B$
c.  $C$         d.  ($A$ or $B$)
e.  ($A$ or $C$)    f.  ($B$ or $C$)
g.  ($A$ and $B$)    h.  ($B$ and $C$)

**8.** In his garage, Gus Triandos has a box containing nine wood screws. Four are No. 12 flat heads ($f_1, f_2, f_3, f_4$), three are No. 8 round heads ($r_1, r_2, r_3$), and two are $\frac{1}{4}$-inch lag screws ($l_1, l_2$). Two screws are randomly selected without replacement. There are 36 outcomes in a sample space. The following events are defined:

> $A$:   Both screws are No. 12 flat heads.
>
> $B$:   Both screws are of the same type.
>
> $C$:   One is a No. 8 round head and one is a $\frac{1}{4}$-inch lag screw.

List the outcomes in the following events:

a.  $A$         b.  $B$
c.  $C$         d.  ($A$ and $B$)
e.  ($A$ or $B$)    f.  ($A$ and $C$)
g.  ($A$ or $C$)    h.  ($B$ or $C$)

*In Exercises 9–16, find the indicated probabilities.*

**9.** A card is randomly selected from a deck of playing cards. Find the following probabilities:
a.  $P$(the card is a king or queen)
b.  $P$(the card is a red card or an ace)
c.  $P$(the card is a face card or a heart)
d.  $P$(the card is even-numbered or a spade)
e.  $P$(the card is odd-numbered or less than 5)

**10.** A pair of dice is rolled. Find the following probabilities:
a.  $P$(7 or 11 is thrown)
b.  $P$(2 or 12 is thrown)
c.  $P$(an even number or a number less than 6)
d.  $P$(an odd number or a number greater than 9)
e.  $P$(1 or 13 is thrown)

**11.** The ten letters in the word "statistics" are each painted on disks of the same size. A disk is randomly selected and the outcome is the letter on the disk. Find the following probabilities.
a.  $P$(i or t)
b.  $P$(s or a vowel)
c.  $P$(letter in "algebra" or "geometry")
d.  $P$(letter in "probability" or "certainty")
e.  $P$(e or b)

**12.** The 11 letters in the word "probability" are each painted on disks of the same size. A disk is randomly selected and the outcome is the letter on the disk. Find the following probabilities:
a.  $P$(a or t)
b.  $P$(b or i)
c.  $P$(letter in "mutually" or "exclusive")
d.  $P$(letter in "union" or "intersection")
e.  $P$(e or d)

**13.** One hundred students at Seminole Community College were interviewed regarding their attitudes toward eating breakfast. The results of the interviews are in Table 3-5. Suppose another student at the college is randomly selected. Use the results of the survey to estimate the following probabilities:
   a. $P$(the student is a female)
   b. $P$(the student regularly eats breakfast)
   c. $P$(a female or regularly eats breakfast)
   d. $P$(a male or regularly skips breakfast)

TABLE 3-5   Results of Survey on Breakfast

|         | Regularly Eat Breakfast | Regularly Skip Breakfast |
|---------|-------------------------|--------------------------|
| Female  | 32                      | 26                       |
| Male    | 30                      | 12                       |

**14.** Two hundred and fifty employees at Woolen Products Manufacturing were interviewed regarding their opinions on required union membership for all employees. The results of the interviews are in Table 3-6. Suppose another employee at the company is randomly selected. Use the results of the survey to estimate the following probabilities:
   a. $P$(the employee is a female)
   b. $P$(the employee rejects required membership)
   c. $P$(a male or has no opinion on membership)
   d. $P$(a female or approves of required membership)

TABLE 3-6   Results of Union Membership Survey

|                 | Approves of Required Membership | Rejects Required Membership | No Opinion |
|-----------------|----------------------------------|------------------------------|------------|
| Female employee | 80                               | 88                           | 10         |
| Male employee   | 43                               | 15                           | 14         |

**15.** Four hundred fifty licensed drivers in a New England state were interviewed regarding their numbers of moving violation convictions within the past 3 years. The results of the interviews are in Table 3-7. Suppose a licensed driver in this state is randomly selected. Use the results of the survey to estimate the following probabilities:
   a. $P$(the driver is younger than 25 years of age)
   b. $P$(between 25 and 40 years or has had 0 violations)
   c. $P$(older than 40 years or has two or more violations)

TABLE 3-7   Results of Moving Violations Survey

|                              | 0 Violations | 1 Violation | 2 or More Violations |
|------------------------------|--------------|-------------|----------------------|
| Drivers younger than 25 years | 8           | 86          | 31                   |
| Drivers between 25 and 40 years | 137       | 31          | 12                   |
| Drivers older than 40 years  | 125          | 15          | 5                    |

**16.** The admissions office at a state college in Pennsylvania examined the current enrollment records of 600 sophomore, junior, and senior students. They noted the number of units in which each student was enrolled for the current term. The results are displayed in Table 3-8.

TABLE 3-8   Results of Enrollment Study

|           | Less Than 10 Units | From 10 to 15 Units | Fifteen Units or More |
|-----------|--------------------|--------------------|-----------------------|
| Sophomore | 45                 | 125                | 55                    |
| Junior    | 70                 | 60                 | 70                    |
| Senior    | 80                 | 25                 | 70                    |

Suppose the record of one of these students is randomly selected. Use the data in the table to determine the following probabilities:
   a. $P$(the student is a sophomore)
   b. $P$(the student is taking 15 or more units)
   c. $P$(a Junior or is taking from 10 to 15 units)

*In Exercises 17–20, list the outcomes in each sample space. Then use the sample space to find the indicated probabilities.*

17. A bag contains eight tulip bulbs. Four bulbs grow red flowers, three grow yellow flowers, and one grows a purple flower. Two bulbs are randomly selected and planted. There are 28 outcomes in a sample space. Find the following probabilities:
    a. $P$(both are red)
    b. $P$(one yellow and one red)
    c. $P$(different colors)
    d. $P$(same colors)
    e. $P$(both are purple)

18. A bag contains eight pieces of chocolate-covered candy. Three are caramels ($c_1, c_2, c_3$), three are peanut clusters ($p_1, p_2, p_3$), and two are marshmallow cremes ($m_1, m_2$). Suppose two pieces of candy are selected from the bag. There are 28 outcomes in a sample space. Find the following probabilities:
    a. $P$(both are caramels)
    b. $P$(one caramel and one peanut cluster)
    c. $P$(different candies)
    d. $P$(same candies)
    e. $P$(two malted-milk balls)

19. There are nine tomatoes in a refrigerator. Four of the tomatoes weigh less than 3 ounces ($s_1, s_2, s_3, s_4$), three weigh between 3 and 4 ounces ($m_1, m_2,$

$m_3$), and two weigh more than 4 ounces ($l_1, l_2$). Two of these tomatoes are selected. There are 36 outcomes in a sample space. Find the following probabilities:
    a. $P$(both weigh less than 3 ounces)
    b. $P$(one weighs less than 3 ounces and one weighs more than 4 ounces)
    c. $P$(both weigh less than 3 ounces or both weigh more than 4 ounces)
    d. $P$(the two together weigh more than 7 ounces)
    e. $P$(the two together weigh less than 5 ounces)

20. Nine employees in the research and development division of V and W Electronics were discussing eating habits. Of these nine employees, four had weights that were appropriate for their size and age ($a_1, a_2, a_3, a_4$), three were overweight ($o_1, o_2, o_3$), and two were underweight ($u_1, u_2$). Two of these employees are randomly selected to participate in a company sponsored physical fitness program. There are 36 outcomes in a sample space. Find the following probabilities regarding the weights of the individuals selected:
    a. $P$(both employees have appropriate weights)
    b. $P$(one employee is underweight and one is overweight)
    c. $P$(both are underweight or both are overweight)
    d. $P$(both are not overweight)

*In Exercises 21–24, use the Venn diagrams and compute the indicated probabilities*

21. a. $P(A)$
    b. $P(B)$
    c. $P(A \text{ or } B)$
    d. $P(A')$
    e. $P(A' \text{ or } B')$

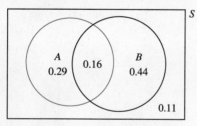

**Figure 3-31.**

22. a. $P(A)$
    b. $P(B)$
    c. $P(A \text{ or } B)$
    d. $P(B')$
    e. $P(A' \text{ or } B')$

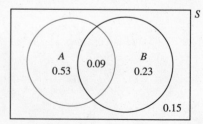

**Figure 3-32.**

**23.** a. $P(A)$
 b. $P(B)$
 c. $P(C)$
 d. $P(A')$
 e. $P(A$ or $B)$
 f. $P(A$ or $C)$
 g. $P(B$ or $C)$
 h. $P(A$ or $B$ or $C)$
 i. $P(A'$ or $B')$
 j. $P(B'$ or $C')$

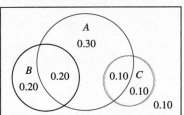

**Figure 3-33.**

**24.** a. $P(A)$
 b. $P(B)$
 c. $P(C)$
 d. $P(A')$
 e. $P(A$ or $B)$
 f. $P(A$ or $C)$
 g. $P(B$ or $C)$
 h. $P(A$ or $B$ or $C)$
 i. $P(A'$ or $B')$
 j. $P(B'$ or $C')$

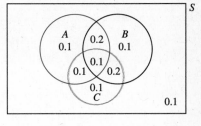

**Figure 3-34.**

*In Exercises 25 and 26, compute the probabilities using the addition rule for mutually exclusive events.*

**25.** A large display of bulbs for Christmas tree lights at Sims Hardware contains 18% blue bulbs, 27% red bulbs, and 13% white bulbs. One bulb is randomly selected and the color of the bulb is noted. Find the probability that the bulb is blue, red, or white.

**26.** Records from the registrar's office at a major college in the South shows that, for the current student body, 18% of the students are taking 12 units, 21% are taking 13 units, and 23% are taking 14 units. One student is randomly selected and the number of units the student is taking is noted. Find the probability that the student is taking 12, 13, or 14 units.

*In Exercises 27–30, find the indicated probability.*

**27.** Bill Cornett has just listed his condominium for sale. He feels there is a 60% chance that the first offer will come within the first month. Based on recent sales of similar condominiums, he feels there is a 30% chance that the first offer will be for the full asking price. If the probability is 70% that either the first offer is within the first month or for the full asking price, find Bill's assessment of the probability that both events will occur.

**28.** Phyllis Klarner is the regional manager of 48 outlets of a national chain of women's clothing. She judges that there is a 45% chance of an employee strike in the coming fiscal year. She also foresees a 60% chance of either a strike or a major layoff. Finally, she believes that there is only a 10%

chance of both a major layoff and a strike. Compute Phyllis's assessment of the likelihood of a major layoff.

**29.** Arnold Parker's and James Marshall's families are close friends. The two families are both shopping for VCRs (video cassette recorders). The following events are defined:

 $A$: The Parker family buys a Sony.

 $B$: The Marshall family buys a Sony.

By their own estimates:

$$P(A) = 0.65 \quad \text{and} \quad P(B) = 0.50$$

If the probability is 0.85 that one, or the other, or both buy a Sony, what is the probability that they both buy a Sony?

**30.** The city council of a large eastern city needs to increase the city's revenue. It is considering raising the fee for collecting garbage and raising the sewer rates. The following events are defined:

 $A$: The city council raises the fee for collecting garbage.

 $B$: The city council raises the sewer rates.

Sharon Noble, the council's public relations person, estimates the following probabilities:

$$P(A) = 0.75 \quad \text{and} \quad P(A \text{ or } B) = 0.92$$

If the probability is 0.59 that the city council raises both fees, what is the probability that they raise the sewer rates?

## Exercises 3-3    Set B

1. Use a table of random numbers to select 50 individuals from Data Set I, Appendix B. For each individual obtain the information on smoker (yes/no) and cancer (yes/no).
   a. Use the sample data to develop counts to place in the following grid:

|  |  | Cancer | | |
| --- | --- | --- | --- | --- |
|  |  | *Yes* | *No* | *Totals* |
| **Smoker** | Yes |  |  |  |
|  | No |  |  |  |
|  | Totals |  |  | 50 |

   b. Use the sample data to estimate the following indicated probabilities for the next randomly selected person from this data set:
   i. *P*(smoker)
   ii. *P*(cancer)
   iii. *P*(smoker or cancer)

2. Use a table of random numbers to select 50 stocks from Data Set II, Appendix B. For each data point obtain the cost of the stock and whether the stock's price increased (+), decreased (−), or was unchanged (0).
   a. Use the sample data to develop counts to place in the following grid:

|  |  | Direction of Stock's Price | | | |
| --- | --- | --- | --- | --- | --- |
|  |  | *Increased ( + )* | *Decreased ( − )* | *Unchanged (0)* | *Totals* |
| **Cost of Stock** | Less than or equal to $10.00 |  |  |  |  |
|  | Between $10.00 and $50.00 |  |  | — |  |
|  | $50.00 or more |  |  |  |  |
|  | Totals |  |  |  |  |

   b. Use the sample data to estimate the following indicated probabilities for the next randomly selected stock from this data set:
   i. *P*(cost is $10.00 or less)
   ii. *P*(stock's price increased)
   iii. *P*(cost is between $10.00 and $50.00 or declined)
   iv. *P*(cost is $50.00 or more or remained unchanged)

*In Exercises 3 and 4, sample 40 students at the school that you are attending.*

3. From each student, obtain the information on sex (male/female) and whether he/she is currently working in addition to being a student (yes/no).

a. Use the sample data to develop counts to place in the following grid:

| Sex | | Working | | |
|-----|-----|-----|-----|-----|
| | | Yes | No | Totals |
| | Male | | | |
| | Female | | | |
| | Totals | | | 40 |

b. Use the sample data to estimate the indicated probabilities for another student at your school:
   i. $P$(female)
   ii. $P$(student is working)
   iii. $P$(male or not working)
   iv. $P$(female or working)

4. From each student, obtain the information on political party affiliation (republican/democrat/independent/none) and whether he/she is receiving some form of government financial assistance for their education costs (yes/no).
   a. Use the sample data to develop counts to place in the following grid:

| Government Assistance | | Political Party Affiliation | | | | |
|-----|-----|-----|-----|-----|-----|-----|
| | | Republican | Democrat | Independent | None | Totals |
| | Yes | | | | | |
| | No | | | | | |
| | Totals | | | | | 40 |

b. Use the sample data to estimate the indicated probabilities for another student at your school:
   i. $P$(Republican)
   ii. $P$(student is getting government assistance)
   iii. $P$(Democrat or getting government assistance)
   iv. $P$(Independent or not getting government assistance)

*In Exercises 5–8, use the given probabilities and label the Venn diagrams with the correct probabilities.*

5. $P(A) = 0.58$
   $P(B) = 0.72$
   $P(A \text{ or } B)$
   $= 0.90$

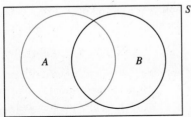

**Figure 3-35.**

6. $P(A) = 0.43$
   $P(A \text{ and } B)$
   $= 0.15$
   $P(A \text{ or } B)$
   $= 0.78$

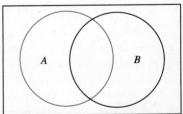

**Figure 3-36.**

**7.** $P(A) = 0.36$    $P(B) = 0.44$
$P(C) = 0.49$    $P(A \text{ and } B) = 0.13$
$P(A \text{ and } C) = 0.15$    $P(B \text{ and } C) = 0.18$
$P(A \text{ and } B \text{ and } C) = 0.08$

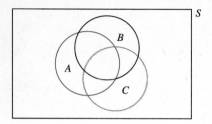

**Figure 3-37.**

**8.** $P(A) = 0.46$    $P(B) = 0.49$
$P(C) = 0.46$    $P(A \text{ and } B) = 0.26$
$P(A \text{ and } C) = 0.18$    $P(B \text{ and } C) = 0.31$
$P(A \text{ and } B \text{ and } C) = 0.18$

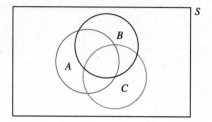

**Figure 3-38.**

# 3-4  Conditional Probability and Independence of Events

**Key Topics**

1. *Conditional probability*

2. *Independent events*

3. *Modeling A and B as independent*

S T A T I S T I C S    I N    A C T I O N

Laura Burnside is a loan officer at SWCU (State Workers Credit Union). Kevin Conors came to her office today to get a new car loan. Laura first had Kevin fill out an application form that included answers to items (1) through (5).

1. What is your current gross annual income?
2. How long have you been employed at your present occupation?
3. Are you married or single?
4. Are you renting or buying a home?
5. List all your current debts and amounts of the monthly payments.

You might wonder why such data are needed to determine whether an individual qualifies for a loan. Keep in mind that loan officers like Laura are trying to assess a probability when someone like Kevin applies for a loan. Specifically Laura wishes to evaluate:

*P*(Kevin fully repays the loan)

Lending institutions have gathered empirical evidence that such probabilities should change, up or down, depending on how a prospective borrower answers these questions. As a consequence, Laura will use Kevin's responses to help her decide whether to recommend to the loan committee that the loan be granted.

On the other hand, lending institutions feel that answers to items like (6) through (8) provide no useful information when evaluating the probability that an individual will repay a loan.

6. What is the color of your eyes?
7. Are you primarily right-handed or left-handed?

8. To which political party are you currently affiliated?

Laura feels that the probability of Kevin repaying the proposed car loan is independent of Kevin's eye color, hand preference, and political party affiliation. The topics of conditional probability and independent events are studied in this section.

## Conditional probability

If $A$ and $B$ are two events defined for some sample space, then the conditional probability of $B$ given $A$, is written $P(B|A)$. The equation in Definition 3-11 defines this conditional probability of $B$, given the partial information that $A$ has occurred in terms of $P(A \text{ and } B)$ and $P(A)$.

### DEFINITION 3-11 The Conditional Probability of B given A

If $P(A) \neq 0$, then:

$$P(B|A) = \frac{P(A \text{ and } B)}{P(A)}$$

The probability the outcome is in ($A$ and $B$)
The probability the outcome is in $A$

▶ **Example 1**  Home Products Manufacturing makes small appliances. A given shift of the employees of this company produced $N = 500$ items. An inventory of the products yielded the data in Table 3-9. Suppose that one of these items is randomly selected. Find the following conditional probabilities:

**a.** $P(\text{defective}|\text{toaster})$
**b.** $P(\text{mixer}|\text{not defective})$

TABLE 3-9  An Inventory of 500 Items

| Product | Not Defective | Defective |
|---------|---------------|-----------|
| Hand mixer | 120 | 10 |
| Electric knife | 130 | 15 |
| Electric toaster | 200 | 25 |

**Solution**  **a.** $P(\text{defective}|\text{toaster}) = \dfrac{P(\text{defective and toaster})}{P(\text{toaster})}$

From Table 3-9, 25 toasters are defective.

$$P(\text{defective and toaster}) = \frac{25}{500} = 0.05$$

Also from the table, 225 items are toasters.

$$P(\text{toaster}) = \frac{225}{500} = 0.45$$

Thus

$$P(\text{defective}|\text{toaster}) = \frac{0.05}{0.45}$$

$$= 0.1111\ldots \approx 11\%$$

The probability is about 11 percent that the item is defective, given that a toaster has been selected.

**b.** $P(\text{mixer}|\text{not defective}) = \dfrac{P(\text{mixer and not defective})}{P(\text{not defective})}$

From Table 3-9, 120 hand mixers are not defective.

$$P(\text{mixer and not defective}) = \frac{120}{500} = 0.24$$

Also from the table, 450 items are not defective.

$$P(\text{not defective}) = \frac{450}{500} = 0.90$$

Thus

$$P(\text{mixer}|\text{not defective}) = \frac{0.24}{0.90}$$

$$= 0.266\ldots \approx 27\%$$

The probability is about 27 percent that the item is a mixer, given that a nondefective item has been selected.   ◄

▶ **Example 2**   Use the probabilities shown in Figure 3-39 to compute the following conditional probabilities:

**a.** $P(B|A)$
**b.** $P(A|B)$

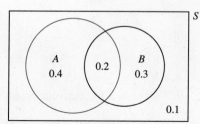

**Figure 3-39.**   Probabilities of $A$ and $B$.

**Solution** **Discussion.** The probabilities follow from the figure:

$$P(A) = 0.6 \quad P(B) = 0.5 \quad P(A \text{ and } B) = 0.2$$

**a.** $P(B|A) = \dfrac{0.2}{0.6}$ $\qquad\qquad P(B|A) = \dfrac{P(A \text{ and } B)}{P(A)}$

$$= \frac{1}{3} \quad \text{or} \quad 33\tfrac{1}{3}\%$$

The conditional probability of $B$ given $A$, is $33\tfrac{1}{3}$ percent.

**b.** $P(A|B) = \dfrac{0.2}{0.5}$ $\qquad\qquad P(A|B) = \dfrac{P(A \text{ and } B)}{P(B)}$

$$= \frac{2}{5} \quad \text{or} \quad 40\%$$

The conditional probability of $A$ given $B$, is 40 percent. ◄

## Independent events

In Example 2, $P(B) = 0.50$ and $P(B|A) = 0.33\tfrac{1}{3}$. Obviously $P(B) \neq P(B|A)$. Thus the partial information that the outcome is in $A$ in some sense changed the probability of $B$ from 50 to $33\tfrac{1}{3}$ percent.

However, for some situations, conditional probabilities are the same as unconditional probabilities of the corresponding events. To illustrate this possibility, consider the probabilities shown in Figure 3-40. Based on the numbers in the diagram:

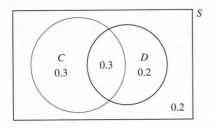

$$P(C) = 0.6 \quad P(D) = 0.5 \quad P(C \text{ and } D) = 0.3$$

$$P(C|D) = \frac{0.3}{0.5} = 0.6 = P(C) \qquad \text{Thus } P(C|D) = P(C)$$

$$P(D|C) = \frac{0.3}{0.6} = 0.5 = P(D) \qquad \text{Thus } P(D|C) = P(D)$$

We therefore say that $C$ and $D$ are *independent events*.

**Figure 3-40.** Probabilities of $C$ and $D$.

### DEFINITION 3-12 Independent Events A and B

The phrase "events $A$ and $B$ are (modeled as) **independent**" means that probabilities have been assigned so that:

$P(B|A) = P(B)$ The conditional probability of $B$ given $A$ equals the probability of $B$.

$P(A|B) = P(A)$ The conditional probability of $A$ given $B$ equals the probability of $A$.

▶ **Example 3**   Garber's Lawn Furniture Manufacturing ships some of its orders with Ace Truck-ing and some with King Transport Services. Some orders are damaged in transit. Garber's records of the last 1000 orders shipped by Ace and King are shown in Table 3-10. A single order is randomly selected from these 1000 orders. Compute the following:

**a.** $P$(an Ace order)
**b.** $P$(Ace and damaged order)
**c.** $P$(damaged order)
**d.** $P$(Ace order|damaged order)
**e.** Are the events "Ace order" and "damaged order" independent?

**TABLE 3-10   Shipping Records**

|  | Damage | No Damage |
|------|--------|-----------|
| Ace  | 40     | 360       |
| King | 60     | 540       |

Solution   **a.** $40 + 360 = 400$ orders are for Ace Trucking:

$$P(\text{an Ace order}) = \frac{400}{1000} = 40\%$$

**b.** 40 orders for Ace Trucking are damaged:

$$P(\text{Ace and damaged order}) = \frac{40}{1000} = 4\%$$

**c.** $40 + 60 = 100$ orders are damaged:

$$P(\text{damaged order}) = \frac{100}{1000} = 10\%$$

**d.** $P(\text{Ace order}|\text{damaged order}) = \dfrac{4\%}{10\%} \quad \dfrac{P(\text{Ace and damaged})}{P(\text{damaged})}$

$\qquad\qquad\qquad = 40\%$

**e.** $\qquad\qquad P(\text{Ace order}) = 40\%$
$P(\text{Ace order}|\text{damaged order}) = 40\%$

The probabilities are the same; therefore the events are independent. This means that the probability a given order was shipped by Ace Trucking is 40 percent, regardless of whether the order was damaged. In the context of the problem, the rates at which Ace Trucking and King Transport Services damage products in shipping are the same. Specifically they both damage orders at a rate of 10 percent.   ◀

▶ **Example 4**   An opinion poll was taken of adult women and men regarding their TV viewing preferences regarding football and baseball. The results of the poll are shown in Table 3-11.

   One person in the group is randomly selected. For the experiment, the fol-lowing events are defined:

$\qquad A$:   The person prefers watching football.
$\qquad B$:   The person is a woman.

TABLE 3-11  Results of Sports Preference Poll

|  | Prefer Watching Football | Prefer Watching Baseball | No Preference | Totals |
|---|---|---|---|---|
| Women | 123 | 82 | 45 | 250 |
| Men | 157 | 78 | 15 | 250 |
| Totals | 280 | 160 | 60 | 500 |

**a.** Compute $P(A)$.
**b.** Compute $P(A|B)$.
**c.** Based on parts (a) and (b), are events $A$ and $B$ independent?

**Solution**  **a.** $P(A) = \dfrac{280}{500}$  or  56%    $\dfrac{280 \text{ prefer watching football}}{500 \text{ people surveyed}}$

**b.** $P(A|B) = \dfrac{123}{250}$  or  49.2%    $\dfrac{123 \text{ women prefer watching football}}{250 \text{ women surveyed}}$

**c.** Because 56% ≠ 49.2%, the events are not independent.    ◄

## Modeling *A* and *B* as independent

For some experiments, the determination of the independence of $A$ and $B$ cannot be made by a direct application of the equations in Definition 3-12. Many times $A$ and $B$ are modeled as (at least approximately) independent because of the nature of the random experiment on which the events are defined. The following examples illustrate some of these circumstances:

**Case 1** **The experiment consists of unrelated but physically similar trials.**
For example, suppose two coins are flipped.

$A$:  A head is obtained on the first coin.

$B$:  A head is obtained on the second coin.

This experiment consists of two unrelated trials:

**Trial 1**  Flip the first coin.

**Trial 2**  Flip the second coin.

Intuitively we feel that the outcome on the second coin is unrelated to the outcome on the first coin. Thus it is common to model $A$ and $B$ as independent.

**Case 2** **The experiment involves sampling with replacement.**
For example, consider an experiment in which two cards are selected, with replacement, from a deck of 52 playing cards.

$A$:  The first card selected is a heart.

$B$:  The second card selected is a heart.

This experiment consists of two trials, but the first card is replaced before the second card is drawn. As a consequence, any effect of the first draw is negated by replacing the card.

**Case 3**  **The experiment involves selecting a relatively small simple random sample from a large population.**

To illustrate such an experiment, suppose two items are randomly selected (without replacement) from a production run of 100,000 items in which 5,000 are known to have a minor defect. The following events are defined:

$A$:   The first item selected has a minor defect.

$B$:   The second item has a minor defect.

$P$(second item is defective) = 0.05.

$P$(defect on second item | defect on first item)

$$= \frac{4,999}{99,999} \qquad \frac{\text{one less defective}}{\text{one less item}}$$

$$= 0.0499905 \ldots$$

$$\approx 0.05 \text{ to two decimal places.}$$

Thus $P(B|A)$ is approximately equal to $P(B)$. It is common practice to model the results of successive selections as independent when the experiment involves taking a small sample from a large population.

**Case 4**  **There is no evidence to support a claim that $A$ and $B$ are dependent events.**

For example, suppose that Cynthia Biron is a salesperson for Mutual Life Insurance Corporation. Tonight Cynthia has appointments to propose TSAs (tax-sheltered annuities) to two clients. For this activity, the following events are defined:

$A$:   Cynthia sells a TSA to the first client.

$B$:   Cynthia sells a TSA to the second client.

It appears, from an intuitive point of view, that $A$ and $B$ should be modeled as independent. That is, any value that Cynthia might give to $P(B)$ would not change, given the partial information of the results of event $A$.

For events such as $A$ and $B$, independence is usually assumed until empirical evidence is obtained to show otherwise. In many areas of science, medicine, business, and so on, the search for empirical evidence to help assess the probabilities and conditional probabilities is very important.

The Statistics in Action part of this section was used to illustrate Case 4. Answers to Items (1) through (5) on the loan application compose events that might produce conditional probabilities different from the unconditional probability

$$P(\text{Kevin repays the car loan})$$

By contrast, answers to items (6) through (8) compose events that are independent of the event of interest. The guidelines in Cases 1 through 4 are intended to help us decide when it is reasonable to model two or more events of interest as independent. We will see in the next section how independence of events can often be useful in the calculation of probabilities of interest.

## Exercises 3-4   Set A

*In Exercises 1–6, find the indicated probabilities.*

1. The class lists of students at Stanislaus College yielded the data shown in Table 3-12. A student in this group is randomly selected and the following events are defined:

   A:   The student is enrolled in Anatomy/Physiology.

   A':   The student is not enrolled in Anatomy/Physiology.

   Find the following conditional probabilities:
   a. $P(\text{Female}|A)$     b. $P(\text{Male}|A')$
   c. $P(A'|\text{Female})$     d. $P(A|\text{Male})$

   TABLE 3-12   A Survey of Stanislaus College Students

   |        | A   | A'  |
   |--------|-----|-----|
   | Female | 290 | 90  |
   | Male   | 210 | 160 |

2. At a school counselors' conference a breathing test was given to 450 volunteers. For all those tested, the results are recorded in Table 3-13.

   TABLE 3-13   Results of a Breathing Test

   |                          | A   | A'  |
   |--------------------------|-----|-----|
   | Person smokes cigarettes | 36  | 84  |
   | Person does not smoke    | 264 | 66  |

   A:   An adequate amount of air is supplied to the lungs.

   A':   An inadequate amount of air is supplied to the lungs.

   Suppose one of the volunteers is randomly selected and the breathing test is given to the person. Compute the following conditional probabilities:
   a. $P(\text{Person smokes cigarettes}|A')$     b. $P(\text{Person does not smoke}|A)$
   c. $P(A'|\text{Person does not smoke})$     d. $P(A|\text{Person smokes cigarettes})$

3. A lot made up of 1000 machine parts is checked piece by piece for Brinell Hardness (BNH) and diameter, with counts resulting as shown in Table 3-14.

TABLE 3-14   Results of Hardness Test and Diameter Measurement

| | | Diameter | | |
|---|---|---|---|---|
| | | Less Than 1.00 in. | Between 1.00 and 1.05 in. | More Than 1.05 in. |
| Brinell Hardness (BNH) | Less than 190 | 150 | 100 | 50 |
| | Between 190 and 210 | 100 | 300 | 100 |
| | More than 210 | 30 | 70 | 100 |

A single part is randomly selected from the lot. Compute the following conditional probabilities:

a. $P$(diameter more than 1.05 in.|BNH more than 210)
b. $P$(diameter less than 1.00 in.|BNH between 190 and 210)
c. $P$(BNH more than 210|diameter less than 1.00 in.)
d. $P$(BNH less than 190|diameter between 1.00 and 1.05 in.)

4. The car inventory at Woodson's Auto World was separated into categories based on manufacturer and miles per gallon (mpg) ratings. The results of the inventory are shown in Table 3-15. Suppose that one of these cars is randomly selected. Find the following conditional probabilities to three decimal places:

a. $P$(less than 20 mpg|American made)
b. $P$(between 20 and 30 mpg|Import)
c. $P$(Import|more than 30 mpg)
d. $P$(American made|between 20 and 30 mpg)

TABLE 3-15   Results of Car Survey

| MPG Rating | American Made | Import |
|---|---|---|
| Less than 20 mpg | 300 | 130 |
| Between 20 and 30 mpg | 340 | 260 |
| More than 30 mpg | 210 | 260 |

5. The employee records at National Syndicated Inc. were surveyed. The age and amount of sick leave each employee took last year were recorded. The results are in Table 3-16.

TABLE 3-16   Results of Employee Survey

| Ages of Employees | 0–3 Sick Days | 4–6 Sick Days | 7 or More Sick Days |
|---|---|---|---|
| Less than 25 yrs old | 180 | 405 | 315 |
| Between 25 and 40 yrs old | 1755 | 540 | 405 |
| More than 40 yrs old | 720 | 540 | 540 |

Suppose that one of these employees is randomly selected. For this experiment, the following events are defined:

> A:   The employee took 0–3 sick days leave last year.
>
> B:   The employee took 4–6 sick days leave last year.
>
> C:   The employee took 7 or more sick days leave last year.

For this experiment, calculate the following conditional probabilities to three decimal places:

a. $P(A|$Employee is between 25 and 40 years old)

b. $P(C|$Employee is less than 25 years old)

c. $P($Employee is more than 40 years old$|B)$

d. $P($Employee is between 25 and 40 years old$|C)$

6. An independent research organization subjected 75 light bulbs from each of three manufacturers, corporations G, S, and I, to a stress test. The result of each test placed a bulb in one of the following three categories:

> A:   The highest quality rating
>
> B:   The middle quality rating
>
> C:   The lowest quality rating

The results of the test are shown in Table 3-17. Suppose a bulb is randomly selected from the 225 tested bulbs. Evaluate:

TABLE 3-17   Results of Light Bulb Test

|   | A | B | C |
|---|---|---|---|
| G | 40 | 25 | 10 |
| S | 30 | 40 | 5 |
| I | 15 | 30 | 30 |

a. $P($A rating$|$Bulb by G)      b. $P($B rating$|$Bulb by I)

c. $P($Bulb by S$|$C rating)      d. $P($Bulb by G$|$B rating)

*In Exercises 7–10, use the probabilities in the Venn diagrams and compute the indicated conditional probabilities.*

7. a. $P(B|A)$
   b. $P(A|B)$

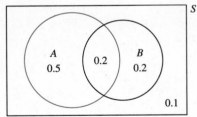

Figure 3-41.

8. a. $P(B|A)$
   b. $P(A|B)$

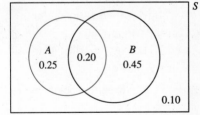

Figure 3-42.

9. a. $P(B|A)$
   b. $P(A|B)$
   c. $P(C|B)$
   d. $P(B|C)$
   e. $P(A|C)$
   f. $P(C|A)$

Figure 3-43.

10. a. $P(B|A)$
    b. $P(A|B)$
    c. $P(C|B)$
    d. $P(B|C)$
    e. $P(A|C)$
    f. $P(C|A)$

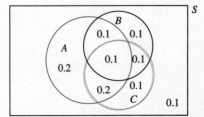

Figure 3-44.

*In Exercises 11–16, find the indicated probabilities and conditional probabilities.*

**11.** An experiment consists of randomly selecting a letter from:

$$S = \{a, b, c, d, e, f, g, h\}.$$

The following events are defined:

      *A*:   The letter is a, b, c, or d.

      *B*:   The letter is a, b, or g.

Find the indicated values:
a.  $P(A)$    b.  $P(A|B)$
c.  $P(B)$    d.  $P(B|A)$
e.  Are *A* and *B* independent?

**12.** An experiment consists of randomly selecting a number from:

$$S = \{1, 2, 3, 4, 5, 6, 7, 8, 9\}.$$

The following events are defined:

      *A*:   The number is 1, 2, 3, 4, or 5.

      *B*:   The number is 2, 4, 6, or 8.

Find the indicated values:
a.  $P(A)$    b.  $P(A|B)$
c.  $P(B)$    d.  $P(B|A)$
e.  Are *A* and *B* independent?

**13.** An experiment consists of rolling a fair red die and a fair green die. The following events are defined:

      *A*:   The outcome on the red die is a 3.

      *B*:   The outcome on the green die is a 4.

Find the indicated values:
a.  $P(A)$
b.  $P(B)$
c.  Are *A* and *B* independent?
d.  $P(A|B)$
e.  $P(B|A)$

**14.** An experiment consists of selecting two cards from a shuffled deck of playing cards. The first card is replaced and the deck shuffled before the second card is selected.

      *A*:   The first card selected is a spade.

      *B*:   The second card selected is an ace.

Find the indicated values:
a.  $P(A)$
b.  $P(B)$
c.  Are *A* and *B* independent?
d.  $P(B|A)$
e.  $P(A|B)$

**15.** An experiment consists of selecting two cars at an intersection in a large metropolitan area in which it is known that 30% of the cars have ineffective emission controls.

      *A*:   The first car has an ineffective emission control.

      *B*:   The second car has an ineffective emission control.

Find the indicated values:
a.  $P(A)$
b.  $P(B)$
c.  Are *A* and *B* independent events?
d.  $P(B|A)$
e.  $P(A|B)$

**16.** In a large batch of iron pipes, it is found that of two types of defects:
  i. 4% of all the pipes have defect *A*.
 ii. 8% of all the pipes have defect *B*.
iii. 2% of all the pipes have defects *A* and *B*.
A pipe is randomly selected. Find the following probabilities:
a.  *P*(pipe has defect *A*).
b.  *P*(pipe has defect *A*|pipe has defect *B*).
c.  Are the events that the pipe has a defect of type *A* and that the pipe has a defect of type *B* independent?

*In Exercises 17—25, discuss whether A and B can be modeled as independent events. Base your decision on whether you would change your assessment of the likelihood of B, given the partial information that the outcome is in event A.*

**17.** Randomly select a coed currently enrolled at Florida State University.

> *A:*  The coed has red hair.
>
> *B:*  The coed is more than 5 feet, 8 inches tall.

**18.** Randomly select a car from the student parking lot.

> *A:*  The car is a Ford product.
>
> *B:*  The car is painted blue.

**19.** Randomly select a male student enrolled at Notre Dame University.

> *A:*  The student is a member of the varsity football team.
>
> *B:*  The student weighs more than 200 pounds.

**20.** Randomly select a card from a deck of 52 playing cards.

> *A:*  The card is a red card.
>
> *B:*  The card is a diamond.

**21.** A professional basketball player is about to shoot two free throws.

> *A:*  The player makes good on the first shot.
>
> *B:*  The player makes good on the second shot.

**22.** A test in Political Science 103 has 25 true-false questions.

> *A:*  The correct answer to Question 5 is true.
>
> *B:*  The correct answer to Question 8 is true.

**23.** Randomly select a credit application for a charge card at J.C. Penney stores.

> *A:*  The applicant is a woman.
>
> *B:*  The applicant is approved for a credit card.

**24.** Randomly select a dog from an animal shelter in a large eastern city.

> *A:*  The dog weighs more than 50 pounds.
>
> *B:*  The dog has long hair.

**25.** Randomly select a state senator from a midwestern state.

> *A:*  The senator was elected from the Democratic party.
>
> *B:*  The senator favors spending state funds to clean up toxic waste sites.

## Exercises 3-4    Set B

1. Use a table of random numbers to select 40 individuals from Data Set I, Appendix B. Use the information from these data points to fill in the following table:

|  | | Smoker | | |
| --- | --- | --- | --- | --- |
|  | | Yes | No | Totals |
|  | Male | | | |
| Sex | Female | | | |
|  | Totals | | | 40 |

Use the results of your sample to estimate the following probabilities for an individual sampled from this population:
a. $P$(male and smoker)
b. $P$(male or smoker)
c. $P$(male|smoker)
d. $P$(smoker|male)
e. Based on your sample, does it appear that the events male and smoker are independent?

2. Use a table of random numbers to select 50 stocks from Data Set II, Appendix B. Use the information from these data points to fill in the following table:

|  | | Direction of Movement of Stock's Price | | | |
| --- | --- | --- | --- | --- | --- |
|  | | Increased ( + ) | Decreased ( − ) | Unchanged (0) | Totals |
|  | Less than or equal to $10.00 | | | | |
| Cost of Stock | Between $10.00 and $50.00 | | | | |
|  | $50.00 or more | | | | |
|  | Totals | | | | |

Use the results of your sample to estimate the following probabilities for a stock price sampled from this population:
a. $P$($50.00 or more and increased)
b. $P$($50.00 or more or increased)
c. $P$($50.00 or more|increased)
d. $P$(increased|$50.00 or more)
e. Based on the sample does it appear that the events "$50.00 or more" and "increased" independent?

**Key Topics**

## 3-5 The Multiplication Rule

*1. A formula for computing P(A and B)*

*2. P(A and B) when A and B are independent events*

*3. The multiplication rule for three or more independent events*

*4. Mutually exclusive events and independent events*

### S T A T I S T I C S   I N   A C T I O N

Ron and Sue Hatton recently vacationed in Reno, Nevada. They stopped in a small casino that had games where a gambler can bet $1.00 and get an apparently large return for winning.

Ron played a game where the activity involved rolling a pair of dice two times. To win the game, Ron had to get a sum of 7 on both rolls of the dice. If either roll resulted in anything but a 7, then Ron would lose. The prize for winning was $25.00. The probability of winning can be written as follows:

Probability of winning =
P(7 on first roll and 7 on second roll)

Sue played a game that involved selecting three cards from a deck of 52 cards. The cards were drawn one after the other without replacement. To win the game, Sue had to get three spade cards. If any of the three cards was not a spade, then Sue would lose. The prize for winning was $50.00. The probability of winning can be written as follows:

Probability of winning =
P(a spade and a spade and a spade)

### A formula for computing *P(A and B)*

The following equation was given in Definition 3 = 11:

$$P(B \mid A) = \frac{P(A \text{ and } B)}{P(A)}, \text{ provided } P(A) \neq 0$$

When both sides of this equation are multiplied by $P(A)$, an equation is obtained that is called *the multiplication rule of probability*.

## The Multiplication Rule of Probability

If $A$ and $B$ are two events defined in a sample space $S$, then:

$$P(A \text{ and } B) = P(A) \cdot P(B|A)$$

The usefulness of this rule is that when two of the three probabilities in the equation are known, then the third probability can be determined. Furthermore, the value of the third probability will be consistent with the values of the two known probabilities.

▶ **Example 1**   Two cards are drawn without replacement from a deck of playing cards. Compute $P(\text{two aces})$.

Solution   $P(\text{first is ace and second is ace}) = P(\text{first is ace}) \cdot P(\text{second is ace}|\text{first is ace})$

$$= \frac{4}{52} \cdot \frac{3}{51} \qquad P(\text{2nd is ace}|\text{1st is ace}) = \frac{3 \text{ aces left}}{51 \text{ cards left}}$$

$$= \frac{1}{221} \qquad \text{Simplify indicated product}$$

The probability is about 0.5 percent that both cards will be aces.   ◀

In the Statistics in Action section, Sue Hatton had to select three spade cards to win the game she played. Recall that the cards were drawn without replacement.

$$P(\text{first is a spade}) = \frac{13}{52} \qquad \frac{13 \text{ spades}}{52 \text{ cards}}$$

$$P(\text{second is a spade}|\text{first is a spade}) = \frac{12}{51} \qquad \frac{12 \text{ spades left}}{51 \text{ cards}}$$

$$P(\text{first and second are spades}) = \frac{13}{52} \cdot \frac{12}{51} \qquad \begin{array}{l} P(A \text{ and } B) \\ = P(A) \cdot P(B|A) \end{array}$$

$$P(\text{third is a spade}|\text{first and second are spades}) = \frac{11}{50} \qquad \frac{11 \text{ spades left}}{50 \text{ cards}}$$

$$P(\text{spade first and spade second and spade third}) = \frac{13}{52} \cdot \frac{12}{51} \cdot \frac{11}{50} \qquad \begin{array}{l} \text{Multiplication} \\ \text{Rule} \end{array}$$

$$= 0.0129\ldots \qquad \text{Simplify product}$$

$$\approx 1.3\%$$

Thus Sue had approximately a 1.3 percent chance of winning the $50.00.

▶ **Example 2**  John Young is an engineer for Dynamo Electric Corporation. He has designed a new electrical transformer and is interested in the following events:

      *A*:  Management will accept the new design.

      *B*:  The new transformer will be more cost effective than the current model.

John feels the following values are reasonable:

$$P(A) = 0.4 \quad \text{and} \quad P(B|A) = 0.9$$

Use these probabilities to estimate the probability that the new transformer will be accepted by management and be more cost effective than the current model.

Solution    **Discussion.** The desired probability can be stated as:

$P(A \text{ and } B)$, where *A* and *B* are the events defined above.

$P(A \text{ and } B) = (0.4)(0.9)$         $P(A) = 0.4 \quad \text{and} \quad P(B|A) = 0.9$

          $= 0.36 \text{ or } 36\%$       Simplify the indicated product.

Based on the given probabilities, John would estimate the probability of the new transformer being accepted by management and being more cost effective as 36 percent.   ◀

## *P*(*A* and *B*) when *A* and *B* are independent events

If *A* and *B* are independent, then $P(B|A) = P(B)$. For independent events, the $P(B|A)$ term can thus be replaced by $P(B)$ in the multiplication rule of probability.

### Multiplication Rule of Probability for Independent Events

If *A* and *B* are independent events, then:

$$P(A \text{ and } B) = P(A) \cdot P(B)$$

▶ **Example 3**  An experiment consists of rolling a fair die and flipping a fair coin. For this experiment, the following events are defined:

           *A*:  The die yields an even number of spots.

           *B*:  The coin yields a tail.

Compute:

**a.** $P(A \text{ and } B)$

**b.** $P(A \text{ or } B)$

**Solution**    **Discussion.** Because the experiment consists of unrelated trials, $A$ and $B$ can be modeled as independent.

$$P(A) = \frac{3}{6} \qquad \frac{\text{3 even numbers}}{\text{6 outcomes}}$$

$$P(B) = \frac{1}{2} \qquad \frac{\text{1 tail}}{\text{2 outcomes}}$$

a. $P(A \text{ and } B) = P(A) \cdot P(B)$     Use the modified equation.

$$= \frac{3}{6} \cdot \frac{1}{2} = \frac{1}{4} \quad \text{or} \quad 25\%$$

The probability is 25 percent that the experiment yields an even number on the die and a tail on the coin.

b. $P(A \text{ or } B) = \frac{3}{6} + \frac{1}{2} - \frac{1}{4}$     $P(A \text{ or } B) = P(A) + P(B) - P(A \text{ and } B)$

$$= \frac{3}{4} \quad \text{or} \quad 75\%$$

The probability is 75 percent that the experiment yields an even number on the die, or a tail on the coin, or both.    ◀

In the Statistics in Action section, Ron Hatton had to roll 7s on two consecutive rolls of the dice. These are unrelated activities, so the events are independent. With fair dice:

$$P(\text{7 on first roll}) = P(\text{7 on second roll}) = \frac{6}{36} \quad \text{or} \quad \frac{1}{6}$$

$$P(\text{7 on first roll and 7 on second roll}) = \frac{1}{6} \cdot \frac{1}{6}$$

$$= 0.02777\ldots$$

$$\approx 2.8\%$$

Thus Ron had an approximately 2.8 percent chance of winning the $25.00.

▶ **Example 4**    Joan Spaulding is a member of the girl's intercollegiate basketball squad. Records of Joan's free throws indicate that she is successful on 82 percent of her attempts. Use 0.82 as the probability that Joan is successful on a given free throw attempt. For the next two free throws that Joan attempts, define the following events:

   $A$:   Joan is successful on the first attempt.

   $B$:   Joan is successful on the second attempt.

Compute the following probabilities:

a. $P(A \text{ and } B)$     b. $P(A \text{ or } B)$

Solution    **Discussion.** Depending on whether Joan tends to shoot in streaks, it may be reasonable to model $A$ and $B$ as independent, and to estimate both $P(A)$ and $P(B)$ to be 0.82.

**a.** $P(A \text{ and } B) = (0.82)(0.82)$
$= 0.6724$ or about 67%     $P(A \text{ and } B) = P(A) \cdot P(B)$

Assuming independence of shots, the probability is about 67 percent that Joan is successful on both shots.

**b.** Using the addition rule:

$$P(A \text{ or } B) = 0.82 + 0.82 - 0.6724$$
$$= 0.9676 \quad \text{or about} \quad 97\%$$

Assuming independence of shots, the probability is about 97 percent that Joan is successful on the first shot, or the second shot, or on both shots.     ◄

▶ **Example 5**    In Figure 3-45, events $A$ and $B$ are shown with probabilities indicated in the four regions. Find:

**a.** $P(A)$
**b.** $P(B)$
**c.** $P(A \text{ and } B)$
**d.** $P(A|B)$
**e.** Are $A$ and $B$ independent?

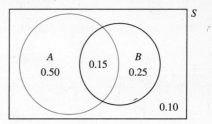

**Figure 3-45.** Events $A$ and $B$, and assigned probabilities.

Solution    **a.** $P(A) = 0.50 + 0.15 = 0.65$    65% of the outcomes are in $A$.
**b.** $P(B) = 0.15 + 0.25 = 0.40$    40% of the outcomes are in $B$.
**c.** $P(A \text{ and } B) = 0.15$    15% of the outcomes are in $A$ and $B$.

**d.** $P(A|B) = \dfrac{0.15}{0.40} = 0.375$    37.5% of the outcomes in $B$ are also in $A$.

**e.** $P(A) = 0.65$ and $P(A|B) = 0.375$

Because these numbers are not the same, $A$ and $B$ are not independent.     ◄

## The multiplication rule for three or more independent events

The concept of independence can be defined for more than two events. For example, let $A$, $B$, and $C$ be three events defined for some experiment. In *qualitative terms*, if $A$, $B$, and $C$ are independent, then knowing whether or not two of the three events have occurred would not change the conditional probability assigned to the third. The modified multiplication rule of probability can be extended to the event $(A \text{ and } B \text{ and } C)$, where $A$, $B$, and $C$ are independent events.

### The Multiplication Rule for $m$ Independent Events

If $A_1, A_2, \ldots, A_m$ are $m$ independent events, then:

$$P(A_1 \text{ and } A_2 \text{ and } \ldots \text{ and } A_m) = P(A_1) \cdot P(A_2) \cdot \cdots \cdot P(A_m)$$

▶ **Example 6**   The residents in a large metropolitan area in the West are considering issuing bonds to finance increased regional transit services. Opinion polls taken in the area indicate that 62 percent of adults favor the issue. A sample is taken of these adults and the following events are defined:

$A$:   The first adult favors the issue.

$B$:   The second adult favors the issue.

$C$:   The third adult favors the issue.

Compute $P(A \text{ and } B \text{ and } C)$.

**Solution**   **Discussion.** Only three adults are sampled in the large metropolitan area. Because a sample of three is small when compared with a large population, it is reasonable to treat $A$, $B$, and $C$ as independent. Using 0.62 for $P(A)$, $P(B)$, and $P(C)$:

$$
\begin{aligned}
P(A \text{ and } B \text{ and } C) &= (0.62)(0.62)(0.62) \qquad \text{\small $P(A \text{ and } B \text{ and } C) = P(A) \cdot P(B) \cdot P(C)$}\\
&= 0.238328 \qquad\qquad\quad \text{\small Simplify the indicated product.}\\
&\approx 24\%
\end{aligned}
$$

The probability is about 24 percent that all three of the residents surveyed will favor the bond issue.   ◀

## Mutually exclusive events and independent events

Many students have difficulty distinguishing between the concepts of mutually exclusive and independent events. We will therefore quickly review the concepts and the related addition and multiplication rules of probability.

**To assess $P(A \text{ or } B)$, we can frequently use the addition rule.** The word *or* is the clue to use the addition rule.

$$P(A \text{ or } B) = P(A) + P(B) - P(A \text{ and } B) \qquad \text{\small For any events $A$ and $B$}$$
$$P(A \text{ or } B) = P(A) + P(B) \qquad\qquad\qquad\quad \text{\small For mutually exclusive events}$$

To determine whether $A$ and $B$ are mutually exclusive, ask:

"Can $A$ and $B$ both occur at the same time?"

If the answer is yes, then $A$ and $B$ are not mutually exclusive. If the answer is no, then $A$ and $B$ are mutually exclusive.

To assess **$P(A$ and $B)$, use the multiplication rule.** The word *and* is the clue to use the multiplication rule.

$$P(A \text{ and } B) = P(A) \cdot P(B|A) \qquad \text{For any events } A \text{ and } B$$

$$P(A \text{ and } B) = P(A) \cdot P(B) \qquad \text{For independent events}$$

To determine whether $A$ and $B$ are independent events, ask:

$$\text{Is } P(B|A) = P(B)?$$

If the answer is no, then $A$ and $B$ are not independent. $\qquad P(B) \neq P(B|A)$

If the answer is yes, then $A$ and $B$ are independent. $\qquad P(B) = P(B|A)$

Notice, if $A$ and $B$ are mutually exclusive events, then $P(A \text{ and } B) = 0$. As a consequence, for such events, $P(B|A) = 0$. For most situations that we study, $P(B) \neq 0$, and $P(B|A)$ is therefore different from $P(B)$ when $A$ and $B$ are mutually exclusive. **We may therefore conclude that mutually exclusive events are, typically not independent.**

## Exercises 3-5    Set A

*In Exercises 1–12, compute the indicated probabilities using the rules of probability.*

1. Two cards are drawn without replacement from a deck of playing cards.
   a. $P$(two kings)
   b. $P$(two spades)
   c. $P$(a queen and a jack)
   Compute $P[$(queen, then jack) *or* (jack, then queen)$]$

2. Two chips are drawn without replacement from a box containing four red chips and three blue chips.
   a. $P$(two red chips)
   b. $P$(two blue chips)
   c. $P$(one of each color)
   Compute $P[$(red, then blue) *or* (blue, then red)$]$

3. A die is biased so that each time it is rolled the probability is 0.30 that six spots will show on the upturned face. The die is rolled twice.

   a. $P$(two 6s)
   b. $P$(neither roll is a 6)
   c. $P$(a 6 and not a 6)
   Compute $P[$(6, then not a 6) *or* (not a 6, then 6)$]$

4. A disk is painted red on one side and blue on the other. Each time the disk is tossed, the probability is 0.6 that the red side will be on top. The disk is tossed twice.
   a. $P$(red and red)
   b. $P$(blue and blue)
   c. $P$(one of each color)
   Compute $P[$(red, then blue) *or* (blue, then red)$]$

5. Dr. Jane Gunton is a surgeon at St. Paul's Methodist Hospital. Tomorrow she is going to operate on a patient suffering from back pain. Jane has defined the following events:

   $A$: The patient has a tumor on the spine.

   $B$: The tumor is benign.

Jane feels the following values are reasonable:

$$P(A) = 0.75 \quad \text{and} \quad P(B|A) = 0.60$$

Use these probabilities to assess the probability that the patient will have a tumor on the spine and it will be benign.

6. Earl Karn is preparing for the final examination in Math 132. He has defined the following events:

   A:   Earl gets a grade of B or better on the final exam.

   B:   Earl gets at least a C for a semester grade.

   Earl feels the following values are reasonable:

   $$P(A) = 0.45 \quad \text{and} \quad P(B|A) = 0.60$$

   Use these probabilities to assess the probability that Earl will get a grade of B or better on the final exam and will get at least a C for the semester grade.

7. A crew is fighting a forest fire in one of the western states. The following events are defined for this activity:

   A:   The wind will die down tonight.

   B:   The crew will surround the fire by tomorrow.

   The crew feels the following values are reasonable:

   $$P(A) = 0.80 \quad \text{and} \quad P(B|A) = 0.35$$

   Use these probabilities to assess the probability that the wind will die down tonight and the crew will surround the fire by tomorrow.

8. The tennis team from Sweetwater College will be playing a powerful squad from Northeastern College on Saturday. The following events are defined for this activity:

   A:   The top seeded player on the Sweetwater team will fully recover from an ankle injury.

   B:   The Sweetwater team will defeat the team from Northeastern.

The tennis team feels the following values are reasonable:

$$P(A) = 0.55 \quad \text{and} \quad P(B|A) = 0.82$$

Use these probabilities to assess the probability that the top seeded player will fully recover and the Sweetwater team will win on Saturday.

9. A student is currently enrolled in Psychology 80 and Ceramics 2A. She has defined the following events:

   A:   She will get a semester grade of A in Psychology 80.

   B:   She will get a semester grade of A in Ceramics 2A.

   By her own estimates:

   $$P(A) = 0.50 \quad \text{and} \quad P(B) = 0.80$$

   a. Based on these estimates what is the probability that she will get an A in both classes?
   b. What are you assuming, to arrive at an answer?

10. A geologist recommends to investors where exploratory gas and oil wells should be drilled. He currently has a test well being drilled in Nevada and one in Utah. He has defined the following events:

    A:   The well in Nevada will be financially successful.

    B:   The well in Utah will be financially successful.

    By his own estimates:

    $$P(A) = 0.90 \quad \text{and} \quad P(B) = 0.65$$

    a. Based on these estimates, what is the probability that both wells will be financially successful?
    b. What are you assuming, to arrive at an answer?

11. An avid bowler estimates that her probability of a strike (all ten pins are toppled on the first roll of the ball) is about 0.45.
    a. Based on this estimate, what is the probability that she will get strikes in the first two frames of the next game she bowls?
    b. What are you assuming, to arrive at an answer?

**12.** A salesperson for a national cosmetics firm estimates that the probability of a sale on a given contact is about 0.67.

    a. Based on this estimate, what is the probability that he will make sales on the next two contacts?

    b. What are you assuming, to arrive at an answer?

*In Exercises 13–16, use the data displayed in the bar charts. In each exercise, an experiment consists of randomly selecting two data points from each finite population shown. For each exercise, find the indicated probabilities, supposing:*

*1. The first data point is replaced before the second one is selected.*

*2. The first data point is not replaced before the second one is selected.*

**13.** Thirty chickens in a shipment to Foster Farms Packing were carefully weighed. The chickens were placed in one of the following three classes based on their weights:

    Class 1:   The chicken weighed less than 3.5 pounds.

    Class 2:   The chicken weighed between 3.5 and 4.5 pounds.

    Class 3:   The chicken weighed more than 4.5 pounds.

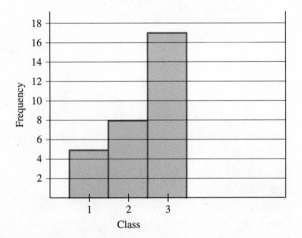

**Figure 3-46.** Weights (coded).

Two of these chickens are randomly selected. Use Figure 3-46 and find the indicated probabilities:

    a. $P$(both are class 1 chickens)

    b. $P$(both are class 3 chickens)

    c. $P$(one class 1 and one class 2 chicken)

    d. $P$(one class 2 and one class 3 chicken)

**14.** The forty children in Katie Naselli's third grade class were asked the question: "How many brothers and sisters do you have?" Two of these children are selected. Use Figure 3-47 and find the indicated probabilities:

    a. $P$(both children have one brother or sister)

    b. $P$(both children have three brothers or sisters)

    c. $P$(one child has none and one child has two)

    d. $P$(one child has one and one child has three)

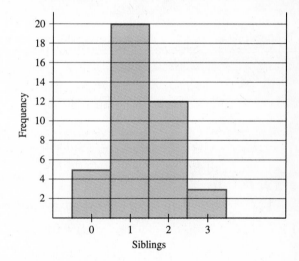

**Figure 3-47.** Numbers of siblings.

**15.** The grade distribution of the 35 students that just completed Ms. Nancy Devereaux's statistics class is shown in Figure 3-48. Two students are selected from this class. Use Figure 3-48 and find the indicated probabilities:

    a. $P$(both grades are A)

    b. $P$(both grades are C)

    c. $P$(one grade is B and one is D)

    d. $P$(one grade is A and one is F)

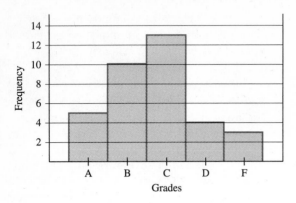

Figure 3-48.   Grade distribution.

**16.** A survey of 45 adults was taken in a shopping mall. Each person was placed in one of the following four classes based on political party affiliation:

Class 1:   Independent
Class 2:   Republican
Class 3:   Democrat
Class 4:   No affiliation

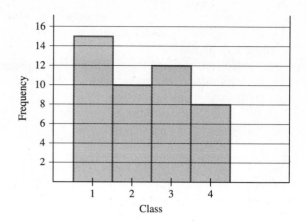

Figure 3-49.   Political party affiliation.

Two of these people are randomly selected. Use Figure 3-49 and find the indicated probabilities:

a. $P$(both are Independents)
b. $P$(both are Democrats)
c. $P$(one Independent and one Republican)
d. $P$(one Democrat and one nonaffiliated)

*In Exercises 17–20, use the Venn diagrams and compute the indicated probabilities.*

**17.** a. $P(A)$
   b. $P(A \text{ and } B)$
   c. $P(A|B)$
   d. Are $A$ and $B$ independent events?

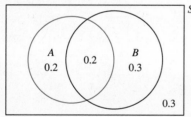

Figure 3-50.

**18.** a. $P(B)$
   b. $P(A \text{ and } B)$
   c. $P(B|A)$
   d. Are $A$ and $B$ independent events?

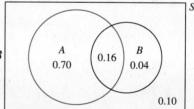

Figure 3-51.

**19.** a. $P(A)$       b. $P(A \text{ and } B)$       c. $P(A|B)$       d. Are $A$ and $B$ independent events?
   e. $P(C)$       f. $P(B \text{ and } C)$       g. $P(C|B)$       h. Are $B$ and $C$ independent events?
   i. $P(A \text{ and } C)$       j. $P(A|C)$       k. Are $A$ and $C$ independent events?

**20.** a. $P(A)$       b. $P(A \text{ and } B)$       c. $P(A|B)$       d. Are $A$ and $B$ independent events?
   e. $P(C)$       f. $P(B \text{ and } C)$       g. $P(C|B)$       h. Are $B$ and $C$ independent events?
   i. $P(A \text{ and } C)$       j. $P(A|C)$       k. Are $A$ and $C$ independent events?

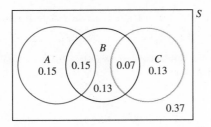

**Figure 3-52.**   Exercise 19.

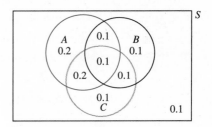

**Figure 3-53.**   Exercise 20.

*In Exercises 21–24, find the indicated probabilities.*

**21.** At a large western university, it is known that 20% of the student body are working on postgraduate degrees. A random selection of three students from this school is made. What is the probability that all three students are working on postgraduate degrees?

**22.** In a large midwestern state, the records at the Department of Motor Vehicles indicate that 65% of the registered vehicles were manufactured in the United States. Three vehicles from that state are randomly selected. What is the probability that the three vehicles were manufactured in the United States?

**23.** A solid is made with 12 faces of the same size and shape. Four of the faces are painted red and eight are painted green. The solid is tossed four times. What is the probability a green face will show on the top face all four times?

**24.** A coin is biased so that the probability of a head on the upturned face is 0.6 for each toss of the coin. The coin is tossed four times. What is the probability that a head will show on all four trials?

## Exercises 3-5   Set B

▶ **Example**   Melissa Gulley and Jim Camp are both shopping for new cars. They live in a large city in the East and do not know each other. For these activities, the following events are defined:

$A$:   Melissa buys a Ford product.

$B$:   Jim buys a Ford product.

Melissa and Jim have both visited Ford dealers and by their own estimates:

$$P(A) = 0.75 \quad \text{and} \quad P(B) = 0.50$$

**a.** Are $A$ and $B$ mutually exclusive events?
**b.** Compute $P(A \text{ and } B)$ with $A$ and $B$ modeled as independent.
**c.** Compute $P(A \text{ or } B)$.

**Solution**      **Discussion.** Melissa and Jim do not know each other. It is therefore reasonable to model $A$ and $B$ as independent.

**a.** Melissa and Jim live in a large city. Therefore it is possible for both to buy Ford products, and the events are not mutually exclusive.

**b.** With $P(A) = 0.75$ and $P(B) = 0.50$:

$$P(A \text{ and } B) = (0.75)(0.50) = 0.375$$

There is about a 38 percent probability that Melissa and Jim both buy Ford products.

**c.** With $P(A) = 0.75$, $P(B) = 0.50$, and $P(A \text{ and } B) = 0.375$:

$$P(A \text{ or } B) = 0.75 + 0.50 - 0.375 = 0.875$$

Based on the given probabilities, there is about an 88 percent probability that Melissa, or Jim, or both will buy a Ford product. ◀

1. Theresa De Palo is working with a travel agent on a vacation package. She is currently giving consideration to the following events:

   $A$:   Theresa selects a riding holiday in Ireland.

   $B$:   Theresa selects a Caribbean cruise.

   Theresa can only afford to buy one of these packages. By her own estimates:

   $$P(A) = 0.60 \quad \text{and} \quad P(B) = 0.25$$

   a. Are $A$ and $B$ mutually exclusive events? Justify your answer.
   b. Are $A$ and $B$ independent events? Justify your answer.
   c. Compute $P(A \text{ and } B)$.
   d. Compute $P(A \text{ or } B)$.
   e. What is the probability that Theresa selects neither of the two packages?

2. Donald Cole, a sportscaster, is reporting on an LPGA tournament scheduled for a local golf course next week. He defines the following events.

   $A$:   Beth H. wins the tournament.

   $B$:   Janis C. wins the tournament.

   Donald makes the following estimates:

   $$P(A) = 0.40 \quad \text{and} \quad P(B) = 0.25$$

   a. Are $A$ and $B$ mutually exclusive events? Justify your answer.
   b. Are $A$ and $B$ independent events? Justify your answer.
   c. Compute $P(A \text{ and } B)$.
   d. Compute $P(A \text{ or } B)$.

3. An employee at a large rocket propulsion plant is randomly selected from the work force of more than 3000 employees. The following events are defined:

   $A$:   The employee is female.

   $B$:   The employee commutes more than 15 miles to work.

   Employee records provide the following:

   $$P(A) = 0.55 \quad \text{and} \quad P(B) = 0.20$$

   a. Are $A$ and $B$ mutually exclusive events? Justify your answer.
   b. Is it reasonable to model $A$ and $B$ as independent events? Justify your answer.
   c. Compute $P(A \text{ and } B)$.
   d. Compute $P(A \text{ or } B)$.

4. At a large college in the Midwest, records indicate that in any semester 40% of the students in the college are enrolled in some course in the Math/Sci/Eng Division. An experiment consists of randomly

selecting two students at this college. The following events are defined for this experiment:

*A:*    The first student is enrolled in a course in the Math/Sci/Eng Division.

*B:*    The second student is enrolled in a course in the Math/Sci/Eng Division.

a. Are *A* and *B* mutually exclusive events? Justify your answer.
b. Is it reasonable to model *A* and *B* as independent? Justify your answer.
c. Compute *P(A and B)*.
d. Compute *P(A or B)*.
e. What is the probability that neither student is enrolled in a course in the Math/Sci/Eng Division?

# Chapter 3 Summary

## Definitions

1. A chance **experiment** is any activity whose result is not predetermined but rather involves chance.
2. An **outcome** is the result of an experiment.
3. A **sample space** is the set of all possible outcomes of an experiment.
4. An **event** is any subset of a sample space.
5. *A* and *B* are **mutually exclusive** if no outcome can simultaneously belong to both *A* and *B*.
6. The outcomes in *S* that do not satisfy the conditions specifying an event *A* are contained in the **complement of *A***, denoted *A'*.
7. The term **equally likely** means that each outcome of an experiment is assigned the same probability of occurring.
8. An outcome is in **the event (*A* or *B*)** if the outcome is in *A* only, or in *B* only, or in both *A* and *B*.
9. An outcome is in **the event (*A* and *B*)** if the outcome is simultaneously in *A* and *B*.
10. If $P(A) \neq 0$, then the **conditional probability of *B* given *A*** is:

$$P(B|A) = \frac{P(A \text{ and } B)}{P(A)}$$

11. Events *A* and *B* are (modeled as) **independent** means that probabilities have been assigned so that:

$$P(B|A) = P(B) \quad \text{and} \quad P(A|B) = P(A)$$

## Rules and Equations

### Methods of Representing a Sample Space

List method   All the possible outcomes in a sample space are written out in some tabular form.

Tree diagram   The possible outcomes of a sample space follow a network of paths.

Venn diagram   A sample space is represented with a rectangle and events by closed regions.

If the outcomes of $S$ are equally likely, then:

$$P(A) = \frac{\text{the number of outcomes in } A}{\text{the number of outcomes in } S} = \frac{n(A)}{n(S)}$$

### Four rules of probability

1. If $A$ contains no outcomes of $S$, then $P(A) = 0$.
2. If $A$ contains all the outcomes in $S$, then $P(A) = 1$.
3. For any event $A$, $0 \leq P(A) \leq 1$.
4. For complementary events $A$ and $A'$, $P(A) + P(A') = 1$.

### The addition rules of probability

1. If $A$ and $B$ are two events defined on a sample space $S$, then:
$$P(A \text{ or } B) = P(A) + P(B) - P(A \text{ and } B)$$

2. If $A$ and $B$ are mutually exclusive, then:
$$P(A \text{ or } B) = P(A) + P(B)$$

3. If $A_1, A_2, \ldots, A_m$ are $m$ mutually exclusive events, then:
$$P(A_1 \text{ or } A_2 \text{ or } \ldots \text{ or } A_m) = P(A_1) + P(A_2) + \cdots + P(A_m)$$

### The multiplication rules of probability

1. If $A$ and $B$ are two events defined on a sample space $S$, then:

$$P(A \text{ and } B) = P(A) \cdot P(B|A)$$

2. If $A$ and $B$ are independent events, then:

$$P(A \text{ and } B) = P(A) \cdot P(B)$$

3. If $A_1, A_2, \ldots, A_m$ are $m$ independent events, then:

$$P(A_1 \text{ and } A_2 \text{ and } \ldots \text{ and } A_m) = P(A_1) \cdot P(A_2) \cdot \cdots \cdot P(A_m)$$

## Symbols

| | | | |
|---|---|---|---|
| $S$ | the sample space of an experiment | $P(A \text{ or } B)$ | the probability of ($A$ or $B$) |
| $A$, $B$, and $C$ | events | $P(B|A)$ | the conditional probability of $B$ given $A$ |
| $A'$ | the complement of $A$ | | |
| $P(A)$ | the probability of $A$ | $P(A \text{ and } B)$ | the probability of $A$ and $B$ |
| $P(A')$ | the probability of the complement of $A$ | | |

## Comparing Mutually Exclusive Events Versus Independent Events

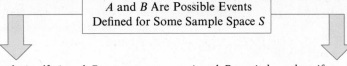

A and B Are Possible Events
Defined for Some Sample Space $S$

$A$ and $B$ are *mutually exclusive* if $A$ and $B$ cannot occur at the same time.

$A$ and $B$ are *independent if*

$$P(B) = P(B|A)$$

where $P(B|A)$ is the conditional probability of $B$, given that $A$ has occurred.

$P(A \text{ and } B) = 0$

$P(A \text{ or } B) = P(A) + P(B)$

Mutually exclusive events are not independent.

$P(A \text{ and } B) \neq 0$

$P(A \text{ and } B) = P(A) \cdot P(B)$

If $A$ and $B$ are events related to different "draws" in an experiment involving sampling, then $A$ and $B$ are independent if:

1. the sampling is done with replacement
2. the sample size is small compared with the size of the population being sampled.

# Chapter 3 Review Exercises

1. At a recent dinner party, a total of eight people were present. While discussing politics, it was revealed that four of the group were Democrats $(d_1, d_2, d_3, d_4)$, three were Republicans $(r_1, r_2, r_3)$, and one was an independent $(i)$. Suppose that two people from the group are randomly selected and the party affiliations of the pair is noted.
   a. List the 28 different possible outcomes in the sample space.
   b. How many of the outcomes contain two Democrats?
   c. How many of the outcomes contain at least one Republican?

2. Four-Square Construction has a staff that includes three plumbers $(p_1, p_2, p_3)$, three electricians $(e_1, e_2, e_3)$, and three carpenters $(c_1, c_2, c_3)$. The foreman randomly selected one of each skill to do a certain job.
   a. Use a tree diagram to show the 27 possible work crews.
   b. How many of the outcomes include $p_1$?
   c. How many of the outcomes include $c_1$?

3. A woman softball player is randomly selected and the position she plays is noted. Name three events that can be defined for this activity.

*In Exercises 4 and 5, determine whether the pairs of events defined for the corresponding random experiment are mutually exclusive.*

4. A box has a collection of nickels, dimes, and quarters. Two coins are randomly selected.

       $A$:   One of the coins is a dime.

       $B$:   The total value of the two coins is less than 20 cents.

5. A solid (called an "octahedron") has eight faces. The digits 1 through 8 are painted on the faces. The solid is rolled once and the outcome is the digit on the upper face.

       $A$:   The digit is an even number.

       $B$:   The digit is 3 or 5.

6. A box contains three disks painted gold $(g_1, g_2, g_3)$, three disks painted silver $(s_1, s_2, s_3)$, and one disk painted black $(b)$. The disks are all the same size and shape. Two disks are randomly selected from the box.
   a. List the 21 outcomes in the sample space.
   b. Find $P$(two gold-painted disks).
   c. Find $P$(at least one silver-painted disk).
   d. Find $P$(both disks are painted the same color).
   e. Find $P$(two disks are painted black).

7. A university in the Midwest has 28,972 students enrolled at the present time. A sample of 120 of these students yielded the following data:

   > 22 were graduate students
   >
   > 17 were seniors
   >
   > 18 were juniors
   >
   > 26 were sophomores
   >
   > 37 were freshmen

   A student from the current student body is randomly selected. Estimate the following probabilities based on the sample:
   a. $P$(the student is a senior)
   b. $P$(the student is a freshman or a sophomore)
   c. $P$(the student is not a graduate student)

8. Nine hikers in a national park happened to meet at one of the many campsites provided by the Park Service. A conversation showed that four were from eastern states, three were from southern states, and two were from western states. Two of these hikers are randomly selected and an outcome is the pair of home states of the hikers.
   a. List the 36 outcomes in the sample space.
   b. Find $P$(both are from the East)
   c. Find $P$(both are from different states)
   d. Find $P$(1 from the East, 1 from the South)
   e. Find $P$(neither is from the West)

9. Use Figure 3-54 and compute the following:
   a. $P(A)$     b. $P(B)$     c. $P(C)$
   d. $P(A')$     e. $P(B')$     f. $P(C')$
   g. $P(A$ or $C)$     h. $P(A$ or $B)$
   i. $P(B$ or $C)$     j. $P(A$ and $B)$
   k. $P(A$ and $B$ and $C)$

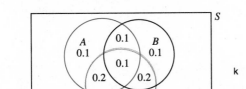

**Figure 3-54.**

10. Use Figure 3-55 and compute the following:
    a. $P(A)$     b. $P(B)$     c. $P(C)$
    d. $P(A')$     e. $P(B')$     f. $P(C')$
    g. $P(A$ or $B)$     h. $P(A$ or $C)$
    i. $P(B$ or $C)$     j. $P(A$ and $B)$
    k. $P(A$ and $C)$     l. $P(B$ and $C)$

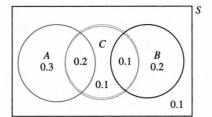

**Figure 3-55.**

11. Five hundred cars and small trucks in a large metropolitan area were randomly sampled and the number of cylinders in the vehicles' engines were recorded. The results of the experiment are in Table 3-18. Suppose an additional vehicle in this area is randomly sampled. Use the results of the sample to estimate the following probabilities to two decimal places:
    a. $P$(Domestic)
    b. $P$(6 cylinders)

Table 3-18    Results of Car and Truck Survey

|  |  | Number of Cylinders | | |
|---|---|---|---|---|
|  |  | 4 | 6 | 8 |
| Type of Vehicle | Domestic | 64 | 160 | 96 |
|  | Import | 108 | 54 | 18 |

   c. $P$(Domestic|8 cylinders)
   d. $P$(4 cylinders|Import)
   e. $P$(Domestic or 6 cylinders)
   f. $P$(Import or 8 cylinders)
   g. $P$(Domestic and 4 cylinders)
   h. $P$(Import and 6 cylinders)

12. Four hundred residents in a large city were randomly sampled and the frequencies with which they exercise were recorded. The results are in Table 3-19. Suppose another resident in this city is randomly sampled. Use the results of the sample to estimate the following probabilities, for this resident, to two decimal places:
    a. $P$(Daily exercise)
    b. $P$(Never exercise)
    c. $P$(Occasional exercise|Female)
    d. $P$(Male|Never exercise)
    e. $P$(Female or Daily exercise)
    f. $P$(Male and Occasional exercise)

TABLE 3-19    Results of Exercise Survey

|  |  | Frequency of Exercise | | |
|---|---|---|---|---|
|  |  | Daily | Occasionally | Never |
| Sex | Female | 50 | 140 | 60 |
|  | Male | 20 | 60 | 70 |

13. Employee records for the Sentinel Life and Casualty Insurance Corporation indicate that 18% of employees are college graduates. The corporation employs 5498 people nationwide. The records of two employees are randomly sampled and the following events are defined:

   A:  The first employee is a college graduate.

   B:  The second employee is a college graduate.

   a. Are A and B mutually exclusive? Justify your answer.
   b. Are A and B independent? Justify your answer.
   c. Compute P(A and B).
   d. Compute P(A or B).

14. The National Safety Council recently released figures that indicate 56 percent of all accidents are caused by men. Assume that 0.56 is a reasonable estimate for the proportion of accidents caused by men in Cleveland, Ohio. Suppose a sample of two recent automobile accidents in this city is examined, and the following events are defined:

   A:  The first accident was caused by a man.

   B:  The second accident was caused by a man.

   a. Are A and B mutually exclusive? Justify your answer.
   b. Are A and B independent? Justify your answer.
   c. Compute P(A and B).
   d. Compute P(A or B).

## Simulation of Experiments

**Key Topics**

1. *Simulation using a computer*

2. *Simulating coin tossing*

## Simulation using a computer

The term *simulation* describes a computer's ability to mimic a theoretical experiment. The Minitab command RANDOM is used with several different subcommands to simulate chance experiments. As an example, suppose that we wish to do an experiment that consists of randomly drawing a disk from a set of ten disks bearing the integers from 1 to 10, recording the number on the disk, replacing the disk, and after thoroughly mixing the disks, drawing again. This experiment is to be repeated until 20 numbers have been recorded. A frequency table is then to be made summarizing the results of the experiment. Figure 3-56 shows the commands and the table resulting from a Minitab simulation of this experiment.

```
MTB>   RANDOM 20 C1;
SUBC>  INTEGER 1:10.
MTB>   PRINT C1
C1
     2    9    2   10    9    7    8    3    6    7   10    3    8
     5    6    1    3    4    2    1
MTB>  TALLY C1
      C1     COUNT
       1       2
       2       3
       3       3
       4       1
       5       1
       6       2
       7       2
       8       2
       9       2
      10       2
      N=      20
```

**Figure 3-56.** Simulation of the disk experiment.

## Simulating coin tossing

Let us consider an experiment in which a fair coin is tossed ten times and the number of heads counted. Figure 3-57 shows the simulated result when this experiment is repeated 50 times.

```
MTB>   RANDOM 50 C2;
SUBC>  BINOMIAL 10 .5.
MTB>   PRINT C2
C1
     7    5    5    6    8    4    7    5    5    4    4    5    4    4    2
     6    5    6    3    4    6    5    4    4    5    4    6    3    4    9
     5    6    3    5    3    4    6    5    3    4    3    3    6    4    5
     7    6    6    5    3
MTB> TALLY C1
      C1     COUNT
       2        1
       3        8
       4       13
       5       13
       6       10
       7        3
       8        1
       9        1
      N=       50
```

**Figure 3-57.**   Simulation of coin toss experiment.

The Minitab command TALLY has been used to make a frequency table of the counts of heads in the 50 trials. Note that two heads occurred only once in the 50 trials, but that there were four and five heads 13 times each.

If we divide each of the counts by 10, the number of tosses in each trial, we obtain the fraction of heads for that trial. When the fractions are plotted against the trial numbers, the result is the graph in Figure 3-58.

The Minitab commands LET and NAME were used to place the relative frequencies and the names of the variables on the worksheet.

The horizontal line on the graph was drawn, by hand, through 0.5 on the RFEQ (relative frequency) axis. Note that in only a few of the 50 trials was the relative frequency close to 0.5. Can you explain why this does not conflict with the fact that the probability that a fair coin will show heads is 0.5?

Suppose that we now simulate an experiment that consists of 50 tosses of 30 fair coins. Again we count the number of heads in each trial and compute a corresponding relative frequency. The graph of the relative frequencies versus trial numbers for one such simulation is shown in Figure 3-59.

```
MTB>  SET C1
DATA> 1:50
MTB> LET C4=C2/10
MTB> NAME C1 'TRIAL' C4 'RFREQ'
MTB> PLOT C4 C1;
SUBC> YSTART O END 1.00.
```

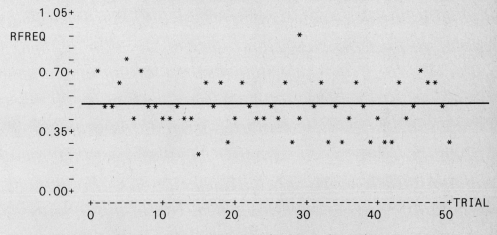

**Figure 3-58.** Relative frequencies of heads in 50 trials of a 10 toss coin experiment.

```
MTB>  RANDOM 5O C2;
SUBC> BINOMIAL 30 .5.

MTB>  LET C4 = C2/30
MTB>  PLOT C4 C1;
SUBC> YSTART O END 1.00.
```

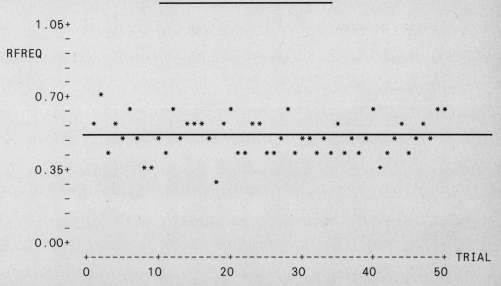

**Figure 3-59.** Relative frequencies of heads in 50 trials of a 30 toss coin experiment.

Figure 3-60 shows the results of a third simulation, 50 tosses of 50 fair coins.

```
MTB>   RANDOM 50 C2;
SUBC>  BINOMIAL 50 .5.
MTB>   LET C4 = C2/50
MTB>   PLOT C4 C1;
SUBC>  YSTART 0 END 1.00.
```

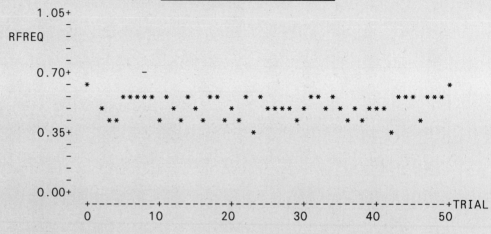

**Figure 3-60.**   Relative frequencies of heads in 50 trials of a 50 toss coin experiment.

## Exercises

1. Examine the graphs shown in Figures 3-58 to 3-60. Does changing the number of coins tossed in each trial seem to have any effect on:
   a. the mean relative frequency?
   (Compare the intersections of the solid horizontal lines with the vertical axes.)
   b. the variance of the relative frequencies? Explain.
   (On each graph, imagine there are two horizontal lines that enclose most of the graphed points. The closer together these lines are, the less the variance in the relative frequencies.)

2. Figure 3-61 shows the result of a simulation in which a pair of dice are tossed 360 times and each time the sum of the spots showing is recorded. Stanley James claims that at least one of the dice must be "crooked." Do you agree with Stanley? Why or why not?

```
MTB>  TALLY C4
   C4    COUNT
    2      15
    3      24
    4      29
    5      45
    6      38
    7      68
    8      54
    9      31
   10      26
   11      21
   12       9
  N=      360
```

**Figure 3-61.**   Results of a dice toss experiment.

## G R A P H I N G

Consider the following chance experiment: A box contains 9 disks of the same size and shape. On one disk a 1 is painted, on another disk a 2 is painted, and so on with a 9 painted on the last disk.

$$S = \{1, 2, 3, 4, 5, 6, 7, 8, 9\}$$

Without looking, one disk is selected, the number on it is recorded, and the disk is replaced in the box. The box is shaken and the process is repeated. Suppose we needed to carry out this chance experiment $n = 60$ times to get 60 data points. The results of the 60 trials can be *simulated* with a graphics calculator.

The RAN # key can be used to generate a three-digit random number between 0 and 1, such as 0.805. To get a random number between 0 and 10, we can multiply the given random number by 10 to get 8.05. If we want to use the calculator to generate only random numbers between 0 and 10, we can fix the calculator to round the random numbers to whole numbers.

Once the calculator has been programmed, repeated pressing of the appropriate key will generate a random number that accurately simulates the activity of selecting (with replacement) the disks from the box. Only two possible outcomes mar the results. Specifically, because of rounding, the calculator can show a 0 or 10. Since 0 and 10 are not in $S$, these outcomes will be ignored.

The Casio Graphics has a feature that keeps a record of the activity, and then displays the frequencies of the outcomes in a histogram. Because each of the digits 1 through 9 occur once in the box, they all have the same probability of being selected. As a consequence, doing this experiment a large number of times should yield a relatively flat histogram. (Keep in mind we will disregard the end bars because they represent the frequencies of 0 and 10, two unacceptable results for this activity.)

▶ **Example 1**     Generate 60 outcomes for the disk experiment.

**Solution**

| Key Sequence | Comments |
|---|---|
| MODE 7 0 EXE | This fixes the calculator to display only whole numbers |
| MODE SHIFT X | Accesses statistics with graphics program (SD 2) |
| SHIFT Scl EXE | Clears memories |
| SHIFT Cls EXE | Clears graphics screen |
| MODE · 11 EXE | Sets graphics display for 11 bars |
| RANGE (−) 0.5 EXE | Sets $x$ and $y$ scales on histogram |
| 10.5 EXE | |
| 1 EXE | |
| 0 EXE | |
| 30 EXE | |
| 1 EXE | |
| SHIFT RAN # × 10 DT | Generates and records a whole number between 0 and 10, inclusive |

◀

Repeated pressing of DT key generates additional random numbers.

**Note 1**  To obtain a display of the histogram based on the stored random numbers use the following sequence.

   GRAPH EXE      The histogram is displayed.

**Note 2**  Suppose you lose track of the number of data points stored in the calculator. To get a record of the number, use the following sequence.

   ALPHA *n* EXE      Display shows the number of stored data points.
(The *n* is below the 3 key)

**Note 3**  After getting a record of the number of stored data points, you must again use the following key sequence to enter more random numbers in the calculator.

The Texas Instrument TI-81 can also simulate the disk experiment. However, there is no graphics capability for the histogram. The following key sequence will generate random numbers between 0 and 10, inclusive.

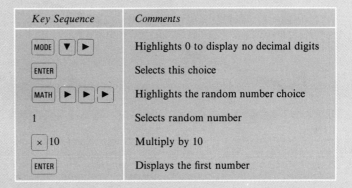

| Key Sequence | Comments |
|---|---|
| MODE ▼ ► | Highlights 0 to display no decimal digits |
| ENTER | Selects this choice |
| MATH ► ► ► | Highlights the random number choice |
| 1 | Selects random number |
| × 10 | Multiply by 10 |
| ENTER | Displays the first number |

Repeated pressing of ENTER key displays another random number.

**Name:** Craig Barbarino
**Occupation:** Statistical and Historical Research
**Employer:** Major League Baseball Office of the Commissioner New York, New York

**A** career in statistics for a professional sports organization is a unique occupation. Though the concepts and methods of gathering and analyzing information are similar to those in other statistical fields, the results of your work can not only show the bottom line, but can stir the loyalties and emotions of millions of people across America. In the sports field, statistics are not just numbers, they represent a player, a time, a memory. These statistics have a place in history. They can make a senior citizen feel like a teenager and a little boy dream of being Willie Mays. The data you produce has a lasting effect on millions of people one way or another.

My introduction to statistics came, as it probably did for so many others, when I started collecting baseball cards as a kid. I was more interested in analyzing and recording batting averages and win-loss records than memorizing them because finding out how and why things work rather than absorbing information was always more appealing to me. By my senior year in college, I really did not know what field I wanted to go into, but I felt I had a lot of opportunities from which to choose. My business major offered an internship and, since I always liked sports, I thought I'd try for one with a sports organization. I landed an internship at the Commissioner's Office and was hired soon after in the public relations department.

As the Commissioner's Office took on new responsibilities, I saw a need for gathering and analyzing statistical and historical information. Slowly I began to shed my public relations responsibilities as more and more statistical work became available. Today, my job involves doing almost the same things I did for fun when I collected baseball cards as a kid, except it's more high-tech and I get paid for it. My responsibilities include collecting and analyzing statistical and historical baseball information for use in office publications, media guides, programs, newsletters, baseball products, a computer service, and the news media, and lastly, for the fans. Baseball facts and figures have exploded in the last five years; however, most of the work being done is so trivial to the average person that only the hardcore fans appreciate it. The statistics I produce are basic and consist of player, club, and league batting, pitching, and fielding statistics. I take that simple data and analyze it to show trends, accomplishments, and/or predictions. Many times, in-depth analysis is needed.

The most satisfying part of my job is knowing my work is being used throughout the baseball community (clubs, newspapers, TV, fans, and so on). The most exciting part of my job is the projects that break new ground. On occasion, these projects have taken close to a year to collect the information, analyze it, and present a significant product in a way never done before.

Statistics is a very analytic, creative, and interesting field. It gets the brain thinking in new and different ways. The usual reason for going into statistics is to make an original contribution to the understanding of a particular subject. But, like every other job, class, or activity, it is only as good, or bad, as you make it. If you put the time and energy into everything you do, whether it's your favorite subject or not, you will be amazed at how much you can accomplish.

## GEOGRAPHY
# A Lost Generation

Is it possible that almost two-thirds of Americans 18 to 24 cannot point to France on an outline map? That 75% of adult Americans are unable to locate the Persian Gulf, where 39 American sailors have died in the past two years? That a few put the U.S. inside Australia or Botswana?

Yes, alas. These are among the revelations in a Gallup study released last week on more than 10,000 adults from nine nations, including Italy, Mexico and Japan. The lowest scorers were young Americans, with an average of just 6.9 correct answers to 16 easy questions (Swedes were tops, with 11.6).

**Figure 4-1.** "A Lost Generation," *Time*, August 8, 1988, p. 19. Copyright © 1988 Time Warner Inc. Reprinted by permission.

# Pall of pessimism hangs over cities
## Poll shows 22% of leaders glum about job prospects in their towns

**Associated Press**

WASHINGTON — You're a young person thinking of starting your career. Where should you go?

Well, don't come to their town, said officials of more than one-fifth of the nation's larger cities.

The National League of Cities' sixth annual survey drew 269 confidential responses from elected municipal officials in cities with a 1980 population of more than 10,000. League officials wouldn't name the cities that responded.

One question in the survey asked about the prospects of a young person finding a job and beginning a career in the official's town.

Twenty-two percent saw the opportunities in their communities as poor for such a person, as contrasted with 19 percent who thought chances were "great" and 59 percent "fair."

"Local elected officials are, on the whole, perpetual boosters of their home towns, so when more than one in five is worried enough to say things are bad, there are some real problems facing our cities and our nation," said Donald J. Borut, executive director of the National League of Cities.

The league's sixth annual survey said there was a 95 percent degree of confidence in the results and the answers would vary by no more than 6 percentage points from the results that would be obtained if all elected officials were polled.

The survey showed these results:

■ Thirty-six percent thought overall economic conditions had worsened, 37 percent saw no change, and 27 percent thought they had improved.

■ The cost of living was seen as worsening by 60 percent, while 37 percent saw no change and 3 percent thought it had improved.

Asked which three conditions are creating the most important problems in the community, 34 percent named economic conditions, 31 percent drugs, 28 percent solid-waste disposal, 23 percent city fiscal conditions and 22 percent crime.

**Figure 4-2.** "Pall of pessimism hangs over cities," *The Sacramento Bee*, January 15, 1991, p. A-8. Reprinted by permission of Associated Press.

# Random Variables and Their Probability Distributions

Figures 4-1 and 4-2 are examples of reports seen almost daily in newspapers and magazines. Such reports are frequently based on statistics computed based on random samples from large populations. These statistics are then reported as indicators, estimates, of corresponding unknown parameters of the populations.

To illustrate, in the Gallup study report in Figure 4-1, we read that "two-thirds of Americans 18 to 24 cannot point to France on an outline map." This statistic is based on the study of more than 10,000 adults from nine nations. Because about two-thirds of the Americans in the study were unable to locate France, we may reasonably assume that roughly two-thirds of all Americans 18 to 24 would similarly fail in this task.

In Figure 4-2, the National League of Cities claims "a 95 percent degree of confidence in the results" of their survey. We will learn a precise definition of *degree of confidence* in Chapter 5. In this chapter, however, we will first need to learn about *random variables* and *probability distributions* associated with them. We will also study two theoretical probability distributions, namely the *binomial* and *normal* distributions, since many random variables encountered in statistical studies can be described by such distributions.

# 4-1 Random Variables

**Key Topics**

1. *The definition of a random variable*

2. *Possible values for random variables*

3. *Random variables and qualitative data*

4. *Probability distributions for discrete random variables*

5. *Probability histograms*

6. *Two basic properties of probability distributions*

## STATISTICS IN ACTION

**A**    Clare Lynch is a consultant with a major financial management corporation. To contact prospective clients, Clare offers 2-hour seminars on the benefits of tax-sheltered annuities. She has such a seminar scheduled for next Saturday, and 40 people have signed up to attend.

Clare wishes she could predict how many of the 40 in attendance will become clients. Based on past seminars, she knows that about 20 percent of those that attend become clients. Therefore Clare knows she can expect about eight clients from this seminar. However, because of variability in the success rate from one seminar to another, she feels the number could range from 3 through 13. The only way she will know for certain the number of clients generated by this seminar is to "wait and see."

**B**    Matt Harr works for the Iowa Department of Transportation System. His primary duty is to weigh large truck and trailer rigs at a state weigh station. Based on the number of axles on a given truck and trailer, the trucker may be hauling a load whose weight exceeds the legal limit. In such cases, the truck is detained until the load can be reduced to the legal limit.

Matt notices that the next truck and trailer to be weighed is loaded with alfalfa hay. From his years of experience, Matt guesses that the hay on the load will weigh about 52,000 pounds. However, because of variation in moisture content and hay composition, he knows the weight can vary between 48,000 and 56,000 pounds. The only way he will know for certain the weight of this particular load of hay is to wait until the truck drives onto the scale.

## The definition of a random variable

Frequently the outcome of a random experiment is a number. For such an experiment, the sample space is a set of numbers. These outcomes are values of a so-called random variable.

### DEFINITION 4-1 Random Variable

A **random variable** is a quantity whose value depends upon the outcome of a chance experiment.

Capital letters such as $V$, $W$, $X$, and $Y$ are used to represent random variables, whereas lowercase letters such as $v$, $w$, $x$, and $y$ represent specific possible values of the random variables.

## Possible values for random variables

In Chapter 1, a definition separates quantitative data into discrete and continuous types. In a similar way, Definition 4-2 separates random variables into the same types.

### DEFINITION 4-2 Discrete and Continuous Random Variables

A random variable is **discrete** when only certain separated numbers are possible as values. If an activity involves *counting*, then the random variable is discrete.

A random variable is **continuous** when any value on an interval of numbers is possible. If an activity involves *measuring*, then the random variable is continuous.

For Clare Lynch, the number of clients she will gain from Saturday's seminar is a discrete random variable. For this particular activity, the possible values for the random variable are integers from 0 through 30 inclusive.

For Matt Harr, the weight of hay that will be weighed next at his station is a continuous random variable. For this particular activity, we might guess that the possible values for the random variable are all numbers from 40,000 through 64,000 pounds.

▶ **Example 1**    On a six-faced cube, a 1 is painted on three faces, a 2 is painted on two faces, and a 3 is painted on one face. The cube is rolled twice and $X$ is the sum of the numbers on the upturned faces. Identify the possible values for $X$.

**Solution**    **Discussion.** *X* is a discrete random variable because only certain separated numbers are possible values for *X*.

| Number on First Roll (1) | Number on Second Roll (2) | Possible Value of Random Variable (1) + (2) |
|:---:|:---:|:---:|
| 1 | 1 | 2 |
| 1 | 2 | 3 |
| 1 | 3 | 4 |
| 2 | 1 | 3 |
| 2 | 2 | 4 |
| 2 | 3 | 5 |
| 3 | 1 | 4 |
| 3 | 2 | 5 |
| 3 | 3 | 6 |

The possible values of the random variable *X* are 2, 3, 4, 5, and 6.    ◀

▶ **Example 2**    A tomato is selected from a shipment delivered to a warehouse in Detroit and the weight in ounces is measured. The supplier has sorted the shipment so that all the tomatoes weigh between 4 and 8 ounces. Identify the possible values for *Y*, the weight of the randomly sampled tomato.

**Solution**    **Discussion.** Weight is measured. Therefore *Y* is a continuous random variable.

Based on the information from the supplier, *Y* can be any number between 4 and 8. This interval can be written using parentheses as follows:

$(4, 8)$    The parentheses indicate a *range* from 4 through 8.

The possible values for *Y* can also be shown on a number line as in Figure 4-3.    ◀

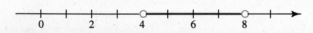

**Figure 4-3.** The interval from 4 to 8.

## Random variables and qualitative data

As defined, the value assumed by a random variable is a number. Therefore, a chance experiment in which some qualitative feature of an object is observed does not directly produce a random variable.

To illustrate, suppose a box contains disks of the same size and weight. The disks are colored blue, red, or white. An experiment consists of selecting one of these disks and noting the color.

The color of a disk would not qualify as a random variable because color is not numerical. The outcomes could, however, be adjusted to make them appear to be numerical. One way that is frequently used is to identify only one of the qualitative outcomes as being of interest. To this outcome, a 1 is assigned. To any other outcome, a 0 is assigned. With this interpretation, the only possible outcomes are 1 and 0. Because 1 and 0 are numbers, the outcomes of the activity can be considered values of a random variable.

▶ **Example 3**     An activity involves checking a randomly selected radial tire from a large production run. The check will determine whether the tire meets minimum quality standards. Define a random variable in the context of this activity.

**Solution**     **Discussion.** An outcome is qualitative in that the tire is either defective or not defective. Let:

$$X = 0, \text{ if the tire is defective, and}$$

$$X = 1, \text{ if the tire is not defective.}$$

The possible values for the random variable $X$ are 0 and 1.     ◀

## Probability distributions for discrete random variables

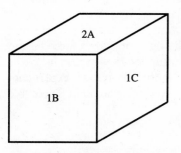

**Figure 4-4.**   A six-faced die.

Consider again the six-faced cube in Example 1. Recall that three faces are labeled 1, two faces are labeled 2, and one face is labeled 3. Figure 4-4 illustrates the cube, and as shown, the faces are labeled:

$$1A, \quad 1B, \quad 1C, \quad 2A, \quad 2B, \quad \text{and} \quad 3A.$$

The die is rolled twice and $X$ is the sum of the numbers on the upturned faces. The outcomes for the activity are shown in Table 4-1. From this table, we see that the only possible values for $X$ are:

$$2, \quad 3, \quad 4, \quad 5, \quad \text{and} \quad 6$$

TABLE 4-1   The 36 Possible Outcomes and Corresponding Values of $X$

|  |  | First Toss | | | | | |
|---|---|---|---|---|---|---|---|
|  |  | *1A* | *1B* | *1C* | *2A* | *2B* | *3A* |
| | 1A | 2 | 2 | 2 | 3 | 3 | 4 |
| | 1B | 2 | 2 | 2 | 3 | 3 | 4 |
| Second Toss | 1C | 2 | 2 | 2 | 3 | 3 | 4 |
| | 2A | 3 | 3 | 3 | 4 | 4 | 5 |
| | 2B | 3 | 3 | 3 | 4 | 4 | 5 |
| | 3A | 4 | 4 | 4 | 5 | 5 | 6 |

Because the relative frequencies of these values are different, we know the probabilities for each of these five possible values of $X$ are different.

$$P(X = 2) = \frac{9}{36} = 0.250 \qquad \frac{9 \text{ outcomes of } 2}{36 \text{ outcomes}}$$

$$P(X = 3) = \frac{12}{36} \approx 0.333 \qquad \frac{12 \text{ outcomes of } 3}{36 \text{ outcomes}}$$

$$P(X = 4) = \frac{10}{36} \approx 0.278 \qquad \frac{10 \text{ outcomes of } 4}{36 \text{ outcomes}}$$

$$P(X = 5) = \frac{4}{36} \approx 0.111 \qquad \frac{4 \text{ outcomes of } 5}{36 \text{ outcomes}}$$

$$P(X = 6) = \frac{1}{36} \approx 0.028 \qquad \frac{1 \text{ outcome of } 6}{36 \text{ outcomes}}$$

The symbol $P(X = 2)$ can be read "the probability that $X$ takes on the value 2," or simply "the probability that $X$ equals 2." When it is understood that a list of probabilities are all for the same random variable, then the "$X = $" is frequently omitted, and we write $P(2)$ for $P(X = 2)$.

Specifying a probability for each of the possible values for $X$ is usually referred to as giving a probability distribution for, or specifying a probability model for, $X$. Thus the probabilities 0.250, 0.333, 0.278, 0.111, and 0.028, together with the values 2, 3, 4, 5, and 6, are a probability model for the die rolling experiment described previously. Such a probability model can be displayed as in Table 4-2.

TABLE 4-2   The Probability Distribution for the Dice Experiment

| $x$ | 2 | 3 | 4 | 5 | 6 |
|------|-------|-------|-------|-------|-------|
| $P(x)$ | 0.250 | 0.333 | 0.278 | 0.111 | 0.028 |

▶ **Example 4**   A computer technician at Metropolitan Airport is concerned with major breakdowns in a computer that assists air traffic controllers. Specifically she is interested in the random variable $Y$, where:

$Y = $ the number of breakdowns of the computer next week.

The data in Table 4-3 have been accumulated for the past 100 weeks.

**a.** Based on the data, identify possible values for $Y$.
**b.** Approximate the probabilities for the possible values of $Y$.

TABLE 4-3   Computer Breakdown Data

| Number of Breakdowns During the Week | Number of Weeks This Happened |
|:---:|:---:|
| 0 | 50 |
| 1 | 30 |
| 2 | 16 |
| 3 | 4 |

**Solution**   **a.** The possible values are 0, 1, 2, and 3.

**b.** Using the following relative frequencies, Table 4-4 can be constructed.

$$P(0) = \frac{50}{100} = 0.50$$   50% of the 100 weeks had 0 breakdowns.

$$P(1) = \frac{30}{100} = 0.30$$   30% of the 100 weeks had 1 breakdown.

$$P(2) = \frac{16}{100} = 0.16$$   16% of the 100 weeks had 2 breakdowns.

$$P(3) = \frac{4}{100} = 0.04$$   4% of the 100 weeks had 3 breakdowns.

TABLE 4-4   The Probability Distribution for the Computer Problem

| $y$ | 0 | 1 | 2 | 3 |
|:---:|:---:|:---:|:---:|:---:|
| $P(y)$ | 0.50 | 0.30 | 0.16 | 0.04 |

◀

Continuous random variables also have probability distributions. To illustrate, consider $Y$, the weight of a tomato from the shipment discussed in Example 2. Recall that the possible values for $Y$ are all numbers between 4 and 8 ounces. Because $Y$ is a continuous random variable, probabilities for $Y$ are associated with *intervals* of possible values. Examples (a) through (c) illustrate three such probabilities.

a. $P(Y < 4.5)$ is the probability the tomato weighs less than 4.5 ounces.
b. $P(5.0 < Y < 6.0)$ is the probability the tomato weighs between 5.0 and 6.0 ounces.
c. $P(Y > 7.0)$ is the probability the tomato weighs more than 7.0 ounces.

The full use of continuous distributions requires techniques learned in calculus. Thus, while we will give a general introduction to discrete distributions in this chapter, we will study only one type of continuous distribution here: bell-shaped continuous distributions are discussed in Section 4-5. Other specific continuous distributions will be introduced in later chapters as they are needed in statistical inference.

## Probability histograms

In Chapter 2, histograms were used to display the relative frequencies of data points in a population or sample. Similar histograms can also be used to display the probability distributions of discrete random variables. Such graphs are called *probability histograms.*

---

### To construct a probability histogram for a discrete random variable X

**Step 1** On a horizontal axis draw a tick mark for each of the possible values of $X$.

         To show uniformly spaced possible values of $X$, make the distance between the marks the same and label them with the possible values of $X$.

**Step 2** Draw a vertical axis at the left end of the horizontal one.

         Mark off an appropriate scale on the vertical axis, using the largest probability in the distribution as a guideline.

**Step 3** Construct a bar centered on each $x$ to a height that shows $P(x)$.

**Step 4** Label the horizontal axis "$x$" and the vertical axis "$P(x)$."

---

To illustrate, probability histograms for the dice and computer breakdown distributions are shown in Figures 4-5 and 4-6 respectively. Notice that the height of each bar reflects the probability associated with the corresponding value of $X$ or $Y$.

For such a probability histogram, the fractional part of the area inside each bar is the same as the probability associated with the corresponding value of the random variable. Using Figure 4-5 to illustrate:

> 25.0% of the area is above 2 and $P(X = 2) = 0.250$.
>
> 33.3% of the area is above 3 and $P(X = 3) = 0.333$.
>
> 27.8% of the area is above 4 and $P(X = 4) = 0.278$.
>
> 11.1% of the area is above 5 and $P(X = 5) = 0.111$.
>
> 2.8% of the area is above 6 and $P(X = 6) = 0.028$.

This kind of relationship between area and probability suggests a way of specifying a probability model for a continuous random variable. To illustrate, consider again the continuous random variable $Y$, where:

$Y =$ the weight of a tomato from the shipment in Example 2.

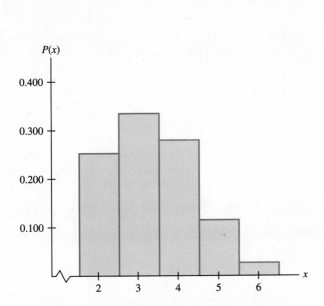

**Figure 4-5.** Probability histogram for the dice experiment.

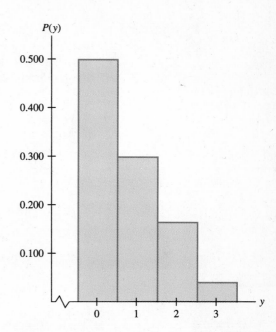

**Figure 4-6.** Probability histogram for the computer breakdown problem.

In Figure 4-7, an *idealized probability histogram* for the values of $Y$ is drawn. Such a histogram is a smooth curve. We specify the probability of $Y$ taking on some value in an interval to be the fraction of the area under the histogram above the interval.

For example, the shaded area in Figure 4-7 is the probability that $Y$ is in the interval 6.5 to 7.0. Suppose that approximately 22% of the area under the curve is between $y_1 = 6.5$ and $y_2 = 7.0$. The following probability equation can be written:

$$P(6.5 < Y < 7.0) = 0.22$$

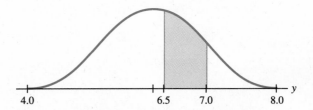

**Figure 4-7.** An idealized probability histogram for $Y$.

This equation states the probability that a randomly selected tomato weighs between 6.5 and 7.0 ounces is 0.22. In Section 4-5, a technique will be studied for finding areas under bell shaped idealized histograms.

▶ **Example 5**   One face of a coin is painted red and the other face is painted green. The coin is biased so that each time the coin is tossed, the probability that the red face will be showing is 0.6. The coin is tossed three times, and $X =$ the number of times the *red face* is showing.

**a.** List the possible values of $X$.
**b.** Compute to three places the probability distribution for $X$.
**c.** Make a probability histogram.

Solution   **a.** The coin is tossed three times. Thus the possible values for $X$ are 0, 1, 2, and 3.
**b.** Let $r$ stand for "red face shows" and $g$ stand for "green face shows." For this experiment, $P(r) = 0.6$ and $P(g) = 0.4$. The possible outcomes for the three tosses are shown in Table 4-5.

1 outcome of 0 red and 3 green

3 outcomes of 1 red and 2 green

3 outcomes of 2 red and 1 green

1 outcome of 3 red and 0 green

**TABLE 4-5   Possible Outcomes of Three Tosses**

| 1st | 2nd | 3rd | Results |
|-----|-----|-----|---------|
| $r$ | $r$ | $r$ | 3 red |
| $r$ | $r$ | $g$ | 2 red and 1 green |
| $r$ | $g$ | $r$ | 2 red and 1 green |
| $r$ | $g$ | $g$ | 1 red and 2 green |
| $g$ | $r$ | $r$ | 2 red and 1 green |
| $g$ | $r$ | $g$ | 1 red and 2 green |
| $g$ | $g$ | $r$ | 1 red and 2 green |
| $g$ | $g$ | $g$ | 3 green |

$$P(0) = P(0 \text{ red and 3 green}) = 1(0.4)(0.4)(0.4) = 0.064$$
$$P(1) = P(1 \text{ red and 2 green}) = 3(0.6)(0.4)(0.4) = 0.288$$
$$P(2) = P(2 \text{ red and 1 green}) = 3(0.6)(0.6)(0.4) = 0.432$$
$$P(3) = P(3 \text{ red and 0 green}) = 1(0.6)(0.6)(0.6) = 0.216$$

See Table 4-6.

**TABLE 4-6   The Probability Distribution for Biased Coin Experiment**

| $x$ | 0 | 1 | 2 | 3 |
|-----|-----|-----|-----|-----|
| $P(x)$ | 0.064 | 0.288 | 0.432 | 0.216 |

**c.** A probability histogram is shown in Figure 4-8.   ◀

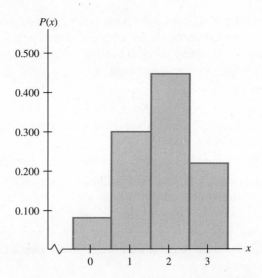

**Figure 4-8.**  A probability histogram for the biased coin toss experiment.

## Two basic properties of probability distributions

Regardless of whether a random variable is discrete or continuous, its probability distribution must conform to the basic rules of probability theory. Two such rules are identified below for both discrete and continuous random variables.

| If X Is a Discrete Random Variable with Possible Values $x_1, x_2, x_3, \ldots$, then: | If Y Is a Continuous Random Variable with Possible Values Between m and M, where m < M, then: |
|---|---|
| 1. $0 \leq P(x_1) \leq 1$ for each value x<br>2. $\sum P(x_i) = 1$ | 1. $0 \leq P(y_1 < Y < y_2) \leq 1$ for each interval $(y_1, y_2)$<br>2. $P(m < Y < M) = 1$ |

The first rule states that the probability associated with any value of $X$ or interval of values of $Y$ must be between 0 and 1 inclusive. The second rule states that whenever the experiment associated with $X$ or $Y$ is carried out, one of the possible values of $X$ or $Y$ will be obtained.

## Exercises 4-1    Set A

*In Exercises 1–10, in the context of each chance experiment, list the possible values for a discrete random variable or specify an interval of possible values for a continuous random variable.*

**1.** A coin is selected from a supply of commonly used coins in the United States and the value of the coin is noted.

**2.** Six coins are simultaneously tossed and the number of heads showing on the upturned faces is counted.

3. The temperature data for a January day in Chicago are checked and the highest temperature for the day is noted. Records indicate the lowest and highest temperatures for this period are 8° and 58° Fahrenheit respectively.

4. The weight of a student from Mark Twain Elementary School is measured. Records indicate that the lightest and heaviest students in this school are currently 57 pounds and 183 pounds, respectively.

5. Ten students at a large community college are selected and the number that are attending their first semester at the school are counted.

6. Three dice are rolled and the number of dots showing on the upturned faces is counted.

7. The height of a tree delivered to the Ladies Auxiliary Christmas Tree Lot is measured. The supplier claims the shortest and tallest trees in the shipment are 3 feet and 7 feet, respectively.

8. The diameter of a ball bearing from a large production run is measured. Quality control records of similar production runs suggest that the smallest and largest diameters are 2.50 and 2.60 centimeters, respectively.

9. A solid has 12 faces of equal area. On four faces a 1 is painted, on three faces a 2 is painted, on two faces a 3 is painted, on two faces a 4 is painted, and on one face a 5 is painted. The solid is rolled twice and $X$ is the sum of the numbers on the top face.

10. A box contains five disks of similar size. The digits 1, 2, 3, 4, and 5 are each painted on a disk. An experiment consists of selecting without replacement two disks. The number on the first disk is used as the tens digit and the number on the second disk is used as the ones digit to form a two-digit number.

*In Exercises 11–16, define a random variable in the context of each activity.*

11. The model of a randomly selected vehicle in the employee parking lot of Sentinel Agriculture Products Corporation is checked. If the vehicle is a pickup truck, then the outcome is considered a success.

12. The manufacturer of woolen sweaters sold in a major department store chain is checked. If the sweater is the product of a U.S. firm, then the outcome is considered a success.

13. The sex of a randomly selected doctor registered in New York State is checked. If the doctor is a woman, then the outcome is considered a success.

14. The race of a randomly selected lawyer in the state of California is checked. If the lawyer is Hispanic, then the outcome is considered a success.

15. The position played by a randomly selected player in the National Football League last year is checked. If the player is a linebacker, then the outcome is considered a success.

16. The class standing of a randomly selected student at Empire State College is checked. If the student is a senior, then the outcome is considered a success.

17. In a supply of poker chips, 40 percent are colored blue. Three chips are selected with replacement and $X$ is the number of blue chips observed.
    a. List the possible values for $X$.
    b. Complete a probability distribution for $X$.
    c. Construct a probability histogram.

**18.** A six-faced die has a 1 painted on three faces, a 3 painted on two faces, and a 5 painted on one face. The die is rolled twice and $X$ is the sum of the numbers on the upturned faces.
   a. List the possible values for $X$.
   b. Complete a probability distribution for $X$.
   c. Construct a probability histogram.

**19.** Gloria Troy is an assistant manager at a Big Bear Supermarket. She is concerned with customer dissatisfaction caused by waiting in a checkout line. Specifically Gloria is interested in the random variable:

$Y =$ the number of customers in the next checkout line.

A summary of the data accumulated by Gloria based on 100 observations at checkout counters is:

   Zero customers at the counter    15 times
   One customer at the counter    30 times
   Two customers at the counter    35 times
   Three customers at the counter    12 times
   Four customers at the counter    8 times

   a. List the possible values for $Y$.
   b. Specify a probability distribution for $Y$.
   c. Construct a probability histogram.

**20.** Kristin Heimstra plays shortstop for the Tri Valley Wildcats softball team. She is concerned with the number of hits she gets in any game in which she bats exactly three times. Specifically Kristin is interested in the random variable:

$Y =$ the number of hits in the next game in which she bats three times.

A summary of the data accumulated by Kristin based on 80 games in which she batted exactly three times is:

   Zero hits    12 games
   One hit    36 games
   Two hits    24 games
   Three hits    8 games

   a. List the possible values for $Y$.
   b. Specify a probability distribution for $Y$.
   c. Construct a probability histogram.

**21.** Sandra Rainer is an inspector in a plant in Canton, Ohio. One of her responsibilities is to check for defectives in production runs of type 3W907H circular saws. Table 4-7 is a summary of past records of defective saws in lots of the same size.
   Let $U$ be the number of defectives in the next lot checked by Sandra.
   a. List the possible values for $U$.
   b. Specify a probability distribution for $U$.
   c. Construct a probability histogram.

TABLE 4-7   Record of Defective Saws in Lots of 5000

| Number of Defectives | Number of Lots |
| --- | --- |
| 0 | 28 |
| 1 | 32 |
| 2 | 20 |
| 3 | 16 |
| 4 | 4 |

**22.** The number of units carried by each of the 5000 students at Marymount College is shown in Table 4-8. Suppose one of these students is randomly sampled and $U$ is the number of units in which the student is enrolled.

a. List the possible values for $U$.
b. Specify a probability distribution for $U$.
c. Construct a probability histogram.

TABLE 4-8   Units Carried by Marymount College Students

| Number of Units | Number of Students |
|---|---|
| 10 | 310 |
| 11 | 390 |
| 12 | 550 |
| 13 | 850 |
| 14 | 900 |
| 15 | 1250 |
| 16 | 750 |

**23.** A solid has 12 faces and each face has the same area. On five faces a 1 is painted, on four faces a 2 is painted, and on three faces a 3 is painted. The solid is rolled twice, and

$$X = \text{the sum of the numbers on the upturned faces.}$$

a. List the possible values for $X$.
b. Specify a probability distribution for $X$.
c. Construct a probability histogram.

**24.** A solid has 12 faces and each face has the same area. On four faces a 1 is painted, on three faces a 2 is painted, on three faces a 3 is painted, and on two faces a 4 is painted. The solid is rolled twice, and

$$X = \text{the sum of the numbers on the upturned faces.}$$

a. List the possible values for $X$.
b. Specify a probability distribution for $X$.
c. Construct a probability histogram.

# Exercises 4-1   Set B

*In Exercises 1 and 2, consider a container that has in it three disks of the same size and shape, and labeled 1, 2, and 3.*

**1.** Two disks are sampled from the container and the first one is replaced before the second one is drawn.

$$X = \text{the sum of the numbers on the disks.}$$

a. List the possible values for $X$.
b. Specify the probability distribution for $X$.
c. Construct a probability histogram.

**2.** Two disks are sampled from the container and the first one is replaced before the second one is drawn.

$$\bar{X} = \text{the mean of the numbers on the disks.}$$

a. List the possible values for $\bar{X}$.
b. Specify the probability distribution for $\bar{X}$.
c. Construct a probability histogram.
d. Compare the histograms in Exercises 1 and 2 to see whether they are essentially the same.

*In Exercises 3 and 4, consider a container that has four disks in it of the same size and shape, labeled 1, 2, 3, and 4.*

**3.** Two disks are sampled from the container and the first one is replaced before the second one is drawn.

$$Y = \text{the sum of the numbers on the disks.}$$

a. List the possible values for $Y$.
b. Specify the probability distribution for $Y$.
c. Construct a probability histogram.

**4.** Two disks are sampled from the container and the first one is replaced before the second one is drawn.

$$\bar{Y} = \text{the mean of the numbers on the disks.}$$

a. List the possible values for $\bar{Y}$.
b. Specify the probability distribution for $\bar{Y}$.
c. Construct a probability histogram.
d. Compare the histograms in Exercises 3 and 4 to see whether they are essentially the same.

# 4-2  The Mean and Variance of a Discrete Random Variable

**Key Topics**

1. *The mean of a discrete random variable*

2. *The variance of a discrete random variable*

S T A T I S T I C S    I N    A C T I O N

Juan Martinez is a counselor at Northern Pointe Community College (NPCC). He recently got a report from an agency of the federal government claiming that 38 percent of students who got educational grants last year dropped out of school before the year was completed. Juan helped establish the process for screening grant applicants at NPCC. The process specifically focused on trying to minimize the kind of negative results summarized in the report.

Juan therefore felt that the dropout rate of grant recipients at NPCC was considerably less than the 38 percent figure in the report. However, he wanted to check the records of last year's recipients to see whether the records would support his intuition. With the help of a computer program, Juan got a sample of 17 records of students who got grants last year. The sample included records of three students who did not complete the school year.

Before taking the sample, Juan knew that the number of dropouts would be only one of the possible values of a random variable:

X = the number of students in the sample who
dropped out during the school year.

Because the sample contained 17 student records, the possible values for X were 0, 1, 2, 3, 4, . . . , 17. Furthermore based on an assumed 38-percent dropout rate at NPCC, Juan knew that on the average the sample should yield between six and seven dropouts. He was also able to determine that the value of x he actually observed in the sample was 1.75 standard deviations below the expected value. As a result of the data Juan collected and the subsequent statistical analysis of the data, he reported to the staff at the college that he had "strong evidence" that fewer than 38 percent of the grant recipients at the school dropped out last year.

## The mean of a discrete random variable

From Chapter 2, we know that the mean of a population of quantitative data is the sum of the data points divided by the number of data points. Using summation notation:

$$\mu = \frac{\sum x_i}{N}$$

If $X$ is the value of a randomly selected data point from this set, then $\mu_X$ is the mean (or expected value) of the random variable $X$. The subscript $X$ on $\mu$ is used to distinguish between the mean of the population of data and the mean of the random variable $X$. Read $\mu_X$ as "mu sub $X$."

### DEFINITION 4-3 The Mean of a Discrete Random Variable X

If $X$ is a discrete random variable, then the **mean** of $X$ is

$$\mu_X = \sum x_i \cdot P(x_i)$$

Based on the equation, we need to know two related sets of values to compute $\mu_X$:

Set 1   The possible values of the random variable

Set 2   The probabilities that correspond to each of these values

The sum of the products of the paired values from sets 1 and 2 is $\mu_X$.

▶ Example 1    A box contains ten disks. On five of the disks a 1 is painted, on three a 3 is painted, and on two a 5 is painted. One disk is randomly sampled, and $X$ = the number on the disk. Compute $\mu_X$.

Solution    Table 4-9 contains the possible values of $X$ and corresponding probabilities. The third column contains the products of each $x_i$ and corresponding $P(x_i)$. For the given experiment, the mean of $X$ is 2.4.    ◀

TABLE 4-9   Calculations for $\mu_X$

| $x_i$ | $P(x_i)$ | $x_i \cdot P(x_i)$ |
|-------|----------|--------------------|
| 1 | $\dfrac{5}{10} = 0.5$ | 0.5 |
| 3 | $\dfrac{3}{10} = 0.3$ | 0.9 |
| 5 | $\dfrac{2}{10} = 0.2$ | 1.0 |

$$\mu_X = \sum x_i \cdot P(x_i)$$

$$\mu_X = 0.5 + 0.9 + 1.0$$
$$= 2.4$$

▶ **Example 2**   Recall the computer technician in Example 4 of Section 4-1 at Metropolitan Airport and the issue of major breakdowns in a computer at the facility.

**a.** Compute $\mu_Y$.
**b.** Draw a probability histogram and locate $\mu_Y$ on the diagram.

**Solution**   **Discussion.** The probabilities for $Y$ were computed in Section 4-1. The product of each $y_i$ and corresponding $P(y_i)$ is given in Table 4-10.

**a.**

TABLE 4-10   Calculations for $\mu_{\hat{Y}}$

| $y$ | $P(y)$ | $y \cdot P(y)$ |
|-----|--------|----------------|
| 0 | 0.50 | 0.00 |
| 1 | 0.30 | 0.30 |
| 2 | 0.16 | 0.32 |
| 3 | 0.04 | 0.12 |

$$\mu_Y = \sum y_i \cdot P(y_i)$$

$$\mu_Y = 0.00 + 0.30 + 0.32 + 0.12$$
$$= 0.74$$

Using probabilities based on the historical data, the number of major breakdowns for the computer expected next week is 0.74.

**b.** A probability histogram for $Y$ is shown in Figure 4-9. The location for $\mu_Y$ is included on the graph. A fulcrum is drawn at 0.74 to illustrate the *balance point* interpretation of $\mu_Y$.   ◀

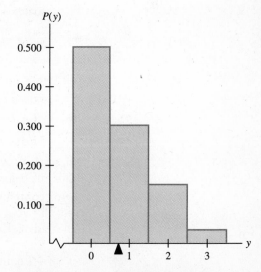

**Figure 4-9.**   A probability histogram for the computer breakdown problem.

## The variance of a discrete random variable

From Chapter 2, we know that the variance of a population of quantitative data can be calculated using the following formula:

$$\sigma^2 = \frac{\sum(x_i - \mu)^2}{N}$$

If $X$ is the value of a randomly selected data point from this set, then $\sigma_X^2$ describes the variability among the possible values of $X$ in much the same way that $\sigma^2$ describes the variability among the data points. The subscript $X$ on $\sigma_X^2$ is used to distinguish between the variance of the population of data and the variance of the random variable $X$. Read $\sigma_X^2$ as "sigma squared sub $X$."

### DEFINITION 4-5 The Variance and Standard Deviation of a Discrete Random Variable $X$

If $X$ is a discrete random variable, then the variance of $X$ is $\sigma_X^2$, and

$$\sigma_X^2 = \sum(x_i - \mu_X)^2 \cdot P(x_i)$$

or equivalently, as shown by expanding the summation notation:

$$\sigma_X^2 = \left[\sum x_i^2 \cdot P(x_i)\right] - (\mu_X)^2$$

The positive square root of $\sigma_X^2$ is called **the standard deviation of** $X$, as shown by:

$$\sigma_X = \sqrt{\sigma_X^2} = \sqrt{\sum x_i^2 \cdot P(x_i) - (\mu_X)^2}$$

▶ **Example 3**    The box in Example 1 has ten disks. A 1 is painted on five disks, a 3 is painted on three, and a 5 is painted on two. One disk is randomly selected, and $X =$ the number on the disk.

a. Compute $\sigma_X^2$ to two decimal places.
b. Compute $\sigma_X$ to one decimal place.
c. Construct a probability histogram and show $\mu_X$ on the graph.

**Solution**    **Discussion.** We will use the alternate formula for computing $\sigma_X^2$. Table 4-11 contains the calculations.

**TABLE 4-11   Calculations for $\sigma_X^2$**

| $x$ | $x^2$ | $P(x)$ | $x^2 \cdot P(x)$ |
|-----|-------|--------|------------------|
| 1   | 1     | 0.5    | 0.50             |
| 3   | 9     | 0.3    | 2.70             |
| 5   | 25    | 0.2    | 5.00             |

**a.** $\sigma_X^2 = (0.50 + 2.70 + 5.00) - 2.4^2$    Recall $\mu_X = 2.4$.
$\quad\ = 8.20 - 5.76$
$\quad\ = 2.44$    The variance is 2.44.

**b.** $\sigma_X = \sqrt{2.44}$    $\sigma_X = \sqrt{\sigma_X^2}$
$\quad\ = 1.562\ldots$    The unrounded value
$\quad\ \approx 1.6$    $\sigma_X$ to one decimal place

**c.** Use the values of $x$ and $P(x)$ in the table to draw the probability histogram in Figure 4-10. ◄

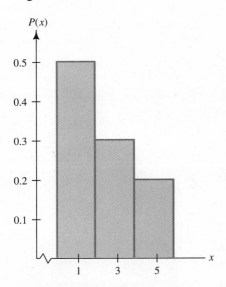

**Figure 4-10.** A probability histogram for the disk experiment.

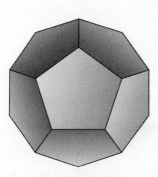

**Figure 4-11.** A top view of a dodecahedron.

▶**Example 4**   A **dodecahedron**, shown in Figure 4-11, has 12 faces of equal area. A 1 is painted on five of the faces, a 2 is painted on four faces, and a 3 is painted on three faces. The solid is rolled twice and $X$ is the sum of the numbers on the topmost faces.

**a.** List the possible values for $X$.
**b.** Compute a probability distribution for $X$. Round each probability to three places.
**c.** Compute $\mu_X$ to one decimal place.
**d.** Compute $\sigma_X$ to one decimal place.
**e.** Construct a probability histogram for $X$, and indicate $\mu_X$ on the histogram.

**Solution**   Label the five 1s as: 1A, 1B, 1C, 1D, and 1E. Label the four 2s as: 2A, 2B, 2C, and 2D. Label the three 3s as: 3A, 3B, and 3C. The possible outcomes are shown in Table 4-12.

TABLE 4-12    The 144 Possible Outcomes and Corresponding Values of $X$

| | | First Toss | | | | | | | | | | | |
|---|---|---|---|---|---|---|---|---|---|---|---|---|---|
| | | 1A | 1B | 1C | 1D | 1E | 2A | 2B | 2C | 2D | 3A | 3B | 3C |
| | 1A | 2 | 2 | 2 | 2 | 2 | 3 | 3 | 3 | 3 | 4 | 4 | 4 |
| | 1B | 2 | 2 | 2 | 2 | 2 | 3 | 3 | 3 | 3 | 4 | 4 | 4 |
| | 1C | 2 | 2 | 2 | 2 | 2 | 3 | 3 | 3 | 3 | 4 | 4 | 4 |
| | 1D | 2 | 2 | 2 | 2 | 2 | 3 | 3 | 3 | 3 | 4 | 4 | 4 |
| | 1E | 2 | 2 | 2 | 2 | 2 | 3 | 3 | 3 | 3 | 4 | 4 | 4 |
| | 2A | 3 | 3 | 3 | 3 | 3 | 4 | 4 | 4 | 4 | 5 | 5 | 5 |
| Second | 2B | 3 | 3 | 3 | 3 | 3 | 4 | 4 | 4 | 4 | 5 | 5 | 5 |
| Toss | 2C | 3 | 3 | 3 | 3 | 3 | 4 | 4 | 4 | 4 | 5 | 5 | 5 |
| | 2D | 3 | 3 | 3 | 3 | 3 | 4 | 4 | 4 | 4 | 5 | 5 | 5 |
| | 3A | 4 | 4 | 4 | 4 | 4 | 5 | 5 | 5 | 5 | 6 | 6 | 6 |
| | 3B | 4 | 4 | 4 | 4 | 4 | 5 | 5 | 5 | 5 | 6 | 6 | 6 |
| | 3C | 4 | 4 | 4 | 4 | 4 | 5 | 5 | 5 | 5 | 6 | 6 | 6 |

**a.** The possible values of $X$ are 2, 3, 4, 5, and 6.

**b.** $P(2) = \dfrac{25}{144} \approx 0.174$    $\dfrac{\text{25 outcomes of 2}}{\text{144 outcomes}}$

$P(3) = \dfrac{40}{144} \approx 0.278$    $\dfrac{\text{40 outcomes of 3}}{\text{144 outcomes}}$

$P(4) = \dfrac{46}{144} \approx 0.319$    $\dfrac{\text{46 outcomes of 4}}{\text{144 outcomes}}$

$P(5) = \dfrac{24}{144} \approx 0.167$    $\dfrac{\text{24 outcomes of 5}}{\text{144 outcomes}}$

$P(6) = \dfrac{9}{144} \approx 0.063$    $\dfrac{\text{9 outcomes of 5}}{\text{144 outcomes}}$

The calculations for $\mu_X$ and $\sigma_X$ are in Table 4-13.

$$\sum x_i \cdot P(x_i) = 3.671$$
$$\sum x_i^2 \cdot P(x_i) = 14.745$$

TABLE 4-13    Calculations for $\mu_X$ and $\sigma_X$

| $x$ | $P(x)$ | $x \cdot P(x)$ | $x^2$ | $x^2 \cdot P(x)$ |
|---|---|---|---|---|
| 2 | 0.174 | 0.348 | 4 | 0.696 |
| 3 | 0.278 | 0.834 | 9 | 2.502 |
| 4 | 0.319 | 1.276 | 16 | 5.104 |
| 5 | 0.167 | 0.835 | 25 | 4.175 |
| 6 | 0.063 | 0.378 | 36 | 2.268 |

c. $\mu_X = 3.671$
   $\approx 3.7$, to one decimal place

d. $\sigma_X = \sqrt{14.745 - 3.671^2}$
   $\approx 1.1$, to one decimal place

e. A probability histogram is shown in Figure 4-12.   ◀

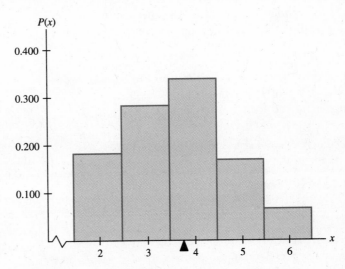

**Figure 4-12.**   A probability histogram for the 12-faced solid problem.

## Exercises 4-2   Set A

1. As part of a grand opening celebration, the management of a new supermarket put 100 bills of paper money in a large container.

   > Five were $100 bills.
   > Ten were $50 bills.
   > Fifteen were $20 bills.
   > Thirty were $10 bills.
   > Forty were $5 bills.

   Each hour a shopper in the market was chosen to randomly select a bill from the container. The denomination of the bill chosen was replaced by management to maintain the above frequencies. Let $X$ be the value of the bill selected by the next shopper.

   a. List the possible values for $X$.
   b. Compute the probability distribution for $X$.
   c. Construct a probability histogram.
   d. Compute $\mu_X$ and locate it on the histogram.
   e. Compute $\sigma_X$ to one decimal place.

2. A box contains 25 disks of the same size. A number is painted on each disk.

   > One disk is labeled 100.
   > Three disks are labeled 50.
   > Five disks are labeled 25.
   > Ten disks are labeled 10.
   > Six disks are labeled 5.

   Let $X$ be the number of one disk randomly selected from the box.

a. List the possible values for $X$.
b. Compute the probability distribution for $X$.
c. Construct a probability histogram.
d. Compute $\mu_X$ and locate it on the histogram.
e. Compute $\sigma_X$ to one decimal place.

3. A six-faced solid has a 1 painted on three faces, a 2 on two faces, and a 3 on one face. An experiment consists of rolling the die twice, and $Y$ is the sum of the numbers on the topmost faces.
a. List the possible values for $Y$.
b. Compute the probability distribution for $Y$.
c. Construct a probability histogram.
d. Compute $\mu_Y$ and locate it on the histogram.
e. Compute $\sigma_Y$ to one decimal place.

4. Ten cards of the same size and shape have numbers marked on them.

   A 1 is marked on five cards.
   A 2 is marked on three cards.
   A 3 is marked on two cards.

Two cards are selected with replacement, and $Y$ is the sum of the numbers on the cards.
a. List the possible values for $Y$.
b. Compute the probability distribution for $Y$.
c. Construct a probability histogram.
d. Compute $\mu_Y$ and locate it on the histogram.
e. Compute $\sigma_Y$ to one decimal place.

5. Lucinda Turley inspects small circuit boards for faulty connections. Each inspection unit consists of the same number of randomly selected circuit boards. Past records indicate that if $W$ is the number of defects in the next inspection unit, then:

$$P(W = 0) = 0.30$$
$$P(W = 1) = 0.40$$
$$P(W = 2) = 0.20$$
$$P(W = 3) = 0.10$$

a. Construct a probability histogram for $W$.
b. Compute $\mu_W$ to one decimal place.
c. Compute $\sigma_W$ to one decimal place.

6. Dr. Edison Diest is an optometrist who has volunteered to give free eye examinations to elementary school children. Edison regularly schedules a block of ten children at a time. If $W$ is the number of children in a given block with an undetected vision problem, past records indicate that:

$$P(W = 0) = 0.12$$
$$P(W = 1) = 0.32$$
$$P(W = 2) = 0.26$$
$$P(W = 3) = 0.18$$
$$P(W = 4) = 0.12$$

a. Construct a probability histogram for $W$.
b. Compute $\mu_W$ to one decimal place.
c. Compute $\sigma_W$ to one decimal place.

7. When a pair of dice is tossed, the number of dots on the upturned faces is a random variable $X$ with possible values of 2 through 12 inclusive. The probabilities for $X$ to three decimal places are:

$P(2) = 0.028$     $P(3) = 0.056$     $P(4) = 0.083$
$P(5) = 0.111$     $P(6) = 0.139$     $P(7) = 0.167$
$P(8) = 0.139$     $P(9) = 0.111$     $P(10) = 0.083$
$P(11) = 0.056$     $P(12) = 0.028$

a. Compute $\mu_X$ to one decimal place.
b. State the meaning of $\mu_X$.
c. Compute $\sigma_X$ to one decimal place.
d. What outcomes are within one standard deviation of $\mu_X$?

8. When five coins are tossed simultaneously, the number of heads showing on the upturned faces is a random variable $X$ with possible values 0 through 5 inclusive. Assuming the coins are balanced, the probabilities for $X$ to three decimal places are:

$P(0) = 0.031$     $P(1) = 0.156$     $P(2) = 0.313$
$P(3) = 0.313$     $P(4) = 0.156$     $P(5) = 0.031$

a. Compute $\mu_X$ to one decimal place.
b. State the meaning of $\mu_X$.
c. Compute $\sigma_X$ to one decimal place.
d. What outcomes are within one standard deviation of $\mu_X$?

9. A study was made in a large eastern state on residents of the state who do not have health insurance. Based on the study, 59 percent of the residents with no such insurance are women. If eight residents of this state with no health insurance are randomly sampled and $Y$ is the number in the sample that are women, then the possible values of $Y$ are 0 through 8 inclusive. The probabilities for $Y$ to three decimal places are:

$P(0) = 0.001$     $P(1) = 0.009$     $P(2) = 0.046$

$P(3) = 0.133$     $P(4) = 0.240$     $P(5) = 0.276$

$P(6) = 0.199$     $P(7) = 0.082$     $P(8) = 0.015$

a. Compute $\mu_Y$ to one decimal place.
b. Interpret the meaning of $\mu_Y$.
c. Compute $\sigma_Y$ to one decimal place.
d. What outcomes are within two standard deviations of $\mu_Y$?

10. According to a recent survey, 40 percent of high school seniors tested in Kansas City could not name three countries in South America. If seven high school seniors in the city are randomly sampled and $Y$ is the number in the sample that cannot name three countries in South America, then the possible values of $Y$ are 0 through 7 inclusive. The probabilities for $Y$ to three decimal places are:

$P(0) = 0.028$     $P(1) = 0.131$     $P(2) = 0.261$

$P(3) = 0.290$     $P(4) = 0.194$     $P(5) = 0.077$

$P(6) = 0.017$     $P(7) = 0.002$

a. Compute $\mu_Y$ to one decimal place.
b. Interpret the meaning of $\mu_Y$.

c. Compute $\sigma_Y$ to one decimal place.
d. What outcomes are within two standard deviations of $\mu_Y$?

11. Records of the student population at a private liberal arts college in the East produced the data in Table 4-14 on the number of units in which the 5000 students are currently enrolled.

Suppose one student from this population is randomly selected and $X$ is the number of units the student is taking.
a. List the possible values of $X$.
b. Compute a probability distribution for $X$.
c. Compute $\mu_X$ to one decimal place.
d. Construct a probability histogram and locate $\mu_X$ on the graph.
e. Compute $\sigma_X$ to one decimal place.
f. What values of $X$ are within one standard deviation of the mean?

12. An eastern state recently gave a banquet to honor state employees with 25 or more years of service. Table 4-15 shows the years of service distribution for those honored.

Suppose one employee from this population is randomly selected and $X$ is the number of years of state service for the employee.
a. List the possible values of $X$.
b. Compute a probability distribution for $X$.
c. Compute $\mu_X$ to one decimal place.
d. Construct a probability histogram and locate $\mu_X$ on the graph.
e. Compute $\sigma_X$ to one decimal place.
f. What values of $X$ are within one standard deviation of the mean?

TABLE 4-14   Units for 5000 College Students

| Number of Units | 10 | 11 | 12 | 13 | 14 | 15 | 16 |
|---|---|---|---|---|---|---|---|
| Number of Students | 310 | 390 | 550 | 850 | 900 | 1250 | 750 |

TABLE 4-15   Years of Service for 500 State Employees

| Number of Years of Service | 25 | 26 | 27 | 28 | 29 | 30 | 31 |
|---|---|---|---|---|---|---|---|
| Number of Employees | 110 | 90 | 75 | 80 | 65 | 50 | 30 |

**13.** Figure 4-13(a) and (b) show probability histograms for random variables $X$ and $Y$. Both $\mu_X$ and $\mu_Y$ are 3.5. From the shapes of the histograms, it appears that the probability distribution in (a) is "more spread out" than the one in (b).
   a. Compute $\sigma_X$ to one decimal place.
   b. Compute $\sigma_Y$ to one decimal place.

**14.** Figure 4-14(a) and (b) show probability histograms for random variables $X$ and $Y$. Both $\mu_X$ and $\mu_Y$ are 2.5. From the shapes of the histograms, it appears that the probability distribution in (a) is more spread out than the one in (b).
   a. Compute $\sigma_X$ to one decimal place.
   b. Compute $\sigma_Y$ to one decimal place.

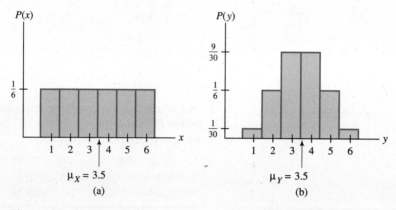

**Figure 4-13.**  (a) Probability histogram for $X$. (b) Probability histogram for $Y$.

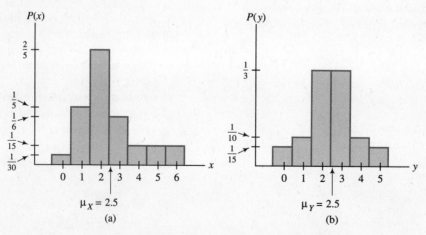

**Figure 4-14.**  (a) Probability histogram for $X$. (b) Probability histogram for $Y$.

# Exercises 4-2   Set B

*In Exercises 1 and 2, use the information in Figure 4-15.*

1. According to the study by Bumpass, Sweet, and Castro, about 40 percent of women in their 30s who separated recently will never remarry. Assume that the 40 percent figure applies to such women in the Dallas/Fort Worth area. Suppose $n = 2$ women in their 30s who recently separated are sampled and $Y =$ the number who will never remarry.
   a. List the possible values for $Y$.
   b. Compute the probability distribution for $Y$.
   c. Compute $\mu_Y$ to one decimal place.
   d. Compute $\sigma_Y^2$ to one decimal place.

### By Malcolm Ritter
**Associated Press**

NEW YORK — About 40 percent of women who separated recently while in their 30s will never remarry, nor will about 70 percent of women who separated when older than 40, a new study projects.

And while 72 percent of recently separated women will eventually remarry, half will still be single seven years after the split, the projections suggest.

The estimates, for women who separated in the early 1980s, emphasize that separation can mean long periods of single life and economic hardship for some women and their children, researchers said.

"For many of the children, it's the rest of their childhood years," said study co-author Larry Bumpass. "For many of these women, it's for the rest of their lives."

Bumpass, a sociology professor at the University of Wisconsin in Madison, developed the projections with colleagues James Sweet and Teresa Castro. He spoke in a telephone interview before presenting the study Sunday in Atlanta at the annual meeting of the American Sociological Association.

**Figure 4-15.** "Study on marriage and divorced women," *The Sacramento Bee*, Monday August 29, 1988, p. A6. Reprinted by permission of Associated Press.

2. According to the study, about 70 percent of women over 40 who recently separated will never remarry. Assume that the 70 percent figure applies to such women in the New Orleans metropolitan area. Suppose $n = 3$ women over 40 who recently separated are sampled and $W =$ the number in the sample who will never remarry.
   a. List the possible values for $W$.
   b. Compute the probability distribution for $W$.
   c. Compute $\mu_W$ to one decimal place.
   d. Compute $\sigma_W^2$ to one decimal place.
   e. Construct a probability histogram.

*In Exercises 3 and 4, use the information in Figure 4-16 on the following page.*

3. According to recent figures, 23 percent of preschool children of working mothers are placed in day-care centers. Suppose that the 23 percent figure applies to such children in the Atlanta metropolitan area and $n = 2$ children with working mothers are sampled. Let:

   $X =$ the number in the sample that are placed in day-care centers.

   a. List the possible values for $X$.
   b. Compute the probability distribution for $X$.
   c. Compute $\mu_X$ to one decimal place.
   d. Compute $\sigma_X^2$ to one decimal place.
   e. Construct a probability histogram.

4. According to the numbers in Figure 4-16, 48 percent of preschool children are cared for by relatives. Suppose the 48 percent figure applies to such children in the Denver metropolitan area and $n = 3$ children are randomly sampled. Let:

   $Y =$ the number in the sample that are cared for by relatives.

   a. List the possible values for $Y$.
   b. Compute the probability distribution for $Y$.
   c. Compute $\mu_Y$ to one decimal place.
   d. Compute $\sigma_Y^2$ to one decimal place.

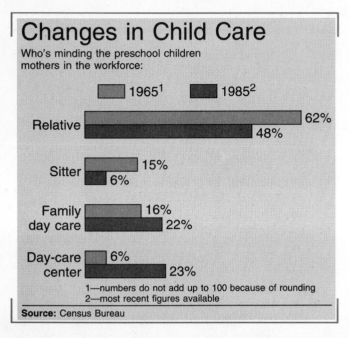

**Figure 4-16.**   Changes in child care.

*An interesting application of expected value is the concept of a fair game. If in a game of chance there is the possibility of positive monetary gain (winnings) and the possibility of negative monetary gain (losses), the game is called fair if the expected monetary gain is 0. One who plays a fair game expects to break even in the long run.*

5. Sam Donnelly claims his game is a fair game. He has three similar boxes. In one box he has $3, in a second $6, and in the third $21. If you pay Sam $10, you may select any of the boxes and keep the contents. Is the game fair?

6. Clark Kent has a game similar to Sam's. In one box he has $3, in a second $12, and in the third $14. If it costs $10 to play Clark's game, is the game fair?

## 4-3  Binomial Distributions: Basic Assumptions and a Formula for $P(x)$

**Key Topics**

1. *Binomial random variables*

2. *A formula for binomial probabilities*

3. *Factor 1 in the formula*

4. *Factor 2 in the formula*

5. *Factor 3 in the formula*

6. *Computing the probability distribution for a binomial random variable*

7. *Using Appendix Table 3 to find binomial probabilities*

## STATISTICS IN ACTION

Barbara Burnett is a motorcycle officer for the Georgia Highway Patrol (GHP). A campaign has recently been initiated by the GHP to get vehicles with faulty smog controls off the highway. To help with the campaign, all patrol officers have been issued a small electronic device to place inside a vehicle's exhaust pipe. In 30 seconds, the device will analyze the exhaust and indicate whether the smog controls on that vehicle are operating properly. Each officer has been instructed to make a few "random" checks each day while on patrol.

Before leaving the station on this day, Barbara decided to make five "random" checks of vehicles for faulty smog controls. She knows that GHP records indicate that about 30 percent of all vehicles currently on the road would fail the smog check. However, Barbara knows that the actual number that will fail her checks today is a random variable:

$X$ = the number of vehicles that fail the smog inspection.

Because Barbara will check five vehicles, the possible values for $X$ are 0, 1, 2, 3, 4, and 5.

This experiment is one that statisticians call a *binomial experiment*. Using the formula for computing probabilities associated with a random variable modeled as binomial, the following probabilities for $X$ can be determined:

| | |
|---|---|
| $P(X = 0) = 0.168$ | the probability no vehicles fail the check |
| $P(X = 1) = 0.360$ | the probability one vehicle fails the check |
| $P(X = 2) = 0.309$ | the probability two vehicles fail the check |
| $P(X = 3) = 0.132$ | the probability three vehicles fail the check |
| $P(X = 4) = 0.028$ | the probability four vehicles fail the check |
| $P(X = 5) = 0.002$ | the probability all five vehicles fail the check |

## Binomial random variables

Frequently investigations are made in which the investigator is interested in one of two possible outcomes. These possibilities can be called *success* and *failure*. As an example, consider the success-failure experiment conducted by Officer Barbara Burnett:

Success   The investigated vehicle does not pass the smog check.

Failure   The investigated vehicle does pass the smog check.

When such an experiment is repeated $n$ times, interest usually centers on the number of successes in the $n$ trials of the experiment. In the example, five vehicles are checked, therefore $n = 5$. Barbara is primarily interested only in the number that fail. Because successes are *counted*, the random variable associated with such an experiment is discrete. If certain assumptions are appropriate for describing such an experiment, then the random variable involved is called **binomial**.

## Criteria for a random variable to be modeled as binomial

If $X$ is the number of successes obtained in $n$ repetitions of a success-failure experiment and it is reasonable to make the assumptions:

1. The probability of success (represented by $p$) on each of the $n$ trials is the same,
2. The events "success on trial 1," "success on trial 2," ..., "success on trial $n$," may all be described as independent,

then we say $X$ **is modeled as binomial** or $X$ **is binomial**.

▶ **Example 1**    An experiment consists of randomly sampling ten electronic components from a production run of 30,000 and checking for defects. If $X$ is the number of defective components in the sample, discuss whether $X$ is binomial.

Solution    **Discussion.** To be able to model $X$ as binomial, we need to show that $n$ success-failure experiments are involved and assumptions 1 and 2 are reasonable.

a. For each of the ten trials (10 items selected), the component will be either defective (a success) or not defective (a failure).
b. The relative frequency of defective components in the production run is the success probability $p$. This value of $p$ pertains to each of the components examined.
c. Because only ten components are sampled from a production run of 30,000, it is reasonable to model as independent the events "the first component is defective," "the second component is defective," ..., "the tenth component is defective." Therefore, for this experiment, we can model $X$ as binomial.    ◀

## A formula for binomial probabilities

The usefulness of assumptions 1 and 2 for a random variable $X$ associated with $n$ repetitions of a success-failure experiment is that they provide a convenient formula for computing $P(x)$. Example 2 will be used to motivate this formula.

▶ **Example 2**    In a large metropolitan area in the South, studies have shown that approximately 30 percent of the residents regularly take medication for allergies. A random sample of three residents in this area is taken, and $W$ is the number in the sample that regularly take medication for allergies.

a. List the possible values for $W$.
b. Complete a probability distribution for $W$.

**Solution**    **Discussion.** $W$ can be modeled as binomial:

1. On each of $n = 3$ trials, only two potential outcomes are of interest, namely:

   Success    The resident takes medication for allergies.

   Failure    The resident does not take medication.

   $W$ = the count of successes in these trials
2. $p = 0.30$ can be used as the probability of success for each trial.

   $(1 - p) = 0.70$ can be used as the probability of failure for each trial.

3. The trials can be modeled as approximately independent because a small sample is taken from a large population.

**a.** 0, 1, 2, and 3    From 0 through 3 successes
**b.** Table 4-16 shows possible sequences of successes ($s$) and failures ($f$).

TABLE 4-16    Sequences of Successes and Failures in Three Trials

| 1st Resident | 2nd Resident | 3rd Resident | Result of Trial |
| --- | --- | --- | --- |
| $s$ | $s$ | $s$ | 3 successes, 0 failures |
| $s$ | $s$ | $f$ | 2 successes, 1 failure |
| $s$ | $f$ | $s$ | 2 successes, 1 failure |
| $s$ | $f$ | $f$ | 1 success, 2 failures |
| $f$ | $s$ | $s$ | 2 successes, 1 failure |
| $f$ | $s$ | $f$ | 1 success, 2 failures |
| $f$ | $f$ | $s$ | 1 success, 2 failures |
| $f$ | $f$ | $f$ | 0 successes, 3 failures |

The probability distribution is shown in Table 4-17 on the next page.

$P(0) = P(0 \text{ successes})$      0 successes and 3 failures

$\quad = 1(0.7)(0.7)(0.7)$      $P(\text{failure}) = 0.7$

$\quad = 0.343$      Only 1 such sequence

$P(1) = P(1 \text{ success})$      1 success and 2 failures

$\quad = 3(0.3)(0.7)(0.7)$      $P(\text{success}) = 0.3;\ P(\text{failure}) = 0.7$

$\quad = 0.441$      3 such sequences of $s$ and $f$s

$P(2) = P(2 \text{ successes})$      2 successes and 1 failure

$\quad = 3(0.3)(0.3)(0.7)$      $P(\text{success}) = 0.3;\ P(\text{failure}) = 0.7$

$\quad = 0.189$      3 such sequences of $s$ and $f$

$P(3) = P(3 \text{ successes})$      3 successes and 0 failures

$\quad = 1(0.3)(0.3)(0.3)$      $P(\text{success}) = 0.3$

$\quad = 0.027$      Only 1 such sequence    ◀

TABLE 4-17   The Probability Distribution for $X$

| $x$ | 0 | 1 | 2 | 3 |
|-----|-----|-----|-----|-----|
| $P(x)$ | 0.343 | 0.441 | 0.189 | 0.027 |

A pattern can be seen when the probabilities are written together. Recall from algebra that $a^1 \cdot 1 = a$, for any real number, and $a^0 = 1$, if $a \neq 0$. Remembering these facts, we can write $P(0)$ through $P(3)$ as the products of three factors.

| $x_i$ | $P(x_i)$ |
|-----|-----|
| 0 | $1(0.3)^0(0.7)^3$ |
| 1 | $3(0.3)^1(0.7)^2$ |
| 2 | $3(0.3)^2(0.7)^1$ |
| 3 | $1(0.3)^3(0.7)^0$ |

**Factor 1**   The numbers 1, 3, 3, and 1 are called binomial coefficients.

**Factor 2**   A power of 0.3, the *probability of success* on each trial

**Factor 3**   A power of 0.7, the *probability of failure* on each trial

The entries in the previous table fit a pattern given by the general formula for binomial probabilities. The formula will be stated below, and then each of the three factors in the formula will be discussed.

### The general formula for computing binomial probabilities

If $X$ is the number of successes obtained in $n$ trials of a success-failure experiment and assumptions 1 and 2 are appropriate, then:

$$P(x) = \binom{n}{x} p^x (1 - p)^{n - x}$$

for $x = 0, 1, 2, \ldots, n$.

$\binom{n}{x}$      stands for the binomial coefficient.

$p$      stands for the probability of success on each trial

$1 - p$   stands for the probability of failure on each trial

$x$      stands for the number of successes in the $n$ trials

$n - x$   stands for the number of failures in the $n$ trials

## Factor 1 in the formula

The binomial coefficient (factor 1) represents the number of different ways in which exactly $x$ successes can occur in $n$ trials of a success-failure activity. The symbol $\binom{n}{x}$ represents the binomial coefficient and is read "the combinations of $n$ things taken $x$ at a time." Combinations were presented in an informal way in Exercises 3-1, Set B. For further discussion of combinations, consult that exercise set.

### A Formula for Computing $\binom{n}{x}$, for $x = 0, 1, 2, \ldots, n$

$$\binom{n}{x} = \frac{n!}{x!(n-x)!}$$

Read $n!$ as "$n$-factorial."
Read $x!$ as "$x$-factorial."
Read $(n - x)!$ as "$n$ minus $x$ factorial"

By definition,

$$n! = n(n-1)(n-2)\ldots(3)(2)(1)$$
$$x! = x(x-1)(x-2)\ldots(3)(2)(1)$$
$$0! = 1$$

▶ **Example 3**   Compute the number of ways in which four successes can occur in ten trials.

**Solution**   With $n = 10$ and $x = 4$:

$$\binom{10}{4} = \frac{10!}{4!(10-4)!}$$    $\binom{n}{x} = \frac{n!}{x!(n-x)!}$

$$= \frac{10 \cdot 9 \cdot 8 \cdot 7 \cdot 6!}{4 \cdot 3 \cdot 2 \cdot 1 \cdot 6!}$$    Write 10! as shown. There is a $(10 - 4)! = 6!$ in denominator.

$$= 210$$    Simplify.

There are 210 different ways of getting four successes and six failures in ten trials.  ◀

Appendix Table 2 is 15 rows of an array of numbers known as *Pascal's Triangle*. The first six rows of this triangle are shown in Figure 4-17.

```
                1
             1     1
          1     2     1
       1     3     3     1
    1     4     6     4     1
 1     5    10    10     5     1
```

**Figure 4-17.**  The first six rows of Appendix Table 2.

### Features of Pascal's Triangle

1. The first and last number in each row is 1.
2. Beginning with row 3, the numbers in each row between the first and last are the sums of the numbers immediately to the left and right in the previous row. Using rows 5 and 6 to illustrate:

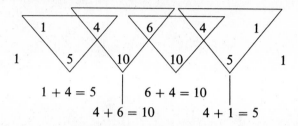

$$1 + 4 = 5 \qquad 6 + 4 = 10$$
$$4 + 6 = 10 \qquad 4 + 1 = 5$$

The relevance of Pascal's Triangle is that the binomial coefficients $\binom{n}{x}$ for $n \leq 15$ are the numbers in one of the rows of Appendix Table 2. **To determine which row, look at the second number from the left in a given row. This number corresponds to $n$.**

▶ **Example 4**   Use the formula for $\binom{n}{x}$ to verify that the entries in the seventh row of Appendix Table 2 are the binomial coefficients for $n = 6$.

**Solution**   **Discussion.** Notice that the second number from the left is 6. In this row, the numbers are:

$$1 \quad 6 \quad 15 \quad 20 \quad 15 \quad 6 \quad 1$$

$$\binom{6}{0} = \frac{6!}{0!(6-0)!} = \frac{6!}{1 \cdot 6!} = 1 \qquad \text{Only 1 sequence with all 6 failures}$$

$$\binom{6}{1} = \frac{6!}{1!(6-1)!} = \frac{6 \cdot 5!}{1 \cdot 5!} = 6 \qquad \text{6 different sequences with exactly 1 success}$$

$$\binom{6}{2} = \frac{6!}{2!(6-2)!} = \frac{6 \cdot 5 \cdot 4}{2 \cdot 1 \cdot 4!} = 15 \qquad \text{15 different sequences with exactly 2 successes}$$

$$\binom{6}{3} = \frac{6!}{3!(6-3)!} = \frac{6 \cdot 5 \cdot 4 \cdot 3!}{3 \cdot 2 \cdot 1 \cdot 3!} = 20 \qquad \text{20 different sequences with exactly 3 successes}$$

$$\binom{6}{4} = \frac{6!}{4!(6-4)!} = \frac{6 \cdot 5 \cdot 4!}{4! \cdot 2 \cdot 1} = 15 \qquad \text{15 different sequences with exactly 4 successes}$$

$$\binom{6}{5} = \frac{6!}{5!(6-5)!} = \frac{6 \cdot 5!}{5! \cdot 1} = 6 \qquad \text{6 different sequences with exactly 5 successes}$$

$$\binom{6}{6} = \frac{6!}{6!(6-6)!} = \frac{6!}{6! \cdot 1} = 1 \qquad \text{Only 1 sequence with all 6 successes} \qquad ◀$$

## Factor 2 in the formula

The second factor in the formula is $p^x$, where $p$ stands for the probability of success on each trial. $P(x)$ represents the probability of $x$ successes in the $n$ trials. There must be exactly $x$ factors of $p$ in the probability corresponding to each sequence that has exactly $x$ successes. The $x$ factors of $p$ can be written $p^x$.

▶ **Example 5**   Let $X$ be a binomial random variable. Evaluate $p^x$ for the following:

$$P(4), \text{ if } n = 10 \quad \text{and} \quad p = 0.2$$

**Solution**   $P(4)$ means $x$ is 4. With $p = 0.2$ and $x = 4$, $p^x$ becomes

$$(0.2)^4 = 0.0016 \quad \blacktriangleleft$$

## Factor 3 in the formula

The third factor in the formula is $(1 - p)^{n-x}$. If $p$ stands for the probability of success on each trial, then $(1 - p)$ stands for the probability of failure. (Remember the rule for probabilities of complements discussed in Chapter 3.)

   If $x$ specifies the number of successes in the $n$ trials, then $n - x$ trials must be failures. Thus, in the probability corresponding to each sequence of $n$ trials that has $x$ successes, there must be $(n - x)$ factors of $(1 - p)$. The product of $(n - x)$ factors of $(1 - p)$ can be written $(1 - p)^{n-x}$.

▶ **Example 6**   Let $X$ be a binomial random variable. Evaluate the $(1 - p)^{n-x}$ factor for the following:

$$P(X = 5), \text{ if } n = 9 \text{ and } p = 0.7$$

**Solution**   If $p = 0.7$, then $(1 - p) = 0.3$. If $n$ is 9 and $x$ is 5, then $n - x = 4$. $(1 - p)^{n-x}$ becomes

$$0.3^4 = 0.0081 \quad \blacktriangleleft$$

## Computing the probability distribution for a binomial random variable

Having studied separately the three factors in the general formula, it is time now to put the pieces together and compute a probability distribution for a binomial random variable.

▶ **Example 7**   A box contains 100 disks of the same size and shape. Sixty-five of the disks are silver and 35 are brass. An experiment consists of selecting with replacement seven disks. Success on a trial is getting a silver disk. Let $X$ be the number of successes in the seven trials.

a. Justify modeling $X$ as a binomial random variable.
b. Compute the probability distribution for $X$ and round each probability to three places.
c. State in words the meaning of $P(4)$.

**Solution**   a. Because 65 percent of the disks are silver and the disks are selected with replacement, the value 0.65 will be used as the probability of success on each trial. The trials can be modeled as independent because the sampling is done with replacement.

b. Using row 8 in Appendix Table 2, the binomial coefficients are:

$$1 \quad 7 \quad 21 \quad 35 \quad 35 \quad 21 \quad 7 \quad 1$$

$$P(0) = 1(0.65)^0(0.35)^7 = 0.001$$
$$P(1) = 7(0.65)^1(0.35)^6 = 0.008 \qquad p = 0.65$$
$$P(2) = 21(0.65)^2(0.35)^5 = 0.047 \qquad 1 - p = 0.35$$
$$P(3) = 35(0.65)^3(0.35)^4 = 0.144 \qquad n = 7$$
$$P(4) = 35(0.65)^4(0.35)^3 = 0.268$$
$$P(5) = 21(0.65)^5(0.35)^2 = 0.298$$
$$P(6) = 7(0.65)^6(0.35)^1 = 0.185$$
$$P(7) = 1(0.65)^7(0.35)^0 = 0.049$$

The probabilities are shown in Table 4-18.

**TABLE 4-18   The Probability Distribution for $X$**

| $x$ | 0 | 1 | 2 | 3 | 4 | 5 | 6 | 7 |
|---|---|---|---|---|---|---|---|---|
| $P(x)$ | 0.001 | 0.008 | 0.047 | 0.144 | 0.268 | 0.298 | 0.185 | 0.049 |

c. $P(4) = 0.268$ means "the probability is about 27% that four of the selected discs will be silver and three will be brass."   ◄

## Using Appendix Table 3 to find binomial probabilities

The physical details of an experiment that yield a binomial random variable are important, as far as the calculation of $P(x)$ is concerned, only to such extent that they specify the values of $n$ and $p$. To illustrate, the experiments described below are all different, but the random variables associated with them can all be modeled as binomial with $n = 5$ and $p = 0.5$.

| Random Experiment | The Associated Random Variable |
|---|---|
| Flip a legal coin 5 times. | $W$ is the number of heads. |
| Select, with replacement, five cards from a deck of playing cards. | $X$ is the number of red cards. |
| Roll a pair of dice five times. | $Y$ is the number of even numbers. |

| $n$ | $x$ | $p = 0.50$ |
|-----|-----|-----------|
| 5   | 0   | 0.031 |
|     | 1   | 0.156 |
|     | 2   | 0.312 |
|     | 3   | 0.312 |
|     | 4   | 0.156 |
|     | 5   | 0.031 |

**Figure 4-18.** $n = 5$ and $p = 0.50$.

The probability distributions for $W$, $X$, and $Y$ are the same, namely the binomial distribution with $n = 5$ and $p = 0.5$. Appendix Table 3 contains the binomial probability distributions for a few selected values of $n$ and $p$. Specifically the table contains probabilities for values of $n$ between 2 and 20 inclusive, and for $p = 0.10$, $0.20, 0.30, \ldots, 0.90$.

The portion of Appendix Table 3 for $n = 5$ and $p = 0.50$ is shown in Figure 4-18. From the table: $P(3) = 0.312 \approx 31$ percent. Relating this value to the examples above "The probability is about 31 percent that 3 heads are obtained in the 5 flips." "The probability is about 31 percent that 3 of the 5 cards are red." "The probability is about 31 percent that 3 of the 5 outcomes are even numbers."

▶ **Example 8**

The mayor of a large metropolitan area in the East formed a panel to study the crime rate in the area. In the panel's report it was stated that 40 percent of the residents in a large section of the midtown district had suffered some kind of loss of personal property within the past five years. Subsequent to this report a sample of twelve residents in this district was taken. $X$ is the number in the sample that suffered a loss of personal property within the past five years. Compute:

**a.** $P(5)$
**b.** $P(X \leq 3)$
**c.** $P(X > 6)$

**Solution**

| $n$ | $x$ | $p = 0.40$ |
|-----|-----|-----------|
| 12  | 0   | 0.002 |
|     | 1   | 0.017 |
|     | 2   | 0.064 |
|     | 3   | 0.142 |
|     | 4   | 0.213 |
|     | 5   | 0.227 |
|     | 6   | 0.177 |
|     | 7   | 0.101 |
|     | 8   | 0.042 |
|     | 9   | 0.012 |
|     | 10  | 0.002 |
|     | 11  | 0.0+ |
|     | 12  | 0.0+ |

**Figure 4-19.** Binomial probabilities for $n = 12$ and $p = 0.40$.

The portion of Appendix Table 3 for $n = 12$ and $p = 0.40$ is given in Figure 4-19.

**a.** $P(5) = 0.227$
The probability is about 23 percent that five of these residents have suffered a loss of personal property.

**b.** $P(X \leq 3) = P(0) + P(1) + P(2) + P(3)$
$$= 0.002 + 0.017 + 0.064 + 0.142$$
$$= 0.225$$

The probability is about 23 percent that three or less of these residents have suffered a loss of personal property.

**c.** $P(X > 6) = P(7) + P(8) + \cdots + P(12)$
$$= 0.101 + 0.042 + \cdots + 0.0+$$
$$= 0.157$$

The probability is about 16 percent that more than six have suffered a loss of personal property.    ◀

Notice that $P(11)$ and $P(12)$ are $0.0+$. Such entries in the table mean that these probabilities are 0 to three decimal places. For example, with $n = 12$ and $p = 0.40$, $P(11) = 11(0.4)^{11}(0.6)^1 = 0.0002768\ldots$. To three decimal places, this number rounds to 0.000, and a $0.0+$ is shown in the table.

## Exercises 4-3    Set A

*In Exercises 1–6, argue that X, the number of successes, can be modeled as binomial. Show that assumptions 1 and 2 are acceptable for the experiment on which X is based.*

**1.** Karen LaBonte subjects ten electronic components to extreme temperatures to investigate the durability of such components. She feels that each component has an 80 percent chance of surviving the testing.

   X = the number of components that survive the testing.

**2.** Pat Kelley knows that a stamping machine has a history of correctly cutting 95 percent of its output of cardboard boxes. He checks 100 boxes cut by the machine and counts the number that are correctly cut.

   X = the number of boxes correctly cut.

**3.** A study is made of obesity using the residents in a metropolitan area in the South. A random sample of 30 residents is taken and the number of people that weigh more than their recommended weight as determined by the Council on Weight Control is counted.

   X = the number of people in the sample classified as obese.

**4.** A study is made on manual dexterity using students in the Nantuckett Elementary School District. A random sample of 50 students is taken and the number that perform a prescribed activity in less than 30 seconds is counted.

   X = the number of students that perform the activity in less than 30 seconds.

**5.** A test consists of 25 multiple-choice questions with four possible answers to each question. The answer to each question is determined by selecting (with replacement) a card from a deck of playing cards. If a spade is drawn, choice (a) is selected; if a heart is drawn, choice (b) is selected; if a club is drawn, choice (c) is selected; and if a diamond is drawn, choice (d) is selected.

   X = the number of correct test answers.

**6.** As a result of a study in a large midwestern city, the claim is made that 30 percent of the vehicles in the city have at least one faulty tire. A police inspection station stops and inspects the tires of 40 vehicles in the city.

   X = the number of vehicles that have at least one faulty tire.

*In Exercises 7–12, compute the number of ways in which x successes can occur in n trials.*

**7.** Three successes in 12 trials

**8.** Six successes in 10 trials

**9.** Eight successes in 20 trials

**10.** Twelve successes in 15 trials

**11.** Two successes in 30 trials

**12.** Twenty-three successes in 25 trials

*In Exercises 13–16, use the formula for $\binom{n}{x}$ and compare to the entries in Appendix Table 2 for the indicated values of n and x.*

**13.** $n = 5$
   $x = 1$ and $2$

**14.** $n = 6$
   $x = 2$ and $5$

**15.** $n = 9$
   $x = 0$ and $3$

**16.** $n = 12$
   $x = 4$ and $9$

*In Exercises 17–20, let X be a binomial random variable. Evaluate the $p^x$ factor for the following probabilities:*

**17.** $P(3)$, if $n = 6$ and $p = 0.3$

**18.** $P(5)$, if $n = 8$ and $p = 0.5$

**19.** $P(2)$, if $n = 10$ and $p = 0.72$

**20.** $P(4)$, if $n = 20$ and $p = 0.22$

*In Exercises 21–24, let X be a binomial random variable. Evaluate the $(1 - p)^{n-x}$ factor for the following probabilities:*

**21.** $P(8)$, if $n = 12$ and $p = 0.6$

**22.** $P(7)$, if $n = 10$ and $p = 0.2$

**23.** $P(3)$, if $n = 6$ and $p = 0.37$

**24.** $P(2)$, if $n = 4$ and $p = 0.07$

*In Exercises 25–30, complete the following for the binomial random variables described in these exercises.*
*a. List the possible values for the random variable.*
*b. Compute a probability distribution.*
*c. In the context of the experiment, state the meaning of the indicated probability.*

**25.** A six-faced cube has two faces painted red and four faces painted white. When the cube is rolled, a success is obtained if a red face is showing on top. An experiment consists of rolling the cube $n = 3$ times, and $X$ is the number of successes. State the meaning of $P(1)$.

**26.** One side of a coin is painted red and the other side is painted white. The coin is biased so that when it is flipped, the probability the red face will show on top is 0.65. The coin is flipped $n = 4$ times and $X$ is the number of tosses that yield red faces. State the meaning of $P(3)$.

**27.** A large box contains candy coated chocolates. Twenty-five percent of them have yellow shells. Let $Y$ be the number of yellow candies in a random sample of five of these candies. State the meaning of $P(2)$.

**28.** Past production records indicate that the relative frequency of substandard batteries is about 12 percent for 2R3007-type dry cell batteries. Let $Y$ be the number of substandard batteries in the next five batteries that are produced. State the meaning of $P(1)$.

**29.** According to Dean Kent Merrill, 28 percent of the students at Southern Normal University are post-graduates. Six randomly sampled students are surveyed. Let $W$ be the number of students in the survey that are postgraduates. State the meaning of $P(3)$.

**30.** Approximately 58 percent of commercially sold pain relief medications contain compound J. A sample is taken of eight randomly selected brands of pain relief medications. Let $W$ be the number of medications in the sample that contain compound J. State the meaning of $P(5)$.

**31.** A national poll suggests that 32 percent of American households do not have any pets. A survey is made of 10 residents of Evanston and $X$ is the number in the sample that have no pets.
a. Argue that $X$ can be modeled as binomial.
b. Compute $P(3)$ to three decimal places, using $p = 0.32$.
c. State the meaning of $P(3)$.

**32.** Diane Gray is a field worker for a federal social services department. The Washington headquarters reported that 35 percent of U.S. citizens older than 65 years do not have insurance that covers the cost of medication. Diane sampled 12 names from her files of citizens older than 65 and $X$ was the number in the sample that did not have insurance that covered the cost of medications.
a. Argue that $X$ can be modeled as binomial.
b. Compute $P(4)$ to three decimal places, using $p = 0.35$.
c. State the meaning of $P(4)$.

**33.** A survey on religion in America indicates that 60 percent of Americans consider themselves to be Protestants. Assume that 60 percent of the adults in St. Louis are Protestants. If 10 residents are randomly selected in St. Louis, let $Y$ be the number in the sample that are Protestants.
a. Argue that $Y$ can be modeled as binomial.
b. Compute $P(6)$ to three decimal places, using $p = 0.60$.
c. State the meaning of $P(6)$.

**34.** Vicky Valerin manages Garbo's Sandwich Shop. To help her accurately order bread for the shop, she kept a record of the proportions of orders for wheat, rye, white, and French. Based on the data, she concluded that 64 percent of the sandwich orders were for wheat bread. If 12 sandwich orders are sampled from tomorrow's business, let $Y$ be the number in the sample that specify wheat bread.
a. Argue that $Y$ can be modeled as binomial.
b. Compute $P(8)$ to three decimal places, using $p = 0.64$.
c. State the meaning of $P(8)$.

*In Exercises 35–40, use Appendix Table 3 to compute the indicated probabilities for binomial random variables.*

**35.** With $n = 4$ and $p = 0.40$, find:
a. $P(3)$    b. $P(X \geq 2)$    c. $P(X \leq 1)$    d. $P(X < 1)$

**36.** With $n = 5$ and $p = 0.30$, find:
a. $P(3)$    b. $P(X \geq 2)$    c. $P(X \leq 3)$    d. $P(X < 3)$

**37.** With $n = 7$ and $p = 0.60$, find:
a. $P(5)$    b. $P(X > 4)$    c. $P(X < 3)$    d. $P(X \geq 4)$

**38.** With $n = 10$ and $p = 0.70$, find:
a. $P(7)$    b. $P(X > 8)$    c. $P(X < 4)$    d. $P(X \geq 8)$

**39.** With $n = 15$ and $p = 0.50$, find:
a. $P(8)$    b. $P(X > 10)$    c. $P(4 \leq X \leq 9)$

**40.** With $n = 20$ and $p = 0.40$, find:
a. $P(8)$    b. $P(X < 6)$    c. $P(6 \leq X \leq 10)$

# Exercises 4-3    Set B

*In Exercises 1 and 2, use Data Set I in Appendix B.*

**1.** In Data Set I, approximately 49% of the subjects are female. Use 0.49 as the probability that a randomly sampled subject is female (labeled 2 in the sex column).
a. Suppose $n = 5$ subjects are randomly sampled and $X$ is the number that are female (success). Compute the probability distribution for $X$.
b. Use a table of random numbers to select ten different samples, each of size $n = 5$ from the data set, and record the number of females in each sample.

| Sample number | 1 | 2 | 3 | 4 | 5 | 6 | 7 | 8 | 9 | 10 |
|---|---|---|---|---|---|---|---|---|---|---|
| Number of females in sample | | | | | | | | | | |

c. Divide the number of samples that had exactly two females by 10 and compare the quotient with $P(2)$ in part (a).

d. Repeat part (c) using samples that had exactly three females and compare the quotient with $P(3)$ in part (a).

2. In Data Set I, approximately 15% of the subjects are individuals with cancer. Use 0.15 as the probability that a randomly sampled individual has cancer (labeled 1 in the cancer column).

a. Suppose $n = 6$ subjects are randomly sampled and $Y$ is the number that have cancer (success). Compute, to three decimal places, the probability distribution for $Y$.

b. Use a table of random numbers to select ten different samples, each of size $n = 6$ from the data set, and record the number of individuals with cancer in each sample.

| Sample number | 1 | 2 | 3 | 4 | 5 | 6 | 7 | 8 | 9 | 10 |
|---|---|---|---|---|---|---|---|---|---|---|
| Number of individuals in sample with cancer | | | | | | | | | | |

c. Compare the proportion of samples that had exactly one individual with cancer with $P(1)$.

d. Compare the proportion of the samples that had exactly 0 individuals with cancer with $P(0)$.

*In Exercises 3 and 4, use the information in Figure 4-20.*

3. According to the data summarized in the figure, 26.3 percent of vacation homes are owned by individuals from 35 through 44 years of age. Assume the 26.3 percent figure applies to vacation homes in the Fort Lauderdale area. Suppose $n = 8$ owners of these homes are randomly selected and

$$X = \text{the number in the sample that are from 35 through 44 years old.}$$

a. Argue that $X$ can be modeled as binomial.

b. To three decimal places, compute the probability distribution for $X$.

c. State the meaning of $P(4)$.

4. According to the data in the figure, 20.5 percent of vacation homes are owned by individuals from 55 through 64 years old. Assume this figure applies to vacation homes in the Santa Cruz area. Suppose $n = 9$ owners of these homes are randomly selected and let

$$Y = \text{the number in the sample that are from 55 through 64 years old.}$$

# Who owns vacation homes

Percentage owning a vacation home

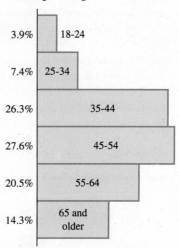

**Figure 4-20.**  Data on ownership of vacation homes. *Resort Research and Marketing Inc.*

a. Argue that $Y$ can be modeled as binomial.

b. To three decimal places, compute the probability distribution for $Y$.

c. State the meaning of $P(2)$.

## 4-4 Binomial Distributions: Probability Histograms, Means, and Variances

**Key Topics**

1. *The probability histogram for a binomial random variable*

2. *The mean of a binomial random variable*

3. *The variance of a binomial random variable*

## S T A T I S T I C S   I N   A C T I O N

Katie Konradt has a PC that she uses in her work as vice president of personnel for a corporation that has 467 department stores nationwide. The number of employees in the corporation is approximately 312,000. Katie frequently needs a table of random numbers to help her take random samples from the computer records of employees. She therefore devised a program to have her PC generate random numbers in which each number consisted of a block of six digits.

Katie wanted to check the validity of the program and decided to check the number of even digits (0, 2, 4, 6, or 8) that occurred in 36 randomly selected locations on a printout page from the computer. She knew that the number of even digits she would get in the sample of 36 was a value of a random variable $X$ and if the program were operating correctly, $X$ could be modeled as binomial. Furthermore

because the relative frequencies of even and odd digits on a page should be approximately the same, for this variable:

$p = 0.5$     The probability a given digit is even is 0.5.

$n = 36$      A total of 36 digits would be checked.

For this experiment, Katie also knew that:

$\mu_X = 18$     The expected number of even digits is 18.

$\sigma_X = 3$     The standard deviation for the distribution of $X$ is 3.

Katie therefore decided that if she observed between 12 and 24 even digits in the sample of 36, she would conclude that the program was operating correctly.

### The probability histogram for a binomial random variable

In Section 4-1, a procedure was introduced for constructing probability histograms for discrete random variables. This procedure can be used to construct a probability histogram for a binomial random variable.

▶ **Example 1**    Employee records at Titan Electronics Corporation indicate 58 percent of the total work force are women. A random sample of five employees is to be selected to work with management to improve working conditions. Let $X$ be the number of women selected to serve on the employee part of the committee.

**a.** Compute to three decimal places the probability distribution for $X$.

**b.** Construct a probability histogram.

**c.** Give the meaning of $P(3)$ in the context of the problem.

**Solution**   **Discussion.** Assuming Titan Electronics Corporation has a large work force, it is reasonable to model $X$ as binomial, with $n = 5$ and $p = 0.58$.

**a.** $P(0) = 1(0.58)^0(0.42)^5 = 0.013$    To three decimal places
$P(1) = 5(0.58)^1(0.42)^4 = 0.090$
$P(2) = 10(0.58)^2(0.42)^3 = 0.249$
$P(3) = 10(0.58)^3(0.42)^2 = 0.344$
$P(4) = 5(0.58)^4(0.42)^1 = 0.238$
$P(5) = 1(0.58)^5(0.42)^0 = 0.066$

**b.** The probability histogram for this distribution is shown in Figure 4-21.

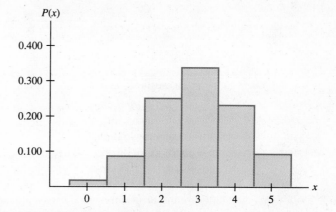

**Figure 4-21.**   A probability histogram for the committee problem.

**c.** From part (a), $P(3) = 0.344$. There is about a 34.4 percent chance that three of the five employees selected will be women.   ◄

## The mean of a binomial random variable

Recall from Section 4-2 the formula for computing $\mu_X$:

$$\mu_X = \sum x_i \cdot P(x_i)$$

There is a much simpler version of this formula for binomial random variables. The experiment described in Example 2 will be used to motivate this simpler version.

▶ **Example 2**   A fair coin is flipped six times and $X$ is the number of heads observed.

**a.** Compute a probability distribution for $X$.

**b.** Compute $\mu_X$.

**Solution**    **Discussion.** The experiment qualifies as binomial with $n = 6$ and $p = 0.5$.

**a.** $P(0) = 1(0.5)^0(0.5)^6 = 0.015625$

$P(1) = 6(0.5)^1(0.5)^5 = 0.093750$

$P(2) = 15(0.5)^2(0.5)^4 = 0.234375$

$P(3) = 20(0.5)^3(0.5)^3 = 0.312500$

$P(4) = 15(0.5)^4(0.5)^2 = 0.234375$

$P(5) = 6(0.5)^5(0.5)^1 = 0.093750$

$P(6) = 1(0.5)^6(0.5)^0 = 0.015625$

These probabilities will be used to compute $\mu_X$ with the formula from Section 4-2.

**b.** The work is organized in Table 4-19.

$$\mu_X = 0 + 0.093750 + \cdots + 0.093750$$

$$= 3.000000$$

Thus the expected value of $X$ is 3.    ◄

TABLE 4-19    Calculations for $\mu_X$

| $x$ | $P(x)$ | $x \cdot P(x)$ |
|---|---|---|
| 0 | 0.015625 | 0.000000 |
| 1 | 0.093750 | 0.093750 |
| 2 | 0.234375 | 0.468750 |
| 3 | 0.312500 | 0.937500 |
| 4 | 0.234375 | 0.937500 |
| 5 | 0.093750 | 0.468750 |
| 6 | 0.015625 | 0.093750 |

In Example 2, $n = 6$ and $p = 0.5$.    The product of $n$ and $p$ is $\mu_X$.

$$\mu_X = (6)(0.5)    \qquad \mu_X = np$$

$$= 3.0$$

This relationship is not accidental and can be shown to hold for any binomial random variable.

## $\mu_X$ for a Binomial Random Variable

If $X$ is a binomial random variable for $n$ trials and probability of success on every trial equal to $p$, then:

$$\mu_X = np$$

Examples (a) through (c) illustrate the equation for three binomial random variables.

|  | Number of Trials, $n$ | Probability of Success, $p$ | $np = \mu_X$ |
|---|---|---|---|
| **a.** | 15 | 0.7 | $15(0.7) = 10.5$ |
| **b.** | 32 | 0.15 | $32(0.15) = 4.8$ |
| **c.** | 140 | 0.58 | $140(0.58) = 81.2$ |

## The variance of a binomial random variable

Recall from Section 4-2 the formula for computing $\sigma_X^2$, the variance of a discrete random variable $X$:

$$\sigma_X^2 = \sum x_i^2 \cdot P(x_i) - (\mu_X)^2$$

There is a much simpler version of this formula for binomial random variables.

▶ **Example 3**  A legal coin is flipped six times and $X$ is the number of heads observed. Compute $\sigma_X^2$.

**Solution**  **Discussion.** The probabilities for the seven possible values of $X$ were computed in Example 2 and are repeated in Table 4-20.

$$\sigma_X^2 = 0 + 0.09375 + \cdots + 0.5625 - (3.0)^2$$
$$= 10.5 - 9$$
$$= 1.5 \quad ◀$$

TABLE 4-20   Calculations for $\sigma_X^2$

| $x$ | $P(x)$ | $x^2$ | $x^2 \cdot P(x)$ |
|---|---|---|---|
| 0 | 0.015625 | 0 | 0.000000 |
| 1 | 0.093750 | 1 | 0.093750 |
| 2 | 0.234375 | 4 | 0.937500 |
| 3 | 0.312500 | 9 | 2.812500 |
| 4 | 0.234375 | 16 | 3.750000 |
| 5 | 0.093750 | 25 | 2.343750 |
| 6 | 0.015625 | 36 | 0.562500 |

In Example 3, $n = 6$, $p = 0.5$, and $1 - p = 0.5$. The product of $n$, $p$, and $1 - p$ is $\sigma_X^2$. That is,

$$\sigma_X^2 = 6(0.5)(0.5) = 1.5$$

This relationship, like the one for $\mu_X$, is not accidental and can be shown to hold for any binomial random variable.

### $\sigma_X^2$ for a Binomial Random Variable

If $X$ is a binomial random variable for $n$ trials and probability of success on every trial equal to $p$, then:

$$\sigma_X^2 = np(1 - p)$$

Furthermore: $\sigma_X = \sqrt{np(1 - p)}$     Standard deviation = $\sqrt{\text{Variance}}$

Examples (a)* through (c)* illustrate the use of these equations for the random variables in examples (a) through (c). For each example, $\sigma_X$ is approximated to one decimal place.

| | $n$ | $p$ | $1-p$ | $np(1-p) = \sigma_X^2$ | $\sqrt{np(1-p)} = \sigma_X$ |
|---|---|---|---|---|---|
| **a\*.** | 15 | 0.7 | 0.3 | $15(0.7)(0.3) = 3.15$ | $\sqrt{3.15} \approx 1.8$ |
| **b\*.** | 32 | 0.15 | 0.85 | $32(0.15)(0.85) = 4.08$ | $\sqrt{4.08} \approx 2.0$ |
| **c\*.** | 140 | 0.58 | 0.42 | $140(0.58)(0.42) = 34.104$ | $\sqrt{34.104} \approx 5.8$ |

▶ **Example 4**    According to a report from the Bureau of Labor Statistics, 42 percent of the current labor force prefers a work week consisting of four 10-hour days. Assume that 0.42 is the probability that a randomly selected employee of Titanic Chemical Corporation prefers a four 10-hour-day work week. Suppose eight of these employees are randomly selected, and $Y$ is the number in the sample that have this opinion.

a. List the possible values for $Y$.
b. Compute the probability distribution for $Y$.
c. Compute $\mu_Y$ to one decimal place and interpret it in the context of the problem.
d. Compute $\sigma_Y$ to one decimal place.
e. Construct a probability histogram. On the graph, locate $\mu_Y$, $\mu_Y + \sigma_Y$, and $\mu_Y - \sigma_Y$.

**Solution**    a. With $n = 8$ trials, the possible values of $Y$ are:

$$0, \quad 1, \quad 2, \quad 3, \quad 4, \quad 5, \quad 6, \quad 7, \quad \text{and} \quad 8$$

b. With $n = 8$, $p = 0.42$, $1 - p = 0.58$, and the binomial coefficients from Pascal's Triangle:

$$1 \quad 8 \quad 28 \quad 56 \quad 70 \quad 56 \quad 28 \quad 8 \quad 1$$

$$P(0) = 1(0.42)^0(0.58)^8 = 0.013$$
$$P(1) = 8(0.42)^1(0.58)^7 = 0.074$$
$$P(2) = 28(0.42)^2(0.58)^6 = 0.188$$
$$P(3) = 56(0.42)^3(0.58)^5 = 0.272$$
$$P(4) = 70(0.42)^4(0.58)^4 = 0.246$$
$$P(5) = 56(0.42)^5(0.58)^3 = 0.143$$
$$P(6) = 28(0.42)^6(0.58)^2 = 0.052$$
$$P(7) = 8(0.42)^7(0.58)^1 = 0.011$$
$$P(8) = 1(0.42)^8(0.58)^0 = 0.001$$

Each probability is rounded to three decimal places.

c. $\mu_Y = 8(0.42) \approx 3.4$        $\mu_Y = np$
An average of about 3.4 employees in samples of size 8 will prefer a four 10-hour-day work week.

**d.** $\sigma_Y = \sqrt{8(0.42)(0.58)}$    $\sigma_Y = \sqrt{np(1-p)}$

$= \sqrt{1.9488} \approx 1.4$

**e.** A probability histogram is shown in Figure 4-22. With $\mu_Y = 3.4$ and $\sigma_Y = 1.4$:

$$\mu_Y + \sigma_Y = 3.4 + 1.4 = 4.8$$

$$\mu_Y - \sigma_Y = 3.4 - 1.4 = 2.0 \quad \blacktriangleleft$$

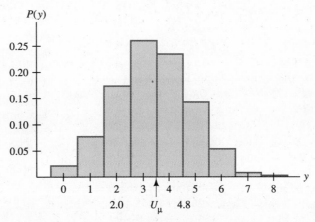

**Figure 4-22.**   A probability histogram for the employee survey problem.

## Exercises 4-4    Set A

*In Exercises 1–4, X is a binomial random variable.*
*For each exercise:*

a. *Compute to three decimal places the probability distribution for X.*

b. *Compute $\mu_X$ to one decimal place.*

c. *Compute $\sigma_X$ to one decimal place.*

d. *Construct a probability histogram.*

e. *Locate $\mu_X$, $\mu_X + \sigma_X$, and $\mu_X - \sigma_X$ on the histogram.*

**1.** The experiment consists of $n = 5$ trials and $p = 0.36$.

**2.** The experiment consists of $n = 6$ trials and $p = 0.57$.

**3.** The experiment consists of $n = 7$ trials and $p = 0.44$.

**4.** The experiment consists of $n = 8$ trials and $p = 0.68$.

**5.** A box contains five disks of the same size. The digits 1, 2, 3, 4, and 5 are each painted on a disk. An experiment consists of $n = 4$ trials. On each trial, a disk is selected, the digit is recorded, and the disk is replaced. Let $X$ be the number of even digits recorded in the experiment.

a. Argue that $X$ can be modeled as binomial.

b. Compute the probability distribution for $X$.

c. Compute $\mu_X$ to one decimal place.

d. Interpret the meaning of $\mu_X$.

e. Compute $\sigma_X$ to one decimal place.

f. Construct a probability histogram and locate $\mu_X$ on the graph.

**6.** Cecilia Lause has a coin that she biased so that when the coin is tossed the probability is approximately 0.65 that the head will show on top. (The approximation of $P$(head) was determined using a large number of tosses.) Let $X$ be the number of heads obtained in the next five tosses of the coin.

a. Argue that $X$ can be modeled as binomial.
b. To three decimal places compute the probability distribution for $X$.
c. Compute $\mu_X$ to one decimal place.
d. Interpret the meaning of $\mu_X$.
e. Compute $\sigma_X$ to one decimal place.
f. Construct a probability histogram and locate $\mu_X$ on the graph.

7. A report from the Pennsylvania Department of Motor Vehicles (DMV) claims that 58 percent of citations issued for moving violations are to drivers younger than the age of 21 years. Lillie Chalmers works in the records section of the Pennsylvania DMV. She will sample six of the moving citations that will be issued tomorrow. Let $Y$ be the number in the sample that are to drivers younger than 21 years of age.

a. To three decimal places, compute the probability distribution for $Y$.
b. Compute $\mu_Y$ to one decimal place.
c. Interpret the meaning of $\mu_Y$.
d. Compute $\sigma_Y$ to one decimal place.
e. Construct a probability histogram.
f. Locate $\mu_Y$, $\mu_Y + \sigma_Y$, and $\mu_Y - \sigma_Y$ on the histogram.

8. A recent epidemic caused by virus D infected 72 percent of all newborn infants in the hospitals of a large metropolitan area. Carol Nielsen, a district supervisor of infant care facilities of these hospitals, will study the records of seven infants born during this period. Let $Y$ be the number in the sample that were afflicted by virus D.

a. To three decimal places compute the probability distribution for $Y$.
b. Compute $\mu_Y$ to one decimal place.
c. State in words the meaning of $\mu_Y$.
d. Compute $\sigma_Y$ to one decimal place.
e. Construct a probability histogram.
f. Locate $\mu_Y$, $\mu_Y + \sigma_Y$, and $\mu_Y - \sigma_Y$ on the histogram.

9. At a major university in the Midwest, the records at the registrar's office provided the following:

32 percent of the students are freshmen.
25 percent of the students are sophomores.
20 percent of the students are juniors.
15 percent of the students are seniors.
 8 percent of the students are graduate students.

Suppose eight of the students at this school are randomly sampled, and $W$ is the number in the sample that are freshmen.

a. Argue that $W$ can be modeled as binomial.
b. Compute the probability distribution for $W$.
c. State the meaning of $P(3)$.
d. Compute $\mu_W$ to one decimal place.
e. Interpret the meaning of $\mu_W$.
f. Compute $\sigma_W$ to one decimal place.
g. Construct a probability histogram and locate $\mu_W$ on it.

10. A large study of the residents in a southern state was made to determine the public's attitude on mandatory testing of individuals for dangerous social diseases. Specifically the question was asked whether such tests should be mandated for all public school teachers and administrators. The results of the study showed the following:

15 percent strongly favor such tests.
28 percent favor such tests.
36 percent have no opinion on such tests.
15 percent oppose such tests.
 6 percent strongly oppose such tests.

Suppose nine of the residents of this state are randomly sampled, and $W$ is the number in the sample that "favor such tests."

a. Argue that $W$ can be modeled as binomial.
b. Compute the probability distribution for $W$.
c. State the meaning of $P(2)$.
d. Compute $\mu_W$ to one decimal place.
e. State the meaning of $\mu_W$.
f. Compute $\sigma_W$ to one decimal place.
g. Construct a probability histogram and locate $\mu_W$ on it.

11. A large load of sweet corn in the husk was delivered to Swenson's Supermarket. It was reported that about 15 percent of the ears had some type of blemish. To build a large display of the corn,

the produce manager put 12 randomly selected ears in each of several hundred paper bags. Let $V$ be the number of ears of corn in a particular bag of 12 that have some type of blemish.

a. Compute $P(3)$ to three decimal places.

b. Interpret the meaning of $P(3)$.

c. Compute $\mu_V$ to one decimal place.

d. State the meaning of $\mu_V$.

e. Compute $\sigma_V$ to one decimal place.

12. A bottling machine underfills 12-ounce cans of soda approximately 8 percent of the time. During the production of several thousand cans of this soda, Joe Thompkins selects 40 cans and checks whether the cans are underfilled. Let $V$ be the number of underfilled cans of soda in the next sample of 40 cans.

a. Compute $P(4)$ to three decimal places.

b. Interpret the meaning of $P(4)$.

c. Compute $\mu_V$ to one decimal place.

d. State the meaning of $\mu_V$.

e. Compute $\sigma_V$ to one decimal place.

13. Figure 4-23 is a probability histogram for a discrete random variable $X$. Use the probabilities from the histogram.

a. Compute $\mu_X$ to one decimal place.

b. Compute $\sigma_X$ to one decimal place.

14. Figure 4-24 is a probability histogram for a discrete random variable $X$. Use the probabilities from the histogram.

a. Compute $\mu_X$ to one decimal place.

b. Compute $\sigma_X$ to one decimal place.

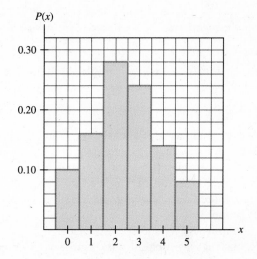

**Figure 4-23.** Probability histogram for $X$.

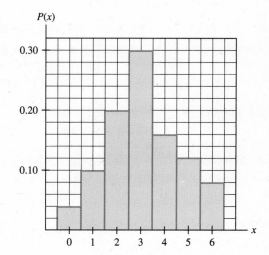

**Figure 4-24.** Probability histogram for $X$.

15. You and a fellow worker flip coins on a daily basis to see which of you must pay for the morning coffee. After losing ten mornings in a row, you ask to see the "favorite coin" your friend has been using. You flip the coin ten times and get two heads. You flip the coin again ten times and get two heads. What conclusion, based on your knowledge of probability, might you make about your friend's favorite coin? What quantitative analysis can you supply to support your conclusion?

16. In a game of dice, one of the players introduces a new pair of dice. She then proceeds to get a total of seven on each of the next five rolls of the dice. What conclusion, based on your knowledge of probability, might you make about the new pair of dice? What quantitative analysis can you supply to support your conclusion?

## Exercises 4-4   Set B

*In Exercises 1 and 2, use Data Set I, Appendix B.*

**1.** Use random numbers to sample $n = 10$ subjects. Let:

  $X$ = the number in the sample that smoke.

These individuals are labeled (1) in the "Smoker" column of the data. For the entire data set, the proportion of smokers is approximately 53 percent.
a. Argue that $X$ can be modeled as binomial.
b. Compute the probability distribution for $X$.
c. Compute $\mu_X$ to one decimal place.
d. Compute $\sigma_X$ to one decimal place.
e. Construct a probability histogram.

**2.** Use random numbers to sample $n = 12$ subjects. Let:

  $Y$ = the number in the sample that have cancer.

These individuals are labeled (1) in the "Cancer" column of the data. For the entire data set, the proportion of individuals that have cancer is approximately 15 percent.
a. Argue that $Y$ can be modeled as binomial.
b. Compute the probability distribution for $Y$.
c. Compute $\mu_Y$ to one decimal place.
d. Compute $\sigma_Y$ to one decimal place.
e. Construct a probability histogram.

*In Exercises 3 and 4, use Data Set II, Appendix B.*

**3.** Use random numbers to sample $n = 9$ stock prices. Let:

  $Y$ = the number of stock prices that advanced.

These stock prices are labeled ($+$) in the "Net Chg" column, and approximately 46 percent advanced on the day reported.
a. Argue that $Y$ can be modeled as binomial.
b. Compute the probability distribution for $Y$.
c. Compute $\mu_Y$ to one decimal place.
d. Compute $\sigma_Y$ to one decimal place.
e. Construct a probability histogram.

**4.** Use random numbers to sample $n = 10$ stock prices. Let:

  $Y$ = the number of stock prices that remained unchanged.

These stock prices are labeled (0) in the "Net Chg" column, and approximately 26 percent were unchanged on the day reported.
a. Argue that $Y$ can be modeled as binomial.
b. Compute the probability distribution for $Y$.
c. Compute $\mu_Y$ to one decimal place.
d. Compute $\sigma_Y$ to one decimal place.
e. Construct a probability histogram.

# 4-5 The Normal Probability Distribution

**Key Topics**

1. *Continuous random variables modeled as normal*

2. *The standard normal distribution*

3. *Finding a value of z corresponding to a given tabled probability*

4. *Computing probabilities for normally distributed random variables*

S T A T I S T I C S   I N   A C T I O N*

Don Cole works for the agriculture department in California. He is currently gathering data on the yields of tomatoes grown for processing as soups, catsups, and so on. Today he is investigating the yields from the Takemori Ranch in Sacramento County. Don measured off a 1-acre region from a 7500-acre field that is currently being harvested. With the assistance of the ranch foreman, one of the ten harvesters working the field was picked to harvest this sampled acre. The trucks that will haul the tomatoes from this acre have been designated, and the weights from these trucks will be given to Don.

Based on data already collected by the agriculture department, the projected state average yield this year is 25.0 tons per acre, with a standard deviation of 2.3 tons

per acre. Furthermore the data suggest that the distribution of weights per acre is approximately bell shaped. For the yield from this particular acre, Don knows the weight is a value of a continuous random variable $X$. Furthermore using the 25.0 mean, the 2.3 standard deviation, and a bell-shaped distribution, Don calculates that for this sampled acre:

$$P(22.7 < X < 27.3) \approx 0.68$$

That is, the probability is about 68 percent that the weight of tomatoes from this acre will be within one standard deviation of the mean of the distribution.

---

* Data for this example were supplied by Takemori Farms, Inc., Elk Grove, California. Courtesy of Mr. and Mrs. Takemori.

## Continuous random variables modeled as normal

The random variable $X$ for the tomato yield experiment is an example of a continuous random variable that might be modeled as normal. This means the *idealized histogram* for the random variable is approximately bell shaped. An idealized histogram can be thought of as the result of drawing a relative frequency histogram in which the number of classes is infinitely large and the class widths are infinitely small. Thus instead of the histogram having well-defined rectangles, the widths of the rectangles are imagined as being so narrow that a smooth curve can be drawn for a corresponding frequency polygon. An idealized histogram for the tomato yield example is given in Figure 4-25. Probabilities associated with normal random variables can be found using the so-called standard normal table. The standard normal table is Appendix Table 4.

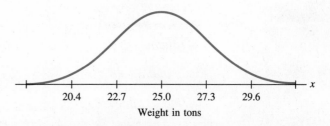

**Figure 4-25.**   An idealized histogram for the tomato yield example.

## The standard normal distribution

Figure 4-26 is a graph of the idealized histogram for *the standard normal distribution,* or Z-*distribution.* There is a special mathematical equation that generates this very useful idealized histogram. There are several features of this distribution that should be noted.

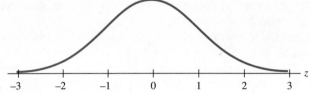

**Figure 4-26.**   The Z-distribution.

1. The corresponding random variable $Z$ is continuous.
2. The probability histogram for $Z$ is bell shaped.
3.* $\mu_Z = 0$. The subscript $Z$ is used to indicate the standard normal, or $Z$ distribution.
4.* $\sigma_Z = 1$. Again the subscript $Z$ is used for the $Z$ distribution.
5. About 68 percent of the area under the curve is between $-1.00$ and $1.00$.
6. About 95 percent of the area under the curve is between $-2.00$ and $2.00$.
7. About 99.8 percent of the area under the curve is between $-3.00$ and $3.00$.
8. The total area under the curve is 1.

Appendix Table 4 is a table of probabilities assigned to the intervals $(-\infty, z)$ by the standard normal distribution. The first 15 lines of this table are given in Figure 4-27.

Appendix Table 4 gives probabilities for values of $z$ with two decimals beyond the units place. To read from this table, first locate the units and tenths digits of the given $z$ in the left column. Then locate the hundredths digit in the top row of the table. To illustrate, suppose $z = -2.05$. Locate $-2.0$ in the left column of the table. Now move to the right under the column headed by 0.05. Read in the table 0.0202.

| *Using the Area Interpretation* | *Using the Probability Interpretation* |
| --- | --- |
| The area of the idealized histogram to the left of $z = -2.05$ is 0.0202. | The probability is about 0.02 that a standard normal random variable $Z$ is less than $-2.05$. |

---

\* The points on the curve are of the form $(z, y)$, where:

$$y = \frac{1}{\sqrt{2\pi}} \, l^{-z^2/2}, \quad \pi \approx 3.14159, \quad \text{and} \quad e \approx 2.71828$$

To establish 3 and 4 above would require using calculus and the above equation. Such mathematical techniques are beyond the scope of this text, and we simply accept these values for the mean and standard deviation of the $Z$ distribution.

| | | | | | $P(Z < z)$ | | | | | |
|---|---|---|---|---|---|---|---|---|---|---|
| $z$ | 0.00 | 0.01 | 0.02 | 0.03 | 0.04 | 0.05 | 0.06 | 0.07 | 0.08 | 0.09 |
| −3.4 | 0.0003 | 0.0003 | 0.0003 | 0.0003 | 0.0003 | 0.0003 | 0.0003 | 0.0003 | 0.0003 | 0.0002 |
| −3.3 | 0.0005 | 0.0005 | 0.0005 | 0.0004 | 0.0004 | 0.0004 | 0.0004 | 0.0004 | 0.0004 | 0.0003 |
| −3.2 | 0.0007 | 0.0007 | 0.0006 | 0.0006 | 0.0006 | 0.0006 | 0.0006 | 0.0005 | 0.0005 | 0.0005 |
| −3.1 | 0.0010 | 0.0009 | 0.0009 | 0.0009 | 0.0008 | 0.0008 | 0.0008 | 0.0008 | 0.0007 | 0.0007 |
| −3.0 | 0.0013 | 0.0013 | 0.0013 | 0.0012 | 0.0012 | 0.0011 | 0.0011 | 0.0011 | 0.0010 | 0.0010 |
| −2.9 | 0.0019 | 0.0018 | 0.0017 | 0.0017 | 0.0016 | 0.0016 | 0.0015 | 0.0015 | 0.0014 | 0.0014 |
| −2.8 | 0.0026 | 0.0025 | 0.0024 | 0.0023 | 0.0023 | 0.0022 | 0.0021 | 0.0021 | 0.0020 | 0.0019 |
| −2.7 | 0.0035 | 0.0034 | 0.0033 | 0.0032 | 0.0031 | 0.0030 | 0.0029 | 0.0028 | 0.0027 | 0.0026 |
| −2.6 | 0.0047 | 0.0045 | 0.0044 | 0.0043 | 0.0041 | 0.0040 | 0.0039 | 0.0038 | 0.0037 | 0.0036 |
| −2.5 | 0.0062 | 0.0060 | 0.0059 | 0.0057 | 0.0055 | 0.0054 | 0.0052 | 0.0051 | 0.0049 | 0.0048 |
| −2.4 | 0.0082 | 0.0080 | 0.0078 | 0.0075 | 0.0073 | 0.0071 | 0.0069 | 0.0068 | 0.0066 | 0.0064 |
| −2.3 | 0.0107 | 0.0104 | 0.0102 | 0.0099 | 0.0096 | 0.0094 | 0.0091 | 0.0089 | 0.0087 | 0.0084 |
| −2.2 | 0.0139 | 0.0136 | 0.0132 | 0.0129 | 0.0125 | 0.0122 | 0.0119 | 0.0116 | 0.0113 | 0.0110 |
| −2.1 | 0.0179 | 0.0174 | 0.0170 | 0.0166 | 0.0162 | 0.0158 | 0.0154 | 0.0150 | 0.0146 | 0.0143 |
| −2.0 | 0.0228 | 0.0222 | 0.0217 | 0.0212 | 0.0207 | 0.0202 | 0.0197 | 0.0192 | 0.0188 | 0.0183 |

**Figure 4-27.**  A portion of Appendix Table 4.

▶ **Example 1**   Find $P(Z < 1.72)$.

**Solution**   From Appendix Table 4, the table value for $z = 1.72$ is 0.9573. Thus:

$$P(Z < 1.72) = 0.9573 \qquad ◀$$

To find $P(Z > z)$ with Appendix Table 4, we use the fact that the total area under the curve is 1.0000. Therefore for a given $z$:

$$P(Z > z) = 1 - P(Z \leq z)$$

For any continuous random variable $X$:

$$P(X = x) = 0, \text{ for any } x \text{ in the sample space of } X.$$

(The area under the idealized histogram that is above the point $x$ is 0). Because $P(Z \leq z)$ can be written as $P(Z < z) + P(Z = z)$ and $Z$ is a continuous random variable, we can simply write:

$$P(Z > z) = 1 - P(Z < z)$$

As a matter of convenience, we will write $<$ instead of $\leq$ and $>$ instead of $\geq$ when computing probabilities involving the $Z$-distribution. This simplification can only be used for continuous random variables, however, and not for discrete random variables.

▶ **Example 2**   Find $P(Z > -0.65)$.

Solution   **Discussion.** The area of the shaded portion of the graph in Figure 4-28 is the desired probability.

From Appendix Table 4, for $z = -0.65$, the table value is 0.2578. Therefore:

$$P(Z > -0.65) = 1 - P(Z < -0.65)$$
$$= 1.0000 - 0.2578$$
$$= 0.7422 \quad ◀$$

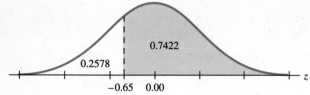

**Figure 4-28.**   $P(Z > -0.65)$.

If $z_1 < z_2$, then we can use Appendix Table 4 to find $P(z_1 < Z < z_2)$. The indicated probability is the area between $z_1$ and $z_2$.

▶ **Example 3**   Find $P(-0.72 < Z < 1.50)$.

Solution   **Discussion.** The area of the shaded portion in Figure 4-29 is the desired probability.

$$P(Z < 1.50) = P(Z < -0.72) + P(-0.72 < Z < 1.50)$$

Therefore:

$$P(-0.72 < Z < 1.50) = P(Z < 1.50) - P(Z < -0.72)$$
$$= 0.9332 - 0.2358$$
$$= 0.6974 \quad ◀$$

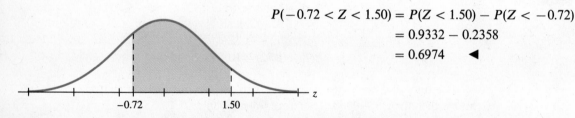

**Figure 4-29.**   $P(-0.72 < Z < 1.50)$.

## Finding a value of z corresponding to a given tabled probability

Sometimes a tabled probability that corresponds to some value of $z$ is given, and the task then is to find the value of $z$. For example:

$$P(Z < z) \approx 0.2000$$

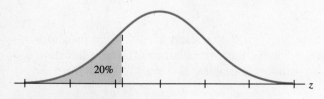

**Figure 4-30.**   $P(Z < z) \approx 0.2000$.

states that for some $z$, the area to the left is 0.2000 (or 20%), as shown in Figure 4-30.

By searching the body of Appendix Table 4, the value closest to 0.2000 is found to be 0.2005. The corresponding value of $z$ is $-0.84$.

Because no value of $z$ produces a probability closer to 0.2000, we will say $z = -0.84$ (rather than trying to interpolate in the table). Thus:

$$P(Z < -0.84) \approx 0.2000.$$

▶ **Example 4**   If $P(Z > z) = 0.3500$, find $z$.

**Solution**   Because Appendix Table 4 gives $P(Z < z)$, we subtract the given probability from 1.0000 before entering the table.

$$P(Z > z) = 0.3500 \quad \text{is equivalent to} \quad P(Z < z) = 0.6500$$

$P(Z < 0.39) = 0.6517$ and 0.6517 is the table value closest to 0.6500. Thus, $z = 0.39$.   ◀

## Computing probabilities for normally distributed random variables

A large number of phenomena exhibit distributions that are bell shaped. For such distributions, the standard normal distribution and a little algebra can be used to produce at least roughly approximate probabilities. To illustrate, consider again the tomato yield problem studied by Don Cole in the Statistics in Action example. The distribution of yields in tons is assumed bell shaped, with mean 25.0 and standard deviation 2.3.

Let $X$ be the weight, in tons, of the tomatoes harvested at the Takemora Ranch from the acre sampled by Don Cole. Suppose that at least approximately $X$ is a normal random variable, with $\mu_X = 25.0$ tons and $\sigma_X = 2.3$ tons. To find, for example, $P(X < 23.5)$, we can use the $Z$-distribution and probabilities in Appendix Table 4.

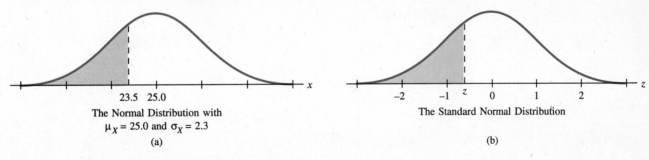

The Normal Distribution with $\mu_X = 25.0$ and $\sigma_X = 2.3$
(a)

The Standard Normal Distribution
(b)

**Figure 4-31.**   (a) $P(X < 23.5)$.   (b) $P(Z < z)$.

In Figure 4-31(a) and (b), the shaded areas represent the desired probability. To find the $z$ that cuts off the area in (b), we use the equation:

$$z = \frac{x - \mu_X}{\sigma_X}$$

With $x = 23.5$, $\mu_X = 25.0$, and $\sigma_X = 2.3$:

$$z = \frac{23.5 - 25.0}{2.3} = -0.652 \approx -0.65 \qquad \text{Express as a two-place decimal.}$$

Now:

$$P(X < 23.5) = P(Z < -0.65) \qquad \text{Using Appendix Table 4}$$
$$= 0.2578$$

The probability is about 26 percent that the acre will yield less than 23.5 tons of tomatoes.

## Changing x from a Normal Distribution to z

If $X$ is a normally distributed random variable with mean $\mu_X$ and standard deviation $\sigma_X$, then the random variable $Z = \dfrac{x - \mu_X}{\sigma_X}$ is a standard normal random variable. If $x$ is a possible value of $X$, then the corresponding value of $Z$ is:

$$z = \frac{x - \mu_X}{\sigma_X}$$

In addition to using $z$ in the above equation in conjunction with Appendix Table 4, **z can also be used to express the number of standard deviations that x is above $\mu_X$.** For example, $x = 23.5$ tons of tomatoes has a corresponding $z = -0.65$. Thus 23.5 is 0.65 standard deviations below (because $z$ is negative) the mean of the distribution.

▶ **Example 5**   Suppose $X$ is a normal random variable, and $\mu_X = 65.0$ and $\sigma_X = 3.5$. Find:

**a.** $P(X > 68.0)$
**b.** $P(62.0 < X < 66.0)$

Solution   **Discussion.** It is recommended that a sketch showing the area of interest be used for problems of this type.

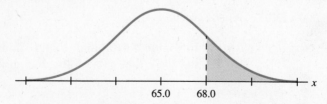

**Figure 4-32.** $P(X > 68.0)$.

**a.** The area shaded in Figure 4-32 is the desired probability. With $x = 68.0$, $\mu_X = 65.0$, and $\sigma_X = 3.5$:

$$z = \frac{68.0 - 65.0}{3.5} \qquad\qquad z = \frac{x - \mu_X}{\sigma_X}$$

| | |
|---|---|
| $= 0.86$, to two places | To agree with Appendix Table 4 |
| $P(X > 68.0) = P(Z > 0.86)$ | The area to the right |
| $= 1 - P(Z < 0.86)$ | One minus the area to the left |
| $= 1.0000 - 0.8051$ | Using Appendix Table 4 |
| $= 0.1949$ | $0.1949 \approx 19\%$ |

The probability is about 19 percent that $X$ will be more than 68.0.

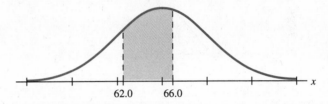

**Figure 4-33.**   $P(62.0 < X < 66.0)$.

**b.** The area shaded in Figure 4-33 is the desired probability.

| For $x_1 = 62.0$ | For $x_2 = 66.0$ |
|---|---|
| $z_1 = \dfrac{62.0 - 65.0}{3.5}$ | $z_2 = \dfrac{66.0 - 65.0}{3.5}$ |
| $= -0.86$, to two places | $= 0.29$, to two places |

| | |
|---|---|
| $P(62.0 < X < 66.0)$ | The area between |
| $= P(-0.86 < Z < 0.29)$ | In terms of the Z-distribution |
| $= P(Z < 0.29) - P(Z < -0.86)$ | Subtract the table values |
| $= 0.6141 - 0.1949$ | Using Appendix Table 4 |
| $= 0.4192$ | $0.4192 \approx 42\%$ |

The probability is about 42 percent that $X$ will be between 62.0 and 66.0.                    ◄

▶ **Example 6**   The warehouse of Savemore Discount Drugs received a large shipment of dispensers of transparent tape. The relative frequency distribution of tape lengths per dispenser is bell shaped with mean 11.5 meters and standard deviation 0.2 meters. Let $Y$ be the length of tape in a randomly selected dispenser.

**a.** Sketch and label an idealized probability histogram for $Y$.
**b.** Compute $P(Y < 11.2)$.
**c.** State the meaning of the answer in (b).
**d.** Find the value of $y$, such that $P(Y < y) = 0.9000$.
**e.** State the meaning of the answer in (d).

**Solution**    **Discussion.** The relative frequency distribution of tape lengths per dispenser is bell shaped. Therefore a normal idealized histogram will be used.

**a.** With $\mu_Y = 11.5$ and $\sigma_Y = 0.2$:

$$\mu_Y - 2\sigma_Y = 11.5 - 0.4 = 11.1$$
$$\mu_Y - \sigma_Y = 11.5 - 0.2 = 11.3$$
$$\mu_Y + \sigma_Y = 11.5 + 0.2 = 11.7$$
$$\mu_Y + 2\sigma_Y = 11.5 + 0.4 = 11.9$$

The labeled histogram is shown in Figure 4-34.
**b.** With $y = 11.2$,   $\mu_Y = 11.5$,   and   $\sigma_Y = 0.2$:

$$z = \frac{11.2 - 11.5}{0.2} = -1.50, \text{ to two decimal places}$$

$$P(Y < 11.2) = P(Z < -1.50)$$
$$= 0.0668$$

**c.** The probability is about 7 percent that a randomly selected dispenser has less than 11.2 meters of tape in it.
**d.** If $P(Y < y) = 0.9000$, then 90 percent of the area lies to the left of $y$, as shown in Figure 4-35. From Appendix Table 4: $P(Z < 1.28) = 0.8997$

$$1.28 = \frac{y - 11.5}{0.2} \qquad z = \frac{y - \mu_Y}{\sigma_Y}$$

$0.256 = y - 11.5$    Multiply both sides by 0.2.

$y = 11.756$       Add 11.5 to both sides.

$y \approx 11.8$       Round to one decimal place.

**e.** Approximately 90 percent of the tape dispensers have less than 11.8 meters of tape.    ◄

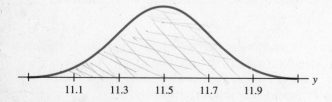

**Figure 4-34.**   A probability histogram for $Y$.

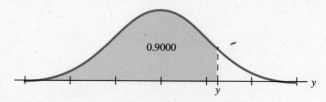

**Figure 4-35.**   $P(Y < y) = 0.9000$

## Exercises 4-5    Set A

*In Exercises 1–4, find the indicated probabilities for a standard variable Z. Use Appendix Table 4.*

**1.** a. $P(Z < -0.85)$     b. $P(Z < 1.33)$

**2.** a. $P(Z < -1.25)$     b. $P(Z < 0.33)$

**3.** a. $P(-1.00 < Z < -0.25)$
   b. $P(0.50 < Z < 1.67)$

**4.** a. $P(-0.33 < Z < 0.33)$
   b. $P(-1.00 < Z < 1.00)$

*In Exercises 5 and 6, use Appendix Table 4 to approximate z for the indicated probabilities.*

**5.** a. $P(Z < z) = 0.1000$     b. $P(Z > z) = 0.2743$

**6.** a. $P(Z < z) = 0.9861$
   b. $P(0 < Z < z) = 0.0500$

**7.** If $X$ is a continuous random variable that can be modeled as normal, with $\mu_X = 70.0$ and $\sigma_X = 5.0$:
   a. Compute $z$ for $x = 62.0$.
   b. Interpret the meaning of the $z$ in part (a).
   c. Sketch and shade an area that corresponds to $P(X < 62.0)$.
   d. Use Appendix Table 4 to find $P(X < 62.0)$.

**8.** If $X$ is a continuous random variable that can be modeled as normal, with $\mu_X = 50.0$ and $\sigma_X = 15.0$:
   a. Compute $z$ for $x = 58.0$.
   b. Interpret the meaning of the $z$ in part (a).
   c. Sketch and shade an area that corresponds to $P(X < 58.0)$.
   d. Use Appendix Table 4 to find $P(X < 58.0)$.

**9.** If $X$ is a continuous random variable that can be modeled as normal, with $\mu_X = 25.0$ and $\sigma_X = 2.8$:
   a. Find $z_1$ and $z_2$ respectively for $x_1 = 21.5$ and $x_2 = 27.1$.
   b. Sketch and shade an area that corresponds to $P(21.5 < X < 27.1)$.
   c. Use Appendix Table 4 to find the probability in (b).

**10.** If $X$ is a continuous random variable that can be modeled as normal, with $\mu_X = 128.0$ and $\sigma_X = 4.5$:
   a. Find $z_1$ and $z_2$ respectively for $x_1 = 126.5$ and $x_2 = 134.0$.
   b. Sketch and shade the area that corresponds to $P(126.5 < Z < 134.0)$.
   c. Use Appendix Table 4 to find the probability in (b).

**11.** The warehouse of Guthrie Electrical Contractors has a large supply of 50 foot rolls of No. 12 Romex Wire. The mean resistance per roll for this population is 0.68 ohms with a standard deviation of 0.08 ohms, and the distribution of resistances is approximately normal. Let $Y$ be the resistance (in ohms) of a randomly selected roll of the wire.
   a. Sketch and label an idealized histogram for $Y$.
   b. Find $z$ for $y = 0.62$.
   c. State in words the meaning of the $z$ in part (b).
   d. Compute $P(Y < 0.62)$.
   e. Interpret the probability in part (d).

**12.** The label from a can of hot dog sauce states that the net weight is 283 grams. The distribution of net weights of a large supply of cans of this hot dog sauce in a warehouse is normal, with $\mu = 285.0$ grams and $\sigma = 3.0$ grams. Let $Y$ be the weight of a can selected from this warehouse.
   a. Sketch and label an idealized histogram for $Y$.
   b. Find $z$ for $y = 283.0$.
   c. Interpret the meaning of the $z$ in part (b).
   d. Compute $P(Y < 283.0)$.
   e. Interpret the meaning of the probability in part (d).

**13.** A General Motors plant in Flint Michigan uses valves that have a mean lip-opening pressure (l. o. p.) of 10.5 pounds per square inch (psi). The standard deviation for the distribution of l. o. p. is 0.4 psi and the distribution is approximately normal. Let $X$ be the l. o. p. of a selected valve in this plant.
   a. Sketch and label an idealized histogram for $X$.
   b. Compute $P(X < 10.0)$.
   c. Compute $P(X > 10.8)$.
   d. Compute $P(10.2 < X < 11.0)$.
   e. Interpret the answer to part (d).

14. All students in grades 10 through 12 in the San Juan school district are required each year to take the APT (All Purpose Test). The relative frequency distribution of these test scores is bell shaped, with mean 80.0 and standard deviation 12.0. Let $X$ be the APT score of a randomly selected student in this school district.
    a. Sketch and label an idealized probability histogram for $X$.
    b. Compute $P(X < 65.0)$.
    c. Compute $P(X > 88.0)$.
    d. Compute $P(90.0 < X < 100.0)$.
    e. State the meaning of the answer to part (d).

15. For a National Weather Service station in Ohio, the distribution of high temperature readings for the months of July and August is bell shaped with mean 87.5° and standard deviation 6.8°. Let $X$ be the high temperature for a randomly selected day in July or August at this station.
    a. Sketch and label an idealized probability histogram for $X$.
    b. Compute $P(X < 85.0°)$.
    c. Compute $P(80.0° < X < 90.0°)$.
    d. Interpret the answer in part (c).

16. Thunder Electronics Corporation manufactures nominal 35-watt stereo receivers. Quality control records indicate that for a typical run of these receivers, the mean rating is 35.1 watts, the standard deviation of ratings is 1.5 watts, and the relative frequency distribution of wattages is reasonably bell shaped. Let $X$ be the wattage of a randomly selected receiver.
    a. Sketch and label an idealized histogram for $X$.
    b. Compute $P(X < 36.6)$.
    c. Compute $P(34.0 < X < 36.0)$.
    d. Interpret the answer in part (c).

17. Newman's Chicken Ranch raises chickens for a major supermarket chain. The weights of chickens ready for market have a mean of 4.8 pounds and a standard deviation of 0.6 pounds. The distribution of weights for a large flock ready for market is approximately bell shaped. Let $X$ be the weight of a randomly selected bird from the flock ready for market.

a. Sketch and label an idealized histogram for $X$.
b. Compute $P(4.0 < X < 5.0)$.
c. Shade the area on the idealized histogram that corresponds to the probability in part (b).
d. State the meaning of the answer in part (b).

18. Whitehouse Corporation produces 100-watt light bulbs. Quality control records indicate that the mean life of a bulb from a large production run of these bulbs is 560 hours with a standard deviation of 20.5 hours. The relative frequency distribution of the lives of these bulbs is approximately bell shaped. Let $X$ be the life of a randomly selected bulb from this production run.
    a. Sketch and label an idealized histogram for $X$.
    b. Compute $P(530 < X < 575)$.
    c. Shade the area on the idealized histogram that corresponds to the probability in part (b).
    d. State the meaning of the answer to part (b).

19. According to an Environmental Protection Agency report, a model C automobile averages 35.0 miles per gallon (mpg) on the highway. Tests on this model indicate that the standard deviation for the distribution of mpgs for cars of this model is 2.4 mpg and the distribution is bell shaped. Let $X$ be the highway mpg rating of a randomly selected model C car.
    a. Sketch and label an idealized probability histogram for $X$.
    b. Find the values of $X$ that are one standard deviation below and above $\mu$.
    c. What is the probability that a randomly selected model C car will have a highway mpg between the values obtained in part (b)?

20. According to the Association of Cosmetologists in a large city in the West, the average cost of a wash and haircut in salons in the city is \$20.25. For the distribution of costs, the standard deviation is \$3.75, and the distribution is approximately bell shaped. Let $X$ be the cost of a wash and haircut in a randomly selected salon in this city.
    a. Sketch and label an idealized probability histogram for $X$.
    b. Find the values of $X$ that are one standard deviation below and above $\mu$.

c. What is the probability that a randomly selected salon will have a cost for wash and haircut between the values obtained in part (b)?

21. One part of a vocational test administered to adults is an exercise in finger dexterity. The average time required to complete the test is 165 seconds, with a standard deviation of 21 seconds. Assume the relative frequency distribution of the times needed to complete the test is approximately bell shaped. What proportion of the adult population complete the test:
   a. in less than 180 seconds?
   b. in more than 190 seconds?
   c. in 140 to 160 seconds?

d. Sketch and label an idealized histogram of the distribution of times.

22. Sandersohn Farmers Association represents a large section of the Midwest. Last year the average corn yield produced by the members of the association was 112 bushels per acre (bu/ac) with a standard deviation of 6.3 bu/ac. Assume the relative frequency distribution of yields for members is approximately bell shaped. What proportion of the association's members had yields:
   a. less than 103 bu/ac?
   b. more than 120 bu/ac?
   c. between 105 and 115 bu/ac?
   d. Sketch and label the member's corn yield distribution.

## Exercises 4-5   Set B

1. Assume that the relative frequency distribution of heights for a population of male students at a large university is approximately normal, with $\mu = 5$ feet, 10 inches (70.0 inches), and $\sigma = 3.2$ inches. What height cuts off:
   a. the bottom 20 percent of the population?
   b. the top 30 percent of the population?
   c. the 50 percent of the population closest to the mean?
   d. Sketch and label an idealized histogram of the distribution of heights.

2. Assume that the relative frequency distribution of weights of football players that play college football is approximately normal, with $\mu = 228$ pounds and $\sigma = 27$ pounds. What weight cuts off:
   a. the bottom 30 percent of the population?
   b. the top 20 percent of the population?
   c. the 40 percent of the population closest to the mean?
   d. Sketch and label an idealized histogram of the distribution of weights.

*In Exercises 3 and 4, use the information in Figure 4-36 on the next page.*

3. Bridgestone S371 tires have a 48,000-mile rating. Assume a large production run of these tires has a

bell-shaped relative frequency distribution of tire life. Furthermore for this population, $\mu = 45,500$ miles and $\sigma = 3,750$ miles. Let:

$X =$ the mileage a randomly selected tire from this population wears.

   a. Sketch and label an idealized histogram for $X$.
   b. Compute $P(X < 48,000)$.
   c. Compute $P(42,000 < X < 46,000)$.

4. Goodrich Advantage tires have a 60,000-mile rating. Assume a large production run of these tires has a bell-shaped relative frequency distribution of tire wear. Furthermore suppose that for this population, $\mu = 54,250$ miles and $\sigma = 4,800$ miles. Let:

$Y =$ the mileage a randomly selected tire from this population wears.

   a. Sketch and label an idealized probability histogram for $Y$.
   b. Compute $P(Y < 57,500)$.
   c. Compute $P(55,000 < Y < 60,000)$.

*In Exercises 5 and 6, use the information in Figure 4-37 on the next page.*

5. According to data in Figure 4-37, in 1988 the average workweek for American workers was 41.1

# Tire ratings list

**Associated Press**

Following is a partial list of the radial tire-wear ratings released Sunday. After each brand name is the model, tire diameter and the mileage rating achieved in Transportation Department tests.

**ARMSTRONG**
Coronet Ultra Trac, 13  40,000.
Tredloc, 13/14, 56,000.
Tredloc, 15, 60,000.
Tru Trac A/S, 13, 30,000.

**BRIDGESTONE**
S371, all diameters, 48,000.
S375, all, 56,000.
137V, all, 28,000.

**CBI**
Classic, all, 60,000.
Pacemaker, all, 64,000.
Styleline, 13, 36,000.

**EXXON**
Signature, 13, 46,000.
Signature, 15, 56,000.
Steel, all, 36,000.

**FIRESTONE**
721 M/S, 13, 48,000.
721 M/S, 15, 54,000.
Firehawk, all, 34,000.
Widetrack/Daytona, 13, 68,000.
Widetrack/Daytona, 15, 64,000.

**GENERAL**
Ameri Classic, 14, 68,000.
Ameri Classic, 15, 70,000.
XP 2000V V/G, all, 20,000.
XP 2000Z G, all, 10,000.

**GOODRICH**
Advantage, 14/15, 60,000.
T/A 60, 65 and 70 S, 14/15, 62,000.
T/A R & R1, all, 10,000.

**GOODYEAR**
Arriva, 13, 36,000.
Double Eagle, all, 52,000.
Eagle GT S, all, 20,000.
Invicta, 14/15, 56,000.
Vector, 14/15, 56,000.

**K-MART**
Avanti Plus, all, 30,000.
Olympian, all, 48,000.
Olympian II, all, 40,000.

**KELLY SPRINGFIELD**
Charger 65 HR, all, 32,000.
Charger SR, all, 64,000.
Voyager 1000, all, 60,000.

**MICHELIN**
X, MXL, all, 56,000.
XA4, all, 62,000.
XH, all, 66,000.
XWX, all, 20,000

**MONTGOMERY WARD**
Gas Miser, 14/15, 48,000.
Gas Miser 70, all, 64,000.
Runabout/All Season, 13, 30,000.

**NATIONAL**
Renegade, all, 44,000.
XT 40 and 50, all, 30,000
XT 50 A/S, all, 42,000.

**PIRELLI**
P3/185, all, 28,000.
P44, 14, 56,000.
P5/205, all, 26,000.
P6, 15, 40,000.

**SEARS**
All Season, all, 58,000.
Guardsman/Weatherman, all, 28,000.
Roadhandler, all, 62,000.
Superguard SBR, all, 30,000.

**UNIROYAL**
Rallye 340, all, 28,000.
Steeler, all, 44,000.
Tiger Paw XTM Gold, all, 62,000.

**YOKOHAMA**
60 A001R/A008R, all, 10,000.
70-Y382-14, all, 44,000
P60-A008R, all, 10,000.
P75-Y371, 15, 54,000.

**Figure 4-36.** "Tire ratings list," *The Sacramento Bee*, August 29, 1988, p. A-7. Reprinted by permission of Associated Press.

hours. Assume the distribution of hours per week for workers in Detroit, Michigan, has a mean of 41.1 hours and a standard deviation of 3.5 hours, and the relative frequency distribution of hours is bell shaped. Identify the hours per week that cut off:

a. the top 15 percent of hours per workweek.

b. the bottom 25 percent of hours per workweek.

c. the 40 percent of hours per workweek closest to the mean.

6. According to data in Figure 4-37, the average duration of commute to work (1979) was 22.5 minutes. Assume the distribution of commute times for workers in Los Angeles, California, has a mean of 22.5 minutes and a standard deviation of 9.7 minutes, and the relative frequency distribution of times is bell shaped. Identify the commute times that cut off:

a. the top 5 percent.

b. the bottom 20 percent.

c. the 50 percent closest to the mean.

**Social indicators**

*Educational attainment* (1987). Percent of population age 25 and over having: less than full primary education 6.9%; primary 5.8%; less than full secondary 11.7%; secondary 38.7%; some postsecondary 17.1%; 4-year higher degree and more 19.8%, of which postgraduate 8.5%. Number of earned degrees (1987–88): bachelor's degree 989,000; master's degree 290,-000; doctor's degree 33,600; first-professional degrees (in fields such as medicine, theology, and law) 74,400.

**Distribution of income (1986)**

percent of national household income by quintile

| 1 | 2 | 3 | 4 | 5 (highest) |
|-----|-----|------|------|------|
| 3.7 | 9.7 | 16.2 | 24.3 | 46.1 |

*Quality of working life* (1988). Average workweek: 41.1 hours (9.5% overtime). Annual rate per 100,000 workers for (1987): injury or accident 1,800; death 10.0. Proportion of labour force insured for damages or income loss resulting from: injury, permanent disability, and death (1986) 49.0%. Average days lost to labour stoppages per 1,000 workdays (1988): 0.8. Average duration of journey to work (1979): 22.5 minutes (85.7% private automobile, 5.9% public transportation, 1.3% bicycle or motorcycle, 3.9% foot, 2.3% work at home, 0.9% other). Rate per 1,000 workers of discouraged (unemployed no longer seeking work; 1983): 53.5.

*Access to services* (1985). Proportion of dwellings having access to: electricity virtually 100.0%; safe public water supply 98.6%; public sewage collection 99.2%; public fire protection, n.a.

**Figure 4-37.** "Social Indicators," *1990 Britannica World Data*, p. 740. Copyright © 1990 by Encyclopaedia Britannica, Inc. Reprinted by permission.

**7.** Find the value of $x_1$ in Figure 4-38.

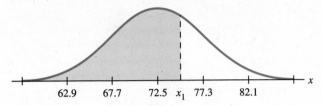

**Figure 4-38.**  $P(X < x_1) \approx 0.60.$

**8.** Find the value of $x_2$ in Figure 4-39.

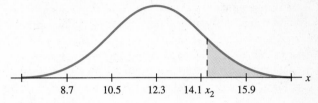

**Figure 4-39.**  $P(X > x_2) \approx 0.15.$

# 4-6  Approximating Binomial Probabilities for Large *n* and Moderate *p*

| Key Topics | |
| --- | --- |
| | *1. Binomial random variables based on many trials* |
| | *2. Selecting the normal curve that gives the best approximation* |

## STATISTICS IN ACTION

Chalon Bridges is the coordinator of the mathematics tutoring center at Chancellor Community College. She has been studying difficulties exhibited by freshmen that come to the center for help. Based on the evidence she has, Chalon feels reasonably certain that about 30 percent of these students would benefit from a minicourse on how to use a pocket calculator. Harold Levinson, the spokesperson for the mathematics department, does not share Chalon's opinion; however, he has agreed to help her devise an assessment test to reveal weaknesses in mathematics comprehension possibly correctable by the course proposed by Chalon.

To test whether the fraction of students that need this course is 30 percent, as Chalon believes, they plan to give the test to 60 freshmen. Based on an agreed upon cutoff test score, each of the 60 students will be either a success or a failure.

Success   A student fails the test and needs the calculator course.

Failure   A student passes the test and does not need the course.

Although the number of successes in the 60 trials is a random variable X, Chalon and Harold agree that they will accept the 30 percent rate if the test yields from 15 to 20 successes inclusive.

Chalon wants to calculate the probability that the assessment test will come out in her favor. She therefore uses a binomial model for $X$, with $p = 0.30$ and $n = 60$. The area in the partial histogram in Figure 4-40 is the desired probability. Because 0.30 is a moderate value for $p$ and 60 is a large number of trials, Chalon uses a normal curve with $\mu_X = 18$ and $\sigma_X = \sqrt{12.6}$ to obtain the desired probability. By changing $x_1 = 14.5$ to $z_1 = -0.99$, and $x_2 = 20.5$ to $z_2 = 0.70$, she is able to approximate the area using Appendix Table 4. The results of these calculations give Chalon a probability of about 60 percent that the test will result in Harold accepting her 30 percent estimate of the rate of deficiencies in freshmen. She realizes, of course, that this 60 percent figure is based on the accuracy of the 0.30 figure she used to calculate the probability.

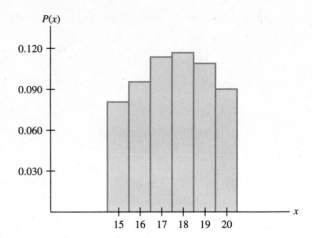

**Figure 4-40.**   A partial histogram for $X$.

## Binomial random variables based on many trials

If a binomial random variable is based on a "moderate" $p$ and number of trials $n$ that is "large enough," then the probability histogram for the associated probability distribution will be approximately bell shaped. By *moderate* and *large enough*, we mean that the inequalities below are satisfied:

### Conditions Under Which a Normal Curve Approximation is Appropriate

As a general rule, most statisticians are willing to use the normal curve approximation method for binomial probabilities when:

1. $np \geq 5$   and
2. $n(1 - p) \geq 5.$

Inequalities 1 and 2 must *both* be satisfied before the approximation method we are about to discuss should be used. The rule ensures us that the histogram of the binomial random variable will be reasonably symmetrical, and not strongly skewed to the left or right. The rule also effectively requires increasing $n$ as $p$ differs more from a moderate value of 0.5. A few selected values of $n$ and $p$ are used to illustrate this fact in Table 4-21.

TABLE 4-21   Some Selected Values of $n$, $p$, and $1 - p$

| $p$ | $1 - p$ | $n$ | $np$ | $n(1 - p)$ |
|-----|---------|-----|------|------------|
| 0.1 | 0.9 | 50 | 5.0 | 45.0 |
| 0.2 | 0.8 | 25 | 5.0 | 20.0 |
| 0.3 | 0.7 | 17 | 5.1 | 11.9 |
| 0.4 | 0.6 | 13 | 5.2 | 7.8 |
| 0.5 | 0.5 | 10 | 5.0 | 5.0 |
| 0.6 | 0.4 | 13 | 7.8 | 5.2 |
| 0.7 | 0.3 | 17 | 11.9 | 5.1 |
| 0.8 | 0.2 | 25 | 20.0 | 5.0 |
| 0.9 | 0.1 | 50 | 45.0 | 5.0 |

▶ **Example 1**   Verify that the mathematics assessment test experiment satisfies the conditions necessary for a normal curve approximation to the distribution of $X$.

**Solution**   $X$ = the number of students who fail the test. With $n = 60$, $p = 0.3$, and $1 - p = 0.7$:

$$60(0.3) = 18, \quad \text{and} \quad 18 > 5 \qquad \text{Inequality 1 is satisfied.}$$
$$60(0.7) = 42, \quad \text{and} \quad 42 > 5 \qquad \text{Inequality 2 is satisfied.} \quad ◀$$

## Selecting the normal curve that gives the best approximation

In the Statistics in Action section, Chalon Bridges used a normal curve to approximate $P(15 \leq X \leq 20)$ for the binomial random variable $X$. Of all the possible normal distributions she could have used, it was reasonable to choose the one with the same mean and variance as the binomial histogram.

### The Mean and Variance for the Approximating Normal Curve

| | Binomial Distribution | | Approximating Normal Curve |
|---|---|---|---|
| **Mean** | $np$ | = | $\mu$ |
| **Variance** | $np(1 - p)$ | = | $\sigma^2$ |

Consider again the partial histogram in Figure 4-40. Chalon wanted to include the total areas of the bars centered on 15 and 20, and all the bars between. However, she wanted to exclude the areas of the bars centered on 14 and 21. As a consequence, she used the *boundary values* as the $x$-values that separated these areas. She therefore changed $x_1 = 14.5$ and $x_2 = 20.5$ to $z$s before using Appendix Table 4. Such use of boundary values when approximating binomial probabilities is sometimes termed use of the *continuity correction factor*.

### Converting a Boundary Value (*bv*) to a *z*

If $X$ is a binomial random variable, then to use Appendix Table 4 to approximate probabilities associated with $X$, use the equation:

$$z = \frac{bv - np}{\sqrt{np(1 - p)}}$$

$bv$ is a **boundary value** of $X$.

$np$ is the **expected value** of $X$.

$\sqrt{np(1 - p)}$ is the **standard deviation** of $X$.

▶ **Example 2**    Sandy Smith coaches a girls' tennis team in a school in South Carolina. An article in *Physical Education Instructors Magazine* gave some statistics on the percentage of women coaches of high school girls' teams in selected states. According to the article, only 46 percent of high school girls' teams in South Carolina are coached by women. Motivated by the information, Sandy will sample high schools in South Carolina and obtain the sex of the coaches of 80 girls' athletic teams. Let $Y$ be the number in the sample that are women. Use $p = 0.46$ as the probability that a given coach is a woman, and model $Y$ as binomial.

**a.** Verify that a normal curve approximation is appropriate for probabilities associated with $Y$.

**b.** Approximate $P(Y \leq 30)$.

**c.** State the meaning of the answer to (b).

**Solution**    **a.** With $n = 80$, $p = 0.46$, and $1 - p = 0.54$:

(1)  $80(0.46) = 36.8$ and $36.8 \geq 5$

(2)  $80(0.54) = 43.2$ and $43.2 \geq 5$

**b.** A partial probability histogram for $Y$ is shown in Figure 4-41.

$$P(Y \leq 30) = P(0) + \cdots + P(29) + P(30)$$

Therefore the upper boundary value for the bar centered on 30 is used, and $bv = 30.5$.

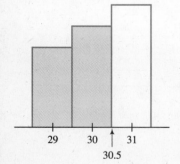

$$z = \frac{30.5 - 80(0.46)}{\sqrt{80(0.46)(0.54)}} \qquad z = \frac{bv - np}{\sqrt{np(1 - p)}}$$

$$= \frac{-6.3}{4.4578 \ldots} \qquad \text{Do not round the denominator.}$$

$$= -1.4132 \ldots \qquad \text{Do the indicated division.}$$

$$\approx -1.41 \qquad \text{Rounded to two decimal places.}$$

$$P(Y \leq 30) \approx P(Z < -1.41)$$

$$= 0.0793 \qquad \text{Appendix Table 4}$$

**Figure 4-41.**  A partial histogram for $Y$.

**c.** Assuming the 46 percent figure in the article is correct, the probability is about 8 percent that 30 or fewer of the coaches in the sample that Sandy takes will be women.  ◀

▶ **Example 3**     Philip Howrigan is personnel director of Allied Manufacturing Corporation. To help in job placement, all Allied applicants are given a manual dexterity test. Past results show that 85 percent of applicants pass the minimum requirements of the test. On Saturday 50 applicants will take the test. Let $X$ be the number who will pass the test.

**a.** Verify that a normal curve approximation is appropriate for probabilities associated with $X$.

**b.** Approximate $P(40 < X < 45)$.

**c.** State the meaning of the answer to (b).

*Binomial-*
*yes, no*
*true, false*

**Solution**     **a.** With $n = 50$, $p = 0.85$, and $1 - p = 0.15$:

(1)  $50(0.85) = 42.5$ and $42.5 \geq 5$

(2)  $50(0.15) = 7.5$ and $7.5 \geq 5$

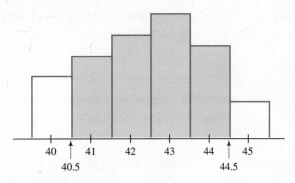

**Figure 4-42.**  A partial histogram for $X$.

**b.** A partial probability histogram for $X$ is shown in Figure 4-42.

$$P(40 < X < 45) = P(41) + P(42) + P(43) + P(44)$$

The area of the bars centered on 41, 42, 43, and 44 is needed; therefore 40.5 and 44.5 are used as the boundary values.

$$z_1 = \frac{40.5 - 50(0.85)}{\sqrt{50(0.85)(0.15)}} \qquad z_2 = \frac{44.5 - 50(0.85)}{\sqrt{50(0.85)(0.15)}}$$

$$= -0.7921\ldots \qquad\qquad = 0.7921\ldots$$

$$\approx -0.79 \qquad\qquad\qquad \approx 0.79$$

$$P(40 < X < 45) \approx P(-0.79 < Z < 0.79)$$

$$= P(Z < 0.79) - P(Z < -0.79)$$

$$= 0.7852 - 0.2148$$

$$= 0.5704$$

**c.** The probability is about 57 percent that more than 40 and less than 45 of the applicants tested will pass the manual dexterity test.    ◀

# Exercises 4-6    Set A

*In Exercises 1–6, do the following:*

*a. Justify using a normal curve to approximate probabilities associated with the given binomial random variable.*

*b. Compute the indicated probability.*

*c. Sketch a partial histogram showing the area that represents the probability.*

*d. In the context of the problem, state the meaning of the probability in part (b).*

1. At a large community college in Arkansas, it is reported that 30 percent of the students transfer after graduation to a state college to pursue a bachelor's degree. Suppose 50 students at this school are interviewed, and $X$ is the number in the sample planning to attend a state college after graduation. Approximate $P(X < 14)$.

2. Highway patrol records indicate that 60 percent of motorcycle accidents in Ohio involve males younger than 25 years of age. Suppose 35 reports of motorcycle accidents in Ohio are randomly selected, and $X$ is the number of reports that involve males younger than 25 years of age. Approximate $P(X < 23)$.

3. A quarter has been biased so that the probability is about 75 percent that a given toss of the coin will yield an upturned head. Suppose the coin is tossed 25 times, and $Y$ is the number of heads observed. Approximate $P(Y \geq 19)$.

4. A six-faced die has been "loaded" so that the probability is about 40 percent that the face with the six spots will be showing on top when the die is tossed. Suppose the die is tossed 30 times, and $Y$ is the number of 6s observed. Approximate $P(Y > 15)$.

5. The union officers of a large national labor organization want to assess the attitude of the rank-and-file members on a strike in favor of job security measures. Suppose a random sample of 150 members is taken, and $W$ is the number in the sample that would strike for such a cause. Furthermore assume that 58 percent of the membership supports a strike. Compute $P(80 \leq W \leq 90)$.

6. The University of California at Davis is known for the large number of bikes used by students. Records kept by campus police indicate that a student who attends the college for 4 years has about a 64 percent chance of having one or more accidents on a bicycle while being a student at the school. Suppose 75 students who attended this campus for 4 years are surveyed, and $W$ is the number that had at least one bicycle accident while a student. Approximate $P(45 < W < 50)$ using a normal curve.

7. The National Fire Protection Association reported that 14 percent of home fires resulting in at least one fatality are caused by faulty heaters. A regional manager for Far Western Insurance Company sampled 60 reports of home fires resulting in at least one fatality. Let $X$ be the number in the sample caused by a faulty heater. Model $X$ as binomial and use $p = 0.14$.
   a. Approximate $P(X > 10)$.
   b. Approximate $P(8 \leq X \leq 12)$.

8. The mayor of Gardnerville proposed a use fee for city parks as a way of obtaining revenue. A reporter for a local newspaper sampled 50 residents and asked them their attitude toward the fee. Suppose 68 percent of all residents oppose the fee. Let $X$ be the number in the survey that oppose the fee. Model $X$ as binomial and use $p = 0.68$.
   a. Approximate $P(X \leq 30)$.
   b. Approximate $P(32 \leq X \leq 36)$.

9. Middlecrest is a large state college. Records at the school indicate that about 15 percent of the students are Asian. The dean of the college's research and development office obtained a random sample of records of 180 students. Let $Y$ be the number in the sample that are Asian. Model $Y$ as binomial and use $p = 0.15$.
   a. Approximate $P(Y \leq 30)$.
   b. Approximate $P(20 < Y < 30)$.

10. The Department of Motor Vehicles (DMV) in a state in the Midwest reported that 28 percent of new vehicle registrations are for pickup trucks. A

supervisor in the vehicle registration division of the DMV randomly selected 240 new vehicle registration applications. Let $Y$ be the number in the sample for pickup trucks. Model $Y$ as binomial and use $p = 0.28$.

a. Approximate $P(Y \geq 75)$.

b. Approximate $P(60 < Y \leq 70)$.

**11.** A recent national survey showed that approximately 10 percent of registered nurses (RNs) in the United States are male. Suppose 250 names are randomly selected from the list of RN graduates last year from nursing schools throughout the country. Assume that 0.10 is the probability that a given name is that of a male. Let $W$ be the number of males in the sample.

a. Approximate $P(26 \leq W \leq 30)$ using a normal curve.

b. State the meaning of the probability in part (a).

**12.** A display placed in a prime location at a supermarket is supposed to generate "impulse buying." Studies have indicated that the probability is about 12 percent that a customer who looks at the display will choose an item from it. Let $W$ be

the number out of the next 237 customers who look at the display and buy.

a. Approximate $P(23 \leq W \leq 33)$.

b. State the meaning of the probability in part (a).

**13.** Let $X$ be a random variable modeled as binomial with $n = 40$ and $p = 0.55$.

a. Compute $P(22)$ using the binomial probability formula.

b. Approximate $P(21 < X < 23)$ using a normal curve.

c. Round the probabilities in parts (a) and (b) to two decimal places and compare the rounded values.

**14.** Let $X$ be a random variable modeled as binomial, with $n = 35$ and $p = 0.42$.

a. Compute $P(15)$ using the binomial probability formula.

b. Approximate $P(14 < X < 16)$ using a normal curve.

c. Round the probabilities in parts (a) and (b) to two decimal places and compare the rounded values.

## Exercises 4-6   Set B

*In Exercises 1 and 2, use Data Set I, Appendix B.*

**1.** In Data Set I, Appendix B, 47 percent of the population are non-smokers. These individuals are identified by a 2 in the "Smoker" column. Suppose $n = 50$ individuals are randomly sampled from this set and

$X = $ the number of non-smokers in the sample.

a. Use a normal curve and approximate $P(20 \leq X \leq 27)$.

b. Use random numbers to select three different samples, each of size 50 from the set, and count the number of non-smokers in each sample.

| Sample Number | 1 | 2 | 3 |
|---|---|---|---|
| Value of X | | | |

c. How many of your samples had from 20 through 27 non-smokers?

d. Is the answer you got for (c) consistent or inconsistent with the probability you got for (a)? Discuss.

**2.** In Data Set I, Appendix B, 15 percent of the population have cancer. These individuals are identified by a 1 in the "Cancer" column. Suppose $n = 60$ individuals are randomly sampled from this set and

$X = $ the number with cancer in the sample.

a. Use a normal curve and approximate $P(6 \leq X \leq 12)$.

b. Use random numbers to select three different samples, each of size 60 from the set, and count the number with cancer in each sample.

| Sample Number | 1 | 2 | 3 |
|---|---|---|---|
| Value of X | | | |

c. How many of your samples had from six through 12 individuals with cancer?

d. Is the answer you got for (c) consistent or inconsistent with the probability you got for (a)? Discuss.

*In Exercises 3 and 4, use the information in Figure 4-43.*

3. According to the study by Kraus, Riggins, and Franti, 33.8 percent of serious motorcycle accidents occur when traveling at 33 to 49 km/hr. Suppose $n = 40$ motorcycle accidents are investigated and

$X = $ the number in which the motorcycle was traveling
$X$    at 33 to 49 km/hr.

Approximate $P(10 \leq X \leq 15)$.

4. According to the study by Kraus, Riggins, and Franti, 16.4 percent of serious motorcycle accidents occur when traveling at 65 to 80 km/hr. Suppose $n = 50$ motorcycle accidents are investigated and

$Y = $ the number in which the motorcycle was traveling at 65 to 80 km/hr.

Approximate $P(8 \leq Y \leq 12)$.

**Number and Percentage of Serious and Not Serious Injuries for Male Drivers According to Estimated Speed of Motorcycle at Time of Impact, Sacramento County, California, 1970**

| Speed of Motorcycle at Impact (km/hr) | Severity of Injury | | | | Ratio Serious/ Not Serious |
|---|---|---|---|---|---|
| | Serious | | Not Serious | | |
| | No. | % | No. | % | |
| ≤16 | 9 | 4.6 | 10 | 5.0 | 0.9 |
| 17–32 | 22 | 11.3 | 31 | 15.6 | 0.7 |
| 33–49 | 66 | 33.8 | 95 | 47.7 | 0.7 |
| 50–64 | 54 | 27.7 | 42 | 21.1 | 1.3 |
| 65–80 | 32 | 16.4 | 13 | 6.5 | 2.5 |
| 81–96 | 8 | 4.1 | 5 | 2.5 | 1.6 |
| 97–113 | 4 | 2.1 | 3 | 1.5 | 1.3 |
| All speeds | 195 | | 199 | | 1.0 |

**Figure 4-43.** Epidemiology of Motorcycle Collision Injuries, II. *Study by Kraus, Riggins and Franti, University of California Medical School, Davis, page 103.*

The **Poisson Probability Distributions** were first introduced by the French mathematician Poisson (1781–1840). The Poisson distributions can be used to approximate binomial probabilities when $p$, the probability of success on each trial, is small.

The probabilities for a particular Poisson distribution depend on a parameter $\lambda$ (read "lambda"), which is the mean of the distribution. The probabilities for Poisson distributions are calculated with the formula

$$P(x) = \frac{e^{-\lambda} \cdot \lambda^x}{x!}, \text{ for } x = 0, 1, 2, 3, \ldots$$

In the formula, e is the irrational number with decimal approximation 2.7182 . . . .

If $n \geq 20$ and $p \leq 0.05$, then a Poisson distribution with $\lambda = np$ can be used to approximate the corresponding Binomial distribution. For $n \geq 100$ and $p \leq 0.01$, the approximations are very good. The following set of exercises provide an opportunity to compare the probabilities of a few binomial experiments using the binomial probability formula and the Poisson probability formula.

*In Exercises 5–8, use a calculator to approximate the following probabilities to five decimal places. Then compare the probabilities obtained in parts (a) and (b).*
*a. Use the binomial probability formula.*
*b. Use the Poisson probability formula.*

**5.** $P(X = 3)$, for $n = 30$ and $p = 0.05$. (Use $\lambda = 1.5$.)

**6.** $P(X = 2)$, for $n = 40$ and $p = 0.04$. (Use $\lambda = 1.6$.)

**7.** $P(X = 1)$, for $n = 50$ and $p = 0.03$. (Use $\lambda = 1.5$.)

**8.** $P(X = 4)$, for $n = 60$ and $p = 0.02$. (Use $\lambda = 1.2$.)

# 4-7  Independence of Random Variables

**Key Topics**

1. *Independence of random variables $X$ and $Y$*

2. *Probability distributions for the random variable $\bar{X}$*

3. *Computing $\mu_{\bar{X}}$ and $\sigma_{\bar{X}}$ for the random variable $\bar{X}$*

$$\text{S T A T I S T I C S   I N   A C T I O N}$$

Last night Kelly Bell and Steve Sidwell were contestants on a TV game show. By correctly answering five questions they earned the opportunity to play the show's bonus game called "Ball Four." The rules of this bonus game required the contestants to select a ball from two different boxes. A number 1, 2, or 3 was painted on each ball. If the two contestants selected balls so that the sum of the numbers was 4, then they would share a cash prize of $10,000.

The boxes were labeled A and B. Kelly and Steve agreed that Kelly would select the ball from box A, and then Steve would select the ball from box B. The frequencies of numbered balls in the two boxes is shown in Table 4-22.

**TABLE 4-22   Frequencies of Numbered Balls in Boxes A and B**

| Box A | | Box B | |
|:---:|:---:|:---:|:---:|
| *Numbers* | *Frequencies* | *Numbers* | *Frequencies* |
| 1 | 7 | 1 | 7 |
| 2 | 2 | 2 | 2 |
| 3 | 1 | 3 | 1 |

Based on the way the game is played and the relative frequencies of numbers in the two boxes, it can be shown that Kelly and Steve had about an 18 percent chance of winning the $10,000. This probability is based on the following calculations:

As it turned out, Kelly got the 3 on her draw from box A, and Steve succeeded in getting a 1 on his draw from box B. Thus they got numbers that totaled 4 and won the $10,000.

| Kelly and Steve Win If: | Box A | Box B | P(Win) |
|---|---|---|---|
| a. Kelly gets a 1 and Steve gets a 3 | $P(1) = 0.7$ | $P(3) = 0.1$ | $P(1 \text{ and } 3) = 0.07$ |
| or | | | |
| b. Kelly gets a 2 and Steve gets a 2 | $P(2) = 0.2$ | $P(2) = 0.2$ | $P(2 \text{ and } 2) = 0.04$ |
| or | | | |
| c. Kelly gets a 3 and Steve gets a 1 | $P(3) = 0.1$ | $P(1) = 0.7$ | $P(3 \text{ and } 1) = 0.07$ |
| | | Sum of winning combinations $= 0.18$ | |

## Independence of random variables $X$ and $Y$

In the Statistics in Action section, we determined that Kelly and Steve had an 18 percent chance of winning "Ball Four" and the $10,000 prize.

A key to finding the probability was the idea of independence discussed in Chapter 3. Recall that we can use $P(A)$ and $P(B)$ to compute $P(A \text{ and } B)$ by multiplication if it is reasonable to model $A$ and $B$ as independent events.

### Multiplication Rule for Independent Events

If $A$ and $B$ are independent events, then $P(A \text{ and } B) = P(A) \cdot P(B)$.

Consider the two activities in the game that Kelly and Steve played. Both activities involved discrete random variables, respectively $X$ and $Y$:

$X$ is the number on the ball from box A with possible values 1, 2, and 3.

$Y$ is the number on the ball from box B with possible values 1, 2, and 3.

From the description of the game, it seems reasonable to describe $X$ and $Y$ as unrelated. That is, once Kelly has selected a ball from box A, there is no reason to suspect that Steve is more likely to draw a particular numbered ball from box B. Thus we can model events related to the random variables $X$ and $Y$ as independent.

▶ **Example 1**   Recompute the probability that Kelly and Steve win the game.

**Solution**   The relative frequencies of the numbered balls in the two boxes yield the following probabilities:

$$P(X = 1) = 0.7 \qquad P(Y = 1) = 0.7$$
$$P(X = 2) = 0.2 \qquad P(Y = 2) = 0.2$$
$$P(X = 3) = 0.1 \qquad P(Y = 3) = 0.1$$

Now with the two draws modeled as independent, the probabilities for all the possible combinations of values of $X$ and $Y$ are in Table 4-23.

TABLE 4-23  Probabilities for $X = x$ and $Y = y$

| y \ x | 1 | 2 | 3 |
|---|---|---|---|
| 1 | 0.49 | 0.14 | 0.07* |
| 2 | 0.14 | 0.04* | 0.02 |
| 3 | 0.07* | 0.02 | 0.01 |

$$P(X = 1 \text{ and } Y = 1) = (0.7)(0.7) = 0.49$$
$$P(X = 1 \text{ and } Y = 2) = (0.7)(0.2) = 0.14$$
$$\vdots \qquad\qquad \vdots \qquad \vdots$$
$$P(X = 3 \text{ and } Y = 3) = (0.1)(0.1) = 0.01$$

The probabilities in the table marked with asterisks correspond to outcomes where $X + Y = 4$. Using the addition rule for mutually exclusive events, the probability of interest is:

$$P(X + Y = 4) = P(X = 1 \text{ and } Y = 3) + P(X = 2 \text{ and } Y = 2)$$
$$+ P(X = 3 \text{ and } Y = 1)$$
$$= 0.07 + 0.04 + 0.07$$
$$= 0.18 \quad \blacktriangleleft$$

The notion of assigning each $P(X = x \text{ and } Y = y)$ as the product of $P(x)$ and $P(y)$ is important enough to merit its own terminology.

## Multiplication Rule for Independent Discrete Random Variables

The phrase "discrete random variables $X_1, X_2, \ldots, X_n$ are modeled as independent" means that each

$$P(X_1 = x_1 \text{ and } X_2 = x_2 \text{ and} \ldots \text{ and } X_n = x_n)$$

is assigned the value:

$$P(x_1) \cdot P(x_2) \cdot \ldots \cdot P(x_n)$$

▶ **Example 2**   A supervisor of the Cosmetics and Fine Jewelry Department in one of the stores of a national chain is responsible for making out the work schedules for the employees in these departments. She is currently working on the schedule for the Memorial Day weekend. She is specifically interested in the random variables $X$ and $Y$, where:

$X$ = the number of employees willing to work overtime.

$Y$ = the number of employees needed this weekend.

Based on past experience, she has in mind the probability distributions shown in Table 4-24 on the following page for $X$ and $Y$.

TABLE 4-24   Probabilities for $X$ and $Y$

| $x$ | $P(x)$ | | $y$ | $P(y)$ |
|---|---|---|---|---|
| 3 | 0.1 | | 3 | 0.2 |
| 4 | 0.2 | | 4 | 0.5 |
| 5 | 0.5 | | 5 | 0.3 |
| 6 | 0.2 | | | |

**a.** Make a table giving probabilities for the events "$X = x$ and $Y = y$" for the supervisor, supposing each pair of events "$X = x$" and "$Y = y$" to be independent.

**b.** Compute the probability that the number of employees who will want to work this weekend will exactly match the number who will be needed.

**Solution**   **Discussion.** There are four possible values for $X$ and three for $Y$. Thus there are $4 \cdot 3 = 12$ possible pairs of values for $X$ and $Y$.

**a.** $P(X = 3 \text{ and } Y = 3) = (0.1)(0.2) = 0.02$
$P(X = 3 \text{ and } Y = 4) = (0.1)(0.5) = 0.05$
$P(X = 3 \text{ and } Y = 5) = (0.1)(0.3) = 0.03$
$P(X = 4 \text{ and } Y = 3) = (0.2)(0.2) = 0.04$
$$\vdots \qquad \vdots \qquad \vdots$$
$P(X = 6 \text{ and } Y = 5) = (0.2)(0.3) = 0.06$

**b.** The events "$X = 3$ and $Y = 3$," "$X = 4$ and $Y = 4$," and "$X = 5$ and $Y = 5$" are the ones in which the number of employees who will want to work exactly matches the number that will be needed. These probabilities are marked with asterisks in Table 4-25.

$$P(X = Y) = 0.02 + 0.10 + 0.15 = 0.27 \quad \blacktriangleleft$$

TABLE 4-25   Probabilities for $X = x$ and $Y = y$

| $y$ \ $x$ | 3 | 4 | 5 | 6 |
|---|---|---|---|---|
| 3 | 0.02* | 0.04 | 0.10 | 0.04 |
| 4 | 0.05 | 0.10* | 0.25 | 0.10 |
| 5 | 0.03 | 0.06 | 0.15* | 0.06 |

Suppose the supervisor wants to assess the probability that there will be enough overtime slots for all the employees who want to work overtime on the Memorial Day weekend. In terms of $X$ and $Y$, she is trying to assess $P(X \le Y)$.

The event "$X \le Y$" can be made up from the mutually exclusive possibilities in (a) through (f). The corresponding probabilities are found in Table 4-25.

a. $X = 3$ and $Y = 3$    $P(X = 3 \text{ and } Y = 3) = 0.02$
b. $X = 3$ and $Y = 4$    $P(X = 3 \text{ and } Y = 4) = 0.05$
c. $X = 3$ and $Y = 5$    $P(X = 3 \text{ and } Y = 5) = 0.03$
d. $X = 4$ and $Y = 4$    $P(X = 4 \text{ and } Y = 4) = 0.10$
e. $X = 4$ and $Y = 5$    $P(X = 4 \text{ and } Y = 5) = 0.06$
f. $X = 5$ and $Y = 5$    $P(X = 5 \text{ and } Y = 5) = 0.15$

Summing the probabilities in (a) through (f):

$$P(X \leq Y) = 0.02 + 0.05 + \cdots + 0.15 = 0.41$$

Thus the supervisor might estimate that there is a 41 percent chance that every employee who wants to work the Memorial Day weekend will have the opportunity to do so.

## Probability distributions for the random variable $\bar{X}$

If $X_1, X_2, \ldots, X_n$ are random variables, then $\bar{X}$, the mean of the $X$s, is also a random variable. In the next two examples, we will illustrate how independence can be used to obtain a probability distribution for the sample mean $\bar{X}$.

▶ **Example 3**  Susan Siegrist is the present state coordinator for this year's Girl Scout cookie sale. Susan is planning on taking a simple random sample of $n = 2$ orders from the 800,000 orders received last year. In particular, she is interested in the mean number of boxes of chocolate mint cookies indicated on the orders.

The (approximate) relative frequency distribution for chocolate mint orders last year is shown in Table 4-26. Use these values to find a probability distribution for $\bar{X}$, the sample mean number of boxes of mint cookies ordered.

TABLE 4-26  Relative Frequency Distribution of Cookie Orders

| Number of Boxes of Mints | Relative Frequency |
| --- | --- |
| 0 | 0.5 |
| 1 | 0.3 |
| 2 | 0.2 |

**Solution**  **Discussion.** We first define $X_1$ and $X_2$:

$X_1 =$ the number of boxes of mints requested on the first order.

$X_2 =$ the number of boxes of mints requested on the second order.

The possible values for both $X_1$ and $X_2$ are 0, 1, and 2.

If $\bar{X}$ is the mean of $X_1$ and $X_2$, then $\bar{X} = \dfrac{1}{2}(X_1 + X_2)$.

The possible values for $\bar{X}$ are obtained by replacing $X_1$ and $X_2$ in the equation by 0, 1, and 2, and simplifying the expression. To illustrate, three of the nine possible combinations are as follows:

$$\text{For } X_1 = 0 \text{ and } X_2 = 1: \quad \bar{X} = \frac{1}{2}(0 + 1) = \frac{1}{2}(1) = \frac{1}{2}$$

$$\text{For } X_1 = 1 \text{ and } X_2 = 1: \quad \bar{X} = \frac{1}{2}(1 + 1) = \frac{1}{2}(2) = 1$$

$$\text{For } X_1 = 2 \text{ and } X_2 = 1: \quad \bar{X} = \frac{1}{2}(2 + 1) = \frac{1}{2}(3) = \frac{3}{2}$$

Following this reasoning, the possible values for $\bar{X}$ are 0, $\frac{1}{2}$, 1, $\frac{3}{2}$, and 2.

We realize that $n = 2$ is a small fraction of the population size $N = 800,000$. We therefore reason that $X_1$ and $X_2$ can be approximately described as independent. Probabilities for the events "$X_1 = x_1$ and $X_2 = x_2$" are obtained by multiplying the relative frequency values for $x_1$ and $x_2$ in the population. These products yield the probabilities in Table 4-27. The probability distribution for $\bar{X}$ can now be obtained by adding the probabilities of the mutually exclusive events in Table 4-27 that yield each of the five possible values for $\bar{X}$. These probabilities are shown in Table 4-28.

**TABLE 4-27   Probabilities for "$X_1 = x_1$ and $X_2 = x_2$"**

|  | $x_1$ | | |
|---|---|---|---|
| $x_2$ | *0* | *1* | *2* |
| 0 | 0.25 | 0.15 | 0.10 |
| 1 | 0.15 | 0.09 | 0.06 |
| 2 | 0.10 | 0.06 | 0.04 |

**TABLE 4-28   Values for $P(\bar{X} = \bar{x})$, Where $\bar{X} = \frac{1}{2}(X_1 + X_2)$**

| $\bar{x}$ | $P(\bar{X} = \bar{x})$ |
|---|---|
| 0 | 0.25 |
| $\frac{1}{2}$ | 0.30 (=0.15 + 0.15) |
| 1 | 0.29 (=0.10 + 0.09 + 0.10) |
| $\frac{3}{2}$ | 0.12 (=0.06 + 0.06) |
| 2 | 0.04 |

For example, we see that the probability that the mean number of boxes of chocolate mint cookies in the two orders sampled will be 1 is about 29 percent. A mean of 1 for the two orders can be obtained in the following three different ways:

$$X_1 = 2 \text{ and } X_2 = 0: \quad P(X_1 = 2 \text{ and } X_2 = 0) = 0.10$$
$$X_1 = 1 \text{ and } X_2 = 1: \quad P(X_1 = 1 \text{ and } X_2 = 1) = 0.09$$
$$X_1 = 0 \text{ and } X_2 = 2: \quad P(X_1 = 0 \text{ and } X_2 = 2) = 0.10$$

The sum of the probabilities for these mutually exclusive events is 0.29, as shown.

◀

## Computing $\mu_{\bar{X}}$ and $\sigma_{\bar{X}}$ for the random variable $\bar{X}$

In Section 4-2, the mean and variance of the random variable $X$ were calculated using the following equations:

$$\mu_X = \sum x_i \cdot P(x_i)$$
$$\sigma_X^2 = \sum x_i^2 \cdot P(x_i) - (\mu_X)^2$$

These equations can also be used to find the mean ($\mu_{\bar{X}}$) and variance ($\sigma_{\bar{X}}^2$) for the random variable $\bar{X}$. (Notice that here the subscript on both $\mu_{\bar{X}}$ and $\sigma_{\bar{X}}^2$ is $\bar{X}$, because these are summary measures for the probability distribution of $\bar{X}$.) To use these equations, we simply replace $X$ by $\bar{X}$ as follows:

$$\mu_{\bar{X}} = \sum \bar{x}_i \cdot P(\bar{x}_i)$$
$$\sigma_{\bar{X}}^2 = \sum \bar{x}_i^2 \cdot P(\bar{x}_i) - (\mu_{\bar{X}})^2$$

The values in Example 3 are used to illustrate the equations. The calculations are shown in Table 4-29.

TABLE 4-29    Calculations for $\mu_{\bar{X}}$ and $\sigma_{\bar{X}}$

| $\bar{x}$ | $P(\bar{x})$ | $\bar{x} \cdot P(\bar{x})$ | $\bar{x}^2$ | $\bar{x}^2 \cdot P(\bar{x})$ |
|---|---|---|---|---|
| 0 | 0.25 | 0.00 | 0 | $(0)(0.25) = 0.000$ |
| $\frac{1}{2}$ | 0.30 | 0.15 | $\frac{1}{4}$ | $(0.25)(0.30) = 0.075$ |
| 1 | 0.29 | 0.29 | 1 | $(1)(0.29) = 0.290$ |
| $\frac{3}{2}$ | 0.12 | 0.18 | $\frac{9}{4}$ | $(2.25)(0.12) = 0.270$ |
| 2 | 0.04 | 0.08 | 4 | $(4)(0.04) = 0.160$ |

$$\sum \bar{x}_i \cdot P(x_i) = 0.70$$
$$\sum \bar{x}_i^2 \cdot P(\bar{x}_i) = 0.795$$

$\mu_{\bar{X}} = 0.7$      The expected mean number of boxes of chocolate mint cookies in the sample of two orders is 0.7 boxes.

$\sigma_{\bar{X}}^2 = 0.795 - 0.7^2$      The variance of the mean is 0.305.

$\quad\quad = 0.305$

The positive square root of $\sigma_{\bar{X}}^2$ is the standard deviation $\sigma_{\bar{X}}$:

$$\sigma_{\bar{X}} = \sqrt{0.305} \approx 0.6, \text{ to one decimal place.}$$

▶ **Example 4**      The Chamber of Commerce of Richland, Indiana, wants to get information on the household furnishings of residents in the area. One item of interest is the number of television sets per household. Assume the relative frequency distribution for the numbers of television sets in households is as shown in Table 4-30 on the following page.

Suppose a random sample of $n = 2$ households from this area is selected to provide information on television sets present.

**a.** Find a probability distribution for $\bar{X}$, the sample mean.

**b.** Compute $\mu_{\bar{x}}$ to one decimal place.

**c.** Compute $\sigma_{\bar{x}}$ to one decimal place.

TABLE 4-30    Relative Frequency Distribution of Television Sets

| Number of Television Sets | Relative Frequency |
|---|---|
| 1 | 0.35 |
| 2 | 0.60 |
| 3 | 0.05 |

**Solution**    **Discussion.** Let:

$X_1$ = the number of television sets in the first household sampled.

$X_2$ = the number of television sets in the second household sampled.

Assuming the number of households in Richland, Indiana, is large compared to the sample size $n = 2$, $X_1$ and $X_2$ can be modeled as approximately independent.

The probabilities for events "$X_1 = x_1$ and $X_2 = x_2$" for each of the nine combinations of values for $x_1$ and $x_2$ are shown in Table 4-31. These values are obtained using the relative frequencies and the multiplication rule for independent random variables. To illustrate:

$$P(X_1 = 2 \text{ and } X_2 = 1) = (0.60)(0.35) = 0.2100$$

which is the entry in row 1 and column 2 of the table.

TABLE 4-31    Probabilities for "$X_1 = x_1$ and $X_2 = x_2$"

| $x_2$ \ $x_1$ | 1 | 2 | 3 |
|---|---|---|---|
| 1 | 0.1225 | 0.2100 | 0.0175 |
| 2 | 0.2100 | 0.3600 | 0.0300 |
| 3 | 0.0175 | 0.0300 | 0.0025 |

Because $\bar{X} = \frac{1}{2}(X_1 + X_2)$, the possible values for $\bar{X}$ are 1, $\frac{3}{2}$, 2, $\frac{5}{2}$, and 3.

**a.** Collecting the appropriate entries of Table 4-31, one can obtain the probability distribution for $\bar{X}$ in Table 4-32. The calculations needed to determine $\mu_{\bar{x}}$ and $\sigma_{\bar{x}}$ are organized in Table 4-33.

TABLE 4-32    Values for $P(\bar{X} = \bar{x})$, Where $\bar{X} = \frac{1}{2}(X_1 + X_2)$

| $\bar{x}$ | $P(\bar{X} = \bar{x})$ |
|---|---|
| 1 | 0.1225 |
| $\frac{3}{2}$ | 0.4200 ($= 0.2100 + 0.2100$) |
| 2 | 0.3950 ($= 0.0175 + 0.3600 + 0.0175$) |
| $\frac{5}{2}$ | 0.0600 ($= 0.0300 + 0.0300$) |
| 3 | 0.0025 |

TABLE 4-33    Calculations for $\mu_{\bar{x}}$ and $\sigma_{\bar{x}}$

| $\bar{x}$ | $P(\bar{x})$ | $\bar{x} \cdot P(\bar{x})$ | $\bar{x}^2$ | $\bar{x}^2 \cdot P(\bar{x})$ |
|---|---|---|---|---|
| 1 | 0.1225 | 0.1225 | 1 | 0.1225 |
| $\frac{3}{2}$ | 0.4200 | 0.6300 | $\frac{9}{4}$ | 0.9450 |
| 2 | 0.3950 | 0.7900 | 4 | 1.5800 |
| $\frac{5}{2}$ | 0.0600 | 0.1500 | $\frac{25}{4}$ | 0.3750 |
| 3 | 0.0025 | 0.0075 | 9 | 0.0225 |

**b.** $\mu_{\bar{x}} = 1.7$     The expected mean number of television sets in the sample of two households is 1.7.

**c.** $\sigma_{\bar{x}} = \sqrt{3.0450 - 1.7^2}$     $\sigma_{\bar{x}} = \sqrt{\sum \bar{x}_i^2 \cdot P(\bar{x}_i) - (\mu_{\bar{x}})^2}$

$\phantom{\text{c.} \sigma_{\bar{x}}} = \sqrt{0.1550}$

$\phantom{\text{c.} \sigma_{\bar{x}}} \approx 0.4$, rounded to 1 decimal place.

Thus the standard deviation for the distribution of $\bar{X}$ is about 0.4.     ◄

There are several important connections between the probability distribution of $\bar{X}$ and the relative frequency distribution of the sampled population. These connections form the basis of statistical inference for a population mean and will be discussed in detail in the next chapter.

# Exercises 4-7    Set A

*In Exercises 1–4, do the following:*
*a. Make a table giving probabilities for the events "$X = x$ and $Y = y$." Use the given probability distributions for $X$ and $Y$, and assume $X$ and $Y$ are independent.*
*b. Using the values in the table, state which event "$X = x$ and $Y = y$" is most likely to occur.*
*c. Using the values in the table, state which event "$X = x$ and $Y = y$" is least likely to occur.*

**1.**

| $x$ | $P(x)$ | $y$ | $P(y)$ |
|-----|--------|-----|--------|
| 0 | 0.2 | 0 | 0.4 |
| 1 | 0.4 | 1 | 0.5 |
| 2 | 0.3 | 2 | 0.1 |
| 3 | 0.1 |   |   |

**2.**

| $x$ | $P(x)$ | $y$ | $P(y)$ |
|-----|--------|-----|--------|
| 1 | 0.1 | 1 | 0.2 |
| 2 | 0.2 | 2 | 0.3 |
| 3 | 0.3 | 3 | 0.4 |
| 4 | 0.4 | 4 | 0.1 |

**3.**

| $x$ | $P(x)$ | $y$ | $P(y)$ |
|-----|--------|-----|--------|
| 0 | 0.10 | 1 | 0.20 |
| 1 | 0.20 | 3 | 0.30 |
| 2 | 0.25 | 5 | 0.50 |
| 3 | 0.30 |   |   |
| 4 | 0.15 |   |   |

**4.**

| $x$ | $P(x)$ | $y$ | $P(y)$ |
|-----|--------|-----|--------|
| 1 | 0.20 | 0 | 0.05 |
| 3 | 0.25 | 2 | 0.10 |
| 5 | 0.40 | 4 | 0.35 |
| 7 | 0.15 | 6 | 0.30 |
|   |   | 8 | 0.20 |

**5.** Suppose the supervisor of the cosmetics and fine jewelry departments had the probability distributions in Table 4-34 in mind for $X$ and $Y$. (See Example 2.)
   a. Suppose $X$ and $Y$ are independent and compute the probabilities for "$X = x$ and $Y = y$."
   b. Compute $P(X < Y)$.

**TABLE 4-34    Probability Distributions for $X$ and $Y$**

| $x$ | $P(x)$ | $y$ | $P(y)$ |
|-----|--------|-----|--------|
| 3 | 0.2 | 3 | 0.3 |
| 4 | 0.3 | 4 | 0.6 |
| 5 | 0.4 | 5 | 0.1 |
| 6 | 0.1 |   |   |

6. Craig Gagstetter is manager of Hi Line Sports Emporium. Occasionally he needs additional help on Saturdays. At such times, he must rely on some of the regular employees to supply the additional help. Define the random variables:

$X$ = the number of regular employees willing to work on Saturday.

$Y$ = the number of additional employees needed to work on Saturday.

Craig's experience has enabled him to give the probability distributions in Table 4-35.

a. Suppose $X$ and $Y$ are independent and compute the probabilities for the events "$X = x$ and $Y = y$."

b. Compute $P(X = Y)$.

TABLE 4-35    Probability Distributions for $X$ and $Y$

| $x$ | $P(x)$ | $y$ | $P(y)$ |
|---|---|---|---|
| 0 | 0.2 | 0 | 0.3 |
| 1 | 0.4 | 1 | 0.4 |
| 2 | 0.3 | 2 | 0.2 |
| 3 | 0.1 | 3 | 0.1 |

7. To win the game in the Statistics in Action example, Kelly and Steve had to get two balls with numbers that added up to 4. Suppose box A and box B each contain five balls numbered 1, three balls numbered 2, and two balls numbered 3. Compute the probability that Kelly and Steve will win with these boxes.

8. In a large manufacturing plant, two machines (call them A and B) produce No. 14 wood screws. Periodically ten screws from each machine are randomly selected and checked for defects. Careful records provide the probability distributions for $X$ and $Y$ in Table 4-36, where $X$ is the number of defective screws in the sample from A and $Y$ is the number of defective screws in the sample from B. Compute the probability that a given pair of samples of ten from each machine contains a total of exactly two defective screws.

TABLE 4-36    Probability Distributions for $X$ and $Y$

| $x$ | $P(x)$ | $y$ | $P(y)$ |
|---|---|---|---|
| 0 | 0.7 | 0 | 0.4 |
| 1 | 0.2 | 1 | 0.3 |
| 2 | 0.1 | 2 | 0.2 |
|  |  | 3 | 0.1 |

9. A school district in a large eastern city has approximately $N = 47,500$ children in the elementary schools. A mobile "sight unit" is provided by the district to check the eyesight of each of these children. Suppose Table 4-37 contains the relative frequency distribution for the number of sight defects per child observed.

a. Find the probability distribution for $\bar{X}$, where $\bar{X}$ is the sample mean number of sight defects in $n = 2$ children selected from the district.

b. Compute $\mu_{\bar{x}}$.

c. Compute $\sigma_{\bar{x}}$.

TABLE 4-37    Relative Frequencies of Sight Defects in Children Checked

| Number of Sight Defects | Relative Frequency |
|---|---|
| 0 | 0.6 |
| 1 | 0.3 |
| 2 | 0.1 |

**10.** A baseball player has played several seasons in the major leagues. Suppose Table 4-38 contains the relative frequency distribution for the number of base hits he got in games in which he batted exactly four times.

   a. Find a probability distribution for $\bar{X}$, where $\bar{X}$ is the sample mean number of base hits for $n = 2$ randomly selected games from this record.

   b. State the meaning of $P(\bar{X} = 1.5)$.

   c. Compute $\mu_{\bar{X}}$ to one decimal place.

   d. State the meaning of $\mu_{\bar{X}}$.

   e. Compute $\sigma_{\bar{X}}$ to one decimal place.

TABLE 4-38   Relative Frequencies of Base Hits in Four At-Bats

| Number of Base Hits | Relative Frequency |
|---|---|
| 0 | 0.20 |
| 1 | 0.45 |
| 2 | 0.25 |
| 3 | 0.08 |
| 4 | 0.02 |

**11.** The (approximate) relative frequency distribution of grades for students at Marymount College that completed Statistics 101 is shown in Table 4-39. Suppose $n = 2$ grades are randomly sampled from these records, and $\bar{X}$ is the sample mean of the two grades.

   a. Find a probability distribution for $\bar{X}$. Round each probability to three decimal places.

   b. State the meaning of $P(\bar{X} = 2.5)$.

   c. Compute $\mu_{\bar{X}}$ to one decimal place.

   d. State the meaning of $\mu_{\bar{X}}$.

   e. Compute $\sigma_{\bar{X}}$ to one decimal place.

TABLE 4-39   Relative Frequencies of Grades of Students who Completed Statistics 101

| Grade | Relative Frequency |
|---|---|
| F = 0 | 0.08 |
| D = 1 | 0.10 |
| C = 2 | 0.40 |
| B = 3 | 0.27 |
| A = 4 | 0.15 |

**12.** The (approximate) relative frequency distribution of the number of motorized vehicles (cars, motorcycles, and trucks) belonging to students at Marymount College is shown in Table 4-40. Suppose two students are randomly selected and $\bar{X}$ is the sample mean number of vehicles owned by the students.

   a. Find a probability distribution for $\bar{X}$. Round each probability to three decimal places.

   b. State the meaning of $P(\bar{X} = 1.5)$

   c. Compute $\mu_{\bar{X}}$ to one decimal place.

   d. State the meaning of $\mu_{\bar{X}}$.

   e. Compute $\sigma_{\bar{X}}$ to one decimal place.

TABLE 4-40   Relative Frequencies of Numbers of Student Vehicles

| Number of Vehicles | Relative Frequency |
|---|---|
| 0 | 0.22 |
| 1 | 0.55 |
| 2 | 0.15 |
| 3 | 0.08 |

# Exercises 4-7    Set B

1. Each of three boxes, A, B, and C, have ten balls with the numbers 1 and 2 painted on them. The frequencies of numbers in the boxes are shown in Table 4-41.

TABLE 4-41    Frequencies of Numbered Balls in Boxes A, B, and C

| Number | Frequency | | |
| | Box A | Box B | Box C |
|---|---|---|---|
| 1 | 8 | 7 | 5 |
| 2 | 2 | 3 | 5 |

An experiment consists of selecting one ball from each of the boxes. The following random variables are identified:

$W$ = the number on the ball from box A.

$X$ = the number on the ball from box B.

$Y$ = the number on the ball from box C.

Let $U$ be the sum of the random variables $W$, $X$, and $Y$.
a. List the eight possible combinations of values of the random variables $W$, $X$, and $Y$.
b. Using (a), list the possible values of $U$.
c. Use the relative frequencies in the table and independence of the random variables $W$, $X$, and $Y$ to complete a probability distribution for $U$.

2. Repeat Exercise 1 using the frequencies of numbers in the boxes shown in Table 4-42.

TABLE 4-42    Frequencies of Numbered Balls in Boxes A, B, and C

| Number | Frequency | | |
| | Box A | Box B | Box C |
|---|---|---|---|
| 1 | 8 | 6 | 4 |
| 2 | 2 | 4 | 6 |

3. Each of the three boxes, A, B, and C, have ten balls with the numbers 1, 2, and 3 painted on them. The frequencies of numbers in the boxes are shown in Table 4-43.
An experiment consists of selecting one ball from each of the boxes. The following random variables are identified:

$W$ = the number on the ball from box A.

$X$ = the number on the ball from box B.

$Y$ = the number on the ball from box C.

Let $U$ be the sum of the random variables.

TABLE 4-43   Frequencies of Numbered Balls in Boxes A, B, and C

| Number | Box A | Frequency Box B | Box C |
|--------|-------|-----------------|-------|
| 1 | 7 | 6 | 5 |
| 2 | 2 | 2 | 3 |
| 3 | 1 | 2 | 2 |

TABLE 4-44   Frequencies of Numbered Balls in Boxes A, B, and C

| Number | Box A | Frequency Box B | Box C |
|--------|-------|-----------------|-------|
| 1 | 8 | 5 | 4 |
| 2 | 1 | 3 | 2 |
| 3 | 1 | 2 | 4 |

a. List the 27 possible combinations of values of the random variables $W$, $X$, and $Y$.
b. Using (a), list the possible values of $U$.
c. Use the relative frequencies in the table and independence of the random variables $W$, $X$, and $Y$ to complete a probability distribution for $U$.

4. Repeat Exercise 3 using the frequencies of numbers in the boxes shown in Table 4-44.

# Chapter 4 Summary

## Definitions

1. A **random variable** is a quantity whose value depends upon the outcome of a chance experiment.
2. a. A random variable is **discrete** when only certain separated numbers are possible as values.
   b. A random variable is **continuous** when any value on an interval of numbers is possible.
3. If $X$ is a discrete random variable, then the **mean of $X$** is

$$\mu_X = \sum x_i \cdot P(x_i)$$

4. If $X$ is a discrete random variable, then the **variance of $X$** is

$$\sigma_X^2 = \sum (x_i - \mu_X)^2 \cdot P(x_i)$$

or

$$\sigma_X^2 = \sum x_i^2 \cdot P(x_i) - (\mu_X)^2$$

## Rules and Equations

The following steps are recommended for constructing a probability histogram for a discrete random variable $X$

Step 1 On a horizontal axis draw a tick mark for each of the possible values of $X$.

Step 2 Draw a vertical axis at the left end of the horizontal one.

Step 3 Construct a bar centered on each $x_i$ to a height that shows $P(x_i)$.

Step 4 Label the horizontal axis $x$ and the vertical axis $P(x)$.

If $X$ is a discrete random variable with set of possible values $\{x_1, x_2, \ldots, \}$ then:

1. $0 \le P(x_i) \le 1$   for each $x_i$
2. $\sum P(x_i) = 1$

If $Y$ is a continuous random variable with set of possible values $(m, M)$, then:

1. $0 \le P(y_1 < Y < y_2) \le 1$   for each interval $(y_1, y_2)$
2. $P(m < Y < M) = 1$

If $X$ is a binomial random variable, then:

$$P(x) = \binom{n}{x} p^x (1 - p)^{n-x} \quad \text{for } x = 0, \quad 1, \quad 2, \quad \ldots, \quad n$$

$$\binom{n}{x} = \frac{n!}{x!(n - x)!}$$

If $X$ is a binomial random variable for $n$ trials and $p$ is the probability of success on every trial, then:

$$\mu_X = np \quad \text{and} \quad \sigma_X^2 = np(1 - p)$$

If $Z$ is a standard normal random variable, then:

$$\mu_Z = 0 \quad \text{and} \quad \sigma_Z = 1$$

If $x$ is a possible value of a normally distributed random variable $X$ with mean $\mu_X$ and standard deviation $\sigma_X$, then:

$$z = \frac{x - \mu_X}{\sigma_X}$$

To approximate probabilities for a binomial random variable $X$ using a normal curve, choose the normal curve with:

$$\mu = np \quad \text{and} \quad \sigma^2 = np(1 - p)$$

To approximate probabilities associated with a binomial random variable $X$, use a $z$ score

$$z = \frac{bv - np}{\sqrt{np(1 - p)}} \quad \text{where } bv \text{ is a } \textbf{boundary value} \text{ for } X$$

Binomial probabilities can be approximated using a normal curve whenever both:

$$np \ge 5 \quad \text{and} \quad n(1 - p) \ge 5$$

For the sample mean $\bar{X}$:

$$\mu_{\bar{x}} = \sum \bar{x}_i \cdot P(\bar{x}_i)$$
$$\sigma_{\bar{x}}^2 = \sum \bar{x}_i^2 \cdot P(\bar{x}_i) - (\mu_{\bar{x}})^2$$

## Symbols

| | |
|---|---|
| $X$, $Y$, and $Z$ | capital letters representing random variable |
| $\mu_X$ | the **mean** of a random variable $X$ |
| $\sigma_X^2$ | the **variance** of a random variable $X$ |
| $\sigma_X$ | the **standard deviation** of a random variable $X$ |
| $\binom{n}{x}$ | the binomial coefficient $n! = n(n-1)(n-2)\ldots(3)(2)(1)$ |
| $Z$ | the symbol used for a random variable having the standard normal distribution |
| $z$ | a possible value of the random variable $Z$ |
| $n$ | the number of trials in a binomial experiment |
| $p$ | the probability of success on every trial in a binomial experiment |
| $\bar{X}$ | the sample **mean** of $n$ random variables $X_1, X_2, \ldots, X_n$ |
| $\mu_{\bar{X}}$ | the mean of the probability distribution of $\bar{X}$ |
| $\sigma_{\bar{X}}^2$ | the variance of probability distribution of $\bar{X}$ |

## Comparing Binomial Random Variables and Normal Random Variables

| If $X$ is a Binomial Random Variable, Then $X$ is *Discrete* |
|---|

| If $X$ is a Normal Random Variable, Then $X$ is *Continuous* |
|---|

The parameters for the probability distribution of $X$ are $n$, the number of trials, and $p$, the probability of success on each trial.

If $x$ is a possible value of $X$, then

$$P(X = x) = \binom{n}{x} p^x (1-p)^{n-x}$$

For given $n$ and $p$:

$$\mu_X = np \qquad \sigma_X = \sqrt{np(1-p)}$$

If $np \geq 5$ and $n(1-p) \geq 5$, then a normal curve with

$$\mu = np \qquad \sigma = \sqrt{np(1-p)}$$

and probabilities from the standard normal distribution can be used to approximate the binomial probabilities. Use *boundary values* to obtain the best approximation.

The parameters for the probability distribution of $X$ are $\mu$ and $\sigma$, the mean and standard deviation of the distribution.

$P(x_1 < X < x_2) = P(z_1 < Z < z_2)$, where $z_1$ and $z_2$ are the z scores respectively for $x_1$ and $x_2$.

$P(z_1 < Z < z_2)$ is found using the standard normal distribution in Appendix Table 4.

Approximately 68 percent of a normal distribution is within one standard deviation of the mean.

Approximately 95 percent of a normal distribution is within two standard deviations of the mean.

# Chapter 4 Review Exercises

*In Exercises 1 and 2, do each part:*
*a. List the possible values for X.*
*b. Compute a probability distribution for X.*
*c. Construct a probability histogram.*
*d. Compute $\mu_X$. State the meaning of $\mu_X$ in the context of the problem.*
*e. Compute $\sigma_X$ to one decimal place.*
*f. Mark on the histogram the values $\mu_X - \sigma_X$ and $\mu_X + \sigma_X$.*

1. Connie Juarez is an avid bowler. In bowling, a "frame" consists of, at most, two attempts by a a bowler to knock down ten pins. A "strike" means the bowler has knocked down all ten pins on the first roll of the ball in a frame. Connie estimates her probability of getting a strike in any frame at 0.35. A type of practice session she sometimes uses has the following two rules for its termination:

   Rule 1 Stop the session when any frame yields a strike.

   Rule 2 Stop the session at the end of five frames if Rule 1 isn't activated.

   Let X be the number of frames in the next practice session of this type.

2. A solid has eight faces of equal area. A 1 is painted on four faces, a 3 is painted on three faces, and a 5 is painted on one face. The solid is rolled twice, and X is the sum of the numbers on the upturned faces.

3. A company that manufactures modular homes has offices in seven states. They spend a large amount of money listing in their toll-free telephone numbers in every Yellow Pages directory in the territories covered by their company. An extensive study showed that 70 percent of the calls their offices receive were made as a result of this advertising. Let Y be the number of individuals in the next four calls who get the telephone number from the Yellow Pages. Using 0.70 as the probability that a given individual gets the number from the Yellow Pages, we can compute the following probabilities.

$$P(0) = 0.008 \quad P(1) = 0.076$$
$$P(2) = 0.265 \quad P(3) = 0.412$$
$$P(4) = 0.240$$

   a. Compute $\mu_Y$ to one decimal place.
   b. Compute $\sigma_Y^2$ to two decimal places.
   c. Compute $\sigma_Y$ to one decimal place.
   d. Construct a probability histogram and locate $\mu_Y$ on the graph.

4. An intersection in the south area of a small western city was the site of three accidents within a short time span. According to state law, a motorist should be prepared to stop when approaching such an intersection. Extensive studies by the highway patrol show that 68 percent of all drivers in the state approach such intersections in the prescribed manner. Let W be the number of drivers in the next five crossing that intersection who approach it at the proper speed. Using 0.68 as the probability that a given driver approaches the intersection properly, the probabilities for W are:

$$P(0) = 0.003 \quad P(1) = 0.036$$
$$P(2) = 0.152 \quad P(3) = 0.322$$
$$P(4) = 0.342 \quad P(5) = 0.145$$

   a. Compute $\mu_W$ to the one decimal place.
   b. Compute $\sigma_W^2$ to two decimal places.
   c. Compute $\sigma_W$ to one decimal place.
   d. Construct a probability histogram and locate $\mu_W$ on the graph.

5. A vehicle inspection site in a large state in the East is staffed by state employees. Records at this site show that approximately 23 percent of the vehicles stopped do not have insurance. For the next six vehicles inspected at this site, let X be the number that do not have insurance.
   a. Argue that X can be modeled as binomial.
   b. List the possible values for X.

c. Compute the probability distribution for $X$.

d. Construct a probability histogram.

e. Compute $\mu_X$ and locate it on the histogram.

f. Compute $\sigma_X$ to one decimal place.

g. What possible values of $X$ are within one standard deviation of $\mu_X$?

6. According to the information in Figure 4-44, on a certain Friday 1,966 stock issues were traded. Of these issues, 937, or about 48 percent advanced in price. Suppose seven stock prices are randomly selected for this day of trading and $X$ is the number that advanced on this day of trading.

a. Argue that $X$ can be modeled as binomial.

b. List the possible values for $X$.

c. Compute the probability distribution for $X$, using 0.48 for $p$.

d. Construct a probability histogram.

e. Compute $\mu_X$ and locate it on the histogram.

f. Compute $\sigma_X$ to one decimal place.

g. What possible values of $X$ are within one standard deviation of $\mu_X$?

7. A study reports that 40 percent of today's teenagers feel that a religious ceremony is important to a marriage. Suppose the 40 percent figure applies to teenagers in Dallas and ten teenagers from the city are interviewed. Let $X$ be the number in the sample that feel a religious ceremony is important. Use Appendix Table 3 to compute:

a. $P(X = 4)$  b. $P(X \le 3)$

c. $P(X > 6)$  d. $P(4 \le X \le 7)$

8. A study by the Federal Aeronautics Administration suggests that 70 percent of female flight attendents have been sexually harassed by male passengers at least once in their career. Suppose that 12 female flight attendants from major airlines are questioned, and $Y$ is the number in the sample that have experienced at least one such incident. Use Appendix Table 3 to compute:

a. $P(Y = 8)$  b. $P(Y < 5)$

c. $P(Y \ge 10)$  d. $P(6 \le Y \le 9)$

9. The scores on one version of the USET (United States Educational Test) have a bell-shaped distribution, with mean 80.0 and standard deviation 12.0. If $X$ is the score of a randomly selected individual that took the test, find:

a. $P(X < 65.0)$  b. $P(X < 100.0)$

c. $P(X > 105.0)$  d. $P(90.0 < X < 96.0)$

10. The Corona Bean Corporation packages and sells navy beans in bags marked 32 ounces. The distribution of net weights of a large production run of these bean bags is bell shaped, with $\mu = 33.5$ ounces and $\sigma = 0.8$ ounces. Suppose a bag from this production run is randomly selected and $Y$ is the weight of the contents. Compute:

a. $P(Y < 32.0)$  b. $P(Y > 34.0)$

c. $P(Y > 31.6)$  d. $P(33.0 < Y < 35.0)$

| WHAT THE MARKET DID | | | | | | |
|---|---|---|---|---|---|---|
| | Fri. | Thur. | Wed. | Tue. | Mon. | Wk. Ago |
| Volume*............... | 177,190 | 200,767 | 187,980 | 162,350 | 168,850 | 161,310 |
| Issues ................. | 1,966 | 1,980 | 1,979 | 1,995 | 1,979 | 1,975 |
| Advances ............. | 937 | 804 | 1,219 | 975 | 584 | 782 |
| Declines............... | 556 | 717 | 354 | 539 | 963 | 737 |
| Unchanged........... | 473 | 459 | 406 | 481 | 432 | 456 |
| New Highs............ | 6 | 7 | 8 | 6 | 8 | 9 |
| New Lows............. | 5 | 6 | 5 | 7 | 16 | 10 |
| * NYSE Only: Thousands of Shares | | | | | | |

**Figure 4-44.**  What the market did.

11. A pentagon official claims that approximately 20 percent of military aircraft cannot be flown because of missing parts. Suppose a sample of 40 military aircraft are inspected and $X$ is the number that cannot be flown because of missing parts.
    a. Justify using a normal curve to approximate probabilities associated with $X$.
    b. Approximate $P(6 \leq X \leq 10)$.
    c. State the meaning of the answer to part (b).

12. The magazine, *Medical Economics*, reported that a person has a 30 percent chance of avoiding a fine on a traffic citation if he or she takes the issue to court. Forty-three people who fight traffic citations by taking their tickets to court are surveyed, and $X$ is the number that avoid a fine.
    a. Justify using a normal curve to approximate probabilities associated with $X$.
    b. Approximate $P(10 \leq X \leq 15)$.
    c. State the meaning of the answer to part (b).

13. a. Make a table giving probabilities for the events "$X = x$ and $Y = y$" based on the probability distributions for $X$ and $Y$ given in Table 4-45. Suppose $X$ and $Y$ are independent variables.
    b. Using the values in the table, state which event "$X = x$ and $Y = y$" is most likely to occur.
    c. Which event is least likely to occur?
    d. Find $P(X < Y)$.

TABLE 4-45    Probability Distributions for $X$ and $Y$

| $x$ | $P(x)$ | | $y$ | $P(y)$ |
|-----|--------|---|-----|--------|
| 1 | 0.2 | | 2 | 0.5 |
| 2 | 0.3 | | 4 | 0.3 |
| 3 | 0.4 | | 6 | 0.2 |
| 4 | 0.1 | | | |

## Computer Generated Distributions

**Key Topics**

1. *Probability distributions*

2. *Comparing binomial and normal distributions*

### Probability distributions

Minitab has a command PDF (Probability Density Function) that may be used to compute probability distributions. Suppose the distribution for a binomial random variable $X$ with $n = 10$ and $p = 0.45$ is needed. First the possible values of $X$ are placed in one column of the Minitab worksheet using the SET command. Then PDF is used with the subcommand BINOMIAL to place the related probabilities into another column. Figure 4-45 shows the Minitab commands and the resulting table.

```
MTB> SET C1
DATA> 0:10
DATA> END
MTB> PDF C1 C2;
SUBC> BINOMIAL N = 10 P = .45.
MTB> NAME C1 = 'X' C2 = 'P(X)'
MTB> PRINT C1 C2
   ROW    X       P(X)
     1    0     0.002533
     2    1     0.020724
     3    2     0.076303
     4    3     0.166478
     5    4     0.238367
     6    5     0.234033
     7    6     0.159568
     8    7     0.074603
     9    8     0.022890
    10    9     0.004162
    11   10     0.000341
```

**Figure 4-45.** Binomial probabilities.

313

The command NAME C1 = 'X'   C2 = 'P(X)' was included so that the table would have column headings.

The PDF command may be used with several other subcommands besides BINOMIAL. Some of these are NORMAL, UNIFORM, T, F, and CHISQUARE.

A cumulative probability table may be constructed for any of the above random variables by using the CDF (Cumulative Density Function) command. Such a table for the binomial variable $X$ with $n = 10$ and $p = 0.45$ is shown in Figure 4-46. From this table, we see that $P(X \leq 3) = 0.26604$ and $P(X \leq 9) = 0.99550$.

```
MTB> CDF C1 C3;
SUBC> BINOMIAL N = 10 P = .45.
MTB> NAME C3 = 'CUMPROB'
MTB> PRINT C1 C3

 ROW    X    CUMPROB
  1     0    0.00253
  2     1    0.02326
  3     2    0.09956
  4     3    0.26604
  5     4    0.50440
  6     5    0.73844
  7     6    0.89801
  8     7    0.97261
  9     8    0.99550
 10     9    0.99966
 11    10    1.00000
```

```
MTB> SET C1
DATA> -4:5
DATA> END
MTB> CDF C1 C2;
SUBC> NORMAL MU = 0 SIGMA = 1.
MTB> NAME C1 = 'Z' C2 = 'P(Z)'
MTB> PRINT C1 C2

 ROW    Z      P(Z)
  1    -4    0.00003
  2    -3    0.00135
  3    -2    0.02275
  4    -1    0.15866
  5     0    0.50000
  6     1    0.84134
  7     2    0.97725
  8     3    0.99865
  9     4    0.99997
 10     5    1.00000
```

**Figure 4-46.** Cumulative binomial probabilities.   **Figure 4-47.** A normal probability distribution.

Figure 4-47 shows part of a table of cumulative probabilities for the standard normal distribution.

## Comparing binomial and normal distributions

You have learned that when $n$ is sufficiently large, a binomial distribution is approximately the same as a normal distribution with the same mean and variance. The graph in Figure 4-48 shows the locations of the midpoints of the tops of the bars of a binomial probability histogram, when $n = 50$ and $p = 0.3$

For this distribution, $\mu_X = 15$ and $\sigma_X$ is approximately 3.24. Figure 4-49 shows points on the normal curve that has these same parameters.

315

```
MTB> SET C1
DATA> 0:50
DATA> END
MTB> PDF data in C1 into C2;
SUBC> BINOMIAL N = 50 P = .3.
MTB> PLOT C2 C1
```

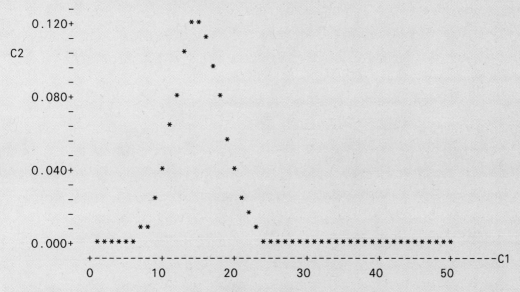

**Figure 4-48.** Graph of binomial probability distribution with $n = 50$ and $p = 0.3$.

```
MTB> PDF C1 C3;
SUBC> NORMAL MU = 15 SIGMA = 3.24.
MTB> PLOT C3 C1
```

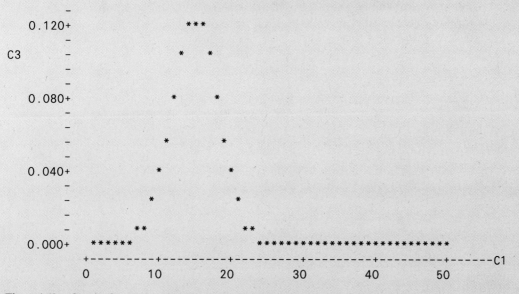

**Figure 4-49.** Graph of normal probability distribution with $\mu = 15$ and $\sigma = 3.24$.

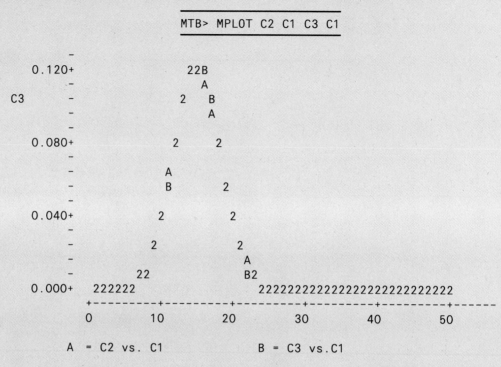

**Figure 4-50.** Combined plots of binomial and approximately normal probability distributions.

To show how closely these distributions match, we may place both graphs on the same axis by using the MPLOT command. The graph in Figure 4-50 was made in this way. In this graph, the As mark the locations of ordinates on the binomial probability histogram, whereas the Bs mark the location of corresponding ordinates for the normal idealized histogram. The 2s indicate two points so close together that they could not be plotted as separate points.

# Exercises

1. Use the table in Figure 4-46. Find:
   a. $P(X \leq 5)$      b. $P(X > 6)$     c. $P(2 \leq X \leq 6)$    d. $P(X \leq 2 \text{ or } X \geq 8)$
2. Use the table in Figure 4-47. Find:
   a. $P(Z < -1)$    b. $P(Z > 2)$     c. $P(Z < -2)$     d. $P(-2 < Z < 1)$

## SCIENTIFIC
### AND
## GRAPHING

If $X$ is a binomial random variable, then the possible values for $X$ are 0, 1, 2, 3, . . . , $n$, where $n$ is the number of trials in the experiment. Letting $p$ represent the probability of success on every trial and $(1 - p)$ represent the probability of failure, the probability that $X$ takes on anyone of the $n + 1$ possible values of $X$ can be calculated using the general binomial probability formula.

$$P(X = x) = \binom{n}{x} p^x (1 - p)^{n - x}$$

The right side of the equation contains three factors:

1. The coefficient $\binom{n}{x}$, that indicates the number of different sequences in which $x$-successes can occur in $n$-trials.
2. The power $p^x = \underbrace{p \cdot p \cdots p}_{x\text{-factors}}$
3. The power $(1 - p)^{n-x} = \underbrace{(1 - p)(1 - p) \ldots (1 - p)}_{(n-x)\text{-factors}}$

To illustrate how a scientific calculator or TI-81 graphics calculator can be used to make the 3 calculations, we will use the following values:

$$n = 5 \qquad x = 3 \qquad (n - x) = 2 \qquad p = 0.25 \qquad (1 - p) = 0.75$$

Thus, $P(X = 3) = \binom{5}{3}(0.25)^3(0.75)^2$

**To calculate** $\binom{5}{3}$

A. A scientific calculator

*Press* 5 [SHIFT] [nCr] 3 [=]

*Display shows*  **10**

B. The TI-81 calculator

*Press* 5 [MATH] [▶] [▶] [▶] [▼] [▼]

[ENTER] 3 [ENTER]

*Display shows*  **10**

**To calculate $(0.25)^3$**

A. A scientific calculator

Press 0.25 $x^y$ 3 $=$

Display shows　0.015625

B. The TI-81 calculator

Press 0.25 $\wedge$ 3 ENTER

Display shows　.015625

**To calculate $(1 - 0.25)^{5-3}$**

A. A scientific calculator

Press $(1 - 0.25)$ $x^y$ $(5 - 3)$ $=$

Displays shows　0.5625

B. The TI-81 calculator

Press $(1 - 0.25)$ $\wedge$ $(5 - 3)$ ENTER

Display shows　.5625

The product of the three factors is $P(x = 3)$. The complete sequences of key strokes are given for both calculators.

A. A scientific calculator

| Press | Display shows | Press | Display shows |
|---|---|---|---|
| 5 | 5. | $-$ | 1. |
| SHIFT nCr | 5. | 0.25 | 0.25 |
| 3 | 3. | $)$ | 0.75 |
| $\times$ | 10 | $x^y$ | 0.75 |
| 0.25 | 0.25 | $($ | [01　　0. * |
| $x^y$ | 0.25 | 5 | 5. |
| 3 | 3. | $-$ | 5. |
| $\times$ | 0.15625 | 3 | 3. |
| $($ | [01　　0. * | $)$ | 2. |
| 1 | 1. | $=$ | 0.087890625 |

---

* The display for this step depends on the type of calculator being used.

B. The TI-81 calculator

| Press | Display shows |
|---|---|

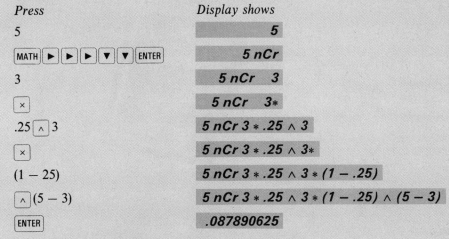

Thus, $P(x = 3) = 0.088$ to three decimal places.

**Name:** Charles B. Sampson
**Occupation:** Consulting Statistician
**Employer:** Eli Lilly and Company, Indianapolis, Indiana

**A**s I planned my college career, I was continually encouraged to study engineering. This was because I seemed to have an aptitude for mathematics and science, and many of the high school advisors of that era (Sputnik) were of the opinion that the most important discipline for overtaking Russia's lead in the space program was engineering. However, I soon discovered that engineering was only one of my interests and that mathematics and physics were the most exciting courses I was exposed to during my first year of college. So I changed my majors to mathematics and physics and continued my studies with little knowledge of the job market or of other areas of study that most suited my interests in science and mathematics and my lack of aptitude for laboratory work. If there were counselors who were aware of the science and profession of statistics, I certainly did not find any. I entered graduate school in mathematics with the goal of becoming a teacher or working in industry as some sort of applied mathematician. However, at the University of Iowa, two professors of statistics had fantastic reputations as teachers and, accordingly, I chose an elective in probability and statistics. These courses and the exposure to statistics were major events in my life although I didn't realize it for some time.

As I completed my M.S. degree in mathematics, I decided that it made sense for me to work for a while before continuing my education. The placement office advised me that my best chances for employment were as a computer programmer or as a statistician. I did not know for sure what a statistician did (the courses I had in statistics and probability were theory courses), but I pursued a position anyway and began my career as a statistician in quality engineering for IBM in upstate New York. I soon learned that statistics was an exciting career and that I did not have enough education to be effective as a consulting statistician, so I attended Iowa State University and received a Ph.D. in statistics.

I am now the Director of Statistical and Mathematical Sciences for Eli Lilly and Company, a research intensive pharmaceutical firm. As a consulting statistician, I have the opportunity to work with scientists, physicians, engineers, psychologists, and businesspersons. I have helped design experiments, and subsequently analyzed and reported on the results of these experiments in basic research, clinical trials, process and product development, quality engineering, marketing, manufacturing, and human resources. Consulting in these different and varied areas of application is most exciting and provides for a stimulating and rewarding career. The field of statistics is increasingly being recognized as one that should play a larger role in decision- and policy-making in the face of uncertainty and variability. The job market for M.S. and Ph.D. statisticians has doubled in the last five years and is expected to double again in the next five years, resulting in extremely rewarding and interesting (and well-paid) careers for those who choose statistics as a profession.

For those of you interested in a career in statistics, I suggest getting an excellent background in mathematics, taking as many statistics courses as possible in undergraduate school. Then go to graduate school in statistics and seek experience by using summer employment and/or internship opportunities in statistics.

**Birth Statistics.** The National Center for Health Statistics provisionally estimated U.S. births in 1988 at 3,913,-000, nearly 3% more than the 3,809,394 registered in 1987. This was the largest number reported since 1964, the last year of the "baby boom" that began in 1947. The estimated birthrate was 15.9 births per 1,000 population, 1% higher than the 1987 rate of 15.7, and the fertility rate was 67.3 births per 1,000 women aged 15–44, 2% above the 1987 figure of 66.1.

**Figure 5-1.** "Birth Statistics," *1990 Britannica Book of the Year*, p. 297. Copyright © 1990 by Encyclopaedia Britannica, Inc. Reprinted by permission.

# Populations and Population Movements

## DEMOGRAPHY

World population stood at 5,234,000,000 as of July 1, 1989, according to Population Reference Bureau estimates, and was growing by about 93 million, or 1.78%, a year, up from 1.73% in 1988. These figures were higher than United Nations medium ("most probable") projections for 1989 prepared in the early 1980s, largely because of an unexpected rise in China's birthrate since 1985 and slower fertility decline and faster death rate decline than projected in India. The estimated mid-1989 populations of China (1,103,900,000) and India (835 million) together made up 37% of the world total. According to revised UN medium projections released in 1988, world population would pass the six billion mark late in 1997, ten years after reaching five billion, and would be 6,251,000,000 in 2000 and 8,467,000,000 in 2025.

**Figure 5-2.** "Population and Population Movements," *1990 Britannica Book of the Year*, pp. 296 & 297. Copyright © 1990 by Encyclopaedia Britannica, Inc. Reprinted by permission.

# 5

# Confidence Intervals for $\mu$ and $P$

Figures 5-1 and 5-2 are examples of the kind of data based reports that are seen almost daily in newspapers. Figure 5-1 contains information from the National Center for Health Statistics that includes a *provisional estimate* of U.S. births in 1988 of 3,913,000. Figure 5-2 contains information from the Population Reference Bureau that includes an estimate of 5,234,000,000 for the world's population. These reports illustrate the way in which population parameters may be estimated. In this chapter, we will study how sample statistics are related to population parameters and the way in which they may be used to estimate unknown population parameters.

## 5-1 The Distribution of the Sample Mean

**Key Topics**

1. The statistic $\bar{X}$ is a random variable

2. $\mu_{\bar{X}} = \mu$  and  $\sigma_{\bar{X}}^2 = \dfrac{\sigma^2}{n}$

3. The sampling distribution of $\bar{X}$ is approximately normal for large n

4. Changing $\bar{X}$, a value of $\bar{X}$, to a z

5. Conditions under which the sampling distribution of $\bar{X}$ is normal for all n

---

## S T A T I S T I C S   I N   A C T I O N

Gladys Washington is store manager of Savemore Foods No. 2107. She wanted an estimate of the average amount spent per customer. Gladys was only interested in an estimate of the average spent per customer for the "primary hours of shopping," namely 9 AM to 7 PM. Although the market opened at 7 AM and closed at 10 PM all seven days of the week, she knew that the first 2 hours and last 3 hours were dominated by "minipurchases" of five items or less. Gladys sampled register receipts several times a day during the primary shopping period until she had collected information on the purchases of 40 customers. The totals (in dollars and cents) indicated on the sampled receipts are listed in Table 5-1.

For these 40 data points:

$$\bar{x} \approx \$57.11 \quad \text{and} \quad s \approx \$44.40$$

Gladys knew that the sample mean of these 40 data points was the value of a random variable because another sample of 40 receipts would undoubtedly yield a different value of $\bar{x}$. However, these data points gave Gladys an indication of the mean dollar amount spent per customer during the hours targeted by the survey. At the national sales meeting in Dallas, Gladys intends to use this value as a *point estimate* for the mean dollar purchase figure for her store.

She also intends to mention in her report that the large sample standard deviation (namely $44.40) reflects the large variability in the amounts of groceries purchased by customers. She will use this fact to make some suggestions on how markets might structure their checkout counters to reduce the time small purchase customers must wait in line when "trapped" behind someone making a large grocery purchase.

**TABLE 5-1  Amounts on Sampled Register Receipts**

| | | | | | | | |
|---|---|---|---|---|---|---|---|
| 73.15 | 35.80 | 88.40 | 142.19 | 43.80 | 173.80 | 9.15 | 34.50 |
| 19.82 | 12.85 | 22.68 | 38.88 | 10.92 | 84.21 | 39.40 | 13.20 |
| 137.32 | 49.60 | 7.49 | 94.10 | 24.63 | 63.60 | 72.75 | 49.25 |
| 64.90 | 42.72 | 56.15 | 101.17 | 97.10 | 44.78 | 83.10 | 26.04 |
| 53.32 | 10.21 | 79.42 | 48.80 | 16.20 | 184.10 | 30.17 | 4.76 |

## The statistic $\bar{X}$ is a random variable

When a simple random sample of size $n$ is selected from a population of quantitative data, the following quantities:

$$X_1 = \text{the first value selected}$$
$$X_2 = \text{the second value selected}$$
$$\vdots$$
$$X_n = \text{the } n\text{th value selected}$$

are all random variables. As a consequence, $\bar{X}$, the mean of the $n$ values, is also a random variable. Because it is the process of sampling that introduces the element of chance, the probability distributions of such sample statistics are often termed *sampling distributions*. For example, the probability distribution of $\bar{X}$ is frequently called the *sampling distribution of $\bar{X}$*.

$$\mu_{\bar{X}} = \mu \quad \text{and} \quad \sigma_{\bar{X}}^2 = \frac{\sigma^2}{n}$$

The first task we want to accomplish in our discussion of the sampling distribution of $\bar{X}$ is to explain the relationships that exist between $\mu_{\bar{X}}$ and $\mu$ and between $\sigma_{\bar{X}}^2$ and $\sigma^2$. The symbols $\mu_{\bar{X}}$ and $\sigma_{\bar{X}}^2$ represent respectively the mean and variance of the probability distribution of $\bar{X}$. As before, $\mu$ and $\sigma^2$ represent respectively the mean and variance of the population being sampled.

In Set B of this section, four exercises are provided with guided instructions intended to make the following pair of equations intuitively agreeable. In each exercise, a different-sized sample from a large population is considered. The probability distribution for the resulting random variable $\bar{X}$ is determined, and the methods for computing $\mu_{\bar{X}}$ and $\sigma_{\bar{X}}^2$ studied in Section 4-7 are used to compute the corresponding mean and variance.

### The Mean and Variance of the Probability Distribution of $\bar{X}$

If a population is large enough so that a model of independence can be used for $X_1, X_2, X_3, \ldots, X_n$, then the sampling distribution of $\bar{X}$ has:

$$\mu_{\bar{X}} = \mu \tag{1}$$

$$\sigma_{\bar{X}}^2 = \frac{\sigma^2}{n} \tag{2}$$

where $\mu$ and $\sigma^2$ are respectively the mean and variance of the sampled population.

▶ **Example 1**   At Diablo Valley Community College, student profile records indicate that the mean distance that students travel to get to school is 5.8 miles. The variance for the distribution of distances is 13.35 (miles)$^2$. Counselor Ed Stupka randomly samples these records and obtains the commuting distances of 30 students. For the sampling distribution of $\bar{X}$ based on a sample of size 30:

**a.** Compute $\mu_{\bar{X}}$.
**b.** Compute $\sigma_{\bar{X}}^2$.
**c.** Compute $\sigma_{\bar{X}}$ to one decimal place.

**Solution**   **a.** $\mu_{\bar{X}} = 5.8$ miles        $\mu_{\bar{X}} = \mu$

**b.** $\sigma_{\bar{X}}^2 = \dfrac{13.35}{30}$        $\sigma_{\bar{X}}^2 = \dfrac{\sigma^2}{n}$

  $= 0.445$

**c.** $\sigma_{\bar{X}} = \sqrt{0.445}$        $\sigma_{\bar{X}} = \sqrt{\sigma_{\bar{X}}^2}$

  $\approx 0.7$        Rounded to one place   ◀

## The sampling distribution of $\bar{X}$ is approximately normal for large $n$

Equations (1) and (2) are important for statistical purposes because for large $n$ they guarantee that the sampling distribution of $\bar{X}$ will be tightly grouped about $\mu$. As a consequence, if the sample size is "large," one is practically assured of obtaining a value of $\bar{X}$ near the population mean. Furthermore, the larger the sample size $n$, the more likely an observed value of $\bar{X}$ is to be close to the population mean.

However, knowing the mean and variance for the probability distribution of $\bar{X}$ does not give a complete description of the sampling distribution. That is, even though $\mu_{\bar{X}}$ and $\sigma_{\bar{X}}^2$ are known in terms of $\mu$ and $\sigma^2$, it is not obvious how to find probabilities for $\bar{X}$. The techniques used in Section 4-7 require lengthy (and very tedious) calculations. One method for avoiding these difficulties is relevant when sample sizes are large and is described in the following important theorem:

### The Central Limit Theorem (CLT)

If $X_1, X_2, X_3, \ldots, X_n$ can be described as independent random variables with a common distribution, then for large $n$ the probability distribution of $\bar{X}$ is approximately normal.

The CLT implies that if $n$ is large (and yet a small enough fraction of $N$ so that a model of independence for $X_1, X_2, X_3, \ldots, X_n$ is reasonable), we may obtain approximate probabilities for $\bar{X}$ using the $Z$-distribution. There is no really satisfactory answer to the question of *how large n* should be to justify normal approximations for probabilities associated with $\bar{X}$. The problem is that how large $n$ should be actually depends to some degree on the shape of the population sampled. But in this text, we will use the rule of thumb that $n \geq 30$ is typically large enough to justify using a normal curve approximation for computing probabilities for $\bar{X}$. For some populations, $n = 30$ is really not large enough to produce good approximations. For other populations, a smaller $n$ is sufficient. However, conventional statistical wisdom is that for most real populations, a sample size of 30 is large enough.

## Changing $\bar{x}$, a value of $\bar{X}$, to a $z$

If $n$ is large enough that the probability distribution of $\bar{X}$ is approximately normal, then the standard normal distribution can be used to compute probabilities associated with $\bar{X}$. Equation 3 below can be used to change a given $\bar{x}$ to a $z$ score.

### To Find a $z$ Score Corresponding to $\bar{x}$

If $\bar{x}$ is the observed mean of a random sample of size $n$ selected from a population with mean $\mu$ and standard deviation $\sigma$, then a $z$ score for $\bar{x}$ is:

$$z = \frac{\bar{x} - \mu}{\dfrac{\sigma}{\sqrt{n}}} \tag{3}$$

Notice that equation (3) is a form of the general relationship:

$$z = \frac{x - \mu}{\sigma}$$

where $x$ is now $\bar{x}$

$\mu$ is now $\mu_{\bar{x}}$ and $\mu_{\bar{x}} = \mu$

$\sigma$ is now $\sigma_{\bar{x}}$ and $\sigma_{\bar{x}} = \dfrac{\sigma}{\sqrt{n}}$

In this text, when a $z$ is computed, no number will be rounded before the last calculation. Then the result will be rounded to two decimal places.

▶ **Example 2**    Records from the telephone company serving Knoxville indicate that an average of $4.30 per day is deposited in each public pay telephone in the area. The records further indicate that the standard deviation for the distribution of amounts is $1.47. A telephone field supervisor intends to check 49 such pay telephones next Wednesday. Let $\bar{X}$ be the mean amount of cash deposited in these phones on that day.

**a.** Compute $P(\bar{X} > \$4.50)$.

**b.** Sketch the area of an idealized probability histogram that represents the probability in part (a).

**c.** State, in the context of the activity, what the answer to part (a) means.

**Solution**   **Discussion.** Because $49 > 30$ we will assume the sampling distribution of $\bar{X}$ is approximately normal,

**a.** with $\bar{x} = 4.50$, $\mu = 4.30$, $\sigma = 1.47$, and $n = 49$:

$$z = \frac{4.50 - 4.30}{\dfrac{1.47}{\sqrt{49}}} \qquad z = \frac{\bar{x} - \mu}{\dfrac{\sigma}{\sqrt{n}}}$$

$$= 0.95 \qquad\qquad \text{Rounded to two decimal places}$$

$$P(\bar{X} > \$4.50) \approx P(Z > 0.95) \qquad \text{Writing the probability in terms of } Z$$

$$= 1 - 0.8289 \qquad \text{Using Appendix Table 4}$$

$$= 0.1711 \qquad\quad \text{The probability is about 17 percent.}$$

**b.** Because $\bar{x} = 4.50$ converts to $z = 0.95$, we know that $\$4.50$ is about 0.95 standard deviations above $\mu$. The area under a normal curve to the right of this value is equal to the indicated probability, and is shaded in the idealized histogram in Figure 5-3.

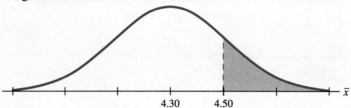

**Figure 5-3.**   Shaded area $= P(\bar{X} > \$4.50)$.

**c.** The probability is approximately 17 percent that the supervisor will get a mean of more than $\$4.50$ from the 49 telephones.   ◄

## Conditions under which the sampling distribution of $\bar{X}$ is normal for all $n$

Earlier we said the techniques of Section 4-7 require tedious calculations. One way to avoid these calculations is to use the central limit theorem when the sample size $n$ is large, (usually $n \geq 30$). If $n$ is small (usually $n < 30$), then a normal distribution can still be used for $\bar{X}$ if we have evidence, or a strong reason, to believe that the population being sampled is itself bell shaped.

## The Distribution of $\bar{X}$ When the Population Sampled Is Bell Shaped

> If $X_1, X_2, X_3, \ldots, X_n$ can be modeled as *independent normal random variables* (that is, if they are the results of $n$ random selections from a large population that has an approximately bell-shaped relative frequency distribution), then the sampling distribution of $\bar{X}$ will be normal for any $n$, large or small.

▶ **Example 3**   The engineering school at Marymount College gives an admissions test to potential engineering students. Each year the grading is adjusted so that the population of test scores has a bell-shaped distribution. Furthermore the mean and standard deviation are respectively 500 and 100. Suppose $n = 4$ scores are randomly selected from last year's scores.

Let $\bar{X}$ be the mean score of the sampled tests.

**a.** Compute $P(460 < \bar{X} < 540)$

**b.** Sketch the area on an idealized histogram that represents the probability of part (a).

**c.** State, in the context of the activity, what the answer to part (a) means.

**Solution**   **Discussion.** The relative frequency distribution of the population of test scores is bell shaped. Therefore the sampling distribution of $\bar{X}$ will also be bell shaped for any sample size $n$ and, in particular, for $n = 4$.

**a.** With $\mu = 500$, $\sigma = 100$, and $n = 4$:

$$\textbf{For } \bar{x}_1 = \textbf{460:} \qquad \textbf{For } \bar{x}_2 = \textbf{540:}$$

$$z_1 = \frac{460 - 500}{\dfrac{100}{\sqrt{4}}} \qquad z_2 = \frac{540 - 500}{\dfrac{100}{\sqrt{4}}}$$

$$= -0.80 \qquad\qquad = 0.80$$

Thus,

$$P(460 < \bar{X} < 540) \approx P(-0.80 < Z < 0.80)$$
$$= 0.7881 - 0.2119$$
$$= 0.5762$$

**b.** Because $\bar{x}_1$ and $\bar{x}_2$ convert to $z$ scores $-0.80$ and $0.80$ respectively, we know these values are 0.80 standard deviations below and above $\mu$. The indicated probability is the shaded area in Figure 5-4 on the next page.

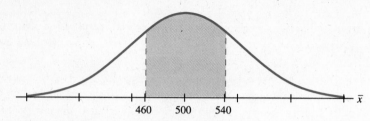

**Figure 5-4.**   Shaded area $= P(460 < \bar{X} < 540)$.

c. The probability is approximately 58 percent that the mean of the four sampled test scores will be between 460 and 540.   ◀

Notice that the solution to Example 3 was similar to the solution of Example 2. The new element in this example is *conceptual* rather than *operational*. Specifically for small $n$, it must first be reasonable to assume that the population sampled has an approximately bell-shaped distribution before treating $\bar{X}$ as normally distributed.

# Exercises 5-1    Set A

*In Exercises 1–4, compute $\mu_{\bar{x}}$ and $\sigma_{\bar{x}}$ for each sampling distribution of $\bar{X}$. Round each answer to one decimal place.*

1. The mean and standard deviation of the sampled population are respectively 100.0 and 20.0.
   a. $n = 64$     b. $n = 225$

2. The mean and standard deviation of the sampled population are respectively 240.0 and 36.0.
   a. $n = 36$     b. $n = 256$

3. The mean and standard deviation of the sampled population are respectively 45.3 and 4.2.
   a. $n = 49$     b. $n = 196$

4. The mean and standard deviation of the sampled population are respectively 82.1 and 6.3.
   a. $n = 81$     b. $n = 441$

5. In a union publication, it was stated that the mean salary of workers in the union is $1800 per month. Assume this value applies to members of this union in Detroit. For this population of workers, $\sigma = \$240$. Suppose 36 workers in this union in Detroit are randomly selected, and $\bar{X}$ is the mean monthly salary for these workers.

   a. Compute $\mu_{\bar{x}}$.
   b. Compute $\sigma_{\bar{x}}$.
   c. Sketch and label a graph of the sampling distribution of $\bar{X}$, showing values for $\mu_{\bar{x}} - 2\sigma_{\bar{x}}$, $\mu_{\bar{x}} - \sigma_{\bar{x}}$, $\mu_{\bar{x}}$, $\mu_{\bar{x}} + \sigma_{\bar{x}}$, and $\mu_{\bar{x}} + 2\sigma_{\bar{x}}$.
   d. Compute the $z$ score for $\bar{x} = \$1825$.
   e. State the meaning of the $z$ score of part (d).

f. Compute $P(\bar{X} < \$1825)$.

g. State the meaning of the answer to part (f).

6. Shanley's Bar and Grill is a favorite restaurant in a large city in the South. Management at Shanley's has kept statistics on the dollar amount spent per customer for dinner. Records indicate that the population mean amount spent per customer is $13.26 and the standard deviation is $4.20. Suppose the amounts spent by 36 customers are randomly sampled, and $\bar{X}$ is the sample mean amount spent on dinner by these customers.

a. Compute $\mu_{\bar{x}}$.

b. Compute $\sigma_{\bar{x}}$.

c. Sketch and label a graph of the sampling distribution of $\bar{X}$, showing values for $\mu_{\bar{x}} - 2\sigma_{\bar{x}}$, $\mu_{\bar{x}} - \sigma_{\bar{x}}$, $\mu_{\bar{x}}$, $\mu_{\bar{x}} + \sigma_{\bar{x}}$, and $\mu_{\bar{x}} + 2\sigma_{\bar{x}}$.

d. Compute the $z$ score for $\bar{x} = \$12.75$.

e. State the meaning of the $z$ score of part (d).

f. Compute $P(\bar{X} < \$12.75)$.

g. State the meaning of the answer to part (f).

7. The mean butterfat content of a carton of low-fat milk is 2.00 percent. A large production run of half-gallon cartons is in a milk cooler at Sinclair's dairy. For this production run, $\mu = 2.00$ percent and $\sigma = 0.12$ percent. A quality control person randomly samples 36 cartons in this production run to check the percent butterfat. Let $\bar{X}$ be the mean butterfat in the sampled cartons.

a. Compute $\mu_{\bar{x}}$.

b. Compute $\sigma_{\bar{x}}$.

c. Sketch and label a graph of the sampling distribution of $\bar{X}$, showing values for $\mu_{\bar{x}} - 2\sigma_{\bar{x}}$, $\mu_{\bar{x}} - \sigma_{\bar{x}}$, $\mu_{\bar{x}}$, $\mu_{\bar{x}} + \sigma_{\bar{x}}$, and $\mu_{\bar{x}} + 2\sigma_{\bar{x}}$.

d. Compute the $z$ score for $\bar{x} = 2.03$.

e. State the meaning of the $z$ score of part (d).

f. Compute $P(\bar{X} > 2.03)$.

g. State the meaning of the answer to part (f).

8. According to the U.S. Census Bureau, the average number of children per married couple is 1.88. Furthermore, records indicate that the standard deviation for the distribution of children per married couple is 0.72. Suppose these figures apply to married couples in Omaha. A field worker for the Nebraska Bureau of Vital Statistics samples the records of 64 married couples in Omaha, and $\bar{X}$ is the mean number of children for the couples.

a. Compute $\mu_{\bar{x}}$.

b. Compute $\sigma_{\bar{x}}$.

c. Sketch and label a graph of the sampling distribution of $\bar{X}$, showing values for $\mu_{\bar{x}} - 2\sigma_{\bar{x}}$, $\mu_{\bar{x}} - \sigma_{\bar{x}}$, $\mu_{\bar{x}}$, $\mu_{\bar{x}} + \sigma_{\bar{x}}$, and $\mu_{\bar{x}} + 2\sigma_{\bar{x}}$.

d. Compute the $z$ score for $\bar{x} = 2.00$.

e. State the meaning of the $z$ score of part (d).

f. Compute $P(\bar{X} > 2.00)$.

g. State the meaning of the answer to part (f).

9. Captain Elizabeth Hayes is an officer in the Central Contracting Division for the Air Force. She studies purchase contracts between the Air Force and suppliers. A survey of the population of contracts suggests that the mean dollar amount of these contracts is $70,500 and the standard deviation for the distribution of dollar amounts is $30,000. Suppose Captain Hayes randomly samples 36 contracts from the files, and $\bar{X}$ is the sample mean dollar amount of the contracts.

a. Compute $\mu_{\bar{x}}$.

b. Compute $\sigma_{\bar{x}}$.

c. Compute $P(\bar{X} > 76,000)$.

d. Shade the area of a graph of the sampling distribution of $\bar{X}$ that represents the probability in part (c).

10. Marja Strutz is a Licensed Vocational Nurse who works in the intensive care unit of a large metropolitan hospital. Records indicate that the mean age of female patients in the unit for the past 10 years has been stable at about 70.5 years. The records also indicate that at any time the standard deviation for the distribution of ages is approximately 13.2 years. Suppose Marja samples 36 female patients' records from this time period, and $\bar{X}$ is the sample mean age of the patients.

a. Compute $\mu_{\bar{x}}$.

b. Compute $\sigma_{\bar{x}}$.

c. Compute $P(\bar{X} > 69.0)$.

d. Shade the area of a graph of the sampling distribution of $\bar{X}$ that represents the probability in part (c).

11. Wellington Paper Products Corporation maintains a farm for growing trees for pulp. The company keeps records of circumference of the trees at a height of 6 feet above the ground. A unit of the farm has trees for which $\mu = 12.6$ inches and $\sigma = 1.2$ inches. Suppose 64 trees from this unit are randomly selected, and $\bar{X}$ is the sample mean circumference of these trees at a height of six feet above the ground.
    a. Compute $\mu_{\bar{x}}$.
    b. Compute $\sigma_{\bar{x}}$.
    c. Compute $P(12.5 < \bar{X} < 12.7)$.
    d. Shade the area of a graph of the sampling distribution of $\bar{X}$ that represents the probability in part (c).

12. The counseling staff at Edith Power Elementary School has been studying the leisure time of children in grades K through 6. The study indicates that children in these grades average 4.8 hours each day watching television. The standard deviation for the distribution of times is 2.1 hours. Suppose the number of hours of television watching are obtained for 49 randomly selected students for a given day, and $\bar{X}$ is the mean of the sample.
    a. Compute $\mu_{\bar{x}}$.
    b. Compute $\sigma_{\bar{x}}$.
    c. Compute $P(4.5 < \bar{X} < 5.0)$.
    d. Shade the area of a graph of the sampling distribution of $\bar{X}$ that represents the probability of part (c).

13. Sandersohn's Electrical Supply Company has a large supply of resistors rated at 560 ohms (a measure of electrical resistance). For this supply of resistors, $\sigma = 18$ ohms and the distribution of resistances is reasonably bell shaped. An employee randomly selects 16 of these resistors, and $\bar{X}$ is the mean resistance of the sampled resistors.
    a. Compute $P(554.0 < \bar{X} < 566.0)$.
    b. Shade the area of a graph of the sampling distribution of $\bar{X}$ that corresponds to the probability of part (a).
    c. State the meaning of the probability.

14. Delma Eckman is a golfing enthusiast. For many years, she has played at the same golf course. For the distribution of her scores, $\mu = 78.5$ and $\sigma = 3.6$. Delma feels the distribution is approximately bell shaped. Suppose nine of these scores are randomly selected, and $\bar{X}$ is the mean of these scores.
    a. Compute $P(77.0 < \bar{X} < 80.0)$.
    b. Shade the area of a graph of the sampling distribution of $\bar{X}$ that corresponds to the probability of part (a).
    c. State the meaning of the probability in part (a).

15. The research and development unit of Universal Tire Corporation has extensive data on the new "All-Star Radial Tire." The data suggest that for any large production run of these tires, the mean life is 33,500 miles with a standard deviation of 2400 miles. Furthermore the distribution of tire life is approximately bell shaped. Suppose 12 such tires are randomly selected, and $\bar{X}$ is the sample mean number of miles in the lives of these tires.
    a. Compute $P(\bar{X} > 34,000)$.
    b. Shade the area of a graph of the sampling distribution of $\bar{X}$ that corresponds to the probability of part (a).
    c. State the meaning of the probability.

16. The mean relief pressure of a large production run of 6KT309 relief valves is 6150 pounds per square inch (psi) with a standard deviation of 225 psi. Records on these valves indicate the distribution of relief pressures is approximately bell shaped. Suppose ten of these valves are randomly sampled, and $\bar{X}$ is the mean relief pressure for the valves.
    a. Compute $P(\bar{X} < 6100)$.
    b. Shade the area of a graph of the sampling distribution of $\bar{X}$ that corresponds to the probability of part (a).
    c. State the meaning of the probability in part (a).

17. The Medina Hogs Cooperative Association raises hogs for market. The coop currently has a large shipment of hogs ready for delivery. The mean weight of these hogs is 202 pounds with a standard

deviation of 18 pounds. Suppose a sample of size $n$ of these hogs is taken, and $\bar{X}$ is the sample mean weight.

a. Compute $P(\bar{X} < 200)$, if $n = 36$.

b. Compute $P(\bar{X} < 200)$, if $n = 81$.

c. By how much was the probability decreased by increasing the sample size from 36 to 81?

18. The members of Wasatch Farm Cooperative grow several thousand acres of wheat each year. For last year's crop, the mean yield per acre was 45.8 bushels with a standard deviation of 5.4 bushels per acre. Suppose a sample of size $n$ of these acres is taken, and $\bar{X}$ is the sample mean yield in bushels per acre.

a. Compute $P(\bar{X} > 47)$, if $n = 36$.

b. Compute $P(\bar{X} > 47)$, if $n = 81$.

c. By how much was the probability decreased by increasing the sample size from 36 to 81?

19. If $\mu = 30.0$ and $\sigma = 9.0$, find $\bar{x}$ such that $P(\bar{X} < \bar{x}) \approx 0.10$, as shown in Figure 5-5. Let $n = 36$.

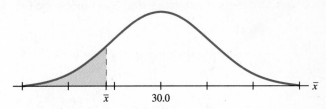

**Figure 5-5.** $P(\bar{X} < \bar{x}) = \approx 0.10$.

20. If $\mu = 50.0$ and $\sigma = 7.2$, find $\bar{x}$ such that $P(\bar{X} > \bar{x}) \approx 0.25$, as shown in Figure 5-6. Let $n = 81$.

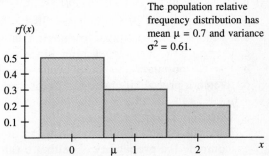

**Figure 5-6.** $P(\bar{X} > \bar{x}) \approx 0.25$.

## Exercises 5-1  Set B

*In Exercises 1–4, suppose there exists a "large" population of disks of similar size and shape. A 0, 1, or 2 is painted on each disk. A relative frequency distribution of these numbers is shown in Figure 5-7. Use the formulas from Section 4-7 to compute $\mu_{\bar{x}}$ and $\sigma_{\bar{x}}$.*

50 percent are numbered 0.
30 percent are numbered 1.
20 percent are numbered 2.

1. Suppose a single disk ($n = 1$) is randomly selected, and $X$ is the number on the disk.

a. List the possible values of $X$.

b. Compute the probability distribution of $X$.

c. Compute $\mu_X$ and verify that $\mu_X = \mu$.

d. Compute $\sigma_X^2$ and verify that $\sigma_{\bar{X}}^2 = \dfrac{\sigma^2}{1}$.

e. Construct a probability histogram for $X$.

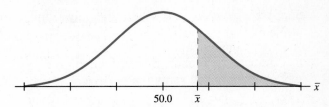

The population relative frequency distribution has mean $\mu = 0.7$ and variance $\sigma^2 = 0.61$.

**Figure 5-7.** A relative frequency distribution of a population of numbered disks.

2. Suppose two disks ($n = 2$) are randomly sampled, and $\bar{X}$ is the mean of the numbers on the disks.
   a. Verify that the possible values of $\bar{X}$ are $0, \frac{1}{2}, 1, \frac{3}{2}$, and 2.
   b. Compute the probability distribution of $\bar{X}$.
   c. Compute $\mu_{\bar{x}}$ and verify that $\mu_{\bar{x}} = \mu$.
   d. Compute $\sigma_{\bar{X}}^2$ and verify that $\sigma_{\bar{X}}^2 = \dfrac{\sigma^2}{2}$
   e. Construct a probability histogram for $\bar{X}$.

3. Suppose four disks ($n = 4$) are randomly sampled, and $\bar{X}$ is the mean of the numbers on the disks.
   a. Verify that the possible values of $\bar{X}$ are $0, \frac{1}{4}, \frac{1}{2}, \frac{3}{4}, 1, \frac{5}{4}, \frac{3}{2}, \frac{7}{4}$, and 2. (Hint: Compute one-half the sums of pairs of values of $\bar{X}$ based on $n = 2$.)
   b. Compute the probability distribution of $\bar{X}$.
   c. Compute $\mu_{\bar{x}}$ and verify that $\mu_{\bar{x}} = \mu$.
   d. Compute $\sigma_{\bar{X}}^2$ and verify that $\sigma_{\bar{X}}^2 = \dfrac{\sigma^2}{4}$.
   e. Construct a probability histogram for $\bar{X}$.

4. Suppose eight disks ($n = 8$) are randomly sampled, and $\bar{X}$ is the mean of the numbers on the disks.
   a. Verify that the possible values of $\bar{X}$ are $0, \frac{1}{8}, \frac{1}{4}, \frac{3}{8}, \frac{1}{2}, \frac{5}{8}, \frac{3}{4}, \frac{7}{8}, 1, \frac{9}{8}, \frac{5}{4}, \frac{11}{8}, \frac{3}{2}, \frac{13}{8}, \frac{7}{4}, \frac{15}{8}$, and 2.
      (Hint: Compute one-half the sums of pairs of values of $\bar{X}$ based on $n = 4$.)
   b. Compute the probability distribution of $\bar{X}$.
   c. Compute $\mu_{\bar{x}}$ and verify that $\mu_{\bar{x}} = \mu$.
   d. Compute $\sigma_{\bar{X}}^2$ and verify that $\sigma_{\bar{X}}^2 = \dfrac{\sigma^2}{8}$.
   e. Construct a probability histogram for $\bar{X}$.

5. Use random numbers to sample the ages of 36 individuals in Appendix B, Data Set I.
   a. Compute the mean age in the sample.
   b. Use the population variance given with the data set to compute $\sigma_{\bar{x}}$ based on $n = 36$.
   c. Sketch a probability histogram of $\bar{X}$ with $n = 36$. Mark on the $\bar{x}$-axis the values $\mu_{\bar{x}}, \mu_{\bar{x}} \pm \sigma_{\bar{x}}$, and $\mu_{\bar{x}} \pm 2\sigma_{\bar{x}}$.
   d. Does the mean of your sample fall in the interval $\mu_{\bar{x}} \pm 2\sigma_{\bar{x}}$?
   e. Does the mean of your sample fall in the interval $\mu_{\bar{x}} \pm \sigma_{\bar{x}}$?
   f. Compute the probability of getting a value of $\bar{X}$ as low, or lower than the sample mean you obtained.

6. Use random numbers table to sample the closing prices of 36 stocks in Appendix B, Data Set II.
   a. Compute the mean stock price in the sample.
   b. Use the population variance given with the data set to compute $\sigma_{\bar{x}}$ based on $n = 36$.
   c. Sketch a probability histogram of $\bar{X}$ with $n = 36$. Mark on the $\bar{x}$-axis the values $\mu_{\bar{x}}, \mu_{\bar{x}} \pm \sigma_{\bar{x}}$, and $\mu_{\bar{x}} \pm 2\sigma_{\bar{x}}$.
   d. Does the mean of your sample fall in the interval $\mu_{\bar{x}} \pm 2\sigma_{\bar{x}}$?
   e. Does the mean of your sample fall in the interval $\mu_{\bar{x}} \pm \sigma_{\bar{x}}$?
   f. Compute the probability of getting a value of $\bar{X}$ as low, or lower than the sample mean you obtained.

# 5-2  Large *n* Confidence Intervals for a Mean

**Key Topics**

1. *The definition of a point estimate*

2. *The definition of an interval estimate*

3. *Tabled z's for four standard confidence coefficients*

4. *Constructing approximate confidence intervals for μ*

5. *Using s to construct confidence intervals*

6. *Choosing a sample size n*

7. *A recommended format for reporting confidence intervals*

---

S T A T I S T I C S   I N   A C T I O N*

Two years ago, the California Department of Fish and Game planted several thousand fingerling trout in a high Sierra lake. Now, 2 years later, Susan Thompson, a field worker for the department, was instructed to sample the lake's fish population to estimate the mean length of the trout in the lake at this time. Susan sampled several locations on the lake and obtained the lengths (to the nearest $\frac{1}{10}$ inch) of $n = 40$ trout. The sample values are listed in Table 5-2.

Using a 95 percent confidence coefficient and the equations for constructing a large *n* confidence interval for an unknown $\mu$, Susan turned in a report that included the following statement:

"Based on the data collected, a 95 percent confidence interval for the mean length of trout in this lake is from 10.9 inches to 12.3 inches."

TABLE 5-2   Lengths (in inches) of the Sampled Trout (For these data $x = 11.6$ inches, and $s \approx 2.2$ inches.)

| | | | | | | | |
|------|------|------|------|------|------|------|------|
| 10.2 | 9.8  | 8.4  | 10.3 | 12.3 | 11.9 | 14.7 | 9.5  |
| 7.9  | 12.4 | 13.0 | 11.4 | 13.3 | 7.6  | 10.5 | 8.7  |
| 14.1 | 14.5 | 12.1 | 15.3 | 11.8 | 10.8 | 15.0 | 12.6 |
| 11.3 | 9.2  | 10.6 | 9.6  | 12.9 | 15.1 | 11.2 | 14.1 |
| 9.9  | 11.9 | 12.8 | 11.0 | 7.7  | 10.1 | 15.6 | 13.0 |

---

* Data for this example were supplied by California Department of Fish and Game, Rancho Cordova, California. Courtesy of Jim Ryan.

## The definition of a point estimate

In many situations, the mean of a population of data of interest is unknown. However, it is usually possible to obtain a sample from the population. The mean of these data can then be used as a *point estimate* of the population mean.

### DEFINITION 5-1 A Point Estimate of a Parameter

A **point estimate** of a parameter is a single value, usually derived from a sample, that is offered as a guess at the unknown value of the parameter.

Susan Thompson could use the 11.6-inch mean length of the sampled trout as a point estimate of the mean length of the population of trout in the lake.

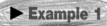 ► **Example 1** The administration of a Regional Transit District (RTD) in a large eastern city received criticism from citizens that buses were not getting to their appointed stops at the times stated in the bus schedules. Melissa Gulley, an RTD field supervisor, was instructed to obtain data to estimate the average difference between *scheduled times* and *actual times* of arrival of buses at various sites. Differences for 40 arrivals were rounded to the nearest second, and both early and late arrivals were recorded as positive numbers. The differences are listed in Table 5-3.

TABLE 5-3　Absolute Differences in Scheduled and Actual Times

| 40 | 180 | 75 | 308 | 82 | 380 | 850 | 50 | 80 | 95 |
|----|-----|-----|-----|-----|-----|-----|-----|-----|-----|
| 450 | 130 | 496 | 140 | 940 | 502 | 436 | 172 | 223 | 5 |
| 16 | 105 | 26 | 155 | 16 | 108 | 144 | 40 | 60 | 248 |
| 100 | 48 | 190 | 30 | 170 | 35 | 110 | 40 | 180 | 45 |

**a.** Compute $\bar{x}$.

**b.** Use the given data to make a point estimate for the mean absolute difference between *scheduled time* and *actual time* for bus arrivals in this RT district.

**Solution**　**a.** $\bar{x} = \dfrac{40 + 180 + 75 + \cdots + 45}{40}$　　$\bar{x} = \dfrac{\sum x_i}{n}$

$\quad\quad = \dfrac{7500}{40}$

$\quad\quad = 187.5$ seconds

**b.** Based on the given data, the mean absolute difference between the scheduled and actual times of arrival of buses to appointed stops is about 3 minutes.　　◄

## The definition of an interval estimate

A shortcoming of the use of only point estimates is that such estimates give no information about "how good" the values are likely to be. To illustrate, consider the 3-minute point estimate of the mean absolute difference between scheduled times and actual times of arrival of buses. We have no idea whether to expect this value to be correct to within 30 seconds, or 30 minutes, or any other time interval we might choose. With this in mind, an interval of values thought likely to contain the unknown mean would be useful. Such an interval is called a *confidence interval* for $\mu$.

### DEFINITION 5-2 A Confidence Interval for $\mu$

A **confidence interval for $\mu$** is an interval of numbers, derived from a sample, that is thought to contain $\mu$. The **confidence coefficient** associated with such an interval is the frequency with which the method used to create it could be expected to successfully bracket $\mu$ if repeated applications of the method were used.

   To give an intuitive introduction to confidence intervals, consider the trout data that Susan Thompson collected. For the fish sampled:

$$\bar{x} = 11.6 \text{ inches}$$

   Because 36 is a large sample size, Susan knows that the sampling distribution of $\bar{X}$ is approximately normal. Furthermore, 95 percent of the normal distribution for $\bar{X}$ is within 1.96 standard deviations of its mean. This fact is shown in Figure 5-8.

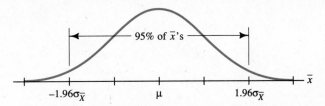

**Figure 5-8.**   95% of the distribution of $\bar{X}$ is within 1.96 $\sigma_{\bar{x}}$ of $\mu$.

   Suppose for the distribution of lengths of trout currently in the lake that Susan sampled:

$$\sigma = 2.2 \text{ inches}$$

For $n = 40$:

$$\sigma_{\bar{x}} = \frac{2.2}{\sqrt{40}} = 0.347 \ldots \text{ inches}$$

and

$$1.96(0.347 \ldots) = 0.681 \ldots \approx 0.7 \text{ inches}.$$

Therefore about 95 percent of all samples based on $n = 40$ produce $\bar{x}$s within 0.7 inches of the mean length of all trout. Knowing this, Susan might well be led to think of $\pm 0.7$ inches as a kind of "precision" figure to attach to the sample mean of 11.6 inches. If the particular sample Susan obtained was in fact one of the "lucky" 95 percent of all possible samples giving an $\bar{x}$ within 0.7 inches of $\mu$, then the interval with endpoints:

$$11.6 - 0.7 = 10.9 \text{ inches} \qquad \text{Lower endpoint of interval}$$

and

$$11.6 + 0.7 = 12.3 \text{ inches} \qquad \text{Upper endpoint of interval}$$

will bracket $\mu$.

Figure 5-9 illustrates what we mean by Susan being lucky and getting one of the 95 percent of all possible samples giving an $\bar{x}$ within 0.7 inches of $\mu$. Keeping in mind that although Susan doesn't know the value of $\mu$, there are numbers $\mu - 0.7$ and $\mu + 0.7$ that bracket the 95 percent of the $\bar{x}$'s within 1.96 standard deviations of $\mu$. These numbers divide the $\bar{x}$-axis into three regions, labeled 1, 2, and 3 in Figure 5-9.

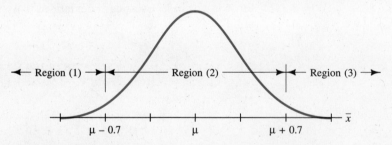

**Figure 5-9.** The possible values of $\bar{X}$ divided into three regions.

If the $\bar{x}$ Susan got is in Region 2, then the interval from 10.9 to 12.3 inches contains $\mu$. If the $\bar{x}$ Susan got is in Region 1 or Region 3, then the interval does not contain $\mu$.

## Tabled *z*'s for four standard confidence coefficients

Consider again the discussion concerning Susan Thompson's confidence interval for the mean length of the trout in the lake. To calculate the lower and upper limits for the interval, we multiplied $\sigma_{\bar{x}}$ by 1.96. The resulting interval has a 95 percent confidence coefficient because the area under the standard normal curve between $z = -1.96$ and $z = 1.96$ is 0.95.

We frequently use 90%, 95%, 98%, and 99% as confidence coefficients for confidence intervals. Figures 5-10 through 5-13 show the *z*'s that correspond to these confidence coefficients as locating areas under standard normal curves.

  a. $P(-1.65 < Z < 1.65) = 0.9505 - 0.0495$
$$= 0.9010$$
$$\approx 90\%$$

Approximately 90 percent of the standard normal distribution in Figure 5-10 is between $-1.65$ and 1.65.

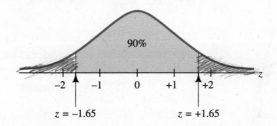

**Figure 5-10.**   About 90% of the area is between $-1.65$ and 1.65.

  b. $P(-1.96 < Z < 1.96) = 0.9750 - 0.0250$
$$= 0.9500$$
$$= 95\%$$

Approximately 95 percent of the standard normal distribution in Figure 5-11 is between $-1.96$ and 1.96.

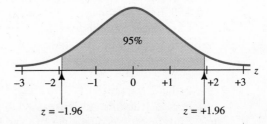

**Figure 5-11.**   About 95% of the area is between $-1.96$ and 1.96.

c.  $P(-2.33 < Z < 2.33) = 0.9901 - 0.0099$
$$= 0.9802$$
$$\approx 98\%$$

Approximately 98 percent of the standard normal distribution in Figure 5-12 is between $-2.33$ and $2.33$.

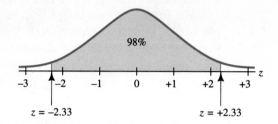

**Figure 5-12.**   About 98% of the area is between $-2.33$ and $2.33$.

d.  $P(-2.58 < Z < 2.58) = 0.9951 - 0.0049$
$$= 0.9902$$
$$\approx 99\%$$

Approximately 99 percent of the standard normal distribution in Figure 5-13 is between $-2.58$ and $2.58$.

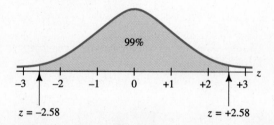

**Figure 5-13.**   About 99% of the area is between $-2.58$ and $2.58$.

In Figures 5-10 through 5-13 the shaded areas are respectively 0.90, 0.95, 0.98 and 0.99. Because the total area under the idealized histogram is 1.00, the unshaded areas are correspondingly 0.10, 0.05, 0.02, and 0.01. We use the symbol $\alpha$ (read "alpha") to represent the **unshaded areas** in the figures. Therefore $(1 - \alpha)$ represents the shaded areas in the figures. To illustrate, in Figure 5-10:

$$\alpha = 0.10 \quad \text{and} \quad 1 - \alpha = 0.90.$$

Because the unshaded areas are divided into two equal parts, the area in each part is $\alpha/2$. People often use the symbol $z_{\alpha/2}$ to represent the $z$ that cuts off an area equal to $\alpha/2$ under the upper part of the standard normal curve. A corresponding negative $z$ cuts off an area equal to $\alpha/2$ under the lower part of the curve. Table 5-4 summarizes these facts for the four commonly used confidence coefficients.

TABLE 5-4   The Relationships Between α, the Confidence Coefficients, and *z*

| Confidence Coefficient | $1 - \alpha$ | $\alpha/2$ | $z_{\alpha/2}$ |
|---|---|---|---|
| 90% | 0.900 | 0.050 | 1.65 and $-1.65$ |
| 95% | 0.950 | 0.025 | 1.96 and $-1.96$ |
| 98% | 0.098 | 0.010 | 2.33 and $-2.33$ |
| 99% | 0.099 | 0.005 | 2.58 and $-2.58$ |

## Constructing approximate confidence intervals for $\mu$

If the standard deviation of a population is known and *n* is large, then equations (4) and (5) can be used to construct an approximate confidence interval for $\mu$.

### Large *n* Confidence Interval for $\mu$ When $\sigma$ is Known

If $n > 30$ and *z* is a positive number, then the interval with:

$$\text{lower limit} = \bar{x} - z\left(\frac{\sigma}{\sqrt{n}}\right) \qquad (4)$$

and

$$\text{upper limit} = \bar{x} + z\left(\frac{\sigma}{\sqrt{n}}\right) \qquad (5)$$

can be used as an approximate confidence interval for $\mu$. The associated confidence coefficient is $P(-z < Z < z)$.

▶ **Example 2**   Kimberly Knop manages a ranch that supplies chickens to a large supermarket chain. Kim needs to estimate the mean weight of 150,000 chickens that are ready for distribution. She samples 50 birds and obtains $\bar{x} = 4.6$ pounds. Past records indicate that the standard deviation of weights of such birds is 0.5 pounds. Construct a 90 percent confidence interval for the mean weight of the entire flock.

**Solution**   **Discussion.** For a 90 percent confidence interval, $z = 1.65$ is used. With $\bar{x} = 4.6$ pounds, $\sigma = 0.5$ pounds, $n = 50$, and $z = 1.65$:

| Lower Limit | Upper Limit |
|---|---|
| $4.6 - 1.65\left(\dfrac{0.5}{\sqrt{50}}\right)$ | $4.6 + 1.65\left(\dfrac{0.5}{\sqrt{50}}\right)$ |
| $= 4.6 - 0.116\ldots$ | $= 4.6 + 0.116\ldots$ |
| $= 4.483\ldots$ | $= 4.716\ldots$ |
| $\approx 4.5$, to one decimal place | $\approx 4.7$, to one decimal place |

Thus a 90 percent confidence interval for the mean weight of the 150,000 chickens is from 4.5 to 4.7 pounds.   ◀

## Using s to construct confidence intervals

Confidence intervals for $\mu$ can be constructed using the population standard deviation $\sigma$, but there are many situations in which $\sigma$ is not known, and yet there is a need to estimate $\mu$. The natural solution to this problem is to use $s$ as a substitute for $\sigma$, hoping that $s$ will provide a good estimate for $\sigma$. As it turns out, this kind of thinking can be made precise, and as a result equations (6) and (7) can be used to construct confidence intervals for $\mu$ based on large samples.

### Large n Confidence Interval for Unknown $\mu$ Using s Instead of $\sigma$

If $n > 30$ and $z$ is a positive number, then the interval with:

$$\text{lower limit} = \bar{x} - z\left(\frac{s}{\sqrt{n}}\right) \qquad (6)$$

and

$$\text{upper limit} = \bar{x} + z\left(\frac{s}{\sqrt{n}}\right) \qquad (7)$$

can be used as a confidence interval for $\mu$. The associated confidence coefficient is $P(-z < Z < z)$.

▶ **Example 3**    Gay Yee works for Hiyashi Medical Supplies Inc. In a laboratory, she checked the amounts of compound V in a large production run of capsules of a pain relieving drug made by the company. A sample of 40 capsules yielded $\bar{x} = 42.70$ milligrams (mg) of compound V and $s = 5.40$ mg. Construct confidence intervals for the mean amount of compound V in capsules from this production run, using:

**a.** a 90% confidence coefficient
**b.** a 98% confidence coefficient

Solution    **a.** For a 90% confidence coefficient, $z = 1.65$.

$$\text{lower limit} = 42.70 - 1.65\left(\frac{5.40}{\sqrt{40}}\right) \qquad \text{lower limit} = \bar{x} - z\left(\frac{s}{\sqrt{n}}\right)$$

$$= 41.29$$

$$\text{upper limit} = 42.70 + 1.65\left(\frac{5.40}{\sqrt{40}}\right) \qquad \text{upper limit} = \bar{x} + z\left(\frac{s}{\sqrt{n}}\right)$$

$$= 44.11$$

A 90% confidence interval for the mean amount of compound V per capsule is from 41.29 to 44.11 mg.

**b.** For a 98% confidence coefficient, $z = 2.33$.

$$\text{lower limit} = 42.70 - 2.33\left(\frac{5.40}{\sqrt{40}}\right) = 40.71, \text{ to two decimal places}$$

$$\text{upper limit} = 42.70 + 2.33\left(\frac{5.40}{\sqrt{40}}\right) = 44.69, \text{ to two decimal places}$$

A 98% confidence interval for the mean amount of compound V per capsule is from 40.71 to 44.69 mg.   ◄

Notice that the length of the interval in part (b) of Example 3 is greater than the length of the interval in part (a). To increase the confidence coefficient from 90 to 98 percent (without increasing the sample size), the length of the interval must be increased.

▶ **Example 4**   Gladys Cooper and Henry White are staff members of the Sunrise Shopping Center Corporation. They were given the task of estimating the mean distance shoppers live from the mall. Gladys and Henry each sampled a total of 100 shoppers. The samples yielded the following summary values:

| Gladys | Henry |
|---|---|
| $\bar{x}_1 = 7.1$ miles | $\bar{x}_2 = 5.6$ miles |
| $s_1 = 3.6$ miles | $s_2 = 4.0$ miles |
| $n_1 = 100$ | $n_2 = 100$ |

Construct 90 percent confidence intervals for the mean distance that shoppers live from the mall based on the data obtained by both Gladys and Henry. Round the limits to the nearest $\frac{1}{10}$ mile.

**Solution**

| Based on Data Collected by Gladys | Based on Data Collected by Henry |
|---|---|
| $\text{lower limit} = 7.1 - 1.65\left(\frac{3.6}{10}\right)$ | $\text{lower limit} = 5.6 - 1.65\left(\frac{4.0}{10}\right)$ |
| $= 6.5$ miles | $= 4.9$ miles |
| $\text{upper limit} = 7.1 + 1.65\left(\frac{3.6}{10}\right)$ | $\text{upper limit} = 5.6 + 1.65\left(\frac{4.0}{10}\right)$ |
| $= 7.7$ miles | $= 6.3$ miles |

Ninety percent confidence intervals for the mean distance are:

According to Gladys's data, from 6.5 to 7.7 miles.
According to Henry's data, from 4.9 to 6.3 miles.   ◄

Example 4 illustrates two important points concerning the confidence interval method for estimating an unknown $\mu$. First, the lengths of the two intervals are different:

Based on Gladys's data:   1.2 miles ($= 7.7 - 6.5$ miles).

Based on Henry's data:   1.4 miles ($= 6.3 - 4.9$ miles).

The reason for the different lengths is the difference in the sample standard deviations, 3.6 for Gladys and 4.0 for Henry. The larger standard deviation yields the longer interval.

Second, the two intervals do not overlap, as can be seen in Figure 5-14. Because in this case the intervals do not overlap, there is no chance that both intervals bracket the population mean. *Therefore at least one of the two 90 percent confidence intervals must be wrong.* This fact helps explain what a 90 percent confidence coefficient really means.

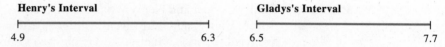

**Henry's Interval**                    **Gladys's Interval**

4.9                           6.3       6.5                           7.7

**Figure 5-14.**   The intervals do not overlap.

## The Meaning of a Confidence Interval

Someone who many times in a lifetime correctly models real situations and applies standard methods to obtain 90 percent confidence intervals expects that about 0.90 of them will successfully bracket their parameters. However, no guarantee can be made about the correctness of any specific one of the intervals she or he makes.

The interpretation of a confidence interval is an endorsement of the method used to construct the interval. It is not a statement of probability of correctness for the interval in hand. The long-run percentage of intervals that are successful in containing $\mu$ depends on the confidence coefficient through the corresponding values of $z$, as shown in Table 5-5.

**TABLE 5-5**   Relating Confidence Coefficients to Long-run Percentages of Intervals that Contain $\mu$

| Confidence Coefficient | Corresponding z Score | Percentage of Intervals Containing $\mu$ | Percentage of Intervals not Containing $\mu$ |
|---|---|---|---|
| 90% | $-1.65$ and $1.65$ | 90% | 10% |
| 95% | $-1.96$ and $1.96$ | 95% | 5% |
| 98% | $-2.33$ and $2.33$ | 98% | 2% |
| 99% | $-2.58$ and $2.58$ | 99% | 1% |

## Choosing a sample size _n_

If $\sigma$ is known for a population we are studying, then we can use equations (4) and (5) to construct a confidence interval for the mean of the population.

$$\text{lower limit} = \bar{x} - z \cdot \frac{\sigma}{\sqrt{n}} \qquad \text{(4)}$$

$$\text{upper limit} = \bar{x} + z \cdot \frac{\sigma}{\sqrt{n}} \qquad \text{(5)}$$

To use these equations, we can somewhat arbitrarily decide on the sample size, and then simply "live with" the length that it produces for the interval. However, we can alternatively decide before selecting the sample to take one of sufficient size to get an interval with a desired length. To illustrate, suppose Kimberly Knop in Example 2 decided to sample enough chickens so that the length of the confidence interval was 0.1 pounds, and not the 0.2 pound length she got from her sample of 50 chickens. As shown in Figure 5-15, this problem amounts to determining determining a value of _n_ so that $z \cdot \dfrac{\sigma}{\sqrt{n}}$ is equal to 0.05.

$$1.65 \cdot \frac{0.5}{\sqrt{n}} = 0.05 \qquad \text{Replace } z \text{ with 1.65 and } \sigma \text{ with 0.5}$$

$$1.65 \cdot \frac{0.5}{0.05} = \sqrt{n} \qquad \text{Solve for } \sqrt{n}$$

$$16.50 = \sqrt{n} \qquad \text{Simplify the left side}$$

$$272.25 = n \qquad \text{Square both sides}$$

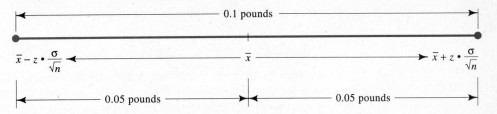

**Figure 5-15.**   A confidence interval with length 0.1 pounds.

Thus, if Kimberly samples 272 chickens, then the confidence interval will have a length of 0.1 pounds. To write a general equation for choosing a sample size, we let 2_e_ represent the desired length of a confidence interval. In the Kimberly Knop example, 2_e_ is 0.1 pounds. The value of _z_ depends on the confidence coefficient associated with the confidence interval.

## Choosing a Sample Size $n$ for a Confidence Interval for $\mu$

Let $e$ be one-half the desired length of a confidence interval for $\mu$ with confidence coefficient corresponding to $z$, then the required sample size is

$$n = \left(\frac{z \cdot \sigma}{e}\right)^2 \tag{8}$$

where $\sigma$ is the (known) standard deviation of the population.

If $\sigma$ in equation 8 is not known, then Stein's two stage procedure (discussed in the Set B exercises of Section 5-3) can be used to determine the value of $n$ on the basis of a preliminary sample.

▶ **Example 5**   Otzen Electronics makes trimmer potentiometers. Katherine Steinbacher measures the diameters of pin heads for these potentiometers to construct confidence intervals for the mean diameter of the pin heads in large production runs. Records of past production runs suggest a standard deviation for the distribution of measurements is 0.0012 inches. Katherine routinely uses a 98 percent confidence coefficient and a confidence interval with length 0.0008 inches. Calculate to the nearest whole number the size of Katherine's samples.

**Solution**   With $\sigma = 0.0012$ inches, $z = 2.33$, and $e = \frac{1}{2}(0.0008) = 0.0004$ inches:

$$n = \left(\frac{2.33(0.0012)}{0.0004}\right)^2 = 48.8601 \approx 49$$

Thus, Katherine uses measurements of 49 pin heads to construct a 98 percent confidence interval for the mean diameter of a given production run.   ◀

## A recommended format for reporting confidence intervals

Confidence intervals can easily be constructed for $\mu$. However, an important part in the practical use of confidence intervals is making an appropriate explanatory statement. We recommend the following wording be used as a general format for such a statement:

### Format for Reporting a Confidence Interval

Based on the given data, a *(state the confidence coefficient)* confidence interval for (name the unknown parameter) is from *(state the lower limit)* to *(state the upper limit)*.

▶ **Example 6**
A simple random sample of 50 automotive seals from a large production run has a mean lip opening pressure of 12.4 pounds per square inch (psi), with a standard deviation of 1.3 psi. Construct a 95 percent confidence interval for the mean lip opening pressure for the entire production run and report the interval in the recommended format.

**Solution**
**Discussion.**   Because $\bar{x}$ and $s$ are given to one decimal place, the limits will also be rounded to one decimal place.

With $\bar{x} = 12.4$, $s = 1.3$, $n = 50$, and $z = 1.96$:

$$\text{lower limit} = 12.4 - 1.96\left(\frac{1.3}{\sqrt{50}}\right)$$

$$= 12.4 - 0.36034\ldots$$

$$\approx 12.0, \text{ to one decimal place}$$

$$\text{upper limit} = 12.4 + 1.96\left(\frac{1.3}{\sqrt{50}}\right)$$

$$= 12.4 + 0.36034\ldots$$

$$\approx 12.8, \text{ to one decimal place}$$

Based on the given data, a 95 percent confidence interval for the mean lip opening pressure of the seals is from 12.0 to 12.8 psi.   ◀

## Exercises 5-2   Set A

*In Exercises 1–4, answer parts (a) through (d).*
a. *Compute $\bar{x}$ to one decimal place.*
b. *Make a point estimate of the mean of the sampled population.*
c. *Construct a confidence interval for $\mu$ with the indicated confidence coefficient.*
d. *State the meaning of the answer to part (c) in the context of the problem.*

1. A sample of 36 couples applying for marriage licenses in Denver yielded the ages for the 36 women:

| | | | | | | | | | | | |
|---|---|---|---|---|---|---|---|---|---|---|---|
| 26 | 28 | 21 | 23 | 28 | 19 | 23 | 33 | 22 | 19 | 25 | 45 |
| 27 | 30 | 41 | 28 | 17 | 27 | 40 | 26 | 24 | 32 | 20 | 37 |
| 22 | 28 | 35 | 30 | 26 | 18 | 25 | 30 | 42 | 23 | 28 | 20 |

Construct a 90 percent confidence interval for $\mu$.

2. A representative of a national corporation that markets pet foods and supplies wanted to estimate the mean number of household pets per household in her territory. She sampled 36 households and the number of pets in each is listed below.

| | | | | | | | | | | |
|---|---|---|---|---|---|---|---|---|---|---|
| 3 | 3 | 2 | 3 | 0 | 0 | 3 | 2 | 10 | 2 | 2 | 2 |
| 2 | 0 | 0 | 1 | 4 | 2 | 0 | 0 | 1 | 3 | 0 | 0 |
| 5 | 2 | 1 | 0 | 0 | 6 | 2 | 4 | 0 | 0 | 3 | 4 |

Construct a 95 percent confidence interval for $\mu$.

3. A major automobile corporation recently produced a new model car. To estimate the model's mean gas mileage rating, the quality control staff carefully tested and recorded the miles per gallon obtained from 40 different cars of this particular model. The numbers, to the nearest $\frac{1}{10}$ mile per gallon are listed below.

| | | | | | | | | | |
|---|---|---|---|---|---|---|---|---|---|
| 31.7 | 30.3 | 29.4 | 29.6 | 30.8 | 27.4 | 28.9 | 30.0 | 29.8 | 26.4 |
| 30.6 | 28.8 | 32.1 | 29.8 | 30.3 | 27.6 | 31.2 | 26.9 | 30.9 | 31.5 |
| 29.4 | 28.5 | 30.6 | 27.9 | 31.0 | 32.8 | 29.0 | 27.9 | 28.8 | 30.6 |
| 30.2 | 29.4 | 26.9 | 30.1 | 31.4 | 28.3 | 27.5 | 31.7 | 30.4 | 28.5 |

Construct a 98 percent confidence interval for $\mu$.

4. A union representative for the United Fellowship of Department Store Workers wanted to estimate the average number of hours worked by member employees classified as part-time. By means of a questionnaire, he got the number of hours worked last week by 45 employees in several different department stores.

| | | | | | | | | | | | | | | |
|---|---|---|---|---|---|---|---|---|---|---|---|---|---|---|
| 23 | 20 | 31 | 27 | 26 | 17 | 19 | 41 | 36 | 32 | 40 | 24 | 30 | 35 | 38 |
| 22 | 33 | 29 | 28 | 34 | 32 | 45 | 42 | 21 | 22 | 23 | 31 | 32 | 21 | 24 |
| 33 | 28 | 30 | 15 | 36 | 34 | 22 | 28 | 25 | 32 | 26 | 23 | 27 | 19 | 32 |

Construct a 99 percent confidence interval for $\mu$.

5. A survey of 32 automotive supply centers in New York City yielded $\bar{x} = 112.4$ cents for the price of a quart of 10W50 motor oil. For the sample data, $s = 15.2$ cents. Construct a 90 percent confidence interval for the mean price of a quart of this motor oil in New York City stores.

6. A survey of 40 markets in San Jose yielded prices with $\bar{x} = \$4.69$ for a 1-pound can of Brewer's Premium Coffee Blend. For the sample data, $s = \$0.32$. Construct a 95 percent confidence interval for the mean price of a 1-pound can of this coffee in San Jose stores.

7. A sample of small bags of peanuts for vending machines was taken from a production run at the Carter Peanut Corporation plant in Atlanta. The numbers of peanuts in the 50 bags sampled gave $\bar{x} = 48.4$ peanuts and $s = 2.4$ peanuts. Construct a 98 percent confidence interval for the mean number of peanuts per bag for this production run.

8. A sample of 75 adults was taken in a large shopping center to estimate the average number of credit cards (Visa, Mastercard, department store, gas company, etc.) by area shoppers. For the shop-

pers in the sample, $\bar{x} = 4.3$ cards and $s = 2.1$ cards. Construct a 99 percent confidence interval for the mean number of credit cards per shopper in the area sampled.

9. The counselor for international students at a community college in New York wanted to estimate the mean grade point average (gpa) for his advisees last semester. He sampled the records of 36 of these students. For the sample, the gpas had $\bar{x} = 2.89$ and $s = 0.33$.

a. Construct a 90 percent confidence interval for the mean gpa for international students last semester.

b. State the meaning of the answer to part (a) in the context of the problem.

10. Melissa Martin is an engineer for the El Dorado Water District. Ripple Creek is one of the sources of water for the district. To estimate the average water flow in cubic feet per second (cfs) during the months of April through October, Melissa has taken 225 readings during these months. For the sample, $\bar{x} = 230$ cfs and $s = 95$ cfs.

a. Construct a 95 percent confidence interval for the mean cfs flow of Ripple Creek for the period checked.

b. State the meaning of the answer to part (a) in the context of the problem.

**11.** Ray Kelley is the manager of the Big Bear Supermarket in Lodi. A directive from corporate headquarters required that the market remain open from 8 PM to 12 PM, rather than closing at 8 PM. Ray wanted to estimate the mean dollar amount of sales to shoppers during the 8 to 12 o'clock period. He recorded the sales of 64 shoppers during these hours. The data yielded $\bar{x} = \$14.60$ and $s = \$17.60$.

a. Construct a 98 percent confidence interval for the unknown mean.

b. State the meaning of the answer to part (a) in the context of the problem.

**12.** Collene McHugh is an employee of Governor's Realty Agency in Buffalo. Management asked Collene to collect some data on the cost of renting a two-bedroom, unfurnished apartment in the midtown section of the city. Collene surveyed the area and got the cost of 1 month's rent for 40 such apartments. For these data, $\bar{x} = \$387$ and $s = \$35$ to the nearest dollar.

a. Construct a 90 percent confidence interval for the mean cost of 1 month's rent of such an apartment in this area.

b. State the meaning of the answer to part (a) in the context of the problem.

**13.** An employee of Ashland Rubber Corporation is responsible for checking balloons made by the company. Specifically she uses a mechanical air pump to inflate a balloon until it bursts. She counts the number of strokes of the pump needed to burst each balloon. Yesterday she checked 50 balloons from a large production run. The data yielded $\bar{x} = 20.9$ strokes with $s = 3.9$ strokes.

a. Construct a 95 percent confidence interval for the mean number of strokes of the pump needed to burst a balloon.

b. State the meaning of the answer to part (a) in the context of the problem.

**14.** A woman frequently goes crabbing off the piers in San Francisco. She checks the crab traps every 10 minutes and removes any crabs. She then records the number of crabs captured before returning the trap into the water. A sample of 45 counts from her records yielded $\bar{x} = 1.53$ crabs per trap for each 10-minute interval. The standard deviation for the distribution of counts was 1.30.

a. Construct a 95 percent confidence interval for the mean number of crabs per trap for a 10-minute time interval.

b. State the meaning of the answer to part (a) in the context of the problem.

**15.** A sample of quantitative data yielded $\bar{x} = 52.6$ with a standard deviation of 8.4.

a. Construct a 90 percent confidence interval for $\mu$, if $n = 36$.

b. Construct a 90 percent confidence interval for $\mu$, if $n = 144$.

c. The sample size in part (b) is four times the sample size in part (a). By what factor was the length of the interval in part (a) reduced by the larger sample in part (b)?

**16.** A sample of quantitative data yielded $\bar{x} = 73.8$ with a standard deviation of 12.6.

a. Construct a 95 percent confidence interval for $\mu$, if $n = 36$.

b. Construct a 95 percent confidence interval for $\mu$, if $n = 324$.

c. The sample size in part (b) is nine times the sample size in part (a). By what factor was the length of the interval in part (a) reduced by the larger sample in part (b)?

*In Exercises 17–24 calculate the value of n necessary to construct confidence intervals with the indicated lengths for the stated confidence coefficients. Round values of n to the nearest whole number.*

**17.** 90% confidence coefficient, $\sigma = 12$, and $2e = 6.6$

**18.** 90% confidence coefficient, $\sigma = 8.4$, and $2e = 4.0$

**19.** 95% confidence coefficient, $\sigma = 6.5$, and $2e = 4.0$

**20.** 95% confidence coefficient, $\sigma = 4.8$, and $2e = 2.2$

**21.** 98% confidence coefficient, $\sigma = 1.3$, and $2e = 0.8$

**22.** 98% confidence coefficient, $\sigma = 3.4$, and $2e = 2.8$

**23.** 99% confidence coefficient, $\sigma = 0.75$, and $2e = 0.32$

**24.** 99% confidence coefficient, $\sigma = 0.042$, and $2e = 0.014$

# Exercises 5-2   Set B

**1.** Use random numbers to sample ages of individuals in Data Set I, Appendix B.
   a. Obtain ten different samples of $n = 30$ data points each.
   b. Compute the sample means $\bar{x}_1, \bar{x}_2, \ldots, \bar{x}_{10}$.
   c. Compute the sample variances $s_1^2, s_2^2, \ldots, s_{10}^2$.
   d. Construct confidence intervals for $\mu$ using a 90 percent confidence coefficient, the population variance $\sigma^2 = 408.04$, and the ten corresponding sample means.
   e. How many of the intervals in part (d) contain the population mean $\mu = 56.1$?
   f. Now construct confidence intervals for $\mu$ using a 90 percent confidence coefficient, the sample variances, and the ten corresponding sample means.
   g. How many of the intervals in part (f) contain the population mean?
   h. Is there any difference between your answers to parts (e) and (g)?

**2.** Use random numbers to sample closing stock prices in Data Set II, Appendix B.
   a. Obtain ten different samples of $n = 30$ data points each.
   b. Compute the sample means $\bar{x}_1, \bar{x}_2, \ldots, \bar{x}_{10}$.
   c. Compute the variances $s_1^2, s_2^2, \ldots, s_{10}^2$.
   d. Construct confidence intervals for $\mu$ using a 90 percent confidence coefficient, the population variance $\sigma^2 = 1751.42$, and each of the ten sample means.
   e. How many of the intervals in part (d) contain $\mu = 36.48$?
   f. Now construct confidence intervals for $\mu$ using a 90 percent confidence coefficient, the sample variances, and the ten corresponding sample means.
   g. How many of the intervals in part (f) contain the population mean?
   h. Is there any difference between your answers to parts (e) and (g)?

*Exercises 3–6 are four suggested projects that can be carried out to practice using confidence intervals to estimate population means.*

**3.** Obtain the ages of $n = 40$ students currently enrolled at your school.
   a. Compute the mean of your sample.
   b. Compute the variance for your sample.
   c. Construct a 95 percent confidence interval for the mean age of the current student body at your school.
   d. Compare your interval with intervals obtained by others in your class.

**4.** Obtain the page count of $n = 40$ casebound texts in your college store. Use only the pages of actual text

material, and do not count the pages of end-matter such as preface, appendices, and tables.
   a. Compute the mean page count of the texts in your sample.
   b. Compute the variance for your sample.
   c. Construct a 90 percent confidence interval for the mean number of pages in casebound texts currently in your college store.
   d. Compare your interval with intervals obtained by others in your class.

**5.** If you live in a large city, obtain current prices per gallon for unleaded gasoline at $n = 30$ service stations in your area.
   a. Compute the mean of your sample.

b. Compute the variance for your sample.

c. Construct a 98 percent confidence interval for the current mean price of unleaded gasoline in your area.

d. Compare your results with the results of others in your class.

6. Obtain current monthly costs of rent for $n = 35$ one-bedroom, furnished apartments within a 5-mile radius of your campus.

a. Compute the mean of your sample.

b. Compute the variance for your sample.

c. Construct a 99 percent confidence interval for the mean rent of a one-bedroom, furnished apartment within 5 miles of your campus.

d. Compare your results with the results of others in your class.

# 5-3 Small *n* Confidence Intervals for the Mean of a Bell-Shaped Population

**Key Topics**

1. *The t-distributions*

2. *Constructing confidence intervals for μ based on small samples*

## S T A T I S T I C S   I N   A C T I O N[*]

In recent weeks, several references were made in local newspapers and television newscasts to the water quality in a midwestern city. The reports focused on the unsafe levels of lead in some locations tested.

Andrew Howell is manager of administrative services and operations for all of the K through 12 schools in the city's district. In response to the media attention on the water, he instructed a member of his staff, Mark Spangler, to sample, analyze, and report on the lead content of drinking water at every school in the district.

At Mount Whitney High school, Mark obtained 21 samples of 100 milliliters. The analyses of these samples were reported in milligrams (mg) of lead per liter (L) of water, and are listed in Table 5-6.

**TABLE 5-6   Milligrams of Lead per Liter of Water in Samples (For these data: $\bar{x} \approx 0.037$ mg/L and $s \approx 0.0061$ mg/L.)**

| | | | | | | |
|---|---|---|---|---|---|---|
| 0.038 | 0.042 | 0.045 | 0.030 | 0.028 | 0.042 | 0.048 |
| 0.040 | 0.036 | 0.041 | 0.045 | 0.038 | 0.026 | 0.041 |
| 0.039 | 0.036 | 0.035 | 0.034 | 0.039 | 0.028 | 0.030 |

---

* Data for this example were supplied by Sacramento City College. Courtesy of Andrew Howell, who graciously volunteered his help.

Mark constructed the stem-and-leaf diagram in Figure 5-16 using the data in Table 5-6. Based on the shape of the diagram, he felt the distribution of lead content for water samples at this school was reasonably bell shaped.

Using a 95 percent confidence coefficient and the equations for constructing confidence intervals for an unknown $\mu$, Mark turned in a report that included the following statement: "Based on the data collected, a 95 percent confidence interval estimate for the mean lead content in water at Mount Whitney High school is from 0.034 to 0.040 mg/L. Both limits are below the 0.050 mg/L limit set by federal guidelines."

```
(0.025-0.029) 0.02 | 8  6  8
(0.030-0.034) 0.03 | 0  4  0
(0.035-0.039) 0.03 | 8  6  8  9  6  5  9
(0.040-0.044) 0.04 | 2  2  0  1  1
(0.045-0.049) 0.04 | 5  8  5
```

**Figure 5-16.** A stem-and-leaf diagram of the water sample data.

## The *t*-distributions

In Section 5-2, we found confidence intervals for unknown means based on large simple random samples. Such intervals are possible because it is a mathematical fact that the probability distribution of

$$Z = \frac{\bar{X} - \mu}{\dfrac{S}{\sqrt{n}}}$$

$\bar{X}$ is the sample mean
$\mu$ is the mean of the population being sampled
$S$ is the sample standard deviation
$n$ is the sample size

is approximately *standard normal* for large $n$ (that is, $n \geq 30$). As a consequence, values from the standard normal distribution were used to construct the confidence intervals.

If $n$ is small, then it is not possible to assert that the quantity we have called $Z$ above is even approximately standard normal. However, it *is* possible to make a different assertion regarding the probability distribution if the population sampled has an approximately bell-shaped distribution. The specifics are stated below.

### Sampling Distributions from Bell-Shaped Populations

If $X_1, X_2, X_3, \ldots, X_n$ can be modeled as **independent normal random variables** with mean $\mu$ and standard deviation $\sigma$, then the quantity

$$\frac{\bar{X} - \mu}{S/\sqrt{n}}$$

has the so-called "*t*-distribution with $n - 1$ degrees of freedom."

The *t*-distributions are bell-shaped probability distributions that have been studied by mathematicians and for which tables of probabilities are available. Appendix Table 5 contains probabilities for *t*-distributions that will be used in this text.

There are many *t*-distributions, one for each possible number of "degrees of freedom." Each *t*-distribution is symmetric about a mean of 0. Taken as a group, the *t*-distributions are in general less peaked and more spread out than the standard normal distribution. However, for large numbers of degrees of freedom (at least 30), the *t*-distributions are almost indistinguishable from the standard normal distribution. In Figure 5-17, three *t*-distributions and the standard normal distribution are sketched. Notice that *df* is used as a symbol for degrees of freedom. Also note that the curve for $df = 20$ is closer to the shape of the standard normal distribution than either of the curves for $df = 10$ or $df = 4$. Because all four curves represent idealized histograms, the areas under the curves are all one.

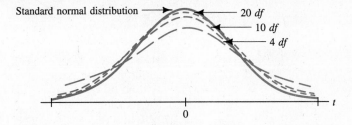

**Figure 5-17.**  Idealized probability histograms for *t*-distributions with $df = 4$, 10, and 20, and the Z-distribution.

A portion of Appendix Table 5 is shown in Figure 5-18. The left column is labeled *df* for degrees of freedom. The headings of the next four columns are $t_{0.050}$, $t_{0.025}$, $t_{0.010}$, and $t_{0.005}$. The subscripts correspond to 5%, 2.5%, 1%, and 0.5%. As indicated by the figure to the left of the table, a given tabled value is the number that will cut off a specified area in the right tail of the distribution.

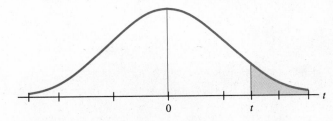

**Figure 5-18.**  A portion of Appendix Table 5.

| d.f. | $t_{0.050}$ | $t_{0.025}$ | $t_{0.010}$ | $t_{0.005}$ | d.f. |
|------|--------|--------|--------|--------|------|
| 1 | 6.314 | 12.706 | 31.821 | 63.657 | 1 |
| 2 | 2.920 | 4.303 | 6.965 | 9.925 | 2 |
| 3 | 2.353 | 3.182 | 4.541 | 5.841 | 3 |
| 4 | 2.132 | 2.776 | 3.747 | 4.604 | 4 |
| 5 | 2.015 | 2.571 | 3.365 | 4.032 | 5 |
| 6 | 1.943 | 2.447 | 3.143 | 3.707 | 6 |
| 7 | 1.895 | 2.365 | 2.998 | 3.499 | 7 |
| 8 | 1.860 | 2.306 | 2.896 | 3.355 | 8 |
| 9 | 1.833 | 2.262 | 2.821 | 3.250 | 9 |
| 10 | 1.812 | 2.228 | 2.764 | 3.169 | 10 |

▶ **Example 1**     Suppose a random variable $T$ has a $t$-distribution. Use Appendix Table 5 to find values of $t$ such that:

**a.** $P(T > t) = 0.025$, if $df = 15$
**b.** $P(-t < T < t) = 0.90$, if $df = 20$

**Solution**     **a.** Because $P(T > t) = 0.025$, we look under the column headed $t_{0.025}$ in the row with $df = 15$. In Appendix Table 5, read 2.131. Thus $t = 2.131$ and $P(T > 2.131) = 0.025$. This probability is represented by the area shaded in Figure 5-19.

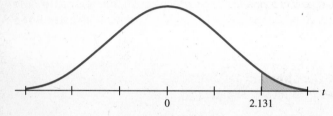

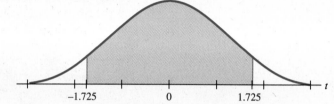

**Figure 5-19.**  Area shaded equals $P(T > 2.131)$.          **Figure 5-20.**  Area shaded equals $P(-1.725 < T < 1.725)$.

**b.** Because $t$ and $-t$ cut off equal areas, together they cut off $(100 - 90)$ percent $= 10\%$, and each cuts off $5\%$ $(=0.050)$. In Appendix Table 5, for $df = 20$ and $t_{0.050}$, $t = 1.725$. Thus $P(-1.725 < T < 1.725) = 0.90$ as in Figure 5-20.  ◀

## Constructing confidence intervals for $\mu$ based on small samples

To construct a confidence interval for an unknown $\mu$ based on a small sample size ($n < 30$), the $z$ in equations (6) and (7) must be replaced by a value from a $t$-distribution. The appropriate $t$-distribution has $n - 1$ degrees of freedom. The examples in the following table identify the appropriate $t$-distribution for three samples sizes.

| Sample Size, n | Degrees of Freedom for the t-Distribution |
|---|---|
| 10 | $df = 10 - 1 = 9$ |
| 18 | $df = 18 - 1 = 17$ |
| 25 | $df = 25 - 1 = 24$ |

In Appendix Table 5, the entries are values that cut off fixed percents of a given $t$-distribution in the right tail. The negatives of these values similarly cut off the same percents in the left tail. When the positive and negative values are taken together, then naturally the percents are doubled. Frequently the sum of the areas in these tails is called $\alpha$ (read "alpha"). The area between the numbers $t$ and $-t$

is $1 - \alpha$. *The value $1 - \alpha$ is the confidence coefficient for an interval constructed using t from Appendix Table 5.*

| Confidence Coefficient, $(1 - \alpha)$ | Desired Tail Area | Notation for Corresponding t-Value, $t_{\alpha/2}$ |
|---|---|---|
| 90% | 0.050 | $t_{0.050}$ |
| 95% | 0.025 | $t_{0.025}$ |
| 98% | 0.010 | $t_{0.010}$ |
| 99% | 0.005 | $t_{0.005}$ |

Equations (9) and (10) below can be used to determine the limits of a confidence interval for the mean of an approximately bell-shaped population of data based on a sample size less than 30.

## Equations for Constructing Confidence Intervals for $\mu$ Based on Small Samples ($n < 30$)

If a population has an approximately bell-shaped distribution, then the interval with:

$$\text{lower limit} = \bar{x} - t\left(\frac{s}{\sqrt{n}}\right) \qquad \textbf{(9)}$$

and

$$\text{upper limit} = \bar{x} + t\left(\frac{s}{\sqrt{n}}\right) \qquad \textbf{(10)}$$

for a positive number $t$, can be used as a confidence interval for $\mu$. The confidence coefficient for the interval is $P(-t < T < t)$ and for a sample of size $n$ is calculated using $n - 1$ degrees of freedom.

In the Statistics in Action section, Mark Spangler constructed a 95 percent confidence interval estimate for the mean lead content in milligrams per liter (mg/L) for samples of the water at Mount Whitney High school. Notice that he asserted the data gave evidence that the distribution of lead content was approximately bell shaped. This shape is necessary before a confidence interval for $\mu$ can be constructed using a sample of size less than 30.

Furthermore, to construct the interval, Mark must have used a value from the $t$-distribution with $21 - 1 = 20$ degrees of freedom. Checking Appendix Table 5, for $df = 20$, $t_{0.025} = 2.086$. Using equations (9) and (10) with $\bar{x} = 0.037$, $s = 0.0061 \ldots$, and $t = 2.086$, the limits of 0.034 and 0.040 mg/L are obtained.

▶ **Example 2**    Dan Cooper is student body president at Apollo Community College. In response to criticism he had been hearing from faculty regarding the lack of student class preparation, he decided to obtain data to estimate the mean time spent studying by students at Apollo. He obtained a sample of 16 students who agreed to keep accurate records of the amount of time spent studying during the week from March 3 through March 9. For the selected time period, these students provided the following summary statistics:

$$\bar{x} = 23.8, \text{ to the nearest } \tfrac{1}{10} \text{ of an hour.}$$
$$s = 3.2, \text{ to the nearest } \tfrac{1}{10} \text{ of an hour.}$$

**a.** Construct a 90 percent confidence interval for $\mu$.
**b.** State the meaning of the answer in part (a).

Solution    **Discussion.** Dan feels that the distribution of hours studied by all students at Apollo for the week under investigation is approximately bell shaped.

**a.** With $n = 16$, $df = 16 - 1 = 15$. From Appendix Table 5, $t_{0.050} = 1.753$. With $\bar{x} = 23.8$, $s = 3.2$, and $n = 16$:

$$\text{lower limit} = 23.8 - 1.753\left(\frac{3.2}{\sqrt{16}}\right)$$
$$= 23.8 - 1.4024$$
$$= 22.4, \text{ to one place}$$

$$\text{upper limit} = 23.8 + 1.753\left(\frac{3.2}{\sqrt{16}}\right)$$
$$= 25.2, \text{ to one place}$$

**b.** Based on the given data, a 90 percent confidence interval for the mean number of hours spent studying during the week under investigation by the students at Apollo College is from 22.4 to 25.2 hours.    ◀

   Notice that the inference in Example 2 is applied only to the week during which the data were collected. (If we further assume that this week is representative of any school week, then the estimate might be extended to other weeks as well.)

▶ **Example 3**    The U.S. Department of Agriculture must make yearly estimates of the annual wheat crop. The management specialist in a large region of South Dakota needed to obtain an estimate of this year's wheat crop in her region. She located 20 test plots that gave $\bar{x} = 44.3$ bushels per acre (bu/acre), with a standard deviation of $s = 3.9$. Assume that the distribution of yields on plots of this size in her region is bell shaped. Construct to one decimal place a 95 percent confidence interval for the mean yield in the region sampled.

**Solution**   With $n = 20$, $df = 20 - 1 = 19$. From Appendix Table 5, $t_{2.025} = 2.093$. With $\bar{x} = 44.3$, $s = 3.9$, and $n = 20$:

$$\text{lower limit} = 44.3 - 2.093\left(\frac{3.9}{\sqrt{20}}\right)$$

$$= 44.3 - 1.825\ldots$$

$$= 42.474\ldots$$

$$= 42.5, \text{ to one place}$$

$$\text{upper limit} = 44.3 + 1.825\ldots$$

$$= 46.125\ldots$$

$$= 46.1, \text{ to one place}$$

Based on the given data, a 95 percent confidence interval for the mean wheat yield this year in the sampled region is from 42.5 to 46.1 bushels per acre.   ◄

Before leaving this section we should review the differences between confidence intervals based on small samples and those based on large samples. The most *obvious* difference is that values from the *t*-distributions replace the values from the *Z*-distribution in the formulas. The most *important* difference deals with the issue of when the two methods are applicable.

Equations (6) and (7) can be used to give a valid estimation method for any population, if *n* is large ($n \geq 30$), but *n* is still a small fraction of *N*. *Equations (9) and (10) can be used to give a valid estimation method for only populations with relative frequency distributions that are bell shaped.*

If the histogram of the sample is approximately bell shaped, then use equations (9) and (10). If the histogram of the sample is not bell shaped, then use the material in Chapter 10. The estimation methods used there are applicable to distributions of any shape.

## Exercises 5-3   Set A

*In Exercises 1–8, suppose a random variable T has a t-distribution. Use Appendix Table 5 to find the indicated positive values of t. For each exercise, also sketch an idealized histogram and shade the area that represents the probability.*

**1.** $P(T > t) = 0.050$, if $df = 8$

**2.** $P(T > t) = 0.010$, if $df = 12$

**3.** $P(-t < T < t) = 0.950$, if $df = 8$

**4.** $P(-t < T < t) = 0.900$, if $df = 12$

**5.** $P(T < -t) = 0.025$, if $df = 17$

**6.** $P(T < -t) = 0.005$, if $df = 22$

**7.** $P(-t < T < t) = 0.950$, if $df = 17$

**8.** $P(-t < T < t) = 0.980$, if $df = 22$

*In Exercises 9–12, answer parts (a), (b), and (c).*
*a. Calculate $\bar{x}$ and s.*
*b. Construct a confidence interval for $\mu$ with the indicated confidence coefficient.*
*c. State, in the context of the problem, the meaning of the interval.*

9. Johnson's Turkey Ranch has a flock of 450,000 birds ready for market. The ranch manager sampled 15 of these birds and obtained their weights to the nearest $\frac{1}{10}$ pound.

    | 12.6 | 14.2 | 10.3 | 8.6 | 13.3 | 10.9 | 9.5 | 15.1 |
    | 16.3 | 13.0 | 11.4 | 13.6 | 14.8 | 15.7 | 12.7 | |

    Assume the distribution of weights is relatively bell shaped. Determine a 90 percent confidence interval for $\mu$.

10. Klein's Medical Supplies Corporation has 3,700 employees. The personnel director sampled 20 employees and obtained the numbers of miles each one had to drive to work.

    | 10.2 | 7.4 | 3.3 | 15.6 | 1.9 | 6.5 | 2.4 | 4.7 | 7.8 | 24.3 |
    | 4.6 | 12.8 | 8.2 | 13.7 | 2.6 | 5.5 | 13.9 | 3.4 | 11.0 | 6.2 |

    Assume the distribution of miles is relatively bell shaped. Determine a 95 percent confidence interval for $\mu$.

11. A waiter at Mama Rosa's Italian Restaurant sampled 24 dinner tabs from his receipts over the past several months. For each tab he determined to the nearest $\frac{1}{10}$ percent the rate at which he was tipped. (Each data point was calculated by dividing his tip by the amount of the tab.)

    | 14.7% | 12.9% | 16.0% | 17.8% | 22.8% | 12.4% | 16.8% | 12.5% |
    | 20.0% | 12.5% | 15.1% | 12.4% | 17.2% | 16.6% | 18.6% | 14.2% |
    | 12.6% | 22.2% | 16.6% | 17.4% | 17.8% | 15.9% | 16.1% | 23.5% |

    Assume the distribution of percents is bell shaped. Determine a 98 percent confidence interval for $\mu$.

12. The owner of the Happiness I.S. novelty shop in the Southland Shopping Mall sampled the amounts spent by 24 customers from the shop receipts over the past several months.

    | $3.38 | 6.25 | 4.76 | 4.19 | 4.85 | 8.52 | 2.70 | 3.76 |
    | $0.95 | 2.02 | 3.18 | 1.33 | 21.20 | 2.92 | 3.18 | 0.20 |
    | $4.19 | 1.01 | 4.24 | 0.32 | 2.92 | 10.49 | 7.16 | 2.12 |

    Assume the distribution of amounts spent is bell shaped. Construct a 99 percent confidence interval for $\mu$.

13. A study was made of the penetration power of .22 caliber bullets fired from 100 feet at targets consisting of stacks of computer paper sandwiched between pine boards. Sixteen tests were made and $\bar{x} = 126$ sheets of paper were penetrated with $s = 11.6$ sheets. Assume the distribution of sheets penetrated is approximately bell shaped.

    a. Construct a 90 percent confidence interval for the mean number of sheets penetrated by .22 caliber bullets fired from 100 feet.

    b. State the meaning of the answer to part (a).

14. A civil engineer wishes to determine the soluble chemical oxygen demand (SCOD) for a new type of sludge culture. She makes a SCOD determination on each of 20 culture samples and obtains $\bar{x} = 17.8$ mg/L and $s = 6.3$ mg/L. The data suggest that the distribution of SCOD is approximately normal.
    a. Construct a 90 percent confidence interval for the mean SCOD for the new type of sludge culture.
    b. State the meaning of the answer to part (a).

15. A sample of 16 Brand X Model M918 dry cell batteries was taken from a large production run in the company's plant in Chicago. The mean output of these batteries was 6.24 volts with a standard deviation of 0.26 volts. Assume the distribution of voltages for this production run is bell shaped.
    a. Construct a 95 percent confidence interval for the mean voltage of these batteries.
    b. State the meaning of the answer to part (a).

16. A sample of 20 cans of New England clam chowder was selected from a production run in the company's plant in Boston. The mean net weight of the contents of these cans was 312 grams with a standard deviation of 8 grams. Assume the distribution of weights in this production run is bell shaped.
    a. Construct a 95 percent confidence interval for the mean net weight of cans in this production run.
    b. State the meaning of the answer to part (a).

17. A sample of 20 jars of Italian seasoning was selected from a company's warehouse in Cleveland. The mean net weight of the contents of these jars was 43.6 grams with a standard deviation of 1.5 grams. Assume the distribution of weights of jars of Italian seasoning is bell shaped.
    a. Construct a 98 percent confidence interval for the mean net weight of jars of this seasoning in the warehouse.
    b. State the meaning of the answer to part (a).

18. General Synthetics Corporation manufactures nylon rope. The rope is periodically tested in the laboratory to determine how much force is needed to break it. Tests on a recent production run of a certain diameter of nylon rope yielded $\bar{x} = 45.2$ newtons (a unit of force) and $s = 1.5$ newtons. The tests were made on fifteen 1-meter pieces of this rope. Assume the distribution of forces needed to break 1-meter pieces of this rope is approximately bell shaped.
    a. Construct a 98 percent confidence interval for the mean force needed to break 1-meter pieces of this rope.
    b. State the meaning of the answer to part (a).

19. Charles Dawkins has been a jogger for several years. His regular routine includes a particular 10-mile course at least once a week. A record of the times it took Charles to run the 10-mile course for 18 trials during the past 5 months yielded $\bar{x} = 71.5$ minutes with $s = 3.6$ minutes. Assume the distribution of times for Charles to run this course is bell shaped.
    a. Construct a 99 percent confidence interval for the mean time it takes Charles to run this course.
    b. State the meaning of the answer to part (a).

20. Universal Brakes Corporation conducted tests on the effectiveness of a new model of disc brake. The engineers used a midsize car for the test. At 60 mph, it took an average of $\bar{x} = 210.5$ feet to stop the car, with $s = 15.6$ feet, and the test was conducted 21 times. Assume the distribution of stopping distances is bell shaped.
    a. Construct a 99 percent confidence interval for the mean distance required to stop the test car at 60 mph.
    b. State the meaning of the answer to part (a).

21. Construct a 90 percent confidence interval for the mean of a bell-shaped distribution, if $\bar{x} = 75.0$ and $s = 4.2$:
    a. If $n = 6$
    b. If $n = 24$

c. Compute the ratio of the lengths of the intervals in parts (a) and (b) by dividing the length of (a) by the length of (b). Round the quotient to one decimal place.

22. Construct a 95 percent confidence interval for the mean of a bell-shaped distribution, if $\bar{x} = 148.0$ and $s = 8.6$:

   a. If $n = 10$

   b. If $n = 20$

   c. Compute the ratio of the lengths of the intervals in parts (a) and (b) by dividing the length of (a) by the length of (b). Round the quotient to one decimal place.

## Exercises 5-3   Set B

*In Exercises 1 and 2 use Data Set I, Appendix B.*

1. Use random numbers to sample the ages of 16 females from the data set. For this set of data.
   a. Compute $\bar{x}$. ·
   b. Compute $s$.
   c. The distribution of female ages is approximately bell shaped. Use your $\bar{x}$ and $s$ to construct a 95 percent confidence interval for the mean age of the female participants.
   d. The mean age of the females is 57.1 years. Does your interval contain the mean?

2. Use random numbers to sample the ages of 20 males in the data set. For this set of data:
   a. Compute $\bar{x}$.
   b. Compute $s$.
   c. The distribution of male ages is approximately bell shaped. Use your $\bar{x}$ and $s$ to construct a 98 percent confidence interval for the mean age of the male participants.
   d. The mean age of the males is 54.9 years. Does your interval contain the mean?

*In Exercises 3 and 4 use Data Set II, Appendix B.*

3. Use random numbers to sample 18 prices from the "52 weeks High column." For this set of data:
   a. Compute $\bar{x}$.
   b. Compute $s$.
   c. The distribution of 52 weeks High data is reasonably bell shaped. Use your $\bar{x}$ and $s$ to construct a 90 percent confidence interval for the mean of this distribution.

4. Use random numbers to sample 25 prices from the "52 weeks Low column." For this set of data:
   a. Compute $\bar{x}$.
   b. Compute $s$.
   c. The distribution of 52 weeks Low data is reasonably bell shaped. Use your $\bar{x}$ and $s$ to construct a 99 percent confidence interval for the mean of this distribution.

*In Exercises 5–10, discuss whether or not you would be willing to assume these relative frequency distributions are bell shaped in order to make a t confidence interval for $\mu$.*

5. The distribution of grade point averages earned last semester by 30,000 undergraduate students at a major university in the Midwest.

6. The distribution of times (in hours) spent sleeping last Saturday by the 23,500 students at a large southern university.

7. The distribution of heights (in inches) of the 3520 male students in grade 6 in a large eastern city school system.

8. The distribution of weights (in pounds) of babies born last year at Kaiser Permanente Hospital in Oakland, California.

9. The distribution of lifetimes (in hours) of a large production run of 40-watt fluorescent tubes manufactured by the Sylvania Corporation.

10. The distribution of ages at time of adoption of the animals at a large animal shelter in Boston last month.

## Stein's 2 stage procedure*

In Section 5-2 we used equation 8 to determine the sample size *n* required to give a confidence interval with length 2*e*. To use this formula, we need to know $\sigma$, the standard deviation of the population whose mean we want to estimate. If $\sigma$ is unknown, then the following steps provide an honest sample size determination method for estimating a population mean $\mu$.

**Step 1.** Select an initial sample of size *m* and based on it compute *s* (the sample standard deviation).

**Step 2.** For estimation $\mu$ with confidence coefficient $(1 - \alpha)$ and an interval of length no more than 2*e*, find the smallest *n* such that

$$t \cdot \frac{s}{\sqrt{n}} \le e \qquad \text{or} \qquad n \ge \left(\frac{t \cdot s}{e}\right)^2$$

where *t* is the upper $\alpha/2$ point of the *t*-distribution with $m - 1$ degrees of freedom.

**Step 3.** If in Step 2 the *n* is greater than *m* (that is, $n > m$), then take an additional $n - m$ observation. If $n \le m$, then no second sample is required.

**Step 4.** Use all observations to compute $\bar{x}$. However, do not use any additional observations to recompute *s*.

**Step 5.** Estimate $\mu$ using end points:

$$\bar{x} \pm t \cdot \frac{s}{\sqrt{n}}, \text{ if } n > m$$

$$\bar{x} \pm t \cdot \frac{s}{\sqrt{m}}, \text{ if } n \le m$$

Additional notes on the equations in Step 5:

**Note 1.** The *t*-distribution used to determine the value of *t* (*t* is the same number used in Step 2) has degrees of freedom $m - 1$, the size of the sample taken in Step 1 minus one.

---

* This procedure is traceable to Charles Stein, "A two-sample test for a linear hypothesis whose power is independent of the variance." *Annals of Mathematical Statistics*, Vol. 16, 243–258 (1945).

**Note 2.**   The $s$ in the equations is the one computed in Step 1 from the $m$ values in the first sample only.

**Note 3.**   The $\bar{x}$ in the equations is based on all the observations, not exclusively the ones obtained in the first sample.

*Exercises 11–14 refer to Stein's two stage procedure.*

11. The first sample of size $m = 20$ from a population with (unknown) mean $\mu$ yielded $\bar{x} = 12.3$ and $s = 1.6$. The desired length of a 90 percent confidence interval is 1.3.
    a. Verify that a second sample is not needed to construct a confidence interval with the required maximum length.
    b. Construct a 90 percent confidence interval using the statistics from the first sample.

12. The first sample of size $m = 16$ from a population with (unknown) mean $\mu$ yielded $\bar{x} = 47.6$ and $s = 3.2$. The desired length of a 95 percent confidence interval is 3.5.
    a. Verify that a second sample is not needed to construct a confidence interval with the required maximum length.
    b. Construct a 95 percent confidence interval using the statistics from the first sample.

13. The first sample of size $m = 12$ from a population with (unknown) mean $\mu$ yielded $\bar{x} = 78.3$ and $s = 9.1$. The desired length of a 98 percent confidence interval is 9.0.
    a. Verify that a second sample is needed to construct a confidence interval with the required maximum length.
    b. Determine the minimum number of additional observations needed to construct the confidence interval.
    c. Assume the mean of the $n$ observations is $\bar{x} = 80.4$. Construct a 98 percent confidence interval for $\mu$.

14. The first sample of size $m = 9$ from a population with (unknown) mean $\mu$ yielded $\bar{x} = 28.1$ and $s = 4.2$. The desired length of a 99 percent confidence interval is 5.0.
    a. Verify that a second sample is needed to construct a confidence interval with the required maximum length.
    b. Determine the minimum number of additional observations needed to construct the confidence interval.
    c. Assume the mean of the $n$ observations is 26.7. Construct a 99 percent confidence interval for $\mu$.

# 5-4  The Distribution of the Sample Proportion

**Key Topics**

1. *The sampling distribution of $\hat{P}$*

2. *$\mu_{\hat{P}} = P$ and $\sigma_{\hat{P}} = \sqrt{\dfrac{P(1 - P)}{n}}$*

3. *Changing $\hat{p}$, a value of $\hat{P}$, to a z score*

4. *Computing probabilities*

S  T  A  T  I  S  T  I  C  S     I  N     A  C  T  I  O  N

Five years ago, the civic and political leaders of a large metropolitan area in the Southwest initiated a campaign titled "Buy American." The campaign was designed to raise the consciousness of consumers in the area to supporting a vigorous economy in the United States by buying products made in America whenever possible.

At the end of 5 years of the campaign, the chamber of commerce announced success, evidenced by the reduced proportion of imported vehicles in the region. They further claimed that "fewer than 25 percent of vehicles less than 5 years old were manufactured in a foreign country."

Cindy Yates is an employee of the state department of motor vehicles (DMV). Based on her experience registering new vehicles in the region targeted by the "Buy American" campaign, she thought the 25 percent figure of the Chamber of Commerce to be too low. She therefore accessed the DMV computers for a random sample of 60 registrations of new vehicles registered for the first time within the last 5 years in the area of interest.

Cindy decided that if more than 30 percent of the sampled registrations were for vehicles manufactured outside the United States, then she would have strong evidence that the 25 percent figure of the Chamber of Commerce was too low. She based this conclusion on the fact that a sample of size 60 from a population in which 25 percent of the items have some characteristic of interest will yield a sample proportion of more than 30 percent less than 19 percent of the time. In symbols:

$P$ = the proportion of vehicles less than 5 years old manufactured outside the United States. (The Chamber of Commerce claims $P = 0.25$.)

$\hat{P}$ = the proportion of vehicles in Cindy's sample manufactured outside the United States.

Based on a probability distribution for $\hat{P}$ centered at the supposed value of $P = 0.25$, Cindy knows:

$$P(\hat{P} > 0.30) \approx 0.19$$

## The sampling distribution of $\hat{P}$

Recall from Chapter 2 that $\hat{p}$ (read "p-hat") was used as a symbol for the proportion of a sample with some characteristic of interest. This symbol will now be used to describe the proportion of successes in a binomial experiment. Because chance is involved in such an experiment, we use a capital letter and write $\hat{P}$ for the random variable. In many applications of statistical methods, $\hat{P}$ will be the result of taking a sample from a population in which $P$ is the proportion with some characteristic of interest. Thus *the probability distribution of $\hat{P}$ is frequently called a sampling distribution.*

To use the random variable $\hat{P}$ to make inferences about a population parameter $P$, we need to know the following three things about the sampling distribution of $\hat{P}$:

     a. The *shape* of the sampling distribution of $\hat{P}$.
     b. The *mean*, $\mu_{\hat{P}}$, of the sampling distribution.
     c. The *standard deviation*, $\sigma_{\hat{P}}$, of the sampling distribution.

Recall from Chapter 4 that a binomial random variable will have an approximately normal probability distribution whenever $np \geq 5$ and $n(1 - p) \geq 5$, where $p$ is the probability of success on each trial of the basic experiment. For experiments involving sampling from a population in which $P$ is the proportion with some characteristic of interest, $P$ can be used as a "probability of success." As a consequence, the conditions for approximate normality of a binomial random variable $X$ (and also the sample proportion $\hat{P} = X/n$) can be stated in terms of $P$.

### Conditions for the Sampling Distribution of $\hat{P}$ to be Approximately Normal

If $\hat{P}$ is the proportion of a sample of size $n$ with a characteristic of interest taken from a population in which a fraction $P$ of the objects have that characteristic, and $n$ is small compared with the population size, then the **sampling distribution of $\hat{P}$ will be approximately normal** if:

$$nP \geq 5 \quad \text{and} \quad n(1 - P) \geq 5$$

In the experiment performed by Cindy Yates in the Statistics in Action section, supposedly 25 percent of the sampled population were vehicles manufactured outside the United States, and so $P = 0.25$. Cindy sampled 60 registrations from the population, and with $n = 60$:

$$nP \text{ becomes } 60(0.25) = 15 \quad \text{and} \quad 15 \geq 5$$
$$n(1 - P) \text{ becomes } 60(0.75) = 45 \quad \text{and} \quad 45 \geq 5$$

Thus the sampling distribution of $\hat{P}$ for Cindy's experiment is approximately normal. As will be seen, the probability Cindy computed can be found using the standard normal distribution and Appendix Table 4.

$$\mu_{\hat{P}} = P \text{ and } \sigma_{\hat{P}} = \sqrt{\frac{P(1 - P)}{n}}$$

The relationship between $X = n\hat{P}$ and the random variable $\hat{P}$ has been used to specify conditions under which the sampling distribution of $\hat{P}$ is approximately normal. This relationship can also be used to find formulas for computing $\mu_{\hat{P}}$ and $\sigma_{\hat{P}}$. The symbol $\mu_{\hat{P}}$ stands for the *mean* (or expected value) of the random variable $\hat{P}$. The symbol $\sigma_{\hat{P}}$ stands for the *standard deviation* of the sampling distribution of $\hat{P}$.

Recall from Chapter 4 that for a binomial random variable $X$:

$$\mu_X = np \quad \text{and} \quad \sigma_X = \sqrt{np(1 - p)}$$

Therefore, substituting $P$ for $p$:

$$\mu_X = nP \quad \text{and} \quad \sigma_X = \sqrt{nP(1 - P)}$$

We can change $\mu_X$ to $\mu_{\hat{P}}$ by dividing $nP$ by $n$. Similarly we can change $\sigma_X$ to $\sigma_{\hat{P}}$ by dividing $\sqrt{nP(1 - P)}$ by $n$:

$$\mu_{\hat{P}} = \frac{\mu_X}{n} \quad \text{and} \quad \sigma_{\hat{P}} = \frac{\sigma_X}{n}$$

$$= \frac{nP}{n} \qquad\qquad = \frac{\sqrt{nP(1 - P)}}{n}$$

$$= P \qquad\qquad = \sqrt{\frac{nP(1 - P)}{n^2}}$$

$$= \sqrt{\frac{P(1 - P)}{n}}$$

## The Mean and Standard Deviation for the Sampling Distribution of $\hat{P}$

$$\mu_{\hat{P}} = P \tag{11}$$

and

$$\sigma_{\hat{P}} = \sqrt{\frac{P(1 - P)}{n}} \tag{12}$$

► **Example 1**   In a large metropolitan school district, records indicate that 38 percent of the elementary school students have single parents. Dr. Meg Weant, a psychologist for the school district, randomly sampled 60 elementary school students in this district. Let $\hat{P}$ be the proportion of students in the sample that have single parents.

    **a.** Verify that the sampling distribution of $\hat{P}$ is approximately normal.
    **b.** Compute $\mu_{\hat{p}}$.
    **c.** Compute $\sigma_{\hat{p}}$ to two decimal places.
    **d.** Draw and label a probability histogram for $\hat{P}$.

**Solution**   **a.** With $n = 60$, $P = 0.38$, and $1 - P = 0.62$:

$$nP = 60(0.38) = 22.8 \qquad nP \geq 5$$
$$n(1 - P) = 60(0.62) = 37.2 \qquad n(1 - P) \geq 5$$

Because both products are greater than 5, the sampling distribution of $\hat{P}$ is approximately normal.

    **b.** For the sampling distribution of $\hat{P}$:

$$\mu_{\hat{p}} = 0.38 \qquad \mu_{\hat{p}} = P$$

    **c.** With $P = 0.38$, $1 - P = 0.62$, and $n = 60$:

$$\sigma_{\hat{p}} = \sqrt{\frac{(0.38)(0.62)}{60}} \qquad \sigma_{\hat{p}} = \sqrt{\frac{P(1-P)}{n}}$$
$$= \sqrt{0.00393\ldots}$$
$$\approx 0.06 \qquad \text{To two decimal places}$$

    **d.** Using $\mu_{\hat{p}} = 0.38$ and $\sigma_{\hat{p}} = 0.06$:

$$\mu_{\hat{p}} - 2\sigma_{\hat{p}} = 0.38 - 2(0.06) = 0.26$$
$$\mu_{\hat{p}} - \sigma_{\hat{p}} = 0.38 - 0.06 = 0.32$$
$$\mu_{\hat{p}} + \sigma_{\hat{p}} = 0.38 + 0.06 = 0.44$$
$$\mu_{\hat{p}} + 2\sigma_{\hat{p}} = 0.38 + 2(0.06) = 0.50$$

An approximate normal distribution for $\hat{P}$ is shown in Figure 5-21.   ◄

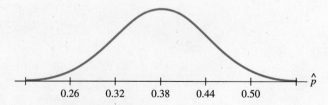

**Figure 5-21.**   A histogram for $\hat{P}$.

BOSTON (AP) — Americans have never been healthier, but they feel worse and worse, a doctor says.

Dr. Arthur J. Barsky calls this "the paradox of health." The nation's collective health has improved dramatically over the past 30 years, yet people report increasingly high rates of disability and a general dissatisfaction with their health.

Barsky, a psychiatrist at Massachusetts General Hospital, blames this on several factors, including people's obsession with diet and exercise and their unrealistic expectations of what medicine can do for them.

Writing in today's New England Journal of Medicine, Barsky noted that the life expectancy of a child born in 1984 was 75 years. In 1900, it was 47. Since 1970 alone, the life expectancy has increased by four years.

However, surveys show that people report twice as many episodes of disabling illness now than they did in the 1920s, before the introduction of antibiotics. In the 1970s, 61 percent of Americans said they were satisfied with their health. A decade later, this had fallen to 55 percent.

"Our subjective feeling of healthiness and physical well-being has decreased, even though there have been major advances in our actual, objective health status," Barsky wrote.

One reason is medicine's conquest of diseases that kill quickly. It lets people grow old so they die with diseases that linger. People who are saved from pneumonia and tuberculosis live long enough to be afflicted with arthritis and diabetes.

**Figure 5-22.**  "America's attitude toward health," *The Sacramento Bee*, February 18, 1988, p. A1. Reprinted by permission of Associated Press.

▶ **Example 2**   In Figure 5-22, Dr. Arthur J. Barsky states that 55 percent of Americans are "satisfied with their health." A reporter for a newspaper in Dallas intends to sample 250 residents. Let $\hat{P}$ be the proportion of her sample that say they are satisfied with their health. Use 55 percent for the proportion of all Dallas residents who are satisfied with their health.

**a.** Verify that the sampling distribution of $\hat{P}$ is approximately normal.
**b.** Compute $\mu_{\hat{P}}$.
**c.** Compute $\sigma_{\hat{P}}$ to two decimal places.
**d.** Draw and label a histogram for $\hat{P}$.

**Solution**   **Discussion.** For the given survey, $n = 250$ and $P = 0.55$.

**a.** $nP = 250(0.55) = 137.5$      $137.5 \geq 5$
$n(1 - P) = 250(0.45) = 112.5$    $112.5 \geq 5$

**b.** $\mu_{\hat{P}} = 0.55$     $\mu_{\hat{P}} = P$

**c.** $\sigma_{\hat{P}} = \sqrt{\dfrac{(0.55)(0.45)}{250}}$     $\sigma_{\hat{P}} = \sqrt{\dfrac{P(1 - P)}{n}}$

$= \sqrt{0.00099}$     The exact value

$\approx 0.03$     Rounded to two places

**d.** Using $\mu_{\hat{p}} = 0.55$ and $\sigma_{\hat{p}} = 0.03$:

$\mu_{\hat{p}} - 2\sigma_{\hat{p}} = 0.55 - 0.06 = 0.49$
$\mu_{\hat{p}} - \sigma_{\hat{p}} = 0.55 - 0.03 = 0.52$
$\mu_{\hat{p}} + \sigma_{\hat{p}} = 0.55 + 0.03 = 0.58$
$\mu_{\hat{p}} + 2\sigma_{\hat{p}} = 0.55 + 0.06 = 0.61$

A graph of the approximate probability distribution for $\hat{P}$ is shown in Figure 5-23.  ◄

**Figure 5-23.**   A probability histogram for $\hat{P}$.

## Changing $\hat{p}$, a value of $\hat{P}$, to a $z$ score

When a normal curve is used to approximate probabilities for $\hat{P}$, then equation (13) can be used to change an observed $\hat{p}$ to a $z$ score.

### To Change $\hat{p}$, a Value of $\hat{P}$, to a $z$ Score

If $P$ is the proportion of a population with some characteristic of interest and $\hat{p}$ is the observed proportion of a random sample from that population with the same characteristic, then the $z$ score for $\hat{p}$ is:

$$z = \frac{\hat{p} - P}{\sqrt{\dfrac{P(1 - P)}{n}}} \qquad (13)$$

Notice that equation (13) is a form of the equation:

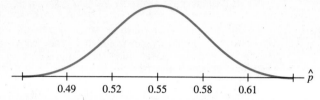

$z = \dfrac{x - \mu}{\sigma}$

$x$ is now $\hat{p}$

$\mu$ is now $\mu_{\hat{p}}$  and  $\mu_{\hat{p}} = P$

$\sigma$ is now $\sigma_{\hat{p}}$  and  $\sigma_{\hat{p}} = \sqrt{\dfrac{P(1 - P)}{n}}$

## Computing probabilities

If $\hat{P}$ is based on a sample size $n$ such that $nP > 5$ and $n(1 - P) > 5$, then Appendix Table 4 can be used to compute probabilities associated with it. The procedure requires finding the $z$ value corresponding to $\hat{p}$.

For the experiment performed by Cindy Yates in Statistics in Action, we previously showed that the probability distribution of $\hat{P}$ is approximately normal. For $\hat{p} = 0.30$:

$$z = \frac{0.30 - 0.25}{\sqrt{\dfrac{(0.25)(0.75)}{60}}} \qquad P = 0.25, \ (1 - P) = 0.75, \text{ and } n = 60$$

$$= \frac{0.05}{0.0559\ldots}$$

$$\approx 0.89$$

$$P(\hat{P} > 0.30) = P(Z > 0.89)$$
$$= 1 - 0.8133$$
$$\approx 0.19$$

Thus Cindy's calculation of about a 19 percent chance of getting $\hat{P} > 0.30$ is correct.

▶ **Example 4**   Joan McKee is a volunteer worker at the blood donor center cited in Figure 5-24. According to the article, 46 people of 100 (or 46 percent) have blood type O. Next Saturday, Joan intends to sample the records at the center. She will check the blood types of 155 randomly selected blood donations obtained over the past year. Let $\hat{P}$ be the proportion of O blood types in the sample.

**a.** Compute $P(0.40 < \hat{P} < 0.50)$.

**b.** Sketch the area of a normal distribution that represents the probability of part (a).

**c.** State, in the context of the study, what the answer to part (a) means.

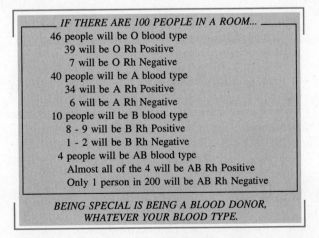

**Figure 5-24.**   A poster from a blood donor center.

**Solution**    **Discussion.** With $P = 0.46$ and $n = 155$, both $nP$ and $n(1 - P)$ are greater than 5. Thus the sampling distribution of $\hat{P}$ is approximately normal.

**a.** With $P = 0.46$, $1 - P = 0.54$, and $n = 155$:

For $\hat{p}_1 = 0.40$:

$$z_1 = \frac{0.40 - 0.46}{\sqrt{\dfrac{(0.46)(0.54)}{155}}}$$

$\approx -1.50$, to two places

For $\hat{p}_2 = 0.50$:

$$z_2 = \frac{0.50 - 0.46}{\sqrt{\dfrac{(0.46)(0.54)}{155}}}$$

$\approx 1.00$, to two places

$$P(0.40 < \hat{P} < 0.50) = P(-1.50 < Z < 1.00)$$
$$= 0.8413 - 0.0668$$
$$= 0.7745$$

**b.** The area shaded in Figure 5-25 represents the desired probability.

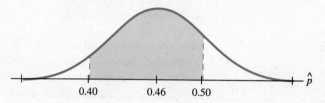

**Figure 5-25.**   Shaded area $= P(0.40 < \hat{P} < 0.50)$.

**c.** The probability is about 77 percent that between 40 and 50 percent of the blood donations sampled will be of blood type O.    ◄

## Exercises 5-4    Set A

*In Exercises 1–10:*

*a. Verify that the sampling distribution of $\hat{P}$ is approximately normal.*

*b. Compute $\mu_{\hat{P}}$ to two decimal places.*

*c. Compute $\sigma_{\hat{P}}$ to two decimal places.*

*d. Draw and label an approximately normal histogram for $\hat{P}$.*

**1.** The Edison Utility Company reports that 18 percent of its customers are past due in paying their utility bills. A random sample of 41 customers is selected, and $\hat{P}$ is the proportion of the accounts in the sample that are past due.

**2.** The Parent Teachers Association (PTA) of Kansas City surveyed the parents of high school students in the city. The results of the survey suggest that 62 percent of all parents feel their children are not being given enough school work to do at home. Suppose a random sample of 68 parents of high school students in this city is taken, and $\hat{P}$ is the proportion of the parents in the sample that feel their children are not being given enough school work to do at home.

**3.** The Right To Life Organization (RTLO) in Pittsburgh surveyed adults in the city. The survey suggests that 72 percent of the residents support the right of families to terminate life-support for terminally ill patients. Suppose a sample of 56 adults in Pittsburgh is taken, and $\hat{P}$ is the proportion of those sampled who feel families have a right to terminate life-support.

4. A member of the Public Relations Office of the regional transit system in San Francisco studied the commuting habits of students at a large community college in the city. Her study suggests that 24 percent of the students regularly use some form of public transportation to get to school. Suppose 30 students from this school are randomly selected, and $\hat{P}$ is the proportion of those sampled that regularly use some form of public transportation to get to school.

5. Matsuii Nursery has a large display of dahlia bulbs. The supplier claims that 30 percent of the bulbs will produce yellow flowers. Suppose 25 of the bulbs are selected, and $\hat{P}$ is the proportion of the bulbs that produce yellow flowers.

6. The 345 − POOL group in Indianapolis promotes car pooling as a means of reducing highway congestion in the city during commuting hours. A study of the problem indicates that 65 percent of vehicles on the city's streets during commuting hours are carrying only one person. Suppose a check is made of 30 vehicles on the city's streets during commuting hours, and $\hat{P}$ = the proportion of these vehicles with only one occupant.

7. A study of adults in Great Britain with titles of nobility showed that 15 percent of them would object to one of their children marrying a commoner. Suppose a sample of 50 of these members of British nobility is taken, and $\hat{P}$ is the proportion of those surveyed that would object to one of their children marrying a commoner.

8. A consumer magazine reported that 72 percent of grocery orders bought at supermarkets are paid for by check or ATM (automatic teller machine) card. Suppose the 0.72 figure applies to people who shop at supermarkets in Denver. A sample

of 150 of these shoppers is taken, and $\hat{P}$ is the proportion of these shoppers who pay for their groceries by check or ATM cards.

9. The SPCA (Society for the Prevention of Cruelty of Animals) in a moderate-sized city in Ohio claims that its program to reduce unwanted pregnancies in house pets has been effective in getting 38 percent of the dog and cat population in the city neutered. A sample of 48 dogs and cats in the city is taken, and $\hat{P}$ is the fraction of dogs and cats in the sample that have been neutered.

10. According to the Edison Electric Institute, 49 percent of single-family homes under construction are installing electric heating. Suppose the 0.49 figure applies to homes under construction in Cincinnati, and a sample of 75 such homes in the area is taken. Let $\hat{P}$ equal the proportion of these homes that are installing electric heating.

*In Exercises 11–16:*
a. *Approximate the indicated probability.*
b. *Sketch the area of an approximately normal distribution that represents the probability of part (a).*
c. *State, in the context of the experiment, what the probability of part (a) means.*

11. Capital Nursery in Camden guarantees that 80 percent of their large stock of bare root trees will grow. A random sample of 48 of these trees is taken and planted. Let $\hat{P}$ equal the fraction of the sampled trees that grow. Approximate $P(\hat{P} < 0.75)$.

12. Records maintained by the Department of Motor Vehicles in a large eastern state indicate that 38 percent of vehicles being registered for the first time are trucks. Suppose a sample of 65 new vehicle registration applications is taken, and $\hat{P}$ is the fraction of the sample that are for trucks. Approximate $P(\hat{P} < 0.40)$.

13. The records from the state highway police in a large southern state show that 72 percent of motorcycle accidents in which the speed of the motorcycle was 40 to 48 mph resulted in death or serious injury. Suppose a sample of 56 accident reports

in this state involving motorcycles traveling between 40 and 48 mph is taken, and $\hat{P}$ is the fraction of the sample that report death or serious injury. Approximate $P(\hat{P} > 0.75)$.

14. According to Youthpoll America, only 13 percent of high school seniors in the United States regularly smoke cigarettes. Suppose the 0.13 figure applies to seniors currently enrolled in the high schools in Seattle. Suppose a random sample of 126 seniors in this city is taken, and $\hat{P}$ is the fraction of the sample that regularly smoke cigarettes. Approximate $P(\hat{P} < 0.15)$.

15. A large study of leisure time activities of high school juniors and seniors was recently conducted in Portland, Maine. The study indicates that 38 percent of all high school juniors and seniors never read anything required school work. Suppose a random sample of 138 high school juniors and seniors is taken in Portland, and $\hat{P}$ is the fraction of the sample that never read anything required school work. Approximate $P(0.30 < \hat{P} < 0.35)$.

16. Records at the state capital suggest that 28 percent of the households in Raleigh own one or more handguns. Suppose a sample of 224 households in this city is surveyed, and $\hat{P}$ is the fraction of the sample that owns at least one handgun. Approximate $P(0.25 < \hat{P} < 0.35)$.

17. Figure 5-26 is a representation of the sampling distribution of $\hat{P}$, if $P = 0.650$ and $n = 91$. Find the value of $\hat{p}_1$ such that $P(\hat{P} > \hat{p}_1) \approx 0.10$.

18. Figure 5-27 is a representation of the sampling distribution of $\hat{P}$, if $P = 0.400$ and $n = 150$. Find the value of $\hat{p}_2$ such that $P(\hat{P} < \hat{p}_2) \approx 0.05$.

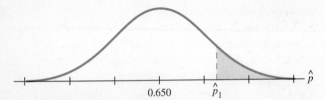

Figure 5-26.  $P(\hat{P} > \hat{p}_1) \approx 0.10$.

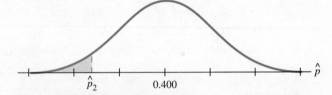

Figure 5-27.  $P(\hat{P} < \hat{p}_2) \approx 0.05$.

19. Figure 5-28 is a representation of the sampling distribution of $\hat{P}$, if $P = 0.500$ and $n = 100$. If the area of the histogram to the left of $\hat{p}_3$ is the same as the area to the right of $\hat{p}_4$, find $\hat{p}_3$ and $\hat{p}_4$ such that $P(\hat{p}_3 < \hat{P} < \hat{p}_4) \approx 0.90$.

20. Figure 5-29 is a representation of the sampling distribution of $\hat{P}$, if $P = 0.270$ and $n = 219$. If the area of the histogram to the left of $\hat{p}_5$ is the same as the area to the right of $\hat{p}_6$, find $\hat{p}_5$ and $\hat{p}_6$ such that $P(\hat{p}_5 < \hat{P} < \hat{p}_6) \approx 0.95$.

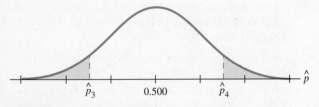

Figure 5-28.  $P(\hat{p}_3 < P < \hat{P}_4) \approx 0.90$.

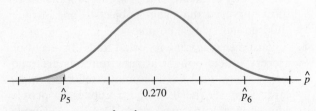

Figure 5-29.  $P(\hat{p}_5 < \hat{P} < \hat{P}_6) \approx 0.95$.

**21.** For a population, $P = 0.37$. Compute the sample size $n$ that will yield $\sigma_{\hat{P}}$ that is approximately 0.05 for the probability distribution of $\hat{P}$ for samples taken from that population.

**22.** For a given population, $P = 0.72$. Compute the sample size $n$ that will yield $\sigma_{\hat{P}}$ that is approximately 0.03 for the probability distribution of $\hat{P}$ for samples taken from that population.

**23.** In a large population, $P$ is the proportion of the population with characteristic A and $P = 0.45$. Suppose 99 items in this population are randomly selected, and $\hat{P}$ is the proportion of the sample with characteristic A. For the sampling distribution of $\hat{P}$, compute to two decimal places:

a. $\sigma_{\hat{P}}$, using $P = 0.45$.

b. $\sigma_{\hat{P}}$, using 0.40 in place of $P$.

c. $\sigma_{\hat{P}}$, using 0.35 in place of $P$.

d. Compare the values of $\sigma_{\hat{P}}$ obtained in parts (a), (b), and (c).

**24.** In a large population, $P$ is the proportion of the population with characteristic B and $P = 0.68$. Suppose 136 items in this population are randomly selected, and $\hat{P}$ is the proportion in the sample with characteristic B. For the sampling distribution of $\hat{P}$, compute to two decimal places:

a. $\sigma_{\hat{P}}$, using $P = 0.68$.

b. $\sigma_{\hat{P}}$, using 0.65 in place of $P$.

c. $\sigma_{\hat{P}}$, using 0.73 in place of $P$.

d. Compare the values of $\sigma_{\hat{P}}$ obtained in parts (a), (b), and (c).

**25.** In a large population, $P$ is the proportion of the population with characteristic C and $P = 0.37$. Suppose $n$ items in this population are randomly selected, and $\hat{P}$ is the proportion of the sample with characteristic C. For the sampling distribution of $\hat{P}$, compute to the nearest whole number the value of $n$ that would yield:

a. $\sigma_{\hat{P}} \approx 0.06$.    b. $\sigma_{\hat{P}} \approx 0.04$.    c. $\sigma_{\hat{P}} \approx 0.02$.

**26.** In a large population, $P$ is the proportion of the population with characteristic D and $P = 0.76$. Suppose $n$ items in the population are randomly selected, and $\hat{P}$ is the proportion in the sample with characteristic D. For the sampling of $\hat{P}$, compute to the nearest whole number the value of $n$ that would yield:

a. $\sigma_{\hat{P}} \approx 0.05$.    b. $\sigma_{\hat{P}} \approx 0.03$.    c. $\sigma_{\hat{P}} \approx 0.01$.

## Exercises 5-4    Set B

**1.** In a large city, the proportion of residents that regularly use the Regional Transit (RT) is 0.32. Suppose $n = 6$ residents in this city are sampled.

a. Let $X$ be the number in the sample that regularly use RT. List the possible values for $X$, compute the probability distribution for $X$ using the binomial probability formula and construct a probability histogram for $X$.

b. Let $\hat{P}$ be the proportion of successes in the six trials, where success is taken to be the event that a sampled resident regularly uses RT. List the possible values for $\hat{P}$, compute the probability distribution for $\hat{P}$, and construct a probability histogram for $\hat{P}$.

c. Compare the probability distributions and histogram in parts (a) and (b).

2. Kirklin Enterprises offers its employees a profit sharing plan. Membership in the plan is voluntary. Records indicate that 56 percent of the eligible employees are participating in the plan. Suppose $n = 7$ employees of Kirklin Enterprises are sampled, and $n$ is small compared with the total number of employees.

   a. Let $Y$ be the number in the sample that are members of the profit sharing plan. List the possible values for $Y$, compute the probability distribution for $Y$ using the binomial probability formula, and construct a probability histogram for $Y$.

   b. Let $\hat{P}$ be the proportion of successes in the seven trials, where success is taken to be the event that a sampled employee is a member of the profit sharing plan. List the possible values for $\hat{P}$, compute the probability distribution for $\hat{P}$, and construct a probability histogram for $\hat{P}$

   c. Compare the probability distributions and histograms in parts (a) and (b).

*In Exercises 3 and 4, use Data Set II, Appendix B.*

3. In Data Set II, 0.458 of the stock prices increased (labeled with "+" in "Net Chg" column).

   a. Use random numbers to sample $n = 20$ stocks.

   b. Count the number of ($+$)-stocks in the sample and compute $\hat{p} = \dfrac{x}{20}$.

   c. Compute $\mu_{\hat{p}}$ for the probability distribution of $\hat{P}$ based on samples of size 20.

   d. Verify that the probability distribution of $\hat{P}$ is approximately normal.

   e. Compute $\sigma_{\hat{p}}$.

   f. Compute $P(\hat{P} < \hat{p})$, where $\hat{p}$ is your observed value from part (b).

4. In Data Set II, 0.262 of the stock prices remained the same (labeled with a "0" in "Net Chg" column).

   a. Use random numbers to sample $n = 30$ stocks.

   b. Count the number of (0)-stocks in the sample and compute $\hat{p} = \dfrac{x}{30}$.

   c. Complete $\mu_{\hat{p}}$ for the probability distribution of $\hat{P}$ based on samples of size 30.

   d. Verify that the probability distribution of $\hat{P}$ is approximately normal.

   e. Compute $\sigma_{\hat{p}}$.

   f. Compute $P(\hat{P} > \hat{p})$, where $\hat{p}$ is your observed value from part (b).

*In Exercises 5 and 6, use the data in Figure 5-30, and do parts (a) and (b).*

*a. Construct and label a histogram for $\hat{P}$.*

*b. Compute the indicated probability.*

5. Assume that 80 percent of managers and professionals in Kansas City disapprove of kissing in the workplace. Suppose $n = 45$ managers and professionals in Kansas City are randomly sampled, and $\hat{P}$ is the proportion in the sample that disapprove of kissing in the workplace.

   Compute $P(0.76 < \hat{P} < 0.84)$.

---

**Kissing taboo; hugging OK**

Managers and professionals do have their pet peeves.

At the top of the list of unacceptable practices in the workplace were kissing, and ethnic and sexual jokes, mentioned by more than 80 percent of those who participated in a survey by Working Smart, a management newsletter.

A solid majority said it was turned off by short skirts, sandals and smoking.

In contrast, asked what was acceptable in the workplace, 98 percent said family pictures. Perfume got the OK from 90 percent; long hair on men, 70 percent; hugging, 60 percent; and physical touching, 58 percent.

**Figure 5-30.** "Kissing Taboo; Hugging OK," *The Sacramento Bee*, 1/27/91, p. G8. Copyright © 1991 *The Sacramento Bee*. Reprinted by permission.

**6.** Assume that 60 percent of managers and professionals in New Orleans approve of hugging in the workplace. Suppose $n = 25$ managers and professionals in New Orleans are randomly sampled, and

$\hat{P}$ is the proportion in the sample that approve of hugging in the workplace.
Compute $P(0.55 < \hat{P} < 0.65)$.

## 5-5 Large *n* Confidence Intervals for a Proportion, *P*

*Key Topics*

1. $\hat{P}$ *is a point estimate for an unknown P*

2. *Constructing confidence intervals*

3. *Choosing the sample size n*

4. *Interpreting a confidence interval for P*

### S T A T I S T I C S   I N   A C T I O N

Bill Tobey is a steward for the Service Employees Union in the northeast region of the United States. At a recent executive council meeting there was a discussion of the salaries paid to restaurant personnel. Several members of the council felt that salaries paid to such personnel by restaurant owners are notoriously low because employers assume that employees get gratuities at a rate of 15 percent of customers' bills. These members felt that the 15 percent rate was not being honored by a growing proportion of restaurant patrons. The decision of the council was to have each steward get some hard data to justify such a claim before taking any action.

Bill contacted union members in his region to enlist their help in collecting data. He instructed each one to obtain copies of restaurant tabs for meals and drinks, and to record the corresponding tips received. Within 2 weeks,

Bill had several thousand data points. From this set, he randomly sampled 80 data points, and for each one he calculated the rate at which the tip was paid, as follows:

$$\text{Rate of tip} = \frac{\text{Amount of tip}}{\text{Amount of restaurant charges}}$$

Each rate that was less than 15 percent was counted a success, and each rate that was 15 percent or more was counted a failure. Bill got 28 successes in the sample of 80. As a consequence, in his subsequent report to the executive council, Bill made the following comment:

"A 90 percent confidence interval for the proportion of restaurant patrons that use less than the 15 percent rate for tipping is between 26 and 44 percent."

## $\hat{p}$ is a point estimate for an unknown $P$

In Section 5-2, we used $\bar{x}$ from a random sample as a point estimate for $\mu$. In a similar way, $\hat{p}$ from a random sample can be used as a point estimate for $P$.

▶ **Example 1**   Brenda Lepley is dean of students at Sun Valley Community College. She wanted to estimate the proportion of students at the college that are getting some form of government assistance to finance their educations. She decided to consider grants, loans, and government-sponsored work-assistance programs to be government assistance. To estimate the unknown proportion, Brenda randomly sampled 60 students and obtained the data in Table 5-7.

**TABLE 5-7**   Results of Government Assistance Poll

| Student | Response | Student | Response | Student | Response | Student | Response |
|---------|----------|---------|----------|---------|----------|---------|----------|
| 1 | No | 16 | Yes | 31 | No | 46 | No |
| 2 | No | 17 | No | 32 | No | 47 | No |
| 3 | Yes | 18 | No | 33 | No | 48 | No |
| 4 | No | 19 | Yes | 34 | No | 49 | No |
| 5 | No | 20 | Yes | 35 | Yes | 50 | No |
| 6 | No | 21 | Yes | 36 | Yes | 51 | Yes |
| 7 | Yes | 22 | No | 37 | No | 52 | Yes |
| 8 | Yes | 23 | No | 38 | Yes | 53 | Yes |
| 9 | No | 24 | No | 39 | Yes | 54 | No |
| 10 | No | 25 | Yes | 40 | No | 55 | No |
| 11 | No | 26 | No | 41 | No | 56 | Yes |
| 12 | Yes | 27 | No | 42 | No | 57 | No |
| 13 | No | 28 | Yes | 43 | Yes | 58 | No |
| 14 | No | 29 | No | 44 | No | 59 | No |
| 15 | Yes | 30 | No | 45 | Yes | 60 | Yes |

Let $\hat{p}$ be the observed fraction of the sample getting some form of government assistance.

**a.** Compute $\hat{p}$ to two decimal places.
**b.** Use $\hat{p}$ as a point estimate for $P$, the proportion of all Sun Valley College students getting some form of government assistance.

**Solution**   **a.** In the table, there are 22 Yes responses.

$$\hat{p} = \frac{22}{60} \qquad \hat{p} = \frac{x}{n}$$

$$= 0.37 \qquad \text{Rounded to two decimal places}$$

**b.** A point estimate of the proportion is 37 percent.   ◀

## Constructing Confidence Intervals for *P*

If $\hat{P}$ is based on a sufficiently large $n$ (such that $nP \geq 5$ and $n(1 - P) \geq 5$), then the sampling distribution of $\hat{P}$ will be approximately normal. For such cases, $\hat{p}$ can be changed to a $z$ score using equation (13):

$$z = \frac{\hat{p} - P}{\sqrt{\dfrac{P(1 - P)}{n}}} \tag{13}$$

and probabilities calculated using the standard normal table. We then know, for example, that the probability is 95 percent that $\hat{P}$ takes on a value $\hat{p}$ with a $z$ score between $-1.96$ and $1.96$. A similar statement about $\bar{X}$ led to a preliminary form for a large $n$ confidence interval for $\mu$. Using the same line of reasoning, we can arrive at the conclusion that approximate 95 percent confidence limits for $P$ are:

$$\text{lower limit} = \hat{p} - 1.96 \sqrt{\frac{P(1 - P)}{n}}$$

and

$$\text{upper limit} = \hat{p} + 1.96 \sqrt{\frac{P(1 - P)}{n}}$$

From a *mathematical point of view* these limits are perfectly fine confidence limits. But from a *practical point of view* they are of little use because we are not going to know the value of $\sqrt{\dfrac{P(1 - P)}{n}}$ when we are trying to estimate $P$.

It turns out that when $n$ is large (but still a small fraction of the population size), the preceding upper and lower limit formulas can be used with $P(1 - P)$ replaced by $\hat{p}(1 - \hat{p})$. The resulting formulas for confidence limits are given in (14) and (15).

### Large *n* Approximate Confidence Intervals for *P*

If $z$ is a positive number, then the interval with:

$$\text{lower limit} = \hat{p} - z \sqrt{\frac{\hat{p}(1 - \hat{p})}{n}} \tag{14}$$

and

$$\text{upper limit} = \hat{p} + z \sqrt{\frac{\hat{p}(1 - \hat{p})}{n}} \tag{15}$$

can be used as a confidence interval for $P$. The associated confidence coefficient is $P(-z < Z < z)$.

▶ **Example 2**    In a study of the lifetimes of a large production run of 40-watt fluorescent tubes, 400 tubes were put to a test and 80 failed before the end of the present guarantee period. Give a 98 percent confidence interval for the fraction of such tubes in this production run that would fail before the end of the present guarantee period.

**Solution**    **Discussion.** Change the count data to a sample proportion:

$$\hat{p} = \frac{80}{400} = 0.20 \qquad \frac{80 \text{ tubes failed}}{400 \text{ tubes tested}}$$

With $\hat{p} = 0.20$, $(1 - \hat{p}) = 0.80$, $n = 400$, $z = 2.33$, and equations (14) and (15):

$$\text{lower limit} = 0.20 - 2.33 \sqrt{\frac{(0.20)(0.80)}{400}}$$

$$= 0.20 - 0.0466$$

$$= 0.1534$$

$$\approx 15\%$$

$$\text{upper limit} = 0.20 + 0.0466$$

$$= 0.2466$$

$$\approx 25\%$$

Thus an approximate 98 percent confidence interval for the proportion of these tubes that would fail before the end of the present guarantee period is from 15 to 25 percent.    ◀

Example 2 serves to illustrate the fact that unless $n$ is very large, equations (14) and (15) can produce fairly wide confidence intervals for $P$. In this example, the sample size was 400, and the 98 percent confidence interval width was:

$$25\% - 15\% = 10\%$$

When estimating $P$, a large sample size is frequently needed to produce an informative interval (one that locates $P$ to an acceptably small interval with an acceptably high associated confidence).

▶ **Example 3**    In a recent issue of a national magazine, an editorial cited data from a large public opinion poll. The poll showed that 72 percent of the adults surveyed named drug abuse as the greatest social problem in the United States. The poll was taken in a large eastern state and involved 1485 adults. Construct a 95 percent confidence interval for $P$, where $P$ is the proportion of adults in this state that would name drug abuse as the greatest social problem.

**Solution**    With $\hat{p} = 0.72$, $1 - \hat{p} = 0.28$, and $n = 1485$:

$$\text{lower limit} = 0.72 - 1.96 \sqrt{\frac{(0.72)(0.28)}{1485}}$$

$$= 0.72 - 0.0228\ldots$$

$$= 0.69716\ldots$$

$$\approx 0.70, \text{ to two decimal places}$$

$$\text{upper limit} = 0.72 + 0.0228\ldots$$

$$= 0.7428\ldots$$

$$\approx 0.74, \text{ to two decimal places}$$

Thus an approximate 95 percent confidence interval for the proportion of adults in this eastern state that feel drug abuse is the greatest social problem in the United States is from 70 to 74 percent.    ◄

## Choosing a sample size *n*

If we let 2*e* represent the length of a confidence interval for *P*, then *e* is the distance between $\hat{p}$ and the end-points of the interval in equations (14) and (15). That is

$$e = z \sqrt{\frac{\hat{p}(1 - \hat{p})}{n}}$$

When this equation is solved for *n*, we get equation (16).

$$n = \frac{z^2 \cdot \hat{p}(1 - \hat{p})}{e^2} \tag{16}$$

To determine a value for *n* using this formula, we need replacements for *z*, $\hat{p}$, $(1 - \hat{p})$, and *e*. Replacements can easily be found for *z* and *e*. Once we identify the confidence coefficient for the confidence interval, we have the corresponding value for *z*. The *e* can be replaced by the maximum difference we want to have between the $\hat{p}$ of our sample (and $\hat{p}$ is yet to be determined) and the end-points of the interval.

The remaining obstacles are replacements for $\hat{p}$ and $(1 - \hat{p})$. The problem is that the sample must be taken before we can calculate $\hat{p}$ and the corresponding $(1 - \hat{p})$. To take this sample we need to decide on *n*, the sample size. But this is the value that the formula is supposed to determine for us!

To get around this dilemma, we simply replace $\hat{p}$ and $(1 - \hat{p})$ with values that yield the *maximum possible product*. The *n* calculated with these replacements is then the *maximum sample size* needed to yield the desired value of *e*. As shown in the following display, the maximum product is obtained when both $\hat{p}$ and $(1 - \hat{p})$ are 0.5.

## Values of $\hat{p}$, Corresponding Values of $(1 - \hat{p})$, and Their Products

| $\hat{p}$ | $(1 - \hat{p})$ | $\hat{p}(1 - \hat{p})$ |
|-----------|-----------------|------------------------|
| 0.2 | 0.8 | 0.16 |
| 0.3 | 0.7 | 0.21 |
| 0.4 | 0.6 | 0.24 |
| 0.5 | 0.5 | 0.25 |
| 0.6 | 0.4 | 0.24 |
| 0.7 | 0.3 | 0.21 |
| 0.8 | 0.2 | 0.16 |

When $\hat{p}$ and $(1 - \hat{p})$ are replaced by 0.5 in equation (16), we get the formula in equation (17).

To get a confidence interval for $P$ with length not greater than $2e$, choose a sample of size $n$, where

$$n = \frac{z^2 \cdot (0.25)}{e^2} \qquad (17)$$

We need to point out that the $\hat{p}$ actually obtained from the sample will very likely be different from 0.5. As a consequence, the product $\hat{p}(1 - \hat{p})$ will be less than the supposed product 0.25. The result will be a confidence interval that is actually shorter than the $2e$ we initially specified. However, the preliminary determination of $n$ using equation (17) provided us with the necessary information on how to get a confidence interval with a maximum length of $2e$.

▶ **Example 4**    Cynthia Wolfe is a registered nurse who is working on a masters degree in health care management services. For her masters thesis, Cynthia is studying cholesterol levels of adults in the Tampa Bay area of Florida. As part of the thesis she wants to estimate the proportion of adults over 25 years of age who have cholesterol levels in the "High" category according to standards established by the American Heart Association. Cynthia will use a 95 percent confidence coefficient, and she wants the interval to be no longer than 8 percent.

**Solution**    With $2e = 0.08$, $e = 0.04$.

$$n = \frac{1.96^2 \cdot 0.25}{0.04^2} \qquad \text{Using equation (17)}$$

$$n = 600.25 \qquad \text{Simplify}$$

$$n \approx 600 \qquad \text{Rounded to the nearest whole number}$$

If Cynthia samples 600 adults, then the confidence interval will have a length of no more than 0.08, or 8 percent.    ◀

## Interpreting a confidence interval for *P*

Regardless of what parameter is being estimated, the confidence coefficient of an interval has the same meaning introduced in Section 5-2. To illustrate, in Example 3, the method that was used for sample selection and calculation based on the sample will produce intervals bracketing *P* for 95 percent of repeated applications. It is impossible (short of surveying every adult in the large eastern state) to know whether the particular interval from 0.70 to 0.74 is one of the 95 percent that successfully bracket *P*, or one of the 5 percent that fail to bracket *P*.

▶ **Example 5**   An Air Force recruitment officer wanted to estimate the proportion of eligible first-term airmen who intend to pursue an Air Force career. An eligible first-term airman is one who is recommended by his or her commander for an Air Force career during the member's third year of a 4-year commitment. The officer instructed several members of his staff to visit Air Force bases and survey such personnel about their intentions regarding an Air Force career. The surveys yielded 805 of 2025 personnel interviewed who intended to pursue an Air Force career. Let *P* be the proportion of all eligible first-term airmen intending to pursue an Air Force career.

**a.** Compute a point estimate of *P*.
**b.** Construct a 98 percent confidence interval for *P*.
**c.** State the meaning of the answer to part (b).

**Solution**   **a.** With $x = 805$ and $n = 2025$:

$$\hat{p} = \frac{805}{2025}$$

$$= 0.39753\ldots$$

$$\approx 0.40, \text{ to two decimal places}$$

Thus a point estimate of *P* is 40 percent.

**b.** With $\hat{p} = \dfrac{805}{2025}$, $1 - \hat{p} = \dfrac{1220}{2025}$, $n = 2025$, and $z = 2.33$:

$$\text{lower limit} = \frac{805}{2025} - 2.33\sqrt{\frac{\frac{805}{2025} \cdot \frac{1220}{2025}}{2025}}$$

$$= 0.3975\ldots - 0.0253\ldots$$

$$\approx 0.37, \text{ to two decimal places}$$

$$\text{upper limit} = 0.3975\ldots + 0.0253\ldots$$

$$\approx 0.42, \text{ to two decimal places}$$

**c.** An approximate 98 percent confidence interval for the proportion of eligible first-term airmen planning to pursue an Air Force career is from 37 to 42 percent.   ◀

# Exercises 5-5    Set A

*In Exercises 1–4:*

*a. Compute $\hat{p}$.*

*b. Use $\hat{p}$ as a point estimate for P.*

*c. Use the data to construct a 95 percent confidence interval for P.*

1. A member of the mayor's staff in a large southern city wanted to estimate the proportion of adults (*P*) who support a proposed city ordinance that would prohibit smoking in all public buildings. She surveyed 75 adults and recorded the opinions of all adults who support such an ordinance, Yes (Y), and those that oppose it, No (N). The results of her survey are in Table 5-8.

TABLE 5-8    Results of a Survey on the Proposed Smoking Ordinance

| Adult | Response | Adult | Response | Adult | Response | Adult | Response |
|-------|----------|-------|----------|-------|----------|-------|----------|
| 1 | N | 20 | Y | 39 | Y | 58 | N |
| 2 | Y | 21 | Y | 40 | N | 59 | N |
| 3 | Y | 22 | N | 41 | Y | 60 | N |
| 4 | Y | 23 | N | 42 | Y | 61 | Y |
| 5 | N | 24 | Y | 43 | Y | 62 | Y |
| 6 | N | 25 | Y | 44 | N | 63 | Y |
| 7 | Y | 26 | Y | 45 | N | 64 | N |
| 8 | Y | 27 | Y | 46 | N | 65 | Y |
| 9 | N | 28 | Y | 47 | Y | 66 | Y |
| 10 | Y | 29 | N | 48 | Y | 67 | Y |
| 11 | Y | 30 | Y | 49 | Y | 68 | Y |
| 12 | Y | 31 | N | 50 | Y | 69 | N |
| 13 | N | 32 | Y | 51 | Y | 70 | Y |
| 14 | Y | 33 | Y | 52 | N | 71 | Y |
| 15 | Y | 34 | Y | 53 | Y | 72 | Y |
| 16 | Y | 35 | N | 54 | Y | 73 | N |
| 17 | N | 36 | N | 55 | N | 74 | Y |
| 18 | Y | 37 | Y | 56 | Y | 75 | Y |
| 19 | Y | 38 | Y | 57 | Y | | |

2. For her thesis, a doctoral candidate in psychology has been investigating alcoholism in the family. One aspect of her study has been the background of non-alcoholic wives who have alcoholic husbands. In particular, she wants to estimate the proportion (*P*) of non-alcoholic wives who had alcoholic fathers. She attended several Al-Anon family group meetings and surveyed 75 non-alcoholic wives. She asked each woman, "Did you have an alcoholic father?." Each woman surveyed answered Yes (Y) or No (N). The results of the survey are in Table 5-9.

3. Brad Rogers is a member of a State Postsecondary Education Commission. He visited one of the campuses of the state's university system and surveyed 80 students. He noted the ethnic group to which each student belongs, as follows: A, Asian; B, black; H, Hispanic; W, white. From the data, Brad wanted to estimate the proportion (*P*) of students at this campus who are Asian. The results of the survey are in Table 5-10.

4. Recent studies of party politics in the United States indicate that the Democratic Party is composed of groups with the following four distinct types of political attitudes.

**TABLE 5-9 Results of a Survey on Alcoholism**

| Wife | Response | Wife | Response | Wife | Response | Wife | Response | Wife | Response |
|------|----------|------|----------|------|----------|------|----------|------|----------|
| 1 | Y | 16 | N | 31 | N | 46 | N | 61 | Y |
| 2 | N | 17 | N | 32 | Y | 47 | Y | 62 | Y |
| 3 | N | 18 | Y | 33 | Y | 48 | N | 63 | N |
| 4 | N | 19 | N | 34 | Y | 49 | Y | 64 | N |
| 5 | Y | 20 | Y | 35 | N | 50 | N | 65 | N |
| 6 | Y | 21 | Y | 36 | N | 51 | N | 66 | Y |
| 7 | N | 22 | N | 37 | N | 52 | Y | 67 | N |
| 8 | Y | 23 | N | 38 | Y | 53 | Y | 68 | Y |
| 9 | N | 24 | Y | 39 | Y | 54 | Y | 69 | N |
| 10 | N | 25 | Y | 40 | N | 55 | N | 70 | N |
| 11 | Y | 26 | N | 41 | N | 56 | N | 71 | Y |
| 12 | Y | 27 | N | 42 | N | 57 | N | 72 | N |
| 13 | Y | 28 | N | 43 | Y | 58 | Y | 73 | N |
| 14 | N | 29 | Y | 44 | N | 59 | N | 74 | N |
| 15 | N | 30 | Y | 45 | N | 60 | N | 75 | Y |

**TABLE 5-10 Results of a Student Survey on Ethnicity**

| Student | Ethnic Group | Student | Ethnic Group | Student | Ethnic Group | Student | Ethnic Group | Student | Ethnic Group |
|---------|--------------|---------|--------------|---------|--------------|---------|--------------|---------|--------------|
| 1 | H | 17 | W | 33 | A | 49 | W | 65 | H |
| 2 | A | 18 | W | 34 | W | 50 | A | 66 | A |
| 3 | W | 19 | B | 35 | A | 51 | W | 67 | A |
| 4 | W | 20 | A | 36 | W | 52 | B | 68 | W |
| 5 | B | 21 | H | 37 | W | 53 | W | 69 | W |
| 6 | A | 22 | W | 38 | B | 54 | A | 70 | B |
| 7 | A | 23 | W | 39 | A | 55 | A | 71 | W |
| 8 | H | 24 | A | 40 | W | 56 | H | 72 | W |
| 9 | W | 25 | B | 41 | W | 57 | W | 73 | A |
| 10 | W | 26 | W | 42 | B | 58 | W | 74 | H |
| 11 | B | 27 | B | 43 | B | 59 | W | 75 | A |
| 12 | W | 28 | W | 44 | A | 60 | W | 76 | W |
| 13 | A | 29 | W | 45 | W | 61 | A | 77 | W |
| 14 | A | 30 | W | 46 | W | 62 | B | 78 | B |
| 15 | B | 31 | A | 47 | A | 63 | A | 79 | A |
| 16 | W | 32 | H | 48 | A | 64 | W | 80 | W |

**New Dealers (D)** older, more traditional, less tolerant, and more anticommunist

**Sixties Democrats (S)** upper middle class, tolerant, and driven by the issues of peace and social justice

**Partisan Poor (P)** financially pressured and highly politicized, who have an extraordinary faith in the Democratic Party

**Passive Poor (R)** low income, less critical of American institutions

Diane Gray wanted to estimate the proportion (*P*) of Sixties Democrats (S) in the Democratic Party in the Los Angeles area. She conducted a survey of a sample of 96 registered Democrats in this area. The results of her survey are in Table 5-11 on the next page.

TABLE 5-11    Results of Survey of Registered Democrats

| Demo | Type | Demo | Type | Demo | Type | Demo | Type | Demo | Type | Demo | Type |
|------|------|------|------|------|------|------|------|------|------|------|------|
| 1 | D | 17 | D | 33 | P | 49 | S | 65 | D | 81 | D |
| 2 | S | 18 | D | 34 | D | 50 | P | 66 | D | 82 | D |
| 3 | P | 19 | P | 35 | D | 51 | D | 67 | S | 83 | R |
| 4 | D | 20 | R | 36 | D | 52 | D | 68 | S | 84 | P |
| 5 | S | 21 | D | 37 | S | 53 | D | 69 | D | 85 | D |
| 6 | S | 22 | D | 38 | R | 54 | D | 70 | R | 86 | D |
| 7 | R | 23 | S | 39 | P | 55 | R | 71 | P | 87 | D |
| 8 | P | 24 | S | 40 | D | 56 | S | 72 | D | 88 | R |
| 9 | D | 25 | S | 41 | D | 57 | R | 73 | P | 89 | S |
| 10 | D | 26 | P | 42 | P | 58 | D | 74 | S | 90 | S |
| 11 | D | 27 | D | 43 | S | 59 | S | 75 | D | 91 | P |
| 12 | P | 28 | D | 44 | R | 60 | S | 76 | P | 92 | D |
| 13 | S | 29 | D | 45 | D | 61 | D | 77 | S | 93 | D |
| 14 | R | 30 | D | 46 | D | 62 | P | 78 | D | 94 | P |
| 15 | D | 31 | R | 47 | R | 63 | D | 79 | D | 95 | D |
| 16 | P | 32 | S | 48 | D | 64 | S | 80 | S | 96 | S |

*In Exercises 5–12:*
a. *Compute a point estimate for P.*
b. *Construct the indicated confidence interval for P.*
c. *State the meaning of the answer to part (b).*

5. The city council of Des Moines wanted to estimate the proportion of inner city workers who would use public transportation to commute if services were expanded. A sample of 493 workers currently driving to work was polled, and 360 said they would use public transportation if better services were available. Let $P$ be the proportion of workers currently driving to work that would use public transportation if services were expanded. Construct a 90 percent confidence interval for $P$.

6. The County Supervisors of Tuscaloosa County wanted to estimate the proportion of county residents in favor of a 10 PM curfew at all the county parks and recreation areas. A sample of 650 residents yielded 468 who favor such a curfew. Let $P$ be the proportion of county residents who favor the 10 PM curfew. Construct a 90 percent confidence interval for $P$.

7. A sample of 280 tomatoes from a truck load for a local cannery yielded 28 that were unfit for processing. Let $P$ be the proportion of tomatoes in this truck that are unfit for processing. Construct a 95 percent confidence interval for $P$.

8. A sample of 160 D-size flashlight batteries was randomly selected from a large production run and 24 had some flaw in construction or failed to pass a voltage check. Let $P$ be the proportion of defective batteries in this production run. Construct a 95 percent confidence interval for $P$.

9. To estimate the proportion of imported cars in a large western city, a newsperson sampled 466 cars at a busy intersection in the city. She counted 140 imported cars in the sample. Let $P$ be the proportion of cars in this city that are imported. Construct a 98 percent confidence interval for $P$.

10. Lactose intolerance is a condition found in some humans that inhibits the digestion of lactose found in milk and milk products. This condition is especially common to blacks, American Indians, and Asians. To estimate the proportion of Chicago's black population with lactose intolerance, a graduate student at the University of Chicago tested 760 black residents. He found 165 displaying symptoms related to this condition. Let $P$ be the proportion of black residents who have symptoms of lactose intolerance. Construct a 98 percent confidence interval for $P$.

11. It is commonly accepted that most babies are born at night. By definition, any baby born between 6 PM and 6 AM is counted as a "night baby." A

sample of 380 Los Angeles birth records are checked for the time of birth. The sample yields 220 night babies. Let $P$ be the proportion of night babies born in the Los Angeles area. Construct a 99 percent confidence interval for $P$.

12. Many people eat at least one meal a day at a fast-food restaurant. To estimate the proportion of the residents of Houston that do, a survey of 1340 residents is conducted. The survey yields 375 that regularly eat at least one meal a day in a fast-food restaurant. Let $P$ be the proportion of Houston residents who regularly eat at least one meal a day in a fast-food restaurant. Construct a 99 percent confidence interval for $P$.

13. A random sample yields $\hat{p} = 0.45$.
    a. Construct a 90 percent confidence interval for $P$, if $n = 50$.
    b. Construct a 90 percent confidence interval for $P$, if $n = 500$.
    c. Compute the ratio of the lengths of the intervals obtained in parts (a) and (b) by dividing the length in (a) by the length in (b).

14. A random sample yields $\hat{p} = 0.32$.
    a. Construct a 95 percent confidence interval for $P$, if $n = 100$.
    b. Construct a 95 percent confidence interval for $P$, if $n = 1000$.
    c. Compute the ratio of the lengths of the intervals obtained in parts (a) and (b) by dividing the length in (a) by the length in (b).

15. A group of business operators in a large southeastern city is promoting a plan that would levy a sales tax to build a large sports complex. The goal would be to attract a major sports team to the city. A poll of 1425 residents yields

    171 strongly favor such a plan (s)
    228 favor such a plan (f)
    513 are indifferent toward such a plan (i)
    399 oppose such a plan (o)
    114 strongly oppose such a plan (n)

    a. Construct to 3 decimal places a 90 percent confidence interval for $P_s$.

    b. Construct to 3 decimal places a 95 percent confidence interval for $P_f$.
    c. Construct to 3 decimal places a 98 percent confidence interval for $P_o$.
    d. Construct to 3 decimal places a 99 percent confidence interval for $P_n$.

16. A statewide group of physicians in an eastern state is interested in public opinion on an issue regarding malpractice suits involving doctors. Specifically, they want to know how the public feels about a proposed state law that would place an upper limit on the amount of money a jury could award in a malpractice suit. The group sponsored a statewide survey of 2450 adults. The responses were placed in five categories with the results:

    441 strongly favor such a law (s)
    833 favor such a law (f)
    588 are indifferent to such a law (i)
    392 oppose such a law (o)
    196 strongly oppose such a law (n)

    a. Construct to 3 decimal places a 90 percent confidence interval for $P_i$.
    b. Construct to 3 decimal places a 95 percent confidence interval for $P_f$.
    c. Construct to 3 decimal places a 98 percent confidence interval for $P_n$.
    d. Construct to 3 decimal places a 99 percent confidence interval for $P_s$.

*In Exercises 17–24, use equation (17) to determine n for the specified confidence coefficients and values of 2e. Round answers to the nearest whole number.*

17. A 90% confidence coefficient and $2e = 0.12$
18. A 90% confidence coefficient and $2e = 0.10$
19. A 95% confidence coefficient and $2e = 0.16$
20. A 95% confidence coefficient and $2e = 0.10$
21. A 98% confidence coefficient and $2e = 0.20$
22. A 98% confidence coefficient and $2e = 0.14$
23. A 99% confidence coefficient and $2e = 0.15$
24. A 99% confidence coefficient and $2e = 0.12$

**25.** A television newsperson frequently used polls of 2000 adults to make inferences regarding the public's opinion on topics of current interest. The value of $\hat{p}$ was usually followed with the statement "plus or minus 3 percent." To three decimal places, find $\sqrt{\dfrac{\hat{p}(1 - \hat{p})}{n}}$ for:

a. $\hat{p} = 0.50$     b. $\hat{p} = 0.40$
c. $\hat{p} = 0.70$     d. $\hat{p} = 0.20$
e. Are the values for parts (a) through (d) approximately the same?

**26.** Refer to Exercise 25 and use the fact that:

$$\sqrt{\frac{\hat{p}(1 - \hat{p})}{n}} \approx 0.01 \text{ for } 0.2 < \hat{p} < 0.8$$

The newsperson used intervals of plus or minus 0.03. Thus:

$$z\sqrt{\frac{\hat{p}(1 - \hat{p})}{n}} \approx 0.03 \quad \text{and} \quad z \approx 3.00$$

Use Appendix Table 4 to approximate the confidence coefficient the newsperson was using when he reported values of $\hat{p}$ between 0.2 and 0.8.

## Exercises 5-5    Set B

*In Exercises 1 and 2, use the data in Data Set I, Appendix B.*

**1.** Use random numbers to sample 50 individuals.
a. Compute $\hat{p}$, where $\hat{p}$ is the proportion of individuals in the sample that smoke.
b. Use the value of $\hat{p}$ from the sample to construct a 95 percent confidence interval for $P$, the proportion of the population that smokes.
c. Check whether the interval of part (b) contains $P = 0.53$.

**2.** Use random numbers to sample 60 individuals.
a. Compute $\hat{p}$, where $\hat{p}$ is the proportion of individuals in the sample that have cancer.
b. Use the value of $\hat{p}$ from the sample to construct a 98 percent confidence interval for $P$, the proportion of the population that has cancer.
c. Check whether the interval of part (b) contains $P = 0.15$.

*In Exercises 3 and 4, use Data Set II, Appendix B.*

**3.** Use random numbers to sample 40 stocks.
a. Compute $\hat{p}$, where $\hat{p}$ is the proportion of stock prices that declined $(-)$.
b. Use the value of $\hat{p}$ to construct a 90 percent confidence interval for the proportion of all stock prices in the population that declined.
c. Does your interval contain $P = 0.277$?

**4.** Use random numbers to sample 50 stocks.
a. Compute $\hat{p}$, where $\hat{p}$ is the proportion of stock prices that remained the same $(0)$.
b. Use the value of $\hat{p}$ to construct a 95 percent confidence interval for the proportion of all stock prices in the population that remained the same.
c. Does your interval contain $P = 0.255$?

*In Exercises 5 and 6, refer to the data in Figure 5-31, a part of an article written about football mania and Super Bowl XXV.*

**5.** Assume a random sample of 360 homes in the Seattle, Washington, area during the time the Super Bowl game was televised yielded 270 that were watching the game. Use these data to construct a 90 percent confidence interval for the proportion of all homes in this area that watched the game.

**6.** Assume a random sample of 265 homes in the Buffalo, New York, area during the time the Super Bowl game was televised yielded 238 that were watching the game. Use these data to construct a 95 percent confidence interval for the proportion of all homes in this area that watched the game.

*Exercises 7 and 8 are suggested student projects for constructing confidence intervals for unknown Ps.*

**7.** Sample 75 students at your college or university. For each student in the sample, obtain his or her eye color. Let $\hat{p}$ be the proportion of students in the sample with brown eyes. Use the results of your sample to construct a 95 percent confidence interval for $P$, the proportion of brown eyed students at your college or university.

> **By Jim Carnes**
> Bee Staff Writer
>
> **F**OR MANY Americans, today doesn't have a date. It has a name: Super Bowl Sunday. No matter what is going on in the rest of the world, the thing that really matters is that game in Tampa, Fla.
>
> In the 24 years the Super Bowl has been played, from 62 percent to 78 percent of the nation's TV-owning homes have tuned in to see the game. ABC estimates that 750 million people worldwide will watch today's edition, either live or on tape delay.

**Figure 5-31.** "Super Bowl Sunday" by Jim Carnes, *The Sacramento Bee*, 1/27/91, p. E4. Copyright © 1991 *The Sacramento Bee*. Reprinted by permission.

**8.** Sample 50 students at your college or university. For each student in the sample obtain his or her political party preference. Let $\hat{p}$ be the proportion of students in the sample that are Democrats. Use the results of your sample to construct a 98 percent confidence interval for $P$, the proportion of students at your school that are Democrats.

# Chapter 5 Summary

## Definitions

**1.** A **point estimate** of a parameter is a single value, usually derived from a sample, that is offered as a guess at the unknown value of the parameter.

**2.** A **confidence interval** for an unknown parameter is an interval of numbers, derived from a sample, that is thought to contain the parameter. The **confidence coefficient** associated with such an interval is the frequency with which the method used to create it could be expected to successfully bracket the unknown parameter if repeated applications of the method were used.

## Rules and Equations

If a population is large enough that a model of independence can be used for $X_1, X_2, \ldots, X_n$, then for the sampling distribution of $\bar{X}$:

$$\mu_{\bar{x}} = \mu \tag{1}$$

$$\sigma_{\bar{x}}^2 = \frac{\sigma^2}{n} \tag{2}$$

where $\mu$ and $\sigma^2$ are respectively the mean and variance of the sampled population.

### The central limit theorem

When $X_1, X_2, \ldots, X_n$ can be described as independent random variables with a common distribution and $n$ is large, the sampling distribution of $\bar{X}$ is approximately normal.

If $\bar{x}$ is the observed mean of a random sample of size $n$ selected from a population with mean $\mu$ and standard deviation $\sigma$, then a $z$ score for $\bar{x}$ is:

$$z = \frac{\bar{x} - \mu}{\dfrac{\sigma}{\sqrt{n}}} \tag{3}$$

If $X_1, X_2, \ldots, X_n$ can be modeled as independent random selections from a large population that has an approximately bell-shaped relative frequency distribution, then the sampling distribution of $\bar{X}$ will be normal for any $n$, large or small.

For large sample confidence intervals based on the normal distribution of $\bar{X}$ or $\hat{P}$:

| Confidence Coefficient | Corresponding $z$'s |
|---|---|
| 90% | $-z = -1.65$ and $z = 1.65$ |
| 95% | $-z = -1.96$ and $z = 1.96$ |
| 98% | $-z = -2.33$ and $z = 2.33$ |
| 99% | $-z = -2.58$ and $z = 2.58$ |

### Large $n$ confidence intervals for $\mu$ when $\sigma$ is known

If $z$ is a positive number and $n$ is large, then the interval with:

$$\text{lower limit} = \bar{x} - z\left(\frac{\sigma}{\sqrt{n}}\right) \tag{4}$$

and

$$\text{upper limit} = \bar{x} + z\left(\frac{\sigma}{\sqrt{n}}\right) \tag{5}$$

can be used as a confidence interval for $\mu$. The associated confidence coefficient is $P(-z < Z < z)$.

**Large $n$ confidence intervals for $\mu$**

If $z$ is a positive number and $n$ is large, then the interval with:

$$\text{lower limit} = \bar{x} - z\left(\frac{s}{\sqrt{n}}\right) \tag{6}$$

$$\text{upper limit} = \bar{x} + z\left(\frac{s}{\sqrt{n}}\right) \tag{7}$$

can be used as a confidence interval for $\mu$. The associated confidence coefficient is $P(-z < Z < z)$.

**Choosing a sample size $n$ for a confidence interval for $\mu$**

Let $e$ be one-half the desired length of a confidence interval for $\mu$ with confidence coefficient corresponding to $z$, then the required sample size is

$$n = \left(\frac{z \cdot \sigma}{e}\right)^2 \tag{8}$$

where $\sigma$ is the (known) standard deviation of the population.

**Small $n$ confidence intervals for $\mu$**

If a sampled population is bell shaped and $t$ is a positive number, then the interval with:

$$\text{lower limit} = x - t\left(\frac{s}{\sqrt{n}}\right) \tag{9}$$

$$\text{upper limit} = x + t\left(\frac{s}{\sqrt{n}}\right) \tag{10}$$

can be used as a confidence interval for $\mu$. The associated confidence coefficient is $P(-t < T < t)$ and is computed using the $t$-distribution with $df = n - 1$.

If $\hat{P}$ is the proportion of a sample of size $n$ with a characteristic of interest taken from a population in which a fraction $P$ of the objects have that characteristic, and $n$ is small compared with the population size, then the sampling distribution of $\hat{P}$ will be approximately normal provided:

$$nP \geq 5 \quad \text{and} \quad n(1 - P) \geq 5$$

The sampling distribution of $\hat{P}$ has:

$$\mu_{\hat{P}} = P \tag{11}$$

and

$$\sigma_{\hat{P}} = \sqrt{\frac{P(1-P)}{n}} \tag{12}$$

To compute a $z$-score corresponding to $\hat{p}$ use:

$$z = \frac{\hat{p} - P}{\sqrt{\dfrac{P(1-P)}{n}}} \tag{13}$$

**Large $n$ confidence intervals for $P$**

If $z$ is a positive number, then the interval with:

$$\text{lower limit} = \hat{p} - z\sqrt{\frac{\hat{p}(1-\hat{p})}{n}} \tag{14}$$

and

$$\text{upper limit} = \hat{p} + z\sqrt{\frac{\hat{p}(1-\hat{p})}{n}} \tag{15}$$

can be used as a confidence interval for $P$. The associated confidence coefficient is $P(-z < Z < z)$.

**Choosing a sample size $n$ for a confidence interval for $P$**

If $2e$ represents the length of a confidence interval for $P$, then $e$ is the distance between $\hat{p}$ and the end points of the interval, and the sample size $n$ associated with a given $e$ is

$$n = \frac{z^2 \cdot \hat{p}(1-\hat{p})}{e^2}. \tag{16}$$

To get a confidence interval for $P$ with length not greater than $2e$, choose a sample of size $n$, where

$$n = \frac{z^2 \cdot (0.25)}{e^2}. \tag{17}$$

## Symbols

$\bar{X}$   the **random variable** that is the mean of a random sample of quantitative data

$\mu_{\bar{x}}$   the **mean** of the **sampling distribution of** $\bar{X}$

$\sigma_{\bar{x}}$   the **standard deviation** of the **sampling distribution of** $\bar{X}$

CLT   the **central limit theorem**

$\hat{P}$   the **random variable** that is the proportion of a random sample of qualitative data belonging to a category of interest

$\mu_{\hat{p}}$   the **mean** of the **sampling distribution of** $\hat{P}$

$\sigma_{\hat{p}}$   the **standard deviation** of the **sampling distribution of** $\hat{P}$

# Comparing the $\bar{X}$ and $\hat{P}$ Random Variables

| | |
|---|---|
| The random variable $\bar{X}$ represents the mean of a sample from a population of quantitative data in which the mean is $\mu$. | The random variable $\hat{P}$ represents the proportion of a sample with a characteristic of interest from a population of qualitative data in which $P$ is the proportion with the characteristic. |
| The probability distribution of $\bar{X}$ is approximately normal if:<br><br>(1) $n \geq 30$, or<br>(2) The population sampled is reasonably bell shaped. | The probability distribution of $\hat{P}$ is approximately normal if:<br><br>(1) $nP \geq 5$, and<br>(2) $n(1 - P) \geq 5$ |
| If $\bar{x}$ is an observed value of $\bar{X}$ and $n$ is the sample size, then $$z = \frac{\bar{x} - \mu}{\frac{\sigma}{\sqrt{n}}}$$ | If $\hat{p}$ is an observed value of $\hat{P}$ and $n$ is the sample size, then $$z = \frac{\hat{p} - P}{\sqrt{\frac{P(1 - P)}{n}}}$$ |
| If $\bar{x}$ and $s$ are based on a large sample, then a confidence interval for $\mu$ has $$\text{lower limit} = \bar{x} - z \cdot \frac{s}{\sqrt{n}}$$ $$\text{upper limit} = \bar{x} + z \cdot \frac{s}{\sqrt{n}}$$ where the value of $z$ depends on the confidence coefficient. | If $\hat{P}$ is based on a large sample, then a confidence interval for $P$ has $$\text{lower limit} = \hat{p} - z \cdot \sqrt{\frac{\hat{p}(1 - \hat{p})}{n}}$$ $$\text{upper limit} = \hat{p} + z \cdot \sqrt{\frac{\hat{p}(1 - \hat{p})}{n}}$$ where the value of $z$ depends on the confidence coefficient. |

If $\bar{x}$ and $s$ are based on a small sample from a population with a bell-shaped frequency distribution, then a confidence interval for $\mu$ has

$$\text{lower limit} = \bar{x} - t \cdot \frac{s}{\sqrt{n}}$$

$$\text{upper limit} = \bar{x} + t \cdot \frac{s}{\sqrt{n}}$$

where the value of $t$ is from a $t$-distribution with $(n - 1)$ degrees of freedom.

# Chapter 5 Review Exercises

1. The Better Foods Corporation processes and cans tomato products at its plant in Stockton. A large production run of No. 7093 cans of peeled tomatoes has a population mean net contents of 454.0 grams with a standard deviation of 3.0 grams. A quality control technician randomly samples 36 cans from this production run, and $\bar{X}$ is the sample mean weight of the contents.
   a. Argue that the sampling distribution of $\bar{X}$ is approximately normal.
   b. Compute $\mu_{\bar{x}}$.
   c. Compute $\sigma_{\bar{x}}$.
   d. Sketch and label a graph of the sampling distribution of $\bar{X}$.

2. Records for the Texas based Mega-Tech Corporation show that the mean age of all employees is currently 38.8 years with a standard deviation of 10.5 years. A member of the personnel office staff randomly samples 49 employee records, and $\bar{X}$ is the sample mean age of those selected.
   a. Argue that the sampling distribution of $\bar{X}$ is approximately normal.
   b. Compute $\mu_{\bar{x}}$.
   c. Compute $\sigma_{\bar{x}}$.
   d. Sketch and label a graph of the sampling distribution of $\bar{X}$.

3. The manager of an Employment Opportunities Office at Midwestern University coordinates part-time job placement for local employers and students at the university. Records indicate that $108.30 is the mean weekly income from jobs handled through his office. The standard deviation for the distribution of weekly incomes is $14.20. Suppose 40 jobs are randomly sampled from the records, and $\bar{X}$ is the sample mean weekly income from these jobs.
   a. Sketch and label a graph of the sampling distribution of $\bar{X}$.
   b. Compute the $z$ score for $\bar{x} = \$105.00$.
   c. State the meaning of the $z$ score in part (b).
   d. Compute $P(\bar{X} < 105.00)$.
   e. State the meaning of the answer to part (d).

4. Gardner Machine Company manufactures roller bearings used on large road construction equipment. The mean diameter of these bearings is 1.860 inches with a standard deviation of 0.030 inches. Suppose a sample of 36 of these bearings from a production run is selected, and $\bar{X}$ is the mean diameter of the bearings.
   a. Sketch and label a graph of the sampling distribution of $\bar{X}$.
   b. Compute the $z$ score for $\bar{x} = 1.867$ inches.
   c. State the meaning of the $z$ score in part (b).
   d. Compute $P(\bar{X} > 1.867)$.
   e. State the meaning of the answer to part (d).

**5.** Laura Lance uses the left turn lane at a traffic signal at Hurley Way and Howe Avenue to get home from work. She has petitioned the County Traffic Control Department to change the signal's mechanical instructions because the first car in the turn lane has to wait too long before the signal changes. To justify her claim, she checked the number of seconds the first car in the left turn lane had to wait for $n = 36$ vehicles. The times below were recorded to the nearest second.

| 67 | 65 | 69 | 81 | 83 | 52 | 46 | 64 | 59 | 94 | 64 | 62 |
|----|----|----|----|----|----|----|----|----|----|----|----|
| 81 | 78 | 63 | 77 | 71 | 62 | 50 | 53 | 57 | 64 | 89 | 56 |
| 72 | 64 | 75 | 53 | 59 | 67 | 49 | 71 | 86 | 44 | 97 | 72 |

  a. Compute $\bar{x}$ to the nearest second.
  b. Give a point estimate of the mean waiting time for cars first to enter the left turn lane at this intersection.
  c. Construct a 95 percent confidence interval for the mean waiting time.
  d. Interpret your answer to part (c).

**6.** Susan Hessny is an employee of a major soup manufacturer. As part of a quality control process, she randomly selects 40 cans of turkey noodle soup from a large production run. The weights (in grams) of the can contents were:

| 301 | 309 | 307 | 315 | 320 | 308 | 309 | 310 | 307 | 309 |
|-----|-----|-----|-----|-----|-----|-----|-----|-----|-----|
| 309 | 305 | 301 | 309 | 306 | 302 | 313 | 305 | 312 | 306 |
| 314 | 307 | 315 | 310 | 308 | 303 | 311 | 309 | 307 | 304 |
| 307 | 315 | 306 | 301 | 315 | 307 | 301 | 309 | 303 | 311 |

  a. Compute $\bar{x}$ to the nearest gram.
  b. Make a point estimate of the mean net weight of a can from this production run.
  c. Construct a 90 percent confidence interval estimate for the mean net weight.
  d. Interpret your answer to part (c).

**7.** The American College Test (ACT) is given to high school students throughout the United States. For the students in the Tampa Bay area that took the test last year, a sample of their test scores on the English section yielded $\bar{x} = 17.4$ and $s = 5.4$. The sample size was $n = 81$.
  a. Construct a 95 percent confidence interval for the mean score on this portion of the ACT for all students in the Tampa Bay area.
  b. State the meaning of the answer to part (a).

**8.** A random sample of 40 stereo speakers was taken from a large production run of Model ZK 709 speakers manufactured by Sonic Boom Corporation. For the speakers sampled, a mean output of 48.6 watts was measured along with a standard deviation of 2.1 watts.

  a. Construct a 98 percent confidence interval for the mean output for the speakers in this production run.
  b. State the meaning of the interval in part (a).

*In Exercises 9 and 10, let T be a random variable with a t-distribution.*

**9.** a. Find $t$, such that $P(T > t) = 0.050$, if $df = 7$.
  b. Find $t$, such that $P(T < t) = 0.010$, if $df = 11$.

**10.** Find $t$, such that $P(-t < T < t) = 0.950$, if $df = 15$.

**11.** Sharon Hollobow, a state worker in New York, uses her economy car solely for commuting to and from work. She drives the same route each day. To estimate the mean gas mileage of her car, Sharon gathered data for $n = 16$ tanks of gas. Her data yielded $\bar{x} = 28.3$ miles per gallon (mpg) and $s = 5.6$ mpg. Assuming the distribution of mileages for tanks of gas is bell shaped:

a. Construct a 95 percent confidence interval for the mean mpg for her car in this kind of driving.

b. State the meaning of the interval in part (a).

12. The Commissioner's Office of Major League Baseball keeps records on all games played by teams in both the American and National Leagues. A sample of the numbers of runs scored by winning teams in 20 games played last season yielded $\bar{x} = 5.8$ runs with $s = 2.9$ runs. The distribution of runs scored per game is reasonably bell shaped.

a. Construct a 98 percent confidence interval for $\mu$, the mean number of runs scored by winning teams in games played last year.

b. State the meaning of the answer to part (a).

13. The publication, *Medical Aspects of Human Sexuality*, surveyed a large number of psychiatrists. The survey indicated that 54 percent of psychiatrists think that companionship is a sounder reason for couples to get married than passionate love. Assume that the fraction of all psychiatrists who feel this way on this issue is $P = 0.54$. Suppose $n = 39$ psychiatrists in San Francisco county are randomly sampled, and $\hat{P}$ is the fraction of the sample that feels companionship is a sounder reason for couples to get married than passionate love.

a. Compute $\mu_{\hat{p}}$.

b. Compute $\sigma_{\hat{p}}$.

c. Sketch and label an approximately normal distribution for $\hat{P}$.

d. Approximate $P(\hat{P} < 0.50)$.

e. State the meaning of the answer to part (d).

14. Based on recently collected data, 54.8 percent of college women's tennis coaches are women. Suppose $n = 40$ colleges in the United States are checked, and $\hat{P}$ is the fraction of these schools that have women coaches for their women's tennis team.

a. Compute $\mu_{\hat{p}}$.

b. Compute $\sigma_{\hat{p}}$.

c. Construct and label an approximately normal distribution for $\hat{P}$.

d. Approximate $P(\hat{P} > 0.60)$.

e. State the meaning of the answer to part (d).

15. A program, funded by the National Heart, Lung, and Blood Institute, treated 2142 habitual cigarette smokers. The program was designed to help these individuals quit smoking. Four years after the program, 45 percent of the participants were still nonsmokers. Let $P$ be the proportion of all habitual cigarette smokers who would complete the program and still be non-smokers 4 years later.

a. Construct a 95 percent confidence interval for $P$.

b. State the meaning of the answer to part (a).

16. A sample of 150 Tau Beta Pi (a national engineering honorary fraternity) members contained 36 percent women. Let $P$ be the proportion of women in Tau Beta Pi throughout the United States.

a. Construct a 90 percent confidence interval for $P$.

b. State the meaning of the answer to part (a).

## Confidence Intervals

**Key Topics**

1. *Central limit theorem*

2. *Z-Intervals*

3. *t-Intervals*

### Central limit theorem

According to the central limit theorem, when *n* is large, means of samples of size *n* drawn from any population should be almost normally distributed regardless of the distributional shape of the population from which they were drawn. We can use Minitab to investigate this phenomenon.

If 40 numbers are drawn randomly with replacement from the integers from 21 to 30, we should expect the frequency distribution to be fairly flat because the probability of drawing each integer is the same, 0.1. A simulation of this experiment, Figure 5-32, shows that our expectations are well founded. The distribution is far from bell shaped.

Let us repeat this experiment a total of 50 times and then look at the distribution of the *means* of these samples of 40 numbers. To make computations easier, we put each sample in a row rather than in a column. Figure 5-33 shows the commands used to do this and the distribution of the sample means. As you can see, the simulated distribution of the 50 means is almost bell shaped, just as the CLT leads us to expect.

### Z-Intervals

In Exercise 1 of the Set A exercises for Section 5-2, you were asked to estimate the mean age of women applying for marriage licenses in Denver. Assume that the ages of the 36 women in the sample have been loaded into column 1 of the Minitab worksheet. Because a sample of 36 is large enough to use a normal distribution and the sample standard deviation to make a confidence interval for $\mu$, we first find *s* using the DESCRIBE command as in Figure 5-34.

Now we use the ZINTERVAL command as shown in Figure 5-35. In this case, we see that a 90 percent confidence interval for the mean age of Denver women applying for a marriage license is from 25.52 years to 29.37 years.

```
MTB> RANDOM 40 C1;
SUBC> INTEGERS 21:30.
MTB> HIST C1

Histogram of C1    N = 40
Midpoint    Count
    21        3      ***
    22        5      *****
    23        3      ***
    24        2      **
    25        2      **
    26        3      ***
    27        5      *****
    28        4      ****
    29        5      *****
    30        8      ********
```

**Figure 5-32.**   A simulated distribution of digits from a uniform distribution.

```
MTB> RANDOM 50 C1-C40;
SUBC> INTEGER 21:30.
MTB> RMEAN C1-C40 into C41
MTB> HIST C41

Histogram of C41    N = 50
Midpoint    Count
   24.6       1      *
   24.8       4      ****
   25.0       4      ****
   25.2       8      ********
   25.4       7      *******
   25.6       9      *********
   25.8       7      *******
   26.0       4      ****
   26.2       5      *****
   26.4       1      *
```

**Figure 5-33.**   Simulated distribution of means of samples from a uniform distribution.

```
MTB> DESCRIBE C1

            N       MEAN     MEDIAN    TRMEAN     STDEV    SEMEAN
AGES       36      27.44      26.50     27.06      7.02      1.17

           MIN       MAX        Q1        Q3
AGES     17.00     45.00     22.25     30.00
```

**Figure 5-34.**   Summary statistics for a sample of ages of women from Denver.

```
MTB> ZINTERVAL 90% C.I. SIGMA = 7.02 DATA IN C1
THE ASSUMED SIGMA = 7.02

            N       MEAN      STDEV    SE MEAN      90.0 PERCENT C.I.
AGES       36      27.44       7.02       1.17    (   25.52,    29.37)
```

**Figure 5-35.**   A 90% confidence interval for $\mu$.

## t-Intervals

If the underlying population is normal, but we do not know sigma, the population variance, we may use the $t$-distribution and the sample standard deviation to estimate the mean of the population.

Suppose the numbers on the next page are known to come from a normal population, and they have been loaded into column 1 of the worksheet.

|          |          |          |          |          |          |
|----------|----------|----------|----------|----------|----------|
| 33.8731  | 39.1790  | 39.0889  | 42.2894  | 36.5092  | 42.7890  |
| 41.0172  | 30.1478  | 37.0391  | 29.5779  | 42.0867  | 41.7376  |
| 48.0280  | 47.9540  | 38.2945  | 27.2928  |          |          |

A 95 percent confidence interval for the population mean and the commands used to produce it are shown in Figure 5-36.

To change the confidence level for either the Z-interval or the t-interval, type the desired confidence coefficient in the command. Thus TINTERVAL 98 C1 will produce a 98 percent confidence interval for the population mean based on the data in column 1.

```
MTB> TINTERVAL 95% DATA IN C1
         N      MEAN    STDEV   SE MEAN    95.0 PERCENT C.I.
C1      16     38.56    6.03      1.51    (   35.34,    41.77)
```

**Figure 5-36.**  A 95% confidence interval for $\mu$.

```
MTB> RANDOM 100 C1-C20;
SUBC> NORMAL MU=40 SIGMA=12.
MTB> ZINT 90% SIGMA=12 DATA IN C1-C20.

THE ASSUMED SIGMA=12.0
         N      MEAN    STDEV   SE MEAN    90.0 PERCENT C.I.
C1      100    41.48   11.45      1.20    (   39.50,    43.45)
C2      100    39.94   10.92      1.20    (   37.96,    41.92)
C3      100    41.12   11.38      1.20    (   39.14,    43.10)
C4      100    42.05   12.28      1.20    (   40.07,    44.02)
C5      100    38.88   12.85      1.20    (   36.90,    40.85)
C6      100    40.96   12.78      1.20    (   38.98,    42.93)
C7      100    39.61   12.43      1.20    (   37.63,    41.58)
C8      100    39.61   11.70      1.20    (   37.64,    41.59)
C9      100    38.39   11.41      1.20    (   36.41,    40.36)
C10     100    39.61   12.52      1.20    (   37.64,    41.59)
C11     100    37.45   12.35      1.20    (   35.47,    39.42)
C12     100    39.80   10.99      1.20    (   37.82,    41.77)
C13     100    37.86   12.88      1.20    (   35.88,    39.83)
C14     100    39.32   13.02      1.20    (   37.34,    41.29)
C15     100    40.17   12.38      1.20    (   38.19,    42.15)
C16     100    40.95   12.03      1.20    (   38.97,    42.92)
C17     100    39.70   12.24      1.20    (   37.72,    41.68)
C18     100    39.16   11.83      1.20    (   37.18,    41.14)
C19     100    40.94   11.16      1.20    (   38.97,    42.92)
C20     100    37.69   11.75      1.20    (   35.72,    39.67)
```

**Figure 5-37.**  90% confidence intervals for $\mu$ based on $n = 100$.

## Exercises

1. Figure 5-37 on the preceding page shows the results of simulating 20 samples, each with size 100, from a normal population with $\mu = 40$ and $\sigma = 12$, and then making 90 percent confidence interval estimates of the population mean using the data in each sample.
   a. In how many of the 20 trials would you expect to find the true population mean in the interval?
   b. In how many of the trials was the population mean actually in the interval?
   c. Do the results in parts (a) and (b) seem contradictory? Why or why not?
   d. Each interval has approximately the same width. Why is this?

2. Figure 5-38 shows twenty 90 percent confidence intervals for the population mean resulting from simulations in which samples of size 16 were randomly selected from a normal population with $\mu = 25$ and $\sigma = 5$.
   a–c. Answer the same questions asked in parts (a–c) of Exercise 1.
   d. This time the intervals all have different widths. Why is this?

```
MTB> RANDOM 16 C1-C20;
SUBC> NORMAL 25 5.
MTB> TINT 90% C1-C20
```

|      | N  | MEAN   | STDEV | SE MEAN | 90.0 PERCENT C.I. |         |
|------|----|--------|-------|---------|-------------------|---------|
| C1   | 16 | 24.34  | 6.11  | 1.53    | ( 21.66,          | 27.01)  |
| C2   | 16 | 24.46  | 5.04  | 1.26    | ( 22.25,          | 26.67)  |
| C3   | 16 | 25.957 | 3.858 | 0.964   | ( 24.266,         | 27.648) |
| C4   | 16 | 24.88  | 4.51  | 1.13    | ( 22.90,          | 26.86)  |
| C5   | 16 | 26.11  | 4.75  | 1.19    | ( 24.03,          | 28.20)  |
| C6   | 16 | 24.17  | 6.04  | 1.51    | ( 21.52,          | 26.82)  |
| C7   | 16 | 24.659 | 3.814 | 0.954   | ( 22.987,         | 26.331) |
| C8   | 16 | 26.22  | 5.03  | 1.26    | ( 24.02,          | 28.42)  |
| C9   | 16 | 25.20  | 4.96  | 1.24    | ( 23.02,          | 27.37)  |
| C10  | 16 | 23.64  | 5.23  | 1.31    | ( 21.34,          | 25.93)  |
| C11  | 16 | 27.13  | 5.23  | 1.31    | ( 24.83,          | 29.42)  |
| C12  | 16 | 23.16  | 6.22  | 1.55    | ( 20.44,          | 25.89)  |
| C13  | 16 | 25.18  | 5.25  | 1.31    | ( 22.88,          | 27.48)  |
| C14  | 16 | 26.45  | 6.88  | 1.72    | ( 23.43,          | 29.46)  |
| C15  | 16 | 23.67  | 6.56  | 1.64    | ( 20.79,          | 26.54)  |
| C16  | 16 | 25.06  | 4.63  | 1.16    | ( 23.03,          | 27.09)  |
| C17  | 16 | 26.77  | 4.29  | 1.07    | ( 24.89,          | 28.66)  |
| C18  | 16 | 24.46  | 4.89  | 1.22    | ( 22.31,          | 26.60)  |
| C19  | 16 | 25.66  | 4.23  | 1.06    | ( 23.80,          | 27.51)  |
| C20  | 16 | 23.70  | 5.49  | 1.37    | ( 21.30,          | 26.11)  |

**Figure 5-38.** 90% confidence intervals for $\mu$ based on $n = 16$.

## A. Confidence intervals for $\mu$

The lower and upper limits of a confidence interval for $\mu$ are obtained by subtracting and adding respectively $z \cdot \dfrac{s}{\sqrt{n}}$ to an observed $\bar{x}$. If a calculator is used, we can first calculate $z \cdot \dfrac{s}{\sqrt{n}}$, and then use the calculator's memory to store that value. Example 1 provides an illustration of the key sequence that can be used on many scientific calculators. For possible variations in sequence consult your calculator's manual, or ask your instructor.

▶ **Example 1**  Construct a 90 percent confidence interval for $\mu$ if a sample of 64 yielded $\bar{x} = \$34.63$ and $s = \$17.60$.

**Solution**

| Key Sequence | Display Shows | Comments |
|---|---|---|
| 1.65 | 1.65 | The $z$ value for 90% |
| ×  | 1.65 | Indicates a multiplication |
| 17.60 | 17.60 | The value of $s$ |
| ÷  | 29.04 | Indicates a division |
| 64 | 64. | The value of $n$ |
| √  | 8 | Indicates a square root |
| =  | 3.63 | The value of $z \cdot \dfrac{s}{\sqrt{n}}$ |
| MIN | 3.63 | Stores the value in memory |
| 34.63 | 34.63 | The value of $\bar{x}$ |
| −  | 34.63 | Indicates a subtraction |
| MR | 3.63 | Recalls the value from memory |
| =  | 31. | The lower limit |
| 34.63 | 34.63 | The value of $\bar{x}$ |
| +  | 34.63 | Indicates an addition |
| MR | 3.63 | Recalls the value from memory |
| =  | 38.26 | The upper limit |

A 90 percent confidence interval is from \$31.00 to \$38.26.  ◄

Example 2 provides an illustration of the key sequence for computing the limits of the confidence interval of Example 1 using a Texas Instrument graphics calculator.

► **Example 2**   Construct a 90 percent confidence interval for $\mu$ if a sample of 64 yielded $\bar{x} = \$34.63$ and $s = \$17.60$.

**Solution**

| Key Sequence | Comments |
|---|---|
| 1.65 | The $z$ value |
| ×  17.60 | The value of $s$ |
| ÷  2nd  √  64 | The square root of $n$ |
| STO  ►  A | Stores the value in A |
| ENTER | Calculates and stores the value |
| **The display shows:** | |
| 1.65 * 17.60 / √ 64 → A<br>                   3.63 | |
| (Now continuing the key sequence) | |
| 34.63 | The $\bar{x}$ value |
| −  ALPHA  A | Subtracts the number in A from $\bar{x}$ |
| ENTER | and calculates the difference. |
| **The display shows:** | |
| 34.63 − A<br>            31 | The lower limit is \$31.00 |
| (Now continuing the key sequence) | |
| 34.63 | The $\bar{x}$ value |
| +  ALPHA  A | Adds the number in A to $\bar{x}$ and |
| ENTER | calculates the sum. |
| **The display shows:** | |
| 34.63 + A<br>           38.26 | The upper limit is \$38.26 |

◄

Example 3 provides an illustration of the key sequence for computing the limits of the confidence interval of Example 1 using a Casio graphics calculator.

▶ **Example 3** Construct a 90 percent confidence interval for $\mu$ if a sample of 64 yielded $\bar{x} =$ \$34.63 and $s =$ \$17.60.

**Solution**

| Key Sequence | Comments |
|---|---|
| 1.65 | The $z$ value for 90% |
| [×] 17.60 | The value of $s$ |
| [÷] [√] 64 | The value of $n$ |
| [➡] | The key between (−) and ( |
| [ALPHA] [A] | Store in memory A, $x^{-1}$ key |
| [EXE] | Computes and stores |

**The display shows:**

$1.65 \times 17.60 \div \sqrt{\phantom{x}} 64 \rightarrow A$
                                  $3.63$

(Now continuing the key sequence)

| 34.63 | The $\bar{x}$ value |
|---|---|
| [−] [ALPHA] [A] | Subtracts the number in A |
| [EXE] | Calculates the lower limit |

**The display shows:**

$34.63 - A$
                $31.$

(Now continuing the key sequence)          The lower limit is \$31.00

| 34.63 | The $\bar{x}$ value |
|---|---|
| [+] [ALPHA] [A] | Adds the number in A |
| [EXE] | Calculates the upper limit |

**The display shows:**

$34.63 + A$
                $38.26$          The upper limit is \$38.26

◀

**B. Confidence intervals for $P$**

The lower and upper limits of a confidence interval for $P$ are obtained by subtracting and adding respectively $z \cdot \sqrt{\dfrac{\hat{p}(1 - \hat{p})}{n}}$ to an observed $\hat{p}$. To use a calculator we write the expression as

$$z\sqrt{((\hat{p}(1 - \hat{p})) \div n)}$$

▶ **Example 4**  Using a scientific calculator, construct a 95 percent confidence interval for $P$ if a sample of 4000 yielded $\hat{p} = 0.380$.

Solution

| Key sequence | Display Shows | Comments |
|---|---|---|
| 1.96 | 1.96 | The $z$ value |
| $\times$ | 1.96 | |
| ( | [01     0. | Beginning the radicand |
| ( | [02     0. | Beginning the numerator |
| 0.380 | 0.380 | The $\hat{p}$ value |
| $\times$ | 0.38 | |
| ( | [03     0. | Beginning the $(1 - \hat{p})$ |
| 1 | 1. | |
| $-$ | 1. | |
| 0.380 | 0.380 | The $\hat{p}$ value |
| ) | 0.62 | The $(1 - \hat{p})$ value |
| ) | 0.2356 | The numerator value |
| $\div$ | 0.2356 | |
| 4000 | 4000. | The value of $n$ |
| ) | $5.89^{-05}$ | The value of the radical |
| $\sqrt{\ }$ | $7.674633542^{-03}$ | |
| $=$ | 0.015042281 | The value of the term |
| MIN | 0.015042281 | Stored in memory |
| 0.380 | 0.380 | The $\hat{p}$ value |
| $-$ | 0.38 | |
| MR | 0.015042281 | Recall the memory |
| $=$ | 0.364957718 | The lower limit is 0.365 |
| 0.380 | 0.380 | The $\hat{p}$ value |
| $+$ | 0.38 | |
| MR | 0.015042281 | Recall the memory |
| $=$ | 0.395042281 | The upper limit is 0.395 |

◀

▶ **Example 5**   Construct the confidence interval of Example 4 using a Texas Instrument graphics calculator.

**Solution**

| Key Sequence | Comments |
|---|---|
| 1.96 | The $z$ value |
| × 2nd √ ( ( .380 | The $\hat{p}$ value |
| × ( 1 − .380 ) ) | The $(1 - \hat{p})$ value |
| ÷ 4000 ) | The value of $n$ |
| STO▶ A ENTER | Stores value in memory A |
| **The display shows:** <br> *1.96 * √ ((.380 * (1 − .380))/4000) → A* <br>                          *.0150422817* <br> (Now continuing the key sequence) | The calculator value |
| .380 − ALPHA A ENTER | Calculates the lower limit |
| **The display shows:** <br> *.380 − A* <br>         *.3649577183* <br> (Now continuing the key sequence) | The lower limit is 0.365 |
| .380 + ALPHA A ENTER | Calculates the upper limit |
| **The display shows:** <br> *.380 + A* <br>         *.3950422817* | The upper limit is 0.395 |

◀

▶ **Example 6**    Calculate the confidence interval of Example 4 using a Casio graphics calculator.

**Solution**

| Key Sequence | Comments |
|---|---|
| 1.96 | The $z$ value |
| ×  √  (  ( .380 | The $\hat{p}$ value |
| ×  ( 1 − .380 )  ) | The $(1 - \hat{p})$ value |
| ÷ 4000 ) | The value of $n$ |
| ➡ ALPHA A EXE | Store the calculation in memory A |

**The display shows:**

$$1.96 \times \sqrt{\ } \ ((.380 \times (1 - .380)) \div 4000) \rightarrow A$$
$$0.01504228174$$

(Now continuing the key sequence)

| .380 − ALPHA A EXE | Calculates the lower limit |
|---|---|

**The display shows:**

.380 − A
          0.3649577183

(Now continuing the key sequence)

The lower limit is 0.365

| .380 + ALPHA A EXE | Calculates the upper limit |
|---|---|

**The display shows:**

.380 + A
          0.3950422817

The upper limit is 0.395

Originally, I went to college to study botany because I loved the outdoors and flowers and plants. During college, I worked as a research assistant for a botanist with the New York Botanical Gardens. The job involved measuring different characteristics from specimens of a particular tree species collected from all over the world. After entering the data into a computer, I used a statistical software package to examine the data for relationships between the location of the tree and the specimen characteristics. I had taken an introductory statistics class as part of my course requirements. This was when I first understood how powerful a tool statistics could be, and how exciting it was to see answers to questions emerge from thousands of numbers. I decided to take as many statistics courses as my schedule allowed. Ultimately, I completed a masters degree in biostatistics.

Today, as a biostatistician, I am responsible for the statistical design, management, and analysis of clinical trials in all phases of cancer research. My job involves interacting closely with the cancer specialists at the institute. The goal of a clinical trial is to compare the outcome of two groups of patients when one group is treated with a new drug and the other with a standard drug. Hopefully, the new treatment will increase the length of survival by an amount that is clinically important. This amount is determined by the specific disease and patient population. We do not want to bias this comparison by selecting patients with a favorable prognosis to receive the new drug. For this reason, patients are randomly assigned to the treatment groups. Another aspect of the experimental design is deciding how many patients must be in each group to have a reasonable chance of measuring this increase in survival, allowing for the variability in each group.

Although my role as a statistician must be unbiased, methodical, and scientific, it is immensely rewarding to be part of the research effort to understand and cure cancer. Statistics, however, may be wed to many disciplines, from accounting to zoology. Statistics is a way of looking at evidence to make a reasonable conclusion to whatever question interests you.

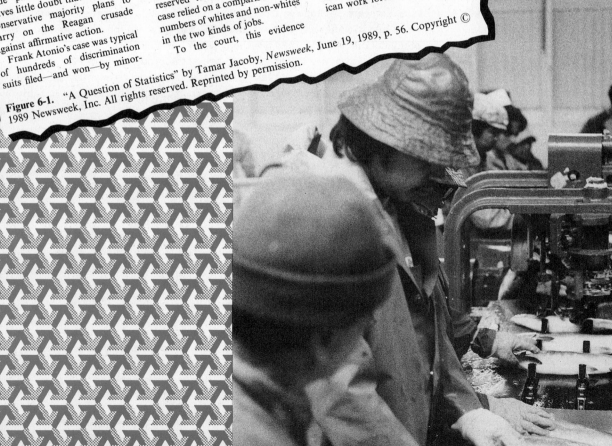

## A Question Of Statistics

### New limits on bias suits

The issues at stake seemed technical and lawyerly: arcane rules about what kinds of statistics count in court and which side in a lawsuit bears the burden of proof. Yet last week's U.S. Supreme Court decision on racial discrimination at Alaskan fish canneries has changed the face of civil-rights law, making it much harder for workers to sue employers. Following a landmark ruling early this year against minority set-aside plans, this opinion also leaves little doubt that the court's conservative majority plans to carry on the Reagan crusade against affirmative action.

Frank Atonio's case was typical of hundreds of discrimination suits filed—and won—by minority workers in the past two decades. A Samoan, he and nine nonwhite co-workers charged that hiring and promotion practices at the Wards Cove Packing Co. had prevented them from rising to better jobs. The law did not require the workers to prove the cannery had *intentionally* discriminated. All they had to show was statistical evidence of racial disparities that served no business purpose.

To Atonio, the evidence was plain: most of the cannery's unskilled work went to minorities, while better-paying slots were filled by whites. The two types of workers ate and bunked separately. Jobs were even tagged with racial labels. Butchering was set aside for a "chink" and certain badge numbers were reserved for "natives." Atonio's case relied on a comparison of the numbers of whites and non-whites in the two kinds of jobs.

To the court, this evidence proved nothing. According to Justice Byron R. White's opinion, Atonio had chosen the wrong statistical comparison. To prove bias, White said, employees must compare the racial mix of job holders with the racial mix among qualified applicants. The court also set new requirements for such suits. From now on, workers must pinpoint exactly which company policy—hiring, promoting, or the like—caused the imbalance. Then *they* must prove it served no business end.

Together, these new rules will discourage many suits. What's more, by reducing the fear of statistical claims, the decision removes a powerful incentive for affirmative-action quotas. In effect, says Prof. Paul Gewirtz of Yale University, "the Supreme Court has dealt body blows to two of the most important mechanism for integrating the American work force."

**Figure 6-1.** "A Question of Statistics" by Tamar Jacoby, *Newsweek*, June 19, 1989, p. 56. Copyright © 1989 Newsweek, Inc. All rights reserved. Reprinted by permission.

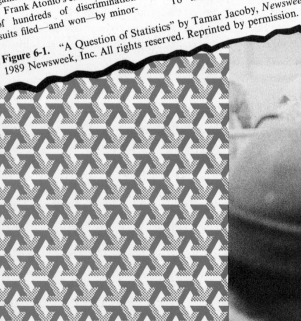

# Hypothesis Testing: One Sample

Figure 6-1 is a copy of an article regarding a recent decision by the U.S. Supreme Court. A suit was brought before the court by employees of Alaskan fish canneries claiming racial discrimination "in hiring and promotion practices." The employees presented data (evidence) to the court that they felt supported their claim. In the middle column, we read that "all they had to show was statistical evidence of racial disparities that served no business purpose." In the right column, we read that the court felt the evidence presented "proved nothing." According to Justice Byron R. White's opinion, "Antonio had chosen the wrong statistical comparison."

This article is placed here at the beginning of a chapter on "hypothesis testing," because it illustrates the use of data to test some claim. The workers claimed racial discrimination. The court reviewed the data and determined the evidence was insufficient to justify the claim of racial discrimination.

The statistical process analogous to this example is called *hypothesis testing*. There are several steps in hypothesis testing that ultimately lead to a conclusion, but the conclusion is always one of two possible decisions:

1. Reject the claim.
2. Do not reject the claim.

The details of how to conduct a hypothesis test, and what kinds of errors can be made, are studied in this chapter.

# 6-1 Hypothesis Testing Generalities

**Key Topics**

1. *The null hypothesis and alternative hypothesis*

2. *Critical values for a hypothesis test*

3. *Definition of the rejection region*

4. *Definition of the level of significance*

5. *Critical values for standard levels of significance*

---

## S T A T I S T I C S   I N   A C T I O N

**A**   Don Hadd is a chemical engineer employed by Gladstone Petroleum Products Corporation. Don had been working on additives that might be added to gasoline to increase mileages for automobiles and small trucks. The research division at Gladstone has a test engine that is used to compare the times the engine can operate on a specific amount of fuel. Current data show that the engine operates an average of 8.35 minutes for the allotted amount of commercially sold unleaded gasoline with $\sigma = 0.32$. Don

has perfected a compound he calls K59MH3 that, when added to the gasoline, he feels will increase the mean operating time. To test his hypothesis, he measured the operating times (in minutes) of 35 samples of gasoline with the compound added. The results are recorded in Table 6-1.

Based on the results of the tests, Don reported that his additive "improved the operating times of the test engine at a 1% level of significance."

TABLE 6-1   Operating Times (in Minutes) for Gasoline Plus Additive Tests

| | | | | | | | | |
|---|---|---|---|---|---|---|---|---|
| 8.17 | 8.08 | 8.40 | 8.93 | 8.88 | 9.04 | 8.36 | 8.77 | 8.62 |
| 8.45 | 9.07 | 8.29 | 8.28 | 9.01 | 8.32 | 8.47 | 8.64 | 8.19 |
| 8.16 | 8.53 | 8.70 | 7.94 | 8.29 | 8.69 | 8.38 | 8.90 | 8.30 |
| 9.12 | 8.72 | 8.13 | 8.36 | 8.09 | 8.58 | 8.43 | 8.21 | |

**B**   Fontaine Evaldo is a member of the Truth in Advertising Commission in the western region of the United States. She was given the responsibility of studying a recent advertising campaign by a national producer of flower and vegetable seeds. The company claims in their advertisement that modern production techniques have increased the germination rates of their seeds to over 90 percent. Fontaine obtained samples of several kinds of their

seeds and subsequently planted 400 seeds in a controlled environment. Later she determined that 352, or 88 percent of the seeds germinated. In a report to the commission on her findings, Fontaine included the following statement:

"Based on the results of my experiment, I was unable to reject the 90% germination rate claim of the company at the 5% level of significance."

## The null hypothesis and alternative hypothesis

A hypothesis test can be helpful when there is a question regarding the correctness of a stated value for a parameter of some population of interest. In this chapter, only values of $\mu$, $P$, and $\sigma$ will be tested. The symbols $\mu_0$ (read "mu naught"), $P_0$ (read "$P$ naught"), and $\sigma_0$ (read "sigma naught") will be used for the given values of the parameters. That is:

> $\mu_0$ represents a stated *mean* of a population of quantitative data.
>
> $P_0$ represents a stated *proportion* of a population with some characteristic of interest.
>
> $\sigma_0$ represents a stated *standard deviation* of a population of quantitative data.

As it turns out, testing for $\mu$ and $P$ have similarities that facilitate their discussion in parallel. Thus, in the first four sections we will concentrate on testing for $\mu$ and $P$. After we have discussed testing for $\mu$ and $P$, we will study testing for $\sigma$ in Section 6-5.

In the Statistics in Action section, Don Hadd tested the mean running time of the test engine with the additive in the gasoline. For this experiment, $\mu_0 = 8.35$ minutes, the assumed (or claimed) value for $\mu$. Fontaine Evaldo tested the germination rate for flower and vegetable seeds. For this experiment, $P_0 = 0.90$, the germination rate claimed by the producer.

The first step in applying a hypothesis-testing method is to write *two opposing statements* concerning the stated value of the parameter. These statements are called the *null hypothesis* and the *alternative hypothesis*.

---

### DEFINITION 6-1 The Null and Alternative Hypotheses

The **null hypothesis** states that there is no difference between the actual unknown value of the parameter and the stated value. The symbol $H_0$ (read "$H$ naught") is used for the null hypothesis.

The **alternative hypothesis** states that there is a difference of a particular type between the actual unknown value of the parameter and the stated value. The symbol $H_a$ (read "$H$ sub-$a$") is used for the alternative hypothesis.

---

Parts (a) through (c) in Figure 6-2 (on the following page) illustrate three possible pairings of $H_0$ and $H_a$ for tests on $\mu$. Parts (a)* through (c)* illustrate three similar pairings of $H_0$ and $H_a$ for tests on $P$. The form of $H_a$ chosen in an application usually reflects the suspicion or opinion of the person investigating the correctness of $H_0$.

| Comparing a Stated Value of the Mean to the Actual Unknown Value of the Mean | Comparing a Stated Value of the Proportion to the Actual Unknown Value of the Proportion | Meaning of the Form of $H_a$ |
|---|---|---|
| a. $H_0$:  $\mu = \mu_0$<br>   $H_a$:  $\mu \neq \mu_0$ | a*. $H_0$:  $P = P_0$<br>    $H_a$:  $P \neq P_0$ | $\mu$ (or $P$) is higher ($>$), or lower ($<$) than the stated value. |
| b. $H_0$:  $\mu = \mu_0$<br>   $H_a$:  $\mu < \mu_0$ | b*. $H_0$:  $P = P_0$<br>    $H_a$:  $P < P_0$ | $\mu$ (or $P$) is lower ($<$) than the stated value. |
| c. $H_0$:  $\mu = \mu_0$<br>   $H_a$:  $\mu > \mu_0$ | c*. $H_0$:  $P = P_0$<br>    $H_a$:  $P > P_0$ | $\mu$ (or $P$) is higher ($>$) than the stated value. |

**Figure 6-2.**   Three pairings of $H_0$ and $H_a$.

The form of $H_a$ in parts (a) and (a)* make these *nondirectional*, or *two-tail alternative hypotheses*. The form of $H_a$ in parts (b), (b)*, (c), and (c)* make these *directional*, or *one-tail alternative hypotheses*.

The alternative hypothesis $H_a$ is sometimes called the **motivated hypothesis** because it usually indicates the motivation for conducting the test.

▶ **Example 1**   Write $H_0$ and $H_a$ for the hypothesis test that Don Hadd conducted in the gasoline additive problem.

Solution   **Discussion.** Don wanted to perfect an additive to increase mileage in cars and small trucks. He therefore hoped to increase the mean running time of the test engine by putting the additive in the gasoline. His motivation was therefore directional with respect to $\mu$ when the engine was run on fuel with the additive included.

$H_0$:   $\mu = 8.35$ minutes          The mean running time is still the same.

$H_a$:   $\mu > 8.35$ minutes          The additive increases the running time.   ◀

▶ **Example 2**   Write $H_0$ and $H_a$ for the hypothesis test that Fontaine Evaldo conducted in the seed germination problem.

Solution   **Discussion.** Fontaine probably had the interest of the consumer in mind when she conducted the seed experiment. Specifically she was concerned about the possibility that less than 90 percent of the seeds that consumers buy will germinate. She would therefore use a directional alternative hypothesis that states that the proportion of seeds that germinate is less than 90 percent.

$H_0$:   $P = 0.90$        The germination rate is 90%.

$H_a$:   $P < 0.90$        The germination rate is less than 90%.   ◀

## Critical values for a hypothesis test

With an $H_0$ and $H_a$ stated, now consider what kinds of samples will lead to the conclusion that $H_0$ can be rejected in favor of $H_a$. On an intuitive basis, it is reasonable to reject $H_0$ only if the observed values of $\bar{X}$ or $\hat{P}$ are "far enough" away from $\mu_0$ (or $P_0$) and in the direction indicated by $H_a$. The values of $\bar{X}$ or $\hat{P}$ that separate values of the statistics that are "far enough," and those that are not, are called *critical values*.

---

### DEFINITION 6-2  A Critical Value for a Hypothesis Test on $\mu$ or $P$

A **critical value** is a boundary point between those possible values of $\bar{X}$ or $\hat{P}$ that will result in rejection of $H_0$ and those that will not result in rejection of $H_0$. The symbols $\bar{x}_c$ and $\hat{p}_c$ will stand for critical values of $\bar{X}$ and $\hat{P}$, respectively.

---

Nondirectional tests have two critical values, a lower one and an upper one, as shown in part (a) of Figure 6-3. Directional tests have only one critical value. Whether the critical value is greater than or less than the value of $\mu_0$ or $P_0$ depends on the form of $H_a$. Parts (b) and (c) of Figure 6-3 illustrate this fact for a test of $H_0: \mu = \mu_0$. Similar statements and diagrams could be given for a test of $H_0: P = P_0$.

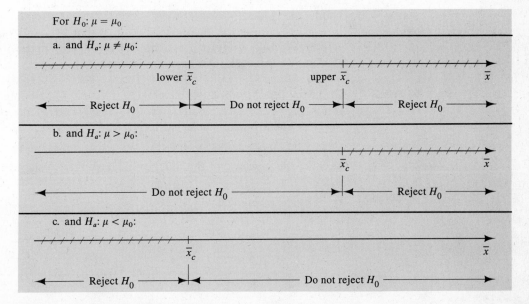

**Figure 6-3.**  Critical values on a number line.

## A definition of the rejection region

The one interval $\bar{x}$ or $\hat{p}$ values in a directional test or two intervals of values in a nondirectional test that cause one to reject $H_0$ are called rejection regions.

### DEFINITION 6-3 Rejection Regions for Hypothesis Tests

The interval (or intervals) of possible values of $\bar{X}$ or $\hat{P}$ that are more extreme than the critical value (or values) is called the **rejection region**.

▶ **Example 3**   In the gasoline additive experiment, Don Hadd wanted to show an increase in the mean running time over 8.35 minutes for the test engine. Suppose he arbitrarily selected $\bar{x}_c = 8.43$ for the test.

**a.** Draw a diagram showing $\bar{x}_c$ and the rejection region.
**b.** State the meaning of the diagram as it applies to the test.

Solution   **a.** In Figure 6-4, $\mu_0 = 8.35$ and $\bar{x}_c = 8.43$ are plotted. The rejection region is cross-hatched and the potential decisions for the test based on $\bar{x}_c$ are labeled.

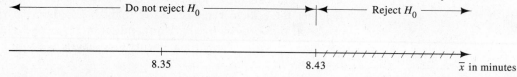

**Figure 6-4.**   Rejection region based on $\bar{x}_c = 8.43$.

**b.** If $\bar{x}$ is the observed mean of the 35 tests that Fred conducts using the additive and $\bar{x} > 8.43$ minutes, then he will reject $H_0$ in favor of $H_a$. If, however, $\bar{x} < 8.43$ minutes, then he will not reject $H_0$.   ◀

▶ **Example 4**   In the seed germination experiment, Fontaine Evaldo wanted to detect a decrease in the germination rate compared to that claimed by the producer of the seeds. Suppose she arbitrarily selected $\hat{p}_c = 0.87$ for the test.

**a.** Draw a diagram showing $\hat{p}_c$ and the rejection region.
**b.** State the meaning of the diagram as it applies to the test.

Solution   **a.** In Figure 6-5, $P_0 = 0.90$ and $\hat{p}_c = 0.87$ are plotted. The rejection region is cross-hatched and the potential decisions for the test based on $\hat{p}_c$ are labeled.

```
◀── Reject H₀ ──▶◀──────── Do not reject H₀ ────────▶

/////////////////
            0.87                    0.90                    p̂
```

**Figure 6-5.**   Rejection region based on $\hat{p}_c = 0.87$.

**b.** If $\hat{p}$ is the observed proportion of the 400 seeds that germinate and $\hat{p} < 0.87$, then Fontaine will reject $H_0$ in favor of $H_a$. If, however, $\hat{p} > 0.87$, then she will not reject $H_0$.   ◄

## Definition of the level of significance

In the hypothesis tests performed by Don and Fontaine, the decisions to reject or not reject $H_0$ depend on values of random variables $\bar{X}$ and $\hat{P}$, respectively. As a consequence, even though the stated values of $\mu$ and $P$ in $H_0$ might be correct, it is possible that the samples studied by Don and Fontaine could yield values of $\bar{X}$ and $\hat{P}$ that would force rejection of $H_0$. The probability of such an event happening is called the level of significance of the test.

---

### DEFINITION 6-4 Level of Significance for a Hypothesis Test on $\mu$ or $P$

The probability that the sampling distribution of $\bar{X}$ or $\hat{P}$ (if $\mu$ or $P$ has the value stated in $H_0$) assigns to the rejection region is called the **level of significance** of the test.

---

▶ **Example 5**   If Don used $\bar{x}_c = 8.43$, then what is the level of significance for the test?

**Solution**   **Discussion.** Consider the situation before collection of the $n = 35$ engine running times. The mean of the 35 run times is $\bar{X}$, and based on the central limit theorem, the probability distribution of $\bar{X}$ is approximately normal. Therefore, after converting to a $z$ score, we can use the standard normal distribution to calculate the probability that $\bar{X} > 8.43$ minutes. (In computing the $z$ score we suppose that $H_0: \mu = 8.35$ is true.)

In Figure 6-6, the shaded area represents $P(\bar{X} > 8.43)$. With $\mu_0 = 8.35$, $\sigma = 0.32$, and $n = 35$:

$$z = \frac{8.43 - 8.35}{\dfrac{0.32}{\sqrt{35}}}$$

$$= 1.4790\ldots$$

$$\approx 1.48$$

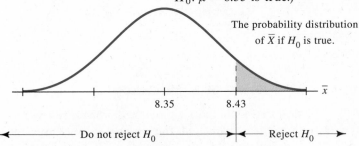

The probability distribution of $\bar{X}$ if $H_0$ is true.

**Figure 6-6.**   The shaded area represents the level of significance.

$$P(\bar{X} > 8.43) = P(Z > 1.48)$$
$$= 1 - 0.9306 \qquad \text{Use Appendix Table 4.}$$
$$= 0.0694 \qquad \text{Level of significance is about 7\%.}   ◄$$

▶ **Example 6**   If Fontaine used $\hat{p}_c = 0.87$, then what is the level of significance for the test?

Solution   **Discussion.** Recall that Fontaine selected and planted $n = 400$ seeds. If $P = 0.90$ as the producer claimed, then $400(0.9) = 360$ and $400(0.1) = 40$. Because both products are greater than 5, the probability distribution of $\hat{P}$ is approximately normal. We can therefore use the standard normal distribution to compute the level of significance (after computing an appropriate $z$ score).

In Figure 6-7, the shaded area represents $P(\hat{P} < 0.87)$. With $P_0 = 0.90$, $1 - P_0 = 0.10$, and $n = 400$:

$$z = \frac{0.87 - 0.90}{\sqrt{\dfrac{(0.9)(0.1)}{400}}}$$

$$= -2.00$$

$$P(\hat{P} < 0.87) = P(Z < -2.00)$$

$$= 0.0228 \qquad \text{Use Appendix Table 4.}$$

Thus the level of significance is about 2 percent when $\hat{p}_c = 0.87$ is used.   ◀

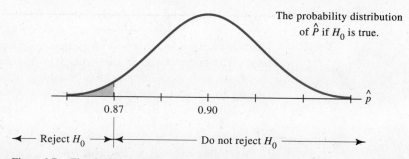

**Figure 6-7.**   The shaded area represents the level of significance.

## Critical values for standard levels of significance

In Examples 5 and 6, we supposed that Don and Fontaine had arbitrarily selected critical values for the tests. We then calculated the corresponding levels of significance.

A more commonly used approach is to first select the level of significance and then from it, calculate the critical value or critical values. Equations (1) and (2) can be used for large sample tests on $\mu$ or $P$.

### Equations for Calculating the Critical Values for Large Sample Hypothesis Tests of $H_0: \mu = \mu_0$ and $H_0: P = P_0$

$$\bar{x}_c = \mu_0 + z_c\left(\frac{\sigma}{\sqrt{n}}\right)$$ ⟶ $\mu_0$ is the stated value for $\mu$.
⟶ $z_c$ is the critical z value.     **(1)**

$$\hat{p}_c = P_0 + z_c\sqrt{\frac{P_0(1 - P_0)}{n}}$$ ⟶ $P_0$ is the stated value for $P$.
⟶ $z_c$ is the critical z value.     **(2)**

A *critical z value*, $z_c$, depends on the level of significance of the test. Many tests are conducted at the 5 and 1 percent levels of significance. The corresponding values of $z_c$ for one- and two-tail tests are given in Table 6-2.

**TABLE 6-2**   $z_c$ Values for 5 Percent and 1 Percent Levels of Significance

| Level of Significance | Two-Tail Test | One-Tail Test |
|---|---|---|
| 5% | 1.96 and −1.96 | 1.65 or −1.65 |
| 1% | 2.58 and −2.58 | 2.33 or −2.33 |

▶ **Example 7**   At a factory of the Dalli Spice Corporation in Calcutta, minced garlic is packaged in small plastic bags. The reported mean weight per bag is 50.0 grams with a population standard deviation of 3.6 grams. Suppose that 36 bags are randomly selected and weighed to check whether $\mu = 50.0$ for a large production run being shipped to the United States.

**a.** State $H_0$ and $H_a$ appropriate for this test.
**b.** Calculate the $\bar{x}_c$s, if a 5 percent level of significance is used.
**c.** Sketch and label an idealized probability histogram for $\bar{X}$ based on $H_0$ being true.

**Solution**   **a.** A nondirectional test would be appropriate.

$H_0$:   $\mu = 50.0$     The mean weight is 50.0 grams.
$H_a$:   $\mu \neq 50.0$     The mean weight is not 50.0 grams.

**b.** With $\mu_0 = 50.0$, $\sigma = 3.6$, $n = 36$, $z_c = -1.96$ and 1.96:

$$\text{lower } \bar{x}_c = 50.0 - 1.96\left(\frac{3.6}{\sqrt{36}}\right)$$

$$= 50.0 - 1.1760$$

$$= 48.8, \text{ to one decimal place}$$

$$\text{upper } \bar{x}_c = 50.0 + 1.96\left(\frac{3.6}{\sqrt{36}}\right)$$

$$= 51.2, \text{ to one decimal place}$$

**c.** An idealized probability histogram centered at 50.0 is shown in Figure 6-8. The total area shaded in the two tails is 0.05. Thus the probability is about 5 percent that $\bar{X}$ will fall in the rejection region and result in the rejection of $H_0$, when in fact $H_0$ is true.   ◀

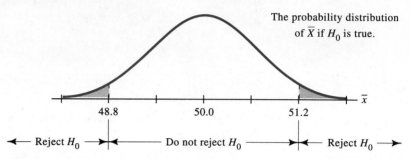

Figure 6-8.   The shaded area is approximately 0.05.

▶ **Example 8**   A report claims that 68 percent of American adults support a proposed international arms agreement. Senator Cindy Yates feels that the percent of her constituents who support the agreement is greater than 0.68. She decides to test this claim using a 1 percent level of significance. She has her staff sample the opinions of 500 residents in her district.

**a.** State $H_0$ and $H_a$ for the test.
**b.** Calculate $\hat{p}_c$.
**c.** Sketch and label an approximate probability histogram for $\hat{P}$ based on $H_0$ being true.

Solution   **a.** Senator Yates feels that $P$ is greater than 0.68.

$$H_0: \quad P = 0.68$$
$$H_a: \quad P > 0.68$$

**b.** With $P_0 = 0.68$, $1 - P_0 = 0.32$, $n = 500$, and $z_c = 2.33$:

$$\hat{p}_c = 0.68 + 2.33 \sqrt{\frac{(0.68)(0.32)}{500}}$$

$$= 0.68 + 0.048607\ldots$$

$$= 0.73, \text{ to two decimal places}$$

**c.** A sketch of an approximate probability histogram for $\hat{P}$ centered at 0.68 is shown in Figure 6-9. The area shaded is approximately 0.01. If $\hat{P} > 0.73$, then $H_0$ will be rejected in favor of $H_a$.   ◀

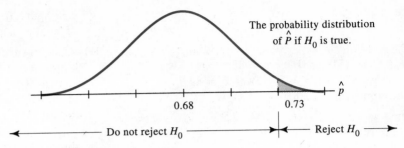

Figure 6-9. The shaded area is approximately 0.01.

# Exercises 6-1   Set A

*In Exercises 1–8:*
*a. State whether the problem involves a test on μ or P.*
*b. Write $H_0$ and $H_a$ that would be appropriate for each test.*
*c. Construct a line graph similar to Figure 6-3 for each test.*

1. A major foreign car manufacturer claims its new economy car averages 41 miles per gallon on the highway. A team from a consumer advocate group wants to check whether the mileage is actually below this claim.

2. The management of a national chain of fast-food restaurants claims its part-time employees make $5.25 per hour on the average. The executive council of a restaurant employees union wants to check whether the hourly rate figure is overstated.

3. A nationwide poll yields data that suggest that 69 percent of the American population considers discarded aluminum beverage cans an environmental problem. An activist group lobbying for a law requiring a deposit on all such cans feels the true percentage is higher.

4. The department of motor vehicles in a large eastern state claims that 35 percent of vehicles in the state have faulty emission controls. A council of automobile mechanics feels that records of automobile repairs by members indicate the percent is actually greater than the one reported.

5. Polls taken in a large southern state suggest that 43 percent of the eligible voters in the state support the re-election of Governor Sandra Nossiter. The governor feels the poll was not properly taken and doubts the accuracy of the 43 percent figure.

6. Studies conducted by a state department of education suggest that 27 percent of students in the state's K through 12 system have a disability that reduces the student's ability to learn. Byron Hopkins, a member of the governor's task force to study educational programs in the K through 12 system, feels this value should be checked for accuracy.

7. A major educational magazine reported recently that the average age of professors in 4-year colleges and universities reached 48.7 years, an all time high. The National Council of Professors feels the figure is not correct and should be tested for accuracy.

8. A manufacturer of light bulbs wants the average life of its 60-watt bulbs to be at least 750 hours. Bill Ball, a quality control engineer, has been given the task of obtaining a sample of these bulbs and devising a test to detect the possibility that the mean is below the minimum standards of the company.

*In Exercises 9–16, compute the level of significance for each indicated test. Round answers to two decimal places.*

9. $H_0$:  $\mu = 85.0$     $H_a$:  $\mu \neq 85.0$
   Suppose $\sigma = 3.4$, $n = 56$, lower $\bar{x}_c = 84.0$, and upper $\bar{x}_c = 86.0$

10. $H_0$:  $\mu = 27.5$     $H_a$:  $\mu \neq 27.5$
    Suppose $\sigma = 8.2$, $n = 75$, lower $\bar{x}_c = 25.8$, and upper $\bar{x}_c = 29.2$

11. $H_0$:  $\mu = 5.90$     $H_a$:  $\mu < 5.90$
    Suppose $\sigma = 1.29$, $n = 40$, and $\bar{x}_c = 5.52$

12. $H_0$:  $\mu = 260$     $H_a$:  $\mu > 260$
    Suppose $\sigma = 18$, $n = 64$, and $\bar{x}_c = 263$

13. $H_0$:  $P = 0.72$     $H_a$:  $P > 0.72$
    Suppose $n = 130$ and $\hat{p}_c = 0.79$

14. $H_0$:  $P = 0.36$     $H_a$:  $P < 0.36$
    Suppose $n = 240$ and $\hat{p}_c = 0.32$

15. $H_0$:  $P = 0.46$     $H_a$:  $P \neq 0.46$
    Suppose $n = 62$, lower $\hat{p}_c = 0.32$, and upper $\hat{p}_c = 0.60$

16. $H_0$:  $P = 0.850$     $H_a$:  $P > 0.850$
    Suppose $n = 260$ and $\hat{p}_c = 0.890$

*In Exercises 17–24, find the critical value (or values) for the appropriate hypothesis tests with the stated levels of significance.*

17. The sponsors that paid for advertising on last year's Super Bowl game were told that during the game 68 percent of TV viewers were watching the game. In an attempt to show that the percent was higher for viewers in a large eastern city, a sports reporter surveyed 175 individuals who were watching TV during the time the game was aired. He used a 5 percent level of significance.

18. A task force commissioned by the state legislature in a western state to study the quality of education in the high schools in the state reported that 32 percent of the 17-year-olds do not know the purpose of the Declaration of Independence. The Association of High School History Teachers feels the figure is too high. They subsequently survey 250 individuals who are 17 years old to test the

0.32 figure and they intend to use a 5 percent level of significance.

19. A major drug company manufacturers a pain-killing tablet that is reported to contain an average of 80.0 mg of a particular drug. A sample of 40 tablets is checked to test the accuracy of the 80.0 mg figure. The test is conducted at a 5 percent level of significance, and records indicate that $\sigma = 8.0$ mg.

20. A large shipment of "ready-mix concrete" contains bags labeled 94 pounds. A sample of 32 bags is weighed to test the 94-pound figure. The test is conducted at a 1 percent level of significance, and production records indicate that $\sigma = 4$ pounds.

21. A large financial consulting firm reported that clients of the firm realized a per account profit of 8.9 percent last year on investments arranged through the firm. A conglomerate of its investors called for an audit of the accounts because the members felt the figure was overstated. If $\sigma = 3.4$ percent and 40 accounts are audited, find $\bar{x}_c$ for a test using a 1 percent level of significance.

22. A large, egg producing company tries to ship no more than eight "flawed" (such as cracked or dirty) eggs per case of 360 eggs. A sample of 30 cases is carefully checked to test the eight-per-case figure. For the distribution of numbers of flawed eggs per case, records indicate that $\sigma = 3.5$ and the test is conducted at a 1 percent level of significance.

23. A study commissioned by a National Law Officer's Organization suggests that 74 percent of Americans think taking the law into one's own hands is sometimes justified. The members of the sheriff's department in a large southwestern state feel the percent is higher for residents in their county. They subsequently sample the opinions of 325 residents and conduct a test at the 1 percent level of significance.

**24.** A report by the registrar's office of a major university in the East suggests that typically approximately 28 percent of the graduate students at the school are Asian. The Asian Club on campus feels the figure is overstated, and decides to test the figure by randomly sampling the records of 275 current graduate students. Members intend to use a 5 percent level of significance.

## Exercises 6-1    Set B

*In Exercises 1 and 2, refer to Data Set I in Appendix B.*

**1.** Consider the data on ages. For this set of data, $\mu = 56.1$ years and $\sigma = 20.2$ years.
   a. Formulate $H_0$ and $H_a$ for someone interested in checking the accuracy of $\mu$.
   b. If a sample of 40 data points is obtained and the level of significance is 5 percent, calculate the values of $\bar{x}_c$ for the test.

**2.** Consider the data on smokers. For the proportion of individuals in the population that smoke, $P = 0.53$.
   a. Formulate $H_0$ and $H_a$ for someone interested in showing that $P$ is lower than the stated value.
   b. If a sample of 50 data points is obtained and a 1 percent level of significance is used, compute the value of $\hat{p}_c$ for the test.

*In Exercises 3 and 4, refer to Data Set II in Appendix B.*

**3.** Consider the data on the stock prices that increased $(+)$, decreased $(-)$, or remained the same $(0)$ on this day of trading. For the proportion of stock prices that increased, $P = 0.468$.
   a. Formulate $H_0$ and $H_a$ for someone who feels the figure is actually higher.
   b. If a sample of 35 data points is obtained and a 1 percent level of significance is used, compute the value of $\hat{p}_c$ for the test.

**4.** Consider the data on the "52 weeks High" stock prices. For this set of data, $\mu = \$36.48$ and $\sigma = \$29.90$.
   a. Formulate $H_0$ and $H_a$ for someone interested in checking the accuracy of $\mu$.
   b. If a sample of 60 data points is obtained and a 5 percent level of significance is used, calculate the values of $\bar{x}_c$ for the test.

*In Exercises 5 and 6, refer to the information in Figure 6-10.*

**5.** The Gallup Poll suggests that 40 percent of the American population is highly concerned about household garbage and trash as an environmental problem. Suppose a hypothesis test is conducted in a particular section of the country to establish $H_a: P > 0.40$. Furthermore suppose a 5 percent level of significance is used for the test.
   a. Compute $\hat{p}_c$, if $n = 30$.
   b. Compute $\hat{p}_c$, if $n = 300$.
   c. Compute $\hat{p}_c$, if $n = 3000$.
   d. Comment on the effect of increasing sample size on the value of $\hat{p}_c$.

# Americans View Waste Disposal As Serious Issue

*By George Gallup Jr. and Alec Gallup*

Princeton, N.J.
   As many as four Americans in 10 are highly concerned about household garbage and trash, which the Environmental Protection Agency has called a "staggering" national crisis. To help cope with the waste-disposal problem, more than one-third are sorting their refuse for recycling.

**Figure 6-10.** "Americans View Waste Disposal As Serious Issue" by George Gallup, Jr. and Alec Gallup, *Gallup Poll*, 1988. Copyright © 1988 American Institute of Public Opinion. Reprinted by permission.

6. The Gallup Poll suggests that approximately 34 percent of the American population is sorting its refuse for recycling. Suppose a hypothesis test is conducted in a particular section of the country to establish $H_a$: $P < 0.34$.

   a. Compute the level of significance, if $\hat{p}_c = 0.30$ and $n = 30$.

   b. Compute the level of significance, if $\hat{p}_c = 0.30$ and $n = 300$.

   c. Compute the level of significance, if $\hat{p}_c = 0.30$ and $n = 3000$.

   d. Comment on the effect of increasing sample size on the magnitude of the level of significance.

*Exercises 7–10 are suggested student projects. For each exercise, do the following tasks:*

a. *State $H_0$ and $H_a$ for a test on the indicated parameter.*

b. *Collect a sample of data points to use in a hypothesis test.*

c. *Use a 5 percent level of significance and compute $\bar{x}_c$ or $\hat{p}_c$.*

d. *Sketch and label a probability distribution showing the decision regions.*

e. *Use the $\bar{x}$ or $\hat{p}$ from your sample to decide whether to reject $H_0$.*

7. The U.S. Department of Vital Statistics reported that the mean age of postsecondary students is 27.0 years with a standard deviation of 6.4 years. Obtain the ages of 36 students to test this figure for your college.

8. The Academy of College Textbook Publishers reported that the mean cost of new casebound college texts is $42.75 with a standard deviation of $10.80. Obtain the prices of 36 texts in the college store on your campus to test this figure for your college.

9. The National Congress of Associated Students reported that 62 percent of all college students earn some money to help pay their college expenses. Sample 60 students on your campus and determine the proportion of these students who earn some money to help pay their college expenses, to test whether the 62 percent figure is too low for your school.

10. The National Alliance of Pizza Makers reported that 80 percent of all college students eat at least one pizza per month. Sample 45 students on your campus and determine the proportion of these students who feel they eat at least one pizza per month, to test whether the 80 percent figure is too high for your school.

# 6-2  Type I and Type II Errors

**Key Topics**

1. *Two types of possible errors in any hypothesis test*

2. $\alpha = P(Type\ I\ error)$

3. $\beta = P(Type\ II\ error)$

4. *Some general comments regarding Type I and Type II error probabilities*

## S T A T I S T I C S   I N   A C T I O N

**A**   Lynn Butler has a pharmaceutical degree from a university in the South. She is currently working for a corporation that manufactures medications for humans. Her duty for today is to check the accuracy of the composition of drugs in a large production run of a medication she helped perfect for alleviating arthritis pain. The key ingredient in the compound is a drug that reportedly averages 200 milligrams (mg) per capsule. This drug is ineffective if the amount per capsule is significantly less than 200 mg, but it causes unacceptable side effects if the amount per capsule is significantly more than 200 mg. For her test Lynn will use:

$$H_0: \quad \mu = 200 \text{ mg}$$
$$H_a: \quad \mu \neq 200 \text{ mg}$$

To make a decision, Lynn will obtain a large random sample ($n \geq 30$) of capsules and carefully analyze the composition of each one in the laboratory. Because humans will use the product, she will conduct the test at a high level of significance. That is, it is better that a production run for which the 200-mg figure is appropriate be rejected, than to fail to reject a production run for which the 200-mg figure is inappropriate. The first situation would cause financial loss to the corporation, but the latter could pose a health hazard to individuals who take the drug.

**B**   James Ertl is currently employed by a corporation that makes cleaning products. James regularly analyzes the composition of large production runs of a granular product made for dishwashers. In particular, he checks the amounts of a phosphorous ingredient. The product reportedly has an average of 1.1 grams of this ingredient in each sample of 15 grams of the granules. James will carefully analyze a large number ($n \geq 30$) of 15-gram samples. The results of the analyses will be used to test:

$$H_0: \quad \mu = 1.1 \text{ grams}$$
$$H_a: \quad \mu \neq 1.1 \text{ grams}$$

For this test, James will use a low level of significance. The accuracy of the 1.1-gram figure is only important in obtaining a product that effectively removes stubborn particles from dishes. Otherwise it poses no health or environmental problems unless the amount is quite different from the amount claimed in $H_0$. Thus an observed $\bar{x}$ quite different from 1.1 would be needed to reject $H_0$ in favor of $H_a$, and produce the accompanying loss of the entire production run of the product. However, $\bar{x}$ close to either $\bar{x}_c$ might indicate that a careful examination should be made of the mechanism that blends the ingredients.

## Two types of possible errors in any hypothesis test

Whenever a hypothesis $H_0: \mu = \mu_0$ or $H_0: P = P_0$ is tested and an $\bar{x}$ or $\hat{p}$ is obtained that falls in the region of rejection, then the decision is made to reject $H_0$ in favor of $H_a$. However, if an $\bar{x}$ or $\hat{p}$ is obtained that does not fall in the region of rejection, then the decision is made not to reject $H_0$. In either case, there is a possibility that the decision made is wrong. We call the two possible errors Type I and Type II errors.

## DEFINITION 6-5 Type I and Type II Errors

A **Type I error** is made when the state of affairs symbolized by the null hypothesis is true, but the decision is made to reject $H_0$ in favor of $H_a$.

A **Type II error** is made when the state of affairs symbolized by the alternate hypothesis is true, but the decision is made to not reject $H_0$.

There are four possibilities regarding the correctness of the decision in any hypothesis test. The following grid identifies the four possibilities. Two possibilities yield correct decisions. One possibility results in a Type I error and one results in a Type II error.

**Four Possibilities for the Decision in a Hypothesis Test**

| | | True State of Affairs | |
|---|---|---|---|
| | | $H_0$ Is True | $H_a$ is True |
| The Decision is Made to: | Reject $H_0$ | Type I error | Correct decision |
| | Not reject $H_0$ | Correct decision | Type II error |

An analogy that is frequently helpful in understanding the four possibilities in a hypothesis test is the situation of a defendant on trial before a jury. In any such trial, the defendant is either guilty or not guilty. The jury examines the evidence presented during the trial. Then based on the collective opinions of the jurors regarding the evidence, the jury makes one of the following decisions:

"We the jury find the defendant guilty."

"We the jury find the defendant not guilty."

Statements (1) through (4) suggest the parallels that exist between a hypothesis test and a jury trial.

| *For a Defendant on Trial* | *For a Hypothesis Test* |
|---|---|
| (*Top left box*)<br>1. The defendant is not guilty, but the jury "finds the defendant guilty." (An innocent person is convicted.) | (*Top left box*)<br>1.* The actual value of the parameter is described by $H_0$, but $H_0$ is rejected in favor of $H_a$ (Type I error). |
| (*Bottom left box*)<br>2. The defendant is not guilty and the jury "finds the defendant not guilty." (An innocent person is not convicted.) | (*Bottom left box*)<br>2.* The actual value of the parameter is described by $H_0$ and $H_0$ is not rejected (correct decision). |
| (*Top right box*)<br>3. The defendant is guilty and the jury "finds the defendant guilty." (A guilty person is found guilty.) | (*Top right box*)<br>3.* The actual value of the parameter is described by $H_a$ and $H_0$ is rejected (correct decision). |
| (*Bottom right box*)<br>4. The defendant is guilty and the jury "finds the defendant not guilty." (A guilty person is found not guilty.) | (*Bottom right box*)<br>4.* The actual value of the parameter is described by $H_a$ and $H_0$ is not rejected (Type II error). |

▶ **Example 1**   Kelly Bell is an employee of Gold's Fitness Gym. Management claims that members spend an average of 50 minutes at each workout session. Kelly feels the figure is overstated. She intends to collect the times spent on 36 randomly selected workouts over the next 3 weeks. She will use these data to test $H_0$: $\mu = 50$ versus $H_a$: $\mu < 50$. In terms of this problem:

**a.** State the conditions under which Kelly would make a Type I error.
**b.** State the conditions under which Kelly would make a Type II error.

**Solution**   **a.** A Type I error is made when $H_0$ is true and $H_0$ is rejected.
     If the 50-minute average is correct, and Kelly obtains $\bar{x} < \bar{x}_c$ causing her to reject $\mu = 50$, then she will make a Type I error.
**b.** A Type II error is made when $H_0$ is not true and $H_0$ is not rejected.
     If the 50-minute average is not correct, and Kelly obtains $\bar{x} > \bar{x}_c$ causing her to not reject $\mu = 50$, then she will make a Type II error.   ◀

## $\alpha = P(\text{Type I error})$

It is standard practice to minimize, or guard against, the possibility of making Type I errors in hypothesis-testing situations. This is done by choosing critical values in such a way that the chance is small of rejecting $H_0$ when $H_0$ is true.

---

### The Probability of a Type I Error

The probability of making a Type I error is symbolized by $\alpha$ (read "alpha"). That is:

$$\alpha = P(\text{Type I error})$$

---

In Section 6-1, the level of significance for a hypothesis test was defined as "the probability that a sampling distribution of $\bar{X}$ or $\hat{P}$ assigns to the rejection region if $\mu$ or $P$ has the value stated in $H_0$." Thus level of significance, $\alpha$, and $P$(Type I error) are three different names for the same quantity.

▶ **Example 2**   The manufacturers of a famous brand of blue jeans claim that "a majority of college students prefer ours to any others." Dorothy Duerr decides to test the validity of this claim so far as the students at a large university in the Midwest are concerned. She uses a student directory to randomly sample 278 names. She will contact each student and obtain the name of the brand of blue jeans each one prefers. The sample will yield $\hat{p}$, an observed value of $\hat{P}$, the proportion of students in the sample that prefer this brand of blue jeans. Dorothy intends to reject the manufacturer's claim if she obtains $\hat{p}$ less than 0.45. Find $\alpha$ for the test she is planning to use.

Solution    **Discussion.** With $P_0 = 0.50$ and $n = 278$, the probability distribution of $\hat{P}$ is approximately normal. See Figure 6-11, which shows the probability distribution of $\hat{P}$ if $H_0$ is true.

$$H_0: \quad P = 0.50$$
$$H_a: \quad P < 0.50$$

Because $\hat{p}_c = 0.45$:

$$P(\text{Type I error}) = P(\hat{P} < 0.45)$$

With $P_0 = 0.50$, $1 - P_0 = 0.50$, $\hat{p}_c = 0.45$, and $n = 278$:

$$z = \frac{0.45 - 0.50}{\sqrt{\dfrac{(0.50)(0.50)}{278}}} \qquad z = \frac{\hat{p}_c - P_0}{\sqrt{\dfrac{P_0(1 - P_0)}{n}}}$$

$$= -1.66733 \ldots \qquad \text{The unrounded value}$$

$$= -1.67 \qquad \text{Rounded to two decimal places}$$

$$P(\hat{P} < 0.45) = P(Z < -1.67)$$

$$\approx 0.0475 \qquad \text{An approximate 5\% level of significance} \qquad \blacktriangleleft$$

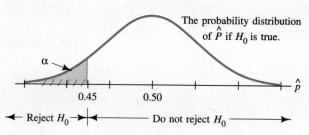

**Figure 6-11.**   $\alpha = P(\hat{P} < 0.45)$.

▶ **Example 3**    The Nevada Alfalfa Producers Association annually sells about 600,000 tons of alfalfa hay. The Association claims that its machines average 100 pounds of hay per bale with a standard deviation of 6 pounds per bale for the distribution of weights. Castle Rock Ranch intends to buy 80 thousand tons of hay from the Association. To test the claim that $\mu = 100.0$ pounds, the ranch foreman will sample and weigh 40 bales. He intends to reject $H_0$ if he gets $\bar{x} < 98.5$ or $\bar{x} > 101.5$. Compute the probability of a Type I error for this test.

Solution    **Discussion.** Because $n = 40$, the probability distribution of $\bar{x}$ is approximately normal. See Figure 6-12, which shows the probability distribution of $\bar{X}$ if $H_0$ is true.

$$H_0: \quad \mu = 100.0$$
$$H_a: \quad \mu \neq 100.0$$

With $\mu_0 = 100.0$, $\sigma = 6.0$, $n = 40$, lower $\bar{x}_c = 98.5$, and upper $\bar{x}_c = 101.5$:

$$\text{lower } z_c = \frac{98.5 - 100.0}{\dfrac{6.0}{\sqrt{40}}} \qquad\qquad \text{upper } z_c = \frac{101.5 - 100.0}{\dfrac{6.0}{\sqrt{40}}}$$

$$= -1.58, \text{ to two decimal places} \qquad\qquad = 1.58, \text{ to two decimal places}$$

$\alpha = P(Z < -1.58 \text{ or } Z > 1.58)$

$\quad = 0.0571 + 0.0571$

$\quad = 0.1142 \qquad\qquad\qquad\qquad$ An approximate 11% level of significance ◄

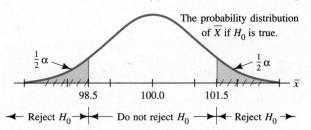

**Figure 6-12.**   $\alpha = P(\bar{X} < 98.5 \text{ or } \bar{X} > 101.5)$.

## $\beta = P$(Type II error)

If the only consideration when setting up a hypothesis test were to protect an investigator from making a Type I error, it would be a relatively simple matter to choose critical values so that $\alpha$ is as small as required. Unfortunately things are not quite that simple. Recall that there are two circumstances that can result in making an error. The second kind of error (called a Type II error) is made when $H_0$ is not true and the decision is made to not reject $H_0$.

### The Probability of a Type II Error

The probability of failing to reject $H_0$ when $H_0$ is not true is symbolized by $\beta$ (read "beta"). Thus:

$$\beta = P(\text{Type II error})$$

If $H_0$ is not true, then $\mu_0$ is not the mean of the population under investigation. Therefore there must be another value, different from $\mu_0$, that is the actual value. If $H_0$ is not true, then the value of $\beta$ depends on which of the many possible means described by $H_a$ is actually correct.

There exists a different $\beta$ for every possible value of $\mu$ described by $H_a$. In the Set B exercises in this section, four problems are given in which $\beta$ is computed supposing different values for $\mu$. These examples illustrate the fact that $\beta$ decreases as the difference between $\mu_0$ and $\mu$ increases. That is, if the true value of a parameter is quite different from the hypothesized value, then one is more likely to detect that the hypothesized value is wrong. (Similar statements could be made regarding a test on $P$ in which $H_0$ is not true.)

▶ **Example 4**  A research assistant for the Kline Hybrid Seed Corporation is currently working on a project to develop a variety of wheat that will average 38 inches in height when fully grown. She intends to sample 75 stalks and test:

$$H_0: \quad \mu = 38.0$$
$$H_a: \quad \mu < 38.0$$

If the research assistant uses a 5 percent level of significance, compute $\beta$, if the actual value of $\mu = 37.7$. Assume $\sigma = 1.4$ inches for the distribution of stalk heights for this variety of wheat.

**Solution**  **Discussion.** First use 38.0 and the level of significance to compute $\bar{x}_c$. Then use 37.7 and $\bar{x}_c$ to compute the probability that $\bar{X}$ will fall in the region labeled "Do not reject $H_0$." Be sure to use 37.7 as the mean of the sampling distribution of $\bar{X}$. With $\mu_0 = 38.0$, $z_c = -1.65$, $\sigma = 1.4$, and $n = 75$:

$$\bar{x}_c = 38.0 - 1.65\left(\frac{1.4}{\sqrt{75}}\right)$$

$$= 37.733\ldots, \text{ to three decimal places}$$

A probability histogram for $\bar{X}$ is shown in Figure 6-13. The shaded area is $\beta$. With $\mu = 37.7$, $\sigma = 1.4$, $\bar{x}_c = 37.733$, and $n = 75$:

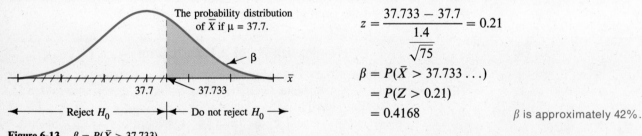

$$z = \frac{37.733 - 37.7}{\dfrac{1.4}{\sqrt{75}}} = 0.21$$

$$\beta = P(\bar{X} > 37.733\ldots)$$
$$= P(Z > 0.21)$$
$$= 0.4168 \qquad \beta \text{ is approximately 42\%.}$$

**Figure 6-13.**  $\beta = P(\bar{X} > 37.733)$.

Thus the probability is about 42 percent that the research assistant will get a sample mean larger than 37.733 inches and therefore fail to reject the false $H_0$.

◀

▶ **Example 5**  A quality control worker for Union Manufacturing Corporation is going to check a large production run of machine parts. If the lot has 5 percent or fewer defectives, then the lot is of acceptable quality. A 1 percent level of significance is used to test:

$$H_0: \quad P = 0.05$$
$$H_a: \quad P > 0.05$$

He will inspect 100 randomly sampled machine parts for defects. If the actual proportion of defectives for this production run is 0.120, calculate $\beta$.

**Solution**   With $P_0 = 0.05$, $1 - P_0 = 0.95$, $z_c = 2.33$, and $n = 100$:

$$\hat{p}_c = 0.05 + 2.33 \sqrt{\frac{(0.05)(0.95)}{100}}$$

$$= 0.101, \text{ to three decimal places}$$

Because $P$ is actually 0.120, a probability histogram for $\hat{P}$ is shown in Figure 6-14. The shaded area is $\beta$.

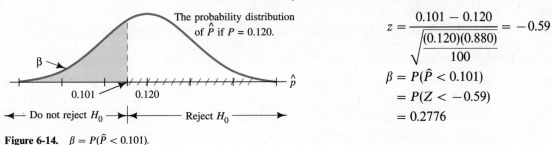

The probability distribution of $\hat{P}$ if $P = 0.120$.

$$z = \frac{0.101 - 0.120}{\sqrt{\dfrac{(0.120)(0.880)}{100}}} = -0.59$$

$$\beta = P(\hat{P} < 0.101)$$
$$= P(Z < -0.59)$$
$$= 0.2776$$

**Figure 6-14.**   $\beta = P(\hat{P} < 0.101)$.

The probability is about 28 percent that the quality control worker will fail to detect that the proportion of defectives for this production run is actually larger than 0.05.   ◄

## Some general comments regarding Type I and Type II error probabilities

When $H_0$ and $H_a$, and the test statistic are specified for a hypothesis test, $n$, $\alpha$, and $\beta$ are related so that only two of them can be chosen independently. In practice, $n$ is frequently dictated by financial constraints and $\alpha$ is then chosen to be small, to protect against the possibility of a Type I error.

On the other hand, it would seem ideal to minimize both $\alpha$ and $\beta$. However, a decrease in either $\alpha$ or $\beta$ is accompanied by an increase in the other when $n$ remains constant. It is possible to achieve both small $\alpha$ and $\beta$ only by a suitably large choice of $n$. (This follows from the fact that the standard deviations of the sampling distributions of $\bar{X}$ and $\hat{P}$ decrease as $n$ increases.)

There are systematic methods of choosing $n$ to meet both small $\alpha$ and $\beta$ requirements for many testing situations. However, these methods are slightly beyond the scope of this text.

# Exercises 6-2    Set A

*In Exercises 1–8, state:*
*a. The conditions under which a Type I error is made.*
*b. The conditions under which a Type II error is made.*

1. The management of Laguna Lake Homesites claims the average commute time to the inner city is 25 minutes. Test $H_0$: $\mu = 25$ minutes versus $H_a$: $\mu > 25$ minutes.

2. The average fill level for a large production run of rubber cement bottles is reported to be 118 milliliters (ml) per bottle. Test $H_0$: $\mu = 118$ ml versus $H_a$: $\mu < 118$ ml.

3. The mean age of the faculty at the nine campuses of a state university system is reported to be 48.6 years. Test $H_0$: $\mu = 48.6$ years versus $H_a$: $\mu \neq 48.6$ years.

4. The gas mileage for the new Model T car manufactured by Mega Manufacturing is reported to be 58.5 miles per gallon (mpg). Test $H_0$: $\mu = 58.5$ mpg versus $H_a$: $\mu \neq 58.5$ mpg.

5. The proportion of voters in a major metropolitan area that favors an ordinance prohibiting local lawyers from using TV advertisements to promote personal injury lawsuits is reported to be 0.54. Test $H_0$: $P = 0.54$ versus $H_a$: $P > 0.54$.

6. The proportion of defective 15.80-centimeter (cm) roller bearings in a large production run at Tittle Roller Bearing Manufacturers is reported to be 0.06. Test $H_0$: $P = 0.06$ versus $H_a$: $P > 0.06$.

7. The proportion of Florida citrus production acres affected by a killer frost is reported to be 0.67. Test $H_0$: $P = 0.67$ versus $H_a$: $P \neq 0.67$.

8. In a large metropolitan school district the proportion of high school students that regularly uses some form of illegal drug is reported to be 0.23. Test $H_0$: $P = 0.23$ versus $H_a$: $P < 0.23$.

*In Exercises 9–12, sketch an idealized probability histogram for $\bar{X}$ showing the decision regions.*
*For each exercise, suppose $\sigma = 6.0$ and the null hypothesis is $H_0$: $\mu = 20.0$.*

9. For $H_a$: $\mu > 20.0$ and $n = 36$:
   a. Determine $\alpha$, if $\bar{x}_c = 22.0$.
   b. Determine $\bar{x}_c$, if $\alpha = 0.05$.

10. For $H_a$: $\mu \neq 20.0$ and $n = 64$:
    a. Determine $\alpha$, if lower and upper $\bar{x}_c$ are 18.5 and 21.5, respectively.
    b. Determine upper and lower $\bar{x}_c$, if $\alpha = 0.01$.

11. For $H_a$: $\mu < 20.0$ and $n = 100$:
    a. Determine $\alpha$, if $\bar{x}_c = 19.0$.
    b. Determine $\bar{x}_c$, if $\alpha = 0.01$.

12. For $H_a$: $\mu \neq 20.0$ and $n = 144$:
    a. Determine $\alpha$, if lower and upper $\bar{x}_c$ are 19.2 and 20.8, respectively.
    b. Determine upper and lower $\bar{x}_c$, if $\alpha = 0.05$.

*In Exercises 13–16, sketch idealized probability histograms for $\hat{P}$ showing the decision regions. For each exercise, use the null hypothesis, $H_0$: $P = 0.30$.*

13. For $H_a$: $P > 0.30$ and $n = 33$:
    a. Determine $\alpha$, if $\hat{p}_c = 0.40$.
    b. Determine $\hat{p}_c$, if $\alpha = 0.05$.

14. For $H_a$: $P \neq 0.30$ and $n = 59$:
    a. Determine $\alpha$, if lower and upper $\hat{p}_c$ are 0.22 and 0.38, respectively.
    b. Determine upper and lower $\hat{p}_c$, if $\alpha = 0.05$.

15. For $H_a$: $P < 0.30$ and $n = 85$:
    a. Determine $\alpha$, if $\hat{p}_c = 0.23$.
    b. Determine $\hat{p}_c$, if $\alpha = 0.01$.

16. For $H_a$: $P \neq 0.30$ and $n = 131$:
    a. Determine $\alpha$, if lower and upper $\hat{p}_c$ are 0.24 and 0.36, respectively.
    b. Determine upper and lower $\hat{p}_c$, if $\alpha = 0.01$.

17. For a hypothesis test of $H_0$: $\mu = 50.0$ versus $H_a$: $\mu > 50.0$, suppose $\sigma = 9.0$.
    a. Calculate $\alpha$, if $\bar{x}_c = 52.0$ and $n = 36$.
    b. Calculate $\alpha$, if $\bar{x}_c = 52.0$ and $n = 81$.
    c. Calculate $\alpha$, if $\bar{x}_c = 52.0$ and $n = 144$.
    d. What happens to the size of $\alpha$ for a fixed $\bar{x}_c$ as the sample size increases?

18. For a hypothesis test of $H_0$: $\mu = 120.0$ versus $H_a$: $\mu < 120.0$, suppose $\sigma = 18.0$.
    a. Calculate $\alpha$, if $\bar{x}_c = 117.0$ and $n = 36$.
    b. Calculate $\alpha$, if $\bar{x}_c = 117.0$ and $n = 81$.
    c. Calculate $\alpha$, if $\bar{x}_c = 117.0$ and $n = 144$.
    d. What happens to the size of $\alpha$ for a fixed $\bar{x}_c$ as the sample size increases?

*In Exercises 19 and 20, sketch idealized histograms for the probability distribution of $\bar{X}$ centered on $\mu$. For each exercise, suppose $\sigma = 6.0$ and the null hypothesis is $H_0$: $\mu = 20.0$. Use $\alpha = 0.05$.*

19. For $H_a$: $\mu > 20.0$ and $n = 36$:
    a. Determine $\beta$, if $\mu = 21.0$.
    b. Determine $\beta$, if $\mu = 22.0$.
    c. What caused $\beta$ to decrease in part (b) as compared to part (a)?

20. For $H_a$: $\mu < 20.0$ and $n = 64$:
    a. Determine $\beta$, if $\mu = 19.5$.
    b. Determine $\beta$, if $\mu = 18.5$.
    c. What caused $\beta$ to decrease in part (b) as compared to part (a)?

*In Exercises 21 and 22, sketch idealized histograms for the probability distribution of $\hat{P}$ centered on P. For each exercise, suppose the null hypothesis is $H_0$: $P = 0.30$. Use $\alpha = 0.05$.*

21. For $H_a$: $P < 0.30$ and $n = 79$:
    a. Determine $\beta$, if $P = 0.25$.
    b. Determine $\beta$, if $P = 0.20$.
    c. What caused $\beta$ to decrease in part (b) as compared to part (a)?

**22.** For $H_a$: $P > 0.30$ and $n = 136$:
   a. Determine $\beta$, if $P = 0.32$.
   b. Determine $\beta$, if $P = 0.35$.
   c. What caused $\beta$ to decrease in part (b) as compared to part (a)?

**23.** For a hypothesis test with $H_0$: $\mu = 50.0$ and $H_a$: $\mu < 50.0$ and $\alpha = 0.05$, using $\sigma = 9.0$:
   a. Calculate $\beta$, if $\mu = 48.0$ and $n = 36$.
   b. Calculate $\beta$, if $\mu = 48.0$ and $n = 81$.
   c. What caused $\beta$ to decrease in part (b) as compared to part (a)?

**24.** For a hypothesis test with $H_0$: $P = 0.75$ and $H_a$: $P > 0.75$ and $\alpha = 0.05$:
   a. Calculate $\beta$, if $P = 0.78$ and $n = 69$.
   b. Calculate $\beta$, if $P = 0.78$ and $n = 191$.
   c. What caused $\beta$ to decrease in part (b) as compared to part (a)?

## Exercises 6-2    Set B

*In Exercises 1 and 2, answer all parts regarding $\beta$.*

**1.** For testing the hypothesis, $H_0$: $\mu = 50.0$ versus $H_a$: $\mu < 50.0$, use $\sigma = 12.0$, $n = 36$, and $\bar{x}_c = 46.7$.
   a. Compute $\beta$, if $\mu = 49.0$.
   b. Compute $\beta$, if $\mu = 48.0$.
   c. Compute $\beta$, if $\mu = 47.0$.
   d. Compute $\beta$, if $\mu = 46.0$.
   e. Plot the four values for $\beta$ on axes like those below and connect the points with a smooth curve.

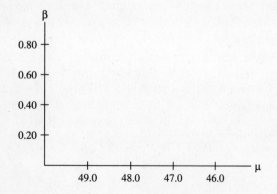

**2.** For testing the hypothesis $H_0$: $P = 0.60$ versus $H_a$: $P > 0.60$, use $n = 267$, and $\hat{p}_c = 0.67$.
   a. Compute $\beta$, if $P = 0.63$.
   b. Compute $\beta$, if $P = 0.66$.
   c. Compute $\beta$, if $P = 0.69$.
   d. Compute $\beta$, if $P = 0.72$.
   e. Plot the four values for $\beta$ on axes like those below and connect the points with a smooth curve.

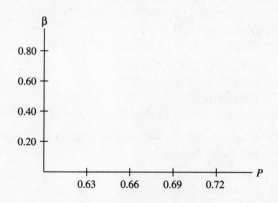

*In Exercises 3 and 4, use Data Set I in Appendix B.*

3. Use random number to obtain a sample of the ages of 36 individuals.
   a. Compute $\bar{x}$ for the sample.
   b. Use $\sigma = 20.2$ years and a 5 percent level of significance to test $H_0: \mu = 56.1$ versus $H_a: \mu \neq 56.1$.
   c. The stated value of $\mu$ is in fact correct. Did your sample mean enable you to make a correct decision or did you commit a Type I error?

4. Use random numbers to obtain a sample of 75 individuals.
   a. Compute $\hat{p}$, where $\hat{p}$ is the proportion of smokers in the sample.
   b. Use a 1 percent level of significance to test $H_0: P = 0.53$ versus $H_a: P \neq 0.53$.
   c. The stated value of $P$ is in fact correct. Did your sample proportion enable you to make a correct decision or did you commit a Type I error?

*In Exercises 5 and 6, use Data Set II in Appendix B.*

5. Use random numbers to obtain a sample of the "high" prices of 40 stocks.
   a. Compute $\bar{x}$ for the sample.
   b. Use $\sigma = \$29.90$ and a 1 percent level of significance and test $H_0: \mu = \$36.48$ versus $H_a: \mu \neq \$36.48$.
   c. The stated value of $\mu$ is in fact correct. Did your $\bar{x}$ enable you to make a correct decision or did you commit a Type I error?

6. Use random numbers to obtain a sample of 50 stock prices. Let $\hat{p}$ be the proportion of stock prices that increased ($+$).
   a. Compute $\hat{p}$ for the sample.
   b. Use $\alpha = 0.01$ and test $H_0: P = 0.468$ versus $H_a: P \neq 0.468$.
   c. The stated value of $P$ is in fact correct. Did your $\hat{p}$ enable you to make a correct decision or did you commit a Type I error?

*For Exercises 7–10, write a paragraph defending your position that it would be better to run a large chance of making a Type I or Type II (you decide) error in the given circumstance.*

7. An automotive engineering team is testing whether a new brake design will stop the new model FA92ST car within the required 75 feet. If $H_0$ is rejected, the team will examine again their brake design.

8. A state water resource employee is testing whether a new filtration system effectively reduces contamination in reclaimed sewage water that will be used for irrigation on local farms. If $H_0$ is rejected, the filtration system will be studied for ways to improve the filtering process.

9. A dairy employee is checking the butterfat content in 1.7 million gallons of recently bottled milk. The advertised butterfat content in the milk is not less than 2%. If $H_0$ is rejected, the 3.4 million cartons will have to be opened, reprocessed, and re-bottled.

10. An electronics corporation employee is checking the mean time until failure for a production run of 5.6 million 100-watt bulbs. The reported mean life of these bulbs is at least 500 hours, and the entire production run will have to be sold at discount if $H_0$ is rejected.

**Key Topics**

# 6-3 Tests Concerning a Single Mean

*1. The test statistic for large n tests on μ*

*2. A six-step format for hypothesis tests*

*3. The test statistic for small n tests on μ*

*4. Observed levels of significance (p-values)*

# STATISTICS IN ACTION

Recall the gasoline additive test conducted by Don Hadd in the Statistics in Action portion of Section 6-1. This test was performed by Don to show that the additive he developed increased the operating time of the test engine. The following six-step format was used in the report he submitted to the executive board summarizing the results of his test. The items specified in the report will be studied in this section.

**Findings of The K59MH3 Additive Experiment**
Submitted by Don Hadd

**Step 1** **State $H_0$ and $H_a$.**

$H_0$:   $=8.35$ minutes     The additive does not increase the mean operating time.

$H_a$:   $>8.35$ minutes     The additive increases the mean operating time.

**Step 2** **State the level of significance of the test.**

For this test, a 1 percent level of significance was used. There was only a 0.01 probability of rejecting $H_0$, if in fact $H_0$ was true.

**Step 3** **Identify the test statistic.**

The test statistic for this test was the one commonly used for large $n$ tests on $\mu$:

$$Z = \frac{\bar{X} - \mu_0}{\frac{S}{\sqrt{n}}}$$     $Z$ is a standard normal random variable, if $H_0$ is true.

**Step 4** **State the decision rule in terms of the rejection region.**

For a 1 percent level of significance and a directional test, $z_c$ is 2.33 (see Figure 6-15), if $z > z_c \, (=2.33)$, then reject $H_0$. Otherwise, do not reject $H_0$.

The approximate probability distribution

of $Z = \frac{\bar{X} - \mu_0}{\frac{S}{\sqrt{n}}}$, if $H_0$ is true.

**Figure 6-15.**   $\alpha = P(Z > 2.33)$.

**Step 5**  **Calculate the observed value of the test statistic.**
For the $n = 35$ data points in Table 6-1,

$$s = \sqrt{\frac{35(2532.2196) - (297.5)^2}{35(34)}}$$

$$= 0.3194 \ldots$$

$$\bar{x} = \frac{297.5}{35} = 8.50$$

$$z = \frac{8.50 - 8.35}{\dfrac{0.3194 \ldots}{\sqrt{35}}}$$

$$= 2.7779 \ldots$$

$$\approx 2.78, \text{ to two decimal places}$$

**Step 6**  **Make a decision about the parameter tested that is consistent with steps 4 and 5.**

Because $2.78 > 2.33$, the mean of the running times of the 35 tests falls in the rejection region shown in Figure 6-15. Thus there is sufficient evidence to conclude that the mean running time of the test engine is increased when K59MH3 is added to the gasoline. This test suggests that the additive will correspondingly increase the miles-per-gallon ratings of cars and trucks using gasoline enriched with the additive.

## The test statistic for large $n$ tests on $\mu$

In Section 6-1, we used an $\bar{x}_c$ for directional tests and $\bar{x}_c$s for non-directional tests as boundaries for the decision regions in tests on $\mu$. The computation of these values depended on knowing $\sigma$, the standard deviation of the population under investigation. However, in applied problems, it is rare to know $\sigma$ when one is testing $H_0$: $\mu = \mu_0$.

To get around this dilemma, we can use a slight extension of the central limit theorem that states that for large $n$ the probability distributions of both

$$Z = \frac{\bar{X} - \mu}{\dfrac{\sigma}{\sqrt{n}}} \tag{1}$$

and

$$Z = \frac{\bar{X} - \mu}{\dfrac{S}{\sqrt{n}}} \tag{2}$$

are approximately standard normal. Notice in the Statistics in Action section that Don used equation (2) with $\mu = \mu_0$ to define the random variable that was used to come to his decision about $H_0$.

### The Test Statistic for Large $n$ Tests on $\mu$

If $\bar{X}$ is the mean and $S$ is the standard deviation of a large sample, then the **test statistic** for testing $H_0$: $\mu = \mu_0$ is:

$$Z = \frac{\bar{X} - \mu_0}{\dfrac{S}{\sqrt{n}}}$$

$\bar{X}$ is the mean of the sample.

$\mu_0$ is the hypothesized value of $\mu$.

$S$ is the sample standard deviation.

$n$ is the sample size.

In step 5 of the report in the Statistics in Action section, the mean of $z = 8.50$ and standard deviation $s = 0.319$ that Don got for his 35 tests converted to an observed value of 2.78 for the test statistic.

▶ **Example 1**    In a test of $H_0$: $\mu = 320$, a sample of 42 data points yielded $\bar{x} = 316$, and $s = 15$. To two decimal places, find the corresponding value of the test statistic.

**Solution**    With $\bar{x} = 316$, $\mu_0 = 320$, $s = 15$, and $n = 42$:

$$z = \frac{316 - 320}{\dfrac{15}{\sqrt{42}}}$$

$$= \frac{-4}{2.31\ldots} \qquad \text{Do not round the denominator.}$$

$$\approx -1.73 \qquad \text{To two decimal places.} \qquad ◀$$

## A six-step format for hypothesis tests

Throughout this section, we will assume that the values of measurements

$$X_1, \quad X_2, \quad X_3, \quad X_4, \quad X_5, \quad \ldots, \quad X_n$$

are reasonably described as independent random variables with a common distribution with mean $\mu$ and standard deviation $\sigma$. As discussed in Chapter 5, such a probability model is appropriate if the values are obtained as a random sample from a large (compared with $n$) population. All the inferences made using the material of this section depend for their validity on the appropriateness of this probability model.

In the Statistics in Action part of this section, Don Hadd used a six-step format to report the results of his test. The following is a general statement of this format. In Examples 2 and 3, the format is applied to two large $n$ tests of $H_0$: $\mu = \mu_0$.

> ## A six-step format for summarizing hypothesis tests

**Step 1** State $H_0$ and $H_a$.

**Step 2** State the level of significance for the test.

**Step 3** State the appropriate formula for the test statistic.

**Step 4** State the decision rule in terms of the rejection region.

**Step 5** Calculate the observed value of the test statistic.

**Step 6** Make a decision about the parameter(s) being tested consistent with steps 4 and 5. Write a concluding statement based on the context of the problem.

In step 2, the level of significance (that is, $\alpha$) to be used is identified. The discussion in Section 6-2 was meant to make us aware that the most rational choices of $\alpha$ involve consideration of the $\beta$s they imply. But because of the mathematical complexity of finding $\beta$s in all but the simplest cases, we can have little more to say about the choice of $\alpha$ in this text. Let us simply agree that each $\alpha$ used here was somehow selected with primary concern for protecting against Type I errors.

▶ **Example 2**

In a recent report published by the National Association of Automotive Mechanics (NAAM), the average cost of replacing the braking devices (shoes or disks) on an automobile is $127.00. A Consumer Alert Group in the Midwest felt the figure published by NAAM was too low. They sampled 40 automotive repair facilities in their region, hoping to obtain sufficient evidence to establish that the real mean cost exceeds $127.00. Use $\alpha = 5\%$ to conduct a hypothesis test, given that the sample yielded $\bar{x} = \$132.09$ with $s = \$13.58$.

**Solution**

**Step 1** $H_0$: $\mu = \$127.00$
$H_a$: $\mu > \$127.00$     A directional test

**Step 2** As indicated, $\alpha = 5\%$     $P$(Type I error) = 0.05

**Step 3** The test statistic will be:

$$Z = \frac{\bar{X} - \$127.00}{\dfrac{S}{\sqrt{n}}}$$     $\mu_0 = 127.00$ and $n$ is greater than 30.

**Step 4** If $z > z_c$ ($= 1.65$), then reject $H_0$.     With $\alpha = 0.05$ and a one-tail test, $z_c = 1.65$.

**Step 5** With $\bar{x} = \$132.09$, $s = \$13.58$, and $n = 40$:

$$z = \frac{\$132.09 - \$127.00}{\dfrac{\$13.58}{\sqrt{40}}}$$     Do not round any numbers until the final calculation.

$$\approx 2.37, \text{ to two decimal places}$$

**Step 6** Because $2.37 > 1.65$, reject $H_0$. That is, there is sufficient evidence, with $\alpha = 0.05$, to conclude that the average cost of a brake job on an automobile is higher than the figure published by NAAM.

Notice that 2.37 falls in the rejection region, as shown in Figure 6-16.   ◀

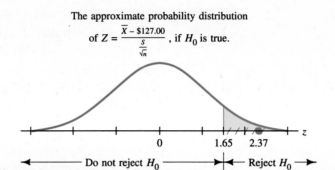

The approximate probability distribution

of $Z = \dfrac{\overline{X} - \$127.00}{\frac{S}{\sqrt{n}}}$ , if $H_0$ is true.

Do not reject $H_0$          Reject $H_0$ ⟶

**Figure 6-16.**   If $z > z_c$ ($=1.65$), then reject $H_0$.

▶ **Example 3**      Jim Camp is a quality control inspector for the Sanford Ink Company. He recently sampled 32 bottles of a large production run of bottles of rubber cement. The mean volume per bottle is reported to be 120.0 milliliters (ml). Jim is concerned with the possibility that the contents of the bottles are, on the average, too low or too high. For the sampled bottles, $\overline{x} = 121.2$ ml and $s = 4.3$ ml. Is the evidence sufficient for Jim to reject the 120.0 figure for this production run? Use $\alpha = 0.05$.

**Solution**      **Step 1** $H_0$:   $\mu = 120.0$
                             $H_a$:   $\mu \neq 120.0$        A non-directional test

**Step 2** As indicated, $\alpha = 5\%$      $P$(Type I error) $= 0.05$

**Step 3** The test statistic will be:

$$Z = \frac{\overline{X} - 120.0}{\dfrac{S}{\sqrt{n}}}$$      $\mu_0 = 120.0$ and $n$ is greater than 30.

**Step 4** If $z <$ lower $z_c$ ($= -1.96$), or $z >$ upper $z_c$ ($=1.96$), then reject $H_0$.

**Step 5** With $\overline{x} = 121.2$ ml, $s = 4.3$ ml, and $n = 32$:

$$z = \frac{121.2 \text{ ml} - 120.0 \text{ ml}}{\dfrac{4.3 \text{ ml}}{\sqrt{32}}}$$      Do not round any numbers until the final calculation.

$\approx 1.58$, to two decimal places

**Step 6** Because $1.58 > -1.96$ and $1.58 < 1.96$, do not reject $H_0$. That is, there is not sufficient evidence, with $\alpha = 0.05$, to conclude that the mean volume of the contents of rubber cement in this production run is different from 120.0 ml.

Notice that 1.58 falls in the region labeled "Do not reject" in Figure 6-17. ◀

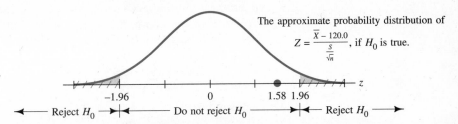

**Figure 6-17.** If $z < -1.96$ or $z > 1.96$, then reject $H_0$.

## The test statistic for small *n* tests on $\mu$

In Chapter 5, we constructed confidence interval estimates based on both large samples ($n \geq 30$) and small samples ($n < 30$). Recall that to construct a confidence interval based on a small sample, it was necessary to:

1. Assume that the relative frequency distribution of the sampled population is reasonably bell shaped.
2. Use a $t$-distribution with $n - 1$ degrees of freedom.

To use a sample of size $n < 30$ to test $H_0: \mu = \mu_0$, we must do the same. Specifically we must be willing to assume a "relatively bell-shaped" frequency distribution and use a $t$-distribution with $n - 1$ degrees of freedom.

### The Test Statistic for Small *n* Tests on $\mu$

If $\bar{X}$ is the mean and $S$ is the standard deviation of a small sample taken from a population with an approximately bell-shaped distribution, then the test statistic for testing $H_0: \mu = \mu_0$ is:

$$T = \frac{\bar{X} - \mu_0}{\dfrac{S}{\sqrt{n}}}$$

→ $\bar{X}$ is the mean of the sample,

→ $\mu_0$ is the mean in $H_0$, and

→ $S$ is the sample standard deviation.

If $H_0$ is true $T$ has a $t$-distribution with $n - 1$ degrees of freedom.

▶ **Example 4**    A new brand of oatmeal flake cereal claims that a 1.5-ounce serving of the cereal has 140 calories. The company that makes the cereal says that records indicate the distribution of caloric contents of 1.5-ounce portions is bell shaped. A member of the Truth-in-Advertising Commission hired a local laboratory to analyze the cereal to test the 140-calorie claim. The staff of the laboratory analyzed 12 different servings of 1.5 ounces each. The results yielded $\bar{x} = 153$ calories with $s = 21$ calories. Can the company's claim of 140 calories be rejected (as too low or too high) based on the data collected? Use $\alpha = 0.01$.

**Solution**    **Step 1**  $H_0$:   $\mu = 140$ calories
             $H_a$:   $\mu \neq 140$ calories          A non-directional test

**Step 2**  As indicated, $\alpha = 0.01$.      $P(\text{Type I error}) = 0.01$

**Step 3**  The test statistic will be:

$$T = \frac{\bar{X} - 140 \text{ calories}}{\dfrac{S}{\sqrt{n}}}$$          Because $n < 30$, a $t$-distribution is used.

**Step 4**  If $t <$ lower $t_c$ ($= -3.106$)          If $n = 12$, then $df = 11$.
             or $t >$ upper $t_c$ ($= 3.106$),          Use the $t_{0.005}$ column.
             then reject $H_0$.

**Step 5**  With $\bar{x} = 153$ calories, $s = 21$ calories, and $n = 12$:

$$t = \frac{153 - 140}{\dfrac{21}{\sqrt{12}}}$$

             $\approx 2.144$, to three decimal places          Do not round any number until the final calculation.

**Step 6**  Because $2.144 > -3.106$ and $2.144 < 3.106$, do not reject $H_0$. That is, there is not sufficient evidence, with $\alpha = 0.01$, to conclude that the mean calories in a 1.5-ounce serving of this cereal is different from 140.

Notice that the observed value of $T$ in the hypothesis test in Example 4 falls in the region labeled "Do not reject," shown in Figure 6-18.    ◀

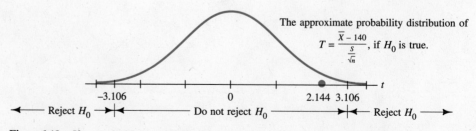

The approximate probability distribution of $T = \dfrac{\bar{X} - 140}{\frac{S}{\sqrt{n}}}$, if $H_0$ is true.

**Figure 6-18.**   If $t < -3.106$ or $t > 3.106$, then reject $H_0$.

## Observed levels of significance (*p*-values)

In Examples 2, 3, and 4, we compared $z$ and $t$ with $z_c$ and $t_c$ respectively. Another approach to hypothesis testing is to compute and report the *p*-value for the observed sample.

### DEFINITION 6-6 The *p*-Value in a Hypothesis Test

The ***p*-value** in a hypothesis test is the smallest level of significance ($\alpha$) for which the observed sample would cause rejection of $H_0$.

For example, in testing $H_0: \mu = \mu_0$ with large $n$, to find a *p*-value one first converts $\bar{x}$ from the sample to a $z$ score. Depending on the form of $H_a$, we then use Appendix Table 4 to compute one of the following probabilities:

a. If $H_a$:   $\mu < \mu_0$, compute $P(Z < z)$.
b. If $H_a$:   $\mu > \mu_0$, compute $P(Z > z)$.
$c_1$. If $H_a$:   $\mu \neq \mu_0$ and $z > 0$, compute $P(Z < -z \text{ or } Z > z)$.
$c_2$. If $H_a$:   $\mu \neq \mu_0$ and $z < 0$, compute $P(Z < z \text{ or } Z > -z)$.

The probabilities in parts ($c_1$) and ($c_2$) can both be computed as $2P(Z > |z|)$.

The probability obtained in parts (a), (b), ($c_1$), or ($c_2$) is then compared to the stated level of significance. If the probability is less than $\alpha$, then $H_0$ is rejected. If the probability is greater than $\alpha$, then $H_0$ is not rejected.

▶ **Example 5**   A study by the Health and Human Services Department suggests that patients in the United States who see doctors at their offices spend an average of 12.0 minutes with the doctor. Terry Depalo is a director of Midland Health Plan. She feels that doctors in her health plan spend more time with their patients than is indicated by the nationwide study. She therefore has her staff study the reports of 50 office calls. The sample yields $\bar{x} = 13.6$ minutes and $s = 8.2$ minutes. Use $\alpha = 0.05$ to test Terry's claim that $\mu > 12.0$ minutes for the members of Midland Health Plan, and then compute the *p*-value and discuss its meaning.

**Solution**   **Step 1**  $H_0$:   $\mu = 12.0$ minutes
                              $H_a$:   $\mu > 12.0$ minutes     A directional test
**Step 2**  As indicated, $\alpha = 0.05$     $P(\text{Type I error}) = 0.05$
**Step 3**  The test statistic will be:

$$Z = \frac{\bar{X} - 12.0 \text{ min}}{\dfrac{S}{\sqrt{n}}}$$

**Step 4**  If $z > z_c \ (= 1.65)$, then reject $H_0$.

**Step 5**   With $\bar{x} = 13.6$ minutes, $s = 8.2$ minutes, and $n = 50$:

$$z = \frac{13.6 \text{ min} - 12.0 \text{ min}}{\dfrac{8.2 \text{ min}}{\sqrt{50}}}$$

$$\approx 1.38, \text{ to two decimal places}$$

**Step 6**   Because $1.38 < 1.65$, do not reject $H_0$. That is, there is not sufficient evidence, with $\alpha = 0.05$, to conclude that Midland Health Plan members spend more than 12.0 minutes with a doctor during an office call.

The observed level of significance ($p$-value) here is $P(Z > 1.38) = 0.0838$. Because $0.0838 > 0.0500$, the $p$-value exceeds the stated $\alpha$ and $H_0$ is not rejected. If, however, a level of significance larger than 0.0838 had been selected, then $H_0$ would be rejected.   ◄

The virtue of a $p$-value computation is that one can simply report the $p$-value and allow others to use their own significance levels. To illustrate, in Example 5, $H_0$ is not rejected because $0.0838 > 0.0500$. However, if a 10 percent level of significance had been used, then $H_0$ would be rejected.

$$p\text{-value} = 0.0838 \begin{cases} \rightarrow \alpha = 10\%: 0.0838 < 0.1000; \text{ reject } H_0. \\ \rightarrow \alpha = 5\%: \ \ 0.0838 > 0.0500; \text{ do not reject } H_0. \end{cases}$$

The geometric representation of this can be seen in Figure 6-19.

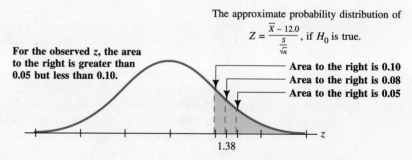

**Figure 6-19.**   Reject $H_0$ for $\alpha = 0.10$, but do not reject $H_0$ for $\alpha = 0.05$.

## Exercises 6-3    Set A

*In Exercises 1–4, for each indicated test, find the value of the test statistic to two decimal places.*

**1.** $H_0: \mu = 73.5$, $\bar{x} = 68.4$, $s = 15.7$, and $n = 40$.

**2.** $H_0: \mu = 128$, $\bar{x} = 123$, $s = 17$, and $n = 58$.

**3.** $H_0: \mu = 8.45$, $\bar{x} = 9.03$, $s = 1.94$, and $n = 38$.

**4.** $H_0: \mu = 13.38$, $\bar{x} = 14.20$, $s = 4.25$, and $n = 75$.

*In Exercises 5–12, practice the six-step hypothesis test format.*

5. $H_0$: $\mu = 50.0$    $H_a$: $\mu \neq 50.0$
Given a sample of 36 with $\bar{x} = 52.5$ and $s = 9.0$, conduct the test at the 5% level of significance.

6. $H_0$: $\mu = 32.0$    $H_a$: $\mu \neq 32.0$
Given a sample of 100 with $\bar{x} = 33.2$ and $s = 4.0$, conduct the test at the 1% level of significance.

7. $H_0$: $\mu = 128.0$    $H_a$: $\mu < 128.0$
Given a sample of 64 with $\bar{x} = 126.5$ and $s = 6.0$, conduct the test at the 5% level of significance.

8. $H_0$: $\mu = 210.0$    $H_a$: $\mu < 210.0$
Given a sample of 49 with $\bar{x} = 202.0$ and $s = 21.0$, conduct the test at the 1% level of significance.

9. $H_0$: $\mu = 340.0$    $H_a$: $\mu > 340.0$
Given a sample of 49 with $\bar{x} = 344.0$ and $s = 18.0$, use $\alpha = 0.01$.

10. $H_0$: $\mu = 12.0$    $H_a$: $\mu > 12.0$
Given a sample of 36 with $\bar{x} = 13.4$ and $s = 3.0$, use $\alpha = 0.05$.

11. $H_0$: $\mu = 5.75$    $H_a$: $\mu \neq 5.75$
Given a sample of 40 with $\bar{x} = 5.68$ and $s = 0.12$, conduct the test at the 1% level of significance.

12. $H_0$: $\mu = 0.095$    $H_a$: $\mu \neq 0.095$
Given a sample of 60 with $\bar{x} = 0.100$ and $s = 0.006$, conduct the test at the 5% level of significance.

13. An article in a recent medical journal claimed that better nutrition has increased the mean weight of adult men in the United States to 175 pounds. The director of a weight control institute felt the figure was too high for the males in the region served by her institute. A sample of 45 men yielded $\bar{x} = 168.7$ pounds with $s = 25.9$ pounds. Conduct a test of $H_0$: $\mu = 175$ versus $H_a$: $\mu < 175$ at the 5 percent level of significance.

14. A national sportscaster recently stated that records over the last 40 years indicate that the mean playing time for a major league baseball game is 2 hours and 25 minutes (that is, 145 minutes). John Miller, the sportscaster at KTXL, feels the mean for baseball games played over the past 5 years is less than 145. He sampled the records of 34 games played within the past 5 years. For these games, he got $\bar{x} = 138$ minutes with $s = 16$ minutes. Use these data to conduct a one-sided test with $\alpha = 0.05$.

15. The mean grade point average (gpa) of the student body at Midwestern University is reported to be 2.83 (based on a four-point scale). The student senate feels that the mean gpa is actually greater than 2.83. They have a computer randomly sample the records of 75 students and record the cumulative gpas of these students. If $\bar{x} = 2.92$ and $s = 0.45$, can the 2.83 figure be rejected at a 1 percent level of significance?

16. The labels on rolls of a brand of twine claim that the rolls contain 285 feet of twine. An employee needs to check a large production run of this product to make sure the mean length of twine is not significantly different from the 285 feet printed on the label. A random sample of 32 rolls yields $\bar{x} = 287$ feet and $s = 7$ feet. Use a 1 percent level of significance to test the 285 feet figure.

17. The manufacturer of safety matches produces a box that is supposed to hold 34 matches. A periodic check of production runs is made to ensure that the "count machine" is putting the correct number of matches in each box. For the last production run, a sample of 30 boxes yielded $\bar{x} = 35.7$ and $s = 2.4$. Use these results and a 1 percent level of significance to test the 34-matches-per-box claim.

18. Joan McKee is a substitute teacher for several large school districts in northern California. Joan has been unable to get full-time teaching status because the districts are not planning to add more classes. Officials in these districts claim that the average class size is 32.0 students. Joan feels the mean is higher. She sampled 38 classes throughout the districts and got $\bar{x} = 34.3$ students per class with $s = 5.6$ students. Test the 32.0-student-per-class claim using a 1 percent level of significance.

19. The label on a package of candy from a vending machine claims that each bag has a net weight of 28.5 grams. Assuming the distribution of weights

from a large production run of these bags of candy is bell shaped, test the 28.5-gram claim if a sample of 15 bags yields $\bar{x} = 30.2$ grams and $s = 3.4$ grams. Use $\alpha = 0.05$.

20. A national chain of fast-food restaurants claims that the mean pay rate for its hourly employees is \$4.89. Assuming the distribution of pay rates for these employees is bell shaped, test the \$4.89 claim if a sample of 20 employees yields $\bar{x} = \$4.48$ and $s = \$0.78$. Use $\alpha = 0.05$.

21. According to the National Weather Service, the mean daily high temperature for July in a large southwestern city is 96.3° Fahrenheit (F). Assuming the distribution of temperatures for this city for July is bell shaped, test the 96.3° F claim if a sample of 16 high temperatures for July over the past 4 years yields $\bar{x} = 91.9°$ F and $s = 7.5°$ F. Use $\alpha = 0.01$.

22. The syllabus of instruction for officers attending Undergraduate Navigator Training (UNT) claims that a total of 80 flying hours are programmed for each officer to justify awarding the aeronautical rating of navigator. Major Glenn Russell, the current chief of UNT, feels that students require more than 80 flying hours. He orders a sample of the records of 20 recent graduates be surveyed and the number of flying hours be recorded. As-

suming the distribution of flying hours is bell shaped, can the 80-hour figure be rejected as too low if the sample yields $\bar{x} = 83.9$ hours and $s = 5.6$ hours? Use $\alpha = 0.01$.

23. The Happy Glen Fishing Resort claims that fishermen at the resort catch an average of seven fish per day. An officer of the Riverdale Sporting Club tested this claim by sampling 50 people over a 2-month period. The sample yielded $\bar{x} = 5.9$ fish with $s = 3.6$ fish. Use these data to test whether the resort's claim is too high.
a. Use a 5% level of significance.
b. Use a 1% level of significance.

24. The plant breeding department at a major university developed a new hybrid strawberry plant called Sequoia II. Based on research data, the claim was made that from the time shoots are planted, 130 days on the average are required to obtain the first berry. A corporation that was interested in marketing the product tested 70 shoots by planting them and recording the number of days before each plant produced its first berry. For the sampled plants, $\bar{x} = 133.1$ days with $s = 12.6$ days. Use these data to test whether the department's claim was too low.
a. Use a 5% level of significance.
b. Use a 1% level of significance.

# Exercises 6-3   Set B

1. An article in a magazine on financial affairs claimed that adults who are heads-of-households have access to an average of 6.0 credit cards. Included are bank, gasoline, telephone, and department store cards. A sample was taken of shoppers in major shopping malls in Detroit. The number of credit cards each of the heads-of-households in the sample had access to is listed in Table 6-3.

TABLE 6-3   The Numbers of Credit Cards of Heads-of-Household

| | | | | | | | | | | | |
|---|---|---|---|---|---|---|---|---|---|---|---|
| 5 | 2 | 5 | 6 | 5 | 3 | 9 | 6 | 3 | 8 | 6 | 9 |
| 8 | 5 | 9 | 0 | 4 | 6 | 5 | 2 | 6 | 7 | 4 | 0 |
| 7 | 4 | 6 | 8 | 6 | 7 | 9 | 4 | 3 | 7 | 8 | 6 |

Test $H_0: \mu = 6.0$ versus $H_a: \mu \neq 6.0$. Use $\alpha = 0.05$.

2. A national home magazine recently reported that the national average number of pets (not counting fish and birds) per household has increased to 2.9. A telephone survey of 36 homes and apartments in the Milwaukee metropolitan area produced the numbers of pets in households given in Table 6.4.

TABLE 6-4   The Numbers of Pets in the 36 Homes Surveyed

| 1 | 0 | 2 | 3 | 1 | 2 | 9 | 0 | 4 | 5 | 2 | 3 |
| 7 | 3 | 9 | 6 | 4 | 7 | 3 | 5 | 1 | 0 | 4 | 6 |
| 5 | 3 | 4 | 0 | 8 | 0 | 5 | 2 | 3 | 6 | 8 | 2 |

Test $H_0$: $\mu = 2.9$ versus $H_a$: $\mu > 2.9$. Use $\alpha = 0.05$.

*In Exercises 3 and 4, use the information in Figure 6-20 and assume the figure represents per hotel averages of "base rates" for all city hotels.*

3. In the figure the average daily base rate of a hotel room in New York is $120.91. Suppose a sample of 32 hotels in New York yields $\bar{x} = \$114.65$ with $s = \$24.60$. Test $H_0$: $\mu = \$120.91$ versus $H_a$: $\mu < \$120.91$. Use $\alpha = 0.05$.

4. In the figure the average daily base rate of a hotel room in San Francisco is $106.50. Suppose a sample of 40 hotels in San Francisco yields $\bar{x} = \$119.36$ and $s = \$31.72$. Test $H_0$: $\mu = \$106.50$ versus $H_a$: $\mu > \$106.50$. Use $\alpha = 0.01$.

## Most expensive places to stay

Hotel rooms in U.S. cities averaged a daily rate of $73.96 in the first half of '88, up 4.6% from '87. Cities with the highest average daily rate:

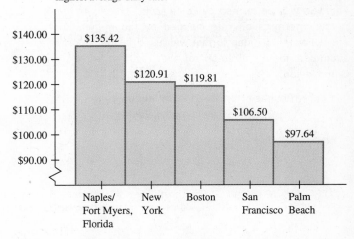

**Figure 6-20.** Quarterly Trends in the Hotel Industry, Pannell Kerr Forster; worldwide survey of 1537 properties.

*In Exercises 5 and 6, use random numbers to sample the indicated number of stocks from Data Set II in Appendix B.*

5. Use the mean and standard deviation of 40 data points to test $H_0$: $\mu = \$36.48$ for the population of "High" stock prices. Use a non-directional test with $\alpha = 0.05$.

6. Use the mean and standard deviation of 50 data points to test $H_0$: $\mu = \$21.47$ for the population of "Low" stock prices. Use a non-directional test with $\alpha = 0.01$.

# 6-4 Large Sample Tests for a Single Proportion

**Key Topics**

1. *The test statistic for large n tests on P*

2. *Using the standard format to test $H_0: P = P_0$*

3. *Statistical significance and practical importance.*

---

## S T A T I S T I C S   I N   A C T I O N

---

Recall the seed germination test conducted by Fontaine Evaldo in the Statistics in Action portion of Section 6-1. This test was performed by Fontaine to investigate the 90 percent germination rate claimed by the manufacturer. The following six-step format was used in the report she submitted to the directors of the Truth in Advertising Commission:

### Results of the Seed Germination Test
Submitted by Fontaine Evaldo

**Step 1** **State $H_0$ and $H_a$.**

$H_0$:   $P = 0.90$     The proportion of seeds that germinate is 0.90.

$H_a$:   $P < 0.90$     The proportion that germinate is less than 0.90.

**Step 2** **State the level of significance of the test.**

For this test, a 5 percent level of significance was used. There was only a 0.05 probability of rejecting $H_0$, if in fact $H_0$ was true.

**Step 3** **Identify the test statistic.**

The test statistic for this test was the one commonly used for large $n$ tests on $P$:

$$Z = \frac{\hat{P} - P_0}{\sqrt{\dfrac{P_0(1 - P_0)}{n}}}$$

$Z$ is a standard normal random variable, if $H_0$ is true.

**Step 4** **State the decision rule in terms of the rejection region.**

For a 5 percent level of significance and a left-tail directional test, $z_c$ is $-1.65$ (see Figure 6-21). If $z < z_c$ ($= -1.65$), then reject $H_0$. Otherwise, do not reject $H_0$.

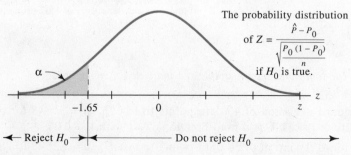

The probability distribution

of $Z = \dfrac{\hat{P} - P_0}{\sqrt{\dfrac{P_0(1 - P_0)}{n}}}$

if $H_0$ is true.

← Reject $H_0$ →|← ———— Do not reject $H_0$ ————→

**Figure 6-21.**  $\alpha = P(Z < -1.65)$.

Step 5   **Calculate the observed value of the test statistic.**

Of the 400 seeds planted in the test, 352 germinated. Thus $\hat{p} = \dfrac{352}{400} = 0.88$. With $\hat{p} = 0.88$, $P_0 = 0.90$, and $n = 400$:

$$z = \frac{0.88 - 0.90}{\sqrt{\dfrac{(0.9)(0.1)}{400}}}$$

$$= -1.333\ldots$$

$$\approx -1.33, \text{ to two decimal places}$$

Step 6   **Make a decision about the parameter tested that is consistent with steps 4 and 5.**

Because $-1.33 > -1.65$, do not reject $H_0$. That is, there is not sufficient evidence, with $\alpha = 0.05$, to conclude that the proportion of seeds that germinate is less than 0.90. The difference between the hypothesized value of 0.90 and the 0.88 figure observed in the test can be attributed to random sampling variation.

## The test statistic for large $n$ tests on $P$

In the Statistics in Action section, Fontaine Evaldo used the value of $\hat{P}$ from her sample to test the claim of the manufacturer. To use the standard normal distribution and corresponding values of $z_c$ to set desired levels of significance, we must be able to claim that $\hat{P}$ has an approximately normal distribution. In Chapter 4, we stated that such a claim is justified provided both:

    1. $nP \geq 5$,           $P$ is the proportion of the sampled population with the characteristic of interest and $n$ is the sample size.

and

    2. $n(1 - P) \geq 5$.

If $H_0: P = P_0$ is true and $\hat{P}$ is based on a sample size such that inequalities 1 and 2 are satisfied, then $\hat{p}$ can be changed to a $z$ value and the standard normal distribution used. The observed $z$ value can then be compared to values of $z_c$ for given levels of significance.

### The Test Statistic for Tests of $H_0: P = P_0$

If $nP_0 \geq 5$ and $n(1 - P_0) \geq 5$, then when $H_0: P = P_0$ is true:

$$Z = \frac{\hat{P} - P_0}{\sqrt{\dfrac{P_0(1 - P_0)}{n}}}$$

is approximately standard normal. Thus $Z$ can be used as a test statistic for $H_0: P = P_0$.

When $n$ and $P_0$ are such that a normal approximation is not appropriate, then exact binomial probabilities (either taken from a table or computed with the binomial probability formula) must be used for choosing critical values. Such calculations will not be covered in this text.

▶ **Example 1**   A report from the national forest agency estimates that as many as 15 percent of trees in national forest lands are infected with a bark beetle. In a large national forest in the Southwest, a sample of 600 trees was studied for beetle infestation, and 78 infected trees were discovered. Compute the observed value of the test statistic for a test of $H_0: P = 0.15$.

**Solution**   For this sample, $\hat{p} = \dfrac{78}{600} = 0.13$. With $P_0 = 0.15$, $1 - P_0 = 0.85$, and $n = 600$:

$$z = \frac{0.13 - 0.15}{\sqrt{\dfrac{(0.15)(0.85)}{600}}}$$

$$= -1.3719\ldots$$

$$\approx -1.37, \text{ to two decimal places} \qquad ◀$$

## Using the standard format to test $H_0: P = P_0$

Examples 2 and 3 illustrate how the six-step hypothesis testing format first discussed in Section 6-3 can be used to summarize tests on $P$.

▶ **Example 2**   A recent article in a medical journal claims that fewer than 24 percent of American adults exercise as part of a daily routine. Sandra Rainer is the director of the Parks and Recreation Division of a moderate-size midwestern district. She feels that residents in her district are much more aware of the value of a regular exercise routine for maintaining a healthy body. Her staff sampled adults from the district regarding their personal exercise plans. The survey yielded 126 of 420 adults who have daily exercise built into their routines. Do these results justify Sandra's claim that the proportion in her district is greater than the 24 percent figure stated in the article? Use $\alpha = 0.05$.

**Solution**   **Step 1**  $H_0$:   $P = 0.24$
             $H_a$:   $P > 0.24$        A directional test

**Step 2**  As indicated, $\alpha = 0.05$        $P(\text{Type I error}) = 0.05$

**Step 3**  The test statistic to be used is:

$$Z = \frac{\hat{P} - 0.24}{\sqrt{\dfrac{(0.24)(0.76)}{n}}} \qquad \text{From } H_0, P_0 = 0.24$$

**Step 4** If $z > z_c$ ($= 1.65$), then reject $H_0$ (see Figure 6-22).

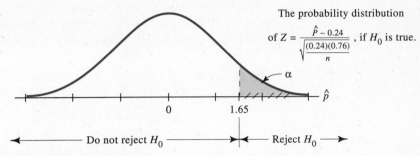

The probability distribution

of $Z = \dfrac{\hat{P} - 0.24}{\sqrt{\dfrac{(0.24)(0.76)}{n}}}$ , if $H_0$ is true.

Do not reject $H_0$     Reject $H_0$

**Figure 6-22.** $\alpha = P(Z > 1.65)$.

**Step 5** With $\hat{p} = \dfrac{126}{420} = 0.30$ and $n = 420$:

$$z = \frac{0.30 - 0.24}{\sqrt{\dfrac{(0.24)(0.76)}{420}}}$$

$$= 2.879 \ldots$$

$$\approx 2.88, \text{ to two decimal places}$$

**Step 6** Because $2.88 > 1.65$, reject $H_0$. That is, with $\alpha = 0.05$, there is sufficient evidence to conclude that the proportion of adults in the district that exercise daily is greater than 24 percent. ◀

Notice that the conclusion of Ms. Rainer's survey was applied only to the population that was sampled. For example, unless a sample of all adults in the United States is taken, there is no way to legitimately draw a conclusion about the proportion of all American adults who exercise daily.

▶ **Example 3**    Vince Straub is the manager of a Las Vegas hotel-casino. One of the slot machines in the casino is supposed to pay off on 20 percent of the quarters played in the machine. Vince wonders whether the machine is out of adjustment. He therefore monitors the next 300 plays of the machine and records 69 payoffs. If Vince uses these data to test $H_0: P = 0.20$ versus $H_a: P \neq 0.20$ using a 1 percent level of significance, can he conclude that the machine is out of adjustment?

**Solution**    **Step 1** $H_0$:  $P = 0.20$

$H_a$:  $P \neq 0.20$     A non-directional test

**Step 2** As indicated, $\alpha = 0.01$     $P(\text{Type I error}) = 0.01$

**Step 3** The test statistic will be:

$$Z = \frac{\hat{P} - 0.20}{\sqrt{\dfrac{(0.20)(0.80)}{n}}} \qquad \text{From } H_0, \ P_0 = 0.20$$

**Step 4** If $z <$ lower $z_c \,(= -2.58)$ or $z >$ upper $z_c \,(= 2.58)$, then reject $H_0$ (see Figure 6-23).

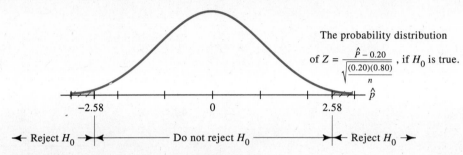

Figure 6-23. $\alpha = P(Z < -2.58 \text{ or } Z > 2.58)$.

**Step 5** With $\hat{p} = \dfrac{69}{300} = 0.23$ and $n = 300$:

$$z = \frac{0.23 - 0.20}{\sqrt{\dfrac{(0.20)(0.80)}{300}}}$$

$$= 1.299 \ldots$$

$$\approx 1.30, \text{ to two decimal places}$$

**Step 6** Because $1.30 > -2.58$ and $1.30 < 2.58$, do not reject $H_0$. That is, with $\alpha = 0.01$, there is not sufficient evidence to conclude that the proportion of payoffs on this machine is other than 0.20. ◀

Notice that Example 3 does not really involve random sampling. The 300 plays of the machine that Mr. Straud watches are in no sense "a random sample of plays." But if the outcomes are assumed to be independent, it is reasonable to model the number of payoffs in the 300 plays as binomial with $n = 300$ and

$$p = P = \text{the machine's current propensity for payoffs.}$$

Such a definition for $P$ is clearly less concrete than "the fraction of a population possessing a certain attribute," but one is commonly called on to use $\hat{P}$ to estimate and test hypotheses for such *theoretical proportions*.

## EPA endorses gasoline gadget

WASHINGTON (AP)—A gadget that cuts off a car's air conditioner when the vehicle accelerates has become the first product aimed at cutting gasoline consumption to win government endorsement.

The device, marketed under the name "Pass Master," can provide a "small but real fuel economy benefit," the Environmental Protection Agency said Wednesday.

Motorists could realize up to 4 percent fuel reduction while using their air conditioners on cars equipped with the device, the agency said. That would translate into .8-miles-per-gallon improvement for a car that normally gets 20 miles to the gallon with the air conditioner on.

The agency cautioned that the 4 percent figure was a maximum amount and could be less depending on a motorist's driving habits, the type of car and the type of air conditioner.

But still the Pass Master, which sells for less than $15, is the first of 40 products to pass the EPA's tests as making any "statistically significant" improvement in a car's mileage.

## Statistical significance and practical importance

The term "statistically significant" is frequently used to describe the results of hypothesis tests. As consumers, we sometimes hear this phrase in connection with new products that are being introduced to the buying public. The products may be presented as "significantly better" than products the manufacturer hopes to replace. We are sometimes subtly encouraged to interpret the phrase "a statistically significant result" to mean "a practically important result."

It may be for this reason that the Environmental Protection Agency (EPA) was cautious in reporting on a gas-saving device it tested. A news article summarizing the report is shown in Figure 6-24. The article is several years old, but it is an excellent example of the EPAs care in noting that the results were "statistically significant," but the savings in miles per gallon was of limited practical importance.

Testing $H_0: \mu = \mu_0$ or $H_0: P = P_0$ can be thought of as a method of deciding whether the evidence against $H_0$ is strong enough to justify its rejection. But testing is not a method for assessing by how much $\mu$ or $P$ differs from the hypothesized value. And it is the size of a possible difference that determines whether it is important in a practical sense. To assess the size of a difference, one must estimate $\mu$ or $P$. For this reason, many statisticians find methods of estimation like those in Chapter 5 to be more valuable than methods of testing like those discussed in the present chapter.

**Figure 6-24.** "EPA endorses gasoline gadget," *Lafayette Journal and Courier*, August 28, 1980. p. D-3. Reprinted by permission of Associated Press.

## Exercises 6-4 Set A

*In Exercises 1–4, compute the value of the test statistic for each indicated test.*

**1.** $H_0: P = 0.35$ versus $H_a: P > 0.35$
$\hat{p} = 0.39$ and $n = 280$

**2.** $H_0: P = 0.72$ versus $H_a: P < 0.72$
$\hat{p} = 0.67$ and $n = 335$

**3.** $H_0: P = 0.42$ versus $H_a: P \neq 0.42$
$\hat{p} = 0.51$ and $n = 72$

**4.** $H_0: P = 0.84$ versus $H_a: P \neq 0.84$
$\hat{p} = 0.77$ and $n = 103$

*In Exercises 5–10, practice the six-step hypothesis test format on the indicated null and alternative hypotheses, using the stated levels of significance.*

**5.** $H_0: \quad P = 0.60$ versus $H_a: \quad P < 0.60$
if a sample of 64 yields $\hat{p} = 0.52$. Conduct the test at the 5% level of significance.

**6.** $H_0: \quad P = 0.60$ versus $H_a: \quad P > 0.60$
if a sample of 96 yields $\hat{p} = 0.70$. Conduct the test at the 1% level of significance.

7. $H_0$:	$P = 0.75$ versus $H_a$:	$P \neq 0.75$
   if a sample of 38 yields $\hat{p} = 0.60$. Conduct the test at the 5% level of significance.

8. $H_0$:	$P = 0.75$ versus $H_a$:	$P \neq 0.75$
   if a sample of 460 yields $\hat{p} = 0.82$. Conduct the test at the 1% level of significance.

9. $H_0$:	$P = 0.32$ versus $H_a$:	$P > 0.32$
   if a sample of 240 yields $\hat{p} = 0.40$. Conduct the test at the 1% level of significance.

10. $H_0$:	$P = 0.23$ versus $H_a$:	$P < 0.23$
    if a sample of 329 yields $\hat{p} = 0.20$. Conduct the test at the 5% level of significance.

11. A report claims that at least one-half the patients with back pain who receive acupuncture treatments obtain relief. The doctors at a major hospital in New York City feel the estimate of 0.50 is too high. They check the records of 225 patients at their hospital that received similar treatment for back pain. If 105 of these patients got relief, can the 0.50 figure be rejected as too high for patients at this hospital? Use $\alpha = 0.05$.

12. A nationwide study of American households revealed that 73 percent have one or more live plants. An organization of florists in Portland, Oregon, feels the estimate is too low for households in that city. Can the value 0.73 be rejected if a survey of 500 homes in Portland yields 385 with one or more live plants? Use $\alpha = 0.05$.

13. A women's glamour magazine reported that 10 percent of American adults bite their fingernails. Lynn Butler is a cosmetologist at a large salon in Boston. She felt the estimate was too high for adults in the Boston metropolitan area. She therefore sampled 210 adults and obtained 15 that were nail-biters. Can the value 0.10 be rejected as too high for adults in Boston? Use $\alpha = 0.05$.

14. A report in a national magazine estimated that 39 percent of American husbands do not know the color of their wives' eyes. Bill Cornett is a marriage counselor in a clinic in New Orleans. He feels the figure is higher for husbands in the greater New Orleans area. A sample of 264 husbands in this area yielded 111 who did not know the color of their wives' eyes. Can the value of 0.39 be rejected as too low for husbands in New Orleans? Use $\alpha = 0.05$.

15. The United States Census Bureau reported that 31.8 percent of all Americans changed their residence at least once in the last 3 years. Jane Gunton lives in a moderate-size community in Arkansas. She wanted to check the accuracy of this figure in her area. Of 270 adults that Jane surveyed, 124 had changed their residence at least once in the last 3 years. Use these data and conduct a hypothesis test using $\alpha = 0.01$.

16. A seed packet of peas had "germination rate 88%" printed on it. Arnold Parker, a member of a consumer advocate group, sampled several packets of these seeds. He planted 250 seeds from them according to instructions. He found 193 that germinated. Can the distributor's claim of 88 percent germination be rejected as too high? Use $\alpha = 0.05$.

*In Exercises 17–20, the statistics are from the book,* 100% American, *by Daniel Evan Weiss, published by Simon and Schuster. An article, entitled "America by the Numbers" and based on his book, appeared in the Sunday Chicago Tribune Magazine, July 3, 1988, Section 10.*

17. It is reported that 47 percent of American women think a man reaches his prime during his 30s. A survey of 375 women students at Washington State University yielded 161 that think a man reaches his prime during his 30s. Use these data and test $H_0: P = 0.47$ versus $H_a: P \neq 0.47$ for female students at this school. Use $\alpha = 0.05$.

18. It is reported that 38 percent of Americans die of heart disease. A survey of 320 death certificates from the past 3 years in the District of Columbia

yielded 190 that listed heart disease as the cause of death. Use these data and test $H_0: P = 0.38$ versus $H_a: P \neq 0.38$ for residents of the District of Columbia. Use $\alpha = 0.01$.

19. It is reported that 74 percent of American adults have a high school diploma. A survey of 260 adults in the Atlanta metropolitan area yielded 182 that have a high school diploma. Use these data to test $H_0: P = 0.74$ versus $H_a: P < 0.74$ for adults in the Atlanta metropolitan area. Use $\alpha = 0.05$

20. It is reported that 63 percent of American car thefts occur at night. A sample of 437 car theft reports in the Cleveland metropolitan area yielded 293 thefts that occurred at night. Use these data to test $H_0: P = 0.63$ versus $H_a: P > 0.63$ for car thefts in the Cleveland metropolitan area. Use $\alpha = 0.01$.

21. An oil company airs a carefully worded television commercial stating, "In tests against the leading motor oil advertising increased gas mileage, ours was unsurpassed."
   a. State the null hypothesis involving mean mileages for two oils the company probably used.
   b. If "ours was unsurpassed" should be interpreted to mean "$H_0$ was not rejected," should the viewer conclude that the difference in gas mileages for the two oils is necessarily very small? Why?

22. A TV commercial on a major brand of trucks states that "our trucks get the best mileage."
   a. State the null hypothesis and alternate hypothesis (both involving two means) for a mileage comparison of this truck with any one of its competitors.
   b. If "best mileage" should be interpreted to mean $H_0$ was rejected, should the viewer conclude that the gas savings on this model truck would be substantial? What questions, as statisticians, would you like to ask concerning the test?

23. Many TV commercials for laundry detergents make claims about a detergent being "more effective at getting clothes whiter and brighter." To justify such a claim, a company will ask individuals to compare washings of similar clothing washed with two different soap products.
   a. In terms of the proportion of those sampled that select clothes washed in the sponsor's detergent, state $H_0$ and $H_a$ for such a test.
   b. If rejection of $H_0$ implies more than half of those sampled prefer the clothes washed in the sponsor's detergent, what additional facts might cause you to change to their brand of detergent? Why?

24. Supermarket chain A claims that "the everyday prices of our groceries are lower than those of supermarket chain B."
   a. Discuss a procedure that might be used to test whether the prices of groceries at the supermarket A stores are lower than those at the supermarket B stores.
   b. As consumers, what factors other than price should we consider when making such a comparison?

25. Find an article from a newspaper or magazine that reports evidence of a "significant breakthrough" in solving some problem that is plaguing our society. Such an article may cite the solution to a social problem, a health problem, a financial dilemma, or some other problem.
   a. Subject the evidence to careful study and decide whether the evidence convinces you, as a skeptical statistically literate consumer of the importance of the solution.
   b. Phrase at least three questions you might ask the investigators about their method of study.

26. Frequently the news media will cite evidence of statistical significance in studies being conducted in the country. State, as best you can, the details of such a report you have recently seen on TV.
   a. Do you believe this statement was the result of a sound statistical study?
   b. Phrase at least three question that you might ask to increase the credibility of the report in your own mind?

## Exercises 6-4   Set B

*In Exercises 1 and 2, the data are the results of samples taken from large production runs of candy-coated chocolate candies. Use these data to test the proportions claimed for this production run. For these data use the following symbols:*

Y is yellow     G is green     B is brown     T is tan     R is red

1. Let $P$ be the proportion of red-colored candies in the production run. $H_0$: $P = 0.28$.

```
T  Y  R  B  B  G  Y  Y  R  T  B  B  G  Y  G
B  R  Y  Y  T  B  R  G  G  B  R  Y  T  R  B
R  T  G  G  B  Y  Y  Y  T  R  G  G  B  B  Y
G  B  B  R  R  G  Y  G  G  B  T  R  Y  T  Y
Y  G  G  T  Y  Y  T  R  B  B  B  B  R  G  T
```

Conduct the test using $H_a$: $P \neq 0.28$ and $\alpha = 0.05$.

2. Let $P$ be the proportion of green-colored candies in the production run. $H_0$: $P = 0.21$.

```
G  Y  Y  B  T  R  R  G  G  B  T  T  G  R  G
B  B  Y  R  G  Y  Y  G  G  Y  Y  R  G  B  R
Y  R  T  T  G  Y  T  G  B  B  R  G  Y  Y  T
G  G  B  R  R  T  R  Y  G  G  G  Y  T  G  G
Y  T  G  Y  B  B  G  T  T  R  R  T  B  Y  Y
```

Conduct the test using $H_a$: $P \neq 0.21$ and $\alpha = 0.05$.

3. Use random numbers to sample $n = 40$ individuals from Data Set I, Appendix B. For each data point, determine whether the individual was a smoker. A 1 indicates a "yes" and a 2 indicates a "no".
   a. Count the number of smokers in your sample.
   b. Compute $\hat{p}$ for your sample.
   c. Use your $\hat{p}$ to test, with $\alpha = 0.05$:

$$H_0: \quad P = 0.53 \quad\quad \text{versus} \quad\quad H_a: \quad P \neq 0.53.$$

4. Use random numbers to sample $n = 50$ individuals from Data Set I Appendix B. For each data point, determine whether the individual is a female or male.
   a. Count the number of females in your sample.
   b. Compute $\hat{p}$ for your sample.
   c. Use your $\hat{p}$ to test, with $\alpha = 0.05$:

$$H_0: \quad P = 0.49 \quad\quad \text{versus} \quad\quad H_a: \quad P < 0.49.$$

# More federal workers women, minorities

By Judith Havemann
Washington Post

WASHINGTON — When President Reagan took his last helicopter turn around Washington and headed for retirement in California in January, he left behind a government bigger than he had inherited eight years before but dramatically altered in mission, race, gender, pay and grade.

This week, Philip A.D. Schneider, head of work force statistics for the Office of Personnel Management (OPM), drew a demographic picture of the federal work force at a research conference.

Schneider reported that while the Defense Department had grown by 7 percent — employing 34 percent of all federal workers, not including those in uniform — the domestic side of the government had declined 5 percent overall. The biggest cutbacks were in some of the smallest agencies.

The General Services Administration shrank by nearly a half, the Tennessee Valley Authority and the Education Department went down by about a third, OPM and the Department of Health and Human Services declined by nearly a quarter, and workers at other agencies, including Housing and Urban Development, Labor, Interior, Energy and Agriculture decreased by 10 percent or more.

The shift in mission from emphasizing domestic to military concerns was precisely what Reagan said he would do when he ran for president in 1981, but some of the other changes represent longer-term trends.

Minority group employment has grown to 27 percent of the federal work force, and virtually every minority group — Indian, Asian, Hispanic and black — is present in proportionally larger numbers than in private employment, Schneider said.

Days before leaving office, President Carter discarded the

standard government entrance examination that had prevented many blacks from getting federal jobs. Minority hiring has since increased at a relatively rapid pace.

Today, nearly half of federal workers are women, and they are slowly creeping into the higher grades. Since 1981, the percentage of women in professional jobs has gone up 34 percent. But the overwhelming majority — 70 percent — of all professional jobs are held by men.

By and large, young women hired recently by the government have gotten better jobs than women did in the past. Ten years ago, 22 percent of women under age 35 were in GS grades 9 through 12, and 9.5 percent held grades of 13 through 15. Today, 40 percent of women under age 35 are in GS grades 9 through 12, and 27.4 percent in grades 13 through 15.

Women older than 35 have moved up in grade levels in the

past 10 years, but not to the extent of the younger group.

Real salaries for all civil servants declined — when adjusted for inflation — between 1969 and 1988. A GS-5 has lost 22 percent of purchasing power, a GS-15, 14 percent, and the first step on the federal executive pay scale has lost 40 percent.

About one-sixth of all government employees leave each year, but virtually all of these comings and goings occur in the relatively low grades of federal employment. Above GS-13, virtually nobody quits, Schneider said.

The work of the government has become more technical, administrative, investigative, legal and complicated in the last 10 years. As a result, the average grade of the federal worker has crept up, he said. Where 34 percent of workers were in grades 9 to 12 in 1978, the figure is now 38 percent.

**Figure 6-25.** "More Federal Workers Women, Minorities" by Judith Havemann, *Washington Post*, 8/19/89. Copyright © 1989 *The Washington Post*. Reprinted by permission.

*Exercises 5 and 6 are based on the report by Philip A. D. Schneider in Figure 6-25.*

**5.** Based on the report, "nearly half of federal workers are women." A sample of 60 employee records at the National Parks and Recreation Department was taken to test:

$$H_0: \quad P = 0.50 \qquad \text{versus} \qquad H_a: P < 0.50$$

where $P$ is the proportion of women employees in this department. Can $H_0$ be rejected at the 5 percent level of significance if the sample yielded records of only 25 women employees?

**6.** Based on the report, "the overwhelming majority— 70 percent—of all professional jobs are held by men." A sample of 75 employee records of lawyers in the Justice Department was taken to test:

$$H_0: \quad P = 0.70 \qquad \text{versus} \qquad H_a: \quad P \neq 0.70$$

where $P$ is the proportion of male lawyers in the Justice Department. Can $H_0$ be rejected at the 1 percent level of significance if the sample yielded records of 40 male lawyers?

*Exercises 7 and 8 are based on the data in Figure 6-26.*

7. Robert Spence, a member of the Sacramento County Board of Supervisors, had his staff contact 340 constituents in his region. In the sample, 156 said they had "hardly any optimism at all" that environmental problems will be under control in the future. Use these results to test:

$$H_0: \quad P = 0.40 \qquad \text{versus} \qquad H_a: \quad P > 0.40$$

for $P$, the fraction of the constituents in Spence's region who had hardly any optimism regarding environmental problems. Use $\alpha = 0.05$.

8. Deborah Harris is current president of Women for Political Progress (WPP) in the North Bay Region of San Francisco. The WPP members surveyed 542 residents and got 145 that expressed little optimism at all regarding environmental problems in the future. Use these results to test:

$$H_0: \quad P = 0.30 \qquad \text{versus} \qquad H_a: \quad P < 0.30$$

for $P$, the fraction of residents in this area that expressed little optimism at all regarding environmental problems. Use $\alpha = 0.05$.

# Ecological fears vary by location
## Sacramentans more gloomy than most

**By Deborah Blum**
**Bee Science Writer**

Residents of Sacramento County are unusually pessimistic about the fate of the environment, according to a new survey.

Forty percent of the people questioned said they have "hardly any optimism at all" that environmental problems will be under control in the future. Nationwide, only about 22 percent of those surveyed expressed a similar lack of hope. Even throughout the state, there is more hope than in Sacramento: Overall, just 30 percent of Californians lacked environmental optimism concerning the future.

**Figure 6-26.** "Ecological Fears Vary by Location" by Deborah Blum, *The Sacramento Bee*, 2/18/91, p. A-23. Copyright © 1991 by *The Sacramento Bee*. Reprinted by permission.

# 6-5 Tests and Confidence Intervals Concerning a Single Variance

**Key Topics**

1. The forms of $H_0$ and $H_a$ for a test on $\sigma^2$

2. The test statistic for a test on $\sigma^2$

3. The $\chi^2$ probability distributions

4. The six-step format and tests on $\sigma^2$

5. Confidence intervals for $\sigma^2$

# STATISTICS IN ACTION

Diane Gray is an employee of Taggart Scientific Instruments Corporation. One of her primary responsibilities is to monitor the quality of instruments purchased by educational institutions. The Howrigan Unified School District recently purchased 250 Model 3B702D analytic balances. Taggart guarantees that this balance model has a precision measured by a variance of at most 5 $(mg)^2$, based on repeated measurements of a 1000 milligram test weight.

To check the precision of the balances scheduled for the Howrigan schools, Diane selected one scale and made 20 measurements using the 1000 milligram (mg) test weight. The weights of this test mass recorded by Diane are in Table 6-5.

For these data points, Diane made the following calculation:

$$s^2 = \frac{20(20,012,122) - (20,006)^2}{20(19)} \approx 6.326 \ldots$$

In a report that Diane subsequently sent to the executive board of Taggart Corporation, she included the following summary statement:

The sample data yielded a $\chi^2$ value of 24.040. Using a 5 percent level of significance and 19 degrees of freedom, the corresponding $\chi_c^2$ was 30.144. Since 24.040 is less than 30.144, I found no statistically significant evidence that the precision of the balance scheduled for delivery to the Howrigan school district was substandard.

The details of how Diane arrived at this conclusion will be covered in this section.

**TABLE 6-5   Recorded Weights of a 1000 mg Mass.**

| Trial Number | Recorded Weight | Trial Number | Recorded Weight | Trial Number | Recorded Weight |
|---|---|---|---|---|---|
| 1 | 1003 | 8 | 999 | 15 | 998 |
| 2 | 998 | 9 | 1000 | 16 | 1001 |
| 3 | 1005 | 10 | 1002 | 17 | 999 |
| 4 | 1001 | 11 | 998 | 18 | 1000 |
| 5 | 1003 | 12 | 1004 | 19 | 1002 |
| 6 | 1002 | 13 | 1001 | 20 | 997 |
| 7 | 997 | 14 | 996 | | |

## The forms of $H_0$ and $H_a$ for a test on $\sigma^2$

In Chapter 2 we studied variance and standard deviation as descriptive measures of variability for quantitative data.

$\sigma^2$ is the variance of a population.
$s^2$ is the variance of a sample.
$\sigma$ is the standard deviation of a population, $\sigma = \sqrt{\sigma^2}$.
$s$ is the standard deviation of a sample, $s = \sqrt{s^2}$.

In the world of mass-produced items, we have come to accept a certain amount of variability in the sizes, shapes, weights, and other measurable features of items we buy. However, there are certain acceptable limits in this variability. For example, we accept a nut and bolt combination that is not perfect so long as it fits well enough to do the job.

As a consequence of the need to control variability, manufacturers need to check the variability of populations of quantitative data almost as regularly as they check the corresponding population locations. In the case of mass-produced items, variability is evaluated on the basis of data regularly sampled from production runs. There are other situations in which variability also needs to be checked, but the production of large quantities of nominally identical items is one that we can easily comprehend.

The three forms of $H_0$ and $H_a$ that will be used in tests for variability are:

**Form 1.**      $H_0$:   $\sigma^2 = \sigma_0^2$      Actual variance $\leq$ hypothesized variance
          **With $H_a$:**   $\sigma^2 > \sigma_0^2$      Actual variance $>$ hypothesized variance

**Form 2.**      $H_0$:   $\sigma^2 = \sigma_0^2$      Actual variance $=$ hypothesized variance
          **With $H_a$:**   $\sigma^2 \neq \sigma_0^2$      Actual variance $\neq$ hypothesized variance

**Form 3.**      $H_0$:   $\sigma^2 = \sigma_0^2$      Actual variance $\geq$ hypothesized variance
          **With $H_a$:**   $\sigma^2 < \sigma_0^2$      Actual variance $<$ hypothesized variance

Form 1 of $H_0$ and $H_a$ was used by Diane Gray in the Statistics in Action part of this section. Recall that she used a sample variance of 20 measurements to test the variance of a single balance.

Before proceeding, we should mention that $H_0$ and $H_a$ in Forms 1, 2, and 3 could also be stated in terms of $\sigma$ and $\sigma_0$. That is, tests on variances are, in effect, also tests on standard deviations. Thus, rejection of $\sigma^2 = \sigma_0^2$ is the same as rejection of $\sigma = \sigma_0$.

## The test statistic for a test on $\sigma^2$

Before identifying the test statistic we will use to conduct hypothesis tests on $\sigma^2$, we first need to identify an important restriction on the population under investigation. *Specifically, we must require that the population has a bell-shaped frequency distribution.* The hypothesis testing procedures studied at this time only produce reliable results for bell-shaped frequency distributions.

### The Test Statistic for Tests on $\sigma^2$

If $S^2$ is the variance of a sample taken from a bell-shaped distribution, then the test statistic for testing $H_0$: $\sigma^2 = \sigma_0^2$ ($H_0$: $\sigma = \sigma_0$) is:

$$X^2 = \frac{(n-1)S^2}{\sigma_0^2}$$

$S^2$ is the variance of the sample.

$n$ is the sample size.

$\sigma_0^2$ is the variance in $H_0$.

If $H_0$ is true, then $X^2$ has a so-called $\chi^2$ distribution with $n - 1$ degrees of freedom.

The symbol $\chi^2$ is read "ky-square." The chi-square probability distributions will be used to set critical values when testing $H_0: \sigma^2 = \sigma_0^2$. Tables for these distributions are given in Appendix Table 6. However, before discussing these distributions, let us first consider the form of the ratio identified as a test statistic for tests about $\sigma^2$.

$S^2$ is a random variable, as we are reminded by the capital letter. If $H_0$ is true, then $S^2$ should be reasonably close to the value of $\sigma_0^2$. Therefore $S^2/\sigma_0^2$ should be fairly close to 1, and $X^2$ will have a value close to $(n-1)$.

Suppose now that $H_0$ is not true. There is a good chance then that $S^2$ will be different from $\sigma_0^2$, and $X^2$ will correspondingly be quite different from the value $(n-1)$. Notice that $X^2$ can be less than $(n-1)$ if $S^2 < \sigma_0^2$, and greater than $(n-1)$ if $S^2 > \sigma_0^2$.

Intuitively, then, large values of $X^2$ or small values of $X^2$ would indicate that $H_0$ is not true. How large or how small a particular value of $X^2$ must be to reject $H_0$ in favor of $H_a$ depends on the level of significance of the test and the sample size $n$. Notice that $(n-1) > 0$ for $n > 1$, and $S^2$ and $\sigma^2$ are nonnegative. As a consequence, $X^2 \geq 0$.

## The $\chi^2$ probability distributions

The question of how large or how small $X^2$ must be to reject $H_0$ depends on the probability distribution of $X^2$. If $S^2$ is the sample variance from a bell-shaped frequency distribution, then the probability distribution of $X^2$ (supposing $H_0$ to be true) is the so-called $\chi^2$ probability distribution.

Just as there are many binomial distributions, many normal distributions, and many $t$ distributions, there are many $\chi^2$ distributions. The various $\chi^2$ probability distributions are distinguished by a so-called "degrees of freedom" parameter and their graphs have the right-skewed general appearance shown in Figure 6-27.

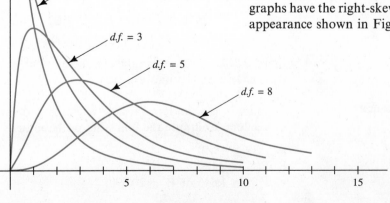

**Figure 6-27.**   Graphs of five $\chi^2$ probability distributions.

| Degrees of Freedom | The Chi-Square ($\chi^2$) Distribution | | | | | | | | | |
|---|---|---|---|---|---|---|---|---|---|---|
| | Right-Tail Area | | | | | | | | | |
| | 0.995 | 0.990 | 0.975 | 0.950 | 0.900 | 0.100 | 0.050 | 0.025 | 0.010 | 0.005 |
| 1 | — | — | 0.001 | 0.004 | 0.016 | 2.706 | 3.841 | 5.024 | 6.635 | 7.879 |
| 2 | 0.010 | 0.020 | 0.051 | 0.103 | 0.211 | 4.605 | 5.991 | 7.378 | 9.210 | 10.597 |
| 3 | 0.072 | 0.115 | 0.216 | 0.352 | 0.584 | 6.251 | 7.815 | 9.348 | 11.345 | 12.838 |
| 4 | 0.207 | 0.297 | 0.484 | 0.711 | 1.064 | 7.779 | 9.488 | 11.143 | 13.277 | 14.860 |
| 5 | 0.412 | 0.554 | 0.831 | 1.145 | 1.610 | 9.236 | 11.071 | 12.833 | 15.086 | 16.750 |
| 6 | 0.676 | 0.872 | 1.237 | 1.635 | 2.204 | 10.645 | 12.592 | 14.449 | 16.812 | 18.548 |
| 7 | 0.989 | 1.239 | 1.690 | 2.167 | 2.833 | 12.017 | 14.067 | 16.013 | 18.475 | 20.278 |
| 8 | 1.344 | 1.646 | 2.180 | 2.733 | 3.490 | 13.362 | 15.507 | 17.535 | 20.090 | 21.955 |
| 9 | 1.735 | 2.088 | 2.700 | 3.325 | 4.168 | 14.684 | 16.919 | 19.023 | 21.666 | 23.589 |
| 10 | 2.156 | 2.558 | 3.247 | 3.940 | 4.865 | 15.987 | 18.307 | 20.483 | 23.209 | 25.188 |

**Figure 6-28.**   Appendix Table 6 for $1 \leq df \leq 10$

Appendix Table 6 has percentiles for $\chi^2$ distributions. A portion of the table is shown in Figure 6-28. The number headings on the ten columns give the areas to the right of particular tabled values. To see the symmetry between the headings of the five columns on the left with the five columns on the right, we need to subtract the headings on the left from 1.000.

| Areas of the $\chi^2$ Distribution with 10 $df$ Corresponding to Table Values | | |
|---|---|---|
| Tabled Value | Area to the Right | Area to the Left |
| 2.156 | 0.995 | 1.000 − 0.995 = 0.005 |
| 2.558 | 0.990 | 1.000 − 0.990 = 0.010 |
| 3.247 | 0.975 | 1.000 − 0.975 = 0.025 |
| 3.940 | 0.950 | 1.000 − 0.950 = 0.050 |
| 4.865 | 0.900 | 1.000 − 0.900 = 0.100 |

Notice that the differences designating areas to the left are the headings of the five columns on the right side of Appendix Table 6. Again using the $\chi^2$ distribution with 10 degrees of freedom, we can write the following probability statements for a $\chi^2$ random variable $W$:

$$P(W \leq 2.156) = 0.005 \quad \text{and} \quad P(W \geq 25.188) = 0.005$$
$$P(W \leq 2.156 \text{ or } W \geq 25.188) = 0.005 + 0.005 = 0.010$$

Since $0.010 = 1\%$, we would use these values of $\chi_c^2$ for a two-tail test and a 1 percent level of significance.

▶ **Example 1**    Suppose $W$ is a random variable with a $\chi^2$ probability distribution with 8 degrees of freedom. Compute the following probabilities:

**a.** $P(W \geq 15.507)$
**b.** $P(W \leq 2.180)$
**c.** $P(W \leq 1.646 \text{ or } W \geq 20.090)$

**Solution**   Using the probabilities in Figure 6-28:

**a.** $P(W \geq 15.507) = 0.050$     Area to the right is 0.050.

**b.** $P(W \leq 2.180) = 1 - P(W \geq 2.180)$
$= 1 - 0.975$     Area to the right is 0.975.
$= 0.025$

**c.** $P(W \leq 1.646 \text{ or } W \geq 20.090) = (1 - 0.990) + 0.010$
$= 0.010 + 0.010 = 0.020$   ◄

## The six-step format and tests on $\sigma^2$

As previously stated, the random variable $(n - 1)S^2/\sigma^2$ has a $\chi^2$ probability distribution. To carry out a hypothesis test on a purported value of $\sigma^2$, we use a critical value $\chi_c^2$ from the $\chi^2$ probability distribution with $(n - 1)$ degrees of freedom. If the test is two-tailed then we need two values for $\chi_c^2$.

▶ **Example 2**   Use the six-step format to carry out the hypothesis test conducted by Diane Gray on the analytic balance in the Statistics in Action section.

**Solution**   **Step 1** $H_0$:  $\sigma^2 = 5$     Diane wanted to show the variance
$H_a$:  $\sigma^2 > 5$     was more than 5 (mg)$^2$.

**Step 2** $\alpha = 0.05$     In her report, Diane stated she used a 5% level of significance.

**Step 3** The test statistic used is

$$X^2 = \frac{(n - 1)S^2}{5}$$     $\sigma_0^2$ is 5 (mg)$^2$.

**Step 4** If $x^2 > \chi_c^2 \ (= 30.144)$, then reject $H_0$.     Appendix Table 6, using $df = 19$ and column 0.050

**Step 5** With $n = 20$ and $s^2 = 6.326\ldots$

$$x^2 = \frac{(20 - 1)6.326\ldots}{5}$$

$$= 24.040$$

**Step 6** Since $24.040 < 30.144$, do not reject $H_0$. That is, there is not sufficient evidence, with $\alpha = 0.05$, to conclude that the variance in the distribution of measurements is more than 5 (mg)$^2$.   ◄

## Confidence intervals for $\sigma^2$

Tabled percentiles for $\chi^2$ probability distributions can also be used to make confidence intervals for $\sigma^2$. If we let $\alpha$ represent the total area in the two tails of a

$\chi^2$ distribution used to construct a particular interval, then $\alpha/2$ is the area in each tail, and $1 - \alpha$ is the confidence coefficient. (This is the same representation we used for $\alpha$ when making confidence intervals for $\mu$ and $P$.)

For large $n$ confidence intervals for $\mu$ and $P$ we used the standard normal distribution to calculate the limits of the intervals. Specifically numbers $z$ and $-z$ were used in computing the upper and lower limits. For small $n$ confidence intervals for means of bell-shaped populations we used numbers $t$ and $-t$ from a $t$-distribution to find the upper and lower limits.

Because the $\chi^2$ distributions are not symmetric, we will have to use the $\chi^2$ table twice in order to make a confidence interval for $\sigma^2$. We need to find two different tabled values that cut off equal areas in the left tail and right tail of the distribution.

As an illustration, Table 6-6 contains tabled values for the $\chi^2$ distribution with 10 degrees of freedom. The tabled values in Example 1 can be used to make $100\% - 10\% = 90\%$ confidence intervals. The tabled values in Example 2 can be used to make $100\% - 2\% = 98\%$ confidence intervals.

**TABLE 6-6   Areas in Left and Right Tails of the $\chi^2$ Distribution with $df = 10$**

| Tabled Value | Area and Location of Tail |
|---|---|
| Example 1 $\longrightarrow$ 3.940  18.307 | Area in left tail: $1.000 - 0.950 = 0.050$  Area in right tail: $0.050$ |
| Example 2 $\longrightarrow$ 2.558  23.209 | Area in left tail: $1.000 - 0.990 = 0.010$  Area in right tail: $0.010$ |

Because two different tabled values are needed for making confidence intervals for $\sigma^2$, we will use subscripts on $\chi^2$ to distinguish between them. Specifically, $\chi_l^2$ will represent the *smaller tabled value* and $\chi_u^2$ will represent the *larger tabled value.* For the tabled values in Table 6-6:

**Example 1**   $\chi_l^2$ is 3.940   and   $\chi_u^2$ is 18.307
**Example 2**   $\chi_l^2$ is 2.558   and   $\chi_u^2$ is 23.209

## Equations for Constructing a Confidence Interval for $\sigma^2$

If $s^2$ is the variance of a sample of size $n$ from a population with a bell-shaped frequency distribution, then a $1 - \alpha$ confidence interval for $\sigma^2$ has:

$$\text{lower limit} = \frac{(n-1)s^2}{\chi_u^2} \quad and \quad \text{upper limit} = \frac{(n-1)s^2}{\chi_l^2}$$

where $\chi_u^2$ and $\chi_l^2$ cut off $\alpha/2$ in the right and left tails, respectively, of the $\chi^2$ distribution with $n - 1$ degrees of freedom.

The square roots of the limits for the variance are the confidence limits for the standard deviation.

▶ **Example 3**  The mathematics department at a major southern university developed an assessment test for students wanting to enroll in the calculus sequence. A sample of 20 scores on the test were:

$$12 \quad 18 \quad 15 \quad 15 \quad 9 \quad 17 \quad 14 \quad 12 \quad 15 \quad 16$$
$$19 \quad 13 \quad 15 \quad 6 \quad 11 \quad 16 \quad 15 \quad 10 \quad 16 \quad 13$$

**a.** Calculate $s^2$ to one decimal place.

**b.** Construct a 90 percent confidence interval for the population variance.

**Solution**  **a.** $s^2 = \dfrac{20(4{,}027) - 277^2}{20(19)}$

$= 10.0289 \ldots$

$\approx 10.0$, to one decimal place

**b.** lower limit $= \dfrac{(20 - 1)(10.028 \ldots)}{30.144}$    $\chi_u^2 = 30.144$

$\approx 6.3$ points$^2$    To one decimal place

upper limit $= \dfrac{(20 - 1)(10.028 \ldots)}{10.117}$    $\chi_l^2 = 10.117$

$\approx 18.8$ points$^2$    To one decimal place

Thus, a 90 percent confidence interval for the population variance is from 6.3 to 18.8.

Notice that limits for the population standard deviation are:

lower limit $= \sqrt{6.32 \ldots} \approx 2.5$ points

upper limit $= \sqrt{18.83 \ldots} \approx 4.3$ points  ◀

## Exercises 6-5  Set A

*In Exercises 1–10, assume W is a random variable with a $\chi^2$ probability distribution. Use Appendix Table 6 to determine the indicated probabilities for W using the stated degrees of freedom.*

**1.** $P(W \geq 27.488)$; $df = 15$

**2.** $P(W \geq 46.963)$; $df = 27$

**3.** $P(W \leq 11.591)$; $df = 21$

**4.** $P(W \leq 1.239)$; $df = 7$

**5.** $P(W \leq 5.142$ or $W \geq 34.267)$; $df = 16$

**6.** $P(W \leq 9.591$ or $W \geq 34.170)$; $df = 20$

**7.** $P(W \geq 7.042)$; $df = 13$

**8.** $P(W \geq 3.571)$; $df = 12$

**9.** $P(W \leq 30.144)$; $df = 19$

**10.** $P(W \leq 21.955)$; $df = 8$

*In Exercises 11–20, determine $\chi_c^2$ (or $\chi_c^2 s$) for each indicated hypothesis test.*

**11.** $H_a$: $\sigma^2 > 3.5$
$\alpha = 0.05$
$n = 16$

**12.** $H_a$: $\sigma^2 > 22.9$
$\alpha = 0.01$
$n = 25$

**13.** $H_a$: $\sigma^2 \neq 25$
$\alpha = 0.01$
$n = 9$

**14.** $H_a$: $\sigma^2 \neq 70$
$\alpha = 0.05$
$n = 20$

**15.** $H_a$: $\sigma^2 < 0.49$
$\alpha = 0.10$
$n = 26$

**16.** $H_a$: $\sigma^2 < 0.04$
$\alpha = 0.025$
$n = 12$

**17.** $H_a$: $\sigma^2 \neq 5.75$
$\alpha = 0.10$
$n = 13$

**18.** $H_a$: $\sigma^2 \neq 9.36$
$\alpha = 0.01$
$n = 6$

**19.** $H_a$: $\sigma^2 > 30.0$
$\alpha = 0.025$
$n = 18$

**20.** $H_a$: $\sigma^2 > 150.0$
$\alpha = 0.010$
$n = 10$

*In Exercises 21–26, conduct a hypothesis test using the six-step format and the given information.*

**21.** $H_0$: $\sigma^2 = 45$
$H_a$: $\sigma^2 > 45$
$\alpha = 0.05$, $n = 14$, and $s^2 = 79$

**22.** $H_0$: $\sigma^2 = 10.1$
$H_a$: $\sigma^2 > 10.1$
$\alpha = 0.01$, $n = 22$, and $s^2 = 19.3$

**23.** $H_0$: $\sigma^2 = 8.75$
$H_a$: $\sigma^2 \neq 8.75$
$\alpha = 0.01$, $n = 7$, and $s^2 = 2.73$

**24.** $H_0$: $\sigma^2 = 340$
$H_a$: $\sigma^2 \neq 340$
$\alpha = 0.05$, $n = 17$, and $s^2 = 582$

**25.** $H_0$: $\sigma^2 = 65$
$H_a$: $\sigma^2 > 65$
$\alpha = 0.10$, $n = 6$, and $s^2 = 124$

**26.** $H_0$: $\sigma^2 = 3.60$
$H_a$: $\sigma^2 > 3.60$
$\alpha = 0.01$, $n = 18$, and $s^2 = 7.29$

*In Exercises 27–34, construct confidence intervals for $\sigma^2$ with the indicated confidence coefficients.*

**27.** $s^2 = 8.0$, $n = 12$, and a 90% confidence coefficient

**28.** $s^2 = 10.5$, $n = 14$, and a 90% confidence coefficient

**29.** $s^2 = 20.25$, $n = 10$, and a 95% confidence coefficient

**30.** $s^2 = 16.75$, $n = 20$, and a 95% confidence coefficient

**31.** $s^2 = 2.42$, $n = 13$, and a 98% confidence coefficient

**32.** $s^2 = 5.30$, $n = 17$, and a 98% confidence coefficient

**33.** $s^2 = 38.5$, $n = 26$, and a 99% confidence coefficient

**34.** $s^2 = 49.3$, $n = 30$, and a 99% confidence coefficient

*In Exercises 35–40, conduct a hypothesis test on $\sigma^2$ using the six-step format.*

**35.** Earl Karn monitors the production line of $\frac{1}{2}$-inch galvanized elbows manufactured by Kessler Plumbing Products, Inc. The company standard for the variance in weights for this product is 0.010 (ounces)$^2$. A sample of 14 elbows from a recent production run yielded a sample variance of 0.014 (ounces)$^2$.

   a. Use $\alpha = 0.05$ and test $H_0: \sigma^2 = 0.010$ versus $H_a: \sigma^2 > 0.010$.

   b. Would Earl ever be concerned about the variance being too small? Discuss.

   c. Suppose $\sigma^2$ is unknown and use the given information to make a 90% confidence interval for $\sigma^2$.

**36.** Valerie Alger periodically checks the variance in the amounts of salad dressing in the 473-milliliter bottles of Bridges Blue Cheese Dressing. The company standard for the variance is 2.56 (ml)$^2$. A sample of 21 fills from a recent production run yielded a sample variance of 3.89 (ml)$^2$.

   a. Use $\alpha = 0.05$ and test $H_0: \sigma^2 = 2.56$ versus $H_a: \sigma^2 > 2.56$.

   b. Would Valerie ever be concerned about the variance being too small? Discuss.

   c. Suppose $\sigma^2$ is unknown and use the given information to make a 90% confidence interval for $\sigma^2$.

**37.** Lillie Chalmers constructs tests designed to assess the educational development of sixth-grade students in mathematics. The several forms of the test have variances of approximately 100 points$^2$ when used in testing math ready sixth graders. Lillie recently developed a new form for the test, and when she administered it to 30 sixth graders, the variance in the scores was 62 points$^2$.

   a. For this new form, use $\alpha = 10\%$ and test $H_0: \sigma^2 = 100$ versus $H_a: \sigma^2 \neq 100$.

   b. Identify at least one reason why a small variance would be unacceptable in this context.

   c. Suppose $\sigma^2$ is not known and use the given information to make a 95% confidence interval for $\sigma^2$.

**38.** Ray Kelley is an air controller at Bay View City airport. A computer at the facility is connected to a sensor at the end of the runway. The computer records the times between the arrivals of aircraft landing on the runway. For the distribution of inter arrival times, the variance is reported to be 56.25 (min)$^2$. A sample of 15 times recorded by the computer yielded $s^2 = 28.75$ (min)$^2$.

   a. Use $\alpha = 10\%$ and test $H_0: \sigma^2 = 56.25$ versus $H_a: \sigma^2 > 56.25$.

   b. Suppose $\sigma^2$ is not known and use the given information to make a 95% confidence interval for $\sigma^2$.

**39.** Sonya Neal is marketing manager for Salladay Tire and Rubber Corporation. She is developing advertising material for the new model 3XK7009 sport car tire. Sonya wants to highlight a uniform wear property of the tire by citing data from the "wear machine," which rotates tires for a specified period of time and then records the amount of tread worn off each tire. Can Sonya claim that the variance in tread wear is clearly less than 0.30 mm$^2$ if 16 tested tires yielded $s^2 = 0.21$ mm$^2$? Use $\alpha = 2.5\%$.

**40.** Renay York is the manager of York and Associates Real Estate Enterprises. The motto of the firm is "We have residential property that will fit anyone's pocketbook." To justify this motto, the firm maintains a listing of residential properties with a large variability in prices. Renay feels that a variance of $625 (thousand dollars)$^2$ is a reasonable guideline for this goal. A recent sample of 12 listings yielded $s^2 = \$210$ (thousand dollars)$^2$ in the prices. Is this strong evidence that the population variance has dropped below (thousand dollars)$^2$? Use $\alpha = 5\%$.

# Exercises 6-5    Set B

1. The weights (in pounds) of potatoes from 20 plots of 0.1 acre each were obtained by researchers at Anderson Seeds Corporation. These are given below:

| | | | | |
|---|---|---|---|---|
| 1420 | 1415 | 1438 | 1429 | 1432 |
| 1447 | 1452 | 1436 | 1428 | 1442 |
| 1432 | 1449 | 1417 | 1442 | 1435 |
| 1439 | 1444 | 1438 | 1426 | 1458 |

   a. Calculate $s^2$ for the given sample.
   b. Test $H_0: \sigma^2 = 100.0$ (lb)$^2$ at the 5% level of significance.
   c. What assumptions were made to conduct the test?
   d. Construct a stem-and-leaf diagram of the data using 141, 142, 143, 144, and 145 for the stem values. Based on the shape of the diagram, are the assumptions of part (c) reasonable?

2. Michelle Beese is president of Beese Batteries Incorporated. The company manufactures a rechargeable flashlight that reportedly provides 3 hours of usable light from a fully charged battery. The quality control staff uses a light meter to determine the minutes of "usable light" provided by the flashlights. Michelle does not want the variance of the distribution of minutes of usable light to exceed 15 (min)$^2$. A sample of 24 fully charged flashlights were tested using the light meter. The periods of usable light in minutes, for these flashlights are recorded below:

| | | | | | | | |
|---|---|---|---|---|---|---|---|
| 179 | 199 | 172 | 185 | 208 | 174 | 186 | 180 |
| 185 | 172 | 193 | 168 | 199 | 186 | 161 | 197 |
| 202 | 181 | 170 | 183 | 204 | 190 | 195 | 188 |

   a. Calculate $s^2$ for the given sample.
   b. Test $H_0: \sigma^2 = 15$ (min)$^2$ at the 10% level of significance.
   c. What assumptions were made to conduct the test?
   d. Construct a stem-and-leaf diagram of the data using 16, 17, 18, 19, and 20 for the stem values. Based on the shape of the diagram, are the assumptions of part (c) reasonable?

*Exercises 3 and 4 refer to the data in Data Set I, Appendix B.*

3. Use random numbers to sample the ages of 20 females.
   a. Compute $s^2$ for your sample data to two decimal places.
   b. Use $\alpha = 0.05$ and test $H_0: \sigma^2 = 420.25$ versus $H_a: \sigma^2 \neq 420.25$.
   c. The stated variance is correct. Did your sample data permit you to make the correct decision?
   d. Compare your results with those of other students in the class.

4. Use random numbers to sample the ages of 24 males.
   a. Compute $s^2$ for your sample data to two decimal places.
   b. Use $\alpha = 0.10$ and test $H_0: \sigma^2 = 380.25$ versus $H_a: \sigma^2 \neq 380.25$.
   c. The stated variance is correct. Did your sample data permit you to make the correct decision?
   d. Compare your results with those of other students in the class.

# Chapter 6 Summary

## Definitions

1. The **null hypothesis** states that there is no difference between the actual unknown value of a parameter and a stated value. The **alternative hypothesis** states that there is a difference between the actual unknown value of the parameter and the stated value.

2. A **critical value** is a boundary point between those values of $\bar{X}$, $\hat{P}$, $Z$, $T$, or $X^2$ that will result in rejection of $H_0$ and those that will not result in rejection of $H_0$.

3. The interval (or intervals) of possible values of $\bar{X}$, $\hat{P}$, $Z$, $T$, or $X^2$ more extreme than the critical value (or values) is called **the rejection region**.

4. The probability that a sampling distribution of $\bar{X}$, $\hat{P}$, $Z$, $T$, or $X^2$ assigns to the rejection region if $\mu$, $P$, or $\sigma^2$ has the value stated in $H_0$ is called **the level of significance** of a test.

5. A **Type I error** is made when the state of affairs symbolized by the null hypothesis is true, but the decision is made to reject $H_0$ in favor of $H_a$. A **Type II error** is made when the state of affairs symbolized by the alternative hypothesis is true, but the decision is made to not reject $H_0$.

6. The ***p*-value** of a hypothesis test is the smallest level of significance for which the observed sample information causes rejection of $H_0$.

## Rules and Equations

a. Calculating the critical values for a hypothesis test
   1. Using $\bar{X}$ as a test statistic for $H_0: \mu = \mu_0$ and knowing $\sigma$

$$\bar{x}_c = \mu_0 + z_c\left(\frac{\sigma}{\sqrt{n}}\right) \quad \begin{array}{l} \longrightarrow \mu_0 \text{ is the mean from } H_0 \\ \longrightarrow z_c \text{ is the critical } z \text{ value} \end{array}$$

   2. Using $\hat{P}$ as the test statistic for $H_0: P = P_0$

$$\hat{p}_c = P_0 + z_c\sqrt{\frac{P_0(1 - P_0)}{n}} \quad \begin{array}{l} \longrightarrow P_0 \text{ is the proportion from } H_0 \\ \longrightarrow z_c \text{ is the critical } z \text{ value} \end{array}$$

b. Four possibilities for the decision in a hypothesis test

|  |  | True State of Affairs | |
|---|---|---|---|
|  |  | $H_0$ **Is True** | $H_a$ **Is True** |
| **The Decision is Made to:** | **Reject $H_0$** | Type I error | Correct decision |
|  | **Not reject $H_0$** | Correct decision | Type II error |

c. Six step format for a hypothesis test

**Step 1** State $H_0$ and $H_a$.

**Step 2** State $\alpha$, the level of significance.

**Step 3** Show the appropriate formula for the test statistic.

**Step 4** State the decision rule in terms of the rejection region.

**Step 5** Calculate the observed value of the test statistic.

**Step 6** State a decision about the parameter(s) being tested that is consistent with a comparison of steps 4 and 5.

d. The test statistic for a large $n$ test on $\mu$ is:

$$Z = \frac{\bar{X} - \mu_0}{\frac{S}{\sqrt{n}}}$$

If $H_0$ is true, $Z$ has a standard normal distribution.

e. The test statistic for a small $n$ test on $\mu$, where $\mu$ is the mean of a bell-shaped population of quantitative data, is:

$$T = \frac{\bar{X} - \mu_0}{\frac{S}{\sqrt{n}}}$$

If $H_0$ is true, $T$ has a $t$-distribution with $n - 1$ degrees of freedom.

f. The test statistic for a large $n$ test on $P$ is:

$$Z = \frac{\hat{P} - P_0}{\sqrt{\frac{P_0(1 - P_0)}{n}}}$$

If $H_0$ is true and $n$ is large, $Z$ has approximately a standard normal distribution.

g. The test statistic for a test on $\sigma^2$, where $\sigma^2$ is the variance of a bell-shaped population is:

$$X^2 = \frac{(n - 1)S^2}{\sigma_0^2},$$

If $H_0$ is true, $X^2$ has a $\chi^2$ distribution with $n - 1$ degrees of freedom.

h. If $s^2$ is the variance of a sample of size $n$ from a population with a bell-shaped distribution, then a $1 - \alpha$ confidence interval for $\sigma^2$ has:

$$\text{lower limit} = \frac{(n - 1)s^2}{\chi_u^2} \quad \text{and} \quad \text{upper limit} = \frac{(n - 1)s^2}{\chi_l^2}$$

where $\chi_u^2$ and $\chi_l^2$ cut off $\alpha/2$ in the right and left tails respectively of the $\chi^2$ distribution with $n - 1$ degrees of freedom.

## Symbols

$H_0$    the **null hypothesis**

$H_a$    the **alternative hypothesis**

$\mu_0$    the **mean** from the null hypothesis

$P_0$    the **proportion** from the null hypothesis

$\sigma_0^2$    the **variance** from the null hypothesis

$\bar{x}_c$    a **critical value** of $\bar{x}$

$\hat{p}_c$    a **critical value** of $\hat{p}$

$z_c$    a **critical value** of $z$

$t_c$    a **critical value** of $t$

$\chi_c^2$    a **critical value** of $x^2$

$\alpha$    $P(\text{Type I error})$

$\beta$    $P(\text{Type II error})$

## Comparing Test Statistics for $H_0\colon \mu = \mu_0$, $H_0\colon P = P_0$, and $H_0\colon \sigma^2 = \sigma_0^2$

**Hypothesis**

**Test statistic**

Large $n$ test:

$$H_0\colon \quad \mu = \mu_0$$

$$Z = \frac{\bar{X} - \mu_0}{\dfrac{S}{\sqrt{n}}}$$

Small $n$ test for bell-shaped populations:

$$H_0\colon \quad \mu = \mu_0$$

$$T = \frac{\bar{X} - \mu_0}{\dfrac{S}{\sqrt{n}}}, \text{ degrees of freedom } n - 1 \text{ for } t\text{-distribution}$$

Large $n$ test:

$$H_0\colon \quad P = P_0$$

$$Z = \frac{\hat{P} - P_0}{\sqrt{\dfrac{P_0(1 - P_0)}{n}}}$$

Test for bell-shaped populations:

$$H_0\colon \quad \sigma^2 = \sigma_0^2$$

$$X^2 = \frac{(n - 1)S^2}{\sigma_0^2}, \text{ degrees of freedom } n - 1 \text{ for } \chi^2\text{-distribution}$$

# Chapter 6 Review Exercises

*In Exercises 1 and 2, write $H_0$ and $H_a$ that would be appropriate for each test.*

1. The superintendent of a large school district reported that the mean score of senior students on the recently administered achievement test was 110.3. Dr. Patricia Quinlin, the principal at Random High School, felt that the senior students in her school scored higher than the district average.

2. A report from the county animal control office suggests that 42 percent of the dogs in the county do not have the proper tags. Mark Weber, an animal control field representative, feels the figure is exaggerated to put pressure on employees that work in the field.

*In Exercises 3 and 4, compute the level of significance for each test. Round answers to two decimal places.*

3. $H_0$:  $P = 0.630$
   $H_a$:  $P > 0.630$
   for $n = 150$ and $\hat{p}_c = 0.705$

4. $H_0$:  $\mu = \$52.75$
   $H_a$:  $\mu < \$52.75$
   for $\sigma = \$3.75$, $n = 38$, and $\bar{x}_c = \$51.87$

*In Exercises 5 and 6, find the critical value (or values) for the stated levels of significance.*

5. A study by the Regional Transit Board in a large metropolitan area in the South suggests that workers in the area drive an average of 135 miles per week to work and home again. A committee of concerned taxpayers feels the figure is overstated. They intend to survey 175 workers to obtain data for a test. Assume that the population standard deviation for the distribution of distances driven by workers in this area is 45 miles per week and a 5 percent level of significance is used for the test.

6. A committee on nutrition reported that 28 percent of high school students do not eat any breakfast on a regular basis. The nurse at Jefferson High School has been giving lectures on the value of proper nutrition to the students at her school. She feels strongly that the figure is now lower at Jefferson High School. She plans to survey 80 students to obtain data for a test and will use a 1 percent level of significance.

*In Exercises 7 and 8 state:*
a. *The conditions that would accompany a Type I error.*
b. *The conditions that would accompany a Type II error.*

7. The mean weight of a large shipment of hogs from the Medina County Pork Cooperative is reported to be 200 pounds. The weights of $n = 36$ hogs are obtained to test $H_0$: $\mu = 200$ pounds versus $H_a$: $\mu < 200$ pounds.

8. The mean score on a standardized achievement test for the 10th grade students of a large metropolitan school district is reported to be 137.6. The scores of $n = 80$ randomly selected 10th grade students in this district are obtained to test $H_0$: $\mu = 137.6$ versus $H_a$: $\mu \neq 137.6$.

9. For the indicated test, suppose $\sigma = 16.0$ and $n = 45$. $H_0$: $\mu = 140.0$ versus $H_a$: $\mu > 140.0$
   a. Calculate $\bar{x}_c$, if $\alpha = 5\%$.
   b. Calculate $\alpha$, if $\bar{x}_c$ is 144.5.

10. For the indicated test, use $n = 125$. $H_0$: $P = 0.42$ versus $H_a$: $P \neq 0.42$
    a. Calculate lower and upper $\hat{p}_c$, if $\alpha = 1\%$.
    b. Calculate $\alpha$, if lower $\hat{p}_c = 0.35$ and upper $\hat{p}_c = 0.49$ are used.

11. For the indicated test, suppose $n = 84$. $H_0$: $P = 0.78$ versus $H_a$: $P < 0.78$. Calculate $\beta$, if $P = 0.73$ and $\hat{p}_c = 0.71$.

12. For the indicated test, suppose $\sigma = 4.50$ and $n = 50$. $H_0$: $\mu = 87.50$ versus $H_a$: $\mu > 87.50$. Calculate $\beta$, if $\mu = 88.75$ and $\bar{x}_c = 89.00$.

13. The management for a large restaurant claimed the waiters and waitresses average $75 a day per employee in tips. The staff felt the figure was too high and collected 270 data points to conduct a test using a 1 percent level of significance. Can the $75 figure be rejected as too high, if the sample yielded $\bar{x} = \$72.33$ with $s = \$15.48$?

14. A major bulb manufacturer claims that its 40-watt fluorescent bulbs average 20,000 hours of life per bulb. A competitor tests 36 bulbs and obtains $\bar{x} = 19,730$ hours with $s = 900$ hours. Can the first manufacturer's claim be rejected as too high at the 5 percent level of significance?

15. An article in a metropolitan newspaper claimed that a dozen red roses could be bought at one of the more than 1000 florists in the city for $20 on the average. An organization of local nurseries felt the figure was too low and sampled 16 florists in the city regarding their prices for a dozen red roses. Assuming the distribution of rose prices in the city is approximately normal, can the $20 figure be rejected as too low, if $\bar{x} = \$24.17$ and $s = \$4.60$? Use $\alpha = 0.01$.

16. The developers of a suburban community outside a large city in the West claim that the mean commute time to the inner city is 30 minutes. To check the accuracy of the 30-minute commute time, the staff of the regional transit system sampled 18 residents already living there. For these residents, the mean commute time was 32.8 minutes with $s = 4.2$ minutes. Assuming the distribution of commuting times for residents in this community is bell shaped, can the 30-minute average claim of the developers be rejected? Use $\alpha = 0.05$.

17. A survey by the Department of Health and Human Services suggests that 34 percent of adults in the U.S. have not had an alcoholic drink in the past year. The manager of a wine distribution company in a large region in the South conducted a poll of 175 shoppers in a mall. He obtained 55 adults in the sample who vowed they had not had one alcoholic drink in the past year. Can these results be used to reject the 34 percent figure as too high for adults living in this region? Use $\alpha = 0.05$.

18. A large shipment of Christmas tree light bulbs arrived at a warehouse in Chicago. According to the manufacturer, the stock contained less than 10 percent defective bulbs. The foreman of the warehouse had 56 bulbs checked for defects. Can the 10 percent figure be rejected as too low if the sample yielded 9 bulbs with defects? Use $\alpha = 0.01$.

*In Exercises 19 and 20, assume W is a random variable with a $\chi^2$ probability distribution. Use Appendix Table 6 to determine the indicated probabilities for W using the stated degrees of freedom.*

19. $P(W \leq 8.907)$; $df = 19$

20. $P(W \leq 5.226$ or $W \geq 21.026)$; $df = 12$

*In Exercises 21 and 22, determine $\chi_c^2$ (or $\chi_c^2$ s) for each indicated hypothesis test.*

21. $H_0$:  $\sigma^2 = 8.45$          22. $H_0$:  $\sigma^2 = 21.7$
    $H_a$:  $\sigma^2 > 8.45$              $H_a$:  $\sigma^2 \neq 21.7$
    $\alpha = 0.10$                        $\alpha = 0.05$
    $n = 15$                              $n = 24$

*In Exercises 23 and 24, conduct a hypothesis test using the six-step format and the given information.*

23. $H_0$:  $\sigma^2 = 490$
    $H_a$:  $\sigma^2 \neq 490$
    $\alpha = 0.10$, $n = 18$, and $s^2 = 804$

24. $H_0$:  $\sigma^2 = 72.5$
    $H_a$:  $\sigma^2 < 72.5$
    $\alpha = 0.05$, $n = 23$, and $s^2 = 38.9$

## Hypothesis Tests

| Key Topics | 1. *Large samples* |
| --- | --- |
| | 2. *Small samples* |

### Large samples

For large samples, a null hypothesis about $\mu$ can be tested versus any of the three alternative hypotheses using the Minitab command ZTEST. The population standard deviation is estimated by $s$, the sample standard deviation.

Suppose that the data of Exercise 1 in Set B of Section 6-3 is in column 1 of the Minitab worksheet and we wish to test:

$$H_0: \quad \mu = 6.0$$

against

$$H_a: \quad \mu \neq 6.0$$

The command STDEV C1 reveals that the sample standard deviation is approximately 2.649 for this data.

We then give the command

```
ZTEST MU = 6.0 SIGMA = 2.649 FOR DATA IN C1
```

Minitab returns the output in Figure 6-29.

```
MTB> ZTEST MU = 6 SIGMA = 2.469 C1
TEST OF MU = 6.000 VS MU N.E.  6.000
THE ASSUMED SIGMA = 2.65
        N     MEAN    STDEV    SE MEAN       Z    P VALUE
C1     36    3.694    2.649      0.441    -5.22    0.0000
```

**Figure 6-29.** Minitab output.

The first line of output, after the ZTEST command gives the null and alternative hypotheses of the test (NE stands for "not equal"). The next line gives the value used for sigma. The final line gives the statistics from the sample, the computed value of $z$ and the $p$-value corresponding to that $z$ value. Because the $p$-value is 0.0000, we may safely conclude that the population mean is different from 6.0.

To make a one-tailed hypothesis test, the ALTERNATIVE = K subcommand is used with K = 1 for an upper-tail test and K = −1 for a lower-tail test. As an example:

```
ZTEST 6.0 2.649 C1;
ALTERNATIVE -1.
```

would be used with the data in column 1 to test:

$$H_0: \quad \mu = 6.0$$
$$H_a: \quad \mu < 6.0$$

## Small samples

When the population is normal the TTEST command is used to test hypotheses about $\mu$. This command may also be used to conduct large sample tests (without having to separately compute the sample standard deviation before issuing a command to perform the test).

Suppose the following data are in column 2 of the Minitab worksheet.

$$21.5 \quad 18.73 \quad 16.49 \quad 25.71 \quad 22.15 \quad 12.32 \quad 21.99$$
$$17.62 \quad 25.02 \quad 16.87 \quad 16.61 \quad 28.53$$

The Minitab output for a test of the hypotheses

$$H_0: \quad \mu = 20.0$$
$$H_a: \quad \mu > 20.0$$

is given in Figure 6-30. Note that the computer output is very similar to that for the ZTEST command. This time, however, the $p$-value is 0.42. Because the $p$-value is so large, there is no significant evidence that the population mean is greater than 20.0.

```
MTB> TTEST MU = 20.0 C2;
SUBC> ALTERNATIVE 1.
TEST OF MU = 20.000 VS MU G.T. 20.000
```

|    | N  | MEAN   | STDEV | SE MEAN | T    | P VALUE |
|----|----|--------|-------|---------|------|---------|
| C2 | 12 | 20.000 | 4.666 | 1.347   | 0.20 | 0.42    |

**Figure 6-30.** Minitab output.

## Exercises

*Four Minitab hypothesis tests are given in Figures 6-31 through 6-34. For each:*
*a. State the null and alternate hypotheses.*
*b. State the proper conclusion, if the test is conducted at the 5% level.*
*c. What is the smallest level of significance at which $H_0$ would be rejected?*

**1.**
```
MTB> ZTEST MU = 11.5 SIGMA = 3 C1;
SUBC> ALTE 1.
TEST OF MU = 11.500 VS MU G.T. 11.500
THE ASSUMED SIGMA = 3.00
          N      MEAN    STDEV    SE MEAN      Z    P VALUE
C1       40    12.625    3.354      0.474    2.37    0.0090
```

**Figure 6-31.** Minitab output.

**2.**
```
MTB> ZTEST 120 7.66 C2
TEST OF MU = 120.000 VS MU N.E. 120.000
THE ASSUMED SIGMA = 7.66
          N      MEAN    STDEV    SE MEAN      Z    P VALUE
C2       60   119.735    7.637      0.989   -0.26     0.79
```

**Figure 6-32.** Minitab output.

**3.**
```
MTB> TTEST 3.0 C3;
SUBC> ALTE 1.
TEST OF MU = 3.000 VS MU G.T. 3.000
          N      MEAN    STDEV    SE MEAN      T    P VALUE
C3       12     2.928    1.018      0.294   -0.24     0.59
```

**Figure 6-33.** Minitab output.

**4.**
```
MTB> TTEST 34 C4;
SUBC> ALTE -1.
TEST OF MU = 34.000 VS MU L.T. 34.000
          N      MEAN    STDEV    SE MEAN      T    P VALUE
C4       20    32.714    4.463      0.998   -1.79    0.045
```

**Figure 6-34.** Minitab output.

## SCIENTIFIC

There are two equations in this chapter for calculating the observed value of the test statistics for hypothesis tests on $\mu$ and $P$:

$$z = \frac{\bar{x} - \mu_0}{\frac{s}{\sqrt{n}}} \quad \text{and} \quad z = \frac{\hat{p} - P_0}{\sqrt{\frac{P_0(1 - P_0)}{n}}}$$

These formulas can be written on one line by using parentheses and the $\div$ symbol:

$$z = (\bar{x} - \mu_0) \div (s \div \sqrt{n}) \quad \text{and} \quad z = (\hat{p} - P_0) \div \sqrt{P_0(1 - P_0) \div n}$$

The parentheses keys ( and ) can be used to make one continuous series of key strokes to calculate $z$ for a given set of data. The following example illustrates the process for a test on $\mu$.

▶ **Example**    Calculate $z$, for $\bar{x} = 2.92$, $\mu_0 = 2.83$, $s = 0.45$, and $n = 225$.

**Solution**

| Press | Display |
|-------|---------|
| ( | [01        0. |
| 2.92 | 2.92 |
| − | 2.92 |
| 2.83 | 2.83 |
| ) | 0.09 |
| ÷ | 0.09 |
| ( | [01        0. |
| .45 | 0.45 |
| ÷ | 0.45 |
| 225 | 225 |
| √ | 15 |
| ) | 0.03 |
| = | 3. |

Therefore $z = 3.00$.    ◀

A similar example and set of values could be used to demonstrate the equation for a test on $P$.

## GRAPHING

Parentheses can also be used to simplify the key sequence for a graphics calculator. The following example illustrates the proper key strokes for a Casio model using the same values for $\bar{x}$, $\mu_0$, $s$, and $n$ in the scientific calculator supplement.

▶ **Example**    Calculate $z$, for $\bar{x} = 2.92$, $\mu_0 = 2.83$, $s = 0.45$, and $n = 225$.

**Solution**

| Press | Display |
|-------|---------|
| ( | ( |
| 2.92 | (2.92 |
| − | (2.92 − |
| 2.83 | (2.92 − 2.83 |
| ) | (2.92 − 2.83) |
| ÷ | (2.92 − 2.83) ÷ |
| ( | (2.92 − 2.83) ÷ ( |
| .45 | (2.92 − 2.83) ÷ (.45 |
| ÷ | (2.92 − 2.83) ÷ (.45  ÷ |
| √ 225 | (2.92 − 2.83) ÷ (.45  ÷√ 225 |
| ) | (2.92 − 2.83) ÷ (.45  ÷√ 225) |
| EXE | 3 |

◀

An advantage of the graphics calculator over the scientific one is seen when any of the values, $\bar{x}$, $\mu_0$, $s$, or $n$, must be changed. Specifically only the value that must be changed is replaced with the revised value and the [EXE] command will give the new value for $z$.

To illustrate, suppose $\mu_0 = 2.89$ is used in the previous example, instead of the given value of 2.83. First press the key marked [⇨]. The old answer disappears and the cursor will blink under the first parenthesis. Now press the [⇨] key and hold it down until the 3 in the 2.83 blinks. Key in 9 as the new digit and press [EXE] to obtain the new value for $z$, namely $z = 1.00$.

Notice that any of the four variables can be similarly changed by locating the cursor under the digits to be replaced. Once the replacements have been made, the single command [EXE] will give the new value for $z$.

**Name:** Roger M. Sauter
**Occupation:** Statistical Consultant
**Employer:** Boeing Commercial
Airplanes Group
Everett, Washington

**M**y introduction to a career of a statistician came in the ninth grade. In a guidance counseling class, we were assigned to pick a career that looked interesting. I chose to be a statistician because it was described as an applied form of mathematics and because of the huge salary (listed as a whopping $8000 in 1972). By the end of my senior year of college, I didn't know what I wanted to do with a major in mathematics. I didn't think I wanted to teach high school math or be a computer programmer, which seemed to be the only options open to me as a math major. I liked the two courses I had taken in statistics. The first was an introductory course in statistics taught by the psychology department. The professor challenged us with a final project that involved obtaining some real data and analyzing it. The other was an introduction to probability course from the math department. During the spring break of my senior year, I checked out the graduate program in statistics at a nearby university. I liked what the professor had to say and enrolled that fall.

Currently my job involves advising people how to collect and analyze data, and answering questions important to them. Many times analyzing the data is as simple as calculating summary statistics, like means and standard deviations, or plotting simple graphs like histograms, scatter plots, and control charts. Other times more in-depth analysis is needed. A current project involves forecasting the status of a performance/business measure based on its past behavior. A previous project involved designing an experiment to isolate a problem and finding the right combination of materials to help a system function properly. Because the human brain does not randomize, some combinations of variables are often overlooked. A designed experiment can aid tremendously in situations like this. The whole idea of statistical analysis is to use objective data to get to the facts rather than relying on subjective opinions and preconceived ideas.

I find that the most gratifying part of my job is helping clients arrive at solutions that solve their problems. The solution may appear obvious after considering the data (just using simple statistical techniques), but because clients are close to their problems they need the statistical evidence to lead them down the proper pathway. Learning about the area to which I am applying statistical techniques is one of the exciting parts about being a statistician. I chose Boeing Commercial Airplanes partly because I wanted to learn about the Boeing 747 airplane. I had never been in a 747 before working at Boeing.

Being a statistician is a rewarding and interesting occupation. Thinking "statistically" will aid you in whatever career you choose, even if you don't become a full-time statistician. It is one of the most important tools you can obtain from school, not just memorizing the formulas, but understanding the thought process that provides the reasoning behind the formulas. Even if math and/or statistics doesn't come easy for you, stick with it and you will benefit from it in the real world of work.

# 'Weaker sex' wins

Men appear to be losing another skirmish in the battle of the sexes.

A new study published in the May issue of Journal of Clinical Psychology suggests men suffer memory loss earlier than women.

The research, led by neurologist Dr. Margit Bleecker of Johns Hopkins Hospital in Baltimore, revealed dramatic differences between what men and women ages 60 to 90 could remember. While both sexes showed a decline with age, men could recall 20 percent fewer words from a list of common nouns than could women.

Unlike studies that have focused on those suffering memory loss due to diseases such as Alzheimer's, Bleecker's study tested healthy patients.

"These are normal, age-related changes that show the boundaries of aging in healthy people," Bleecker said. "They are built into our genes."

**Figure 7-1.** "Weaker sex wins" from *AARP News Bulletin*, June 1988, p. 11. Reprinted by permission of AARP, Washington, DC, 20049.

# Men do more housework

## By RANDOLPH E. SCHMID
ASSOCIATED PRESS.

WASHINGTON — American women still do most of the cooking, laundering and housecleaning, but men have accepted an increased share of other household chores such as pet care, gardening and dealing with bills, a university study shows.

Overall, American women do two hours of housework for every one hour put in by a man, according to the report by sociologist John P. Robinson of the University of Maryland.

Robinson's findings, based on his 1985 survey of 5,000 men and women, compare to earlier, similar studies at the University of Michigan which said that women performed a 3-to-1 share of housework in 1975 and nearly 6 to 1 in 1965.

All three analyses found four primarily female tasks — cooking

meals, cleaning up after meals, housecleaning and laundry.

Men dominated two areas, yard work and home repairs, Robinson found.

And two areas where women previously did the majority of the work — physically dealing with bills and the combined category of gardening and pet care — were about evenly divided in the most recent report.

"Several important trends account for shifts in who does how much housework," Robinson said. These include declines in the share of households with children, a smaller share of married-couple households and the increasing number of working women.

Robinson's study is reported in the December edition of American Demographics magazine, which concentrates on population and statistics issues.

**Figure 7-2.** "Men do more housework" by Randolph E. Schmid, *The Sacramento Union*, December 1, 1988, p. 25. Reprinted by permission of Associated Press.

# 7

# Comparisons Based on Two Independent Samples and Paired Differences

Figures 7-1 and 7-2 contain some of the findings of studies carried out by Dr. Margit Bleecker of Johns Hopkins Hospital in Baltimore and John P. Robinson of the University of Maryland. Both studies compared certain aspects of two different populations. Dr. Bleecker's research involved women (Population I) and men (Population II) between the ages of 60 and 90 years. Robinson's study involved 5000 women (Population I) and men (Population II).

Such statistical studies are frequently performed to compare values of a parameter of interest for two different populations. To conduct such a study, samples of data are independently collected from the populations and then summary statistics are computed for the samples. These measures qualify as random variables. Once appropriate probability models for differences in the sample statistics are determined, analogs of the methods we have used to construct confidence intervals for $\mu$, $P$, and $\sigma^2$ can be used to produce confidence intervals for the differences in the population parameters. Furthermore, test statistics can be formulated to determine whether statistically detectable differences exist between the parameters of the populations.

## 7-1 Large Sample Inferences Based on the Difference Between Independent Sample Means

**Key Topics**

1. *Defining independent sample means*

2. *The probability distribution of $\bar{X} - \bar{Y}$ based on large samples*

3. $\mu_{\bar{X} - \bar{Y}} = \mu_1 - \mu_2$ and $\sigma_{\bar{X} - \bar{Y}} = \sqrt{\dfrac{\sigma_1^2}{n_1} + \dfrac{\sigma_2^2}{n_2}}$

4. *Converting $\bar{x} - \bar{y}$ to a z score*

5. *Constructing confidence intervals for $\mu_1 - \mu_2$*

6. *Testing $H_0$: $\mu_1 - \mu_2 = 0$*

---

### STATISTICS IN ACTION

Joan Spaulding is an epidemiologist working with the World Health Care for Children Organization (WHCCO). She has been studying the effects of maternal cigarette smoking on the weights at birth of babies carried to term. Joan's staff enlisted the support of several hospitals to obtain data on the weights of such babies of non-smoking mothers (Population 1) and women that, on average, smoke at least one pack of cigarettes a day (Population 2). Tables 7-1 and 7-2 contain the birth weights, to the nearest one-tenth of 1 pound, of the babies in the samples.

**TABLE 7-1    Weights of 40 Babies of Non-smoking Mothers**

| | | | | |
|---|---|---|---|---|
| 8.3 | 7.9 | 9.6 | 7.1 | 6.8 |
| 10.2 | 7.3 | 8.8 | 8.0 | 9.5 |
| 5.9 | 10.1 | 8.2 | 8.7 | 9.6 |
| 12.3 | 8.1 | 7.3 | 7.8 | 8.5 |
| 9.1 | 7.4 | 6.8 | 7.5 | 8.2 |
| 6.6 | 7.9 | 8.4 | 8.9 | 10.4 |
| 9.0 | 7.5 | 8.2 | 8.7 | 7.0 |
| 10.8 | 9.9 | 8.8 | 12.3 | 6.6 |

**TABLE 7-2    Weights of 40 Babies of Mothers that Smoke**

| | | | | |
|---|---|---|---|---|
| 8.1 | 6.5 | 7.3 | 6.8 | 7.9 |
| 8.4 | 6.2 | 7.8 | 9.1 | 6.7 |
| 8.8 | 7.5 | 7.0 | 7.3 | 9.6 |
| 5.6 | 8.0 | 6.9 | 7.1 | 7.9 |
| 10.3 | 7.4 | 4.9 | 7.3 | 8.1 |
| 6.2 | 9.9 | 5.7 | 8.6 | 7.4 |
| 8.2 | 10.8 | 6.8 | 7.4 | 8.9 |
| 5.9 | 7.2 | 7.9 | 8.0 | 6.6 |

Joan calculated the sample means and standard deviations for the two samples of weights and obtained the following:

| *Sample 1* | *Sample 2* |
|---|---|
| Mean weight is 8.5 pounds | Mean weight is 7.6 pounds |
| Standard deviation is 1.45 pounds | Standard deviation is 1.27 pounds |

In a report to WHCCO, Joan reported that the data proved at a 1 percent level of significance that the mean weight of babies at birth, of mothers who smoke cigarettes, is less than the mean weight of babies at birth of mothers who do not smoke. The details of the hypothesis test that accompanied the report are given in this section.

## Defining independent sample means

In Figure 7-3, two populations of quantitative data are represented by rectangles. As indicated, the means and standard deviations of the populations are $\mu_1$, $\mu_2$, $\sigma_1$, and $\sigma_2$.

| Population 1 | Population 2 |
|---|---|
| Mean is $\mu_1$ <br> Standard deviation is $\sigma_1$ <br> $X_1, X_2, X_3, \ldots, X_{n_1}$ | Mean is $\mu_2$ <br> Standard deviation is $\sigma_2$ <br> $Y_1, Y_2, Y_3, \ldots, Y_{n_2}$ |

**Figure 7-3.** Two populations of quantitative data.

If separate random samples of sizes $n_1$ and $n_2$ are selected respectively from Populations 1 and 2, the data can then be used to calculate the following statistics:

$$\bar{X} = \frac{\sum X_i}{n_1} \quad \text{and} \quad \bar{Y} = \frac{\sum Y_i}{n_2}$$

### DEFINITION 7-1 Independent Sample Means $\bar{X}$ and $\bar{Y}$

The sample means $\bar{X}$ and $\bar{Y}$ can be modeled as independent when, on an intuitive basis, they can be described as *unrelated* or *based on separately chosen samples*. (In such cases, knowing that a particular value of $x$ appears in the first sample does not change the likelihood that a particular value of $y$ appears in the second sample, and vice versa.)

In the Statistics in Action section, Joan Spaulding obtained independent sample means of the weights at birth of the babies in populations she labeled as 1 and 2. Specifically, knowing the weight of a baby of a mother who does not smoke (or does smoke) did not affect the likelihood of a particular weight appearing in the other sample. As a consequence, in the birth weight example $\bar{X}$ and $\bar{Y}$ can be called independent sample means.

# The probability distribution of $\bar{X} - \bar{Y}$ based on large samples

If $\bar{X}$ and $\bar{Y}$ are both based on large samples (greater than or equal to 30), then by the central limit theorem the probability distributions of $\bar{X}$ and $\bar{Y}$ are both approximately normal. Furthermore it can be shown that the probability distribution of $\bar{X} - \bar{Y}$ is also approximately normal. In Figure 7-4, graphs of the probability distributions for $\bar{X}$, $\bar{Y}$, and $\bar{X} - \bar{Y}$ are shown. These graphs are valid, if $\bar{X}$ and $\bar{Y}$ are based on large samples.

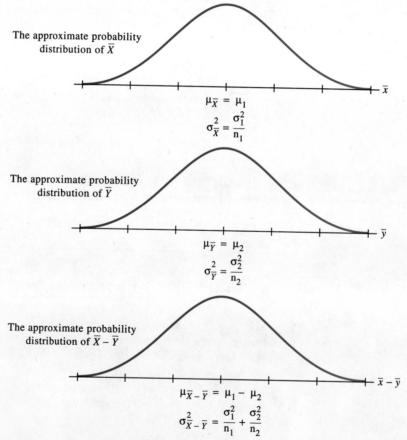

The approximate probability distribution of $\bar{X}$

$$\mu_{\bar{X}} = \mu_1$$
$$\sigma_{\bar{X}}^2 = \frac{\sigma_1^2}{n_1}$$

The approximate probability distribution of $\bar{Y}$

$$\mu_{\bar{Y}} = \mu_2$$
$$\sigma_{\bar{Y}}^2 = \frac{\sigma_2^2}{n_2}$$

The approximate probability distribution of $\bar{X} - \bar{Y}$

$$\mu_{\bar{X} - \bar{Y}} = \mu_1 - \mu_2$$
$$\sigma_{\bar{X} - \bar{Y}}^2 = \frac{\sigma_1^2}{n_1} + \frac{\sigma_2^2}{n_2}$$

**Figure 7-4.**   Probability distributions for $\bar{X}$, $\bar{Y}$, and $\bar{X} - \bar{Y}$.

Joan Spaulding obtained samples of size $n_1 = n_2 = 40$ from Populations 1 and 2. As a consequence, the means of the samples are approximately normal. Furthermore, the difference in the means $\bar{X} - \bar{Y}$ also has an approximately normal shape.

$$\mu_{\bar{X}-\bar{Y}} = \mu_1 - \mu_2 \text{ and } \sigma_{\bar{X}-\bar{Y}} = \sqrt{\frac{\sigma_1^2}{n_1} + \frac{\sigma_2^2}{n_2}}$$

In Figure 7-4, there are five means indicated:

1. $\mu_1$ and $\mu_2$, the means of the populations being sampled.
2. $\mu_{\bar{X}}$ and $\mu_{\bar{Y}}$, the means of the probability distributions of $\bar{X}$ and $\bar{Y}$, respectively.
3. $\mu_{\bar{X}-\bar{Y}}$, the mean of the probability distribution of $\bar{X} - \bar{Y}$.

It can be shown that $\mu_{\bar{X}-\bar{Y}}$ is the difference in the means of the probability distributions of $\bar{X}$ and $\bar{Y}$ (and therefore also the difference in the means of the populations being sampled).

### The Mean of the Probability Distribution of $\bar{X} - \bar{Y}$

$$\mu_{\bar{X}-\bar{Y}} = \mu_{\bar{X}} - \mu_{\bar{Y}}$$
$$= \mu_1 - \mu_2 \tag{1}$$

In Figure 7-4, there are five variances indicated:

1. $\sigma_1^2$ and $\sigma_2^2$, the variances of the populations being sampled.
2. $\sigma_{\bar{X}}^2$ and $\sigma_{\bar{Y}}^2$, the variances of the probability distributions of $\bar{X}$ and $\bar{Y}$, respectively.
3. $\sigma_{\bar{X}-\bar{Y}}^2$, the variance of the probability distribution of $\bar{X} - \bar{Y}$.

It can be shown that $\sigma_{\bar{X}-\bar{Y}}^2$ is the sum of the variances of the probability distributions of $\bar{X}$ and $\bar{Y}$ (and can therefore also be written in terms of the variances of the populations being sampled).

### The Variance of the Probability Distribution of $\bar{X} - \bar{Y}$

$$\sigma_{\bar{X}-\bar{Y}}^2 = \sigma_{\bar{X}}^2 + \sigma_{\bar{Y}}^2$$
$$= \frac{\sigma_1^2}{n_1} + \frac{\sigma_2^2}{n_2} \tag{2}$$

▶ **Example 1** State University is a large public school in the South. Magna College is a large private college located in the same city as State University. Records from the registrar's offices of the two schools yielded the following parameters for the grade point averages (gpas) of the two current student bodies.

| State University | Magna College |
|---|---|
| $\mu_1 = 2.98$ | $\mu_2 = 2.85$ |
| $\sigma_1 = 0.36$ | $\sigma_2 = 0.48$ |

Suppose random samples of sizes $n_1 = 36$ and $n_2 = 36$ are taken from the school's records, and $\bar{X}$ and $\bar{Y}$ are respectively the mean gpas of the students sampled. For the random variable $\bar{X} - \bar{Y}$:

**a.** Compute $\mu_{\bar{X} - \bar{Y}}$.
**b.** Compute $\sigma_{\bar{X} - \bar{Y}}$.
**c.** Sketch and label a graph of the distribution of $\bar{X} - \bar{Y}$.

**Solution**   **Discussion.** Because $n_1$ and $n_2$ are both greater than 30, the probability distribution of $\bar{X} - \bar{Y}$ is approximately normal.

**a.** With $\mu_1 = 2.98$ and $\mu_2 = 2.85$:

$$\mu_{\bar{X} - \bar{Y}} = 2.98 - 2.85 = 0.13$$

**b.** With $\sigma_1 = 0.36$, $n_1 = 36$, $\sigma_2 = 0.48$, and $n_2 = 36$:

$$\sigma_{\bar{X} - \bar{Y}} = \sqrt{\frac{(0.36)^2}{36} + \frac{(0.48)^2}{36}}$$
$$= 0.0036 + 0.0064$$
$$= 0.10$$

**c.** With $\mu_{\bar{X} - \bar{Y}} = 0.13$ and $\sigma_{\bar{X} - \bar{Y}} = 0.10$, a graph of the probability distribution of $\bar{X} - \bar{Y}$ is shown in Figure 7-5.   ◀

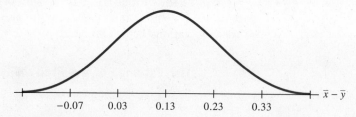

**Figure 7-5.**   A graph of the probability distribution of $\bar{X} - \bar{Y}$ for comparing gpas.

## Converting $\bar{x} - \bar{y}$ to a z score

If the mean and standard deviation of the probability distribution of $\bar{X} - \bar{Y}$ are known and the distribution is approximately normal, then probabilities that $\bar{X} - \bar{Y}$ will take on values of interest can be calculated. To use Appendix Table 4, we must first change an observed value of $\bar{X} - \bar{Y}$ to a z score. The appropriate formula is given in equation (3).

## Formula for the z-score corresponding to $\bar{x} - \bar{y}$

Comparing equation (3) to $z = \dfrac{x - \mu}{\sigma}$

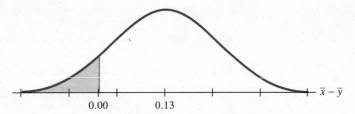

$$z = \frac{(\bar{x} - \bar{y}) - (\mu_1 - \mu_2)}{\sqrt{\dfrac{\sigma_1^2}{n_1} + \dfrac{\sigma_2^2}{n_2}}}$$

$x$ is replaced by $(\bar{x} - \bar{y})$, the difference in the sample means.

$\mu$ is replaced by $(\mu_1 - \mu_2)$, the difference in the population means.

$\sigma$ is replaced by $\sigma_{\bar{x} - \bar{y}}$.

(3)

▶ **Example 2**   Using the distribution in Example 1, find the probability that $\bar{X} - \bar{Y}$ will be negative.

**Solution**   **Discussion.** The area shaded in Figure 7-6 represents the desired probability.

```
                    0.00      0.13                       x̄ − ȳ
```

**Figure 7-6.**   Shaded area represents $P(\bar{X} - \bar{Y} < 0.00)$.

With $\bar{x} - \bar{y} = 0.00$, $\mu_1 - \mu_2 = 0.13$,   and   $\sqrt{\dfrac{\sigma_1^2}{n_1} + \dfrac{\sigma_2^2}{n_2}} = 0.10$:

$$z = \frac{0.00 - 0.13}{0.10} = -1.30 \qquad \text{The boundary is 1.30 standard deviations below the mean.}$$

$$P(\bar{X} - \bar{Y} < 0.00) = P(Z < -1.30)$$

$$= 0.0968 \qquad \text{From Appendix Table 4.}$$

$$\approx 10\%$$

Thus the probability is approximately 10 percent that the difference in the means of the sampled gpas will be negative.   ◀

## Constructing confidence intervals for $\mu_1 - \mu_2$

Recall that an unknown mean of a population can be estimated by a confidence interval centered at $\bar{x}$. It was possible to make such an interval because for large $n$ the probability distribution of $\bar{X}$ is approximately normal and centered at $\mu$. Similarly, if $\bar{X} - \bar{Y}$ has an approximately normal probability distribution, then we can construct confidence interval estimates for an unknown difference $\mu_1 - \mu_2$.

Reasoning as we did in Section 5-2, lower and upper limits of an approximate confidence interval for $\mu_1 - \mu_2$ are:

$$\text{lower limit} = (\bar{x} - \bar{y}) - z \sqrt{\frac{\sigma_1^2}{n_1} + \frac{\sigma_2^2}{n_2}}$$

$$\text{upper limit} = (\bar{x} - \bar{y}) + z \sqrt{\frac{\sigma_1^2}{n_1} + \frac{\sigma_2^2}{n_2}}$$

*The value of z in these equations depends on the confidence coefficient of the interval.*

In most real situations, $\sigma_1$ and $\sigma_2$ are unknown. However, if both $n_1$ and $n_2$ are large, it is common practice (and statistically valid) to estimate $\sigma_1$ and $\sigma_2$ with $s_1$ and $s_2$. A confidence interval can then be constructed using these values in equations (4) and (5).

### Equations for Constructing Confidence Intervals for $\mu_1 - \mu_2$

If $z$ is a positive number and $n_1$ and $n_2$ are large, then the interval with:

$$\text{lower limit} = (\bar{x} - \bar{y}) - z \sqrt{\frac{s_1^2}{n_1} + \frac{s_2^2}{n_2}} \tag{4}$$

and

$$\text{upper limit} = (\bar{x} - \bar{y}) + z \sqrt{\frac{s_1^2}{n_1} + \frac{s_2^2}{n_2}} \tag{5}$$

can be used as a confidence interval for $\mu_1 - \mu_2$. The associated confidence coefficient is $P(-z < Z < z)$.

Suppose Joan Spaulding in the Statistics in Action section wanted to estimate the difference in the mean weights of babies born to mothers who are not smokers and those who are smokers using the data in Tables 7-1 and 7-2, and a 95 percent confidence coefficient. With $\bar{x} = 8.5$ pounds, $s_1^2 = 2.093\ldots$, $n_1 = 40$, $\bar{y} = 7.6$ pounds, $s_2^2 = 1.615\ldots$, $n_2 = 40$, and $z = 1.96$:

$$\text{lower limit} = (8.5 - 7.6) - 1.96 \sqrt{\frac{2.093\ldots}{40} + \frac{1.615\ldots}{40}}$$

$$= 0.9 - 0.596\ldots$$

$$\approx 0.3$$

$$\text{upper limit} = 0.9 + 0.596\ldots$$

$$\approx 1.5$$

Therefore, based on the samples, a 90 percent confidence interval for the difference in mean birth weights of babies of mothers who are not smokers and those who are smokers is between 0.3 and 1.5, to the nearest one-tenth of 1 pound.

▶ **Example 3**   Michaelene Kelly is the personnel manager of NTS Electronics Corporation. As part of an analysis of employee profiles, she wanted to estimate the difference in the mean number of sick days used last year by employees with less than 5 years' service (Population 1) and employees with more than 5 years' service (Population 2). She therefore sampled records of employees from Populations 1 and 2. The samples yielded the following:

| Population 1 | Population 2 |
|---|---|
| $n_1 = 30$ employees | $n_2 = 40$ employees |
| $\bar{x} = 6.3$ days sick leave | $\bar{y} = 3.9$ days sick leave |
| $s_1 = 3.2$ days | $s_2 = 1.5$ days |

**a.** Use these statistics to construct a 90 percent confidence interval for $\mu_1 - \mu_2$.
**b.** Discuss the meaning of the interval in the context of the problem.

**Solution**   **Discussion.** For a 90 percent interval, use:

$$z = 1.65$$

**a.** With $\bar{x} = 6.3$, $s_1 = 3.2$, $n_1 = 30$, $\bar{y} = 3.9$, $s_2 = 1.5$, $n_2 = 40$, and equations (4) and (5):

$$\text{lower limit} = (6.3 - 3.9) - 1.65\sqrt{\frac{3.2^2}{30} + \frac{1.5^2}{40}}$$

$$= 2.4 - (1.65)(0.6305\ldots)$$

$$= 2.4 - 1.040\ldots$$

$$= 1.4, \text{ to one decimal place}$$

$$\text{upper limit} = 2.4 + 1.040\ldots$$

$$= 3.4, \text{ to one decimal place}$$

**b.** Based on the given data, a 90 percent confidence interval for the difference in the mean numbers of days sick leave used by employees with less than 5 years' service and those with more than 5 years' service is between 1.4 and 3.4 days.   ◀

## Testing $H_0$: $\mu_1 - \mu_2 = 0$

The random variable $\bar{X} - \bar{Y}$ can also be used in a hypothesis test that compares the means of two populations of quantitative data. The form of $H_0$ in such tests is usually

$H_0$:   $\mu_1 - \mu_2 = 0$      $H_0$ claims that $\mu_1 = \mu_2$. That is, there is no difference in the means of the two populations.

The purpose of a hypothesis test is to find whether there is sufficient evidence to reject $H_0$ and assert that some type of difference exists between the means. Depending on the motivation for the test, the form of $H_a$ will be one of the following:

$H_a$:   $\mu_1 - \mu_2 \neq 0$     A nondirectional alternative claiming a difference exists between $\mu_1$ and $\mu_2$

$H_a$:   $\mu_1 - \mu_2 < 0$     A directional alternative claiming $\mu_1 - \mu_2 < 0$ and therefore, $\mu_1 < \mu_2$

$H_a$:   $\mu_1 - \mu_2 > 0$     A directional alternative claiming $\mu_1 - \mu_2 > 0$ and therefore, $\mu_1 > \mu_2$

The null hypothesis is rejected when samples from the two populations yield $\bar{x} - \bar{y}$ with an associated $z$-value that falls in a rejection region. The boundary for the rejection region is $z_c$, where $z_c$ depends on the level of significance of the test and whether the test is directional or nondirectional.

## The Test Statistic for $H_0$: $\mu_1 - \mu_2 = 0$

In equation (6), $\bar{X}$ and $\bar{Y}$ are means of large independent samples of sizes $n_1$ and $n_2$ taken from Populations 1 and 2, respectively.

$$Z = \frac{(\bar{X} - \bar{Y}) - (\mu_1 - \mu_2)}{\sqrt{\dfrac{S_1^2}{n_1} + \dfrac{S_2^2}{n_2}}} = \frac{\bar{X} - \bar{Y}}{\sqrt{\dfrac{S_1^2}{n_1} + \dfrac{S_2^2}{n_2}}} \tag{6}$$

The simplified form of equation (6) is based on $H_0$: $\mu_1 - \mu_2 = 0$.

If $H_0$ is true, the random variable in equation (6) has an approximately normal probability distribution with mean 0. Consistent with the hypothesis testing technique studied in Chapter 6, we change an observed value of this random variable to a $z$ score. The resulting value of $z$ is then compared with a $z_c$ for a directional test or values of $z_c$ for a nondirectional test. If $z$ falls in the rejection region, then $H_0$ is rejected in favor of $H_a$. If it does not fall in the rejection region, then $H_0$ cannot be rejected.

Joan Spaulding used the six-step format to prepare the following report on the results of the study on weights of babies at birth of non-smoking and smoking mothers:

**Step 1**  $H_0$:   $\mu_1 - \mu_2 = 0$     No difference in the mean weights
          $H_a$:   $\mu_1 - \mu_2 > 0$     The mean weight of babies born to non-smoking mothers is greater than the mean weight of those born to smoking mothers.

**Step 2**  The level of significance is 0.01.

**Step 3** The test statistic that will be used is:

$$Z = \frac{\bar{X} - \bar{Y}}{\sqrt{\dfrac{S_1^2}{n_1} + \dfrac{S_2^2}{n_2}}}$$

**Step 4** If $z > z_c \ (= 2.33)$, then reject $H_0$. See Figure 7-7.

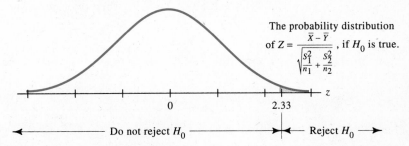

The probability distribution of $Z = \dfrac{\bar{X} - \bar{Y}}{\sqrt{\dfrac{s_1^2}{n_1} + \dfrac{s_2^2}{n_2}}}$, if $H_0$ is true.

0          2.33          $z$

Do not reject $H_0$     Reject $H_0$ →

**Figure 7-7.** $\alpha = P(Z > 2.33) = 0.01$.

**Step 5** With $\bar{x} = 8.5$, $s_1^2 = 2.093\ldots$, $n_1 = 40$, $\bar{y} = 7.6$, $s_2^2 = 1.615\ldots$, and $n_2 = 40$:

$$z = \frac{8.5 - 7.6}{\sqrt{\dfrac{2.09\ldots}{40} + \dfrac{1.61\ldots}{40}}} \approx 2.96$$

**Step 6** Because $2.96 > 2.33$, reject $H_0$ and accept $H_a$. That is, with $\alpha = 0.01$, there is sufficient evidence to conclude that the mean weight of babies born to non-smoking mothers is greater than the mean weight of those born to mothers who smoke.

▶ **Example 4** Earl Karn is a biologist who works at a Colorado State Fish Hatchery. He was interested in testing a new diet formula that will hopefully improve the growth of newly hatched trout over that experienced on the traditional diet. He fed $n_1 = 100$ newly hatched trout the new formula diet for a specified time period. For these fish, the mean length was $\bar{x} = 5.85$ cm with $s_1 = 1.60$ cm. He also fed $n_2 = 80$ newly hatched trout the traditional diet for the same period of time. For these fish, the mean length was $\bar{y} = 5.30$ cm and $s_2 = 1.20$ cm. Use these statistics to conduct a hypothesis test to determine whether the new diet is better than the traditional one for growing newly hatched trout. Use a 5 percent level of significance.

**Solution** Let: $\mu_1 =$ the mean length of trout fed the new diet, and
$\mu_2 =$ the mean length of trout fed the traditional diet.

**Step 1** $H_0$:   $\mu_1 - \mu_2 = 0$
$H_a$:   $\mu_1 - \mu_2 > 0$     Earl wants to show that $\mu_1 > \mu_2$; thus $\mu_1 - \mu_2 > 0$.

**Step 2** $\alpha = 0.05$     $P(\text{Type I error}) = 0.05$.

**Step 3** The test statistic will be

$$Z = \frac{\bar{X} - \bar{Y}}{\sqrt{\dfrac{S_1^2}{n_1} + \dfrac{S_2^2}{n_2}}}$$    Both $n_1$ and $n_2$ qualify as large sample sizes.

**Step 4** If $z > z_c \ (=1.65)$, then reject $H_0$. See Figure 7-8.

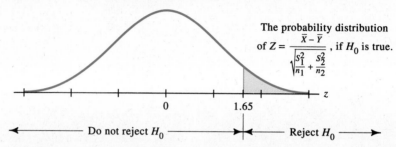

The probability distribution of $Z = \dfrac{\bar{X} - \bar{Y}}{\sqrt{\dfrac{S_1^2}{n_1} + \dfrac{S_2^2}{n_2}}}$, if $H_0$ is true.

Do not reject $H_0$     Reject $H_0$

**Figure 7-8.**   $\alpha = P(Z > 1.65) = 0.05$.

**Step 5** With $\bar{x} = 5.85$, $s_1 = 1.60$, $n_1 = 100$, $\bar{y} = 5.30$, $s_2 = 1.20$, and $n_2 = 80$:

$$z = \frac{5.85 - 5.30}{\sqrt{\dfrac{1.60^2}{100} + \dfrac{1.20^2}{80}}}$$

$$= \frac{0.55}{0.2088\ldots}$$

$$= 2.63, \text{ to two decimal places}$$

**Step 6** Because $2.63 > 1.65$, reject $H_0$ and accept $H_a$. That is, there is sufficient evidence, with $\alpha = 0.05$, to conclude that the new formula diet is more effective than the traditional diet.    ◄

## Exercises 7-1    Set A

*In Exercises 1–4, the means and standard deviations of two populations are given. For each exercise, suppose independent samples are taken from the populations and do the following:*

*a. Compute $\mu_{\bar{X} - \bar{Y}}$.*

*b. Compute $\sigma_{\bar{X} - \bar{Y}}$ to one decimal place.*

*c. Sketch and label a graph for the probability distribution of $\bar{X} - \bar{Y}$.*

*d. Compute the indicated probability.*

*e. State the meaning of the answer to part (d).*

**1.**

| Population 1 | Population 2 |
|---|---|
| $\mu_1 = 60.0$ | $\mu_2 = 50.0$ |
| $\sigma_1 = 8.0$ | $\sigma_2 = 6.0$ |

For sample 1, $n_1 = 64$ and $\bar{X}$ is the mean. For sample 2, $n_2 = 52$ and $\bar{Y}$ is the mean. Compute $P(\bar{X} - \bar{Y} < 8.0)$.

**2.**

| Population 1 | Population 2 |
|---|---|
| $\mu_1 = 108.5$ | $\mu_2 = 102.8$ |
| $\sigma_1 = 7.4$ | $\sigma_2 = 12.6$ |

For sample 1, $n_1 = 46$ and $\bar{X}$ is the mean. For sample 2, $n_2 = 56$ and $\bar{Y}$ is the mean. Compute $P(\bar{X} - \bar{Y} > 8.0)$.

**3.**

| Population 1 | Population 2 |
|---|---|
| $\mu_1 = 26.3$ | $\mu_2 = 19.8$ |
| $\sigma_1 = 2.90$ | $\sigma_2 = 2.10$ |

For sample 1, $n_1 = 38$ and $\bar{X}$ is the mean. For sample 2, $n_2 = 32$ and $\bar{Y}$ is the mean. Compute $P(6.0 < \bar{X} - \bar{Y} < 7.0)$.

**4.**

| Population 1 | Population 2 |
|---|---|
| $\mu_1 = 1450$ | $\mu_2 = 1390$ |
| $\sigma_1 = 78.0$ | $\sigma_2 = 109.0$ |

For sample 1, $n_1 = 90$ and $\bar{X}$ is the mean. For sample 2, $n_2 = 75$ and $\bar{Y}$ is the mean. Compute $P(50.0 < \bar{X} - \bar{Y} < 70.0)$.

**5.** A computer printout of the profile of the current student body at a large university in the Midwest shows the mean height of the women students is 65.0 inches with a standard deviation of 2.8 inches. The same printout shows the mean height of the men students is 70.3 inches with a standard deviation of 3.2 inches. Samples of sizes $n_1 = 75$ and $n_2 = 60$ women and men students, respectively, are taken from this student body. Let $\bar{X}$ and $\bar{Y}$ be the sample mean heights of the women and men, respectively.

a. Compute $\mu_{\bar{X} - \bar{Y}}$.
b. Compute $\sigma_{\bar{X} - \bar{Y}}$ to one decimal place.
c. Sketch and label a graph for the probability distribution of $\bar{X} - \bar{Y}$.
d. Compute $P(\bar{X} - \bar{Y} > -4.5)$.
e. Compute $P(-6.3 < \bar{X} - \bar{Y} < 4.3)$.
f. State the meaning of the answer to part (e).

**6.** A Building Trades Magazine stated that the mean hourly pay rate of carpenters is $14.30 with a standard deviation of $2.10. The mean hourly rate of sheetmetal workers is $13.60 with a standard deviation of $1.60. Samples of sizes $n_1 = n_2 = 43$ are taken from the national carpenters' and sheetmetal workers' unions. Let $\bar{X}$ and $\bar{Y}$ be the sample mean hourly pay rates of the samples of carpenters and sheetmetal workers, respectively.

a. Compute $\mu_{\bar{X} - \bar{Y}}$.
b. Compute $\sigma_{\bar{X} - \bar{Y}}$ to the nearest 1 cent.
c. Sketch and label a graph for the probability distribution of $\bar{X} - \bar{Y}$.
d. Compute $P(\bar{X} - \bar{Y} < 0.00)$.
e. State the meaning of the answer to part (d).

**7.** A national organization of teachers asserted recently that the mean age of college instructors is 42.5 years with a standard deviation of 10.5 years. The mean age of elementary school teachers is 30.5 years with a standard deviation of 8.4 years. Samples of sizes $n_1 = n_2 = 45$ are taken from colleges and elementary schools in Florida. Let $\bar{X}$ and $\bar{Y}$ be the sample mean ages, respectively, of the samples of college instructors and elementary school teachers and suppose the national means and standard deviations hold in Florida.

a. Compute $\mu_{\bar{X} - \bar{Y}}$.
b. Compute $\sigma_{\bar{X} - \bar{Y}}$ to one decimal place.
c. Sketch and label a graph for the probability distribution of $\bar{X} - \bar{Y}$.
d. Compute $P(\bar{X} - \bar{Y} > 13.0)$.
e. State the meaning of the answer to part (d).

**8.** In a large midwestern metropolitan area, there are two large corporations that are major sources of employment for workers in the area. A study was made of the distances the employees of the two corporations must drive to work. The study yielded the following statistics:

| Employees of Corporation A | Employees of Corporation B |
|---|---|
| $\mu_1 = 9.7$ miles | $\mu_2 = 7.2$ miles |
| $\sigma_1 = 5.8$ miles | $\sigma_2 = 4.1$ miles |

Samples of sizes $n_1 = 60$ and $n_2 = 66$ are taken from employee files at the two corporation offices. Let $\bar{X}$ and $\bar{Y}$ be the mean miles the employees in the samples drive to work.

a. Compute $\mu_{\bar{X} - \bar{Y}}$.

b. Compute $\sigma_{\bar{X} - \bar{Y}}$ to one decimal place.

c. Sketch and label a graph for the probability distribution of $\bar{X} - \bar{Y}$.

d. Compute $P(\bar{X} - \bar{Y} < 1.5)$.

e. State the meaning of the answer to part (d).

9. Brand C and brand T are two popular brands of sugarless gum. To estimate the difference in the mean weights of sticks of these brands of gum, random samples of 45 sticks of each kind are taken from a jobbers warehouse and weighed on laboratory scales. For the brand C gum, $\bar{x} = 2.72$ grams and $s_1 = 0.05$ grams. For the brand T gum, $\bar{y} = 1.81$ grams and $s_2 = 0.08$ grams. Use these statistics to construct a 90 percent confidence interval for the difference in the mean weights of sticks of these brands of gum.

10. Model C and model F are two popular models of midsize automobiles. To estimate the differences in acceleration abilities of these cars, samples of 40 model C cars and 45 model F cars are taken from dealers' lots in a large metropolitan area. All the cars are tested by professional drivers on a test course, and the times required to cover a measured distance beginning from a stopped position recorded. The tests yield the following statistics:

| Model C Cars | Model F Cars |
|---|---|
| $\bar{x} = 17.5$ seconds | $\bar{y} = 14.9$ seconds |
| $s_1 = 2.8$ seconds | $s_2 = 3.2$ seconds |

Use these statistics to construct a 95 percent confidence interval for the difference in the mean times required for model C and model F cars to cover the measured distance beginning from a standing start.

11. The management of a record outlet conducted a survey of customers in the store. The purpose of the survey was to estimate the difference between the mean dollar amount of purchases made by customers more than 30 years old, and that of customers less than 30 years old. The results of the survey are stated below.

| More than 30 Years Old | Less than 30 Years Old |
|---|---|
| $\bar{x} = \$12.15$ | $\bar{y} = \$4.20$ |
| $s_1 = \$10.60$ | $s_2 = \$3.80$ |
| $n_1 = 35$ | $n_2 = 46$ |

Use these statistics to construct a 98 percent confidence interval for the difference in the mean amounts spent by customers more than 30 years old and those less than 30 years old.

12. For a class statistics project, a student sampled shoppers in a downtown shopping mall regarding the dollar amounts of their purchases on a given Saturday. (The shoppers in this mall are primarily apartment dwellers living near the mall.) A classmate sampled shoppers in a suburban mall regarding the dollar amounts of their purchases on the same Saturday. (The shoppers in this mall are primarily home owners living in the area of the mall.) The following information was collected:

| Downtown Mall | Suburban Mall |
|---|---|
| $\bar{x} = \$83.40$ | $\bar{y} = \$59.10$ |
| $s_1 = \$22.50$ | $s_2 = \$31.40$ |
| $n_1 = 43$ | $n_2 = 37$ |

Use these statistics to construct a 99 percent confidence interval for the difference in the mean amounts spent by downtown mall shoppers and suburban mall shoppers.

**13.** A comment regarding the life expectancies of women and men motivated Elizabeth Linton, a reporter at a newspaper in a large southern city to do a research project. She determined the ages at death of 61 women and 48 men who were randomly sampled from the obituary section of the newspaper for the past year. The results are summarized below.

| Women's Ages | Men's Ages |
|---|---|
| $\bar{x} = 79.6$ years | $\bar{y} = 72.4$ years |
| $s_1 = 14.6$ years | $s_2 = 15.6$ years |

a. Use these statistics to construct a 90 percent confidence interval for the difference in mean ages of women and men at the time of death for residents in this city.

b. State the meaning of the answer to part (a) in the context of the problem.

**14.** Craig Gagstetter works at a large military facility. He wanted to estimate the difference between the mean numbers of years of government service for people with a GS-11 rating and those with a GS-9 rating. From a directory of about 40,000 civil service employees, Craig sampled 40 employees from each rating. The results are listed below.

| GS-11 Rating | GS-9 Rating |
|---|---|
| $\bar{x} = 10.2$ years service | $\bar{y} = 5.9$ years service |
| $s_1 = 2.3$ years | $s_2 = 1.4$ years |

Use these statistics to construct a 95 percent confidence interval for the difference in mean years' service for employees with GS-11 ratings and GS-9 ratings.

**15.** A standardized aptitude test is given to male and female candidates for entrance into the California Highway Patrol academy. Test the hypothesis that there is no difference between the average scores achieved by male candidates ($\mu_1$) and female candidates ($\mu_2$) at this school, if samples of sizes $n_1 = 64$ and $n_2 = 49$ yield the following: $\bar{x} = 50.0$, with $s_1 = 12.0$; and $\bar{y} = 54.0$, with $s_2 =$

14.0. Conduct the test at a 5 percent level of significance.

**16.** In his annual report to Washington, the campus representative of the Veterans Administration at Ohio State University wants to include a statement to the effect that veteran students have a combined grade point average higher than that of non-veteran students on campus. Samples of 36 veteran and 36 non-veteran students yield a mean gpa of 2.83 and a standard deviation of 0.12 for veterans and a mean gpa of 2.74 and a standard deviation of 0.21 for non-veteran students. Can the representative make the proposed claim at a 5 percent level of significance?

**17.** A Federal Communications Commission representative believes that there is a difference between the amounts of commercial time on two of the major networks. She samples 40 hours of broadcast time on network A and records an average of 12.6 min/hr of commercials, with a standard deviation of 1.6 minutes. She samples 35 hours of broadcast time on network C and records an average of 13.2 min/hr of commercials, with a standard deviation of 1.7 minutes. For $\alpha = 0.05$ and a two-tail test, can the claim be made that a difference exists in the mean amounts of commercial time? (Note: $n_1 = 40$ and $n_2 = 35$.)

**18.** The Jurgensohn Seed Company developed a new barley seed, brand X. It wants to show that the average yield of brand X is better than that of the conventional brand Y. Each brand was planted on 36 different 1 acre plots. For the brand X plots, yields had $\bar{x} = 42.3$ bushels per acre with $s_1 = 3.6$ bushels. For the brand Y plots, yields had $\bar{y} = 40.8$ bushels per acre with $s_2 = 4.8$ bushels. Can the Jurgensohn Seed Company claim that on average brand X seed produces more barley per acre than brand Y? Use $\alpha = 5\%$.

**19.** Doug Gilbert is dean of students at Western National University. He feels that female students at the school are enrolled in more units than male students. He has his staff sample the records of the current student body. The survey yields the following.

| Female Students | Male Students |
|---|---|
| $\bar{x} = 15.2$ units | $\bar{y} = 14.0$ units |
| $s_1 = 4.2$ units | $s_2 = 3.8$ units |
| $n_1 = 40$ | $n_2 = 39$ |

Test, with $\alpha = 0.05$, Doug's claim that female students are enrolled in more units than male students.

20. Sandra Nordstrom is a physical therapist who has been comparing agility in left-handed youngsters to that of right-handed youngsters. From a large city school district, she sampled students to per-

form a manual assembly task that she developed. The times it took the youngsters to complete the tasks were measured. The results of the study are listed below.

| Left-handed Youngsters | Right-handed Youngsters |
|---|---|
| $\bar{x} = 62.5$ seconds | $\bar{y} = 66.8$ seconds |
| $s_1 = 8.7$ seconds | $s_2 = 10.3$ seconds |
| $n_1 = 30$ | $n_2 = 38$ |

Test, with $\alpha = 0.01$, Sandra's claim that left-handed youngsters can complete the assembly task in less time on the average than right-handed youngsters.

## Exercises 7-1    Set B

In Exercises 1 and 2, use the following information. A red box (Population 1) contains disks of the same size and shape. On each disk, a digit is painted.

| | |
|---|---|
| A 0 is painted on 200 disks. | A 1 is painted on 400 disks. |
| A 2 is painted on 1000 disks. | A 3 is painted on 1000 disks. |
| A 4 is painted on 400 disks. | A 5 is painted on 200 disks. |

1. For the disks in the red box:
   a. Compute $\mu_1$.
   b. Compute $\sigma_1^2$.
   c. Use $\bar{x} = 2.40$, $s_1 = 1.40$, and $n_1 = 35$, and $\alpha = 0.05$ to conduct a hypothesis test on the value of $\mu_1$ obtained in part (a). Use a "less than" directional hypothesis for $H_a$.

2. a. Use $\bar{x} = 2.40$, $s_1 = 1.40$, and $n_1 = 35$ to construct a 95 percent confidence interval for $\mu_1$.
   b. Does the interval of part (a) contain the mean of Population 1?

In Exercises 3 and 4, use the following information. A blue box contains disks (Population 2) of the same size and shape. On each disk, a digit is painted.

| | |
|---|---|
| A 0 is painted on 100 disks. | A 1 is painted on 200 disks. |
| A 2 is painted on 500 disks. | A 3 is painted on 1300 disks. |
| A 4 is painted on 1000 disks. | A 5 is painted on 100 disks. |

3. For the disks in the blue box:
   a. Compute $\mu_2$.
   b. Compute $\sigma_2^2$.
   c. Use $\bar{y} = 3.15$, $s_2 = 1.12$, and $n_2 = 40$ and $\alpha = 0.05$ to conduct a hypothesis test on the value of $\mu_2$ obtained in part (a). Use a "greater than" directional hypothesis for $H_a$.

4. a. Use $\bar{y} = 3.15$, $s_2 = 1.12$, and $n_2 = 40$ to construct a 90 percent confidence interval for $\mu_2$.
   b. Does the interval of part (a) contain the mean of Population 2?

*In Exercises 5 and 6, use the distributions of digits in both the red box (Population 1) and the blue box (Population 2).*

5. Use $\bar{x} = 2.40$, $s_1 = 1.40$, $n_1 = 35$, $\bar{y} = 3.15$, $s_2 = 1.12$, and $n_2 = 40$ to test $H_0: \mu_1 - \mu_2 = 0$ versus $H_a: \mu_1 - \mu_2 \neq 0$. Use $\alpha = 0.05$.

6. a. Use $\bar{x} = 2.40$, $s_1 = 1.40$, $n_1 = 35$, $\bar{y} = 3.15$, $s_2 = 1.12$, and $n_2 = 40$ to construct a 98 percent confidence interval estimate for $\mu_1 - \mu_2$.
   b. Does the interval of part (a) contain $\mu_1 - \mu_2$. (From Exercise 1(a), $\mu_1 = 2.5$, and from Exercise 3(a), $\mu_2 = 3.0$. Thus $\mu_1 - \mu_2 = -0.5$.)

7. Based on values in Figure 7-9, the following populations are defined for a large city in the state of Florida:

   $$\text{Population 1} = \text{homes with thermostats}$$
   $$\text{set at } 72°\,F$$

   $$\text{Population 2} = \text{homes with thermostats}$$
   $$\text{set at } 78°\,F$$

   Assume that it has been determined that the mean yearly energy needed to cool a home in this city in Population 1 is $\mu_1 = 3200$ kWH (kilowatt hours of electricity), and the figure for Population 2 is $\mu_2 = 2100$ kWH. In this city, independent samples of 43 homes each are taken from Populations 1 and 2. Suppose $\sigma_1 = 600$ kWH and $\sigma_2 = 400$ kWH. Let $\bar{X}$ and $\bar{Y}$ be the means of the samples selected from Populations 1 and 2, respectively.
   a. Compute $\mu_{\bar{X} - \bar{Y}}$.
   b. Compute $\sigma_{\bar{X} - \bar{Y}}$ to the nearest whole number.
   c. Sketch and label a graph of the probability distribution of $\bar{X} - \bar{Y}$.
   d. Compute $P(1000 < \bar{X} - \bar{Y} < 1300)$.
   e. State the meaning of the answer to part (d).

---

**HOT WEATHER ENERGY SAVERS**

*Some special summer, or warm climate saving tips:* **Set air-conditioning thermostats no lower than 78 degrees.** The 78 degree temperature is judged to be reasonably comfortable and energy efficient. One authority estimates that if this setting raises the temperature 6 degrees (78 degrees vs 72 degrees) home cooling costs should drop about 47 percent. (The Federal Government is enforcing a strict 78-80 degree temperature in all its buildings during the summer.)

If everyone raised cooling thermostats 6 degrees during the summer, the Nation would save more than the equivalent of 36 billion kilowatt hours of electricity, or 2 percent of the Nation's total electricity consumption for a year.

**Figure 7-9.** From the pamphlet "Tips for Energy Savers: In and Around the Home," Energy Conservation and Environment, Federal Energy Administration, Washington, D.C.

**8.** Based on the values in Figure 7-10, the following populations are defined for homes in a large city in the state of New York:

$$\text{Population 1} = \text{homes with thermostats}$$
$$\text{set at } 74° \text{ and } 66° \text{ F}$$

$$\text{Population 2} = \text{homes with thermostats}$$
$$\text{set at } 68° \text{ and } 60° \text{ F}$$

Assume that it has been determined that the mean yearly cost of heating a home in Population 1 is $\mu_1 = \$670$. The mean yearly cost of heating a home in Population 2 is $\mu_2 = \$580$. Independent samples of 51 each are taken from Populations 1 and 2. Suppose $\sigma_1 = \$95$ and $\sigma_2 = \$75$. Let $\bar{X}$ and $\bar{Y}$ be the sample mean costs from Populations 1 and 2, respectively.

a. Compute $\mu_{\bar{X} - \bar{Y}}$.

b. Compute $\sigma_{\bar{X} - \bar{Y}}$ to the nearest dollar.

c. Sketch and label a graph of the probability distribution of $\bar{X} - \bar{Y}$.

d. Compute $P(\$70 < \bar{X} - \bar{Y} < \$100)$

e. State the meaning of the answer to part (d).

*Exercises 9 and 10 are suggested student projects.*

**9.** Sample $n_1 = 35$ female students (Population 1) and $n_2 = 35$ male students (Population 2) at your college. From each student, obtain the number of units she/he is currently taking this term. Use these data to construct a 90 percent confidence

**COLD WEATHER ENERGY SAVERS**

*To save on heating energy and heating costs:* **Lower thermostats to 68 degrees during the day and 60 degrees at night.** If these settings reduce the temperature an average of 6 degrees, heating costs should run about 15 percent less.

If every household in the United States lowered heating temperatures 6 degrees, the demand for fuel would drop by more than 570,000 barrels of oil per day (enough to heat over 9 million homes during the winter season).

**Figure 7-10.** From the pamphlet "Tips for Energy Savers: In and Around the Home," Energy Conservation and Environment, Federal Energy Administration, Washington, D.C.

interval for the difference in the means of the numbers of units being taken this term by female and male students.

**10.** Sample $n_1 = 35$ female students (Population 1) and $n_2 = 35$ male students (Population 2) at your college. From each student, obtain the cumulative grade point average (gpa). Use these data to test:

$$H_0: \quad \mu_1 - \mu_2 = 0 \quad \text{versus} \quad H_a: \quad \mu_1 - \mu_2 \neq 0$$

# 7-2 Inferences Based on Paired Differences

**Key Topics**

1. *A definition of paired data*

2. *The probability distribution of $\bar{D}$ for large $n$*

3. *Large sample confidence intervals for $\mu_d$*

4. *Small sample confidence intervals for $\mu_d$*

5. *Large sample tests of $H_0: \mu_d = 0$*

6. *Small sample tests of $H_0: \mu_d = 0$*

# STATISTICS IN ACTION

Sandy Fong is a nutritionist who works with the National Institute on Weight Control (NIWC). She has recently been working on the weight problems of individuals who have difficulty maintaining proper weight because of low-activity jobs. These individuals have jobs that typically require many hours each day, and they usually work 6 or 7 days a week. Furthermore their work schedules are not regular, in that they do not work the same hours every day. As a consequence, attempts to build an exercise program into a daily routine are hampered by time constraints of the job.

To help such individuals reduce caloric intake while maintaining proper nutrition, Sandy has developed a food supplement. She feels that this supplement, when taken according to her instructions, will cause weight loss to occur, while maintaining an acceptable level of energy in these people. To test her claim, she sampled 40 individuals who had the kind of occupation profile she wanted from the files of NIWC. Each of these individuals expressed a desire to lose weight, but cited long hours on the job as an obstacle to achieving success.

At the first meeting, Sandy told the group about her supplement and the program they would be following during the study. Each member of the group was weighed and the weights are shown as values of x in Table 7-3. The group met periodically over the several weeks the group was monitored to discuss problems. At the end of the study, each member of the group was again weighed, and the weights are shown as values of y in Table 7-3.

To make an inference based on these data, Sandy calculated the loss in weight of each individual. The differences in the 40 pairs of numbers are recorded in Table 7-4.

For the $n = 40$ data points, Sandy calculated the following:

$\bar{d} = 11.8$  The sample mean of the differences

$s_d = 5.96$  The sample standard deviation of the differences

In her report to the executive board on the results of the supplement study, Sandy made the following statement:

"If the supplement is used according to instructions, then based on the data obtained, a 95% confidence interval for the mean weight loss over the time period of the test program for individuals with job profiles similar to those targeted in the study is between 10.0 and 13.6 pounds."

The details of how the confidence interval was constructed are studied in this section.

**TABLE 7-3    The Before (x) and After (y) Weights of 40 Individuals**

| x/y | x/y | x/y | x/y | x/y |
|-----|-----|-----|-----|-----|
| 185/176 | 143/137 | 233/218 | 164/152 | 155/138 |
| 190/180 | 205/189 | 175/165 | 142/138 | 164/150 |
| 139/128 | 173/165 | 186/170 | 143/133 | 139/126 |
| 127/120 | 164/155 | 243/218 | 151/143 | 137/135 |
| 215/190 | 145/140 | 159/156 | 125/113 | 210/195 |
| 173/162 | 158/143 | 178/163 | 138/130 | 149/136 |
| 185/176 | 203/181 | 163/150 | 266/238 | 118/112 |
| 258/240 | 125/116 | 133/128 | 175/167 | 138/128 |

**TABLE 7-4    The Differences in Weights of 40 Individuals**

| d = x − y | | | | |
|-----|-----|-----|-----|-----|
| 9 | 6 | 15 | 12 | 17 |
| 10 | 16 | 10 | 4 | 14 |
| 11 | 8 | 16 | 10 | 13 |
| 7 | 9 | 25 | 8 | 2 |
| 25 | 5 | 3 | 12 | 15 |
| 11 | 15 | 15 | 8 | 13 |
| 9 | 22 | 13 | 28 | 6 |
| 18 | 9 | 5 | 8 | 10 |

## A definition of paired data

When samples from two populations are not chosen separately, then the method of statistical analysis discussed in Section 7-1 is not applicable. But consider the following,

$$\{x_1, x_2, x_3, \ldots, x_N\} \text{ are } N \text{ data points in Population 1}$$
$$\{y_1, y_2, y_3, \ldots, y_N\} \text{ are } N \text{ data points in Population 2}$$

Now suppose that each $y_i$ corresponds naturally to the $x_i$ with the same subscript. For such a correspondence, the data points are said to be *paired*.

### DEFINITION 7-2 A Sample of Paired Data

A sample from Populations 1 and 2 contains *paired data*, if for each $x_i$ from Population 1 there exists exactly one data point $y_i$ from Population 2 that *naturally corresponds* to it in some way.

Suppose a single sample $n$ of data pairs is randomly sampled from Populations 1 and 2. For each pair of data points, the difference $d$ can be computed.

$$d_1 = x_1 - y_1$$
$$d_2 = x_2 - y_2$$
$$\vdots \quad \vdots \quad \vdots$$
$$d_n = x_n - y_n$$

For the $n$ data pairs, the sample mean is $\bar{d}$ and the standard deviation for the $d$'s is $s_d$. That is,

$$\bar{d} = \frac{\sum d_i}{n} \tag{7}$$

and

$$s_d = \sqrt{\frac{n \sum d_i^2 - (\sum d_i)^2}{n(n-1)}} \tag{8}$$

In the Statistics in Action portion of this section, Sandy Fong used equations (7) and (8) to calculate $\bar{d}$ and $s_d$ for the 40 differences in Table 7-4. Specifically:

$$\bar{d} = \frac{472}{40} = 11.8 \quad \text{and} \quad s_d = \sqrt{\frac{40(6956) - 472^2}{40(39)}} \approx 5.96$$

▶ **Example 1**    To determine the effect of hunger on the test performance of elementary school children, 36 fifth grade students in a large school district are given a general intelligence test on a morning that they eat breakfast. The same children are given another form of the test on a day that they do not eat breakfast. The results are stated in the form $(x_i, y_i)$, where $x_i$ and $y_i$ are the scores of each student on the test taken *with* and *without* breakfast, respectively.

$$(17, 14)\quad (20, 16)\quad (12, 13)\quad (24, 23)\quad (10, 8)\quad (14, 12)$$
$$(16, 15)\quad (14, 17)\quad (19, 18)\quad (20, 18)\quad (15, 11)\quad (17, 12)$$
$$(21, 18)\quad (13, 15)\quad (14, 10)\quad (18, 13)\quad (17, 13)\quad (19, 18)$$
$$(20, 15)\quad (10, 9)\quad (15, 15)\quad (18, 16)\quad (24, 24)\quad (16, 19)$$
$$(21, 16)\quad (23, 21)\quad (18, 15)\quad (11, 7)\quad (13, 12)\quad (18, 16)$$
$$(19, 15)\quad (21, 18)\quad (13, 13)\quad (19, 15)\quad (15, 14)\quad (24, 22)$$

For these data compute:

**a.** $\bar{d}$

**b.** $s_d$

**Solution**    **Discussion.** We must first compute $d_i$ for each pair of data points. For example, $d_1 = 17 - 14 = 3$ and $d_2 = 20 - 16 = 4$. The 36 values of $d_i$ are:

$$\begin{array}{cccccccccccc}
3 & 4 & -1 & 1 & 2 & 2 & 1 & -3 & 1 & 2 & 4 & 5 \\
3 & -2 & 4 & 5 & 4 & 1 & 5 & 1 & 0 & 2 & 0 & -3 \\
5 & 2 & 3 & 4 & 1 & 2 & 4 & 3 & 0 & 4 & 1 & 2
\end{array}$$

**a.** $\bar{d} = \dfrac{72}{36} = 2.0$    For these data, $\sum d_i = 72$.

**b.** $s_d = \sqrt{\dfrac{36(306) - 72^2}{36(35)}}$    For these data, $\sum d_i^2 = 306$, $n = 36$, and $n - 1 = 35$.

$$= \sqrt{\dfrac{5832}{1260}}$$

$$= 2.1514\ldots$$

$$= 2.2, \text{ to one decimal place}$$

The experiment yielded a mean increase of 2.0 points on the test scores with a standard deviation of 2.2.    ◀

## The probability distribution of $\bar{D}$ for large $n$

The random variable $\bar{D}$ has the same relation to a population of $d$'s that $\bar{X}$ has to a population of $x$'s. Therefore the discussion in Chapter 5 that led to a conclusion regarding the shape of the probability distribution of $\bar{X}$ can be used to make similar conclusions regarding the shape of the probability distribution of $\bar{D}$.

Specifically suppose $d_1, d_2, d_3, \ldots, d_n$ are values sampled from a large population of $d$'s. If $n$ is large, then by the central limit theorem, the probability distribution of $\bar{D}$ is approximately normal. The mean and standard deviation of the probability distribution can be found using equations (9) and (10).

## The Mean and Standard Deviation of $\bar{D}$

If $\bar{D}$ is the mean of a sample from a population of differences with mean $\mu_d$ and standard deviation $\sigma_d$, then:

$$\mu_{\bar{D}} = \mu_d \tag{9}$$

$$\sigma_{\bar{D}} = \frac{\sigma_d}{\sqrt{n}} \tag{10}$$

If $n$ is large, the sample standard deviation $s_d$ can be used as an estimate of $\sigma_d$, and $\sigma_{\bar{D}}$ can be estimated by:

$$\frac{s_d}{\sqrt{n}} \tag{11}$$

If $\bar{D}$ is based on a large $n$, then the standard normal distribution can be used to compute probabilities for $\bar{D}$. To make such calculations, it is necessary to change a value of $\bar{d}$ to a $z$ score using equation (12).

## Formula for the z-Score Corresponding to $\bar{d}$

Comparing equation (12) to $z = \dfrac{x - \mu}{\sigma}$

$$z = \frac{\bar{d} - \mu_d}{\dfrac{\sigma_d}{\sqrt{n}}} \tag{12}$$

Replaced by $x$ by $\bar{d}$

Replace $\mu$ by $\mu_d$

Replace $\sigma$ by $\dfrac{\sigma_d}{\sqrt{n}}$

▶ **Example 2**    The effect of *gentle acceleration* versus *jerky acceleration* on the gas mileage of automobiles has been studied by major car producers. The study indicated that drivers can improve the miles per gallon (mpg) rating of their cars an average of 2.6 mpg by using gentle acceleration instead of jerky acceleration. Furthermore the standard deviation for the distribution of differences is approximately 3.0 mpg. Suppose 36 cars are tested and $\bar{d}$ is the mean of the sample differences in mpgs.

   **a.** Compute $\mu_{\bar{D}}$.
   **b.** Compute $\sigma_{\bar{D}}$.
   **c.** Sketch and label a graph of the probability distribution of $\bar{D}$.
   **d.** Compute $P(\bar{D} < 1.5)$.

**Solution**

**a.** $\mu_{\bar{D}} = 2.6$ mpg    $\mu_{\bar{D}} = \mu_d$

**b.** $\sigma_{\bar{D}} = \dfrac{3.0}{\sqrt{36}} = 0.5$ mpg    $\sigma_{\bar{D}} = \dfrac{\sigma_d}{\sqrt{n}}$

**c.** Because 36 is a large sample size, the probability distribution of $\bar{D}$ is approximately normal. A graph is shown in Figure 7-11.

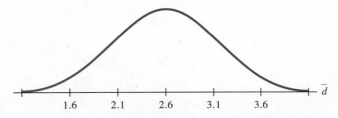

**Figure 7-11.** A graph of the probability distribution of $\bar{D}$.

**d.** With $\bar{d} = 1.5$, $n = 36$, $\mu_d = 2.6$, and $\sigma_d = 3.0$:

$$z = \dfrac{1.5 - 2.6}{\dfrac{3.0}{\sqrt{36}}} \qquad z = \dfrac{\bar{d} - \mu_d}{\dfrac{\sigma_d}{\sqrt{n}}}$$

$$= -2.20$$

$$P(\bar{D} < 1.5) = P(Z < -2.20)$$

$$= 0.0139$$

Thus the probability is about 1 percent that the mean difference in mpgs for the 36 sampled cars is less than 1.5 mpg.    ◄

## Large sample confidence intervals for $\mu_d$

If $\bar{D}$ is based on a large sample size, then values of the standard normal distribution can be used to construct confidence intervals for $\mu_d$. That is, large $n$ confidence limits for a population mean difference $\mu_d$ are:

$$\text{lower limit} = \bar{d} - z\left(\dfrac{\sigma_d}{\sqrt{n}}\right)$$

$$\text{upper limit} = \bar{d} + z\left(\dfrac{\sigma_d}{\sqrt{n}}\right)$$

The value of $z$ in these equations depends on the confidence coefficient of the interval. If $\sigma_d$ is unknown (and it nearly always is), then common practice is to estimate $\sigma_d$ by $s_d$, and therefore estimate $\sigma_{\bar{D}}$ with the expression in (11). When this substitution is made, we obtain practical confidence limits for $\mu_d$.

## Large Sample Equations for Constructing Confidence Intervals for $\mu_d$

If $z$ is a positive number and $n$ is large, then the interval with:

$$\text{lower limit} = \bar{d} - z\left(\frac{s_d}{\sqrt{n}}\right) \tag{13}$$

and

$$\text{upper limit} = \bar{d} + z\left(\frac{s_d}{\sqrt{n}}\right) \tag{14}$$

can be used as a confidence interval for $\mu_d$. The associated confidence coefficient is $P(-z < Z < z)$.

▶ **Example 3**   Sandra Broom manages the Tip-Top Fitness Club. She recently installed a machine that is advertised to reduce the waists of individuals who exercise with it. To estimate the effectiveness of the machine, she closely monitored the measurements of 64 members of the club. For the period of the study, Sandra obtained the following statistics for the 64 differences:

$$\bar{d} = 12.5 \text{ cm} \quad \text{and} \quad s_d = 2.4 \text{ cm}$$

Use these to construct a 95 percent confidence interval for the mean difference in waist measurements for individuals that exercise with this machine for a period of time comparable to the one in the study.

**Solution**   With $\bar{d} = 12.5$, $s_d = 2.4$, $n = 64$, and $z = 1.96$:

$$\text{lower limit} = 12.5 - 1.96\left(\frac{2.4}{8}\right)$$

$$= 12.5 - 0.588$$

$$= 11.9, \text{ to one decimal place}$$

$$\text{upper limit} = 12.5 + 0.588$$

$$= 13.1, \text{ to one decimal place}$$

Based on the given data, a 95 percent confidence interval for the mean difference in waist measurements of individuals who exercise with the machine for a period of time comparable to the one in Sandra's study is between 11.9 cm and 13.1 cm. ◀

## Small sample confidence intervals for $\mu_d$

Equations (13) and (14) can be used to estimate $\mu_d$ if $n$ is large. To estimate $\mu_d$ with a confidence interval based on a small sample size ($n < 30$), one must add the restriction that the population of differences has a relative frequency distribution that is approximately bell shaped. *That is, the differences in the sample must be modeled as independent normal random variables.*

When such an assumption can be made, then the random variable

$$\frac{\bar{d} - \mu_d}{\dfrac{s_d}{\sqrt{n}}}$$

has a *t*-distribution with $n - 1$ degrees of freedom.

---

### Small Sample Equations for Constructing Confidence Intervals for $\mu_d$

If the values in a sample of differences can be described as independent normal random variables with mean $\mu_d$ and standard deviation $\sigma_d$, then confidence limits for $\mu_d$ are:

$$\text{lower limit} = \bar{d} - t\left(\frac{s_d}{\sqrt{n}}\right) \tag{15}$$

and

$$\text{upper limit} = \bar{d} + t\left(\frac{s_d}{\sqrt{n}}\right) \tag{16}$$

where $t$ is a percentage point for the *t*-distribution with $n - 1$ degrees of freedom chosen for the desired confidence coefficient.

---

▶ **Example 4**    Hector Rodriquez owns a company that installs insulation in homes. He wanted to estimate the mean reduction in natural gas consumption in homes that his company has insulated over the past couple of years. He therefore sampled eight homes from his large list of former customers. From the owners of these homes, he got the natural gas consumption in cubic feet for the January immediately preceding and the January immediately following installation of additional insulation. Hector compared the weather conditions for the two Januarys and found them to be essentially the same. The pairs of before and after figures obtained were:

$$(900, 820) \quad (1100, 930) \quad (1060, 790) \quad (880, 780)$$
$$(1020, 875) \quad (1370, 1020) \quad (680, 615) \quad (1460, 1120)$$

Hector feels the population of differences has a relative frequency distribution that is approximately bell shaped. Use the eight data points to construct a 90 percent confidence interval for the mean decrease in gas consumption for the months studied in homes insulated by Hector's company.

**Solution**

| Natural Gas Consumption Before Insulation (1) | Natural Gas Consumption After Insulation (2) | Difference in Gas Consumption (1) − (2) |
|---|---|---|
| 900 | 820 | 80 |
| 1100 | 930 | 170 |
| 1060 | 790 | 270 |
| 880 | 780 | 100 |
| 1020 | 875 | 145 |
| 1370 | 1020 | 350 |
| 680 | 615 | 65 |
| 1460 | 1120 | 340 |

For the sampled differences:

$$\bar{d} = 190 \text{ cubic feet} \quad \text{and} \quad s_d = 115 \text{ cubic feet.}$$

From Appendix Table 5, with $n - 1 = 7$ and $t = 1.895$ for a 90 percent confidence coefficient:

$$\text{lower limit} = 190 - 1.895\left(\frac{115}{\sqrt{8}}\right)$$

$$= 190 - 77.048 \ldots$$

$$= 113, \text{ to the nearest whole number}$$

$$\text{upper limit} = 190 + 77.048 \ldots$$

$$= 267, \text{ to the nearest whole number}$$

Based on the given data, a 90 percent confidence interval for the mean reduction in natural gas consumption in homes insulated by Rodriquez's company during the time compared is between 113 and 267 cubic feet of gas.  ◀

## Large sample tests of $H_0$: $\mu_d = 0$

The random variable $\bar{D}$ can also be used in a hypothesis test for the mean difference. The form of $H_0$ for such tests is usually

$$H_0: \quad \mu_d = 0 \qquad \text{\small $H_0$ claims there is on average no difference.}$$

A hypothesis test determines whether a sample gives sufficient evidence to reject $H_0$ and assert that some kind of difference exists between the $x$ and $y$ values. Depending on the motivation for the test, the form of $H_a$ will be one of the following:

$H_a$:  $\mu_d \neq 0$    A nondirectional alternative that claims a difference does exist between the $x$ and $y$ values.

$H_a$:  $\mu_d < 0$    A directional alternative that claims the mean difference is negative, that is, $x$ values tend to be smaller than $y$ values.

$H_a$:  $\mu_d > 0$    A directional alternative that claims the mean difference is positive, that is, $x$ values tend to be larger than $y$ values.

### The Large Sample Test Statistic for $H_0$: $\mu_d = 0$

For large $n$, the hypothesis $H_0$: $\mu_d = 0$ can be tested using

$$Z = \frac{\bar{D} - \mu_d}{\frac{S_d}{\sqrt{n}}} = \frac{\bar{D}}{\frac{S_d}{\sqrt{n}}} \quad \text{or} \quad Z = \frac{\bar{D}\sqrt{n}}{S_d} \qquad (17)$$

The simplified form is based on $H_0$: $\mu_d = 0$. If $H_0$ is true, $Z$ has an approximately standard normal distribution.

As is by now standard, to conduct such a test, the computed value of $z$ is compared to $z_c$, the critical $z$ for the test. The $z_c$ depends on the $\alpha$ of the test, and whether the test is directional or nondirectional. If the $z$ obtained by using the equation (17) falls in the rejection region, then $H_0$ is rejected in favor of $H_a$. If $z$ does not fall in the rejection region, then $H_0$ cannot be rejected.

▶ **Example 5**   Programmable Incorporated, a Chicago software company, has developed a typing tutor program that helps increase the number of words per minute (wpm) a typist can type. To test whether the program is effective, the company sampled employees in the area that type as part of their job responsibilities. Those sampled possessed a wide variety of typing skills. Typing proficiencies were determined using pre and post tests, and differences were measured by subtracting pretest scores from posttest scores. For the 81 differences:

$$\bar{d} = 3.7 \text{ wpm}$$

$$s_d = 21.6 \text{ wpm}$$

Use these results to test the alternative hypothesis that the typing tutor program is effective in increasing typing speed. Use $\alpha = 0.05$.

**Solution**   **Step 1**  $H_0$:  $\mu_d = 0$
    $H_a$:  $\mu_d > 0$    The company wants to prove that posttest-pretest scores
                    are, on the average, positive numbers.

**Step 2**  $\alpha = 0.05$    $P$(Type I error) $= 0.05$.

**Step 3**  The test statistic used is:

$$Z = \frac{\bar{D}\sqrt{n}}{S_d}$$

**Step 4**  If $z > z_c$ ($= 1.65$), then reject $H_0$. See Figure 7-12 on the next page.

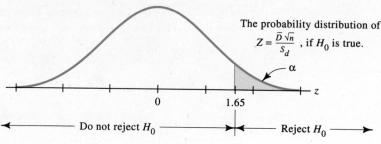

**Figure 7-12.**   $\alpha = P(Z > 1.65)$.

**Step 5**   With $\bar{d} = 3.7$, $s_d = 21.6$, and $n = 81$:

$$z = \frac{3.7(9)}{21.6} = 1.54166\ldots$$
$$= 1.54, \text{ to two decimal places}$$

**Step 6**   Because $1.54 < 1.65$, do not reject $H_0$. That is, there is not sufficient evidence, with $\alpha = 0.05$, to conclude that the tutor program is effective in increasing typing speed in terms of results on a words-per-minute typing test.   ◀

## Small sample tests of $H_0$: $\mu_d = 0$

Recall that we had to assume that a population of $d$'s has a bell-shaped relative frequency distribution to construct a small sample confidence interval for $\mu_d$. Similarly this added assumption is needed before we can test $H_0$: $\mu_d = 0$ based on a small sample. If $H_0$ is true, when the assumption is made that the differences in the sample are independent normal random variables with mean $\mu_d$ and standard deviation $\sigma_d$, then the test statistic can be shown to have a $t$-distribution with $n - 1$ degrees of freedom.

### The Small Sample Test Statistic for $H_0$: $\mu_d = 0$

If the differences in a sample can be described as independent normal random variables, then the test statistic for $H_0$: $\mu_d = 0$ is:

$$T = \frac{\bar{D} - \mu_d}{\dfrac{S_d}{\sqrt{n}}} \quad \text{or} \quad T = \frac{\bar{D}\sqrt{n}}{S_d} \tag{18}$$

If $H_0$ is true, then $T$ has a $t$-distribution with $n - 1$ degrees of freedom.

▶ **Example 6**  In an experiment intended to study the effect of altitude on running times, nine professional football players were asked to run 880 yards in Oakland, California, and then again in Denver, Colorado. The times, to the nearest 1 second, for the nine athletes are given below with the Oakland time listed first.

(210, 225)  (130, 135)  (190, 202)  (198, 201)  (150, 156)
(175, 183)  (280, 289)  (252, 265)  (186, 193)

Assuming the distribution of differences in times for running 880 yards is approximately bell shaped for professional football players, does the sample provide sufficient data to conclude $\mu_d \neq 0$? Use $\alpha = 0.05$.

**Solution**  Because the Denver times are greater, we will subtract from them the Oakland times to keep the differences positive. (This arbitrary choice has no effect on the final conclusion of the analysis.)

| Denver Times (1) | Oakland Times (2) | Differences in Times (1) − (2) |
|---|---|---|
| 225 | 210 | 15 |
| 135 | 130 | 5 |
| 202 | 190 | 12 |
| 201 | 198 | 3 |
| 156 | 150 | 6 |
| 183 | 175 | 8 |
| 289 | 280 | 9 |
| 265 | 252 | 13 |
| 193 | 186 | 7 |

For these data, $\bar{d} = 8.666\ldots$ seconds and $s_d = 3.968\ldots$ seconds.

**Step 1**  $H_0$:  $\mu_d = 0$
$H_a$:  $\mu_d \neq 0$      A nondirectional test

**Step 2**  $\alpha = 0.05$      $P(\text{Type I error}) = 0.05$.

**Step 3**  The test statistic used is:

$$T = \frac{\bar{D}\sqrt{n}}{S_d}$$

**Step 4**  If $t <$ lower $t_c$ ($= -2.306$) or $t >$ upper $t_c$ ($=2.306$), then reject $H_0$. (With $n = 9$, the $t$-distribution with 8 degrees of freedom is used.)

**Step 5**  With $\bar{d} = 8.66\ldots$, $s_d = 3.96\ldots$, and $n = 9$:

$$t = \frac{(8.66\ldots)(3)}{3.96\ldots} = 6.551\ldots$$

$$= 6.551, \text{ to three decimal places}$$

**Step 6** Because $6.551 > 2.306$, reject $H_0$. That is, there is sufficient evidence, with $\alpha = 0.05$, to conclude that a difference exists between the mean times it takes professional football players to run 880 yards in Oakland and in Denver.  ◄

If $n$ is small, then the assumption regarding the shape of the relative frequency distribution of differences is essential to using $\bar{d}$ to make a confidence interval or conduct a hypothesis test. If it seems unlikely that a population of differences has a bell-shaped distribution, then the procedures discussed in Chapter 10 must be used.

# Exercises 7-2  Set A

*In Exercises 1–4, consider the sample mean difference of n paired data points. The mean and standard deviation of the population of differences are given. In each exercise, do the following:*
*a. Compute $\mu_{\bar{D}}$.*
*b. Compute $\sigma_{\bar{D}}$ to one decimal place.*
*c. Sketch and label a graph of the probability distribution of $\bar{D}$.*
*d. Compute the indicated probability.*
*e. State the meaning of the probability in part (d).*

1. For the population of differences:

$$\mu_d = 25.0 \quad \text{and} \quad \sigma_d = 4.8$$

A sample of $n = 64$ data pairs is obtained. Compute $P(\bar{D} < 24.0)$.

2. For the population of differences:

$$\mu_d = 3.75 \quad \text{and} \quad \sigma_d = 0.84.$$

A sample of $n = 36$ data pairs is obtained. Compute $P(\bar{D} > 3.90)$.

3. For the population of differences:

$$\mu_d = 138 \quad \text{and} \quad \sigma_d = 16.1.$$

A sample of $n = 49$ data pairs is obtained. Compute $P(136 < \bar{D} < 140)$.

4. For the population of differences:

$$\mu_d = 0.075 \quad \text{and} \quad \sigma_d = 0.016$$

A sample of $n = 64$ data pairs is obtained. Compute $P(0.072 < \bar{D} < 0.077)$.

5. Mary Ellen Murnin-Heise saw advertisements that claimed supermarket A had prices that were less than a competing supermarket B. She was interested enough in the claim to do a mini comparison of several items that she knew both markets carried in their stock.

The following ordered pairs contain the price of a given item at supermarket A ($x$) and the price of the same item at supermarket B ($y$). The ordered pairs are listed as ($x$, $y$). For this sample, compute:

a. $\bar{d}$ to the nearest 1 cent, where $d_i = x_i - y_i$.

b. $s_d$ to the nearest 1 cent.

| | | | | |
|---|---|---|---|---|
| (1.79, 2.09) | (1.86, 1.75) | (2.62, 2.69) | (1.49, 1.59) | (0.89, 0.98) |
| (1.39, 1.61) | (1.89, 2.31) | (1.48, 1.75) | (5.19, 5.25) | (2.56, 2.29) |
| (1.23, 1.25) | (1.49, 1.84) | (3.89, 3.59) | (0.49, 0.73) | (1.85, 2.09) |
| (4.39, 4.39) | (6.98, 6.75) | (3.19, 3.35) | (0.19, 0.21) | (1.32, 1.48) |

**6.** Donald Kozlowski was the superintendent on a project to test the effect of a smog-control device on the gas mileage of 30 cars. The following pairs of numbers are the mpgs of the cars with the device ($x$) and the mpgs of the same cars before the device was installed ($y$). For this sample of data, compute:

a. $\bar{d}$ to one decimal place, where $d_i = x_i - y_i$.

b. $s_d$ to one decimal place.

| | | | | |
|---|---|---|---|---|
| (18.9, 17.8) | (20.3, 19.5) | (19.1, 18.6) | (21.5, 21.0) | (25.5, 24.1) |
| (16.3, 16.1) | (18.5, 17.8) | (28.4, 26.2) | (20.3, 19.1) | (22.6, 22.0) |
| (22.2, 21.8) | (20.1, 19.7) | (23.4, 20.8) | (20.6, 19.4) | (14.3, 12.8) |
| (18.3, 18.0) | (19.6, 19.0) | (25.0, 23.5) | (21.7, 20.9) | (30.5, 27.9) |
| (21.7, 21.2) | (22.3, 20.8) | (20.7, 19.5) | (17.8, 15.9) | (17.4, 17.2) |
| (20.3, 18.7) | (16.2, 15.9) | (14.8, 12.5) | (17.5, 16.3) | (16.7, 15.5) |

**7.** Fifty adults completed a speed-writing course. Let $d$, measured in words per minute (wpm), represent the difference in the writing speed of a participant (after − before the course). The following summary statistics were obtained:

$$n = 50 \qquad \sum d_i = 2000 \qquad \sum d_i^2 = 83{,}540$$

a. Calculate $\bar{d}$ to one decimal place.

d. Calculate $s_d$ to one decimal place.

**8.** The girths of 60 seedling trees were measured at the time of planting and the girths are measured again 5 years later. If $d$, measured in centimeters (cm), is the difference in the girth measurements, then the following summary statistics are obtained:

$$n = 60 \qquad \sum d_i = 756 \qquad \sum d_i^2 = 9{,}660$$

a. Calculate $\bar{d}$ to one decimal place.

b. Calculate $s_d$ to one decimal place.

**9.** A machine measures in kilograms (kg) the mass a person can lift. Thirty people participated in a body-building program designed to increase lifting ability. The mass each person could lift was measured before and after the program. The following summary statistics were obtained:

$$d = \text{ability after − ability before}$$
$$n = 30 \qquad \sum d_i = 465.0 \qquad \sum d_i^2 = 7625.0$$

a. Calculate $\bar{d}$ to one decimal place.

b. Calculate $s_d$ to one decimal place.

**10.** A special diet is developed to reduce the blood pressure of a person with high blood pressure. Forty volunteers participate in a well-supervised program in which only the individual's diet is changed. The blood pressure of each volunteer is measured before and after the program. The following summary statistics are obtained:

$$n = 40 \qquad \sum d_i = 740.0 \qquad \sum d_i^2 = 14{,}900.0$$

a. Calculate $\bar{d}$ to one decimal place.

b. Calculate $s_d$ to one decimal place.

11. A quality control inspection for one of the types of springs produced by Sutter Springs Corporation consists of stretching the springs by a force A, and then again by a different force B. The difference in the amount of stretch produced by the two forces is measured for each spring. For the population of differences for these springs, assume $\mu_d =$ 13.8 cm and $\sigma_d = 1.2$ cm. Suppose a sample of 36 springs is taken from a large production run of these springs and $\bar{D}$ is the mean difference in the amounts of stretch for this sample.

    a. Compute $\mu_{\bar{D}}$.
    b. Compute $\sigma_{\bar{D}}$ to one decimal place.
    c. Sketch and label a graph of the probability distribution of $\bar{D}$.
    d. Compute $P(\bar{D} < 13.5)$.

12. A new formula feed developed by Snowdon Feed and Supply Company is specifically designed to increase milk production in Holstein cows. Records indicate that the new formula feed increases the milk production of such cows an average of 3.2 pounds of milk per day over the traditional feed. For the distribution of differences in milk production, the standard deviation is 0.9 pounds. Suppose a sample of 36 Holstein cows is taken from a major milk producing area in Wisconsin. The cows are fed the traditional formula feed for a period of time and the milk productions are measured. The cows are then fed the new formula feed and the milk productions are again measured. Let $\bar{D}$ be the mean weight increase in pounds per day in the milk productions of these cows.

    a. Compute $\mu_{\bar{D}}$.
    b. Compute $\sigma_{\bar{D}}$ to the nearest one-tenth of 1 pound.
    c. Sketch and label a graph of the probability distribution of $\bar{D}$.
    d. Compute $P(\bar{D} < 9.0)$.

13. A mild sedative developed by Dearinger Drug Corporation is reported to reduce the time it takes someone to fall asleep an average of 8 to 10 minutes. Suppose that $\mu_d = 9.0$ minutes and $\sigma_d = 3.5$ minutes for the population of differences in

times. A sample of 49 employees of the legislative branch of government of a central state is selected to test the effectiveness of the drug. Let $\bar{D}$ be the mean difference in the times it takes these individuals to fall asleep without, and then with, the sedative.

    a. Compute $\mu_{\bar{D}}$.
    b. Compute $\sigma_{\bar{D}}$.
    c. Sketch and label a graph of the probability distribution of $\bar{D}$.
    d. Compute $P(8.5 < \bar{D} < 9.5)$.

14. The effect of the addition of an air conditioner on the mpg rating of vehicles of all types (cars, trucks, tractors, and so on) has been studied by an independent research corporation. The study suggests that the mean difference in ratings with, and without, an air conditioner is 4.6 mpg with differences having a standard deviation of 3.1 mpg. Suppose a sample of 40 vehicles is taken in Tuscon, Arizona. Let $\bar{D}$ be the sample mean difference in the mpgs of the 40 vehicles when operated without versus with an air conditioner running.

    a. Compute $\mu_{\bar{D}}$.
    b. Compute $\sigma_{\bar{D}}$ to one decimal place.
    c. Sketch and label a graph of the probability distribution of $\bar{D}$.
    d. Compute $P(4.0 < \bar{D} < 5.0)$.

15. An advertisement promoting car pooling for commuters in the Los Angeles area claims a participant can save $30.00 per month in commuting costs. Suppose the estimate is based on a study of 250 commuters that changed to car pooling, and for the sample $s_d = \$7.50$.

    a. Construct a 90 percent confidence interval for $\mu_d$.
    b. State what the interval means.

16. A major tire manufacturer claims that gas mileage can be improved an average of 2.5 mpg by switching to steel-belted radial tires. Suppose the estimate is based on a study of the mileages of 40 automobiles and for the sample $s_d = 1.3$ mpg.

    a. Construct a 90 percent confidence interval for $\mu_d$.
    b. State what the interval means.

17. Susan Wilcox-Garner, a reading instructor at a large community college, developed a program to increase reading speeds. Based on the results of 400 participants who completed the program, the reading speeds measured in words per minute (wpm) were increased by 120 wpm with a standard deviation of 60 wpm.
    a. Construct a 95 percent confidence interval for $\mu_d$.
    b. State what the interval means.

18. Jeffrey Stephens, a basic skills educator, developed a program that teaches basic skills in mathematics to adult learners. A sample of 144 adults completed the program. A standardized mathematics test of 100 problems was given to these adults before, and a similar one again after, completion of the program. For these adults, the mean increase in the number of correct answers was 35 answers with a standard deviation of 18.
    a. Construct a 98 percent confidence interval for $\mu_d$.
    b. State what the interval means.

*In Exercises 19–22, construct confidence intervals for $\mu_d$. In each exercise assume the population of differences is approximately bell shaped.*

19. An electronic sensing device is developed by a company that builds carburetors for automobiles. The device electronically controls the air and gas mixture flowing to a car's carburetor, and is supposed to improve the gas mileage of a car. The device is tested on eight cars selected from car dealers in the city. The cars were first driven with the device and then without the device. The mpgs are listed below as ordered pairs (with, without).

    (18.8, 16.5)   (20.1, 18.3)   (24.9, 22.6)   (25.8, 22.3)
    (23.1, 21.5)   (27.5, 23.1)   (31.8, 27.5)   (27.4, 23.8)

    Construct a 90 percent confidence interval for $\mu_d$.

20. A "Members Only Supply Store" claims its prices on building tools and equipment are lower than any other supply outlet in the area. Ten tools were priced at a store of a national supply outlet chain ($x$). The same ten items were priced at Members Only ($y$). The data are listed below in the form ($x, y$):

    (7.98, 7.59)    (1.79, 1.69)   (3.29, 3.19)    (5.49, 5.19)    (10.98, 9.89)
    (23.49, 21.89)  (2.69, 2.43)   (19.89, 19.29)  (9.49, 8.75)    (6.99, 6.19)

    Construct a 95 percent confidence interval for $\mu_d$.

21. A cement manufacturing company has developed a silicone-based brick mortar that reportedly enables construction workers to set more bricks in a given time period than they can with the standard mortar mix. To test the claim, 12 construction workers laying bricks at different construction sites are observed. Each worker records the number of bricks he or she can set in a specified time using the new formula mortar ($x$). The same worker then records the number of bricks he or she can set in the same time using the standard mortar mix ($y$). The data are listed below in the form ($x, y$).

    (63, 61)   (72, 67)   (55, 49)   (70, 62)   (65, 62)   (89, 81)
    (76, 71)   (67, 60)   (59, 55)   (74, 66)   (58, 56)   (66, 60)

    Construct a 98 percent confidence interval for $\mu_d$.

22. Test Anxiety Inc. is an organization of behavioral scientists studying the extreme physiological responses that students exhibit while taking examinations. At a large university in the Northeast, they randomly selected 16 students. They monitored the heart rates of these students while taking a final examination ($x$). They then obtained the heart rates of these same students while in a relaxed setting ($y$). Their data are listed below in the form ($x$, $y$).

    (98, 78)   (112, 76)  (85, 80)   (89, 76)   (106, 82)  (110, 85)  (92, 75)   (86, 76)
    (102, 80)  (105, 74)  (120, 86)  (83, 78)   (97, 74)   (90, 80)   (101, 87)  (88, 72)

    Construct a 99 percent confidence interval for $\mu_d$.

*In Exercises 23–28, use the given data to test $H_0$: $\mu_d = 0$.*

23. A finger dexterity exercise was developed by a typing instructor to increase students' typing speed, measured in words per minute (wpm). To test the exercise, the teacher obtained the typing speeds ($x$) of 49 students. The students used the exercise for 2 weeks and their typing speeds were again measured ($y$). If $d_i = x_i - y_i$, and for the 49 students, $\bar{d} = -3.10$ wpm and $s_d = 14.00$, determine whether the exercise increases typing speed. Use $\alpha = 0.05$.

24. A device is supposed to reduce the carbon monoxide emissions in cars. A sample of 36 cars is obtained, and the carbon monoxide emissions ($x$), measured in grams per mile (gpm), are obtained. The device is installed and the carbon monoxide emissions ($y$) are measured again. If $d_i = x_i - y_i$, and for the 36 cars, $\bar{d} = 4.73$ gpm and $s_d = 8.40$, determine whether the device is effective in reducing emissions. Use $\alpha = 0.01$.

25. A doctor thinks that the number of migraine headaches in chronic sufferers can be reduced by eliminating certain foods from the diet. The number ($x$) of headaches is counted for a 6-month period for 36 volunteers. The suspected foods are eliminated from the volunteers' diets and the number ($y$) of headaches is again counted for a 6-month period. If $d_i = x_i - y_i$, and for the 36 volunteers, $\bar{d} = 2.3$ migraines and $s_d = 4.8$, can the claim be made that the elimination of these foods is effective in reducing the number of migraine headaches for chronic sufferers? Use $\alpha = 0.01$.

26. An experiment is conducted to determine whether music can affect the egg laying productivity of chickens. Forty-nine hens are selected and the number ($x$) of eggs laid over a 3-month period without music is determined for each hen. With music, the number ($y$) of eggs laid over a 3-month period by the same hens is counted. If $d_i = x_i - y_i$, and for the 49 hens, $\bar{d} = -1.3$ eggs and $s_d = 4.2$, can the claim be made that music affects the egg laying productivity of chickens? Use $\alpha = 0.05$.

27. A test is conducted to determine whether the amount of food eaten by laboratory rats changes when they live under colored lighting. Thirty-six rats are selected for the study. The amount of food ($x$) each rat eats for 2 weeks is measured in grams while the rat lives continuously under white lights. The amount of food ($y$) each rat eats for 2 weeks is measured in grams while the rat lives continuously under blue lights. If $d_i = x_i - y_i$ and $\bar{d} = 2.3$ grams and $s_d = 5.4$, can the claim be made that rats eat different mean amounts of food living under white and blue lights? Use $\alpha = 0.01$.

28. A test was conducted on 36 volunteers in Syracuse, New York, to see whether the frequency of common colds can be reduced by taking massive doses of vitamin C. From December 1 through March 31 one-half of the volunteers took no vitamin C tablets, whereas the other one-half took massive doses of the vitamin. The number of colds contracted per individual was recorded. From December 1 through March 31 of the next year, the same individuals were tested. Now the ones who took no vitamin C and those who did

were reversed. The number of colds contracted per individual was again recorded. With $x$ = the number of colds without vitamin C, $y$ = the number of colds with vitamin C, and $d_i = x_i - y_i$, can $H_0: \mu_d = 0$ be rejected in favor of $H_a: \mu_d > 0$ for residents of Syracuse with $\alpha = 0.01$, if $\bar{d} = 1.6$ colds and $s_d = 2.4$ for the 36 individuals tested?

*In Exercises 29 and 30, assume the population of differences is approximately bell shaped, and test $H_0: \mu_d = 0$ at the indicated level of significance.*

29. A commercial for brand X pain reliever claims the product reduces the pain of a headache faster than any other brand available without a doctor's prescription. Nancy lives in a large mobile home park and enlists the help of 15 residents to test this claim as part of a science project for high school. Each volunteer agrees to use brand X next time they get a headache and carefully record the time it takes to get noticeable relief. They also agree to use brand Y (another popular brand pain reliever) the following time they get a headache and again carefully record the time to get similar relief. From

the $n = 15$ volunteers, Nancy gets $\bar{d} = -2.8$ minutes and $s_d = 7.3$ minutes, where:

$$d = \text{time to relief with brand X} - \text{time to relief with brand Y}$$

and the times are measured in minutes. Use $\alpha = 0.05$.

30. In an attempt to improve "small car" driving safety, a major U.S. car manufacturer designed a new braking system to reduce stopping distances for its small car products. The engineers selected ten cars with the standard braking design from the company's inventory. Cars were driven by a test driver who, while driving 50 mph, stopped using maximum braking power. The distances ($x$) needed to stop were measured. The same cars were then equipped with the new braking systems. With the same drivers and the same 50 mph speed, the distances ($y$) needed to stop were measured. Let:

$$d_i = x_i - y_i$$

and for the ten cars, $\sum d_i = 24.8$ feet and $s_d = 1.3$ feet. Use $\alpha = 0.01$.

## Exercises 7-2   Set B

*In Exercises 1 and 2, use $\bar{d}$ and $s_d$ from the 40 differences in Table 7-4 in the Statistics in Action part of this section.*

1. Use the values to construct a 98 percent confidence interval for $\mu_d$, the mean weight loss for all the individuals in the program.

2. Using $\alpha = 0.01$, test:

$$H_0: \quad \mu_d = 0$$
$$H_a: \quad \mu_d > 0$$

*In Exercises 3 and 4, refer to the report in Figure 7-13 on the next page.*

3. Suppose a sample of 36 homes in the Cincinnati, Ohio, area that have inadequate insulation is taken. A record of heating costs for 1 week in January is obtained. The same homes are insulated and a record of heating costs for 1 week of comparable weather

conditions is again obtained. If $d$ is the difference in the heating costs of each home for the weeks tested,

$$d = \text{cost before insulation} - \text{cost after insulation}$$

Suppose $\bar{d} = \$8.73$ with $s_d = \$10.80$. Do these data substantiate the claim that insulation reduces heating costs in the Cincinnati area? Use $\alpha = 0.01$.

4. Suppose a sample of 40 homes in the St. Petersburg, Florida, area that have inadequate insulation is taken. A record of heating costs for 1 week in January is obtained. The same homes are insulated and a record of heating costs for 1 week of comparable weather conditions is again obtained. If $d$ is the difference in the heating costs of each home for the weeks tested,

$$d = \text{cost before insulation} - \text{cost after insulation}$$

Suppose $\bar{d} = \$4.92$ and $s_d = \$10.76$. Do these data substantiate the claim that insulation reduces heating costs in the St. Petersburg area? Use $\alpha = 0.01$.

*In Exercises 5 and 6, use random numbers to sample the indicated number of stocks from Data Set II, Appendix B. For these stocks compute:*

$$d = 52 \text{ week high} - 52 \text{ week low}$$

a. *Compute $\bar{d}$.*
b. *Compute $s_d$.*
c. *Use the statistics from parts (a) and (b) to construct a 90 percent confidence interval for $\mu_d$.*

5. Sample 36 stocks.

6. Sample 50 stocks.

**INSULATE THE ATTIC AND THE WALLS.**

Install mineral wool, glass fiber, or cellulose insulation to a depth of 6 inches in the attic. Heating costs should drop about 20 percent.

If 15 million homes with inadequate attic insulation were upgraded, about 400,000 barrels of heating oil would be saved each winter day — reducing the Nation's demand for residential heating fuels by 4 percent. Installation of insulation in the walls also yields a large energy saving but requires special equipment and professional help in existing homes.

**Figure 7-13.** From the pamphlet "Tips for Energy Savers: In and Around the Home," Energy Conservation and Environment, Federal Energy Administration, Washington, D.C.

# 7-3 Inferences Based on the Difference Between Independent Sample Proportions—Large Samples

**Key Topics**

1. *The probability distribution of $\hat{P}_1 - \hat{P}_2$ for large $n_1$ and $n_2$*

2. $\mu_{\hat{P}_1 - \hat{P}_2} = P_1 - P_2$ *and* $\sigma_{\hat{P}_1 - \hat{P}_2} = \sqrt{\dfrac{P_1(1 - P_1)}{n_1} + \dfrac{P_2(1 - P_2)}{n_2}}$

3. *Converting $\hat{p}_1 - \hat{p}_2$ to a z score*

4. *Large sample confidence intervals for $P_1 - P_2$*

5. *Large sample tests of $H_0: P_1 - P_2 = 0$*

S T A T I S T I C S      I N      A C T I O N

Velma Stein is a psychologist working for a large metropolitan school district in the East. She has recently been studying teenage relationships with an emphasis on one that she calls "best friend." The following statement describes her meaning of a best friend:

"A best friend is a person of the same sex and approximate age whose friendship is valued above anyone else's. Such a friendship is exhibited by activities such as personal contacts with this person every day, a commitment to the happiness of this person, a feeling of protection for this person's safety. This is someone from whom you can keep no secrets."

Velma wondered whether a difference exists in the proportions of female and male teenagers that feel they have a best friend based on this definition. To obtain data for a comparison, she prepared the following questionnaire:

| A Survey on Best Friends | | |
|---|---|---|
| 1. Please check the appropriate box. | Female ☐ | Male ☐ |
| 2. After reading the statement that describes a best friend, do you feel that you currently have such a person in your life?<br>If you answered no to question 2, then please answer 3 and 4. | Yes ☐ | No ☐ |
| 3. Do you feel you have a best friend, but she/he is of the opposite sex? | Yes ☐ | No ☐ |
| 4. Based on the description of a best friend and your "no" answer to question 2, do you wish you had such a person in your life? | Yes ☐ | No ☐ |

With the help of four other staff members, Velma and her assistants visited the junior high and high school campuses in the district. Each survey was personally handed to a student who read, marked, and returned it at the time of contact. The results were then tabulated and are summarized below.

| Female Students | Male Students |
|---|---|
| 1. $n_1 = 435$ | 1. $n_2 = 396$ |
| 2. Yes, 139; No, 296 | 2. Yes, 87; No, 309 |
| 3. Yes, 74; No, 222 | 3. Yes, 117; No, 192 |
| 4. Yes, 118; No, 178<br>For questions 3 and 4, $n_1 = 296$. | 4. Yes, 108; No, 201<br>For questions 3 and 4, $n_2 = 309$. |

Velma was asked to speak at faculty meetings throughout the district on her perception of current trends in the social and personal development of students in the district. She included the results of the "best friend" survey as part of her presentation. Based on the data gathered in the survey, Velma made the following claims:

"Question 2 showed that significantly more female students give best friend status to another female, than male students give to another male. However, question 3 showed that significantly more male students give best friend status to a female, than female students give to a male. Finally, question 4 showed that there is no significant difference between the proportions of female and male students that do not have a best friend of the same sex, but would like to have one."

The details of how these conclusions were obtained are studied in this section.

## The probability distribution of $\hat{P}_1 - \hat{P}_2$ for large $n_1$ and $n_2$

In Figure 7-14, two populations are represented as rectangles. Suppose a characteristic is defined that is present in some of the members in each population. For Populations 1 and 2, $P_1$ and $P_2$ are the proportions, respectively, of the members with the characteristic.

Parameter $P_1$                Parameter $P_2$

**Figure 7-14.**   Two populations in which some of the members have a specified characteristic.

In the Statistics in Action part of this section, Populations 1 and 2 were, respectively, the female and male students in the district. Velma Stein investigated the proportions of students in these populations who claimed they had a best friend that fit Velma's description.

If samples are taken from the populations, then each sample will yield a value of $\hat{P}_1$ or $\hat{P}_2$, where

$$\hat{P}_1 = \frac{X}{n_1} \quad \text{and} \quad \hat{P}_2 = \frac{Y}{n_2}$$

and $X$ and $Y$ are the numbers of items in the samples with the characteristic of interest. If $n_1$ and $n_2$ are sufficiently large, then the probability distributions of $\hat{P}_1$ and $\hat{P}_2$ will be approximately normal.

### DEFINITION 7-3 Independent Sample Proportions $\hat{P}_1$ and $\hat{P}_2$

If $\hat{P}_1$ and $\hat{P}_2$ are calculated from unrelated samples, then they may be modeled as independent. Furthermore, if the sample sizes are large, they have probability distributions that are approximately normal. In such a case, $\hat{P}_1 - \hat{P}_2$ is a random variable that also has an approximately normal probability distribution.

The samples of female and male students that Velma surveyed were sufficiently large so that the observed sample proportions were values of random variables with approximately normal probability distributions. Based on Definition 7-3, the probability distribution of $\hat{P}_1 - \hat{P}_2$ is also approximately normal.

$$\mu_{\hat{P}_1 - \hat{P}_2} = P_1 - P_2 \quad \text{and}$$

$$\sigma_{\hat{P}_1 - \hat{P}_2} = \sqrt{\frac{P_1(1 - P_1)}{n_1} + \frac{P_2(1 - P_2)}{n_2}}$$

Equations (19) and (20) relate the mean and standard deviation of the probability distribution of $\hat{P}_1 - \hat{P}_2$ to $P_1$ and $P_2$.

### The Mean and Standard Deviation for the Probability Distribution of $\hat{P}_1 - \hat{P}_2$

For the probability of distribution of $\hat{P}_1 - \hat{P}_2$:

$$\mu_{\hat{P}_1 - \hat{P}_2} = P_1 - P_2 \tag{19}$$

$$\sigma_{\hat{P}_1 - \hat{P}_2} = \sqrt{\frac{P_1(1 - P_1)}{n_1} + \frac{P_2(1 - P_2)}{n_2}} \tag{20}$$

▶ **Example 1**   A statewide study by the Department of Motor Vehicles (DMV) was conducted to check on compliance with a new "seat belt law." The law requires all passengers in a moving vehicle wear a seat belt. The study suggests that 49 percent of female passengers (Population 1) and 38 percent of male passengers (Population 2) obey the law. Suppose samples are taken of 120 female and 120 male passengers in the state. Let $P_1$ and $P_2$ be the proportions of female and male passengers, respectively, who are wearing seat belts.

**a.** Compute $\mu_{\hat{P}_1 - \hat{P}_2}$.
**b.** Compute $\sigma_{\hat{P}_1 - \hat{P}_2}$ to two decimal places.
**c.** Sketch a graph of the probability distribution of $\hat{P}_1 - \hat{P}_2$.

**Solution**   **a.** With $P_1 = 0.49$ and $P_2 = 0.38$:

$$\mu_{\hat{P}_1 - \hat{P}_2} = 0.49 - 0.38 = 0.11$$

**b.** With $P_1 = 0.49$, $1 - P_1 = 0.51$, $n_1 = 120$, $P_2 = 0.38$, $1 - P_2 = 0.62$, and $n_2 = 120$:

$$\sigma_{\hat{P}_1 - \hat{P}_2} = \sqrt{\frac{(0.49)(0.51)}{120} + \frac{(0.38)(0.62)}{120}}$$

$$= \sqrt{0.004045\ldots}$$

$$= 0.06, \text{ to two decimal places.}$$

**c.** With the mean of 0.11 and a standard deviation of 0.06, a sketch of the probability distribution of $\hat{P}_1 - \hat{P}_2$ is shown in Figure 7-15 on the following page.   ◀

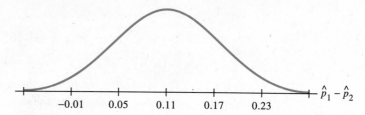

**Figure 7-15.**   A graph of the probability distribution of $\hat{P}_1 - \hat{P}_2$.

## Converting $\hat{p}_1 - \hat{p}_2$ to a z score

Once the mean and standard deviation of the probability distribution of $\hat{P}_1 - \hat{P}_2$ are known, probabilities for $\hat{P}_1 - \hat{P}_2$ can be calculated using Appendix Table 4. However, to use this table, we must first change values of $\hat{p}_1 - \hat{p}_2$ to z scores.

### Formula for the z-Score Corresponding to $\hat{p}_1 - \bar{p}_2$

$$z = \frac{(\hat{p}_1 - \hat{p}_2) - (P_1 - P_2)}{\sqrt{\dfrac{P_1(1 - P_1)}{n_1} + \dfrac{P_2(1 - P_2)}{n_2}}} \quad (21)$$

Comparing Equation (21) to $z = \dfrac{x - \mu}{\sigma}$

$x$ is replaced by $\hat{p}_1 - \hat{p}_2$, the difference in the sample proportions.

$\mu$ is replaced by $P_1 - P_2$, the difference in the population proportions.

$\sigma$ is replaced by $\sigma_{\hat{p}_1 - \hat{p}_2}$.

▶ **Example 2**   For the probability distribution of $\hat{P}_1 - \hat{P}_2$ in Example 1, compute:

$$P(\hat{P}_1 - \hat{P}_2 > 0.20)$$

**Solution**   Using the values from Example 1:

$$z = \frac{0.20 - 0.11}{\sqrt{0.04045\ldots}}$$

$P_1 - P_2 = 0.11.$
Use the unrounded value for $\sigma_{\hat{p}_1 - \hat{p}_2}$.

$$= 1.41494\ldots$$

$$= 1.41, \text{ to two decimal places}$$

$$P(\hat{P}_1 - \hat{P}_2 > 0.20) = P(Z > 1.41)$$

$$= 0.0793$$

Thus the probability is about 8 percent that such samples will yield a difference of more than 20 percent in the proportions of female and male passengers who are wearing seat belts.   ◀

## Large sample confidence intervals for $P_1 - P_2$

If the values of $\hat{P}_1$ and $\hat{P}_2$ are based on reasonably large sample sizes, then $\hat{P}_1$ and $\hat{P}_2$ can be used to construct confidence intervals for an unknown $P_1 - P_2$. Because $P_1$, $(1 - P_1)$, $P_2$, and $(1 - P_2)$ will not be known, it is common practice (and theoretically sound) to replace these by natural estimates based on $\hat{P}_1$ and $\hat{P}_2$. That is, $\sigma_{\hat{P}_1 - \hat{P}_2}$ is estimated by the square roots given in equations (22) and (23).

### Equations for Constructing Confidence Intervals for $P_1 - P_2$

If $z$ is a positive number, and $n_1$ and $n_2$ are large, then the interval with:

$$\text{lower limit} = (\hat{p}_1 - \hat{p}_2) - z\sqrt{\frac{\hat{p}_1(1 - \hat{p}_1)}{n_1} + \frac{\hat{p}_2(1 - \hat{p}_2)}{n_2}} \tag{22}$$

and

$$\text{upper limit} = (\hat{p}_1 - \hat{p}_2) + z\sqrt{\frac{\hat{p}_1(1 - \hat{p}_1)}{n_1} + \frac{\hat{p}_2(1 - \hat{p}_2)}{n_2}} \tag{23}$$

can be used as a confidence interval for $P_1 - P_2$. The associated confidence coefficient is $P(-z < Z < z)$.

▶ **Example 3**   Dr. Delma Lowell is a specialist in education working with the state school superintendent's office. She recently completed a study in a large district in the state. She surveyed 280 high school seniors in the lower one-half of their classes (Population 1), and found 53 percent do no reading other than that required for school. She also surveyed 230 high school seniors in the top one-half of their classes (Population 2), and found 30 percent do no reading other than that required for school. Use this information to construct a 95 percent confidence interval for the difference in proportions of high school seniors in the lower one-half and upper one-half of their classes in this district who do no reading other than that required for school.

**Solution**   With $\hat{p}_1 = 0.53$, $1 - \hat{p}_1 = 0.47$, $n_1 = 280$, $\hat{p}_2 = 0.30$, $1 - \hat{p}_2 = 0.70$, $n_2 = 230$, and equations (22) and (23):

$$\text{lower limit} = (0.53 - 0.30) - 1.96\sqrt{\frac{(0.53)(0.47)}{280} + \frac{(0.30)(0.70)}{230}}$$

$$= 0.23 - 1.96(0.04245\ldots)$$

$$= 0.23 - 0.0832\ldots$$

$$= 0.15, \text{ to two decimal places}$$

$$\text{upper limit} = 0.23 + 0.0832\ldots$$

$$= 0.31, \text{ to two decimal places}$$

Based on the given data, a 95 percent confidence interval for the difference in proportions of high school seniors in the district in the lower one-half and upper one-half of their classes who do no reading other than that required for school is from 0.15 to 0.31.  ◀

## Large sample tests of $H_0$: $P_1 - P_2 = 0$

In Section 7-1, we used hypothesis tests for $\mu_1 - \mu_2$ to determine whether there was a detectable difference between the means of two populations. We shall now use hypothesis tests on $P_1 - P_2$ to decide whether a detectable difference exists between the proportions of two populations with a particular characteristic. The form of $H_0$ in such tests is usually:

$$H_0: \quad P_1 - P_2 = 0$$

$H_0$ claims that $P_1 = P_2$, that is, there is *no difference* in the proportions of the two populations with the characteristic of interest.

The purpose of the test is to find if there is sufficient evidence to reject $H_0$ and assert that some type of difference exists between the proportions. Depending on the motivation for the test, the form of $H_a$ will be one of the following:

$$H_a: \quad P_1 - P_2 \neq 0$$

A nondirectional alternative claiming *a difference exists* between $P_1$ and $P_2$.

$$H_a: \quad P_1 - P_2 < 0$$

A directional alternative claiming $P_1$ is smaller than $P_2$; that is, $P_1 - P_2 < 0$.

$$H_a: \quad P_1 - P_2 > 0$$

A directional alternative claiming $P_1$ is greater than $P_2$; that is, $P_1 - P_2 > 0$.

If independent random samples from Populations 1 and 2 yield $\hat{p}_1 - \hat{p}_2$ that converts to a $z$ value that falls in a rejection region, then $H_0: P_1 - P_2 = 0$ is rejected. There is a slight (perhaps unexpected) wrinkle in converting $\hat{p}_1 - \hat{p}_2$ to a $z$ value. This wrinkle arises because the calculation is made supposing $H_0$ is true. If $H_0$ is true (that is, $P_1 = P_2$), then the samples are being taken from two populations with the same proportion. We temporarily call the (unknown) *common value P*. As a consequence of the fact that $P_1 = P_2 = P$, the standard deviation for the probability distribution of $\hat{P}_1 - \hat{P}_2$ is:

$$\sigma_{\hat{P}_1 - \hat{P}_2} = \sqrt{\frac{P(1-P)}{n_1} + \frac{P(1-P)}{n_2}}$$

Replace both $P_1$ and $P_2$ by the common value $P$

$$= \sqrt{P(1-P)} \sqrt{\frac{1}{n_1} + \frac{1}{n_2}}$$

Factor out the $P(1-P)$.

This standard deviation is estimated by "pooling" the data from both samples to get an estimate of $P$. We will use $\hat{p}$ to represent this estimate of $P$, and $\hat{p}$ is then further used to calculate the estimate for the standard deviation of $\hat{P}_1 - \hat{P}_2$.

Depending on the way the data from the samples are given, the value of $\hat{p}$ can be calculated by one of the following formulas:

1. If $x$ and $y$ are *the numbers of data points* in the samples of size $n_1$ and $n_2$ with the specified characteristic, then

$$\hat{p} = \frac{x + y}{n_1 + n_2} \left( = \frac{\text{total successes}}{\text{total sample size}} \right)$$

2. If $\hat{p}_1$ (based on $n_1$) and $\hat{p}_2$ (based on $n_2$) are *the sample proportions*, then $x = n_1 p_1$ and $y = n_2 p_2$.

$$\hat{p} = \frac{n_1 \hat{p}_1 + n_2 \hat{p}_2}{n_1 + n_2}$$

When testing $H_0: P_1 - P_2 = 0$:

$$\sigma_{\hat{P}_1 - \hat{P}_2} \text{ is estimated by } \sqrt{\hat{p}(1 - \hat{p})} \sqrt{\frac{1}{n_1} + \frac{1}{n_2}}$$

### The Test Statistic for a Test of $H_0: P_1 - P_2 = 0$

If large independent samples of sizes $n_1$ and $n_2$ are taken from Populations 1 and 2 respectively, then the test statistic for $H_0: P_1 - P_2 = 0$ is:

$$Z = \frac{(\hat{P}_1 - \hat{P}_2) - (P_1 - P_2)}{\sqrt{\hat{P}(1 - \hat{P})} \sqrt{\frac{1}{n_1} + \frac{1}{n_2}}} = \frac{\hat{P}_1 - \hat{P}_2}{\sqrt{\hat{P}(1 - \hat{P})} \sqrt{\frac{1}{n_1} + \frac{1}{n_2}}} \tag{24}$$

The simplified form of equation (24) is based on $H_0: P_1 - P_2 = 0$, and if $H_0$ is true, the test statistic has an approximately standard normal distribution.

▶ **Example 4**   Dr. Steven Sidwell is a psychology instructor at Marymount College. He read a paper, by Dr. Phillip G. Zimbardo, a Stanford University professor, on "shyness" in the American population. Steve thinks that changing values in the United States are making young people more self-assertive and less shy. To test his hypothesis, he surveyed 110 individuals in Boston younger than 21 years of age (Population 1) and found 39 that qualified as shy based on his criteria. He also surveyed 125 individuals in Boston older than 21 years of age (Population 2) and found 60 that qualified as shy. Do the data support Steve's suspicion at a 5 percent level of significance?

**Solution**   **Discussion.** In the pooled sample, $39 + 60 = 99$ individuals qualify as shy. The total of the two sample sizes is $110 + 125 = 235$. Thus:

$$\hat{p} = \frac{99}{235} = 0.42127 \ldots \qquad \text{Do not round.}$$

**Step 1** $H_0$:   $P_1 - P_2 = 0$      Steven wants to show that the proportion of
        $H_a$:   $P_1 - P_2 < 0$      individuals under 21 that are shy is *less than*
                                   the proportion of individuals over 21.

**Step 2** $\alpha = 0.05$

**Step 3** The test statistic used is:

$$Z = \frac{\hat{P}_1 - \hat{P}_2}{\sqrt{\hat{P}(1 - \hat{P})}\sqrt{\dfrac{1}{n_1} + \dfrac{1}{n_2}}}$$

**Step 4** If $z < z_c \, (= -1.65)$, then reject $H_0$. See Figure 7-16.

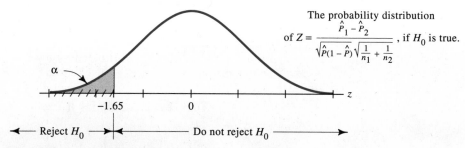

The probability distribution of $Z = \dfrac{\hat{P}_1 - \hat{P}_2}{\sqrt{\hat{P}(1 - \hat{P})}\sqrt{\frac{1}{n_1} + \frac{1}{n_2}}}$ , if $H_0$ is true.

Reject $H_0$            Do not reject $H_0$

**Figure 7-16.**   $\alpha = P(Z < -1.65)$.

**Step 5** With $\hat{p}_1 = \frac{39}{110}$, $\hat{p}_2 = \frac{60}{125}$, and $\hat{p} = 0.42127\ldots$

$$z = \frac{\frac{39}{110} - \frac{60}{125}}{\sqrt{(0.42\ldots)(0.57\ldots)}\sqrt{\frac{1}{110} + \frac{1}{125}}}$$

$$= \frac{-0.1254\ldots}{0.0645\ldots}$$

$$= -1.94, \text{ to two decimal places}$$

**Step 6** Because $-1.94 < -1.65$, reject $H_0$. That is, there is sufficient evidence, with $\alpha = 0.05$, to conclude that in Boston the proportion of individuals younger than 21 years who are shy is less than the proportion older than 21 years who are shy.   ◄

# Exercises 7-3    Set A

*In Exercises 1–4, assume Populations 1 and 2 have some members with a specified characteristic of interest. Furthermore $n_1$ and $n_2$ are the sizes of independent random samples taken from Populations 1 and 2 respectively. For each exercise, do the following:*

*a. Compute $\mu_{\hat{p}_1 - \hat{p}_2}$.*

*b. Compute $\sigma_{\hat{p}_1 - \hat{p}_2}$.*

c. Sketch a graph of the probability distribution of $\hat{P}_1 - \hat{P}_2$.

d. Compute the indicated probability.

e. State the meaning of the answer to part (d).

1. $P_1 = 0.60$, $P_2 = 0.52$, $n_1 = 120$, and $n_2 = 156$. Compute $P(\hat{P}_1 - \hat{P}_2 < 0.00)$.

2. $P_1 = 0.78$, $P_2 = 0.65$, $n_1 = 240$, and $n_2 = 257$. Compute $P(\hat{P}_1 - \hat{P}_2 > 0.18)$.

3. $P_1 = 0.36$, $P_2 = 0.28$, $n_1 = 40$, and $n_2 = 86$. Compute $P(0.00 < \hat{P}_1 - \hat{P}_2 < 0.20)$.

4. $P_1 = 0.483$, $P_2 = 0.459$, $n_1 = 530$, and $n_2 = 515$. Compute $P(0.000 < \hat{P}_1 - \hat{P}_2 < 0.050)$.

5. The U.S. Labor Department reports that 55 percent of workers in jobs related to the building industry (Population 1) earn more than the national median income. It also reports that 35 percent of workers in service oriented jobs (Population 2) earn more than the median income. Samples of sizes $n_1 = 165$ and $n_2 = 220$ are selected from the two populations and the proportions of each sample that earn more than the national median income are computed.

a. Compute $\mu_{\hat{P}_1 - \hat{P}_2}$.

b. Compute $\sigma_{\hat{P}_1 - \hat{P}_2}$.

c. Sketch a graph of the probability distribution of $\hat{P}_1 - \hat{P}_2$.

d. Compute $P(0.16 < \hat{P}_1 - \hat{P}_2 < 0.24)$.

e. State what the answer to part (d) means.

6. Machines A and B at Ohio Screw and Bolt Manufacturers put threads on carriage bolts. Records indicate that 8 percent of machine A bolts and 5 percent of machine B bolts have some type of defect. A quality control program requires periodic sampling of 120 bolts threaded by these machines. Let $\hat{P}_1$ and $\hat{P}_2$ be the proportions of defective bolts, respectively, from machines A and B in the samples.

a. Compute $\mu_{\hat{P}_1 - \hat{P}_2}$.

b. Compute $\sigma_{\hat{P}_1 - \hat{P}_2}$.

c. Sketch a graph of the probability distribution of $\hat{P}_1 - \hat{P}_2$.

d. Compute $P(0.00 < \hat{P}_1 - \hat{P}_2 < 0.05)$.

e. State what the answer to part (d) means.

7. A large agency called Hire a Musician (HAM) promotes live musical entertainment. Its current records indicate that 60 percent of union musicians and 45 percent of nonunion musicians are currently working. Suppose independent random samples of $n_1 = 80$ union musicians (Population 1) and $n_2 = 130$ nonunion musicians (Population 2) represented by HAM are taken. Let $\hat{P}_1$ and $\hat{P}_2$ be the proportions in the respective samples that are currently working.

a. Compute $\mu_{\hat{P}_1 - \hat{P}_2}$.

b. Compute $\sigma_{\hat{P}_1 - \hat{P}_2}$.

c. Sketch a graph of the probability distribution of $\hat{P}_1 - \hat{P}_2$.

d. Compute $P(\hat{P}_1 - \hat{P}_2 > 0.10)$.

e. State what the answer to part (d) means.

8. A nationwide poll suggests that 73 percent of women and 58 percent of men currently rate toxic waste disposal as the most critical problem in the United States. Suppose that in Orlando, Florida, samples of $n_1 = 540$ women (Population 1) and $n_2 = 455$ men (Population 2) are taken and $\hat{P}_1$ and $\hat{P}_2$ are the proportions of women and men in the samples that rate toxic waste disposal as the most critical problem.

a. Compute $\mu_{\hat{P}_1 - \hat{P}_2}$.

b. Compute $\sigma_{\hat{P}_1 - \hat{P}_2}$.

c. Sketch a graph of the probability distribution of $\hat{P}_1 - \hat{P}_2$.

d. Compute $P(0.10 < \hat{P}_1 - \hat{P}_2 < 0.16)$.

e. State what the answer to part (d) means.

9. The need for water in California has divided the state into southern Californians (Population 1) and northern Californians (Population 2). A member of the state water resources board surveyed 340 residents in southern California and 285 residents in northern California. Based on a series of questions asked, she found 204 southern Californians and 136 northern Californians who

actively practice water conservation in their homes. Use these data to construct a 90 percent confidence interval for the difference in proportions of southern and northern California residents that actively practice water conservation.

10. A reporter at a large metropolitan area newspaper wondered whether there was a difference related to educational attainment in attitudes toward mandatory testing for detection of serious diseases. He subsequently conducted a survey in which 105 of 150 individuals with, at most, a high school diploma (Population 1) felt that such mandatory testing is necessary. The survey also yielded 28 of 80 individuals with a college degree (Population 2) who felt such mandatory testing is necessary. Use these data to construct a 90 percent confidence interval for the difference in proportions of nongraduates and college graduates who favor mandatory testing for detection of serious diseases.

11. Recently the issue of a single national primary was studied by the League of American Voters (LAV). The league sponsored a survey of Democratic (Population 1) and Republican (Population 2) voters in a ten-state region in the East. The survey yielded 70 percent of the 1220 Democratic voters and 65 percent of the 940 Republican voters who favor a single national primary. Let $P_1$ and $P_2$ be the proportions, respectively, of Democratic and Republican voters in this area who favor a single national primary.
    a. Construct a 95 percent confidence interval for $P_1 - P_2$.
    b. State the meaning of the interval found in part (a).

12. The issue of imported goods was being studied by an economics class at a university in Michigan. To gather information on one aspect of the issue, the students surveyed the vehicles in the student parking lots (Population 1) and the vehicles in the faculty parking lots (Population 2). The surveys yielded 56 percent imported vehicles of the 75 checked in Population 1 and 45 percent imported vehicles of the 80 checked in Population 2. Let $P_1$

and $P_2$ be the proportions of imported vehicles in the student and faculty lots, respectively, at this university.
    a. Construct a 95 percent confidence interval for $P_1 - P_2$.
    b. State the meaning of the interval found in part (a).

13. Katie Konradt is the personnel manager for a large tire manufacturing corporation in the Akron, Ohio, area. The company's officers have been discussing the issue of changing from the traditional 5-day/8-hour shifts to a new 4-day/10-hour shift plan. The change could result in a substantial savings in company operating expenses. Katie was instructed to poll male and female opinions on the possible change. A survey of 140 men (Population 1) and 140 women (Population 2) employees yielded 112 men and 80 women who favor the proposed new work schedule. Let $P_1$ and $P_2$ be the proportions of male and female workers, respectively, who favor the 4-day/10-hour shift work schedule.
    a. Construct a 98 percent confidence interval for $P_1 - P_2$.
    b. State the meaning of the interval in part (a).

14. Bud Granger is the president of a large chapter of a BMW motorcycle club. His club members are strong advocates of a state law requiring motorcyclists to wear helmets. A member of the club has access to accident reports throughout the state. A sample of motorcycle accidents over the past few years yielded 66 serious head injuries of 150 involving motorcyclists not wearing helmets (Population 1). Another sample yielded 33 serious head injuries of 132 accidents involving motorcyclists wearing helmets (Population 2). Let $P_1$ and $P_2$ be the proportions of serious head injuries in accidents involving motorcyclists not wearing helmets and wearing helmets, respectively.
    a. Construct a 98 percent confidence interval for $P_1 - P_2$.
    b. State the meaning of the interval in part (a).

*In Exercises 15–22, test* $H_0$: $P_1 - P_2 = 0$ *at the indicated level of significance.*

15. A study by Dr. John A. Benvenuto, Jr., of the National Institute of Drug Abuse in Rockville, Maryland, suggests that one-third of all U.S. adults use some form of a tranquilizer at least once a year. A sociologist conducts a study expecting to show that a greater proportion of residents of large cities (more than 100,000 residents) use tranquilizers than residents of small cities (fewer than 100,000 residents). She samples 250 large city residents (Population 1) and finds 105 that take tranquilizers. A sample of 180 small city residents (Population 2) yields 52 who take tranquilizers. Do the results support the sociologist's claim at the 5 percent level of significance?

16. A recent issue of *Field & Stream* magazine stated that 17 percent of the registered voters in the United States support a ban on handguns. The article was based on findings in a poll taken by the Decision Making Information (DMI) firm. A member of a Women for Political Action group in a midwestern city feels the proportion of women in the city (Population 1) who favor a ban on handguns is greater than the proportion of men (Population 2) who favor a ban on handguns. With the help of other members, 140 women voters and 93 men voters were contacted. The results yielded sample proportions $\hat{p}_1 = 0.33$ and $\hat{p}_2 = 0.25$, respectively, of the women and men voters who support a ban on handguns. Do the data support the expectation of the member of the Action group regarding the residents of this city? Use $\alpha = 0.05$.

17. An organization of clergy involved in premarital counseling wants to evaluate its effectiveness in reducing divorce in teenage marriages. A survey of 60 couples who were married in their teens without counseling (Population 1) yielded 35 divorces within the first 7 years of marriage. A survey of 48 couples who were married in their teens after counseling (Population 2) yielded 21 divorces within the first 7 years. Can the members of the clergy claim that they are effective in reducing divorces among teenagers? Use $\alpha = 0.05$.

18. A community college district has an urban and a suburban campus. The central administration wants to know whether the Hispanic students in the district attend the campuses in equal proportions. A sample of 180 urban and 160 suburban campus students includes 27 and 26 Hispanic students, respectively. Do the data contradict the hypothesis that the proportions of students at the two campuses who are Hispanic are equal? Use $\alpha = 0.05$.

19. A recently conducted poll indicated that the opinion of the vast majority of workers in the United States is that "ability to do the job" should be the principal consideration for hiring. Caralee Woods is the director of an unemployment office in a region in the Midwest. She sampled the opinions of 82 unemployed minority workers (Population 1) and 74 unemployed non-minority workers (Population 2) that reported to her office regarding the "ability to do the job" principle. Of those questioned, 67 minority and 68 non-minority workers agreed that ability should be the main consideration for hiring. Do these results support the claim that a difference exists between the proportions of these groups regarding ability as the main consideration for hiring? Use $\alpha = 0.01$.

20. Mayors Pratt and Gulley of large eastern and western cities, respectively, met at a mayors' convention in St. Louis. The topic of air pollution as a major concern of large cities was discussed at the meeting. Pratt and Gulley wondered whether the proportions of residents in their cities who rate air pollution as a major concern were the same. They agreed to conduct similar polls and share the results. Pratt obtained 172 of 220 people (Population 1) who named air pollution as a major concern. Gulley got 185 of 210 people (Population 2) who similarly named air pollution. Do these results show that the proportions in the two cities are different? Use $\alpha = 0.01$.

21. Luwanna Johnson is a candidate for the state senate. She sampled union workers (Population 1) and nonunion workers (Population 2) regarding their intentions to vote for her. Of the 125 union workers, 52 percent expressed support for Ms. Johnson. Of the 130 nonunion workers, 40 percent expressed support for Ms. Johnson. Do the data support the claim that a difference exists in the proportions of union and nonunion workers who support Luwanna? Use $\alpha = 0.05$.

22. Herman Gladstone works at a lunch counter near several government buildings. He recorded 54 men of 146 surveyed who drink their coffee black. He similarly recorded 65 women of 144 surveyed who drink their coffee black. Do the data support Herman's claim that more women than men drink black coffee? Use $\alpha = 0.05$.

## Exercises 7-3    Set B

*In Exercises 1–3, use the data obtained by Velma Stein in the Statistics in Action part of this section. Let Population 1 be the female students in the district and let Population 2 be the male students.*

1. Use the results of question 2 on the survey:
   $x = 139$, $n_1 = 435$, $y = 87$, and $n_2 = 396$.
   a. Compute $\hat{p}_1$.
   b. Compute $\hat{p}_2$.
   c. Estimate $\hat{P}_1 - \hat{P}_2$ with a 90 percent confidence interval.
   d. Test $H_0: P_1 - P_2 = 0$ versus $H_0: P_1 - P_2 \neq 0$, using $\alpha = 0.01$.

2. Use the results of question 3 on the survey:
   $x = 74$, $n_1 = 296$, $y = 117$, and $n_2 = 309$.
   a. Compute $\hat{p}_1$.
   b. Compute $\hat{p}_2$.
   c. Estimate $P_1 - P_2$ with a 95 percent confidence interval.
   d. Test $H_0: P_1 - P_2 = 0$ versus $H_a: P_1 - P_2 < 0$, using $\alpha = 0.01$.

3. Use the results of question 4 on the survey:
   $x = 118$, $n_1 = 296$, $y = 108$, and $n_2 = 309$.
   a. Compute $\hat{p}_1$.
   b. Compute $\hat{p}_2$.
   c. Estimate $P_1 - P_2$ with a 98 percent confidence interval.
   d. Test $H_0: P_1 - P_2 = 0$ versus $H_a: P_1 - P_2 \neq 0$, using $\alpha = 0.05$.

4. Use random numbers to obtain $n_1 = 40$ males and $n_2 = 40$ females from Data Set I in Appendix B.
   a. Compute $\hat{p}_1$ and $\hat{p}_2$, the proportions of males and females respectively in the samples who have cancer.

b. Use $\hat{p}_1$ and $\hat{p}_2$ from part (a) to test, using $\alpha = 0.05$:

$$H_0: \quad P_1 - P_2 = 0 \quad\quad \text{versus} \quad\quad H_a: \quad P_1 - P_2 > 0$$

c. Use the values of $\hat{p}_1$ and $\hat{p}_2$ from part (a) to construct a 95 percent confidence interval for the difference in the proportions of males and females in the population who have cancer.

5. Use random numbers to obtain $n_1 = 50$ males and $n_2 = 50$ females from Data Set I in Appendix B.

a. Compute $\hat{p}_1$ and $\hat{p}_2$, the proportions of males and females respectively, who smoke.

b. Use $\hat{p}_1$ and $\hat{p}_2$ from part (a) to test, using $\alpha = 0.05$:

$$H_0: \quad P_1 - P_2 = 0 \quad\quad \text{versus} \quad\quad H_a: \quad P_1 - P_2 \neq 0$$

c. Use $\hat{p}_1$ and $\hat{p}_2$ from part (a) to construct a 90 percent confidence interval for the difference in the proportions of males and females in the population who smoke.

*In Exercises 6–8, refer to the information in Figure 7-17.*

6. A study was made of people who recently used an emergency room (ER) in an urban area (Population 1) and a rural area (Population 2). Of the 260 cases studied in the urban area ER, 169 required urgent or routine care. Of the 240 cases studied in the rural area ER, 120 required urgent or routine care. Use these data to test, with $\alpha = 0.01$, whether a difference exists in the proportions of urban and rural residents who use emergency rooms for non-emergency cases.

7. A study was made of the ages of people who recently used the emergency room (ER) in a large metropolitan area. Of the 310 cases of individuals younger than 25 years of age (Population 1) studied, 217 were non-emergency. Of the 250 cases of individuals 25 years of age or older (Population 2) studied, 110 were non-emergency. Use these data to construct a 95 percent confidence interval for the difference in proportions of non-emergency uses of the ER for individuals younger than 25 years and individuals 25 and older.

8. A study was made of the sex of the people who recently used the emergency department in a large

> • **Misused emergency departments** – A local study reveals that of people using emergency departments, 10% are in trauma, 30% are emergencies, and the remaining 60% require urgent or routine care that should have been handled in a physician's office.

**Figure 7-17.** "Misused emergency departments" by Sacramento-Sierra Hospital Conference. Reprinted by permission.

metropolitan area. Of the 280 cases of female patients (Population 1) studied, 154 were non-emergency. Of the 260 cases of male patients (Population 2) studied, 156 were non-emergency. Use these data to test, with $\alpha = 0.05$, whether a difference exists in the proportions of female and male patients who use emergency rooms for non-emergency cases.

## Chapter 7 Summary

### Definitions

1. The sample means $\bar{X}$ and $\bar{Y}$ can be modeled as **independent** when, on an intuitive basis, they can be described as unrelated, or based on separately chosen samples.

2. Populations 1 and 2 contains **paired (dependent) data**, if for each $x_i$ from Population 1 there exists exactly one data point $y_i$ from Population 2 that naturally corresponds to it in some way.

3. If $\hat{P}_1$ and $\hat{P}_2$ are calculated from unrelated samples, then they may be modeled as independent. Furthermore, if the sample sizes are large, they have probability distributions that are approximately normal. In such a case, $\hat{P}_1 - \hat{P}_2$ is a random variable that also has an approximately normal probability distribution.

### Rules and Equations

a. $\mu_{\bar{X} - \bar{Y}} = \mu_1 - \mu_2$          **(1)**

b. $\sigma_{\bar{X} - \bar{Y}} = \sqrt{\dfrac{\sigma_1^2}{n_1} + \dfrac{\sigma_2^2}{n_2}}$          **(2)**

c. $z = \dfrac{(\bar{x} - \bar{y}) - (\mu_1 - \mu_2)}{\sqrt{\dfrac{\sigma_1^2}{n_1} + \dfrac{\sigma_2^2}{n_2}}}$          **(3)**

d. The lower and upper limits of a large sample confidence interval for $\mu_1 - \mu_2$ are:

$$\text{lower limit} = (\bar{x} - \bar{y}) - z \sqrt{\frac{s_1^2}{n_1} + \frac{s_2^2}{n_2}} \qquad \textbf{(4)}$$

$$\text{upper limit} = (\bar{x} - \bar{y}) + z \sqrt{\frac{s_1^2}{n_1} + \frac{s_2^2}{n_2}} \qquad \textbf{(5)}$$

where $z$ is based on the confidence coefficient.

e. The large sample test statistic for a test of $H_0 : \mu_1 - \mu_2 = 0$ based on independent samples of sizes $n_1$ and $n_2$ is:

$$Z = \frac{\bar{X} - \bar{Y}}{\sqrt{\dfrac{S_1^2}{n_1} + \dfrac{S_2^2}{n_2}}} \qquad \textbf{(6)}$$

f. Let $d_1, d_2, \ldots, d_n$ be the $n$ differences of paired data from Populations 1 and 2:

$$\bar{d} = \frac{\sum d_i}{n} \tag{7}$$

$$s_d = \sqrt{\frac{n\sum d_i^2 - (\sum d_i)^2}{n(n-1)}} \tag{8}$$

g. If $\bar{D}$ is the mean of a sample from a population of differences with mean $\mu_d$ and standard deviation $\sigma_d$, then:

$$\mu_{\bar{D}} = \mu_d \tag{9}$$

$$\sigma_{\bar{D}} = \frac{\sigma_d}{\sqrt{n}} \tag{10}$$

h. The sample standard deviation $s_d$ can be used as an estimate for $\sigma_d$, and $\sigma_{\bar{D}}$ can be estimated by:

$$\frac{s_d}{\sqrt{n}} \tag{11}$$

$$z = \frac{\bar{d} - \mu_d}{\dfrac{\sigma_d}{\sqrt{n}}} \tag{12}$$

i. If $z$ is a positive number and $n$ is large, then the interval with:

$$\text{lower limit} = \bar{d} - z\left(\frac{s_d}{\sqrt{n}}\right) \tag{13}$$

and

$$\text{upper limit} = \bar{d} + z\left(\frac{s_d}{\sqrt{n}}\right) \tag{14}$$

can be used as a confidence interval for $\mu_d$. The associated confidence coefficient is $P(-z < Z < z)$.

j. If the differences in a sample can be described as independent normal random variables with mean $\mu_d$ and standard deviation $\sigma_d$, then confidence limits for $\mu_d$ are:

$$\text{lower limit} = \bar{d} - t\left(\frac{s_d}{\sqrt{n}}\right) \tag{15}$$

and

$$\text{upper limit} = \bar{d} + t\left(\frac{s_d}{\sqrt{n}}\right) \tag{16}$$

where $t$ is a percentage point for the $t$-distribution with $n - 1$ degrees of freedom chosen to produce a desired confidence coefficient.

k. The test statistic for a test of $H_0: \mu_d = 0$ based on a large sample of paired data is:

$$Z = \frac{\bar{D} - \mu_d}{\frac{S_d}{\sqrt{n}}} = \frac{\bar{D}\sqrt{n}}{S_d} \tag{17}$$

l. For the probability distribution of $\hat{P}_1 - \hat{P}_2$:

$$\mu_{\hat{P}_1 - \hat{P}_2} = P_1 - P_2 \tag{18}$$

$$\sigma_{\hat{P}_1 - \hat{P}_2} = \sqrt{\frac{P_1(1 - P_1)}{n_1} + \frac{P_2(1 - P_2)}{n_2}} \tag{19}$$

m. $z = \dfrac{(\hat{p}_1 - \hat{p}_2) - (P_1 - P_2)}{\sqrt{\dfrac{P_1(1 - P_1)}{n_1} + \dfrac{P_2(1 - P_2)}{n_2}}}$ \hfill (20)

n. If $z$ is a positive number, and $n_1$ and $n_2$ are large, then the interval with:

$$\text{lower limit} = (\hat{p}_1 - \hat{p}_2) - z\sqrt{\frac{\hat{p}_1(1 - \hat{p}_1)}{n_1} + \frac{\hat{p}_2(1 - \hat{p}_2)}{n_2}} \tag{21}$$

and

$$\text{upper limit} = (\hat{p}_1 - \hat{p}_2) + z\sqrt{\frac{\hat{p}_1(1 - \hat{p}_1)}{n_1} + \frac{\hat{p}_2(1 - \hat{p}_2)}{n_2}} \tag{22}$$

can be used as a confidence interval for $P_1 - P_2$. The associated confidence coefficient is $P(-z < Z < z)$.

o. The test statistic for a test of $H_0: P_1 - P_2 = 0$ based on large independent samples $n_1$ and $n_2$ taken from Populations 1 and 2, respectively, is:

$$Z = \frac{(\hat{P}_1 - \hat{P}_2) - (P_1 - P_2)}{\sqrt{\hat{P}(1 - \hat{P})}\sqrt{\dfrac{1}{n_1} + \dfrac{1}{n_2}}}$$

$$= \frac{\hat{P}_1 - \hat{P}_2}{\sqrt{\hat{P}(1 - \hat{P})}\sqrt{\dfrac{1}{n_1} + \dfrac{1}{n_2}}} \tag{23}$$

where $\hat{P} = \dfrac{n_1\hat{P}_1 + n_2\hat{P}_2}{n_1 + n_2}$

## Symbols

$\bar{X} - \bar{Y}$    the **difference of the means** of independent random samples from two populations

$\mu_{\bar{X}-\bar{Y}}$    the **mean** of the probability distribution of $\bar{X} - \bar{Y}$

$\sigma_{\bar{X}-\bar{Y}}$    the **standard deviation** of the probability distribution of $\bar{X} - \bar{Y}$

$\bar{D}$    the **mean** of a sample of differences of paired data

$S_d$    the **standard deviation** of a sample of differences of paired data

$\mu_d$    the **mean** of a population of differences of paired data

$\mu_{\bar{D}}$    the **mean** of the probability distribution of $\bar{D}$

$\sigma_d$    the **standard deviation** of a population of differences

$\sigma_{\bar{D}}$    the **standard deviation** of the probability distribution of $\bar{D}$

$\hat{P}_1 - \hat{P}_2$    a random variable based on the **difference of the proportions** of independent samples from two populations

$\mu_{\hat{P}_1-\hat{P}_2}$    the **mean** of the probability distribution of $\hat{P}_1 - \hat{P}_2$

$\sigma_{\hat{P}_1-\hat{P}_2}$    the **standard deviation** of the probability distribution of $\hat{P}_1 - \hat{P}_2$

## Random Variables $\bar{X} - \bar{Y}$, $\bar{D}$, and $\hat{P}_1 - \hat{P}_2$

1. An observed value of $\bar{X} - \bar{Y}$ is obtained by calculating the difference in the means of independent samples from two populations with means $\mu_1$ and $\mu_2$.

2. The probability distribution of $\bar{X} - \bar{Y}$ will be approximately normal if $n_1 \geq 30$ and $n_2 \geq 30$, and:

$$\mu_{\bar{X}-\bar{Y}} = \mu_1 - \mu_2$$

$$\sigma_{\bar{X}-\bar{Y}} = \sqrt{\frac{\sigma_1^2}{n_1} + \frac{\sigma_2^2}{n_2}}$$

3. A $z$-score for $\bar{x} - \bar{y}$ is:

$$z = \frac{(\bar{x} - \bar{y}) - (\mu_1 - \mu_2)}{\sqrt{\dfrac{\sigma_1^2}{n_1} + \dfrac{\sigma_2^2}{n_2}}}$$

1. An observed value of $\bar{D}$ is obtained by calculating the mean of differences of paired dependent data from two populations.

2. The probability distribution of $\bar{D}$ will be approximately normal if $n$ is large, or if the distribution of $d$'s is bell shaped, and:

$$\mu_{\bar{D}} = \mu_d$$

$$\sigma_{\bar{D}} = \frac{\sigma_d}{\sqrt{n}}$$

3. A $z$-score for $\bar{d}$ is:

$$z = \frac{\bar{d} - \mu_d}{\dfrac{\sigma_d}{\sqrt{n}}}$$

1. An observed value of $\hat{P}_1 - \hat{P}_2$ is obtained by calculating the difference in the proportions of independent samples from two populations with proportions $P_1$ and $P_2$.

2. The probability distribution of $\hat{P}_1 - \hat{P}_2$ will be approximately normal if $n_1 P_1 \geq 5$, $n_1(1 - P_1) \geq 5$, $n_2 P_2 \geq 5$, and $n_2(1 - P_2) \geq 5$, and:

$$\mu_{\hat{P}_1-\hat{P}_2} = P_1 - P_2$$

$$\sigma_{\hat{P}_1-\hat{P}_2} = \sqrt{\frac{P_1(1 - P_1)}{n_1} + \frac{P_2(1 - P_2)}{n_2}}$$

3. A $z$-score for $\hat{p}_1 - \hat{p}_2$ is:

$$z = \frac{(\hat{p}_1 - \hat{p}_2) - (P_1 - P_2)}{\sqrt{\dfrac{P_1(1 - P_2)}{n_1} + \dfrac{P_2(1 - P_2)}{n_2}}}$$

4. The large sample limits of a confidence interval for $\mu_1 - \mu_2$ are

$$\text{lower limit} = (\bar{x} - \bar{y}) - z\sqrt{\frac{s_1^2}{n_1} + \frac{s_2^2}{n_2}}$$

$$\text{upper limit} = (\bar{x} - \bar{y}) + z\sqrt{\frac{s_1^2}{n_1} + \frac{s_2^2}{n_2}}$$

5. The test statistic for
$H_0: \mu_1 - \mu_2 = 0$

$$Z = \frac{\bar{X} - \bar{Y}}{\sqrt{\dfrac{S_1^2}{n_1} + \dfrac{S_2^2}{n_2}}}$$

4. The large sample limits of a confidence interval for $\mu_d$ has

$$\text{lower limit} = \bar{d} - z \cdot \frac{s_d}{\sqrt{n}}$$

$$\text{upper limit} = \bar{d} + z \cdot \frac{s_d}{\sqrt{n}}$$

5. The test statistic for
$H_0: \mu_d = 0$

$$Z = \frac{\bar{D}\sqrt{n}}{S_d}$$

where $n \geq 30$, or

$$T = \frac{\bar{D}\sqrt{n}}{S_d}$$

where $n < 30$

4. The large sample limits of a confidence interval for $P_1 - P_2$ has

$$\text{lower limit} = (\hat{p}_1 - \hat{p}_2)$$
$$- z\sqrt{\frac{\hat{p}_1(1 - \hat{p}_1)}{n_1} + \frac{\hat{p}_2(1 - \hat{p}_2)}{n_2}}$$

$$\text{upper limit} = (\hat{p}_1 - \hat{p}_2)$$
$$+ z\sqrt{\frac{\hat{p}_1(1 - \hat{p}_1)}{n_1} + \frac{\hat{p}_2(1 - \hat{p}_2)}{n_2}}$$

5. The test statistic for
$H_0: P_1 - P_2 = 0$

$$Z = \frac{\hat{P}_1 - \hat{P}_2}{\sqrt{\hat{P}(1 - \hat{P})}\sqrt{\dfrac{1}{n_1} + \dfrac{1}{n_2}}}$$

where $\hat{P} = \dfrac{n_1\hat{P}_1 + n_2\hat{P}_2}{n_1 + n_2}$

# Chapter 7 Review Exercises

*In Exercises 1 and 2, do the following:*
*a. Compute $\mu_{\bar{X} - \bar{Y}}$.*
*b. Compute $\sigma_{\bar{X} - \bar{Y}}$.*
*c. Sketch and label a graph of the probability distribution of $\bar{X} - \bar{Y}$.*
*d. Compute the indicated probability.*
*e. State the meaning of the answer to part (d).*

1. The Wechsler Adult Intelligence Scale (WAIS) is a test for which standard means and variances for various age groups have been developed. For this test, the following populations are defined:

   Population 1 = {test scores for
                  individuals ages 20–24 years}
   Population 2 = {test scores for
                  individuals ages 35–44 years}

   For these populations, the following parameters are given:

   $\mu_1 = 110.10$      $\mu_2 = 106.30$
   $\sigma_1 = 25.69$      $\sigma_2 = 24.77$

   Suppose $n_1 = 33$ and $n_2 = 40$ test scores are sampled from Populations 1 and 2, respectively. Let $\bar{X}$ and $\bar{Y}$ be the sample means. Compute $P(\bar{X} - \bar{Y} > 10.00)$.

2. The Green Valley Chicken Ranch has a computerized system that records the daily egg production of its chickens. Each hen is housed in a separate cage and the eggs laid by each hen are recorded. The ranch has two types of chickens for egg production. One type (Population 1) is lighter in body weight, lays more eggs, but is susceptible to chicken diseases. The other type (Population 2) is heavier in body weight, lays fewer eggs, and is less susceptible to chicken diseases. An analysis is made of the records for these two populations in terms of the numbers of eggs laid in one year. For these populations

$$\mu_1 = 336 \text{ eggs} \qquad \mu_2 = 275 \text{ eggs}$$
$$\sigma_1 = 15 \text{ eggs} \qquad \sigma_2 = 10 \text{ eggs}$$

Suppose $n_1 = 50$ and $n_2 = 40$ records of these hens are selected from Populations 1 and 2, respectively. Let $\bar{X}$ and $\bar{Y}$ be the corresponding sample means. Compute $P(\bar{X} - \bar{Y} < 60)$.

3. Hank Petrowski is the personnel manager for a large department store chain. He wants to estimate the difference between the mean dollar sales per day in the "Women's Casual Clothes" department for employees that work the evening shift (Population 1) and employees that work the early shift (Population 2). Samples of one shift sales performances of 88 evening shift and 92 early shift employees that work this department were taken from the chain's records. The samples yield the following data:

$$\bar{x} = \$330 \qquad \bar{y} = \$272$$
$$s_1 = \$68 \qquad s_2 = \$42$$

Construct a 90 percent confidence interval for the difference in the means of one day sales for evening shift and early shift employees in this department.

4. Nancy Klein is a research analyst for an independent research company. She developed a device to test the time it takes to drain the charge on dry cell batteries. She recently used the device to estimate the difference in the mean lives of a particular size dry cell battery manufactured by corporation E and corporation D. Samples of 40 each of these batteries yielded the following results:

| Dry cells from Corporation E | Dry cells from Corporation D |
| --- | --- |
| $\bar{x} = 485$ seconds | $\bar{y} = 364$ seconds |
| $s_1 = 21$ seconds | $s_2 = 27$ seconds |
| $n_1 = 40$ | $n_2 = 40$ |

Construct a 95 percent confidence interval for the difference in the mean lives for these two brands of dry cell batteries when tested with Ms. Klein's device.

5. Two different work area configurations for an assembly process are under consideration for adoption in a plant. A time and motion study shows that 30 workers can perform the assembly job using configuration X in a mean time of 425 seconds with a standard deviation of 35 seconds. A sample of 30 different workers can perform the assembly job using configuration Y in a mean time of 384 seconds with a standard deviation of 48 seconds. Test, using $\alpha = 0.01$, that there is a difference in the mean times for the assembling process using the two configurations.

6. A famous professional football player recently made the statement that "football players are getting bigger." A woman sportscaster felt the statement was not true. She obtained rosters of college football teams from 1958 and selected the weights of 36 players. For this sample $\bar{x} = 190.8$ pounds and $s_1 = 24.6$ pounds. She then sampled the weights of 36 players from the rosters of college football players for the current year. For this sample $\bar{y} = 198.2$ pounds and $s_2 = 21.6$ pounds. With $\alpha = 0.05$, do these data verify the claim that $\mu_1 - \mu_2 < 0$ for the college football players in 1958 and the current year?

*In Exercises 7 and 8, do the following:*
*a. Compute $\mu_{\bar{D}}$.*
*b. Compute $\sigma_{\bar{D}}$.*
*c. Sketch and label a graph of the probability distribution of $\bar{D}$.*
*d. Compute the indicated probability.*

7. An advertisement for a speed reading course claims that "it is possible to double your reading rate in words per minute (wpm) with one free lesson." Let:

$$d = \text{reading rate after first lesson}$$
$$- \text{reading rate before the lesson.}$$

Records of the company offering the course indicate that in words per minute (wpm):

$$\mu_d = 30 \text{ wpm} \quad \text{and} \quad \sigma_d = 12 \text{ wpm}$$

Let $\bar{D}$ be the mean difference in reading rates of 36 individuals that take the free lesson. Compute $P(28 < \bar{D} < 32)$.

8. A solar unit sold by Sunburst Energy Corporation is designed to preheat water that goes into hot water tanks. Records kept by Sunburst on units installed on all types and sizes of tanks indicate the average decrease in costs for heating water is $12.80 per month with a standard deviation of $4.62. Let $\bar{D}$ be her mean difference in heating costs of 40 installations of solar units by Sunburst. Compute $P(\bar{D} > \$13.50)$.

9. An elementary school PE teacher developed a training program designed to reduce the time it takes a 12-year-old to run a specified distance. She selected 48 students from within the district. She obtained the times $x$, in seconds, that it took each student to run the distance before the training program. Each student then participated in the program and the times $y$, in seconds, to run the distance were again recorded. Letting

$$d_i = x_i - y_i$$

for each student,

$$\sum d_i = 590.4 \text{ seconds}$$

and

$$\sum d_i^2 = 8091.6 \text{ seconds}^2.$$

a. Compute $\bar{d}$ to one decimal place.
b. Compute $s_d$ to one decimal place.
c. Construct a 95 percent confidence interval for $\mu_d$.
d. State the meaning of the answer to part (c).

10. The manufacturers of a wrist support for bowlers claim that the device will improve bowling scores. A sample of 64 bowlers is selected in Akron, Ohio. Each bowler has an established average of over 170 with the American Bowling Congress. Each volunteer rolls two complete games, the first one with the support and the second one without the support. Letting

$$d = \text{score with the support}$$
$$- \text{score without the support}$$

for the 64 bowlers:

$$\bar{d} = 8.45 \text{ pins} \quad \text{and} \quad s_d = 12.80 \text{ pins}$$

a. Construct a 90 percent confidence interval for $\mu_d$.
b. State the meaning of the answer to part (a).

11. In Toledo, Ohio, the management of a membership supermarket chain (MSC) has been advertising that its prices are lower than those of a national supermarket chain (NSC) with stores in the area. A sample of 30 items sold by both outlets was taken, and:

$$d = \text{cost of item at NSC}$$
$$- \text{cost of item at MSC}$$

For the 30 items, $\bar{d} = \$0.303$ and $s_d = \$0.089$. Use these results to test $H_0: \mu_d = 0$ versus $H_a: \mu_d > 0$. Use $\alpha = 0.01$.

12. A new tool was designed to reduce the time needed to complete a standard assembly task in an electronics plant. A sample of 35 employees was timed doing the task without the tool. The same employees were timed again using the tool. For the 35 employees, $\bar{d} = 1.2$ minutes with $s_d = 2.3$ minutes. Use $\alpha = 0.05$ to test the hypothesis that the tool is effective in reducing the time required to do the assembly task.

13. According to the National Center for Health Statistics, a sister agency of the Office on Smoking and Health, the proportions of males and females who smoke are 36.7 percent and 28.9 percent, respectively. Assume these figures apply to the residents in Detroit, Michigan. Sup-

pose $n_1 = 135$ males and $n_2 = 109$ females are selected from the residents in Detroit, and:

$\hat{P}_1 =$ the proportion of the men in the sample who smoke

$\hat{P}_2 =$ the proportion of the women in the sample who smoke

a. Find $\mu_{\hat{p}_1 - \hat{p}_2}$ to three decimal places.
b. Find $\sigma_{\hat{p}_1 - \hat{p}_2}$ to three decimal places.
c. Sketch and label a graph of the probability distribution of $\hat{P}_1 - \hat{P}_2$.
d. Find $P(0.100 < \hat{P}_1 - \hat{P}_2 < 0.200)$.

14. The legislature of a midwestern state has been debating whether to increase the sales tax by 1 cent to raise revenues to improve mass transit systems in the major cities in the state. Surveys throughout the state indicate that 67 percent of urban citizens (Population 1) and 45 percent of rural citizens (Population 2) favor such a tax increase. Suppose $n_1 = 110$ urban and $n_2 = 85$ rural citizens are sampled and $\hat{P}_1$ and $\hat{P}_2$ are the proportions, respectively, of urban and rural citizens that favor the tax increase.
a. Compute $\mu_{\hat{p}_1 - \hat{p}_2}$.
b. Compute $\sigma_{\hat{p}_1 - \hat{p}_2}$ to two decimal places.
c. Sketch and label a graph of the probability distribution of $\hat{P}_1 - \hat{P}_2$.
d. Compute $P(\hat{P}_1 - \hat{P}_2 < 0.15)$.

15. A research worker feels that the proportion of adult women who regularly suffer migraine headaches is greater than the proportion of adult men. She sampled $n_1 = 130$ women and $n_2 = 110$ men. Of those sampled, 34 women and 22 men regularly suffer migraine headaches. Construct a 95 percent confidence interval for the difference in proportions of women and men that regularly suffer migraine headaches.

16. A member of the Society for the Prevention of Cruelty to Animals wanted to estimate the difference in proportions of dogs (Population 1) and cats (Population 2) that have been neutered. She surveyed pet owners in a large area around the city and found 72 of 180 dogs and 78 of 234 cats in the survey were neutered. Construct a 98 percent confidence interval for the difference in the proportions of neutered pet dogs and cats in the area surveyed.

17. A motor magazine claims that the proportion of foreign-made cars driven on the West Coast is greater than the proportion of foreign-made cars driven on the East Coast. A sample of registration records is made in a major West Coast city and a major East Coast city. For the $n_1 = 340$ West Coast records, $\hat{p}_1 = 0.38$ foreign-made cars. For the $n_2 = 1014$ East Coast records, $\hat{p}_2 = 0.30$ foreign-made cars. Do these results substantiate the claim of the motor magazine? Use $\alpha = 0.01$.

18. A study was conducted in Mansfield, Ohio, to evaluate the effectiveness of a bicycle safety program for youngsters less than 15 years old. One year after the program was conducted, 100 youngsters who were in the program (Population 1) and 100 youngsters who were not in the program (Population 2) were interviewed. The results of the interviews showed that eight members of sample 1 and 16 members of sample 2 suffered some form of bicycle-related injury in the past year. Do the data support the claim that the program is effective in reducing injuries for youngsters younger than 15 years old in this city? Use $\alpha = 0.05$.

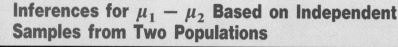

## Inferences for $\mu_1 - \mu_2$ Based on Independent Samples from Two Populations

**Key Topics**

1. *Small sample inference for $\mu_1 - \mu_2$ with population variances unequal*

2. *Small sample inference for $\mu_1 - \mu_2$ with population variances equal*

In this lab we will use Minitab to do small sample hypothesis testing and make confidence intervals for $\mu_1 - \mu_2$. These methods are not covered in Sections 7-1 through 7-3, but represent an extension of the material discussed thus far. This text's formal discussion of small sample inferences for $\mu_1 - \mu_2$ can be found in Section 8-2.

### Small sample inference for $\mu_1 - \mu_2$ with population variances unequal

The Minitab command TWOSAMPLE is used to test the hypothesis of equality of means for two populations. The same command gives a confidence interval for the difference between the population means.

The TWOSAMPLE test is based on the *t*-distribution. If this procedure is used with samples in which one or both contain fewer than 30 data points, it must be assumed that both populations are normally distributed.

When this test is made, the computer uses:

$$\sqrt{\frac{s_1^2}{n_1} + \frac{s_2^2}{n_2}}$$

to estimate the standard deviation of the difference of the means. An approximation not discussed in this chapter is used to develop appropriate degrees of freedom for use in an approximate *t*-interval for $\mu_1 - \mu_2$.

Suppose the following data have been put into columns 1 and 2 of the worksheet:

| | | | | | | | | |
|---|---|---|---|---|---|---|---|---|
| Sample 1: | 20.6 | 20.8 | 19.6 | 18.5 | 19.3 | 22.0 | 22.1 | 18.8 |
| Sample 2: | 18.4 | 18.6 | 17.4 | 20.2 | 21.5 | 17.8 | | |

To test:

$H_0$:   $\mu_1 - \mu_2 = 0$

$H_a$:   $\mu_1 - \mu_2 \neq 0$

the command TWOSAMPLE 90 C1 C2 is given. The output in Figure 7-18 is the result

```
MTB> TWOSAMPLE 90% C1 C2
TWOSAMPLE T FOR C1 VS C2
        N     MEAN    STDEV    SE MEAN
C1      8     20.21   1.38     0.49
C2      6     18.98   1.56     0.64
90 PCT CI FOR MU C1 - MU C2: (-0.23, 2.69)
TTEST MU C1 = MU C2 (VS NE): T = 1.53 P = 0.16 DF = 10
```

**Figure 7-18.** Minitab output.

The first four lines of this display show what type test has been made and some statistics for the two samples. The next line gives a 90% confidence interval estimate of the true difference in population means. (For a different confidence level, type the desired value in place of 90 when the TWOSAMPLE command is given. If no level is specified, Minitab automatically chooses 95%.)

The last line of the output lists the null and alternative hypotheses, the $t$-value, the corresponding $p$-value, and the number of degrees of freedom.

## Small sample inference for $\mu_1 - \mu_2$ with population variances equal

If it can be assumed that both populations have the same variance, the POOLED subcommand may be used with TWOSAMPLE to estimate the standard deviation of the difference of the means. This estimate is made by "averaging" the sample standard deviations and using:

$$\sqrt{\frac{(n_1 - 1)s_1^2 + (n_2 - 1)s_2^2}{n_1 + n_2 - 2}\left(\frac{1}{n_1} + \frac{1}{n_2}\right)}$$

to estimate $\sigma_{\bar{X} - \bar{Y}}$.

Assuming that the populations of the previous example have equal variances, the computer run in Figure 7-19 was made.

```
MTB> TWOSAMPLE 90% C1 C2;
SUBC> POOLED.
TWOSAMPLE T FOR C1 VS C2
        N     MEAN    STDEV    SE MEAN
C1      8     20.21   1.38     0.49
C2      6     18.98   1.56     0.64
90 PCT CI FOR MU C1 - MU C2: (-0.18, 2.64)
TTEST MU C1 = MU C2 (VS NE): T = 1.56 P = 0.15 DF = 12
POOLED STDEV =       1.46
```

**Figure 7-19.** Minitab output.

Note that this output contains, at the bottom, an additional bit of information, namely the value of the pooled sample standard deviation. Also, because the estimate of the standard deviation of the difference of means is different from the unpooled estimate, the width of the 90% confidence interval and the p-value are slightly different than those obtained in the first test.

# Exercises

1. Two drugs, X and Y, were compared for their effects on the ovulation of rabbits. The data consists of the numbers of mature eggs found upon dissection of 13 rabbits. The computer run in Figure 7-20 was made.

   a. How many rabbits were given drug X?

   b. Which sample had the largest mean?

   c. Compare the standard deviations of the two samples. Do you think the researcher was correct in his decision to use the unpooled variance? (At this time, this will have to be a subjective judgment on your part.)

   d. Give the null and alternative hypotheses of the test.

   e. On the basis of this test, what should be concluded?

   f. What is the p-value?

   g. Give an estimate of the true difference between the mean numbers of mature eggs from the two populations. What is the confidence level of this estimate?

```
MTB> TWOSAMPLE 'X' 'Y';
SUBC> POOLED;
SUBC> ALTERNATIVE -1.
TWOSAMPLE T FOR X VS Y
     N    MEAN   STDEV   SE MEAN
X    7    5.86   1.35    0.51
Y    6    7.83   1.94    0.79
95 PCT CI FOR MU X - MU Y: (-3.99, 0.04)
TTEST MU X = MU Y (VS LT): T = -2.16 P = 0.027 DF = 11
POOLED STDEV =       1.64
```

**Figure 7-20.**  Minitab output.

```
MTB> TWOSAMPLE 99% C3 C4;
SUBC> ALTERNATIVE 1.
TWOSAMPLE T FOR C3 VS C4
      N    MEAN   STDEV   SE MEAN
C3   10   14.40   2.55    0.81
C4   10   12.80   2.25    0.71
99 PCT CI FOR MU C3 - MU C4 (-1.52, 4.72)
TTEST MU C3 = MU C4 (VS GT): T = 1.49 P = 0.077 DF = 17

MTB> TWOSAMPLE 99% C3 C4;
SUBC> POOLED;
SUBC> ALTERNATIVE 1.
TWOSAMPLE T FOR C3 VS C4
      N    MEAN   STDEV   SE MEAN
C3   10   14.40   2.55    0.81
C4   10   12.80   2.25    0.71
99 PCT CI FOR MU C3 - MU C4 (-1.49, 4.69)
TTEST MU C3 = MU C4 (VS GT): T = 1.49 P = 0.077 DF = 18
POOLED STDEV =       2.40
```

**Figure 7-21.**  Minitab output.

**2.** Figures 7-21 through 7-23 are parts of pairs of computer outputs. In each pair, a single set of data was analyzed, first using the assumption that the population variances were unequal, then using the assumption that the variances were equal.

a. For each pair, give the null and alternative hypotheses.

b. Compare the $p$-values for the tests in each pair. Which assumption, variances unequal or variances equal, leads to a smaller $p$-value?

c. Compare the confidence intervals in each pair. Which assumption, variances unequal or variances equal, gives the widest confidence interval?

```
MTB> TWOSAMPLE C5 C6;
SUBC> ALTERNATIVE -1.

TWOSAMPLE T FOR C5 VS C6
          N      MEAN     STDEV    SE MEAN
C5       10    101.15      3.16     0.88
C6       10    105.10      1.79     0.57

99 PCT CI FOR MU C5 - MU C6 (-6.13, -1.76)
TTEST MU C5 = MU C6 (VS LT): T = -3.78 P = 0.0006 DF = 19


MTB> TWOSAMPLE C5 C6;
SUBC> POOLED;
SUBC> ALTERNATIVE -1.

TWOSAMPLE T FOR C5 VS C6
          N      MEAN     STDEV    SE MEAN
C5       10    101.15      3.16     0.88
C6       10    105.10      1.79     0.57

95 PCT CI FOR MU C5 - MU C6 (-6.27, -1.62)
TTEST MU C5 = MU C6 (VS LT): T = -3.53 P = 0.0010 DF = 21
POOLED STDEV =        2.66
```

**Figure 7-22.** Minitab output.

```
MTB> TWOSAMPLE 90% C7 C8

TWOSAMPLE T FOR C7 VS C8
          N      MEAN     STDEV    SE MEAN
C7       16    1.1662    0.0944     0.024
C8       24    1.2313    0.0709     0.014

90 PCT CI FOR MU C7 - MU C8 (-0.112, -0.018)
TTEST MU C7 = MU C8 (VS NE): T = -2.35 P = 0.027 DF = 26


MTB> TWOSAMPLE 90% C7 C8;
POOLED.

TWOSAMPLE T FOR C7 VS C8
          N      MEAN     STDEV    SE MEAN
C7       16    1.1662    0.0944     0.024
C8       24    1.2313    0.0709     0.014

90 PCT CI FOR MU C7 - MU C8 (-0.118, -0.012)
TTEST MU C7 = MU C8 (VS NE): T = -2.49 P = 0.017 DF = 38
POOLED STDEV =        0.0810
```

**Figure 7-23.** Minitab output.

## SCIENTIFIC

A scientific calculator can be used to calculate the observed value of the test statistic for a test of:

$$H_0: \quad \mu_1 - \mu_2 = 0$$

Parentheses can be used to write the equation on one line as follows:

$$z = (\bar{x} - \bar{y}) \div \sqrt{s_1^2 \div n_1 + s_2^2 \div n_2}$$

The following example demonstrates the key strokes and resulting calculator displays.

▶ **Example** Calculate $z$ for $\bar{x} = 2.78$, $\bar{y} = 2.82$, $s_1 = 0.24$, $n_1 = 144$, $s_2 = 0.22$, and $n_2 = 97$.

**Solution**

| Press | Display |
|---|---|
| ( | [01        0. |
| 2.78 | 2.78 |
| − | 2.78 |
| 2.82 | 2.82 |
| ) | −0.04 |
| ÷ | −0.04 |
| ( | [01        0. |
| 0.24 | 0.24 |
| $x^2$ | 0.0576 |
| ÷ | 0.0576 |
| 144 | 144 |
| + | 4. $^{-04}$ |
| 0.22 | 0.22 |
| $x^2$ | 0.0484 |
| ÷ | 0.0484 |
| 97 | 97 |
| ) | 8.989690722 $^{-04}$ |
| √ | 0.029982812 |
| = | −1.33409764 |

Therefore $z = -1.33$.  ◀

In a similar manner, the scientific calculator can be used to calculate the observed value of the test statistic for a test of:

$$H_0: \quad P_1 - P_2 = 0$$

Parentheses can be used to write the equation on one line as follows:

$$z = (\hat{p}_1 - \hat{p}_2) \div \sqrt{\hat{p}(1 - \hat{p})(1 \div n_1 + 1 \div n_2)}$$

In this equation, keep in mind that $\hat{p}$ is the "pooled proportion" and:

$$\hat{p} = \frac{x + y}{n_1 + n_2} \quad \text{or} \quad \hat{p} = \frac{n_1\hat{p}_1 + n_2\hat{p}_2}{n_1 + n_2}$$

# GRAPHING

To use Casio or Texas Instrument (TI) graphics calculators, the test statistic for a test of:

$$H_0: \quad \mu_1 - \mu_2 = 0$$

is written on one line as follows:

$$z = (\bar{x} - \bar{y}) \div \sqrt{s_1^2 \div n_1 + s_2^2 \div n_2}$$

The following example illustrates the key strokes for the Casio model. The keys for the TI are identical with only one exception. A "÷" symbol is used to indicate division on the Casio model, and a "/" symbol is used on the TI model.

▶ **Example**  Compute $z$ for $\bar{x} = 2.78$, $\bar{y} = 2.82$, $s_1 = 0.24$, $n_1 = 144$, $s_2 = 0.22$, and $n_2 = 97$.

**Solution**

| Press | Display |
|---|---|
| ( | *(* |
| 2.78 | *(2.78* |
| − | *(2.78 −* |
| 2.82 | *(2.78 − 2.82* |
| ) | *(2.78 − 2.82)* |
| ÷ | *(2.78 − 2.82) ÷* |
| √ | *(2.78 − 2.82) ÷ √* |
| ( | *(2.78 − 2.82) ÷ √ (* |
| .24 | *(2.78 − 2.82) ÷ √ (.24* |
| $x^2$ | *(2.78 − 2.82) ÷ √ (.24²* |
| ÷ | *(2.78 − 2.82) ÷ √ (.24² ÷* |
| 144 | *(2.78 − 2.82) ÷ √ (.24² ÷ 144* |
| + | *(2.78 − 2.82) ÷ √ (.24² ÷ 144 +* |
| .22 | *(2.78 − 2.82) ÷ √ (.24² ÷ 144 + .22* |
| $x^2$ | *(2.78 − 2.82) ÷ √ (.24² ÷ 144 + .22²* |
| ÷ | *(2.78 − 2.82) ÷ √ (.24² ÷ 144 + .22² ÷* |
| 97 | *(2.78 − 2.82) ÷ √ (.24² ÷ 144 + .22² ÷ 97* |
| ) | *(2.78 − 2.82) ÷ √ (.24² ÷ 144 + .22² ÷ 97)* |
| EXE | *−1.33409764* |

Therefore $z = -1.33$.  ◀

An advantage the graphics calculator has over the scientific calculator is that a given value of $\bar{x}$, $\bar{y}$, $s_1$, $s_2$, $n_1$, or $n_2$ can be changed and the revised value of $z$ obtained with a single key stroke. Pressing the ➡ key deletes the value of $z$ and positions the cursor at the beginning of the expression. By using the arrow key, any value in the expression can be changed. Pressing the EXE key will display the new value of $z$. (Press the ▲ key for the TI calculator.)

**Name:** James Inglis

**Occupation:** Head of the Business Analysis and Advanced Computing Department

**Employer:** AT & T Bell Laboratories, Murray Hill, New Jersey

I became interested in statistics as a mathematics major in college in the mid-60s. The problems in the probability and mathematical statistics courses I took had a complexity and an *association with the real world* that was fascinating. In graduate school, I became more interested in applied statistics, seeing how the statistical way of thinking and statistical techniques could make sense out of seemingly confused and messy, real-world data.

In my career, I've helped the chaplain of my Army battalion in Vietnam, who was dealing with a complaint about racial prejudice, to understand the racial mix of our unit. I've helped a magazine interpret data on the "power" of Broadway theater newspaper critics to influence theater attendance with their praising and panning of shows. I've developed an equation to estimate the weight of babies, while still in the uterus, to aid in the management of high-risk pregnancies. I've helped a telephone company and a state public utilities commission understand the effects of rate changes on the demand for telephone service. The variety of opportunities for meaningful data analysis are limitless.

In the course of my career, I've been an Army officer and a professor of statistics and biostatistics. Since 1978, I've been at AT & T Bell Laboratories, starting as an applied statistical researcher and consultant, and then working through two promotions as a technical manager.

I'm presently Head of the Business Analysis and Advanced Computing Department (about 40 people) at Bell Laboratories' Murray Hill location. I lead and support projects involving large-scale data analysis of some billing and marketing data for AT & T. Being involved with projects where data analysis and the software systems for data analysis have a real, measurable impact on company operations is very satisfying.

In all my jobs, statistical thinking and an understanding of data analysis have been important. In my current job, I see many kinds of data—both managerial (budget, personnel) and technical project data. The data are in many forms: tables, histograms, graphs. Understanding how to look at data and how to draw conclusions from it are part of my job.

Statistics is a fascinating subject with many uses in many fields. A few useful suggestions: learn about computers and their abilities in data analysis—they are already heavily used for computations, and their role will only increase. Learn the principles of statistical inference and the difference between statistical correlation and physical causality. Most importantly, understand the fundamental role variability plays in any analysis—from surveys reported in the newspaper to complex medical studies, the inherent variability of people and nature is what applied statistics strives to help us understand.

# Survey shows perception of bias follows racial lines

By Lee May
Los Angeles Times

WASHINGTON — In a vivid illustration of the gulf still dividing American society, a new national survey released Wednesday found dramatic disagreement between blacks and whites about the severity of racial problems in such basic areas as education, housing, job opportunities and the law.

For example, two-thirds of the whites surveyed believe black people "get equal pay for equal work," while two-thirds of blacks believe just the opposite.

Similarly, half of the black people surveyed say local police "keep blacks down," while an equal percentage of white people say police are more helpful than harmful to blacks, according to the study, which was conducted by Louis Harris for the NAACP Legal Defense and Educational Fund.

"These gaps in perception underscore the need for continuing educational efforts to help white Americans appreciate that race continues to be a factor in influencing our daily lives," the study's authors said.

James O. Gibson, who directs the Rockefeller Foundation's national programs on equal opportunity, said the gap stems from the differences in experiences among blacks and whites. "If people aren't touched by (racial bias) then they have no informed basis for an attitude," he said. "It's the out-of-sight, out-of-mind factor."

On housing, 55 percent of whites said blacks are not worse off than other groups with comparable education and income, but 64 percent of blacks believe they are worse off.

At the same time, the study found significant areas of agreement. About eight of 10 respondents of both races, for instance, agreed that more money should be spent attacking the causes of crime — such as poverty and lack of motivation — than on sentencing criminals to long jail terms.

Nine of every 10 blacks and whites also believe in the need for special school programs to motivate poor youngsters to stay in school. And nine of every 10 members of both races support efforts to locate more jobs in areas where poor people live, the study found.

**Figure 8-1** "The Unfinished Agenda on Race in America," by Eddie Lee May, *Los Angeles Times*. Copyright © 1989 Los Angeles Times. Reprinted by permission.

# Comparing Several Populations

In Chapters 5 and 6, the parameters $\mu$, $P$, and $\sigma$ of a single population were estimated and tested using $\bar{X}$, $\hat{P}$, and $S$ respectively. In Chapter 7, the quantities $\mu_1 - \mu_2$, $\mu_d$, and $P_1 - P_2$, all relating to two populations, were estimated and tested using $\bar{X} - \bar{Y}$, $\bar{D}$, and $\hat{P}_1 - \hat{P}_2$ respectively. Now, in Chapter 8, the parameters of several populations will be simultaneously examined. The content of the article in Figure 8-1 illustrates a study of this type. The survey conducted by Louis Harris for the NAACP Legal Defense and Education Fund, included 3123 interviews with people of all races. Thus the study collected data on the attitudes of not only blacks and whites, but also on the attitudes of Asians, Hispanics, and others. In this chapter, we will study the probability models and statistical methods that can be used to make comparisons of several populations based on means and proportions. This material is an extension of the material in the past few chapters.

# 8-1 The $\chi^2$ Test for Equality of $r$ Distributions

**Key Topics**

1. *A general version of the two-way table*

2. *The fraction of the ith sample falling into category j*

3. *The null hypothesis for the test*

4. *A probability model for testing equality of r distributions*

5. *The $\chi^2$ probability distributions (Review)*

6. *Testing the equality of r distributions*

7. *Constructing confidence intervals for the difference in 2 proportions*

## S T A T I S T I C S   I N   A C T I O N

Pat Crowley is the education editor for the *Daily Express*, a newspaper in the Northeast that has statewide circulation. Recently the voters in the state passed a referendum instituting a state lottery. A major portion of the proceeds were directed toward education. A controversy in the state legislature arose over priorities for these funds in the education program. Josh Steinberg, a state senator, in an open address to the lawmakers, claimed that "teachers themselves cannot agree on where the money should be spent."

To challenge this claim, Pat decided to gather data on the opinions of teachers within the state regarding what areas of the education program should get the major portion of the additional money generated by the lottery. For the purpose of this study, she divided all teachers in state supported schools into the following three populations:

Population 1    The elementary school teachers (grades K through 8)

Population 2    The high school teachers (grades 9 through 12)

Population 3    The postsecondary teachers (community and four-year colleges)

Pat wanted to focus on four broad areas of the educational program. She therefore defined the following four categories:

Category 1    Improvement in teachers' salaries

Category 2    Reduction in class sizes

Category 3    Improvement in staff development activities

Category 4    Improvement in equipment and facilities

Pat contacted newspaper personnel who work in several large cities in the state. With the help of these volunteers, Pat surveyed 100 teachers in each of the three populations. Each teacher was personnally interviewed and the purpose of the survey was explained. Each teacher was then asked the following question:

"In your opinion, taking into consideration all levels of education, into which of these four categories should the state direct most of the additional funds for education?"

After the surveys were taken and the results returned to Pat, she separated the data into the 12-cell display in Table 8-1.

With the help of Glenn Fleming, the staff statistician, Pat used a $\chi^2$ (read "ky-squared") test of equality of $r$ distributions to compare the opinions of teachers in the three populations. In a subsequent edition of the Daily Express, Pat published an article that included the following statement:

"Based on the results of the survey, there is sufficient evidence to conclude that a difference of opinions does exit among elementary, high school, and college teachers as to where the lottery money should be spent. Further analysis of the results of this survey will be the topic of future articles in this column."

The $\chi^2$ test of equality of $r$ distributions will be studied in this section.

**TABLE 8-1   Results of Teacher Opinion Survey**

|  | Improvement in Teachers' Salaries | Reduction in Class Sizes | Improvement in Staff Development Activities | Improvement in Equipment and Facilities | Row Totals |
|---|---|---|---|---|---|
| Elementary Teachers | 24 | 42 | 12 | 22 | 100 |
| High School Teachers | 26 | 26 | 20 | 28 | 100 |
| Postsecondary Teachers | 30 | 10 | 34 | 26 | 100 |
| Column Totals | 80 | 78 | 66 | 76 | 300 |

Adapted table from *Nonparametrical Statistical Methods* by Myles Hollander and Douglas Wolfe. Copyright © 1973 by John Wiley & Sons, Inc. Reprinted by permission.

## A general version of the two-way table

The study conducted by Pat Crowley in the Statistics in Action section is typical of ones in which we simultaneously study three or more samples of qualitative data with similar categories. In the teacher opinion survey, the three populations are the elementary, high school, and college teachers in the state. The categories are preferences for the four general areas of the education program specified in the survey.

When independent random samples are taken from the populations of interest, the data are usually recorded in a table divided into cells, such as the one in Table 8-1. Table 8-2 (on the following page) is a general version of such a table. Notice the following features of this table:

**Feature 1**   The table has $r$ rows. One sample is taken from each population under investigation. Therefore $r$ represents both the number of populations and the number of samples.

**Feature 2**   The table has $c$ columns. The number of columns represents the number of categories into which each population is divided.

**Feature 3**   The table is divided into $(r)(c)$ cells. In Table 8-1, the 3 rows and 4 columns divide the table into $(3)(4) = 12$ cells.

**Feature 4**  The symbol $n_i$ stands for the number of items in the sample from the $i$th population. In Table 8-1, $n_1 = n_2 = n_3 = 100$. That is, 100 teachers were sampled from each population. The value for each $n_i$ is in the column headed by "Row totals."

**Feature 5**  The symbol $m_j$ stands for the number of items observed in the $j$th category. Using Table 8-1 as an example:

$m_1$ counts the 80 preferences for "Improvement in teachers salaries."
$m_2$ counts the 78 preferences for "Reduction in class size."
$m_3$ counts the 66 preferences for "Improvement in staff development activities."
$m_4$ counts the 76 preferences for "Improvement in equipment and facilities."

**Feature 6**  The symbol $n$ stands for the total of the sample sizes. That is:

$$n = n_1 + n_2 + \cdots + n_r \quad \text{or} \quad n = m_1 + m_2 + \cdots + m_c$$

In Table 8-1:

$$n = 100 + 100 + 100 = 300 \qquad \text{The sum of the row totals}$$
$$n = 80 + 78 + 66 + 76 = 300 \qquad \text{The sum of the column totals}$$

**Feature 7**  The symbol $o_{ij}$ (read "oh sub eye-jay") stands for the number of items "observed" in the $i$th row and $j$th column.
In Table 8-1:

$$o_{12} = 42 \qquad \text{1st row and 2nd column}$$
$$o_{34} = 26 \qquad \text{3rd row and 4th column}$$

**TABLE 8-2   A Table with $r$ Rows and $c$ Columns**

| | | | | Category | | |
|---|---|---|---|---|---|---|
| | | *1* | *2* | $\cdots$ | *c* | *Total* |
| | *1* | $o_{11}$ | $o_{12}$ | $\cdots$ | $o_{1c}$ | $n_1$ |
| | *2* | $o_{21}$ | $o_{22}$ | $\cdots$ | $o_{2c}$ | $n_2$ |
| **Sample** | $\vdots$ | $\vdots$ | $\vdots$ | $\cdots$ | $\vdots$ | $\vdots$ |
| | *r* | $o_{r1}$ | $o_{r2}$ | $\cdots$ | $o_{rc}$ | $n_r$ |
| | *Total* | $m_1$ | $m_2$ | $\cdots$ | $m_c$ | $n$ |

Adapted table from *Nonparametrical Statistical Methods* by Myles Hollander and Douglas Wolfe. Copyright © 1973 by John Wiley & Sons, Inc. Reprinted by permission.

▶ **Example 1**   In Exercises (a) through (e), identify each indicated value.

**a.** $n_2$     **b.** $m_4$     **c.** $o_{23}$     **d.** $o_{15}$     **e.** $n$

| | | Category | | | | | |
|---|---|---|---|---|---|---|---|
| | | *A* | *B* | *C* | *D* | *E* | *Totals* |
| | *1* | 10 | 12 | 8 | 15 | 11 | 56 |
| | *2* | 9 | 13 | 20 | 11 | 14 | 67 |
| **Sample** | *3* | 13 | 16 | 18 | 17 | 10 | 74 |
| | *4* | 7 | 13 | 19 | 5 | 9 | 53 |
| | *Totals* | 39 | 54 | 65 | 48 | 44 | 250 |

**Solution**   **a.** $n_2 = 67$     The row total for sample 2 is 67.

**b.** $m_4 = 48$     The column total for category *D* is 48.

**c.** $o_{23} = 20$     The number of items in the cell where row 2 intersects column 3 is 20.

**d.** $o_{15} = 11$     The number of items in the cell where row 1 intersects column 5 is 11.

**e.** $n = 250$     The total number of items in all samples is 250.   ◀

## The fraction of the *i*th sample falling into category *j*

The symbol $o_{ij}$ stands for the number of items observed in the cell that is the intersection of the *i*th row and the *j*th column. This number can be converted to a sample proportion by dividing it by $n_i$, the size of the *i*th sample.

### The Proportion of the *i*th Sample Falling into Category *j*

The ratio $\dfrac{o_{ij}}{n_i}$ stands for the *proportion* of the *i*th sample that falls into category *j*.

▶ **Example 2**   Convert the value of each $o_{ij}$ in Table 8-1 to a sample proportion.

**Solution**   For simplicity, the four opinion categories are labeled 1, 2, 3, and 4.   ◀

As a result of the computations shown in Table 8-3 on the next page, we can make statements such as the following.

"24% of the elementary teachers surveyed chose the improvement in salaries response."

"26% of the high school teachers surveyed chose the reduction in class size response."

**TABLE 8-3    The Values of $o_{ij}$ Written as Sample Proportions**

|          | Category 1 | Category 2 | Category 3 | Category 4 |
|----------|------------|------------|------------|------------|
| Sample 1 | $\frac{24}{100} = 0.24$ | $\frac{42}{100} = 0.42$ | $\frac{12}{100} = 0.12$ | $\frac{22}{100} = 0.22$ |
| Sample 2 | $\frac{26}{100} = 0.26$ | $\frac{26}{100} = 0.26$ | $\frac{20}{100} = 0.20$ | $\frac{28}{100} = 0.28$ |
| Sample 3 | $\frac{30}{100} = 0.30$ | $\frac{10}{100} = 0.10$ | $\frac{34}{100} = 0.34$ | $\frac{26}{100} = 0.26$ |

Corresponding to the sample proportion $\frac{o_{ij}}{n_i}$, the symbol $P_{ij}$ will stand for the parameter that describes the proportion of population $i$ that belongs to category $j$. Using Table 8-1 to illustrate, let us consider the following proportions of all high school teachers in the state (Population 2):

$P_{21}$   stands for the proportion of high school teachers in category 1

$P_{22}$   stands for the proportion of high school teachers in category 2

$P_{23}$   stands for the proportion of high school teachers in category 3

$P_{24}$   stands for the proportion of high school teachers in category 4

## The null hypothesis for the test

Recall part of the question that Pat Crowley wanted each teacher to answer:

"... into which of these four categories should the state direct most of the additional funds ..."

The intent of the survey was to consider whether the pattern of teachers' preferences at the elementary school, high school, and college levels would be approximately the same. This possibility can be written as a series of equations using $P_{ij}$'s.

1. If the proportions of preferences for category 1 in the three populations are the same

$$P_{11} = P_{21} = P_{31} \Big\langle \begin{array}{l} \text{Populations 1, 2, and 3} \\ \text{Category 1} \end{array}$$

2. If the proportions of preferences for category 2 in the three populations are the same

$$P_{12} = P_{22} = P_{32} \Big\langle \begin{array}{l} \text{Populations 1, 2, and 3} \\ \text{Category 2} \end{array}$$

3. If the proportions of preferences for category 3 in the three populations are the same

$$P_{13} = P_{23} = P_{33} \begin{cases} \nearrow \text{Populations 1, 2, and 3} \\ \searrow \text{Category 3} \end{cases}$$

4. If the proportions of preferences for category 4 in the three populations are the same

$$P_{14} = P_{24} = P_{34} \begin{cases} \nearrow \text{Populations 1, 2, and 3} \\ \searrow \text{Category 4} \end{cases}$$

When these four extended equalities are joined, they form a null hypothesis that asserts "there is no difference between the population distributions." The alternative hypothesis that is most often used in such a test is "any departure from the null hypothesis," and we simply write $H_a$: not $H_0$.

Therefore a suitable statement of $H_0$ and $H_a$ for the problem posed by Pat Crowley would be:

$$\begin{aligned} H_0: \quad & P_{11} = P_{21} = P_{31} \\ & P_{12} = P_{22} = P_{32} \\ & P_{13} = P_{23} = P_{33} \\ & P_{14} = P_{24} = P_{34} \\ H_a: \quad & \text{not } H_0 \end{aligned}$$

### $H_0$ and $H_a$ for a Test of Equality of $r$ Distributions

$$\begin{aligned} H_0: \quad & P_{11} = P_{21} = \cdots = P_{r1} \\ & P_{12} = P_{22} = \cdots = P_{r2} \\ & \quad \vdots \qquad \vdots \qquad \qquad \vdots \\ & P_{1c} = P_{2c} = \cdots = P_{rc} \\ H_a: \quad & \text{not } H_0 \end{aligned}$$

## A probability model for testing equality of $r$ distributions

As is always the case, we need a probability model to do any inferential statistics. To obtain such a model, we first assume that each population of interest is large compared with the size of the sample taken from it. Furthermore we suppose that separate simple random samples are selected from populations. Under these conditions, the following assumptions are appropriate.

> **Assumption 1** The $r$ samples are independent.
>
> **Assumption 2** Within a particular sample (say, sample $i$), the category values corresponding to the $n_i$ items in the sample are independent variables with probabilities of the possible values $P_{i1}, P_{i2}, \ldots, P_{ic}$.

Based on such model assumptions, a theoretically sound test of equality of the $r$ distributions specified by the $P_{ij}$ is possible.

The formula for the test statistic that will eventually be used to test the equality of several distributions is more complicated than any seen thus far in this text. There is, however, some simple intuitive appeal attached to the statistic. We will use the survey of teachers' opinions in Table 8-1 to illustrate.

Recall that 100 teachers from each of the populations of elementary schools, high schools, and colleges in the state were surveyed. Their opinions on which area of the educational program should get more funding were separated into four categories. Suppose that $H_0$ is true; that is, the patterns of opinions of teachers are the same in all three samples. We would then expect approximately the same proportion of teachers from each population to occur in a given opinion category. Consider, for example, the counts in Table 8-4 for categories 1 and 4.

**TABLE 8-4   Only Categories 1 and 4 (from Table 8-1)**

|          | Improvement in Teachers' Salaries | Improvement in Equipment and Facilities |
|----------|:---------------------------------:|:---------------------------------------:|
| Sample 1 | 24 | 22 |
| Sample 2 | 26 | 28 |
| Sample 3 | 30 | 26 |

The numbers 24, 26, and 30 for the "Improvement in teachers' salaries" category are quite believable as observed values of random variables with means equal to 100 times a common proportion of teachers in this category from each of the three populations. Similarly the 22, 28, and 26 for the "Improvement in equipment and facilities" category are also quite believable as observed values of random variables with means equal to 100 times a common proportion of teachers in this category from each of the three populations.

But consider the counts in Table 8-5 that occurred in categories 2 and 3. The numbers 42, 26, and 10 for the "Reduction in class size" category are quite different in the three samples. (There is a difference of 32 in the numbers of elementary school and college teachers that marked this category.) Similarly, in the "Improvement in staff development activities" category, there are large differences in the numbers observed from the three samples. If the proportions of teachers from the elementary schools, high schools, and colleges were the same in these two categories, then it would be quite unlikely that such large variations would occur in these observed values. Possible, but not likely.

TABLE 8-5  Only Categories 2 and 3 (from Table 8-1)

|  | Reduction in Class Sizes | Improvement in Staff Development Activities |
|---|---|---|
| Sample 1 | 42 | 12 |
| Sample 2 | 26 | 20 |
| Sample 3 | 10 | 34 |

This discussion suggests we look at the number of teachers that "should" occur in each cell, supposing $H_0$ to be true. For example, let us consider only category 1. The first line in $H_0$ states:

$$P_{11} = P_{21} = P_{31}$$

The proportion of preferences for "Improvement in teachers' salaries" is the same in all three populations.

From Table 8-1, 80 teachers out of 300 were counted in column 1. Thus $\frac{80}{300} = 0.27$ is a reasonable estimate for a common proportion of teachers from each of the three populations that are in this category. Because 100 teachers were surveyed from each population, $100(0.27) = 27$ is approximately the *expected number* of teachers that we should see in each cell of category 1.

| Elementary Teachers | 24 observed | 27 expected | a difference of 3 |
| High School Teachers | 26 observed | 27 expected | a difference of 1 |
| College Teachers | 30 observed | 27 expected | a difference of 3 |

The 24, 26, and 30 are the *observed values*. The 27 is an estimate of the so-called *expected value* for these cells, based on the estimated common proportion of teachers in this category. Similar calculations can be used to obtain estimates for the expected counts in the other cells in Table 8-1 if $H_0$ is true.

## An Expression for Calculating the Estimated Expected Count in Any Cell

The expected count for the cell that is the intersection of the $i$th row and $j$th column is estimated by:

$$e_{ij} = \frac{n_i m_j}{n} \qquad \frac{\text{(total for } i\text{th row)(total for } j\text{th column)}}{\text{total sample size}}$$

Large differences between the values of the $o_{ij}$ and corresponding $e_{ij}$ would therefore seem to be evidence that $H_0$ is not true. Table 8-6 contains the values of the $o_{ij}$ and $e_{ij}$ for the sampled teachers in Table 8-1. In each cell, the $o_{ij}$ is above the diagonal and the corresponding $e_{ij}$ is below the diagonal.

TABLE 8-6   $o_{ij}$ and $e_{ij}$ Values for Teacher Opinion Survey

|  | Category 1 | Category 2 | Category 3 | Category 4 | Row Totals |
|---|---|---|---|---|---|
| Elementary | 24 / 27 | 42 / 26 | 12 / 22 | 22 / 25 | 100 |
| High | 26 / 27 | 26 / 26 | 20 / 22 | 28 / 25 | 100 |
| College | 30 / 27 | 10 / 26 | 34 / 22 | 26 / 25 | 100 |
| Column Totals | 80 | 78 | 66 | 76 | 300 |

## The test statistic

There are some large individual differences between the $o_{ij}$ and $e_{ij}$ values in Table 8-6. However, we still need to use these individual differences to calculate a single collective *test statistic*. This test statistic will then be used to determine whether the differences are extreme enough to force rejection of $H_0$. The test statistic we will use is:

### A Formula to Summarize the Difference in the $O_{ij}$ and $E_{ij}$ Values

$$X^2 = \sum \frac{(O_{ij} - E_{ij})^2}{E_{ij}} \qquad (1)$$

Or equivalently:

$$X^2 = \sum \left( \frac{O_{ij}^2}{E_{ij}} \right) - n \qquad (2)$$

The numerator of each term in equation (1) is the indicated square of the difference between an observed count and the estimated value of the expected count supposing $H_0$ is true. By squaring, the negative differences are changed to positive numbers. The denominator of each term in the sum "weights" the squares of any difference. To illustrate, consider the following possibilities.

| $o_{ij}$ | $e_{ij}$ | $(o_{ij} - e_{ij})^2$ | $\dfrac{(o_{ij} - e_{ij})^2}{e_{ij}}$ |
|---|---|---|---|
| a.  9 | 10 | $(-1)^2 = 1$ | $\frac{1}{10} = 0.10$ |
| b. 99 | 100 | $(-1)^2 = 1$ | $\frac{1}{100} = 0.01$ |

In both (a) and (b), the square of the difference is 1. However, a difference of 1 when the estimated expected value is 10 is much more significant than a difference of 1 when the estimated expected value is 100. By dividing the square of the difference by the estimated expected value, a 0.10 is added to the computation of $X^2$ in (a), but only 0.01 is added in (b). In this way, the more important difference in (a) is given greater weight than the same difference in (b).

▶ **Example 3**   Calculate the observed value of $X^2$ for the data in Table 8-6.

**Solution**   **Discussion.** Equation (2) requires only one subtraction. Therefore it is usually easier to use than equation (1). To calculate $x^2$, use unrounded values for $e_{ij}$.

$$x^2 = \frac{24^2}{26.6\ldots} + \frac{26^2}{26.6\ldots} + \frac{30^2}{26.6\ldots} + \cdots + \frac{26^2}{25.3\ldots} - 300$$

$$= 21.60 + 25.35 + 33.75 + \cdots + 26.68\ldots - 300$$

$$= 332.40\ldots - 300$$

$$\approx 32.402, \text{ to three decimal places}$$

Thus the sum of the weighted squares of the differences between the observed frequencies and the estimated expected frequencies is 32.402.   ◀

## The $\chi^2$ probability distributions (Review)

The question of "how large" $X^2$ must be to reject $H_0$ depends on the probability distribution of $X^2$ supposing $H_0$ to be true. It turns out that *if all the sample sizes $n_i$ are large*, then the probability distribution of $X^2$ can be approximated using a standard $\chi^2$ probability distribution. Recall that the $\chi^2$ distributions were first introduced in Section 6-5 in the context of inference for the variance of bell-shaped populations. Remember that just as there are many binomial distributions, many normal distributions, and many $t$-distributions, there are many $\chi^2$-distributions. The various $\chi^2$ probability distributions are distinguished by a so-called degrees-of-freedom parameter, and their graphs have the general right-skewed appearance shown in Figure 8-2 on the following page.

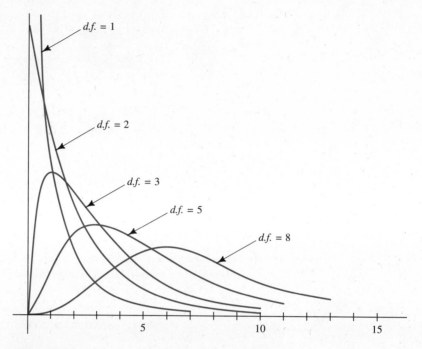

**Figure 8-2.**   Graphs of five $\chi^2$ probability distributions.

Appendix Table 6 has the values of the upper and lower percentage points of several $\chi^2$-distributions with various degrees of freedom ($df$). In this section we will use only the table values for the 0.050 and 0.010 percentage points. Thus, for example, if $df = 5$:

| $df$ | $\chi^2_{0.050}$ | $\chi^2_{0.010}$ |
|------|------------------|------------------|
| 5    | 11.071           | 15.086           |

Therefore if $W$ is a random variable with a $\chi^2$ probability distribution with 5 degrees of freedom, then:

$$P(W \geq 11.071) = 0.050 \quad \text{and} \quad P(W \geq 15.086) = 0.010$$

## Testing the equality of *r* distributions

The $\chi^2$ probability distribution to use in a test of equality of $r$ distributions depends on the number of populations compared and the number of categories used. Remember again that these values are symbolized by $r$, the number of *rows* in the table, and $c$, the number of *columns*.

## The $\chi^2$-Distribution to Use in a Test of Equality of $r$ Distributions

If $H_0$ is true, then the $\chi^2$-distribution that can be used to approximate the distribution of the test statistic is the one characterized by:

$$\text{degrees of freedom} = (r - 1)(c - 1)$$

For the survey of teachers' opinions that Pat Crowley carried out, there were three populations of teachers ($r = 3$) and four categories of opinions ($c = 4$). Therefore for the test of $H_0$, the appropriate probability distribution has

$$df = (3 - 1)(4 - 1) = 6$$

Using Appendix Table 6:

$$\text{if } \alpha = 0.05, \quad \chi_c^2 = 12.592$$
$$\text{if } \alpha = 0.01, \quad \chi_c^2 = 16.812$$

▶ **Example 4**   Using the data in Table 8-1, test the null hypothesis that the patterns of teachers' preferences in the three populations are the same. Use $\alpha = 0.05$.

**Solution**   **Discussion.** The six-step format for conducting hypothesis tests will be used.

**Step 1** $H_0$:  $P_{11} = P_{21} = P_{31}$
$\phantom{H_0: }P_{12} = P_{22} = P_{32}$
$\phantom{H_0: }P_{13} = P_{23} = P_{33}$
$\phantom{H_0: }P_{14} = P_{24} = P_{34}$

$\phantom{..}H_a$:  not $H_0$

**Step 2** $\alpha = 0.05$     $P(\text{Type I error}) = 0.05$.

**Step 3** The test statistic will be:

$$X^2 = \sum \frac{O_{ij}^2}{E_{ij}} - n$$

**Step 4** If $x^2 > \chi_c^2 (= 12.592)$, then reject $H_0$. The value of $\chi_c^2$ is taken from Appendix Table 6 using:

$$df = (3 - 1)(4 - 1) = 6$$

**Step 5** From Example 3:

$$x^2 = 32.402$$

**Step 6** Because $32.402 > 12.592$, reject $H_0$. That is, there is sufficient evidence, with $\alpha = 0.05$, to conclude that the patterns of elementary school, high school, and college teachers preferences are not the same.    ◀

The approximate nature of the $\chi^2$ test above deserves some additional comments. The mathematical result that justifies using a critical point from the $\chi^2$ table is applicable only when all $n_i$ are large. One famous statistician's partial answer to the question "how large" is known as Cochran's rule.

### Cochran's Rule

A $\chi^2$ test for equality of $r$ distributions can be used, if the $n_i$ are large enough so that all $o_{ij}$ are at least one and at least 80 percent of the $o_{ij}$ are at least five.

The rule is intended to alert us to the fact that small $o_{ij}$'s indicate some $P_{ij}$'s are small. If some $P_{ij}$'s are small, then very large $n_i$'s are needed to make the $\chi^2$ approximation valid. Notice that the $n_i$'s in Example 4 were large enough to produce $o_{ij}$'s satisfying the conditions stated in Cochran's rule.

▶ **Example 5**    A study was recently completed by a federal labor commission on the impact of imported goods on the sale of American made products in the United States. The officials of a nationwide labor union wondered whether there are regional differences in the attitudes of workers on this issue. As a consequence, they directed the officers in the regional offices in the East, Midwest, South, and West to take opinion polls of workers in their regions. The possible responses in the polls were "no impact," "some impact," and "a large impact" of imported goods on the sale of American made products. The results of the polls are listed in Table 8-7.

TABLE 8-7    The $o_{ij}$'s for the Four Polls on Worker's Attitudes Toward Imports

|  | **Attitudes** | | | |
|---|---|---|---|---|
|  | No Impact | Some Impact | Large Impact | Totals |
| East | 13 | 31 | 81 | 125 |
| Midwest | 52 | 53 | 70 | 175 |
| South | 38 | 75 | 37 | 150 |
| West | 30 | 35 | 35 | 100 |
| Totals | 133 | 194 | 223 | 550 |

Use a $\chi^2$ test of equality of distributions to compare distributions of worker's attitudes in the four regions. Use $\alpha = 0.01$.

**Solution**    **Step 1**  $H_0$:  $P_{11} = P_{21} = P_{31} = P_{41}$
$P_{12} = P_{22} = P_{32} = P_{42}$
$P_{13} = P_{23} = P_{33} = P_{43}$

$H_a$:  not $H_0$

**Step 2** $\alpha = 0.01$

**Step 3** The test statistic will be:

$$X^2 = \sum\left(\frac{O_{ij}^2}{E_{ij}}\right) - n$$

**Step 4** If $x^2 > \chi_c^2$ ($=16.812$), then reject $H_0$. The value of $\chi_c^2$ is taken from Appendix Table 6 using:

$$df = (4 - 1)(3 - 1) = 6$$

**Step 5** Using equation (2):

$$x^2 = \frac{13^2}{30.227\ldots} + \frac{52^2}{42.318\ldots} + \cdots + \frac{35^2}{40.545\ldots} - 550$$

$$= 5.590\ldots + 63.896\ldots + \cdots + 30.213\ldots - 550$$

$$= 606.097\ldots - 550$$

$$= 56.097, \text{ to three decimal places}$$

**Step 6** Because $56.097 > 16.812$, reject $H_0$ and accept $H_a$. That is, there is sufficient evidence, with $\alpha = 0.01$, to conclude that there are regional differences in the patterns of opinions of workers regarding the impact of imported goods on the sale of American made products.    ◄

## Constructing confidence intervals for differences in proportions

The use of a $\chi^2$ test of equality of $r$ distributions helps answer the question "Are there differences in the distributions?" It does not help to assess the size of any differences among the $P_{ij}$'s for a given category $j$. To estimate the size of a particular difference, a confidence interval for the difference in two proportions can be used. Remember the following formula from Section 7-3.

### Equations for Constructing Confidence Intervals for the Difference in Two Proportions

$$\text{lower limit} = (\hat{p}_1 - \hat{p}_2) - z\sqrt{\frac{\hat{p}_1(1 - \hat{p}_1)}{n_1} + \frac{\hat{p}_2(1 - \hat{p}_2)}{n_{21}}} \qquad (3)$$

$$\text{upper limit} = (\hat{p}_1 - \hat{p}_2) + z\sqrt{\frac{\hat{p}_1(1 - \hat{p}_1)}{n_1} + \frac{\hat{p}_2(1 - \hat{p}_2)}{n_2}} \qquad (4)$$

▶ **Example 6** Use the results of the opinion polls in Table 8-7 to estimate the difference in proportions of workers in the East and Midwest who feel imports have a "large impact" on the sale of American made products in the United States with a 90 percent confidence interval.

Solution From Table 8-7:

$$\hat{p}_1 = \frac{81}{125} = 0.648$$

$$\hat{p}_2 = \frac{70}{175} = 0.400$$

For a 90 percent confidence interval, $z = 1.65$.

$$\text{lower limit} = (0.648 - 0.400) - 1.65\sqrt{\frac{(0.648)(0.352)}{125} + \frac{(0.400)(0.600)}{175}}$$

$$= 0.248 - 0.0932\ldots$$

$$= 0.155, \text{ to three decimal places}$$

$$\text{upper limit} = 0.248 + 0.0932\ldots$$

$$= 0.341, \text{ to three places}$$

Based on the given data, a 90 percent confidence interval for the difference in proportions of American workers in the East and Midwest who feel imports have a large impact on the sale of American products is from 15.5 to 34.1 percent. ◀

# Exercises 8-1 Set A

*In Exercises 1 and 2, use the data in the following table of observed counts and identify each indicated value:*

|        |        | Category |     |     |     |        |
|--------|--------|----------|-----|-----|-----|--------|
|        |        | W        | X   | Y   | Z   | Totals |
|        | A      | 10       | 20  | 30  | 40  | 100    |
| Sample | B      | 80       | 50  | 60  | 70  | 260    |
|        | C      | 90       | 100 | 110 | 120 | 420    |
|        | D      | 160      | 150 | 130 | 140 | 580    |
|        | Totals | 340      | 320 | 330 | 370 | 1360   |

**1.** a. $n_2$    b. $m_3$    c. $o_{33}$    d. $o_{42}$    e. $n$

**2.** a. $n_4$    b. $m_1$    c. $o_{14}$    d. $o_{41}$    e. $o_{22}$

*In Exercises 3 and 4, use the data in the following table of observed counts and identify each indicated value:*

|  | Category | | | | | |
|---|---|---|---|---|---|---|
|  | V | W | X | Y | Z | Totals |
| **A** | 10 | 20 | 25 | 15 | 30 | 100 |
| **B** | 12 | 15 | 13 | 40 | 40 | 120 |
| **Sample**   **C** | 14 | 25 | 30 | 51 | 20 | 140 |
| **D** | 14 | 20 | 42 | 24 | 60 | 160 |
| Totals | 50 | 80 | 110 | 130 | 150 | 520 |

**3.** a. $n_4$   b. $m_2$   c. $o_{24}$   d. $o_{31}$   e. $n$

**4.** a. $n_2$   b. $m_5$   c. $o_{44}$   d. $o_{15}$   e. $o_{23}$

*In Exercises 5–8, convert the value of each $o_{ij}$ to a sample proportion.*

**5.**

|  | Category | | |
|---|---|---|---|
|  | X | Y | Totals |
| **A** | 5 | 5 | 10 |
| **Sample**   **B** | 12 | 8 | 20 |
| **C** | 8 | 22 | 30 |
| Totals | 25 | 35 | 60 |

**6.**

|  | Category | | | |
|---|---|---|---|---|
|  | X | Y | Z | Totals |
| **A** | 6 | 7 | 7 | 20 |
| **B** | 7 | 8 | 10 | 25 |
| **Sample**   **C** | 15 | 12 | 13 | 40 |
| **D** | 18 | 7 | 10 | 35 |
| **E** | 14 | 6 | 10 | 30 |
| Totals | 60 | 40 | 50 | 150 |

**7.**

|  | Category | | | |
|---|---|---|---|---|
|  | X | Y | Z | Totals |
| **Sample**   **A** | 65 | 95 | 40 | 200 |
| **B** | 85 | 155 | 60 | 300 |
| Totals | 150 | 250 | 100 | 500 |

**8.**

|  | Category | | | | | |
|---|---|---|---|---|---|---|
|  | V | W | X | Y | Z | Totals |
| **A** | 40 | 50 | 30 | 35 | 45 | 200 |
| **Sample**   **B** | 60 | 70 | 25 | 30 | 15 | 200 |
| **C** | 50 | 80 | 45 | 15 | 10 | 200 |
| Totals | 150 | 200 | 100 | 80 | 70 | 600 |

*In Exercises 9–12, conduct a $\chi^2$ test for equality of distribution for each situation. In each cell, the upper left value is $o_{ij}$ and the lower right value is $e_{ij}$.*

**9.** Use $\alpha = 0.05$.

| | | **Category** | | |
|---|---|---|---|---|
| | | *X* | *Y* | *Totals* |
| **Sample** | *A* | 25 / 20 | 25 / 30 | 50 |
| | *B* | 15 / 20 | 35 / 30 | 50 |
| | *Totals* | 40 | 60 | 100 |

**10.** Use $\alpha = 0.01$.

| | | **Category** | | | |
|---|---|---|---|---|---|
| | | *X* | *Y* | *Z* | *Totals* |
| **Sample** | *A* | 18 / 15 | 16 / 15 | 6 / 10 | 40 |
| | *B* | 12 / 15 | 14 / 15 | 14 / 10 | 40 |
| | *Totals* | 30 | 30 | 20 | 80 |

**11.** Use $\alpha = 0.01$.

| | | **Category** | | | |
|---|---|---|---|---|---|
| | | *X* | *Y* | *Z* | *Totals* |
| **Sample** | *A* | 32 / 36 | 39 / 32 | 19 / 22 | 90 |
| | *B* | 30 / 24 | 17 / 21 | 13 / 15 | 60 |
| | *C* | 18 / 20 | 14 / 17 | 18 / 13 | 50 |
| | *Totals* | 80 | 70 | 50 | 200 |

**12.** Use $\alpha = 0.05$.

| Sample | Category | | | |
|---|---|---|---|---|
| | W | X | Y | Totals |
| A | 32 / 40 | 28 / 24 | 20 / 16 | 80 |
| B | 60 / 50 | 24 / 30 | 16 / 20 | 100 |
| C | 70 / 65 | 29 / 39 | 31 / 26 | 130 |
| D | 38 / 45 | 39 / 27 | 13 / 18 | 90 |
| Totals | 200 | 120 | 80 | 400 |

*In Exercises 13 and 14, the data refer to a recent nationwide study of individuals regarding attitudes toward "walking in the neighborhood at night."*

**13.** Geraldine White is a member of a federal task force on violence in the country. She wonders whether community size is a factor in an individual's fear for personal safety. She therefore conducts four surveys. Each individual surveyed is asked whether she/he is afraid to walk in the neighborhood at night.

Sample *A*   Residents in a *large eastern city*

Sample *B*   Residents in a *moderate-size eastern city*

Sample *C*   Residents in a *small city in the East*

Sample *D*   Residents in a *"farming town" in the East*

| Sample | Category | | |
|---|---|---|---|
| | Yes | No | Totals |
| A | 112 | 88 | 200 |
| B | 75 | 75 | 150 |
| C | 45 | 55 | 100 |
| D | 14 | 36 | 50 |
| Totals | 246 | 254 | 500 |

a. Use a $\chi^2$ test of equality of distribution to compare attitudes of residents in the four cities polled. Use $\alpha = 0.01$.

b. Estimate the difference in proportions of residents in *A* and *D* who are afraid to walk in their neighborhoods at night, with a 90 percent confidence interval.

14. Frank E. Lee is the current president of Retired Individuals Who Travel (RIWT). He wonders whether there are regional differences in the attitudes of RIWT members regarding fear of walking in their neighborhoods at night. He therefore contacts regional offices in the East, Midwest, South, and West and instructs them to poll members in these regions.

| | | Category | | | |
|---|---|---|---|---|---|
| | | *None* | *Some* | *Much* | *Totals* |
| **Sample** | *East* | 10 | 25 | 65 | 100 |
| | *Midwest* | 30 | 30 | 40 | 100 |
| | *South* | 25 | 50 | 25 | 100 |
| | *West* | 30 | 35 | 35 | 100 |
| | *Totals* | 95 | 140 | 165 | 400 |

a. Use a $\chi^2$ test of equality of distributions to compare attitudes of RIWT members in the four regions polled. Use $\alpha = 0.01$.

b. Estimate the difference in population proportions of residents in the East and West who have no fear of walking in their neighborhoods at night, with a 95 percent confidence interval.

15. The student government of a large college formed a committee to study grading patterns of instructors. With the permission of the faculty senate, the registrars office supplied the committee with randomly selected grades from files covering the past few years. Professors Henry and Davis share teaching the elementary statistics classes at the school. The following table shows the grade distributions given to the committee for these instructors:

| | | Category | | | | |
|---|---|---|---|---|---|---|
| | | *A* | *B* | *C* | *No Credit* | *Totals* |
| **Sample** | *Henry* | 10 | 15 | 45 | 10 | 80 |
| | *Davis* | 15 | 25 | 40 | 10 | 90 |
| | *Totals* | 25 | 40 | 85 | 20 | 170 |

a. Use a $\chi^2$ test of equality of distribution to compare the grade distributions for Henry and Davis. Use $\alpha = 0.05$.

b. Estimate the difference in population proportions in *C* grades, with a 98 percent confidence interval.

16. A member of the state board of trustees of community colleges wonders whether attitude patterns toward "the right to strike" are the same for various teachers' organizations. The three principal teachers' organizations are:

      *A*   A *conservative* professional group

      *B*   A *moderate* statewide teachers' association

      *C*   A *liberal* teachers' union

The trustee polls members of each organization through a random selection of its membership rolls. The results are:

| Sample | | Favor | Oppose | Undecided | Totals |
|---|---|---|---|---|---|
| | | **Category** | | | |
| | A | 8 | 28 | 14 | 50 |
| | B | 50 | 44 | 26 | 120 |
| | C | 62 | 8 | 10 | 80 |
| | Totals | 120 | 80 | 50 | 250 |

a. Use a $\chi^2$ test of equality of distribution to compare the attitude patterns of the three groups regarding the right-to-strike issue. Use $\alpha = 0.05$.
b. Estimate the difference in population proportions favoring the right-to-strike for groups $A$ and $C$ with a 99 percent confidence interval.

17. A political science major at a state college is interested in knowing whether the area of the city in which a voter lives is a factor that determines the political party of the individual. She divides the city into the following three areas:

   Area $A$   A section with predominantly *low-cost* housing

   Area $B$   A section with predominantly *moderately priced* housing

   Area $C$   A section with predominantly *high-cost* housing

She randomly selects 100 voters from each area. The results are:

| Sample | | Republican | Democrat | Independent | Totals |
|---|---|---|---|---|---|
| | | **Category** | | | |
| | A | 18 | 36 | 46 | 100 |
| | B | 18 | 40 | 42 | 100 |
| | C | 27 | 56 | 17 | 100 |
| | Totals | 63 | 132 | 105 | 300 |

Use a $\chi^2$ test of equality of distribution to compare the party affiliation patterns of the voters from the three areas. Use $\alpha = 0.01$.

18. A national association of lawyers wonders whether there are regional differences in public attitudes on the issue of capital punishment. They have local organizations in the East, Midwest, South, and West obtain samples of 200 peoples' opinions each in their sections of the country. The results are shown in the table on the next page.

|  | Category | | | |
|---|---|---|---|---|
|  | For | Against | No Opinion | Totals |
| **Sample** East | 120 | 55 | 25 | 200 |
| Midwest | 115 | 50 | 35 | 200 |
| South | 130 | 30 | 40 | 200 |
| West | 95 | 85 | 20 | 200 |
| Totals | 460 | 220 | 120 | 800 |

Use a $\chi^2$ test of equality of distribution to compare the opinion patterns in the four designated regions of the country. Use $\alpha = 0.05$.

# Exercises 8-1   Set B

*In Exercises 1–4, suppose the data were obtained in supplemental surveys taken in response to the report in Figure 8-3.*

**1.** The following groups were surveyed regarding the problem of controlling medical costs:

     A   Medical doctors and nurses

     B   Individuals not employed in medical professions

|  | Favor Control | Oppose Control | No Opinion | Totals |
|---|---|---|---|---|
| A | 105 | 70 | 35 | 210 |
| B | 295 | 80 | 65 | 440 |
| Totals | 400 | 150 | 100 | 650 |

a. Use a $\chi^2$ test of equality of distribution to compare the patterns of opinion for medical and non-medical individuals on the issue of controlling medical costs.

b. Estimate the difference in proportions of individuals in populations $A$ and $B$ who oppose control, with a 90 percent confidence interval.

**2.** The following groups were surveyed regarding the problem of transportation needs and a possible solution involving the raising the gas tax:

     A   Employees of oil companies

     B   Employees of mass transportation systems

     C   Individuals not in groups $A$ or $B$

|  | Favor Tax Increase | Oppose Tax Increase | No Opinion | Totals |
|---|---|---|---|---|
| A | 46 | 76 | 10 | 132 |
| B | 61 | 37 | 8 | 106 |
| C | 131 | 70 | 18 | 219 |
| Totals | 238 | 183 | 36 | 457 |

# YOUR VIEWS–COUNTED UP

*What follows is a summary report addressing major issues raised in the voter survey mailed out a few months ago.\**

**The top five responses you offered to the question about rating the State's most pressing problems were:**

| | |
|---|---|
| Controlling Medical Costs | 67% |
| Controlling Insurance Rates | 61% |
| Crime and Punishment | 61% |
| Drug/Alcohol Addiction | 60% |
| Honesty in Government | 57% |

As solutions to California's transportation needs most of you favored raising the gas tax (**60%**) and investing more in mass transit (**63%**). The least popular solutions were more bond financing (**92%**) and establishing toll roads (**75%**).

Most of you (**82%**) felt that the State should begin regulating insurance companies as a way of curbing rising insurance rates.

An overwhelming majority of you (**93%**) felt there should be strong legal protections for San Francisco Bay, the Delta, and fish and wildlife in any water transfer legislation.

Most of you (**68%**) said you would favor a Constitutional Amendment to allow the State to increase spending on programs like education, health, transportation, and AIDS.

Issues which you felt the State should be paying more attention to included: cleaning up toxic waste, helping the homeless, and crime prevention.

You felt it would be a good idea (**76%**) to lease surplus State property for use as homeless shelters, food closets or community kitchens to address the State's homeless crisis.

Most respondents (**62%**) also indicated that state government needs to increase the amount of money that it sends to local government.

Most of you (**90%**) favored reducing the first prize amounts in the California Lottery in favor of a system that had more winners receiving smaller prizes.

A large majority of you (**62%**) favored simplification of the State's tax laws.

With regard to the question about how you receive the majority of your information about state government, the following chart shows your responses.

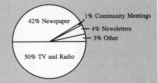

*Thank you for taking time to complete the voter issue survey. Your views count and your suggestions provide helpful input for my dections in the State Senate.*

**Figure 8-3.** "Your Views Counted Up" by Senator John Garamendi. Reprinted by permission.

a. Use a $\chi^2$ test of equality of distribution to compare the opinion patterns of individuals in groups $A$, $B$, and $C$ on the issue of increasing the tax on gasoline.

b. Estimate the difference in proportions of individuals in populations $A$ and $B$ who oppose a tax increase, with a 95 percent confidence interval.

3. The following groups were surveyed regarding the issue of how important crime and punishment is in the state:

     *A*    Individuals who earn less than $10,000 a year

     *B*    Individuals who earn between $10,000 and $25,000 a year

     *C*    Individuals who earn between $25,000 and $50,000 a year

     *D*    Individuals who earn more than $50,000 a year

|  | Very Important | Important | Not Very Important | Totals |
|---|---|---|---|---|
| *A* | 54 | 46 | 43 | 143 |
| *B* | 117 | 73 | 22 | 212 |
| *C* | 146 | 34 | 15 | 195 |
| *D* | 66 | 9 | 3 | 78 |
| Totals | 383 | 162 | 83 | 628 |

a. Use a $\chi^2$ test of equality of distribution to compare the opinion patterns of individuals in groups *A*, *B*, *C*, and *D* on the issue of crime and punishment.

b. Estimate the difference in proportions of individuals in populations *A* and *B* who feel the issue is of little importance, with a 98 percent confidence interval.

4. The political party affiliations of individuals were thought to influence their opinions on the issue of a constitutional amendment to allow the state to increase spending on programs like education, health, transportation, and AIDS research. Samples of Democrats, Republicans, and Independents were taken with the following results:

|  | Strongly Favor | Moderately Favor | Moderately Oppose | Strongly Oppose | Totals |
|---|---|---|---|---|---|
| Democrats | 85 | 111 | 42 | 27 | 265 |
| Independents | 30 | 54 | 20 | 8 | 112 |
| Republicans | 29 | 68 | 64 | 33 | 194 |
| Totals | 144 | 233 | 126 | 68 | 571 |

a. Use a $\chi^2$ test of equality of distribution to compare the opinion patterns of Democrats, Independents, and Republicans on the issue of the constitutional amendment.

b. Estimate the difference in proportions of Democrats and Republicans who strongly favor the amendment, with a 90 percent confidence interval.

*Exercises 5 and 6 are suggested student projects.*

5. Survey college students (Population 1) and residents of the community (Population 2) in which your college is located. Summarize the responses to the following question in the table below:

"Should college students be banned from parking vehicles on city streets near the college campus?"

Sample from both populations until at least five responses are recorded in each cell. Then conduct a $\chi^2$ test to determine whether a difference exists in the opinion patterns in the two populations on this issue. Use $\alpha = 0.05$.

|  | Oppose | Favor | No Opinion | Totals |
|---|---|---|---|---|
| College Students |  |  |  |  |
| Residents |  |  |  |  |
| Totals |  |  |  |  |

**6.** Survey college students (Population 1) and residents of the community (Population 2) in which your college is located. Summarize the responses to the following question in the table below:

"Should students who live in off-campus housing be issued a card that qualifies them for a 5% discount on all purchases made at local businesses?"

Sample from both populations until at least five responses are recorded in each cell. Then conduct a $\chi^2$ test to determine whether a difference exists in the two population opinion patterns on this issue. Use $\alpha = 0.05$.

|  | Favor | Oppose | No Opinion | Totals |
|---|---|---|---|---|
| College Students |  |  |  |  |
| Residents |  |  |  |  |
| Totals |  |  |  |  |

# 8-2  A Test for Equality of *r* Means Analysis of Variance (ANOVA)

**Key Topics**

1. *The model assumptions necessary for the ANOVA test*

2. *Computing a pooled sample variance*

3. *Computing the "between" sample variance*

4. *The test statistic for testing* $H_0: \mu_1 = \mu_2 = \cdots = \mu_r$

5. *The probability distribution of* $\dfrac{S_b^2}{S_p^2}$

6. *The F-distribution for testing* $H_0: \mu_1 = \mu_2 = \cdots = \mu_r$

7. *Constructing small sample confidence intervals for* $\mu_1 - \mu_2$

# S T A T I S T I C S   I N   A C T I O N

Claudine Young is a sociology professor at a large university. The Sociology 235 class she is teaching this term has been studying the attitudes that Americans have had about marriage. Discussions have shown that most students agree that current attitudes are quite different from those held in the recent past. Claudine asked for suggestions on what data could be collected to test this hypothesis. Ray Hernandez, a student in his 50s, suggested that one measure would be the average age at which adults get married the first time. He said that his recollection of the decades of the 1950s, 1960s, 1970s, and 1980s would lead him to assume that average ages of individuals at the time of their first marriage increased across these decades. He mentioned several historical events during these years that somehow influenced adults' attitudes toward getting married and establishing a family.

The class agreed that Ray's suggestion was a good one and decided to obtain data to test his hypothesis. Five students volunteered to contact Cathy Parker, a friend of Claudine's who works in the State Department of Vital Statistics. Cathy can access computer records of events such as births, deaths, and marriages within the state. With her help, the students obtained samples of the ages of 12 females at the time of their first marriage in each of the four decades being studied. As a result of this activity, the volunteers presented the class with the data in Table 8-8.

**TABLE 8-8   Ages of Females at the Time of Their First Marriage**

| 1950–1959 | 1960–1969 | 1970–1979 | 1980–1989 |
|-----------|-----------|-----------|-----------|
| 23 | 18 | 23 | 17 |
| 18 | 22 | 20 | 32 |
| 25 | 26 | 16 | 21 |
| 16 | 19 | 27 | 26 |
| 28 | 15 | 22 | 19 |
| 20 | 20 | 19 | 29 |
| 19 | 24 | 23 | 25 |
| 21 | 29 | 18 | 30 |
| 17 | 18 | 24 | 27 |
| 21 | 25 | 17 | 24 |
| 18 | 21 | 29 | 28 |
| 20 | 19 | 26 | 31 |

With a calculator, the following descriptive measures were determined:

| Sample 1: 1950s | Sample 2: 1960s | Sample 3: 1970s | Sample 4: 1980s |
|-----------------|-----------------|-----------------|-----------------|
| $\bar{x}_1 = 20.5$ | $\bar{x}_2 = 21.33\ldots$ | $\bar{x}_3 = 22.0$ | $\bar{x}_4 = 25.75$ |
| $s_1^2 = 11.90\ldots$ | $s_2^2 = 16.06\ldots$ | $s_3^2 = 16.90\ldots$ | $s_4^2 = 22.75$ |

To summarize these results in the form of a hypothesis test using ANOVA and the standard format first studied in Chapter 6, the following populations and parameters were defined. The details of the hypothesis test are studied in this section.

**Population 1**  The ages of all females at the time of their first marriage in the years 1950–59. The mean is $\mu_1$.

**Population 2**  The ages of all females at the time of their first marriage in the years 1960–69. The mean is $\mu_2$.

**Population 3**  The ages of all females at the time of their first marriage in the years 1970–79. The mean is $\mu_3$.

**Population 4**  The ages of all females at the time of their first marriage in the years 1980–89. The mean is $\mu_4$.

**Step 1**  $H_0$:  $\mu_1 = \mu_2 = \mu_3 = \mu_4$    The means are the same.

$H_a$:  not $H_0$    At least one mean is different from the others.

**Step 2**  $\alpha = 0.05$    $P(\text{Type I error}) = 0.05$.

**Step 3**  The test statistic used is:

$$F = \frac{S_b^2}{S_p^2}$$

**Step 4**  If $f > f_c$ (=2.84), then reject $H_0$.

**Step 5**  Using the data in Table 8-8:

$s_p^2 = 16.907\ldots$    The "pooled" variance

$s_b^2 = 64.520\ldots$    The "betweens" variance

$f = \dfrac{64.520\ldots}{16.907\ldots} \approx 3.82$, to two decimal places

**Step 6**  Because $3.82 > 2.84$, reject $H_0$. That is, there is sufficient evidence, with $\alpha = 0.05$, to conclude that the means of the four populations are not the same.

## The model assumptions necessary for the ANOVA test

In the Statistics in Action part of this section, the means from the four samples are, to one decimal place:

$$\bar{x}_1 = 20.5 \text{ years}, \quad \bar{x}_2 = 21.3 \text{ years}, \quad \bar{x}_3 = 22.0 \text{ years}, \quad \text{and} \quad \bar{x}_4 = 25.8 \text{ years}$$

On studying the four means, it is clear that some variation exists *between the sample means*. However, it is not obvious whether there is enough variation to support the conclusion that variation exists *among the four population means*. Using the $\bar{x}_i$'s and $s_i^2$'s as tools, we can calculate a statistic that is used to make such a decision. Before proceeding to develop that statistic, we need to identify the model assumptions under which it is possible to decide whether to reject

$$H_0: \mu_1 = \mu_2 = \mu_3 = \cdots = \mu_r$$

### Model Assumptions for the ANOVA Test

To test $H_0: \mu_1 = \mu_2 = \cdots = \mu_r$ at a particular level of significance, the following assumptions must be made:

1. The observed values are all independent random variables.
2. Each population sampled has a normal distribution.
3. The variances of all the populations are the same. The symbol $\sigma^2$ is used for the common value.
4. The mean of the *i*th population is $\mu_i$.

The most unappealing of these assumptions are 2 and 3, namely the normal probability distributions and the equal variances assumptions. These are restrictive in the sense that not all populations have normal distributions. Furthermore in a particular study, all population variances need not necessarily be equal. However, these assumptions are at least approximately appropriate if the $r$ samples are separate simple random samples from large populations possessing bell-shaped relative frequency distributions and similar standard deviations.

## Computing a pooled sample variance

In view of assumption 3, the sample variances $s_1^2, s_2^2, \ldots, s_r^2$ are all estimates of a common population variance $\sigma^2$. (Notice the subscript $r$ on the last variance indicates that $r$ populations have been sampled.) To get the best estimate for $\sigma^2$ based on the sample data, we "pool" these estimates together to get a single value.

### The Pooled Estimate of $\sigma^2$

$$s_p^2 = \frac{\sum(n_i - 1)s_i^2}{n - r}$$

where $n = n_1 + n_2 + \cdots + n_r$, the total number of observations. (The small $p$ subscript stands for "pooled.")

Because $s_p^2$ is a weighted average of the sample variances, it will always be somewhere between the largest and smallest $s_i^2$.

▶ **Example 1**   Compute $s_p^2$ for the samples from the four populations of ages.

**Solution**   The samples were all of size 12.
That is $n_1 = n_2 = n_3 = n_4 = 12$. Thus each $(n_i - 1) = (12 - 1) = 11$. With $s_1^2 = 11.90 \ldots, s_2^2 = 16.06 \ldots, s_3^2 = 16.90 \ldots, s_4^2 = 22.75, n = 48,$ and $r = 4$:

$$s_p^2 = \frac{11(11.90 \ldots) + 11(16.06 \ldots) + 11(16.09 \ldots) + 11(22.75)}{48 - 4}$$

$$= \frac{11(67.62 \ldots)}{44}$$

$$\approx 16.907, \text{ to three decimal places} \qquad ◀$$

It is the value of $s_p^2$ that is used as a basis of comparison for judging whether the observed variability in the $\bar{x}_i$ values is large enough to force rejection of $H_0$.

## Computing the "between" sample variance

There is another descriptive measure that we need to calculate for the sample data. This measure describes the overall variability in the sample means. Because it measures the differences *between* the $\bar{x}_i$'s, (where $1 \le i \le r$), the subscript "$b$" will be used.

## The "Between" Variance

$$s_b^2 = \frac{\sum n_i(\bar{x}_i - \bar{x})^2}{r - 1}$$

where $\bar{x}$ is the *grand mean*,

$$\bar{x} = \frac{\text{grand total of the observations}}{\text{total number of observations}} = \frac{1}{n}\sum n_i \bar{x}_i$$

Notice that $s_b^2$ will tend to be large when the $\bar{x}_i$'s are very different.

▶ **Example 2**    Calculate $s_b^2$ for the $\bar{x}_i$'s from the four samples of ages that Cathy Parker recorded in Table 8-8.

**Solution**    The values from the samples needed to calculate $s_b^2$ are:

$$n_1 = 12 \qquad n_2 = 12 \qquad n_3 = 12 \qquad n_4 = 12$$
$$\bar{x}_1 = 20.5 \qquad \bar{x}_2 = 21.3\ldots \qquad \bar{x}_3 = 22.0 \qquad \bar{x}_4 = 25.75$$

$$\bar{x} = \frac{12(20.5) + 12(21.3\ldots) + 12(22.0) + 12(25.75)}{48}$$

$$= \frac{1075}{48} = 22.39\ldots \qquad \qquad \text{The mean of the 48 ages}$$

With $r = 4$, $\bar{x} = 22.39\ldots$, and the values of $\bar{x}_i$ and $n_i$ above:

$$s_b^2 = \frac{12(20.5 - 22.39\ldots)^2 + \cdots + 12(25.75 - 22.39\ldots)^2}{4 - 1}$$

$$= \frac{43.13\ldots + 13.54\ldots + 1.88\ldots + 135.00\ldots}{3}$$

$$= \frac{193.5\ldots}{3} \approx 64.521, \text{ to three decimal places} \qquad ◀$$

## The test statistic for testing $H_0: \mu_1 = \mu_2 = \cdots = \mu_r$

It is possible to show mathematically that if

$$H_0: \quad \mu_1 = \mu_2 = \cdots = \mu_r$$

is true, then $s_b^2$ and $s_p^2$ are both estimates of $\sigma^2$. If, however, $H_0$ is not true (at least two means are different), then $s_b^2$ estimates a value larger than $\sigma^2$. Comparison of $s_b^2$ and $s_p^2$ is then an intuitively appealing way of deciding whether to reject, or not reject, $H_0$. The exact method used is to form the ratio

$$F = \frac{S_b^2}{S_p^2}$$

and reject $H_0$ for large observed values of $F$.

### The Test Statistic for Testing the Equality of $r$ Means

To test $H_0$: $\mu_1 = \mu_2 = \cdots = \mu_r$ versus $H_a$: not $H_0$, use

$$F = \frac{S_b^2}{S_p^2} \tag{3}$$

If $H_0$ is true, then $S_b^2$ and $S_p^2$ are both random variables estimating the unknown population variance $\sigma^2$. Therefore the ratio of the "between" and "pooled" sample variances will tend to be close to 1. However, if $H_0$ is not true, then $S_b^2$ will tend to be larger than $S_p^2$, and the corresponding ratio will be greater than 1. To conclude that a given ratio is significantly greater than 1, and hence $H_0$ is not true, we compare the observed value with a critical $f$-value from Appendix Table 7A or 7B. The use of this table is studied later in this section.

▶ **Example 3**     Compute the observed value of $F$ for the hypothesis test on the means of the four populations of ages in the Statistics in Action section.

**Solution**     From Example 1: $s_p^2 \approx 16.907$   and   from Example 2: $s_b^2 \approx 64.521$.
Using equation (3):

$$f = \frac{64.521}{16.907} \approx 3.82, \text{ to two decimal places} \quad ◀$$

## The probability distribution of $\frac{S_b^2}{S_p^2}$

If $H_0$: $\mu_1 = \mu_2 = \cdots = \mu_r$ is true, then the distribution of $\frac{S_b^2}{S_p^2}$ is a standard probability distribution called an $F$-distribution. There are many $F$-distributions and each one is determined by two so-called degrees-of-freedom parameters.

## The Degrees-of-Freedom Parameters
## for the *F* Probability Distributions

The two degrees-of-freedom parameters are:

1. The degrees of freedom for the *numerator*.
2. The degrees of freedom for the *denominator*.

The various *F* probability distributions have idealized histograms with generally right-skewed appearances, as shown in Figure 8-4.

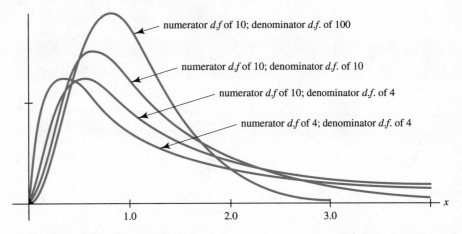

**Figure 8-4.**   Graphs of four *F*-distributions.

Appendix Table 7A contains some upper 5% points for the *F*-distributions. Appendix Table 7B contains some upper 1% points for the same *F*-distributions.

In Figure 8-5 (on the next page), a portion of Appendix Table 7A is shown. To use this table, locate the degrees of freedom for denominator along the left and the degrees of freedom for numerator along the top. The number in the table in the corresponding row and column is the upper percentage point of that *F*-distribution. To illustrate, the upper 5% point of the *F*-distribution with numerator degrees of freedom 10 and denominator degrees of freedom 4 is 5.96.

Therefore if *W* is a random variable with an *F* probability distribution with 10 and 4 *df* (the numerator degrees of freedom are always listed first), then:

$$P(W \geq 5.96) = 0.05$$

This probability statement is illustrated in Figure 8-6 on the following page.

**Degrees of Freedom for Numerator**

| | | 1 | 2 | 3 | 4 | 5 | 6 | 7 | 8 | 9 | 10 |
|---|---|---|---|---|---|---|---|---|---|---|---|
| | 1 | 161 | 200 | 216 | 225 | 230 | 234 | 237 | 239 | 241 | 242 |
| | 2 | 18.5 | 19.0 | 19.2 | 19.2 | 19.3 | 19.3 | 19.4 | 19.4 | 19.4 | 19.4 |
| | 3 | 10.1 | 9.55 | 9.28 | 9.12 | 9.01 | 8.94 | 8.89 | 8.85 | 8.81 | 8.79 |
| | 4 | 7.71 | 6.94 | 6.59 | 6.39 | 6.26 | 6.16 | 6.09 | 6.04 | 6.00 | 5.96 |
| | 5 | 6.61 | 5.79 | 5.41 | 5.19 | 5.05 | 4.95 | 4.88 | 4.82 | 4.77 | 4.74 |
| | 6 | 5.99 | 5.14 | 4.76 | 4.53 | 4.39 | 4.28 | 4.21 | 4.15 | 4.10 | 4.06 |
| | 7 | 5.59 | 4.74 | 4.35 | 4.12 | 3.97 | 3.87 | 3.79 | 3.73 | 3.68 | 3.64 |
| | 8 | 5.32 | 4.46 | 4.07 | 3.84 | 3.69 | 3.58 | 3.50 | 3.44 | 3.39 | 3.35 |
| | 9 | 5.12 | 4.26 | 3.86 | 3.63 | 3.48 | 3.37 | 3.29 | 3.23 | 3.18 | 3.14 |
| | 10 | 4.96 | 4.10 | 3.71 | 3.48 | 3.33 | 3.22 | 3.14 | 3.07 | 3.02 | 2.98 |
| | 11 | 4.84 | 3.98 | 3.59 | 3.36 | 3.20 | 3.09 | 3.01 | 2.95 | 2.90 | 2.85 |
| | 12 | 4.75 | 3.89 | 3.49 | 3.26 | 3.11 | 3.00 | 2.91 | 2.85 | 2.80 | 2.75 |
| | 13 | 4.67 | 3.81 | 3.41 | 3.18 | 3.03 | 2.92 | 2.83 | 2.77 | 2.71 | 2.67 |
| | 14 | 4.60 | 3.74 | 3.34 | 3.11 | 2.96 | 2.85 | 2.76 | 2.70 | 2.65 | 2.60 |
| | 15 | 4.54 | 3.68 | 3.29 | 3.06 | 2.90 | 2.79 | 2.71 | 2.64 | 2.59 | 2.54 |
| | 16 | 4.49 | 3.63 | 3.24 | 3.01 | 2.85 | 2.74 | 2.66 | 2.59 | 2.54 | 2.49 |
| | 17 | 4.45 | 3.59 | 3.20 | 2.96 | 2.81 | 2.70 | 2.61 | 2.55 | 2.49 | 2.45 |
| | 18 | 4.41 | 3.55 | 3.16 | 2.93 | 2.77 | 2.66 | 2.58 | 2.51 | 2.46 | 2.41 |
| | 19 | 4.38 | 3.52 | 3.13 | 2.90 | 2.74 | 2.63 | 2.54 | 2.48 | 2.42 | 2.38 |
| | 20 | 4.35 | 3.49 | 3.10 | 2.87 | 2.71 | 2.60 | 2.51 | 2.45 | 2.39 | 2.35 |

*Degrees of Freedom for Denominator*

**Figure 8-5.**    A portion of Appendix Table 7A.

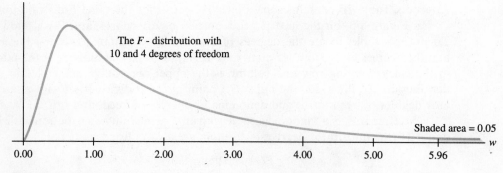

The $F$ - distribution with
10 and 4 degrees of freedom

Shaded area = 0.05

**Figure 8-6.**    $P(W \geq 5.96) = 0.05$.

## The *F*-distribution for testing $H_0$: $\mu_1 = \mu_2 = \cdots = \mu_r$

The $F$ probability distribution to use in a test of equality of $r$ means depends on degrees of freedom associated with $s_b^2$ and $s_p^2$.

### The *F*-Distribution for Testing $H_0$: $\mu_1 = \mu_2 = \cdots = \mu_r$

The $F$-distribution to use in testing the equality of $r$ means has:

1. Numerator degrees of freedom $= r - 1$
2. Denominator degrees of freedom $= n - r$

where $n$ is the total number of data points in all of the $r$ samples.

Notice that the degrees of freedom for the appropriate $F$-distribution are, respectively, the denominators of the formulas for calculating $s_b^2$ and $s_p^2$.

▶ **Example 4**   Allenton Plastics Corporation has four units that manufacture fittings for $\frac{1}{2}$-in. PVC pipe. Periodic samples are taken from those fittings produced by the four units. A critical dimension of the fittings is measured. Table 8-9 contains the data on the last ten fittings inspected from each unit. The units are 0.001 inches above the nominal (blueprint) dimension. Using $\alpha = 0.01$ and the data in Table 8-9, test whether the mean dimension on fittings produced by units $A$, $B$, $C$, and $D$ is the same.

TABLE 8-9   0.001 Inches Above Nominal for a Critical Dimension of Fittings

| Unit *A* | | | | | Unit *B* | | | | | Unit *C* | | | | | Unit *D* | | | | |
|---|---|---|---|---|---|---|---|---|---|---|---|---|---|---|---|---|---|---|---|
| 2 | 3 | 2 | 4 | 0 | 3 | 5 | 4 | 2 | 2 | 1 | 0 | 3 | 3 | 2 | 2 | 2 | 3 | 2 | 1 |
| 4 | 3 | 2 | 1 | 2 | 4 | 3 | 1 | 5 | 3 | 2 | 5 | 4 | 3 | 1 | 3 | 2 | 4 | 2 | 0 |
| $\bar{x}_1 = 2.3$ | | | | | $\bar{x}_2 = 3.2$ | | | | | $\bar{x}_3 = 2.4$ | | | | | $\bar{x}_4 = 2.1$ | | | | |
| $s_1^2 = 1.56\ldots$ | | | | | $s_2^2 = 1.73\ldots$ | | | | | $s_3^2 = 2.26\ldots$ | | | | | $s_4^2 = 1.21\ldots$ | | | | |

**Solution**   **Step 1**   $H_0$: $\mu_1 = \mu_2 = \mu_3 = \mu_4$     The mean dimension on fittings
$H_a$:   not $H_0$     from $A$, $B$, $C$, and $D$ is the same.

**Step 2**   $\alpha = 0.01$

**Step 3**   The test statistic used is:

$$F = \frac{S_b^2}{S_p^2}$$

**Step 4**   If $f > f_c$ ($=4.31$), then reject $H_0$.
(With $n = 40$ and $r = 4$; $r - 1 = 3$ and $n - r = 36$. From Appendix Table 7B, $f_c = 4.31$.)

**Step 5**  $$s_p^2 = \frac{9(1.56\ldots) + \cdots + 9(1.21\ldots)}{40 - 4} = 1.69\ldots$$

$$\bar{x} = \frac{10(2.3) + \cdots + 10(2.1)}{40} = 2.5$$

$$s_b^2 = \frac{10(2.3 - 2.5)^2 + \cdots + 10(2.1 - 2.5)^2}{4 - 1} = 2.33\ldots$$

$$f = \frac{2.33\ldots}{1.69\ldots} = 1.38, \text{ to two decimal places}$$

**Step 6**  Because $1.38 < 4.31$, do not reject $H_0$. That is, there is not sufficient evidence, with $\alpha = 0.01$, to conclude that the means of the dimensions on fittings from $A$, $B$, $C$, and $D$ differ.   ◄

The name analysis of variance (ANOVA) associated with the test of equality of $r$ means is based on the fact that the sample variance of all $n$ observations (calculated ignoring the fact that $r$ different samples are involved), can be broken into two parts. The parts represent variation "within" samples and variation "between" samples as follows:

$$(n - 1)\left(\begin{array}{l}\text{overall sample variance}\\ \text{calculated from } n \text{ observations}\end{array}\right) = (r - 1)s_b^2 + (n - r)s_p^2$$

The calculations of $s_b^2$ and $s_p^2$ (and then $F$, based on them) can thus be thought of in terms of making an **analysis of the raw variability** present in the $n$ observations.

► **Example 5**  Dr. Maureen Kennedy is an economics professor at a large university. She is currently studying the consumer behavior of college students. She sampled from the records of the current student body the names of ten freshmen, five sophomores, seven juniors, eight seniors, and five graduate students. After contacting these students, she obtained copies of their September telephone bills. The data collected yielded the following summaries:

| Class | Sample Size | Sample Mean | Sample Variance |
|-------|-------------|-------------|-----------------|
| Freshman | 10 | $30.20 | 36.00 |
| Sophomore | 5 | 14.20 | 10.24 |
| Junior | 7 | 12.30 | 16.81 |
| Senior | 8 | 15.80 | 14.44 |
| Graduate | 5 | 21.50 | 28.09 |

Test the null hypothesis that the mean September phone bills for the five categories of students are the same. Use $\alpha = 0.01$.

**Solution**   Before starting the hypothesis test, calculate the values for $\bar{x}$, $s_p^2$, and $s_b^2$.

$$\bar{x} = \frac{10(30.20) + 5(14.20) + 7(12.30) + 8(15.80) + 5(21.50)}{35}$$

$$= \frac{693.00}{35} = 19.80$$

$$s_p^2 = \frac{9(36.00) + 4(10.24) + 6(16.81) + 7(14.44) + 4(28.09)}{35 - 5}$$

$$= \frac{679.26}{30} = 22.642$$

$$s_b^2 = \frac{10(30.2 - 19.8)^2 + 5(14.2 - 19.8)^2 + \cdots + 5(21.5 - 19.8)^2}{5 - 1}$$

$$= \frac{1774.60}{4} = 443.65$$

**Step 1**  $H_0$:  $\mu_1 = \mu_2 = \mu_3 = \mu_4 = \mu_5$
   $H_a$:  not $H_0$

**Step 2**  $\alpha = 0.01$

**Step 3**  The test statistic will be:

$$F = \frac{S_b^2}{S_p^2}$$

**Step 4**  If $f > f_c$ ($=4.02$), then reject $H_0$. (The numerator degrees of freedom is $5 - 1 = 4$, and the denominator degrees of freedom is $35 - 5 = 30$. From Appendix Table 7B, $f_c = 4.02$.)

**Step 5**  With $s_b^2 = 443.65$ and $s_p^2 = 22.642$:

$$f = \frac{443.65}{22.642} = 19.59, \text{ to two decimal places}$$

**Step 6**  Because $19.59 > 4.02$, reject $H_0$. That is, there is sufficient evidence, with $\alpha = 0.01$, to conclude that the mean September phone bills for the five categories of students at this university are not all the same.   ◀

## Constructing small sample confidence intervals for $\mu_1 - \mu_2$

As is always the case with testing methods, the ANOVA technique can be useful in answering the question "Is the null hypothesis true?" However, testing is not helpful in answering the important question "By how much do the means of the populations differ?" Beginning with the four assumptions necessary for the ANOVA test, it is possible to show that if $\mu_1$, $\mu_2$, $\bar{X}_1$, and $\bar{X}_2$ are the population and sample means for two of the *r* groups, then

$$T = \frac{(\bar{X}_1 - \bar{X}_2) - (\mu_1 - \mu_2)}{\sqrt{S_p^2}\sqrt{\dfrac{1}{n_1} + \dfrac{1}{n_2}}}$$

has a $t$-distribution with $n - r$ degrees of freedom. This fact yields equations (4) and (5) that can be used to construct confidence intervals for $\mu_1 - \mu_2$. The equations specify interval estimates for $\mu_1 - \mu_2$, but they can obviously also be used for comparing any other pair of means by simply renaming the samples.

## Equations for Confidence Interval Estimates of $\mu_1 - \mu_2$

Under the model assumptions for an ANOVA test of equality of $r$ means, lower and upper confidence limits for the difference in a particular pair of means (say, $\mu_1 - \mu_2$) are:

$$\text{lower limit} = (\bar{x}_1 - \bar{x}_2) - t\sqrt{s_p^2}\sqrt{\frac{1}{n_1} + \frac{1}{n_2}} \tag{4}$$

$$\text{upper limit} = (\bar{x}_1 - \bar{x}_2) + t\sqrt{s_p^2}\sqrt{\frac{1}{n_1} + \frac{1}{n_2}} \tag{5}$$

where $t$ is a percentage point of the $t$-distribution with $df = n - r$.

▶ **Example 6**   Use the data in Table 8-8.

    **a.** Construct a 90% confidence interval for $\mu_4 - \mu_1$.
    **b.** Construct a 95% confidence interval for $\mu_3 - \mu_1$.

**Solution**   **a.** From Example 1, $s_p^2 = 16.90\ldots$. From Table 8-8, $\bar{x}_4 = 25.75$ and $\bar{x}_1 = 20.5$. With $n - r = 48 - 4 = 44$, $t = 1.65$.

$$\text{lower limit} = (25.75 - 20.5) - 1.65\sqrt{16.90\ldots}\sqrt{\frac{1}{12} + \frac{1}{12}}$$

$$= 5.25 - 2.769\ldots$$

$$= 2.5, \text{ to one decimal place}$$

$$\text{upper limit} = 5.25 + 2.769\ldots$$

$$= 8.0, \text{ to one decimal place}$$

Thus a 90 percent confidence interval for $\mu_4 - \mu_1$ is from 2.5 to 8.0 years.
**b.** From Table 8-8, $\bar{x}_3 = 22.0$ and $\bar{x}_1 = 20.5$. For $n - r = 44$, $t = 1.96$.

$$\text{lower limit} = (22.0 - 20.5) - 1.96\sqrt{16.90\ldots}\sqrt{\frac{1}{12} + \frac{1}{12}}$$

$$= 1.5 - 3.29\ldots$$

$$= -1.8, \text{ to one decimal place}$$

$$\text{upper limit} = 1.5 + 3.29\ldots$$

$$= 4.8, \text{ to one decimal place}$$

Thus a 95 percent confidence interval for $\mu_3 - \mu_1$ is from $-1.8$ to $4.8$ years.

◀

The confidence intervals constructed in Example 6 have confidence levels associated only with the intervals individually. That is, in this example, the 90 percent confidence level is applicable only to the interval for $\mu_4 - \mu_1$. Similarly the 95 percent confidence level applies only to the interval for $\mu_3 - \mu_1$. The confidence level for the simultaneous correctness of both intervals is less than the confidence levels of either one of the intervals separately.

There are statistical methods that are slightly beyond the scope of this text that can be used to make many confidence intervals comparing means with a desired *simultaneous confidence*. As fas as what is done here, though, we shall be content to make intervals with *individual confidence coefficients* and be careful to interpret them as such.

## Exercises 8-2   Set A

*In Exercises 1–4, use Appendix Table 7A and B to find values of $f_c$ for tests of*
$H_0: \mu_1 = \mu_2 = \cdots \mu_r$ *for the following values of n, r, and indicated d's.*

**1.** For $r = 4$ and $n = 20$,
   a. $\alpha = 0.05$
   b. $\alpha = 0.01$

**2.** For $r = 8$ and $n = 32$,
   a. $\alpha = 0.05$
   b. $\alpha = 0.01$

**3.** For $r = 5$ and $n = 30$,
   a. $\alpha = 0.05$
   b. $\alpha = 0.01$

**4.** For $r = 10$ and $n = 50$,
   a. $\alpha = 0.05$
   b. $\alpha = 0.01$

*In Exercises 5–8, calculate the following for each set of data. Round each value to, at most, one decimal place. For sample A, identify the statistics as $\bar{x}_1$ and $s_1^2$. For sample B, identify the statistics as $\bar{x}_2$ and $s_2^2$. Similarly identify the other sample statistics.*

a. $\bar{x}$ for each sample     b. $s^2$ for each sample     c. $s_p^2$
d. $\bar{x}$                     e. $s_b^2$                   f. $f$

**5.**

| Sample | | |
|---|---|---|
| A | B | C |
| 3 | 2 | 4 |
| 4 | 5 | 3 |
| 3 | 8 | 10 |
| 6 | 7 | 6 |
| 5 | 6 | 2 |
| 3 | 8 | 5 |

**6.**

| Sample | | | |
|---|---|---|---|
| A | B | C | D |
| 8 | 4 | 9 | 6 |
| 5 | 9 | 14 | 9 |
| 9 | 10 | 10 | 8 |
| 6 | 2 | 10 | 7 |
| 7 | 5 | 7 | 10 |

**7.**

| | Sample | | |
|---|---|---|---|
| A | B | C | D |
| 19 | 13 | 8 | 15 |
| 9 | 20 | 13 | 20 |
| 15 | 8 | 10 | 18 |
| 12 | 21 | 15 | 16 |
| 10 | 16 | 9 | 19 |
| 7 | 12 | 11 | 14 |

**8.**

| | | Sample | | |
|---|---|---|---|---|
| A | B | C | D | E |
| 20 | 17 | 9 | 18 | 10 |
| 30 | 24 | 23 | 14 | 19 |
| 25 | 20 | 15 | 20 | 28 |
| 25 | 19 | 11 | 20 | 23 |
| 22 | 19 | 19 | 17 | 15 |
| 18 | 23 | 21 | 16 | 27 |
| 21 | 18 | 14 | 21 | 25 |

**9.** Three students, Linda, Nancy, and Tom, are given five laboratory rats each for a nutrition experiment. Each rat's weight is recorded in grams. Linda feeds her rats a special diet, formula $A$. Nancy feeds her rats formula $B$. Tom feeds his rats formula $C$. At the end of a specified time period, each rat is again weighed and the net gain is recorded. The results are in the following table.

a. Using $\alpha = 0.05$ and the weight gain data, test the hypothesis that the three formulas produce the same mean weight gain.

b. Construct a 90% confidence interval for $\mu_1 - \mu_2$.

c. Construct a 95% confidence interval for $\mu_2 - \mu_3$.

| Linda's Rats | Nancy's Rats | Tom's Rats |
|---|---|---|
| 43.5 | 48.0 | 51.2 |
| 38.8 | 41.5 | 43.8 |
| 40.2 | 39.8 | 38.6 |
| 44.0 | 46.7 | 45.0 |
| 41.5 | 44.5 | 47.4 |

**10.** An engine, which is used to test the quality of three major brands of gasoline (brand $A$, brand $S$, and brand $T$), is secured to a stationary platform. The quality of the gasoline is determined by the number of seconds the engine operates on a measured quantity of gasoline. The test is conducted six times for each brand of gasoline. The operation times are in the following table.

a. Using $\alpha = 0.01$ and the operation times, test the hypothesis that the three brands produce the same mean operation time.

b. Construct a 98% confidence interval estimate for $\mu_1 - \mu_3$.

c. Construct a 99% confidence interval estimate for $\mu_1 - \mu_2$.

| Brand A | Brand S | Brand T |
|---|---|---|
| 43.6 | 46.4 | 42.5 |
| 49.2 | 48.2 | 39.8 |
| 46.4 | 43.9 | 41.3 |
| 38.9 | 47.5 | 45.0 |
| 44.0 | 46.0 | 38.6 |
| 40.1 | 48.8 | 41.2 |

**11.** Four major brands of dry cell batteries are compared. A laboratory lamp is attached to each battery used in the test, and the time the battery keeps the bulb lit is measured. The time is measured to the nearest one-tenth of 1 minute. The results of the test are in the following table.

a. Using $\alpha = 0.01$ and the times data, test the hypothesis that the four brands have the same mean operation time.

b. Construct a 90% confidence interval for $\mu_1 - \mu_2$.

c. Construct a 98% confidence interval for $\mu_3 - \mu_4$.

| Brand E | Brand D | Brand S | Brand J |
|---|---|---|---|
| 13.8 | 16.2 | 15.5 | 11.5 |
| 11.3 | 14.5 | 13.2 | 13.8 |
| 13.2 | 15.8 | 18.0 | 12.7 |
| 15.1 | 14.9 | 16.2 | 13.0 |
| 14.0 | 17.5 | 12.8 | 14.0 |

**12.** So-called economy model cars from four automobile manufacturers are tested on a standard test track to obtain gas mileage figures. Five different cars from each manufacturer are tested. Each car is driven by the same driver at the same speed under similar conditions. The miles-per-gallon (mpg) figures are in the following table.

   a. Using $\alpha = 0.05$ and the mpg data, test the hypothesis that the four models have the same mean mileage.

   b. Construct a 95% confidence interval for comparing mean mileages for car *G* and car *F*.

   c. Construct a 99% confidence interval for comparing mean mileages for car *C* and car *A*.

| Car G | Car F | Car C | Car A |
|-------|-------|-------|-------|
| 28.5  | 26.5  | 32.5  | 23.5  |
| 30.2  | 27.2  | 34.8  | 28.4  |
| 27.1  | 25.9  | 33.6  | 30.2  |
| 29.4  | 26.8  | 32.9  | 25.6  |
| 27.8  | 27.5  | 31.4  | 27.3  |

**13.** An air traffic controller at a large air force base is interested in knowing whether the average number of military aircraft that contact the base air traffic control tower is the same for each day of the week. She obtains samples of the numbers of contacts made on six Sundays, Wednesdays, and Fridays from the records of the past several years. The results of the survey are in the following table.

   a. Use $\alpha = 0.01$ and the contact data to test the hypothesis that the mean numbers of military aircraft that contact this control tower on Sundays, Wednesdays, and Fridays are the same.

   b. Construct a 95% confidence interval for the difference in the mean numbers of contacts for Sundays and Wednesdays.

   c. Construct a 98% confidence interval for the difference in the mean numbers of contacts for Wednesdays and Fridays.

| Sunday | Wednesday | Friday |
|--------|-----------|--------|
| 183    | 383       | 551    |
| 86     | 325       | 540    |
| 121    | 155       | 395    |
| 74     | 479       | 668    |
| 32     | 183       | 779    |
| 139    | 434       | 416    |

**14.** An agriculture school wants to test the effects of three different fertilizer formulas on the yield of wheat. They select 21 test plots, seven for each fertilizer, in the school's test acreage. They use the same seed and the soils in plots are essentially the same. Thus they feel the primary variable involved in the study is the effectiveness of the fertilizer. The yield data, measured in bushels per test plot, are given in the following table.

   a. Using $\alpha = 0.05$, and the data, test the hypothesis that the mean wheat yields using the three different fertilizer formulas are the same.

   b. Construct a 90% confidence interval for the difference in mean yields for formulas *A* and *B*.

   c. Construct a 95% confidence interval for the difference in mean yields for formulas *B* and *C*.

| Formula A | Formula B | Formula C |
|-----------|-----------|-----------|
| 123.5     | 130.5     | 120.5     |
| 118.9     | 126.4     | 116.8     |
| 121.0     | 129.2     | 123.3     |
| 125.6     | 128.3     | 124.0     |
| 120.3     | 135.0     | 118.3     |
| 124.5     | 127.3     | 119.6     |
| 122.7     | 132.4     | 121.8     |

# Exercises 8-2   Set B

*When the means of two populations are compared based on small samples, and the alternative hypothesis is two sided (that is, $H_a$: $\mu_1 - \mu_2 \neq 0$), then either an F-test or t-test may be used. The connection between the two techniques is that*

$$T^2 = F$$

*The critical values for a non-directional t-test at a given level of significance when squared give the critical value of the F-test, which is a basically two-sided test.*

| For a t-test, the Test Statistic Is: | For an F-test, the Test Statistic Is: |
|---|---|
| $$T = \dfrac{\bar{X}_1 - \bar{X}_2}{\sqrt{S_p^2}\sqrt{\dfrac{1}{n_1} + \dfrac{1}{n_2}}}$$ | $$F = \dfrac{S_b^2}{S_p^2}$$ |
| The degrees of freedom for the appropriate $T$-distribution is $n_1 + n_2 - 2 = n - 2$. | The degrees of freedom for the numerator is $r - 1 = 2 - 1 = 1$, and the degrees of freedom for the denominator is $n - r = n - 2$. |

1. Use the following data from independent random samples from Population 1 with mean $\mu_1$ and Population 2 with mean $\mu_2$:

$$\bar{x}_1 = 10.3 \qquad \bar{x}_2 = 12.7$$
$$s_1^2 = 9.61 \qquad s_2^2 = 26.01$$
$$n_1 = 6 \qquad n_2 = 8$$

Test with $\alpha = 0.05$:

$$H_0: \quad \mu_1 = \mu_2$$
$$H_a: \quad \mu_1 \neq \mu_2$$

a. Use an $F$-test.
b. Use a $t$-test.
c. Verify that the sample value of $F$ is the square of the sample value for $T$.
d. Verify that $f_c$ is the square of $t_c$.

2. Use the following data from independent random samples from Population 1 with mean $\mu_1$ and Population 2 with mean $\mu_2$:

$$\bar{x}_1 = 14.2 \qquad \bar{x}_2 = 13.8$$
$$s_1^2 = 0.020 \qquad s_2^2 = 0.015$$
$$n_1 = 8 \qquad n_2 = 8$$

Test with $\alpha = 0.01$:

$$H_0: \quad \mu_1 = \mu_2$$
$$H_a: \quad \mu_1 \neq \mu_2$$

   a. Use an $F$-test.
   b. Use a $t$-test.
   c. Verify that the sample value of $F$ is the square of the sample value for $T$.
   d. Verify that $f_c$ is the square of $t_c$.

3. What are the model assumptions that must be made whenever an ANOVA test is used?

4. What is $H_0$ for any test using ANOVA?

5. What are the two parameters for the $F$-distributions?

6. If $r$ means are compared and $m$ items are sampled from each population, then the $F$-distribution appropriate to describing the ratio $\dfrac{S_b^2}{S_p^2}$ is the one with:

   a. What degrees of freedom for the numerator?
   b. What degrees of freedom for the denominator?

*Exercises 7 and 8 are suggested student projects. For each project do parts (a) through (g).*
*a. Calculate $\bar{x}$ for each sample.*
*b. Calculate $s^2$ for each sample.*
*c. Calculate $s_p^2$ for the data.*
*d. Calculate $\bar{x}$ for the data.*
*e. Calculate $s_b^2$ for the data.*
*f. Determine f for an ANOVA test on the means of the populations.*
*g. Perform an ANOVA test comparing the means of the populations, including a well-written statement on what the results mean in the context of the problem.*

7. For this ANOVA test, the following populations are defined:

   Population 1   The current freshmen at your college

   Population 2   The current sophomores at your college

   Population 3   The current juniors at your college

   Population 4   The current seniors at your college

Sample ten students from each population and obtain the *number of units* in which each student is currently enrolled.

8. For this ANOVA test, the following populations are defined:

   Population 1   The current freshmen and sophomores at your college

   Population 2   The current juniors and seniors at your college

   Population 3   The current graduate students at your college

Sample 15 students from each population and obtain the *cumulative grade point average* (gpa) of each student. For the graduate students, use only the gpa in graduate school.

## Chapter 8 Summary

### Rules and Equations

a. The test for equality of $r$ distributions has:

$$H_0: \quad P_{11} = P_{21} = \cdots = P_{r1}$$
$$P_{12} = P_{22} = \cdots = P_{r2}$$
$$\vdots \qquad \vdots \qquad\qquad \vdots$$
$$P_{1c} = P_{2c} = \cdots = P_{rc}$$
$$H_a: \quad \text{not } H_0$$

b. The test statistic for a test of equality of $r$ distributions will be:

$$X^2 = \sum \frac{(O_{ij} - E_{ij})^2}{E_{ij}} \quad \text{or} \quad X^2 = \sum \frac{O_{ij}^2}{E_{ij}} - n$$

where $E_{ij} = \dfrac{n_i m_j}{n}$.

c. The test for equality of $r$ means has:

$$H_0: \quad \mu_1 = \mu_2 = \mu_3 = \cdots = \mu_r$$
$$H_a: \quad \text{not } H_0$$

d. The "pooled" estimate of $\sigma^2$ is obtained using the formula:

$$s_p^2 = \frac{\sum (n_i - 1)s_i^2}{n - r}$$

where $n$ is the total number of observations.

e. The "between" variance is obtained using the formula:

$$s_b^2 = \frac{\sum n_i(\bar{x}_i - \bar{x})^2}{r - 1}$$

where

$$\bar{x} = \frac{1}{n} \sum n_i \bar{x}_i$$

f. The test statistic for a test of equality of $r$ means is:

$$F = \frac{S_b^2}{S_p^2}$$

If $H_0$ is true, then $F$ has an $F$-distribution with:
1. Numerator degrees of freedom $= r - 1$
2. Denominator degrees of freedom $= n - r$

g. Equations for confidence interval estimates of $\mu_1 - \mu_2$ are:

$$\text{lower limit} = (\bar{x}_1 - \bar{x}_2) - t\sqrt{s_p^2}\sqrt{\frac{1}{n_1} + \frac{1}{n_2}}$$

$$\text{upper limit} = (\bar{x}_1 - \bar{x}_2) + t\sqrt{s_p^2}\sqrt{\frac{1}{n_1} + \frac{1}{n_2}}$$

where $t$ is a percentage point of the $t$-distribution with $df = n - r$.

## Symbols

A general table for a test of equality of distributions has $r$ rows and $c$ columns.

| | |
|---|---|
| $n_i$ | the number of items in the $i$th row |
| $m_j$ | the number of items in the $j$th column |
| $n$ | the total of the sample sizes |
| $o_{ij}$ | the number of items in the $i$th sample that falls in the $j$th category |
| $\dfrac{o_{ij}}{n}$ | the proportion of the $i$th sample that falls into the $j$th category |
| $P_{ij}$ | the parameter that describes the proportion of population $i$ that belongs to category $j$ |
| $\dfrac{n_i m_j}{n}$ | an estimate for the expected count for the cell that is the intersection of the $i$th row and $j$th column |
| $e_{ij}$ | $\dfrac{n_i m_j}{n}$ |
| $\chi^2$ | the probability distribution that can be used in a test of the equality of $r$ distributions |
| ANOVA | analysis of variance |
| $s_p^2$ | the "pooled" sample variance |
| $\bar{x}$ | $\dfrac{\text{grand total of the observations}}{\text{total number of observations}} = \dfrac{1}{n}\sum n_i \bar{x}_i$ |
| $s_b^2$ | the "between" sample variance |
| $r$ | the number of samples |

# Comparing the $\chi^2$ and ANOVA Tests

A $\chi^2$ test is used to test the equality of proportions of items in $r$ populations that fall into $c$ specified categories of interest. (Qualitative data)

ANOVA is used to test the equality of means in $r$ populations. (Quantitative data)

For both tests obtain independent random samples from the populations under investigation.

Organize counts in a table of $r$ rows and $c$ columns, where $r$ is the number of populations investigated and $c$ is the number of categories of interest.

List the data points in a table of $r$ columns, where $r$ is the number of populations under investigation.

Compute the following for each cell in the table:

$$o_{ij}^2 \quad \text{and} \quad e_{ij}$$

Compute the following for the data in the table:

$$\bar{x}_i, \, s_i^2, \, s_p^2, \, \bar{x}, \, s_b^2$$

Use the statistics determined above, a $\chi_c^2$ from Appendix Table 6, and the six-step hypothesis testing format to complete a $\chi^2$ test of equality of $r$ distributions.

Use the statistics determined above, an $f_c$ from Appendix Table 7, and the six-step hypothesis testing format to complete an ANOVA test of equality of $r$ means.

# Chapter 8 Review Exercises

1. A nationwide survey was taken of citizens in the United States regarding the issue of being "fearful in the home at night." The staff of an animal shelter in a large city wondered whether the proportions of people who are fearful in the home in this city are the same for people with and without one or more dogs in the home. They sample the reactions of 100 people in each of the following populations:

$$A = \{\text{people with at least one dog in the home}\}$$
$$B = \{\text{people with no dogs in the home}\}$$

The results are shown in the following table:

|        |        | Category |      |        |
|--------|--------|----------|------|--------|
|        |        | Yes      | No   | Totals |
| Sample | A      | 15       | 85   | 100    |
|        | B      | 29       | 71   | 100    |
|        | Totals | 44       | 156  | 200    |

   a. For each cell, compute $e_{ij}$.

   b. Compute $x^2$ from the four values of $o_{ij}$ and $e_{ij}$.

   c. Use a $\chi^2$ test of equality of distribution to compare the proportions of people with dogs and those without dogs who are afraid in the home at night. Use $\alpha = 0.01$.

   d. Use the data and estimate the difference between the proportions of people with dogs and those without dogs who are fearful in the home at night, with a 90 percent confidence interval.

2. The mathematics and English departments of a large community college met to discuss plans to coordinate their efforts to offer a comprehensive "basic skills program." The current student body is composed of approximately equal proportions of Asian, black, Hispanic, and white students. The staff wondered whether there were any differences in the proportions of students with these ethnic backgrounds who need basic skills classes. They subsequently sampled assessment scores of students in each of the four populations. Based on the test scores, they rated individual needs for basic skills remediation as "severe," "moderate," or "minor." The results of the samples are given in the following table:

|  |  | Category | | | |
|---|---|---|---|---|---|
|  |  | Severe | Moderate | Minor | Totals |
|  | Asian | 13 | 27 | 40 | 80 |
|  | Black | 20 | 31 | 59 | 110 |
| Sample | Hispanic | 18 | 29 | 48 | 95 |
|  | White | 29 | 33 | 53 | 115 |
|  | Totals | 80 | 120 | 200 | 400 |

   a. For each cell, compute $e_{ij}$.

   b. Compute $x^2$ from the 12 values of $o_{ij}$ and $e_{ij}$.

   c. Use a $\chi^2$ test of equality of distribution to compare the proportions of Asian, black, Hispanic, and white students at this school that exhibit severe, moderate, and minor deficiencies in basic skills. Use $\alpha = 0.05$.

   d. Use the data and estimate the difference between the proportions of Asian and black students at this school who exhibit moderate deficiencies in basic skills.

3. A large western university provides off-campus programs at three military bases near the main campus. To compare the effectiveness of the off-campus programs, five randomly selected graduating students from each location were asked to take a general achievement test. The resulting test scores are given in the following table:

| On Campus | Base M | Base B | Base T |
|---|---|---|---|
| 65 | 96 | 60 | 72 |
| 91 | 72 | 81 | 85 |
| 88 | 80 | 83 | 59 |
| 79 | 63 | 64 | 68 |
| 83 | 75 | 74 | 73 |

Consider the *On Campus* test scores as sample 1, *Base M* test scores as sample 2, *Base B* test scores as sample 3, and *Base T* test scores as sample 4.
   a. Calculate $\bar{x}$ for each sample.
   b. Calculate $s^2$ for each sample.
   c. Calculate $s_p^2$.
   d. Calculate $\bar{\bar{x}}$.
   e. Calculate $s_b^2$.
   f. Calculate $f$.
   g. Using $\alpha = 0.05$ and the test scores, test the hypothesis that the four populations of test scores have equal means.
   h. Construct a 90% confidence interval estimate for $\mu_1 - \mu_2$.
   i. Construct a 95% confidence interval estimate for $\mu_3 - \mu_4$.

4. Four new brake designs ($A$, $B$, $C$, and $D$) were tested seven times each on a new model car prototype. Each test consisted of driving the car at a specified speed, and then measuring the distance (in meters) required to stop. The results of the tests are in the following table:

| Design A | Design B | Design C | Design D |
|---|---|---|---|
| 12 | 9  | 13 | 7 |
| 10 | 10 | 9  | 5 |
| 11 | 10 | 12 | 6 |
| 10 | 11 | 12 | 7 |
| 10 | 9  | 11 | 6 |
| 9  | 8  | 9  | 5 |
| 12 | 11 | 10 | 6 |

   a. Calculate $\bar{x}$ for each sample to one decimal place.
   b. Calculate $s^2$ for each sample to one decimal place.
   c. Calculate $s_p^2$ to two decimal places.
   d. Calculate $\bar{\bar{x}}$ to one decimal place.
   e. Calculate $s_b^2$ to two decimal places.
   f. Calculate $f$ to two decimal places.
   g. Use the data to test, with $\alpha = 0.01$, the claim that the mean braking distances for the four designs are the same.
   h. Construct a 90% confidence interval estimate for $\mu_1 - \mu_4$.
   i. Construct a 95% confidence interval estimate for $\mu_2 - \mu_4$.

## ANOVA

1. *The ANOVA command*

2. *ANOVA and confidence intervals*

### The ANOVA command

After performing one or two ANOVA tests by hand, or even with the aid of a calculator, one rapidly becomes convinced that here, if ever, a computer is needed. The test studied in Section 8-2 is known as the one-way analysis of variance test. The Minitab command used to perform the test is AOVONEWAY abbreviated as AOVO. Let us consider four samples that are in the worksheet as in Figure 8-7:

```
MTB> PRINT C1-C4
ROW    SAMP1    SAMP2    SAMP3    SAMP4
  1    12.6      9.0     10.3      9.6
  2    14.1     14.2      5.9      9.3
  3    10.1      9.9      9.8      8.2
  4    12.7     11.1     10.0     11.3
  5    10.9      9.0     10.9      7.1
  6    11.8      6.9      9.8     14.5
  7     6.8      7.0      9.8     11.3
  8    10.3     10.3      6.7     14.0
  9    15.0      9.1
 10     7.6     10.1
 11              7.8
 12             10.7
```

**Figure 8-7.** Minitab output.

The command

AOVO of data in C1–C4

is given and Minitab responds with the output in Figure 8-8:

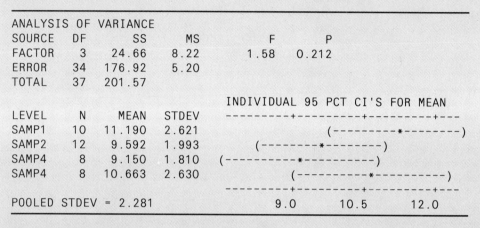

```
ANALYSIS OF VARIANCE
SOURCE   DF       SS       MS         F       P
FACTOR    3    24.66     8.22      1.58    0.212
ERROR    34   176.92     5.20
TOTAL    37   201.57

                                 INDIVIDUAL 95 PCT CI'S FOR MEAN
LEVEL     N     MEAN   STDEV    ---------+---------+---------+---
SAMP1    10   11.190   2.621                   (---------*--------)
SAMP2    12    9.592   1.993            (--------*--------)
SAMP4     8    9.150   1.810      (-----------*----------)
SAMP4     8   10.663   2.630              (----------*----------)
                                 ---------+---------+---------+---
POOLED STDEV = 2.281                  9.0       10.5      12.0
```

Figure 8-8.    Minitab output.

This output comes in four parts:

1. An ANOVA table that includes the ANOVA $F$-test and its $p$-value.
2. A descriptive summary of the various samples.
3. A diagram of 95 percent confidence intervals for the means of the populations using the pooled standard deviation. For each sample the * shows the location of the sample mean and the ( )s show the limits of each confidence interval.
4. The pooled standard deviation.

## ANOVA and confidence intervals

Minitab produces confidence intervals for the means of the sampled populations as a guide to possible further investigation. For instance, the four 95 percent confidence intervals in Figure 8-8 suggest that the means of Populations 2, 3, and 4 may not be statistically distinguishable. There is some hint, however, that the mean of Population 1 may be larger than the mean of either Population 2 or Population 3. The two-sample test of:

$$H_0: \quad \mu_1 = \mu_2$$
$$H_a: \quad \mu_1 > \mu_2$$

in Figure 8-9 shows that, at the 6.6% level, we could conclude that the mean of Population 1 is larger than the mean of Population 2.

```
MTB> TWOSAMPLE C1 C2;
SUBC> ALTE 1.
TWOSAMPLE T FOR SAMP1 VS SAMP2
            N     MEAN    STDEV    SE MEAN
SAMP1      10    11.19     2.62      0.83
SAMP2      12     9.59     1.99      0.58
95 PCT CI FOR MU SAMP1 - MU SAMP2: (-0.54, 3.74)
TTEST MU SAMP1 = MU SAMP2 (VS GT): T = 1.58 P = 0.066 DF = 16
```

**Figure 8-9.** Minitab output.

# Exercises

*Figure 8-10 summarizes a one-way analysis of variance test for data in columns 1 through 4 of the Minitab worksheet.*

**1.** Should the hypothesis of equal means be accepted or rejected at the 5% level?

```
ANALYSIS OF VARIANCE
SOURCE    DF      SS      MS        F       P
FACTOR     3    7.00    2.33     1.38    0.265
ERROR     36   61.00    1.69
TOTAL     39   68.00

                                INDIVIDUAL 95 PCT CI'S FOR MEAN
LEVEL      N    MEAN   STDEV    -----+--------+---------+---------
C1        10   2.300   1.252    (---------*--------)
C2        10   3.200   1.317               (--------*--------)
C3        10   2.400   1.506      (----------*----------)
C4        10   2.100   1.101    (----------*----------)
                                ---------+---------+---------+----
POOLED STDEV = 1.302                  1.60      2.40      3.20
```

**Figure 8-10.** Minitab output.

## SCIENTIFIC
## AND
## GRAPHING

### The $\chi^2$ test for equality of $r$ distributions

The number of multiplications, divisions, and additions needed to calculate $x^2$ for a set of tabled values can be very tedious. To use a calculator to carry out these computations it is helpful to change the formula to a form that uses the row and column totals in the table. The change is explained in the following:

Since

$$x^2 = \sum \frac{o_{ij}^2}{e_{ij}} - n$$

where

$$e_{ij} = \frac{n_i m_j}{n}$$

The formula for $x^2$ can be written as:

$$x^2 = \sum \frac{o_{ij}^2 n}{n_i m_j} - n$$

▶ **Example 1**   Find $x^2$ for the data in the table below.

|  | A | B | Totals |
|---|---|---|---|
| Sample 1 | 30 | 70 | 100 |
| Sample 2 | 30 | 20 | 50 |
| Totals | 60 | 90 | 150 |

**Solution**   To solve this with a scientific calculator, do the following:

| Key Sequence | Display | Comments |
|---|---|---|
| 0 | 0. | |
| MIN | 0. | Clears memory |
| 30 | 30. | $o_{11}$ |
| $x^2$ | 900. | $o_{11}^2$ |
| $x$ | 900. | Indicates multiplication |
| 150 | 150. | $n$ |
| ÷ | 135000. | Indicates division |
| 100 | 100. | $n_1$ |
| ÷ | 1350. | Indicates division |
| 60 | 60. | $m_1$ |
| M+ | 22.5 | Calculates and adds to memory |
| 70 | 70. | $o_{12}$ |
| $x^2$ | 4900. | $o_{12}^2$ |
| $x$ | 4900. | Indicates a multiplication |
| 150 | 150. | $n$ |
| ÷ | 735000. | Indicates division |
| 100 | 100. | $n_1$ |
| ÷ | 7350. | Indicates division |
| 90 | 90. | $m_2$ |
| M+ | 81.66666667 | Calculates and adds to memory |
| 30 | 30. | $o_{21}$ |
| $x^2$ | 900. | $o_{21}^2$ |
| $x$ | 900. | Indicates multiplication |
| 150 | 150. | $n$ |
| ÷ | 135000. | Indicates division |

*table continued on p. 596*

| Key Sequence | Display | Comments |
|---|---|---|
| 50 | 50. | $n_2$ |
| $\div$ | 2700. | Indicates division |
| 60 | 60. | $m_1$ |
| M+ | 45. | Calculates and adds to memory |
| 20 | 20. | $o_{22}$ |
| $x^2$ | 400. | $o_{22}^2$ |
| $\times$ | 400. | Indicates multiplication |
| 150 | 150. | $n$ |
| $\div$ | 60000. | Indicates division |
| 50 | 50. | $n_2$ |
| $\div$ | 1200. | Indicates division |
| 90 | 90. | $m_2$ |
| M+ | 13.33333333 | Calculates and adds to memory |
| MR | 162.5. | Recalls sum from memory |
| $-$ | 162.5 | Indicates subtraction |
| 150 | 150. | $n$ |
| $=$ | 12.5 | The value of $x^2$ |

◀

## Casio Graphics Calculator

The Casio Graphics Calculator can use the same sequence of key strokes as the scientific calculator if the following substitutions are made:

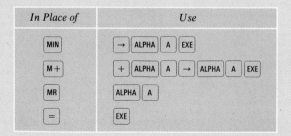

| In Place of | Use |
|---|---|
| MIN | $\rightarrow$ ALPHA A EXE |
| M+ | $+$ ALPHA A $\rightarrow$ ALPHA A EXE |
| MR | ALPHA A |
| $=$ | EXE |

Note that each time the EXE key is pressed, the subtotal found in memory A is displayed.

## Texas Instruments TI-81

The Texas Instruments TI-81 Calculator can use the same sequence of key strokes as the scientific calculator if the following substitutions are made:

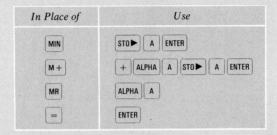

| In Place of | Use |
|---|---|
| MIN | STO▶ A ENTER |
| M+ | + ALPHA A STO▶ A ENTER |
| MR | ALPHA A |
| = | ENTER |

Note that each time the ENTER key is pressed, the subtotal found in memory A is displayed.

**Name:** Kirk Wolter
**Occupation:** Vice President and Director of Statistical Design Worldwide
**Employer:** AC Nielsen Company Northbrook, Illinois

**M**y introduction to statistics came when I was a senior math major at St. Olaf College in Minnesota wondering which courses to take my final year and what to do following graduation. By chance, I enrolled in a two-semester course in statistics, and as that year progressed, I became inspired by the opportunities in statistics for intellectual challenge and practical utility as well as personal enjoyment. This chance decision, made late in my college career, was to reshape the rest of my life.

I attended graduate school at Iowa State University and received as MS and a PhD in statistics. I served as a teaching and research assistant helping students appreciate how statistics could be useful in almost all fields of human endeavor.

My career began in Washington, DC, at the Census Bureau. Contrary to some public opinion, the Census Bureau collects information about America, its people, and its industry evey month of every year, not just once every ten years as part of the population and housing census. One of my most challenging and rewarding areas of work was the development of statistical methods to measure the number of people missed in a census.

After fourteen years, I left the Census Bureau to accept a new position as vice president with the AC Nielsen Company. Nielsen operates in two fundamental businesses, both in the United States and in 27 other countries worldwide. Television ratings is our most widely known business, measuring the number of Americans watching network programming, cable programming, and home video. The second component of our business is market research. We collect data on sales volumes, prices, and promotions from large samples of stores carrying consumer packaged goods. The data help businessmen decide what products to manufacture, how many to make, how to market them, and where to market them. We produce data from a large national sample of households regarding the characteristics of the individual people buying consumer packaged goods.

My current work involves designing samples of stores and households, and estimation procedures for projecting data from a sample to an entire universe of stores or households. My colleagues and I study the relationships between the sales volume of a given product, the price of that product, and the promotional conditions (coupons, in-store displays, and advertising) under which the product is sold. Such relationships are helpful in formulating optimal strategies for marketing products. In addition, I work toward exporting statistical methods and innovations to other Nielsen countries around the world.

Quantitative literacy has become important for all Americans, especially the young who are beginning careers. It is a means to maintaining or improving our competitive position in an emerging global marketplace. Statistics is a rewarding career because it is the essence of both quantitative literacy and scientific method. It teaches how to approach thinking in confronting important national and global issues. It helps formulate the right questions to ask and how to solve or answer them. A good grounding in statistics and the scientific method gives you flexibility to work in many different areas of science, business, and government.

It's one of society's "most blatant prejudices": taller men land better jobs, make more money and get more votes. And short women have their problems, too

Condensed from "THE HEIGHT OF YOUR LIFE"
RALPH KEYES

# The Height Report

SOME YEARS AGO, Leland P. Deck, then the director of personnel at the University of Pittsburgh, was waiting in front of the city's exclusive Duquesne Club. He amused himself by watching members of Pittsburgh's establishment enter the building. One thing about them was striking: their height. "They were," recalls Deck, "uniformly tall."

This observation intrigued Deck so much that he decided to survey a sample of Pitt's graduates and compare height with starting salary. His results were revealing: Among 91 graduates of one class, those under 6' averaged $701 a month in starting salary, followed by 6-footers who were paid $719; 6'1"-ers, $723; and 6'2"-ers, $788. The overall salary bonus for being 6'2" rather than 5'11" was 12 percent. (The bonus for being academically above average was only 4 percent.) Deck's findings, compiled in 1967, were the first real confirmation of something suspected for a long time: The rewards for being tall in this society include money.

RALPH KEYES is 5 feet, 7.62 inches.

**Figure 9-1.** From *The Height of Your Life* by Ralph Keyes. Copyright © 1980 by Ralph Keyes. Reprinted by permission of Sterling Lord Literistic, Inc.

# 9

# Using Bivariate Data

The article in Figure 9-1 appeared in *Reader's Digest* several years ago. The article describes an interesting relationship that apparently exists between the quantitative variables *height* (*x*) and *starting salary* (*y*). (At least such a relationship seems to have held for the graduates at one university during a particular time period.) Researchers are frequently involved in projects in which they try to find evidence that some type of *relationship* exists between two or more variables. The existence of a relationship between variables *x* and *y* means that certain values of *y* are more likely to be paired with particular values of *x*, than other possible values of *y*. Some of the probability models and statistical methods that are used to establish relationships between variables will be studied in this chapter.

# 9-1 Testing Independence—Another Use of $\chi^2$

**Key Topics**

1. *A general version of the two-way table*

2. *The null hypothesis for a test of independence*

3. *Motivation for the testing method*

4. *Two examples of tests of independence*

S T A T I S T I C S   I N   A C T I O N

Cecelia Adams is currently completing training at the New Jersey Highway Patrol Academy prior to becoming a high-way patrol officer. Recently, she studied the topic of moving violations in one of her classes. The issue of how age is related to the numbers of moving violations came up. The instructor, Officer Marie Hall, claimed that the number of moving violations an individual accumulates over a speci-fied period of time depends on the age of the driver.

Cecelia asked Officer Hall what she meant by the claim relating age to number of moving violations. Marie replied that it is normal operating procedure to check the record of any driver who has been stopped for a moving violation. She said her experience has been that the age of a driver is a good indicator of the number of moving violations that will show up on the driver's record.

Cecelia decided to gather evidence to investigate Officer Hall's claim. With the help of a friend at the New Jersey Department of Motor Vehicles, Cecelia obtained a sample of the records of 1000 New Jersey drivers. For each driver in the sample, she recorded the values of 2 variables, namely:

Variable 1   The current age of the driver

Variable 2   The number of moving violations the driver had over the past 3 years

Cecelia put each data point in one of the 12 cells shown in Table 9-1. The age variable is labeled along the left edge of the table and the number of moving violations is labeled along the top.

**TABLE 9-1   Frequencies of Ages and Moving Violations for the Past 3 Years**

| Variable 1 | | *0 Violations* | *1 Violation* | *More Than 1 Violation* | *Totals* |
|---|---|---|---|---|---|
| | *18–25 Years* | 220 | 35 | 25 | 280 |
| | *26–34 Years* | 240 | 70 | 10 | 320 |
| | *35–50 Years* | 215 | 27 | 8 | 250 |
| | *More Than 50 Years* | 136 | 9 | 5 | 150 |
| | *Totals* | 811 | 141 | 48 | 1000 |

*(Variable 2 spans the 0 Violations, 1 Violation, and More Than 1 Violation columns)*

Cecelia then used these data to conduct a $\chi^2$ test of independence of the variables age of driver and number of moving violations. Rejection of $H_0$ would enable her to agree with Officer Hall's claim that the number of moving violations a New Jersey driver accumulated over the past 3 years depends (at least in part) on the age of the driver. Cecelia used the standard six-step format to summarize her hypothesis test. The symbols and procedure used in the following test will be studied in this section.

**Step 1** $H_0$: $P_{ij} = P_{i \cdot} P_{\cdot j}$ for all $i$ and $j$.

$H_a$: not $H_0$.

**Step 2** $\alpha = 0.01$

**Step 3** The test statistic will be: $\quad X^2 = \sum \left( \dfrac{O_{ij}^2}{E_{ij}} \right) - n$

**Step 4** If $x^2 > \chi_c^2$ ($= 16.812$), then reject $H_0$.

(The value of $\chi_c^2$ was obtained using Appendix Table 6 and a distribution with $(4-1)(3-1) = 6$ degrees of freedom.)

**Step 5** $x^2 = \dfrac{48{,}400}{227.08} + \dfrac{57{,}600}{259.52} + \cdots + \dfrac{25}{7.2} - 1000$

$= 213.140 \ldots + \cdots + 3.472 \ldots - 1000$

$= 1041.079 \ldots - 1000$

$= 41.079$, to three decimal places

**Step 6** Because $41.301 > 16.812$, reject $H_0$. That is, there is sufficient evidence, with $\alpha = 0.01$, to conclude that the number of moving violations a New Jersey driver got over the last 3 years is not independent of the driver's age.

## A general version of the two-way table

In the Statistics in Action account in this section, Cecelia Adams established the existence of a relationship between the age and number of moving violations variables for New Jersey drivers. In Step 6 of the hypothesis testing format, this conclusion was phrased in terms of the variables being "dependent" (not independent). Remember that we studied in Chapter 3 the notions of independence and dependence of events $A$ and $B$. In fact, recall that:

If $P(B|A) = P(B)$, then $A$ and $B$ are independent events.

If $P(B|A) \neq P(B)$, then $A$ and $B$ are not independent events.

In the age versus number of moving violations example, let us specify that event $B$ is obtaining the record of a New Jersey driver with more than one moving violation over the past 3 years. Given no other information, we could estimate $P(B)$ using the data in Table 9-1 as:

$$P(B) \approx \frac{48}{1000} \approx 0.5\% \qquad \frac{48 \text{ drivers with more than 1 violation}}{1000 \text{ drivers in the sample}}$$

Let us now specify that event $A$ is obtaining a driver between 18 and 25 years old. Based on the results of the hypothesis test performed by Cecelia Adams, we might suspect that $P(B|A)$ appears to be substantially different from the 0.5% value we estimated for $P(B)$. Our suspicion is verified by the data in Table 9-1.

$$P(B|A) \approx \frac{25}{280} \approx 9\% \qquad \frac{25 \text{ drivers between 18 and 25 with more than 1 violation}}{280 \text{ drivers between 18 and 25 in the sample}}$$

Notice, $P(B|A)$ appears to be about 18 times the size of $P(B)$. As a consequence of this kind of difference, Officer Marie Hall's belief that the driving record of a

New Jersey driver depends on the age of the driver is verified by the $\chi^2$ test of independence. Specifically rejection of $H_0$ enables one to conclude that the variables tested are not independent, but rather that some form of dependence exists.

If a simple random sample is selected from a large population, then the model assumptions needed to test a null hypothesis of independence of two categorical variables are relatively few. One only has to assume that the individual observations are independent. The probability that an observation falls into the $i$th category of the first variable and $j$th category of the second variable is $P_{ij}$. Because the $P_{ij}$ represent probabilities, the sum of all the $P_{ij}$ values must be 1.

To illustrate the meaning of the $P_{ij}$ symbol, the cells in the age versus moving violations example are shown in Table 9-2. The meanings of $P_{11}$ and $P_{42}$, respectively are then:

a. $P_{11}$ stands for the probability that a randomly selected New Jersey driver is between 18 and 25 years old and had no (zero) moving violations over the past 3 years.

b. $P_{42}$ stands for the probability that a randomly selected New Jersey driver is more than 50 years old and had one moving violation over the past 3 years.

**TABLE 9-2**   $P_{ij}$ **for Each Cell of Data in Table 9-1**

|  | Variable 2 | | | |
|---|---|---|---|---|
|  | *0 Violations* | *1 Violation* | *More Than 1 Violation* | *Totals* |
| *18–25 Years* | $P_{11}$ | $P_{12}$ | $P_{13}$ | $P_{1.}$ |
| *26–34 Years* | $P_{21}$ | $P_{22}$ | $P_{23}$ | $P_{2.}$ |
| *35–50 Years* | $P_{31}$ | $P_{32}$ | $P_{33}$ | $P_{3.}$ |
| *More Than 50 Years* | $P_{41}$ | $P_{42}$ | $P_{43}$ | $P_{4.}$ |
| *Totals* | $P_{.1}$ | $P_{.2}$ | $P_{.3}$ |  |

To motivate the use of the $\chi^2$ statistic for a two-way table to test independence, let us recall that notation used to describe the entries in a table for a test of equality of $r$ distributions. This notation is displayed in Table 9-3.

As before:

$$n = n_1 + n_2 + \cdots + n_r \qquad \text{There are } r \text{ rows.}$$

and

$$n = m_1 + m_2 + \cdots + m_c \qquad \text{There are } c \text{ columns.}$$

Notice, however, that unlike the table for a test of equality for $r$ distributions, the row totals $n_i$ are observed values of random variables, rather than fixed sample sizes. That is, the individual row totals will not be known until the entire sample has been cross-categorized.

TABLE 9-3   A General Version of the Two-Way Table

| | | 2nd Variable | | | |
|---|---|---|---|---|---|
| | | Category 1 | Category 2 | $\cdots$ | Category c | Totals |
| 1st Variable | Category 1 | $o_{11}$ | $o_{12}$ | $\cdots$ | $o_{1c}$ | $n_1$ |
| | Category 2 | $o_{21}$ | $o_{22}$ | $\cdots$ | $o_{2c}$ | $n_2$ |
| | $\vdots$ | $\vdots$ | $\vdots$ | | $\vdots$ | $\vdots$ |
| | Category r | $o_{r1}$ | $o_{r2}$ | $\cdots$ | $o_{rc}$ | $n_r$ |
| | Totals | $m_1$ | $m_2$ | $\cdots$ | $m_c$ | $n$ |

## The null hypothesis for a test of independence

The purpose of such a test is to determine whether the two variables are related. It is therefore reasonable to adopt a null hypothesis that the $P_{ij}$'s express *independence of the variables.*

To write such a hypothesis in symbols, the following notation will be used:

$$P_{i\cdot} \text{ will stand for } P_{i1} + P_{i2} + \cdots + P_{ic}$$

(Thus $P_{i\cdot}$ is the *row total* for the population proportions in the $i$th category for the first variable)

$$P_{\cdot j} \text{ will stand for } P_{1j} + P_{2j} + \cdots + P_{rj}$$

(Thus $P_{\cdot j}$ is the *column total* for the population proportions in the $j$th category for the second variable)

Independence means that each $P_{ij}$ is the product of its $P_{i\cdot}$ and $P_{\cdot j}$.
Thus one can use the hypotheses:

$$H_0: \quad P_{ij} = P_{i\cdot}P_{\cdot j} \text{ for all } i \text{ and } j.$$
$$H_a: \quad \text{not } H_0.$$

## Motivation for the testing method

Referring to the general table in Table 9-3, we can make the following observations:

1. $\dfrac{n_i}{n}$ is the fraction of the sample falling into the $i$th category of the first variable.

   This ratio is a natural estimate of $P_{i\cdot}$. Using Table 9-1 to illustrate, the following estimates can be given:

   $$\frac{280}{1000} \approx 28\% \text{ of New Jersey drivers are between 18 and 25 years old.}$$

$$\frac{320}{1000} \approx 32\% \text{ of New Jersey drivers are between 26 and 34 years old.}$$

$$\frac{250}{1000} \approx 25\% \text{ of New Jersey drivers are between 35 and 50 years old.}$$

$$\frac{150}{1000} \approx 15\% \text{ of New Jersey drivers are more than 50 years old.}$$

2. $\frac{m_j}{n}$ is the fraction of the sample falling into the $j$th category of the second variable.

   This ratio is a natural estimate of $P_{.j}$. Using Table 9-1 to illustrate, the following estimates can be given:

$$\frac{811}{1000} \approx 81\% \text{ of New Jersey drivers had no violations the past 3 years.}$$

$$\frac{141}{1000} \approx 14\% \text{ of New Jersey drivers had one violation the past 3 years.}$$

$$\frac{48}{1000} \approx 5\% \text{ of New Jersey drivers had more than one violation.}$$

3. If $H_0$ is true, and $P_{ij} = P_i.P_{.j}$ for all $i$ and $j$, then $\left(\frac{n_i}{n}\right)\left(\frac{m_j}{n}\right)$ is a natural estimate of $P_{ij}$.

4. Finally, if $H_0$ is true, then a fraction of roughly $\left(\frac{n_i}{n}\right)\left(\frac{m_j}{n}\right)$ of the $n$ observations in the sample is expected to fall in the $i$th category of the first variable and the $j$th category of the second variable. That is, if $P_{ij} = P_i.P_{.j}$ for all $i$ and $j$, then one expects to see:

$$o_{ij} \approx n\left(\frac{n_i}{n}\right)\left(\frac{m_j}{n}\right) = \frac{n_i m_j}{n}$$

We will again use $e_{ij}$ to represent $\frac{n_i m_j}{n}$.

Large differences between the $o_{ij}$ and $e_{ij}$ would then seem to be evidence against the possibility that the null hypothesis of independence is true. We have therefore come to the point of comparing $o_{ij}$ and $e_{ij}$ values, which is the same situation we faced in the test of equality of $r$ distributions in Section 8-1 and will use the same test statistic. In fact, we may even use the same critical values.

### The Test Statistic for a Test of Independence

If $H_0: P_{ij} = P_i.P._j$ for all $i$ and $j$ is true and $n$ is large, then it is possible to show that the probability distribution of:

$$X^2 = \sum \frac{(O_{ij} - E_{ij})^2}{E_{ij}} \tag{1}$$

(or equivalently)

$$X^2 = \sum \left(\frac{O_{ij}^2}{E_{ij}}\right) - n \tag{2}$$

is approximately $\chi^2$ with degrees of freedom $(r - 1)(c - 1)$.

Thus the null hypothesis of independence can be tested using exactly the same procedure as that used to test the equality of $r$ distributions.

## Two examples of tests of independence

The following examples illustrate tests of independence.

▶ **Example 1**  The issue of whether to allow drilling for oil in the Santa Barbara Channel has been vigorously debated. Elizabeth Murry, a television reporter, conducted an opinion poll of residents who live along the coast near the debated drilling site. She asked each individual whether the state should intervene in the matter, and how much their yearly income was. The results are shown in Table 9-4. Test the null hypothesis that the variables "income" and "attitude toward state intervention" are independent in this community. Use $\alpha = 0.05$.

TABLE 9-4   Public Opinion Poll Results on State Intervention

|  | *Under $20,000* | *$20,000–$50,000* | *Over $50,000* | *Totals* |
|---|---|---|---|---|
| *State Should Intervene* | 85 | 65 | 20 | 170 |
| *State Should Not Intervene* | 115 | 55 | 65 | 235 |
| *Undecided* | 50 | 30 | 15 | 95 |
| *Totals* | 250 | 150 | 100 | 500 |

**Solution**   **Step 1**  $H_0$:  $P_{ij} = P_i.P._j$ for all $i$ and $j$.

  $H_a$:  not $H_0$.

**Step 2**  $\alpha = 0.05$

**Step 3** The test statistic will be:

$$X^2 = \sum \left( \frac{O_{ij}^2}{E_{ij}} \right) - n$$

**Step 4** If $x^2 > \chi_c^2$ ($=9.488$), then reject $H_0$. (The value of $\chi_c^2$ was obtained using Appendix Table 6 and a distribution with $(3-1)(3-1) = 4$ degrees of freedom.)

**Step 5** $x^2 = \dfrac{7225}{85.0} + \dfrac{13,225}{117.5} + \cdots + \dfrac{225}{19.0} - 500$

$\quad\quad = 85.0 + 112.55\ldots + \cdots + 11.84\ldots - 500$

$\quad\quad = 521.015\ldots - 500$

$\quad\quad = 21.015$, to three decimal places

**Step 6** Because $21.015 > 9.488$ reject $H_0$. That is, there is sufficient evidence, with $\alpha = 0.05$, to conclude that the variables "income" and "attitude toward state intervention" are dependent for residents that live along the Santa Barbara Channel. ◄

▶ **Example 2** Juan Hernandez is a counselor at a southwestern university. He has been supervising a program designed to assist minority students complete degree programs. A recent concern has been the number of units completed each term by students from minority groups. To test whether the number of units a student at this university completed last term is independent of race, Juan took a random sample of 500 student records for last term. The results are shown in Table 9-5. Use $\alpha = 0.01$.

TABLE 9-5   Race and Completed Units Last Term for 500 Students

|  | Less Than 9 Units | Between 9 and 12 Units | Between 12 and 16 Units | More Than 16 Units | Totals |
|---|---|---|---|---|---|
| Asian | 19 | 30 | 57 | 24 | 130 |
| Black | 8 | 15 | 40 | 7 | 70 |
| Hispanic | 10 | 24 | 38 | 13 | 85 |
| White | 26 | 41 | 67 | 16 | 150 |
| Other | 12 | 17 | 28 | 8 | 65 |
| Totals | 75 | 127 | 230 | 68 | 500 |

**Solution** **Step 1** $H_0$:   $P_{ij} = P_i.P._j$ for all $i$ and $j$.
$\quad\quad\quad\quad\quad H_a$:   not $H_0$.

**Step 2** $\alpha = 0.01$

**Step 3** The test statistic used is:

$$X^2 = \sum \left( \frac{O_{ij}^2}{E_{ij}} \right) - n$$

**Step 4** If $x^2 > \chi_c^2 \, (=26.217)$, then reject $H_0$. (The $\chi_c^2$ was obtained using Appendix Table 6 and 12 degrees of freedom.)

**Step 5** $x^2 = \dfrac{361}{19.50} + \dfrac{64}{10.50} + \cdots + \dfrac{64}{8.84} - 500$

$= 9.845$, to three decimal places

**Step 6** Because $9.845 < 26.217$, do not reject $H_0$. That is, there is not sufficient evidence, with $\alpha = 0.01$, to conclude that the number of units completed last term and race are dependent variables at this university.   ◀

# Exercises 9-1   Set A

In Exercises 1–4, use the given values of $n_i$, $m_j$, and $n$ to compute $e_{ij}$ for each cell in the table. Round each value to one decimal place. In these tables, "Cat." is an abbreviation for category.

**1.**

| 1st Variable | | 2nd Variable | | | |
|---|---|---|---|---|---|
| | | Cat. 1 | Cat. 2 | Cat. 3 | Totals |
| | Cat. 1 | | | | 40 |
| | Cat. 2 | | | | 25 |
| | Cat. 3 | | | | 35 |
| | Totals | 20 | 30 | 50 | 100 |

**2.**

| 1st Variable | | 2nd Variable | | | | |
|---|---|---|---|---|---|---|
| | | Cat. 1 | Cat. 2 | Cat. 3 | Cat. 4 | Totals |
| | Cat. 1 | | | | | 80 |
| | Cat. 2 | | | | | 120 |
| | Totals | 60 | 80 | 25 | 35 | 200 |

**3.**

| 1st Variable | | 2nd Variable | | | | |
|---|---|---|---|---|---|---|
| | | Cat. 1 | Cat. 2 | Cat. 3 | Cat. 4 | Totals |
| | Cat. 1 | | | | | 80 |
| | Cat. 2 | | | | | 100 |
| | Cat. 3 | | | | | 70 |
| | Totals | 90 | 50 | 70 | 40 | 250 |

**4.**

| | | 2nd Variable | | | | |
|---|---|---|---|---|---|---|
| | | Cat. 1 | Cat. 2 | Cat. 3 | Cat. 4 | Totals |
| **1st Variable** | Cat. 1 | | | | | 60 |
| | Cat. 2 | | | | | 100 |
| | Cat. 3 | | | | | 90 |
| | Cat. 4 | | | | | 150 |
| | Totals | 80 | 90 | 110 | 120 | 400 |

In Exercises 5–8, use the given values of $o_{ij}$ (upper left corner of cell) and $e_{ij}$ (lower right corner of cell) to perform a test of:

$$H_0: \quad P_{ij} = (P_{i\cdot})(P_{\cdot j}) \text{ for all } i \text{ and } j.$$
$$H_a: \quad \text{not } H_0.$$

**5.** Use $\alpha = 0.05$.

| | | 2nd Variable | | |
|---|---|---|---|---|
| | | Cat. 1 | Cat. 2 | Cat. 1 |
| **1st Variable** | Cat. 1 | 32   36 | 39   32 | 19   22 |
| | Cat. 2 | 30   24 | 17   21 | 13   15 |
| | Cat. 3 | 18   20 | 14   17 | 18   13 |

**6.** Use $\alpha = 0.01$.

| | | 2nd Variable | | |
|---|---|---|---|---|
| | | Cat. 1 | Cat. 2 | Cat. 3 |
| **1st Variable** | Cat. 1 | 32   40 | 28   24 | 20   16 |
| | Cat. 2 | 60   50 | 24   30 | 16   20 |
| | Cat. 3 | 70   65 | 29   39 | 31   26 |
| | Cat. 4 | 38   45 | 39   27 | 13   18 |

**7.** Use $\alpha = 0.01$.

| 1st Variable | | 2nd Variable | | | | |
|---|---|---|---|---|---|---|
| | | Cat. 1 | Cat. 2 | Cat. 3 | Cat. 4 | Cat. 5 |
| | Cat. 1 | 15 / 24 | 27 / 20 | 18 / 12 | 15 / 10 | 5 / 14 |
| | Cat. 2 | 45 / 36 | 23 / 30 | 12 / 18 | 10 / 15 | 30 / 21 |

**8.** Use $\alpha = 0.05$.

| 1st Variable | | 2nd Variable | | | | |
|---|---|---|---|---|---|---|
| | | Cat. 1 | Cat. 2 | Cat. 3 | Cat. 4 | Cat. 5 |
| | Cat. 1 | 19 / 27 | 24 / 30 | 33 / 24 | 38 / 36 | 36 / 33 |
| | Cat. 2 | 15 / 23 | 20 / 26 | 30 / 21 | 35 / 31 | 30 / 29 |
| | Cat. 3 | 28 / 22 | 30 / 24 | 10 / 19 | 32 / 29 | 20 / 26 |
| | Cat. 4 | 28 / 18 | 26 / 20 | 7 / 16 | 15 / 24 | 24 / 22 |

**9.** A survey was taken at a large naval station to obtain the opinions of navy personnel regarding a service career. The marital status of each person interviewed was also recorded. Do the results of the survey, summarized in the following table, show that attitude toward a career in the navy and marital status are dependent? Use $\alpha = 0.01$.

| 1st Variable | | 2nd Variable | | |
|---|---|---|---|---|
| | | Married | Single | Totals |
| | Favors a Career | 89 | 21 | 110 |
| | Against a Career | 83 | 58 | 141 |
| | Undecided | 28 | 21 | 49 |
| | Totals | 200 | 100 | 300 |

**10.** A survey was taken at a large medical facility to obtain people's opinions regarding abortion on request. The sex of each individual surveyed was also noted. Do the results of the survey, given in the following table, show that opinions on abortion in this facility are dependent on the sex of the person surveyed? Use $\alpha = 0.05$.

| | | 2nd Variable | | |
|---|---|---|---|---|
| | | Female | Male | Totals |
| | Favors Abortion | 42 | 22 | 64 |
| 1st Variable | Against Abortion | 20 | 9 | 29 |
| | No Opinion | 8 | 9 | 17 |
| | Totals | 70 | 40 | 110 |

11. The state legislature of a midwestern state passed a bill that would give an incurably ill patient the right to refuse medication for prolonging life. A field project was conducted in the capital city to determine whether the variables age and opinion on the "Right to Die" bill are dependent. Use the table of $o_{ij}$'s to conduct a test of independence at a 1 percent level of significance.

| | | 2nd Variable | | | | |
|---|---|---|---|---|---|---|
| | | Less than 25 Years Old | Between 25 and 39 | Between 40 and 59 | More Than 60 Years Old | Totals |
| | Favors Bill | 88 | 84 | 90 | 68 | 330 |
| 1st Variable | Opposes Bill | 6 | 14 | 18 | 6 | 44 |
| | No Opinion | 6 | 22 | 12 | 26 | 66 |
| | Totals | 100 | 120 | 120 | 100 | 440 |

12. A Threat Analysis Center (TAC) is a collection center for threat messages sent to the U.S. Department of Justice in an eastern state. Each message is rated $A$, $B$, $C$, or $D$ with an $A$ given to the most dangerous and $D$ to the least dangerous threats. Because half the state is politically more liberal than the other half, a test is conducted to determine whether the type of message received is dependent on the area of the state from which it comes. Use the table of $o_{ij}$'s to conduct a test of independence for these variables. Use $\alpha = 0.01$.

| | | 2nd Variable | | |
|---|---|---|---|---|
| | | Liberal Area | Conservative Area | Totals |
| | A Message | 10 | 5 | 15 |
| | B Message | 25 | 20 | 45 |
| 1st Variable | C Message | 25 | 40 | 65 |
| | D Message | 15 | 60 | 75 |
| | Totals | 75 | 125 | 200 |

*In Exercises 13–15, use the information obtained by a graduate student in a large western state. She surveyed 250 faculty members on the nine campuses of the state's university system. The survey was part of a 1-year study that involved face-to-face interviews. The study enabled her to test the independence of several pairs of variables related to college teaching.*

**13.** One portion of the survey divided the instructors on the basis of the variables age and job satisfaction. The results of the survey are summarized in the following table. Test the independence of the variables age and job satisfaction. Use $\alpha = 0.05$.

| | | 2nd Variable | | | |
|---|---|---|---|---|---|
| | | *Very Satisfied* | *Moderately Satisfied* | *Little Satisfaction* | *Totals* |
| | *Less Than 35 Years Old* | 10 | 14 | 8 | 32 |
| **1st Variable** | *Between 35 and 50* | 98 | 54 | 10 | 162 |
| | *More Than 50 Years Old* | 20 | 28 | 8 | 56 |
| | *Totals* | 128 | 96 | 26 | 250 |

**14.** One portion of the survey divided instructors on the basis of the variables, sex and opinion on why most students are going to college. For the opinion variable, the survey offered the following choices:

   *A*   Most students are genuinely trying to learn and better themselves.

   *B*   Most students are appeasing their parents.

   *C*   Most students go to college for lack of something better to do.

   *D*   Most students do not know why they are going to college.

The results are shown in the following table. Test the independence of these variables using a 1 percent level of significance.

| | | 2nd Variable | | |
|---|---|---|---|---|
| | | *Female* | *Male* | *Totals* |
| | *A* | 43 | 109 | 152 |
| | *B* | 19 | 7 | 26 |
| **1st Variable** | *C* | 15 | 17 | 32 |
| | *D* | 13 | 27 | 40 |
| | *Totals* | 90 | 160 | 250 |

**15.** Another portion of the survey divided instructors on the basis of the variables years of college education and attitude toward student qualifications in college. For the attitude variable, the survey offered the following choices:

  *A*   Students are very qualified.

  *B*   Students are moderately qualified.

  *C*   Students are generally lacking in some basic areas.

The results are shown in the following table. Test the independence of these variables using a 5 percent level of significance.

|  | | 2nd Variable | | |
|---|---|---|---|---|
|  |  | *5 to 6 Years* | *7 or More Years* | *Totals* |
| **1st Variable** | *A* | 60 | 57 | 117 |
|  | *B* | 21 | 60 | 81 |
|  | *C* | 9 | 43 | 53 |
|  | *Totals* | 90 | 160 | 250 |

**16.** The director of a state employment agency in a multiethnic region studied the ethnic mix of workers in three job categories. The study used the following categories:

  *A*   Black workers      *X*   Retail sales

  *B*   White workers      *Y*   Blue collar jobs

  *C*   Hispanic workers   *Z*   Service occupations

  *D*   Asian workers

Use the data from this study and test the independence of the variables, "race" and "job category." Use a 5 percent level of significance.

|  | | 2nd Variable | | | |
|---|---|---|---|---|---|
|  |  | *X* | *Y* | *Z* | *Totals* |
| **1st Variable** | *A* | 46 | 72 | 42 | 160 |
|  | *B* | 38 | 57 | 55 | 150 |
|  | *C* | 31 | 67 | 42 | 140 |
|  | *D* | 45 | 54 | 51 | 150 |
|  | *Totals* | 160 | 250 | 190 | 600 |

# Exercises 9-1 Set B

1. Use random numbers to obtain a sample of 100 individuals from Data Set I in Appendix B. Place each individual in a cell of the following table based on the variables, "sex" and "cancer." With $\alpha = 0.05$, test the independence of these variables in this population.

| | | 2nd Variable | | |
|---|---|---|---|---|
| | | Has Cancer | Does Not Have Cancer | Totals |
| 1st Variable | Female | | | |
| | Male | | | |
| | Totals | | | 100 |

2. Use random numbers to obtain a sample of 100 individuals from Data Set I in Appendix B. Place each individual in a cell of the following table based on the variables "smoker" and "cancer." With $\alpha = 0.01$, test the independence of these variables in this population.

| | | 2nd Variable | | |
|---|---|---|---|---|
| | | Has Cancer | Does Not Have Cancer | Totals |
| 1st Variable | Smoker | | | |
| | Not a Smoker | | | |
| | Totals | | | 100 |

*The magnitude of $x^2$ in a test of independence is an indication of the strength of the evidence against $H_0$. It is not an indication of the size of any dependence present in the $P_{ij}$'s. Statisticians have invented many measures of the strength of association in a two-way table. For example, the value of $\dfrac{x^2}{n}$ is sometimes used as an indicator of the strength of association between the two variables. It can be shown that $\dfrac{x^2}{n}$ is an estimate of $\sum \dfrac{[P_{ij} - (P_{i\cdot})(P_{\cdot j})]^2}{(P_{i\cdot})(P_{\cdot j})}$, a quantity that can be viewed as an index of dependence between two categorical variables.*

3. Compute $\dfrac{x^2}{n}$ for Exercise 9, Set A.

4. Compute $\dfrac{x^2}{n}$ for Exercise 10, Set A.

5. Compute $\dfrac{x^2}{n}$ for Exercise 11, Set A.

6. Compute $\dfrac{x^2}{n}$ for Exercise 12, Set A.

*Exercises 7 and 8 are suggested student projects.*

7. Sample the opinions of 75 students. Put each data point in a cell of the following table based on the variables, "sex" and "opinion." The opinion variable is an individual's response to the question:

> "Should the college offer grants for academic excellence equal to the amount of grants the college gives annually for sport scholarships?"

|  | Opinion | | | |
|---|---|---|---|---|
|  | *Favor* | *Oppose* | *No Opinion* | *Totals* |
| *Female* |  |  |  |  |
| *Male* |  |  |  |  |
| *Totals* |  |  |  |  |

**Sex** labels the rows Female, Male, Totals.

Use the data from your sample and test the independence of the variables "Sex" and "Opinion." Use $\alpha = 0.05$.

**8.** Sample the opinions of 100 students. Put each individual in a cell of the following table based on the variables "Class Standing" and "Opinion." The opinion variable is an individual's response to the question:

> "Taking a course from which of the following departments would cause you the most anxiety: mathematics, English, or science?"

|  | Opinion | | | |
|---|---|---|---|---|
|  | *Mathematics* | *English* | *Science* | *Totals* |
| *Freshman* |  |  |  |  |
| *Sophomore* |  |  |  |  |
| *Junior* |  |  |  |  |
| *Senior* |  |  |  |  |
| *Totals* |  |  |  |  |

**Class Standing** labels the rows Freshman, Sophomore, Junior, Senior, Totals.

Use the data from your sample and test the independence of the variables "Class Standing" and "Opinion." Use $\alpha = 0.05$.

# 9-2 Descriptive Statistics for Bivariate Quantitative Data—Correlation and the Least Squares Line

**Key Topics**

1. *Correlation, a measure of strength of linear relationship*

2. *Taking a look at what correlation means*

3. *Some cautions regarding the use of correlation*

4. *The least squares line*

$\mathbb{S}\ \mathbb{T}\ \mathbb{A}\ \mathbb{T}\ \mathbb{I}\ \mathbb{S}\ \mathbb{T}\ \mathbb{I}\ \mathbb{C}\ \mathbb{S}\qquad\mathbb{I}\ \mathbb{N}\qquad\mathbb{A}\ \mathbb{C}\ \mathbb{T}\ \mathbb{I}\ \mathbb{O}\ \mathbb{N}$

Susan Thompson is a dance choreographer who also teaches several dance classes for physical fitness. Students in these classes have frequently asked her how to determine ideal weights they should try to maintain. Susan felt that female and male students needed different predictive models, so she concentrated on first developing one for females.

She first sampled the records of eight women in the advanced dance classes. She felt these students were more likely to have achieved their ideal weights than students in beginning classes. Each woman was the source of the following two items of quantitative data:

Datum number 1   The student's height, measured to the nearest 1 inch

Datum number 2   The student's weight, measured to the nearest 1 pound

Susan listed the eight data pairs in Table 9-6.

**TABLE 9-6   The Heights and Weights of Eight Women Students**

| Individual | Height (inches) | Weight (pounds) |
|---|---|---|
| 1 | 62 | 105 |
| 2 | 66 | 120 |
| 3 | 70 | 156 |
| 4 | 58 | 85 |
| 5 | 64 | 130 |
| 6 | 61 | 112 |
| 7 | 63 | 142 |
| 8 | 68 | 126 |

To get a sense of the relationship between height and weight, Susan graphed the data on a coordinate system like the one in Figure 9-2. The horizontal axis was the height axis and the vertical axis was the weight axis. The coordinates of each point are listed on Figure 9-2 as ordered pairs having the form (height, weight).

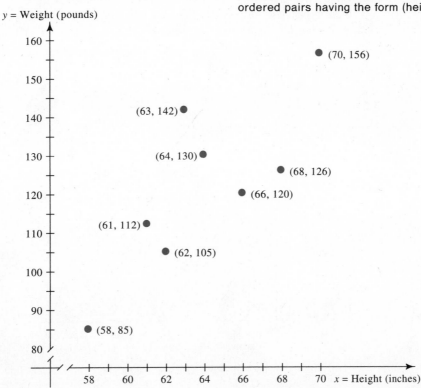

**Figure 9-2.**
A scatterplot of the height-weight data.

The graph, called a *scatterplot*, showed an approximate "straight-line" relationship. Specifically large weight values are paired with large height values and small weight values with small height values. Susan perceived that a straight line could be drawn as a description of the entire data set. She used the data to calculate two descriptive measures. These measures were included in a newsletter that she gave to her students. The following is a portion of that newsletter:

"The ideal-weight formula below was calculated using sample data that yielded a correlation of 0.80. Because 1.00 is perfect correlation, the 0.80 figure is quite good. For more details of what this figure means, please see me personally.

To compute your ideal weight (represented by $y$ in the equation), replace $x$ in the following formula with your height to the nearest 1 inch:

$$y = -169.0 + 4.5x$$

(First multiply your height by 4.5, then subtract 169.0.)"

## Correlation, a measure of strength of linear relationship

The way the points in Figure 9-2 are arrayed suggests that a straight line can be easily drawn through the array of dots that will be "fairly close" to all of the dots. Scatterplots of other sets of paired quantitative data may suggest that other types of curves might be used to "stay close" to the dots, or that no simple curve can be easily drawn through the array of dots. The scatterplots in Figure 9-3 show an apparent *quadratic relationship* (with corresponding equation: $y = ax^2 + bx + c$), a *periodic relationship* (with corresponding equation: $y = a \sin bx$), an *exponential relationship* (with corresponding equation: $y = b^x$), and no apparent relationship.

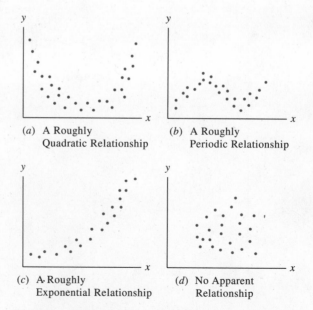

(a)  A Roughly
     Quadratic Relationship

(b)  A Roughly
     Periodic Relationship

(c)  A Roughly
     Exponential Relationship

(d)  No Apparent
     Relationship

**Figure 9-3.**   Four nonlinear scatterplots.

Although an approximate straight-line relationship is not the only one possible, it is the simplest type that will adequately describe many real data sets and is easy to work within statistical settings. It is therefore desirable to have a measure of the strength of straight-line association between two quantitative variables. This measure can then be used to assess how linear a scatterplot appears.

### The Correlation Coefficient, a Measure of the Strength of Linear Association

For a set of bivariate quantitative data

$$\{(x_1, y_1), (x_2, y_2), (x_3, y_3), \ldots, (x_n, y_n)\}$$

the sample correlation is:

$$r = \frac{\sum(x_i - \bar{x})(y_i - \bar{y})}{\sqrt{\sum(x_i - \bar{x})^2 \sum(y_i - \bar{y})^2}} \tag{3}$$

In equation (3), $x_i$ represents the values listed first in each data pair. The $x_i$'s usually stand for the horizontal positions of points on a scatterplot. The $y_i$'s represent the values listed second in each data pair. The $y_i$'s stand for vertical positions of points on a scatterplot.

▶ **Example 1**   Find the correlation between the height and weight values in Table 9-6.

**Solution**   **Discussion.** The calculations necessary to find $r$ are conveniently summarized in Table 9-7. Columns I and II contain the raw data. Columns III and IV contain $(x_i - \bar{x})$'s and $(y_i - \bar{y})$'s, respectively. Column V contains the products of the entries in columns III and IV. Columns VI and VII contain, respectively, the squares of the entries in columns III and IV. From column I,

$$\sum x_i = 512 \quad \text{and} \quad \bar{x} = \frac{512}{8} = 64$$

TABLE 9-7   Calculations for the Height and Weight Data

| | $I$ $x_i$ | $II$ $y_i$ | $III$ $x_i - \bar{x}$ | $IV$ $y_i - \bar{y}$ | $V$ $(x_i - \bar{x})(y_i - \bar{y})$ | $VI$ $(x_i - \bar{x})^2$ | $VII$ $(y_i - \bar{y})^2$ |
|---|---|---|---|---|---|---|---|
| | 62 | 105 | −2 | −17 | 34 | 4 | 289 |
| | 66 | 120 | 2 | −2 | −4 | 4 | 4 |
| | 70 | 156 | 6 | 34 | 204 | 36 | 1156 |
| | 58 | 85 | −6 | −37 | 222 | 36 | 1369 |
| | 64 | 130 | 0 | 8 | 0 | 0 | 64 |
| | 61 | 112 | −3 | −10 | 30 | 9 | 100 |
| | 63 | 142 | −1 | 20 | −20 | 1 | 400 |
| | 68 | 126 | 4 | 4 | 16 | 16 | 16 |
| Totals | 512 | 976 | 0 | 0 | 482 | 106 | 3398 |

From column II,

$$\sum y_i = 976 \quad \text{and} \quad \bar{y} = \frac{976}{8} = 122$$

$$r = \frac{482}{\sqrt{(106)(3398)}}$$

$$r = \frac{\sum(x_i - \bar{x})(y_i - \bar{y})}{\sqrt{\sum(x_i - \bar{x})^2 \sum(y_i - \bar{y})^2}}$$

$$= \frac{482}{600.156\ldots}$$

Do not round the denominator.

$$= 0.8031\ldots$$

The unrounded quotient

$$= 0.80, \text{ to two decimal places}$$

This is the correlation coefficient Susan Thompson reported in the newsletter to her students.  ◄

There are several alternate (but equivalent) formulas for $r$ that may be more convenient to use than equation (3). The following formula will be used in Example 2:

## An Alternate Formula for Computing $r$

$$r = \frac{\sum x_i y_i - \frac{(\sum x_i)(\sum y_i)}{n}}{\sqrt{\left(\sum x_i^2 - \frac{(\sum x_i)^2}{n}\right)\left(\sum y_i^2 - \frac{(\sum y_i)^2}{n}\right)}} \tag{4}$$

## Taking a look at what correlation means

To understand what a value of $r$ means, it is important to know that it is always the case that:

$$-1 \leq r \leq 1$$

Furthermore $r$ is $+1$ or $-1$ only when all the points of a scatterplot fall perfectly on a single straight line. An $r = +1$ indicates a perfect straight-line relationship with all the dots on a line inclined upward to the right, as in Figure 9-4(a). An $r = -1$ indicates a perfect straight-line relationship with all the dots on a line inclined downward to the right, as in Figure 9-4(b). As a consequence, we interpret values of $r$ near $+1$ or $-1$ as indicating a strong linear relationship between $x$ and $y$.

Values of $r$ close to zero are interpreted as indicating a lack of a strong linear relationship between $x$ and $y$. However, such values of $r$ do not necessarily mean that no relationship exists, but only that *no linear relationship exists*. The data that produced the graphs in parts (a), (b), and (d) of Figure 9-3 would all yield an $r$ close to zero, but in Figure 9-3(a) and (b) there is clearly *some* relationship present.

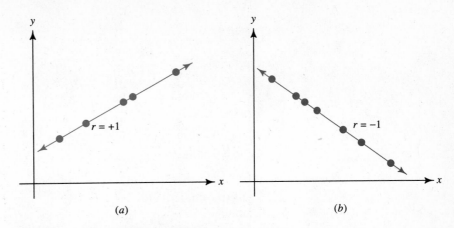

Figure 9-4. Perfect straight-line relationships.

▶ **Example 2** Joan Sanford is a graduate student at a major university in the West. For her masters degree, she studied a species of American shad that spawns only in the Yuba River. With the help of the U.S. Forest Service, she obtained the data in Table 9-8 on the lengths in millimeters (mm) and weights in grams (g) of a sample of 34 fish.

**a.** Make a scatterplot. **b.** Calculate $r$.

TABLE 9-8 Lengths and Weights of 34 American Shad

| Body Length (x) | Weight (y) | Body Length (x) | Weight (y) |
|---|---|---|---|
| 438 | 2128 | 455 | 2464 |
| 438 | 2240 | 458 | 2156 |
| 442 | 2044 | 460 | 2660 |
| 442 | 2212 | 460 | 2086 |
| 442 | 2408 | 460 | 2464 |
| 445 | 2044 | 460 | 2478 |
| 445 | 2380 | 463 | 2156 |
| 447 | 2576 | 465 | 2520 |
| 448 | 2184 | 470 | 2472 |
| 448 | 2072 | 470 | 2520 |
| 450 | 2086 | 470 | 2520 |
| 450 | 2408 | 470 | 2408 |
| 450 | 2380 | 475 | 2520 |
| 450 | 2408 | 475 | 2240 |
| 450 | 2492 | 475 | 2576 |
| 452 | 2380 | 475 | 2408 |
| 455 | 2408 | 480 | 2579 |

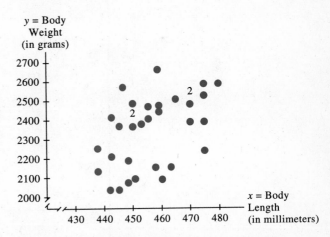

Figure 9-5. A scatterplot of body length versus weight for American shad. A numeral 2 indicates two data points.

**Solution** **a.** Figure 9-5 is a scatterplot of the shad data (shown in Table 9-9 on the next page). Although it does not look overwhelmingly linear, it is typical of real bivariate data sets to which the methods of this section and the next are often successfully applied.

TABLE 9-9   Calculating $r$ for the Shad Data

| | $x_i$ | $y_i$ | $x_iy_i$ | $x_i^2$ | $y_i^2$ |
|---|---|---|---|---|---|
| | 438 | 2128 | 932,064 | 191,844 | 4,528,384 |
| | 438 | 2240 | 981,120 | 191,844 | 5,017,600 |
| | 442 | 2044 | 903,448 | 195,364 | 4,177,936 |
| | 442 | 2212 | 977,704 | 195,364 | 4,892,944 |
| | 442 | 2408 | 1,064,336 | 195,364 | 5,798,464 |
| | 445 | 2044 | 909,580 | 198,025 | 4,177,936 |
| | 445 | 2380 | 1,059,100 | 198,025 | 5,664,400 |
| | 447 | 2576 | 1,151,472 | 199,809 | 6,635,776 |
| | 448 | 2184 | 978,432 | 200,704 | 4,769,856 |
| | 448 | 2072 | 928,256 | 200,704 | 4,293,184 |
| | 450 | 2086 | 938,700 | 202,500 | 4,351,396 |
| | 450 | 2408 | 1,083,600 | 202,500 | 5,798,464 |
| | 450 | 2380 | 1,071,000 | 202,500 | 5,664,400 |
| | 450 | 2408 | 1,083,600 | 202,500 | 5,798,464 |
| | 450 | 2492 | 1,121,400 | 202,500 | 6,210,064 |
| | 452 | 2380 | 1,075,760 | 204,304 | 5,664,400 |
| | 455 | 2408 | 1,095,640 | 207,025 | 5,798,464 |
| | 455 | 2464 | 1,121,120 | 207,025 | 6,071,296 |
| | 458 | 2156 | 987,448 | 209,764 | 4,648,336 |
| | 460 | 2660 | 1,223,600 | 211,600 | 7,075,600 |
| | 460 | 2086 | 959,560 | 211,600 | 4,351,396 |
| | 460 | 2464 | 1,133,440 | 211,600 | 6,071,296 |
| | 460 | 2478 | 1,139,880 | 211,600 | 6,140,484 |
| | 463 | 2156 | 998,228 | 214,369 | 4,648,336 |
| | 465 | 2520 | 1,171,800 | 216,225 | 6,350,400 |
| | 470 | 2472 | 1,161,840 | 220,900 | 6,110,784 |
| | 470 | 2520 | 1,184,400 | 220,900 | 6,350,400 |
| | 470 | 2520 | 1,184,400 | 220,900 | 6,350,400 |
| | 470 | 2408 | 1,131,760 | 220,900 | 5,798,464 |
| | 475 | 2520 | 1,197,000 | 225,625 | 6,350,400 |
| | 475 | 2240 | 1,064,000 | 225,625 | 5,017,600 |
| | 475 | 2576 | 1,223,600 | 225,625 | 6,635,776 |
| | 475 | 2408 | 1,143,800 | 225,625 | 5,798,464 |
| | 480 | 2579 | 1,237,920 | 230,400 | 6,651,241 |
| Totals | 15,533 | 80,077 | 36,619,008 | 7,101,159 | 189,662,805 |

b. Now using equation (4):

$$r = \frac{36,619,008 - \dfrac{(15,533)(80,077)}{34}}{\sqrt{\left(7,101,159 - \dfrac{(15,533)^2}{34}\right)\left(189,662,805 - \dfrac{(80,077)^2}{34}\right)}}$$

$$= \frac{35,595.03}{71,959.936\ldots}$$

$$= 0.4946\ldots$$

$$\approx 0.49, \text{ to two decimal places}$$

This moderate but positive correlation is about what one would expect from looking at the scatterplot.   ◄

## Some cautions regarding the use of correlation

The concept of correlation is one of the most frequently misinterpreted statistical ideas. It is therefore important, before leaving this topic, to add several cautions concerning the use of $r$.

### Caution Number 1

Correlation measures only the degree of *linear relationship* present in a set of $(x, y)$ points. It is quite possible for a perfect (but not linear) relationship to exist between $x$ and $y$ and at the same time to have a correlation near zero.

As illustrations of this caution, see again scatterplots (a) and (b) in Figure 9-3. In Exercises 9-2 Set B, there are four exercises in which $x$ and $y$ values in sets of bivariate data are perfectly related by a mathematical equation. But for each of these sets of data, the correlation between $x$ and $y$ is close to zero.

### Caution Number 2

Even when $r$ indicates that a strong linear relationship exists between $x$ and $y$, *causality* is *not* necessarily established.

If a strong linear relationship is found between two variables, then it is tempting to conclude that one of the variables "causes" the other. A value of $r$ can be used to establish the existence of an apparent relationship between two quantitative variables. But it is not useful in revealing the mechanism or cause of the relationship. In some cases, a third factor causes both $x$ and $y$. In other situations, even after careful consideration, there may be no obvious useful explanation of why certain types of $x$- and $y$-values tend to occur together. This leaves open the possibility that an apparent relationship between two variables is completely spurious.

### Caution Number 3

Correlation is highly sensitive to one or more *extreme data pairs*.

A single "wild" data point can cause $r$ to be close to $+1$ or $-1$ when the corresponding scatterplot indicates "no linear relationship." To illustrate this fact, Figure 9-6 (on the following page) is a scatterplot of the ages and heights of 37 students in a class taught by one of the authors of this text. Notice the circled data point at the upper right of the plot. If $r$ is calculated excluding that data point, then $r = 0.03$. If, however, that data point is included, then $r = 0.73$! The very large difference in the two values of $r$ is caused by one student that happened to be unusually tall and also much older than the other students in the class.

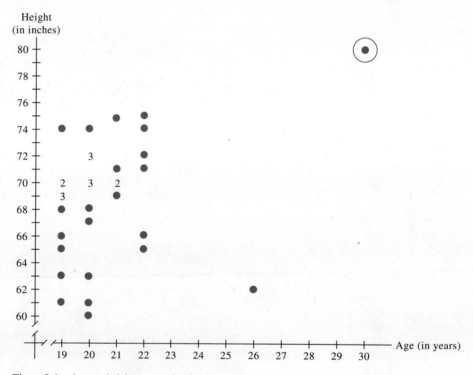

**Figure 9-6.**    An age-height scatterplot for 37 students.

## The least squares line

Suppose a scatterplot of a set of bivariate data and the corresponding value of $r$ indicate that the data might be described as approximately linear. The question that then needs answering is "What line best describes the data?"

In an algebra course, we learn that a line plotted in an $xy$-coordinate system can be described by the *slope* and *y-intercept* of the line. The so-called *slope-intercept equation* of a line has the general form:

$$y = mx + b$$

$m$ is the slope
$b$ is the $y$-intercept

The slope describes the inclination of the line relative to the positively directed $x$-axis. If the line inclines upward to the right, then $m$ is positive. If the line inclines downward to the right, then $m$ is negative. If the line is horizontal, then $m$ is 0. The value of $b$ is the $y$-coordinate of the point where the line intersects the $y$-axis. Thus $(0, b)$ is a point on the line.

## The Slope of a Nonvertical Line

If $(x_1, y_1)$ and $(x_2, y_2)$ are two different points on a nonvertical line, then the slope of the line through the points can be found using equation (5):

$$\text{slope} = \frac{y_2 - y_1}{x_2 - x_1} \qquad \text{slope} = \frac{\text{change in } y}{\text{change in } x} \qquad \textbf{(5)}$$

Four different lines and their slopes and $y$-intercepts are shown in Figure 9-7.

If the slope and $y$-intercept of a line are known, then a graph of the line can easily be drawn.

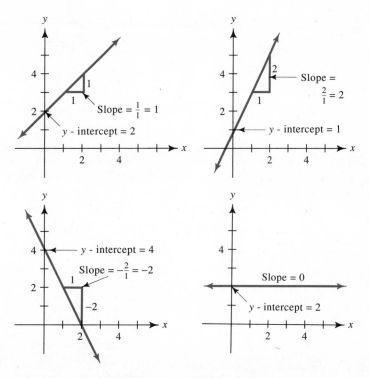

**Figure 9-7.**  Four lines with slopes and $y$-intercepts indicated.

▶**Example 3**    Graph: $y = \dfrac{3}{5}x + 2$

Solution    **Discussion.** The $y$-intercept is 2, therefore $(0, 2)$ is a point on the line. The slope is $\frac{3}{5}$, therefore the vertical rise is 3 for a corresponding horizontal run of 5. The slope is positive, thus the line is inclined upward to the right. A graph of the line is shown in Figure 9-8.    ◀

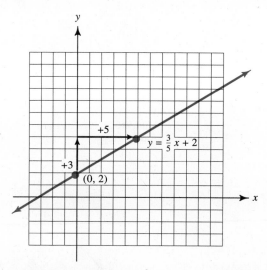

**Figure 9-8.**    A graph of $y = \dfrac{3}{5}x + 2$.

---

### The Slope-Intercept Equation of a Line Using $\beta_0$ and $\beta_1$

If a line has $y$-intercept $\beta_0$ and slope $\beta_1$, then the $(x, y)$ coordinates of every point on the line satisfy:

$$y = \beta_0 + \beta_1 x \tag{6}$$

Notice that equation (6) is the same as the $y = mx + b$ form that we study in algebra. The only differences in the two equations are in the symbols used for the slope and the $y$-intercept. Specifically, $m$ and $\beta_1$ represent the slope, and $b$ and $\beta_0$ represent the $y$-intercept. The notation in equation (6) is the one most commonly used by statisticians and we will use it in this text.

The problem of choosing a line to describe a set of bivariate data amounts to choosing an appropriate $y$-intercept $\beta_0$ and an appropriate slope $\beta_1$. An objective and mathematically convenient way of choosing a line is to apply the principle of least squares.

## The Principle of Least Squares

When choosing among the different possible lines to represent a data set, the line that makes the sum of squared vertical distances from the data points to the line as small as possible should be used.

To illustrate the principle of least squares, Figure 9-9(a) and (b) shows two scatterplots of the following set of paired data:

$$\{(1, 1), (2, 3), (3, 2), (4, 4)\}$$

Also shown are two lines that might be used for describing the scatterplot, and the signed vertical distances from the lines to the points.

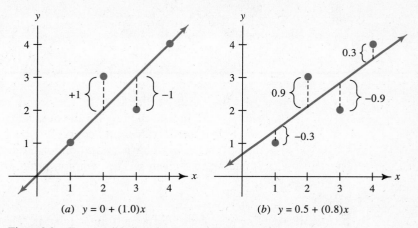

(a) $y = 0 + (1.0)x$  (b) $y = 0.5 + (0.8)x$

**Figure 9-9.** Two possible lines for the set of bivariate data.

The equation of the line in 9(a) is $y = 0 + (1)x$. Two points are on the line. One point is vertically one unit above the line $(+1)$, and the last point is vertically one unit below the line $(-1)$. The sum of the squared vertical distances of the four points from this line is:

$$0^2 + (+1)^2 + (-1)^2 + 0^2 = 2.0 \text{ units}^2$$

The equation of the line in 9(b) is $y = 0.5 + 0.8x$. As shown in the figure, the four points are (from left to right), $(-0.3)$, $(+0.9)$, $(-0.9)$, and $(+0.3)$ units vertically from the line. The sum of the squared vertical distances of the four points from this line is:

$$(-0.3)^2 + (+0.9)^2 + (-0.9)^2 + (0.3)^2 = 1.8 \text{ units}^2$$

Because $1.8 < 2.0$, based on the principle of least squares, the equation in 9(b) is a better description of the data than the one in 9(a).

The mathematical solution to the problem of finding the "best" line according to the least squares criterion is obtained by using calculus. Here we simply state the formulas that result from that application.

### The Least Squares Line

For a set of $(x_i, y_i)$ pairs, the least squares line has:

1. slope $b_1 = \dfrac{\sum(x_i - \bar{x})(y_i - \bar{y})}{\sum(x_i - \bar{x})^2}$    (7)

2. $y$-intercept $b_0 = \bar{y} - b_1\bar{x}$    (8)

An alternate formula for $b_1$ is:

$$b_1 = \frac{\sum x_i y_i - \dfrac{(\sum x_i)(\sum y_i)}{n}}{\sum x_i^2 - \dfrac{(\sum x_i)^2}{n}}$$    (9)

When fitting a line to data, the notations $b_0$ and $b_1$ are used in place of $\beta_0$ and $\beta_1$ used in equation (6). (In the next section we will use $\beta_0$ and $\beta_1$ for population or parameter values and continue to use $b_0$ and $b_1$ for values computed from data.) It is useful to have a notation for the $y$-value on the least squares line corresponding to a given $x$. The common notation for this "predicted" value of $y$ is:

$$\hat{y} = b_0 + b_1 x$$

▶ **Example 4**    For the height and weight data that Ms. Thompson collected in Table 9-10:

**a.** Find the least squares line.
**b.** Draw the least squares line on the scatterplot.

**TABLE 9-10**   Computations for $b_0$ and $b_1$

| | Heights, $x_i$ | Weights, $y_i$ | $x_i y_i$ | $x_i^2$ |
|---|---|---|---|---|
| | 62 | 105 | 6,510 | 3,844 |
| | 66 | 120 | 7,920 | 4,356 |
| | 70 | 156 | 10,920 | 4,900 |
| | 58 | 85 | 4,930 | 3,364 |
| | 64 | 130 | 8,320 | 4,096 |
| | 61 | 112 | 6,832 | 3,721 |
| | 63 | 142 | 8,946 | 3,969 |
| | 68 | 126 | 8,568 | 4,624 |
| *Totals* | 512 | 976 | 62,946 | 32,874 |

**Solution**
$$b_1 = \frac{62{,}946 - \dfrac{(512)(976)}{8}}{32{,}874 - \dfrac{512^2}{8}}$$   Using equation (9)

$$= 4.547\ldots$$

$$\approx 4.5, \text{ to one decimal place}$$

$$b_0 = \frac{976}{8} - (4.547\ldots)\frac{512}{8}$$   $b_0 = \bar{y} - b_1\bar{x}$, and

$$= 122 - 291.018\ldots$$   use unrounded value for $b_1$.

$$= -169.0, \text{ to one decimal place}$$

**a.** Thus the least squares line is, as Susan stated in the newsletter to her students:

$$\hat{y} = -169.0 + 4.5x$$

**b.** The scatterplot and least squares line are both shown in Figure 9-10.

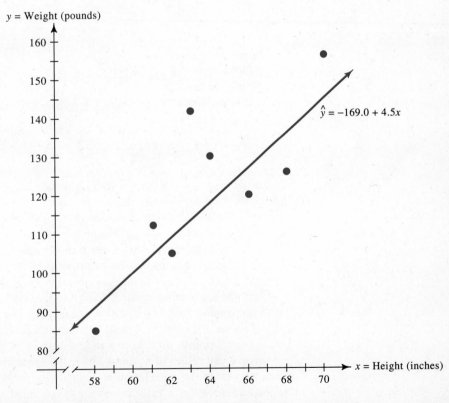

**Figure 9-10.**   A scatterplot and the least squares line for the height-weight data.

If we use the least squares line to describe the scatterplot, then the slope $b_1 = 4.5$ represents an increase of 4.5 pounds in weight for each added inch of height. The $y$-intercept $b_0 = -169.0$ has little practical meaning in and of itself because it represents a weight of $-169.0$ pounds for a woman of height 0 inches!   ◀

▶ **Example 5**   Find the least squares line for the American Shad data.

**Solution**   From Table 9-9:

$$\sum x_i = 15{,}533, \quad \sum y_i = 80{,}077, \quad \sum x_i y_i = 36{,}619{,}008,$$
$$\sum x_i^2 = 7{,}101{,}159, \quad \text{and} \quad n = 34.$$

$$b_1 = \frac{36{,}619{,}008 - \dfrac{(15{,}533)(80{,}077)}{34}}{7{,}101{,}159 - \dfrac{15{,}533^2}{34}}$$

$$= \frac{35{,}595.03}{4{,}862.265}$$

$$= 7.3206 \ldots$$

$$= 7.3 \text{ g/mm, to one decimal place}$$

$$b_0 = \frac{80{,}077}{34} - (7.3206 \ldots)\frac{15{,}533}{34}$$

$$= -989.3 \text{ g, to one decimal place}$$

Thus the least squares line for the length and weight data is:

$$\hat{y} = -989.3 \text{ g} + 7.3x \text{ g}   ◀$$

Today, users of statistics usually have access to computer programs that can find the least squares or *estimated regression line* for bivariate data sets. Even many hand-held electronic calculators are preprogrammed to do the required calculations (as shown in the calculator supplements at the end of this chapter). This easy availability of computational power has helped make describing a bivariate data set with its least squares line quite popular. This popularity makes it important to mention two cautions before leaving this section.

First, **there are no safeguards built into the formulas for $b_0$ and $b_1$ to stop them from being applied to data sets that are nonlinear.** That is, only by using a scatterplot and the value of $r$ habitually and intelligently, can inappropriate use of a line to describe $(x, y)$ points that do not in any way look like a straight line be avoided.

Second, **the values of $b_0$ and $b_1$ have the same kind of sensitivity to "wild" data points we saw in the correlation $r$.** A single extreme data pair is sufficient to give the least squares line a relatively poor fit to an otherwise linear appearing collection of $(x, y)$ pairs. With this in mind, it can at times make sense to develop a least squares line using only the part of a data set that "looks" like a straight line.

## Exercises 9-2 Set A

In Exercises 1–4, for each set of ordered pairs written in the form $(x_i, y_i)$, find the following:

a. $\sum x_i$    b. $\sum x_i^2$    c. $\sum y_i$

d. $\sum y_i^2$    e. $\sum x_i y_i$    f. $r$

1. $\{(-1, -3), (1, 2), (3, 5), (5, 8), (7, 13)\}$

2. $\{(-2, 10), (1, 2), (2, -2), (4, -5), (5, -10), (7, -13)\}$

3. $\{(-2, -4), (1, -1), (4, 1), (5, 1), (6, -1), (8, 2), (10, 3)\}$

4. $\{(1, 1), (2, 5), (3, 11), (4, 11), (5, 12), (6, 18), (7, 18), (3, 20)\}$

In Exercises 5–10, graph each equation.

5. $y = \dfrac{1}{4}x + 5$        6. $y = \dfrac{3}{7}x - 3$

7. $y = \dfrac{-4}{3}x - 2$        8. $y = \dfrac{-3}{8}x + 6$

9. $y = \dfrac{7}{2}x - 4$        10. $y = \dfrac{-6}{5}x + 8$

In Exercises 11–22, do each of the following:

a. Make a scatterplot of the data.

b. Find $r$ for the data.

c. Find the least squares line through the data.

11. Many machines have springs that stretch when a force is applied to them, and then return to their original length when the force is removed. A spring has a "spring constant" that is based in part on the type of metal used to make the spring and the dimensions of the spring. The spring constants of seven 20D94T-type springs made by Collings Spring Manufacturers were measured using test weights $A$ and $B$. The results of the two determinations of the spring constants are given below. The first number in each ordered pair is the spring constant determined using test weight $A$, and the second number is the spring constant of the same spring determined using test weight $B$.

$\{(1.84, 1.65), (1.91, 2.00), (1.84, 1.72), (1.79, 1.70), (1.90, 1.84), (1.73, 1.56), (2.15, 2.41)\}$

12. Reagent 4UD900 was developed to cause a rapid increase in the temperature of water when it is added. In a laboratory, a series of tests were performed in which various quantities of the reagent were added to a fixed quantity of water at approximately the same initial temperature. The temperatures of the water were measured exactly 60 seconds after the reagent was added, and the differences in temperature readings (final − initial) were determined. The amounts of reagent added (in milligrams) and the corresponding changes in temperatures (in degrees Celsius) are listed below.

$\{(27, 57), (45, 64), (41, 80), (19, 46), (35, 62), (39, 72), (19, 52), (49, 77), (15, 57), (31, 68)\}$

13. An inertia welding experiment was performed to establish how the *weld breaking strength* measured in standard units (su) varied with the *velocity* of the rotating workpiece measured in feet per second (fps). In this study, the rotating workpiece was brought to a standstill by forcing it into contact with the stationary part. The heat generated by friction at the interface produced a hot-pressure weld. The data are given in the form (velocity, weld breaking strength).

$$\{(3.3, 89), (4.2, 97), (4.2, 91), (4.6, 98), (5.0, 100), (5.0, 104), (5.0, 97)\}$$

14. Eight volunteers are given a test in manual dexterity, and the times $x$ (measured in minutes) to do the test are recorded. The volunteers are then given "anxiety" tests, and the scores $y$ are noted. The following set of $(x, y)$ pairs shows the results of the two tests.

$$\{(1.0, 14), (1.4, 13), (2.0, 9), (2.0, 6), (3.6, 3), (4.2, 0), (4.2, -2), (4.5, 1)\}$$

15. The silicone based compound SS707 was developed specifically to reduce the friction on surfaces of planks made of 2-in. by 6-in. white pine boards. A test was performed in which a fixed force was applied to a 2-kilogram disk, propelling it over such a surface. Between each test, the surface was given another application of SS707 and the distances (in meters) the disk traveled with each new application were measured. The results are given below.

| x-Number of Applications | 1 | 2 | 3 | 4 | 5 | 6 | 7 | 8 |
|---|---|---|---|---|---|---|---|---|
| y-Distance Disk Traveled (m) | 0.58 | 0.67 | 0.88 | 1.20 | 1.45 | 1.64 | 1.70 | 1.74 |

16. Helen Petrowski keeps records for the staff that manages a large area of national forest land in Montana. She recently compiled some data on annual rainfall and corresponding number of forest fires for that area for 8 years. The following set of $(x, y)$ ordered pairs shows the annual rainfalls $x$ (in inches) and numbers of forest fires for the year $y$.

$$\{(65, 23), (83, 12), (70, 21), (48, 30), (89, 10), (70, 18), (80, 10), (55, 25)\}$$

17. A recent study suggests that 42 percent of all Americans consider themselves shy. A test was developed by two experimental psychologists to measure the shyness of volunteers. The test was used to determine whether a linear relationship exists between the order of birth $x$ of an individual (first born, second born, and so on) and the individual's shyness. The following set of $(x, y)$ ordered pairs shows the birth positions $x$ of nine volunteers with their scores $y$ on the shyness test.

$$\{(1, 5), (1, 10), (2, 3), (2, 12), (2, 8), (3, 5), (3, 13), (4, 8), (6, 12)\}$$

18. According to a report published by the Federal Energy Administration, proper insulation can increase temperature control by as much as 20 to 30 percent. A study was conducted to determine the strength of linear relationship between the thickness of insulation in walls $x$ (measured in inches) and the corresponding increase in temperature control $y$ (measured as a fraction). The following set of $(x, y)$ ordered pairs is from a laboratory study of the relationship of $y$ to $x$.

$$\{(1.0, 0.02), (2.0, 0.05), (3.0, 0.08), (4.0, 0.14), (5.0, 0.24), (6.0, 0.28)\}$$

*Exercises 19 and 20 contain data from a study of stomata. Stomata are small pores in the epidermis of a plant. Water vapor is diffused to the atmosphere and carbon dioxide is taken in through the stomata. A study was conducted to determine whether a linear relationship exists between the rates of water vapor diffusion (measured in grams per unit of time) through stomata and either the areas of the stomata (measured in square millimeters) or the circumferences (measured in millimeters).*

**19.** The following set of $(x, y)$ pairs gives the results of the test for eight stomata, in which $x$ is the area and $y$ is the amount of water vapor diffused through a stoma during the test period:

$$\{(5.5, 2.7), (2.0, 1.6), (0.7, 0.9), (0.5, 0.8), (0.4, 0.7), (0.3, 0.5), (0.2, 0.5), (0.1, 0.4)\}$$

**20.** The following set of $(x, y)$ pairs gives the results of the test for the eight stomata used in Exercise 19, and now $x$ is the circumference and $y$ is the amount of water vapor diffused through a stoma during the test period:

$$\{(8.3, 2.7), (5.0, 1.6), (3.0, 0.9), (2.5, 0.8), (2.3, 0.7), (1.8, 0.5), (1.5, 0.5), (1.1, 0.4)\}$$

**21.** The following set of $(x, y)$ pairs shows the total losses in weight $y$ (measured in pounds) for time intervals $x$ (measured in weeks) during a particular weight loss program:

$$\{(1, 2.5), (2, 3.0), (3, 4.5), (4, 7.2), (5, 10.5), (6, 12.5)\}$$

**22.** At the same time of day, and on the same latitude, the temperatures are measured at six different altitudes in the Sierra mountain range. The following set of $(x, y)$ pairs shows the altitudes $x$ (measured in thousands of feet) and the temperatures $y$ (measured in degrees Celsius):

$$\{(1, 13), (2, 12), (4, 10), (6, 6), (8, 2), (9, -4)\}$$

## Exercises 9-2   Set B

*In Exercises 1–4, for each set of ordered pairs written in the form $(x_i, y_i)$:*
a. *Make a scatterplot of the data.*
b. *Compute r.*
c. *Verify that each value of y can be found using the given equation and the corresponding value of x.*

**1.** $\{(-3, 7), (-2, 2), (-1, -1), (0, -2), (1, -1), (2, 2), (3, 7)\}$; $y = x^2 - 2$

**2.** $\{(-2, 16), (-1, 1), (0, 0), (1, 1), (2, 16)\}$; $y = x^4$

**3.** $\{(0, 0), (1, 5), (2, 8), (3, 9), (4, 8), (5, 5), (6, 0)\}$; $y = 6x - x^2$

**4.** $\{(0, 0), (1, 7), (2, 12), (3, 15), (4, 16), (5, 15), (6, 12), (7, 7), (8, 0)\}$; $y = 8x - x^2$

*In Exercises 5–8, for each exercise:*

a. *For equations (A) and (B), find the value of y for each x appearing in one of the ordered pairs.*

b. *For equations (A) and (B), compute the differences between the actual values of y in the ordered pairs and the predicted values found in part (a).*

c. *Find the squares of the differences computed in part (b) and sum the squares separately for equations (A) and (B).*

d. *Which equation, (A) or (B), is the better choice to represent the data, based on a comparison of the sum of the squares of the differences?*

**5.** $\{(1, 1), (2, 3), (3, 3), (4, 5)\}$

         (A)    $y = 0 + 1.2x$

         (B)    $y = 0.5 + 1.3x$

**6.** $\{(1, 6), (2, 2), (3, 4), (5, 1)\}$

         (A)    $y = 7.2 - 1.4x$

         (B)    $y = 6.5 - 1.3x$

**7.** $\{(1, 8), (2, 5), (3, 0), (4, 1)\}$

         (A)    $y = 9.5 - 2.5x$

         (B)    $y = 10 - 2.6x$

**8.** $\{(1, -2), (2, 3), (3, 3), (4, 6)\}$

         (A)    $y = -3.5 + 2.4x$

         (B)    $y = -3.6 + 2.5x$

**9.** Using random numbers, sample $n = 10$ stocks from Data Set II, Appendix B. Let $x$ be the *highest stock price* for the past 52 weeks and $y$ be the corresponding *lowest stock price*. For the ten pairs of data:

     a. Make a scatterplot of the paired data.

     b. Compute $r$.

     c. Find the equation of the least squares line.

**10.** Using another set of ten random numbers, repeat the tasks described in Exercise 9.

## 9-3 Inferences Using the Least Squares Line

**Key Topics**

1. The probability model used in simple linear regression analysis

2. Using $b_0$ and $b_1$ as point estimates for $\beta_0$ and $\beta_1$

3. Using $s_e^2$ as a point estimate for $\sigma^2$

4. Confidence interval estimates for $\beta_1$

5. Testing $H_0: \beta_1 = 0$

6. Point estimates for $\mu_{y|x*}$

7. Confidence interval estimates for $\mu_{y|x*}$

8. Testing $H_0: \mu_{y|x*} = \mu_0$

9. Predicting the next value of $y$

---

### STATISTICS IN ACTION

Geraldine Smith, Juanita Rivera, and Tami Fong are three students in a modern dance class conducted by Susan Thompson. Each woman is 65 inches tall. The regression equation in Susan's newsletter was:

$$\hat{y} = -169.0 + 4.5x$$

If $x$ is replaced by 65 inches:

$$\hat{y} = -169.0 \text{ pounds} + 4.5 \frac{\text{pounds}}{\text{inch}} (65 \text{ inches})$$

$$\approx 124 \text{ pounds}$$

Geraldine, Juanita, and Tami approached Susan to discuss the fact that each one weighed something differ-

ent from the predicted weight. At the morning weigh-in session, Geraldine weighed 130 pounds, Juanita weighed 126 pounds, and Tami weighed 116 pounds. They wondered whether the deviations in weight from 124 pounds indicated a need to get their weights closer to the predicted weight.

Ms. Thompson had anticipated such questions, and had contacted Nick Millstein, a biometrician, for assistance in answering them. Nick developed a table of prediction intervals for various values of $x$. He used a prediction confidence coefficient of 90 percent for each one. Susan showed Geraldine, Juanita, and Tami the 90 percent prediction interval for $x = 65$ inches.

For $x = 65$ inches: 94.8 pounds to 153.2 pounds.

The weights of all three women were easily within the limits of this interval. Susan therefore suggested that no immediate changes be made in their weight maintenance programs, but that each could meet with her individually to look at factors other than height that affect weight. The details of how to construct such prediction intervals, and other types of inference based on a regression line, are discussed in this section.

## The probability model used in simple linear regression analysis

Before beginning a discussion on the types of inferences that can be based on $n$ data pairs $(x_1, y_1), (x_2, y_2), (x_3, y_3), \ldots, (x_n, y_n)$, we first need to introduce the probability model that is used in simple linear regression analysis. (For reasons that are not so obvious, inference based on equations fitted using the least squares principle is called "regression" analysis.)

The simple linear regression model is intended to express an *approximately linear relation* between the variables $x$ and $y$, where any deviation from a perfect straight-line relationship is to be attributed to chance or random variation. In the data pairs $(x_i, y_i)$, we think of the $x$-values as *fixed* or *known*. The $y$-values are then described as *observed values of normal random variables* with the same (unknown) standard deviation $\sigma$.

For each $Y_i$, we will suppose the mean of the probability distribution of $Y_i$ is $\beta_0 + \beta_1 x_i$, where $x_i$ is the corresponding $i$th value of $x$. The values of the parameters $\beta_1$, and $\beta_0$ will also be unknown. These model assumptions are often expressed in compact form as equation (10).

### Model Assumptions for Simple Linear Regression Analysis

$$Y_i = \beta_0 + \beta_1 x_i + \epsilon_i \qquad (10)$$

Read $\epsilon_i$ as "epsilon sub $i$." The $\epsilon_i$'s are *independent normal random variables* with mean 0 and constant variance $\sigma^2$.

In equation (10), the $\epsilon_i$'s are the deviations of the corresponding $Y_i$ above or below the line $y = \beta_0 + \beta_1 x$. That is, for each $x_i$, the corresponding observed $y_i$ can vary from ordered pair to ordered pair. The amount by which a given $y_i$ deviates from the mean $\beta_0 + \beta_1 x_i$ is the value taken on by $\epsilon_i$. For a fixed $x$, $\epsilon$ measures the deviation of $Y$ from its mean.

Figure 9-11(a) and (b) show two pictures that may be helpful in understanding the simple linear regression model. In Figure 9-11(a), $x_1$, $x_2$, $x_3$, and $x_4$ are four known values of $x$. The bell-shaped curves are each centered on the regression line $y = \beta_0 + \beta_1 x$. (Ideally the curves should be above the page of the text to demonstrate the intended three dimensional nature of the graph.) Thus we view the possible $y$-values for a particular $x$ as producing a distribution with mean $\mu_{y|x} = \beta_0 + \beta_1 x$ and variance $\sigma^2$. Read $\mu_{y|x}$ as "the mean $y$ for a given $x$."

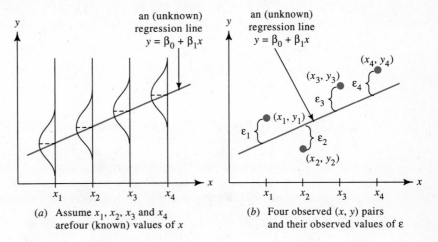

(a)  Assume $x_1$, $x_2$, $x_3$ and $x_4$ arefour (known) values of $x$

(b)  Four observed $(x, y)$ pairs and their observed values of $\varepsilon$

**Figure 9-11.**   A regression line with four known values of $x$ indicated.

In Figure 9-11(b), four observed values are plotted, one for each of the values of $x$. The points are plotted off the regression line by an amount labeled $\epsilon$. For example, $y_1$ is above the regression line by an amount $\epsilon_1$. Thus:

$$y_1 = \beta_0 + \beta_1 x_1 + \epsilon_1 \quad \text{for} \quad \epsilon_1 > 0 \qquad (y_1 \text{ is } above \text{ the line.})$$

Similarly $\quad y_2 = \beta_0 + \beta_1 x_2 + \epsilon_2 \quad \text{for} \quad \epsilon_2 < 0 \qquad (y_2 \text{ is } below \text{ the line.})$

Similarly $\quad y_3 = \beta_0 + \beta_1 x_3 + \epsilon_3 \quad \text{for} \quad \epsilon_3 > 0 \qquad (y_3 \text{ is } above \text{ the line.})$

Similarly $\quad y_4 = \beta_0 + \beta_1 x_4 + \epsilon_4 \quad \text{for} \quad \epsilon_4 > 0 \qquad (y_4 \text{ is } above \text{ the line.})$

### The Parameters for Simple Linear Regression Model

$\beta_0$ is the unknown *y-intercept* of the regression line.
$\beta_1$ is the unknown *slope* of the regression line.
$\sigma^2$ is the unknown *variance* of the $y$-values about the regression line, for any given $x$.

## Using $b_0$ and $b_1$ as point estimates for $\beta_0$ and $\beta_1$

Once the simple linear regression model is adopted as an appropriate description of a set of $n$ data pairs, it is important to be able to estimate the unknown parameters of the model. The descriptive measures from the previous section can be used to make such inferences. The point estimates of $\beta_0$ and $\beta_1$ are, respectively, $b_0$ and $b_1$, computed from the observed data pairs.

▶ **Example 1**     Assuming the simple linear regression model for the height-weight data of Table 9-6 in Section 9-2, estimate $\beta_0$ and $\beta_1$.

Solution     From Example 4 of Section 9-2, the least squares line is:

$$\hat{y} = -169.0 + 4.5x$$

The point estimate for $\beta_0$ is $b_0 = -169.0$. The point estimate for $\beta_1$ is $b_1 = 4.5$.     ◀

## Using $s_e^2$ as a point estimate for $\sigma^2$

In Chapter 5, a sample variance $s^2$ from a single sample was used as a point estimate for $\sigma^2$. In a similar way, a kind of sample variance based on observed data pairs can be used as a point estimate for $\sigma^2$ in the simple linear regression model. Remember that $\sigma^2$ measures the variability of the $y$-values about the line

$$y = \beta_0 + \beta_1 x$$

One can estimate each value $\beta_0 + \beta_1 x$ by

$$\hat{y} = b_0 + b_1 x$$

and use differences between observed $y$-values and $\hat{y}$-values to make an estimate of $\sigma^2$. The symbol $s_e^2$ is used to stand for a kind of sample variance based on the set of $y_i$'s and corresponding $\hat{y}_i$'s. The subscript $e$ stands for "error" and refers to the fact that $s_e^2$ *estimates* $\sigma^2$, the variance of the "errors" $\epsilon_i$. The formula for computing $s_e^2$ is equation (11), with equation (12) offered as an equivalent but computationally more efficient formula.

---

### Equations for Calculating $s_e^2$

A point estimate for $\sigma^2$ in the simple linear regression model is $s_e^2$, where:

$$s_e^2 = \frac{\sum(y_i - \hat{y}_i)^2}{n - 2}$$

or:

$$s_e^2 = \frac{1}{n - 2}\left[\sum y_i^2 - \frac{(\sum y_i)^2}{n} - b_1^2\left(\sum x_i^2 - \frac{(\sum x_i)^2}{n}\right)\right] \qquad \textbf{(12)}$$

---

▶ **Example 2**     Estimate $\sigma^2$ to the nearest whole number for the simple linear regression model and the height-weight data in Table 9-6.

Solution     From Example 4 of Section 9-2, the unrounded value of $b_1$ is $\dfrac{482}{106} = 4.547\ldots$.

The other values needed for use of equation (12) are given in Table 9-11.

TABLE 9-11   Calculations for $s_e^2$

| $x_i$ | $y_i$ | $x_i^2$ | $y_i^2$ |
|---|---|---|---|
| 62 | 105 | 3,844 | 11,025 |
| 66 | 120 | 4,356 | 14,400 |
| 70 | 156 | 4,900 | 24,336 |
| 58 | 85 | 3,364 | 7,225 |
| 64 | 130 | 4,096 | 16,900 |
| 61 | 112 | 3,721 | 12,544 |
| 63 | 142 | 3,969 | 20,164 |
| 68 | 126 | 4,624 | 15,876 |
| *Totals* 512 | 976 | 32,874 | 122,470 |

Now using equation (12):

$$s_e^2 = \frac{1}{8-2}\left[122{,}470 - \frac{976^2}{8} - \left(\frac{482}{106}\right)^2\left(32{,}874 - \frac{512^2}{8}\right)\right]$$

$$= \frac{1}{6}(122{,}470 - 119{,}072 - 2{,}191.7\ldots)$$

$$= 201.044\ldots$$

$$= 201, \text{ to the nearest whole number.} \quad \blacktriangleleft$$

▶ **Example 3**   Assuming the simple linear regression model for the American shad length-weight data of Table 9-8 in Section 9-2:

**a.** Estimate $\beta_0$ and $\beta_1$ to one decimal place.
**b.** Estimate $\sigma^2$ to the nearest whole number.

**Solution**   The following values are in Table 9-9 and the values of $b_0$ and $b_1$ are given in Example 5 of Section 9-2:

$$\sum x_i = 15{,}533, \quad \sum y_i = 80{,}077, \quad \sum x_i^2 = 7{,}101{,}159, \quad \sum y_i^2 = 189{,}662{,}805, \quad n = 34,$$
$$b_0 = -989.26\ldots, \quad \text{and} \quad b_1 = 7.3206\ldots$$

**a.** To one decimal place, $\beta_0$ is estimated to be $-989.3$. To one decimal place, $\beta_1$ is estimated to be 7.3.

**b.** $s_e^2 = \dfrac{1}{34-2}\left[189{,}662{,}805 - \dfrac{(80{,}077)^2}{34} - (7.32\ldots)^2\left(7{,}101{,}159 - \dfrac{(15{,}533)^2}{34}\right)\right]$

$$= \frac{1}{32}\left[189{,}662{,}805 - (188{,}597{,}821.4\ldots) - (260{,}579.41\ldots)\right]$$

$$= 25{,}137.63\ldots$$

$$= 25{,}138$$

$\sigma^2$ is estimated to be 25,138. This means, of course, that the standard deviation of weights of American shad of fixed length is estimated to be

$$s_e = \sqrt{s_e^2} = \sqrt{25{,}137.63\ldots} \approx 159 \text{ g.} \quad \blacktriangleleft$$

## Confidence interval estimates for $\beta_1$

If the simple linear regression model is adopted as an appropriate description of a set of $n$ data pairs, then it is the case that:

$$T = \frac{B_1 - \beta_1}{\sqrt{\dfrac{S_e^2}{\sum(x_i - \bar{x})^2}}} \tag{13}$$

has a $t$-distribution with $n - 2$ degrees of freedom. (The slope $\beta_1$ is in fact the mean of the probability distribution of $B_1$. Furthermore the expression in the denominator of the right side of equation (13) is an estimate of the standard deviation of the probability distribution of $B_1$.) This fact leads to the following confidence limits for $\beta_1$.

### Equations to Construct Confidence Intervals for $\beta_1$

Confidence limits for $\beta_1$ are:

$$\text{lower limit} = b_1 - t\sqrt{\frac{s_e^2}{\sum(x_i - \bar{x})^2}} \tag{14}$$

and

$$\text{upper limit} = b_1 + t\sqrt{\frac{s_e^2}{\sum(x_i - \bar{x})^2}} \tag{15}$$

where $t$ is an upper percentage point of the $t$-distribution with $n - 2$ degrees of freedom. Remember that $\sum(x_i - \bar{x})^2$ can be computed as $\sum x_i^2 - \dfrac{(\sum x_i)^2}{n}$.

▶ **Example 4**     For the height-weight data from Ms. Thompson's class, construct a 90 percent confidence interval for the increase in mean weight that accompanies a 1-inch increase in height.

**Solution**     From Example 4 of Section 9-2:

$$b_1 = \frac{482}{106} = 4.547\ldots$$

$$\sum x_i = 512 \quad \text{and} \quad \sum x_i^2 = 32{,}874$$

From Example 2 of this section:

$$s_e^2 = \frac{1206.2642}{6} = 201.044\ldots$$

With $n = 8$, $n - 2 = 6$, and Appendix Table 5 shows that $t_{0.050} = 1.943$. Thus,

$$\text{lower limit} = (4.547\ldots) - 1.943 \sqrt{\frac{201.044\ldots}{32{,}874 - \dfrac{512^2}{8}}}$$

$$= (4.547\ldots) - (2.675\ldots)$$
$$= 1.9, \text{ to one decimal place}$$
$$\text{upper limit} = (4.547\ldots) + (2.675\ldots)$$
$$= 7.2, \text{ to one decimal place}$$

Based on the given data, a 90 percent confidence interval estimate for the increase in weight that accompanies a 1-inch increase in height is from 1.9 pounds to 7.2 pounds. ◀

## Testing $H_0$: $\beta_1 = 0$

A test of $H_0$: $\beta_1 = 0$ is sometimes used as a way of answering the question "Does any significant change in average $y$ accompany a change in $x$?" If $H_0$ is not rejected, then the regression line becomes:

$$y = \beta_0 + (0)x = \beta_0$$

A line defined by the equation $y = \beta_0$ is parallel to the $x$-axis. For such a line, each $(x, y)$ pair on the line has the same $y$-coordinate. Thus a change in $x$ does not yield a change in the corresponding average value of $y$.

The form of $H_a$ for such a test is one of the following:

| | |
|---|---|
| $H_a$: $\beta_1 \neq 0$ | The slope of the regression line is not 0. |
| $H_a$: $\beta_1 > 0$ | The slope of the regression line is positive. |
| $H_a$: $\beta_1 < 0$ | The slope of the regression line is negative. |

### The Test Statistic for $H_0$: $\beta_1 = 0$

The hypothesis $H_0$: $\beta_1 = 0$ can be tested using the statistic

$$T = \frac{B_1}{\sqrt{\dfrac{S_e^2}{\sum(x_i - \bar{x})^2}}} \qquad \text{Notice, this is equation (13) with } \beta_1 \text{ replaced by 0.} \qquad \textbf{(16)}$$

If $H_0$ is true, $T$ has a $t$-distribution with $n - 2$ degrees of freedom.

▶ **Example 5**    With $\alpha = 0.05$, is there sufficient evidence in the American shad data to conclude that an increase in average weight accompanies an increase in length for shad that spawn in the Yuba River?

**Solution**   **Discussion.** Because we are specifically looking for an *increase* in weight to ac-company an *increase in length*, we want to reject $\beta_1 = 0$ in favor of $\beta_1 > 0$, and thus use the appropriate one-sided alternative hypothesis.

**Step 1**  $H_0$:  $\beta_1 = 0$     No change in average weight
$H_a$:  $\beta_1 > 0$     An increase in average weight

**Step 2**  $\alpha = 0.05$     $P$ (Type I error) = 0.05.

**Step 3**  The test statistic used is:

$$T = \frac{B_1}{\sqrt{\dfrac{S_e^2}{\sum(x_i - \bar{x})^2}}}$$

**Step 4**  If $t > t_c \, (= 1.65)$, then reject $H_0$. (With $34 - 2 = 32$ degrees of freedom, $t_c = 1.65$)

**Step 5**  With $b_1 = 7.3206\ldots$, $s_e^2 = 25{,}137.63\ldots$, $n = 34$, $\sum x_i^2 = 7{,}101{,}159$, and $\sum x_i = 15{,}533$:

$$t = \frac{7.3206\ldots}{\sqrt{\dfrac{25{,}137.63\ldots}{7{,}101{,}159 - \dfrac{15{,}533^2}{34}}}}$$

$$= 3.219\ldots$$

$$= 3.22, \text{ to two decimal places}$$

**Step 6**  Because $3.22 > 1.65$, reject $H_0$. That is, there is sufficient evidence, with $\alpha = 0.05$, to conclude that average weight increases with length for the population of American shad that spawn in the Yuba River.   ◀

If $H_0: \beta_1 = 0$ is rejected, then the potentially misleading language "$x$ has an effect on $y$" is sometimes used. Recall the discussion in Section 9-2, that establishing an association between two variables is not the same as establishing a causal relationship between them. As a result, we do not recommend the use of the "$x$ affects $y$" language in conjunction with rejection of $H_0: \beta_1 = 0$.

## Point estimates for $\mu_{y|x*}$

A question that is asked in many applications of regression analysis is "What can be said about the values of $y$ if $x$ is assumed to have a specific value $x*$?" For example, we might ask "What can be said about the weights of shad whose lengths are 450 mm?" One way to answer such questions is to provide inference methods for the mean $y$ if $x$ is assumed to be a fixed value $x*$. The following symbol will help us phrase this concept in a compact way:

$\mu_{y|x*}$ stands for "the mean $y$ for $x = x*$."

If one uses the simple linear regression model in the analysis of bivariate data, then the following relation is one of the basic assumptions:

$$\mu_{y|x^*} = \beta_0 + \beta_1 x^*$$

This equation asserts the mean $y$ has a straight-line relationship with $x$. It is quite natural to estimate the mean $y$ for $x = x^*$ with the $y$-coordinate of the point on the least squares line having $x$-coordinate $x^*$.

### A Point Estimate for $\mu_{y|x^*}$

A point estimate for $\mu_{y|x^*}$ is $\hat{y}$, where:

$$\hat{y} = b_0 + b_1 x^* \qquad \textbf{(17)}$$

▶ **Example 6**  Find a point estimate of the mean weight of 450-mm American shad spawning in the Yuba River.

**Solution**  The unrounded values for $b_0$ and $b_1$ are given in Example 5 of Section 9-2.

$$b_0 = -989.26 \ldots \quad \text{and} \quad b_1 = 7.3206 \ldots$$

With $x^* = 450$:

$$\hat{y} = (-986.26 \ldots) + (7.3206 \ldots)(450)$$

$$= 2305, \text{ to the nearest whole number}$$

Therefore a point estimate for the mean weight of the 450 mm American shad in the Yuba River is 2305 grams. ◀

## Confidence interval estimates for $\mu_{y|x^*}$

Frequently more than a point estimate of $\mu_{y|x^*}$ is needed. To construct confidence interval estimates for $\mu_{y|x^*}$, we need to know more about the probability distribution of $\hat{Y}$. It can be shown that under the assumptions of the simple linear regression model,

$$T = \frac{\hat{Y} - \mu_{y|x^*}}{\sqrt{S_e^2}\sqrt{\dfrac{1}{n} + \dfrac{(x^* - \bar{x})^2}{\sum(x - \bar{x})^2}}} \qquad \textbf{(18)}$$

has a $t$-distribution with $n - 2$ degrees of freedom. This basic probability fact leads to confidence intervals and tests for $\mu_{y|x^*}$.

### Confidence Intervals for $\mu_{y|x^*}$

Confidence limits for the mean $y$ when $x = x^*$ are:

$$\text{lower limit} = \hat{y} - t\sqrt{s_e^2}\,\sqrt{\frac{1}{n} + \frac{(x^* - \bar{x})^2}{\sum(x_i - \bar{x})^2}} \qquad \textbf{(19)}$$

and

$$\text{upper limit} = \hat{y} + t\sqrt{s_e^2}\,\sqrt{\frac{1}{n} + \frac{(x^* - \bar{x})^2}{\sum(x_i - \bar{x})^2}} \qquad \textbf{(20)}$$

where $t$ is an upper percentage point of the $t$-distribution with $n - 2$ degrees of freedom.

▶ **Example 7**　Give 95 percent confidence intervals for the mean weight of American shad that spawn in the Yuba River for:

**a.** $x^* = 457$ mm　　**b.** $x^* = 475$ mm

**Solution**　Because $n = 34$, $n - 2 = 32$, and we can use $t = 1.96$ to construct the confidence intervals.

**a.** $s_e^2 = 25,\ 137.63\ldots,$　$\sum(x_i - \bar{x})^2 = 4{,}862.265,$　$n = 34,$　and　$(x^* - \bar{x})^2$ $= 0.0216\ldots$ :

$$t\sqrt{s_e^2}\,\sqrt{\frac{1}{n} + \frac{(x^* - \bar{x})^2}{\sum(x_i - \bar{x})^2}} \approx 53$$

For $x^* = 457$:
$$\hat{y} = -989.26 + 7.32(457) = 2356,$$
so,　　　　　　　　lower limit $= 2356 - 53 = 2303$
and,　　　　　　　upper limit $= 2356 + 53 = 2409$

Thus a 95 percent confidence interval for the mean weight of the 457-mm shad is from 2303 to 2409 grams.

**b.** With $s_e^2 = 25{,}137.63\ldots,$ $\sum(x_i - \bar{x})^2 = 4{,}862.265,$ $n = 34,$ and $(x^* - \bar{x})^2 = 329.315\ldots$ :

$$t\sqrt{s_e^2}\,\sqrt{\frac{1}{n} + \frac{(x^* - \bar{x})^2}{\sum(x_i - \bar{x})^2}} \approx 97$$

For $x^* = 475$:
$$\hat{y} = -989.26 + 7.32(475) = 2488,$$
so,　　　　　　　　lower limit $= 2488 - 97 = 2391$
and,　　　　　　　upper limit $= 2488 + 97 = 2585$

Thus a 95 percent confidence interval for the mean weight of the 475-mm shad is from 2391 to 2585 grams.　◀

Notice that the interval for the 475-mm shad ($2585 - 2391 = 194$ grams) is longer than the interval with the same confidence coefficient for the 457-mm shad ($2409 - 2303 = 106$ grams). The reason is that for the 34 data points, $\bar{x}$ is approximately 456.9 mm, and 457 is much closer to $\bar{x}$ than is 475. The estimated standard deviation for $\hat{Y}$ is smallest when $x^*$ is near $\bar{x}$. The relative lengths of confidence intervals for $\mu_{y|x^*}$ with various $x^*$ are illustrated in Figure 9-12 for a hypothetical set of $(x_i, y_i)$ pairs.

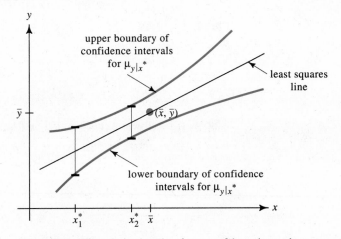

**Figure 9-12.** The relative lengths of two confidence intervals for $\mu_{y|x^*}$ for $x_1^*$ and $x_2^*$.

## Testing $H_0$: $\mu_{y|x^*} = \mu_0$

The fact that the statistic $T$ specified in display (18) has a $t$-distribution with $n - 2$ degrees of freedom provides a method of testing $\mu_{y|x^*} = \mu_0$.

### The Test Statistic for $H_0$: $\mu_{y|x^*} = \mu_0$

The null hypothesis that the mean $y$ when $x = x^*$ is $\mu_0$ can be tested using:

$$T = \frac{\hat{Y} - \mu_0}{\sqrt{S_e^2}\sqrt{\dfrac{1}{n} + \dfrac{(x^* - \bar{x})^2}{\sum(x_i - \bar{x})^2}}}$$

Notice that this is display (18) with $\mu_{y|x^*}$ replaced by $\mu_0$. **(21)**

and critical values taken from the $t$-table, using $n - 2$ degrees of freedom.

▶ **Example 8**    Suppose data collected 10 years before the shad data given in Section 9-2 indicated the mean weight of 450-mm shad at that time was 2400 g. Do the present data represent sufficient evidence (with $\alpha = 0.05$) to support the conclusion that the present mean weight of 450-mm shad is different from 2400 g? Note, $\hat{y} = -989.26 + 7.32(450) = 2305$.

**Solution**    **Step 1**   $H_0$:   $\mu_{y|450} = 2400$
                 $H_a$:   $\mu_{y|450} \neq 2400$

**Step 2**   $\alpha = 0.05$

**Step 3**   The test statistic will be:

$$T = \frac{\hat{Y} - 2400}{\sqrt{S_e^2}\sqrt{\dfrac{1}{n} + \dfrac{(450 - \bar{x})^2}{\sum(x_i - \bar{x})^2}}}$$

**Step 4**   If $t <$ lower $t_c \, (= -1.96)$ or $t >$ upper $t_c \, (= 1.96)$, then reject $H_0$.

**Step 5**   The sample yields:

$$t = \frac{2305 - 2400}{\sqrt{25{,}137}\sqrt{\dfrac{1}{34} + \dfrac{(450 - 456.8)^2}{4862}}}$$

$$= -3.04.$$

**Step 6**   Because $-3.04 < -1.96$, reject $H_0$. That is, there is sufficient evidence, with $\alpha = 0.05$, to conclude that the mean weight of the 450-mm shad today is not the same as the mean weight of 450-mm shad 10 years ago.    ◀

## Predicting the next value of y

Besides giving inference methods for $\mu_{y|x^*}$, there is another type of answer that can be given to the question, "What can be said about values of $y$ if $x$ is assumed to have the value $x^*$?" That is, it is possible to predict the next value of $y$ that would be obtained holding $x$ at $x^*$. For example, the weight of an additional shad not included in the original data set can be predicted based on the supposition that the shad is 450 mm long.

If the symbol $y^*$ is used to stand for an additional $y$ observation that has corresponding $x$ coordinate $x^*$, and the simple linear regression model is assumed to describe the $(n + 1)$ pairs $(x_1, y_1), (x_2, y_2), \ldots, (x_n, y_n)$, and $(x^*, y^*)$, it follows that:

$$T = \frac{Y^* - \hat{Y}}{\sqrt{S_e^2}\sqrt{1 + \dfrac{1}{n} + \dfrac{(x^* - \bar{x})^2}{\sum(x_i - \bar{x})^2}}} \tag{22}$$

has a $t$-distribution with $n - 2$ degrees of freedom. (The values $\hat{Y}$, $\bar{x}$, $S_e^2$, and $\sum(x_i - \bar{x})^2$ refer to the $n$ data points used to obtain the least squares line.) This fact leads to the so-called "prediction limits" for an additional $y$-value.

### Prediction Intervals for y*, the Next Observed Value of y when x = x*

A prediction interval for a single additional $y$-value obtained when $x = x^*$ can be made using:

$$\text{lower limit} = \hat{y} - t\sqrt{s_e^2}\sqrt{1 + \frac{1}{n} + \frac{(x^* - \bar{x})^2}{\sum(x_i - \bar{x})^2}} \qquad (23)$$

and

$$\text{upper limit} = \hat{y} + t\sqrt{s_e^2}\sqrt{1 + \frac{1}{n} + \frac{(x^* - \bar{x})^2}{\sum(x_i - \bar{x})^2}} \qquad (24)$$

where $t$ is chosen from a $t$-distribution with $n - 2$ degrees of freedom to give a desired degree of "certainty" attached to the prediction.

Equations (23) and (24) differ from those used to construct confidence intervals for $\mu_{y|x^*}$ only in that an additional 1 is added under the second square root sign. This additional 1 increases the length of the interval over a corresponding confidence interval for $\mu_{y|x^*}$.

▶ **Example 9**  Give a 90 percent prediction interval for the weight of another shad not included in the original data set, if the length of that fish is known to be 450 mm.

**Solution**  $\hat{y} = -989.26 + 7.32(450) = 2305$ with $s_e^2 = 25{,}137.63\ldots$, $\sum(x_i - \bar{x})^2 = 4{,}862.265$, $n = 34$, $(x^* - \bar{x})^2 = 46.96\ldots$, and $t = 1.65$:

$$t\sqrt{s_e^2}\sqrt{1 + \frac{1}{n} + \frac{(x^* - \bar{x})^2}{\sum(x_i - \bar{x})^2}} \approx 267,$$

so,                     lower limit $= 2305 - 267 = 2038$

and,                    upper limit $= 2305 + 267 = 2572$

Thus a 90 percent prediction interval for the weight of another shad that is 450 mm long is from 2038 to 2572 grams.  ◀

The 90 percent prediction limits mean that the limits have been obtained using a method that, if the whole process of selecting a sample, computing an interval, and seeing whether an additional $y$ falls in the interval is repeated many times, will be successful in approximately 90 percent of those trials. Notice that this interpretation is close to that of a confidence interval. There is, however, a conceptual difference between *estimating a fixed quantity* (such as $\mu_{y|x^*}$) and *predicting a random variable* (such as $Y^*$). We have therefore chosen to use the terminology *prediction limits*.

A note of caution should be made concerning the indiscriminate use of techniques of estimation and prediction discussed here. In an application of regression analysis, the only information obtained on the relation between $x$ and $y$ is for those values of $x$ appearing in the data. In real situations, even if it appears reasonable to assume the mean of $y$ is linearly related to $x$ for those $x$ in the data set, there is no guarantee that such a relationship holds true for other values of $x$. Consider the two graphs in Figure 9-13.

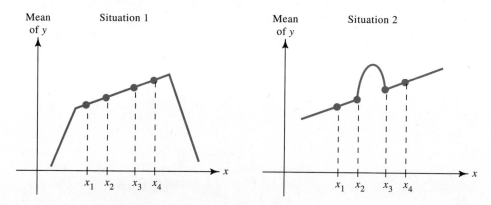

**Figure 9-13.** Two situations in which four observed values of $x$ have a linear relationship with average $y$ that does not hold true for all possible $x$ values.

In both situations depicted in the figure, the mean of $y$ is linearly related to $x$ for those $x$ observed. In situation 1, the relationship is decidedly not linear outside the range of observed $x$. An attempt to predict values of $y$ less than the smallest $x$-value or greater than the largest $x$-value is called *extrapolation*.

In situation 2, the relationship is definitely not straight line between the middle two observed $x$-values. An attempt to predict values of $y$ between the observed values of $x$ is called *interpolation*. Both extrapolation outside and interpolation between observed $x$-values are potentially dangerous. Conventional statistical wisdom is that real problems are more likely to resemble situation 1 in Figure 9-13 than situation 2, so that extrapolation is more likely to cause a problem than interpolation. But unless familiarity with a physical situation under study allows ruling out the possibility of circumstances like those in Figure 9-13, neither extrapolation nor interpolation is a 100 percent safe practice.

## Exercises 9-3    Set A

*In Exercises 1–4, assume the simple linear regression model for each set of $(x_i, y_i)$ data.*
*a. Find a point estimate for $\beta_1$.*
*b. Find a point estimate for $\beta_0$.*
*c. Find a point estimate for $\sigma^2$.*
*d. Use the value of $b_1$ found in part (a) to construct a 90 percent confidence interval for $\beta_1$.*
*e. Find the least squares line.*
*f. For the given $x^*$, find $\hat{y}$ as a point estimate for $\mu_{y|x^*}$.*
*g. Find a 95 percent confidence interval estimate for $\mu_{y|x^*}$ using the $x^*$ and $\hat{y}$ of part (f).*

1. $\{(-2, 10), (1, 2), (2, -2), (4, -5), (5, -10), (7, -13)\}$ and $x^* = 4$.

2. $\{(1, 12), (2, 8), (3, 7), (4, 3), (5, 5), (6, 2)\}$ and $x^* = 3$.

3. $\{(-2, -4), (1, -1), (4, 1)\ (5, 0), (6, -1), (8, 2), (10, 3)\}$ and $x^* = 5$.

4. $\{(1, 3), (2, 5), (3, 11)\ (4, 9), (5, 10), (6, 16), (7, 18), (8, 16)\}$ and $x^* = 4$.

*The U.S. Department of Energy publishes a gas mileage guide. The booklet contains fuel economy estimates of cars, vans, and light trucks. Some of the information in the guide includes engine sizes in liters, numbers of cylinders, and estimated gas mileages in miles per gallon. Eight passenger cars were selected from the guide and the data related to the vehicles were organized in Table 9-12.*

TABLE 9-12    Data on Eight Cars

| Vehicle | A | B | C | D | E | F | G | H |
|---|---|---|---|---|---|---|---|---|
| Engine Size | 2.5 | 4.0 | 4.0 | 2.0 | 5.8 | 3.8 | 5.0 | 2.3 |
| Number of Cylinders | 4 | 6 | 8 | 4 | 8 | 6 | 8 | 4 |
| Estimated Gas Mileage | 24 | 28 | 20 | 34 | 18 | 24 | 20 | 22 |

*For the $(x_i, y_i)$ ordered pairs:*
*a. Make a scatterplot.*
*b. Calculate r.*
*c. Find the least squares line for the data.*

5. Let $x$ be the engine size of a car and $y$ the corresponding estimated gas mileage. In addition to parts (a), (b), and (c):
   d. Estimate to one decimal place the change in the gas mileage estimate for a 1-liter increase in the engine size.
   e. Find a point estimate for the mean gas mileage estimate for an engine size of 3.8 liters.
   f. Construct a 90 percent confidence interval for $\mu_{y|x^*}$ if $x^* = 3.8$.

6. Let $x$ be the number of cylinders in the engine and $y$ the corresponding estimated gas mileage. In addition to parts (a), (b), and (c):
   d. Estimate to one decimal place the change in the gas mileage estimate for a two-cylinder increase in the engine size.
   e. Find a point estimate for the mean gas mileage estimate of an engine with six cylinders.
   f. Construct a 95 percent confidence interval for $\mu_{y|x*}$ if $x* = 6$.

*In Exercises 7–14, refer to the data given in Exercises 11–18 in Exercises 9-2 Set A.*

7. For the spring constant data,

$$\hat{y} = -2.1 + 2.1x, \quad n = 7, \quad \sum x_i = 13.16, \quad \sum y_i = 12.88, \quad \sum x_i y_i = 24.4401,$$
$$\sum x_i^2 = 24.8488 \quad \text{and} \quad \sum y_i^2 = 24.1982.$$

   a. Calculate $s_e^2$.
   b. Construct a 95 percent confidence interval for $\beta_1$ using $b_1 = 0.01$.
   c. Find $\hat{y}$ for $x* = 2.00$.
   d. Test, with $\alpha = 0.05$:

   $$H_0: \quad \mu_{y|2.00} = 1.80$$
   $$H_a: \quad \mu_{y|2.00} > 1.80$$

   e. Construct a 95 percent prediction interval for the next spring constant determined by test weight $B$ when the spring constant determined by $A$ is 2.00.

8. For the reagent 4UD900 temperature data,

$$\hat{y} = -39.05 + 0.76x, \quad n = 10, \quad \sum x_i = 320, \quad \sum y_i = 635, \quad \sum x_i y_i = 21{,}275, \quad \sum x_i^2 = 11{,}490,$$
$$\text{and} \quad \sum y_i^2 = 41{,}395.$$

   a. Calculate $s_e^2$.
   b. Construct a 95 percent confidence interval for $\beta_1$ using $b_1 = 0.76$.
   c. Find $\hat{y}$ for $x* = 50$ mg.
   d. Test, with $\alpha = 0.05$:

   $$H_0: \quad \mu_{y|50} = 75°$$
   $$H_a: \quad \mu_{y|50} \neq 75°$$

   e. Construct a 95 percent prediction interval for the next temperature change when 50 mg of reagent is added.

9. For the velocity-weld breaking strength data,

$$\hat{y} = 65.7 + 6.9x, \quad n = 7, \quad \sum x_i = 31.3, \quad \sum y_i = 676, \quad \sum x_i y_i = 3039.10,$$
$$\sum x_i^2 = 142.33, \quad \text{and} \quad \sum y_i^2 = 65{,}440.$$

   a. Calculate $s_e^2$
   b. Construct a 95 percent confidence interval for $\beta_1$, using $b_1 = 16.1$.
   c. Find $\hat{y}$ for $x* = 4.8$ fps.
   d. Test, with $\alpha = 0.01$:

   $$H_0: \quad \mu_{y|4.8} = 102$$
   $$H_a: \quad \mu_{y|4.8} < 102$$

   e. Construct a 98 percent prediction interval for the next weld breaking strength for a velocity of 4.8 fps.

10. For the manual dexterity-anxiety data,

$$\hat{y} = 17.2 - 4.1x, \quad n = 8, \quad \sum x_i = 22.9, \quad \sum y_i = 44, \quad \sum x_i y_i = 69.1,$$
$$\sum x_i^2 = 79.45, \quad \text{and} \quad \sum y_i^2 = 496.$$

a. Calculate $s_e^2$.
b. Construct a 98 percent confidence interval for $\beta_1$ using $b_1 = -4.1$.
c. Find $\hat{y}$ for a manual dexterity score of $x^* = 3.6$.
d. Test, with $\alpha = 0.01$: $\qquad H_0: \ \mu_{y|3.6} = 0$
$\qquad\qquad\qquad\qquad\qquad\qquad H_a: \ \mu_{y|3.6} > 0$

e. Construct a 98 percent prediction interval for the anxiety test score of an individual who got a score of 3.6 on the manual dexterity test.

11. For the silicone SS707 data,

$$\hat{y} = 0.39 + 0.19x, \quad n = 8, \quad \sum x_i = 36, \quad \sum y_i = 9.86,$$
$$\sum x_i y_i = 52.27, \ \sum x_i^2 = 204, \quad \text{and} \quad \sum y_i^2 = 13.71.$$

a. Test, with $\alpha = 0.05$: $\qquad H_0: \ \beta_1 = 0$
$\qquad\qquad\qquad\qquad\qquad\qquad H_a: \ \beta_1 > 0$

b. Find $\hat{y}$ for $x^* = 9$.
c. Test, with $\alpha = 0.05$: $\qquad H_0: \ \mu_{y|9} = 2.8$
$\qquad\qquad\qquad\qquad\qquad\qquad H_a: \ \mu_{y|9} < 2.8$

12. For the annual rainfall-forest fire data,

$$\hat{y} = 54.8 - 0.5x, \quad n = 8, \quad \sum x_i = 560, \quad \sum y_i = 149, \quad \sum x_i y_i = 9{,}726,$$
$$\sum x_i^2 = 40{,}564, \quad \text{and} \quad \sum y_i^2 = 3{,}163.$$

a. Test, with $\alpha = 0.05$: $\qquad H_0: \ \beta_1 = 0$
$\qquad\qquad\qquad\qquad\qquad\qquad H_a: \ \beta_1 < 0$

b. Find $y$ for $x^* = 70$ inches of rain.
c. Test, with $\alpha = 0.05$: $\qquad H_0: \ \mu_{y|70} = 20.0$
$\qquad\qquad\qquad\qquad\qquad\qquad H_a: \ \mu_{y|70} < 20.0$

13. For the order of birth-shyness study,

$$\hat{y} = 6.3 + 0.8x, \quad n = 9, \quad \sum x_i = 24, \quad \sum y_i = 76, \quad \sum x_i y_i = 219,$$
$$\sum x_i^2 = 84, \quad \text{and} \quad \sum y_i^2 = 744$$

a. Test, with $\alpha = 0.01$: $\qquad H_0: \ \beta_1 = 0$
$\qquad\qquad\qquad\qquad\qquad\qquad H_a: \ \beta_1 > 0$

b. Find $\hat{y}$ for $x^* = 3$.
c. Test, with $\alpha = 0.01$: $\qquad H_0: \ \mu_{y|3} = 9.0$
$\qquad\qquad\qquad\qquad\qquad\qquad H_a: \ \mu_{y|3} < 9.0$

**14.** For the insulation-temperature control study,

$$\hat{y} = -0.06 + 0.06x, \quad n = 6, \quad \sum x_i = 21.0, \quad \sum y_i = 0.81, \quad \sum x_i y_i = 3.80,$$
$$\sum x_i^2 = 91.00, \quad \text{and} \quad \sum y_i^2 = 0.1649.$$

a. Test, with $\alpha = 0.01$:
$$H_0: \quad \beta_1 = 0$$
$$H_a: \quad \beta_1 > 0$$

b. Find $\hat{y}$ for $x^* = 3.5$.

c. Test, with $\alpha = 0.01$:
$$H_0: \quad \mu_{y|3.5} = 0.15$$
$$H_a: \quad \mu_{y|3.5} < 0.15$$

## Exercises 9-3　Set B

*In Exercises 1–3, use the data in Table 9-13. Assume these data are the test scores of eight randomly selected students from the Chicago public school system on a standardized test that is given annually across the United States. (Any inferences made apply only to the students in the system sampled.)*

TABLE 9-13　Scores on a Standardized Test for Eight Students from the Chicago School System

| Student | A | B | C | D | E | F | G | H |
|---------|----|----|----|----|----|----|----|----|
| English | 16 | 26 | 18 | 18 | 12 | 30 | 22 | 27 |
| Math | 22 | 16 | 17 | 10 | 14 | 22 | 20 | 30 |
| Composite | 18 | 20 | 17 | 12 | 13 | 24 | 23 | 29 |

**1.** Let $x$ be the score of a student in English and $y$ the corresponding score of the same student in mathematics.
a. Make a scatterplot.
b. Calculate $r$.
c. Estimate the change in the mean score in mathematics for a one-unit increase in the score in English.
d. Write the equation of the least squares line for the data.
e. Find a point estimate for the mean math score for an English score of 22.
f. Construct a 90 percent confidence interval for $\mu_{y|x^*}$, if $x^* = 22$.

**2.** Let $x$ be the score of a student in English and $y$ the corresponding composite score of the same student.
a. Make a scatterplot.
b. Calculate $r$.
c. Estimate the change in the mean composite score for a one-unit increase in the English score.
d. Write the equation of the least squares line for the data.
e. Find a point estimate for the mean composite score for an English score of 18.
f. Construct a 95 percent confidence interval for $\mu_{y|x^*}$, if $x^* = 18$.

3. Let $x$ be the score of a student in mathematics and $y$ the corresponding composite score of the same student.
   a. Make a scatterplot.
   b. Calculate $r$.
   c. Estimate the change in the mean composite score for a one-unit increase in the mathematics score.
   d. Write the equation of the least squares line for the data.
   e. Find a point estimate for the mean composite score for a math score of 18.
   f. Construct a 98 percent confidence interval for $\mu_{y|x^*}$ and $x^* = 18$.

*Exercises 4 and 5 are suggested student projects.*

4. Use a city map of the area within a 5-mile radius of campus. Sample ten 1-bedroom furnished apartments in this area for rent information. Let $x$ be the distance (to the nearest one-tenth of 1 mile) an apartment is from the campus, and $y$ the corresponding monthly rent.
   a. Plot the ordered pairs $(x_i, y_i)$.
   b. Calculate $r$.
   c. If the scatterplot suggests a linear relationship between $x$ and $y$, determine the least squares line for the data.
   d. Make a concluding statement regarding your results.

5. Sample 12 students (female or male) on your campus. Let $x$ be the height (in inches) of a given student and $y$ the corresponding grade point average (gpa).
   a. Plot the ordered pairs $(x_i, y_i)$.
   b. Calculate $r$.
   c. If the scatterplot suggests a linear relationship between $x$ and $y$, determine the least squares line for the data.
   d. Make a concluding statement regarding your results.

# Chapter 9 Summary

## Rules and Equations

   a. If $H_0$: $P_{ij} = P_i.P._j$ for all $i$ and $j$ is true and $n$ is large, then the probability distribution of

   $$X^2 = \sum \frac{(O_{ij} - E_{ij})^2}{E_{ij}} \tag{1}$$

   (or equivalently)

   $$X^2 = \sum \left(\frac{O_{ij}^2}{E_{ij}}\right) - n \tag{2}$$

   is approximately $\chi^2$ with $(r - 1)(c - 1)$ degrees of freedom.
   b. Correlation, a measure of the strength of linear association:

   $$r = \frac{\sum(x_i - \bar{x})(y_i - \bar{y})}{\sqrt{\sum(x_i - \bar{x})^2 \sum(y_i - \bar{y})^2}} \tag{3}$$

c. An alternate formula for computing $r$

$$r = \frac{\sum x_i y_i - \dfrac{(\sum x_i)(\sum y_i)}{n}}{\sqrt{\left(\sum x_i^2 - \dfrac{(\sum x_i)^2}{n}\right)\left(\sum y_i^2 - \dfrac{(\sum y_i)^2}{n}\right)}} \tag{4}$$

d. If $(x_1, y_1)$ and $(x_2, y_2)$ are coordinates of points on a nonvertical line, then *the slope of the line is*

$$\text{slope} = \frac{y_2 - y_1}{x_2 - x_1} \tag{5}$$

e. *The point-slope equation of a line*
   If a line has $y$-intercept $\beta_0$ and slope $\beta_1$, then:

$$y = \beta_0 + \beta_1 x \tag{6}$$

f. *The least squares line*
   For a collection of $(x, y)$ pairs, the least squares line has:

slope $b_1$, where

$$b_1 = \frac{\sum(x_i - \bar{x})(y_i - \bar{y})}{\sum(x_i - \bar{x})^2} \tag{7}$$

and

$y$-intercept $b_0$, where

$$b_0 = \bar{y} - b_1 \bar{x} \tag{8}$$

An alternate formula for $b_1$ is:

$$b_1 = \frac{\sum x_i y_i - \dfrac{(\sum x_i)(\sum y_i)}{n}}{\sum x_i^2 - \dfrac{(\sum x_i)^2}{n}} \tag{9}$$

g. The probability model for simple linear regression analysis:

$$Y_i = \beta_0 + \beta_1 x_i + \epsilon_i \tag{10}$$

h. A point estimate for $\sigma^2$ is $s_e^2$, where:

$$s_e^2 = \frac{\sum(y_i - \hat{y}_i)^2}{n - 2} \tag{11}$$

or:

$$s_e^2 = \frac{1}{n - 2}\left[\sum y_i^2 - \frac{(\sum y_i)^2}{n} - b_1^2\left(\sum x_i^2 - \frac{(\sum x_i)^2}{n}\right)\right] \tag{12}$$

i. If the simple linear regression model:

$$Y_i = \beta_0 + \beta_1 x_i + \epsilon_i$$

is adopted as an appropriate description of a set of $n$ data pairs, then

$$T = \frac{B_1 - \beta_1}{\sqrt{\dfrac{S_e^2}{\sum(x_i - \bar{x})^2}}} \tag{13}$$

has a $t$-distribution with $n - 2$ degrees of freedom.

j. To construct confidence interval estimates for $\beta_1$ use:

$$\text{lower limit} = b_1 - t \sqrt{\frac{s_e^2}{\sum(x_i - \bar{x})^2}} \tag{14}$$

$$\text{upper limit} = b_1 + t \sqrt{\frac{s_e^2}{\sum(x_i - \bar{x})^2}} \tag{15}$$

where $t$ is chosen from the $t$-distribution with $n - 2$ degrees of freedom to yield the desired confidence level.

k. The test statistic for the test $H_0: \beta_1 = 0$ is:

$$T = \frac{B_1}{\sqrt{\dfrac{S_e^2}{\sum(x_i - \bar{x})^2}}} \tag{16}$$

where the appropriate $t$-distribution has $n - 2$ degrees of freedom.

l. A point estimate for $\mu_{y|x^*}$ for a given $x = x^*$ is $\hat{y}$, where:

$$\hat{y} = b_0 + b_1 x^* \tag{17}$$

m. Under the assumptions of the simple linear regression model:

$$T = \frac{\hat{Y} - \mu_{y|x^*}}{\sqrt{S_e^2}\sqrt{\dfrac{1}{n} + \dfrac{(x^* - \bar{x})^2}{\sum(x_i - \bar{x})^2}}} \tag{18}$$

has a $t$-distribution with $n - 2$ degrees of freedom.

n. To construct confidence intervals for $\mu_{y|x^*}$ the mean $y$ for $x = x^*$:

$$\text{lower limit} = \hat{y} - t\sqrt{s_e^2}\sqrt{\frac{1}{n} + \frac{(x^* - \bar{x})^2}{\sum(x_i - \bar{x})^2}} \tag{19}$$

and

$$\text{upper limit} = \hat{y} + t\sqrt{s_e^2}\sqrt{\frac{1}{n} + \frac{(x^* - \bar{x})^2}{\sum(x_i - \bar{x})^2}} \tag{20}$$

where $t$ is chosen from Appendix Table 5 using $n - 2$ degrees of freedom to yield the desired confidence level.

o. The test statistic for the test $H_0: \mu_{y|x*} = \mu_0$ is:

$$T = \frac{\hat{Y} - \mu_0}{\sqrt{S_e^2}\sqrt{\frac{1}{n} + \frac{(x* - \bar{x})^2}{\sum(x_i - \bar{x})^2}}} \tag{21}$$

and critical values are taken from the $t$-distribution with $n - 2$ degrees of freedom.

p. If the symbol $y*$ is used to stand for future $y$-value that has corresponding $x$-coordinate $x*$, and the simple linear regression model is assumed to describe the $(n + 1)$ pairs $(x_1, y_1), (x_2, y_2), \ldots, (x_n, y_n)$ and $(x*, y*)$, it follows that:

$$T = \frac{Y* - \hat{Y}}{\sqrt{S_e^2}\sqrt{1 + \frac{1}{n} + \frac{(x* - \bar{x})^2}{\sum(x_i - \bar{x})^2}}} \tag{22}$$

has a $t$-distribution with $n - 2$ degrees of freedom.

q. To construct *prediction intervals* for $y*$, the next observed value of $y$ when $x = x*$ use:

$$\text{lower limit} = \hat{y} - t\sqrt{s_e^2}\sqrt{1 + \frac{1}{n} + \frac{(x* - \bar{x})^2}{\sum(x_i - \bar{x})^2}} \tag{23}$$

and

$$\text{upper limit} = \hat{y} + t\sqrt{s_e^2}\sqrt{1 + \frac{1}{n} + \frac{(x* - \bar{x})^2}{\sum(x_i - \bar{x})^2}} \tag{24}$$

where $t$ is chosen from the $t$-distribution with $n - 2$ degrees of freedom.

## Symbols

| | |
|---|---|
| $P_{i \cdot}$ | $P_{i1} + P_{i2} + P_{i3} + \cdots + P_{ic}$ |
| $P_{\cdot j}$ | $P_{1j} + P_{2j} + P_{3j} + \cdots + P_{rj}$ |
| $\dfrac{n_i}{n}$ | the fraction of a sample falling into the $i$th category of the 1st variable |
| $\dfrac{m_j}{n}$ | the fraction of a sample falling into the $j$th category of the 2nd variable |
| $o_{ij}$ | the number of observations in the $i$th category of the 1st variable and the $j$th category of the 2nd variable |
| $e_{ij}$ | an estimate of the expected number of observations in the $i$th category of the 1st variable and the $j$th category of the 2nd variable, if $H_0: P_{ij} = P_{i \cdot}P_{\cdot j}$ is true |

$r$      the correlation coefficient

$(x_i, y_i)$      a bivariate data point

$\beta_0$      the $y$-intercept of the regression line $y = \beta_0 + \beta_1 x$

$\beta_1$      the slope of the regression line $y = \beta_0 + \beta_1 x$

$\epsilon_i$      the amount by which $Y_i$ deviates from its expected value described by $\beta_0 + \beta_1 x$

$\sigma^2$      the variance of $y$ about the regression line for a given value of $x$

$b_0$      an estimate of $\beta_0$

$b_1$      an estimate of $\beta_1$

$s_e^2$      a point estimate of $\sigma^2$

$\mu_{y|x*}$      the mean $y$ for $x = x^*$

$\hat{y}$      a point estimate for $\mu_{y|x*}$ for a given $x = x^*$

$y^*$      an additional $y$-observation that has corresponding $x$-coordinate $x^*$

## Comparing the $\chi^2$ Test of Independence and Regression Analysis

Tests of independence are used with bivariate *qualitative data*. They can be used to determine whether the presence of one type of characteristic in members of a population changes the probability that a second type of characteristic also is present.

Each item sampled is a source of two characteristics of interest. The items are recorded in a contingency table of $r$ rows and $c$ columns.

The number of data points in each cell is the *observed frequency*, $o_{ij}$.

Assuming a model of independence, calculate the estimated *expected frequency*, $e_{ij}$, for each cell:

$$e_{ij} = \frac{n_i m_j}{n} \quad \frac{(\text{row total})(\text{column total})}{\text{grand total}}$$

From Appendix Table 6, find $\chi_c^2$ for a specified level of significance and degrees of freedom.

Least squares lines are used with bivariate *quantitative data*. The value of one variable can be used to predict the corresponding value of the second of two linearly related variables.

Plot the ordered pairs in an $xy$-coordinate system to determine whether the scatterplot suggests that a linear relationship is appropriate for $x$ and $y$.

List the ordered pairs $(x_i, y_i)$ in a table. To calculate $b_0$ and $b_1$ for the equation:

$$y = b_0 + b_1 x$$

include columns in the table for $x_i y_i$, $x_i^2$, and $y_i^2$.

Assuming that $x$ and $y$ are linearly related, calculate the following:

$$b_1 = \frac{\sum x_i y_i - \dfrac{(\sum x_i)(\sum y_i)}{n}}{\sum x_i^2 - \dfrac{(\sum x_i)^2}{n}}$$

$$b_0 = \bar{y} - b_1 \bar{x}$$

If $x^2 > \chi_c^2$, then reject the model of independence for the variables of interest in the sampled population. If $x^2 < \chi_c^2$, then the model of independence cannot be rejected.

To determine the correlation for the set of data, calculate:

$$r = \frac{\sum x_i y_i - \dfrac{(\sum x_i)(\sum y_i)}{n}}{\sqrt{\left(\sum x_i^2 - \dfrac{(\sum x_i)^2}{n}\right)\left(\sum y_i^2 - \dfrac{(\sum y_i)^2}{n}\right)}}$$

# Chapter 9 Review Exercises

1. Use the following table of values of $o_{ij}$ to test $H_0: P_{ij} = P_i.P_{.j}$ for all $i$ and $j$ versus $H_a$: not $H_0$. Use $\alpha = 0.05$.

|  |  | 2nd Variable | | | |
| --- | --- | --- | --- | --- | --- |
|  |  | Cat. 1 | Cat. 2 | Cat. 3 | Totals |
| **1st Variable** | Cat. 1 | 7 | 3 | 10 | 20 |
|  | Cat. 2 | 10 | 15 | 5 | 30 |
|  | Cat. 3 | 8 | 17 | 25 | 50 |
|  | Totals | 25 | 35 | 40 | 100 |

2. A study was performed in Cleveland to determine whether peoples' opinions of state lotteries financing college athletic programs are dependent on political affiliation. Use the following values of $o_{ij}$ obtained in the study to test $H_0: P_{ij} = P_i.P_{.j}$ for all $i$ and $j$ versus $H_a$: not $H_0$. Use $\alpha = 0.05$.

|  |  | 2nd Variable | | | |
| --- | --- | --- | --- | --- | --- |
|  |  | In Favor | Against | No Opinion | Totals |
| **1st Variable** | Democrat | 18 | 39 | 24 | 81 |
|  | Republican | 123 | 15 | 21 | 159 |
|  | Independent | 21 | 24 | 15 | 60 |
|  | Totals | 162 | 78 | 60 | 300 |

*In Exercises 3 and 4, do the following:*
*a. Make a scatterplot of the data.*
*b. Find the least squares line.*
*c. Calculate r.*

3. A small cart is used in a college physics laboratory to conduct certain experiments. One such experiment requires approximating the time and distance a cart moves while undergoing uniform motion. The following set of $(x, y)$ pairs shows the times $x$ (measured in seconds) and distances $y$ (measured in centimeters):

$$\{(0.1, 5.0), (0.3, 7.8), (0.4, 9.1), (0.7, 12.2), (1.0, 14.7), (1.6, 20.5), (2.1, 26.2)\}$$

4. The solubility of a compound in a specified quantity of water is tested at various temperatures. The following set of $(x, y)$ pairs shows the temperatures of the water $x$ (measured in degrees Celsius) and the quantities of the compound $y$ (measured in grams) that dissolve.

$$\{(10, 1.8), (20, 2.0), (35, 2.7), (50, 2.8), (65, 3.6), (70, 3.6), (90, 4.0)\}$$

5. Assuming the simple linear regression model for the given set of ordered pairs $(x, y)$:

$$\{(-3, 10), (-2, 9), (-1, 6), (0, 3), (1, 5), (2, 2)\}$$

Find:
   a. A point estimate for $\beta_1$.
   b. A point estimate for $\beta_0$.
   c. A point estimate for $\sigma^2$.
   d. Using the value of $b_1$ in part (a), give a 95% confidence interval for the increase in average $y$ that accompanies a unit increase in $x$ in the population sampled.

6. For the area of stomata–water vapor diffusion rate data, the least squares line is:

$$\hat{y} = 0.5 + 0.4x$$

Furthermore:

$$\sum x_i = 9.7, \quad \sum y_i = 8.1, \quad \sum x_i y_i = 19.65, \quad \sum x_i^2 = 35.29, \quad \sum y_i^2 = 12.45, \quad \text{and} \quad n = 8.$$

   a. Find $s_e^2$.
   b. Construct a 95% confidence interval for $\beta_1$ using $b_1 = 0.4$.
   c. Find a point estimate for $\mu_{y|x^*}$, if $x^* = 0.5$.
   d. Construct a 95% confidence interval for $\mu_{y|10.5}$.
   e. Give a 95% prediction interval for the rate of water vapor diffusion through a stoma with an area of 0.5 mm$^2$.
   f. Does it seem reasonable to use the least squares line to estimate the rate of water vapor diffusion for a value of $x^*$ between 5.5 and 0.1 mm$^2$?

7. For the circumference of stomata–water vapor diffusion rate data, the least squares line is:

$$\hat{y} = -0.03 + 0.3x$$

Furthermore:

$$\sum x_i = 25.5, \quad \sum y_i = 8.1, \quad \sum x_i y_i = 38.81, \quad \sum x_i^2 = 121.13, \quad \sum y_i^2 = 12.45, \quad \text{and} \quad n = 8.$$

   a. Find $s_e^2$.
   b. Construct a 95% confidence interval for $\beta_1$ using $b_1 = 0.3$.
   c. Find a point estimate for $\mu_{y|x^*}$, if $x^* = 1.8$.
   d. Construct to two decimal places a 95% confidence interval for $\mu_{y|1.8}$.
   e. Find a point estimate for $\mu_{y|3.0}$.
   f. Construct to two decimal places a 95% confidence interval for $\mu_{y|3.0}$.
   g. Why are the confidence intervals in parts (d) and (f) of different lengths, even though they have the same confidence coefficient?

h. Give a 95% prediction interval for the rate of water vapor diffusion through a stoma with a circumference of 3.0 mm.

i. Does it seem reasonable to use the least squares line to estimate the rate of water vapor diffusion for an $x^*$ between 8.3 and 1.1 mm?

8. A study was made of the relationship between the length of a cotyledon (seed leaf) in a particular plant and the number of days after the seed was planted. The following table contains the data on the total increase in length $y$ of the cotyledon and $x$, the number of days the seed was planted. The values of $y$ are in millimeters.

| $x$ | 3 | 4 | 5 | 6 | 7 | 8 | 9 | 10 | 11 | 12 | 13 | 14 |
|-----|-----|-----|------|------|------|------|-------|-------|-------|-------|-------|-------|
| $y$ | 2.4 | 8.1 | 18.9 | 38.0 | 64.0 | 94.6 | 123.0 | 145.0 | 157.3 | 164.0 | 167.4 | 168.6 |

a. Make a scatterplot of the data.
b. Calculate $\sum x_i$.
c. Calculate $\sum y_i$.
d. Calculate $\sum x_i y_i$.
e. Calculate $\sum x_i^2$.
f. Calculate $\sum y_i^2$.
g. Find the least squares line.
h. Compute $r$.

9. For the shad fish of the American River studied by Joan Sanford, another aspect of the study compared the body length $y$ in millimeters and the corresponding scale length $x$ in millimeters. Some of the data are given in the following table:

| $x$ | 82 | 85 | 97 | 85 | 101 | 93 | 80 | 94 | 100 | 80 |
|-----|-----|-----|-----|-----|-----|-----|-----|-----|-----|-----|
| $y$ | 408 | 410 | 420 | 430 | 432 | 428 | 400 | 422 | 438 | 402 |

a. Make a scatterplot of the data.
b. Calculate $\sum x_i$.
c. Calculate $\sum y_i$.
d. Calculate $\sum x_i y_i$.
e. Calculate $\sum x_i^2$.
f. Calculate $\sum y_i^2$.
g. Find the equation of the least squares line.
h. Compute $r$.
i. Compute $s_e^2$.
j. Construct a 98% confidence interval for $\beta_1$ using $b_1$ of part (g).
k. Find a point estimate for $\mu_{y|x^*}$, if $x^* = 93$ mm.
l. Construct a 98% confidence interval for $\mu_{y|93}$.
m. Approximate the mean increase in body length of a shad fish, for a 1-mm increase in the scale length of the fish.
n. Give a 95% prediction interval for the next body length for $x^* = 93$ in the population of shad fish studied.

**Key Topics**

1. *The REGRESSION command*

2. *Prediction using the regression equation*

3. *Multiple regression*

4. *Polynomial regression*

In this section we will study the way Minitab can be used to obtain the equation of a least squares line. The additional topics of multiple regression and polynomial regression are also briefly introduced, even though we did not cover them in the previous sections of this text.

## The REGRESSION command

Let us consider the data on length and weight of shad (Section 9-2, Example 2). When these data have been entered into two columns of the worksheet and these columns are named 'LENGTH' and 'WEIGHT', the command:

```
REGRESS 'WEIGHT' ON 1 VARIABLE 'LENGTH'
```

causes the output in Figure 9-14.

```
THE REGRESSION EQUATION IS
WEIGHT = -989 + 7.32 LENGTH

PREDICTOR     COEF      STDEV     T-RATIO       P
CONSTANT      -989       1039        0.95    0.348
LENGTH       7.321      2.274        3.22    0.003

S = 158.5    R-SQ = 24.5%    R-SQ(ADJ) = 22.1%

ANALYSIS OF VARIANCE

SOURCE         DF         SS         MS        F        P
REGRESSION      1     260579     260579    10.37    0.003
ERROR          32     804404      25138
TOTAL          33    1064983
```

**Figure 9-14.** Minitab output.

The first part of this display is the regression equation. Next comes a block of information that repeats the computed values of the slope and intercept of the line with the $t$- and $p$-values that may be used to test whether each is different from zero. For instance, the $p$-value of 0.003 in this example shows that it may be concluded that the slope is detectably different from zero.

The next line gives the values for $s_e$ (the estimated standard deviation of the $y$-values about the regression line), $R$-sq (the square of the correlation coefficient), and $R$-sq adjusted, a statistic that we have not discussed.

At the end of the display is an ANOVA table that gives a breakdown of the raw variation in the data in terms that are suitable to the context of fitting a line to $(x, y)$ data.

In this particular example, none of the $x$'s in the sample had substantially more effect on the fitted line than any of the others. If this had not been true, Minitab would have included such points in another block of information entitled **UNUSUAL OBSERVATIONS**. In this block, there would also be any data points for which the residuals (differences between $y$-values and corresponding $\hat{y}$-values) are large.

## Prediction using the regression line

To make predictions using the regression line, add the subcommand PREDICT K (where K is a value for the "independent" variable $x$) to the regression command. To predict the weight of a shad whose length is 457 mm, the commands are:

```
REGRESS 'WEIGHT' 1 'LENGTH';
PREDICT 457.
```

The output is the same as before with the additional lines:

| FIT | STDEV. FIT | 95% C.I. | 95% P.I. |
|-----|-----------|----------|----------|
| 2356.3 | 27.2 | (2300.9, 2411.7) | (2028.5, 2684.1) |

The value for "Fit" (2356.3) is a point predictor for the weight of a shad 457 mm long. The "95% C.I." is a confidence interval for the *mean length* of *all* shad in the Yuba River that are 457 mm long. The "95% P.I." is a prediction interval for the length of *one* shad that is 457 mm long.

## Multiple regression

Very often the value of one variable depends on two or more independent variables. (For instance, in geometry we learn that the volume of a cylinder depends on both the area of its base and its height.)

Consider the $(x, z, y)$ data in Figure 9-15.

```
MTB> PRINT C1 C2 C3
ROW     X       Z       Y
  1    0.0     1.3     8.7
  2    2.0     1.9     6.3
  3    2.8     2.7     6.7
  4    4.0     4.0     7.6
  5    5.0     4.1     7.2
  6    6.0     5.6     8.2
```

**Figure 9-15.**   Minitab output.

Suspecting that the y-values may be dependent on both the x- and the z-values, we give the command:

```
REGRESS C3 ON 2 VARIABLES IN C1 AND C2
```

and Minitab returns the output in Figure 9-16:

```
THE REGRESSION EQUATION IS
Y = 5.61 - 1.53 X + 2.11 Z

PREDICTOR      COEF     STDEV    T-RATIO       P
CONSTANT     5.6105    0.6580       8.53   0.003
X           -1.5268    0.4244      -3.60   0.037
Z            2.1055    0.5766       3.65   0.035

S = 0.4999   R-SQ = 81.7%   R-SQ(ADJ) = 69.5%

ANALYSIS OF VARIANCE

SOURCE        DF        SS        MS      F        P
REGRESSION     2    3.3453    1.6727   6.69   0.078
ERROR          3    0.74976   0.2499
TOTAL          5    4.0950

SOURCE        DF     SEQ SS
X              1    0.0129
Z              1    3.3325
```

**Figure 9-16.** Minitab output.

The regression equation is given in the form:

$$y = b_0 + b_1 x + b_2 z$$

and t- and p-values are given for all three coefficients. Note also in the ANOVA table, the Regression SS is broken in two components, one for x and one for z after x.

## Polynomial regression

Consider the following data:

$$
\begin{array}{ccccccccc}
y & 4 & 5 & 9 & 9 & 8 & 7 & 5 & 1 \\
x & 1 & 2 & 3 & 4 & 5 & 6 & 7 & 8
\end{array}
$$

A scattergram showing this data is in Figure 9-17.

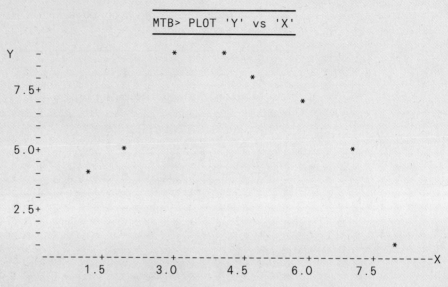

**Figure 9-17.**   Minitab output.

The points on the graph seem more likely to lie on a curve than a straight line. In fact, the points look as though they might lie on a parabola. This prompts us to try a regression equation of the form:

$$y = b_0 + b_1 x + b_2 x^2$$

With the $y$-values in C1 and the $x$-values in C2, the command:

```
LET C3 = C2*C2
NAME C3 'XSQ'
```

puts the squares of the $x$-values into column 3 and names the variable $XSQ$. Then:

```
REGRESS C1 ON TWO VARIABLES IN C2 AND C3
```

causes the output in Figure 9-18.

```
THE REGRESSION EQUATION IS
Y = -0.36 + 4.38 X - 0.524 XSQ

PREDICTOR        COEF      STDEV     T-RATIO        P
CONSTANT       -0.357      1.123       -0.32     0.763
X               4.3810     0.5724       7.65     0.001
XSQ            -0.52381    0.06209     -8.44     0.000

S = 0.8047    R-SQ = 94.0%    R-SQ(ADJ) = 91.6%

ANALYSIS OF VARIANCE

SOURCE          DF        SS          MS        F        P
REGRESSION       2    50.762      25.381    39.19    0.001
ERROR            5     3.238       0.648
TOTAL            7    54.000

SOURCE          DF     SEQ SS
X                1      4.667
Z                1     46.095
```

**Figure 9-18.** Minitab output.

# Exercises

1. Consider the polynomial regression done above.
   a. What is the regression equation?
   b. Are the coefficients of $X$ and $XSQ$ significantly different from 0? Explain.

2. Using the regression information on the shad data given in this section, find the value of *SS Regression/SS Total*. Compare this with *R-sq*. Can you see why *R-sq* is said to give the part of the variation accounted for by the regression equation?

3. The "predict" subcommand may be used with any regression command. When it was used with the data in the example of multiple regression discussed above in this way:

```
REGRESS C3 2 C1 C2;
PREDICT 2.0 3.0;
PREDICT 3.5 6.2.
```

(The first number after "predict" is the value selected to replace $x$, the second is the number for $z$.)

The result was the standard regression output with these additional lines:

```
  FIT    STDEV.FIT        95% C.I.             95% P.I.
 8.873    0.453     (7.432, 10.315)     (6.727, 11.020)
13.321    1.622     (8.159, 18.483)     (7.919, 18.722) XX
```

$X$ denotes a row with $(x, z)$ values away from the center of the $(x, z)$ data. $XX$ denotes a row with very extreme $(x, z)$ values.
   a. Give a point estimate of the mean $y$, when $x^* = 2.0$ and $z^* = 3.0$.
   b. Find a prediction interval for the next value of $y$, when $x^* = 2.0$ and $z^* = 3.0$.
   c. Would you agree that the mean $y$, when $x^* = 3.5$ and $z^* = 6.2$, is probably between 8.159 and 18.483? Why?

## SCIENTIFIC

A scientific calculator with a *linear regression mode* (mode LR) can be used to obtain $b_0$ and $b_1$ in the equation:

$$\hat{y} = b_0 + b_1 x$$

The correlation coefficient $r$ can also be calculated at the same time. The following example illustrates the necessary key strokes and the corresponding displays.

▶ **Example** Find the least squares equation of the line for the following data, and the correlation coefficient:

$$(2, 1), \quad (3, 3), \quad (5, 4), \quad (7, 5)$$

**Solution**

| Display | Key Sequence | Comments |
|---|---|---|
| [MODE] 2 | *0.* | Selects linear regression (LR). |
| [SHIFT] [KAC] | *0.* | Clears statistical memories. |
| 2 [xD, yD] | *2.* | Enters 2 for $x_1$. |
| 1 [DATA] | *1.* | Enters 1 for $y_1$. |
| 3 [xD, yD] | *3.* | Enters 3 for $x_2$. |
| 3 [DATA] | *3.* | Enters 3 for $y_2$. |
| 5 [xD, yD] | *5.* | Enters 5 for $x_3$. |
| 4 [DATA] | *4.* | Enters 4 for $y_3$. |
| 7 [xD, yD] | *7.* | Enters 7 for $x_4$. |
| 5 [DATA] | *5.* | Enters 5 for $y_4$. |
| [SHIFT] [A] | *0.152542372* | Displays value of $b_0$. |
| [SHIFT] [B] | *0.728813559* | Displays value of $b_1$. |
| [SHIFT] [r] | *0.946255523* | Displays value of $r$. |

$\hat{y} = 0.15 + 0.73x$ is the least squares line. 0.95 is the correlation coefficient. ◀

## GRAPHING

The Casio graphics calculator will plot the points of paired data used to determine the least squares line and also sketch the line through the points. At the same time, it will calculate $b_0$ and $b_1$ in the linear regression equation $\hat{y} = b_0 + b_1 x$, as well as the correlation coefficient $r$. The LR2 mode of the calculator is used to do these operations and calculations. The next example illustrates the necessary keystrokes.

In the following key sequence, these symbols will be used:

a. To enter $-1$, press: $\boxed{(-)}1$

b. The $\boxed{A}$ key is: $\boxed{\text{SHIFT}}\,7$

c. The $\boxed{B}$ key is: $\boxed{\text{SHIFT}}\,8$

d. The $\boxed{r}$ key is: $\boxed{\text{SHIFT}}\,9$

e. The $\boxed{,}$ key is: $\boxed{\text{SHIFT}}\,\boxed{(}$

f. The $\boxed{\text{DT}}$ key is the $\boxed{\sqrt{\ }}$ key (no shift)

g. The $\boxed{\text{LINE}}$ key is: $\boxed{\text{SHIFT}}\,\boxed{⇧}$

▶ **Example**    Determine $b_0$, $b_1$, $r$, and the least squares line for the 4 data pairs:

$$(2, 1), \quad (3, 3), \quad (5, 4), \quad (7, 5)$$

**Solution**

| Key Sequence | Comments |
|---|---|
| $\boxed{\text{SHIFT}}\ \boxed{\text{MODE}}\ \boxed{÷}$ | Puts calculator in LR2 mode. |
| $\boxed{\text{RANGE}}$ <br> $-1\ \boxed{\text{EXE}}\ 10\ \boxed{\text{EXE}}$ <br> $1\ \boxed{\text{EXE}}$ <br> $-1\ \boxed{\text{EXE}}\ 10\ \boxed{\text{EXE}}$ <br> $1\ \boxed{\text{EXE}}$ | This sequence of 13 key strokes sets the scales on the horizontal and vertical axes for the graphics display of the points and line. |

*table continues on p. 668*

**667**

| Key Sequence | Comments |
|---|---|
| SHIFT Scl EXE | Clears the STAT memories. |
| SHIFT Cls EXE | Clears the Graphics memories. |
| 2 , 1 DT | Enters and plots (2, 1). |
| 3 , 3 DT | Enters and plots (3, 3). |
| 5 , 4 DT | Enters and plots (5, 4). |
| 7 , 5 DT | Enters and plots (7, 5). |
| GRAPH SHIFT LINE 1 EXE | This sequence of 5 key strokes draws the regression line. |
| SHIFT A EXE | Displays value of $b_0$. |
| SHIFT B EXE | Displays value of $b_1$. |
| SHIFT r EXE | Displays value of $r$. |

From Step 10:   $b_0 \approx 0.15$
From Step 11:   $b_1 \approx 0.73$
From Step 12:   $r \approx 0.95$

Thus the regression line equation is: $\hat{y} = 0.15 + 0.73x$.     ◄

Before a new set of data is entered, the SHIFT Scl EXE key sequence must be used to clear and reset the statistical memories. The range for the scales on the horizontal and vertical axes must be set to include the largest and smallest values in the ordered pairs. Consult the owner's manual for instructions on how to add and delete data points from an existing set, as well as other operations that can be performed.

The TI-81 can also be used to determine $b_0$, $b_1$, and $r$ for a set of $xy$ data. This graphics calculator uses menus displayed on the screen. When specific keys are pressed, a menu of options will appear. The "blue arrow keys" are then used to highlight a choice, and the ENTER key selects the choice. The next example illustrates the necessary keystrokes.

► **Example**    Find $b_0$, $b_1$, and $r$ for:

$$(2, 1), \quad (3, 3), \quad (5, 4), \quad \text{and} \quad (7, 5)$$

**Solution**

| Key Sequence | Comments |
|---|---|
| [RANGE] −1 [ENTER] 10 [ENTER]<br>1 [ENTER] −1 [ENTER] 10 [ENTER]<br>1 [ENTER] 1 [ENTER]<br>[2nd] [QUIT] | This sequence sets the scales on the $x$ and $y$ axes. |
| [2nd] [STAT] | Statistics menu is displayed. |
| Highlight [DATA] | The data menu is displayed. |
| Press 2 [ENTER] | The statistics memories are cleared. |
| [2nd] [STAT] | Statistics menu is displayed. |
| Highlight [DATA] | The data menu is displayed. |
| Press 1 | Starts the data entry. |
| 2 [ENTER] 1 [ENTER] | First data pair is entered. |
| 3 [ENTER] 3 [ENTER] | Second data pair is entered. |
| 5 [ENTER] 4 [ENTER] | Third data pair is entered. |
| 7 [ENTER] 5 [ENTER] | Fourth data pair is entered. |
| [2nd] [STAT] | Statistics menu is displayed. |
| Press 2 [ENTER] | Selects linear regression and displays $a$, $b$, and $r$.<br>$\hat{y} = 0.15 + 0.73x (\hat{y} = a + bx)$ |
| [2nd] [STAT] | Statistics menu is displayed. |
| Highlight [DRAW] | The draw menu is displayed. |
| Press 2 [ENTER] | The scatterplot is displayed. |
| [VARS] | Displays the VARS menu. |
| Highlight [LR] | Displays the LR menu. |
| Press 4 | Displays the equation. |

◄

**Name:** James A. Adams
**Occupation:** Quality Control Engineer
**Employer:** Ford Motor Company, Dearborn, Michigan

After earning my BS in biology from Bowling Green State University and an MBA from Central Michigan University in 1981, which included a graduate statistics course and a course in quantitative business methods, I began my quality control career at the Livonia, Michigan, plant of Chevrolet Motor Division, GMC. Between then and 1987, I took many industrial seminars in quality control applications of statistics, totaling about 1500 contact hours.

I am currently employed as a quality control engineer for the Stamping Division of Ford Motor Company, Body and Assembly General Office, Dearborn, Michigan. My job duties consist primarily of performing quality systems audits on Ford stamping plants and looking at quality improvement through the application of statistical process control (SPC) methods by plant personnel. The primary statistical methods applied in stamping operations are variable control charts (average and range, individual and moving range) and attributes control charts ($c$-charts and $p$-charts); there is some use of designed experiments. These statistical techniques are used to control piece-to-piece variation for metal stamping surface quality, weld integrity, and dimensional fit (checking fixtures and mating vehicle components).

I have worked in the quality control sector of the automotive industry for 11 of my 14 years of employment, as a statistician, using both basic and advanced statistical techniques, histograms, correlation regression, analysis of variance, control charts, lot sampling plans with operating characteristic curves, and designed experiments. The position is gratifying because I have an opportunity to help plant personnel make meaningful improvements to quality systems and vehicle quality through effective application of statistical methods.

In 1986, I began the part-time study of law at the University of Toledo, and I received my JD degree in 1990. I elected to take courses in product liability, environmental law, labor law, and evidence, all of which covered the use of statistical data evidence in trial proceedings. My new career objective is to specialize in product liability law, making extensive use of my background in statistical quality control in the auto industry.

The study of statistics has enabled me to attain a very rewarding and successful career in quality control. The kinds of statistical methods used in industry are indeed some of the most basic, especially quality control. After your basic courses, take as many courses in quality control as possible, especially if you are considering a career in industry. Statistics is a most practical and useful tool in business; learn as much about it as you can.

# 1987 motorcyclist deaths by age group

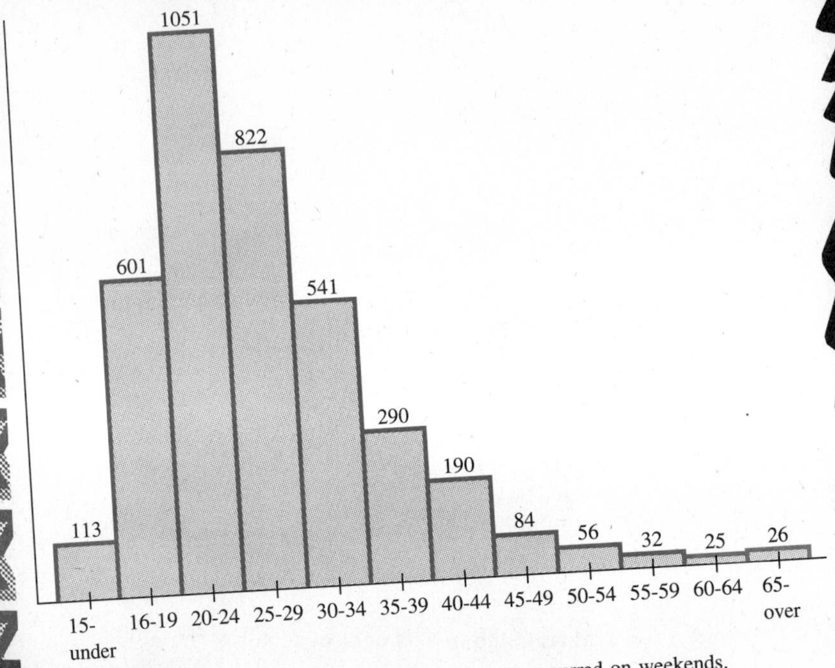

- Of all 1987 fatal motorcycle accidents, 41 percent occurred on weekends.
- The hours between 6 p.m. and 3 a.m. saw more than half — 56 percent — of all fatal motorcycle accidents in 1987.
- Warm-weather months — May through September — account for about two-thirds of motorcycle fatalities.
- Of all fatal single-vehicle cycle accidents, 55 percent involved a driver with a blood alcohol level higher than 0.10 percent — legally drunk in most states.

**Figure 10-1.** "Motorcycle Statistics Highlight Dangers" by Julie Stacey, *USA Today*, Dec. 6, 1988, p. 6A. Copyright © 1988 USA Today. Reprinted with permission.

# 10

# Nonparametric Methods

Figure 10-1 contains some information on motorcycle accidents. From the horizontal histogram, we can see that the distribution of motorcycle deaths by age is not approximately normal. Suppose that a researcher wanted to study this population and needed to make an inference about a measure of central tendency based on a small sample size. The assumption of an approximately normal population relative frequency distribution needed to justify using a *t*-distribution to perform a hypothesis test or construct a confidence interval would be inappropriate.

The situation above illustrates a kind of problem researchers can encounter in "real-world situations." In this final chapter, we will study some statistical techniques and probability models that may be used to make inferences about populations when, in situations like this, normal distribution assumptions are not appropriate. These are some of the statistical methods that are collectively referred to as *nonparametric methods.*

673

# 10-1 Inference for a Median—The Sign Test and Related Confidence Intervals

**Key Topics**

1. *A probability model for n observations*

2. *Testing $H_0$: $\tilde{\mu} = \tilde{\mu}_0$*

3. *Choosing the critical value (or values) $b_c$*

4. *The sign test*

5. *Finding approximate critical values for the sign test*

6. *Testing $H_0$: $\tilde{\mu}_d = 0$*

7. *Estimating a median*

---

## S T A T I S T I C S   I N   A C T I O N

Bill Young works in the research and development unit of an international paper corporation. He has been the leader of a team of five researchers trying to develop a new material for disposable diapers. The primary goal of the project is to develop a material that has the same absorption rate as the current material used, but has a higher decomposition rate to make it ecologically more acceptable. The research team has developed a material they labeled CPF903, that has most of the ingredients in the product currently being manufactured, plus a special fiber with a high decomposition rate. (Note: CPF903 stands for "cotton plus fiber, project number 903.")

A roll of CPF903 100 meters by 2 meters was produced and delivered to the research team. Bill wanted to first test CPF903 to see whether it had the same absorption rate as the material currently being used in the company's diapers.

To make the comparison, the team used the same test that the quality control unit uses to regularly check absorption properties of production runs of this product.

**Step 1** Pieces $5 \text{ cm}^2$ in area are cut from ten randomly selected sites on a roll of the material to be tested.

**Step 2** Each piece is placed on the surface of water in a closed container for a specified time period.

**Step 3** Each piece is then placed in a centrifuge and spun for a specified period of time.

**Step 4** The water collected from each piece is weighed to the nearest one-tenth of 1 gram and recorded.

The weights of the water recovered by Bill Young and his team from the ten test sites on the roll of CPF903 are shown in Table 10-1.

**TABLE 10-1    Weights of Water (in Grams) Recovered from Pieces of CPF903**

| | | | | | |
|---|---|---|---|---|---|
| Piece 1 | 3.4 | Piece 5 | 2.6 | Piece 8 | 5.0 |
| Piece 2 | 2.1 | Piece 6 | 0.5 | Piece 9 | 1.2 |
| Piece 3 | 4.8 | Piece 7 | 4.1 | Piece 10 | 2.3 |
| Piece 4 | 1.7 | | | | |

The data collected by quality control people using the same test on the material currently used by the company in disposable diapers suggests the median weight of water absorbed per piece is 3.0 grams. Bill used the data in Table 10-1 to compare the median absorption rates of CPF903 with the 3.0 grams per piece figure for the material currently used. The following hypothesis test was included in the report regarding CPF903 given to the executive board of the corporation.

In this test, the following symbols are used:

| | |
|---|---|
| $\tilde{\mu}$ | the median weight of water absorbed by 5-cm$^2$ pieces of the material developed by the research team |
| $B^*$ | the random variable that represents the number of pieces of CPF903 in the sample of ten whose collected water weights exceed 3.0 grams |
| $\tilde{\mu}_0 = 3.0$ | the hypothesized median amount of water absorbed per piece |

**Step 1**   $H_0$:   $\tilde{\mu} = 3.0$
            $H_a$:   $\tilde{\mu} < 3.0$

**Step 2**   $\alpha = 0.055$

**Step 3**   The test statistic will be:
         $B^* =$ the number of values in the sample greater than 3.0

**Step 4**   If $b^* \leq b_c \, (=2)$, then reject $H_0$.
         [From the binomial probability distribution, with $n = 10$ and $p = 0.5$, if $H_0$ is true $P(B^* \leq 2) = 0.055$.]

**Step 5**   From Table 10-1, $b^* = 4$ (that is, 3.4, 4.1, 4.8, and 5.0 are all greater than 3.0)

**Step 6**   Because $4 > 2$, do not reject $H_0$. That is, there is not sufficient evidence, with $\alpha = 0.055$, to conclude that the median weight of water absorbed by pieces of CPF903 5-cm$^2$ in area is less than 3.0 grams, the median weight of water absorbed by this size pieces of material currently used in disposable diapers.

## A probability model for *n* observations

If $X_1, X_2, X_3, \ldots, X_n$ represent a simple random sample of $n$ data points from a large population with median $\tilde{\mu}$, then the following is reasonable as an approximate probability model for the observations:

### A Probability Model Used to Make Inferences on $\tilde{\mu}$

Methods of inference for a median are based on a probability model for observations

$$X_1, \quad X_2, \quad X_3, \quad \ldots, \quad X_n$$

that says the $X$'s are *independent random variables* with a common individual (continuous) probability distribution for which:

$$P(X > \tilde{\mu}) = 0.5 \quad \text{and} \quad P(X < \tilde{\mu}) = 0.5$$

The independence assumption is reasonable, if the sample size $n$ is small compared with the population size $N$. The common distribution requirement is appropriate because the data are selected from the same population. Finally, the probability statements are reasonable because approximately one-half the values in a population of quantitative data are greater than $\tilde{\mu}$ (thus $P(X > \tilde{\mu}) = 0.5$) and one-half are less than $\tilde{\mu}$ (thus $P(X < \tilde{\mu}) = 0.5$).

This probability model was appropriate in the context of the test conducted by Bill Young on CPF903 for the following reasons:

1. The ten test pieces represent an area that is small compared with the total area of 200 m$^2$ in the entire roll.
2. All ten pieces in the sample were cut from the same roll.
3. Even though the median weight of water collected for every 5-cm$^2$ piece of the entire roll is unknown, the probability is 0.5 that a given piece in the sample will absorb less than (or more than) the median value $\tilde{\mu}$.

The key feature of this model is that the number of observations in the sample that are larger than $\tilde{\mu}$ becomes an approximately binomial random variable for $n$ trials with $p = 0.5$. The $n$ "trials" are the occurrences or nonoccurrences of the $n$ events:

$$X_i \text{ is larger than } \tilde{\mu}$$

The model states that $P(X_i > \tilde{\mu}) = 0.5$. Because the $X_i$'s are independent, the trials are also independent.

---

### DEFINITION 10-1 The Random Variable $B$

If $X_1, X_2, X_3, \ldots, X_n$ are $n$ independent random variables with a common individual probability distribution, then:

$$B = \text{the number of } X_i \text{ larger than } \tilde{\mu}$$

---

$B$ can be modeled as binomial with $p = 0.5$ and $n$ the number of items observed.

▶ **Example 1**    Assume that the median weight of water absorbed by test pieces of CPF903 is 3.0 grams per piece. Find the probability that eight or more of the pieces in the sample have weights greater than 3.0 grams.

**Solution**    The random variable

$B = $ the number of test pieces of CPF903 with more than 3.0 grams of water

has (approximately) a binomial distribution with $n = 10$ and $p = 0.5$. Using Appendix Table 3:

$$P(B \geq 8) = P(8) + P(9) + P(10)$$
$$= 0.044 + 0.010 + 0.001$$
$$= 0.055$$

Thus the probability is about 6 percent that eight or more test pieces of CPF903 would absorb more than the population median amount per piece.  ◀

## Testing $H_0$: $\tilde{\mu} = \tilde{\mu}_0$

The model above can be used to test a null hypothesis regarding $\tilde{\mu}$. We will use the symbol $\tilde{\mu}_0$ to stand for the hypothesized value of $\tilde{\mu}$.

### The Test Statistic for $H_0$: $\tilde{\mu} = \tilde{\mu}_0$

Because the number of observations with values greater than $\tilde{\mu}$ has a binomial distribution, we use

$B^*$ = the number of observations with values greater than $\tilde{\mu}_0$

as a test statistic for testing $H_0$: $\tilde{\mu} = \tilde{\mu}_0$.

For a test on $\tilde{\mu}$, there are three possibilities regarding the actual value of $\tilde{\mu}$ and $\tilde{\mu}_0$, shown graphically in Figure 10-2.

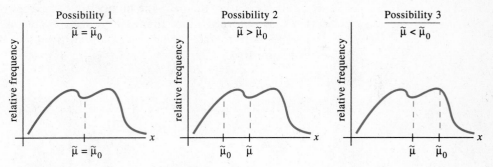

**Figure 10-2.** Population relative frequency distributions showing three possibilities with regard to values of $\tilde{\mu}$ and $\tilde{\mu}_0$.

*Possibility 1 represents the null hypothesis $\tilde{\mu} = \tilde{\mu}_0$.* If this possibility is true, then $B^*$ has a binomial distribution for $n$ trials and $p = 0.5$.

*Possibility 2 represents a case in which more than one-half of the actual population exceeds $\tilde{\mu}_0$, that is, $\tilde{\mu} > \tilde{\mu}_0$.* One should therefore expect $B^*$ to "tend to be larger" than a binomial random variable for $n$ trials and $p = 0.5$. (In other words, there is a *high probability* that more than one-half the sample will be greater than $\tilde{\mu}_0$.)

*Possibility 3 represents a case in which less than one-half the actual population exceeds $\mu_0$, that is, $\tilde{\mu} < \tilde{\mu}_0$.* One should therefore expect $B^*$ to "tend to be smaller" than a binomial random variable for $n$ trials and $p = 0.5$. (In other words, there is a *low probability* that more than one-half the sample will be greater than $\tilde{\mu}_0$.)

## The Rejection Regions for a Test of $H_0$: $\tilde{\mu} = \tilde{\mu}_0$

| If $H_a$: $\tilde{\mu} > \tilde{\mu}_0$, then reject $H_0$ if | If $H_a$: $\tilde{\mu} < \tilde{\mu}_0$, then reject $H_0$ if | If $H_a$: $\tilde{\mu} \neq \tilde{\mu}_0$, then reject $H_0$ if |
|---|---|---|
| $b^* \geq b_c$ | $b^* \leq b_c$ | $b^* \leq$ lower $b_c$,   or $b^* \geq$ upper $b_c$ |

where $b_c$ is a critical value (or values) for the test.

### Choosing the critical value (or values) $b_c$

The test statistic $B^*$ is a discrete random variable. As a consequence, only a few different values corresponding to particular binomial probabilities are possible as significance levels when testing $H_0$: $\tilde{\mu} = \tilde{\mu}_0$. To illustrate, consider the test performed by Bill Young in the Statistics in Action section for the hypotheses:

$$H_0: \quad \tilde{\mu} = 3.0$$
$$H_a: \quad \tilde{\mu} < 3.0$$

If $H_0$ is true, then $B^*$, the number of test pieces of CPF903 in the sample of ten that absorbed more than 3.0 grams of water, is binomial with $n = 10$ and $p = 0.5$. A histogram of the probability distribution of $B^*$ if $H_0$ is true, is shown in Figure 10-3.

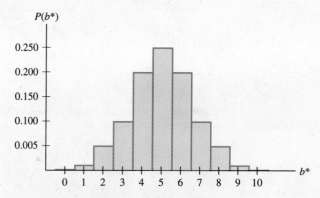

**Figure 10-3.**   Binomial probabilities based on $n = 10$ and $p = 0.5$.

If $H_a$ is true, then more weights less than 3.0 grams and fewer weights greater than 3.0 grams would be expected. The precise number of weights greater than 3.0 grams that would constitute sufficient evidence to reject $H_0$ is the value of $b_c$. The level of significance would then be $P(B^* \leq b_c)$. To illustrate, consider the following possibilities for $b_c$ and the corresponding values of $\alpha$.

**Possibility 1** The researchers require no pieces absorb more than 3.0 grams of water in order to reject $H_0$:

$$b_c = 0 \quad \text{and} \quad P(B^* \leq 0) = 0.001, \quad \alpha = 0.1\%$$

**Possibility 2** The researchers require one or fewer pieces absorb more than 3.0 grams of water in order to reject $H_0$:

$$b_c = 1 \quad \text{and} \quad P(B^* \leq 1) = 0.001 + 0.010$$
$$= 0.011, \quad \alpha = 1.1\%$$

**Possibility 3** The researchers require two or fewer pieces absorb more than 3.0 grams of water in order to reject $H_0$:

$$b_c = 2 \quad \text{and} \quad P(B^* \leq 2) = 0.001 + 0.010 + 0.044$$
$$= 0.055, \quad \alpha = 5.5\%$$

Notice that as $b_c$ increases by ones, the corresponding $\alpha$'s increase by discrete amounts equal to particular binomial probabilities.

▶ **Example 2**  Michelle Thomas is dean of women at a large university in the Midwest. She frequently sends information on housing costs in the area to prospective students. A recent report from the area's board of realtors states that the median rent for a one-bedroom furnished apartment is $385. Michelle feels that the figure is too low, and obtains a sample of monthly rents for such units from the university's directory of apartments. The following list contains the rents of the apartments in the sample:

| $410 | $375 | $480 | $525 | $430 | $368 |
|------|------|------|------|------|------|
| $395 | $460 | $350 | $400 | $500 | $450 |

Use these data, with $\alpha = 0.073$, to test the claim $H_0$: $\tilde{\mu} = \$385$ versus $H_a$: $\tilde{\mu} > \$385$.

**Solution**  **Step 1** $H_0$: $\tilde{\mu} = \$385$
$H_a$: $\tilde{\mu} > \$385$

**Step 2** $\alpha = 0.073$

**Step 3** The test statistic used is:
$B^*$ = the number of rents more than $385

**Step 4** If $b^* \geq b_c \ (= 9)$, then reject $H_0$.
(From Appendix Table 3, with $n = 12$ and $p = 0.50$, $P(B^* \geq 9) = 0.073$.)

**Step 5** In the data collected by Michelle, the following rents are more than $385:

$410   $480   $525   $430   $395   $460   $400   $500   $450

Thus $b^* = 9$.

**Step 6** Because $9 \geq 9$, reject $H_0$. That is, there is sufficient evidence, with $\alpha = 0.073$, to conclude that the median monthly rent of a one-bedroom furnished apartment in this area is more than $385.  ◀

## The sign test

Notice in the above test that we were only interested in the number of data points more than \$385. We were not concerned with the size of any difference between a data point and \$385. It is therefore possible to think of approaching a test of $H_0: \tilde{\mu} = \tilde{\mu}_0$ by changing each data point to a plus $(+)$ or minus $(-)$ sign depending upon whether it is respectively more or less than $\tilde{\mu}_0$. That is, if one forms the difference between each $x$ in a sample and $\tilde{\mu}_0$, then the following set of values is obtained:

$$x_1 - \tilde{\mu}_0, \quad x_2 - \tilde{\mu}_0, \quad x_3 - \tilde{\mu}_0, \quad \ldots, \quad x_n - \tilde{\mu}_0$$

If $x < \tilde{\mu}_0$, then $x - \tilde{\mu}_0 < 0$ and we write a *minus sign*. If $x > \tilde{\mu}_0$, then $x - \tilde{\mu}_0 > 0$ and we write a *plus sign*.

Common statistical practice is to discard any data points with $x = \tilde{\mu}_0$ and reduce $n$ accordingly. That is, if $x = \tilde{\mu}_0$, then $x - \tilde{\mu}_0 = 0$ and the data point is *deleted*.

### The Sign Test for $H_0: \tilde{\mu} = \tilde{\mu}_0$

If the plus $(+)$ or minus $(-)$ signs that result from the subtractions $x_i - \tilde{\mu}_0$ are used to test $H_0: \tilde{\mu} = \tilde{\mu}_0$, then the test is frequently called a *sign test*.

## Finding approximate critical values for the sign test

In Example 2, $n$ is 12. We were therefore able to use Appendix Table 3 to choose an appropriate $b_c$. For samples larger than 20, the binomial tables in this text would not be sufficient to obtain $b_c$. To choose critical values for the sign test for $n > 20$, either more complete tables or a normal approximation to the binomial distribution must be used. That is, if $n \geq 20$, then both $np = n(0.5)$ and $n(1 - p) = n(0.5)$ are at least 10. Therefore, according to the rule of thumb introduced in Chapter 4, the binomial probability distribution for $p = 0.5$ may be approximated using a normal distribution.

### Approximating $b_c$ for $n \geq 20$

If $n \geq 20$, then approximate critical values for the sign test may be obtained with equations 1 and 2:

$$\text{a lower } b_c = n(0.5) - z_c \sqrt{n(0.5)(0.5)} \quad \begin{array}{l} \rightarrow \text{ for } H_a: \tilde{\mu} < \tilde{\mu}_0 \\ \rightarrow \text{ for } H_a: \tilde{\mu} \neq \tilde{\mu}_0 \end{array} \quad \text{(1)}$$
$$\text{an upper } b_c = n(0.5) + z_c \sqrt{n(0.5)(0.5)} \quad \begin{array}{l} \\ \rightarrow \text{ for } H_a: \tilde{\mu} > \tilde{\mu}_0 \end{array} \quad \text{(2)}$$

The value of $z_c$ depends on the level of significance and whether the test is a one- or two-tail test.

▶ **Example 3**     Find approximate critical values for the sign test with $n = 50$ and $\alpha = 0.10$ for:

**a.** $H_a$: $\tilde{\mu} > \tilde{\mu}_0$
**b.** $H_a$: $\tilde{\mu} \neq \tilde{\mu}_0$

**Solution**     **a.** For $\alpha = 0.10$ and a one-tail test, $z_c = 1.28$. Thus:

$$b_c = 50(0.5) + 1.28\sqrt{50(0.5)(0.5)}$$
$$= 25 + 4.5$$
$$= 30$$

**b.** For $\alpha = 0.10$ and a two-tail test, $z_c = 1.65$. Thus:

$$\text{lower } b_c = 50(0.5) - 1.65\sqrt{50(0.5)(0.5)}$$
$$= 25 - 5.8$$
$$= 19$$

and

$$\text{upper } b_c = 50(0.5) + 1.65\sqrt{50(0.5)(0.5)}$$
$$= 25 + 5.8$$
$$= 31$$

Notice that in a conservative fashion, lower approximate critical values are rounded down and upper approximate critical values are rounded up.     ◀

## Testing $H_0$: $\tilde{\mu}_d = 0$

One of the most common uses of the sign test is in cases where the values used to obtain $B^*$ are themselves differences of naturally paired values, which are then used to test whether the population median difference is 0. For example, Table 10-2 (on the following page) contains examination scores of 20 academically deficient high school juniors. These students participated in a special summer program at a campus of a state university. The scores listed are on the same arithmetic test given at the start and again at the end of the 8-week program.

A natural question is whether these scores provide sufficient evidence to conclude that a majority test scores of such academically deficient high school juniors in the state would improve as a result of participation in such a program. Are there enough + differences in the 20 data pairs to show that the program in general helps such students?

Let us suppose that a normal distribution assumption is not appropriate for the distribution of differences. Then, of the statistical tests studied in this text, only the sign test is applicable. That is, if we let $\tilde{\mu}_d$ stand for the median difference (increase in score) that would result from giving all such academically deficient juniors the program, then the 20 data pairs can be used to test:

$$H_0: \quad \tilde{\mu}_d = 0 \qquad \text{The program is not effective.}$$
$$H_a: \quad \tilde{\mu}_d > 0 \qquad \text{The program is effective.}$$

TABLE 10-2    Arithmetic Scores of 20 Academically Deficient High School Juniors that Participated in a Summer Program

| Student | Pretest Score, $y_i$ | Posttest Score, $x_i$ | Difference, $x_i - y_i$ |
|---------|---------------------|----------------------|------------------------|
| 1 | 18 | 21 | +3 |
| 2 | 15 | 17 | +2 |
| 3 | 25 | 24 | −1 |
| 4 | 20 | 19 | −1 |
| 5 | 12 | 18 | +6 |
| 6 | 15 | 19 | +4 |
| 7 | 20 | 21 | +1 |
| 8 | 12 | 11 | −1 |
| 9 | 16 | 21 | +5 |
| 10 | 17 | 19 | +2 |
| 11 | 22 | 21 | −1 |
| 12 | 14 | 18 | +4 |
| 13 | 17 | 20 | +3 |
| 14 | 11 | 10 | −1 |
| 15 | 13 | 19 | +6 |
| 16 | 19 | 21 | +2 |
| 17 | 18 | 20 | +2 |
| 18 | 21 | 20 | −1 |
| 19 | 16 | 17 | +1 |
| 20 | 22 | 25 | +3 |

▶ **Example 4**    Use the sign test with $\alpha$ approximately 0.10 to test:

$$H_0: \quad \tilde{\mu}_d = 0$$
$$H_a: \quad \tilde{\mu}_d > 0$$

**Solution**    **Step 1**  $H_0: \quad \tilde{\mu}_d = 0$
$H_a: \quad \tilde{\mu}_d > 0$

**Step 2**  $\alpha = 0.10$

**Step 3**  The test statistic will be $B^* =$ the number of $d$'s larger than 0.

**Step 4**  $H_0$ should be rejected if $b^* \geq b_c$, where $b_c$ is such that the probability is 0.10 that a binomial random variable with $n = 20$ and $p = 0.5$ is at least $b_c$. For $\alpha = 0.10$ and a one-tail test, $z_c = 1.28$. Thus:

$$b_c = 20(0.5) + 1.28\sqrt{20(0.5)(0.5)}$$
$$= 10 + 2.9$$
$$= 13$$

**Step 5**  From Table 10-2, $b^* = 14$.

**Step 6**  Because $14 > 13$, reject $H_0$. That is, there is sufficient evidence, with $\alpha = 0.10$, to conclude that the median increase in scores for all members of the population is positive.    ◀

## Estimating a median

Examples 2 and 4 illustrate how the sign test can be used to test hypotheses about a median, or median difference. There remains, however, the issue of estimation for a population median or median difference. The first step in the process of estimation is to order the sample values from the smallest to the largest. Once this has been done, then:

$\tilde{x}$, the *sample median*, is a point estimate of $\tilde{\mu}$.

$\tilde{d}$, the *sample median difference*, is a point estimate of $\tilde{\mu}_d$.

▶ **Example 5**  Collene McHugh is a conservation worker at a state fish hatchery. She is given the task of estimating the median length of the trout in a tank that is to be used for stocking Lake Alminor in the Sierra Mountains. The lengths of ten randomly selected fish are measured to the nearest 1 millimeter (mm). The lengths are listed below.

$$\{148, 150, 146, 141, 168, 149, 140, 147, 150, 145\}$$

Give a point estimate of the median length of the trout in this tank.

**Solution**  The data are first ordered from smallest to largest:

$$\{140, 141, 145, 146, 147, 148, 149, 150, 150, 168\}$$

$$\tilde{x} = \frac{147 + 148}{2} = 147.5 \text{ mm}$$

Thus a point estimate of the median length of the trout is 147.5 mm.  ◀

Beyond giving point estimates for a population median, it is also possible to obtain confidence intervals using the ordered array of sample values. To obtain such an interval, use two values in the ordered array of observations that are the same distance from each end of the array. These observations are then used as the lower and upper limits of a confidence interval for $\tilde{\mu}$ (or $\tilde{\mu}_d$). That is, the confidence intervals for a median that we will study are determined by some number $j \in \{1, 2, 3, \ldots, n\}$, that specifies where in the ordered array of data points the end points are located.

## Confidence Limits for a Median

To construct confidence limits for a median, use:

$$\text{lower limit} = \text{the } j\text{th } smallest \ value \text{ in the sample} \qquad \textbf{(3)}$$

and

$$\text{upper limit} = \text{the } j\text{th } largest \ value \text{ in the sample} \qquad \textbf{(4)}$$

Using reasoning that is slightly beyond the scope of this text, it can be established that the confidence associated with such an interval is

$$[1 - 2P(X \le j - 1)] \cdot 100\%$$

for $X$ a binomial random variable for $n$ trials and $p = 0.5$.

Binomial tables can be used to determine how many values to count in from each end of the ordered array of values to obtain a particular confidence.

▶ **Example 6**   If $n = 8$, find the confidence associated with an interval from the second smallest to the second largest values in the sample for estimating a population median.

**Solution**   If the second smallest and largest values are used to determine the interval, then $j = 2$ and $j - 1 = 1$. Thus we need to compute:

$$P(X \leq 1)$$

with $X$ modeled as binomial with $n = 8$ and $p = 0.5$. Using Appendix Table 3, we have:

$$P(X \leq 1) = P(0) + P(1)$$
$$= 0.004 + 0.031$$
$$= 0.035$$

Now using the expression for the confidence coefficient

$$[1 - 2P(X \leq j - 1)] \cdot 100\%$$

becomes

$$[1 - 2(0.035)] \cdot 100\% = (1 - 0.070) \cdot 100\%$$
$$= 93\%$$

Thus employing an interval from the second smallest to the second largest sample values, we get the confidence associated with the interval of about 93 percent.   ◀

▶ **Example 7**   If $n = 9$, find a $j$ so that the interval from the $j$th smallest to the $j$th largest values in the sample has an associated confidence near 95 percent for estimating a population median.

**Solution**   For approximately 95 percent confidence:

$$1 - 2P(X \leq j - 1) \approx 0.95$$

that is:

$$2P(X \leq j - 1) \approx 0.05$$
$$P(X \leq j - 1) \approx 0.025$$

From Appendix Table 3, with $n = 9$ and $p = 0.5$:

$$P(0) + P(1) = 0.002 + 0.018$$

Thus:

$$P(X \leq 1) \approx 0.025$$

and

$$j - 1 = 1$$
$$j = 2$$

That is, an interval from the second smallest to the second largest value in the array of nine data points will have the confidence of about 95 percent associated with it. The exact value of the confidence is:

$$[1 - 2(0.020)] \cdot 100\% = 96\% \quad \blacktriangleleft$$

▶ **Example 8**  Give an interval with confidence close to 90 percent for the median length of the trout sampled by Collene.

**Solution**  **Discussion.** Because the desired confidence is 90 percent, choose $j$ so that:

$$1 - 2P(X \leq j - 1) \approx 0.90$$

where $X$ has a binomial distribution with $n = 10$ and $p = 0.5$. That is, $j$ should be such that:

$$P(X \leq j - 1) \approx 0.05$$

From Appendix Table 3, with $n = 10$ and $p = 0.5$:

$$P(0) = 0.001 \qquad P(X \leq 0) = 0.001$$
$$P(1) = 0.010 \qquad P(X \leq 1) = 0.011$$
$$P(2) = 0.044 \qquad P(X \leq 2) = 0.055$$

Notice that we have computed the cumulative probabilities until we obtained two values closest to the required value (one larger than specified and the other smaller). We then choose the one that is closer to the specified value.

Therefore:

$$j - 1 = 2$$
$$j = 3$$

and $[1 - 2(0.0550)] \cdot 100\% = 89\%$ is the exact confidence associated with an interval with $j = 3$. From Example 5, the ordered array of lengths is:

$$\{140, 141, 145, 146, 147, 148, 149, 150, 150, 168\}$$

Using 145, the third smallest data point as a lower limit, and 150, the third largest data point as an upper limit, we obtain an 89 percent confidence interval with:

$$\text{lower limit} = 145 \text{ mm}$$
$$\text{upper limit} = 150 \text{ mm} \quad \blacktriangleleft$$

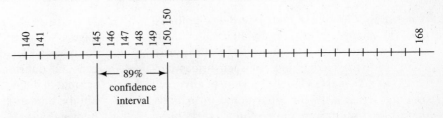

**Figure 10-4.** An 89% confidence interval for median length of trout in the tank.

For small $n$, binomial tables can be used to make a proper choice of $j$ to obtain a particular confidence level. For large $n$, the normal approximation to the binomial distribution yields an approximate choice of $j$.

### Approximating $j$ for Large $n$

If $n \geq 20$, then an approximate choice of $j$ for estimating a median can be made using:

$$j = 1 + n(0.5) - z\sqrt{n(0.5)(0.5)} \qquad (5)$$

▶ **Example 9**    Use the data in Table 10-2 to give a confidence interval estimate for the median improvement in arithmetic scores for the population of students under study. Use a confidence coefficient of about 90 percent.

Solution    **Discussion.** The difference in the test scores from Table 10-2 must first be ordered from smallest to largest. The ordered array is given below

$$\{-1, -1, -1, -1, -1, -1, 1, 1, 2, 2, 2, 2, 3, 3, 3, 4, 4, 5, 6, 6\}$$

Now using equation (5), with $z = 1.65$ for a 90 percent confidence coefficient:

$$j = 1 + 20(0.5) - 1.65\sqrt{20(0.5)(0.5)}$$
$$= 7.31 \ldots$$

or (conservatively) rounding down, $j = 7$ is indicated. As the seventh smallest and seventh largest differences are respectively 1 and 3, the interval with

$$\text{lower limit} = 1 \text{ point}$$

and

$$\text{upper limit} = 3 \text{ points}$$

is an approximately 90 percent confidence interval for $\tilde{\mu}_d$.    ◀

## Exercises 10-1    Set A

1. A newspaper in a large midwestern city reported that the median age of individuals previously divorced who remarry continued a downward trend, with the median age for such brides currently at 32 years. Of ten previously divorced brides married last year, what is the probability that at least seven were more than 32 years old if this reported median is correct?

2. A statistics department at a large state college gave a standardized test to the 847 students that completed Elementary Statistics 120 during a recent fall quarter. The median score was 83. If a sample of eight tests is selected, what is the probability that at least six of the tests have scores higher than 83?

3. A major segment of the industrial sector employs about 280,000 hourly rated workers. A management report states that the median salary of these workers is \$11.25 per hour. A sample of 12 of these employees is selected. What is the probability that at least nine of those sampled earn more than the stated median wage if the reported median is correct?

4. A computer printout of the heights of all children entering the first grade in a large school district in a city in Texas gave a median height of 35.8 in. A random sample of nine of these students is selected. What is the probability that at least seven are taller than the median height?

5. A report from a real estate board appeared in a large eastern city newspaper. The report stated that the median price of a three-bedroom house in the city was \$74,900. A sample of $n = 10$ such houses is taken from the list of those currently for sale. What is the probability that less than three of those sampled are priced over \$74,900 if the reported median is correct?

6. According to a report by the Population Division of the U.S. Census Bureau, the median age of widowed brides recently rose to 53.1 years. A sample of $n = 12$ brides who were widowed was taken from marriage licenses issued during 1 month in a large southern city. If the reported median is correct, what is the probability that fewer than five of these brides are more than 53.1 years old?

7. A famous trainer of hunter and jumper thoroughbred horses claims that the median height of these horses is 66 in. (The height of a horse is measured at the base of the neck and is usually stated in hands. One hand is 4 in., thus 66 in. = 16.2 hands.) A second trainer feels the figure is too high and samples ten hunter and jumper thoroughbreds registered with the Jockey Club. If the sample yields three horses more than 16.2 hands tall, can the claim $H_0$: $\tilde{\mu} = 16.2$ be rejected with $\alpha$ approximately 5%?

8. A teachers' organization reports that the median age of state high school teachers is 42 years. A sample of 12 names randomly selected from the current directory yielded four who were older than 42 years. Test:

$$H_0: \quad \tilde{\mu} = 42$$
$$H_a: \quad \tilde{\mu} < 42$$

Use $\alpha \approx 7\%$.

9. A state employees' organization asserts that the median number of years worked by the 137,000 state workers is 11.4 years. An audit of 11 such workers yielded two who have worked more than 11.4 years. Use the data to test:

$$H_0: \quad \tilde{\mu} = 11.4 \text{ years}$$
$$H_a: \quad \tilde{\mu} \neq 11.4 \text{ years}$$

Use $\alpha \approx 6\%$.

10. An association of nurseries in a large western city reported that a recent study showed that the median number of living plants in city households was six. Feeling that the figure was overstated, a statistics student sampled 14 households in the city. The sample yielded four with more than six living plants. Use the data to test:

$$H_0: \quad \tilde{\mu} = 6 \text{ plants}$$
$$H_a: \quad \tilde{\mu} < 6 \text{ plants}$$

Use $\alpha \approx 9\%$.

*In Exercises 11–14, find approximate critical values for a sign test for each set of n and α. See Example 3 and round each critical value in a conservative manner.*

11. With $n = 48$ and $\alpha = 0.10$, for:
   a. $H_a$: $\tilde{\mu} < \tilde{\mu}_0$
   b. $H_a$: $\tilde{\mu} \neq \tilde{\mu}_0$

12. With $n = 64$ and $\alpha = 0.05$, for:
   a. $H_a$: $\tilde{\mu} \neq \tilde{\mu}_0$
   b. $H_a$: $\tilde{\mu} > \tilde{\mu}_0$

13. With $n = 88$ and $\alpha = 0.02$, for:
   a. $H_a$: $\tilde{\mu} \neq \tilde{\mu}_0$
   b. $H_a$: $\tilde{\mu} < \tilde{\mu}_0$

14. With $n = 100$ and $\alpha = 0.01$, for:
   a. $H_a$: $\tilde{\mu} > \tilde{\mu}_0$
   b. $H_a$: $\tilde{\mu} \neq \tilde{\mu}_0$

*In Exercises 15–18, find the confidence associated with the indicated intervals for estimating a population median.*

**15.** If $n = 9$ and the second smallest and the second largest values in the sample are used, find the confidence associated with the interval.

**16.** If $n = 12$ and the third smallest and the third largest values in the sample are used, find the confidence associated with the interval.

**17.** If $n = 15$ and the fifth smallest and the fifth largest values in the sample are used, find the confidence associated with the interval.

**18.** If $n = 20$ and the fifth smallest and the fifth largest values in the sample are used, find the confidence associated with the interval.

*In Exercises 19–22, find j so that an interval for estimating a population median from the jth smallest to the jth largest observed value has approximately the stated confidence.*

**19.** $n = 10$ and the confidence is near 90%     **20.** $n = 12$ and the confidence is near 95%

**21.** $n = 16$ and the confidence is near 98%     **22.** $n = 20$ and the confidence is near 99%

*In Exercises 23–26, for the following values of n, find j so that the interval for estimating a population median from the jth smallest to the jth largest observed value has approximately the stated confidence.*

**23.** $n = 30$ and the confidence is near 90%     **24.** $n = 48$ and the confidence is near 95%

**25.** $n = 64$ and the confidence is near 98%     **26.** $n = 100$ and the confidence is near 99%

**27.** A sample of $n = 8$ tomatoes randomly selected from several cases delivered to a local supermarket yielded the following weights to the nearest one-tenth of an ounce:

$$2.8 \quad 3.0 \quad 3.1 \quad 3.3 \quad 3.3 \quad 3.4 \quad 3.6 \quad 3.9$$

a. Give a point estimate for the median weight of these tomatoes.

b. State an interval with a confidence of approximately 90 percent for the median weight of these tomatoes.

**28.** The percent moisture content in hard red winter wheat loaded on a ship for export is checked, by using a Motomco Moisture Meter. The following values are obtained:

$$9.1 \quad 9.3 \quad 9.9 \quad 9.9 \quad 10.3 \quad 10.6 \quad 10.6 \quad 10.7 \quad 11.0 \quad 11.2$$
$$11.3 \quad 11.5 \quad 11.7 \quad 11.8 \quad 12.4 \quad 12.8 \quad 13.0 \quad 13.9 \quad 14.0 \quad 14.5$$

a. Give a point estimate for the median moisture content for this shipment of wheat.

b. State an interval with a confidence of approximately 95 percent for the median moisture content.

**29.** A sample of 12 automobile supply stores in a greater metropolitan area in Oregon yielded the following prices (dollars) for a set of brake shoes:

$$16.50 \quad 19.95 \quad 21.95 \quad 22.50 \quad 24.44 \quad 25.00$$
$$26.50 \quad 27.00 \quad 28.11 \quad 28.50 \quad 28.95 \quad 29.46$$

a. Give a point estimate of the median price of a set of brake shoes for this area.

b. State an interval with a confidence of approximately 96 percent for the median price of brake shoes.

**30.** The ages in 36 personnel files selected at a large air force base in California are:

| | | | | | | | | |
|---|---|---|---|---|---|---|---|---|
| 19.2 | 19.6 | 19.8 | 20.2 | 20.3 | 20.8 | 21.9 | 22.3 | 22.4 |
| 22.4 | 22.6 | 22.8 | 22.8 | 23.4 | 23.4 | 23.6 | 23.8 | 24.2 |
| 24.3 | 24.6 | 25.3 | 26.3 | 26.8 | 26.8 | 26.8 | 27.4 | 28.1 |
| 28.6 | 28.8 | 30.1 | 20.6 | 30.7 | 32.8 | 33.5 | 34.3 | 36.3 |

a. Give a point estimate of the median age of the personnel at this base.

b. State an interval with a confidence of approximately 98 percent for the median age of these personnel.

# Exercises 10-1    Set B

*In Exercises 1 and 2, do the following:*

*a. Randomly sample the specified number of stocks from Data Set II, Appendix B.*

*b. Count the number of stock prices that increased in price on this day ( + ).*

*c. Use the results of Part (b) to test $H_0$: $\tilde{\mu}_d = 0$ at the indicated level of significance.*

**1.** Sample $n = 25$ stock prices and use $H_a$: $\tilde{\mu}_d > 0$. Use $\alpha = 0.05$.

**2.** Sample $n = 20$ stock prices and use $H_a$: $\tilde{\mu}_d < 0$. Use $\alpha = 0.01$.

**3.** An instructor at a small college in the East developed a program to improve physical capabilities of certain physically handicapped students. To evaluate the program he gave a pretest and posttest to 16 students in the program. The results of the two tests are in the following table:

| Student | Posttest Score $(x_i)$ | Pretest Score $(y_i)$ | Difference $d_i = x_i - y_i$ |
|---|---|---|---|
| 1 | 18 | 12 | |
| 2 | 10 | 4 | |
| 3 | 20 | 15 | |
| 4 | 8 | 6 | |
| 5 | 8 | 10 | |
| 6 | 6 | 3 | |
| 7 | 14 | 7 | |
| 8 | 4 | 5 | |
| 9 | 13 | 10 | |
| 10 | 16 | 6 | |
| 11 | 12 | 8 | |
| 12 | 13 | 9 | |
| 13 | 19 | 12 | |
| 14 | 11 | 8 | |
| 15 | 17 | 13 | |
| 16 | 12 | 7 | |

a. Calculate $d$ for each of the 16 pairs of scores.

b. Test $H_0$: $\tilde{\mu}_d = 0$ versus $H_a$: $\tilde{\mu}_d > 0$. Use $\alpha = 0.10$.

**4.** To determine whether prices on items at a discount store $A$ are lower than prices on the same items at a similar store $B$ not classified as a discount store, a homemaker sampled 20 items from each of the two stores. The prices of the 20 items are shown in the following table:

| Item | Store A ($x_i$) | Store B ($y_i$) | Difference $d_i = x_i - y_i$ |
|------|-----------------|-----------------|------------------------------|
| 1  | $0.29 | $0.31 | |
| 2  | 1.35 | 1.33 | |
| 3  | 0.99 | 1.09 | |
| 4  | 0.62 | 0.69 | |
| 5  | 0.79 | 0.85 | |
| 6  | 1.85 | 1.79 | |
| 7  | 0.49 | 0.47 | |
| 8  | 0.09 | 0.10 | |
| 9  | 0.19 | 0.21 | |
| 10 | 0.50 | 0.45 | |
| 11 | 2.21 | 1.98 | |
| 12 | 0.89 | 1.05 | |
| 13 | 0.98 | 1.19 | |
| 14 | 1.89 | 1.99 | |
| 15 | 2.29 | 2.09 | |
| 16 | 3.36 | 3.79 | |
| 17 | 1.69 | 1.85 | |
| 18 | 2.85 | 2.79 | |
| 19 | 0.79 | 1.03 | |
| 20 | 2.29 | 2.59 | |

a. Calculate $d$ for each of the 20 pairs of prices.

b. Test $H_0: \tilde{\mu}_d = 0$ versus $H_a: \tilde{\mu}_d < 0$. Use $\alpha = 0.05$.

**5.** The effect of altitude on running ability was tested on 12 individuals who specialized in different sports. The participants were asked to run 880 m (about 0.5 mi) at sea level and then again at an altitude of 4500 feet. Their times are listed in the following table:

| Athlete | Time at Sea Level ($x_i$) (sec) | Time at 4500 ft ($y_i$) (sec) | Difference $d_i = x_i - y_i$ |
|---------|----------------------------------|-------------------------------|------------------------------|
| 1  | 210 | 225 | |
| 2  | 130 | 125 | |
| 3  | 198 | 202 | |
| 4  | 190 | 193 | |
| 5  | 150 | 148 | |
| 6  | 178 | 183 | |
| 7  | 234 | 252 | |
| 8  | 167 | 170 | |
| 9  | 126 | 123 | |
| 10 | 145 | 144 | |
| 11 | 188 | 193 | |
| 12 | 204 | 223 | |

a. Calculate $d$ for each of the 12 pairs of times.

b. Construct an approximately 90 percent confidence interval for the median difference in times.

**6.** The ability of individuals to perform tasks with the hand they normally do not use was tested by a clinical psychologist at a major university. She had each person in the experiment do a series of five manipulative tasks with the hand (right or left) they normally use. She then had each person do the same tasks with the other hand (left or right). The times, in seconds, for the two trials are in the following table:

| Volunteer | Time with Hand Normally Used $(x_i)$ (sec) | Time with Other Hand $(y_i)$ (sec) | Difference $d_i = y_i - x_i$ |
|---|---|---|---|
| 1 | 120 | 195 | |
| 2 | 95 | 128 | |
| 3 | 142 | 203 | |
| 4 | 78 | 92 | |
| 5 | 133 | 140 | |
| 6 | 107 | 162 | |
| 7 | 151 | 220 | |
| 8 | 84 | 92 | |
| 9 | 116 | 173 | |
| 10 | 140 | 189 | |
| 11 | 98 | 127 | |
| 12 | 138 | 177 | |
| 13 | 104 | 145 | |
| 14 | 166 | 219 | |
| 15 | 129 | 140 | |
| 16 | 100 | 134 | |

a. Calculate $d$ for each of the 16 pairs of times.
b. Construct an approximately 95 percent confidence interval for the median difference in times.

## 10-2 Inference for a Difference in Medians— The Wilcoxon Rank Sum Test and Related Confidence Intervals

**Key Topics**

1. *Model assumptions for comparing the medians of two populations*

2. *The Wilcoxon rank sum test for* $H_0: \tilde{\mu}_1 - \tilde{\mu}_2 = 0$

3. *Nonparametric estimation of a difference in location*

4. *Confidence intervals for a difference in location*

### STATISTICS IN ACTION

Buena Vista Community College District has five campuses with an estimated student population of 78,500. Pat Kelly is the central administrator responsible for buying the operating supplies for all five campuses. He is known for his efforts to get ''the most for the money.'' As a consequence, he frequently conducts comparative tests of similar products to determine which one is the best buy.

The district has recently been buying brand *A* and brand *B* light bulbs from two manufacturers. The brand *B* manufacturer contacted Pat with an offer to give an 8% discount, if he purchased all light bulbs for the district from them. When notified of the offer by the brand *B* manufacturer, the maker of brand *A* bulbs refused to match the offer. They contended that brand *A* bulbs were better than brand *B* bulbs, based on a comparison of hours of continuous service before failure.

Pat decided to make a comparison of the service lives of brand *A* and brand *B* bulbs before responding to either manufacturer. He therefore contacted service personnel at the five campuses for help in performing the test. Eleven sites were selected where bulbs remained lit until they burned out. New brand *A* bulbs were placed in five locations and new brand *B* bulbs were placed in six locations. Timers were installed at each site to measure the number of hours each bulb operated from the time it was turned on until it burned out. The life times of the 11 bulbs, measured to the nearest 1 hour, are listed in Table 10-3.

Pat arrayed the data in Table 10-3 from smallest to largest values. The ordered array is shown in Figure 10-5, and the five data points for brand *A* bulbs are circled.

Pat used the *Wilcoxon rank sum test* based on the ordered array of data. He was unable, on the basis of the data in Table 10-3, to reject the null hypothesis that the two brands have the same lifetime distribution. He therefore decided to conduct a more extensive study, involving larger samples, before changing his purchasing policy. The details of the Wilcoxon rank sum test Pat used are studied in this section.

**TABLE 10-3**   **The Numbers of Hours of Service Before Failure of 11 Light Bulbs**

| Brand A | 390 | 840 | 900 | 1070 | 1700 | |
| Brand B | 150 | 260 | 530 | 710 | 890 | 1590 |

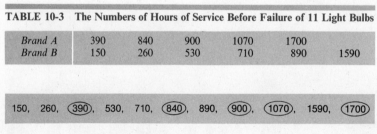

**Figure 10-5.**   The hours of service data arranged in order from smallest to largest, with the brand *A* data circled.

## Model assumptions for comparing the medians of two populations

In Section 8-2, small sample tests and confidence intervals for comparing two means were studied. The tests and intervals were seen to be appropriate under the assumption that the populations are normal with a common variance.

In this section, we shall study nonparametric methods for comparing the locations of two populations. Such methods do not require the populations to possess bell-shaped relative frequency distributions.

## Model Assumptions for Comparing the Medians of Populations

Nonparametric methods of inference for the difference in location of two populations using samples

$$x_1, \quad x_2, \quad x_3, \quad \ldots, \quad x_{n_1}$$
$$y_1, \quad y_2, \quad y_3, \quad \ldots, \quad y_{n_2}$$

are based on a probability model in which the $x_i$'s and $y_j$'s are observed values of continuous random variables. The $X_i$'s are assumed to have a common individual probability distribution that differs from the common individual probability distribution of the $Y_j$'s only in location but not in shape.

To facilitate using tables later in this section, we will name the samples so that $n_1 \leq n_2$. That is, the second sample is at least as large as the first. Notice that in the light bulb study, Pat Kelly used five brand $A$ bulbs and six brand $B$ bulbs.

The model assumption regarding the shapes of the probability distributions of the $X_i$'s and $Y_j$'s is illustrated graphically in Figure 10-6. Notice that the probability distributions have the same general shape, but the locations of the two distributions are different.

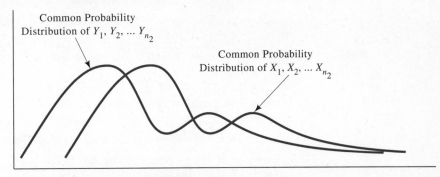

**Figure 10-6.**   Probability distribution of $X_i$'s and $Y_j$'s.

The assumptions do not require a normal probability distribution for the $x$ or $y$ variables. These assumptions are appropriate when the data are simple random samples from large populations that have relative frequency distributions that are similar in shape, but with possibly different medians $\tilde{\mu}_1$ and $\tilde{\mu}_2$. The difference $\tilde{\mu}_1 - \tilde{\mu}_2$ corresponds to the amount by which the $x$-distribution is shifted to the right of the $y$-distribution.

## The Wilcoxon rank sum test for $H_0: \tilde{\mu}_1 - \tilde{\mu}_2 = 0$

The test we will study for

$$H_0: \quad \tilde{\mu}_1 - \tilde{\mu}_2 = 0$$

uses *the ranks of the values in two samples.* To rank the values from two samples, we first combine the data into one set. The values in this set are then ordered from the smallest to the largest.

▶ **Example 1**    Find the ranks for the brand $A$ bulbs.

**Solution**    In Figure 10-5, Pat arrayed the data from the two samples of bulbs and circled the numbers that correspond to brand $A$ bulbs. The ranks of every bulb's lifetime are identified in the following table:

| Observation | Rank |
|:---:|:---:|
| 150 | 1 |
| 260 | 2 |
| (390) | 3 |
| 530 | 4 |
| 710 | 5 |
| (840) | 6 |
| 890 | 7 |
| (900) | 8 |
| (1070) | 9 |
| 1590 | 10 |
| (1700) | 11 |

The ranks for the brand $A$ bulbs are 3, 6, 8, 9, and 11.    ◀

If $H_0: \tilde{\mu}_1 - \tilde{\mu}_2 = 0$ is true, then $\tilde{\mu}_1 = \tilde{\mu}_2$, and there is no difference in the location of the $x$- and $y$-distributions. As a consequence, every collection of $n_1$ of the numbers

$$1, \quad 2, \quad 3, \quad 4, \quad \ldots, \quad (n_1 + n_2 - 1), \quad (n_1 + n_2)$$

is equally likely to compose the set of ranks corresponding to values in the first sample. That is, if the two distributions are the same, then the ranks for the first sample are a simple random sample of size $n_1$ from the numbers:

$$1, \quad 2, \quad 3, \quad 4, \quad \ldots, \quad (n_1 + n_2)$$

To illustrate, consider the 11 brand $A$ and brand $B$ bulbs in the test conducted by Pat Kelly. If the distributions of life times for bulbs from the two manufacturers are the same, then the ranks of the brand $A$ bulbs are equally likely to take on any five of the 11 possible rankings. (In Example 1, the rankings turned out to be 3, 6, 8, 9, and 11.)

## The sum of the ranks random variable W

If the $x$-distribution and $y$-distribution have equal medians (and are therefore the same), then the probability distribution of the random variable

$$W = \text{the sum of the ranks of the first sample}$$

can be obtained by treating the $x$ sample as a simple random sample of size $n_1$ from the integers

$$1, 2, 3, 4, \ldots, n$$

To illustrate how a probability distribution for $W$ can be constructed, the case for $n_1 = 2$ and $n_2 = 3$ is shown in Example 2.

▶ **Example 2**   If $n_1 = 2$ and $n_2 = 3$ and there is no difference in the locations of the $x$- and $y$-distributions, find the probability distribution for the sum of the ranks assigned to the values in the first sample.

**Solution**   With $n_1 + n_2 = 5$ and $n_1 = 2$:   $\binom{5}{2} = \dfrac{5!}{2!(5-2)!} = \dfrac{5 \cdot 4 \cdot 3 \cdot 2 \cdot 1}{2 \cdot 1 \cdot 3 \cdot 2 \cdot 1} = 10$

There are therefore ten different possible ways in which the $x$'s from sample 1 can be ranked. Because each ranking is equally likely, the probability of any one ranking is 0.1.

| Collection of Two of the Values from $\{1, 2, 3, 4, 5\}$ | Probability of the Pair Occurring | Corresponding Value of W, W = Sum of Ranks |
|---|---|---|
| 1 and 2 | 0.1 | $1 + 2 = 3$ |
| 1 and 3 | 0.1 | $1 + 3 = 4$ |
| 1 and 4 | 0.1 | $1 + 4 = 5$ |
| 1 and 5 | 0.1 | $1 + 5 = 6$ |
| 2 and 3 | 0.1 | $2 + 3 = 5$ |
| 2 and 4 | 0.1 | $2 + 4 = 6$ |
| 2 and 5 | 0.1 | $2 + 5 = 7$ |
| 3 and 4 | 0.1 | $3 + 4 = 7$ |
| 3 and 5 | 0.1 | $3 + 5 = 8$ |
| 4 and 5 | 0.1 | $4 + 5 = 9$ |

Collecting the possible values and probabilities for $W$, we obtain the following table for the probability distribution of $W$:

| $w$ | 3 | 4 | 5 | 6 | 7 | 8 | 9 |
|---|---|---|---|---|---|---|---|
| $P(w)$ | 0.1 | 0.1 | 0.2 | 0.2 | 0.2 | 0.1 | 0.1 |

Based on this table, we see that $P(W = 5) = 0.2$. That is, the probability is 0.2 that the sum of the ranks of the values in the first sample is 5.   ◀

The random variable $W$ is a convenient choice of a test statistic for testing:

$$H_0: \quad \tilde{\mu}_1 - \tilde{\mu}_2 = 0$$

The graphs in Figure 10-7 are useful in determining appropriate rejection regions based on $W$.

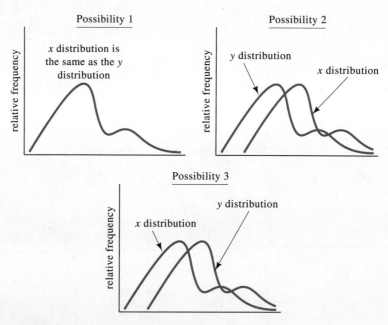

**Figure 10-7.**   Three possibilities for $x$ and $y$-distributions when testing $H_0: \tilde{\mu}_1 - \tilde{\mu}_2 = 0$.

## Three possibilities for *x*- and *y*-distributions

**Possibility 1**  The null hypothesis is true so $\tilde{\mu}_1 = \tilde{\mu}_2$.

If $H_0$ is true, then a probability distribution for $W$ can be constructed as in Example 2.

**Possibility 2**  The null hypothesis is false and $\tilde{\mu}_1 - \tilde{\mu}_2 > 0$.

The median of the $x$-distribution is greater than the median of the $y$-distribution. As a consequence, one can expect $W$ to "tend to be larger than it would be if $H_0$ were true."

**Possibility 3**  The null hypothesis is false and $\tilde{\mu}_1 - \tilde{\mu}_2 < 0$.

The median of the $x$-distribution is less than the median of the $y$-distribution. As a consequence, one can expect $W$ to "tend to be smaller than it would be if $H_0$ were true."

As a consequence, appropriate critical regions for the three possible alternatives to $H_0: \tilde{\mu}_1 - \tilde{\mu}_2 = 0$ are:

a. with $H_a: \tilde{\mu}_1 - \tilde{\mu}_2 > 0$, $H_0$ should be rejected for $W \geq w_c$.
b. with $H_a: \tilde{\mu}_1 - \tilde{\mu}_2 < 0$, $H_0$ should be rejected for $W \leq w_c$.
c. with $H_a: \tilde{\mu}_1 - \tilde{\mu}_2 \neq 0$, $H_0$ should be rejected for $W \leq$ lower $w_c$ or $W \geq$ upper $w_c$.

Two tables giving upper and lower percentage points of the distribution of $W$, if $H_0$ is true, are given in Appendix Table 8 of this text. The format of these tables is illustrated in Figure 10-8 for $n_1 = 5$ and $n_2 = 6$.

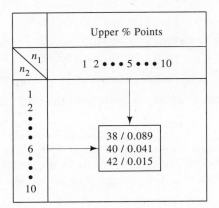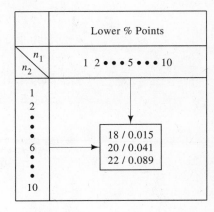

**Figure 10-8.** A portion of Appendix Table 9.

According to the table, if $n_1 = 5$ and $n_2 = 6$, and $H_0$ is true, then calculations like those illustrated in Example 2 show that $P(W \geq 38) = 0.089$. With five values in the first sample, the probability that the sum of the ranks of these values is greater than or equal to 38 is approximately 0.089 or 8.9 percent. Also, for $n_1 = 5$ and $n_2 = 6$, if $H_0$ is true, then $P(W \leq 22)$ is also 0.089. That is, the probability is about 8.9 percent that the sum of the ranks of these five values is less than or equal to 22. The following probabilities are also given in Figure 10-8 for $W$:

$$P(W \geq 40) = 0.041 \text{ or about } 4.1\%$$

$$P(W \geq 42) = 0.015 \text{ or about } 1.5\%$$

$$P(W \leq 18) = 0.015 \text{ or about } 1.5\%$$

$$P(W \leq 20) = 0.041 \text{ or about } 4.1\%$$

## For Testing $H_0 : \tilde{\mu}_1 - \tilde{\mu}_2 = 0$

If $H_a: \tilde{\mu}_1 - \tilde{\mu}_2 < 0$, then use the table of *lower percentage points*.
If $H_a: \tilde{\mu}_1 - \tilde{\mu}_2 \neq 0$, then use both the *lower and upper percentage points*.
If $H_a: \tilde{\mu}_1 - \tilde{\mu}_2 > 0$, then use the table of *upper percentage points*.

▶ **Example 3**     Use the data in Table 10-3 to test:

$H_0$:   $\tilde{\mu}_1 - \tilde{\mu}_2 = 0$     There is no difference in the medians (and therefore distributions as well).

$H_a$:   $\tilde{\mu}_1 - \tilde{\mu}_2 > 0$     The brand A bulbs have a greater median service life than brand B bulbs.

Use $\alpha = 8.9\%$.     This value is in Appendix Table 8 with $n_1 = 5$ and $n_2 = 6$.

**Solution**     **Step 1**  $H_0$:   $\tilde{\mu}_1 - \tilde{\mu}_2 = 0$
$H_a$:   $\tilde{\mu}_1 - \tilde{\mu}_2 > 0$

**Step 2**  $\alpha = 8.9\%$     $P$(Type I error) $\approx 0.089$.

**Step 3**  The test statistic used is:

$W$ = the sum of the ranks of the values from the first sample

**Step 4**  If $w \geq w_c$ (=38), then reject $H_0$. (From Figure 10-8, $P(W \geq 38) = 0.089$.)

**Step 5**  From Example 2, the sum of the ranks is:

$$W = 3 + 6 + 8 + 9 + 11 = 37$$

**Step 6**  Because $37 < 38$, do not reject $H_0$. That is, with $\alpha = 0.089$, there is not sufficient evidence to conclude that brand A bulbs have a greater median service time than brand B bulbs.     ◀

## Approximating $w_c$ for large $n_1$ and $n_2$

If both $n_1$ and $n_2$ are at least 10, then approximate critical points for the Wilcoxon rank sum test can be calculated using equations 6 and 7.

### Approximating $w_c$ for Large $n_1$ and $n_2$

If $n_1 \geq 10$ and $n_2 \geq 10$, then approximately:

$$\text{lower } w_c = \frac{n_1(n_1 + n_2 + 1)}{2} - z_c \sqrt{\frac{n_1 n_2(n_1 + n_2 + 1)}{12}} \tag{6}$$

and/or

$$\text{upper } w_c = \frac{n_1(n_1 + n_2 + 1)}{2} + z_c \sqrt{\frac{n_1 n_2(n_1 + n_2 + 1)}{12}} \tag{7}$$

▶ **Example 4**     A retail outlet chain has two suppliers of dry cell batteries for flashlights. Lucinda Turley, the manager responsible for buying this item for the chain, obtained a test device that drains the charge of a battery in a relatively short period of time. She sampled 12 batteries from shipments received from both suppliers and tested them with the device. The numbers of seconds it took to render each battery useless were recorded and are displayed below.

Supplier $A$    250, 330, 390, 420, 540, 590, 600, 820, 950, 1070, 1160, 1700

Supplier $B$    70, 150, 180, 260, 410, 530, 680, 710, 840, 890, 1510, 1590

Relative frequency distributions of battery lifetimes using the test device are typically skewed (and thus not normal). Use the Wilcoxon rank sum test to test the null hypothesis that the two shipments have the same median battery lifetime. Use $\alpha = 0.05$.

**Solution**    **Discussion.** Before using the six-step hypothesis-testing format, we find the ranks of the values in the first sample (supplier $A$). The circled values in the following ordered array belong to the first sample:

70   150   180   (250)   260   (330)   (390)   410   (420)   530   (540)   (590)   (600)
680   710   (820)   840   890   (950)   (1070)   (1160)   1510   1590   (1700)

Thus the ranks corresponding to the first sample are 4, 6, 7, 9, 11, 12, 13, 16, 19, 20, 21, and 24.

**Step 1**  $H_0$:  $\tilde{\mu}_1 - \tilde{\mu}_2 = 0$
$H_a$:  $\tilde{\mu}_1 - \tilde{\mu}_2 \neq 0$

**Step 2**  $\alpha = 0.05$

**Step 3**  The test statistic used is:
$W$ = the sum of the ranks corresponding to the lifetimes of the supplier $A$ batteries.

**Step 4**  $H_0$ will be rejected for $w \leq$ lower $w_c$ or $w \geq$ upper $w_c$. Approximate values of these critical values for the two-sided alternative are obtained using equations (6) and (7). Recall, for $\alpha = 0.05$ and a two-sided test, $z_c = 1.96$.

$$\text{lower } w_c = \frac{12(12 + 12 + 1)}{2} - 1.96 \sqrt{\frac{(12)(12)(12 + 12 + 1)}{12}}$$

$$= 150 - 33.9 \ldots$$

$$= 116, \text{ to the nearest whole number}$$

and

$$\text{upper } w_c = \frac{12(12 + 12 + 1)}{2} + 1.96 \sqrt{\frac{(12)(12)(12 + 12 + 1)}{12}}$$

$$= 150 + 33.9 \ldots$$

$$= 184, \text{ to the nearest whole number}$$

**Step 5**  $w = 4 + 6 + 7 + 9 + 11 + 12 + 13 + 16 + 19 + 20 + 21 + 24 = 162$

**Step 6** Because $162 > 116$ and $162 < 184$, do not reject $H_0$. That is, there is not sufficient evidence, with $\alpha = 0.05$, to conclude that the median lifetimes of the batteries in these shipments are different.    ◄

One additional matter concerning the Wilcoxon test that needs to be discussed is that of tied values. Suppose two data sets are combined to give the following ordered array of values:

$$16 \;\; ⑰ \;\; 17 \;\; 18 \;\; ⑳ \;\; ㉑ \;\; 21 \;\; 21$$

The circled values in the set are the ones found in the first sample. What ranks are assigned to the values 17 and 21? Common statistical practice is to assign the same *mean of the ranks* to each of a group of identical values.

Because 17 in the preceding set is ranked 2 and 3, we assign a rank of 2.5 to·both values of 17.

Because 21 in the preceding set is ranked 6, 7, and 8, we assign a rank of 7 to each value of 21.

Thus the value of the rank sum statistic would be calculated as:

$$W = 2.5 + 5 + 7 = 14.5$$

## Nonparametric estimation of a difference in location

We now consider the task of estimating the size of any difference in location for the $x$- and $y$-distributions. Nonparametric techniques for estimating $\tilde{\mu}_1 - \tilde{\mu}_2$ use the ordered array of the $n_1 \times n_2$ possible differences of $x_i - y_j$. To illustrate such an ordered array of differences, suppose:

3, 7, and 10 are the $n_1 = 3$ values from an $x$ sample.

2, 5, 6, and 12 are the $n_2 = 4$ values from a $y$ sample.

Because $n_1 \times n_2 = 3 \times 4 = 12$, there are 12 possible differences in the ordered array. To compute each $x_i - y_j$, we subtract 2, 5, 6, and 12 from 3, 7, and 10. The 12 differences are listed as:

$$
\begin{array}{lll}
3 - 2 = 1 & 7 - 2 = 5 & 10 - 2 = 8 \\
3 - 5 = -2 & 7 - 5 = 2 & 10 - 5 = 5 \\
3 - 6 = -3 & 7 - 6 = 1 & 10 - 6 = 4 \\
3 - 12 = -9 & 7 - 12 = -5 & 10 - 12 = -2
\end{array}
$$

The ordered array of these 12 differences is:

$$-9, \;\; -5, \;\; -3, \;\; -2, \;\; -2, \;\; 1, \;\; 1, \;\; 2, \;\; 4, \;\; 5, \;\; 5, \;\; 8$$

▶ **Example 5**    For the data on light bulbs in Table 10-3, produce the ordered array of differences $x_i - y_j$.

**Solution**   From Table 10-3:

$$x \in \{390, 840, 900, 1070, 1700\}$$

$$y \in \{150, 260, 530, 710, 890, 1590\}$$

With $n_1 = 5$ and $n_2 = 6$, there are $5 \times 6 = 30$ differences $x_i - y_j$. These values must then be ordered from the smallest to the largest. Table 10-4 contains the 30 differences.

**TABLE 10-4   The 30 Values of $x_i - y_j$**

| $y$ \ $x$ | 390 | 840 | 900 | 1070 | 1700 |
|---|---|---|---|---|---|
| 150 | 240 | 690 | 750 | 920 | 1550 |
| 260 | 130 | 580 | 640 | 810 | 1440 |
| 530 | −140 | 310 | 370 | 540 | 1170 |
| 710 | −320 | 130 | 190 | 360 | 990 |
| 890 | −500 | −50 | 10 | 180 | 810 |
| 1590 | −1200 | −750 | −690 | −520 | 110 |

The ordered array of the 30 differences is:

$$-1200 \quad -750 \quad -690 \quad -520 \quad -500 \quad -320 \quad -140 \quad -50 \quad 10$$

$$110 \quad 130 \quad 130 \quad 180 \quad 190 \quad 240 \quad 310 \quad 360 \quad 370 \quad 540 \quad 580 \quad 640$$

$$690 \quad 750 \quad 810 \quad 810 \quad 920 \quad 990 \quad 1170 \quad 1440 \quad 1550 \quad \blacktriangleleft$$

## A Point Estimate of $\tilde{\mu}_1 - \tilde{\mu}_2$

A nonparametric *point estimate* of the size of the difference in locations of the x- and y-distributions is the median value of the ordered array of $x_i - y_j$ values.

▶ **Example 6**   Give a point estimate of the difference in the median lifetimes for the brand $A$ and brand $B$ light bulbs.

**Solution**   **Discussion.** In Example 5, the 30 differences are ordered from the smallest value to the largest value. The median of the array is the mean of the fifteenth and sixteenth values in the array. Thus for the ordered differences:

$$\text{median ordered difference} = \frac{240 + 310}{2} = 275 \text{ hours} \quad \blacktriangleleft$$

## Confidence intervals for a difference in location

Beyond giving point estimates for a difference in locations, confidence intervals can also be determined by using the ordered array of differences $x_i - y_j$. The idea behind these intervals is to count in $j$ differences from each end of the ordered array. We then use the resulting values as the lower and upper limits of a confidence interval for $\tilde{\mu}_1 - \tilde{\mu}_2$.

### Confidence Intervals for $\tilde{\mu}_1 - \tilde{\mu}_2$

To construct a confidence interval for $\tilde{\mu}_1 - \tilde{\mu}_2$, use

$$\text{lower limit} = \text{the } j\text{th smallest } x_i - y_j$$
$$\text{upper limit} = \text{the } j\text{th largest } x_i - y_j$$

The confidence associated with such an interval is:

$$\left[ 1 - 2P\left( W \leq j - 1 + \frac{n_1(n_1 + 1)}{2} \right) \right] \cdot 100\% \tag{8}$$

The indicated probability is calculated supposing that the $x$- and $y$-distributions are identical.

▶ **Example 7**   For estimating a difference in location:

a. Find the confidence associated with an interval from the fourth smallest to the fourth largest $x_i - y_j$, if $n_1 = 5$ and $n_2 = 6$.
b. Find a $j$ so that the interval from the $j$th smallest to the $j$th largest $x_i - y_j$ has associated confidence near 95%, if $n_1 = 4$ and $n_2 = 8$.

**Solution**   a. For the given values of $n_1$, $n_2$, and $j$:

$$(j - 1) + \frac{n_1(n_1 + 1)}{2} = (4 - 1) + \frac{5(6)}{2}$$
$$= 3 + 15$$
$$= 18$$

From Appendix Table 8B of lower percentage points for $W$, we find that for $n_1 = 5$ and $n_2 = 6$:

$$P(W \leq 18) = 0.015$$

Thus the desired confidence is:

$$[1 - 2(0.015)] \cdot 100\% = 97\%$$

**b.** For a confidence of approximately 95%:

$$1 - 2P\left(W \le j - 1 + \frac{n_1(n_1 + 1)}{2}\right) \approx 0.95$$

or

$$P\left(W \le j - 1 + \frac{n_1(n_1 + 1)}{2}\right) \approx 0.025$$

Consulting Appendix Table 8B of lower percentage points for $W$, we see that with $n_1 = 4$ and $n_2 = 8$:

$$P(W \le 14) = 0.024$$

Therefore using $w_c = 14$:

$$j - 1 + \frac{4(5)}{2} = 14$$

Now solving for $j$:

$$j = 14 + 1 - 10$$
$$= 5$$

Thus an interval from the fifth smallest to the fifth largest $x_i - y_j$ will have the associated confidence of:

$$95.2\% = [1 - 2(0.024)] \cdot 100\% \qquad \blacktriangleleft$$

▶ **Example 8**    Give an approximately 90 percent confidence interval for the difference in median lifetimes for brand $A$ and brand $B$ light bulbs.

Solution    **Discussion.** With $n_1 = 5$ and $n_2 = 6$, the lower percentage points of the distribution of $W$ in Appendix Table 8B give:

$$P(W \le 20) = 0.041$$

Notice

$$2(0.041) = 0.082, \text{ and } 100\% - 8.2\% \approx 90\%$$

Using $w_c = 20$:

$$j - 1 + \frac{5(6)}{2} = 20 \qquad j - 1 + \frac{n_1(n_1 + 1)}{2} = w_c$$

$$j - 1 = 5 \qquad \text{Subtract 15 from both sides.}$$

$$j = 6 \qquad \text{Add 1 to both sides.}$$

Using the sixth smallest and the sixth largest ordered differences, we get an approximately 90 percent confidence interval. From Example 5 where the ordered differences are listed:

$$\text{the sixth smallest value} = -320$$

$$\text{the sixth largest value} = 810$$

Thus the difference in median hours of service life for brand $A$ and brand $B$ bulbs is between $-320$ hours and 810 hours, with a confidence coefficient of approximately 90 percent.    ◄

## Exercises 10-2    Set A

*In Exercises 1 and 2, find the probability distribution for the sum of ranks assigned to the first sample, assuming that there is no difference in the locations of the x- and y-distributions.*

**1.** For $n_1 = 2$ and $n_2 = 4$

**2.** For $n_1 = 3$ and $n_2 = 3$

*In Exercises 3–6, use Appendix Tables 8A and 8B to find the indicated probabilities for various values of $n_1$ and $n_2$, assuming there is no difference in the locations of the x- and y-distributions.*

**3.** For $n_1 = 3$ and $n_2 = 4$
   a. $P(W \geq 17)$
   b. $P(W \leq 6)$

**4.** For $n_1 = 5$ and $n_2 = 6$
   a. $P(W \geq 40)$
   b. $P(W \leq 22)$

**5.** For $n_1 = 7$ and $n_2 = 7$
   a. $P(W \geq 69)$
   b. $P(W \leq 36)$

**6.** For $n_1 = 8$ and $n_2 = 10$
   a. $P(W \geq 96)$
   b. $P(W \leq 60)$

**7.** A standardized aptitude test was given to 16 college students at a large university in Ohio. Eight of the students were majors in physical sciences and eight were majors in physical education. The results of the test are given in the following table:

| Physical Science Majors | 115 | 128 | 103 | 133 | 110 | 142 | 105 | 145 |
|---|---|---|---|---|---|---|---|---|
| Physical Education Majors | 98 | 125 | 118 | 120 | 135 | 112 | 123 | 108 |

   a. Array the 16 scores from the smallest to the largest. Circle the scores for the students who are physical science majors.
   b. Compute $w$ = sum of the ranks corresponding to the circled values.
   c. Test $H_0$:  $\tilde{\mu}_1 - \tilde{\mu}_2 = 0$
   $\qquad H_a$:  $\tilde{\mu}_1 - \tilde{\mu}_2 \neq 0$
   with $\alpha \approx 0.05$.

**8.** A study was conducted to compare the amount of time (in minutes) spent watching television each day by suburban and urban elementary school children in the greater

Dallas area. On a given Wednesday, a survey was taken of eight urban and nine suburban children randomly selected from school rosters. The results are given in the following table:

| Urban Children | 60 | 240 | 190 | 75 | 30 | 150 | 220 | 90 | |
|---|---|---|---|---|---|---|---|---|---|
| Suburban Children | 140 | 80 | 45 | 210 | 120 | 135 | 30 | 180 | 200 |

a. Array the 17 times from the smallest to the largest. Circle the times for the $n_1 = 8$ urban children.
b. Compute $w = $ sum of the ranks corresponding to the circled values.
c. Test $H_0$:  $\tilde{\mu}_1 - \tilde{\mu}_2 = 0$
   $H_a$:  $\tilde{\mu}_1 - \tilde{\mu}_2 \neq 0$
   with $\alpha \approx 0.05$.

9. Two experimental plots of tomatoes were grown at an agricultural school in the South. A standard brand of commercial fertilizer was used on plot A. A mixture of nutrients developed at the school to increase tomato size was used on plot B. A sample of nine tomatoes from each plot yielded the following weights, in grams:

| Plot A | 109 | 98 | 118 | 105 | 130 | 102 | 126 | 127 | 121 |
|---|---|---|---|---|---|---|---|---|---|
| Plot B | 112 | 138 | 141 | 150 | 105 | 148 | 121 | 142 | 131 |

a. Array the 18 weights from the smallest to the largest. Circle the weights for the $n_1 = 9$ tomatoes from plot $A$.
b. Compute $w = $ sum of the ranks corresponding to the circled values.
c. Test $H_0$:  $\tilde{\mu}_1 - \tilde{\mu}_2 = 0$
   $H_a$:  $\tilde{\mu}_1 - \tilde{\mu}_2 < 0$
   with $\alpha \approx 0.05$.

10. A manufacturer of electronic components needs to find out whether certain assembly tasks involving considerable finger dexterity can be performed more efficiently by women than men. A test of manual dexterity is given to 10 women and 10 men employees. Their times, measured in seconds, for completing the test are given in the following table:

| Times for Women Employees | 96 | 85 | 120 | 112 | 90 | 103 | 112 | 90 | 98 | 87 |
|---|---|---|---|---|---|---|---|---|---|---|
| Times for Men Employees | 93 | 107 | 95 | 125 | 98 | 90 | 173 | 119 | 144 | 101 |

a. Array the 20 times from the smallest to the largest. Circle the times for the $n_1 = 10$ women employees.
b. Compute $w = $ sum of the ranks corresponding to the circled values.
   Test $H_0$:  $\tilde{\mu}_1 - \tilde{\mu}_2 = 0$
   $H_a$:  $\tilde{\mu}_1 - \tilde{\mu}_2 < 0$
   with $\alpha \approx 0.10$.

*In Exercises 11–14, use equation (6), (7), or both, to find approximate values for $w_c$ for each $H_a$.*

**11.** If $n_1 = 16$, $n_2 = 16$, $\alpha = 0.10$, and $H_a$: $\tilde{\mu}_1 - \tilde{\mu}_2 < 0$

**12.** If $n_1 = 20$, $n_2 = 24$, $\alpha = 0.05$, and $H_a$: $\tilde{\mu}_1 - \tilde{\mu}_2 > 0$

**13.** If $n_1 = 25$, $n_2 = 25$, $\alpha = 0.05$, and $H_a$: $\tilde{\mu}_1 - \tilde{\mu}_2 \neq 0$

**14.** If $n_1 = 30$, $n_2 = 30$, $\alpha = 0.01$, and $H_a$: $\tilde{\mu}_1 - \tilde{\mu}_2 \neq 0$

*In Exercises 15–18, for estimating a difference in location, find the confidence associated with an interval for $\tilde{\mu}_1 - \tilde{\mu}_2$ of the type indicated.*

**15.** If $n_1 = 5$, $n_2 = 7$, and the sixth smallest and the sixth largest $x_i - y_j$ are used as endpoints

**16.** If $n_1 = 6$, $n_2 = 8$, and the eleventh smallest and the eleventh largest $x_i - y_j$ are used as endpoints

**17.** If $n_1 = 8$, $n_2 = 9$, and the nineteenth smallest and the nineteenth largest $x_i - y_j$ are used as endpoints

**18.** If $n_1 = 10$, $n_2 = 10$, and the 28th smallest and the 28th largest $x_i - y_j$ are used as endpoints

**19.** In a large class of students taking physical education, four men and four women were randomly selected. Each student was then asked to hurl a disk-shaped object as far as possible, using any manner he or she chose. The distances, measured in feet, that the men ($x$) and women ($y$) hurled the disk are shown in the following table:

| Men (x) | 135 | 89 | 50 | 162 |
|---------|-----|-----|-----|-----|
| Women (y) | 43 | 72 | 95 | 60 |

a. Give a point estimate of the difference in median distances for the men and women in the class.

b. Find the confidence associated with an interval extending from the second smallest to the second largest $x_i - y_j$ for estimating $\tilde{\mu}_1 - \tilde{\mu}_2$.

**20.** A rancher raises chickens that are sold as fryers. Four chicks are randomly selected from a large flock, and fed the usual diet. Five more chicks are randomly selected and fed a new formula diet. At the end of a specified time, the birds are weighed. The weights, measured in pounds, of the birds fed the usual diet ($x$) and those fed the new formula diet ($y$) are given in the following table:

| Usual Diet | 3.6 | 3.9 | 4.2 | 3.8 | |
|------------|-----|-----|-----|-----|-----|
| New Formula | 4.3 | 4.0 | 3.8 | 4.4 | 4.1 |

a. Give a point estimate of the difference in median weights for such chickens fed on the two diets.

b. Find the confidence associated with an interval extending from the third smallest to the third largest $x_i - y_j$ for estimating $\tilde{\mu}_1 - \tilde{\mu}_2$.

## Exercises 10-2    Set B

*In Exercises 1 and 2, use random numbers and Data Set I, Appendix B.*

**1.** a. Obtain the ages of simple random samples of $n_1 = 5$ women and $n_2 = 6$ men.
  b. Find the ranks for the ages of the women.
  c. Use these data to test with $\alpha = 0.05$:
   $H_0$:  $\tilde{\mu}_1 - \tilde{\mu}_2 = 0$
   $H_a$:  $\tilde{\mu}_1 - \tilde{\mu}_2 < 0$

**2.** a. Obtain the ages of simple random samples of $n_1 = 4$ women and $n_2 = 5$ men.
  b. Produce the ordered array of the differences $x_i - y_j$.
  c. Give a point estimate for the difference in the median ages of the women and men in these populations.
  d. Give an approximately 90 percent confidence interval for the difference in the median ages of the women and men.

## 10-3 Making Inferences Based on the Correlation Between Two Sets of Ranks— Using Spearman's $\tilde{r}$

**Key Topics**

1. *The Spearman rank correlation, $\tilde{r}$*

2. *A probability model for $\tilde{R}$ when x and y are unrelated*

3. *Testing $H_0$: Every pairing of ranks is equally likely*

4. *Approximating $\tilde{r}_c$ for large n*

5. *Averaging ranks for identical x- or y-values*

S  T  A  T  I  S  T  I  C  S     I  N     A  C  T  I  O  N

In Section 9-2, Susan Thompson used the data from eight students in dance classes to compute a least squares line for the height versus weight relationship. The data used by Susan are repeated here in Table 10-5.

Using these data, Susan calculated the following descriptive measures:

Least squares line          $\hat{y} = -169.0 + 4.5x$

Sample correlation coefficient   $r = 0.80$

Susan was discussing the questions of some of her students regarding their weights and the weights predicted by the regression equation, with Caralee Woods, a friend and dance instructor. Caralee had recently completed a statistics course as a requirement in a degree program she was pursuing. She suggested that a Spearman rank correlation test could be used to determine whether the data Susan had used indicated that a relationship exists between heights and weights of students in modern dance classes.

The two women obtained the necessary formula from Caralee's statistics book and subsequently calculated 0.76 for the rank correlation for the data pairs in Table 10-5. Because the computed value was greater than $\tilde{r}_c$ at the 5 percent level of significance, Susan and Caralee concluded that the data justified their claim that a relationship between height and weight exists in dance students. The details of the Spearman rank correlation test will be studied in this section.

**TABLE 10-5   The Heights and Weights of Eight Women Students**

| Individual | Height in Inches (x) | Weight in Pounds (y) |
|:---:|:---:|:---:|
| 1 | 62 | 105 |
| 2 | 66 | 120 |
| 3 | 70 | 156 |
| 4 | 58 | 85 |
| 5 | 64 | 130 |
| 6 | 61 | 112 |
| 7 | 63 | 142 |
| 8 | 68 | 126 |

## The Spearman rank correlation, $\tilde{r}$

In our study of bivariate data, one of the descriptive measures we calculated was the sample correlation coefficient $r$. Recall that correlation measures the strength of linear relationship between two variables. However, we did not at that time describe methods of statistical inference based on $r$, because to do so requires discussion of some very complicated probability models. However, using correlation to test for a relationship between two variables is easy, if the original $x$- and $y$-values are converted to ranks before calculating $r$. The correlation between the two sets of ranks is known as the *Spearman rank correlation* and will be symbolized as $\tilde{r}$.

### Spearman Rank Correlation, $\tilde{r}$

For a set of $x$ and $y$ ranked data:

$$\tilde{r} = \frac{\sum(x \text{ rank})(y \text{ rank}) - \dfrac{\left[\sum(x \text{ rank})\right]\left[\sum(y \text{ rank})\right]}{n}}{\sqrt{\left(\sum(x \text{ rank})^2 - \dfrac{\left[\sum(x \text{ rank})\right]^2}{n}\right)\left(\sum(y \text{ rank})^2 - \dfrac{\left[\sum(y \text{ rank})\right]^2}{n}\right)}} \quad (9)$$

▶ **Example 1** Calculate $\tilde{r}$ for the height-weight data in Table 10-5.

**Solution** **Discussion.** The first step is to rank the $x$ and the $y$ data from smallest to largest. For example, individual 4 has the smallest values in both height and weight and is given a ranking of 1 in both categories. The other seven data points in the two categories are ranked in Table 10-6.

**TABLE 10-6 Ranks of $x$ and $y$ Data**

| Individual | Height (x) | x Rank | Weight (y) | y Rank |
|:---:|:---:|:---:|:---:|:---:|
| 1 | 62 | 3 | 105 | 2 |
| 2 | 66 | 6 | 120 | 4 |
| 3 | 70 | 8 | 156 | 8 |
| 4 | 58 | 1 | 85 | 1 |
| 5 | 64 | 5 | 130 | 6 |
| 6 | 61 | 2 | 112 | 3 |
| 7 | 63 | 4 | 142 | 7 |
| 8 | 68 | 7 | 126 | 5 |

These rankings are now used to compute the following quantities:

$$\sum(x \text{ rank}), \quad \sum(y \text{ rank}), \quad \left[\sum(x \text{ rank})\right]^2, \quad \left[\sum(y \text{ rank})\right]^2, \quad \text{and} \quad \sum(x \text{ rank})(y \text{ rank})$$

| Individual | x Rank | y Rank | (x Rank)² | (y Rank)² | (x Rank)(y Rank) |
|:---:|:---:|:---:|:---:|:---:|:---:|
| 1 | 3 | 2 | 9 | 4 | 6 |
| 2 | 6 | 4 | 36 | 16 | 24 |
| 3 | 8 | 8 | 64 | 64 | 64 |
| 4 | 1 | 1 | 1 | 1 | 1 |
| 5 | 5 | 6 | 25 | 36 | 30 |
| 6 | 2 | 3 | 4 | 9 | 6 |
| 7 | 4 | 7 | 16 | 49 | 28 |
| 8 | 7 | 5 | 49 | 25 | 35 |
| Totals | 36 | 36 | 204 | 204 | 194 |

Now with $n = 8$ and the column totals from the table:

$$\tilde{r} = \frac{194 - \dfrac{(36)(36)}{8}}{\sqrt{\left(204 - \dfrac{36^2}{8}\right)\left(204 - \dfrac{36^2}{8}\right)}}$$

$$= \frac{32}{42}$$

$$= 0.76, \text{ to two decimal places.} \quad \blacktriangleleft$$

The calculation of $\tilde{r}$ can be simplified by using the fact that each of the numbers 1 through $n$ will appear once as $x$ and $y$ ranks. The simplified equation is equation (10).

### Alternate Formula for Calculating the Spearman Rank Correlation, $\tilde{r}$

$$\tilde{r} = 1 - \frac{6}{n(n^2 - 1)} \sum d^2 \qquad (10)$$

where the $d$'s are the differences in $x$ and $y$ ranks for the $n$ data pairs.

▶ **Example 2**   Use equation (10) to calculate $\tilde{r}$ for the height-weight data.

**Solution**   The $x$ ranks, $y$ ranks, $d$'s and $d^2$s are in the following table:

| Individual | x Rank | y Rank | d | d² |
|:---:|:---:|:---:|:---:|:---:|
| 1 | 3 | 2 | 1 | 1 |
| 2 | 6 | 4 | 2 | 4 |
| 3 | 8 | 8 | 0 | 0 |
| 4 | 1 | 1 | 0 | 0 |
| 5 | 5 | 6 | −1 | 1 |
| 6 | 2 | 3 | −1 | 1 |
| 7 | 4 | 7 | −3 | 9 |
| 8 | 7 | 5 | 2 | 4 |
| | | | | $\sum d^2 = 20$ |

Now using equation (10):

$$\tilde{r} = 1 - \frac{6}{8(8^2 - 1)} (20)$$

$$= 1 - 0.238 \ldots$$

$$= 0.76, \text{ to two decimal places.} \quad ◀$$

## A probability model for $\widetilde{R}$ when $x$ and $y$ are unrelated

Ms. Thompson, the modern dance instructor, should view an "extreme" value of $\tilde{r}$ as indicating the presence of a relationship between the $x$ and $y$ variables in her population of students. The question that arises is "How extreme must $\tilde{r}$ be before it is safe to conclude that there is a relationship between $x$ and $y$?"

This question can be answered by using a probability model that expresses *the unrelatedness of the variables x and y* and by computing the probability distribution of $\widetilde{R}$ that this model implies. Then, if Susan's observed $\tilde{r}$ is extreme when compared to an appropriate percentage point of this probability distribution, it can be concluded that $x$ and $y$ are related.

Part of the reason the Spearman rank correlation is a much used summary statistic is that many different probability models expressing the unrelatedness of $x$ and $y$ lead to the statement:

Every pairing of $y$ ranks to $x$ ranks is equally likely.

Based on this statement, a probability distribution for $\tilde{R}$ can be derived. In Example 3, such a probability distribution is derived for $n = 3$.

▶ **Example 3** Use the statement "every pairing of $x$ ranks to $y$ ranks is equally likely" to derive a probability distribution for $\tilde{R}$, if $n = 3$.

Solution **Discussion.** There are exactly six ways in which the three $x$ ranks and three $y$ ranks can be paired. If we assume the variables $x$ and $y$ are unrelated, then each of the six pairings has the same probability of occurring, namely $\frac{1}{6}$. The six pairings and corresponding values of $\tilde{r}$ [computed using equation (10)] are listed below.

|  | x Rank | y Rank |  | x Rank | y Rank |  | x Rank | y Rank |
|---|---|---|---|---|---|---|---|---|
| **Case 1** | 1<br>2<br>3 | 1<br>2<br>3 | **Case 3** | 1<br>2<br>3 | 2<br>1<br>3 | **Case 5** | 1<br>2<br>3 | 3<br>1<br>2 |
|  | $\tilde{r} = 1$ | | | $\tilde{r} = 0.5$ | | | $\tilde{r} = -0.5$ | |
| **Case 2** | 1<br>2<br>3 | 1<br>3<br>2 | **Case 4** | 1<br>2<br>3 | 2<br>3<br>1 | **Case 6** | 1<br>2<br>3 | 3<br>2<br>1 |
|  | $\tilde{r} = 0.5$ | | | $\tilde{r} = -0.5$ | | | $\tilde{r} = -1.0$ | |

Consolidating the common values of $\tilde{r}$, namely cases 2 and 3, and also cases 4 and 5, we get the following probability distribution for $\tilde{R}$, in which the probabilities have been rounded to two decimal places:

| $\tilde{r}$ | $P(\tilde{r})$ |
|---|---|
| $-1.0$ | 0.17 |
| $-0.5$ | 0.33 |
| 0.5 | 0.33 |
| 1.0 | 0.17 |

◀

| $n$ | $\tilde{r}_c$ | $P(\tilde{R} \geq \tilde{r}_c)$ |
|---|---|---|
| 5 | 1.000 | 0.008 |
|  | 0.900 | 0.042 |
|  | 0.800 | 0.067 |
|  | 0.700 | 0.117 |
|  | 0.600 | 0.175 |
|  | 0.500 | 0.225 |
|  | 0.400 | 0.258 |
|  | 0.300 | 0.342 |
|  | 0.200 | 0.392 |
|  | 0.100 | 0.475 |
|  | 0.000 | 0.525 |

**Figure 10-9.** A portion of Appendix Table 9 for $n = 5$.

A table for the distribution of $\tilde{R}$ derived from the statement that all pairings of $y$ ranks to $x$ ranks are equally likely is found in Appendix Table 9. This table gives

$$P(\tilde{R} \geq \tilde{r}_c)$$

for various $n$ and positive critical values $\tilde{r}_c$. Figure 10-9 contains the part of the table for $n = 5$.

If the method used in Example 3 is used for $n = 5$, then the probabilities in Figure 10-9 will be obtained. For example:

$$P(\tilde{R} \geq 0.800) = 0.067$$
$$P(\tilde{R} \geq 0.300) = 0.342$$

Lower percentage points for $\tilde{R}$ are found using the fact that if $\tilde{r}_c \leq 0$, then $-\tilde{r}_c \geq 0$. Thus, for example:

$$P(\tilde{R} \leq -0.600) = P(\tilde{R} \geq 0.600) = 0.175$$
$$P(\tilde{R} \leq -0.200) = P(\tilde{R} \geq 0.200) = 0.392$$

▶ **Example 4**    Use Appendix Table 9 to find the following probabilities:

**a.** If $n = 6$, $P(\tilde{R} \geq 0.543)$
**b.** If $n = 6$, $P(\tilde{R} \leq -0.600)$
**c.** If $n = 8$, $P(\tilde{R} \leq -0.905 \text{ or } R \geq 0.905)$

**Solution**    **a.** Using Appendix Table 9 with $n = 6$:

$$P(\tilde{R} \geq 0.543) = 0.149$$

**b.** Because $P(\tilde{R} \leq -0.600) = P(\tilde{R} \geq 0.600)$, the desired value is obtained from Appendix Table 9.

$$P(\tilde{R} \geq 0.600) = 0.121$$

**c.** $P(\tilde{R} \leq -0.905) + P(\tilde{R} \geq 0.905) = 2P(\tilde{R} \geq 0.905)$

Using Appendix Table 9, we find:

$$2(0.002) = 0.004 \quad ◀$$

## Testing $H_0$: Every pairing of ranks is equally likely

In qualitative terms, the possibility that $x$ and $y$ are unrelated should be rejected in favor of a positive relationship if $\tilde{r}$ is large, or in favor of an inverse relationship if $\tilde{r}$ is small, or in favor of some relationship if $\tilde{r}$ is either large or small. But it is difficult to state these ideas in quantitative terms. That is, unlike the situation with testing based on $\bar{X}$ or $\hat{P}$, null and alternative hypotheses for tests based on $\tilde{R}$ are not easily phrased in terms of a population parameter. (In this regard, here the terminology "nonparametric test" is particularly appropriate.) Recognizing this problem, we shall agree to state our null and alternative hypotheses for tests based on $\tilde{R}$ in qualitative terms, as:

$H_0$:    Every pairing of ranks is equally likely

with

$H_a$:    A positive relationship exists

or

$H_a$:    An inverse relationship exists

or

$H_a$:    Some relationship exists.

▶ **Example 5**   Test the null hypothesis of unrelatedness with an alternative hypothesis of some relationship based on the height-weight data from the students in Ms. Thompson's modern dance class. Use $\alpha$ near 0.05.

**Solution**   **Discussion.** The alternative hypothesis suggests $H_0$ should be rejected for $\tilde{r} \leq$ lower $\tilde{r}_c$ or $\tilde{r} \geq$ upper $\tilde{r}_c$. Furthermore the two critical values should be chosen so that supposing $H_0$ to be true

$$P(\tilde{R} \leq \text{lower } \tilde{r}_c) = P(\tilde{R} \geq \text{upper } \tilde{r}_c) \approx \tfrac{1}{2}(0.05)$$
$$\approx 0.025$$

From Appendix Table 9, for $n = 8$:

$$P(\tilde{R} \geq 0.738) = 0.023 \begin{cases} \rightarrow \text{upper } \tilde{r}_c = 0.738 \\ \rightarrow \text{lower } \tilde{r}_c = -0.738 \end{cases}$$

**Step 1**   $H_0$:   Every pairing of ranks is equally likely
        $H_a$:   Some relationship exists

**Step 2**   $\alpha = 0.046$

**Step 3**   The test statistic will be:

$$\tilde{R} = 1 - \frac{6}{n(n^2 - 1)} \sum D^2$$

**Step 4**   If $\tilde{r} \leq$ lower $\tilde{r}_c$ $(= -0.738)$ or if $\tilde{r} \geq$ upper $\tilde{r}_c$ $(= 0.738)$, then reject $H_0$.

**Step 5**   From Example 1:

$$\tilde{r} = 0.76$$

**Step 6**   Because $0.76 > 0.738$, reject $H_0$. That is, there is sufficient evidence, with $\alpha = 0.046$, to conclude that a relationship exists between the heights and weights of students in Ms. Thompson's modern dance class.   ◀

Remember that the tabled distribution for $\tilde{R}$ can be derived from the statement *every pairing of ranks is equally likely*, which implies that inferences can also be drawn based on $\tilde{R}$ in situations where the object is other than to use a sample to establish a relationship between $x$ and $y$ for a population of bivariate data. The following example describes such a situation.

▶ **Example 6**   An ESP expert is given cards numbered 1 through 10 placed face down on a table. Without viewing the faces, he attempts to place them in order from the smallest to the largest with the following results:

| | |
|---|---|
| *Position* | 1, 2, 3, 4, 5, 6, 7, 8, 9, 10 |
| *Card number* | 4, 1, 3, 2, 6, 7, 5, 9, 8, 10 |

Use the correlation between these two sets of values to test whether the expert has shown unusual ability in ordering the cards. Use an $\alpha$ of approximately 0.10.

**Solution**   **Discussion.** If the ESP expert is just guessing, then every arrangement of the cards is equally likely. Furthermore if the expert is guessing and we think of the position and card number variables as types of rankings, then the tabled distribution of $\tilde{R}$ is appropriate for judging the size of a correlation between positions and card numbers generated by the ESP expert.

**Step 1**  $H_0$:   Every pairing of ranks is equally likely
  $H_a$:   A positive relationship exists

**Step 2**  Using Appendix Table 9, with $n = 10$, $\alpha = 0.096$ is possible for a one-sided test based on $\tilde{R}$. Thus we will use $\alpha = 0.096$ because $0.096 \approx 0.10$.

**Step 3**  The test statistic will be:

$$\tilde{R} = 1 - \frac{6}{n(n^2 - 1)}\sum D^2$$

**Step 4**  $H_0$ will be rejected if $\tilde{r} \geq \tilde{r}_c$. Because the table gives $P(\tilde{R} \geq 0.455) = 0.096$, $H_0$ will be rejected if $\tilde{r} \geq 0.455$.

**Step 5**  The expert's ordering of the cards gives the values in Table 10-7. From the table:

$$\tilde{r} = 1 - \frac{6}{10(100 - 1)}(22)$$

$$= 1 - 0.133\ldots$$

$$= 0.867, \text{ to three}$$
$$\text{decimal places}$$

**TABLE 10-7   Ranks of Position and Card Number**

| Position | Card Number | $d$ | $d^2$ |
|---|---|---|---|
| 1 | 4 | −3 | 9 |
| 2 | 1 | 1 | 1 |
| 3 | 3 | 0 | 0 |
| 4 | 2 | 2 | 4 |
| 5 | 6 | −1 | 1 |
| 6 | 7 | −1 | 1 |
| 7 | 5 | 2 | 4 |
| 8 | 9 | −1 | 1 |
| 9 | 8 | 1 | 1 |
| 10 | 10 | 0 | 0 |
| | | | 22 |

**Step 6**  Reject $H_0$. With $\alpha = 0.096$, the expert's pairing shows sufficient evidence of a positive relationship between position and card number to conclude that he is not just guessing.   ◄

## Approximating $\tilde{r}_c$ for large $n$

If

$$H_0:\quad \text{Every pairing of ranks is equally likely}$$

is true, then the probability distribution of $\tilde{R}$ can be approximated using a normal distribution with mean 0 and variance $\dfrac{1}{n - 1}$. This fact allows one to compute approximate critical values for large $n$.

## Approximating $\tilde{r}_c$ for Large $n$

For large $n$, approximate critical values for $\tilde{r}$ can be obtained using equation (11):

$$\tilde{r}_c = \frac{z_c}{\sqrt{n-1}} \qquad (11)$$

▶ **Example 7**     In $n = 20$ consecutive mileage tests with the same car and driver at different speeds, the correlation between ranked mileages observed and speeds used was $\tilde{r} = -0.925$. With $\alpha = 0.05$, is this value of $\tilde{r}$ evidence of an inverse relationship between speed and mileage for this car-driver combination? An inverse relationship means that higher ranked mileages are generally paired with lower ranked speeds, and vice versa.

**Solution**     **Step 1** $H_0$:   Every pairing of ranks is equally likely
              $H_a$:   An inverse relationship exists

**Step 2** $\alpha = 0.05$

**Step 3** The test statistic will be:

$$\tilde{R} = 1 - \frac{6}{n(n^2 - 1)} \sum D^2$$

**Step 4** $H_0$ will be rejected if $\tilde{r} \leq \tilde{r}_c$. With $z_c = -1.65$ and $n = 20$:

$$\tilde{r}_c = \frac{-1.65}{\sqrt{20 - 1}}$$

$$= -0.380, \text{ to three decimal places}$$

$H_0$ will be rejected for $\tilde{r} \leq -0.380$.

**Step 5** The 20 tests gave $\tilde{r} = -0.925$.

**Step 6** Reject $H_0$. With $\alpha = 0.05$, there is sufficient evidence to conclude that mileage is inversely related to speed for the car-driver combination in question.   ◀

## Averaging ranks for identical x- or y-values

Sometimes an $x$-, or a $y$-value may be repeated in a given set of data. For example, consider the following set of $(x, y)$ pairs:

$$\{(2, 2), (6, 2), (6, 6), (6, 7), (3, 8)\}$$

The $x$-value 6 is repeated three times, and the $y$-value 2 is repeated twice. To calculate $\tilde{r}$, the procedure usually recommended is to use *average ranks*. In the following table, the $x$-value of 6 is given an average rank of $\dfrac{3+4+5}{3} = 4$, and the $y$-value of 2 is given an average rank of $\dfrac{1+2}{2} = 1.5$.

| $x$ | $y$ | Average x Rank | Average y Rank | $\left(\dfrac{Average}{x\ Rank}\right)^2$ | $\left(\dfrac{Average}{y\ Rank}\right)^2$ | $\left(\dfrac{Average}{x\ Rank}\right)\left(\dfrac{Average}{y\ Rank}\right)$ |
|---|---|---|---|---|---|---|
| 2 | 2 | 1 | 1.5 | 1 | 2.25 | 1.5 |
| 6 | 2 | 4 | 1.5 | 16 | 2.25 | 6.0 |
| 6 | 6 | 4 | 3 | 16 | 9.00 | 12.0 |
| 6 | 7 | 4 | 4 | 16 | 16.00 | 16.0 |
| 3 | 8 | 2 | 5 | 4 | 25.00 | 10.0 |
| | | 15 | 15 | 53 | 54.50 | 45.5 |

Thus:

$$\tilde{r} = \frac{45.5 - \dfrac{(15)(15)}{5}}{\sqrt{\left(53 - \dfrac{(15)(15)}{5}\right)\left(54.5 - \dfrac{(15)(15)}{5}\right)}}$$

and

$$\tilde{r} = 0.057$$

## Exercises 10-3   Set A

*In Exercises 1–4, suppose every pairing of y ranks to x ranks is equally likely and use Appendix Table 9 to find the following probabilities for $\tilde{R}$:*

1.  a. If $n = 4$, $P(\tilde{R} \geq 0.400)$
    b. If $n = 7$, $P(\tilde{R} \geq 0.143)$
    c. If $n = 5$, $P(\tilde{R} \leq -0.200)$
    d. If $n = 6$, $P(\tilde{R} \leq -0.886$ or $\tilde{R} \geq 0.886)$

2.  a. If $n = 4$, $P(\tilde{R} \geq 1.000)$
    b. If $n = 6$, $P(\tilde{R} \geq 0.714)$
    c. If $n = 8$, $P(\tilde{R} \leq -0.643)$
    d. If $n = 7$, $P(\tilde{R} \leq -0.786$ or $\tilde{R} \geq 0.786)$

3.  a. If $n = 8$, $P(\tilde{R} \geq 0.524)$
    b. If $n = 9$, $P(\tilde{R} \leq -0.600)$
    c. If $n = 10$, $P(\tilde{R} \leq -0.648$ or $\tilde{R} \geq 0.648)$

4.  a. If $n = 10$, $P(\tilde{R} \geq 0.733)$
    b. If $n = 8$, $P(\tilde{R} \leq -0.810)$
    c. If $n = 9$, $P(\tilde{R} \leq -0.817$ or $\tilde{R} \geq 0.817)$

*In Exercises 5–8, use equation (11) to approximate $\tilde{r}_c$ to three decimal places.*

5.  If $P(\tilde{R} \geq \tilde{r}_c) = 0.050$ and $n = 26$
6.  If $P(\tilde{R} \geq \tilde{r}_c) = 0.010$ and $n = 37$
7.  If $P(\tilde{R} \leq -\tilde{r}_c$ or $\tilde{R} \geq \tilde{r}_c) = 0.050$ and $n = 50$
8.  If $P(\tilde{R} \leq -\tilde{r}_c$ or $\tilde{R} \geq \tilde{r}_c) = 0.010$ and $n = 26$

9. The semester test averages of 12 students
enrolled in the same calculus and physics classes
are shown in the following table at the right.

a. Rank the calculus test averages, $x$.
b. Rank the physics test averages, $y$.
c. Calculate $\tilde{r}$ for the data.
d. Test $H_0$: Every pairing of ranks is equally likely
with $H_a$: A positive relationship exists.
Use $\alpha \approx 0.05$.

| Student | Calculus Test Average, x (%) | Physics Test Average, y (%) |
|---------|------------------------------|-----------------------------|
| A | 83 | 76 |
| B | 92 | 89 |
| C | 75 | 83 |
| D | 95 | 96 |
| E | 68 | 73 |
| F | 88 | 88 |
| G | 84 | 77 |
| H | 90 | 85 |
| I | 63 | 60 |
| J | 70 | 90 |
| K | 73 | 75 |
| L | 86 | 68 |

10. Sixteen students at a small southern college completed two semesters of a calculus course.
The students were ranked at the end of each semester using their test averages. The
lowest test average was ranked 1, and the highest test average was ranked 16. The results
of the rankings are given in the following table:

| Student | A | B | C | D | E | F | G | H | I | J | K | L | M | N | O | P |
|---------|---|---|---|---|---|---|---|---|---|---|---|---|---|---|---|---|
| Semester 1 | 1 | 2 | 3 | 4 | 5 | 6 | 7 | 8 | 9 | 10 | 11 | 12 | 13 | 14 | 15 | 16 |
| Semester 2 | 2 | 1 | 8 | 6 | 4 | 3 | 11 | 14 | 9 | 5 | 12 | 7 | 15 | 16 | 10 | 13 |

a. Find $\tilde{r}$ for the data.
b. Test $H_0$: Every pairing of ranks is equally likely with $H_a$: A positive relationship exists.
Use $\alpha \approx 0.01$.

11. A consumer protection agency was asked to rank ten brands of electric knives on the
basis of performance. A ranking of 1 was given to the knife with the worst performance
and 10 to the one with the best. The corresponding suggested manufacturers' list prices
to the nearest 1 dollar for each knife are also given in the following table:

| Electric Knife | A | B | C | D | E | F | G | H | I | J |
|----------------|---|---|---|---|---|---|---|---|---|---|
| Quality Rank | 8 | 7 | 10 | 1 | 3 | 5 | 9 | 6 | 4 | 2 |
| Suggested Price | $42 | 35 | 38 | 25 | 28 | 23 | 41 | 33 | 36 | 24 |

a. Rank the list prices from 1 through 10 with the lowest price ranked 1 and the highest
price ranked 10.
b. Compute $\tilde{r}$ for the ranked data.
c. Test $H_0$: Every pairing of ranks is equally likely with $H_a$: A positive relationship exists.
Use $\alpha \approx 0.05$.

12. A preseason poll is conducted by two press organizations on ten college football teams
that are usually ranked as the top ten teams in the United States. The teams are ranked
1 through 10 from the lowest to the highest by both organizations. The results of the poll
are given in the table on the following page.

| Team | A | B | C | D | E | F | G | H | I | J |
|------|---|---|---|---|---|---|---|---|---|---|
| Poll A | 10 | 9 | 8 | 7 | 6 | 5 | 4 | 3 | 2 | 1 |
| Poll B | 3 | 5 | 4 | 10 | 9 | 6 | 8 | 1 | 7 | 2 |

a. For the ranked data, calculate $\tilde{r}$:
b. Test $H_0$: Every pairing of ranks is equally likely with $H_a$: Some relationship exists.
c. Use $\alpha \approx 0.02$.

13. In the following table, eight cars made by U.S. auto manufacturers are listed. The suggested list prices and the EPA mileage ratings are stated for each car.

| Car | A | B | C | D | E | F | G | H |
|-----|---|---|---|---|---|---|---|---|
| Price | $10,900 | 12,500 | 14,000 | 10,950 | 10,750 | 9,890 | 13,250 | 11,600 |
| Mileage | 28 | 26 | 22 | 33 | 42 | 39 | 21 | 36 |

a. Rank the prices from the lowest to the highest.
b. Rank the mileage ratings from the lowest to the highest.
c. Find $\tilde{r}$ for the ranked data.
d. Test $H_0$: Every pairing of ranks is equally likely with $H_a$: An inverse relationship exists. Use $\alpha \approx 0.05$.

14. In the following table, seven freshmen at Grandiose University who took an entrance examination are represented. The scores on the test and the family gross annual incomes to the nearest thousand dollars are listed in the following table:

| Student | 1 | 2 | 3 | 4 | 5 | 6 | 7 |
|---------|---|---|---|---|---|---|---|
| Family Income (x) | $130 | 115 | 140 | 190 | 145 | 195 | 210 |
| Test Score (y) | 470 | 355 | 460 | 530 | 410 | 645 | 480 |

a. Rank the family incomes from lowest to highest.
b. Rank the test scores from lowest to highest.
c. Find $\tilde{r}$ for the ranked data.
d. Test $H_0$: Every pairing of ranks is equally likely with $H_a$: A positive relationship exists. Use $\alpha \approx 0.05$.

*In Exercises 15–18, conduct the indicated hypothesis tests for the indicated values of $\alpha$, n, and $\tilde{r}$.*

15. Test $H_0$: Every pairing of ranks is equally likely with $H_a$: A positive relationship exists if $n = 7$, $\alpha \approx 0.050$, and $\tilde{r} = 0.620$.

16. Test $H_0$: Every pairing of ranks is equally likely with $H_a$: An inverse relationship exists if $n = 8$, $\alpha \approx 0.100$, and $\tilde{r} = -0.510$.

17. Test $H_0$: Every pairing of ranks is equally likely with $H_a$: Some relationship exists if $n = 10$, $\alpha \approx 0.050$, and $\tilde{r} = -0.696$.

18. Test $H_0$: Every pairing of ranks is equally likely with $H_a$: Some relationship exists if $n = 26$, $\alpha \approx 0.020$, and $\tilde{r} = 0.725$.

# Exercises 10-3    Set B

*In Exercises 1 and 2, use random numbers to obtain a sample of n stocks from Data Set II, Appendix B.*

**1.** Sample $n = 12$ stocks. Let:

  $x$ = the *52 weeks High* price of a selected stock.

  $y$ = the *52 weeks Low* price of the same stock.

  a. Assign ranks 1 through 12 for the $x$-values.
  b. Assign ranks 1 through 12 for the $y$-values.
  c. Compute $d$, the difference in each pair of ranks.
  d. Calculate $\tilde{r}$ for the sample data.
  e. Test $H_0$: Every pairing of ranks is equally likely with $H_a$: A positive relationship exists. Use $\alpha \approx 0.05$.

**2.** Sample $n = 15$ stocks. Let:

  $x$ = the *52 weeks High* price of a selected stock.

  $y$ = the *Close* price of the same stock.

  a. Assign ranks 1 through 15 for the $x$-values.
  b. Assign ranks 1 through 15 for the $y$-values.
  c. Compute $d$, the difference in each pair of ranks.
  d. Calculate $\tilde{r}$ for the sample data.
  e. Test $_2$: Every pairing of ranks is equally likely with $H_a$: A positive relationship exists. Use $\alpha \approx 0.05$.

# Chapter 10 Summary

## Rules and Equations

  a. For the *sign test*:
     If $x - \tilde{\mu}_0 < 0$, write a minus sign.
     If $x - \tilde{\mu}_0 = 0$, the data point is deleted.
     If $x - \tilde{\mu}_0 > 0$, write a plus sign.
  b. If $n \geq 20$, then approximate critical values for the sign test may be obtained with equations 1 and 2:

$$\text{lower } b_c = n(0.5) - z_c\sqrt{n(0.5)(0.5)} \quad\begin{array}{l}\text{for } H_a: \quad \tilde{\mu} < \tilde{\mu}_0 \\ \text{for } H_a: \quad \tilde{\mu} \neq \tilde{\mu}_0\end{array} \quad (1)$$
$$\text{upper } b_c = n(0.5) + z_c\sqrt{n(0.5)(0.5)} \quad\begin{array}{l}\\ \text{for } H_a: \quad \tilde{\mu} > \tilde{\mu}_0\end{array} \quad (2)$$

The value of $z_c$ depends on the level of significance and whether the test is a one- or two-tail test.

  c. Confidence limits for a median

$$\text{lower limit} = \text{the } j\text{th smallest value in the sample} \quad (3)$$
$$\text{upper limit} = \text{the } j\text{th largest value in the sample} \quad (4)$$

The confidence associated with such an interval is:

$$[1 - 2P(X \leq j - 1)] \cdot 100\%$$

for $X$ a binomial random variable for $n$ trials and $p = 0.5$.

d. Approximating $j$ for large $n$
If $n \geq 20$, then approximately:

$$j = 1 + n(0.5) - z\sqrt{n(0.5)(0.5)} \qquad (5)$$

e. Approximating $w_c$ for large $n_1$ and $n_2$
If $n_1 \geq 10$ and $n_2 \geq 10$, then approximately:

$$\text{lower } w_c = \frac{n_1(n_1 + n_2 + 1)}{2} - z_c\sqrt{\frac{n_1 n_2(n_1 + n_2 + 1)}{12}} \qquad (6)$$

and/or

$$\text{upper } w_c = \frac{n_1(n_1 + n_2 + 1)}{2} + z_c\sqrt{\frac{n_1 n_2(n_1 + n_2 + 1)}{12}} \qquad (7)$$

f. Confidence intervals for $\tilde{\mu}_1 - \tilde{\mu}_2$
To construct a confidence interval for $\tilde{\mu}_1 - \tilde{\mu}_2$, use:

$$\text{lower limit} = \text{the } j\text{th smallest } x_i - y_j$$
$$\text{upper limit} = \text{the } j\text{th largest } x_i - y_j$$

The confidence associated with such an interval is:

$$\left[1 - 2P\left(W \leq j - 1 + \frac{n_1(n_1 + 1)}{2}\right)\right] \cdot 100\% \qquad (8)$$

where the probability is computed supposing the $x$ and $y$ distributions are the same.

g. Spearman rank correlation, $\tilde{r}$
For a set of $x$ and $y$ ranked data:

$$\tilde{r} = \frac{\sum(x \text{ rank})(y \text{ rank}) - \dfrac{[\sum(x \text{ rank})][\sum(y \text{ rank})]}{n}}{\sqrt{\left(\sum(x \text{ rank})^2 - \dfrac{[\sum(x \text{ rank})]^2}{n}\right)\left(\sum(y \text{ rank})^2 - \dfrac{[\sum(y \text{ rank})]^2}{n}\right)}} \qquad (9)$$

h. Alternate formula for calculating the Spearman rank correlation, $\tilde{r}$

$$\tilde{r} = 1 - \frac{6}{n(n^2 - 1)}\sum d^2 \qquad (10)$$

where the $d$'s are the differences in $x$ and $y$ ranks for the $n$ data pairs.

i. Approximating $r_c$ for large $n$
For large $n$:

$$\tilde{r}_c \approx \frac{z_c}{\sqrt{n - 1}} \qquad (11)$$

## Symbols

| | |
|---|---|
| $\tilde{\mu}$ | the median of a population of quantitative data |
| $\tilde{x}$ | the median of a sample of quantitative data |
| $B$ | the number of $X_i$ larger than $\tilde{\mu}$ |
| $B^*$ | the test statistic for testing $H_0$: $\tilde{\mu} = \tilde{\mu}_0$, namely the number of $X_i$ larger than $\tilde{\mu}_0$ |
| $b_c$ | the boundary value of the rejection region for testing $H_0$: $\tilde{\mu} = \tilde{\mu}_0$ |
| $\tilde{\mu}_d$ | the median difference in a population of paired data |
| $\tilde{\mu}_1 - \tilde{\mu}_2$ | the difference in the medians of an $x$-distribution and a $y$-distribution |
| $W$ | a random variable equal to the sum of the ranks corresponding to the values in the first sample |
| $w_c$ | the boundary value of the rejection region for testing $H_0$: $\tilde{\mu} - \tilde{\mu}_2 = 0$ |
| $\tilde{r}$ | the symbol for the Spearman rank correlation |
| $\tilde{r}_c$ | the boundary value of the rejection region for testing $H_0$: Every pairing of ranks is equally likely |

# Comparing the Sign Test, the Wilcoxon Rank Sum Test, and the Spearman Rank Correlation Test

### The Sign Test

If $n$ is small and the normal distribution assumption is not appropriate, then a test of the population median is used instead of the population mean.

Array the sample data points from smallest to largest.

Based on $H_a$, and a binomial probability distribution with $p = 0.5$ and the number of data points $n$, determine the critical values for the test of

$$H_0: \quad \tilde{\mu} = \tilde{\mu}_0$$

### Wilcoxon Rank Sum Test

If measures of center of nonnormal Populations 1 and 2 need to be compared and the samples are small, then only a comparison of the population medians can be used.

Combine the data points from both samples into one data set of $n_1 + n_2$ data points.

Array the $n_1 + n_2$ data points from smallest to largest and circle the data points from Population 1.

### Spearman Rank Correlation Test

If a measure of correlation between $x$- and $y$-values is to be tested, then Spearman's $\tilde{r}$ can be used.

To determine $\tilde{r}$ for a set of $(x_i, y_i)$ data points, each $x_i$ is assigned a rank based on the ordered array of $x$ data points. Similarly, each $y_i$ is assigned a rank based on the ordered array of data points.

### The Sign Test

For the sample, count the number of data points that are larger than $\tilde{\mu}_0$. Based on the number of data points larger than $\tilde{\mu}_0$ conclude whether to reject $H_0$.

The test can be carried out by computing $x_i - \mu_0$ for each $x_i$, and recording a $+$ or $-$ for each difference. Delete any data point whose difference is 0. The number of corresponding "plus signs" is then compared with the critical values.

### Wilcoxon Rank Sum Test

Assign a rank to each of the data points and sum the ranks of the circled values. Let $w$ be the sum of the ranks.

To test $H_0: \tilde{\mu}_1 - \tilde{\mu}_2 = 0$, obtain a critical value from Appendix Table 8 based on $n_1$, $n_2$, and the level of significance.

Depending on the form of $H_a$ and $w$, determine whether to reject $H_0$.

### Spearman Rank Correlation Test

The values obtained from the rankings can be used in equation 9 to calculate $\tilde{r}$. The differences in the paired rankings can be used in equation 10 to calculate $\tilde{r}$.

To test $H_0$: Every pairing of ranks is equally likely, obtain an $\tilde{r}_c$ from Appendix Table 9 based on $n$, the form of $H_a$, and the level of significance.

Compare $\tilde{r}$ and $\tilde{r}_c$ to determine whether to reject $H_0$.

## Chapter 10 Review Exercises

1. An editorial in the school paper of a large university reported that the median age of the students enrolled in the school was 23.7 years. A random sample of the student body records yields the ages of nine students. What is the probability that fewer than three of these ages are more than 23.7 years?

2. The statistics department at a southern university gives a standardized test to all students who complete the first course in statistics. It has been determined that the median score for the test is 72.5. A random sample of 15 of these test scores is obtained. What is the probability that more than ten of these scores are greater than 72.5?

3. A national motor magazine reports that the median miles-per-gallon (mpg) estimate for six-cylinder automobiles made in the United States is 20.0. The owner of a foreign car dealership feels the figure is too high. He obtains information on a series of ten new American car registrations in his county and uses the current issue of the EPA gas mileage guide to find their mile-

age estimates. The sample yields four cars with mpg estimates of more than 20.0. Test, with $\alpha = 0.055$:

$$H_0: \quad \tilde{\mu} = 20.0$$

versus

$$H_a: \quad \tilde{\mu} < 20.0$$

4. Find approximate critical values for the sign test, with $n = 36$ and $\alpha = 0.05$, for:
   a. $H_a: \quad \tilde{\mu} > \tilde{\mu}_0$
   b. $H_a: \quad \tilde{\mu} \neq \tilde{\mu}_0$

5. An administrator at a large hospital in Los Angeles stated that the median weight of babies born at the hospital was 7.5 lb. A nurse at the hospital felt that most of the babies she helped to deliver in the past year weighed more than this amount. She randomly selected 27 records of births for the past 12 months. The sample yielded 17 babies with weights of more than 7.5 lb. Use $\alpha = 0.05$ and test:

$$H_0: \quad \tilde{\mu} = 7.5$$
$$H_a: \quad \tilde{\mu} > 7.5$$

6. Find the confidence associated with an interval for estimating a population median if $n = 13$ and the fourth smallest and fourth largest values in the sample are used to construct the interval.

7. Find a $j$ so that the interval for estimating a population median from the $j$th smallest to the $j$th largest value has confidence of approximately 90 percent for $n = 15$.

8. A social worker at a large clinic in Minneapolis noted the ages of 17 expectant mothers pregnant with their first child. The ages in years are listed below.

| 28 | 29 | 21 | 30 | 17 | 26 | 24 | 41 | 20 |
|----|----|----|----|----|----|----|----|----|
| 28 | 27 | 25 | 29 | 31 | 23 | 30 | 27 | |

Give an interval with confidence close to 95 percent for the median age of expectant mothers at this clinic who are pregnant with their first child.

9. A disc jockey at a local radio station became interested in the obvious fact that record albums have different amounts of playing time. She selected $n = 25$ albums copyrighted last year. The ordered array of playing times in seconds is given below.

| 1802 | 1954 | 1975 | 2015 | 2062 |
|------|------|------|------|------|
| 2108 | 2135 | 2136 | 2157 | 2196 |
| 2199 | 2221 | 2231 | 2263 | 2314 |
| 2321 | 2324 | 2365 | 2403 | 2454 |
| 2498 | 2526 | 2534 | 2635 | 2676 |

Give an interval with confidence close to 90 percent for the median playing time of an album recorded last year.

10. A college newspaper reporter felt that female students on campus were currently taking more units per student than male students. She randomly selected six female and six male students and obtained the data shown in the following table:

| | Number of Units | | | | | |
|---|---|---|---|---|---|---|
| Female Students | 17 | 12 | 12 | 14 | 18 | 21 |
| Male Students | 4 | 12 | 11 | 9 | 16 | 17 |

If $\tilde{\mu}_1$ and $\tilde{\mu}_2$ are the median numbers of units currently being taken by female and male students respectively at this school, use these results to test, with $\alpha$ near 0.10:

$$H_0: \quad \tilde{\mu}_1 - \tilde{\mu}_2 = 0$$

versus

$$H_a: \quad \tilde{\mu}_1 - \tilde{\mu}_2 > 0$$

11. A health insurance sales representative wondered whether the median ages of men and women hospitalized for cardiovascular disorders were different. She felt that a large general hospital in the downtown area would have a good representation of city residents. She obtained samples of $n_1 = n_2 = 12$ records of men and women admitted to this hospital during the past year with cardiovascular disorders. The ages (in years) of these patients are listed in the table below.

| | Ages of Patients | | | | | | | | | | | |
|---|---|---|---|---|---|---|---|---|---|---|---|---|
| Men | 71 | 64 | 70 | 58 | 78 | 69 | 47 | 30 | 51 | 33 | 53 | 41 |
| Women | 48 | 70 | 75 | 43 | 87 | 76 | 80 | 65 | 88 | 71 | 64 | 66 |

If $\tilde{\mu}_1$ and $\tilde{\mu}_2$ are the median ages of males and females, respectively, who are admitted to this hospital with diagnosed cardiovascular disorders, with $\alpha = 0.05$, test:

$$H_0: \quad \tilde{\mu}_1 - \tilde{\mu}_2 = 0$$
$$H_a: \quad \tilde{\mu}_1 - \tilde{\mu}_2 \neq 0$$

12. a. For the data in the following table, produce the ordered array of differences $x_i - y_j$:

| x | 3 | 5 | 7 | 9 | 10 |
|---|---|---|---|---|----|
| y | 2 | 3 | 4 | 6 | |

b. Give a point estimate of the difference in locations of the $x$- and $y$-distributions.

13. For estimating a difference in location:
    a. Find the confidence associated with an interval from the ninth smallest to the ninth largest $x_i - y_j$, if $n_1 = 6$ and $n_2 = 7$.
    b. Find a $j$ so that the interval from the $j$th smallest to the $j$th largest $x_i - y_j$ has associated confidence near 85 percent, if $n_1 = 5$ and $n_2 = 7$.

14. At a large factory in Cleveland:

    Population 1 = {workers who work from
    7 AM to 4 PM}

    Population 2 = {workers who work from
    4 PM to 1 AM}

    Random samples of $n_1 = n_2 = 5$ workers yielded the numbers of times each worker was late to work during the last 6 months.

    | Worker from Population 1 (x) | 4 | 2 | 0 | 8 | 3 |
    |---|---|---|---|---|---|
    | Worker from Population 2 (y) | 6 | 1 | 3 | 5 | 10 |

    a. Produce the ordered array of differences $x_i - y_j$.
    b. Give a point estimate of the difference in median times late for the two populations of workers for the past 6 months.
    c. Give an approximate 90 percent confidence interval for the difference in medians.

15. Suppose every pairing of $y$ ranks to $x$ ranks is equally likely, and use Appendix Table 9 to find the following probabilities for $\tilde{R}$:
    a. If $n = 6$, and $P(\tilde{R} \geq 0.257)$
    b. If $n = 8$, and $P(\tilde{R} \geq 0.333)$
    c. If $n = 7$, and $P(\tilde{R} \leq -0.607)$
    d. If $n = 10$, and $P(\tilde{R} \leq -0.564)$

16. Approximate $\tilde{r}_c$ to three decimal places for the indicated alternative hypotheses and sample sizes.
    a. If $\alpha = 0.05$, $H_a$: a positive relationship exists and $n = 26$
    b. If $\alpha = 0.01$, $H_a$: an inverse relationship exists and $n = 37$
    c. If $\alpha = 0.05$, $H_a$: some relationship exists and $n = 65$

17. Seven students in a coeducational physical education class were given the following two tasks to perform:

    A   Throw a softball as far as possible

    B   Run 100 feet as fast as possible

    The results are in the table below. The task A values are distances in feet and the task B values are times in seconds.

    | Student | A | B | C | D | E | F | G |
    |---|---|---|---|---|---|---|---|
    | Task A | 210 | 187 | 110 | 163 | 201 | 151 | 174 |
    | Task B | 4.8 | 5.4 | 5.1 | 5.0 | 4.6 | 4.2 | 5.6 |

    a. Rank the distances from lowest to highest.
    b. Rank the times from lowest to highest.
    c. Find $\tilde{r}$ for the ranked data.
    d. Test $H_0$: Every pairing of ranks is equally likely versus $H_a$: Some relationship exists. Use $\alpha \approx 0.05$.

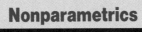

## Nonparametrics

**Key Topics**

*1. The sign test*

*2. The Mann-Whitney test*

### The sign test

Minitab nonparametric tests are very similar to their parametric counterparts. As a first example, let us consider the problem posed in Exercise 28 in Exercise Set 10-1, Set A. There you were asked to make a 90% confidence interval estimate for the median percent moisture content of wheat. To do this with Minitab, after placing the sample data in column 1, the command is:

```
SINTERVAL 95% C1
```

and the output is in Figure 10-10:

```
SIGN CONFIDENCE INTERVAL FOR MEDIAN
                           ACHIEVED
           N    MEDIAN   CONFIDENCE   CONFIDENCE    INTERVAL    POSITION
    C1    20     11.25     0.8847    (   10.60,     11.80)           7
                           0.9500    (   10.60,     12.26)         NLI
                           0.9586    (   10.60,     12.40)           6
```

**Figure 10-10.** Minitab output.

Three intervals are supplied. This is due to the fact that only rarely is it possible to obtain an interval with confidence level exactly equal to the one discussed. The first interval given has confidence slightly smaller than desired (see achieved confidence column). The third interval has confidence slightly larger than required. The middle interval was obtained from the other two by nonlinear interpolation and has associated confidence level approximately 95 percent.

The POSITION values at the right of the printout tells the position of the boundaries. For instance, the 7 for the 88.5% confidence interval tells that the interval starts with the seventh data point (when the data is ordered from smallest to largest) and ends at the seventh data point from the upper end of the data.

Minitab also has a sign test to be used in testing a hypothesis about the median of a population.

Suppose that we wish to use the wheat sample and test:

$$H_0: \quad \tilde{\mu} = 12$$
$$H_a: \quad \tilde{\mu} < 12$$

With the sample in column 1, the command

```
STEST MEDIAN = 12 DATA IN C1;
ALTERNATIVE = -1.
```

is given with the following result in Figure 10-11:

```
SIGN TEST OF MEDIAN = 12.00 VERSUS L.T. 1200
         N    BELOW   EQUAL   ABOVE   P-VALUE   MEDIAN
C1      20      14       0       6    0.0577    11.25
```

**Figure 10-11.** Minitab output.

STEST may be used with the ALTERNATIVE 1 or $-1$ subcommand for a one-tailed test, or with no subcommand for a two-tailed test.

## The Mann-Whitney test

Because there is a one-sample test called the Wilcoxon signed rank test for testing individual medians when the population is symmetrical but not normal, to avoid confusion the procedure that is called the Wilcoxon rank sum method in Section 10-2 of this text is identified as the MANN-WHITNEY test by Minitab. (Mann and Whitney developed a variant of the method discussed in Section 10-2.)

Let us solve Exercise 32 in Exercises 10-2B. First the $x$ values are placed in column 1 and the $y$ values are put into column 2. Next the command

```
MANNWHITNEY C1 C2
```

is given.

The computer output is in Figure 10-12 and has three parts:

1. A point estimate for $\tilde{\mu}_1 - \tilde{\mu}_2$ is given.
2. A confidence interval for $\tilde{\mu}_1 - \tilde{\mu}_2$ is given. The available confidence level closest value to 95 percent.
3. A two-sample rank test of

$$H_0: \quad \tilde{\mu}_1 - \mu_2 = 0$$
$$H_a: \quad \tilde{\mu}_1 - \tilde{\mu}_2 \neq 0$$

is summarized. The Mann-Whitney version of the test statistic and the test $p$-value are given.

```
MTB> MANNWHITNEY C1 C2
MANN-WHITNEY CONFIDENCE INTERVAL AND TEST
  SMOKE      N =  40      MEDIAN =        8.250
  NOSMOKE    N =  40      MEDIAN =        7.450
  POINT ESTIMATE FOR ETA1-ETA2 IS        0.750
  95.1  PCT C.I. FOR ETA1-ETA2 IS (     0.100,     1.300)
  W =   1879.5
  TEST OF ETA1 = ETA2  VS.  ETA1 N.E. ETA2 IS SIGNIFICANT AT 0.0127
```

**Figure 10-12.**   Minitab output.

# Exercises

1. Look at the output for Exercise 28 of Set A, Section 10-1 given in Figure 10-10.
   a. What is the smallest number in the 95.8% confidence interval? The largest?
   b. Can it be concluded that the median moisture content of the wheat sampled is less than 12%?
   c. How many data points from the sample were below the hypothesized median? How many were above it?

2. Consider the output given in Figure 10-12.
   a. What are the median scores for each sample?
   b. With what confidence may it be said that the difference in medians is between .100 and 1.300?

## Calculation of $\tilde{r}$

The value of $\tilde{r}$ can be found by using the formula:

$$\tilde{r} = 1 - \frac{6}{n(n^2 - 1)} \sum d^2$$

where the $d$'s are the differences in the $x$ and $y$ ranks for the $n$ data pairs.
  The formula can also be written:

$$\tilde{r} = 1 - 6 \div (n(n^2 - 1)) \sum d^2$$

It is best to consider this form when using a calculator.

▶ **Example**   Find $\tilde{r}$ for the data pairs at the right.

| x-rank | y-rank |
|--------|--------|
| 2 | 4 |
| 1 | 1 |
| 3 | 3 |
| 5 | 6 |
| 4 | 2 |
| 6 | 7 |
| 7 | 5 |

  In most cases, it is easy to find the differences in ranks and their squares without the use of a calculator.

| x-rank | y-rank | d | d² |
|--------|--------|-----|-----|
| 2 | 4 | −2 | 4 |
| 1 | 1 | 0 | 0 |
| 3 | 3 | 0 | 0 |
| 5 | 6 | −1 | 1 |
| 4 | 2 | 2 | 4 |
| 6 | 7 | −1 | 1 |
| 7 | 5 | 2 | 4 |
|  |  |  | 14 |

**Solution**

$$\sum d^2 = 14 \quad \text{and} \quad n = 7$$

To solve this with a scientific calculator, do the following:

| Key Sequence | Display | Comments |
|--------------|---------|----------|
| 1 | *1.* | |
| $-$ | *1.* | Indicates subtraction |
| 6 | *6.* | |
| $\div$ | *6.* | Indicates division |

| Key Sequence | Display | Comments |
|:---:|:---:|:---|
| ( | [01  0. | First open ( |
| 7 | 7. | $n$ |
| × | 7. | Indicates multiplication |
| ( | [02  0. | Second open ( |
| 7 | 7. | $n$ |
| $x^2$ | 49. | $n^2$ |
| − | 49. | Indicates subtraction |
| 1 | 1. | |
| ) | 48. | First closed ) |
| ) | 336. | Second closed ) |
| × | 0.017857142 | Indicates multiplication |
| 14 | 14. | $\sum d^2$ |
| = | 0.75 | The value of $\tilde{r}$ |

## Casio Graphics Calculator

Press the following keys for the Casio Graphics Calculator:

$$1 \boxed{-} 6 \boxed{\div} \boxed{(} 7 \boxed{\times} \boxed{(} 7 \boxed{x^2} \boxed{-} 1 \boxed{)} \boxed{)} \boxed{\times} 14 \boxed{\text{EXE}}$$

The display will look like the following:

$$1 - 6 \div (7 \times (7^2 - 1)) \times 14$$
$$0.75$$

## Texas Instruments TI-81

Press the following keys for the TI-81 Calculator:

$$1 \boxed{-} 6 \boxed{\div} \boxed{(} 7 \boxed{\times} \boxed{(} 7 \boxed{x^2} \boxed{-} 1 \boxed{)} \boxed{)} \boxed{\times} 14 \boxed{\text{ENTER}}$$

The display will look like the following:

$$1 - 6/(7 * (7^2 - 1)) * 14$$
$$.75$$

The value for $\tilde{r}$ can also be found using the Linear Regression features on the calculators. The correlation coefficient ($r$) found after entering the pairs of ranks is the same as the $\tilde{r}$ calculated above. See the calculator activities for Chapter 9 for examples of Linear Regression operations on the calculator.

**N**umber crunchers have long been in demand by government, medical and academic institutions for their all-important conclusions and projections.

Such analysis is the traditional work of statisticians, the mathematicians who do research, create studies and then interpret the results of their experiments.

But today, these professionals are being recruited by private businesses to work in quality control, production, research and development, tax and pension laws, marketing and sales. Statisticians, once ignored, are now vital members of the business team.

And U.S. businesses now are hiring them, even though most still are private consultants, part of college research teams or work for government agencies.

The U.S. Department of Labor reports that businesses are relying more heavily on statisticians with strong backgrounds in mathematics and computer science to forecast sales, analyze business conditions, modernize accounting procedures and help solve management problems.

The American Statistical Association estimates there are some 30,000 statisticians nationwide. A 1990 Northwestern University survey shows graduates with a bachelor's degree in mathematics or statistics are being hired at $28,900. Those with master's degrees start at $35,000; with a doctorate, $45,000 and more.

"Right now, statistics is a booming field to go into," said Barbara Bailar, executive director of the statistical association based in Alexandria, Va. "Statisticians are actively involved in new applications in every field, designing clinical trials for new drugs

Statisticians like Frank J. Rossi sell statistical software products to Fortune 500 companies.

and new products. They're being hired by industrial firms, especially automobile, food and chemical manufacturers. There's a new emphasis on quality assurance, which is based on conclusions derived from statistical techniques."

Bailar is a statistician with a doctorate from American University; a master's from Virginia Polytechnic Institute; and a bachelor's degree from State University of New York at Albany. Her association has 15,000 members.

"Today, some 20 to 25 percent of statisticians are in business," Bailar said. "Most take the route of getting a job with a bachelor's degree and going on for their master's later. They probably don't need a doctorate, but it gives you more job options."

Business statisticians have to sell the value of their services to colleagues, she says. "You have to keep reminding them you're there and what you can do to help them," Bailar said. "But once you help the firm save money or time, they beat a path to your door."

At Smith Hanley Associates Inc., a New York recruitment firm, demand for statisticians with master's or doctoral degrees doubled over the last five years and is expected to double again by 1995. "We're in an information age, and the need for statisticians is on the rise in business environments," said Tracey Gmoser, vice president of the firm. Four of the firm's 25 consultants specialize in statisticians. Gmoser says businesses are looking for statisticians who "have hands-on problem-solving experience—as opposed to the ability to analyze problems that haven't happened yet."

Statisticians are being hired as full-time employees by pharmaceutical companies, chemical manufacturing firms, consumer package goodsindustries and financial service institutions, she says. "A large pharmaceutical firm could have 40 statisticians on staff, while a smaller manufacturing firm may be hiring their first statistician," Gmoser observed.

The recruiter reports that "job opportunities are excellent. Statisticians never had as much variety in job choices as they do now." She believes that "statisticians in business who develop managerial skills have a greater opportunity to move into a management track."

Twenty years ago, Robert Wilkinson was managing an industrial department that included statisticians. "A statistician left, and I couldn't find an agency to get us a replacement, so I started my own," said Wilkinson, whose Blauvelt, N.Y., search firm specializes in statisticians. "Today, more business are learning that statistical techniques can provide them answers they need."

He says graduates of the top universities such as Iowa State, Michigan, North Carolina, Virginia Polytechnic and Wisconsin are highly sought after. "Nationwide, the trend toward statisticians in business is significantly upward," he noted.

Working with a team that includes engineers, Frank J. Rossi, a statistician, sells statistical software products to Fortune 500 manufacturing companies. "It's a full data analysis and data software package with quality control and statistically designed programs," said Rossi, a technical consultant for BBN Software Products Corp. in Northbrook. The company, which has branches worldwide, is based in Cambridge, Mass.

Rossi has an undergraduate and master's degree in statistics from Penn State University and learned computer skills in college. "We customize our software to individual companies so that the best possible product comes out, but, at the same time, a significant amount of money is saved," he said.

The statistician, who previously worked for the U.S. Census Bureau in Washington and for General Foods Corp. in Tarrytown, N.J., says "industry is the place for statisticians to go. Businesses are getting clobbered by all the data available and have to turn it into information to make decisions."

Statisticians know how to use the new tools automation provides, Rossi says, "because we're the ones who invented them."

# Appendix A  Tables

TABLE 1.  Random Numbers

| 2743 | 9077 | 8944 | 4609 | 6185 | 2620 | 4942 | 1699 | 3422 | 4144 | 8674 | 1871 |
|------|------|------|------|------|------|------|------|------|------|------|------|
| 9204 | 3655 | 2067 | 8413 | 4006 | 5209 | 1297 | 7846 | 8089 | 4565 | 2893 | 6517 |
| 2026 | 7171 | 4327 | 6965 | 1107 | 9385 | 4198 | 2373 | 3563 | 4725 | 1421 | 7662 |
| 1817 | 4595 | 0417 | 7354 | 2708 | 0030 | 9332 | 5018 | 0241 | 4603 | 0806 | 6203 |
| 2288 | 9713 | 4217 | 2244 | 7925 | 8468 | 2363 | 3707 | 9579 | 4127 | 8369 | 8856 |
| 3454 | 0531 | 7569 | 8523 | 8828 | 0289 | 7134 | 4960 | 7874 | 6828 | 6255 | 0976 |
| 3149 | 9716 | 3953 | 1303 | 3440 | 1346 | 2182 | 9337 | 4933 | 8834 | 9442 | 7557 |
| 4054 | 8522 | 1718 | 9208 | 4874 | 5341 | 2766 | 0356 | 2308 | 9157 | 9872 | 5816 |
| 5920 | 5122 | 4080 | 9387 | 1294 | 8963 | 8089 | 7832 | 8294 | 9713 | 1410 | 5332 |
| 5168 | 0795 | 5561 | 3195 | 3871 | 2989 | 5903 | 9057 | 9367 | 5997 | 2807 | 4150 |
| 6569 | 9901 | 7065 | 5968 | 2447 | 0315 | 9180 | 3785 | 0256 | 1863 | 8357 | 1804 |
| 4174 | 2085 | 4266 | 6853 | 1078 | 4325 | 2050 | 1424 | 7342 | 4345 | 8444 | 2695 |
| 1192 | 7641 | 6517 | 7090 | 8141 | 3553 | 4675 | 1410 | 2553 | 6952 | 4649 | 5748 |
| 1001 | 9482 | 5940 | 7736 | 8211 | 1077 | 9261 | 0171 | 7465 | 2379 | 6624 | 8805 |
| 3980 | 7644 | 8910 | 0214 | 5370 | 9569 | 1889 | 0634 | 9782 | 8066 | 0406 | 8680 |
| 7345 | 4772 | 4599 | 5672 | 0160 | 5452 | 7356 | 2710 | 6879 | 0553 | 0363 | 0311 |
| 7689 | 3516 | 7793 | 1559 | 4849 | 8094 | 6704 | 5689 | 3619 | 7051 | 9467 | 0957 |
| 8390 | 9282 | 5635 | 6360 | 4207 | 2478 | 9643 | 5992 | 4099 | 5188 | 5067 | 7867 |
| 6607 | 4267 | 0518 | 5949 | 1481 | 6290 | 6533 | 6208 | 9540 | 4353 | 9862 | 0367 |
| 7022 | 1226 | 3787 | 7460 | 1746 | 4850 | 7135 | 3856 | 6989 | 4572 | 0254 | 9793 |
| 2980 | 4145 | 6395 | 4289 | 1054 | 6295 | 5019 | 0218 | 1614 | 8318 | 9904 | 6933 |
| 3337 | 5788 | 2075 | 9170 | 9391 | 6085 | 9860 | 0906 | 2658 | 9608 | 1515 | 9401 |
| 8306 | 1612 | 1160 | 5998 | 5417 | 2237 | 9295 | 5834 | 6931 | 3371 | 9232 | 2978 |
| 5918 | 4323 | 5140 | 1468 | 5522 | 6118 | 7133 | 0364 | 7954 | 0212 | 7729 | 4185 |
| 0431 | 6066 | 2815 | 3480 | 3121 | 6161 | 2166 | 3047 | 5310 | 5684 | 7450 | 0363 |
| 0531 | 8830 | 4076 | 4783 | 2705 | 0691 | 6361 | 6028 | 8419 | 3941 | 8666 | 2494 |
| 5850 | 9894 | 8574 | 1971 | 7842 | 8920 | 1380 | 0493 | 5215 | 6738 | 9480 | 6193 |
| 8990 | 0745 | 3955 | 5778 | 5321 | 3534 | 0606 | 5585 | 0373 | 3711 | 9959 | 4306 |
| 2749 | 0415 | 0857 | 1088 | 0098 | 0830 | 6369 | 1742 | 0315 | 2402 | 7093 | 7035 |
| 4720 | 6668 | 4349 | 7614 | 8270 | 6823 | 4544 | 1883 | 4639 | 8930 | 7757 | 8050 |
| 3972 | 0973 | 4005 | 4678 | 7724 | 0267 | 1230 | 6386 | 6202 | 2775 | 9371 | 7511 |
| 6976 | 5709 | 7299 | 6035 | 8687 | 6028 | 8680 | 0285 | 4508 | 1595 | 0445 | 5443 |
| 2489 | 4531 | 3504 | 3161 | 8835 | 4752 | 0181 | 3641 | 9612 | 4524 | 5683 | 1396 |
| 7560 | 9334 | 7789 | 7970 | 0164 | 7250 | 6066 | 8505 | 6214 | 4883 | 2863 | 8838 |
| 9366 | 9765 | 7701 | 8372 | 7302 | 4554 | 5540 | 8431 | 0142 | 5736 | 1147 | 5207 |
| 0045 | 3704 | 6709 | 2623 | 9464 | 7072 | 1524 | 8954 | 1897 | 4248 | 3028 | 1067 |
| 4235 | 9659 | 2021 | 0906 | 0714 | 8780 | 8187 | 9990 | 4467 | 3255 | 4603 | 4301 |
| 1488 | 4520 | 6327 | 5067 | 5745 | 4896 | 3362 | 3584 | 5074 | 1012 | 1345 | 7783 |
| 8259 | 2595 | 5691 | 9961 | 0849 | 3966 | 0536 | 2950 | 3064 | 2531 | 9079 | 1464 |
| 6650 | 4372 | 8256 | 2336 | 2275 | 7794 | 7592 | 1233 | 8609 | 8462 | 2333 | 2306 |

**TABLE 1.**   Random Numbers (Continued)

| | | | | | | | | | | | |
|---|---|---|---|---|---|---|---|---|---|---|---|
| 2735 | 1599 | 3878 | 2307 | 5592 | 6900 | 4802 | 8483 | 1476 | 2266 | 4960 | 8558 |
| 8865 | 7173 | 4013 | 5076 | 6518 | 0173 | 2103 | 0328 | 3442 | 7607 | 3970 | 0178 |
| 9636 | 6841 | 7815 | 8999 | 7506 | 2689 | 0038 | 1884 | 1444 | 9343 | 0290 | 0387 |
| 4407 | 9324 | 8730 | 4915 | 2525 | 3080 | 0656 | 0047 | 5651 | 4386 | 0915 | 4994 |
| 5640 | 2157 | 9385 | 9238 | 9442 | 6419 | 6134 | 9463 | 8609 | 5647 | 5476 | 0033 |
| 2099 | 7096 | 0374 | 2607 | 9075 | 2050 | 3610 | 3010 | 8790 | 6383 | 8184 | 0910 |
| 0368 | 0558 | 3926 | 3884 | 8198 | 9571 | 2721 | 4200 | 8398 | 1775 | 8106 | 6602 |
| 1879 | 7548 | 7141 | 5441 | 8676 | 6438 | 8021 | 4921 | 5126 | 8670 | 4934 | 7328 |
| 5131 | 9988 | 1983 | 8179 | 7419 | 3086 | 0964 | 4480 | 0037 | 6924 | 9961 | 6632 |
| 5297 | 2906 | 1726 | 7201 | 4352 | 4046 | 8652 | 8366 | 6273 | 8283 | 4433 | 1319 |
| 5897 | 4415 | 6022 | 1596 | 8065 | 5962 | 8017 | 1847 | 4852 | 4229 | 5295 | 1384 |
| 9733 | 7908 | 9298 | 4257 | 1363 | 5968 | 4632 | 2218 | 1849 | 2663 | 0476 | 8144 |
| 8762 | 7429 | 9654 | 0179 | 1376 | 6348 | 6190 | 5774 | 5237 | 4078 | 0239 | 0583 |
| 5115 | 9640 | 7964 | 4164 | 1426 | 3931 | 2442 | 0064 | 1978 | 8765 | 0195 | 2411 |
| 7918 | 8330 | 9351 | 1388 | 4347 | 5138 | 0614 | 8328 | 7591 | 4466 | 4228 | 0654 |
| 3139 | 9891 | 1740 | 6749 | 9430 | 0142 | 3613 | 9899 | 9934 | 0024 | 8545 | 8483 |
| 6119 | 0726 | 8665 | 1444 | 7693 | 2061 | 6536 | 5648 | 7876 | 4377 | 9686 | 6201 |
| 8988 | 9988 | 6019 | 7710 | 2657 | 9629 | 8708 | 6906 | 9534 | 6854 | 8656 | 3840 |
| 7655 | 3625 | 4945 | 5170 | 6317 | 4280 | 0892 | 8800 | 7263 | 7181 | 5882 | 1291 |
| 5550 | 8142 | 5537 | 4709 | 2500 | 5087 | 9901 | 1437 | 9102 | 6165 | 4178 | 4410 |
| 0565 | 5610 | 9345 | 3408 | 9788 | 1810 | 5273 | 4881 | 8844 | 0473 | 9401 | 8046 |
| 1636 | 6360 | 3952 | 2759 | 0969 | 4321 | 5199 | 6353 | 9431 | 7452 | 0997 | 9411 |
| 4692 | 9173 | 8321 | 4983 | 7054 | 1544 | 1403 | 0602 | 6854 | 8424 | 7109 | 7742 |
| 5342 | 4019 | 2146 | 2169 | 8264 | 3360 | 5154 | 6867 | 4252 | 1395 | 1453 | 8695 |
| 2298 | 9684 | 4305 | 1980 | 8716 | 6095 | 9096 | 8390 | 8389 | 6618 | 1626 | 4395 |
| 0573 | 9173 | 1642 | 6306 | 5479 | 0338 | 2087 | 2892 | 3206 | 0945 | 1061 | 2593 |
| 3003 | 0155 | 2635 | 5258 | 1574 | 6942 | 6733 | 4978 | 7627 | 9401 | 9291 | 0861 |
| 7267 | 8883 | 0634 | 2460 | 5119 | 4609 | 1805 | 3880 | 7311 | 4927 | 7727 | 4261 |
| 8117 | 8530 | 3855 | 0061 | 9273 | 5027 | 4228 | 2499 | 2542 | 9721 | 0632 | 7429 |
| 6974 | 5375 | 1821 | 4415 | 0209 | 3975 | 7823 | 6168 | 6930 | 4124 | 3471 | 4516 |
| 3610 | 8778 | 0434 | 5861 | 2769 | 9348 | 1023 | 2987 | 0964 | 7299 | 1867 | 4323 |
| 7074 | 3384 | 0370 | 8542 | 6009 | 9530 | 7846 | 6481 | 7339 | 1346 | 8956 | 9110 |
| 5576 | 4487 | 5980 | 8702 | 3307 | 8055 | 4076 | 9772 | 0273 | 7877 | 2121 | 5092 |
| 7495 | 9999 | 4388 | 2394 | 4236 | 3001 | 8446 | 3456 | 7584 | 4604 | 4708 | 2778 |
| 8208 | 4958 | 6966 | 6044 | 7880 | 2040 | 7795 | 1943 | 1276 | 9816 | 8959 | 4116 |
| 9933 | 7008 | 7890 | 5799 | 9779 | 6594 | 0259 | 8864 | 6318 | 1026 | 5067 | 0322 |
| 4261 | 3799 | 6943 | 4108 | 7202 | 1033 | 6800 | 7512 | 4313 | 8973 | 4181 | 7873 |
| 9572 | 5737 | 6270 | 4456 | 9920 | 5339 | 8223 | 4766 | 6737 | 2903 | 7539 | 6097 |
| 8023 | 0019 | 3263 | 7716 | 6178 | 2198 | 7403 | 9422 | 4928 | 5592 | 6429 | 2306 |
| 9297 | 4400 | 4266 | 6024 | 6052 | 1930 | 8887 | 8386 | 6537 | 9018 | 0149 | 2905 |
| 7717 | 9936 | 9022 | 6408 | 4696 | 5367 | 0571 | 6643 | 0589 | 4146 | 1665 | 1413 |
| 6120 | 7438 | 7124 | 4024 | 4831 | 9767 | 4631 | 9462 | 5886 | 0743 | 3156 | 6838 |
| 7261 | 4747 | 1755 | 4210 | 0778 | 6153 | 7401 | 7762 | 9905 | 5678 | 6128 | 8342 |
| 4170 | 9567 | 9407 | 8665 | 3930 | 0893 | 0384 | 0081 | 7892 | 0630 | 3278 | 4625 |
| 6105 | 9044 | 3879 | 0287 | 2702 | 7420 | 7084 | 7081 | 4348 | 2649 | 1814 | 8987 |
| 6752 | 0168 | 0063 | 6824 | 6582 | 9060 | 2842 | 7031 | 1571 | 3508 | 0403 | 3473 |
| 0743 | 5216 | 2608 | 9872 | 4141 | 0023 | 7375 | 3207 | 2470 | 6279 | 4437 | 8078 |
| 5680 | 0676 | 4160 | 5163 | 7166 | 7997 | 2188 | 6639 | 7314 | 0075 | 6813 | 3409 |
| 9360 | 0581 | 0359 | 2582 | 5616 | 4278 | 7563 | 0150 | 6621 | 6392 | 3902 | 7778 |
| 4379 | 7693 | 1272 | 6842 | 0584 | 9880 | 6308 | 2118 | 4860 | 2991 | 8903 | 2129 |

**TABLE 1.** Random Numbers (Continued)

| | | | | | | | | | | | |
|---|---|---|---|---|---|---|---|---|---|---|---|
| 5691 | 1638 | 7040 | 2978 | 3110 | 5751 | 5755 | 9217 | 8492 | 3561 | 4046 | 2395 |
| 8094 | 2144 | 1133 | 7620 | 7166 | 3388 | 7310 | 4550 | 1246 | 2789 | 2642 | 1507 |
| 4523 | 8588 | 3358 | 0029 | 1447 | 9769 | 4021 | 6029 | 0728 | 6333 | 4788 | 0486 |
| 6446 | 3364 | 6141 | 4088 | 0786 | 0950 | 9577 | 5629 | 1875 | 6807 | 0371 | 6472 |
| 4059 | 0805 | 1722 | 7385 | 9533 | 2550 | 2170 | 9948 | 1372 | 4701 | 6285 | 3699 |
| 4487 | 7588 | 5928 | 4854 | 5615 | 2584 | 4330 | 5694 | 6208 | 7723 | 3384 | 7653 |
| 8297 | 8178 | 6846 | 7780 | 8541 | 2448 | 5459 | 7080 | 6477 | 7383 | 1750 | 4265 |
| 1530 | 3751 | 3062 | 4082 | 3534 | 9518 | 9251 | 8574 | 2136 | 3733 | 1465 | 2580 |
| 2921 | 5525 | 8650 | 8332 | 6491 | 7279 | 0677 | 8935 | 5891 | 1705 | 8587 | 1846 |
| 8500 | 5953 | 4618 | 2431 | 6942 | 1431 | 9729 | 4980 | 6899 | 4998 | 8505 | 6446 |
| 8274 | 3691 | 8976 | 0389 | 8716 | 2916 | 2667 | 1804 | 3263 | 4153 | 0054 | 3511 |
| 2152 | 0805 | 0137 | 3147 | 0475 | 1290 | 9031 | 6675 | 5507 | 0593 | 9966 | 9831 |
| 4828 | 3137 | 0809 | 1870 | 0833 | 4384 | 0662 | 3765 | 7044 | 1314 | 0562 | 3519 |
| 1790 | 7272 | 2101 | 0659 | 5223 | 2698 | 1917 | 6795 | 6318 | 2152 | 0816 | 8207 |
| 1149 | 0042 | 9996 | 2110 | 4212 | 9450 | 0259 | 8300 | 0957 | 4522 | 3592 | 3968 |
| 8483 | 9014 | 8902 | 1669 | 5451 | 9736 | 9453 | 5035 | 6556 | 4062 | 4710 | 5145 |
| 4367 | 4887 | 9463 | 9205 | 3942 | 4720 | 1081 | 8419 | 0342 | 4636 | 5942 | 6339 |
| 6791 | 9235 | 0307 | 8751 | 7816 | 9309 | 1576 | 8456 | 4416 | 3615 | 4154 | 0002 |
| 3147 | 3555 | 5936 | 9852 | 9302 | 4232 | 4438 | 2068 | 5743 | 6900 | 1257 | 1572 |
| 8975 | 8022 | 5429 | 6440 | 3086 | 5985 | 9838 | 9283 | 2595 | 4593 | 2174 | 0579 |
| 1998 | 8187 | 0987 | 0356 | 3320 | 8088 | 6298 | 3214 | 9626 | 0785 | 0496 | 8667 |
| 2175 | 6619 | 7552 | 8832 | 2123 | 2733 | 1618 | 2253 | 6261 | 3369 | 4029 | 7970 |
| 3830 | 8635 | 9309 | 3893 | 4697 | 5488 | 2757 | 5446 | 5602 | 2337 | 4901 | 5772 |
| 2883 | 3271 | 8764 | 9189 | 6577 | 0298 | 6843 | 4455 | 3521 | 7494 | 1436 | 3902 |
| 5859 | 5875 | 5107 | 1447 | 4690 | 5049 | 8006 | 8066 | 0142 | 9015 | 1466 | 3780 |
| 3085 | 3513 | 2996 | 9118 | 6418 | 9395 | 7211 | 7675 | 8391 | 6798 | 9284 | 0581 |
| 0207 | 5418 | 6221 | 6376 | 2597 | 0750 | 4800 | 4681 | 6799 | 7040 | 0906 | 2422 |
| 5218 | 6905 | 0944 | 8407 | 6654 | 1879 | 7904 | 3240 | 6263 | 6978 | 4854 | 3037 |
| 0682 | 7709 | 5693 | 6423 | 4562 | 6032 | 7430 | 4299 | 2923 | 1234 | 8124 | 8860 |
| 3334 | 8172 | 7804 | 3340 | 2808 | 8052 | 1810 | 9363 | 1875 | 5695 | 9541 | 3964 |

**TABLE 2.** Fifteen Rows of Pascal's Triangle

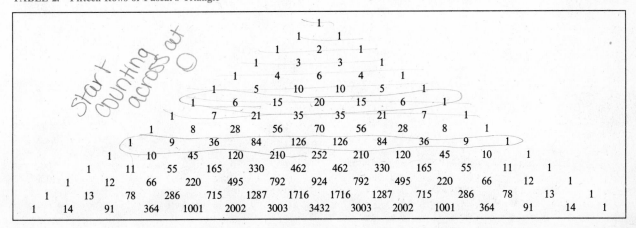

**TABLE 3.**　**Binomial Probabilities**

$$P(x) = \binom{n}{x} p^x (1 - p)^{n-x}$$

| n | x | 0.10 | 0.20 | 0.30 | 0.40 | 0.50 | p<br>0.60 | 0.70 | 0.80 | 0.90 |
|---|---|------|------|------|------|------|------|------|------|------|
| 2 | 0 | 0.810 | 0.640 | 0.490 | 0.360 | 0.250 | 0.160 | 0.090 | 0.040 | 0.010 |
|   | 1 | 0.180 | 0.320 | 0.420 | 0.480 | 0.500 | 0.480 | 0.420 | 0.320 | 0.180 |
|   | 2 | 0.010 | 0.040 | 0.090 | 0.160 | 0.250 | 0.360 | 0.490 | 0.640 | 0.810 |
| 3 | 0 | 0.729 | 0.512 | 0.343 | 0.216 | 0.125 | 0.064 | 0.027 | 0.008 | 0.001 |
|   | 1 | 0.243 | 0.384 | 0.441 | 0.432 | 0.375 | 0.288 | 0.189 | 0.096 | 0.027 |
|   | 2 | 0.027 | 0.096 | 0.189 | 0.288 | 0.375 | 0.432 | 0.441 | 0.384 | 0.243 |
|   | 3 | 0.001 | 0.008 | 0.027 | 0.064 | 0.125 | 0.216 | 0.343 | 0.512 | 0.729 |
| 4 | 0 | 0.656 | 0.410 | 0.240 | 0.130 | 0.062 | 0.026 | 0.008 | 0.002 | 0.0+ |
|   | 1 | 0.292 | 0.410 | 0.412 | 0.346 | 0.250 | 0.154 | 0.076 | 0.026 | 0.004 |
|   | 2 | 0.049 | 0.154 | 0.265 | 0.346 | 0.375 | 0.346 | 0.265 | 0.154 | 0.049 |
|   | 3 | 0.004 | 0.026 | 0.076 | 0.154 | 0.250 | 0.346 | 0.412 | 0.410 | 0.292 |
|   | 4 | 0.0+ | 0.002 | 0.008 | 0.026 | 0.062 | 0.130 | 0.240 | 0.410 | 0.656 |
| 5 | 0 | 0.590 | 0.328 | 0.168 | 0.078 | 0.031 | 0.010 | 0.002 | 0.0+ | 0.0+ |
|   | 1 | 0.328 | 0.410 | 0.360 | 0.259 | 0.156 | 0.077 | 0.028 | 0.006 | 0.0+ |
|   | 2 | 0.073 | 0.205 | 0.309 | 0.346 | 0.312 | 0.230 | 0.132 | 0.051 | 0.008 |
|   | 3 | 0.008 | 0.051 | 0.132 | 0.230 | 0.312 | 0.346 | 0.309 | 0.205 | 0.073 |
|   | 4 | 0.0+ | 0.006 | 0.028 | 0.077 | 0.156 | 0.259 | 0.360 | 0.410 | 0.328 |
|   | 5 | 0.0+ | 0.0+ | 0.002 | 0.010 | 0.031 | 0.078 | 0.168 | 0.328 | 0.590 |
| 6 | 0 | 0.531 | 0.262 | 0.118 | 0.047 | 0.016 | 0.004 | 0.001 | 0.0+ | 0.0+ |
|   | 1 | 0.354 | 0.393 | 0.303 | 0.187 | 0.094 | 0.037 | 0.010 | 0.002 | 0.0+ |
|   | 2 | 0.098 | 0.246 | 0.324 | 0.311 | 0.234 | 0.138 | 0.060 | 0.015 | 0.001 |
|   | 3 | 0.015 | 0.082 | 0.185 | 0.276 | 0.312 | 0.276 | 0.185 | 0.082 | 0.015 |
|   | 4 | 0.001 | 0.015 | 0.060 | 0.138 | 0.234 | 0.311 | 0.324 | 0.246 | 0.098 |
|   | 5 | 0.0+ | 0.002 | 0.010 | 0.037 | 0.094 | 0.187 | 0.303 | 0.393 | 0.354 |
|   | 6 | 0.0+ | 0.0+ | 0.001 | 0.004 | 0.016 | 0.047 | 0.118 | 0.262 | 0.531 |
| 7 | 0 | 0.478 | 0.210 | 0.082 | 0.028 | 0.008 | 0.002 | 0.0+ | 0.0+ | 0.0+ |
|   | 1 | 0.372 | 0.367 | 0.247 | 0.131 | 0.055 | 0.017 | 0.004 | 0.0+ | 0.0+ |
|   | 2 | 0.124 | 0.275 | 0.318 | 0.261 | 0.164 | 0.077 | 0.025 | 0.004 | 0.0+ |
|   | 3 | 0.023 | 0.115 | 0.227 | 0.290 | 0.273 | 0.194 | 0.097 | 0.029 | 0.003 |
|   | 4 | 0.003 | 0.029 | 0.097 | 0.194 | 0.273 | 0.290 | 0.227 | 0.115 | 0.023 |
|   | 5 | 0.0+ | 0.004 | 0.025 | 0.077 | 0.164 | 0.261 | 0.318 | 0.275 | 0.124 |
|   | 6 | 0.0+ | 0.0+ | 0.004 | 0.017 | 0.055 | 0.131 | 0.247 | 0.367 | 0.372 |
|   | 7 | 0.0+ | 0.0+ | 0.0+ | 0.002 | 0.008 | 0.028 | 0.082 | 0.210 | 0.478 |
| 8 | 0 | 0.430 | 0.168 | 0.058 | 0.017 | 0.004 | 0.001 | 0.0+ | 0.0+ | 0.0+ |
|   | 1 | 0.383 | 0.336 | 0.198 | 0.090 | 0.031 | 0.008 | 0.001 | 0.0+ | 0.0+ |
|   | 2 | 0.149 | 0.294 | 0.296 | 0.209 | 0.109 | 0.041 | 0.010 | 0.001 | 0.0+ |
|   | 3 | 0.033 | 0.147 | 0.254 | 0.279 | 0.219 | 0.124 | 0.047 | 0.009 | 0.0+ |
|   | 4 | 0.005 | 0.046 | 0.136 | 0.232 | 0.273 | 0.232 | 0.136 | 0.046 | 0.005 |

**TABLE 3. Binomial Probabilities (Continued)**

$$P(x) = \binom{n}{x} p^x(1 - p)^{n-x}$$

| n | x | 0.10 | 0.20 | 0.30 | 0.40 | 0.50 | $p$ 0.60 | 0.70 | 0.80 | 0.90 |
|---|---|------|------|------|------|------|------|------|------|------|
|   | 5 | 0.0+ | 0.009 | 0.047 | 0.124 | 0.219 | 0.279 | 0.254 | 0.147 | 0.033 |
|   | 6 | 0.0+ | 0.001 | 0.010 | 0.041 | 0.109 | 0.209 | 0.296 | 0.294 | 0.149 |
|   | 7 | 0.0+ | 0.0+ | 0.001 | 0.008 | 0.031 | 0.090 | 0.198 | 0.336 | 0.383 |
|   | 8 | 0.0+ | 0.0+ | 0.0+ | 0.001 | 0.004 | 0.017 | 0.058 | 0.168 | 0.430 |
| 9 | 0 | 0.387 | 0.134 | 0.040 | 0.010 | 0.002 | 0.0+ | 0.0+ | 0.0+ | 0.0+ |
|   | 1 | 0.387 | 0.302 | 0.156 | 0.060 | 0.018 | 0.004 | 0.0+ | 0.0+ | 0.0+ |
|   | 2 | 0.172 | 0.302 | 0.267 | 0.161 | 0.070 | 0.021 | 0.004 | 0.003 | 0.0+ |
|   | 3 | 0.045 | 0.176 | 0.267 | 0.251 | 0.164 | 0.074 | 0.021 | 0.003 | 0.0+ |
|   | 4 | 0.007 | 0.066 | 0.172 | 0.251 | 0.246 | 0.167 | 0.074 | 0.017 | 0.001 |
|   | 5 | 0.001 | 0.017 | 0.074 | 0.167 | 0.246 | 0.251 | 0.172 | 0.066 | 0.007 |
|   | 6 | 0.0+ | 0.003 | 0.021 | 0.074 | 0.164 | 0.251 | 0.267 | 0.176 | 0.045 |
|   | 7 | 0.0+ | 0.0+ | 0.004 | 0.021 | 0.070 | 0.161 | 0.267 | 0.302 | 0.172 |
|   | 8 | 0.0+ | 0.0+ | 0.0+ | 0.004 | 0.018 | 0.060 | 0.156 | 0.302 | 0.387 |
|   | 9 | 0.0+ | 0.0+ | 0.0+ | 0.0+ | 0.002 | 0.010 | 0.040 | 0.134 | 0.387 |
| 10 | 0 | 0.349 | 0.107 | 0.028 | 0.006 | 0.001 | 0.0+ | 0.0+ | 0.0+ | 0.0+ |
|   | 1 | 0.387 | 0.268 | 0.121 | 0.040 | 0.010 | 0.002 | 0.0+ | 0.0+ | 0.0+ |
|   | 2 | 0.194 | 0.302 | 0.233 | 0.121 | 0.044 | 0.011 | 0.001 | 0.0+ | 0.0+ |
|   | 3 | 0.057 | 0.201 | 0.267 | 0.215 | 0.117 | 0.042 | 0.009 | 0.001 | 0.0+ |
|   | 4 | 0.011 | 0.088 | 0.200 | 0.251 | 0.205 | 0.111 | 0.037 | 0.006 | 0.0+ |
|   | 5 | 0.001 | 0.026 | 0.103 | 0.201 | 0.246 | 0.201 | 0.103 | 0.026 | 0.001 |
|   | 6 | 0.0+ | 0.006 | 0.037 | 0.111 | 0.205 | 0.251 | 0.200 | 0.088 | 0.011 |
|   | 7 | 0.0+ | 0.001 | 0.009 | 0.042 | 0.117 | 0.215 | 0.267 | 0.201 | 0.057 |
|   | 8 | 0.0+ | 0.0+ | 0.001 | 0.011 | 0.044 | 0.121 | 0.233 | 0.302 | 0.194 |
|   | 9 | 0.0+ | 0.0+ | 0.0+ | 0.002 | 0.010 | 0.040 | 0.121 | 0.268 | 0.387 |
|   | 10 | 0.0+ | 0.0+ | 0.0+ | 0.0+ | 0.001 | 0.006 | 0.028 | 0.107 | 0.349 |
| 11 | 0 | 0.314 | 0.086 | 0.020 | 0.004 | 0.0+ | 0.0+ | 0.0+ | 0.0+ | 0.0+ |
|   | 1 | 0.384 | 0.236 | 0.093 | 0.027 | 0.005 | 0.001 | 0.0+ | 0.0+ | 0.0+ |
|   | 2 | 0.213 | 0.295 | 0.200 | 0.089 | 0.027 | 0.005 | 0.001 | 0.0+ | 0.0+ |
|   | 3 | 0.071 | 0.221 | 0.257 | 0.177 | 0.081 | 0.023 | 0.004 | 0.0+ | 0.0+ |
|   | 4 | 0.016 | 0.111 | 0.220 | 0.236 | 0.161 | 0.070 | 0.017 | 0.002 | 0.0+ |
|   | 5 | 0.002 | 0.039 | 0.132 | 0.221 | 0.226 | 0.147 | 0.057 | 0.010 | 0.0+ |
|   | 6 | 0.0+ | 0.010 | 0.057 | 0.147 | 0.226 | 0.221 | 0.132 | 0.039 | 0.002 |
|   | 7 | 0.0+ | 0.002 | 0.017 | 0.070 | 0.161 | 0.236 | 0.220 | 0.111 | 0.016 |
|   | 8 | 0.0+ | 0.0+ | 0.004 | 0.023 | 0.081 | 0.177 | 0.257 | 0.221 | 0.071 |
|   | 9 | 0.0+ | 0.0+ | 0.001 | 0.005 | 0.027 | 0.089 | 0.200 | 0.295 | 0.213 |
|   | 10 | 0.0+ | 0.0+ | 0.0+ | 0.001 | 0.005 | 0.027 | 0.093 | 0.236 | 0.384 |
|   | 11 | 0.0+ | 0.0+ | 0.0+ | 0.0+ | 0.0+ | 0.004 | 0.020 | 0.086 | 0.314 |

TABLE 3.   Binomial Probabilities (Continued)

$$P(x) = \binom{n}{x} p^x (1 - p)^{n-x}$$

| n | x | 0.10 | 0.20 | 0.30 | 0.40 | 0.50 | p 0.60 | 0.70 | 0.80 | 0.90 |
|---|---|------|------|------|------|------|------|------|------|------|
| 12 | 0 | 0.282 | 0.069 | 0.014 | 0.002 | 0.0+ | 0.0+ | 0.0+ | 0.0+ | 0.0+ |
| | 1 | 0.377 | 0.206 | 0.071 | 0.017 | 0.003 | 0.0+ | 0.0+ | 0.0+ | 0.0+ |
| | 2 | 0.230 | 0.283 | 0.168 | 0.064 | 0.016 | 0.002 | 0.0+ | 0.0+ | 0.0+ |
| | 3 | 0.085 | 0.236 | 0.240 | 0.142 | 0.054 | 0.012 | 0.001 | 0.0+ | 0.0+ |
| | 4 | 0.021 | 0.133 | 0.231 | 0.213 | 0.121 | 0.042 | 0.008 | 0.001 | 0.0+ |
| | 5 | 0.004 | 0.053 | 0.158 | 0.227 | 0.193 | 0.101 | 0.029 | 0.003 | 0.0+ |
| | 6 | 0.0+ | 0.016 | 0.079 | 0.177 | 0.226 | 0.177 | 0.079 | 0.016 | 0.0+ |
| | 7 | 0.0+ | 0.003 | 0.029 | 0.101 | 0.193 | 0.227 | 0.158 | 0.053 | 0.004 |
| | 8 | 0.0+ | 0.001 | 0.008 | 0.042 | 0.121 | 0.213 | 0.231 | 0.133 | 0.021 |
| | 9 | 0.0+ | 0.0+ | 0.001 | 0.012 | 0.054 | 0.142 | 0.240 | 0.236 | 0.085 |
| | 10 | 0.0+ | 0.0+ | 0.0+ | 0.002 | 0.016 | 0.064 | 0.168 | 0.283 | 0.230 |
| | 11 | 0.0+ | 0.0+ | 0.0+ | 0.0+ | 0.003 | 0.017 | 0.071 | 0.206 | 0.377 |
| | 12 | 0.0+ | 0.0+ | 0.0+ | 0.0+ | 0.0+ | 0.002 | 0.014 | 0.069 | 0.282 |
| 13 | 0 | 0.254 | 0.055 | 0.010 | 0.001 | 0.0+ | 0.0+ | 0.0+ | 0.0+ | 0.0+ |
| | 1 | 0.367 | 0.179 | 0.054 | 0.011 | 0.02 | 0.0+ | 0.0+ | 0.0+ | 0.0+ |
| | 2 | 0.245 | 0.268 | 0.139 | 0.045 | 0.010 | 0.001 | 0.0+ | 0.0+ | 0.0+ |
| | 3 | 0.100 | 0.246 | 0.218 | 0.111 | 0.035 | 0.006 | 0.001 | 0.0+ | 0.0+ |
| | 4 | 0.028 | 0.154 | 0.234 | 0.184 | 0.087 | 0.024 | 0.003 | 0.0+ | 0.0+ |
| | 5 | 0.006 | 0.069 | 0.180 | 0.221 | 0.157 | 0.066 | 0.014 | 0.001 | 0.0+ |
| | 6 | 0.001 | 0.023 | 0.103 | 0.197 | 0.209 | 0.131 | 0.044 | 0.006 | 0.0+ |
| | 7 | 0.0+ | 0.006 | 0.044 | 0.131 | 0.209 | 0.197 | 0.103 | 0.023 | 0.001 |
| | 8 | 0.0+ | 0.001 | 0.014 | 0.066 | 0.157 | 0.221 | 0.180 | 0.069 | 0.006 |
| | 9 | 0.0+ | 0.0+ | 0.003 | 0.024 | 0.087 | 0.184 | 0.234 | 0.154 | 0.028 |
| | 10 | 0.0+ | 0.0+ | 0.001 | 0.006 | 0.035 | 0.111 | 0.218 | 0.246 | 0.100 |
| | 11 | 0.0+ | 0.0+ | 0.0+ | 0.001 | 0.010 | 0.045 | 0.139 | 0.268 | 0.245 |
| | 12 | 0.0+ | 0.0+ | 0.0+ | 0.0+ | 0.002 | 0.011 | 0.054 | 0.179 | 0.367 |
| | 13 | 0.0+ | 0.0+ | 0.0+ | 0.0+ | 0.0+ | 0.001 | 0.010 | 0.055 | 0.254 |
| 14 | 0 | 0.229 | 0.044 | 0.007 | 0.001 | 0.0+ | 0.0+ | 0.0+ | 0.0+ | 0.0+ |
| | 1 | 0.356 | 0.154 | 0.041 | 0.007 | 0.001 | 0.0+ | 0.0+ | 0.0+ | 0.0+ |
| | 2 | 0.257 | 0.250 | 0.113 | 0.032 | 0.006 | 0.001 | 0.0+ | 0.0+ | 0.0+ |
| | 3 | 0.114 | 0.250 | 0.194 | 0.085 | 0.022 | 0.003 | 0.0+ | 0.0+ | 0.0+ |
| | 4 | 0.035 | 0.172 | 0.229 | 0.155 | 0.061 | 0.014 | 0.001 | 0.0+ | 0.0+ |
| | 5 | 0.008 | 0.086 | 0.196 | 0.207 | 0.122 | 0.041 | 0.007 | 0.0+ | 0.0+ |
| | 6 | 0.001 | 0.032 | 0.126 | 0.207 | 0.183 | 0.092 | 0.023 | 0.002 | 0.0+ |
| | 7 | 0.0+ | 0.009 | 0.062 | 0.157 | 0.209 | 0.157 | 0.062 | 0.009 | 0.0+ |
| | 8 | 0.0+ | 0.002 | 0.023 | 0.092 | 0.183 | 0.207 | 0.126 | 0.032 | 0.001 |
| | 9 | 0.0+ | 0.0+ | 0.007 | 0.041 | 0.122 | 0.207 | 0.196 | 0.086 | 0.008 |

**TABLE 3.  Binomial Probabilities (Continued)**

$$P(x) = \binom{n}{x} p^x (1 - p)^{n-x}$$

| n | x | 0.10 | 0.20 | 0.30 | 0.40 | 0.50 | p 0.60 | 0.70 | 0.80 | 0.90 |
|---|---|------|------|------|------|------|--------|------|------|------|
|    | 10 | 0.0+ | 0.0+ | 0.001 | 0.014 | 0.061 | 0.155 | 0.229 | 0.172 | 0.035 |
|    | 11 | 0.0+ | 0.0+ | 0.0+ | 0.003 | 0.022 | 0.085 | 0.194 | 0.250 | 0.114 |
|    | 12 | 0.0+ | 0.0+ | 0.0+ | 0.001 | 0.006 | 0.032 | 0.113 | 0.250 | 0.257 |
|    | 13 | 0.0+ | 0.0+ | 0.0+ | 0.0+ | 0.001 | 0.007 | 0.041 | 0.154 | 0.356 |
|    | 14 | 0.0+ | 0.0+ | 0.0+ | 0.0+ | 0.0+ | 0.001 | 0.007 | 0.044 | 0.229 |
| 15 | 0 | 0.206 | 0.035 | 0.005 | 0.0+ | 0.0+ | 0.0+ | 0.0+ | 0.0+ | 0.0+ |
|    | 1 | 0.343 | 0.132 | 0.031 | 0.005 | 0.0+ | 0.0+ | 0.0+ | 0.0+ | 0.0+ |
|    | 2 | 0.267 | 0.231 | 0.092 | 0.022 | 0.003 | 0.0+ | 0.0+ | 0.0+ | 0.0+ |
|    | 3 | 0.129 | 0.250 | 0.170 | 0.063 | 0.014 | 0.002 | 0.0+ | 0.0+ | 0.0+ |
|    | 4 | 0.043 | 0.188 | 0.219 | 0.127 | 0.042 | 0.007 | 0.001 | 0.0+ | 0.0+ |
|    | 5 | 0.010 | 0.103 | 0.206 | 0.186 | 0.092 | 0.024 | 0.003 | 0.0+ | 0.0+ |
|    | 6 | 0.002 | 0.043 | 0.147 | 0.207 | 0.153 | 0.061 | 0.012 | 0.001 | 0.0+ |
|    | 7 | 0.0+ | 0.014 | 0.081 | 0.177 | 0.196 | 0.118 | 0.035 | 0.003 | 0.0+ |
|    | 8 | 0.0+ | 0.003 | 0.035 | 0.118 | 0.196 | 0.177 | 0.081 | 0.014 | 0.0+ |
|    | 9 | 0.0+ | 0.001 | 0.012 | 0.061 | 0.153 | 0.207 | 0.147 | 0.043 | 0.002 |
|    | 10 | 0.0+ | 0.0+ | 0.003 | 0.024 | 0.092 | 0.186 | 0.206 | 0.103 | 0.010 |
|    | 11 | 0.0+ | 0.0+ | 0.001 | 0.007 | 0.042 | 0.127 | 0.219 | 0.188 | 0.043 |
|    | 12 | 0.0+ | 0.0+ | 0.0+ | 0.002 | 0.014 | 0.063 | 0.170 | 0.250 | 0.129 |
|    | 13 | 0.0+ | 0.0+ | 0.0+ | 0.0+ | 0.003 | 0.022 | 0.092 | 0.231 | 0.267 |
|    | 14 | 0.0+ | 0.0+ | 0.0+ | 0.0+ | 0.0+ | 0.005 | 0.031 | 0.132 | 0.343 |
|    | 15 | 0.0+ | 0.0+ | 0.0+ | 0.0+ | 0.0+ | 0.0+ | 0.005 | 0.035 | 0.206 |
| 16 | 0 | 0.185 | 0.028 | 0.003 | 0.0+ | 0.0+ | 0.0+ | 0.0+ | 0.0+ | 0.0+ |
|    | 1 | 0.329 | 0.113 | 0.023 | 0.003 | 0.0+ | 0.0+ | 0.0+ | 0.0+ | 0.0+ |
|    | 2 | 0.274 | 0.211 | 0.073 | 0.015 | 0.002 | 0.0+ | 0.0+ | 0.0+ | 0.0+ |
|    | 3 | 0.142 | 0.246 | 0.146 | 0.047 | 0.008 | 0.001 | 0.0+ | 0.0+ | 0.0+ |
|    | 4 | 0.051 | 0.200 | 0.204 | 0.101 | 0.028 | 0.004 | 0.0+ | 0.0+ | 0.0+ |
|    | 5 | 0.014 | 0.120 | 0.210 | 0.162 | 0.067 | 0.014 | 0.001 | 0.0+ | 0.0+ |
|    | 6 | 0.003 | 0.055 | 0.165 | 0.198 | 0.122 | 0.039 | 0.006 | 0.0+ | 0.0+ |
|    | 7 | 0.0+ | 0.020 | 0.101 | 0.189 | 0.175 | 0.084 | 0.018 | 0.001 | 0.0+ |
|    | 8 | 0.0+ | 0.006 | 0.049 | 0.142 | 0.196 | 0.142 | 0.049 | 0.006 | 0.0+ |
|    | 9 | 0.0+ | 0.001 | 0.018 | 0.084 | 0.175 | 0.189 | 0.101 | 0.020 | 0.0+ |
|    | 10 | 0.0+ | 0.0+ | 0.006 | 0.039 | 0.122 | 0.198 | 0.165 | 0.055 | 0.003 |
|    | 11 | 0.0+ | 0.0+ | 0.001 | 0.014 | 0.067 | 0.162 | 0.210 | 0.120 | 0.014 |
|    | 12 | 0.0+ | 0.0+ | 0.0+ | 0.004 | 0.028 | 0.101 | 0.204 | 0.200 | 0.051 |
|    | 13 | 0.0+ | 0.0+ | 0.0+ | 0.001 | 0.008 | 0.047 | 0.146 | 0.246 | 0.142 |
|    | 14 | 0.0+ | 0.0+ | 0.0+ | 0.0+ | 0.002 | 0.015 | 0.073 | 0.211 | 0.274 |
|    | 15 | 0.0+ | 0.0+ | 0.0+ | 0.0+ | 0.0+ | 0.003 | 0.023 | 0.113 | 0.329 |
|    | 16 | 0.0+ | 0.0+ | 0.0+ | 0.0+ | 0.0+ | 0.0+ | 0.003 | 0.028 | 0.185 |

TABLE 3.   Binomial Probabilities (Continued)

$$P(x) = \binom{n}{x} p^x (1 - p)^{n-x}$$

| n | x | 0.10 | 0.20 | 0.30 | 0.40 | 0.50 | p 0.60 | 0.70 | 0.80 | 0.90 |
|---|---|------|------|------|------|------|------|------|------|------|
| 17 | 0 | 0.167 | 0.022 | 0.002 | 0.0+ | 0.0+ | 0.0+ | 0.0+ | 0.0+ | 0.0+ |
|  | 1 | 0.315 | 0.096 | 0.017 | 0.002 | 0.0+ | 0.0+ | 0.0+ | 0.0+ | 0.0+ |
|  | 2 | 0.280 | 0.191 | 0.058 | 0.010 | 0.001 | 0.0+ | 0.0+ | 0.0+ | 0.0+ |
|  | 3 | 0.156 | 0.239 | 0.124 | 0.034 | 0.005 | 0.0+ | 0.0+ | 0.0+ | 0.0+ |
|  | 4 | 0.060 | 0.209 | 0.187 | 0.080 | 0.018 | 0.002 | 0.0+ | 0.0+ | 0.0+ |
|  | 5 | 0.018 | 0.136 | 0.208 | 0.138 | 0.047 | 0.008 | 0.001 | 0.0+ | 0.0+ |
|  | 6 | 0.004 | 0.068 | 0.178 | 0.184 | 0.094 | 0.024 | 0.003 | 0.0+ | 0.0+ |
|  | 7 | 0.001 | 0.027 | 0.120 | 0.193 | 0.148 | 0.057 | 0.010 | 0.0+ | 0.0+ |
|  | 8 | 0.0+ | 0.008 | 0.064 | 0.161 | 0.186 | 0.107 | 0.028 | 0.002 | 0.0+ |
|  | 9 | 0.0+ | 0.002 | 0.028 | 0.107 | 0.186 | 0.161 | 0.064 | 0.008 | 0.0+ |
|  | 10 | 0.0+ | 0.0+ | 0.010 | 0.057 | 0.148 | 0.193 | 0.120 | 0.027 | 0.001 |
|  | 11 | 0.0+ | 0.0+ | 0.003 | 0.024 | 0.094 | 0.184 | 0.178 | 0.068 | 0.004 |
|  | 12 | 0.0+ | 0.0+ | 0.001 | 0.008 | 0.047 | 0.138 | 0.208 | 0.136 | 0.018 |
|  | 13 | 0.0+ | 0.0+ | 0.0+ | 0.002 | 0.018 | 0.080 | 0.187 | 0.209 | 0.060 |
|  | 14 | 0.0+ | 0.0+ | 0.0+ | 0.0+ | 0.005 | 0.034 | 0.124 | 0.239 | 0.156 |
|  | 15 | 0.0+ | 0.0+ | 0.0+ | 0.0+ | 0.001 | 0.010 | 0.058 | 0.191 | 0.280 |
|  | 16 | 0.0+ | 0.0+ | 0.0+ | 0.0+ | 0.0+ | 0.002 | 0.017 | 0.096 | 0.315 |
|  | 17 | 0.0+ | 0.0+ | 0.0+ | 0.0+ | 0.0+ | 0.0+ | 0.002 | 0.022 | 0.167 |
| 18 | 0 | 0.150 | 0.018 | 0.002 | 0.0+ | 0.0+ | 0.0+ | 0.0+ | 0.0+ | 0.0+ |
|  | 1 | 0.300 | 0.081 | 0.013 | 0.001 | 0.0+ | 0.0+ | 0.0+ | 0.0+ | 0.0+ |
|  | 2 | 0.284 | 0.172 | 0.046 | 0.007 | 0.001 | 0.0+ | 0.0+ | 0.0+ | 0.0+ |
|  | 3 | 0.168 | 0.230 | 0.105 | 0.025 | 0.003 | 0.0+ | 0.0+ | 0.0+ | 0.0+ |
|  | 4 | 0.070 | 0.215 | 0.168 | 0.061 | 0.012 | 0.001 | 0.0+ | 0.0+ | 0.0+ |
|  | 5 | 0.002 | 0.151 | 0.202 | 0.115 | 0.033 | 0.004 | 0.0+ | 0.0+ | 0.0+ |
|  | 6 | 0.005 | 0.082 | 0.187 | 0.166 | 0.071 | 0.014 | 0.001 | 0.0+ | 0.0+ |
|  | 7 | 0.001 | 0.035 | 0.138 | 0.189 | 0.121 | 0.037 | 0.005 | 0.0+ | 0.0+ |
|  | 8 | 0.0+ | 0.012 | 0.081 | 0.173 | 0.167 | 0.077 | 0.015 | 0.001 | 0.0+ |
|  | 9 | 0.0+ | 0.003 | 0.038 | 0.128 | 0.186 | 0.128 | 0.038 | 0.003 | 0.0+ |
|  | 10 | 0.0+ | 0.001 | 0.015 | 0.077 | 0.167 | 0.173 | 0.081 | 0.012 | 0.0+ |
|  | 11 | 0.0+ | 0.0+ | 0.005 | 0.037 | 0.121 | 0.189 | 0.138 | 0.035 | 0.001 |
|  | 12 | 0.0+ | 0.0+ | 0.001 | 0.014 | 0.071 | 0.166 | 0.187 | 0.082 | 0.005 |
|  | 13 | 0.0+ | 0.0+ | 0.0+ | 0.004 | 0.033 | 0.115 | 0.202 | 0.151 | 0.022 |
|  | 14 | 0.0+ | 0.0+ | 0.0+ | 0.001 | 0.012 | 0.061 | 0.168 | 0.215 | 0.070 |
|  | 15 | 0.0+ | 0.0+ | 0.0+ | 0.0+ | 0.003 | 0.025 | 0.105 | 0.230 | 0.168 |
|  | 16 | 0.0+ | 0.0+ | 0.0+ | 0.0+ | 0.001 | 0.007 | 0.046 | 0.172 | 0.284 |
|  | 17 | 0.0+ | 0.0+ | 0.0+ | 0.0+ | 0.0+ | 0.001 | 0.013 | 0.081 | 0.300 |
|  | 18 | 0.0+ | 0.0+ | 0.0+ | 0.0+ | 0.0+ | 0.0+ | 0.002 | 0.018 | 0.150 |

TABLE 3.   Binomial Probabilities (Continued)

$$P(x) = \binom{n}{x} p^x (1 - p)^{n-x}$$

| n | x | 0.10 | 0.20 | 0.30 | 0.40 | 0.50 | p 0.60 | 0.70 | 0.80 | 0.90 |
|---|---|------|------|------|------|------|------|------|------|------|
| 19 | 0 | 0.135 | 0.014 | 0.001 | 0.0+ | 0.0+ | 0.0+ | 0.0+ | 0.0+ | 0.0+ |
|    | 1 | 0.285 | 0.069 | 0.009 | 0.001 | 0.0+ | 0.0+ | 0.0+ | 0.0+ | 0.0+ |
|    | 2 | 0.285 | 0.154 | 0.036 | 0.005 | 0.0+ | 0.0+ | 0.0+ | 0.0+ | 0.0+ |
|    | 3 | 0.180 | 0.218 | 0.087 | 0.018 | 0.002 | 0.0+ | 0.0+ | 0.0+ | 0.0+ |
|    | 4 | 0.080 | 0.218 | 0.149 | 0.047 | 0.007 | 0.0+ | 0.0+ | 0.0+ | 0.0+ |
|    | 5 | 0.027 | 0.164 | 0.192 | 0.093 | 0.022 | 0.002 | 0.0+ | 0.0+ | 0.0+ |
|    | 6 | 0.007 | 0.096 | 0.192 | 0.145 | 0.052 | 0.008 | 0.0+ | 0.0+ | 0.0+ |
|    | 7 | 0.001 | 0.044 | 0.152 | 0.180 | 0.096 | 0.024 | 0.002 | 0.0+ | 0.0+ |
|    | 8 | 0.0+ | 0.017 | 0.098 | 0.180 | 0.144 | 0.053 | 0.008 | 0.0+ | 0.0+ |
|    | 9 | 0.0+ | 0.005 | 0.051 | 0.146 | 0.176 | 0.098 | 0.022 | 0.001 | 0.0+ |
|    | 10 | 0.0+ | 0.001 | 0.022 | 0.098 | 0.176 | 0.146 | 0.051 | 0.005 | 0.0+ |
|    | 11 | 0.0+ | 0.0+ | 0.008 | 0.053 | 0.144 | 0.180 | 0.098 | 0.017 | 0.0+ |
|    | 12 | 0.0+ | 0.0+ | 0.002 | 0.024 | 0.096 | 0.180 | 0.152 | 0.044 | 0.001 |
|    | 13 | 0.0+ | 0.0+ | 0.0+ | 0.008 | 0.052 | 0.145 | 0.192 | 0.096 | 0.007 |
|    | 14 | 0.0+ | 0.0+ | 0.0+ | 0.002 | 0.022 | 0.093 | 0.192 | 0.164 | 0.027 |
|    | 15 | 0.0+ | 0.0+ | 0.0+ | 0.0+ | 0.007 | 0.047 | 0.149 | 0.218 | 0.080 |
|    | 16 | 0.0+ | 0.0+ | 0.0+ | 0.0+ | 0.002 | 0.018 | 0.087 | 0.218 | 0.180 |
|    | 17 | 0.0+ | 0.0+ | 0.0+ | 0.0+ | 0.0+ | 0.005 | 0.036 | 0.154 | 0.285 |
|    | 18 | 0.0+ | 0.0+ | 0.0+ | 0.0+ | 0.0+ | 0.001 | 0.009 | 0.069 | 0.285 |
|    | 19 | 0.0+ | 0.0+ | 0.0+ | 0.0+ | 0.0+ | 0.0+ | 0.001 | 0.014 | 0.135 |
| 20 | 0 | 0.122 | 0.012 | 0.001 | 0.0+ | 0.0+ | 0.0+ | 0.0+ | 0.0+ | 0.0+ |
|    | 1 | 0.270 | 0.058 | 0.007 | 0.0+ | 0.0+ | 0.0+ | 0.0+ | 0.0+ | 0.0+ |
|    | 2 | 0.285 | 0.137 | 0.028 | 0.003 | 0.0+ | 0.0+ | 0.0+ | 0.0+ | 0.0+ |
|    | 3 | 0.190 | 0.205 | 0.072 | 0.012 | 0.001 | 0.0+ | 0.0+ | 0.0+ | 0.0+ |
|    | 4 | 0.090 | 0.218 | 0.130 | 0.035 | 0.005 | 0.0+ | 0.0+ | 0.0+ | 0.0+ |
|    | 5 | 0.032 | 0.175 | 0.179 | 0.075 | 0.015 | 0.001 | 0.0+ | 0.0+ | 0.0+ |
|    | 6 | 0.009 | 0.109 | 0.192 | 0.124 | 0.037 | 0.005 | 0.0+ | 0.0+ | 0.0+ |
|    | 7 | 0.002 | 0.054 | 0.164 | 0.166 | 0.074 | 0.015 | 0.001 | 0.0+ | 0.0+ |
|    | 8 | 0.0+ | 0.022 | 0.114 | 0.180 | 0.120 | 0.036 | 0.004 | 0.0+ | 0.0+ |
|    | 9 | 0.0+ | 0.007 | 0.065 | 0.160 | 0.160 | 0.071 | 0.012 | 0.0+ | 0.0+ |
|    | 10 | 0.0+ | 0.002 | 0.031 | 0.117 | 0.176 | 0.117 | 0.031 | 0.002 | 0.0+ |
|    | 11 | 0.0+ | 0.0+ | 0.012 | 0.071 | 0.160 | 0.160 | 0.065 | 0.007 | 0.0+ |
|    | 12 | 0.0+ | 0.0+ | 0.004 | 0.036 | 0.120 | 0.180 | 0.114 | 0.022 | 0.0+ |
|    | 13 | 0.0+ | 0.0+ | 0.001 | 0.015 | 0.074 | 0.166 | 0.164 | 0.054 | 0.002 |
|    | 14 | 0.0+ | 0.0+ | 0.0+ | 0.005 | 0.037 | 0.124 | 0.192 | 0.109 | 0.009 |
|    | 15 | 0.0+ | 0.0+ | 0.0+ | 0.001 | 0.015 | 0.075 | 0.179 | 0.175 | 0.032 |
|    | 16 | 0.0+ | 0.0+ | 0.0+ | 0.0+ | 0.005 | 0.035 | 0.130 | 0.218 | 0.090 |
|    | 17 | 0.0+ | 0.0+ | 0.0+ | 0.0+ | 0.001 | 0.012 | 0.072 | 0.205 | 0.190 |
|    | 18 | 0.0+ | 0.0+ | 0.0+ | 0.0+ | 0.0+ | 0.003 | 0.028 | 0.137 | 0.285 |
|    | 19 | 0.0+ | 0.0+ | 0.0+ | 0.0+ | 0.0+ | 0.0+ | 0.007 | 0.058 | 0.270 |
|    | 20 | 0.0+ | 0.0+ | 0.0+ | 0.0+ | 0.0+ | 0.0+ | 0.001 | 0.012 | 0.122 |

**TABLE 4.   The Standard Normal Distribution**

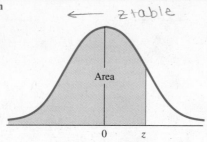

| | | | | | $P(Z < z)$ | | | | | |
|---|---|---|---|---|---|---|---|---|---|---|
| $z$ | 0.00 | 0.01 | 0.02 | 0.03 | 0.04 | 0.05 | 0.06 | 0.07 | 0.08 | 0.09 |
| −3.4 | 0.0003 | 0.0003 | 0.0003 | 0.0003 | 0.0003 | 0.0003 | 0.0003 | 0.0003 | 0.0003 | 0.0002 |
| −3.3 | 0.0005 | 0.0005 | 0.0005 | 0.0004 | 0.0004 | 0.0004 | 0.0004 | 0.0004 | 0.0004 | 0.0003 |
| −3.2 | 0.0007 | 0.0007 | 0.0006 | 0.0006 | 0.0006 | 0.0006 | 0.0006 | 0.0005 | 0.0005 | 0.0005 |
| −3.1 | 0.0010 | 0.0009 | 0.0009 | 0.0009 | 0.0008 | 0.0008 | 0.0008 | 0.0008 | 0.0007 | 0.0007 |
| −3.0 | 0.0013 | 0.0013 | 0.0013 | 0.0012 | 0.0012 | 0.0011 | 0.0011 | 0.0011 | 0.0010 | 0.0010 |
| −2.9 | 0.0019 | 0.0018 | 0.0017 | 0.0017 | 0.0016 | 0.0016 | 0.0015 | 0.0015 | 0.0014 | 0.0014 |
| −2.8 | 0.0026 | 0.0025 | 0.0024 | 0.0023 | 0.0023 | 0.0022 | 0.0021 | 0.0021 | 0.0020 | 0.0019 |
| −2.7 | 0.0035 | 0.0034 | 0.0033 | 0.0032 | 0.0031 | 0.0030 | 0.0029 | 0.0028 | 0.0027 | 0.0026 |
| −2.6 | 0.0047 | 0.0045 | 0.0044 | 0.0043 | 0.0041 | 0.0040 | 0.0039 | 0.0038 | 0.0037 | 0.0036 |
| −2.5 | 0.0062 | 0.0060 | 0.0059 | 0.0057 | 0.0055 | 0.0054 | 0.0052 | 0.0051 | 0.0049 | 0.0048 |
| −2.4 | 0.0082 | 0.0080 | 0.0078 | 0.0075 | 0.0073 | 0.0071 | 0.0069 | 0.0068 | 0.0066 | 0.0064 |
| −2.3 | 0.0107 | 0.0104 | 0.0102 | 0.0099 | 0.0096 | 0.0094 | 0.0091 | 0.0089 | 0.0087 | 0.0084 |
| −2.2 | 0.0139 | 0.0136 | 0.0132 | 0.0129 | 0.0125 | 0.0122 | 0.0119 | 0.0116 | 0.0113 | 0.0110 |
| −2.1 | 0.0179 | 0.0174 | 0.0170 | 0.0166 | 0.0162 | 0.0158 | 0.0154 | 0.0150 | 0.0146 | 0.0143 |
| −2.0 | 0.0228 | 0.0222 | 0.0217 | 0.0212 | 0.0207 | 0.0202 | 0.0197 | 0.0192 | 0.0188 | 0.0183 |
| −1.9 | 0.0287 | 0.0281 | 0.0274 | 0.0268 | 0.0262 | 0.0256 | 0.0250 | 0.0244 | 0.0239 | 0.0233 |
| −1.8 | 0.0359 | 0.0352 | 0.0344 | 0.0336 | 0.0329 | 0.0322 | 0.0314 | 0.0307 | 0.0301 | 0.0294 |
| −1.7 | 0.0446 | 0.0436 | 0.0427 | 0.0418 | 0.0409 | 0.0401 | 0.0392 | 0.0384 | 0.0375 | 0.0367 |
| −1.6 | 0.0548 | 0.0537 | 0.0526 | 0.0516 | 0.0505 | 0.0495 | 0.0485 | 0.0475 | 0.0465 | 0.0455 |
| −1.5 | 0.0668 | 0.0655 | 0.0643 | 0.0630 | 0.0618 | 0.0606 | 0.0594 | 0.0582 | 0.0571 | 0.0559 |
| −1.4 | 0.0808 | 0.0793 | 0.0778 | 0.0764 | 0.0749 | 0.0735 | 0.0722 | 0.0708 | 0.0694 | 0.0681 |
| −1.3 | 0.0968 | 0.0951 | 0.0934 | 0.0918 | 0.0901 | 0.0885 | 0.0869 | 0.0853 | 0.0838 | 0.0823 |
| −1.2 | 0.1151 | 0.1131 | 0.1112 | 0.1093 | 0.1075 | 0.1056 | 0.1038 | 0.1020 | 0.1003 | 0.0985 |
| −1.1 | 0.1357 | 0.1335 | 0.1314 | 0.1292 | 0.1271 | 0.1251 | 0.1230 | 0.1210 | 0.1190 | 0.1170 |
| −1.0 | 0.1587 | 0.1562 | 0.1539 | 0.1515 | 0.1492 | 0.1469 | 0.1446 | 0.1423 | 0.1401 | 0.1379 |
| −0.9 | 0.1841 | 0.1814 | 0.1788 | 0.1762 | 0.1736 | 0.1711 | 0.1685 | 0.1660 | 0.1635 | 0.1611 |
| −0.8 | 0.2119 | 0.2090 | 0.2061 | 0.2033 | 0.2005 | 0.1977 | 0.1949 | 0.1922 | 0.1894 | 0.1867 |
| −0.7 | 0.2420 | 0.2389 | 0.2358 | 0.2327 | 0.2296 | 0.2266 | 0.2236 | 0.2206 | 0.2177 | 0.2148 |
| −0.6 | 0.2743 | 0.2709 | 0.2676 | 0.2643 | 0.2611 | 0.2578 | 0.2546 | 0.2514 | 0.2483 | 0.2451 |
| −0.5 | 0.3085 | 0.3050 | 0.3015 | 0.2981 | 0.2946 | 0.2912 | 0.2877 | 0.2843 | 0.2810 | 0.2776 |

TABLE 4.   The Standard Normal Distribution (Continued)

| z | 0.00 | 0.01 | 0.02 | 0.03 | 0.04 | 0.05 | 0.06 | 0.07 | 0.08 | 0.09 |
|---|------|------|------|------|------|------|------|------|------|------|
| | | | | | $P(Z < z)$ | | | | | |
| −0.4 | 0.3446 | 0.3409 | 0.3372 | 0.3336 | 0.3300 | 0.3264 | 0.3228 | 0.3192 | 0.3156 | 0.3121 |
| −0.3 | 0.3821 | 0.3783 | 0.3745 | 0.3707 | 0.3669 | 0.3632 | 0.3594 | 0.3557 | 0.3520 | 0.3483 |
| −0.2 | 0.4207 | 0.4168 | 0.4129 | 0.4090 | 0.4052 | 0.4013 | 0.3974 | 0.3936 | 0.3897 | 0.3859 |
| −0.1 | 0.4602 | 0.4562 | 0.4522 | 0.4483 | 0.4443 | 0.4404 | 0.4364 | 0.4325 | 0.4286 | 0.4247 |
| −0.0 | 0.5000 | 0.4960 | 0.4920 | 0.4880 | 0.4840 | 0.4801 | 0.4761 | 0.4721 | 0.4681 | 0.4641 |
| 0.0 | 0.5000 | 0.5040 | 0.5080 | 0.5120 | 0.5160 | 0.5199 | 0.5239 | 0.5279 | 0.5319 | 0.5359 |
| 0.1 | 0.5398 | 0.5438 | 0.5478 | 0.5517 | 0.5557 | 0.5596 | 0.5636 | 0.5675 | 0.5714 | 0.5753 |
| 0.2 | 0.5793 | 0.5832 | 0.5871 | 0.5910 | 0.5948 | 0.5987 | 0.6026 | 0.6064 | 0.6103 | 0.6141 |
| 0.3 | 0.6179 | 0.6217 | 0.6255 | 0.6293 | 0.6331 | 0.6368 | 0.6406 | 0.6443 | 0.6480 | 0.6517 |
| 0.4 | 0.6554 | 0.6591 | 0.6628 | 0.6664 | 0.6700 | 0.6736 | 0.6772 | 0.6808 | 0.6844 | 0.6879 |
| 0.5 | 0.6915 | 0.6950 | 0.6985 | 0.7019 | 0.7054 | 0.7088 | 0.7123 | 0.7157 | 0.7190 | 0.7224 |
| 0.6 | 0.7257 | 0.7291 | 0.7324 | 0.7357 | 0.7389 | 0.7422 | 0.7454 | 0.7486 | 0.7517 | 0.7549 |
| 0.7 | 0.7580 | 0.7611 | 0.7642 | 0.7673 | 0.7704 | 0.7734 | 0.7764 | 0.7794 | 0.7823 | 0.7852 |
| 0.8 | 0.7881 | 0.7910 | 0.7939 | 0.7967 | 0.7995 | 0.8023 | 0.8051 | 0.8078 | 0.8106 | 0.8133 |
| 0.9 | 0.8159 | 0.8186 | 0.8212 | 0.8238 | 0.8264 | 0.8289 | 0.8315 | 0.8340 | 0.8365 | 0.8389 |
| 1.0 | 0.8413 | 0.8438 | 0.8461 | 0.8485 | 0.8508 | 0.8531 | 0.8554 | 0.8577 | 0.8599 | 0.8621 |
| 1.1 | 0.8643 | 0.8665 | 0.8686 | 0.8708 | 0.8729 | 0.8749 | 0.8770 | 0.8790 | 0.8810 | 0.8830 |
| 1.2 | 0.8849 | 0.8869 | 0.8888 | 0.8907 | 0.8925 | 0.8944 | 0.8962 | 0.8980 | 0.8997 | 0.9015 |
| 1.3 | 0.9032 | 0.9049 | 0.9066 | 0.9082 | 0.9099 | 0.9115 | 0.9131 | 0.9147 | 0.9162 | 0.9177 |
| 1.4 | 0.9192 | 0.9207 | 0.9222 | 0.9236 | 0.9251 | 0.9265 | 0.9278 | 0.9292 | 0.9306 | 0.9319 |
| 1.5 | 0.9332 | 0.9345 | 0.9357 | 0.9370 | 0.9382 | 0.9394 | 0.9406 | 0.9418 | 0.9429 | 0.9441 |
| 1.6 | 0.9452 | 0.9463 | 0.9474 | 0.9484 | 0.9495 | 0.9505 | 0.9515 | 0.9525 | 0.9535 | 0.9545 |
| 1.7 | 0.9554 | 0.9564 | 0.9573 | 0.9582 | 0.9591 | 0.9599 | 0.9608 | 0.9616 | 0.9625 | 0.9633 |
| 1.8 | 0.9641 | 0.9649 | 0.9656 | 0.9664 | 0.9671 | 0.9678 | 0.9686 | 0.9693 | 0.9699 | 0.9706 |
| 1.9 | 0.9713 | 0.9719 | 0.9726 | 0.9732 | 0.9738 | 0.9744 | 0.9750 | 0.9756 | 0.9761 | 0.9767 |
| 2.0 | 0.9772 | 0.9778 | 0.9783 | 0.9788 | 0.9793 | 0.9798 | 0.9803 | 0.9808 | 0.9812 | 0.9817 |
| 2.1 | 0.9821 | 0.9826 | 0.9830 | 0.9834 | 0.9838 | 0.9842 | 0.9846 | 0.9850 | 0.9854 | 0.9857 |
| 2.2 | 0.9861 | 0.9864 | 0.9868 | 0.9871 | 0.9875 | 0.9878 | 0.9881 | 0.9884 | 0.9887 | 0.9890 |
| 2.3 | 0.9893 | 0.9896 | 0.9898 | 0.9901 | 0.9904 | 0.9906 | 0.9909 | 0.9911 | 0.9913 | 0.9916 |
| 2.4 | 0.9918 | 0.9920 | 0.9922 | 0.9925 | 0.9927 | 0.9929 | 0.9931 | 0.9932 | 0.9934 | 0.9936 |
| 2.5 | 0.9938 | 0.9940 | 0.9941 | 0.9943 | 0.9945 | 0.9946 | 0.9948 | 0.9949 | 0.9951 | 0.9952 |
| 2.6 | 0.9953 | 0.9955 | 0.9956 | 0.9957 | 0.9959 | 0.9960 | 0.9961 | 0.9962 | 0.9963 | 0.9964 |
| 2.7 | 0.9965 | 0.9966 | 0.9967 | 0.9968 | 0.9969 | 0.9970 | 0.9971 | 0.9972 | 0.9973 | 0.9974 |
| 2.8 | 0.9974 | 0.9975 | 0.9976 | 0.9977 | 0.9977 | 0.9978 | 0.9979 | 0.9979 | 0.9980 | 0.9981 |
| 2.9 | 0.9981 | 0.9982 | 0.9982 | 0.9983 | 0.9984 | 0.9984 | 0.9985 | 0.9985 | 0.9986 | 0.9986 |
| 3.0 | 0.9987 | 0.9987 | 0.9987 | 0.9988 | 0.9988 | 0.9989 | 0.9989 | 0.9989 | 0.9990 | 0.9990 |
| 3.1 | 0.9990 | 0.9991 | 0.9991 | 0.9991 | 0.9992 | 0.9992 | 0.9992 | 0.9992 | 0.9993 | 0.9993 |
| 3.2 | 0.9993 | 0.9993 | 0.9994 | 0.9994 | 0.9994 | 0.9994 | 0.9994 | 0.9995 | 0.9995 | 0.9995 |
| 3.3 | 0.9995 | 0.9995 | 0.9995 | 0.9996 | 0.9996 | 0.9996 | 0.9996 | 0.9996 | 0.9996 | 0.9997 |
| 3.4 | 0.9997 | 0.9997 | 0.9997 | 0.9997 | 0.9997 | 0.9997 | 0.9997 | 0.9997 | 0.9997 | 0.9998 |

From *Probability and Statistics*, 3/E by B. W. Lindgren & G. W. McElrath. Copyright © 1969 by B. W. Lindgren. Reprinted by permission of Macmillan Publishing Company.

## TABLE 5.  *t*-Distributions

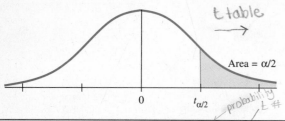

Area = α/2

0        $t_{\alpha/2}$

| d.f. | $t_{0.050}$ | $t_{0.025}$ | $t_{0.010}$ | $t_{0.005}$ | d.f. |
|------|------|------|------|------|------|
| 1 | 6.314 | 12.706 | 31.821 | 63.657 | 1 |
| 2 | 2.920 | 4.303 | 6.965 | 9.925 | 2 |
| 3 | 2.353 | 3.182 | 4.541 | 5.841 | 3 |
| 4 | 2.132 | 2.776 | 3.747 | 4.604 | 4 |
| 5 | 2.015 | 2.571 | 3.365 | 4.032 | 5 |
| 6 | 1.943 | 2.447 | 3.143 | 3.707 | 6 |
| 7 | 1.895 | 2.365 | 2.998 | 3.499 | 7 |
| 8 | 1.860 | 2.306 | 2.896 | 3.355 | 8 |
| 9 | 1.833 | 2.262 | 2.821 | 3.250 | 9 |
| 10 | 1.812 | 2.228 | 2.764 | 3.169 | 10 |
| 11 | 1.796 | 2.201 | 2.718 | 3.106 | 11 |
| 12 | 1.782 | 2.179 | 2.681 | 3.055 | 12 |
| 13 | 1.771 | 2.160 | 2.650 | 3.012 | 13 |
| 14 | 1.761 | 2.145 | 2.624 | 2.977 | 14 |
| 15 | 1.753 | 2.131 | 2.602 | 2.947 | 15 |

| d.f. | $t_{0.050}$ | $t_{0.025}$ | $t_{0.010}$ | $t_{0.005}$ | d.f. |
|------|------|------|------|------|------|
| 16 | 1.746 | 2.120 | 2.583 | 2.921 | 16 |
| 17 | 1.740 | 2.110 | 2.567 | 2.898 | 17 |
| 18 | 1.734 | 2.101 | 2.552 | 2.878 | 18 |
| 19 | 1.729 | 2.093 | 2.539 | 2.861 | 19 |
| 20 | 1.725 | 2.086 | 2.528 | 2.845 | 20 |
| 21 | 1.721 | 2.080 | 2.518 | 2.831 | 21 |
| 22 | 1.717 | 2.074 | 2.508 | 2.819 | 22 |
| 23 | 1.714 | 2.069 | 2.500 | 2.807 | 23 |
| 24 | 1.711 | 2.064 | 2.492 | 2.797 | 24 |
| 25 | 1.708 | 2.060 | 2.485 | 2.787 | 25 |
| 26 | 1.706 | 2.056 | 2.479 | 2.779 | 26 |
| 27 | 1.703 | 2.052 | 2.473 | 2.771 | 27 |
| 28 | 1.701 | 2.048 | 2.467 | 2.763 | 28 |
| 29 | 1.699 | 2.045 | 2.462 | 2.756 | 29 |
| inf. | 1.645 | 1.960 | 2.326 | 2.576 | inf. |

Adapted from *Statistics and Probability in Modern Life*, 2/E by J. Newmark. Copyright © 1977 by Holt, Rinehart and Winston. This table is abridged from Table IV from *Statistical Tables for Biological, Agricultural and Medical Research* by R. A. Fisher & F. Yates, published by Longman Group, Ltd., London (previously published by Oliver & Boyd, Edinburgh). Reprinted by permission of the authors and publishers.

## TABLE 6.  The Chi-Square ($\chi^2$) Distributions

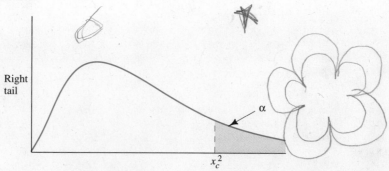

Right tail

α

$x_c^2$

| Degrees of Freedom | Right Tail Area | | | | | | | | | |
|------|------|------|------|------|------|------|------|------|------|------|
| | 0.995 | 0.99 | 0.975 | 0.95 | 0.90 | 0.10 | 0.05 | 0.025 | 0.01 | 0.005 |
| 1 | — | — | 0.001 | 0.004 | 0.016 | 2.706 | 3.841 | 5.024 | 6.635 | 7.879 |
| 2 | 0.010 | 0.020 | 0.051 | 0.103 | 0.211 | 4.605 | 5.991 | 7.378 | 9.210 | 10.597 |
| 3 | 0.072 | 0.115 | 0.216 | 0.352 | 0.584 | 6.251 | 7.815 | 9.348 | 11.345 | 12.838 |
| 4 | 0.207 | 0.297 | 0.484 | 0.711 | 1.064 | 7.779 | 9.488 | 11.143 | 13.277 | 14.860 |
| 5 | 0.412 | 0.554 | 0.831 | 1.145 | 1.610 | 9.236 | 11.071 | 12.833 | 15.086 | 16.750 |

TABLE 6.  The Chi-Square ($\chi^2$) Distributions (Continued)

| Degrees of Freedom | Right Tail Area | | | | | | | | | |
|---|---|---|---|---|---|---|---|---|---|---|
| | 0.995 | 0.99 | 0.975 | 0.95 | 0.90 | 0.10 | 0.05 | 0.025 | 0.01 | 0.005 |
| 6 | 0.676 | 0.872 | 1.237 | 1.635 | 2.204 | 10.645 | 12.592 | 14.449 | 16.812 | 18.548 |
| 7 | 0.989 | 1.239 | 1.690 | 2.167 | 2.833 | 12.017 | 14.067 | 16.013 | 18.475 | 20.278 |
| 8 | 1.344 | 1.646 | 2.180 | 2.733 | 3.490 | 13.362 | 15.507 | 17.535 | 20.090 | 21.955 |
| 9 | 1.735 | 2.088 | 2.700 | 3.325 | 4.168 | 14.684 | 16.919 | 19.023 | 21.666 | 23.589 |
| 10 | 2.156 | 2.558 | 3.247 | 3.940 | 4.865 | 15.987 | 18.307 | 20.483 | 23.209 | 25.188 |
| 11 | 2.603 | 3.053 | 3.816 | 4.575 | 5.578 | 17.275 | 19.675 | 21.920 | 24.725 | 26.757 |
| 12 | 3.074 | 3.571 | 4.404 | 5.226 | 6.304 | 18.549 | 21.026 | 23.337 | 26.217 | 28.299 |
| 13 | 3.565 | 4.107 | 5.009 | 5.892 | 7.042 | 19.812 | 22.362 | 24.736 | 27.688 | 29.819 |
| 14 | 4.075 | 4.660 | 5.629 | 6.571 | 7.790 | 21.064 | 23.685 | 26.119 | 29.141 | 31.319 |
| 15 | 4.601 | 5.229 | 6.262 | 7.261 | 8.547 | 22.307 | 24.996 | 27.488 | 30.578 | 32.801 |
| 16 | 5.142 | 5.812 | 6.908 | 7.962 | 9.312 | 23.542 | 26.296 | 28.845 | 32.000 | 34.267 |
| 17 | 5.697 | 6.408 | 7.564 | 8.672 | 10.085 | 24.769 | 27.587 | 30.191 | 33.409 | 35.718 |
| 18 | 6.265 | 7.015 | 8.231 | 9.390 | 10.865 | 25.989 | 28.869 | 31.526 | 34.805 | 37.156 |
| 19 | 6.844 | 7.633 | 8.907 | 10.117 | 11.651 | 27.204 | 30.144 | 32.852 | 36.191 | 38.582 |
| 20 | 7.434 | 8.260 | 9.591 | 10.851 | 12.443 | 28.412 | 31.410 | 34.170 | 37.566 | 39.997 |
| 21 | 8.034 | 8.897 | 10.283 | 11.591 | 13.240 | 29.615 | 32.671 | 35.479 | 38.932 | 41.401 |
| 22 | 8.643 | 9.542 | 10.982 | 12.338 | 14.042 | 30.813 | 33.924 | 36.781 | 40.289 | 42.796 |
| 23 | 9.260 | 10.196 | 11.689 | 13.091 | 14.848 | 32.007 | 35.172 | 38.076 | 41.638 | 44.181 |
| 24 | 9.886 | 10.856 | 12.401 | 13.848 | 15.659 | 33.196 | 36.415 | 39.364 | 42.980 | 45.559 |
| 25 | 10.520 | 11.524 | 13.120 | 14.611 | 16.473 | 34.382 | 37.652 | 40.646 | 44.314 | 46.928 |
| 26 | 11.160 | 12.198 | 13.844 | 15.379 | 17.292 | 35.563 | 38.885 | 41.923 | 45.642 | 48.290 |
| 27 | 11.808 | 12.879 | 14.573 | 16.151 | 18.114 | 36.741 | 40.113 | 43.194 | 46.963 | 49.645 |
| 28 | 12.461 | 13.565 | 15.308 | 16.928 | 18.939 | 37.916 | 41.337 | 44.461 | 48.278 | 50.993 |
| 29 | 13.121 | 14.257 | 16.047 | 17.708 | 19.768 | 39.087 | 42.557 | 45.772 | 49.588 | 52.336 |
| 30 | 13.787 | 14.954 | 16.791 | 18.493 | 20.599 | 40.256 | 43.773 | 46.979 | 50.892 | 53.672 |
| 40 | 20.707 | 22.164 | 24.433 | 26.509 | 29.051 | 51.805 | 55.758 | 59.342 | 63.691 | 66.766 |
| 50 | 27.991 | 29.707 | 32.357 | 34.764 | 37.689 | 63.167 | 67.505 | 71.420 | 76.154 | 79.490 |
| 60 | 35.534 | 37.485 | 40.482 | 43.188 | 46.459 | 74.397 | 79.082 | 83.298 | 88.379 | 91.952 |
| 70 | 43.275 | 45.442 | 48.758 | 51.739 | 55.329 | 85.527 | 90.531 | 95.023 | 100.425 | 104.215 |
| 80 | 51.172 | 53.540 | 57.153 | 60.391 | 64.278 | 96.578 | 101.879 | 106.629 | 112.329 | 116.321 |
| 90 | 59.196 | 61.754 | 65.647 | 69.126 | 73.291 | 107.565 | 113.145 | 118.136 | 124.116 | 128.299 |
| 100 | 67.328 | 70.065 | 74.222 | 77.929 | 82.358 | 118.498 | 124.342 | 129.561 | 135.807 | 140.169 |

Table from *Handbook of Statistical Tables* by Donald B. Owen. Copyright © 1962 by Addison-Wesley Publishing Company. Reprinted with permission of the publisher.

TABLE 7.  *F*-Distributions

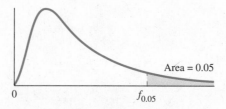

| a. Critical Values of the F-Distributions ($\alpha = 0.05$)* | | | | | | | | | |
|---|---|---|---|---|---|---|---|---|---|
| **Degrees of Freedom for Numerator** | | | | | | | | | |
| *1* | *2* | *3* | *4* | *5* | *6* | *7* | *8* | *9* | *10* |

| | 1 | 2 | 3 | 4 | 5 | 6 | 7 | 8 | 9 | 10 |
|---|---|---|---|---|---|---|---|---|---|---|
| *1* | 161 | 200 | 216 | 225 | 230 | 234 | 237 | 239 | 241 | 242 |
| *2* | 18.5 | 19.0 | 19.2 | 19.2 | 19.3 | 19.3 | 19.4 | 19.4 | 19.4 | 19.4 |
| *3* | 10.1 | 9.55 | 9.28 | 9.12 | 9.01 | 8.94 | 8.89 | 8.85 | 8.81 | 8.79 |
| *4* | 7.71 | 6.94 | 6.59 | 6.39 | 6.26 | 6.16 | 6.09 | 6.04 | 6.00 | 5.96 |
| *5* | 6.61 | 5.79 | 5.41 | 5.19 | 5.05 | 4.95 | 4.88 | 4.82 | 4.77 | 4.74 |
| *6* | 5.99 | 5.14 | 4.76 | 4.53 | 4.39 | 4.28 | 4.21 | 4.15 | 4.10 | 4.06 |
| *7* | 5.59 | 4.74 | 4.35 | 4.12 | 3.97 | 3.87 | 3.79 | 3.73 | 3.68 | 3.64 |
| *8* | 5.32 | 4.46 | 4.07 | 3.84 | 3.69 | 3.58 | 3.50 | 3.44 | 3.39 | 3.35 |
| *9* | 5.12 | 4.26 | 3.86 | 3.63 | 3.48 | 3.37 | 3.29 | 3.23 | 3.18 | 3.14 |
| *10* | 4.96 | 4.10 | 3.71 | 3.48 | 3.33 | 3.22 | 3.14 | 3.07 | 3.02 | 2.98 |
| *11* | 4.84 | 3.98 | 3.59 | 3.36 | 3.20 | 3.09 | 3.01 | 2.95 | 2.90 | 2.85 |
| *12* | 4.75 | 3.89 | 3.49 | 3.26 | 3.11 | 3.00 | 2.91 | 2.85 | 2.80 | 2.75 |
| *13* | 4.67 | 3.81 | 3.41 | 3.18 | 3.03 | 2.92 | 2.83 | 2.77 | 2.71 | 2.67 |
| *14* | 4.60 | 3.74 | 3.34 | 3.11 | 2.96 | 2.85 | 2.76 | 2.70 | 2.65 | 2.60 |
| *15* | 4.54 | 3.68 | 3.29 | 3.06 | 2.90 | 2.79 | 2.71 | 2.64 | 2.59 | 2.54 |
| *16* | 4.49 | 3.63 | 3.24 | 3.01 | 2.85 | 2.74 | 2.66 | 2.59 | 2.54 | 2.49 |
| *17* | 4.45 | 3.59 | 3.20 | 2.96 | 2.81 | 2.70 | 2.61 | 2.55 | 2.49 | 2.45 |
| *18* | 4.41 | 3.55 | 3.16 | 2.93 | 2.77 | 2.66 | 2.58 | 2.51 | 2.46 | 2.41 |
| *19* | 4.38 | 3.52 | 3.13 | 2.90 | 2.74 | 2.63 | 2.54 | 2.48 | 2.42 | 2.38 |
| *20* | 4.35 | 3.49 | 3.10 | 2.87 | 2.71 | 2.60 | 2.51 | 2.45 | 2.39 | 2.35 |
| *21* | 4.32 | 3.47 | 3.07 | 2.84 | 2.68 | 2.57 | 2.49 | 2.42 | 2.37 | 2.32 |
| *22* | 4.30 | 3.44 | 3.05 | 2.82 | 2.66 | 2.55 | 2.46 | 2.40 | 2.34 | 2.30 |
| *23* | 4.28 | 3.42 | 3.03 | 2.80 | 2.64 | 2.53 | 2.44 | 2.37 | 2.32 | 2.27 |
| *24* | 4.26 | 3.40 | 3.01 | 2.78 | 2.62 | 2.51 | 2.42 | 2.36 | 2.30 | 2.25 |
| *25* | 4.24 | 3.39 | 2.99 | 2.76 | 2.60 | 2.49 | 2.40 | 2.34 | 2.28 | 2.24 |
| *30* | 4.17 | 3.32 | 2.92 | 2.69 | 2.53 | 2.42 | 2.33 | 2.27 | 2.21 | 2.16 |
| *40* | 4.08 | 3.23 | 2.84 | 2.61 | 2.45 | 2.34 | 2.25 | 2.18 | 2.12 | 2.08 |
| *60* | 4.00 | 3.15 | 2.76 | 2.53 | 2.37 | 2.25 | 2.17 | 2.10 | 2.04 | 1.99 |
| *120* | 3.92 | 3.07 | 2.68 | 2.45 | 2.29 | 2.18 | 2.09 | 2.02 | 1.96 | 1.91 |
| *∞* | 3.84 | 3.00 | 2.60 | 2.37 | 2.21 | 2.10 | 2.01 | 1.94 | 1.88 | 1.83 |

**Degrees of Freedom for Denominator**

**TABLE 7.   *F*-Distributions (Continued)**

| | | | | | Degrees of Freedom for Numerator | | | | |
|---|---|---|---|---|---|---|---|---|---|
| | 12 | 15 | 20 | 24 | 30 | 40 | 60 | 120 | ∞ |
| 1 | 244 | 246 | 248 | 249 | 250 | 251 | 252 | 253 | 254 |
| 2 | 19.4 | 19.4 | 19.4 | 19.5 | 19.5 | 19.5 | 19.5 | 19.5 | 19.5 |
| 3 | 8.74 | 8.70 | 8.66 | 8.64 | 8.62 | 8.59 | 8.57 | 8.55 | 8.53 |
| 4 | 5.91 | 5.86 | 5.80 | 5.77 | 5.75 | 5.72 | 5.69 | 5.66 | 5.63 |
| 5 | 4.68 | 4.62 | 4.56 | 4.53 | 4.50 | 4.46 | 4.43 | 4.40 | 4.37 |
| 6 | 4.00 | 3.94 | 3.87 | 3.84 | 3.81 | 3.77 | 3.74 | 3.70 | 3.67 |
| 7 | 3.57 | 3.51 | 3.44 | 3.41 | 3.38 | 3.34 | 3.30 | 3.27 | 3.23 |
| 8 | 3.28 | 3.22 | 3.15 | 3.12 | 3.08 | 3.04 | 3.01 | 2.97 | 2.93 |
| 9 | 3.07 | 3.01 | 2.94 | 2.90 | 2.86 | 2.83 | 2.79 | 2.75 | 2.71 |
| 10 | 2.91 | 2.85 | 2.77 | 2.74 | 2.70 | 2.66 | 2.62 | 2.58 | 2.54 |
| 11 | 2.79 | 2.72 | 2.65 | 2.61 | 2.57 | 2.53 | 2.49 | 2.45 | 2.40 |
| 12 | 2.69 | 2.62 | 2.54 | 2.51 | 2.47 | 2.43 | 2.38 | 2.34 | 2.30 |
| 13 | 2.60 | 2.53 | 2.46 | 2.42 | 2.38 | 2.34 | 2.30 | 2.25 | 2.21 |
| 14 | 2.53 | 2.46 | 2.39 | 2.35 | 2.31 | 2.27 | 2.22 | 2.18 | 2.13 |
| 15 | 2.48 | 2.40 | 2.33 | 2.29 | 2.25 | 2.20 | 2.16 | 2.11 | 2.07 |
| 16 | 2.42 | 2.35 | 2.28 | 2.24 | 2.19 | 2.15 | 2.11 | 2.06 | 2.01 |
| 17 | 2.38 | 2.31 | 2.23 | 2.19 | 2.15 | 2.10 | 2.06 | 2.01 | 1.96 |
| 18 | 2.34 | 2.27 | 2.19 | 2.15 | 2.11 | 2.06 | 2.02 | 1.97 | 1.92 |
| 19 | 2.31 | 2.23 | 2.16 | 2.11 | 2.07 | 2.03 | 1.98 | 1.93 | 1.88 |
| 20 | 2.28 | 2.20 | 2.12 | 2.08 | 2.04 | 1.99 | 1.95 | 1.90 | 1.84 |
| 21 | 2.25 | 2.18 | 2.10 | 2.05 | 2.01 | 1.96 | 1.92 | 1.87 | 1.81 |
| 22 | 2.23 | 2.15 | 2.07 | 2.03 | 1.98 | 1.94 | 1.89 | 1.84 | 1.78 |
| 23 | 2.20 | 2.13 | 2.05 | 2.01 | 1.96 | 1.91 | 1.86 | 1.81 | 1.76 |
| 24 | 2.18 | 2.11 | 2.03 | 1.98 | 1.94 | 1.89 | 1.84 | 1.79 | 1.73 |
| 25 | 2.16 | 2.09 | 2.01 | 1.96 | 1.92 | 1.87 | 1.82 | 1.77 | 1.71 |
| 30 | 2.09 | 2.01 | 1.93 | 1.89 | 1.84 | 1.79 | 1.74 | 1.68 | 1.62 |
| 40 | 2.00 | 1.92 | 1.84 | 1.79 | 1.74 | 1.69 | 1.64 | 1.58 | 1.51 |
| 60 | 1.92 | 1.84 | 1.75 | 1.70 | 1.65 | 1.59 | 1.53 | 1.47 | 1.39 |
| 120 | 1.83 | 1.75 | 1.66 | 1.61 | 1.55 | 1.50 | 1.43 | 1.35 | 1.25 |
| ∞ | 1.75 | 1.67 | 1.57 | 1.52 | 1.46 | 1.39 | 1.32 | 1.22 | 1.00 |

*Degrees of Freedom for Denominator* (left axis label)

* The entries in this table are critical values of *F* for which the area under the curve to the right is equal to 0.05.

**TABLE 7.** *F*-Distributions (Continued)

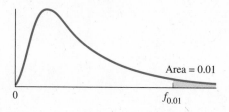

Area = 0.01

0           $f_{0.01}$

### b. Critical Values of the F-Distributions ($\alpha = 0.01$)*

| | **Degrees of Freedom for Numerator** | | | | | | | | | |
|---|---|---|---|---|---|---|---|---|---|---|
| | *1* | *2* | *3* | *4* | *5* | *6* | *7* | *8* | *9* | *10* |
| *1* | 4,052 | 5,000 | 5,403 | 5,625 | 5,764 | 5,859 | 5,928 | 5,982 | 6,023 | 6,056 |
| *2* | 98.5 | 99.0 | 99.2 | 99.2 | 99.3 | 99.3 | 99.4 | 99.4 | 99.4 | 99.4 |
| *3* | 34.1 | 30.8 | 29.5 | 28.7 | 28.2 | 27.9 | 27.7 | 27.5 | 27.3 | 27.2 |
| *4* | 21.2 | 18.0 | 16.7 | 16.0 | 15.5 | 15.2 | 15.0 | 14.8 | 14.7 | 14.5 |
| *5* | 16.3 | 13.3 | 12.1 | 11.4 | 11.0 | 10.7 | 10.5 | 10.3 | 10.2 | 10.1 |
| *6* | 13.7 | 10.9 | 9.78 | 9.15 | 8.75 | 8.47 | 8.26 | 8.10 | 7.98 | 7.87 |
| *7* | 12.2 | 9.55 | 8.45 | 7.85 | 7.46 | 7.19 | 6.99 | 6.84 | 6.72 | 6.62 |
| *8* | 11.3 | 8.65 | 7.59 | 7.01 | 6.63 | 6.37 | 6.18 | 6.03 | 5.91 | 5.81 |
| *9* | 10.6 | 8.02 | 6.99 | 6.42 | 6.06 | 5.80 | 5.61 | 5.47 | 5.35 | 5.26 |
| *10* | 10.0 | 7.56 | 6.55 | 5.99 | 5.64 | 5.39 | 5.20 | 5.06 | 4.94 | 4.85 |
| *11* | 9.65 | 7.21 | 6.22 | 5.67 | 5.32 | 5.07 | 4.89 | 4.74 | 4.63 | 4.54 |
| *12* | 9.33 | 6.93 | 5.95 | 5.41 | 5.06 | 4.82 | 4.64 | 4.50 | 4.39 | 4.30 |
| *13* | 9.07 | 6.70 | 5.74 | 5.21 | 4.86 | 4.62 | 4.44 | 4.30 | 4.19 | 4.10 |
| *14* | 8.86 | 6.51 | 5.56 | 5.04 | 4.70 | 4.46 | 4.28 | 4.14 | 4.03 | 3.94 |
| *15* | 8.68 | 6.36 | 5.42 | 4.89 | 4.56 | 4.32 | 4.14 | 4.00 | 3.89 | 3.80 |
| *16* | 8.53 | 6.23 | 5.29 | 4.77 | 4.44 | 4.20 | 4.03 | 3.89 | 3.78 | 3.69 |
| *17* | 8.40 | 6.11 | 5.19 | 4.67 | 4.34 | 4.10 | 3.93 | 3.79 | 3.68 | 3.59 |
| *18* | 8.29 | 6.01 | 5.09 | 4.58 | 4.25 | 4.01 | 3.84 | 3.71 | 3.60 | 3.51 |
| *19* | 8.19 | 5.93 | 5.01 | 4.50 | 4.17 | 3.94 | 3.77 | 3.63 | 3.52 | 3.43 |
| *20* | 8.10 | 5.85 | 4.94 | 4.43 | 4.10 | 3.87 | 3.70 | 3.56 | 3.46 | 3.37 |
| *21* | 8.02 | 5.78 | 4.87 | 4.37 | 4.04 | 3.81 | 3.64 | 3.51 | 3.40 | 3.31 |
| *22* | 7.95 | 5.72 | 4.82 | 4.31 | 3.99 | 3.76 | 3.59 | 3.45 | 3.35 | 3.26 |
| *23* | 7.88 | 5.66 | 4.76 | 4.26 | 3.94 | 3.71 | 3.54 | 3.41 | 3.30 | 3.21 |
| *24* | 7.82 | 5.61 | 4.72 | 4.22 | 3.90 | 3.67 | 3.50 | 3.36 | 3.26 | 3.17 |
| *25* | 7.77 | 5.57 | 4.68 | 4.18 | 3.86 | 3.63 | 3.46 | 3.32 | 3.22 | 3.13 |
| *30* | 7.56 | 5.39 | 4.51 | 4.02 | 3.70 | 3.47 | 3.30 | 3.17 | 3.07 | 2.98 |
| *40* | 7.31 | 5.18 | 4.31 | 3.83 | 3.51 | 3.29 | 3.12 | 2.99 | 2.89 | 2.80 |
| *60* | 7.08 | 4.98 | 4.13 | 3.65 | 3.34 | 3.12 | 2.95 | 2.82 | 2.72 | 2.63 |
| *120* | 6.85 | 4.79 | 3.95 | 3.48 | 3.17 | 2.96 | 2.79 | 2.66 | 2.56 | 2.47 |
| *∞* | 6.63 | 4.61 | 3.78 | 3.32 | 3.02 | 2.80 | 2.64 | 2.51 | 2.41 | 2.32 |

**Degrees of Freedom for Denominator**

**TABLE 7.** *F*-Distributions (Continued)

| | | Degrees of Freedom for Numerator | | | | | | | | |
|---|---|---|---|---|---|---|---|---|---|---|
| | | 12 | 15 | 20 | 24 | 30 | 40 | 60 | 120 | ∞ |
| Degrees of Freedom for Denominator | 1 | 6,106 | 6,157 | 6,209 | 6,235 | 6,261 | 6,287 | 6,313 | 6,339 | 6,366 |
| | 2 | 99.4 | 99.4 | 99.4 | 99.5 | 99.5 | 99.5 | 99.5 | 99.5 | 99.5 |
| | 3 | 27.1 | 26.9 | 26.7 | 26.6 | 26.5 | 26.4 | 26.3 | 26.2 | 26.1 |
| | 4 | 14.4 | 14.2 | 14.0 | 13.9 | 13.8 | 13.7 | 13.7 | 13.6 | 13.5 |
| | 5 | 9.89 | 9.72 | 9.55 | 9.47 | 9.38 | 9.29 | 9.20 | 9.11 | 9.02 |
| | 6 | 7.72 | 7.56 | 7.40 | 7.31 | 7.23 | 7.14 | 7.06 | 6.97 | 6.88 |
| | 7 | 6.47 | 6.31 | 6.16 | 6.07 | 5.99 | 5.91 | 5.82 | 5.74 | 5.65 |
| | 8 | 5.67 | 5.52 | 5.36 | 5.28 | 5.20 | 5.12 | 5.03 | 4.95 | 4.86 |
| | 9 | 5.11 | 4.96 | 4.81 | 4.73 | 4.65 | 4.57 | 4.48 | 4.40 | 4.31 |
| | 10 | 4.71 | 4.56 | 4.41 | 4.33 | 4.25 | 4.17 | 4.08 | 4.00 | 3.91 |
| | 11 | 4.40 | 4.25 | 4.10 | 4.02 | 3.94 | 3.86 | 3.78 | 3.69 | 3.60 |
| | 12 | 4.16 | 4.01 | 3.86 | 3.78 | 3.70 | 3.62 | 3.54 | 3.45 | 3.36 |
| | 13 | 3.96 | 3.82 | 3.66 | 3.59 | 3.51 | 3.43 | 3.34 | 3.25 | 3.17 |
| | 14 | 3.80 | 3.66 | 3.51 | 3.43 | 3.35 | 3.27 | 3.18 | 3.09 | 3.00 |
| | 15 | 3.67 | 3.52 | 3.37 | 3.29 | 3.21 | 3.13 | 3.05 | 2.96 | 2.87 |
| | 16 | 3.55 | 3.41 | 3.26 | 3.18 | 3.10 | 3.02 | 2.93 | 2.84 | 2.75 |
| | 17 | 3.46 | 3.31 | 3.16 | 3.08 | 3.00 | 2.92 | 2.83 | 2.75 | 2.65 |
| | 18 | 3.37 | 3.23 | 3.08 | 3.00 | 2.92 | 2.84 | 2.75 | 2.66 | 2.57 |
| | 19 | 3.30 | 3.15 | 3.00 | 2.92 | 2.84 | 2.76 | 2.67 | 2.58 | 2.49 |
| | 20 | 3.23 | 3.09 | 2.94 | 2.86 | 2.78 | 2.69 | 2.61 | 2.52 | 2.42 |
| | 21 | 3.17 | 3.03 | 2.88 | 2.80 | 2.72 | 2.64 | 2.55 | 2.46 | 2.36 |
| | 22 | 3.12 | 2.98 | 2.83 | 2.75 | 2.67 | 2.58 | 2.50 | 2.40 | 2.31 |
| | 23 | 3.07 | 2.93 | 2.78 | 2.70 | 2.62 | 2.54 | 2.45 | 2.35 | 2.26 |
| | 24 | 3.03 | 2.89 | 2.74 | 2.66 | 2.58 | 2.49 | 2.40 | 2.31 | 2.21 |
| | 25 | 2.99 | 2.85 | 2.70 | 2.62 | 2.53 | 2.45 | 2.36 | 2.27 | 2.17 |
| | 30 | 2.84 | 2.70 | 2.55 | 2.47 | 2.39 | 2.30 | 2.21 | 2.11 | 2.01 |
| | 40 | 2.66 | 2.52 | 2.37 | 2.29 | 2.20 | 2.11 | 2.02 | 1.92 | 1.80 |
| | 60 | 2.50 | 2.35 | 2.20 | 2.12 | 2.03 | 1.94 | 1.84 | 1.73 | 1.60 |
| | 120 | 2.34 | 2.19 | 2.03 | 1.95 | 1.86 | 1.76 | 1.66 | 1.53 | 1.38 |
| | ∞ | 2.18 | 2.04 | 1.88 | 1.79 | 1.70 | 1.59 | 1.47 | 1.32 | 1.00 |

* The entries in the table are critical values of *F* for which the area under the curve to the right is equal to 0.01.

From Table 18 of the *Biometrika Tables for Statisticians*, Vol. I, 3/E 1966. Reprinted by permission of the Biometrika Trustees.

**TABLE 8. Critical Points of the Wilcoxon Rank Sum Statistic**

*a. Some upper percentage points of the distribution of W if $H_0$ is true. Given are some values of $w_c$ and $P(W \geq w_c)$*

| $n_2$ \ $n_1$ | 1 | 2 | 3 | 4 | 5 | 6 | 7 | 8 | 9 | 10 |
|---|---|---|---|---|---|---|---|---|---|---|
| 1 | 2/0.500 | | | | | | | | | |
| 2 | 3/0.333 | 7/0.167 | | | | | | | | |
| 3 | 4/0.250 | 9/0.100 | 14/0.100<br>15/0.050 | | | | | | | |
| 4 | 5/0.200 | 11/0.067 | 17/0.057<br>18/0.029 | 23/0.100<br>25/0.029<br>26/0.014 | | | | | | |
| 5 | 6/0.167 | 12/0.095<br>13/0.048 | 19/0.071<br>20/0.036<br>21/0.018 | 26/0.095<br>28/0.032<br>29/0.016 | 35/0.075<br>36/0.048<br>38/0.016 | | | | | |
| 6 | 7/0.143 | 14/0.071<br>15/0.036 | 21/0.083<br>22/0.048<br>23/0.024 | 29/0.086<br>31/0.033<br>32/0.019 | 38/0.089<br>40/0.041<br>42/0.015 | 48/0.090<br>50/0.047<br>52/0.021 | | | | |
| 7 | 8/0.125 | 16/0.056<br>17/0.028 | 23/0.092<br>25/0.033<br>26/0.017 | 32/0.082<br>34/0.036<br>35/0.021 | 42/0.074<br>44/0.037<br>45/0.024 | 52/0.090<br>55/0.037<br>57/0.017 | 64/0.082<br>66/0.049<br>69/0.019 | | | |
| 8 | 9/0.111 | 17/0.089<br>18/0.044<br>19/0.022 | 25/0.097<br>27/0.042<br>28/0.024 | 35/0.077<br>37/0.036<br>38/0.024 | 45/0.085<br>47/0.047<br>49/0.023 | 56/0.091<br>59/0.041<br>61/0.021 | 68/0.095<br>71/0.047<br>74/0.020 | 81/0.097<br>85/0.041<br>87/0.025 | | |
| 9 | 10/0.100 | 19/0.073<br>20/0.036<br>21/0.018 | 28/0.073<br>29/0.050<br>31/0.018 | 37/0.099<br>40/0.038<br>41/0.025 | 48/0.095<br>51/0.041<br>53/0.021 | 60/0.091<br>63/0.044<br>65/0.025 | 73/0.087<br>76/0.045<br>79/0.021 | 86/0.100<br>90/0.046<br>93/0.023 | 101/0.095<br>105/0.047<br>108/0.025 | |
| 10 | 11/0.091 | 20/0.091<br>22/0.030<br>23/0.015 | 30/0.080<br>32/0.038<br>33/0.024 | 40/0.094<br>43/0.038<br>45/0.018 | 52/0.082<br>54/0.050<br>57/0.020 | 64/0.090<br>67/0.047<br>70/0.021 | 77/0.097<br>81/0.044<br>84/0.022 | 92/0.086<br>96/0.042<br>99/0.022 | 107/0.091<br>111/0.047<br>115/0.022 | 123/0.095<br>128/0.045<br>132/0.022 |

*b. Some lower percentage points of the distribution of W if $H_0$ is true. Given are some values of $w_c$ and $P(W \geq w_c)$*

| $n_2$ \ $n_1$ | 1 | 2 | 3 | 4 | 5 | 6 | 7 | 8 | 9 | 10 |
|---|---|---|---|---|---|---|---|---|---|---|
| 1 | 1/0.500 | | | | | | | | | |
| 2 | 1/0.333 | 3/0.167 | | | | | | | | |
| 3 | 1/0.250 | 3/0.100 | 6/0.050<br>7/0.100 | | | | | | | |
| 4 | 1/0.200 | 3/0.067 | 6/0.029<br>7/0.057 | 10/0.014<br>11/0.029<br>13/0.100 | | | | | | |

**TABLE 8.** Critical Points of the Wilcoxon Rank Sum Statistic (Continued)

| $n_2$ \ $n_1$ | 1 | 2 | 3 | 4 | 5 | 6 | 7 | 8 | 9 | 10 |
|---|---|---|---|---|---|---|---|---|---|---|
| 5 | 1/0.167 | 3/0.048<br>4/0.095 | 6/0.018<br>7/0.036<br>8/0.071 | 11/0.016<br>12/0.032<br>14/0.095 | 17/0.016<br>19/0.048<br>20/0.075 | | | | | |
| 6 | 1/0.143 | 3/0.036<br>4/0.071 | 7/0.024<br>8/0.048<br>9/0.083 | 12/0.019<br>13/0.033<br>15/0.086 | 18/0.015<br>20/0.041<br>22/0.089 | 26/0.021<br>28/0.047<br>30/0.090 | | | | |
| 7 | 1/0.125 | 3/0.028<br>4/0.056 | 8/0.017<br>9/0.033<br>11/0.092 | 13/0.021<br>14/0.036<br>16/0.082 | 20/0.024<br>21/0.037<br>23/0.074 | 27/0.017<br>29/0.037<br>32/0.090 | 36/0.019<br>39/0.049<br>41/0.082 | | | |
| 8 | 1/0.111 | 3/0.022<br>4/0.044<br>5/0.089 | 8/0.024<br>9/0.042<br>11/0.097 | 14/0.024<br>15/0.036<br>17/0.077 | 21/0.023<br>23/0.047<br>25/0.085 | 29/0.021<br>31/0.041<br>34/0.091 | 38/0.020<br>41/0.047<br>44/0.095 | 49/0.025<br>51/0.041<br>55/0.097 | | |
| 9 | 1/0.100 | 3/0.018<br>4/0.036<br>5/0.073 | 9/0.018<br>11/0.050<br>12/0.073 | 15/0.025<br>16/0.038<br>19/0.099 | 22/0.021<br>24/0.041<br>27/0.095 | 31/0.025<br>33/0.044<br>36/0.091 | 40/0.021<br>43/0.045<br>46/0.087 | 51/0.023<br>54/0.046<br>58/0.100 | 63/0.025<br>66/0.047<br>70/0.095 | |
| 10 | 1/0.091 | 3/0.015<br>4/0.030<br>6/0.091 | 9/0.024<br>10/0.038<br>12/0.080 | 15/0.018<br>17/0.038<br>20/0.094 | 23/0.020<br>26/0.050<br>28/0.082 | 32/0.021<br>35/0.047<br>38/0.090 | 42/0.022<br>45/0.044<br>49/0.097 | 53/0.022<br>56/0.042<br>60/0.086 | 65/0.022<br>69/0.047<br>73/0.091 | 78/0.022<br>82/0.045<br>87/0.095 |

**TABLE 9.** Upper Critical Points for Spearman's $\tilde{r}$

| $n$ | $\tilde{r}_c$ | $P(\tilde{R} \geq \tilde{r}_c)$ | $n$ | $\tilde{r}_c$ | $P(\tilde{R} \geq \tilde{r}_c)$ | $n$ | $\tilde{r}_c$ | $P(\tilde{R} \geq \tilde{r}_c)$ | $n$ | $\tilde{r}_c$ | $P(\tilde{R} \geq \tilde{r}_c)$ |
|---|---|---|---|---|---|---|---|---|---|---|---|
| 3 | 1.000 | 0.167 | | 0.200 | 0.392 | | 0.200 | 0.357 | | 0.536 | 0.118 |
| | 0.500 | 0.500 | | 0.100 | 0.475 | | 0.143 | 0.401 | | 0.500 | 0.133 |
| | | | | 0.000 | 0.525 | | 0.086 | 0.460 | | 0.464 | 0.151 |
| 4 | 1.000 | 0.042 | | | | | 0.029 | 0.500 | | 0.429 | 0.177 |
| | 0.800 | 0.167 | 6 | 1.000 | 0.001 | | | | | 0.393 | 0.198 |
| | 0.600 | 0.208 | | 0.943 | 0.008 | 7 | 1.000 | 0.000 | | 0.357 | 0.222 |
| | 0.400 | 0.375 | | 0.886 | 0.017 | | 0.964 | 0.001 | | 0.321 | 0.249 |
| | 0.200 | 0.458 | | 0.829 | 0.029 | | 0.929 | 0.003 | | 0.286 | 0.278 |
| | 0.000 | 0.542 | | 0.771 | 0.051 | | 0.893 | 0.006 | | 0.250 | 0.297 |
| | | | | 0.714 | 0.068 | | 0.857 | 0.012 | | 0.214 | 0.331 |
| 5 | 1.000 | 0.008 | | 0.657 | 0.088 | | 0.821 | 0.017 | | 0.179 | 0.357 |
| | 0.900 | 0.042 | | 0.600 | 0.121 | | 0.786 | 0.024 | | 0.143 | 0.391 |
| | 0.800 | 0.067 | | 0.543 | 0.149 | | 0.750 | 0.033 | | 0.107 | 0.420 |
| | 0.700 | 0.117 | | 0.486 | 0.178 | | 0.714 | 0.044 | | 0.071 | 0.453 |
| | 0.600 | 0.175 | | 0.429 | 0.210 | | 0.679 | 0.055 | | 0.036 | 0.482 |
| | 0.500 | 0.225 | | 0.371 | 0.249 | | 0.643 | 0.069 | | 0.000 | 0.518 |
| | 0.400 | 0.258 | | 0.314 | 0.282 | | 0.607 | 0.083 | | | |
| | 0.300 | 0.342 | | 0.257 | 0.329 | | 0.571 | 0.100 | | | |

TABLE 9.   Upper Critical Points for Spearman's $\tilde{r}$ (Continued)

| $n$ | $\tilde{r}_c$ | $P(\tilde{R} \geq \tilde{r}_c)$ | $n$ | $\tilde{r}_c$ | $P(\tilde{R} \geq \tilde{r}_c)$ | $n$ | $\tilde{r}_c$ | $P(\tilde{R} \geq \tilde{r}_c)$ | $n$ | $\tilde{r}_c$ | $P(\tilde{R} \geq \tilde{r}_c)$ |
|---|---|---|---|---|---|---|---|---|---|---|---|
| 8 | 1.000 | 0.000 | | 0.933 | 0.000 | | 0.133 | 0.372 | | 0.539 | 0.057 |
| | 0.976 | 0.000 | | 0.917 | 0.001 | | 0.117 | 0.388 | | 0.527 | 0.062 |
| | 0.952 | 0.001 | | 0.900 | 0.001 | | 0.100 | 0.405 | | 0.515 | 0.067 |
| | 0.929 | 0.001 | | 0.883 | 0.002 | | 0.083 | 0.422 | | 0.503 | 0.072 |
| | 0.905 | 0.002 | | 0.867 | 0.002 | | 0.067 | 0.440 | | 0.491 | 0.077 |
| | 0.881 | 0.004 | | 0.850 | 0.003 | | 0.050 | 0.456 | | 0.479 | 0.083 |
| | 0.857 | 0.005 | | 0.833 | 0.004 | | 0.033 | 0.474 | | 0.467 | 0.089 |
| | 0.833 | 0.008 | | 0.817 | 0.005 | | 0.017 | 0.491 | | 0.455 | 0.096 |
| | 0.810 | 0.011 | | 0.800 | 0.007 | | 0.000 | 0.509 | | 0.442 | 0.102 |
| | 0.786 | 0.014 | | 0.783 | 0.009 | | | | | 0.430 | 0.109 |
| | 0.762 | 0.018 | | 0.767 | 0.011 | 10 | 1.000 | 0.000 | | 0.418 | 0.116 |
| | 0.738 | 0.023 | | 0.750 | 0.013 | | 0.988 | 0.000 | | 0.406 | 0.124 |
| | 0.714 | 0.029 | | 0.733 | 0.016 | | 0.976 | 0.000 | | 0.394 | 0.132 |
| | 0.690 | 0.035 | | 0.717 | 0.018 | | 0.964 | 0.000 | | 0.382 | 0.139 |
| | 0.667 | 0.042 | | 0.700 | 0.022 | | 0.952 | 0.000 | | 0.370 | 0.148 |
| | 0.643 | 0.048 | | 0.683 | 0.025 | | 0.939 | 0.000 | | 0.358 | 0.156 |
| | 0.619 | 0.057 | | 0.667 | 0.029 | | 0.927 | 0.000 | | 0.345 | 0.165 |
| | 0.595 | 0.066 | | 0.650 | 0.033 | | 0.915 | 0.000 | | 0.333 | 0.174 |
| | 0.571 | 0.076 | | 0.633 | 0.038 | | 0.903 | 0.000 | | 0.321 | 0.184 |
| | 0.548 | 0.085 | | 0.617 | 0.043 | | 0.891 | 0.001 | | 0.309 | 0.193 |
| | 0.524 | 0.098 | | 0.600 | 0.048 | | 0.879 | 0.001 | | 0.297 | 0.203 |
| | 0.500 | 0.108 | | 0.583 | 0.054 | | 0.867 | 0.001 | | 0.285 | 0.214 |
| | 0.476 | 0.122 | | 0.567 | 0.060 | | 0.855 | 0.001 | | 0.273 | 0.224 |
| | 0.452 | 0.134 | | 0.550 | 0.066 | | 0.842 | 0.002 | | 0.261 | 0.235 |
| | 0.429 | 0.150 | | 0.533 | 0.074 | | 0.830 | 0.002 | | 0.248 | 0.246 |
| | 0.405 | 0.163 | | 0.517 | 0.081 | | 0.818 | 0.003 | | 0.236 | 0.257 |
| | 0.381 | 0.180 | | 0.500 | 0.089 | | 0.806 | 0.004 | | 0.224 | 0.268 |
| | 0.357 | 0.195 | | 0.483 | 0.097 | | 0.794 | 0.004 | | 0.212 | 0.280 |
| | 0.333 | 0.214 | | 0.467 | 0.106 | | 0.782 | 0.005 | | 0.200 | 0.292 |
| | 0.310 | 0.231 | | 0.450 | 0.115 | | 0.770 | 0.007 | | 0.188 | 0.304 |
| | 0.286 | 0.250 | | 0.433 | 0.125 | | 0.758 | 0.008 | | 0.176 | 0.316 |
| | 0.262 | 0.268 | | 0.417 | 0135 | | 0.745 | 0.009 | | 0.164 | 0.328 |
| | 0.238 | 0.291 | | 0.400 | 0.146 | | 0.733 | 0.010 | | 0.152 | 0.341 |
| | 0.214 | 0.310 | | 0.383 | 0.156 | | 0.721 | 0.012 | | 0.139 | 0.354 |
| | 0.190 | 0.332 | | 0.367 | 0.168 | | 0.709 | 0.013 | | 0.127 | 0.367 |
| | 0.167 | 0.352 | | 0.350 | 0.179 | | 0.697 | 0.015 | | 0.115 | 0.379 |
| | 0.143 | 0.376 | | 0.333 | 0.193 | | 0.685 | 0.017 | | 0.103 | 0.393 |
| | 0.119 | 0.397 | | 0.317 | 0.205 | | 0.673 | 0.019 | | 0.091 | 0.406 |
| | 0.095 | 0.420 | | 0.300 | 0.218 | | 0.661 | 0.022 | | 0.079 | 0.419 |
| | 0.071 | 0.441 | | 0.283 | 0.231 | | 0.648 | 0.025 | | 0.067 | 0.433 |
| | 0.048 | 0.467 | | 0.267 | 0.247 | | 0.636 | 0.027 | | 0.055 | 0.446 |
| | 0.024 | 0.488 | | 0.250 | 0.260 | | 0.624 | 0.030 | | 0.042 | 0.459 |
| | 0.000 | 0.512 | | 0.233 | 0.276 | | 0.612 | 0.033 | | 0.030 | 0.473 |
| | | | | 0.217 | 0.290 | | 0.600 | 0.037 | | 0.018 | 0.486 |
| 9 | 1.000 | 0.000 | | 0.200 | 0.307 | | 0.588 | 0.040 | | 0.006 | 0.500 |
| | 0.983 | 0.000 | | 0.183 | 0.322 | | 0.576 | 0.044 | | | |
| | 0.967 | 0.000 | | 0.167 | 0.339 | | 0.564 | 0.048 | | | |
| | 0.950 | 0.000 | | 0.150 | 0.354 | | 0.552 | 0.052 | | | |

# Appendix B    Data Sets

Parameters for Data Set I on Cancer.

"While the risks of lung cancer associated with the use of tobacco products is well established, less than 20% of heavy smokers will develop lung cancer in their lifetime. To examine the role of host factors in the pathogenesis of the disease, a study was conducted in southern Louisiana in the early 1980's to determine if relatives of lung cancer patients demonstrated higher rates of lung cancer than could be expected. The findings of this research, conducted by Dr. Wee Lock Ooi and Dr. Henry Rothschild at Louisiana State University, were published in the Journal of the National Cancer Institute in 1986 (vol 76, pages 217–222). The data suggested that some families demonstrate a predisposition for lung cancer (a 2.4-fold greater risk after allowing for personal smoking habits). Additional analyses by Dr. Thomas A. Sellers demonstrated that this familial risk extended to cancers at other sites (American Journal of Epidemiology, 1987; 126:237–246), and that the pattern of lung cancer in these families was consistent with Mendelian inheritance of a gene that produced an earlier age of onset of lung cancer (Journal of the National Cancer Institute, 1990; 82:1272–1279). The data found in this textbook represent a randomly selected subset of the total study sample, and were kindly provided by Dr. Henry Rothschild."

$N = 2{,}538$ data points

| Age Data | Sex Data | Cancer Data |
|---|---|---|
| Men $\longrightarrow \begin{array}{l}\mu = 54.9 \text{ years} \\ \sigma = 19.5 \text{ years}\end{array}$ | 1292 Males $\qquad P_m = 50.9\%$ <br> 1246 Females $\qquad P_f = 49.1\%$ | 381 have cancer $\qquad P_c = 15.0\%$ <br> 2157 do not $\qquad P_n = 85.0\%$ |
| Women $\longrightarrow \begin{array}{l}\mu = 57.1 \text{ years} \\ \sigma = 20.5 \text{ years}\end{array}$ | Smoker Data | |
| For men and women: | 1349 Smokers $\qquad P_s = 53.2\%$ <br> 1189 Non-smokers $\qquad P_n = 46.8\%$ | |
| $\mu = 56.1$ years <br> $\sigma = 20.2$ years | | |

---

## Data Set I

| | Key | | | | | | |
|---|---|---|---|---|---|---|---|
| | Sex | | | Smoker | | Cancer | |
| | Male | (M) | | Yes | (Y) | Yes | (1) |
| | Female | (F) | | No | (N) | No | (2) |

| Number | Age | Sex | Smoker | Cancer | Number | Age | Sex | Smoker | Cancer | Number | Age | Sex | Smoker | Cancer |
|---|---|---|---|---|---|---|---|---|---|---|---|---|---|---|
| 0001 | 56 | M | Y | 1 | 0006 | 64 | M | Y | 2 | 0011 | 63 | M | Y | 2 |
| 0002 | 31 | M | N | 2 | 0007 | 61 | M | N | 2 | 0012 | 48 | M | Y | 2 |
| 0003 | 90 | F | N | 2 | 0008 | 20 | M | N | 2 | 0013 | 52 | F | Y | 2 |
| 0004 | 69 | M | N | 2 | 0009 | 56 | F | Y | 2 | 0014 | 43 | F | Y | 2 |
| 0005 | 51 | M | Y | 2 | 0010 | 86 | F | N | 2 | 0015 | 38 | F | N | 2 |

**Data Set I (Continued)**

| Number | Age | Sex | Smoker | Cancer | Number | Age | Sex | Smoker | Cancer | Number | Age | Sex | Smoker | Cancer |
|--------|-----|-----|--------|--------|--------|-----|-----|--------|--------|--------|-----|-----|--------|--------|
| 0016 | 35 | M | N | 2 | 0065 | 60 | M | Y | 1 | 0114 | 20 | M | N | 1 |
| 0017 | 33 | F | N | 2 | 0066 | 76 | F | N | 2 | 0115 | 58 | F | N | 2 |
| 0018 | 78 | F | N | 1 | 0067 | 71 | F | Y | 2 | 0116 | 52 | F | Y | 2 |
| 0019 | 48 | M | Y | 1 | 0068 | 67 | F | Y | 2 | 0117 | 48 | F | N | 2 |
| 0020 | 55 | F | N | 2 | 0069 | 61 | F | Y | 2 | 0118 | 50 | M | Y | 2 |
| 0021 | 79 | M | Y | 2 | 0070 | 46 | M | Y | 2 | 0119 | 48 | M | Y | 2 |
| 0022 | 68 | M | Y | 2 | 0071 | 46 | F | Y | 2 | 0120 | 25 | M | N | 2 |
| 0023 | 80 | M | Y | 2 | 0072 | 84 | M | Y | 1 | 0121 | 35 | M | Y | 2 |
| 0024 | 78 | M | Y | 2 | 0073 | 85 | M | Y | 2 | 0122 | 53 | F | N | 2 |
| 0025 | 74 | M | Y | 2 | 0074 | 88 | F | N | 1 | 0123 | 46 | F | Y | 2 |
| 0026 | 67 | M | Y | 2 | 0075 | 70 | M | Y | 2 | 0124 | 40 | F | Y | 1 |
| 0027 | 29 | M | Y | 2 | 0076 | 71 | F | N | 2 | 0125 | 58 | F | Y | 1 |
| 0028 | 62 | M | Y | 2 | 0077 | 83 | F | N | 2 | 0126 | 86 | M | N | 1 |
| 0029 | 57 | M | Y | 2 | 0078 | 60 | F | N | 1 | 0127 | 38 | F | N | 1 |
| 0030 | 73 | F | N | 2 | 0079 | 73 | F | N | 1 | 0128 | 55 | M | Y | 2 |
| 0031 | 72 | F | N | 2 | 0080 | 90 | F | N | 2 | 0129 | 60 | M | N | 1 |
| 0032 | 70 | F | N | 2 | 0081 | 87 | F | N | 1 | 0130 | 50 | M | Y | 2 |
| 0033 | 54 | F | N | 2 | 0082 | 63 | M | Y | 2 | 0131 | 48 | M | Y | 2 |
| 0034 | 44 | M | Y | 2 | 0083 | 66 | F | N | 2 | 0132 | 25 | M | N | 2 |
| 0035 | 56 | F | N | 2 | 0084 | 60 | F | N | 2 | 0133 | 35 | M | Y | 2 |
| 0036 | 54 | F | N | 2 | 0085 | 54 | F | Y | 2 | 0134 | 53 | F | N | 2 |
| 0037 | 65 | M | N | 2 | 0086 | 75 | F | N | 2 | 0135 | 46 | F | Y | 2 |
| 0038 | 76 | F | N | 2 | 0087 | 75 | M | Y | 2 | 0136 | 40 | F | Y | 1 |
| 0039 | 67 | F | N | 2 | 0088 | 60 | F | N | 1 | 0137 | 64 | F | Y | 1 |
| 0040 | 77 | M | Y | 2 | 0089 | 70 | F | N | 2 | 0138 | 50 | M | N | 2 |
| 0041 | 71 | F | N | 2 | 0090 | 63 | M | Y | 2 | 0139 | 61 | F | N | 2 |
| 0042 | 54 | M | N | 2 | 0091 | 66 | F | N | 2 | 0140 | 47 | M | N | 2 |
| 0043 | 41 | M | Y | 2 | 0092 | 60 | F | N | 2 | 0141 | 66 | F | N | 2 |
| 0044 | 60 | F | N | 2 | 0093 | 54 | F | Y | 2 | 0142 | 64 | F | Y | 2 |
| 0045 | 54 | F | N | 2 | 0094 | 75 | M | Y | 1 | 0143 | 61 | F | N | 2 |
| 0046 | 77 | M | Y | 1 | 0095 | 90 | M | N | 2 | 0144 | 60 | F | N | 2 |
| 0047 | 74 | M | Y | 2 | 0096 | 85 | F | Y | 2 | 0145 | 55 | F | N | 2 |
| 0048 | 68 | F | Y | 2 | 0097 | 69 | M | Y | 1 | 0146 | 52 | F | Y | 2 |
| 0049 | 69 | F | Y | 1 | 0098 | 75 | M | Y | 2 | 0147 | 46 | M | Y | 2 |
| 0050 | 91 | M | Y | 2 | 0099 | 81 | M | Y | 2 | 0148 | 44 | F | Y | 2 |
| 0051 | 99 | F | N | 2 | 0100 | 49 | M | N | 2 | 0149 | 40 | F | N | 2 |
| 0052 | 54 | M | N | 2 | 0101 | 71 | F | Y | 2 | 0150 | 78 | F | N | 1 |
| 0053 | 85 | F | N | 2 | 0102 | 92 | M | N | 2 | 0151 | 86 | M | N | 2 |
| 0054 | 82 | F | N | 2 | 0103 | 83 | F | N | 1 | 0152 | 64 | M | Y | 1 |
| 0055 | 81 | F | N | 2 | 0104 | 72 | M | Y | 2 | 0153 | 74 | M | Y | 1 |
| 0056 | 23 | F | N | 2 | 0105 | 17 | M | N | 2 | 0154 | 1 | M | N | 2 |
| 0057 | 46 | M | Y | 2 | 0106 | 69 | M | Y | 2 | 0155 | 72 | M | Y | 2 |
| 0058 | 46 | F | Y | 2 | 0107 | 74 | F | N | 2 | 0156 | 63 | M | Y | 1 |
| 0059 | 59 | M | Y | 1 | 0108 | 49 | M | N | 2 | 0157 | 72 | F | N | 1 |
| 0060 | 65 | F | Y | 1 | 0109 | 48 | M | Y | 1 | 0158 | 81 | F | Y | 2 |
| 0061 | 69 | M | Y | 2 | 0110 | 72 | F | N | 2 | 0159 | 78 | F | Y | 2 |
| 0062 | 47 | M | Y | 2 | 0111 | 72 | M | Y | 1 | 0160 | 76 | F | N | 2 |
| 0063 | 65 | M | Y | 2 | 0112 | 57 | M | N | 2 | 0161 | 69 | F | Y | 2 |
| 0064 | 65 | M | Y | 2 | 0113 | 59 | F | N | 1 | 0162 | 54 | M | Y | 2 |

Data Set I (Continued)

| Number | Age | Sex | Smoker | Cancer | Number | Age | Sex | Smoker | Cancer | Number | Age | Sex | Smoker | Cancer |
|--------|-----|-----|--------|--------|--------|-----|-----|--------|--------|--------|-----|-----|--------|--------|
| 0163 | 46 | M | Y | 2 | 0212 | 28 | M | Y | 2 | 0261 | 02 | M | N | 2 |
| 0164 | 43 | M | Y | 2 | 0213 | 30 | F | Y | 2 | 0262 | 54 | M | Y | 1 |
| 0165 | 61 | F | N | 2 | 0214 | 48 | F | N | 2 | 0263 | 48 | M | Y | 1 |
| 0166 | 60 | F | N | 2 | 0215 | 47 | M | Y | 1 | 0264 | 44 | M | Y | 2 |
| 0167 | 80 | M | Y | 2 | 0216 | 72 | F | N | 2 | 0265 | 72 | F | N | 2 |
| 0168 | 69 | M | Y | 2 | 0217 | 44 | F | Y | 2 | 0266 | 65 | F | N | 2 |
| 0169 | 93 | F | N | 2 | 0218 | 31 | F | Y | 2 | 0267 | 62 | F | N | 2 |
| 0170 | 73 | M | N | 2 | 0219 | 28 | M | Y | 2 | 0268 | 69 | F | N | 1 |
| 0171 | 66 | M | Y | 2 | 0220 | 30 | F | Y | 2 | 0269 | 40 | F | N | 2 |
| 0172 | 92 | F | N | 2 | 0221 | 62 | F | Y | 1 | 0270 | 34 | F | Y | 2 |
| 0173 | 12 | F | N | 2 | 0222 | 71 | M | Y | 2 | 0271 | 33 | M | Y | 2 |
| 0174 | 70 | F | N | 2 | 0223 | 70 | F | N | 1 | 0272 | 63 | F | N | 2 |
| 0175 | 54 | M | Y | 2 | 0224 | 60 | M | Y | 2 | 0273 | 99 | F | N | 2 |
| 0176 | 46 | M | Y | 2 | 0225 | 58 | M | Y | 2 | 0274 | 70 | M | Y | 2 |
| 0177 | 43 | M | Y | 2 | 0226 | 70 | F | N | 2 | 0275 | 56 | M | Y | 1 |
| 0178 | 61 | F | N | 2 | 0227 | 68 | F | N | 2 | 0276 | 54 | M | Y | 2 |
| 0179 | 60 | F | N | 2 | 0228 | 38 | F | N | 2 | 0277 | 83 | F | Y | 2 |
| 0180 | 65 | M | Y | 1 | 0229 | 36 | F | N | 2 | 0278 | 82 | F | N | 2 |
| 0181 | 80 | M | N | 2 | 0230 | 58 | F | Y | 2 | 0279 | 66 | F | N | 1 |
| 0182 | 63 | F | N | 2 | 0231 | 70 | M | Y | 2 | 0280 | 69 | F | Y | 1 |
| 0183 | 44 | M | Y | 2 | 0232 | 80 | M | N | 2 | 0281 | 72 | F | N | 2 |
| 0184 | 40 | M | N | 2 | 0233 | 70 | F | Y | 1 | 0282 | 01 | F | N | 2 |
| 0185 | 14 | M | N | 2 | 0234 | 71 | M | Y | 2 | 0283 | 68 | F | N | 2 |
| 0186 | 17 | F | N | 2 | 0235 | 62 | M | Y | 2 | 0284 | 40 | F | N | 2 |
| 0187 | 64 | F | N | 2 | 0236 | 33 | F | Y | 2 | 0285 | 34 | F | Y | 2 |
| 0188 | 75 | F | N | 2 | 0237 | 60 | F | Y | 2 | 0286 | 33 | M | Y | 2 |
| 0189 | 72 | F | N | 2 | 0238 | 32 | F | Y | 2 | 0287 | 68 | M | Y | 1 |
| 0190 | 30 | M | Y | 2 | 0239 | 32 | M | Y | 2 | 0288 | 76 | M | Y | 2 |
| 0191 | 28 | F | N | 2 | 0240 | 19 | F | Y | 2 | 0289 | 37 | F | N | 2 |
| 0192 | 23 | F | Y | 2 | 0241 | 37 | F | Y | 2 | 0290 | 75 | M | Y | 2 |
| 0193 | 52 | F | N | 1 | 0242 | 86 | M | Y | 1 | 0291 | 73 | M | Y | 2 |
| 0194 | 84 | M | Y | 1 | 0243 | 80 | M | N | 2 | 0292 | 46 | M | Y | 1 |
| 0195 | 66 | F | N | 2 | 0244 | 93 | F | N | 2 | 0293 | 63 | F | Y | 1 |
| 0196 | 57 | M | Y | 2 | 0245 | 80 | M | Y | 2 | 0294 | 41 | F | N | 2 |
| 0197 | 58 | M | N | 2 | 0246 | 77 | F | N | 2 | 0295 | 46 | F | N | 2 |
| 0198 | 55 | M | Y | 2 | 0247 | 55 | M | N | 2 | 0296 | 44 | F | N | 2 |
| 0199 | 49 | M | Y | 2 | 0248 | 52 | F | Y | 2 | 0297 | 43 | F | N | 2 |
| 0200 | 60 | F | N | 2 | 0249 | 49 | F | Y | 2 | 0298 | 72 | F | N | 2 |
| 0201 | 30 | M | Y | 2 | 0250 | 80 | F | Y | 1 | 0299 | 80 | M | N | 2 |
| 0202 | 28 | F | Y | 2 | 0251 | 77 | M | N | 2 | 0300 | 51 | F | N | 1 |
| 0203 | 23 | F | Y | 2 | 0252 | 87 | F | Y | 2 | 0301 | 80 | M | N | 2 |
| 0204 | 52 | M | Y | 1 | 0253 | 77 | F | N | 2 | 0302 | 78 | M | Y | 1 |
| 0205 | 81 | F | N | 2 | 0254 | 58 | F | N | 2 | 0303 | 62 | M | Y | 2 |
| 0206 | 46 | F | Y | 1 | 0255 | 55 | M | Y | 2 | 0304 | 59 | M | Y | 2 |
| 0207 | 60 | M | Y | 1 | 0256 | 52 | F | Y | 2 | 0305 | 78 | F | N | 2 |
| 0208 | 52 | M | Y | 2 | 0257 | 49 | F | Y | 2 | 0306 | 66 | M | Y | 1 |
| 0209 | 46 | M | N | 2 | 0258 | 59 | M | Y | 1 | 0307 | 73 | F | Y | 2 |
| 0210 | 57 | F | N | 2 | 0259 | 76 | F | Y | 2 | 0308 | 71 | F | N | 2 |
| 0211 | 32 | F | N | 2 | 0260 | 62 | M | N | 2 | 0309 | 69 | F | N | 2 |

**Data Set I (Continued)**

| Number | Age | Sex | Smoker | Cancer | Number | Age | Sex | Smoker | Cancer | Number | Age | Sex | Smoker | Cancer |
|--------|-----|-----|--------|--------|--------|-----|-----|--------|--------|--------|-----|-----|--------|--------|
| 0310 | 66 | F | N | 2 | 0359 | 52 | M | Y | 2 | 0408 | 67 | F | N | 2 |
| 0311 | 65 | F | Y | 2 | 0360 | 58 | M | N | 2 | 0409 | 54 | M | Y | 2 |
| 0312 | 56 | M | Y | 2 | 0361 | 43 | M | Y | 2 | 0410 | 49 | F | Y | 2 |
| 0313 | 53 | M | Y | 2 | 0362 | 37 | M | Y | 2 | 0411 | 43 | F | N | 2 |
| 0314 | 48 | M | Y | 2 | 0363 | 80 | F | N | 2 | 0412 | 40 | F | Y | 2 |
| 0315 | 55 | F | N | 2 | 0364 | 66 | F | N | 2 | 0413 | 57 | M | Y | 1 |
| 0316 | 52 | F | Y | 2 | 0365 | 73 | M | Y | 1 | 0414 | 65 | M | Y | 1 |
| 0317 | 49 | F | Y | 2 | 0366 | 64 | F | Y | 2 | 0415 | 65 | F | N | 1 |
| 0318 | 47 | F | Y | 2 | 0367 | 79 | M | Y | 1 | 0416 | 66 | M | Y | 2 |
| 0319 | 73 | F | N | 2 | 0368 | 70 | M | Y | 1 | 0417 | 65 | M | Y | 2 |
| 0320 | 80 | M | Y | 2 | 0369 | 88 | M | Y | 2 | 0418 | 57 | F | N | 2 |
| 0321 | 54 | F | Y | 2 | 0370 | 85 | M | N | 2 | 0419 | 41 | F | N | 2 |
| 0322 | 75 | M | Y | 2 | 0371 | 83 | F | N | 2 | 0420 | 40 | F | Y | 2 |
| 0323 | 66 | M | Y | 2 | 0372 | 78 | F | N | 2 | 0421 | 39 | F | Y | 2 |
| 0324 | 60 | M | Y | 2 | 0373 | 59 | F | Y | 2 | 0422 | 54 | F | Y | 2 |
| 0325 | 62 | F | N | 2 | 0374 | 82 | F | N | 2 | 0423 | 68 | F | N | 2 |
| 0326 | 70 | F | N | 2 | 0375 | 61 | M | Y | 2 | 0424 | 93 | M | Y | 2 |
| 0327 | 58 | F | Y | 2 | 0376 | 58 | F | Y | 2 | 0425 | 68 | M | Y | 2 |
| 0328 | 60 | F | N | 2 | 0377 | 54 | F | N | 2 | 0426 | 88 | F | N | 2 |
| 0329 | 11 | F | N | 2 | 0378 | 85 | F | N | 1 | 0427 | 86 | F | N | 2 |
| 0330 | 56 | M | Y | 2 | 0379 | 48 | M | N | 2 | 0428 | 80 | F | N | 2 |
| 0331 | 53 | M | Y | 2 | 0380 | 51 | F | N | 2 | 0429 | 77 | F | N | 2 |
| 0332 | 48 | M | Y | 2 | 0381 | 81 | M | Y | 1 | 0430 | 69 | F | Y | 2 |
| 0333 | 55 | F | N | 2 | 0382 | 57 | F | Y | 1 | 0431 | 70 | F | N | 2 |
| 0334 | 52 | F | Y | 2 | 0383 | 64 | M | Y | 1 | 0432 | 63 | F | N | 2 |
| 0335 | 49 | F | Y | 2 | 0384 | 16 | F | N | 2 | 0433 | 41 | M | Y | 2 |
| 0336 | 47 | F | Y | 2 | 0385 | 72 | M | Y | 1 | 0434 | 40 | F | Y | 2 |
| 0337 | 43 | F | N | 1 | 0386 | 58 | F | N | 2 | 0435 | 39 | F | Y | 2 |
| 0338 | 49 | M | Y | 2 | 0387 | 72 | M | N | 1 | 0436 | 62 | F | Y | 2 |
| 0339 | 66 | M | Y | 1 | 0388 | 69 | M | Y | 2 | 0437 | 93 | F | N | 2 |
| 0340 | 75 | M | Y | 2 | 0389 | 68 | M | Y | 2 | 0438 | 23 | M | N | 2 |
| 0341 | 52 | F | N | 2 | 0390 | 70 | M | Y | 2 | 0439 | 81 | F | N | 2 |
| 0342 | 50 | M | Y | 2 | 0391 | 67 | M | Y | 2 | 0440 | 42 | M | Y | 2 |
| 0343 | 55 | M | Y | 2 | 0392 | 18 | M | N | 2 | 0441 | 72 | M | Y | 2 |
| 0344 | 72 | M | Y | 2 | 0393 | 72 | F | N | 2 | 0442 | 53 | M | Y | 2 |
| 0345 | 40 | M | Y | 2 | 0394 | 86 | F | N | 2 | 0443 | 46 | M | Y | 2 |
| 0346 | 71 | M | Y | 1 | 0395 | 65 | F | N | 2 | 0444 | 49 | F | N | 2 |
| 0347 | 80 | F | N | 2 | 0396 | 60 | F | N | 2 | 0445 | 55 | F | N | 2 |
| 0348 | 85 | F | N | 2 | 0397 | 54 | M | Y | 2 | 0446 | 67 | M | Y | 2 |
| 0349 | 70 | F | N | 2 | 0398 | 49 | F | Y | 2 | 0447 | 65 | M | Y | 1 |
| 0350 | 79 | F | Y | 1 | 0399 | 48 | F | Y | 2 | 0448 | 77 | F | N | 1 |
| 0351 | 70 | F | Y | 2 | 0400 | 43 | F | N | 2 | 0449 | 65 | F | Y | 2 |
| 0352 | 67 | F | Y | 1 | 0401 | 40 | F | Y | 2 | 0450 | 27 | M | N | 2 |
| 0353 | 76 | F | N | 2 | 0402 | 70 | F | N | 2 | 0451 | 56 | F | Y | 2 |
| 0354 | 72 | F | Y | 2 | 0403 | 71 | M | Y | 2 | 0452 | 85 | M | N | 2 |
| 0355 | 71 | F | N | 1 | 0404 | 68 | F | N | 2 | 0453 | 85 | F | Y | 2 |
| 0356 | 52 | M | Y | 2 | 0405 | 78 | M | Y | 2 | 0454 | 71 | M | N | 2 |
| 0357 | 73 | F | N | 2 | 0406 | 72 | M | Y | 2 | 0455 | 68 | M | Y | 2 |
| 0358 | 40 | M | Y | 2 | 0407 | 63 | M | Y | 2 | 0456 | 64 | F | Y | 2 |

## Data Set I (Continued)

| Number | Age | Sex | Smoker | Cancer | Number | Age | Sex | Smoker | Cancer | Number | Age | Sex | Smoker | Cancer |
|---|---|---|---|---|---|---|---|---|---|---|---|---|---|---|
| 0457 | 62 | F | Y | 2 | 0506 | 54 | F | N | 2 | 0555 | 67 | F | Y | 2 |
| 0458 | 27 | M | N | 2 | 0507 | 76 | M | Y | 2 | 0556 | 42 | M | Y | 2 |
| 0459 | 35 | F | Y | 2 | 0508 | 56 | M | Y | 2 | 0557 | 34 | M | Y | 2 |
| 0460 | 31 | F | Y | 2 | 0509 | 19 | M | N | 2 | 0558 | 52 | F | N | 2 |
| 0461 | 78 | M | Y | 1 | 0510 | 50 | M | Y | 2 | 0559 | 47 | F | N | 2 |
| 0462 | 75 | M | N | 2 | 0511 | 62 | M | Y | 2 | 0560 | 20 | M | Y | 2 |
| 0463 | 80 | F | N | 2 | 0512 | 60 | M | N | 2 | 0561 | 15 | M | Y | 2 |
| 0464 | 84 | M | Y | 2 | 0513 | 71 | F | N | 2 | 0562 | 04 | M | N | 2 |
| 0465 | 83 | M | N | 2 | 0514 | 71 | F | N | 2 | 0563 | 18 | F | Y | 2 |
| 0466 | 73 | F | N | 2 | 0515 | 56 | F | N | 2 | 0564 | 12 | F | N | 2 |
| 0467 | 76 | F | N | 2 | 0516 | 44 | F | N | 2 | 0565 | 07 | F | N | 2 |
| 0468 | 78 | F | N | 2 | 0517 | 65 | F | N | 2 | 0566 | 06 | F | N | 2 |
| 0469 | 27 | F | N | 2 | 0518 | 56 | F | N | 2 | 0567 | 41 | F | Y | 2 |
| 0470 | 49 | F | N | 2 | 0519 | 17 | F | N | 2 | 0568 | 47 | M | Y | 2 |
| 0471 | 52 | M | Y | 2 | 0520 | 33 | M | Y | 2 | 0569 | 78 | F | Y | 2 |
| 0472 | 55 | F | N | 2 | 0521 | 26 | M | Y | 2 | 0570 | 52 | M | Y | 2 |
| 0473 | 46 | F | Y | 2 | 0522 | 35 | F | Y | 2 | 0571 | 38 | M | Y | 2 |
| 0474 | 79 | F | N | 2 | 0523 | 30 | F | Y | 2 | 0572 | 35 | M | Y | 2 |
| 0475 | 78 | M | N | 2 | 0524 | 28 | F | Y | 2 | 0573 | 47 | F | N | 2 |
| 0476 | 49 | F | N | 2 | 0525 | 24 | F | Y | 2 | 0574 | 44 | F | N | 2 |
| 0477 | 72 | M | N | 2 | 0526 | 58 | M | Y | 1 | 0575 | 20 | M | Y | 2 |
| 0478 | 66 | M | N | 2 | 0527 | 79 | M | Y | 1 | 0576 | 15 | M | Y | 2 |
| 0479 | 70 | M | Y | 2 | 0528 | 83 | F | N | 2 | 0577 | 04 | M | N | 2 |
| 0480 | 82 | F | N | 2 | 0529 | 71 | M | Y | 2 | 0578 | 18 | F | Y | 2 |
| 0481 | 72 | F | N | 2 | 0530 | 66 | M | Y | 1 | 0579 | 12 | F | N | 2 |
| 0482 | 77 | F | N | 2 | 0531 | 62 | M | Y | 2 | 0580 | 07 | F | N | 2 |
| 0483 | 75 | F | N | 2 | 0532 | 48 | M | Y | 2 | 0581 | 06 | F | N | 2 |
| 0484 | 12 | F | N | 2 | 0533 | 70 | F | N | 1 | 0582 | 58 | M | N | 2 |
| 0485 | 34 | F | N | 2 | 0534 | 67 | F | N | 2 | 0583 | 62 | F | Y | 1 |
| 0486 | 52 | M | Y | 2 | 0535 | 59 | F | N | 2 | 0584 | 78 | M | Y | 2 |
| 0487 | 55 | F | N | 2 | 0536 | 55 | F | N | 2 | 0585 | 83 | F | N | 2 |
| 0488 | 46 | F | Y | 2 | 0537 | 36 | M | Y | 2 | 0586 | 63 | M | Y | 2 |
| 0489 | 60 | F | Y | 1 | 0538 | 29 | M | Y | 2 | 0587 | 63 | M | Y | 2 |
| 0490 | 88 | M | Y | 1 | 0539 | 25 | M | Y | 2 | 0588 | 61 | M | Y | 2 |
| 0491 | 60 | M | Y | 1 | 0540 | 37 | F | N | 2 | 0589 | 58 | M | Y | 2 |
| 0492 | 68 | M | Y | 1 | 0541 | 32 | F | N | 2 | 0590 | 54 | M | Y | 2 |
| 0493 | 67 | M | N | 2 | 0542 | 61 | F | N | 2 | 0591 | 43 | M | Y | 2 |
| 0494 | 78 | F | N | 2 | 0543 | 86 | M | Y | 1 | 0592 | 47 | F | N | 2 |
| 0495 | 21 | F | N | 2 | 0544 | 75 | F | N | 2 | 0593 | 48 | F | Y | 2 |
| 0496 | 73 | F | Y | 2 | 0545 | 68 | M | N | 2 | 0594 | 45 | F | N | 2 |
| 0497 | 70 | F | Y | 2 | 0546 | 66 | M | Y | 2 | 0595 | 42 | M | Y | 2 |
| 0498 | 65 | F | Y | 2 | 0547 | 65 | M | N | 2 | 0596 | 35 | M | N | 2 |
| 0499 | 58 | F | Y | 2 | 0548 | 58 | F | N | 2 | 0597 | 43 | F | N | 2 |
| 0500 | 33 | M | Y | 2 | 0549 | 36 | M | Y | 2 | 0598 | 54 | F | N | 2 |
| 0501 | 25 | M | Y | 2 | 0550 | 29 | M | Y | 2 | 0599 | 60 | M | N | 2 |
| 0502 | 35 | F | Y | 2 | 0551 | 25 | M | Y | 2 | 0600 | 90 | F | N | 2 |
| 0503 | 30 | F | Y | 2 | 0552 | 37 | F | N | 2 | 0601 | 02 | M | N | 2 |
| 0504 | 28 | F | Y | 2 | 0553 | 32 | F | N | 2 | 0602 | 45 | M | Y | 2 |
| 0505 | 22 | F | Y | 2 | 0554 | 44 | M | Y | 1 | 0603 | 02 | F | N | 2 |

## Data Set I (Continued)

| Number | Age | Sex | Smoker | Cancer | Number | Age | Sex | Smoker | Cancer | Number | Age | Sex | Smoker | Cancer |
|--------|-----|-----|--------|--------|--------|-----|-----|--------|--------|--------|-----|-----|--------|--------|
| 0604 | 49 | F | N | 2 | 0653 | 76 | M | N | 2 | 0702 | 12 | M | N | 2 |
| 0605 | 42 | M | Y | 2 | 0654 | 66 | M | Y | 2 | 0703 | 09 | M | Y | 2 |
| 0606 | 35 | M | N | 2 | 0655 | 63 | F | Y | 2 | 0704 | 05 | M | N | 2 |
| 0607 | 43 | F | N | 2 | 0656 | 55 | M | Y | 1 | 0705 | 63 | M | N | 2 |
| 0608 | 71 | M | Y | 1 | 0657 | 82 | F | N | 2 | 0706 | 73 | F | N | 2 |
| 0609 | 87 | M | Y | 2 | 0658 | 02 | M | N | 2 | 0707 | 60 | F | Y | 2 |
| 0610 | 36 | F | N | 2 | 0659 | 51 | M | Y | 2 | 0708 | 38 | M | Y | 2 |
| 0611 | 67 | M | Y | 2 | 0660 | 19 | M | N | 2 | 0709 | 40 | F | Y | 2 |
| 0612 | 56 | M | Y | 2 | 0661 | 37 | F | Y | 2 | 0710 | 68 | F | Y | 2 |
| 0613 | 61 | M | Y | 2 | 0662 | 58 | F | N | 2 | 0711 | 69 | M | Y | 2 |
| 0614 | 59 | M | Y | 2 | 0663 | 72 | M | Y | 2 | 0712 | 78 | F | N | 2 |
| 0615 | 71 | F | N | 2 | 0664 | 81 | F | N | 2 | 0713 | 35 | M | Y | 2 |
| 0616 | 37 | F | N | 1 | 0665 | 56 | M | Y | 2 | 0714 | 20 | M | Y | 2 |
| 0617 | 62 | F | N | 2 | 0666 | 62 | F | N | 2 | 0715 | 84 | F | N | 2 |
| 0618 | 74 | F | Y | 2 | 0667 | 56 | F | N | 2 | 0716 | 50 | F | Y | 2 |
| 0619 | 87 | F | N | 2 | 0668 | 19 | M | N | 2 | 0717 | 34 | F | N | 2 |
| 0620 | 64 | M | Y | 2 | 0669 | 37 | F | Y | 2 | 0718 | 61 | F | Y | 2 |
| 0621 | 37 | F | Y | 1 | 0670 | 79 | F | N | 1 | 0719 | 38 | M | Y | 2 |
| 0622 | 63 | M | Y | 1 | 0671 | 72 | M | Y | 1 | 0720 | 40 | F | Y | 2 |
| 0623 | 60 | F | Y | 2 | 0672 | 75 | F | Y | 2 | 0721 | 64 | M | Y | 1 |
| 0624 | 45 | M | N | 2 | 0673 | 54 | M | Y | 1 | 0722 | 80 | F | N | 2 |
| 0625 | 40 | F | Y | 1 | 0674 | 61 | F | N | 2 | 0723 | 68 | M | Y | 2 |
| 0626 | 38 | F | Y | 2 | 0675 | 57 | F | N | 2 | 0724 | 76 | F | Y | 1 |
| 0627 | 22 | M | N | 2 | 0676 | 53 | F | N | 1 | 0725 | 74 | F | Y | 2 |
| 0628 | 17 | M | N | 2 | 0677 | 51 | F | Y | 2 | 0726 | 45 | F | N | 2 |
| 0629 | 16 | M | Y | 2 | 0678 | 62 | M | Y | 2 | 0727 | 63 | F | N | 2 |
| 0630 | 09 | M | Y | 2 | 0679 | 76 | M | Y | 1 | 0728 | 33 | F | Y | 2 |
| 0631 | 20 | F | N | 2 | 0680 | 69 | F | N | 1 | 0729 | 67 | F | N | 2 |
| 0632 | 45 | M | Y | 2 | 0681 | 83 | M | Y | 1 | 0730 | 72 | M | Y | 2 |
| 0633 | 70 | M | N | 1 | 0682 | 60 | F | N | 1 | 0731 | 86 | F | N | 2 |
| 0634 | 70 | F | N | 2 | 0683 | 63 | F | Y | 1 | 0732 | 78 | F | N | 2 |
| 0635 | 46 | M | Y | 2 | 0684 | 62 | F | Y | 2 | 0733 | 69 | F | Y | 2 |
| 0636 | 47 | M | Y | 2 | 0685 | 83 | F | N | 2 | 0734 | 40 | M | Y | 2 |
| 0637 | 49 | M | Y | 2 | 0686 | 39 | M | Y | 2 | 0735 | 36 | M | Y | 2 |
| 0638 | 51 | M | Y | 2 | 0687 | 76 | F | N | 2 | 0736 | 33 | F | Y | 2 |
| 0639 | 43 | M | Y | 2 | 0688 | 75 | M | N | 2 | 0737 | 67 | M | Y | 1 |
| 0640 | 45 | M | Y | 2 | 0689 | 71 | M | Y | 2 | 0738 | 68 | F | N | 2 |
| 0641 | 38 | M | Y | 2 | 0690 | 78 | F | N | 2 | 0739 | 45 | M | Y | 2 |
| 0642 | 57 | F | Y | 2 | 0691 | 78 | F | N | 2 | 0740 | 01 | F | N | 2 |
| 0643 | 43 | F | N | 2 | 0692 | 73 | F | N | 2 | 0741 | 47 | M | Y | 2 |
| 0644 | 41 | F | Y | 2 | 0693 | 06 | F | N | 2 | 0742 | 37 | F | Y | 2 |
| 0645 | 22 | M | N | 2 | 0694 | 63 | F | Y | 1 | 0743 | 23 | F | N | 2 |
| 0646 | 17 | M | N | 2 | 0695 | 62 | F | Y | 2 | 0744 | 62 | M | N | 2 |
| 0647 | 16 | M | Y | 2 | 0696 | 66 | M | Y | 1 | 0745 | 60 | M | N | 2 |
| 0648 | 09 | M | Y | 2 | 0697 | 68 | M | Y | 1 | 0746 | 69 | F | N | 2 |
| 0649 | 20 | F | N | 2 | 0698 | 82 | F | N | 2 | 0747 | 67 | M | Y | 1 |
| 0650 | 69 | M | Y | 1 | 0699 | 50 | M | Y | 2 | 0748 | 74 | F | N | 2 |
| 0651 | 56 | F | N | 2 | 0700 | 77 | M | Y | 1 | 0749 | 64 | M | Y | 2 |
| 0652 | 72 | F | N | 2 | 0701 | 67 | M | Y | 1 | 0750 | 51 | M | Y | 2 |

**Data Set I (Continued)**

| Number | Age | Sex | Smoker | Cancer | Number | Age | Sex | Smoker | Cancer | Number | Age | Sex | Smoker | Cancer |
|--------|-----|-----|--------|--------|--------|-----|-----|--------|--------|--------|-----|-----|--------|--------|
| 0751 | 47 | M | Y | 2 | 0800 | 65 | F | N | 2 | 0849 | 45 | M | Y | 2 |
| 0752 | 37 | F | Y | 2 | 0801 | 55 | F | Y | 2 | 0850 | 57 | F | Y | 2 |
| 0753 | 23 | F | N | 2 | 0802 | 31 | M | N | 2 | 0851 | 09 | M | Y | 2 |
| 0754 | 62 | M | Y | 1 | 0803 | 33 | F | Y | 2 | 0852 | 45 | M | Y | 2 |
| 0755 | 88 | F | N | 2 | 0804 | 17 | F | N | 2 | 0853 | 50 | M | Y | 1 |
| 0756 | 63 | M | Y | 2 | 0805 | 63 | M | Y | 1 | 0854 | 76 | F | N | 2 |
| 0757 | 63 | F | Y | 2 | 0806 | 78 | M | Y | 2 | 0855 | 65 | M | Y | 2 |
| 0758 | 59 | F | Y | 2 | 0807 | 69 | F | N | 2 | 0856 | 32 | M | N | 2 |
| 0759 | 37 | M | N | 2 | 0808 | 75 | F | Y | 2 | 0857 | 29 | M | N | 2 |
| 0760 | 40 | F | N | 2 | 0809 | 73 | F | N | 2 | 0858 | 27 | M | N | 2 |
| 0761 | 59 | F | Y | 2 | 0810 | 38 | M | N | 2 | 0859 | 30 | F | N | 2 |
| 0762 | 83 | M | Y | 1 | 0811 | 22 | F | N | 2 | 0860 | 50 | F | N | 2 |
| 0763 | 82 | F | N | 2 | 0812 | 65 | F | Y | 2 | 0861 | 47 | F | N | 1 |
| 0764 | 65 | M | Y | 2 | 0813 | 59 | F | N | 2 | 0862 | 32 | M | N | 2 |
| 0765 | 67 | F | Y | 2 | 0814 | 66 | M | Y | 2 | 0863 | 29 | M | N | 2 |
| 0766 | 37 | M | N | 2 | 0815 | 61 | F | N | 2 | 0864 | 27 | M | N | 2 |
| 0767 | 40 | F | N | 2 | 0816 | 38 | M | N | 2 | 0865 | 30 | F | N | 2 |
| 0768 | 57 | M | Y | 1 | 0817 | 22 | F | N | 2 | 0866 | 37 | M | N | 2 |
| 0769 | 59 | M | N | 2 | 0818 | 63 | M | Y | 1 | 0867 | 44 | F | N | 2 |
| 0770 | 83 | F | N | 2 | 0819 | 74 | M | Y | 2 | 0868 | 42 | F | Y | 2 |
| 0771 | 60 | M | Y | 2 | 0820 | 50 | F | N | 2 | 0869 | 40 | F | Y | 2 |
| 0772 | 29 | M | N | 2 | 0821 | 77 | M | N | 1 | 0870 | 58 | M | Y | 1 |
| 0773 | 49 | M | Y | 1 | 0822 | 02 | M | N | 2 | 0871 | 78 | F | Y | 2 |
| 0774 | 56 | M | N | 2 | 0823 | 65 | M | Y | 1 | 0872 | 62 | M | N | 2 |
| 0775 | 60 | F | Y | 2 | 0824 | 69 | M | N | 2 | 0873 | 69 | M | Y | 2 |
| 0776 | 84 | M | Y | 2 | 0825 | 64 | M | N | 2 | 0874 | 66 | M | N | 2 |
| 0777 | 80 | F | N | 2 | 0826 | 62 | M | N | 1 | 0875 | 59 | M | Y | 2 |
| 0778 | 63 | M | Y | 2 | 0827 | 72 | F | N | 2 | 0876 | 52 | M | N | 2 |
| 0779 | 42 | M | Y | 1 | 0828 | 40 | M | N | 2 | 0877 | 63 | F | N | 1 |
| 0780 | 50 | M | Y | 2 | 0829 | 38 | M | N | 2 | 0878 | 60 | F | N | 2 |
| 0781 | 56 | F | Y | 2 | 0830 | 36 | M | N | 2 | 0879 | 55 | F | Y | 2 |
| 0782 | 36 | M | N | 2 | 0831 | 30 | M | N | 2 | 0880 | 38 | M | Y | 2 |
| 0783 | 32 | M | Y | 2 | 0832 | 26 | M | N | 2 | 0881 | 33 | M | Y | 2 |
| 0784 | 58 | M | Y | 1 | 0833 | 43 | F | N | 2 | 0882 | 48 | M | Y | 2 |
| 0785 | 86 | F | Y | 2 | 0834 | 65 | F | N | 2 | 0883 | 58 | F | Y | 2 |
| 0786 | 50 | M | Y | 2 | 0835 | 63 | M | N | 2 | 0884 | 63 | M | Y | 2 |
| 0787 | 48 | M | Y | 2 | 0836 | 70 | F | N | 2 | 0885 | 85 | F | N | 2 |
| 0788 | 36 | M | Y | 2 | 0837 | 63 | F | N | 2 | 0886 | 57 | F | N | 2 |
| 0789 | 44 | M | Y | 2 | 0838 | 57 | F | N | 1 | 0887 | 55 | F | Y | 2 |
| 0790 | 53 | F | Y | 2 | 0839 | 58 | F | N | 2 | 0888 | 39 | M | Y | 2 |
| 0791 | 51 | F | N | 2 | 0840 | 40 | M | N | 2 | 0889 | 33 | M | Y | 2 |
| 0792 | 31 | M | N | 2 | 0841 | 38 | M | N | 2 | 0890 | 73 | M | Y | 1 |
| 0793 | 33 | F | Y | 2 | 0842 | 36 | M | N | 2 | 0891 | 84 | M | Y | 2 |
| 0794 | 17 | F | N | 2 | 0843 | 30 | M | N | 2 | 0892 | 84 | F | N | 2 |
| 0795 | 53 | F | Y | 2 | 0844 | 26 | M | N | 2 | 0893 | 80 | M | Y | 2 |
| 0796 | 77 | M | Y | 2 | 0845 | 43 | F | N | 2 | 0894 | 79 | M | Y | 2 |
| 0797 | 81 | F | N | 1 | 0846 | 66 | M | N | 1 | 0895 | 72 | M | Y | 2 |
| 0798 | 66 | M | N | 2 | 0847 | 84 | F | N | 2 | 0896 | 69 | M | Y | 2 |
| 0799 | 51 | M | Y | 2 | 0848 | 62 | F | N | 1 | 0897 | 87 | F | N | 2 |

**Data Set I (Continued)**

| Number | Age | Sex | Smoker | Cancer | Number | Age | Sex | Smoker | Cancer | Number | Age | Sex | Smoker | Cancer |
|--------|-----|-----|--------|--------|--------|-----|-----|--------|--------|--------|-----|-----|--------|--------|
| 0898 | 84 | F | N | 2 | 0947 | 56 | M | N | 2 | 0996 | 90 | F | N | 1 |
| 0899 | 76 | F | N | 2 | 0948 | 48 | F | Y | 2 | 0997 | 60 | M | Y | 1 |
| 0900 | 76 | F | N | 2 | 0949 | 38 | F | Y | 2 | 0998 | 60 | M | Y | 2 |
| 0901 | 74 | F | N | 2 | 0950 | 77 | F | N | 2 | 0999 | 61 | M | Y | 1 |
| 0902 | 35 | M | Y | 2 | 0951 | 29 | F | N | 2 | 1000 | 71 | F | N | 1 |
| 0903 | 51 | F | N | 2 | 0952 | 79 | M | Y | 2 | 1001 | 76 | F | N | 2 |
| 0904 | 50 | F | N | 2 | 0953 | 58 | M | N | 2 | 1002 | 69 | F | N | 2 |
| 0905 | 67 | F | N | 2 | 0954 | 56 | M | N | 2 | 1003 | 43 | F | Y | 2 |
| 0906 | 57 | M | Y | 2 | 0955 | 48 | F | Y | 2 | 1004 | 30 | M | Y | 1 |
| 0907 | 80 | F | N | 2 | 0956 | 38 | F | Y | 2 | 1005 | 29 | F | N | 2 |
| 0908 | 73 | M | Y | 1 | 0957 | 56 | F | Y | 1 | 1006 | 59 | F | Y | 2 |
| 0909 | 22 | M | Y | 1 | 0958 | 37 | M | Y | 2 | 1007 | 81 | M | Y | 2 |
| 0910 | 66 | M | Y | 2 | 0959 | 67 | F | Y | 2 | 1008 | 83 | F | N | 1 |
| 0911 | 65 | M | Y | 2 | 0960 | 57 | M | Y | 2 | 1009 | 56 | M | Y | 2 |
| 0912 | 52 | M | Y | 2 | 0961 | 53 | M | Y | 2 | 1010 | 53 | M | Y | 2 |
| 0913 | 77 | F | N | 2 | 0962 | 47 | M | Y | 2 | 1011 | 40 | F | N | 2 |
| 0914 | 55 | F | N | 1 | 0963 | 23 | M | Y | 2 | 1012 | 35 | F | Y | 2 |
| 0915 | 55 | F | Y | 2 | 0964 | 62 | F | Y | 2 | 1012 | 35 | F | Y | 2 |
| 0916 | 58 | F | N | 2 | 0965 | 47 | F | N | 2 | 1014 | 69 | M | Y | 1 |
| 0917 | 35 | M | Y | 2 | 0966 | 37 | M | N | 2 | 1015 | 92 | F | N | 2 |
| 0918 | 51 | F | N | 2 | 0965 | 47 | F | N | 2 | 1016 | 65 | M | Y | 2 |
| 0919 | 50 | F | N | 2 | 0968 | 34 | M | Y | 2 | 1017 | 43 | F | N | 2 |
| 0920 | 78 | M | Y | 2 | 0969 | 28 | M | N | 2 | 1018 | 30 | M | Y | 1 |
| 0921 | 67 | M | Y | 1 | 0970 | 24 | M | N | 2 | 1019 | 29 | F | N | 2 |
| 0922 | 71 | F | Y | 2 | 0971 | 67 | M | Y | 2 | 1020 | 83 | F | N | 2 |
| 0923 | 65 | F | Y | 2 | 0972 | 65 | F | Y | 2 | 1021 | 66 | M | N | 1 |
| 0924 | 23 | M | Y | 2 | 0973 | 37 | M | N | 2 | 1022 | 57 | M | N | 2 |
| 0925 | 22 | M | Y | 2 | 0974 | 35 | M | Y | 2 | 1023 | 71 | F | N | 2 |
| 0926 | 17 | M | N | 2 | 0975 | 34 | M | Y | 2 | 1024 | 36 | F | N | 1 |
| 0927 | 24 | F | N | 2 | 0976 | 28 | M | N | 2 | 1025 | 62 | F | N | 2 |
| 0928 | 13 | F | N | 2 | 0977 | 24 | M | N | 2 | 1026 | 62 | F | N | 2 |
| 0929 | 43 | F | N | 2 | 0978 | 60 | M | Y | 1 | 1027 | 60 | F | N | 2 |
| 0930 | 65 | M | N | 2 | 0979 | 48 | M | Y | 2 | 1028 | 43 | M | N | 1 |
| 0931 | 62 | F | N | 2 | 0980 | 43 | M | Y | 2 | 1029 | 69 | M | N | 2 |
| 0932 | 40 | M | N | 2 | 0981 | 38 | M | N | 2 | 1030 | 83 | F | N | 2 |
| 0933 | 23 | M | Y | 2 | 0982 | 32 | M | N | 2 | 1031 | 70 | M | Y | 2 |
| 0934 | 22 | M | Y | 2 | 0983 | 60 | F | Y | 2 | 1032 | 66 | M | Y | 1 |
| 0935 | 17 | M | N | 2 | 0984 | 77 | M | Y | 1 | 1033 | 75 | M | Y | 2 |
| 0936 | 24 | F | N | 2 | 0985 | 78 | F | N | 2 | 1034 | 90 | F | N | 2 |
| 0937 | 13 | F | N | 2 | 0986 | 65 | M | N | 2 | 1035 | 65 | M | Y | 2 |
| 0938 | 77 | M | Y | 1 | 0987 | 54 | M | Y | 2 | 1036 | 63 | M | Y | 2 |
| 0939 | 74 | M | N | 2 | 0988 | 57 | F | Y | 2 | 1037 | 58 | M | N | 2 |
| 0940 | 69 | F | N | 1 | 0989 | 43 | M | Y | 2 | 1038 | 68 | F | Y | 2 |
| 0941 | 70 | M | Y | 2 | 0990 | 38 | M | Y | 2 | 1039 | 60 | F | N | 1 |
| 0942 | 56 | M | Y | 2 | 0991 | 32 | M | N | 2 | 1040 | 57 | F | N | 2 |
| 0943 | 84 | F | N | 2 | 0992 | 56 | M | Y | 1 | 1041 | 55 | F | N | 2 |
| 0944 | 82 | F | N | 2 | 0993 | 62 | M | Y | 2 | 1042 | 49 | M | Y | 2 |
| 0945 | 78 | F | Y | 2 | 0994 | 56 | F | Y | 2 | 1043 | 49 | M | Y | 2 |
| 0946 | 58 | M | N | 2 | 0995 | 89 | M | Y | 2 | 1044 | 46 | M | Y | 2 |

**Data Set I (Continued)**

| Number | Age | Sex | Smoker | Cancer | Number | Age | Sex | Smoker | Cancer | Number | Age | Sex | Smoker | Cancer |
|---|---|---|---|---|---|---|---|---|---|---|---|---|---|---|
| 1045 | 90 | M | N | 2 | 1094 | 68 | F | Y | 2 | 1143 | 75 | F | N | 2 |
| 1046 | 68 | M | Y | 2 | 1095 | 47 | M | Y | 2 | 1144 | 70 | F | Y | 2 |
| 1047 | 71 | F | Y | 2 | 1096 | 55 | F | Y | 2 | 1145 | 64 | F | Y | 2 |
| 1048 | 49 | M | Y | 2 | 1097 | 51 | F | N | 2 | 1146 | 43 | M | Y | 2 |
| 1049 | 49 | M | Y | 2 | 1098 | 49 | F | Y | 2 | 1147 | 40 | M | N | 2 |
| 1050 | 46 | M | Y | 2 | 1099 | 78 | F | N | 2 | 1148 | 37 | M | Y | 2 |
| 1051 | 73 | M | Y | 1 | 1100 | 60 | M | Y | 2 | 1149 | 33 | M | N | 2 |
| 1052 | 58 | M | N | 2 | 1101 | 82 | F | N | 2 | 1150 | 32 | M | N | 2 |
| 1053 | 82 | F | N | 2 | 1102 | 85 | M | Y | 2 | 1151 | 38 | F | Y | 2 |
| 1054 | 63 | M | Y | 2 | 1103 | 80 | M | Y | 2 | 1152 | 26 | F | Y | 2 |
| 1055 | 60 | M | N | 2 | 1104 | 70 | M | Y | 2 | 1153 | 66 | F | Y | 2 |
| 1056 | 02 | F | N | 2 | 1105 | 59 | M | N | 2 | 1154 | 77 | M | Y | 2 |
| 1057 | 78 | F | N | 1 | 1106 | 43 | M | N | 2 | 1155 | 89 | F | N | 2 |
| 1058 | 76 | F | N | 2 | 1107 | 72 | M | N | 2 | 1155 | 89 | F | N | 2 |
| 1059 | 66 | F | N | 2 | 1108 | 70 | M | N | 2 | 1157 | 64 | M | Y | 2 |
| 1060 | 51 | M | Y | 2 | 1109 | 74 | M | N | 2 | 1158 | 71 | F | N | 2 |
| 1061 | 75 | F | N | 2 | 1110 | 78 | F | N | 2 | 1159 | 65 | F | N | 2 |
| 1062 | 59 | M | N | 2 | 1111 | 84 | F | N | 2 | 1160 | 68 | F | N | 2 |
| 1063 | 86 | F | N | 2 | 1112 | 47 | M | Y | 2 | 1161 | 46 | F | N | 1 |
| 1064 | 66 | M | N | 1 | 1113 | 55 | F | Y | 2 | 1162 | 43 | M | Y | 2 |
| 1065 | 64 | M | Y | 2 | 1114 | 51 | F | N | 2 | 1163 | 40 | M | N | 2 |
| 1066 | 59 | F | N | 2 | 1115 | 49 | F | Y | 2 | 1164 | 37 | M | Y | 2 |
| 1067 | 65 | F | N | 2 | 1116 | 42 | M | Y | 1 | 1165 | 33 | M | N | 2 |
| 1068 | 73 | F | N | 2 | 1117 | 37 | M | N | 2 | 1166 | 32 | M | N | 2 |
| 1069 | 51 | M | Y | 2 | 1118 | 75 | F | N | 2 | 1167 | 38 | F | Y | 2 |
| 1070 | 68 | M | Y | 1 | 1119 | 51 | M | Y | 2 | 1168 | 26 | F | Y | 2 |
| 1071 | 72 | M | N | 2 | 1120 | 32 | M | Y | 2 | 1169 | 65 | M | Y | 1 |
| 1072 | 61 | F | N | 1 | 1121 | 43 | M | Y | 2 | 1170 | 54 | M | N | 2 |
| 1073 | 74 | M | Y | 2 | 1122 | 56 | F | N | 2 | 1171 | 68 | F | N | 2 |
| 1074 | 77 | F | Y | 2 | 1123 | 52 | F | N | 2 | 1172 | 28 | M | N | 2 |
| 1075 | 62 | F | Y | 2 | 1124 | 42 | F | N | 2 | 1173 | 74 | M | Y | 2 |
| 1076 | 57 | F | Y | 2 | 1125 | 40 | F | N | 2 | 1174 | 57 | M | Y | 2 |
| 1077 | 44 | M | Y | 2 | 1126 | 48 | M | Y | 2 | 1175 | 57 | M | Y | 2 |
| 1078 | 47 | F | Y | 2 | 1127 | 11 | F | N | 2 | 1176 | 54 | M | Y | 2 |
| 1079 | 59 | F | Y | 2 | 1128 | 10 | F | N | 2 | 1177 | 72 | F | N | 2 |
| 1080 | 62 | F | N | 2 | 1129 | 19 | F | Y | 2 | 1178 | 65 | F | N | 2 |
| 1081 | 54 | F | N | 2 | 1130 | 25 | F | Y | 2 | 1179 | 41 | M | N | 2 |
| 1082 | 77 | M | Y | 1 | 1131 | 35 | F | N | 2 | 1180 | 36 | M | N | 2 |
| 1083 | 54 | M | Y | 2 | 1132 | 61 | M | Y | 2 | 1181 | 66 | F | Y | 2 |
| 1084 | 70 | F | N | 2 | 1133 | 60 | F | N | 2 | 1182 | 41 | M | N | 2 |
| 1085 | 86 | M | Y | 2 | 1134 | 28 | F | N | 2 | 1183 | 36 | M | N | 2 |
| 1086 | 84 | M | Y | 2 | 1135 | 18 | M | Y | 2 | 1184 | 51 | M | Y | 1 |
| 1087 | 49 | M | Y | 2 | 1136 | 11 | F | N | 2 | 1185 | 16 | F | N | 2 |
| 1088 | 78 | M | N | 2 | 1137 | 10 | F | N | 2 | 1186 | 53 | F | N | 2 |
| 1089 | 76 | M | Y | 2 | 1138 | 09 | F | Y | 2 | 1187 | 16 | F | N | 2 |
| 1090 | 74 | F | Y | 2 | 1139 | 15 | F | Y | 2 | 1188 | 70 | F | N | 1 |
| 1091 | 74 | F | N | 2 | 1140 | 90 | F | N | 2 | 1189 | 50 | M | N | 1 |
| 1092 | 72 | F | Y | 2 | 1141 | 68 | M | N | 2 | 1190 | 88 | F | N | 2 |
| 1093 | 70 | F | N | 2 | 1142 | 82 | F | Y | 2 | 1191 | 05 | M | N | 2 |

**Data Set I (Continued)**

| Number | Age | Sex | Smoker | Cancer | Number | Age | Sex | Smoker | Cancer | Number | Age | Sex | Smoker | Cancer |
|---|---|---|---|---|---|---|---|---|---|---|---|---|---|---|
| 1192 | 22 | M | Y | 2 | 1241 | 50 | M | Y | 2 | 1290 | 65 | M | Y | 2 |
| 1193 | 68 | M | Y | 1 | 1242 | 43 | M | Y | 2 | 1291 | 76 | F | N | 2 |
| 1194 | 46 | M | Y | 1 | 1243 | 35 | M | Y | 2 | 1292 | 73 | F | N | 2 |
| 1195 | 58 | M | Y | 1 | 1244 | 67 | F | Y | 2 | 1293 | 74 | F | Y | 2 |
| 1196 | 61 | M | Y | 2 | 1245 | 82 | F | N | 1 | 1294 | 68 | F | N | 2 |
| 1197 | 37 | M | Y | 2 | 1246 | 65 | M | Y | 1 | 1295 | 63 | F | N | 2 |
| 1198 | 40 | M | N | 2 | 1247 | 18 | M | Y | 2 | 1296 | 34 | M | N | 2 |
| 1199 | 51 | F | Y | 2 | 1248 | 28 | M | Y | 2 | 1297 | 33 | M | Y | 2 |
| 1200 | 76 | M | Y | 2 | 1249 | 70 | M | Y | 2 | 1298 | 31 | M | Y | 2 |
| 1201 | 71 | M | Y | 2 | 1250 | 76 | F | N | 2 | 1299 | 26 | M | N | 2 |
| 1202 | 94 | F | N | 2 | 1251 | 84 | F | N | 2 | 1300 | 24 | M | N | 2 |
| 1203 | 78 | F | N | 2 | 1252 | 83 | F | Y | 2 | 1301 | 36 | F | N | 2 |
| 1204 | 37 | M | Y | 2 | 1253 | 50 | M | Y | 2 | 1302 | 29 | F | Y | 2 |
| 1205 | 40 | M | N | 2 | 1254 | 43 | M | Y | 2 | 1303 | 22 | F | N | 2 |
| 1206 | 51 | F | Y | 2 | 1255 | 35 | M | Y | 2 | 1304 | 68 | F | Y | 1 |
| 1207 | 78 | M | N | 2 | 1256 | 53 | M | Y | 1 | 1305 | 60 | M | Y | 1 |
| 1208 | 92 | F | N | 2 | 1257 | 32 | M | N | 2 | 1306 | 45 | F | N | 2 |
| 1209 | 75 | M | N | 2 | 1258 | 57 | F | Y | 2 | 1307 | 60 | M | Y | 2 |
| 1210 | 27 | F | N | 2 | 1259 | 79 | M | Y | 1 | 1308 | 61 | F | Y | 1 |
| 1211 | 65 | F | N | 2 | 1260 | 55 | M | N | 2 | 1309 | 64 | F | Y | 1 |
| 1212 | 55 | M | Y | 2 | 1261 | 50 | F | N | 2 | 1310 | 83 | M | N | 1 |
| 1213 | 52 | F | Y | 2 | 1262 | 49 | M | Y | 2 | 1311 | 66 | F | N | 2 |
| 1214 | 49 | F | N | 2 | 1263 | 60 | M | Y | 2 | 1312 | 89 | M | Y | 2 |
| 1215 | 46 | F | Y | 2 | 1264 | 18 | M | Y | 2 | 1313 | 80 | M | N | 2 |
| 1216 | 68 | F | Y | 2 | 1265 | 23 | F | Y | 2 | 1314 | 83 | M | N | 2 |
| 1217 | 71 | M | N | 1 | 1266 | 77 | F | N | 2 | 1315 | 93 | F | N | 2 |
| 1218 | 38 | F | N | 2 | 1267 | 50 | F | N | 2 | 1316 | 68 | F | N | 2 |
| 1219 | 22 | M | Y | 2 | 1268 | 55 | F | N | 2 | 1317 | 73 | F | N | 2 |
| 1220 | 58 | M | Y | 2 | 1269 | 62 | M | Y | 2 | 1318 | 74 | F | N | 2 |
| 1221 | 55 | M | Y | 2 | 1250 | 76 | F | N | 2 | 1319 | 65 | M | N | 2 |
| 1222 | 73 | F | N | 2 | 1271 | 33 | M | Y | 2 | 1320 | 62 | M | N | 2 |
| 1223 | 62 | F | Y | 2 | 1272 | 31 | M | Y | 2 | 1321 | 77 | M | Y | 1 |
| 1224 | 45 | F | Y | 2 | 1273 | 26 | M | N | 2 | 1322 | 99 | F | N | 2 |
| 1225 | 66 | F | N | 2 | 1274 | 24 | M | N | 2 | 1323 | 35 | M | Y | 2 |
| 1226 | 62 | F | N | 2 | 1275 | 19 | F | N | 2 | 1324 | 63 | M | N | 2 |
| 1227 | 55 | M | Y | 2 | 1276 | 57 | F | N | 2 | 1325 | 77 | F | N | 2 |
| 1228 | 52 | F | Y | 2 | 1277 | 49 | F | N | 2 | 1326 | 80 | F | Y | 2 |
| 1229 | 49 | F | N | 2 | 1278 | 53 | F | N | 2 | 1327 | 78 | F | N | 2 |
| 1230 | 46 | F | Y | 2 | 1279 | 48 | F | Y | 2 | 1328 | 76 | F | N | 2 |
| 1231 | 77 | M | Y | 1 | 1280 | 32 | F | Y | 2 | 1329 | 73 | F | N | 2 |
| 1232 | 95 | M | Y | 2 | 1281 | 29 | F | Y | 2 | 1330 | 56 | M | Y | 2 |
| 1233 | 75 | F | N | 2 | 1282 | 22 | F | N | 2 | 1331 | 49 | M | Y | 2 |
| 1234 | 65 | M | Y | 2 | 1283 | 74 | M | Y | 2 | 1332 | 46 | M | N | 2 |
| 1235 | 77 | M | N | 2 | 1284 | 60 | F | N | 2 | 1333 | 62 | F | N | 2 |
| 1236 | 65 | M | Y | 2 | 1285 | 70 | M | Y | 2 | 1334 | 60 | F | N | 2 |
| 1237 | 80 | F | N | 2 | 1286 | 71 | F | N | 2 | 1335 | 53 | F | N | 2 |
| 1238 | 17 | F | N | 2 | 1287 | 53 | M | Y | 1 | 1336 | 79 | F | N | 2 |
| 1239 | 72 | F | N | 2 | 1288 | 73 | M | Y | 1 | 1337 | 62 | M | Y | 1 |
| 1240 | 70 | F | N | 2 | 1289 | 58 | M | N | 2 | 1338 | 20 | F | N | 2 |

**Data Set I (Continued)**

| Number | Age | Sex | Smoker | Cancer | Number | Age | Sex | Smoker | Cancer | Number | Age | Sex | Smoker | Cancer |
|--------|-----|-----|--------|--------|--------|-----|-----|--------|--------|--------|-----|-----|--------|--------|
| 1339 | 56 | M | Y | 2 | 1388 | 73 | M | Y | 2 | 1437 | 76 | M | Y | 1 |
| 1340 | 49 | M | Y | 2 | 1389 | 83 | F | N | 2 | 1438 | 95 | M | Y | 2 |
| 1341 | 46 | M | N | 2 | 1390 | 81 | F | N | 2 | 1439 | 92 | F | N | 2 |
| 1342 | 62 | F | N | 2 | 1391 | 64 | F | Y | 1 | 1440 | 70 | M | Y | 1 |
| 1343 | 60 | F | N | 2 | 1392 | 68 | F | Y | 2 | 1441 | 58 | F | Y | 2 |
| 1344 | 53 | F | N | 2 | 1393 | 40 | M | Y | 2 | 1442 | 44 | F | N | 2 |
| 1345 | 32 | M | N | 2 | 1394 | 35 | M | Y | 2 | 1443 | 67 | F | N | 2 |
| 1345 | 32 | M | N | 2 | 1395 | 44 | F | Y | 2 | 1444 | 78 | M | Y | 2 |
| 1347 | 69 | F | N | 2 | 1396 | 66 | M | Y | 1 | 1445 | 74 | F | N | 2 |
| 1348 | 66 | F | N | 2 | 1397 | 64 | F | Y | 2 | 1446 | 65 | M | N | 2 |
| 1349 | 84 | F | N | 1 | 1398 | 72 | F | N | 1 | 1447 | 59 | F | N | 2 |
| 1350 | 53 | F | N | 2 | 1399 | 90 | M | N | 2 | 1448 | 44 | F | N | 2 |
| 1351 | 79 | M | N | 1 | 1400 | 65 | F | N | 1 | 1449 | 76 | M | Y | 1 |
| 1352 | 65 | M | Y | 1 | 1401 | 81 | M | Y | 1 | 1450 | 63 | M | Y | 1 |
| 1353 | 78 | M | Y | 2 | 1402 | 70 | M | Y | 1 | 1451 | 58 | M | Y | 2 |
| 1354 | 82 | M | N | 2 | 1403 | 63 | M | Y | 2 | 1452 | 67 | M | Y | 2 |
| 1355 | 74 | M | Y | 2 | 1404 | 65 | M | Y | 2 | 1453 | 44 | F | Y | 1 |
| 1356 | 77 | F | N | 2 | 1405 | 52 | M | Y | 2 | 1454 | 58 | M | N | 2 |
| 1357 | 85 | M | N | 1 | 1406 | 46 | M | Y | 2 | 1455 | 36 | M | Y | 2 |
| 1358 | 77 | F | N | 2 | 1407 | 56 | M | N | 2 | 1456 | 68 | F | N | 2 |
| 1359 | 80 | F | N | 1 | 1408 | 55 | M | Y | 1 | 1457 | 63 | M | N | 2 |
| 1360 | 29 | F | N | 2 | 1409 | 55 | M | N | 2 | 1458 | 66 | F | N | 2 |
| 1361 | 67 | M | Y | 1 | 1410 | 68 | F | Y | 2 | 1459 | 83 | M | N | 2 |
| 1362 | 87 | M | N | 1 | 1411 | 55 | M | Y | 2 | 1460 | 65 | M | N | 2 |
| 1363 | 75 | F | N | 2 | 1412 | 68 | M | Y | 1 | 1461 | 44 | M | Y | 2 |
| 1364 | 72 | M | N | 2 | 1413 | 45 | M | N | 2 | 1462 | 64 | M | N | 2 |
| 1365 | 62 | M | N | 2 | 1414 | 43 | F | Y | 2 | 1463 | 78 | F | N | 2 |
| 1366 | 76 | M | N | 2 | 1415 | 73 | M | Y | 1 | 1464 | 70 | F | N | 2 |
| 1367 | 74 | M | N | 1 | 1416 | 66 | M | N | 2 | 1465 | 38 | M | N | 2 |
| 1368 | 60 | F | N | 1 | 1417 | 79 | F | N | 2 | 1466 | 36 | M | Y | 2 |
| 1369 | 34 | M | N | 2 | 1418 | 70 | F | N | 2 | 1467 | 74 | F | Y | 1 |
| 1370 | 37 | F | Y | 2 | 1419 | 54 | M | Y | 2 | 1468 | 89 | M | Y | 1 |
| 1371 | 70 | F | Y | 2 | 1420 | 53 | M | N | 2 | 1469 | 82 | F | N | 2 |
| 1372 | 80 | M | Y | 2 | 1421 | 46 | F | N | 2 | 1470 | 34 | M | Y | 1 |
| 1373 | 79 | F | N | 2 | 1422 | 79 | F | N | 1 | 1471 | 74 | M | Y | 2 |
| 1374 | 40 | F | Y | 1 | 1423 | 69 | M | Y | 2 | 1472 | 56 | M | Y | 1 |
| 1375 | 56 | F | Y | 2 | 1424 | 48 | F | N | 2 | 1473 | 69 | M | Y | 2 |
| 1376 | 34 | M | N | 2 | 1425 | 81 | M | N | 2 | 1474 | 61 | F | Y | 1 |
| 1377 | 37 | F | Y | 2 | 1426 | 55 | M | N | 2 | 1475 | 74 | M | Y | 1 |
| 1378 | 80 | M | Y | 1 | 1427 | 75 | M | N | 2 | 1476 | 64 | F | N | 2 |
| 1379 | 90 | M | Y | 2 | 1428 | 73 | M | N | 2 | 1477 | 67 | M | Y | 2 |
| 1380 | 93 | F | N | 2 | 1429 | 71 | M | Y | 2 | 1478 | 57 | M | Y | 1 |
| 1381 | 48 | M | Y | 1 | 1430 | 69 | F | N | 2 | 1479 | 61 | F | N | 2 |
| 1382 | 23 | M | N | 2 | 1431 | 67 | F | N | 2 | 1480 | 59 | M | Y | 2 |
| 1383 | 66 | F | Y | 2 | 1432 | 65 | F | N | 2 | 1481 | 40 | M | N | 2 |
| 1384 | 23 | M | N | 2 | 1433 | 63 | F | N | 2 | 1482 | 24 | M | N | 2 |
| 1385 | 62 | F | Y | 1 | 1434 | 54 | M | Y | 2 | 1483 | 22 | M | N | 2 |
| 1386 | 66 | M | Y | 2 | 1435 | 53 | M | N | 2 | 1484 | 39 | F | N | 2 |
| 1387 | 68 | M | Y | 1 | 1436 | 46 | F | N | 2 | 1485 | 36 | F | N | 2 |

Data Set I (Continued)

| Number | Age | Sex | Smoker | Cancer | Number | Age | Sex | Smoker | Cancer | Number | Age | Sex | Smoker | Cancer |
|--------|-----|-----|--------|--------|--------|-----|-----|--------|--------|--------|-----|-----|--------|--------|
| 1486 | 86 | M | Y | 2 | 1535 | 50 | M | Y | 2 | 1584 | 50 | F | N | 2 |
| 1487 | 60 | M | N | 2 | 1536 | 39 | M | Y | 2 | 1585 | 40 | M | N | 2 |
| 1488 | 90 | F | N | 2 | 1537 | 43 | F | Y | 2 | 1586 | 33 | M | Y | 2 |
| 1489 | 92 | M | N | 2 | 1538 | 41 | F | N | 2 | 1587 | 37 | M | N | 2 |
| 1490 | 80 | M | N | 1 | 1539 | 33 | F | Y | 2 | 1588 | 17 | M | N | 2 |
| 1488 | 90 | F | N | 2 | 1540 | 25 | M | Y | 2 | 1589 | 35 | M | Y | 2 |
| 1492 | 83 | F | N | 2 | 1541 | 22 | M | Y | 2 | 1590 | 38 | F | N | 2 |
| 1493 | 40 | F | N | 1 | 1542 | 20 | M | Y | 2 | 1591 | 71 | M | Y | 2 |
| 1494 | 63 | M | Y | 2 | 1543 | 15 | M | N | 2 | 1592 | 80 | M | N | 2 |
| 1495 | 58 | M | N | 2 | 1544 | 28 | F | Y | 2 | 1593 | 77 | F | N | 1 |
| 1496 | 40 | M | N | 2 | 1545 | 23 | F | Y | 2 | 1594 | 81 | M | Y | 2 |
| 1497 | 24 | M | N | 2 | 1546 | 19 | F | Y | 2 | 1595 | 71 | M | N | 2 |
| 1498 | 22 | M | N | 2 | 1547 | 54 | M | Y | 1 | 1596 | 68 | M | Y | 2 |
| 1499 | 39 | F | N | 2 | 1548 | 60 | M | Y | 2 | 1597 | 62 | M | Y | 1 |
| 1500 | 36 | F | N | 2 | 1549 | 85 | F | N | 1 | 1598 | 75 | F | N | 2 |
| 1501 | 81 | M | Y | 1 | 1550 | 62 | M | Y | 1 | 1599 | 78 | F | N | 2 |
| 1502 | 90 | M | N | 2 | 1551 | 62 | M | Y | 2 | 1600 | 53 | F | N | 2 |
| 1503 | 74 | F | N | 1 | 1552 | 49 | M | Y | 2 | 1601 | 55 | F | N | 2 |
| 1504 | 70 | M | Y | 1 | 1553 | 44 | F | Y | 2 | 1602 | 40 | M | N | 2 |
| 1505 | 63 | M | Y | 2 | 1554 | 42 | M | Y | 2 | 1603 | 37 | M | N | 2 |
| 1506 | 65 | M | Y | 2 | 1555 | 40 | M | Y | 2 | 1604 | 35 | M | Y | 2 |
| 1507 | 52 | M | Y | 2 | 1556 | 54 | F | N | 2 | 1605 | 33 | M | Y | 2 |
| 1508 | 46 | M | Y | 2 | 1557 | 51 | F | N | 1 | 1606 | 17 | M | N | 2 |
| 1509 | 56 | M | N | 2 | 1558 | 47 | F | Y | 2 | 1607 | 38 | F | N | 2 |
| 1510 | 55 | M | Y | 1 | 1559 | 37 | F | Y | 2 | 1608 | 72 | M | Y | 1 |
| 1511 | 55 | M | N | 2 | 1560 | 73 | M | Y | 1 | 1609 | 54 | M | N | 2 |
| 1512 | 72 | F | N | 1 | 1561 | 85 | M | Y | 2 | 1610 | 89 | F | N | 2 |
| 1513 | 69 | F | Y | 2 | 1562 | 56 | F | N | 2 | 1611 | 60 | M | Y | 2 |
| 1514 | 37 | F | Y | 1 | 1563 | 71 | M | Y | 2 | 1612 | 76 | F | Y | 2 |
| 1515 | 15 | M | N | 2 | 1564 | 70 | M | Y | 1 | 1613 | 72 | F | N | 2 |
| 1516 | 48 | M | Y | 2 | 1565 | 69 | M | Y | 2 | 1614 | 62 | F | Y | 1 |
| 1517 | 15 | M | N | 2 | 1566 | 16 | M | N | 2 | 1615 | 57 | F | Y | 2 |
| 1518 | 52 | M | Y | 1 | 1567 | 58 | M | Y | 2 | 1616 | 48 | M | Y | 2 |
| 1519 | 83 | M | N | 2 | 1568 | 78 | F | N | 2 | 1617 | 52 | F | N | 2 |
| 1520 | 83 | F | N | 2 | 1569 | 67 | M | N | 2 | 1618 | 50 | F | N | 2 |
| 1521 | 58 | M | Y | 2 | 1570 | 36 | F | N | 2 | 1619 | 75 | M | N | 2 |
| 1522 | 57 | M | N | 2 | 1571 | 80 | M | Y | 2 | 1620 | 27 | F | N | 1 |
| 1523 | 52 | M | Y | 2 | 1572 | 72 | M | N | 2 | 1621 | 63 | F | N | 2 |
| 1524 | 25 | M | Y | 2 | 1573 | 76 | F | Y | 2 | 1622 | 59 | F | N | 2 |
| 1525 | 22 | M | Y | 2 | 1574 | 65 | M | Y | 1 | 1623 | 48 | M | Y | 2 |
| 1526 | 21 | M | Y | 2 | 1575 | 52 | F | Y | 2 | 1624 | 52 | F | N | 2 |
| 1527 | 15 | M | N | 2 | 1576 | 56 | F | Y | 1 | 1625 | 56 | F | Y | 2 |
| 1528 | 28 | F | Y | 2 | 1577 | 58 | M | N | 2 | 1626 | 60 | F | Y | 1 |
| 1529 | 23 | F | Y | 2 | 1578 | 82 | F | Y | 2 | 1627 | 50 | M | N | 2 |
| 1530 | 19 | F | Y | 2 | 1579 | 71 | M | Y | 2 | 1628 | 85 | F | N | 2 |
| 1531 | 48 | F | Y | 2 | 1580 | 59 | M | Y | 2 | 1629 | 72 | M | Y | 1 |
| 1532 | 73 | M | N | 2 | 1581 | 58 | M | Y | 2 | 1630 | 63 | M | Y | 2 |
| 1533 | 45 | F | N | 1 | 1582 | 69 | F | N | 2 | 1631 | 76 | F | Y | 2 |
| 1534 | 53 | M | Y | 2 | 1583 | 62 | F | N | 2 | 1632 | 72 | F | N | 2 |

**Data Set I (Continued)**

| Number | Age | Sex | Smoker | Cancer | Number | Age | Sex | Smoker | Cancer | Number | Age | Sex | Smoker | Cancer |
|--------|-----|-----|--------|--------|--------|-----|-----|--------|--------|--------|-----|-----|--------|--------|
| 1633 | 57 | F | Y | 2 | 1682 | 67 | F | N | 2 | 1731 | 74 | F | Y | 2 |
| 1634 | 32 | M | Y | 1 | 1683 | 43 | M | Y | 2 | 1732 | 47 | M | Y | 2 |
| 1635 | 39 | M | Y | 2 | 1684 | 63 | F | N | 1 | 1733 | 43 | M | Y | 2 |
| 1636 | 35 | M | Y | 2 | 1685 | 83 | F | Y | 2 | 1734 | 36 | F | N | 2 |
| 1637 | 30 | F | N | 2 | 1686 | 60 | M | Y | 2 | 1735 | 54 | M | Y | 2 |
| 1638 | 67 | M | Y | 2 | 1687 | 58 | M | Y | 2 | 1736 | 46 | F | N | 2 |
| 1639 | 64 | M | Y | 2 | 1688 | 43 | M | Y | 2 | 1737 | 21 | M | Y | 2 |
| 1640 | 50 | F | N | 2 | 1689 | 76 | M | Y | 1 | 1738 | 25 | M | N | 2 |
| 1641 | 46 | M | Y | 2 | 1690 | 85 | M | Y | 2 | 1739 | 33 | F | Y | 2 |
| 1642 | 75 | F | N | 2 | 1691 | 63 | F | N | 2 | 1740 | 27 | F | Y | 2 |
| 1643 | 73 | F | Y | 2 | 1692 | 71 | M | Y | 2 | 1741 | 74 | M | Y | 1 |
| 1644 | 71 | F | Y | 2 | 1693 | 62 | M | Y | 2 | 1742 | 76 | M | Y | 2 |
| 1645 | 62 | F | Y | 2 | 1694 | 75 | F | N | 2 | 1743 | 69 | F | N | 2 |
| 1646 | 62 | F | Y | 2 | 1695 | 75 | M | Y | 2 | 1744 | 79 | M | N | 2 |
| 1647 | 43 | M | N | 2 | 1696 | 69 | F | N | 2 | 1745 | 77 | M | Y | 2 |
| 1648 | 32 | M | Y | 1 | 1697 | 64 | F | N | 2 | 1746 | 67 | M | Y | 2 |
| 1649 | 39 | M | Y | 2 | 1698 | 51 | M | Y | 2 | 1747 | 51 | M | Y | 2 |
| 1650 | 35 | M | Y | 2 | 1699 | 66 | M | Y | 2 | 1748 | 81 | F | N | 2 |
| 1651 | 30 | F | N | 2 | 1700 | 53 | F | N | 2 | 1749 | 69 | F | Y | 2 |
| 1652 | 84 | M | Y | 2 | 1701 | 49 | F | N | 2 | 1750 | 72 | M | Y | 1 |
| 1653 | 84 | M | Y | 2 | 1702 | 47 | F | N | 2 | 1751 | 73 | F | N | 1 |
| 1654 | 20 | M | Y | 2 | 1703 | 73 | F | N | 2 | 1752 | 68 | M | Y | 1 |
| 1655 | 82 | F | N | 2 | 1704 | 64 | M | Y | 2 | 1753 | 65 | M | Y | 2 |
| 1656 | 50 | F | N | 2 | 1705 | 95 | F | N | 2 | 1754 | 56 | M | Y | 1 |
| 1657 | 69 | M | Y | 1 | 1706 | 77 | M | Y | 2 | 1755 | 54 | M | Y | 2 |
| 1658 | 81 | M | N | 2 | 1707 | 58 | M | Y | 2 | 1756 | 57 | F | Y | 1 |
| 1659 | 57 | F | N | 2 | 1708 | 59 | M | Y | 2 | 1757 | 65 | F | Y | 2 |
| 1660 | 70 | M | Y | 2 | 1709 | 53 | M | Y | 2 | 1758 | 74 | M | N | 2 |
| 1661 | 68 | M | Y | 2 | 1710 | 53 | M | Y | 2 | 1759 | 75 | F | N | 2 |
| 1662 | 64 | M | N | 2 | 1711 | 72 | F | N | 2 | 1760 | 69 | M | N | 2 |
| 1663 | 66 | M | Y | 2 | 1712 | 70 | F | N | 2 | 1761 | 24 | M | Y | 2 |
| 1664 | 44 | M | N | 2 | 1713 | 64 | F | N | 2 | 1762 | 72 | F | N | 2 |
| 1665 | 39 | F | Y | 2 | 1714 | 60 | F | N | 2 | 1763 | 66 | F | N | 2 |
| 1666 | 69 | F | Y | 2 | 1715 | 51 | M | Y | 2 | 1764 | 45 | M | Y | 2 |
| 1667 | 24 | M | N | 2 | 1716 | 53 | F | N | 2 | 1765 | 40 | M | Y | 2 |
| 1668 | 79 | F | N | 1 | 1717 | 49 | F | N | 2 | 1766 | 38 | M | Y | 2 |
| 1669 | 68 | F | Y | 2 | 1718 | 47 | F | N | 2 | 1767 | 36 | M | Y | 2 |
| 1670 | 67 | F | Y | 2 | 1719 | 52 | M | Y | 1 | 1768 | 34 | M | Y | 2 |
| 1671 | 44 | M | N | 2 | 1720 | 66 | M | Y | 1 | 1769 | 32 | M | Y | 2 |
| 1672 | 39 | F | Y | 2 | 1721 | 74 | F | Y | 2 | 1770 | 30 | M | Y | 2 |
| 1673 | 66 | M | N | 1 | 1722 | 49 | M | Y | 2 | 1771 | 42 | F | N | 2 |
| 1674 | 67 | F | Smoker | 2 | 1723 | 54 | M | Y | 2 | 1772 | 71 | M | Y | 2 |
| 1675 | 77 | M | Y | 1 | 1724 | 51 | F | N | 2 | 1773 | 70 | F | Y | 2 |
| 1676 | 61 | M | Y | 1 | 1725 | 48 | F | Y | 2 | 1774 | 26 | M | N | 1 |
| 1677 | 65 | M | Y | 2 | 1726 | 38 | F | N | 2 | 1775 | 60 | F | N | 2 |
| 1678 | 73 | M | Y | 2 | 1727 | 25 | M | N | 2 | 1776 | 55 | F | N | 2 |
| 1679 | 45 | F | N | 2 | 1728 | 27 | F | Y | 2 | 1777 | 53 | F | N | 2 |
| 1680 | 65 | F | N | 2 | 1729 | 52 | F | Y | 2 | 1778 | 45 | M | Y | 2 |
| 1681 | 64 | F | N | 2 | 1730 | 72 | M | Y | 2 | 1779 | 40 | M | Y | 2 |

**Data Set I (Continued)**

| Number | Age | Sex | Smoker | Cancer | Number | Age | Sex | Smoker | Cancer | Number | Age | Sex | Smoker | Cancer |
|--------|-----|-----|--------|--------|--------|-----|-----|--------|--------|--------|-----|-----|--------|--------|
| 1780 | 38 | M | Y | 2 | 1829 | 51 | F | Y | 2 | 1878 | 56 | M | Y | 1 |
| 1781 | 36 | M | Y | 2 | 1830 | 49 | M | N | 2 | 1879 | 60 | M | Y | 2 |
| 1782 | 34 | M | Y | 2 | 1831 | 67 | F | N | 2 | 1880 | 69 | F | N | 2 |
| 1783 | 32 | M | Y | 2 | 1832 | 64 | F | N | 2 | 1881 | 36 | M | N | 2 |
| 1784 | 30 | M | Y | 2 | 1833 | 44 | F | N | 2 | 1882 | 31 | M | Y | 2 |
| 1785 | 42 | F | N | 2 | 1834 | 51 | F | Y | 2 | 1883 | 45 | F | N | 2 |
| 1786 | 70 | M | Y | 1 | 1835 | 62 | F | N | 2 | 1884 | 25 | F | N | 2 |
| 1787 | 80 | M | Y | 2 | 1836 | 60 | F | N | 2 | 1885 | 64 | F | N | 2 |
| 1788 | 71 | F | N | 2 | 1837 | 58 | F | N | 2 | 1886 | 76 | M | Y | 2 |
| 1789 | 74 | M | Y | 2 | 1838 | 56 | F | N | 2 | 1887 | 87 | F | N | 2 |
| 1790 | 57 | M | Y | 2 | 1839 | 54 | F | N | 2 | 1888 | 66 | M | N | 2 |
| 1791 | 17 | M | Y | 2 | 1840 | 46 | F | N | 2 | 1889 | 58 | M | Y | 1 |
| 1792 | 67 | M | Y | 2 | 1841 | 28 | M | Y | 2 | 1890 | 15 | M | N | 2 |
| 1793 | 65 | M | Y | 2 | 1842 | 24 | M | N | 2 | 1891 | 48 | M | Y | 2 |
| 1794 | 63 | F | N | 2 | 1843 | 21 | M | N | 2 | 1892 | 42 | M | Y | 2 |
| 1795 | 61 | F | N | 2 | 1844 | 62 | M | Y | 1 | 1893 | 33 | F | N | 2 |
| 1796 | 59 | F | N | 2 | 1845 | 40 | M | Y | 2 | 1894 | 58 | F | N | 2 |
| 1797 | 49 | M | Y | 2 | 1846 | 80 | F | Y | 2 | 1895 | 53 | F | N | 2 |
| 1798 | 41 | M | Y | 2 | 1847 | 29 | M | Y | 2 | 1896 | 36 | M | N | 2 |
| 1799 | 45 | F | Y | 2 | 1848 | 62 | F | N | 2 | 1897 | 31 | M | Y | 2 |
| 1800 | 34 | F | Y | 2 | 1849 | 60 | F | Y | 2 | 1898 | 45 | F | N | 2 |
| 1801 | 33 | F | Y | 2 | 1850 | 58 | F | N | 2 | 1899 | 25 | F | N | 2 |
| 1802 | 69 | F | N | 2 | 1851 | 26 | F | Y | 2 | 1900 | 67 | M | Y | 1 |
| 1803 | 82 | M | N | 2 | 1852 | 36 | M | Y | 2 | 1901 | 89 | M | N | 2 |
| 1804 | 77 | F | N | 2 | 1853 | 34 | M | Y | 2 | 1902 | 82 | F | N | 2 |
| 1805 | 70 | M | Y | 2 | 1854 | 32 | F | N | 2 | 1903 | 69 | M | Y | 1 |
| 1806 | 76 | F | N | 2 | 1855 | 65 | F | Y | 2 | 1904 | 67 | M | Y | 2 |
| 1807 | 74 | F | N | 2 | 1856 | 74 | M | Y | 2 | 1905 | 65 | F | N | 2 |
| 1808 | 62 | F | N | 2 | 1857 | 84 | F | N | 1 | 1906 | 63 | F | N | 2 |
| 1809 | 57 | F | N | 2 | 1858 | 62 | M | Y | 2 | 1907 | 30 | M | Y | 2 |
| 1810 | 49 | M | Y | 2 | 1859 | 49 | M | Y | 1 | 1908 | 23 | M | Y | 2 |
| 1811 | 41 | M | Y | 2 | 1860 | 60 | M | Y | 2 | 1909 | 74 | F | Y | 2 |
| 1812 | 45 | F | Y | 2 | 1861 | 72 | F | N | 2 | 1910 | 80 | M | N | 2 |
| 1813 | 34 | F | Y | 2 | 1862 | 71 | F | Y | 2 | 1911 | 80 | F | N | 2 |
| 1814 | 33 | F | Y | 2 | 1863 | 70 | F | N | 2 | 1912 | 16 | M | N | 2 |
| 1815 | 50 | M | Y | 1 | 1864 | 67 | F | Y | 2 | 1913 | 81 | M | Y | 2 |
| 1816 | 72 | M | N | 2 | 1865 | 36 | M | Y | 2 | 1914 | 18 | F | Y | 2 |
| 1817 | 62 | F | N | 2 | 1866 | 34 | M | Y | 2 | 1915 | 84 | F | N | 2 |
| 1818 | 54 | M | Y | 2 | 1867 | 32 | F | N | 2 | 1916 | 80 | F | N | 2 |
| 1819 | 52 | M | Y | 2 | 1868 | 77 | M | Y | 1 | 1917 | 73 | F | Y | 2 |
| 1820 | 41 | F | N | 2 | 1869 | 71 | M | Y | 1 | 1918 | 70 | F | Y | 1 |
| 1821 | 28 | M | Y | 2 | 1870 | 54 | M | Y | 1 | 1919 | 56 | F | N | 2 |
| 1822 | 24 | M | N | 2 | 1871 | 22 | M | N | 2 | 1920 | 53 | F | Y | 2 |
| 1823 | 21 | M | N | 2 | 1872 | 56 | F | N | 2 | 1921 | 51 | F | Y | 2 |
| 1824 | 53 | F | N | 2 | 1873 | 22 | M | N | 2 | 1922 | 46 | F | N | 2 |
| 1825 | 83 | M | N | 2 | 1874 | 64 | M | Y | 1 | 1923 | 58 | M | Y | 1 |
| 1826 | 68 | F | N | 2 | 1875 | 69 | M | Y | 1 | 1924 | 63 | M | Y | 2 |
| 1827 | 69 | M | Y | 2 | 1876 | 86 | F | N | 1 | 1925 | 53 | F | N | 2 |
| 1828 | 68 | M | Y | 2 | 1877 | 71 | M | Y | 2 | 1926 | 58 | M | N | 2 |

Data Set I (Continued)

| Number | Age | Sex | Smoker | Cancer | Number | Age | Sex | Smoker | Cancer | Number | Age | Sex | Smoker | Cancer |
|--------|-----|-----|--------|--------|--------|-----|-----|--------|--------|--------|-----|-----|--------|--------|
| 1927 | 54 | M | Y | 2 | 1976 | 87 | F | N | 1 | 2025 | 57 | M | N | 2 |
| 1928 | 56 | F | N | 2 | 1977 | 59 | M | Y | 1 | 2026 | 51 | F | N | 2 |
| 1929 | 53 | F | N | 2 | 1978 | 36 | M | N | 2 | 2027 | 76 | F | N | 2 |
| 1930 | 50 | F | N | 2 | 1979 | 73 | F | N | 2 | 2028 | 69 | M | Y | 2 |
| 1931 | 48 | F | Y | 2 | 1980 | 64 | M | Y | 2 | 2029 | 87 | F | N | 2 |
| 1932 | 35 | F | Y | 2 | 1981 | 75 | M | N | 2 | 2030 | 73 | M | Y | 2 |
| 1933 | 31 | F | N | 2 | 1982 | 50 | M | Y | 2 | 2031 | 70 | F | N | 1 |
| 1934 | 28 | F | Y | 2 | 1983 | 61 | M | Y | 1 | 2032 | 68 | F | N | 2 |
| 1935 | 65 | M | Y | 2 | 1984 | 68 | M | Y | 1 | 2033 | 66 | F | N | 2 |
| 1936 | 26 | M | N | 2 | 1985 | 66 | M | Y | 2 | 2034 | 63 | F | N | 2 |
| 1937 | 51 | F | N | 2 | 1986 | 64 | M | N | 2 | 2035 | 62 | M | Y | 1 |
| 1938 | 76 | M | Y | 2 | 1987 | 37 | M | Y | 2 | 2036 | 63 | F | Y | 2 |
| 1939 | 74 | F | N | 2 | 1988 | 34 | M | N | 2 | 2037 | 60 | M | Y | 1 |
| 1940 | 56 | M | Y | 2 | 1989 | 26 | F | Y | 2 | 2038 | 79 | M | Y | 2 |
| 1939 | 74 | F | N | 2 | 1990 | 24 | F | Y | 2 | 2039 | 99 | F | Y | 2 |
| 1942 | 53 | F | N | 2 | 1991 | 59 | F | Y | 2 | 2040 | 66 | M | Y | 2 |
| 1943 | 52 | F | N | 2 | 1992 | 67 | M | Y | 2 | 2041 | 22 | M | Y | 2 |
| 1944 | 31 | F | N | 2 | 1993 | 69 | F | N | 2 | 2042 | 69 | F | Y | 2 |
| 1945 | 28 | F | Y | 2 | 1994 | 46 | M | Y | 2 | 2043 | 57 | F | Y | 2 |
| 1946 | 64 | M | Y | 1 | 1995 | 64 | M | Y | 2 | 2044 | 54 | F | Y | 2 |
| 1947 | 67 | M | Y | 2 | 1996 | 51 | M | Y | 1 | 2045 | 32 | M | N | 2 |
| 1948 | 56 | M | N | 2 | 1997 | 71 | F | N | 2 | 2046 | 23 | M | N | 2 |
| 1949 | 63 | M | Y | 2 | 1998 | 68 | F | Y | 2 | 2047 | 36 | F | Y | 2 |
| 1950 | 59 | M | N | 2 | 1999 | 37 | M | Y | 2 | 2048 | 26 | F | Y | 2 |
| 1951 | 56 | M | N | 2 | 2000 | 34 | M | N | 2 | 2049 | 28 | F | N | 2 |
| 1952 | 51 | M | Y | 2 | 2001 | 26 | F | Y | 2 | 2050 | 25 | F | N | 2 |
| 1953 | 78 | F | N | 2 | 2002 | 24 | F | Y | 2 | 2051 | 56 | F | Y | 2 |
| 1954 | 52 | F | Y | 2 | 2003 | 72 | M | Y | 1 | 2052 | 70 | M | Y | 2 |
| 1955 | 53 | F | N | 2 | 2004 | 76 | F | N | 2 | 2053 | 74 | F | N | 2 |
| 1956 | 48 | F | Y | 2 | 2005 | 59 | M | Y | 1 | 2054 | 67 | M | Y | 2 |
| 1957 | 41 | M | Y | 2 | 2006 | 55 | M | Y | 1 | 2055 | 64 | M | Y | 2 |
| 1958 | 31 | M | Y | 2 | 2007 | 80 | F | N | 1 | 2056 | 62 | M | Y | 2 |
| 1959 | 23 | M | Y | 2 | 2008 | 75 | M | N | 2 | 2057 | 48 | M | N | 2 |
| 1960 | 34 | F | N | 2 | 2009 | 66 | M | N | 2 | 2058 | 55 | F | N | 2 |
| 1961 | 28 | F | N | 2 | 2010 | 59 | M | Y | 2 | 2059 | 50 | F | Y | 2 |
| 1962 | 62 | F | N | 1 | 2011 | 64 | F | N | 2 | 2060 | 43 | F | Y | 2 |
| 1963 | 67 | F | N | 2 | 2012 | 28 | F | Y | 2 | 2061 | 32 | M | N | 2 |
| 1964 | 93 | F | N | 2 | 2013 | 54 | F | N | 2 | 2062 | 23 | M | N | 2 |
| 1965 | 73 | M | N | 2 | 2014 | 49 | F | N | 2 | 2063 | 36 | F | Y | 2 |
| 1966 | 66 | M | N | 2 | 2015 | 23 | M | Y | 2 | 2064 | 26 | F | Y | 2 |
| 1967 | 78 | F | N | 2 | 2016 | 22 | M | Y | 2 | 2065 | 28 | F | N | 2 |
| 1968 | 16 | F | Y | 2 | 2017 | 42 | F | N | 2 | 2066 | 25 | F | N | 2 |
| 1969 | 56 | F | Y | 2 | 2018 | 23 | M | Y | 2 | 2067 | 72 | M | Y | 1 |
| 1970 | 41 | M | Y | 2 | 2019 | 22 | M | Y | 2 | 2068 | 58 | M | N | 1 |
| 1971 | 31 | M | Y | 2 | 2020 | 69 | M | Y | 1 | 2069 | 80 | F | N | 2 |
| 1972 | 23 | M | Y | 2 | 2021 | 65 | M | Y | 2 | 2070 | 68 | M | Y | 2 |
| 1973 | 34 | F | N | 2 | 2022 | 78 | F | N | 1 | 2071 | 45 | F | N | 2 |
| 1974 | 28 | F | N | 2 | 2023 | 44 | M | Y | 1 | 2072 | 68 | F | N | 2 |
| 1975 | 69 | M | Y | 2 | 2024 | 47 | M | Y | 1 | 2073 | 66 | F | N | 2 |

**Data Set I (Continued)**

| Number | Age | Sex | Smoker | Cancer | Number | Age | Sex | Smoker | Cancer | Number | Age | Sex | Smoker | Cancer |
|--------|-----|-----|--------|--------|--------|-----|-----|--------|--------|--------|-----|-----|--------|--------|
| 2074 | 64 | F | N | 2 | 2123 | 52 | M | Y | 1 | 2172 | 31 | M | N | 2 |
| 2075 | 62 | F | N | 2 | 2124 | 76 | F | N | 2 | 2173 | 56 | M | Y | 1 |
| 2076 | 60 | F | N | 2 | 2125 | 60 | F | Y | 2 | 2174 | 53 | M | Y | 1 |
| 2077 | 58 | F | N | 2 | 2126 | 65 | F | Y | 2 | 2175 | 74 | F | N | 1 |
| 2078 | 39 | M | Y | 2 | 2127 | 42 | F | N | 2 | 2176 | 49 | M | Y | 2 |
| 2079 | 37 | F | Y | 2 | 2128 | 77 | F | N | 2 | 2177 | 46 | M | Y | 2 |
| 2080 | 67 | F | N | 2 | 2129 | 33 | M | N | 2 | 2178 | 56 | F | N | 2 |
| 2081 | 56 | M | N | 2 | 2130 | 98 | F | N | 2 | 2179 | 54 | F | Y | 2 |
| 2082 | 77 | F | N | 2 | 2131 | 76 | M | N | 2 | 2180 | 18 | M | N | 2 |
| 2083 | 74 | M | Y | 1 | 2132 | 57 | M | N | 2 | 2181 | 16 | M | N | 2 |
| 2084 | 74 | M | Y | 2 | 2133 | 66 | M | N | 1 | 2182 | 28 | F | N | 2 |
| 2085 | 69 | M | N | 2 | 2134 | 81 | F | N | 2 | 2183 | 17 | F | Y | 1 |
| 2086 | 72 | M | N | 2 | 2135 | 42 | F | N | 2 | 2184 | 57 | F | N | 2 |
| 2087 | 65 | M | Y | 2 | 2136 | 79 | M | Y | 1 | 2185 | 59 | M | Y | 1 |
| 2088 | 58 | M | N | 2 | 2137 | 68 | F | N | 2 | 2186 | 82 | F | N | 2 |
| 2089 | 61 | F | N | 2 | 2138 | 76 | M | N | 2 | 2187 | 54 | M | N | 2 |
| 2090 | 39 | M | Y | 2 | 2139 | 69 | F | N | 2 | 2188 | 62 | F | N | 2 |
| 2091 | 37 | F | Y | 2 | 2140 | 56 | M | N | 2 | 2189 | 60 | F | N | 2 |
| 2092 | 84 | M | N | 1 | 2141 | 52 | M | N | 2 | 2190 | 18 | M | N | 2 |
| 2093 | 90 | M | N | 2 | 2142 | 79 | F | N | 2 | 2191 | 16 | M | N | 2 |
| 2094 | 64 | M | Y | 2 | 2143 | 81 | F | N | 2 | 2192 | 28 | F | N | 2 |
| 2095 | 74 | M | Y | 1 | 2144 | 88 | M | N | 2 | 2193 | 27 | F | Y | 1 |
| 2096 | 96 | F | N | 1 | 2145 | 84 | M | Y | 2 | 2194 | 70 | M | Y | 1 |
| 2097 | 91 | F | N | 1 | 2146 | 73 | M | Y | 2 | 2195 | 51 | F | N | 2 |
| 2098 | 75 | F | Y | 2 | 2147 | 80 | F | N | 2 | 2196 | 70 | M | Y | 1 |
| 2099 | 45 | M | Y | 1 | 2148 | 56 | M | N | 2 | 2197 | 70 | F | Y | 2 |
| 2100 | 42 | F | N | 2 | 2149 | 52 | M | N | 2 | 2198 | 62 | F | N | 2 |
| 2101 | 51 | F | N | 2 | 2150 | 85 | M | N | 2 | 2199 | 47 | M | N | 2 |
| 2102 | 49 | F | Y | 2 | 2151 | 65 | F | N | 2 | 2200 | 44 | M | Y | 2 |
| 2103 | 47 | F | N | 2 | 2152 | 78 | M | Y | 2 | 2201 | 50 | F | Y | 2 |
| 2104 | 33 | F | Y | 2 | 2153 | 81 | M | Y | 2 | 2202 | 74 | F | N | 2 |
| 2105 | 75 | F | N | 2 | 2154 | 92 | F | N | 2 | 2203 | 57 | M | Y | 2 |
| 2106 | 79 | M | Y | 2 | 2155 | 88 | F | N | 2 | 2204 | 73 | F | N | 2 |
| 2107 | 50 | M | Y | 2 | 2156 | 39 | M | N | 2 | 2205 | 58 | M | Y | 2 |
| 2108 | 45 | M | Y | 1 | 2157 | 69 | M | Y | 2 | 2206 | 53 | F | Y | 2 |
| 2109 | 42 | F | N | 2 | 2158 | 72 | F | N | 2 | 2207 | 67 | F | Y | 2 |
| 2110 | 51 | F | N | 2 | 2159 | 67 | F | N | 2 | 2208 | 63 | F | Y | 2 |
| 2111 | 49 | F | Y | 2 | 2160 | 70 | M | Y | 1 | 2209 | 47 | M | N | 2 |
| 2112 | 47 | F | Y | 2 | 2161 | 69 | F | N | 2 | 2210 | 44 | M | Y | 2 |
| 2113 | 33 | F | Y | 2 | 2162 | 59 | F | Y | 1 | 2211 | 50 | F | Y | 2 |
| 2114 | 60 | M | Y | 1 | 2163 | 55 | M | Y | 2 | 2212 | 59 | M | Y | 1 |
| 2115 | 49 | F | N | 2 | 2164 | 61 | F | N | 1 | 2213 | 13 | M | N | 2 |
| 2116 | 66 | M | Y | 1 | 2165 | 63 | F | N | 2 | 2214 | 53 | F | Y | 2 |
| 2117 | 74 | M | N | 1 | 2166 | 58 | F | Y | 2 | 2215 | 64 | M | Y | 1 |
| 2118 | 79 | F | N | 2 | 2167 | 31 | M | N | 2 | 2216 | 46 | M | Y | 2 |
| 2119 | 69 | M | Y | 2 | 2168 | 78 | M | Y | 2 | 2217 | 54 | F | N | 2 |
| 2120 | 70 | M | N | 2 | 2169 | 64 | M | Y | 2 | 2218 | 72 | M | N | 2 |
| 2121 | 53 | M | Y | 2 | 2170 | 60 | M | Y | 2 | 2219 | 56 | M | Y | 1 |
| 2122 | 71 | M | N | 2 | 2171 | 60 | F | Y | 1 | 2220 | 47 | M | N | 2 |

Data Set I (Continued)

| Number | Age | Sex | Smoker | Cancer | Number | Age | Sex | Smoker | Cancer | Number | Age | Sex | Smoker | Cancer |
|--------|-----|-----|--------|--------|--------|-----|-----|--------|--------|--------|-----|-----|--------|--------|
| 2221 | 42 | M | Y | 2 | 2270 | 41 | M | Y | 2 | 2319 | 53 | M | Y | 2 |
| 2222 | 41 | F | Y | 2 | 2271 | 21 | M | Y | 2 | 2320 | 89 | M | Y | 1 |
| 2223 | 39 | F | N | 2 | 2272 | 32 | M | Y | 2 | 2321 | 57 | F | N | 2 |
| 2224 | 35 | F | N | 2 | 2273 | 68 | F | N | 2 | 2322 | 62 | M | Y | 1 |
| 2225 | 32 | F | Y | 2 | 2274 | 78 | M | N | 2 | 2323 | 82 | M | N | 2 |
| 2226 | 65 | M | N | 2 | 2275 | 83 | F | N | 2 | 2324 | 74 | F | N | 2 |
| 2227 | 68 | M | Y | 2 | 2276 | 68 | M | N | 2 | 2325 | 69 | M | Y | 1 |
| 2228 | 75 | F | N | 2 | 2277 | 72 | M | Y | 2 | 2326 | 70 | M | Y | 2 |
| 2229 | 62 | M | Y | 2 | 2278 | 67 | M | Y | 2 | 2327 | 67 | M | Y | 2 |
| 2230 | 48 | M | Y | 2 | 2279 | 65 | M | N | 2 | 2328 | 69 | F | N | 2 |
| 2231 | 53 | M | Y | 2 | 2280 | 60 | M | Y | 2 | 2329 | 75 | F | N | 2 |
| 2232 | 59 | M | Y | 2 | 2281 | 55 | M | N | 2 | 2330 | 73 | F | Y | 2 |
| 2233 | 66 | M | N | 2 | 2282 | 74 | F | N | 2 | 2331 | 39 | M | Y | 2 |
| 2234 | 81 | F | N | 2 | 2283 | 62 | F | N | 2 | 2332 | 38 | M | Y | 2 |
| 2235 | 73 | F | Y | 1 | 2284 | 57 | F | Y | 1 | 2333 | 36 | M | Y | 2 |
| 2236 | 66 | F | Y | 2 | 2285 | 23 | F | Y | 2 | 2334 | 24 | M | N | 2 |
| 2237 | 63 | F | Y | 2 | 2286 | 43 | F | Y | 2 | 2335 | 44 | F | Y | 2 |
| 2238 | 47 | M | N | 2 | 2287 | 41 | F | Y | 2 | 2336 | 42 | F | Y | 2 |
| 2239 | 42 | M | Y | 2 | 2288 | 21 | M | Y | 2 | 2337 | 40 | F | Y | 2 |
| 2240 | 41 | F | Y | 2 | 2289 | 32 | M | Y | 2 | 2338 | 33 | F | N | 2 |
| 2241 | 39 | F | N | 2 | 2290 | 60 | M | Y | 1 | 2339 | 66 | F | Y | 2 |
| 2242 | 35 | F | N | 2 | 2291 | 80 | M | Y | 2 | 2340 | 63 | M | Y | 2 |
| 2243 | 32 | F | Y | 2 | 2292 | 59 | F | N | 2 | 2341 | 44 | F | N | 2 |
| 2244 | 59 | M | Y | 1 | 2293 | 63 | M | Y | 2 | 2342 | 42 | M | Y | 2 |
| 2245 | 25 | F | N | 2 | 2294 | 48 | M | N | 1 | 2343 | 61 | M | Y | 2 |
| 2246 | 69 | F | N | 2 | 2295 | 51 | M | N | 2 | 2344 | 60 | F | Y | 2 |
| 2247 | 59 | F | Y | 2 | 2296 | 45 | M | N | 2 | 2345 | 56 | F | Y | 2 |
| 2248 | 63 | F | N | 2 | 2297 | 62 | F | N | 2 | 2346 | 39 | M | Y | 2 |
| 2249 | 39 | F | N | 2 | 2298 | 41 | F | N | 2 | 2347 | 38 | M | Y | 2 |
| 2250 | 37 | F | Y | 2 | 2299 | 57 | F | N | 2 | 2348 | 36 | M | Y | 2 |
| 2251 | 60 | F | Y | 2 | 2300 | 84 | M | N | 2 | 2349 | 24 | M | N | 2 |
| 2252 | 42 | F | N | 2 | 2301 | 75 | F | N | 2 | 2350 | 44 | F | Y | 2 |
| 2253 | 58 | F | Y | 2 | 2302 | 66 | F | N | 2 | 2351 | 42 | F | Y | 2 |
| 2254 | 39 | F | N | 2 | 2303 | 50 | M | N | 2 | 2352 | 40 | F | Y | 2 |
| 2255 | 37 | F | Y | 2 | 2304 | 62 | F | Y | 1 | 2353 | 33 | F | N | 2 |
| 2256 | 68 | M | Y | 2 | 2305 | 81 | M | Y | 2 | 2354 | 56 | F | Y | 1 |
| 2257 | 66 | M | Y | 2 | 2306 | 83 | F | N | 2 | 2355 | 42 | M | Y | 2 |
| 2258 | 62 | M | Y | 2 | 2307 | 63 | M | Y | 2 | 2356 | 89 | F | N | 2 |
| 2259 | 55 | M | Y | 2 | 2308 | 60 | F | Y | 2 | 2357 | 59 | M | Y | 1 |
| 2260 | 34 | F | N | 1 | 2309 | 43 | M | Y | 2 | 2358 | 61 | M | Y | 2 |
| 2261 | 64 | M | Y | 1 | 2310 | 33 | M | N | 2 | 2359 | 57 | F | Y | 2 |
| 2262 | 64 | M | N | 2 | 2311 | 41 | F | Y | 2 | 2360 | 37 | F | N | 2 |
| 2263 | 75 | F | N | 1 | 2312 | 39 | F | N | 2 | 2361 | 23 | F | N | 2 |
| 2264 | 64 | M | Y | 2 | 2313 | 66 | M | Y | 2 | 2362 | 60 | M | N | 1 |
| 2265 | 62 | M | N | 2 | 2314 | 43 | M | Y | 2 | 2363 | 86 | M | Y | 2 |
| 2266 | 60 | M | Y | 2 | 2315 | 33 | M | N | 2 | 2364 | 62 | F | N | 1 |
| 2267 | 55 | M | Y | 2 | 2316 | 41 | F | Y | 2 | 2365 | 60 | F | Y | 2 |
| 2268 | 52 | F | N | 2 | 2317 | 39 | F | N | 2 | 2366 | 38 | M | N | 2 |
| 2269 | 43 | M | Y | 2 | 2318 | 48 | F | Y | 1 | 2367 | 40 | F | Y | 2 |

Data Set I (Continued)

| Number | Age | Sex | Smoker | Cancer | Number | Age | Sex | Smoker | Cancer | Number | Age | Sex | Smoker | Cancer |
|--------|-----|-----|--------|--------|--------|-----|-----|--------|--------|--------|-----|-----|--------|--------|
| 2368 | 62 | F | N | 2 | 2417 | 68 | F | Y | 2 | 2466 | 88 | F | N | 2 |
| 2369 | 77 | M | Y | 1 | 2418 | 64 | F | Y | 2 | 2467 | 35 | M | Y | 1 |
| 2370 | 83 | F | Y | 2 | 2419 | 56 | F | N | 2 | 2468 | 69 | M | Y | 2 |
| 2371 | 63 | M | Y | 1 | 2420 | 12 | F | N | 2 | 2469 | 67 | M | N | 2 |
| 2372 | 16 | M | N | 2 | 2421 | 48 | F | N | 2 | 2470 | 52 | M | Y | 2 |
| 2373 | 38 | M | N | 2 | 2422 | 45 | M | Y | 2 | 2471 | 17 | M | N | 2 |
| 2374 | 40 | F | Y | 2 | 2423 | 43 | M | Y | 2 | 2472 | 54 | M | Y | 2 |
| 2375 | 56 | F | Y | 1 | 2424 | 44 | F | N | 2 | 2473 | 61 | F | N | 2 |
| 2376 | 58 | M | Y | 1 | 2425 | 69 | M | Y | 2 | 2474 | 57 | F | N | 2 |
| 2377 | 57 | M | Y | 1 | 2426 | 50 | M | Y | 2 | 2475 | 23 | M | Y | 2 |
| 2378 | 57 | F | N | 2 | 2427 | 66 | M | N | 2 | 2476 | 33 | F | N | 2 |
| 2379 | 57 | M | Y | 1 | 2428 | 56 | M | Y | 2 | 2477 | 69 | F | N | 2 |
| 2380 | 34 | F | N | 2 | 2429 | 70 | F | N | 2 | 2478 | 62 | M | Y | 2 |
| 2381 | 60 | F | Y | 2 | 2430 | 68 | F | N | 2 | 2479 | 82 | F | N | 2 |
| 2382 | 47 | M | Y | 2 | 2431 | 64 | F | N | 2 | 2480 | 64 | M | Y | 2 |
| 2383 | 40 | F | N | 2 | 2432 | 62 | F | N | 2 | 2481 | 63 | M | Y | 1 |
| 2384 | 17 | M | N | 2 | 2433 | 60 | F | N | 2 | 2482 | 62 | M | Y | 2 |
| 2385 | 65 | M | Y | 2 | 2434 | 58 | F | N | 2 | 2483 | 55 | M | N | 2 |
| 2386 | 59 | M | N | 2 | 2435 | 31 | F | N | 2 | 2484 | 52 | M | Y | 2 |
| 2387 | 57 | M | N | 2 | 2436 | 28 | F | Y | 2 | 2485 | 28 | M | Y | 2 |
| 2388 | 74 | F | N | 2 | 2437 | 63 | F | N | 1 | 2486 | 18 | M | N | 2 |
| 2389 | 72 | F | N | 2 | 2438 | 68 | M | Y | 2 | 2487 | 70 | F | N | 2 |
| 2390 | 72 | F | N | 2 | 2439 | 69 | F | N | 2 | 2488 | 59 | F | N | 2 |
| 2391 | 24 | F | N | 1 | 2440 | 38 | M | N | 2 | 2489 | 23 | M | Y | 2 |
| 2392 | 70 | F | N | 2 | 2441 | 57 | M | Y | 2 | 2490 | 33 | F | N | 2 |
| 2393 | 56 | F | Y | 2 | 2442 | 49 | M | N | 2 | 2491 | 62 | M | Y | 1 |
| 2394 | 34 | F | N | 2 | 2443 | 51 | M | N | 2 | 2492 | 57 | M | Y | 1 |
| 2395 | 65 | M | Y | 1 | 2444 | 54 | F | Y | 2 | 2493 | 84 | M | Y | 1 |
| 2396 | 71 | M | N | 2 | 2445 | 31 | F | N | 2 | 2494 | 61 | F | N | 2 |
| 2397 | 60 | F | N | 2 | 2446 | 28 | F | Y | 2 | 2495 | 57 | M | Y | 2 |
| 2398 | 79 | M | Y | 2 | 2447 | 69 | M | Y | 1 | 2496 | 47 | M | Y | 2 |
| 2399 | 75 | M | N | 2 | 2448 | 70 | M | Y | 2 | 2497 | 59 | F | N | 2 |
| 2400 | 69 | M | Y | 2 | 2449 | 83 | F | N | 2 | 2498 | 54 | F | N | 2 |
| 2401 | 86 | F | N | 2 | 2450 | 68 | F | N | 2 | 2499 | 49 | F | N | 2 |
| 2402 | 80 | F | N | 2 | 2451 | 53 | M | Y | 2 | 2500 | 73 | M | Y | 1 |
| 2403 | 72 | F | N | 1 | 2452 | 42 | M | Y | 2 | 2501 | 94 | M | Y | 1 |
| 2404 | 60 | F | N | 2 | 2453 | 38 | F | N | 2 | 2502 | 67 | F | N | 2 |
| 2405 | 77 | F | N | 2 | 2454 | 64 | F | N | 2 | 2503 | 67 | M | Y | 2 |
| 2406 | 64 | F | Y | 2 | 2455 | 92 | M | N | 2 | 2504 | 62 | M | Y | 2 |
| 2407 | 45 | M | Y | 2 | 2456 | 72 | F | N | 2 | 2505 | 61 | M | Y | 2 |
| 2408 | 43 | M | Y | 2 | 2457 | 82 | M | N | 2 | 2506 | 82 | F | N | 1 |
| 2409 | 44 | F | N | 2 | 2458 | 79 | F | N | 2 | 2507 | 69 | F | N | 2 |
| 2410 | 62 | F | N | 2 | 2459 | 76 | F | N | 2 | 2508 | 67 | F | Y | 2 |
| 2411 | 77 | M | Y | 2 | 2460 | 75 | F | N | 2 | 2509 | 65 | M | N | 2 |
| 2412 | 66 | F | N | 2 | 2461 | 61 | F | N | 2 | 2510 | 63 | F | N | 2 |
| 2413 | 72 | M | Y | 2 | 2462 | 42 | M | Y | 2 | 2511 | 59 | F | N | 2 |
| 2414 | 59 | M | Y | 2 | 2463 | 38 | F | N | 2 | 2512 | 78 | F | Y | 2 |
| 2415 | 69 | F | N | 2 | 2464 | 69 | M | Y | 1 | 2513 | 42 | M | Y | 2 |
| 2416 | 70 | F | N | 2 | 2465 | 80 | M | Y | 2 | 2514 | 78 | F | N | 2 |

## Data Set I (Continued)

| Number | Age | Sex | Smoker | Cancer | Number | Age | Sex | Smoker | Cancer | Number | Age | Sex | Smoker | Cancer |
|--------|-----|-----|--------|--------|--------|-----|-----|--------|--------|--------|-----|-----|--------|--------|
| 2515 | 31 | M | Y | 2 | 2523 | 84 | F | N | 2 | 2531 | 82 | M | Y | 2 |
| 2516 | 68 | M | Y | 2 | 2524 | 75 | M | Y | 2 | 2532 | 63 | F | N | 1 |
| 2517 | 69 | F | Y | 2 | 2525 | 68 | M | N | 2 | 2533 | 62 | M | Y | 2 |
| 2518 | 43 | F | Y | 2 | 2526 | 28 | M | N | 2 | 2534 | 63 | M | Y | 2 |
| 2519 | 60 | F | Y | 2 | 2527 | 78 | F | Y | 2 | 2535 | 62 | F | N | 1 |
| 2520 | 32 | F | Y | 1 | 2528 | 69 | F | N | 2 | 2536 | 76 | F | Y | 2 |
| 2521 | 60 | F | Y | 1 | 2529 | 36 | F | Y | 2 | 2537 | 65 | F | N | 1 |
| 2522 | 68 | M | N | 2 | 2530 | 70 | M | Y | 2 | 2538 | 61 | F | N | 2 |

## Parameters for Data Set II on New York Stock Exchange

$N = 1,925$ data points

**52 weeks High Data**

$\mu = \$36.48$
$\sigma = \$29.90$

**52 weeks Low Data**

$\mu = \$21.47$
$\sigma = \$21.88$

**Closing Price Data**

$\mu = \$28.05$
$\sigma = \$24.29$

**Net Change Data**

901 increased $(+)$    $P_i = 0.468$
533 decreased $(-)$    $P_d = 0.277$
491 unchanged $(0)$    $P_\mu = 0.255$

*How to read the stock prices*

The stock prices are written as mixed numbers, when (if necessary) the cents part of the price is in one-eighths of a dollar. To illustrate:

| Stock Number 0001 | High | 277/8 | means | $27\frac{7}{8}$ |
|---|---|---|---|---|
| | Low | 14 | means | $14.00 |
| | Close | 251/4 | means | $25\frac{1}{4}$ |

To enter prices in calculator, the following decimal conversions can be used:

$\frac{1}{8} = 0.125$   $\frac{1}{4} = 0.25$   $\frac{3}{8} = 0.375$   $\frac{1}{2} = 0.5$
$\frac{5}{8} = 0.625$   $\frac{3}{4} = 0.75$   $\frac{7}{8} = 0.875$

## Data Set II

| Stock Number | 52 weeks High | 52 weeks Low | Close | Net Chg | Stock Number | 52 weeks High | 52 weeks Low | Close | Net Chg |
|--------------|------|-----|-------|---------|--------------|------|-----|-------|---------|
| 0001 | 277/8 | 14 | 251/4 | + | 0015 | 101/2 | 65/8 | 71/8 | 0 |
| 0002 | 121/8 | 83/4 | 111/2 | − | 0016 | 20 | 141/8 | 16 | + |
| 0003 | 121/8 | 103/4 | 111/4 | + | 0017 | 191/4 | 67/8 | 161/4 | 0 |
| 0004 | 101/8 | 9 | 93/8 | + | 0018 | 247/8 | 71/2 | 13 | + |
| 0005 | 273/4 | 101/2 | 267/8 | 0 | 0019 | 563/4 | 291/4 | 381/2 | − |
| 0006 | 77/8 | 25/8 | 4 | 0 | 0020 | 117/8 | 45/8 | 73/8 | 0 |
| 0007 | 85/8 | 31/4 | 53/4 | + | 0021 | 201/4 | 161/4 | 173/4 | 0 |
| 0008 | 293/8 | 17 | 223/8 | + | 0022 | 213/4 | 171/2 | 21 | − |
| 0009 | 651/2 | 263/4 | 441/2 | + | 0023 | 13 | 57/8 | 83/8 | 0 |
| 0010 | 111/8 | 53/4 | 81/4 | − | 0024 | 641/4 | 391/2 | 445/8 | + |
| 0011 | 727/8 | 40 | 405/8 | + | 0025 | 413/4 | 201/2 | 267/8 | − |
| 0012 | 223/8 | 91/2 | 193/8 | 0 | 0026 | 233/4 | 13 | 155/8 | + |
| 0013 | 663/8 | 40 | 441/4 | + | 0027 | 4 | 13/4 | 31/2 | + |
| 0014 | 163/4 | 81/2 | 105/8 | + | 0028 | 537/8 | 29 | 453/8 | + |

**Data Set II (Continued)**

| Stock Number | 52 weeks High | 52 weeks Low | Close | Net Chg | Stock Number | 52 weeks High | 52 weeks Low | Close | Net Chg |
|---|---|---|---|---|---|---|---|---|---|
| 0029 | 291/8 | 111/8 | 163/4 | + | 0076 | 421/2 | 271/4 | 295/8 | + |
| 0030 | 161/2 | 63/4 | 155/8 | + | 0077 | 183/4 | 43/4 | 53/4 | + |
| 0031 | 20 | 131/2 | 191/4 | + | 0078 | 81/2 | 8 | 81/8 | − |
| 0032 | 93/4 | 75/8 | 91/4 | 0 | 0079 | 181/2 | 12 | 181/2 | 0 |
| 0033 | 98 | 811/2 | 901/8 | + | 0080 | 32 | 23 | 251/2 | + |
| 0034 | 101 | 861/2 | 94 | 0 | 0081 | 125/8 | 43/4 | 111/2 | + |
| 0035 | 237/8 | 121/4 | 177/8 | + | 0082 | 241/4 | 171/4 | 23 | − |
| 0036 | 38 | 143/4 | 36 | + | 0083 | 963/4 | 62 | 733/8 | + |
| 0037 | 281/4 | 125/8 | 261/8 | + | 0084 | 997/8 | 74 | 90 | + |
| 0038 | 34 | 201/4 | 303/4 | − | 0085 | 833/4 | 49 | 591/4 | + |
| 0039 | 377/8 | 18 | 301/4 | + | 0086 | 201/2 | 105/8 | 171/8 | 0 |
| 0040 | 295/8 | 151/4 | 243/4 | + | 0087 | 495/8 | 215/8 | 305/8 | + |
| 0041 | 281/4 | 157/8 | 233/4 | + | 0088 | 871/4 | 46 | 563/8 | + |
| 0042 | 59 | 353/4 | 51 | 0 | 0089 | 171/2 | 121/2 | 161/8 | + |
| 0043 | 921/2 | 611/2 | 703/8 | 0 | 0090 | 6 | 31/4 | 43/4 | 0 |
| 0044 | 19 | 21/8 | 33/8 | + | 0091 | 201/4 | 101/8 | 145/8 | 0 |
| 0045 | 81 | 17 | 193/4 | − | 0092 | 223/4 | 163/8 | 18 | − |
| 0046 | 34 | 151/8 | 305/8 | 0 | 0093 | 77/8 | 31/4 | 43/4 | − |
| 0047 | 411/2 | 313/8 | 37 | + | 0094 | 861/4 | 411/2 | 49 | + |
| 0048 | 175/8 | 51/2 | 12 | + | 0095 | 931/2 | 51 | 57 | + |
| 0049 | 22 | 91/4 | 133/4 | − | 0096 | 357/8 | 23 | 261/4 | + |
| 0050 | 113/8 | 10 | 103/8 | − | 0097 | 201/4 | 137/8 | 161/8 | − |
| 0051 | 285/8 | 123/8 | 191/2 | − | 0098 | 141/8 | 81/4 | 14 | 0 |
| 0052 | 483/4 | 26 | 351/4 | + | 0099 | 82 | 751/8 | 751/4 | − |
| 0053 | 23/8 | 5/8 | 3/4 | − | 0100 | 545/8 | 29 | 41 | + |
| 0054 | 19 | 31/8 | 31/4 | − | 0101 | 1341/2 | 116 | 133 | − |
| 0055 | 103/8 | 87/8 | 10 | − | 0102 | 401/2 | 243/4 | 365/8 | − |
| 0056 | 101/4 | 91/4 | 91/2 | 0 | 0103 | 293/4 | 71/2 | 161/2 | + |
| 0057 | 353/4 | 23 | 333/4 | + | 0104 | 195/8 | 12 | 141/2 | 0 |
| 0058 | 643/4 | 333/4 | 511/8 | + | 0105 | 111/2 | 91/4 | 105/8 | 0 |
| 0059 | 211/4 | 91/4 | 153/4 | − | 0106 | 495/8 | 24 | 467/8 | + |
| 0060 | 291/4 | 121/2 | 227/8 | + | 0107 | 873/8 | 57 | 761/2 | + |
| 0061 | 171/2 | 71/2 | 137/8 | 0 | 0108 | 711/2 | 341/8 | 45 | + |
| 0062 | 415/8 | 211/2 | 281/8 | − | 0109 | 19 | 111/8 | 131/8 | + |
| 0063 | 303/8 | 123/8 | 185/8 | − | 0110 | 221/2 | 63/4 | 207/8 | − |
| 0064 | 60 | 361/2 | 471/8 | + | 0111 | 167/8 | 63/8 | 71/2 | + |
| 0065 | 333/8 | 273/4 | 305/8 | + | 0112 | 325/8 | 201/2 | 24 | 0 |
| 0066 | 253/4 | 15 | 251/8 | − | 0113 | 111/2 | 33/4 | 9 | + |
| 0067 | 301/4 | 151/2 | 26 | 0 | 0114 | 347/8 | 191/4 | 235/8 | + |
| 0068 | 225/8 | 177/8 | 201/2 | 0 | 0115 | 233/4 | 83/8 | 135/8 | − |
| 0069 | 313/4 | 20 | 227/8 | − | 0116 | 293/4 | 113/4 | 151/4 | 0 |
| 0070 | 101/8 | 97/8 | 97/8 | − | 0117 | 343/8 | 181/2 | 255/8 | − |
| 0071 | 19 | 75/8 | 97/8 | − | 0118 | 131/4 | 95/8 | 10 | − |
| 0072 | 57 | 29 | 501/8 | + | 0119 | 401/8 | 253/4 | 295/8 | − |
| 0073 | 293/4 | 231/8 | 28 | + | 0120 | 173/8 | 63/4 | 121/8 | − |
| 0074 | 397/8 | 203/4 | 271/8 | + | 0121 | 171/8 | 73/4 | 161/2 | − |
| 0075 | 181/2 | 97/8 | 127/8 | − | 0122 | 297/8 | 201/2 | 271/8 | + |

Data Set II (Continued)

| Stock Number | 52 weeks | | | Net Chg | Stock Number | 52 weeks | | | Net Chg |
|---|---|---|---|---|---|---|---|---|---|
| | High | Low | Close | | | High | Low | Close | |
| 0123 | 121/2 | 65/8 | 77/8 | 0 | 0170 | 28 | 17 | 257/8 | + |
| 0124 | 83/8 | 25/8 | 23/4 | 0 | 0171 | 467/8 | 255/8 | 285/8 | + |
| 0125 | 273/4 | 261/8 | 261/2 | + | 0172 | 273/8 | 101/2 | 211/2 | + |
| 0126 | 36 | 171/4 | 313/4 | + | 0173 | 193/8 | 93/4 | 133/4 | 0 |
| 0127 | 207/8 | 81/2 | 145/8 | + | 0174 | 34 | 19 | 311/4 | + |
| 0128 | 273/4 | 171/2 | 193/4 | − | 0175 | 60 | 50 | 501/2 | − |
| 0129 | 383/4 | 17 | 331/4 | − | 0176 | 275/8 | 161/8 | 253/8 | + |
| 0130 | 39 | 163/4 | 301/4 | 0 | 0177 | 285/8 | 161/2 | 223/4 | + |
| 0131 | 96 | 79 | 80 | − | 0178 | 64 | 41 | 553/4 | 0 |
| 0132 | 261/2 | 155/8 | 193/8 | + | 0179 | 41/2 | 11/8 | 13/8 | 0 |
| 0133 | 531/2 | 341/2 | 40 | + | 0180 | 683/4 | 42 | 611/4 | + |
| 0134 | 143/4 | 71/8 | 111/8 | + | 0181 | 357/8 | 177/8 | 271/2 | 0 |
| 0135 | 251/2 | 181/4 | 24 | + | 0182 | 523/4 | 431/2 | 47 | + |
| 0136 | 46 | 373/4 | 433/4 | + | 0183 | 527/8 | 371/2 | 443/8 | − |
| 0137 | 473/8 | 221/2 | 363/8 | + | 0184 | 371/4 | 201/4 | 271/2 | + |
| 0138 | 39 | 13 | 381/2 | + | 0185 | 457/8 | 241/2 | 341/8 | + |
| 0139 | 121/8 | 47/8 | 10 | 0 | 0186 | 1011/4 | 100 | 1013/4 | + |
| 0140 | 211/2 | 12 | 181/2 | + | 0187 | 145/8 | 65/8 | 135/8 | + |
| 0141 | 395/8 | 11 | 311/2 | + | 0188 | 367/8 | 241/2 | 361/4 | − |
| 0142 | 37 | 143/4 | 223/8 | 0 | 0189 | 611/2 | 42 | 601/2 | − |
| 0143 | 341/4 | 15 | 255/8 | + | 0190 | 91/8 | 61/8 | 7 | + |
| 0144 | 761/4 | 461/2 | 733/8 | + | 0191 | 541/4 | 261/4 | 353/4 | + |
| 0145 | 12 | 35/8 | 7 | 0 | 0192 | 273/4 | 133/8 | 245/8 | − |
| 0146 | 251/8 | 97/8 | 221/4 | − | 0193 | 363/4 | 253/8 | 281/2 | − |
| 0147 | 28 | 193/8 | 267/8 | + | 0194 | 495/8 | 25 | 44 | + |
| 0148 | 361/2 | 283/4 | 323/4 | + | 0195 | 401/4 | 261/4 | 331/8 | − |
| 0149 | 991/8 | 583/4 | 81 | + | 0196 | 413/4 | 271/8 | 347/8 | + |
| 0150 | 53 | 231/2 | 361/4 | 0 | 0197 | 93/4 | 45/8 | 6 | 0 |
| 0151 | 8 | 3 | 45/8 | 0 | 0198 | 61/2 | 1/2 | 5/8 | − |
| 0152 | 28 | 105/8 | 131/8 | − | 0199 | 273/8 | 123/4 | 161/8 | − |
| 0153 | 367/8 | 10 | 34 | − | 0200 | 48 | 303/4 | 44 | − |
| 0154 | 541/2 | 321/2 | 383/4 | + | 0201 | 291/4 | 151/2 | 197/8 | + |
| 0155 | 8 | 41/8 | 51/8 | 0 | 0202 | 507/8 | 421/2 | 435/8 | − |
| 0156 | 283/4 | 145/8 | 25 | − | 0203 | 93 | 61 | 66 | + |
| 0157 | 291/4 | 151/8 | 241/8 | + | 0204 | 233/4 | 101/4 | 161/2 | 0 |
| 0158 | 391/4 | 181/2 | 23 | + | 0205 | 271/8 | 193/4 | 237/8 | − |
| 0159 | 385/8 | 191/4 | 241/2 | + | 0206 | 211/2 | 8 | 12 | − |
| 0160 | 261/8 | 241/8 | 241/4 | + | 0207 | 417/8 | 201/8 | 39 | − |
| 0161 | 351/4 | 16 | 203/8 | 0 | 0208 | 69 | 421/4 | 501/4 | + |
| 0162 | 32 | 231/2 | 303/8 | + | 0209 | 11/8 | 3/8 | 5/8 | − |
| 0163 | 201/4 | 133/8 | 157/8 | + | 0210 | 27/8 | 1/4 | 13/4 | − |
| 0164 | 95/8 | 31/4 | 73/4 | − | 0211 | 397/8 | 181/4 | 343/4 | 0 |
| 0165 | 317/8 | 231/2 | 307/8 | 0 | 0212 | 793/4 | 601/2 | 687/8 | + |
| 0166 | 195/8 | 141/2 | 185/8 | 0 | 0213 | 233/8 | 111/4 | 153/8 | − |
| 0167 | 421/4 | 20 | 361/2 | + | 0214 | 437/8 | 281/8 | 41 | + |
| 0168 | 273/8 | 111/8 | 15 | − | 0215 | 385/8 | 211/8 | 281/2 | + |
| 0169 | 661/8 | 385/8 | 455/8 | + | 0216 | 251/2 | 123/4 | 231/8 | + |

Data Set II (Continued)

| Stock Number | 52 weeks High | 52 weeks Low | Close | Net Chg | Stock Number | 52 weeks High | 52 weeks Low | Close | Net Chg |
|---|---|---|---|---|---|---|---|---|---|
| 0217 | 623/4 | 281/2 | 431/2 | — | 0264 | 42 | 19 | 27 | — |
| 0218 | 477/8 | 38 | 441/2 | + | 0265 | 261/4 | 173/4 | 211/8 | + |
| 0219 | 271/2 | 23 | 243/4 | — | 0266 | 203/4 | 161/4 | 167/8 | + |
| 0220 | 8 | 21/4 | 41/4 | 0 | 0267 | 18 | 121/4 | 151/4 | + |
| 0221 | 621/8 | 1/2 | 11/4 | + | 0268 | 283/4 | 12 | 19 | — |
| 0222 | 201/2 | 53/4 | 81/4 | + | 0269 | 827/8 | 40 | 651/4 | + |
| 0223 | 15 | 6 | 141/8 | 0 | 0270 | 255/8 | 233/8 | 247/8 | + |
| 0224 | 251/2 | 91/8 | 243/4 | 0 | 0271 | 203/8 | 93/8 | 103/4 | — |
| 0225 | 551/4 | 29 | 531/2 | + | 0272 | 157/8 | 61/2 | 12 | + |
| 0226 | 271/2 | 131/2 | 263/8 | 0 | 0273 | 317/8 | 16 | 291/2 | 0 |
| 0227 | 151/2 | 41/8 | 61/2 | — | 0274 | 541/2 | 377/8 | 52 | + |
| 0228 | 223/8 | 121/4 | 145/8 | — | 0275 | 2261/4 | 1401/2 | 1573/8 | + |
| 0229 | 283/8 | 101/8 | 105/8 | + | 0276 | 53/8 | 25/8 | 33/8 | 0 |
| 0230 | 231/4 | 101/8 | 205/8 | — | 0277 | 691/2 | 411/4 | 451/4 | + |
| 0231 | 261/2 | 13 | 217/8 | + | 0278 | 7 | 2 | 65/8 | — |
| 0232 | 281/8 | 191/2 | 275/8 | — | 0279 | 321/2 | 12 | 197/8 | + |
| 0233 | 10 | 10 | 10 | 0 | 0280 | 221/4 | 101/2 | 205/8 | + |
| 0234 | 343/8 | 20 | 253/8 | — | 0281 | 661/2 | 47 | 543/4 | + |
| 0235 | 223/4 | 171/2 | 173/4 | + | 0282 | 123/8 | 91/4 | 111/4 | 0 |
| 0236 | 9 | 45/8 | 53/4 | 0 | 0283 | 441/2 | 161/4 | 255/8 | 0 |
| 0237 | 595/8 | 335/8 | 595/8 | + | 0284 | 361/2 | 19 | 251/2 | + |
| 0238 | 511/8 | 283/4 | 431/4 | + | 0285 | 581/2 | 26 | 451/2 | + |
| 0239 | 247/8 | 113/4 | 167/8 | 0 | 0286 | 331/4 | 223/8 | 32 | + |
| 0240 | 195/8 | 93/4 | 181/4 | — | 0287 | 215/8 | 161/2 | 183/4 | 0 |
| 0241 | 637/8 | 311/4 | 515/8 | + | 0288 | 193/8 | 14 | 151/4 | 0 |
| 0242 | 21 | 81/4 | 107/8 | — | 0289 | 19 | 143/4 | 15 | — |
| 0243 | 137/8 | 103/8 | 131/8 | 0 | 0290 | 231/4 | 91/2 | 22 | 0 |
| 0244 | 22 | 121/2 | 145/8 | — | 0291 | 413/4 | 221/8 | 257/8 | + |
| 0245 | 97 | 83 | 85 | — | 0292 | 301/8 | 17 | 243/8 | — |
| 0246 | 16 | 131/2 | 143/4 | + | 0293 | 153/4 | 71/4 | 113/4 | + |
| 0247 | 441/2 | 22 | 297/8 | 0 | 0294 | 491/4 | 251/4 | 36 | + |
| 0248 | 141/4 | 83/4 | 9 | — | 0295 | 353/4 | 101/2 | 271/8 | + |
| 0249 | 417/8 | 201/4 | 323/8 | + | 0296 | 87/8 | 6 | 67/8 | 0 |
| 0250 | 541/8 | 281/4 | 421/8 | + | 0297 | 357/8 | 181/2 | 241/4 | — |
| 0251 | 371/2 | 221/8 | 26 | + | 0298 | 65/8 | 41/8 | 5 | 0 |
| 0252 | 341/2 | 301/2 | 32 | + | 0299 | 361/8 | 151/8 | 191/4 | 0 |
| 0253 | 771/4 | 447/8 | 537/8 | + | 0300 | 461/8 | 22 | 321/4 | — |
| 0254 | 19 | 61/2 | 71/4 | 0 | 0301 | 67/8 | 17/8 | 33/4 | 0 |
| 0255 | 187/8 | 12 | 121/4 | 0 | 0302 | 23 | 10 | 14 | + |
| 0256 | 473/4 | 35 | 427/8 | 0 | 0303 | 66 | 353/4 | 451/2 | 0 |
| 0257 | 321/2 | 161/2 | 261/2 | — | 0304 | 31/8 | 11/8 | 11/4 | — |
| 0258 | 267/8 | 185/8 | 237/8 | + | 0305 | 341/2 | 223/4 | 253/4 | + |
| 0259 | 29 | 261/2 | 275/8 | 0 | 0306 | 221/8 | 123/4 | 183/8 | 0 |
| 0260 | 193/4 | 121/2 | 141/2 | 0 | 0307 | 61/2 | 3 | 31/8 | — |
| 0261 | 437/8 | 265/8 | 32 | + | 0308 | 450 | 297 | 3121/2 | + |
| 0262 | 353/4 | 171/2 | 24 | + | 0309 | 351/4 | 241/4 | 315/8 | + |
| 0263 | 301/4 | 103/4 | 213/8 | + | 0310 | 1051/4 | 971/4 | 981/2 | + |

Data Set II (Continued)

| Stock Number | 52 weeks High | 52 weeks Low | Close | Net Chg | Stock Number | 52 weeks High | 52 weeks Low | Close | Net Chg |
|---|---|---|---|---|---|---|---|---|---|
| 0311 | 153/8 | 41/2 | 135/8 | − | 0362 | 533/8 | 333/8 | 387/8 | 0 |
| 0312 | 373/8 | 22 | 333/8 | 0 | 0363 | 365/8 | 191/2 | 213/8 | 0 |
| 0313 | 113/4 | 51/2 | 7 | + | 0364 | 265/8 | 141/4 | 181/4 | + |
| 0314 | 401/4 | 171/4 | 205/8 | − | 0365 | 623/4 | 32 | 461/2 | + |
| 0315 | 375/8 | 301/4 | 327/8 | + | 0366 | 1661/4 | 118 | 1613/4 | + |
| 0316 | 113/4 | 51/2 | 7 | + | 0367 | 77 | 41 | 661/2 | 0 |
| 0317 | 401/4 | 171/4 | 205/8 | − | 0368 | 59 | 22 | 471/8 | + |
| 0318 | 375/8 | 301/4 | 327/8 | + | 0369 | 11 | 53/8 | 9 | − |
| 0319 | 537/8 | 335/8 | 491/2 | 0 | 0370 | 293/8 | 103/4 | 227/8 | − |
| 0320 | 8 | 3 | 53/8 | 0 | 0371 | 48 | 195/8 | 225/8 | + |
| 0321 | 161/2 | 71/4 | 135/8 | + | 0372 | 70 | 507/8 | 541/4 | + |
| 0322 | 19 | 61/2 | 95/8 | 0 | 0373 | 117/8 | 5 | 7 | 0 |
| 0323 | 537/8 | 24 | 363/8 | − | 0374 | 8 | 35/8 | 43/8 | 0 |
| 0324 | 205/8 | 91/2 | 183/8 | 0 | 0375 | 367/8 | 29 | 313/4 | − |
| 0325 | 161/4 | 11 | 151/8 | + | 0376 | 323/4 | 193/8 | 311/8 | + |
| 0326 | 287/8 | 12 | 253/4 | − | 0377 | 29 | 231/8 | 261/2 | + |
| 0327 | 287/8 | 141/4 | 261/4 | 0 | 0378 | 51 | 42 | 463/4 | − |
| 0328 | 211/8 | 33/4 | 107/8 | 0 | 0379 | 771/4 | 651/2 | 72 | 0 |
| 0329 | 743/4 | 411/8 | 613/4 | − | 0380 | 961/8 | 821/2 | 90 | 0 |
| 0330 | 101/4 | 53/4 | 101/8 | + | 0381 | 97 | 851/8 | 91 | − |
| 0331 | 521/4 | 323/4 | 431/4 | + | 0382 | 35 | 14 | 24 | + |
| 0332 | 181/2 | 143/8 | 43/4 | − | 0383 | 155/8 | 73/4 | 91/2 | + |
| 0333 | 293/4 | 151/8 | 173/8 | 0 | 0384 | 185/8 | 7 | 153/8 | 0 |
| 0334 | 343/4 | 27 | 31 | + | 0385 | 39 | 17 | 383/8 | + |
| 0335 | 251/4 | 161/2 | 201/4 | + | 0386 | 343/4 | 171/2 | 301/8 | + |
| 0336 | 251/8 | 191/2 | 213/8 | − | 0387 | 341/8 | 157/8 | 24 | + |
| 0337 | 351/4 | 281/2 | 313/8 | + | 0388 | 1003/4 | 73 | 79 | − |
| 0338 | 18 | 121/2 | 173/4 | − | 0389 | 561/4 | 331/2 | 457/8 | + |
| 0339 | 293/4 | 151/8 | 273/8 | 0 | 0390 | 73/8 | 13/4 | 2 | + |
| 0340 | 343/4 | 27 | 31 | + | 0391 | 113/8 | 25/8 | 33/4 | 0 |
| 0341 | 251/4 | 161/2 | 201/4 | + | 0392 | 357/8 | 171/4 | 333/4 | + |
| 0342 | 351/4 | 281/2 | 313/8 | + | 0393 | 133/8 | 7 | 101/4 | + |
| 0343 | 18 | 121/2 | 173/4 | 0 | 0394 | 95/8 | 43/8 | 73/8 | − |
| 0344 | 253/4 | 205/8 | 241/8 | + | 0395 | 233/4 | 73/4 | 225/8 | − |
| 0345 | 371/2 | 143/4 | 345/8 | 0 | 0396 | 36 | 231/2 | 285/8 | + |
| 0346 | 203/4 | 161/4 | 177/8 | 0 | 0397 | 243/4 | 9 | 151/2 | 0 |
| 0347 | 445/8 | 231/4 | 333/4 | + | 0398 | 135/8 | 65/8 | 113/4 | 0 |
| 0348 | 163/8 | 77/8 | 113/4 | + | 0399 | 9 | 35/8 | 61/2 | 0 |
| 0349 | 155/8 | 141/4 | 14 | − | 0400 | 22 | 123/4 | 157/8 | − |
| 0350 | 61/8 | 13/4 | 4 | 0 | 0401 | 403/8 | 21 | 301/2 | + |
| 0351 | 461/4 | 193/8 | 283/4 | + | 0402 | 381/2 | 241/4 | 301/2 | + |
| 0352 | 55 | 45 | 501/2 | − | 0403 | 531/8 | 29 | 373/4 | + |
| 0353 | 533/8 | 42 | 463/4 | − | 0404 | 211/4 | 101/2 | 14 | + |
| 0354 | 531/4 | 377/8 | 421/4 | − | 0405 | 103/8 | 11/4 | 17/8 | 0 |
| 0355 | 151/4 | 31/2 | 4 | 0 | 0406 | 431/2 | 261/4 | 391/8 | 0 |
| 0356 | 231/8 | 111/4 | 167/8 | + | 0407 | 525/8 | 28 | 43 | + |
| 0357 | 445/8 | 253/4 | 327/8 | + | 0408 | 231/2 | 101/8 | 141/2 | 0 |
| 0358 | 453/8 | 20 | 297/8 | + | 0409 | 10 | 10 | 10 | 0 |
| 0359 | 67/8 | 21/8 | 41/2 | 0 | 0410 | 93/4 | 71/8 | 91/2 | + |
| 0360 | 121/4 | 71/2 | 87/8 | + | 0411 | 561/2 | 267/8 | 377/8 | + |
| 0361 | 53 | 40 | 443/8 | 0 | 0412 | 161/8 | 63/4 | 10 | 0 |

Data Set II (Continued)

| Stock Number | 52 weeks High | Low | Close | Net Chg | Stock Number | 52 weeks High | Low | Close | Net Chg |
|---|---|---|---|---|---|---|---|---|---|
| 0413 | 123/4 | 51/2 | 83/4 | + | 0465 | 221/4 | 93/8 | 143/4 | + |
| 0414 | 123/4 | 51/2 | 9 | − | 0466 | 351/2 | 171/2 | 305/8 | + |
| 0415 | 261/2 | 24 | 26 | − | 0467 | 1184/4 | 47 | 801/4 | + |
| 0416 | 453/8 | 225/8 | 313/4 | + | 0468 | 353/4 | 18 | 315/8 | + |
| 0417 | 361/2 | 12 | 22 | + | 0469 | 171/4 | 71/4 | 141/4 | + |
| 0418 | 345/8 | 17 | 251/2 | 0 | 0470 | 223/8 | 135/8 | 18 | + |
| 0419 | 311/2 | 141/2 | 273/4 | − | 0471 | 1001/2 | 93 | 981/2 | − |
| 0420 | 133/4 | 61/4 | 123/8 | + | 0472 | 1397/8 | 851/2 | 1131/4 | + |
| 0421 | 343/4 | 223/4 | 291/4 | + | 0473 | 285/8 | 121/2 | 227/8 | + |
| 0422 | 201/4 | 161/2 | 183/8 | − | 0474 | 141/8 | 41/8 | 67/8 | + |
| 0423 | 211/4 | 171/4 | 195/8 | + | 0475 | 93 | 403/4 | 523/8 | − |
| 0424 | 851/2 | 75 | 771/4 | − | 0476 | 67 | 37 | 44 | − |
| 0425 | 26 | 233/4 | 243/4 | 0 | 0477 | 123/8 | 101/2 | 12 | − |
| 0426 | 277/8 | 251/4 | 261/8 | − | 0478 | 69 | 435/8 | 491/8 | 0 |
| 0427 | 331/2 | 251/4 | 295/8 | − | 0479 | 125/8 | 57/8 | 97/8 | − |
| 0428 | 85/8 | 31/8 | 45/8 | + | 0480 | 491/2 | 191/4 | 293/4 | + |
| 0429 | 337/8 | 22 | 277/8 | − | 0481 | 233/8 | 151/4 | 221/4 | + |
| 0430 | 321/4 | 19 | 237/8 | + | 0482 | 275/8 | 221/8 | 26 | − |
| 0431 | 781/2 | 34 | 57 | − | 0483 | 153/8 | 75/8 | 123/4 | + |
| 0432 | 151/4 | 51/2 | 91/4 | − | 0484 | 283/4 | 9 | 22 | − |
| 0433 | 371/4 | 151/8 | 275/8 | + | 0485 | 541/4 | 271/2 | 345/8 | + |
| 0434 | 271/4 | 71/2 | 151/2 | + | 0486 | 20 | 6 | 181/2 | + |
| 0435 | 73 | 38 | 451/8 | + | 0487 | 131/8 | 51/8 | 75/8 | 0 |
| 0436 | 161/8 | 91/8 | 125/8 | − | 0488 | 363/4 | 16 | 187/8 | − |
| 0437 | 101/8 | 10 | 10 | 0 | 0489 | 91/4 | 33/8 | 43/4 | − |
| 0438 | 38 | 207/8 | 303/4 | + | 0490 | 307/8 | 151/2 | 25 | + |
| 0439 | 243/4 | 183/4 | 221/4 | 0 | 0491 | 105/8 | 51/4 | 9 | + |
| 0440 | 205/8 | 153/4 | 171/4 | + | 0492 | 177/8 | 71/4 | 111/4 | − |
| 0441 | 151/4 | 7 | 103/8 | − | 0493 | 601/2 | 211/2 | 35 | 0 |
| 0442 | 265/8 | 16 | 171/2 | − | 0494 | 371/4 | 223/4 | 257/8 | 0 |
| 0443 | 471/2 | 371/2 | 431/2 | + | 0495 | 10 | 93/8 | 93/8 | 0 |
| 0444 | 573/4 | 51 | 537/8 | − | 0496 | 501/2 | 223/4 | 447/8 | + |
| 0445 | 411/4 | 223/4 | 325/8 | + | 0497 | 22 | 16 | 167/8 | − |
| 0446 | 465/8 | 281/2 | 351/4 | + | 0498 | 213/8 | 161/4 | 171/2 | 0 |
| 0447 | 407/8 | 197/8 | 313/4 | + | 0499 | 601/4 | 32 | 503/8 | − |
| 0448 | 87/8 | 25/8 | 55/8 | + | 0500 | 65/8 | 31/2 | 61/2 | + |
| 0449 | 28 | 12 | 233/8 | − | 0501 | 373/4 | 20 | 23 | + |
| 0450 | 50 | 401/2 | 451/2 | − | 0502 | 347/8 | 211/4 | 253/8 | + |
| 0451 | 763/8 | 66 | 681/2 | 0 | 0503 | 42 | 21 | 34 | 0 |
| 0452 | 77 | 64 | 701/2 | + | 0504 | 161/8 | 12 | 14 | + |
| 0453 | 791/2 | 67 | 721/2 | − | 0505 | 93 | 80 | 841/2 | + |
| 0454 | 795/8 | 661/2 | 74 | − | 0506 | 787/8 | 65 | 705/8 | 0 |
| 0455 | 395/8 | 25 | 357/8 | − | 0507 | 765/8 | 601/2 | 68 | − |
| 0456 | 505/8 | 301/2 | 387/8 | + | 0508 | 741/2 | 62 | 68 | − |
| 0457 | 57/8 | 21/4 | 51/8 | − | 0509 | 271/4 | 241/2 | 255/8 | 0 |
| 0458 | 473/4 | 33 | 41 | + | 0510 | 291/2 | 25 | 271/8 | − |
| 0459 | 3/8 | 1/8 | 3/8 | 0 | 0511 | 281/2 | 231/4 | 27 | − |
| 0460 | 121/4 | 37/8 | 41/8 | 0 | 0512 | 263/4 | 243/8 | 26 | + |
| 0461 | 381/4 | 175/8 | 221/2 | + | 0513 | 251/4 | 191/4 | 213/4 | + |
| 0462 | 85/8 | 33/8 | 47/8 | 0 | 0514 | 323/8 | 17 | 24 | 0 |
| 0463 | 121/2 | 85/8 | 11 | 0 | 0515 | 451/2 | 201/2 | 413/4 | + |
| 0464 | 123/4 | 51/4 | 61/4 | − | 0516 | 313/4 | 12 | 21 | + |

Data Set II (Continued)

| Stock Number | 52 weeks High | 52 weeks Low | 52 weeks Close | Net Chg | Stock Number | 52 weeks High | 52 weeks Low | 52 weeks Close | Net Chg |
|---|---|---|---|---|---|---|---|---|---|
| 0517 | 201/2 | 141/8 | 161/2 | 0 | 0564 | 145/8 | 11 | 141/4 | + |
| 0518 | 17 | 73/8 | 153/4 | 0 | 0565 | 151/2 | 61/2 | 103/4 | + |
| 0519 | 121/2 | 47/8 | 47/8 | 0 | 0566 | 407/8 | 23 | 301/8 | + |
| 0520 | 563/4 | 311/8 | 363/4 | − | 0567 | 471/4 | 193/8 | 30 | 0 |
| 0521 | 451/2 | 193/4 | 313/8 | − | 0568 | 291/2 | 19 | 237/8 | + |
| 0522 | 1991/2 | 991/4 | 1057/8 | + | 0569 | 323/8 | 211/8 | 245/8 | + |
| 0523 | 261/8 | 125/8 | 153/8 | + | 0570 | 705/8 | 393/4 | 445/8 | + |
| 0524 | 821/2 | 411/4 | 613/4 | + | 0571 | 1073/4 | 551/2 | 81 | + |
| 0525 | 297/8 | 213/4 | 257/8 | + | 0572 | 195/8 | 103/4 | 171/8 | + |
| 0526 | 61/4 | 31/8 | 37/8 | 0 | 0573 | 333/4 | 103/4 | 171/8 | + |
| 0527 | 471/4 | 365/8 | 421/4 | + | 0574 | 401/4 | 217/8 | 261/2 | − |
| 0528 | 171/2 | 83/4 | 113/8 | + | 0575 | 21 | 115/8 | 161/4 | 0 |
| 0529 | 253/4 | 11 | 217/8 | − | 0576 | 321/4 | 141/4 | 187/8 | − |
| 0530 | 453/8 | 251/2 | 351/2 | + | 0577 | 41/8 | 15/8 | 21/4 | + |
| 0531 | 777/8 | 433/4 | 641/4 | − | 0578 | 131/8 | 51/4 | 7 | 0 |
| 0532 | 1095/8 | 593/4 | 845/8 | + | 0579 | 201/8 | 117/8 | 183/8 | − |
| 0533 | 55 | 263/4 | 331/4 | + | 0580 | 41/2 | 21/8 | 27/8 | − |
| 0534 | 211/8 | 10 | 151/2 | 0 | 0581 | 21/8 | 1 | 11/8 | 0 |
| 0535 | 217/8 | 83/4 | 147/8 | 0 | 0582 | 101/8 | 51/4 | 93/8 | − |
| 0536 | 355/8 | 175/8 | 301/8 | + | 0583 | 101/8 | 10 | 10 | 0 |
| 0537 | 223/8 | 17 | 191/4 | + | 0584 | 423/8 | 263/4 | 293/4 | + |
| 0538 | 38 | 16 | 257/8 | − | 0585 | 63/4 | 21/8 | 33/4 | 0 |
| 0539 | 101/8 | 73/8 | 10 | 0 | 0586 | 15 | 33/4 | 43/8 | 0 |
| 0540 | 12 | 111/8 | 113/4 | + | 0587 | 265/8 | 151/2 | 213/8 | 0 |
| 0541 | 131 | 75 | 847/8 | − | 0588 | 317/8 | 273/8 | 291/2 | 0 |
| 0542 | 47 | 391/2 | 423/8 | + | 0589 | 61/4 | 5 | 6 | 0 |
| 0543 | 61 | 501/2 | 521/2 | − | 0590 | 65/8 | 51/4 | 57/8 | 0 |
| 0544 | 91/2 | 7 | 81/4 | 0 | 0591 | 141/8 | 12 | 123/8 | 0 |
| 0545 | 501/4 | 401/8 | 445/8 | + | 0592 | 243/4 | 171/8 | 223/4 | + |
| 0546 | 99 | 86 | 901/8 | + | 0593 | 28 | 143/4 | 183/8 | + |
| 0547 | 96 | 791/4 | 847/8 | + | 0594 | 307/8 | 18 | 255/8 | − |
| 0548 | 90 | 77 | 813/4 | 0 | 0595 | 531/2 | 31 | 38 | 0 |
| 0549 | 951/2 | 811/4 | 861/2 | + | 0596 | 181 | 1261/2 | 139 | + |
| 0550 | 73/4 | 55/8 | 61/2 | + | 0597 | 253/8 | 143/4 | 19 | 0 |
| 0551 | 11/4 | 3/8 | 7/8 | − | 0598 | 171/2 | 83/4 | 101/8 | + |
| 0552 | 713/4 | 441/2 | 481/2 | + | 0599 | 103/4 | 41/4 | 61/2 | − |
| 0553 | 151/4 | 105/8 | 151/4 | 0 | 0600 | 123/4 | 4 | 65/8 | 0 |
| 0554 | 221/2 | 19 | 211/2 | 0 | 0601 | 161/2 | 61/2 | 111/8 | 0 |
| 0555 | 191/2 | 161/2 | 18 | 0 | 0602 | 217/8 | 123/4 | 18 | − |
| 0556 | 201/2 | 171/8 | 193/4 | + | 0603 | 361/4 | 8 | 131/4 | + |
| 0557 | 24 | 193/4 | 221/2 | − | 0604 | 327/8 | 171/2 | 25 | 0 |
| 0558 | 263/8 | 143/4 | 231/4 | 0 | 0605 | 161/2 | 9 | 5 | 0 |
| 0559 | 255/8 | 9 | 203/8 | + | 0606 | 97/8 | 67/8 | 91/8 | 0 |
| 0560 | 14 | 73/4 | 93/4 | 0 | 0607 | 437/8 | 271/4 | 321/4 | − |
| 0561 | 29 | 51/2 | 57/8 | 0 | 0608 | 95/8 | 31/4 | 33/8 | + |
| 0562 | 451/8 | 27 | 32 | + | 0609 | 317/8 | 151/2 | 277/8 | + |
| 0563 | 121/2 | 71/2 | 123/8 | 0 | 0610 | 461/8 | 24 | 33 | − |

Data Set II (Continued)

| Stock Number | 52 weeks | | | Net Chg | Stock Number | 52 weeks | | | Net Chg |
|---|---|---|---|---|---|---|---|---|---|
| | High | Low | Close | | | High | Low | Close | |
| 0611 | 331/2 | 123/4 | 291/4 | — | 0658 | 255/8 | 21 | 221/4 | — |
| 0612 | 21 | 73/4 | 151/2 | + | 0659 | 145/8 | 81/2 | 93/4 | + |
| 0613 | 303/8 | 15 | 225/8 | + | 0660 | 407/8 | 33 | 373/8 | + |
| 0614 | 17 | 131/2 | 147/8 | 0 | 0661 | 81/4 | 41/2 | 73/8 | 0 |
| 0615 | 503/4 | 331/4 | 461/8 | + | 0662 | 623/4 | 35 | 521/4 | + |
| 0616 | 233/4 | 13 | 181/2 | 0 | 0663 | 17/8 | 1/4 | 11/8 | 0 |
| 0617 | 603/8 | 243/8 | 351/4 | + | 0664 | 391/8 | 34 | 38 | — |
| 0618 | 175/8 | 81/2 | 11 | — | 0665 | 191/8 | 91/2 | 163/8 | + |
| 0619 | 34 | 243/8 | 297/8 | + | 0666 | 111/8 | 71/8 | 107/8 | — |
| 0620 | 113/4 | 6 | 71/4 | — | 0667 | 55/8 | 3/4 | 1 | — |
| 0621 | 151/2 | 77/8 | 103/8 | + | 0668 | 5 | 3/8 | 1/2 | — |
| 0622 | 42 | 351/4 | 411/8 | + | 0669 | 97/8 | 53/8 | 51/2 | — |
| 0623 | 12 | 43/8 | 6 | 0 | 0670 | 267/8 | 121/4 | 213/4 | — |
| 0624 | 193/8 | 7 | 14 | — | 0671 | 261/2 | 111/2 | 211/4 | — |
| 0625 | 171/4 | 81/2 | 125/8 | 0 | 0672 | 63/8 | 31/2 | 51/2 | — |
| 0626 | 171/8 | 63/4 | 91/4 | 0 | 0673 | 263/4 | 16 | 20 | — |
| 0627 | 167/8 | 63/4 | 91/4 | 0 | 0674 | 343/4 | 211/4 | 257/8 | 0 |
| 0628 | 91/2 | 43/4 | 85/8 | — | 0675 | 417/8 | 301/2 | 383/4 | — |
| 0629 | 103/4 | 41/2 | 101/8 | + | 0676 | 317/8 | 175/8 | 283/4 | — |
| 0630 | 295/8 | 175/8 | 273/4 | + | 0677 | 175/8 | 111/4 | 157/8 | + |
| 0631 | 751/2 | 351/4 | 413/4 | — | 0678 | 211/4 | 63/8 | 101/4 | 0 |
| 0632 | 1401/2 | 42 | 1203/8 | — | 0679 | 293/4 | 17 | 251/8 | + |
| 0633 | 49 | 291/8 | 42 | 0 | 0680 | 305/8 | 14 | 221/2 | + |
| 0634 | 481/4 | 25 | 457/8 | 0 | 0681 | 427/8 | 22 | 313/4 | + |
| 0635 | 181/2 | 5 | 127/8 | + | 0682 | 437/8 | 25 | 393/8 | 0 |
| 0636 | 27 | 11 | 19 | — | 0683 | 113/4 | 25/8 | 3 | 0 |
| 0637 | 543/4 | 311/2 | 423/4 | + | 0684 | 771/2 | 391/2 | 601/4 | — |
| 0638 | 225/8 | 171/4 | 201/4 | — | 0685 | 39 | 293/8 | 333/4 | + |
| 0639 | 243/8 | 141/4 | 213/8 | + | 0686 | 497/8 | 181/4 | 483/4 | 0 |
| 0640 | 73 | 283/8 | 531/8 | + | 0687 | 81/8 | 27/8 | 61/2 | + |
| 0641 | 403/4 | 171/2 | 351/4 | + | 0688 | 22 | 15 | 167/8 | + |
| 0642 | 333/4 | 131/4 | 231/4 | + | 0689 | 231/4 | 11 | 213/8 | + |
| 0643 | 187/8 | 73/4 | 81/4 | + | 0690 | 143/4 | 53/4 | 12 | 0 |
| 0644 | 33/4 | 11/8 | 17/8 | 0 | 0691 | 343/4 | 191/2 | 267/8 | — |
| 0645 | 187/8 | 11/2 | 21/8 | + | 0692 | 81/4 | 33/4 | 73/8 | + |
| 0646 | 223/8 | 117/8 | 123/4 | 0 | 0693 | 561/4 | 30 | 521/8 | + |
| 0647 | 185/8 | 93/4 | 15 | — | 0694 | 161/2 | 41/2 | 123/4 | + |
| 0648 | 14 | 101/2 | 13 | + | 0695 | 151/2 | 121/8 | 145/8 | + |
| 0649 | 401/8 | 241/8 | 31 | + | 0696 | 62 | 32 | 541/2 | + |
| 0650 | 341/2 | 175/8 | 213/4 | + | 0697 | 25 | 91/2 | 151/8 | + |
| 0651 | 481/2 | 201/2 | 403/4 | + | 0698 | 397/8 | 183/8 | 303/8 | + |
| 0652 | 93/4 | 63/4 | 83/4 | + | 0699 | 133/8 | 6 | 91/8 | — |
| 0653 | 121/8 | 107/8 | 107/8 | 0 | 0700 | 271/4 | 12 | 201/4 | + |
| 0654 | 133/4 | 41/4 | 57/8 | — | 0701 | 151/8 | 103/8 | 131/8 | — |
| 0655 | 27 | 16 | 211/4 | — | 0702 | 191/2 | 101/2 | 133/8 | — |
| 0656 | 345/8 | 165/8 | 331/8 | + | 0703 | 7 | 41/2 | 5 | 0 |
| 0657 | 54 | 403/4 | 533/4 | — | 0704 | 261/2 | 171/4 | 23 | + |

Data Set II (Continued)

| Stock Number | 52 weeks High | Low | Close | Net Chg | Stock Number | 52 weeks High | Low | Close | Net Chg |
|---|---|---|---|---|---|---|---|---|---|
| 0705 | 313/8 | 163/4 | 27 | − | 0752 | 511/4 | 331/4 | 413/4 | + |
| 0706 | 323/4 | 21 | 291/4 | − | 0753 | 321/4 | 231/2 | 281/4 | + |
| 0707 | 271/2 | 135/8 | 26 | + | 0754 | 211/4 | 31/4 | 113/4 | + |
| 0708 | 6 | 17/8 | 27/8 | + | 0755 | 61/2 | 23/4 | 41/2 | + |
| 0709 | 171/4 | 6 | 97/8 | − | 0756 | 185/8 | 61/8 | 85/8 | + |
| 0710 | 383/8 | 191/2 | 275/8 | − | 0757 | 443/8 | 271/4 | 341/4 | 0 |
| 0711 | 103/8 | 7 | 85/8 | 0 | 0758 | 753/4 | 28 | 671/2 | + |
| 0712 | 691/2 | 313/4 | 461/8 | + | 0759 | 483/8 | 223/4 | 37 | + |
| 0713 | 503/4 | 34 | 481/2 | + | 0760 | 267/8 | 223/4 | 247/8 | − |
| 0714 | 1291/4 | 901/8 | 1231/2 | 0 | 0761 | 261/8 | 22 | 245/8 | 0 |
| 0715 | 65/8 | 23/4 | 43/4 | + | 0762 | 251/2 | 197/8 | 233/4 | + |
| 0716 | 6 | 21/4 | 21/4 | 0 | 0763 | 267/8 | 233/4 | 243/4 | 0 |
| 0717 | 443/4 | 293/8 | 391/4 | + | 0764 | 291/4 | 251/4 | 275/8 | − |
| 0718 | 287/8 | 223/4 | 265/8 | + | 0765 | 263/4 | 213/4 | 253/8 | 0 |
| 0719 | 163/8 | 135/8 | 141/2 | − | 0766 | 263/4 | 21 | 253/8 | + |
| 0720 | 107/8 | 61/2 | 93/4 | 0 | 0767 | 281/2 | 231/4 | 26 | − |
| 0721 | 211/2 | 137/8 | 171/4 | 0 | 0768 | 801/4 | 68 | 75 | 0 |
| 0722 | 91/4 | 21/2 | 43/4 | 0 | 0769 | 63 | 221/2 | 451/2 | + |
| 0723 | 47/8 | 11/2 | 15/8 | 0 | 0770 | 241/8 | 121/8 | 201/4 | + |
| 0724 | 561/4 | 26 | 31 | + | 0771 | 121/8 | 55/8 | 7 | + |
| 0725 | 777/8 | 16 | 287/8 | − | 0772 | 213/8 | 121/4 | 181/8 | − |
| 0726 | 35/8 | 5/8 | 11/8 | 0 | 0773 | 251/8 | 131/4 | 163/8 | − |
| 0727 | 171/2 | 91/4 | 117/8 | − | 0774 | 101/2 | 3 | 3 | 0 |
| 0728 | 135/8 | 93/4 | 123/4 | + | 0775 | 49 | 175/8 | 395/8 | + |
| 0729 | 383/8 | 161/2 | 21 | + | 0776 | 301/4 | 153/4 | 165/8 | 0 |
| 0730 | 531/2 | 223/8 | 227/8 | − | 0777 | 191/4 | 81/2 | 121/8 | 0 |
| 0731 | 22 | 111/2 | 151/8 | 0 | 0778 | 321/2 | 167/8 | 211/2 | 0 |
| 0732 | 313/4 | 15 | 213/4 | + | 0779 | 103/8 | 91/8 | 91/2 | 0 |
| 0733 | 91/2 | 3 | 31/2 | + | 0780 | 101/4 | 71/2 | 81/8 | − |
| 0734 | 211/2 | 81/8 | 125/8 | 0 | 0781 | 103/8 | 8 | 95/8 | 0 |
| 0735 | 737/8 | 435/8 | 521/4 | + | 0782 | 41/4 | 1 | 1 | − |
| 0736 | 663/8 | 383/8 | 417/8 | + | 0783 | 83/4 | 17/8 | 17/8 | + |
| 0737 | 47/8 | 2 | 31/8 | − | 0784 | 101/2 | 87/8 | 93/8 | 0 |
| 0738 | 157/8 | 65/8 | 11 | + | 0785 | 151/4 | 75/8 | 125/8 | − |
| 0739 | 113/8 | 53/4 | 9 | + | 0786 | 433/4 | 201/2 | 28 | + |
| 0740 | 477/8 | 211/2 | 333/4 | + | 0787 | 73/4 | 23/4 | 31/4 | 0 |
| 0741 | 621/8 | 403/4 | 47 | + | 0788 | 65 | 273/4 | 49 | 0 |
| 0742 | 941/8 | 50 | 80 | + | 0789 | 623/4 | 36 | 53 | − |
| 0743 | 651/2 | 491/2 | 58 | 0 | 0790 | 761/2 | 35 | 603/4 | + |
| 0744 | 51 | 30 | 403/4 | 0 | 0791 | 187/8 | 97/8 | 175/8 | + |
| 0745 | 405/8 | 201/8 | 295/8 | 0 | 0792 | 251/4 | 8 | 15 | + |
| 0746 | 61/4 | 25/8 | 43/4 | − | 0793 | 371/4 | 191/8 | 253/4 | + |
| 0747 | 355/8 | 213/4 | 343/8 | − | 0794 | 351/2 | 22 | 341/4 | − |
| 0748 | 613/8 | 451/2 | 535/8 | + | 0795 | 713/4 | 45 | 581/8 | + |
| 0749 | 221/4 | 131/2 | 221/4 | + | 0796 | 181/2 | 85/8 | 103/8 | + |
| 0750 | 611/4 | 331/4 | 513/4 | + | 0797 | 467/8 | 29 | 39 | + |
| 0751 | 281/4 | 181/2 | 221/4 | + | 0798 | 77 | 40 | 581/4 | − |

**Data Set II (Continued)**

| Stock Number | 52 weeks | | | Net Chg | Stock Number | 52 weeks | | | Net Chg |
|---|---|---|---|---|---|---|---|---|---|
| | High | Low | Close | | | High | Low | Close | |
| 0799 | 301/8 | 211/2 | 261/2 | + | 0846 | 281/2 | 213/4 | 251/4 | 0 |
| 0800 | 601/2 | 27 | 415/8 | + | 0847 | 43/4 | 13/8 | 15/8 | 0 |
| 0801 | 243/8 | 12 | 141/4 | + | 0848 | 233/4 | 10 | 153/4 | − |
| 0802 | 263/4 | 201/2 | 24 | + | 0849 | 301/4 | 123/8 | 185/8 | − |
| 0803 | 271/4 | 117/8 | 143/4 | 0 | 0850 | 513/4 | 331/2 | 403/4 | + |
| 0804 | 427/8 | 191/4 | 311/4 | − | 0851 | 421/2 | 203/4 | 38 | 0 |
| 0805 | 591/2 | 49 | 52 | 0 | 0852 | 363/8 | 171/2 | 205/8 | + |
| 0806 | 145/8 | 51/2 | 13 | − | 0853 | 147/8 | 73/4 | 91/2 | − |
| 0807 | 10 | 85/8 | 91/8 | 0 | 0854 | 731/2 | 40 | 453/8 | + |
| 0808 | 71/4 | 33/8 | 4 | + | 0855 | 373/4 | 203/4 | 24 | + |
| 0809 | 307/8 | 171/8 | 211/8 | + | 0856 | 735/8 | 391/8 | 487/8 | + |
| 0810 | 277/8 | 243/4 | 27 | − | 0857 | 391/2 | 205/8 | 351/4 | + |
| 0811 | 161/2 | 8 | 113/4 | − | 0858 | 193/4 | 121/4 | 181/4 | + |
| 0812 | 391/2 | 231/8 | 253/4 | + | 0859 | 101/8 | 91/2 | 97/8 | 0 |
| 0813 | 463/4 | 293/4 | 423/8 | + | 0860 | 103/8 | 91/4 | 93/8 | 0 |
| 0814 | 171/8 | 81/2 | 153/8 | + | 0861 | 101/8 | 91/8 | 91/4 | − |
| 0815 | 77/8 | 43/4 | 67/8 | + | 0862 | 353/5 | 191/4 | 343/4 | − |
| 0816 | 36 | 251/4 | 281/2 | + | 0863 | 125/8 | 51/4 | 95/8 | − |
| 0817 | 247/8 | 173/4 | 24 | + | 0864 | 507/8 | 271/2 | 457/8 | + |
| 0818 | 263/4 | 203/8 | 23 | + | 0865 | 541/4 | 23 | 371/8 | − |
| 0819 | 721/2 | 51 | 62 | + | 0866 | 1471/8 | 70 | 1801/8 | + |
| 0820 | 83/4 | 41/4 | 61/2 | − | 0867 | 37 | 17 | 251/8 | + |
| 0821 | 241/4 | 171/4 | 23 | − | 0868 | 451/8 | 243/8 | 39 | 0 |
| 0822 | 101/8 | 3 | 47/8 | + | 0869 | 293/4 | 121/8 | 263/8 | + |
| 0823 | 121/8 | 21/2 | 43/8 | − | 0870 | 347/8 | 187/8 | 243/4 | 0 |
| 0824 | 423/4 | 201/8 | 281/2 | + | 0871 | 21 | 101/2 | 131/4 | − |
| 0825 | 305/8 | 133/8 | 157/8 | + | 0872 | 231/8 | 19 | 221/8 | + |
| 0826 | 203/4 | 111/4 | 173/4 | 0 | 0873 | 30 | 101/2 | 131/2 | − |
| 0827 | 16 | 121/2 | 147/8 | + | 0874 | 24 | 121/2 | 145/8 | 0 |
| 0828 | 233/8 | 161/4 | 201/4 | 0 | 0875 | 9 | 41/8 | 51/8 | − |
| 0829 | 343/8 | 131/2 | 31 | + | 0876 | 9 | 5 | 47/8 | − |
| 0830 | 271/8 | 131/4 | 171/2 | − | 0877 | 1811/2 | 78 | 168 | + |
| 0831 | 333/4 | 17 | 30 | − | 0878 | 901/2 | 49 | 637/8 | + |
| 0832 | 36 | 241/2 | 331/2 | − | 0879 | 183/4 | 81/8 | 83/8 | + |
| 0833 | 543/4 | 271/2 | 45 | 0 | 0880 | 681/2 | 51 | 675/8 | + |
| 0834 | 157/8 | 9 | 123/8 | + | 0881 | 71/2 | 2 | 21/8 | − |
| 0835 | 35/8 | 13/4 | 21/2 | 0 | 0882 | 50 | 235/8 | 341/4 | 0 |
| 0836 | 131/4 | 33/4 | 113/8 | + | 0883 | 221/2 | 111/2 | 151/8 | 0 |
| 0837 | 113/4 | 53/4 | 111/8 | 0 | 0884 | 421/2 | 203/4 | 37 | + |
| 0838 | 281/2 | 163/8 | 201/8 | − | 0885 | 20 | 107/8 | 181/4 | 0 |
| 0839 | 291/4 | 91/4 | 285/8 | + | 0886 | 621/2 | 321/2 | 567/8 | + |
| 0840 | 405/8 | 22 | 273/8 | + | 0887 | 120 | 831/4 | 115 | − |
| 0841 | 393/8 | 231/2 | 321/4 | − | 0888 | 35 | 261/2 | 313/4 | + |
| 0842 | 343/4 | 181/4 | 257/8 | + | 0889 | 41/4 | 11/2 | 15/8 | 0 |
| 0843 | 193/8 | 131/2 | 153/8 | 0 | 0890 | 141/2 | 8 | 101/4 | − |
| 0844 | 335/8 | 221/4 | 281/8 | − | 0891 | 193/4 | 105/8 | 131/2 | 0 |
| 0845 | 97/8 | 63/4 | 83/4 | − | 0892 | 201/8 | 131/8 | 171/8 | − |

Data Set II (Continued)

| Stock Number | 52 weeks | | | Net Chg | Stock Number | 52 weeks | | | Net Chg |
|---|---|---|---|---|---|---|---|---|---|
| | High | Low | Close | | | High | Low | Close | |
| 0893 | 291/2 | 161/8 | 241/2 | 0 | 0940 | 221/2 | 171/2 | 211/4 | − |
| 0894 | 287/8 | 165/8 | 283/8 | + | 0941 | 593/4 | 291/2 | 573/4 | + |
| 0895 | 36 | 22 | 291/8 | 0 | 0942 | 2551/2 | 1261/2 | 192 | + |
| 0896 | 201/4 | 103/4 | 131/4 | 0 | 0943 | 55 | 34 | 413/4 | + |
| 0897 | 411/4 | 223/8 | 34 | + | 0944 | 281/2 | 151/4 | 277/8 | − |
| 0898 | 153/8 | 8 | 101/2 | + | 0945 | 1757/8 | 100 | 1233/4 | + |
| 0899 | 133/4 | 55/8 | 67/8 | 0 | 0946 | 58 | 371/4 | 485/8 | + |
| 0900 | 25 | 207/8 | 23 | + | 0947 | 53 | 28 | 401/4 | + |
| 0901 | 431/4 | 221/2 | 37 | + | 0948 | 67 | 481/2 | 551/2 | + |
| 0902 | 191/2 | 141/8 | 167/8 | − | 0949 | 671/4 | 481/8 | 571/4 | + |
| 0903 | 253/4 | 171/4 | 22 | + | 0950 | 393/4 | 221/2 | 31 | − |
| 0904 | 191/8 | 127/8 | 173/8 | − | 0951 | 56 | 27 | 433/4 | + |
| 0905 | 663/8 | 413/4 | 497/8 | + | 0952 | 121/8 | 43/8 | 67/8 | − |
| 0906 | 1081/8 | 79 | 85 | − | 0953 | 157/8 | 17/8 | 33/8 | 0 |
| 0907 | 1051/2 | 731/2 | 83 | 0 | 0954 | 431/2 | 223/4 | 33 | 0 |
| 0908 | 83 | 541/8 | 641/2 | + | 0955 | 251/2 | 193/8 | 211/8 | 0 |
| 0909 | 261/4 | 19 | 207/8 | + | 0956 | 26 | 201/4 | 23 | 0 |
| 0910 | 4 | 17/8 | 25/8 | + | 0957 | 113/8 | 7 | 83/8 | + |
| 0911 | 27 | 161/2 | 185/8 | − | 0958 | 411/2 | 333/4 | 363/4 | + |
| 0912 | 221/2 | 181/4 | 185/8 | − | 0959 | 221/8 | 151/4 | 155/8 | 0 |
| 0913 | 25 | 203/4 | 211/4 | − | 0960 | 245/8 | 191/2 | 227/8 | 0 |
| 0914 | 993/4 | 891/2 | 90 | + | 0961 | 183/4 | 8 | 9 | + |
| 0915 | 50 | 45 | 465/8 | 0 | 0962 | 79 | 373/4 | 671/2 | − |
| 0916 | 51 | 361/2 | 381/2 | − | 0963 | 523/4 | 361/2 | 423/4 | + |
| 0917 | 42 | 28 | 30 | + | 0964 | 12 | 61/4 | 77/8 | 0 |
| 0918 | 491/2 | 251/4 | 38 | + | 0965 | 247/8 | 12 | 17 | 0 |
| 0919 | 23 | 11 | 207/8 | − | 0966 | 33 | 113/4 | 231/2 | + |
| 0920 | 1081/8 | 661/2 | 717/8 | + | 0967 | 175/8 | 63/4 | 157/8 | − |
| 0921 | 161/4 | 7 | 131/4 | + | 0968 | 381/8 | 181/2 | 241/8 | + |
| 0922 | 351/8 | 123/8 | 307/8 | 0 | 0969 | 581/2 | 371/2 | 443/4 | + |
| 0923 | 815/8 | 65 | 741/4 | − | 0970 | 521/2 | 291/4 | 433/4 | − |
| 0924 | 913/8 | 795/8 | 83 | + | 0971 | 147/8 | 57/8 | 103/4 | + |
| 0925 | 271/2 | 25 | 261/4 | − | 0972 | 421/2 | 23 | 335/8 | + |
| 0926 | 313/4 | 233/4 | 303/4 | 0 | 0973 | 133/4 | 5 | 113/4 | + |
| 0927 | 453/4 | 221/2 | 387/8 | + | 0974 | 231/4 | 191/4 | 213/4 | 0 |
| 0928 | 383/8 | 17 | 373/4 | + | 0975 | 1053/8 | 55 | 771/4 | + |
| 0929 | 571/4 | 45 | 56 | − | 0976 | 40 | 201/2 | 337/8 | + |
| 0930 | 681/2 | 43 | 68 | − | 0977 | 241/2 | 121/8 | 161/4 | 0 |
| 0931 | 251/4 | 14 | 20 | − | 0978 | 143/8 | 61/8 | 63/4 | + |
| 0932 | 101/8 | 33/4 | 7 | 0 | 0979 | 30 | 173/4 | 27 | 0 |
| 0933 | 93/4 | 41/8 | 53/4 | 0 | 0980 | 243/8 | 151/8 | 175/8 | − |
| 0934 | 327/8 | 143/4 | 161/4 | 0 | 0981 | 195/8 | 81/2 | 171/8 | + |
| 0935 | 461/2 | 361/2 | 38 | 0 | 0982 | 273/4 | 131/4 | 183/4 | + |
| 0936 | 447/8 | 30 | 341/2 | + | 0983 | 483/8 | 215/8 | 335/8 | + |
| 0937 | 24 | 141/2 | 151/2 | + | 0984 | 191/8 | 111/2 | 157/8 | − |
| 0938 | 57/8 | 23/8 | 27/8 | 0 | 0985 | 231/8 | 73/4 | 173/4 | 0 |
| 0939 | 143/4 | 63/4 | 93/4 | + | 0986 | 111/2 | 31/4 | 33/8 | 0 |

**Data Set II (Continued)**

| Stock Number | 52 weeks | | | Net Chg | Stock Number | 52 weeks | | | Net Chg |
|---|---|---|---|---|---|---|---|---|---|
| | High | Low | Close | | | High | Low | Close | |
| 0987 | 47/8 | 11/4 | 21/8 | 0 | 1034 | 157/8 | 103/4 | 141/2 | + |
| 0988 | 623/4 | 33 | 621/4 | + | 1035 | 24 | 13 | 155/8 | − |
| 0989 | 31 | 21 | 281/8 | 0 | 1036 | 327/8 | 26 | 29 | + |
| 0990 | 411/2 | 351/2 | 381/2 | − | 1037 | 191/4 | 95/8 | 173/4 | + |
| 0991 | 251/2 | 21 | 24 | 0 | 1038 | 121/8 | 41/4 | 111/8 | − |
| 0992 | 793/4 | 343/8 | 385/8 | − | 1039 | 177/8 | 105/8 | 131/8 | 0 |
| 0993 | 25 | 16 | 191/8 | + | 1040 | 307/8 | 121/4 | 28 | − |
| 0994 | 273/8 | 20 | 24 | + | 1041 | 155/8 | 83/8 | 13 | + |
| 0995 | 215/8 | 105/8 | 153/4 | − | 1042 | 211/4 | 97/8 | 161/4 | + |
| 0996 | 141/2 | 73/4 | 105/8 | − | 1043 | 297/8 | 20 | 283/8 | 0 |
| 0997 | 18 | 9 | 121/2 | + | 1044 | 193/4 | 91/2 | 107/8 | − |
| 0998 | 291/2 | 173/4 | 211/2 | − | 1045 | 351/2 | 20 | 275/8 | − |
| 0999 | 191/2 | 71/2 | 171/4 | 0 | 1046 | 18 | 105/8 | 123/8 | + |
| 1000 | 673/4 | 377/8 | 541/4 | + | 1047 | 93/8 | 3 | 43/4 | − |
| 1001 | 41 | 15 | 27 | 0 | 1048 | 361/2 | 161/4 | 223/8 | + |
| 1002 | 123/8 | 113/4 | 12 | − | 1049 | 281/2 | 121/2 | 181/4 | + |
| 1003 | 101/8 | 10 | 101/8 | 0 | 1050 | 113/8 | 41/4 | 61/2 | − |
| 1004 | 401/4 | 20 | 311/4 | + | 1051 | 17 | 81/2 | 117/8 | + |
| 1005 | 211/8 | 15 | 181/4 | + | 1052 | 101/8 | 51/8 | 63/4 | 0 |
| 1006 | 233/4 | 15 | 187/8 | − | 1053 | 53 | 321/2 | 391/4 | − |
| 1007 | 467/8 | 291/8 | 341/4 | + | 1054 | 121/8 | 5 | 83/8 | + |
| 1008 | 295/8 | 185/8 | 205/8 | + | 1055 | 1073/4 | 573/4 | 811/4 | + |
| 1009 | 28 | 11 | 111/8 | + | 1056 | 47 | 16 | 26 | − |
| 1010 | 24 | 121/2 | 215/8 | + | 1057 | 527/8 | 157/8 | 213/4 | − |
| 1011 | 641/2 | 393/8 | 56 | − | 1058 | 131/8 | 87/8 | 117/8 | 0 |
| 1012 | 331/3 | 13 | 203/4 | + | 1059 | 601/2 | 353/4 | 461/4 | − |
| 1013 | 111/2 | 63/4 | 101/2 | 0 | 1060 | 261/4 | 22 | 251/4 | − |
| 1014 | 611/4 | 331/4 | 413/8 | + | 1061 | 1081/4 | 64 | 741/4 | + |
| 1015 | 273/8 | 111/8 | 137/8 | + | 1062 | 257/8 | 20 | 22 | − |
| 1016 | 283/4 | 201/2 | 267/8 | 0 | 1063 | 611/2 | 283/4 | 431/8 | + |
| 1017 | 193/4 | 87/8 | 151/8 | − | 1064 | 39 | 19 | 32 | + |
| 1018 | 611/8 | 267/8 | 60 | 0 | 1065 | 961/4 | 60 | 691/4 | + |
| 1019 | 893/8 | 407/8 | 717/8 | − | 1066 | 29 | 15 | 231/8 | + |
| 1020 | 62 | 38 | 537/8 | + | 1067 | 341/2 | 15 | 181/4 | − |
| 1021 | 391/2 | 233/8 | 327/8 | + | 1068 | 235/8 | 157/8 | 225/8 | 0 |
| 1022 | 15 | 51/2 | 121/2 | + | 1069 | 251/2 | 16 | 217/8 | − |
| 1023 | 983/4 | 581/2 | 951/2 | + | 1070 | 383/8 | 15 | 311/2 | + |
| 1024 | 231/2 | 101/2 | 217/8 | − | 1071 | 14 | 61/2 | 127/8 | + |
| 1025 | 151/4 | 63/4 | 115/8 | 0 | 1072 | 631/2 | 361/4 | 51 | + |
| 1026 | 241/4 | 165/8 | 227/8 | − | 1073 | 381/4 | 211/8 | 375/8 | 0 |
| 1027 | 95/8 | 41/2 | 57/8 | − | 1074 | 405/8 | 21 | 39 | + |
| 1028 | 51/4 | 21/4 | 35/8 | − | 1075 | 381/8 | 221/2 | 371/2 | − |
| 1029 | 8 | 3 | 63/4 | 0 | 1076 | 421/8 | 233/4 | 41 | 0 |
| 1030 | 251/4 | 101/2 | 173/4 | + | 1077 | 371/2 | 197/8 | 361/8 | 0 |
| 1031 | 63/8 | 23/4 | 41/2 | 0 | 1078 | 301/2 | 161/4 | 287/8 | + |
| 1032 | 85/8 | 23/8 | 27/8 | 0 | 1079 | 305/8 | 183/4 | 293/4 | 0 |
| 1033 | 141/4 | 97/8 | 127/8 | + | 1080 | 41 | 251/8 | 353/8 | + |

Data Set II (Continued)

| Stock Number | 52 weeks | | | Net Chg | Stock Number | 52 weeks | | | Net Chg |
| | High | Low | Close | | | High | Low | Close | |
|---|---|---|---|---|---|---|---|---|---|
| 1081 | 47 | 25 | 361/8 | + | 1128 | 561/2 | 35 | 411/4 | + |
| 1082 | 141/8 | 10 | 121/8 | − | 1129 | 391/4 | 183/4 | 27 | + |
| 1083 | 431/8 | 25 | 31 | + | 1130 | 421/2 | 341/2 | 377/8 | + |
| 1084 | 393/4 | 211/2 | 331/2 | − | 1131 | 228 | 1031/4 | 2271/8 | + |
| 1085 | 323/4 | 29 | 307/8 | 0 | 1132 | 157/8 | 61/8 | 81/4 | + |
| 1086 | 277/8 | 223/8 | 26 | − | 1133 | 103/4 | 71/4 | 91/8 | + |
| 1087 | 365/8 | 301/8 | 321/8 | 0 | 1134 | 16 | 6 | 71/2 | − |
| 1088 | 293/8 | 151/4 | 201/2 | + | 1135 | 463/4 | 321/4 | 371/2 | 0 |
| 1089 | 423/8 | 251/4 | 371/8 | + | 1136 | 497/8 | 221/4 | 357/8 | 0 |
| 1090 | 317/8 | 20 | 233/8 | + | 1137 | 321/4 | 17 | 223/4 | + |
| 1091 | 583/4 | 28 | 441/2 | + | 1138 | 333/8 | 181/2 | 241/2 | + |
| 1092 | 163/8 | 71/4 | 101/2 | 0 | 1139 | 281/2 | 201/2 | 231/2 | 0 |
| 1093 | 211/8 | 83/4 | 193/4 | − | 1140 | 33 | 13 | 195/8 | − |
| 1094 | 22 | 105/8 | 181/2 | − | 1141 | 9 | 15/8 | 23/4 | 0 |
| 1095 | 641/2 | 30 | 415/8 | + | 1142 | 93/4 | 51/2 | 63/4 | + |
| 1096 | 87/8 | 2 | 23/8 | 0 | 1143 | 611/8 | 313/8 | 441/4 | + |
| 1097 | 333/4 | 16 | 171/2 | − | 1144 | 793/8 | 541/2 | 647/8 | + |
| 1098 | 191/8 | 14 | 157/8 | 0 | 1145 | 841/2 | 43 | 66 | + |
| 1099 | 11 | 41/2 | 5 | − | 1146 | 631/4 | 23 | 617/8 | − |
| 1100 | 223/4 | 17 | 185/8 | 0 | 1147 | 387/8 | 231/4 | 337/8 | + |
| 1101 | 91/4 | 33/4 | 51/2 | 0 | 1148 | 11/4 | 1/8 | 1/4 | − |
| 1102 | 103/8 | 93/8 | 95/8 | + | 1149 | 483/8 | 21 | 393/4 | + |
| 1103 | 101/2 | 71/2 | 101/4 | + | 1150 | 357/8 | 191/2 | 32 | 0 |
| 1104 | 97/8 | 71/8 | 93/8 | 0 | 1151 | 213/8 | 14 | 173/4 | + |
| 1105 | 10 | 10 | 10 | 0 | 1152 | 1081/2 | 64 | 781/4 | + |
| 1106 | 105/8 | 71/2 | 101/8 | + | 1153 | 441/8 | 223/4 | 297/8 | + |
| 1107 | 105/8 | 91/2 | 101/8 | 0 | 1154 | 283/4 | 253/8 | 263/4 | + |
| 1108 | 21 | 141/4 | 191/4 | − | 1155 | 84 | 441/4 | 671/8 | + |
| 1109 | 19 | 53/4 | 113/4 | − | 1156 | 517/8 | 305/8 | 391/8 | 0 |
| 1110 | 33/4 | 11/8 | 15/8 | 0 | 1157 | 741/4 | 48 | 545/8 | + |
| 1111 | 831/8 | 455/8 | 83 | + | 1158 | 103/4 | 41/2 | 85/8 | 0 |
| 1112 | 163/4 | 4 | 87/8 | − | 1159 | 431/4 | 23 | 297/8 | + |
| 1113 | 81/8 | 41/2 | 53/8 | 0 | 1160 | 411/2 | 191/2 | 263/8 | + |
| 1114 | 193/8 | 95/8 | 11 | 0 | 1161 | 171/8 | 91/8 | 123/4 | + |
| 1115 | 461/4 | 183/4 | 287/8 | + | 1162 | 151/4 | 9 | 117/8 | + |
| 1116 | 51 | 311/2 | 401/2 | 0 | 1163 | 17/8 | 1 | 11/4 | 0 |
| 1117 | 471/2 | 28 | 361/2 | + | 1164 | 453/4 | 33 | 35 | 0 |
| 1118 | 41/8 | 13/4 | 17/8 | 0 | 1165 | 21/2 | 1 | 11/2 | 0 |
| 1119 | 287/8 | 141/2 | 231/8 | + | 1166 | 93/4 | 5 | 57/8 | − |
| 1120 | 613/4 | 397/8 | 561/2 | − | 1167 | 833/4 | 681/2 | 75 | 0 |
| 1121 | 61/8 | 13/4 | 21/2 | 0 | 1168 | 12 | 7 | 9 | 0 |
| 1122 | 411/2 | 145/8 | 191/8 | + | 1169 | 141/4 | 3 | 5 | 0 |
| 1123 | 97/8 | 6 | 95/8 | 0 | 1170 | 40 | 361/2 | 38 | 0 |
| 1124 | 19 | 87/8 | 117/8 | + | 1171 | 311/2 | 173/4 | 301/2 | 0 |
| 1125 | 413/4 | 24 | 283/8 | − | 1172 | 141/8 | 73/4 | 131/8 | + |
| 1126 | 72 | 433/4 | 517/8 | + | 1173 | 181/2 | 83/4 | 117/8 | − |
| 1127 | 25 | 95/8 | 167/8 | − | 1174 | 211/4 | 15 | 181/2 | − |

Data Set II (Continued)

| Stock Number | 52 weeks | | | Net Chg | Stock Number | 52 weeks | | | Net Chg |
|---|---|---|---|---|---|---|---|---|---|
| | High | Low | Close | | | High | Low | Close | |
| 1175 | 461/2 | 271/8 | 37 | + | 1222 | 227/8 | 161/4 | 18 | − |
| 1176 | 16 | 81/2 | 151/8 | − | 1223 | 101/2 | 33/4 | 47/8 | + |
| 1177 | 831/2 | 45 | 623/4 | + | 1224 | 201/2 | 101/4 | 161/2 | 0 |
| 1178 | 273/4 | 191/2 | 241/2 | 0 | 1225 | 531/2 | 343/4 | 47 | + |
| 1179 | 43/4 | 2 | 23/8 | + | 1226 | 30 | 161/8 | 225/8 | 0 |
| 1180 | 55 | 32 | 441/8 | + | 1227 | 81/4 | 35/8 | 61/4 | + |
| 1181 | 361/8 | 121/4 | 331/4 | + | 1228 | 5 | 13/4 | 31/8 | 0 |
| 1182 | 331/2 | 31 | 331/8 | − | 1229 | 27/8 | 5/8 | 13/8 | + |
| 1183 | 85 | 381/2 | 463/4 | + | 1230 | 33/4 | 11/4 | 2 | 0 |
| 1184 | 241/2 | 117/8 | 181/4 | 0 | 1231 | 251/2 | 113/4 | 191/4 | + |
| 1185 | 1001/4 | 57 | 863/8 | + | 1232 | 52 | 411/2 | 503/4 | − |
| 1186 | 371/4 | 291/2 | 343/8 | + | 1233 | 451/2 | 113/4 | 157/8 | + |
| 1187 | 181/2 | 73/4 | 133/8 | + | 1234 | 23 | 117/8 | 181/8 | − |
| 1188 | 213/4 | 171/4 | 181/4 | 0 | 1235 | 221/4 | 163/8 | 20 | |
| 1189 | 81/4 | 51/2 | 71/4 | − | 1236 | 101/2 | 93/4 | 101/4 | − |
| 1190 | 265/8 | 161/2 | 24 | + | 1237 | 4 | 21/2 | 25/8 | 0 |
| 1191 | 535/8 | 27 | 357/8 | + | 1238 | 183/8 | 111/2 | 163/4 | − |
| 1192 | 93/4 | 3/34 | 8 | 0 | 1239 | 281/8 | 20 | 225/8 | |
| 1193 | 143/8 | 81/4 | 93/8 | 0 | 1240 | 235/8 | 161/8 | 18 | − |
| 1194 | 301/2 | 145/8 | 181/4 | + | 1241 | 175/8 | 103/4 | 141/2 | 0 |
| 1195 | 857/8 | 381/4 | 791/2 | + | 1242 | 281/8 | 191/2 | 23 | + |
| 1196 | 545/8 | 291/2 | 411/2 | + | 1243 | 245/8 | 191/2 | 22 | 0 |
| 1197 | 131/4 | 77/8 | 103/4 | + | 1244 | 463/4 | 211/8 | 431/8 | − |
| 1198 | 74 | 35 | 461/4 | + | 1245 | 353/8 | 24 | 331/8 | − |
| 1199 | 377/8 | 123/4 | 141/2 | − | 1246 | 46 | 213/4 | 395/8 | 0 |
| 1200 | 95/8 | 23/4 | 4 | 0 | 1247 | 83/8 | 21/2 | 27/8 | 0 |
| 1201 | 423/4 | 203/4 | 323/4 | − | 1248 | 73/8 | 5 | 53/4 | + |
| 1202 | 351/2 | 163/4 | 35 | + | 1249 | 483/4 | 271/8 | 391/8 | − |
| 1203 | 163/8 | 13 | 14 | 0 | 1250 | 681/8 | 241/2 | 391/2 | − |
| 1204 | 147/8 | 75/8 | 113/8 | + | 1251 | 355/8 | 121/4 | 173/8 | 0 |
| 1205 | 161/8 | 83/8 | 91/2 | − | 1252 | 155/8 | 111/8 | 131/2 | 0 |
| 1206 | 401/8 | 253/4 | 361/2 | + | 1253 | 36 | 31 | 353/8 | + |
| 1207 | 131/8 | 35/8 | 41/4 | + | 1254 | 221/2 | 191/4 | 207/8 | 0 |
| 1208 | 46 | 25 | 421/4 | − | 1255 | 78 | 65 | 751/4 | + |
| 1209 | 291/8 | 151/2 | 231/8 | + | 1256 | 167/8 | 113/8 | 131/2 | − |
| 1210 | 871/4 | 42 | 573/4 | + | 1257 | 93/8 | 4 | 71/4 | + |
| 1211 | 121/8 | 8 | 107/8 | 0 | 1258 | 247/8 | 8 | 103/4 | − |
| 1212 | 11 | 41/4 | 63/4 | + | 1259 | 311/4 | 20 | 303/4 | 0 |
| 1213 | 161/2 | 97/8 | 151/4 | 0 | 1260 | 183/8 | 10 | 115/8 | 0 |
| 1214 | 211/4 | 157/8 | 183/8 | − | 1261 | 183/4 | 6 | 103/8 | + |
| 1215 | 731/2 | 305/8 | 455/8 | + | 1262 | 381/4 | 21 | 281/4 | + |
| 1216 | 403/4 | 183/4 | 315/8 | + | 1263 | 393/8 | 19 | 33 | − |
| 1217 | 25 | 223/4 | 243/8 | − | 1264 | 145/8 | 5 | 75/8 | 0 |
| 1218 | 11/4 | 7/8 | 1 | − | 1265 | 213/4 | 141/2 | 145/8 | + |
| 1219 | 123/8 | 6 | 91/2 | − | 1266 | 227/8 | 71/2 | 13 | + |
| 1220 | 307/8 | 15 | 28 | − | 1267 | 321/2 | 16 | 211/4 | + |
| 1221 | 4 | 11/8 | 13/8 | − | 1268 | 241/4 | 18 | 183/4 | − |

Data Set II (Continued)

| Stock Number | 52 weeks | | | Net Chg | Stock Number | 52 weeks | | | Net Chg |
|---|---|---|---|---|---|---|---|---|---|
| | High | Low | Close | | | High | Low | Close | |
| 1269 | 421/2 | 373/4 | 421/2 | 0 | 1316 | 191/2 | 51/2 | 143/4 | − |
| 1270 | 34 | 261/4 | 307/8 | + | 1317 | 125/8 | 51/4 | 103/8 | + |
| 1271 | 481/4 | 42 | 44 | − | 1318 | 38 | 161/4 | 303/4 | − |
| 1272 | 543/4 | 43 | 48 | 0 | 1319 | 281/2 | 15 | 231/4 | + |
| 1273 | 91 | 751/2 | 801/2 | − | 1320 | 145/8 | 73/4 | 131/2 | + |
| 1274 | 241/4 | 14 | 177/8 | + | 1321 | 267/8 | 9 | 225/8 | + |
| 1275 | 97/8 | 37/8 | 6 | − | 1322 | 24 | 81/8 | 233/4 | − |
| 1276 | 511/8 | 243/4 | 307/8 | + | 1323 | 19 | 91/8 | 10 | − |
| 1277 | 26 | 24 | 26 | 0 | 1324 | 45 | 253/4 | 321/4 | 0 |
| 1278 | 241/2 | 111/8 | 243/8 | 0 | 1325 | 14 | 61/8 | 8 | + |
| 1279 | 64 | 313/4 | 491/2 | − | 1326 | 51 | 331/4 | 435/8 | + |
| 1280 | 32 | 211/8 | 317/8 | + | 1327 | 371/2 | 281/4 | 353/4 | − |
| 1281 | 121/4 | 91/2 | 105/8 | + | 1328 | 381/4 | 283/8 | 35 | − |
| 1282 | 105/8 | 93/8 | 97/8 | 0 | 1329 | 51 | 271/2 | 435/8 | + |
| 1283 | 101/8 | 83/8 | 97/8 | − | 1330 | 351/8 | 203/4 | 291/2 | 0 |
| 1284 | 12 | 103/8 | 107/8 | + | 1331 | 171/2 | 111/4 | 121/8 | + |
| 1285 | 101/8 | 7 | 93/8 | 0 | 1332 | 161/4 | 131/8 | 151/8 | − |
| 1286 | 151/8 | 15 | 15 | − | 1333 | 611/4 | 435/8 | 445/8 | + |
| 1287 | 783/8 | 58 | 64 | + | 1334 | 207/8 | 14 | 161/2 | + |
| 1288 | 13/4 | 7/8 | 11/8 | 0 | 1335 | 143/4 | 71/2 | 141/2 | 0 |
| 1289 | 421/2 | 243/4 | 411/4 | 0 | 1336 | 241/4 | 19 | 233/4 | − |
| 1290 | 137/8 | 53/4 | 57/8 | 0 | 1337 | 191/2 | 93/4 | 12 | 0 |
| 1291 | 387/8 | 221/4 | 26 | + | 1338 | 333/4 | 24 | 287/8 | + |
| 1292 | 301/8 | 143/8 | 153/4 | + | 1339 | 37 | 263/4 | 34 | + |
| 1293 | 445/8 | 171/2 | 273/8 | + | 1340 | 36 | 131/8 | 157/8 | + |
| 1294 | 231/4 | 161/2 | 181/8 | + | 1341 | 25 | 123/4 | 153/4 | − |
| 1295 | 471/2 | 40 | 43 | − | 1342 | 51/2 | 23/8 | 25/8 | + |
| 1296 | 87 | 69 | 79 | + | 1343 | 17/8 | 7/8 | 15/8 | 0 |
| 1297 | 22 | 111/4 | 181/2 | − | 1344 | 343/4 | 181/4 | 243/4 | + |
| 1298 | 851/2 | 731/2 | 781/2 | − | 1345 | 17 | 5 | 95/8 | − |
| 1299 | 823/4 | 681/2 | 763/8 | − | 1346 | 273/4 | 101/2 | 151/4 | + |
| 1300 | 801/2 | 69 | 763/8 | + | 1347 | 26 | 121/8 | 141/4 | 0 |
| 1301 | 25 | 203/4 | 241/4 | + | 1348 | 103/4 | 43/4 | 91/4 | + |
| 1302 | 841/4 | 681/8 | 771/2 | + | 1349 | 83/4 | 31/2 | 6 | + |
| 1303 | 333/8 | 28 | 301/4 | + | 1350 | 241/2 | 12 | 197/8 | + |
| 1304 | 60 | 325/8 | 467/8 | 0 | 1351 | 81/2 | 21/2 | 41/4 | + |
| 1305 | 113/8 | 51/8 | 81/8 | + | 1352 | 49 | 241/2 | 321/8 | − |
| 1306 | 221/4 | 55/8 | 55/8 | − | 1353 | 5 | 27/8 | 4 | 0 |
| 1307 | 195/8 | 101/2 | 153/4 | − | 1354 | 117/8 | 35/8 | 51/4 | 0 |
| 1308 | 377/8 | 95/8 | 18 | + | 1355 | 271/2 | 95/8 | 263/4 | 0 |
| 1309 | 131/2 | 6 | 12 | − | 1356 | 301/4 | 181/2 | 223/8 | + |
| 1310 | 117/8 | 101/2 | 111/2 | + | 1357 | 66 | 351/4 | 453/4 | + |
| 1311 | 103/4 | 55/8 | 83/4 | 0 | 1358 | 377/8 | 285/8 | 347/8 | + |
| 1312 | 331/2 | 25 | 293/4 | + | 1359 | 52 | 431/2 | 48 | + |
| 1313 | 35/8 | 13/8 | 21/4 | + | 1360 | 93 | 76 | 85 | + |
| 1314 | 25 | 11 | 161/8 | − | 1361 | 91 | 78 | 83 | − |
| 1315 | 25 | 133/4 | 181/2 | 0 | 1362 | 841/2 | 343/4 | 77 | + |

Data Set II (Continued)

| Stock Number | 52 weeks High | 52 weeks Low | 52 weeks Close | Net Chg | Stock Number | 52 weeks High | 52 weeks Low | 52 weeks Close | Net Chg |
|---|---|---|---|---|---|---|---|---|---|
| 1363 | 483/4 | 201/2 | 441/2 | − | 1410 | 271/4 | 93/4 | 191/8 | + |
| 1364 | 811/4 | 381/2 | 713/4 | + | 1411 | 147/8 | 3 | 6 | 0 |
| 1365 | 205/8 | 141/4 | 205/8 | + | 1412 | 27 | 205/8 | 22 | 0 |
| 1366 | 187/8 | 91/2 | 121/8 | + | 1413 | 287/8 | 25 | 263/4 | − |
| 1367 | 421/4 | 251/2 | 35 | + | 1414 | 391/4 | 21 | 311/2 | + |
| 1368 | 12 | 73/8 | 101/2 | 0 | 1415 | 79 | 48 | 65 | + |
| 1369 | 391/4 | 181/2 | 221/2 | + | 1416 | 241/2 | 18 | 205/8 | + |
| 1370 | 81/8 | 51/8 | 51/2 | 0 | 1417 | 361/4 | 181/2 | 331/2 | 0 |
| 1371 | 101/4 | 63/4 | 71/2 | 0 | 1418 | 331/8 | 21 | 291/8 | 0 |
| 1372 | 133/8 | 51/4 | 73/8 | − | 1419 | 63/4 | 43/4 | 55/8 | + |
| 1373 | 40 | 141/4 | 175/8 | 0 | 1420 | 31 | 121/8 | 137/8 | − |
| 1374 | 21 | 153/8 | 201/2 | + | 1421 | 50 | 21 | 361/8 | + |
| 1375 | 33 | 201/2 | 243/8 | + | 1422 | 191/2 | 123/4 | 163/4 | − |
| 1376 | 177/8 | 15 | 161/2 | 0 | 1423 | 493/4 | 213/4 | 245/8 | 0 |
| 1377 | 2 | 3/4 | 11/8 | 0 | 1424 | 1031/2 | 60 | 731/4 | + |
| 1378 | 76 | 40 | 503/8 | + | 1425 | 19 | 123/8 | 171/4 | + |
| 1379 | 56 | 231/2 | 40 | − | 1426 | 347/8 | 251/2 | 311/2 | 0 |
| 1380 | 74 | 37 | 56 | 0 | 1427 | 701/2 | 331/2 | 68 | + |
| 1381 | 233/8 | 163/4 | 181/4 | + | 1428 | 101/8 | 95/8 | 95/8 | 0 |
| 1382 | 461/2 | 361/2 | 393/8 | − | 1429 | 15/8 | 7/8 | 11/4 | 0 |
| 1383 | 13 | 11 | 121/4 | − | 1430 | 7 | 5 | 61/2 | 0 |
| 1384 | 123/8 | 97/8 | 113/8 | 0 | 1431 | 101/4 | 9 | 9 | − |
| 1385 | 79 | 653/8 | 69 | + | 1432 | 231/8 | 171/4 | 215/8 | + |
| 1386 | 117/8 | 93/4 | 111/8 | 0 | 1433 | 231/2 | 191/4 | 22 | − |
| 1387 | 117 | 1061/8 | 112 | + | 1434 | 83 | 681/2 | 70 | + |
| 1388 | 783/4 | 64 | 691/2 | + | 1435 | 89 | 803/4 | 84 | + |
| 1389 | 167/8 | 121/8 | 143/4 | + | 1436 | 61/4 | 21/4 | 45/8 | + |
| 1390 | 1241/2 | 771/8 | 897/8 | + | 1437 | 141/2 | 4 | 73/8 | 0 |
| 1391 | 237/8 | 121/2 | 193/4 | + | 1438 | 173/8 | 31/2 | 81/4 | 0 |
| 1392 | 27 | 125/8 | 143/4 | − | 1439 | 16 | 33/8 | 73/4 | + |
| 1393 | 183/4 | 10 | 173/8 | + | 1440 | 323/8 | 143/4 | 15 | 0 |
| 1394 | 253/8 | 23 | 247/8 | 0 | 1441 | 271/4 | 20 | 231/8 | 0 |
| 1395 | 261/2 | 71/2 | 93/4 | + | 1442 | 461/2 | 40 | 41 | 0 |
| 1396 | 143/8 | 43/4 | 101/2 | + | 1443 | 47 | 40 | 411/2 | 0 |
| 1397 | 83/4 | 51/4 | 8 | 0 | 1444 | 771/4 | 521/2 | 681/8 | + |
| 1398 | 97/8 | 35/8 | 63/8 | 0 | 1445 | 37/8 | 13/8 | 17/8 | 0 |
| 1399 | 487/8 | 28 | 361/2 | + | 1446 | 463/8 | 161/2 | 431/4 | − |
| 1400 | 32 | 223/4 | 233/4 | + | 1447 | 217/8 | 173/4 | 19 | + |
| 1401 | 557/8 | 293/4 | 501/2 | + | 1448 | 93/4 | 33/8 | 9 | + |
| 1402 | 501/4 | 295/8 | 423/4 | + | 1449 | 101/8 | 61/8 | 81/4 | + |
| 1403 | 181/8 | 81/2 | 151/2 | 0 | 1450 | 101/8 | 97/8 | 10 | + |
| 1404 | 213/8 | 107/8 | 131/2 | 0 | 1451 | 101/8 | 9 | 91/2 | + |
| 1405 | 311/8 | 161/4 | 26 | − | 1452 | 101/8 | 91/4 | 95/8 | + |
| 1406 | 24 | 111/2 | 197/8 | − | 1453 | 101/8 | 93/8 | 95/8 | + |
| 1407 | 371/2 | 231/2 | 287/8 | 0 | 1454 | 7 | 35/8 | 57/8 | + |
| 1408 | 83/4 | 21/2 | 41/4 | 0 | 1455 | 261/2 | 71/2 | 171/8 | − |
| 1409 | 433/4 | 161/2 | 431/8 | + | 1456 | 41/8 | 1 | 11/8 | 0 |

Data Set II (Continued)

| Stock Number | 52 weeks | | | Net Chg | Stock Number | 52 weeks | | | Net Chg |
|---|---|---|---|---|---|---|---|---|---|
| | High | Low | Close | | | High | Low | Close | |
| 1457 | 573/8 | 313/4 | 457/8 | + | 1504 | 531/4 | 24 | 34 | + |
| 1458 | 281/2 | 127/8 | 203/8 | 0 | 1505 | 39 | 123/4 | 287/8 | 0 |
| 1459 | 141/4 | 41/4 | 125/8 | − | 1506 | 257/8 | 111/2 | 181/8 | + |
| 1460 | 105 | 49 | 921/4 | + | 1507 | 243/8 | 171/4 | 22 | |
| 1461 | 11 | 71/2 | 101/8 | 0 | 1508 | 391/8 | 193/4 | 371/4 | + |
| 1462 | 107/8 | 47/8 | 73/4 | + | 1509 | 111/4 | 33/4 | 6 | 0 |
| 1463 | 453/4 | 261/2 | 321/8 | + | 1510 | 141 | 943/8 | 1131/2 | + |
| 1464 | 243/4 | 10 | 11 | − | 1511 | 131/2 | 5 | 115/8 | + |
| 1465 | 111/2 | 41/2 | 11 | + | 1512 | 93/4 | 53/4 | 85/8 | 0 |
| 1466 | 711/8 | 341/2 | 493/4 | + | 1513 | 35 | 19 | 221/2 | + |
| 1467 | 115/8 | 45/8 | 105/8 | + | 1514 | 453/4 | 171/2 | 213/4 | 0 |
| 1468 | 153/8 | 7 | 87/8 | − | 1515 | 223/8 | 11 | 131/2 | 0 |
| 1469 | 7 | 3 | 55/8 | − | 1516 | 183/4 | 105/8 | 151/2 | + |
| 1470 | 63/4 | 13/8 | 33/4 | 0 | 1517 | 41 | 20 | 251/4 | + |
| 1471 | 94 | 575/8 | 763/8 | + | 1518 | 321/4 | 173/8 | 281/4 | − |
| 1472 | 93/4 | 4 | 75/8 | 0 | 1519 | 281/2 | 11 | 167/8 | − |
| 1473 | 63/4 | 31/8 | 55/8 | 0 | 1520 | 203/4 | 71/2 | 111/2 | 0 |
| 1474 | 543/4 | 295/8 | 463/4 | − | 1521 | 12 | 8 | 91/4 | 0 |
| 1475 | 201/4 | 85/8 | 93/4 | − | 1522 | 381/4 | 275/8 | 311/8 | + |
| 1476 | 223/4 | 145/8 | 21 | + | 1523 | 17 | 5 | 75/8 | + |
| 1477 | 13 | 45/8 | 47/8 | − | 1524 | 133/8 | 73/8 | 8 | − |
| 1478 | 847/8 | 571/4 | 661/4 | + | 1525 | 461/2 | 207/8 | 401/4 | + |
| 1479 | 65/8 | 11/4 | 13/8 | + | 1526 | 395/8 | 213/4 | 365/8 | + |
| 1480 | 121/2 | 31/4 | 4 | + | 1527 | 313/8 | 13 | 231/4 | − |
| 1481 | 9 | 17/8 | 21/4 | + | 1528 | 333/8 | 161/2 | 183/4 | + |
| 1482 | 201/2 | 141/4 | 173/8 | + | 1529 | 14 | 10 | 111/2 | 0 |
| 1483 | 12 | 51/4 | 77/8 | − | 1530 | 195/8 | 9 | 155/8 | 0 |
| 1484 | 23 | 7 | 14 | 0 | 1531 | 391/2 | 213/4 | 251/2 | 0 |
| 1485 | 16 | 81/4 | 12 | − | 1532 | 9 | 53/4 | 81/4 | 0 |
| 1486 | 21/2 | 3/8 | 13/8 | − | 1533 | 241/2 | 175/8 | 211/4 | − |
| 1487 | 77/8 | 43/4 | 61/8 | 0 | 1534 | 141/4 | 53/4 | 13 | + |
| 1488 | 173/8 | 11 | 15 | − | 1535 | 891/2 | 61 | 775/8 | − |
| 1489 | 111/2 | 41/2 | 6 | 0 | 1536 | 383/4 | 165/8 | 235/8 | + |
| 1490 | 85/8 | 45/8 | 53/4 | 0 | 1537 | 35 | 281/4 | 343/8 | + |
| 1491 | 551/4 | 361/4 | 441/2 | 0 | 1538 | 93/4 | 65/8 | 7 | + |
| 1492 | 613/4 | 28 | 52 | + | 1539 | 141/4 | 81/2 | 133/8 | − |
| 1493 | 223/4 | 81/4 | 201/8 | − | 1540 | 325/8 | 231/2 | 305/8 | 0 |
| 1494 | 461/4 | 281/2 | 343/4 | + | 1541 | 213/4 | 143/8 | 165/8 | − |
| 1495 | 11/4 | 1/2 | 3/4 | + | 1542 | 343/8 | 141/4 | 181/4 | − |
| 1496 | 201/4 | 111/2 | 125/8 | 0 | 1543 | 473/8 | 261/2 | 38 | + |
| 1497 | 295/8 | 121/4 | 257/8 | − | 1544 | 28 | 16 | 273/4 | − |
| 1498 | 181/4 | 141/4 | 175/8 | + | 1545 | 13/4 | 3/8 | 7/8 | − |
| 1499 | 493/4 | 37 | 451/2 | 0 | 1546 | 85/8 | 13/4 | 51/2 | + |
| 1500 | 123/8 | 55/8 | 97/8 | + | 1547 | 71/2 | 53/8 | 6 | 0 |
| 1501 | 20 | 141/4 | 195/8 | + | 1548 | 71/8 | 5 | 53/4 | + |
| 1502 | 297/8 | 141/4 | 207/8 | + | 1549 | 341/2 | 261/2 | 305/8 | − |
| 1503 | 101/2 | 47/8 | 83/4 | + | 1550 | 93/8 | 53/8 | 73/8 | 0 |

**Data Set II (Continued)**

| Stock Number | 52 weeks High | 52 weeks Low | Close | Net Chg | Stock Number | 52 weeks High | 52 weeks Low | Close | Net Chg |
|---|---|---|---|---|---|---|---|---|---|
| 1551 | 573/8 | 311/4 | 52 | + | 1598 | 101/2 | 41/4 | 41/2 | 0 |
| 1552 | 51 | 26 | 341/8 | + | 1599 | 371/4 | 213/4 | 263/4 | + |
| 1553 | 17 | 53/4 | 73/8 | + | 1600 | 521/4 | 251/8 | 505/8 | + |
| 1554 | 201/4 | 81/8 | 141/4 | + | 1601 | 301/2 | 121/4 | 221/4 | + |
| 1555 | 431/2 | 271/2 | 373/8 | + | 1602 | 413/4 | 30 | 373/8 | 0 |
| 1556 | 157/8 | 10 | 131/8 | − | 1603 | 263/8 | 23 | 243/4 | 0 |
| 1557 | 13 | 51/4 | 93/4 | + | 1604 | 22 | 16 | 183/8 | 0 |
| 1558 | 301/2 | 12 | 251/8 | + | 1605 | 251/2 | 141/4 | 233/8 | + |
| 1559 | 171/2 | 133/4 | 165/8 | − | 1606 | 291/2 | 171/4 | 25 | + |
| 1560 | 58 | 371/2 | 50 | + | 1607 | 783/4 | 54 | 70 | 0 |
| 1561 | 823/8 | 49 | 543/4 | + | 1608 | 241/2 | 177/8 | 221/4 | + |
| 1562 | 235/8 | 12 | 13 | 0 | 1609 | 301/8 | 231/4 | 285/8 | 0 |
| 1563 | 501/2 | 281/2 | 43 | + | 1610 | 573/4 | 43 | 501/4 | − |
| 1564 | 591/2 | 293/4 | 36 | + | 1611 | 34 | 28 | 291/4 | 0 |
| 1565 | 437/8 | 201/2 | 363/4 | + | 1612 | 14 | 75/8 | 93/8 | − |
| 1566 | 883/8 | 403/4 | 633/4 | + | 1613 | 101/4 | 21/8 | 31/4 | − |
| 1567 | 911/2 | 423/4 | 661/2 | − | 1614 | 415/8 | 91/8 | 113/4 | 0 |
| 1568 | 303/8 | 18 | 191/2 | + | 1615 | 273/4 | 53/4 | 83/8 | − |
| 1569 | 131/8 | 21/4 | 61/4 | 0 | 1616 | 231/4 | 113/4 | 181/8 | − |
| 1570 | 297/8 | 201/2 | 257/8 | − | 1617 | 251/2 | 181/4 | 201/4 | 0 |
| 1571 | 261/4 | 121/2 | 191/4 | − | 1618 | 451/2 | 281/4 | 371/4 | + |
| 1572 | 27 | 125/8 | 245/8 | − | 1619 | 263/4 | 153/8 | 19 | 0 |
| 1573 | 31 | 121/2 | 215/8 | 0 | 1620 | 29 | 221/8 | 253/4 | 0 |
| 1574 | 237/8 | 111/2 | 125/8 | + | 1621 | 121/8 | 10 | 10 | − |
| 1575 | 961/4 | 67 | 731/8 | + | 1622 | 191/4 | 9 | 13 | + |
| 1576 | 361/8 | 201/8 | 267/8 | − | 1623 | 323/4 | 103/4 | 325/8 | + |
| 1577 | 163/4 | 61/8 | 87/8 | 0 | 1624 | 20 | 71/8 | 127/8 | + |
| 1578 | 241/4 | 18 | 217/8 | 0 | 1625 | 383/4 | 203/4 | 327/8 | + |
| 1579 | 143/4 | 43/4 | 43/4 | − | 1626 | 651/2 | 43 | 501/2 | + |
| 1580 | 383/8 | 211/4 | 311/4 | + | 1627 | 1027/8 | 551/4 | 641/2 | + |
| 1581 | 203/4 | 151/2 | 17 | − | 1628 | 317/8 | 105/8 | 143/8 | 0 |
| 1582 | 183/8 | 111/8 | 141/8 | 0 | 1629 | 291/4 | 17 | 287/8 | − |
| 1583 | 29 | 14 | 281/4 | − | 1630 | 93/4 | 45/8 | 85/8 | − |
| 1584 | 107/8 | 41/2 | 83/4 | − | 1631 | 21 | 111/8 | 127/8 | − |
| 1585 | 723/4 | 393/4 | 451/2 | − | 1632 | 131/4 | 61/2 | 103/8 | + |
| 1586 | 60 | 39 | 541/8 | − | 1633 | 35 | 177/8 | 295/8 | − |
| 1587 | 461/2 | 241/4 | 365/8 | − | 1634 | 24 | 121/2 | 203/4 | + |
| 1588 | 101/4 | 61/2 | 67/8 | 0 | 1635 | 213/4 | 93/4 | 20 | + |
| 1589 | 203/4 | 151/2 | 17 | − | 1636 | 365/8 | 211/4 | 257/8 | + |
| 1590 | 183/8 | 111/8 | 141/8 | 0 | 1637 | 117/8 | 93/8 | 11 | + |
| 1591 | 29 | 14 | 281/4 | − | 1638 | 6 | 21/2 | 5 | 0 |
| 1592 | 107/8 | 41/2 | 83/4 | − | 1639 | 153/8 | 7 | 111/2 | 0 |
| 1593 | 723/4 | 393/4 | 451/2 | − | 1640 | 16 | 53/4 | 63/8 | − |
| 1594 | 60 | 39 | 541/8 | − | 1641 | 871/2 | 551/2 | 67 | − |
| 1595 | 461/2 | 241/4 | 365/8 | − | 1642 | 393/4 | 151/4 | 341/8 | − |
| 1596 | 101/4 | 61/2 | 67/8 | 0 | 1643 | 111/8 | 51/2 | 67/8 | 0 |
| 1597 | 203/4 | 135/8 | 141/4 | − | 1644 | 93/4 | 61/4 | 85/8 | − |

## Data Set II (Continued)

| Stock Number | 52 weeks | | | Net Chg | Stock Number | 52 weeks | | | Net Chg |
|---|---|---|---|---|---|---|---|---|---|
| | High | Low | Close | | | High | Low | Close | |
| 1645 | 181/8 | 65/8 | 167/8 | + | 1692 | 343/8 | 245/8 | 28 | + |
| 1646 | 203/4 | 127/8 | 151/4 | + | 1693 | 91/4 | 4 | 61/2 | 0 |
| 1647 | 673/4 | 34 | 571/2 | + | 1694 | 101/4 | 95/8 | 97,8 | + |
| 1648 | 137 | 98 | 1183/4 | 0 | 1695 | 367/8 | 171/4 | 24 | + |
| 1649 | 64 | 36 | 503/4 | 0 | 1696 | 795/8 | 411/4 | 511/2 | 0 |
| 1650 | 73/4 | 31/4 | 41/4 | − | 1697 | 87/8 | 41/4 | 81/2 | + |
| 1651 | 10 | 71/4 | 83/4 | − | 1698 | 201/4 | 14 | 143/8 | − |
| 1652 | 273/4 | 17 | 213/8 | + | 1699 | 263/4 | 10 | 163/4 | + |
| 1653 | 301/8 | 16 | 211/4 | + | 1700 | 673/4 | 411/2 | 543/4 | + |
| 1654 | 271/8 | 93/4 | 173/8 | − | 1701 | 241/2 | 137/8 | 215/8 | − |
| 1655 | 15 | 77/8 | 115/8 | − | 1702 | 25 | 83/8 | 211/8 | 0 |
| 1656 | 483/4 | 23 | 39 | + | 1703 | 113/4 | 9 | 111/8 | 0 |
| 1657 | 411/2 | 221/2 | 317/8 | + | 1704 | 231/8 | 10 | 133/4 | + |
| 1658 | 91/4 | 31/2 | 73/4 | + | 1705 | 23 | 5 | 51/8 | − |
| 1659 | 83/4 | 53/4 | 77/8 | − | 1706 | 101/2 | 4 | 63/4 | + |
| 1660 | 981/2 | 53 | 801/2 | + | 1707 | 41 | 141/2 | 333/8 | − |
| 1661 | 263/8 | 211/4 | 223/8 | 0 | 1708 | 173/4 | 53/4 | 13 | + |
| 1662 | 95/8 | 51/4 | 83/4 | + | 1709 | 1167/8 | 653/4 | 981/2 | + |
| 1663 | 26 | 111/4 | 173/4 | + | 1710 | 527/8 | 293/8 | 305/8 | + |
| 1664 | 215/8 | 171/4 | 193/4 | − | 1711 | 825/8 | 431/2 | 663/4 | − |
| 1665 | 70 | 37 | 451/2 | + | 1712 | 71/8 | 23/4 | 3 | − |
| 1666 | 23 | 91/4 | 175/8 | + | 1713 | 97/8 | 15/8 | 21/2 | − |
| 1667 | 3 | 3/4 | 15/8 | 0 | 1714 | 351/4 | 137/8 | 21 | − |
| 1668 | 243/8 | 91/2 | 133/4 | + | 1715 | 233/4 | 193/4 | 231/4 | + |
| 1669 | 32 | 131/2 | 191/4 | − | 1716 | 27 | 231/4 | 253/4 | − |
| 1670 | 711/8 | 443/4 | 541/2 | + | 1717 | 22 | 191/4 | 213/8 | − |
| 1671 | 37 | 131/2 | 141/8 | 0 | 1718 | 127/8 | 43/8 | 53/8 | − |
| 1672 | 561/2 | 28 | 423/4 | − | 1719 | 221/2 | 73/8 | 91/2 | − |
| 1673 | 181/2 | 91/4 | 151/8 | − | 1720 | 365/8 | 203/4 | 311/2 | + |
| 1674 | 403/4 | 201/2 | 227/8 | + | 1721 | 363/4 | 213/4 | 311/8 | + |
| 1675 | 37/8 | 15/8 | 13/4 | 0 | 1722 | 1071/4 | 981/2 | 991/2 | 0 |
| 1676 | 20 | 141/4 | 17 | + | 1723 | 247/8 | 111/8 | 19 | 0 |
| 1677 | 382 | 242 | 3231/4 | + | 1724 | 33/8 | 11/8 | 31/4 | − |
| 1678 | 291/2 | 16 | 241/8 | − | 1725 | 44 | 16 | 421/4 | − |
| 1679 | 267/8 | 10 | 151/2 | 0 | 1726 | 423/4 | 22 | 371/8 | + |
| 1680 | 681/2 | 35 | 497/8 | − | 1727 | 121/8 | 71/4 | 91/8 | − |
| 1681 | 101/2 | 97/8 | 97/8 | 0 | 1728 | 361/4 | 14 | 341/4 | + |
| 1682 | 621/2 | 361/8 | 471/4 | + | 1729 | 173/8 | 14 | 17 | − |
| 1683 | 361/2 | 12 | 167/8 | − | 1730 | 511/2 | 225/8 | 321/2 | + |
| 1684 | 161/2 | 73/4 | 97/8 | 0 | 1731 | 261/8 | 21 | 241/2 | − |
| 1685 | 281/4 | 167/8 | 173/4 | − | 1732 | 145/8 | 101/4 | 103/4 | − |
| 1686 | 521/2 | 267/8 | 471/2 | + | 1733 | 117/8 | 45/8 | 71/8 | 0 |
| 1687 | 131/4 | 11/2 | 11/2 | 0 | 1734 | 43 | 18 | 315/8 | + |
| 1688 | 413/8 | 201/2 | 251/2 | + | 1735 | 52 | 373/4 | 471/8 | − |
| 1689 | 441/2 | 253/8 | 367/8 | − | 1736 | 131/2 | 47/8 | 65/8 | + |
| 1690 | 801/4 | 361/4 | 421/4 | + | 1737 | 61/4 | 21/4 | 4 | 0 |
| 1691 | 343/4 | 227/8 | 301/4 | − | 1738 | 261/2 | 24 | 251/4 | − |

**Data Set II (Continued)**

| Stock Number | 52 weeks High | 52 weeks Low | 52 weeks Close | Net Chg | Stock Number | 52 weeks High | 52 weeks Low | 52 weeks Close | Net Chg |
|---|---|---|---|---|---|---|---|---|---|
| 1739 | 307/8 | 141/2 | 183/4 | + | 1786 | 483/8 | 24 | 335/8 | + |
| 1740 | 487/8 | 303/4 | 337/8 | − | 1787 | 815/8 | 48 | 607/8 | + |
| 1741 | 34 | 203/8 | 211/2 | + | 1788 | 41/2 | 11/8 | 21/8 | 0 |
| 1742 | 307/8 | 273/8 | 273/8 | − | 1789 | 187/8 | 87/8 | 131/2 | − |
| 1743 | 493/4 | 291/2 | 365/8 | + | 1790 | 181/4 | 91/4 | 161/4 | 0 |
| 1744 | 377/8 | 141/4 | 311/4 | + | 1791 | 341/4 | 18 | 331/2 | − |
| 1745 | 441/4 | 201/4 | 261/4 | − | 1792 | 275/8 | 191/8 | 235/8 | + |
| 1746 | 193/8 | 63/4 | 15 | 0 | 1793 | 151/4 | 121/2 | 131/4 | − |
| 1747 | 241/4 | 117/8 | 173/8 | + | 1794 | 183/4 | 101/4 | 145/8 | 0 |
| 1748 | 29 | 181/8 | 231/2 | 0 | 1795 | 373/4 | 18 | 243/8 | + |
| 1749 | 611/2 | 491/4 | 535/8 | + | 1796 | 297/8 | 17 | 213/8 | + |
| 1750 | 151/8 | 7 | 83/4 | − | 1797 | 121/8 | 53/4 | 81/2 | + |
| 1751 | 331/2 | 171/4 | 281/8 | − | 1798 | 113/8 | 23/4 | 4 | − |
| 1752 | 361/2 | 153/4 | 323/8 | − | 1799 | 21/4 | 11/8 | 17/8 | 0 |
| 1753 | 173/4 | 91/4 | 163/8 | 0 | 1800 | 531/2 | 26 | 361/8 | 0 |
| 1754 | 1057/8 | 55 | 933/4 | + | 1801 | 53/4 | 15/8 | 21/2 | 0 |
| 1755 | 201/4 | 113/4 | 19 | − | 1802 | 327/8 | 123/4 | 183/8 | − |
| 1756 | 30 | 213/4 | 28 | + | 1803 | 36 | 211/4 | 297/8 | − |
| 1757 | 133/4 | 41/2 | 93/4 | + | 1804 | 601/4 | 421/2 | 433/4 | + |
| 1758 | 273/8 | 153/8 | 231/4 | 0 | 1805 | 155/8 | 53/8 | 53/4 | 0 |
| 1759 | 451/4 | 261/4 | 307/8 | + | 1806 | 11 | 95/8 | 97/8 | 0 |
| 1760 | 563/4 | 411/4 | 447/8 | − | 1807 | 601/2 | 30 | 37 | + |
| 1761 | 73/4 | 63/8 | 71/2 | + | 1808 | 343/8 | 231/2 | 321/4 | + |
| 1762 | 28 | 161/4 | 213/8 | − | 1809 | 23 | 14 | 191/8 | − |
| 1763 | 321/4 | 191/2 | 305/8 | + | 1810 | 14 | 53/4 | 81/8 | − |
| 1764 | 393/8 | 21 | 305/8 | 0 | 1811 | 7 | 37/8 | 53/8 | 0 |
| 1765 | 51 | 437/8 | 501/8 | − | 1812 | 45 | 21 | 351/2 | − |
| 1766 | 1031/2 | 951/2 | 1017/8 | − | 1813 | 533/4 | 231/4 | 313/4 | + |
| 1767 | 63 | 411/2 | 513/8 | − | 1814 | 103/8 | 7 | 83/4 | 0 |
| 1768 | 53/4 | 1/8 | 5/8 | 0 | 1815 | 441/8 | 263/4 | 381/8 | + |
| 1769 | 373/8 | 103/8 | 107/8 | + | 1816 | 107/8 | 81/4 | 9 | + |
| 1770 | 301/4 | 171/2 | 253/8 | − | 1817 | 31 | 203/4 | 291/8 | + |
| 1771 | 46 | 283/8 | 325/8 | 0 | 1818 | 205/8 | 135/8 | 173/4 | + |
| 1772 | 731/2 | 38 | 541/2 | + | 1819 | 481/2 | 22 | 281/2 | + |
| 1773 | 475/8 | 26 | 337/8 | − | 1820 | 145/8 | 61/2 | 93/8 | − |
| 1774 | 321/8 | 151/2 | 211/2 | − | 1821 | 13 | 41/2 | 85/8 | − |
| 1775 | 91/2 | 4 | 91/8 | + | 1822 | 28 | 223/8 | 255/8 | − |
| 1776 | 257/8 | 197/8 | 225/8 | 0 | 1823 | 251/8 | 151/4 | 221/4 | − |
| 1777 | 491/8 | 40 | 441/2 | + | 1824 | 267/8 | 16 | 181/4 | − |
| 1778 | 501/8 | 401/2 | 453/8 | − | 1825 | 4 | 11/2 | 25/8 | − |
| 1779 | 86 | 723/4 | 80 | − | 1826 | 215/8 | 133/4 | 173/4 | − |
| 1780 | 24 | 181/2 | 233/4 | + | 1827 | 95/8 | 25/8 | 31/2 | + |
| 1781 | 83 | 69 | 791/2 | − | 1828 | 393/4 | 181/4 | 283/4 | + |
| 1782 | 22 | 131/8 | 145/8 | + | 1829 | 37/8 | 2 | 31/4 | + |
| 1783 | 865/8 | 451/8 | 583/4 | − | 1830 | 251/2 | 141/2 | 22 | + |
| 1784 | 145/8 | 53/4 | 101/2 | + | 1831 | 231/2 | 67/8 | 191/2 | + |
| 1785 | 181/2 | 9 | 131/4 | − | 1832 | 221/2 | 111/8 | 19 | + |

Data Set II (Continued)

| Stock Number | 52 weeks High | 52 weeks Low | Close | Net Chg | Stock Number | 52 weeks High | 52 weeks Low | Close | Net Chg |
|---|---|---|---|---|---|---|---|---|---|
| 1833 | 135/8 | 111/8 | 127/8 | 0 | 1880 | 371/2 | 23 | 28 | + |
| 1834 | 63/4 | 27/8 | 47/8 | + | 1881 | 381/4 | 197/8 | 243/4 | + |
| 1835 | 873/4 | 74 | 781/4 | + | 1882 | 50 | 321/2 | 351/4 | 0 |
| 1836 | 881/2 | 73 | 80 | − | 1883 | 271/2 | 57/8 | 22 | 0 |
| 1837 | 85 | 711/2 | 763/8 | + | 1884 | 53 | 16 | 50 | + |
| 1838 | 321/8 | 14 | 303/4 | − | 1885 | 46 | 121/2 | 421/2 | + |
| 1839 | 61 | 20 | 537/8 | + | 1886 | 40 | 201/4 | 265/8 | + |
| 1840 | 133/4 | 6 | 9 | − | 1887 | 251/4 | 101/4 | 115/8 | − |
| 1841 | 101 | 74 | 101 | + | 1888 | 391/4 | 221/8 | 315/8 | − |
| 1842 | 164 | 95 | 154 | 0 | 1889 | 211/2 | 71/4 | 93/4 | 0 |
| 1843 | 417/8 | 263/4 | 411/8 | + | 1890 | 107/8 | 21/2 | 31/2 | 0 |
| 1844 | 9 | 23/8 | 71/4 | − | 1891 | 297/8 | 173/8 | 22 | 0 |
| 1845 | 483/4 | 421/2 | 445/8 | 0 | 1892 | 61/4 | 33/4 | 43/4 | − |
| 1846 | 223/4 | 103/4 | 215/8 | + | 1893 | 197/8 | 71/2 | 177/8 | − |
| 1847 | 9 | 4 | 61/8 | 0 | 1894 | 377/8 | 193/4 | 30 | + |
| 1848 | 427/8 | 20 | 317/8 | + | 1895 | 77/8 | 43/8 | 61/4 | 0 |
| 1849 | 447/8 | 243/4 | 321/8 | + | 1896 | 113/4 | 21/8 | 3 | + |
| 1850 | 491/2 | 313/8 | 417/8 | + | 1897 | 507/8 | 371/2 | 39 | + |
| 1851 | 391/4 | 171/2 | 353/8 | − | 1898 | 135/8 | 7 | 103/4 | 0 |
| 1852 | 71 | 381/2 | 643/4 | − | 1899 | 4 | 11/4 | 11/2 | 0 |
| 1853 | 85/8 | 35/8 | 55/8 | 0 | 1900 | 273/8 | 21 | 26 | 0 |
| 1854 | 871/2 | 481/4 | 675/8 | + | 1901 | 253/4 | 187/8 | 213/8 | 0 |
| 1855 | 261/2 | 191/2 | 253/4 | + | 1902 | 42 | 265/8 | 351/2 | − |
| 1856 | 167/8 | 87/8 | 143/8 | + | 1903 | 133/4 | 71/2 | 111/2 | 0 |
| 1857 | 377/8 | 191/4 | 281/4 | − | 1904 | 603/4 | 291/2 | 493/8 | + |
| 1858 | 285/8 | 221/4 | 267/8 | + | 1905 | 93/8 | 51/8 | 51/2 | + |
| 1859 | 481/2 | 28 | 353/8 | + | 1906 | 191/2 | 103/8 | 16 | 0 |
| 1860 | 383/4 | 19 | 251/2 | + | 1907 | 411/4 | 191/2 | 353/8 | + |
| 1861 | 261/2 | 63/4 | 15 | + | 1908 | 23/4 | 3/4 | 13/8 | 0 |
| 1862 | 155/8 | 67/8 | 75/8 | − | 1909 | 171/2 | 7 | 10 | + |
| 1863 | 281/4 | 181/2 | 273/4 | 0 | 1910 | 303/8 | 141/2 | 191/2 | 0 |
| 1864 | 415/8 | 281/4 | 283/8 | 0 | 1911 | 393/4 | 131/4 | 195/8 | − |
| 1865 | 611/4 | 371/2 | 60 | + | 1912 | 85 | 50 | 523/4 | + |
| 1866 | 501/2 | 367/8 | 411/2 | 0 | 1913 | 503/4 | 471/2 | 481/8 | + |
| 1867 | 201/2 | 143/8 | 173/4 | 0 | 1914 | 353/4 | 191/8 | 331/4 | + |
| 1868 | 117/8 | 41/8 | 53/4 | 0 | 1915 | 593/4 | 171/2 | 57 | − |
| 1869 | 217/8 | 121/4 | 141/4 | + | 1916 | 63/4 | 21/4 | 27/8 | 0 |
| 1870 | 54 | 471/2 | 51 | + | 1917 | 37 | 131/2 | 221/4 | + |
| 1871 | 411/2 | 21 | 363/4 | + | 1918 | 161/8 | 83/4 | 113/4 | − |
| 1872 | 31/2 | 1/4 | 1/2 | 0 | 1919 | 335/8 | 10 | 227/8 | 0 |
| 1873 | 113/8 | 4 | 45/8 | 0 | 1920 | 103/4 | 91/2 | 97/8 | 0 |
| 1874 | 163/4 | 131/4 | 143/4 | + | 1921 | 101/8 | 1 | 11/4 | 0 |
| 1875 | 215/8 | 45/8 | 61/8 | + | 1922 | 24 | 131/2 | 20 | − |
| 1876 | 43/8 | 15/8 | 23/4 | 0 | 1923 | 203/8 | 113/8 | 16 | + |
| 1877 | 913/7 | 771/2 | 783/8 | + | 1924 | 303/4 | 15 | 231/8 | + |
| 1878 | 231/2 | 161/4 | 181/4 | + | 1925 | 111/4 | 75/8 | 101/2 | 0 |
| 1879 | 75 | 40 | 523/8 | + | | | | | |

# Answers to Selected Exercises

If you want further help with this course, you may want to obtain a copy of the *Student's Solutions Manual* and the *Statistical Utility Disk*. The first item contains solutions to half the odd-numbered exercises plus more aids, and the second item contains programs for solving the many problems in the text. You college bookstore either has the manual or can order it for you, and you may obtain a copy of the disk through your instructor.

## CHAPTER 1

### EXERCISES 1-1, SET A  (page 12)

**1.** The data are scores of games played on a given Tuesday by teams in the National Basketball Association (NBA). The data are descriptive in that they are a record of historical evidence.
**3.** Some of the data on the microwave oven are descriptive. "More than 12 million of them *were sold* last year. . . ." Furthermore "research has shown. . . ." [emphasis added] Some of the data are inferential in that they site the results of research, such as "the microwave oven also saves nutrients."
**5.** The data are descriptive because they are a record of something that has already occurred. "We ate 923 million gallons. . . ."
**7.** Well defined.   **9.** Not well defined.   **11.** Well defined.   **13.** Not well defined.   **15.** Well defined.   **17.** Well defined.
**19.** Well defined.   **21.** At the time licenses must be renewed.
**23.** At the start (or close) of the school year to get a record of the number of students in the system.
**25.** Every morning (or night) to accurately account for each inmate.
**27.** Assign to each faculty member a block of three (or four) digits, depending on the number of faculty members at Kent State U. Use a table of random numbers to select the 30 individuals for the sample.
**29.** If some type of serial number on the toasters is given on an invoice, then a table of random numbers could be used to select numbers from the list of possible numbers on the invoice. Or supposing the toasters are packaged in large cartons, numbering the cartons could be used to first decide which cartons should be sampled. Another numbering system could then be used to decide which toasters to sample from each carton.
**31.** Use a table of random numbers to obtain twenty-five numbers of six digits each that correspond to numbers on water meters in the area.   **33.** Not everyone has a telephone, and many telephone numbers are not listed in a directory.
**35.** Such a box is not equally accessible to all employees, and not all employees will use such an opportunity, especially as a result of where the box is placed.

### EXERCISES 1-2, SET A  (page 20)

Answers to Exercises 1–19 can vary. Answers given are suggestions for possible answers.

**1.** Republican, Democrat, Independent, Other.   **3.** Categories based on regions, such as Eastern, Southern, etc.
**5.** Categories based on flower color, such as red, yellow, etc. Categories based on life cycle, such as annual, perennial, etc.
**7.** Categories based on type of illness or injury. Categories based on method of payment of hospital expenses.
**9.** Categories based on political party affiliation. Categories based on region of the country represented by the congressperson.
**11. a.** The heights of the junipers.   **b.** The number of branches over a specified length on each juniper.
**13. a.** The weights of the books.   **b.** The number of pages in each book.
**15. a.** The lengths of the diagonals of the screens.   **b.** The number of channels that each set can receive.
**17. a.** The weights of the heads of lettuce.   **b.** The numbers of leaves on each head that are unfit for consumption.
**19. a.** The heights of the students.   **b.** The number of brothers and sisters each student has.
**21.** Discrete. The number of heads is counted.   **23.** Continuous. The quantities of liquids are measured.
**25.** Discrete. The numbers of games won, lost, and tied are counted.   **27.** Continuous. Temperature and rainfall are measured.
**29.** $x_1 = 3, x_2 = 0, x_3 = 1, x_4 = 0, x_5 = 5, x_6 = 2, x_7 = 1, x_8 = 2, x_9 = 3,$ and $x_{10} = 1$.
**31.** $x_1 = 2, x_2 = 5, x_3 = 11, x_4 = 3, x_5 = 4, x_6 = 2, x_7 = 5, x_8 = 1, x_9 = 12, x_{10} = 8, x_{11} = 2, x_{12} = 9, x_{13} = 3, x_{14} = 6,$ and $x_{15} = 2$

The answers to Exercises 33–37 can vary.

**33.** An aspect such as obedience. A dog is "very obedient," "average obedience," . . . , "disobedient."

**35.** Aspects such as "noise level" inside car, "response to bumps" in the road, "ease of turning," and so on.

**37.** Aspects such as time spent on practice field, willingness to play at more than one position, and willingness to help other players during practice sessions.

## EXERCISES 1-2, SET B  (page 23)

**1. a.** continuous   **b.** continuous   **3. a.** continuous   **b.** continuous   **c.** continuous

## EXERCISES 1-3, SET A  (page 29)

**1. a.** $\left(\begin{array}{c}\text{Paint several surfaces}\\\text{with the oxide coating}\end{array}\right) \longrightarrow \left(\begin{array}{c}\text{Study the bleaching}\\\text{effect on the surfaces}\end{array}\right)$   **b.** Thickness of coating and duration of test.

**3. a.** $\left(\begin{array}{c}\text{Record the number of}\\\text{workers that report late}\end{array}\right) \longrightarrow \left(\begin{array}{c}\text{Install the}\\\text{time clock}\end{array}\right) \longrightarrow \left(\begin{array}{c}\text{Record the number of}\\\text{workers that report late}\end{array}\right)$

**b.** Weather conditions during test that may affect commute times and workers' health conditions.

**5. a.** $\left(\begin{array}{c}\text{Play several rounds of}\\\text{golf with regular shaft clubs}\end{array}\right) \longrightarrow \left(\begin{array}{c}\text{Replace the clubs with}\\\text{graphite shaft clubs}\end{array}\right) \longrightarrow \left(\begin{array}{c}\text{Play several rounds}\\\text{with new clubs}\end{array}\right)$

**b.** Playing conditions during tests and variability in players ability.

**7. a.** $\left(\begin{array}{c}\text{Take the weights of}\\\text{several volunteers}\end{array}\right) \longrightarrow \left(\begin{array}{c}\text{Have the volunteers use the}\\\text{belt according to instructions}\end{array}\right) \longrightarrow \left(\begin{array}{c}\text{Take again the weights}\\\text{of the volunteers}\end{array}\right)$

**b.** Eating habits of volunteers and amount of weight loss available on each volunteer.

**9. a.** $\left(\begin{array}{c}\text{Arrange for several volunteers to have morning}\\\text{and evening massages using the restoration gel}\end{array}\right) \longrightarrow \left(\begin{array}{c}\text{Check to see whether the massages}\\\text{are successful in stopping hair loss}\end{array}\right)$

**b.** The benefits derived from the massaging itself without the gel and the possible differences in times spent massaging.

**11. a.** $\left(\begin{array}{c}\text{Use rivets to secure}\\\text{several seams on plows}\end{array}\right) \longrightarrow \left(\begin{array}{c}\text{Count the number that fail}\\\text{when subjected to stress test}\end{array}\right)$

**b.** The quality of steel used in the plows and the different sizes possible for rivits.

**13. a.** $\left(\begin{array}{c}\text{Determine the gasoline}\\\text{mileage without the device}\end{array}\right) \longrightarrow \left(\begin{array}{c}\text{Install}\\\text{the device}\end{array}\right) \longrightarrow \left(\begin{array}{c}\text{Determine the gasoline}\\\text{mileage with the device}\end{array}\right)$

**b.** Differences in types of driving without versus with the device on the car, and differences in the ways in which Connie drove the car during the two tests.

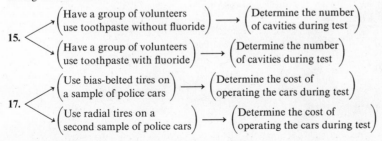

**15.** $\left(\begin{array}{c}\text{Have a group of volunteers}\\\text{use toothpaste without fluoride}\end{array}\right) \longrightarrow \left(\begin{array}{c}\text{Determine the number}\\\text{of cavities during test}\end{array}\right)$
$\left(\begin{array}{c}\text{Have a group of volunteers}\\\text{use toothpaste with fluoride}\end{array}\right) \longrightarrow \left(\begin{array}{c}\text{Determine the number}\\\text{of cavities during test}\end{array}\right)$

**17.** $\left(\begin{array}{c}\text{Use bias-belted tires on}\\\text{a sample of police cars}\end{array}\right) \longrightarrow \left(\begin{array}{c}\text{Determine the cost of}\\\text{operating the cars during test}\end{array}\right)$
$\left(\begin{array}{c}\text{Use radial tires on a}\\\text{second sample of police cars}\end{array}\right) \longrightarrow \left(\begin{array}{c}\text{Determine the cost of}\\\text{operating the cars during test}\end{array}\right)$

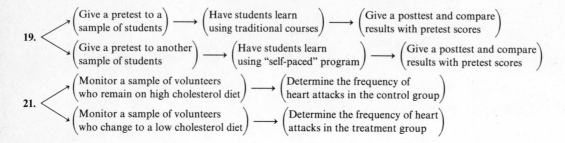

**19.**

**21.**

## CHAPTER 1 REVIEW EXERCISES　(page 33)

**1.** Descriptive　**3.** Not well defined
**5.** Use a table of random numbers to obtain 50 different five-digit numbers from 00001 through 80,000. The chickens in cages with tags whose numbers are the list obtained from the random number table comprise the sample.
**7.** Answers can vary. One possible categorization is based on the sizes of the aircraft, such as small private planes, small commercial aircraft, large commercial aircraft, and so on.
**9.** Answers can vary. As examples:
**a.** The annual consumption of energy by the college or university.
**b.** The annual enrollment determined by the number of full-time students.
**11.** Answers can vary. As examples:
**a.** The contestant's ability to perform very difficult tasks.　**b.** Very poor (1), poor (2), average (3), good (4), very good (5)
**13.** Using design (B)

$$\left(\begin{array}{l}\text{Observe the condition}\\\text{of the plants}\end{array}\right) \longrightarrow \left(\begin{array}{l}\text{Make regular applications}\\\text{of iron supplements}\end{array}\right) \longrightarrow \left(\begin{array}{l}\text{Observe the condition}\\\text{of the plants}\end{array}\right)$$

Two possible compounding factors are:
1. General improvement in weather may also have a positive effect on the general condition of the plants.
2. Insects and/or disease may have a negative effect on the general condition of the plants in spite of the supplement.

## MINITAB SUPPLEMENT　(page 36)

**1.** 320  415  216  127  210  115　**3.** Answers can vary.

## CHAPTER 2

### EXERCISES 2-1, SET A　(page 56)

**1. a.**

| Class Number | Class Limits | Class Boundaries | Class Mark | Tally Marks | Frequency | Relative Frequency |
|---|---|---|---|---|---|---|
| 1 | 1–2 | 0.5–2.5 | 1.5 | \|\| | 2 | 6.7% |
| 2 | 3–4 | 2.5–4.5 | 3.5 | \|\|\|\| | 4 | 13.3% |
| 3 | 5–6 | 4.5–6.5 | 5.5 | ⊤⊢⊢ ⊤⊢⊢ \|\| | 12 | 40.0% |
| 4 | 7–8 | 6.5–8.5 | 7.5 | ⊤⊢⊢ \|\| | 7 | 23.3% |
| 5 | 9–10 | 8.5–10.5 | 9.5 | ⊤⊢⊢ | 5 | 16.7% |

**b.**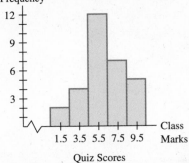

Quiz Scores

**c.**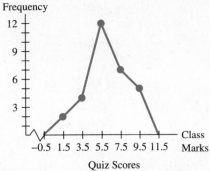

Quiz Scores

**d.** Bell shaped

**3. a.**

| Class Number | Class Limits | Class Boundaries | Class Marks | Tally Marks | Frequency | Relative Frequency |
|---|---|---|---|---|---|---|
| 1 | 19.0–22.9 | 18.95–22.95 | 20.95 | 卌 I | 6 | 15% |
| 2 | 23.0–26.9 | 22.95–26.95 | 24.95 | 卌 卌 III | 13 | 32.5% |
| 3 | 27.0–30.9 | 26.95–30.95 | 28.95 | 卌 卌 | 10 | 25.0% |
| 4 | 31.0–34.9 | 30.95–34.95 | 32.95 | 卌 II | 7 | 17.5% |
| 5 | 35.0–38.9 | 34.95–38.95 | 36.95 | IIII | 4 | 10.0% |

**b.**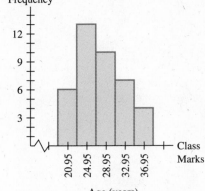

Age (years)

**c.**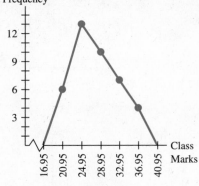

Age (years)

**d.** Bell shaped

**5. a.**

| Class Number | Class Limits | Class Boundaries | Class Mark | Tally Marks | Frequency | Relative Frequency |
|---|---|---|---|---|---|---|
| 1 | 6.20–8.89 | 6.195–8.895 | 7.545 | II | 2 | 5.6% |
| 2 | 8.90–11.59 | 8.895–11.595 | 10.245 | IIII | 4 | 11.1% |
| 3 | 11.60–14.29 | 11.595–14.295 | 12.945 | 卌 卌 | 10 | 27.8% |
| 4 | 14.30–16.99 | 14.295–16.995 | 15.645 | 卌 卌 II | 12 | 33.3% |
| 5 | 17.00–19.69 | 16.995–19.695 | 18.345 | 卌 IIII | 8 | 22.2% |

**b.**

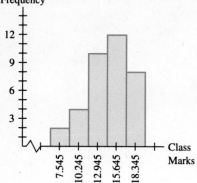

**c.**

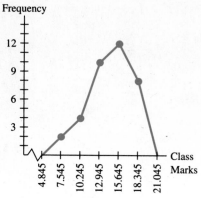

**d.** Left skewed

**7. a.**

| Class Number | Class Limits | Class Boundaries | Class Mark | Tally Marks | Frequency | Relative Frequency |
|---|---|---|---|---|---|---|
| 1 | 3.0–3.3 | 2.95–3.35 | 3.15 | IIII | 4 | 4% |
| 2 | 3.4–3.7 | 3.35–3.75 | 3.55 | ⊞ ⊞ | 10 | 10% |
| 3 | 3.8–4.1 | 3.75–4.15 | 3.95 | ⊞ ⊞ ⊞ I | 16 | 16% |
| 4 | 4.2–4.5 | 4.15–4.55 | 4.35 | ⊞ ⊞ ⊞ ⊞ IIII | 24 | 24% |
| 5 | 4.6–4.9 | 4.55–4.95 | 4.75 | ⊞ ⊞ ⊞ ⊞ ⊞ ⊞ | 30 | 30% |
| 6 | 5.0–5.3 | 4.95–5.35 | 5.15 | ⊞ ⊞ | 10 | 10% |
| 7 | 5.4–5.7 | 5.35–5.75 | 5.55 | ⊞ I | 6 | 6% |

**b.**

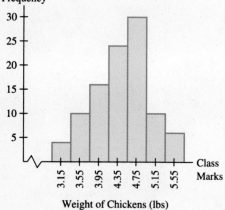

**c.**

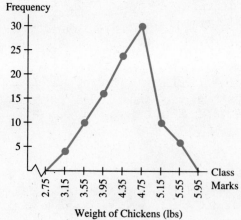

**d.** Left skewed

**9. a.**

```
0 | 6  6  8  8  8
1 | 0  0  0  2  2  2  4  5  5  5  5  6  6  8  8
2 | 0  0  0  0  0  0  0  0  1  2  4  4  5  5  5  5  5  5  5  8
3 | 0  0  0  0  2  2  4  5  5
4 | 0  0
```

**b.** Right skewed

**EXERCISES 2-1, SET B**   (page 60)

**1. a.**

| Class Number | Class Limits | Class Boundaries | Class Mark | Tally Marks | Frequency | Relative Frequency |
|---|---|---|---|---|---|---|
| 1 | 7–11 | 6.5–11.5 | 9 | ⵘⵘ ⅠⅠⅠ | 8 | 30% |
| 2 | 12–16 | 11.5–16.5 | 14 | ⵘⵘ Ⅰ | 6 | 22% |
| 3 | 17–21 | 16.5–21.5 | 19 | ⅠⅠⅠ | 3 | 11% |
| 4 | 22–26 | 21.5–26.5 | 24 | ⵘⵘ ⅠⅠⅠ | 8 | 30% |
| 5 | 27–31 | 26.5–31.5 | 29 | ⅠⅠ | 2 | 7% |

**b.**

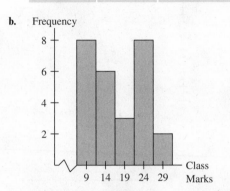

**c.**

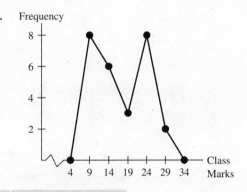

**d.** Bimodal
**e.** Eastern Conference
**f.** Eastern Conference

**7. a.**

| Class Number | Class Boundaries | Frequency | Cumulative Frequency | |
|---|---|---|---|---|
| 1 | 0.5–2.5 | 2 | less than 2.5 | 2 |
| 2 | 2.5–4.5 | 4 | less than 4.5 | 6 |
| 3 | 4.5–6.5 | 12 | less than 6.5 | 18 |
| 4 | 6.5–8.5 | 7 | less than 8.5 | 25 |
| 5 | 8.5–10.5 | 5 | less than 10.5 | 30 |

**b.**

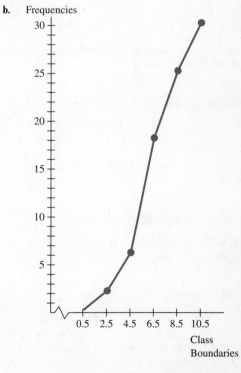

**9. a.**

| Class Number | Class Boundaries | Frequency | Cumulative Frequency | |
|---|---|---|---|---|
| 1 | 18.95–22.95 | 6 | less than 22.95 | 6 |
| 2 | 22.95–26.95 | 13 | less than 26.95 | 19 |
| 3 | 26.95–30.95 | 10 | less than 30.95 | 29 |
| 4 | 30.95–34.95 | 7 | less than 34.95 | 36 |
| 5 | 34.95–38.95 | 4 | less than 38.95 | 40 |

**b.**   Frequencies

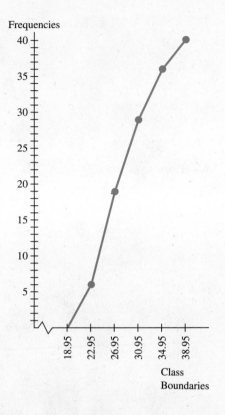

Class Boundaries

**11. a.**
```
0.61 | 2  5  8
0.62 | 2  2  4  6  6  7
0.63 | 0  0  0  1  3  3  9
0.64 | 0  0  3  3  5  6
0.65 | 2  4  7  9
0.66 | 1  8
0.67 | 2  9
```
**b.** Right skewed

**13. a.**
```
(2.0–2.4)2 | .15  .25  .39  .45
(2.5–2.9)2 | .51  .56  .63  .63  .65  .71  .77  .88  .92  .96
(3.0–3.4)3 | .00  .02  .07  .22  .23  .24  .26  .27  .34  .35  .38  .39  .43  .47
(3.5–3.9)3 | .50  .51  .52  .57  .58  .58  .63  .64  .65  .68  .72  .81  .88  .98
(4.0–4.4)4 | .07  .21  .26  .31  .33  .35  .38  .49
```
**b.** Bell shaped

**15.** With low annual wages on the left and high annual wages on the right, a distribution of the data would be skewed right.
**17.** The distribution would be bell shaped with the frequencies decreasing at approximately the same rate on both sides of the mean.
**19.** The distribution is probably bell shaped around the mean sales tax rate for all states.
**21.** With youngest aged driver on the left and oldest aged driver on the right, a distribution would be skewed right.

**EXERCISES 2.2, SET A**   (page 71)

**1. a.** 6 pounds   **b.** 5 pounds   **c.** 4 pounds   **3. a.** 76 grams   **b.** 76 grams   **c.** 73 grams   **5. a.** 211.8   **b.** 211.5   **c.** 209   **7. a.** $45.41   **b.** $34.44
**c.** no mode   **9. a.** 52.7°   **b.** 52.5°   **c.** 52°   **11. a.** 2.63 grams   **b.** 2.5 grams   **c.** 2.5 grams   **13. a.** 26.5 years   **b.** 23.5 years   **c.** 19 years
**15. a.** $37.17   **b.** $38.85   **c.** $38.99 and $39.45   **17.** $0.13; supermarket B   **19.** 28 words per minute
**21.** If the distribution of salaries is skewed to the right, then the $5.35 could be the median and $8.35 the mean of the distribution.
**23. a.** Use the median of 72° F.   **b.** Use the mode of 73° F.

**EXERCISES 2-2, SET B**   (page 75)

**1.** Mean   **3.** Median   **5.** Mode   **7.** Mean   **9.** Median   **13. a.** $95,870,000   **b.** $13,000,000   **c.** $270.00, $470.00, and $1,100.00 are extreme data points

**EXERCISES 2-3, SET A**   (page 88)

**1. a.** 11   **b.** 12.0   **c.** 3.5   **3. a.** 19.3 miles   **b.** 21.14 miles   **c.** 4.6 miles   **5. a.** $100.80   **b.** 532.75   **c.** $23.08
**7. a.** 10.9 inches   **b.** 9.1 inches$^2$   **c.** 3.0 inches   **9. a.** 0.0839 mm   **b.** 0.0818 mm   **c.** 0.0048 mm   **11. a.** $63.5   **b.** $64.0   **c.** $12.0
**13. a.** 4.0 hours   **b.** 2.8 hours   **c.** 4.4 hours   **15. a.** 33.9 years   **b.** 31.5 years   **c.** 9.4 years

**17. a.**

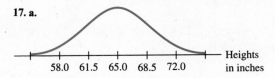

58.0  61.5  65.0  68.5  72.0  Heights in inches

**b.** 61.5 and 68.5 inches  **c.** 58.0 and 72.0 inches

**19. a.**

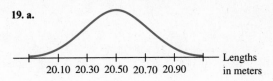

20.10  20.30  20.50  20.70  20.90  Lengths in meters

**b.** 20.30 and 20.70 m  **c.** 20.10 and 20.90 m

**EXERCISES 2-3, SET B**  (page 92)

**5. a.** Approximately 29.4 games and 5.0 games  **b.** Approximately 28.8 games and 7.9 games  **c.** Atlanta Hawks  **d.** Boston Celtics
**7.** 56%  **9.** 80%  **11.** 6.2  **13.** 3.3 miles compared with 4.6 miles

**EXERCISES 2-4, SET A**  (page 103)

**1. a.** 1.18  **b.** 1.46  **c.** Clayton Turner  **3. a.** −1.00  **b.** −0.60  **c.** Wilma  **5. a.** 0.35  **b.** 0.36  **c.** city C  **7. a.** 3.57  **b.** 3.23  **c.** 2.65  **d.** 2.30
**9. a.** 46  50  52  53  56  57  58  59  61  64  65  66  66  **b.** 75  **c.** 82  **d.** 71  **e.** 90  **f.** 66  **g.** 84
69  70  71  72  73  73  73  75  75  75  75  76  76
76  76  78  78  79  80  80  82  83  84  84  84  84
85  85  86  90  90  90  90  92  93  93  96

**11. a.**

| 7 | 8 | 9 | | | | | | | | |
|---|---|---|---|---|---|---|---|---|---|---|
| 8 | 7 | 7 | 8 | 9 | | | | | | |
| 9 | 2 | 3 | 5 | 6 | 6 | 6 | 7 | 8 | 8 | 9 |
| 10 | 3 | 3 | 4 | 6 | 6 | 8 | | | | |
| 11 | 0 | 2 | 9 | | | | | | | |

**b.** 91;  97;  105

**c.**

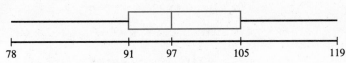

78  91  97  105  119

**13. a.**

| 12. | 7 | 8 | 8 | 9 | | | | | | | |
|---|---|---|---|---|---|---|---|---|---|---|---|
| 13. | 0 | 2 | 2 | 4 | 5 | 5 | 5 | 7 | 8 | 8 | 9 | 9 |
| 14. | 1 | 2 | 3 | 3 | 5 | 6 | 7 | 8 | 9 | | |
| 15. | 0 | 2 | 3 | 4 | 4 | 7 | 7 | 8 | | | |
| 16. | 4 | 5 | 7 | 7 | 9 | | | | | | |
| 17. | 3 | 8 | | | | | | | | | |

**b.** 13.5  14.4  15.6

**c.**

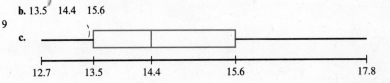

12.7  13.5  14.4  15.6  17.8

**15. a.** 76%  **b.** 48%  **17. a.** 88%  **b.** 31%

**EXERCISES 2-4, SET B**  (page 107)

**1. a.** 45.3, 46.8, 46.9, 48.3, 51.2, 53.9, 54.4, 55.9, 56.3, 56.7, 59.5, 60.0, 62.0, 63.8, 64.0, 65.0, 66.0, 66.0, 66.0, 66.0, 66.3, 66.8, 66.8, 66.8, 67.1, 67.1, 67.5, 67.8, 68.0, 68.0, 68.0, 68.1, 68.7, 69.0, 69.0, 69.2, 69.3, 69.6, 69.7, 69.9, 70.0, 70.0, 70.0, 70.0, 70.0, 70.0, 70.2, 70.2, 70.8, 71.0, 71.0, 71.1, 71.9, 72.0, 72.1, 72.4, 72.5, 73.1, 73.7, 74.2  **b.** 64.5 years  **c.** 68.0 years  **d.** 70.0 years

**e.**

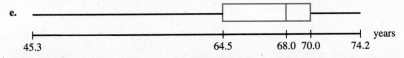

45.3  64.5  68.0  70.0  74.2  years

**f.** 19.2 years; 3.5 years; 2.0 years; 4.2 years  **g.** The interval from $\tilde{\mu}$ to $Q_3$
**3. a.** −0.33  **b.** −0.31  **c.** Krosno  **5. a.** 0.21  **b.** 0.21  **c.** approximately the same

**EXERCISES 2-5, SET A**    (page 112)

**1. a.** 16%   **b.** 44%   **c.** 33%   **d.** 7%   **3. a.** 13%   **b.** 37%   **d.** 49%   **5. a.** 50%   **b.** 17%   **c.** 23%   **d.** 10%   **7. a.** 5.6%   **b.** 1.5%
**9. a.** 26.7%   **b.** 48.9%   **c.** 24.4%   **11. a.** 85%   **b.** 15%   **13. a.** 17.8%   **b.** 17.8%   **c.** 11.5%   **d.** 8.9%   **e.** 7.6%
**15. a.** 40.4%   **b.** 22.3%   **c.** 16.4%   **d.** 11.8%   **17.** Tim. A proportion discounts the numbers of students the instructors had in their classes.

**EXERCISES 2-6, SET A**    (page 120)

**1. a.**

| $f_i x_i$ | $x_i^2$ | $f_i x_i^2$ |
|---|---|---|
| 218.0 | 2970.25 | 11,881.00 |
| 322.5 | 4160.25 | 20,801.25 |
| 1341.0 | 5550.25 | 99,904.50 |
| 591.5 | 7140.25 | 49,981.75 |
| 567.0 | 8930.25 | 53,581.50 |

**b.** 76.0
**c.** 127.75
**d.** 11.3

**3. a.**

| $x_i$ | $f_i x_i$ | $x_i^2$ | $f_i x_i^2$ |
|---|---|---|---|
| 702 | 2,808 | 492,804 | 1,971,216 |
| 707 | 8,484 | 499,849 | 5,998.188 |
| 712 | 15,664 | 506,944 | 11,152,768 |
| 717 | 12,906 | 514,089 | 9,253,602 |
| 722 | 11,552 | 521,284 | 8,340,544 |
| 727 | 2,181 | 528,529 | 1,585,587 |

**b.** 715 g
**c.** 39 g$^2$
**d.** 6 g

**5. a.**

| Class Number | Class Limits | Class Mark, $x_i$ | Frequency, $f_i$ | $f_i x_i$ | $x_i^2$ | $f_i x_i^2$ |
|---|---|---|---|---|---|---|
| 1 | 20–29 | 24.5 | 17 | 416.5 | 600.25 | 10,204.25 |
| 2 | 30–39 | 34.5 | 18 | 621.0 | 1190.25 | 21,424.50 |
| 3 | 40–49 | 44.5 | 11 | 489.5 | 1980.25 | 21,782.75 |
| 4 | 50–59 | 54.5 | 7 | 381.5 | 2970.25 | 20,791.75 |
| 5 | 60–69 | 64.5 | 11 | 709.5 | 4160.25 | 45,762.75 |

**b.** 40.9 years
**c.** 201.15 years$^2$
**d.** 14.2 years
**e.** 41.0 and 13.8 years

**7. a.**

| Class Number | Class Limits | Class Mark, $x_i$ | Frequency, $f_i$ | $f_i x_i$ | $f_i x_i^2$ |
|---|---|---|---|---|---|
| 1 | 9.00–15.99 | 12.495 | 10 | 124.950 | 1,561.25025 |
| 2 | 16.00–22.99 | 19.495 | 12 | 233.940 | 4,560.66030 |
| 3 | 23.00–29.99 | 26.495 | 10 | 264.950 | 7,019.85025 |
| 4 | 30.00–36.99 | 33.495 | 9 | 301.455 | 10,097.23523 |
| 5 | 37.00–43.99 | 40.495 | 10 | 404.950 | 16,398.45025 |
| 6 | 44.00–50.99 | 47.495 | 9 | 427.455 | 20,301.97523 |

**b.** $29.30
**c.** 140.79 cent$^2$
**d.** $11.87
**e.** $29.92 and $12.56

**9. a.**

| Class Number | Class Limits | Class Mark, $x_i$ | Frequency, $f_i$ | $f_i x_i$ | $f_i x_i^2$ |
|---|---|---|---|---|---|
| 1 | 2.00–2.39 | 2.195 | 9 | 19.755 | 43.362225 |
| 2 | 2.40–2.79 | 2.595 | 9 | 23.355 | 60.606225 |
| 3 | 2.80–3.19 | 2.995 | 9 | 26.955 | 80.730225 |
| 4 | 3.20–3.59 | 3.395 | 7 | 23.765 | 80.682175 |
| 5 | 3.60–3.99 | 3.795 | 2 | 7.590 | 28.804050 |

**b.** 2.82
**c.** 0.24
**d.** 0.48
**e.** 2.81 and 0.51

**EXERCISES 2-6, SET B**    (page 124)

**3. a.**

| Class Number | Class Limits | Class Mark, $x_i$ | Frequency $f_i$ | $f_i \cdot x_i$ | $f_i \cdot x_i^2$ |
|---|---|---|---|---|---|
| 1 | −34−−15 | −24.5 | 1 | −24.5 | 600.25 |
| 2 | −14−5 | −4.5 | 0 | 0.0 | 0.00 |
| 3 | 6−25 | 15.5 | 4 | 62.0 | 961.00 |
| 4 | 26−45 | 35.5 | 43 | 1526.5 | 54,190.75 |
| 5 | 46−65 | 55.5 | 35 | 1942.5 | 107,808.75 |
| 6 | 66−85 | 75.5 | 9 | 679.5 | 51,302.25 |

**b.** 46 degrees
**c.** 16 degrees
**d.** 78%
**e.** 97%

## CHAPTER 2 REVIEW EXERCISES (page 129)

**1. a.**

| Class Number | Class Limits | Class Boundaries | Class Mark | Tally Marks | Frequency | Relative Frequency |
|---|---|---|---|---|---|---|
| 1 | 5–24 | 4.50–24.50 | 14.50 | ||||  |||| |||| |||| |||| | 25 | 50% |
| 2 | 25–44 | 24.50–44.50 | 34.50 | |||| |||| | | 11 | 22% |
| 3 | 45–64 | 44.50–64.50 | 54.50 | |||| | 5 | 10% |
| 4 | 65–84 | 64.50–84.50 | 74.50 | |||| || | 7 | 14% |
| 5 | 85–104 | 84.50–104.50 | 94.50 | || | 2 | 4% |

**b.** Frequency

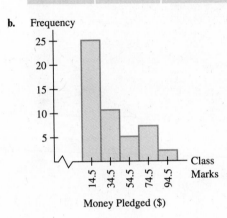

Money Pledged ($)

**c.** Frequency

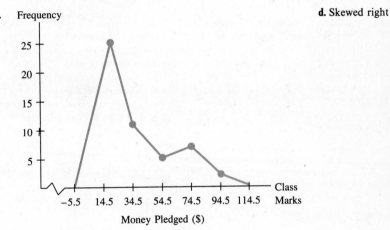

Money Pledged ($)

**d.** Skewed right

**3.**

```
2 | 6 8 8 9 9 9
3 | 3 4 4 5 6 6 7 8 8 8
4 | 0 0 0 2 2 3 3 3 3 4 6 7 7 8
5 | 0 1 1 2 2 3
```

**5. a.** 10.19 **b.** 10.19 **c.** 10.19 and 10.25 **7. a.** 50.7 pounds **b.** 5.4 pounds$^2$ **c.** 2.3 pounds

**9. a.**
```
4  16  20  21  22  23  25  26  27  27  27  28  29  31
31  32  33  33  33  33  35  36  36  36  37  37  37  39
40  40  40  41  41  42  42  42  42  43  44  44  45  45
45  46  47  47  48  48  48  48  49  50  50  51  51  51
53  56  56  57  57  60  60  62  63  64  65  70  73  76
```
**b.** 36 **c.** 57 **d.** 31 **e.** 48 **f.** 33 **g.** 51

**11. a.** 91% **b.** 37% **13. a.** 31.3% **b.** 23.8% **c.** 17.5% **d.** 15.0% **3.** 12.5%

## CHAPTER 3

### EXERCISES 3-1, SET A (page 152)

**1. a.**
$vw$ $wx$ $xy$ $yz$
$vx$ $wy$ $xz$
$vy$ $wz$
$vz$
**b.** 7 **c.** 1

**3. a.**
$x_1x_2$ $x_2x_3$ $x_3x_4$ $x_4y_1$ $y_1y_2$ $y_2z$
$x_1x_3$ $x_2x_4$ $x_3y_1$ $x_4y_2$ $y_1z$
$x_1x_4$ $x_2y_1$ $x_3y_2$ $x_4z$
$x_1y_1$ $x_2y_2$ $x_3z$
$x_1y_2$ $x_2z$
$x_1z$
**b.** 6 **c.** 3

**5. a.**
$n_1n_2n_3$ $n_1n_3d_1$ $n_1d_1d_3$ $n_2n_3d_3$ $n_3d_1d_2$
$n_1n_2d_1$ $n_1n_3d_2$ $n_1d_2d_3$ $n_2d_1d_2$ $n_3d_1d_3$
$n_1n_2d_2$ $n_1n_3d_3$ $n_2n_3d_1$ $n_2d_1d_3$ $n_3d_2d_3$
$n_1n_2d_3$ $n_1d_1d_2$ $n_2n_3d_2$ $n_2d_2d_3$ $d_1d_2d_3$
**b.** 9 **c.** 10

**7. a.**

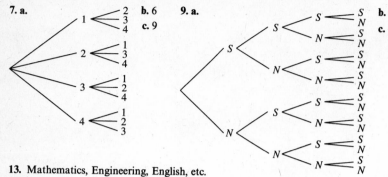

**b.** 6   **c.** 9

**9. a.**

**b.** 6   **c.** 15

**11. a.**

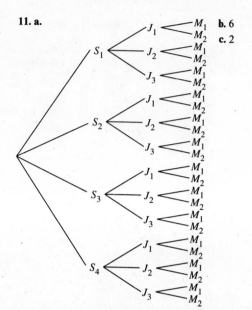

**b.** 6   **c.** 2

**13.** Mathematics, Engineering, English, etc.
**15.** Maine, New Hampshire, Vermont, etc.   **17.** Between $100,000 and $125,000, more than the median cost, less than the median cost, etc.
**19.** The stock price increased, the stock price decreased, the stock price remained the same   **21.** Not mutually exclusive   **23.** Mutually exclusive
**25.** Not mutually exclusive   **27.** Mutually exclusive   **29.** Not mutually exclusive
**31. a.** 67, 68, 76, 78, 86, and 87   **b.** 68, 76, 78, and 86   **c.** 78 and 87   **d.** Venn diagram
**33. a.** 1, 2, 3, 4, 5, 6, 7, and 8   **b.** 1, 3, 5, and 7   **c.** 5, 6, and 8   **d.** Venn diagram

## EXERCISES 3-1, SET B    (page 156)

**1.** 120   **3.** 479,001,600   **5.** 6,720   **7.** 17,160   **9.** 15   **11.** 3,432   **13.** 45   **15.** 35   **17.** 1   **19.** 25
**21. a.** $1.64649 \times 10^{12}$   **b.** $3.535316 \times 10^{18}$

## EXERCISES 3-2, SET A    (page 165)

**1. a.** $\frac{1}{10}$   **b.** $\frac{1}{2}$   **c.** 1   **d.** 0   **e.** Answer depends on current year.   **3. a.** $\frac{2}{7}$   **b.** $\frac{5}{7}$   **c.** $\frac{3}{7}$   **d.** $\frac{2}{7}$   **e.** 0

**5. a.** $q_1q_2$   $q_1n_2$   $q_2d_3$   $d_1d_2$   $d_1n_4$   $d_2n_4$   $n_1n_2$   $n_3n_4$   **b.** $\frac{1}{3}$   **c.** $\frac{5}{12}$   **d.** $\frac{1}{2}$
$q_1d_1$   $q_1n_3$   $q_2n_1$   $d_1d_3$   $d_2d_3$   $d_3n_1$   $n_1n_3$
$q_1d_2$   $q_1n_4$   $q_2n_2$   $d_1n_1$   $d_2n_1$   $d_3n_2$   $n_1n_4$
$q_1d_3$   $q_2d_1$   $q_2n_3$   $d_1n_2$   $d_2n_2$   $d_3n_3$   $n_2n_3$
$q_1n_1$   $q_2d_2$   $q_2n_4$   $d_1n_3$   $d_2n_3$   $d_3n_4$   $n_2n_4$

**7. a.**

| + | 1 | 2 | 3 | 4 | 5 | 6 | 7 | 8 |
|---|---|---|---|---|---|---|---|---|
| 1 | 2 | 3 | 4 | 5 | 6 | 7 | 8 | 9 |
| 2 | 3 | 4 | 5 | 6 | 7 | 8 | 9 | 10 |
| 3 | 4 | 5 | 6 | 7 | 8 | 9 | 10 | 11 |
| 4 | 5 | 6 | 7 | 8 | 9 | 10 | 11 | 12 |
| 5 | 6 | 7 | 8 | 9 | 10 | 11 | 12 | 13 |
| 6 | 7 | 8 | 9 | 10 | 11 | 12 | 13 | 14 |
| 7 | 8 | 9 | 10 | 11 | 12 | 13 | 14 | 15 |
| 8 | 9 | 10 | 11 | 12 | 13 | 14 | 15 | 16 |

**b.** $\frac{7}{64}$

**c.** $\frac{3}{32}$

**d.** $\frac{1}{2}$

**e.** $\frac{5}{32}$

**9. a.** $\frac{1}{6}$   **b.** $\frac{11}{15}$   **11. a.** $\frac{2}{5}$   **b.** $\frac{31}{70}$   **c.** $\frac{31}{35}$   **13. a.** $\frac{1}{8}$   **b.** $\frac{7}{10}$   **c.** $\frac{33}{40}$   **15. a.** $\frac{9}{35}$   **b.** $\frac{1}{7}$   **c.** $\frac{1}{2}$   **17.** Elizabeth's high school gpa; Elizabeth's scores on a standarized entrance test; Elizabeth's scores on some assessment tests given at Grand Central U, etc.   **19.** How tired the starting pitcher is; how successful the starting pitcher has been in getting this batter out; the hitting style of this batter (left-handed or right-handed) and the pitching style of the starting and relief pitchers; etc.   **21.** How extensive the damage is to the knee; the success of past surgeries in similar cases; the opinions of other surgeons; etc.   **23. a.** 0.72   **b.** 0.24   **c.** 0.28   **d.** 0.76   **25. a.** 0.18   **b.** 0.82   **c.** 0.64   **d.** 0.36   **e.** 0.15   **f.** 0.85

### EXERCISES 3-2, SET B   (page 169)

**1.** The sum of the probabilities is greater than 1.00.   **3.** The sum of the probabilities is 0.97, which is less than 1.00.
**5.** The sum of the indicated probabilities is 1.04.   **7. a.** 0.4   **b.** 0.5   **c.** 0.6   **d.** 0.5   **9. a.** 0.4   **b.** 0.6   **c.** 0.3   **d.** 0.7   **e.** 0.3   **f.** 0.7

**11.**
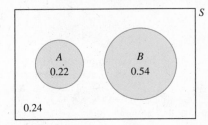

**13.**
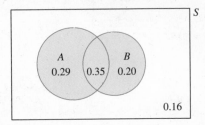

### EXERCISES 3-3, SET A   (page 179)

**1. a.** 66, 77, 88, 99   **b.** 67, 69, 77, 79, 87, 89, 97, 99   **c.** 66, 67, 69, 77, 79, 87, 88, 89, 97, 99   **d.** 77, 99   **3. a.** a, b   **b.** a, d   **c.** a, b, d   **d.** a
**5. a.** $x_1x_2, x_1x_3, x_2x_3$   **b.** $y_1z, y_2z$   **c.** $x_1y_1, x_1y_2, x_1z, x_2y_1, x_2y_2, x_2z, x_3y_1, x_3y_2, x_3z, y_1z, y_2z$   **d.** The 5 outcomes listed in parts (a) and (b)
**e.** The 14 outcomes listed in parts (a) and (c)   **f.** The 11 outcomes listed in part (c)   **g.** None   **h.** The 2 outcomes in part (b)
**7. a.** $q_1d_1, q_1d_2, q_1d_3, q_2d_1, q_2d_2, q_2d_3$   **b.** $d_1n_1, d_1n_2, d_1n_3, d_1n_4, d_2n_1, d_2n_2, d_2n_3, d_2n_4, d_3n_1, d_3n_2, d_3n_3, d_3n_4$
**c.** $d_1n_1, d_1n_2, d_1n_3, d_1n_4, d_2n_1, d_2n_2, d_2n_3, d_2n_4, d_3n_1, d_3n_2, d_3n_3, d_3n_4, n_1n_2, n_1n_3, n_1n_4, n_2n_3, n_2n_4, n_3n_4$
**d.** The 18 outcomes listed in parts (a) and (b)   **e.** The 24 outcomes listed in parts (a) and (c)   **f.** The 18 outcomes listed in part (c)
**g.** None   **h.** The 12 outcomes listed in part (b)   **9. a.** $\frac{2}{13}$   **b.** $\frac{7}{13}$   **c.** $\frac{11}{26}$   **d.** $\frac{7}{13}$   **e.** $\frac{7}{13}$   **11. a.** $\frac{1}{2}$   **b.** $\frac{3}{5}$   **c.** $\frac{2}{5}$   **d.** $\frac{7}{10}$   **e.** 0   **13. a.** 0.58   **b.** 0.62
**c.** 0.88   **d.** 0.68   **15. a.** $\frac{5}{18}$   **b.** $\frac{313}{450}$   **c.** $\frac{94}{225}$   **17. a.** $\frac{3}{14}$   **b.** $\frac{3}{7}$   **c.** $\frac{19}{28}$   **d.** $\frac{9}{28}$   **e.** 0   **19. a.** $\frac{1}{6}$   **b.** $\frac{2}{9}$   **c.** $\frac{7}{36}$   **d.** and **e.** Cannot determine.   **21. a.** 0.45
**b.** 0.60   **c.** 0.89   **d.** 0.55   **e.** 0.84   **23. a.** 0.6   **b.** 0.4   **c.** 0.2   **d.** 0.4   **e.** 0.8   **f.** 0.7   **g.** 0.6   **h.** 0.9   **i.** 0.8   **j.** 1.0   **25.** 0.58   **27.** 0.20   **29.** 0.30

### EXERCISES 3-3, SET B   (page 184)

**5.**
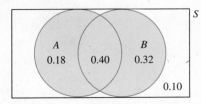

**7.**
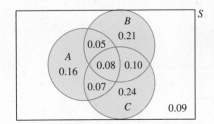

### EXERCISES 3-4, SET A   (page 193)

**1. a.** $\frac{29}{50}$   **b.** $\frac{16}{25}$   **c.** $\frac{9}{38}$   **d.** $\frac{21}{37}$   **3. a.** $\frac{1}{2}$   **b.** $\frac{1}{5}$   **c.** $\frac{3}{28}$   **d.** $\frac{10}{47}$   **5. a.** 0.650   **b.** 0.350   **c.** 0.364   **d.** 0.321   **7. a.** $\frac{2}{7}$   **b.** $\frac{1}{2}$   **9. a.** $\frac{1}{2}$   **b.** $\frac{2}{5}$   **c.** $\frac{2}{5}$   **d.** $\frac{2}{3}$
**e.** 0   **f.** 0   **11. a.** $\frac{1}{2}$   **b.** $\frac{2}{3}$   **c.** $\frac{3}{8}$   **d.** $\frac{1}{2}$   **e.** No   **13. a.** $\frac{1}{6}$   **b.** $\frac{1}{6}$   **c.** Yes   **d.** $\frac{1}{6}$   **e.** $\frac{1}{6}$   **15. a.** 0.30   **b.** 0.30   **c.** Yes   **d.** 0.30   **e.** 0.30   **17.** Yes
**19.** No   **21.** Yes   **23.** Yes   **25.** Yes

### EXERCISES 3-5, SET A   (page 205)

**1. a.** $\frac{1}{221}$   **b.** $\frac{1}{17}$   **c.** $\frac{8}{663}$   **3. a.** 0.09   **b.** 0.49   **c.** 0.42   **5.** 0.45   **7.** 0.28   **9. a.** 0.40   **b.** $A$ and $B$ are independent events.
**11. a.** 0.2025   **b.** The events are independent.   **13. 1. a.** $\frac{1}{36}$   **b.** $\frac{289}{900}$   **c.** $\frac{4}{45}$   **d.** $\frac{68}{225}$   **2. a.** $\frac{2}{87}$   **b.** $\frac{136}{435}$   **c.** $\frac{8}{87}$   **d.** $\frac{136}{435}$
**15. 1. a.** $\frac{1}{49}$   **b.** $\frac{169}{1225}$   **c.** $\frac{16}{245}$   **d.** $\frac{6}{245}$   **2. a.** $\frac{2}{119}$   **b.** $\frac{78}{595}$   **c.** $\frac{8}{119}$   **d.** $\frac{3}{119}$   **17. a.** 0.4   **b.** 0.2   **c.** 0.4   **d.** Yes
**19. a.** 0.30   **b.** 0.15   **c.** $\frac{3}{7}$   **d.** No   **e.** 0.20   **f.** 0.07   **g.** 0.20   **h.** Yes   **i.** 0   **j.** 0   **k.** No   **21.** 0.008   **23.** $\frac{16}{81}$

## EXERCISES 3-5, SET B   (page 209)

**1. a.** Yes. She can afford only one holiday.   **b.** No. Mutually exclusive events are dependent.   **c.** 0   **d.** 0.85   **e.** 0.15
**3. a.** No, a female employee can commute more than 15 miles to work.   **b.** Yes, commute distances are not necessarily related to sex of employee.
**c.** 0.11   **d.** 0.64

## CHAPTER 3 REVIEW EXERCISES   (page 214)

**1. a.** $d_1d_2$   $d_2d_3$   $d_3d_4$   $d_4r_1$   $r_1r_2$   $r_2r_3$   $r_3i$   **b.** 6
     $d_1d_3$   $d_2d_4$   $d_3r_1$   $d_4r_2$   $r_1r_3$   $r_2i$   **c.** 18
     $d_1d_4$   $d_2r_1$   $d_3r_2$   $d_4r_3$   $r_1i$
     $d_1r_1$   $d_2r_2$   $d_3r_3$   $d_4i$
     $d_1r_2$   $d_2r_3$   $d_3i$
     $d_1r_3$   $d_2i$
     $d_1i$

**3.** She plays the infield; she plays the outfield; she is a pitcher; etc.
**5.** Mutually exclusive.   **7. a.** 0.14   **b.** 0.53   **c.** 0.82   **9. a.** 0.5   **b.** 0.5   **c.** 0.6
**d.** 0.5   **e.** 0.5   **f.** 0.4   **g.** 0.8   **h.** 0.8   **i.** 0.8   **j.** 0.2   **k.** 0.1   **11. a.** 0.64   **b.** 0.43
**c.** 0.84   **d.** 0.60   **e.** 0.75   **f.** 0.55   **g.** 0.13   **h.** 0.11   **13. a.** No, both could be
college graduates   **b.** Yes, small sample from large population   **c.** 0.0324
**d.** 0.3276

## CHAPTER 4

## EXERCISES 4-1, SET A   (page 237)

**1.** 1, 5, 10, 25, 50, 100   **3.** $(8°, 58°)$   **5.** 0, 1, 2, 3, 4, 5, 6, 7, 8, 9, 10   **7.** (3 ft, 7 ft)   **9.** 2, 3, 4, 5, 6, 7, 8, 9, 10
**11.** $X = 1$, if the vehicle is a pickup truck; $X = 0$, if the vehicle is not a pickup truck.
**13.** $X = 1$, if the doctor is a woman; $X = 0$, if the doctor is not a woman.
**15.** $X = 1$, if the player is a linebacker; $X = 0$, if the player is not a linebacker.
**17. a.** 0, 1, 2, 3
**b.** 0.216; 0.432; 0.288; 0.064

**19. a.** 0, 1, 2, 3, 4
**b.** 0.15; 0.30; 0.35; 0.12; 0.08

**c.**
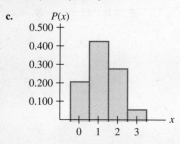
X = number of
blue chips selected

**c.**
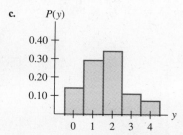
Y = number of customers
in next checkout line

**21. a.** 0, 1, 2, 3, 4
**b.** 0.28; 0.32; 0.20; 0.16; 0.04
**c.**
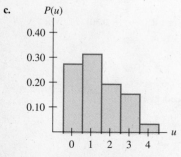
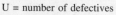
U = number of defectives

**23. a.** 2, 3, 4, 5, 6   **b.** $P(2) = \frac{25}{144}$; $P(3) = \frac{40}{144}$; $P(4) = \frac{46}{144}$; $P(5) = \frac{24}{144}$; $P(6) = \frac{9}{144}$
**c.**
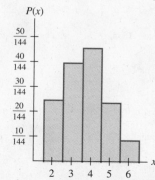
X = sum of numbers on faces

**EXERCISES 4-1, SET B** (page 240)

**1. a.** 2, 3, 4, 5, 6

**b.** $P(2) = \frac{1}{9}$; $P(3) = \frac{2}{9}$; $P(4) = \frac{3}{9}$; $P(5) = \frac{2}{9}$; $P(6) = \frac{1}{9}$

**c.**

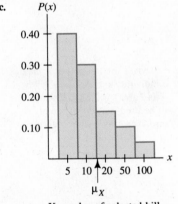

X = sum of two disks

**2. a.** 1, 1.5, 2, 2.5, 3

**b.** $P(1) = \frac{1}{9}$; $P(1.5) = \frac{2}{9}$; $P(2) = \frac{3}{9}$; $P(2.5) = \frac{2}{9}$; $P(3) = \frac{1}{9}$

**c.**

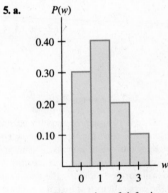

$\overline{X}$ = mean of the total

**d.** They are the same shape.

**EXERCISES 4-2, SET A** (page 247)

**1. a.** 5, 10, 20, 50, 100

**b.** $P(5) = 0.40$; $P(10) = 0.30$; $P(20) = 0.15$; $P(50) = 0.10$; $P(100) = 0.05$

**c.**

X = value of selected bill

**d.** 18   **e.** 22.9

**3. a.** 2, 3, 4, 5, 6

**b.** $P(2) = \frac{1}{4}$; $P(3) = \frac{1}{3}$; $P(4) = \frac{5}{18}$;

$P(5) = \frac{1}{9}$; $P(6) = \frac{1}{36}$

**c.**

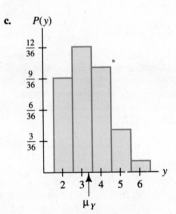

Y = sum of two tosses

**d.** $\frac{10}{3}$   **e.** 1.1

**5. a.**

W = number of defectives

**b.** 1.1   **c.** 0.9

**7. a.** 7.0   **b.** The expected outcome of each roll of the dice is 7.0.   **c.** 2.4   **d.** 5, 6, 7, 8, and 9   **9. a.** 4.7   **b.** The average number of women in a sample of eight is 4.7.   **c.** 1.4   **d.** 2, 3, 4, 5, 6, and 7

**11. a.** 10, 11 ,12 ,13, 14, 15, and 16   **b.** $P(10) = 0.062$,
$P(11) = 0.078$, $P(12) = 0.110$, $P(13) = 0.170$, $P(14) = 0.180$,
$P(15) = 0.250$, and $P(16) = 0.150$   **c.** 13.7

**d.**

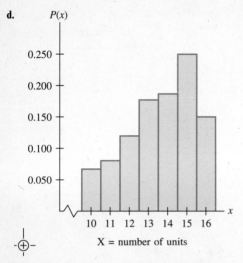

X = number of units

**e.** 1.7   **f.** 12, 13, 14, and 15   **13. a.** 1.7   **b.** 1.1

**1. a.** 0, 1, 2   **b.** $P(0) = 0.36$; $P(1) = 0.48$; $P(2) = 0.16$   **c.** 0.8   **d.** 0.5
**3. a.** 0, 1, 2   **b.** $P(0) = 0.59$; $P(1) = 0.35$; $P(2) = 0.05$   **c.** 0.5   **d.** 0.4

**e.**                                    **5.** Yes, the game is fair

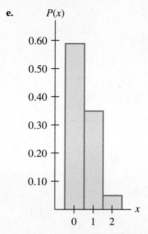

X = number in day-care
centers

**EXERCISES 4-3, SET A**   (page 262)

**1.** Assumption 1. $p = 0.80$ will be used for each component. Assumption 2. Whether a given component survives the test or not is *unrelated* to other components; thus the trials are independent.   **3.** Assumption 1. The relative frequency of obese individuals in this area will be used as $p$ for each individual sampled. Assumption 2. Because only 30 individuals in the "large metropolitan area" will be sampled, it is reasonable to model the trials of the experiment independent.   **5.** Assumption 1. A value of $p = 0.25$ will be used that the randomly selected card will correctly identify the answer of each of the 25 questions. Assumption 2. Because the cards are drawn *with replacement*, the trials of the experiment will be independent.   **7.** 220   **9.** 125, 970   **11.** 435

**13.** $\binom{5}{1} = 5$; $\binom{5}{2} = 10$   **15.** $\binom{9}{0} = 1$; $\binom{9}{3} = 84$   **17.** 0.027   **19.** 0.5184   **21.** 0.0256   **23.** 0.250047

**25. a.** 0, 1, 2, 3   **b.** $P(0) = 0.296$; $P(1) = 0.444$; $P(2) = 0.222$; $P(3) = 0.037$
**c.** The probability is approximately 44.4% that exactly one red face will occur in the three trials.
**27. a.** 0, 1, 2, 3, 4, 5   **b.** $P(0) = 0.237$; $P(1) = 0.396$; $P(2) = 0.264$; $P(3) = 0.088$; $P(4) = 0.015$; $P(5) = 0.001$
**c.** The probability is approximately 26.4% that exactly two yellow candies will be selected in the sample of five.
**29. a.** 0, 1, 2, 3, 4, 5, 6   **b.** $P(0) = 0.139$; $P(1) = 0.325$; $P(2) = 0.316$; $P(3) = 0.164$; $P(4) = 0.048$; $P(5) = 0.007$; $P(6) = 0.000$
**c.** The probability is approximately 16.4% that exactly three of the students in the sample will be postgraduates.
**31. a.** A given household does (a success), or does not (a failure) have pets. The value 0.32 will be used for the probability of success on each trial. Because a small sample will be taken from a large population, the trials can be described as independent.   **b.** 0.264
**c.** The probability is approximately 26.4% that exactly three of the ten residents sampled will live in a household with no pets.
**33. a.** A given resident will consider himself to be Protestant (a success) or not (a failure). The value 0.60 will be used for each resident sampled. Only ten residents in a large city will be sampled; therefore the trials can be described as independent.   **b.** 0.251
**c.** The probability is approximately 25.1% that exactly six of the ten residents sampled will consider themselves Protestants.
**35. a.** 0.154   **b.** 0.526   **c.** 0.476   **d.** 0.130   **37. a.** 0.261   **b.** 0.420   **c.** 0.096   **d.** 0.710   **39. a.** 0.196   **b.** 0.059   **c.** 0.832

**EXERCISES 4-3, SET B** (page 264)

**3. a.** A given owner will either be from 35 to 44 years old (a success), or not (a failure). The 26.3% value will be used for each individual sampled. Only eight homes in a large area will be sampled; thus the trials can be described as independent.
**b.** $P(0) = 0.087$; $P(1) = 0.249$; $P(2) = 0.310$; $P(3) = 0.222$; $P(4) = 0.099$; $P(5) = 0.028$; $P(6) = 0.005$; $P(7) = 0.001$; $P(8) = 0.0+$
**c.** The probability is about 9.9% that exactly four of the eight in the sample will be from 35 to 44 years old.

**EXERCISES 4-4, SET A** (page 271)

**1. a.** 0.107; 0.302; 0.340; 0.191; 0.054; 0.006  **b.** 1.8  **c.** 1.1

**d.** and **e.**

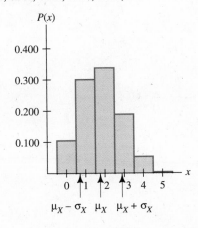

**3. a.** 0.017; 0.095; 0.224; 0.293; 0.230; 0.109; 0.028; 0.003
**b.** 3.1  **c.** 1.3

**d.** and **e.**

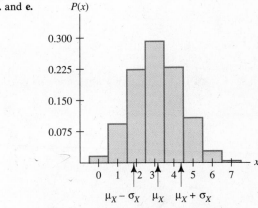

**5. a.** A given disk will have an even number (a success), or an odd number (a failure). The probability of success on each disk will be 0.4. The disks are sampled with replacement; thus the trials can be described as independent.  **b.** 0.130; 0.346; 0.346; 0.154; 0.026  **c.** 1.6  **d.** On the average, 1.6 even-numbered disks are selected  **e.** 1.0

**7. a.** 0.005; 0.045; 0.157; 0.289; 0.299; 0.165; 0.038  **b.** 3.5
**c.** On the average, the six citations will have 3.5 drivers less than the age of 21 years  **d.** 1.2

**e.** and **f.**

**f.**

**9. a.** Each student in the sample is either a freshman (a success), or not a freshman (a failure). The value 0.32 can be used for the probability that a given student is a freshman. Because eight is small compared with the total enrollment of a university, the trials can be described as independent.  **b.** 0.046; 0.172; 0.283; 0.267; 0.157; 0.059; 0.014; 0.002; 0.0+  **c.** The probability is about 26.7 percent that the sample will contain exactly three freshmen  **d.** 2.6  **e.** On the average, a sample of eight students will contain 2.6 freshmen  **f.** 1.3

g.

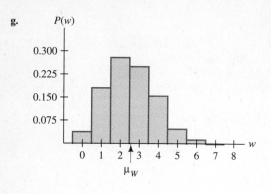

**11. a.** 0.172   **b.** There is an approximate 17.2 percent probability that a randomly selected bag of 12 ears of corn will have exactly 3 with some type of blemish   **c.** 1.8   **d.** On the average, a bag of 12 ears of this corn will have 1.8 ears with some type of blemish   **e.** 1.2   **13. a.** 2.4   **b.** 1.4   **15.** You might conclude that the coin is biased in favor of tails. If $p = 0.5$ (an unbiased coin), then $P(X \leq 2) = 0.055$. Because this happened twice in a row, and $P(X \leq 2)$ twice if the coin is fair is $(0.055)(0.055) \approx 0.003$, there is reason to doubt $p = 0.5$.

## EXERCISES 4-4, SET B   (page 274)

**1. a.** omitted   **b.** 0.001; 0.006; 0.030; 0.091; 0.179; 0.242; 0.227; 0.146; 0.062; 0.016; 0.002   **c.** 5.3   **d.** 1.6   **e.** omitted
**3. a.** omitted   **b.** 0.004; 0.030; 0.102; 0.203; 0.259; 0.221; 0.125; 0.046; 0.010; 0.001   **c.** 4.1   **d.** 1.5   **e.** omitted

## EXERCISES 4-5, SET A   (page 283)

**1. a.** 0.1977   **b.** 0.9082   **3. a.** 0.2426   **b.** 0.2610   **5. a.** $-1.28$   **b.** 0.60
**7. a.** $-1.60$   **b.** $x = 62.0$ is 1.60 standard deviations below $\mu_x$.

**9. a.** $z_1 = -1.25$ and $z_2 = 0.75$

**c.**

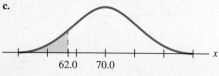

**d.** 0.0548

**b.**

**c.** 0.6678

**11. a.**

**b.** $-0.75$   **c.** 0.62 ohms is 0.75 standard deviations below the mean of 0.68 ohms **d.** 0.2266   **e.** The probability is approximately 23 percent that a randomly selected roll of this wire has a resistance less than 0.62 ohms.

**13. a.**

**b.** 0.1056   **c.** 0.2266   **d.** 0.6678   **e.** The probability is approximately 67 percent that a randomly selected valve has a lip-opening pressure between 10.2 and 11.0 psi.

**15. a.**

**b.** 0.3557   **c.** 0.5086   **d.** The probability is approximately 51 percent that a randomly selected days weather report from this station for July or August indicates a high temperature of from 80.0 to 90.0°.

**17. a. and c.**

**b.** 0.5375   **d.** The probability is approximately 54 percent that a randomly selected chicken from this flock will weigh between 4.0 and 5.0 pounds.

**19. a.**

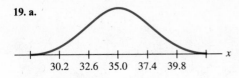

**b.** 32.6 and 37.4 mpg   **c.** 0.6826

## EXERCISES 4-5, SET B   (page 285)

**1. a.** 67.3 inches   **b.** 71.7 inches   **c.** 67.9 to 72.1 inches
**d.**

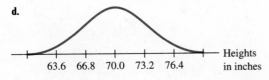

**5. a.** 44.7   **b.** 38.8   **c.** 39.3 to 42.9   **7.** 73.7

## EXERCISES 4-6, SET A   (page 292)

**1. a.** $(50)(0.30) \geq 5$ and $(50)(0.70) \geq 5$   **b.** 0.3228

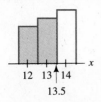

**d.** The probability is approximately 32 percent that fewer than 14 students in the sample are planning to attend a state college after graduation.

**3. a.** $(25)(0.75) \geq 5$ and $(25)(0.25) \geq 5$   **b.** 0.5478
**c.**

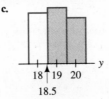

**d.** The probability is approximately 55 percent that 19 or more heads will be observed in the 25 tosses of this coin.

**21. a.** 76%   **b.** 12%   **c.** 29%
**d.**

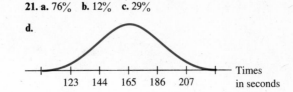

**3. a.**

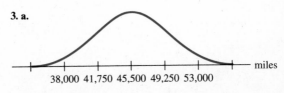

**b.** 0.7486   **c.** 0.3755

**5. a.** $(150)(0.58) \geq 5$ and $(150)(0.42) \geq 5$   **b.** 0.6115
**c.**

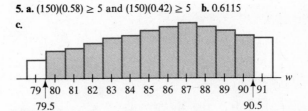

**d.** The probability is approximately 61 percent that between 80 and 90 union members inclusive in the sample of 150 would favor a strike in favor of "job security measures."

**7. a.** 0.2177   **b.** 0.5663   **9. a.** 0.7673   **b.** 0.6116   **11. a.** 0.3332   **b.** There is an approximate 33 percent probability that between 26 and 30, inclusive, of the individuals in the sample of 250 will be males.
**13. a.** 0.1260   **b.** 0.1272   **c.** 0.13 and 0.13; the same.

## EXERCISES 4-6, SET B   (page 293)

**1. a.** 0.7416   **3.** 0.6553   **5. a.** 0.12705   **b.** 0.12551   **7. a.** 0.33721   **b.** 0.33470

## EXERCISES 4-7, SET A　(page 303)

**1. a.**

| Y \ X | 0 | 1 | 2 | 3 |
|---|---|---|---|---|
| 0 | 0.08 | 0.16 | 0.12 | 0.04 |
| 1 | 0.10 | 0.20 | 0.15 | 0.05 |
| 2 | 0.02 | 0.04 | 0.03 | 0.01 |

**b.** $X = 1$ and $Y = 1$
**c.** $X = 3$ and $Y = 2$

**3. a.**

| Y \ X | 0 | 1 | 2 | 3 | 4 |
|---|---|---|---|---|---|
| 1 | 0.02 | 0.04 | 0.05 | 0.06 | 0.03 |
| 3 | 0.03 | 0.06 | 0.075 | 0.09 | 0.045 |
| 5 | 0.05 | 0.10 | 0.125 | 0.15 | 0.075 |

**b.** $X = 3$ and $Y = 5$
**c.** $X = 0$ and $Y = 1$

**5. a.**

| Y \ X | 3 | 4 | 5 | 6 |
|---|---|---|---|---|
| 3 | 0.06 | 0.09 | 0.12 | 0.03 |
| 4 | 0.12 | 0.18 | 0.24 | 0.06 |
| 5 | 0.02 | 0.03 | 0.04 | 0.01 |

**b.** 0.17

**7.** 0.29

**9. a.**

| $\bar{X}_i$ | $P(\bar{X}_i = \bar{x}_i)$ |
|---|---|
| 0 | 0.36 |
| $\frac{1}{2}$ | 0.36 ( = 0.18 + 0.18) |
| 1 | 0.21 ( = 0.06 + 0.09 + 0.06) |
| $\frac{3}{2}$ | 0.06 ( = 0.03 + 0.03) |
| 2 | 0.01 |

**b.** 0.5　**c.** 0.47, to two places.

**11. a.**

| $\bar{X}_i$ | $P(\bar{X}_i = \bar{x}_i)$ |
|---|---|
| 0 | 0.006 |
| $\frac{1}{2}$ | 0.016 ( = 0.008 + 0.008) |
| 1 | 0.074 ( = 0.032 + 0.010 + 0.032) |
| $\frac{3}{2}$ | 0.123 ( = 0.0216 + 0.04 + 0.04 + 0.0216) |
| 2 | 0.238 ( = 0.012 + 0.027 + 0.160 + 0.027 + 0.012) |
| $\frac{5}{2}$ | 0.246 ( = 0.015 + 0.108 + 0.108 + 0.015) |
| 3 | 0.193 ( = 0.060 + 0.0729 + 0.060) |
| $\frac{7}{2}$ | 0.081 ( = 0.405 + 0.405) |
| 4 | 0.023 |

**b.** The probability is approximately 24.6 percent that the mean of the two grades in the sample is 2.5
**c.** 2.3
**d.** On the average, the mean of the two grades is 2.3
**e.** 0.8

## EXERCISES 4-7, SET B (page 306)

**1. a.**

| A | B | C | Sum |
|---|---|---|-----|
| 1 | 1 | 1 | 3 |
| 1 | 1 | 2 | 4 |
| 1 | 2 | 1 | 4 |
| 1 | 2 | 2 | 5 |
| 2 | 1 | 1 | 4 |
| 2 | 1 | 2 | 5 |
| 2 | 2 | 1 | 5 |
| 2 | 2 | 2 | 6 |

**b.** {3, 4, 5, 6}
**c.** $P(3) = 0.28$; $P(4) = 0.47$; $P(5) = 0.22$; $P(6) = 0.03$

**3. a.**

| A | B | C | Sum |   | A | B | C | Sum |
|---|---|---|-----|---|---|---|---|-----|
| 1 | 1 | 1 | 3 |   | 2 | 2 | 3 | 7 |
| 1 | 1 | 2 | 4 |   | 2 | 3 | 1 | 6 |
| 1 | 1 | 3 | 5 |   | 2 | 3 | 2 | 7 |
| 1 | 2 | 1 | 4 |   | 2 | 3 | 3 | 8 |
| 1 | 2 | 2 | 5 |   | 3 | 1 | 1 | 5 |
| 1 | 2 | 3 | 6 |   | 3 | 1 | 2 | 6 |
| 1 | 3 | 1 | 5 |   | 3 | 1 | 3 | 7 |
| 1 | 3 | 2 | 6 |   | 3 | 2 | 1 | 6 |
| 1 | 3 | 3 | 7 |   | 3 | 2 | 2 | 7 |
| 2 | 1 | 1 | 4 |   | 3 | 2 | 3 | 8 |
| 2 | 1 | 2 | 5 |   | 3 | 3 | 1 | 7 |
| 2 | 1 | 3 | 6 |   | 3 | 3 | 2 | 8 |
| 2 | 2 | 1 | 5 |   | 3 | 3 | 3 | 9 |
| 2 | 2 | 2 | 6 |   |   |   |   |   |

**b.** {3, 4, 5, 6, 7, 8, 9}
**c.** $P(3) = 0.210$; $P(4) = 0.256$; $P(5) = 0.282$; $P(6) = 0.154$; $P(7) = 0.076$; $P(8) = 0.018$; $P(9) = 0.004$

## CHAPTER 4 REVIEW EXERCISES (page 310)

**1. a.** {1, 2, 3, 4, 5}   **b.** 0.350; 0.228; 0.148; 0.096; 0.179

**c. and f.**

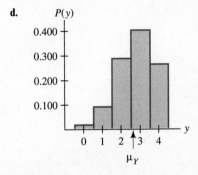

**c.** 2.529   **d.** On the average, Connie bowls about 2.5 frames for a practice session such as the one described.   **e.** 1.5

**3. a.** 2.8   **b.** 0.83   **c.** 0.9

**d.**

**5. a.** Each vehicle sampled either does (a success), or does not (a failure), need repairs on the smog control device. The value 0.23 will be used for the probability of success on each vehicle. Because only six vehicles in the state's population of vehicles will be sampled, the trials can be described as independent.   **b.** {0, 1, 2, 3, 4, 5, 6}   **c.** 0.208; 0.374; 0.279; 0.111; 0.025; 0.003; 0.0+

**d.**

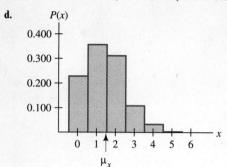

**e.** 1.38   **f.** 1.0   **g.** 1 and 2

**7. a.** 0.251   **b.** 0.382   **c.** 0.055   **d.** 0.605
**9. a.** 0.1056   **b.** 0.9525   **c.** 0.0188   **d.** 0.1115   **11. a.** (40)(0.20) ≥ 5 and (40)(0.80) ≥ 5   **b.** 0.6778
**c.** The probability is approximately 68 percent that between 6 and 10, inclusive, of the aircraft sampled cannot be flown because of missing parts.

**13. a.**

| Y \ X | 1 | 2 | 3 | 4 |
|---|---|---|---|---|
| 2 | 0.10 | 0.15 | 0.20 | 0.05 |
| 4 | 0.06 | 0.09 | 0.12 | 0.03 |
| 6 | 0.04 | 0.06 | 0.08 | 0.02 |

**b.** $X = 3$ and $Y = 2$
**c.** $X = 4$ and $Y = 6$
**d.** 0.57

# CHAPTER 5

## EXERCISES 5-1, SET A   (page 330)

**1. a.** $\mu_{\bar{x}} = 100.0$; $\sigma_{\bar{x}} = 2.5$   **b.** $\mu_{\bar{x}} = 100.0$; $\sigma_{\bar{x}} = 1.3$   **3. a.** $\mu_{\bar{x}} = 45.3$; $\sigma_{\bar{x}} = 0.6$   **b.** $\mu_{\bar{x}} = 45.3$; $\sigma_{\bar{x}} = 0.3$
**5. a.** $1800
**b.** $40

**c.**

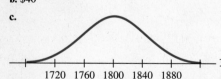

**d.** 0.63   **e.** $1825 is approximately 0.63 standard deviations above $\mu$.
**f.** 0.7357   **g.** The probability is approximately 74 percent that the mean of a sample of 36 will be less than $1825.

**7. a.** 2.00
**b.** 0.02

**c.**

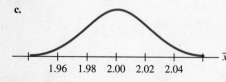

**d.** 1.50   **e.** 2.03 is 1.50 standard deviations above $\mu$.   **f.** 0.0668
**g.** The probability is approximately 7 percent that the mean butterfat content of the sampled cartons will be more than 2.03.

**9. a.** $70,500 **b.** $5,000 **c.** 0.1357

**d.**

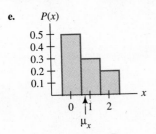

$70,500 $75,000

**13. a.** 0.8164

**b.**

554.0  560.0  566.0

**c.** The probability is approximately 82 percent that the mean of a random sample of 16 resistors is between 554.0 and 566.0 ohms.

**17. a.** 0.2514 **b.** 0.1587 **c.** Approximately 9% **19.** 28.1

**11. a.** 12.6 inches **b.** 0.15 **c.** 0.4972

**d.**

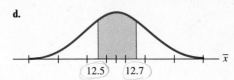

12.5  12.7

**15. a.** 0.2358

**b.**

33,500  34,000

**c.** The probability is approximately 24 percent that the mean of a random sample of 12 tires is more than 34,000 miles.

---

**EXERCISES 5-1, SET B** (page 333)

**1. a.** $\{0, 1, 2\}$
**b.** $P(0) = 0.5$; $P(1) = 0.3$; $P(2) = 0.2$
**c.** 0.7 **d.** 0.61

**e.**

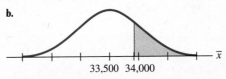

$P(x)$

0.5
0.4
0.3
0.2
0.1

0  1  2  $x$

$\mu_x$

**3. a.** $\frac{0+0}{2} = 0$; $\frac{0+\frac{1}{2}}{2} = \frac{1}{4}$; $\frac{0+1}{2} = \frac{1}{2}$; $\frac{0+\frac{3}{2}}{2} = \frac{3}{4}$; $\frac{0+2}{2} = 1$; $\frac{\frac{1}{2}+1}{2} = \frac{3}{4}$; $\frac{\frac{1}{2}+\frac{3}{2}}{2} = 1$; $\frac{\frac{1}{2}+2}{2} = \frac{5}{4}$; $\frac{1+\frac{3}{2}}{2} = \frac{5}{4}$;

$\frac{1+2}{2} = \frac{3}{2}$; $\frac{\frac{3}{2}+2}{2} = \frac{7}{4}$; $\frac{2+2}{2} = 2$ **b.** $P(0) = 0.0625$; $P(\frac{1}{4}) = 0.1500$; $P(\frac{1}{2}) = 0.2350$;

$P(\frac{3}{4}) = 0.2340$; $P(1) = 0.1761$; $P(\frac{5}{4}) = 0.0936$; $P(\frac{3}{2}) = 0.0376$; $P(\frac{7}{4}) = 0.0096$; $P(2) = 0.0016$
**c.** 0.7 **d.** 0.1525

**e.**

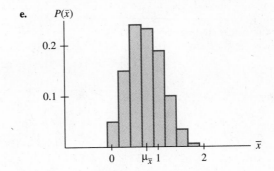

$P(\bar{x})$

0.2

0.1

0  $\mu_{\bar{x}}$  1  2  $\bar{x}$

---

**EXERCISES 5-2, SET A** (page 347)

**1. a.** 27.4 years **b.** 27.4 years **c.** 25.5 to 29.4 years
**d.** A 90 percent confidence interval for the mean age of women that apply for marriage licenses in Denver is between 25.5 and 29.4 years.
**3. a.** 29.6 mpg **b.** 1.6 mpg **c.** 29.0 to 30.2 mpg
**d.** A 98 percent confidence interval for the mean miles per gallon for this new model car is between 29.0 and 30.2 mpg.
**5.** From 108.0 to 116.8 cents **7.** From 47.6 to 49.2 peanuts **9. a.** 2.80 to 2.98 **b.** A 90 percent confidence interval for the mean grade point average of international students at this college is between 2.80 and 2.98.
**11. a.** From $9.47 to $19.73 **b.** A 98 percent confidence interval for the mean dollar amount of sales to customers at the Big Bear Supermarket in Lodi between 8 and 12 PM is between $9.47 and $19.73.
**13. a.** From 19.8 to 22.0 strokes **b.** A 95 percent confidence interval for the mean number of strokes of the pump to burst a balloon from this production run is between 19.8 and 22.0 strokes.
**15. a.** From 50.3 to 54.9 **b.** From 51.4 to 53.8 **c.** By 2.2 ($=4.6 - 2.4$) **17.** 36 **19.** 41 **21.** 57 **23.** 146

## EXERCISES 5-3, SET A  (page 357)

**1.** 1.860

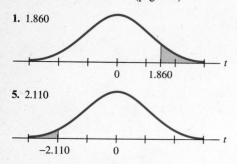

**3.** 2.306

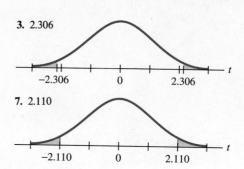

**5.** 2.110

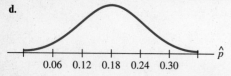

**7.** 2.110

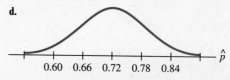

**9. a.** 12.8 pounds; 2.3 pounds, to the nearest tenth   **b.** From 11.8 to 13.8 pounds
**c.** A 90 percent confidence interval for the mean weight of the 450,000 turkeys is from 11.8 to 13.8 pounds.
**11. a.** 16.4%, to the nearest tenth; 3.3%, to the nearest tenth   **b.** From 14.7% to 18.1%
**c.** A 98 percent confidence interval for the rate of tips at Mama Rosa's Italian Restaurant based on the tab is from 14.7% to 18.1%.
**13. a.** From 120.9 to 131.1 sheets   **b.** A 90 percent confidence interval for the number of sheets of computer paper penetrated by a .22 caliber bullet fired from 100 feet is from 120.9 to 131.1 sheets.
**15. a.** From 6.10 to 6.38 volts   **b.** A 95 percent confidence interval for the mean voltage of the brand X Model M918 batteries in this production run is from 6.10 to 6.38 volts.
**17. a.** From 42.7 to 44.5 grams   **b.** A 98 percent confidence interval for the mean weight of the contents in jars of Italian Seasoning in this warehouse is from 42.7 to 44.5 grams.
**19. a.** From 69.0 to 74.0 minutes   **b.** A 99% confidence interval for the mean time it takes Charles to run this 10-mile course is from 69.0 to 74.0 minutes.   **21. a.** From 71.5 to 78.5   **b.** From 73.5 to 76.5   **c.** 2.3 ($=7.0 \div 3.0$)

## EXERCISES 5-3, SET B  (page 360)

**5.** Records indicate gpas at major universities are reasonably bell shaped.
**7.** Heights of males and females in the same age groups are reasonably bell shaped.
**9.** It seems reasonable that the distribution should be relatively bell shaped.

## EXERCISES 5-4, SET A  (page 370)

**1. a.** With $P = 0.18$, $(1 - P) = 0.82$, and $n = 41$:
$(41)(0.18) = 7.38$ ($\geq 5$) and $(41)(0.82) = 33.62$ ($\geq 5$)   **b.** 0.18   **c.** 0.06

**d.**

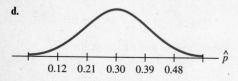

**3. a.** With $P = 0.72$, $(1 - P) = 0.28$, and $n = 56$:
$(56)(0.72) = 40.32$ ($\geq 5$) and $56(0.28) = 15.68$ ($\geq 5$)   **b.** 0.72   **c.** 0.06

**d.**

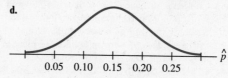

**5. a.** With $P = 0.30$, $(1 - P) = 0.70$, and $n = 25$:
$(25)(0.30) = 7.5$ ($\geq 5$) and $(25)(0.70) = 17.5$ ($\geq 5$)   **b.** 0.30   **c.** 0.09

**d.**

**7. a.** With $P = 0.15$, $(1 - P) = 0.85$, and $n = 50$:
$(50)(0.15) = 7.5$ ($\geq 5$) and $(50)(0.85) = 42.5$ ($\geq 5$)   **b.** 0.15   **c.** 0.05

**d.**

**9. a.** With $P = 0.38$, $(1 - P) = 0.62$, and $n = 48$:
$(48)(0.38) = 18.24$ ($\geq 5$) and $(48)(0.62) = 29.76$ ($\geq 5$)
**b.** 0.38   **c.** 0.07

**d.**

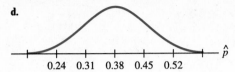

0.24   0.31   0.38   0.45   0.52   $\hat{p}$

**13. a.** 0.3085

**b.**

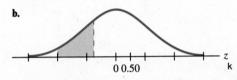

0 0.50   $z$   k

**c.** The probability is approximately 31 percent that more than 75 percent of the accidents involved death or injury.

**11. a.** 0.1922

**b.**

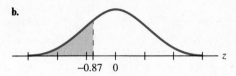

−0.87   0   $z$

**c.** The probability is approximately 19 percent that fewer than 75 percent of the trees will grow.

**15. a.** 0.2065

**b.**

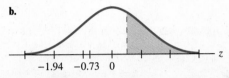

−1.94   −0.73   0   $z$

**c.** The probability is approximately 21 percent that between 30 and 35 percent of the juniors and seniors in the sample never read anything that is not required school work.

**17.** 0.714   **19.** 0.418 and 0.583   **21.** 93   **23. a.** 0.050   **b.** 0.049   **c.** 0.048   **d.** About the same   **25. a.** 65   **b.** 146   **c.** 583

**EXERCISES 5-4, SET B**   (page 373)

**1. a.** $\{0, 1, 2, 3, 4, 5, 6\}$
$P(0) = 0.0989$; $P(1) = 0.2792$; $P(2) = 0.3284$; $P(3) = 0.2164$;
$P(4) = 0.0727$; $P(5) = 0.0137$; $P(6) = 0.0011$

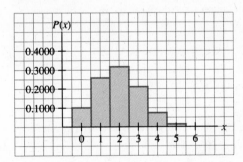

**b.** $\{0, \frac{1}{6}, \frac{1}{3}, \frac{1}{2}, \frac{2}{3}, \frac{5}{6}, 1\}$
$P(0) = 0.0989$; $P(\frac{1}{6}) = 0.2792$; $P(\frac{1}{3}) = 0.3284$; $P(\frac{1}{2}) = 0.2164$;
$P(\frac{2}{3}) = 0.0727$; $P(\frac{5}{6}) = 0.0137$; $P(1) = 0.0011$

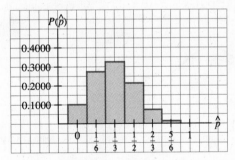

**c.** They are the same.

**5. a.**

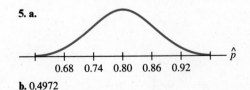

0.68   0.74   0.80   0.86   0.92   $\hat{p}$

**b.** 0.4972

## EXERCISES 5-5, SET A    (page 382)

**1. a.** $\frac{51}{75} = 0.68$  **b.** 68%  **c.** From 57% to 79%  **3. a.** $\frac{24}{80} = 0.30$  **b.** 30%  **c.** From 20% to 40%  **5.** From 70% to 76%  **7.** From 6.5% to 13.5%
**9.** From 25.1% to 35.0%  **11.** From 51.4% to 64.4%  **13. a.** From 0.33 to 0.57  **b.** From 0.41 to 0.49  **c.** 3 to 1 (= 0.24 to 0.08)
**15. a.** From 10.6% to 13.4%  **b.** From 14.1% to 17.9%  **c.** From 25.2% to 30.8%  **d.** From 6.1% to 9.9%  **17.** 189  **19.** 150  **21.** 136  **23.** 296
**25. a.** 0.011  **b.** 0.011  **c.** 0.010  **d.** 0.009  **e.** yes

## EXERCISES 5-5, SET B    (page 386)

**5.** From 71% to 79%

## CHAPTER 5 REVIEW EXERCISES    (page 392)

**1. a.** $n$ is 36 and 36 > 30  **b.** 454.0 grams  **c.** 0.5 grams

**d.**

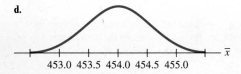

**3. a.**

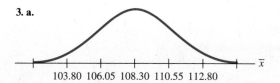

**b.** −1.47  **c.** $105.00 is 1.47 standard deviations below the mean $108.30.  **d.** 0.0708  **e.** The probability is approximately 7 percent that the mean of the sample of 40 weekly incomes is less than $105.00.

**5. a.** 67 seconds  **b.** 67 seconds  **c.** From 63 to 71 seconds
**d.** A 95 percent confidence interval for the mean waiting time of the first car to enter the left turn lane is from 63 to 71 seconds.
**7. a.** From 16.2 to 18.6  **b.** A 95 percent confidence interval estimate of the mean score of this part of the ACT for all students in the Tampa Bay area is from 16.2 to 18.6.
**9. a.** 1.895  **b.** −2.718  **11. a.** From 25.3 to 31.3 miles per gallon  **b.** A 95 percent confidence interval for the mean mileage of Sharon Hollobow's car during the period checked is from 25.3 to 31.3 mpg.
**13. a.** 0.54  **b.** 0.08

**c.**

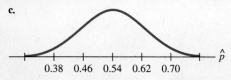

**d.** 0.3085  **e.** The probability is approximately 31 percent that fewer than 50 percent of the 39 psychiatrists surveyed would agree that companionship is a sounder reason for couples to get married than passionate love.
**15. a.** From 0.43 to 0.47
**b.** A 95 percent confidence interval for the proportion of all smokers who were non-smokers 4 years after completing the program is from 43 to 47 percent.

## CHAPTER 6

## EXERCISES 6-1, SET A    (page 417)

**1. a.** A test on $\mu$  **b.** $H_0: \mu = 41$ mpg  **c.**
         $H_a: \mu < 41$ mpg

**3. a.** A test on $P$    **b.** $H_0$: $P = 0.69$    **c.**
$H_a$: $P > 0.69$

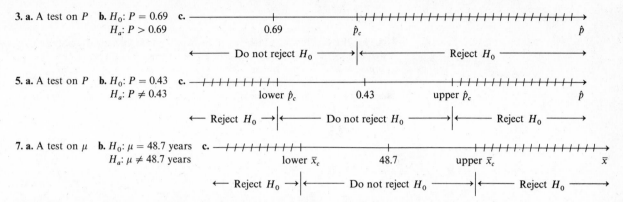

**5. a.** A test on $P$    **b.** $H_0$: $P = 0.43$    **c.**
$H_a$: $P \neq 0.43$

**7. a.** A test on $\mu$    **b.** $H_0$: $\mu = 48.7$ years    **c.**
$H_a$: $\mu \neq 48.7$ years

**9.** 0.03    **11.** 0.03    **13.** 0.04    **15.** 0.03    **17.** 0.74    **19.** 77.5 and 82.5 mg    **21.** 7.6 percent    **23.** 80 percent

## EXERCISES 6-1, SET B    (page 419)

**1. a.** $H_0$: $\mu = 56.1$ years    **b.** lower $\bar{x}_c = 49.8$ and upper $\bar{x}_c = 62.4$    **3. a.** $H_0$: $P = 0.458$    **b.** $\hat{p}_c = 0.654$
$H_a$: $\mu \neq 56.1$ years                                                                    $H_a$: $P > 0.458$
**5. a.** 0.55    **b.** 0.45    **c.** 0.41    **d.** The larger the value of $n$, the smaller the value of $\hat{p}_c$, and thus closer to the value specified by $H_0$.
**7. a.** $H_0$: $\mu = 27.0$    **9. a.** $H_0$: $P = 0.62$
$H_a$: $\mu \neq 27.0$        $H_a$: $P > 0.62$

## EXERCISES 6-2, SET A    (page 428)

**1. a.** If the average commute time is 25 minutes, and the figure is rejected.
**b.** If the average commute time is greater than 25 minutes, and the figure is not rejected.
**3. a.** If the mean age is 48.6 years, and the figure is rejected.
**b.** If the mean age is less than 48.6 years or greater than 48.6 years, and the figure is not rejected.
**5. a.** If the proportion is 0.54, and the figure is rejected.
**b.** If the proportion is greater than 0.54, and the figure is not rejected.
**7. a.** If the proportion is 0.67, and the figure is rejected.
**b.** If the proportion is either less than 0.67 or greater than 0.67, and the figure is not rejected.
**9. a.** 2%    **b.** 21.7    **11. a.** 5%    **b.** 18.6    **13. a.** 11%    **b.** 0.43    **15. a.** 8%    **b.** 0.18    **17. a.** 9%    **b.** 2%    **c.** 0.4%    **d.** Gets smaller

**19. a.** 0.7422                                    **b.** 0.3632                                    **c.** The $\mu$ in part (b) is farther away
from $\mu_0$ than the one in part (a).

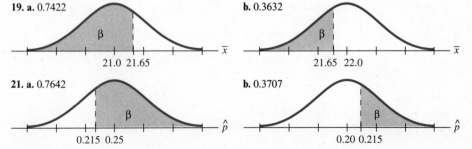

**21. a.** 0.7642                                    **b.** 0.3707                                    **c.** The $P$ in part (b) is farther away
from $P_0$ than the one in part (a).

**23. a.** 0.6255    **b.** 0.3632    **c.** The larger sample size

## EXERCISES 6-2, SET B (page 430)

**1. a.** 0.8749 **b.** 0.7422 **c.** 0.5596 **d.** 0.3632

**e.**

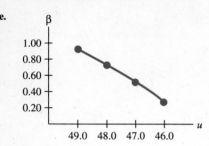

**7.** A type II error. It would be better if the team rejected an adequate braking design and reexamined it than to accept a faulty design that would cause a risk to someone who bought the car. **9.** A type I error. A minor difference in the butterfat content from the hypothesized value of 2 percent is not a major health hazard. The cost involved in repackaging the milk would be a considerable loss to the dairy.

## EXERCISES 6-3, SET A (page 440)

**1.** $-2.05$  **3.** 1.84  **5.** With $Z = 1.67$ and $\alpha = 0.05$, do not reject $H_0$.  **7.** With $Z = -2.00$ and $\alpha = 0.05$, reject $H_0$.
**9.** With $Z = 1.56$ and $\alpha = 0.01$, do not reject $H_0$.  **11.** With $Z = -3.69$ and $\alpha = 0.01$, reject $H_0$.
**13.** With $Z = -1.63$ and $\alpha = 0.05$, do not reject $H_0$.  **15.** With $Z = 1.73$ and $\alpha = 0.01$, do not reject $H_0$.
**17.** With $Z = 3.88$ and $\alpha = 0.01$, reject $H_0$.  **19.** With $T = 1.936$ and $\alpha = 0.05$, do not reject $H_0$.
**21.** With $T = -2.347$ and $\alpha = 0.01$, do not reject $H_0$.  **23. a.** With $Z = -2.16$ and $\alpha = 0.05$, reject $H_0$.
**b.** With $Z = -2.16$ and $\alpha = 0.01$, do not reject $H_0$.

## EXERCISES 6-3, SET B (page 442)

**1.** With $Z = -1.26$ and $\alpha = 0.05$, do not reject $H_0$.  **3.** With $Z = -1.44$ and $\alpha = 0.05$, do not reject $H_0$.

## EXERCISES 6-4, SET A (page 449)

**1.** 1.40  **3.** 1.55  **5.** With $Z = -1.31$ and $\alpha = 0.05$, do not reject $H_0$.  **7.** With $Z = -2.14$ and $\alpha = 0.05$, reject $H_0$.
**9.** With $Z = 2.66$ and $\alpha = 0.01$, reject $H_0$.  **11.** With $Z = -1.00$ and $\alpha = 0.05$, do not reject $H_0$.
**13.** With $Z = -1.38$ and $\alpha = 0.05$, do not reject $H_0$.  **15.** With $Z = 4.98$ and $\alpha = 0.01$, reject $H_0$.
**17.** With $Z = -1.58$ and $\alpha = 0.05$, do not reject $H_0$.  **19.** With $Z = -1.47$ and $\alpha = 0.05$, do not reject $H_0$.
**21. a.** $H_a$: Gas mileage obtained using our oil is the same as (equal to) gas mileage obtained using the competitive motor oil.
**b.** No. A low level of significance or small $n$ could require a large difference to reject $H_0$.
**23. a.** $H_0$: The proportion of people who select clothes washed in our detergent is the same as the proportion of people who select clothes washed in their detergent.
   $H_a$: The proportion of people who select clothes washed in our detergent is greater than the proportion of people who select clothes washed in their detergent.
**b.** Examples of additional facts that might cause one to change to their brand of detergent are:
1. How large was the difference in proportions?
2. How many people were tested?
3. Was the same amount of detergent used in each test?
4. Were the people sampled completely informed as to the criteria of the comparison test?
5. What happened to the results of a test in which the person perceived no difference in the two washings.

## EXERCISES 6-4, SET B (page 452)

**1.** With $Z = -2.06$ and $\alpha = 0.05$, reject $H_0$.  **5.** With $Z = -1.29$ and $\alpha = 0.05$, do not reject $H_0$.
**7.** With $Z = 2.21$ and $\alpha = 0.05$, reject $H_0$.

## EXERCISES 6-5, SET A   (page 461)

**1.** 0.025   **3.** 0.050   **5.** 0.010   **7.** 0.900   **9.** 0.950   **11.** 24.996   **13.** lower $\chi_c^2$ is 1.344; upper $\chi_c^2$ is 21.955   **15.** 16.473
**17.** lower $\chi_c^2$ is 5.226; upper $\chi_c^2$ is 21.026   **19.** 7.564   **21.** With $\chi^2 = 22.822$ and $\alpha = 0.05$, reject $H_0$.
**23.** With $\chi^2 = 1.872$ and $\alpha = 0.01$, do not reject $H_0$.   **25.** With $\chi^2 = 9.538$ and $\alpha = 0.10$, reject $H_0$.
**27.** lower limit = 4.5 and upper limit = 19.2   **29.** lower limit = 9.58 and upper limit = 67.50
**31.** lower limit = 1.11 and upper limit = 89.12   **33.** lower limit = 20.5 and upper limit = 91.5
**35. a.** With $\chi^2 = 18.200$ and $\alpha = 0.05$, do not reject $H_0$.   **b.** Probably not, because a small variance would indicate uniformity in products.
**37.** With $\chi^2 = 17.980$ and $\alpha = 0.10$, reject $H_0$.   **39.** With $\chi^2 = 31.500$ and $\alpha = 0.025$, do not reject $H_0$.

## EXERCISES 6-5, SET B   (page 464)

**1. a.** 130.89 . . .   **b.** With $x^2 = 24.870$ and $\alpha = 0.050$, do not reject $H_0$.
**c.** The distribution of weights from 0.1-acre plots is approximately bell shaped.

**d.** 141 | 5  7
141 | 5  7
142 | 0  9  8  6
143 | 8  2  6  2  5  9  8
144 | 7  2  9  2  4
145 | 2  8

Yes, the assumption seems reasonable.

## CHAPTER 6 REVIEW EXERCISES   (page 468)

**1.** $H_0: \mu = 110.3$ versus $H_a: \mu > 110.3$   **3.** With $Z_c = 1.90$, $\alpha = 0.0287$   **5.** $\bar{x}_c = 129$ miles
**7. a.** If the mean weight is not less than 200 pounds and the figure is rejected.
**b.** If the mean weight is less than 200 pounds and the figure is not rejected.
**9. a.** 143.9   **b.** With $Z_c = 1.89$, $\alpha = 0.0294$   **11.** 0.6591   **13.** With $Z = -2.83$ and $\alpha = 0.01$, reject $H_0$.
**15.** With $T = 3.63$ and $\alpha = 0.01$, reject $H_0$.   **17.** With $Z = -0.72$, do not reject $H_0$.   **19.** 0.025   **21.** 21.064
**23.** With $\chi^2 = 27.894$ and $\alpha = 0.10$, reject $H_0$.

## CHAPTER 7
### EXERCISES 7-1, SET A   (page 490)

**1. a.** 10.0   **b.** 1.4
**c.**

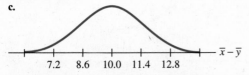

7.2   8.6   10.0   11.4   12.8   $\bar{x} - \bar{y}$

**d.** 0.0793   **e.** The probability is approximately 8 percent that the difference in the sample means is less than 8.0.

**5. a.** $-5.3$ inches   **b.** 0.7 inches
**c.**

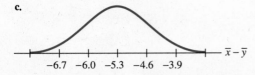

$-6.7$   $-6.0$   $-5.3$   $-4.6$   $-3.9$   $\bar{x} - \bar{y}$

**d.** 0.1271   **e.** 0.8472   **f.** The probability is approximately 85 percent that the difference in the mean heights is between $-6.3$ and $-4.3$ inches.

**3. a.** 6.5   **b.** 0.6
**c.**

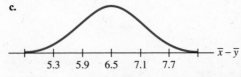

5.3   5.9   6.5   7.1   7.7   $\bar{x} - \bar{y}$

**d.** 0.5934   **e.** The probability is approximately 59 percent that the difference in the sample means is between 6.0 and 7.0.

**7. a.** 12.0 years   **b.** 2.0 years
**c.**

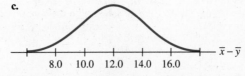

8.0   10.0   12.0   14.0   16.0   $\bar{x} - \bar{y}$

**d.** 0.3085   **e.** The probability is approximately 31 percent that the difference in the sample means is more than 13.0 years.

**9.** From 0.89 to 0.93 grams   **11.** From $3.58 to $12.32   **13. a.** From 2.4 to 12.0 years
**b.** A 90 percent confidence interval for the difference in the mean ages of women and men at the time of death for residents in this city
is between 2.4 and 12.0 years.   **15.** With $z = -1.60$ and $\alpha = 0.05$, do not reject $H_0$.
**17.** With $z = -1.57$ and $\alpha = 0.05$, do not reject $H_0$.   **19.** With $z = 1.33$ and $\alpha = 0.05$, do not reject $H_0$.

## EXERCISES 7-1, SET B   (page 494)

**1. a.** 2.500   **b.** 1.500   **c.** With $Z = -0.48$ and $\alpha = 0.05$, do not reject $H_0$.
**3. a.** 3.000   **b.** 1.125   **c.** With $Z = 0.89$ and $\alpha = 0.05$, do not reject $H_0$.
**5.** With $z = -2.54$ and $\alpha = 0.05$, reject $H_0$.
**7. a.** 1100 kWH   **b.** 110 kWH

**c.**

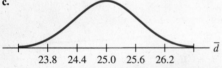

**d.** 0.7842   **e.** The probability is approximately 78 percent that the difference in
the means is between 1000 and 1300 kWH.

## EXERCISES 7-2, SET A   (page 508)

**1. a.** 25.0   **b.** 0.6

**c.**

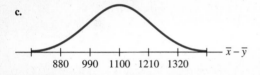

**d.** 0.0475   **e.** The probability is approximately 5 percent that
the mean of the differences is less than 24.0.

**3. a.** 138   **b.** 2.3

**c.**

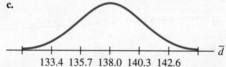

**d.** 0.6156   **e.** The probability is approximately 62 percent that
the mean of the differences is between 136 and 140.

**5. a.** $-0.09   **b.** $0.20   **7. a.** 40 wpm   **b.** 8.5 wpm   **9. a.** 15.5 kg   **b.** 3.8 kg

**11. a.** 13.8 cm   **b.** 0.2 cm

**c.**

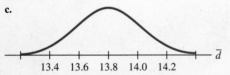

**d.** 0.0668

**13. a.** 9.0 minutes   **b.** 0.5 minutes

**c.**

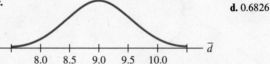

**d.** 0.6826

**15. a.** From $29.22 to $30.78   **b.** A 90 percent confidence interval for the mean amount saved per month by commuters who convert
to car pooling is from $29.22 to $30.78.
**17. a.** From 114 to 126 wpm   **b.** A 95 percent confidence interval for the mean increase in reading speeds for individuals that complete
the program is from 114 to 126 wpm.
**19.** From 2.2 to 3.7 mpg   **21.** From 3.6 to 6.9 bricks   **23.** With $z = -1.55$ and $\alpha = 0.05$, do not reject $H_0$.
**25.** With $z = 2.88$ and $\alpha = 0.01$, reject $H_0$.   **27.** With $z = 3.01$ and $\alpha = 0.01$, reject $H_0$.
**29.** With $t = -1.486$ and $\alpha = 0.05$, do not reject $H_0$.

**EXERCISES 7-2, SET B** (page 513)

**1.** From 9.6 to 14.0 pounds   **3.** With $z = 4.43$ and $\alpha = 0.01$, reject $H_0$.

**EXERCISES 7-3, SET A** (page 522)

**1. a.** 0.08   **b.** 0.06

**c.**

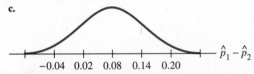

$$-0.04 \quad 0.02 \quad 0.08 \quad 0.14 \quad 0.20$$

**d.** 0.0918   **e.** The probability is approximately 9 percent that the difference in sample proportions is less than 0.00.

**5. a.** 0.20   **b.** 0.05

**c.**

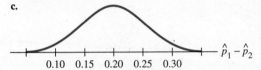

$$0.10 \quad 0.15 \quad 0.20 \quad 0.25 \quad 0.30$$

**d.** 0.5762   **e.** The probability is approximately 58 percent that the difference in sample proportions is between 0.16 and 0.24.

**3. a.** 0.08   **b.** 0.09

**c.**

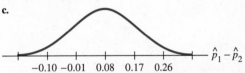

$$-0.10 \quad -0.01 \quad 0.08 \quad 0.17 \quad 0.26$$

**d.** 0.7215   **e.** The probability is approximately 72 percent that the difference in sample proportions is between 0.00 and 0.20.

**7. a.** 0.15   **b.** 0.07

**c.**

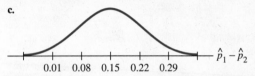

$$0.01 \quad 0.08 \quad 0.15 \quad 0.22 \quad 0.29$$

**d.** 0.7611   **e.** The probability is approximately 76 percent that the difference in sample proportions is greater than 0.10.

**9.** From 5% to 18%   **11. a.** From 1% to 9%   **b.** A 95 percent confidence interval for the difference in proportions of Democratic and Republican voters who favor a single national primary is between 1% and 9%.   **13. a.** From 12% to 38%
**b.** A 98 percent confidence interval for the difference in proportions of male and female employees at this plant who prefer the proposed work schedule is between 12 and 38 percent.   **15.** With $z = 2.79$ and $\alpha = 0.05$, reject $H_0$.
**17.** With $z = 1.51$ and $\alpha = 0.05$, do not reject $H_0$.   **19.** With $z = -1.86$ and $\alpha = 0.01$, do not reject $H_0$.
**21.** With $z = 1.92$ and $\alpha = 0.05$, do not reject $H_0$.

**EXERCISES 7-3, SET B** (page 526)

**1. a.** 0.32   **b.** 0.22   **c.** From 5% to 15%   **d.** With $z = 3.27$ and $\alpha = 0.01$, reject $H_0$.
**3. a.** 0.40   **b.** 0.35   **c.** From $-0.04$ to 0.14   **d.** With $z = 1.25$ and $\alpha = 0.05$, do not reject $H_0$.   **7.** From 18% to 34%

**CHAPTER 7 REVIEW EXERCISES** (page 532)

**1. a.** 3.80   **b.** 5.94

**c.**

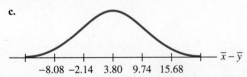

$$-8.08 \quad -2.14 \quad 3.80 \quad 9.74 \quad 15.68$$

**d.** 0.1492   **e.** The probability is approximately 15 percent that the difference in the means of the sampled test scores is greater than 10.00.

**3.** From $44.03 to $71.97
**5.** With $z = 3.78$ and $\alpha = 0.01$, reject $H_0$.
**7. a.** 30 wpm   **b.** 2 wpm

**c.**

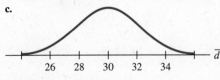

$$26 \quad 28 \quad 30 \quad 32 \quad 34$$

**d.** 0.6826

**9. a.** 12.3 seconds   **b.** 4.2 seconds   **c.** From 11.1 to 13.5 seconds   **d.** A 95 percent confidence interval for the mean difference in times for all 12 year olds to run the specified distance is from 11.1 to 13.5 seconds.   **11.** With $z = 6.34$ and $\alpha = 0.01$, reject $H_0$.
**13. a.** 0.078   **b.** 0.060

**c.**

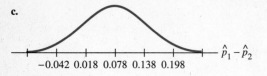

$-0.042 \quad 0.018 \quad 0.078 \quad 0.138 \quad 0.198$

**d.** 0.3345   **15.** From $-0.04$ to 0.17   **17.** With $z = 2.74$ and $\alpha = 0.01$, reject $H_0$.

## CHAPTER 8

### EXERCISES 8-1, SET A   (page 560)

**1. a.** 260   **b.** 330   **c.** 110   **d.** 150   **e.** 1360   **3. a.** 160   **b.** 80   **c.** 40   **d.** 14   **e.** 520

**5.**

|        | Category | | |
|--------|------|------|--------|
|        | X | Y | Totals |
| A      | 4.2  | 5.8  | 10  |
| Sample B | 8.3 | 11.7 | 20 |
| C      | 12.5 | 17.5 | 30  |
| Totals | 25   | 35   | 60  |

**7.**

|        | Category | | | |
|--------|------|-------|------|--------|
|        | X | Y | Z | Totals |
| A      | 60.0 | 100.0 | 40.0 | 200 |
| Sample B | 90.0 | 150.0 | 60.0 | 300 |
| Totals | 150  | 250   | 100  | 500 |

**9.** With $x^2 = 4.167$ and $\alpha = 0.05$, reject $H_0$.
**11.** With $x^2 = 7.566$ and $\alpha = 0.01$, do not reject $H_0$.
**13. a.** With $x^2 = 13.435$ and $\alpha = 0.01$, reject $H_0$.
**b.** From 0.16 to 0.40
**15. a.** With $x^2 = 3.217$, and $\alpha = 0.05$, do not reject $H_0$.   **b.** From $-0.18$ to 0.10
**17.** With $x^2 = 21.777$ and $\alpha = 0.01$, reject $H_0$.

### EXERCISES 8-1, SET B   (page 566)

**1. a.** With $x^2 = 21.185$, $df = 2$, and $\alpha = 0.01$, reject $H_0$.   **b.** From 0.09 to 0.21
**3. a.** With $x^2 = 88.643$, $df = 6$, and $\alpha = 0.01$, reject $H_0$.   **b.** From 0.10 to 0.30

### EXERCISES 8-2, SET A   (page 581)

**1. a.** 3.24   **b.** 5.29   **3. a.** 2.76   **b.** 4.18   **5. a.** 4.0; 6.0; 5.0   **b.** 1.6; 5.2; 8.0   **c.** 4.9   **d.** 5.0   **e.** 6.0   **f.** 1.2
**7. a.** 12.0; 15.0; 11.0; 17.0   **b.** 19.2; 24.8; 6.8; 5.6   **c.** 14.1   **d.** 13.8   **e.** 45.5   **f.** 3.23
**9. a.** With $F = 1.34$ and $\alpha = 0.05$, do not reject $H_0$.   **b.** From $-6.5$ to 1.5 pounds   **c.** From $-6.0$ to 3.8 pounds
**11. a.** With $F = 3.88$, and $\alpha = 0.01$, do not reject $H_0$.   **b.** From $-4.0$ to $-0.6$ minutes   **c.** From $-0.3$ to 4.6 minutes
**13. a.** With $F = 21.91$ and $\alpha = 0.01$, reject $H_0$.   **b.** From 307 to 598 contacts   **c.** From 43 to 399 contacts

**EXERCISES 8-2, SET B**   (page 584)

**3.**  Four assumptions are necessary:
**a.** The observed values are all independent random variables.   **b.** Each observed value has a normal distribution.
**c.** The variances of all the observations are the same.   **d.** The mean of any observation in the $i$th group is $\mu_i$.
**5.**  The two parameters are:
**a.** The degrees of freedom for numerator   **b.** The degrees of freedom for denominator

## CHAPTER 8 REVIEW EXERCISES   (page 588)

**1. a.**

|  |  | Category | | |
|---|---|---|---|---|
|  |  | Yes | No | Totals |
| **Sample** | A | 22 | 78 | 100 |
|  | B | 22 | 78 | 100 |
|  | Totals | 44 | 156 | 200 |

**b.** $x^2 = 5.711$   **c.** With $x^2 = 5.711$ and $\alpha = 0.01$, do not reject $H_0$.
**d.** From 0.045 to 0.235

**3. a.** $\bar{x}_1 = 81.2$; $\bar{x}_2 = 77.2$; $\bar{x}_3 = 72.4$; $\bar{x}_4 = 71.4$   **b.** $s_1^2 = 103.2$; $s_2^2 = 148.7$; $s_3^2 = 103.3$; $s_4^2 = 88.3$   **c.** 110.9   **d.** 75.6   **e.** 103.0   **f.** 0.93
**g.** With $F = 0.93$ and $\alpha = 0.05$, do not reject $H_0$.   **h.** From $-7.0$ to 15.0   **i.** From $-5.8$ to 17.4

## CHAPTER 9
**EXERCISES 9-1, SET A**   (page 609)

**1.**

|  |  | 2nd Variable | | | |
|---|---|---|---|---|---|
|  |  | Cat. 1 | Cat. 2 | Cat. 3 | Totals |
| **1st Variable** | Cat. 1 | 8.0 | 12.0 | 20.0 | 40 |
|  | Cat. 2 | 5.0 | 7.5 | 12.5 | 25 |
|  | Cat. 3 | 7.0 | 10.5 | 17.5 | 35 |
|  | Totals | 20 | 30 | 50 | 100 |

**3.**

|  |  | 2nd Variable | | | | |
|---|---|---|---|---|---|---|
|  |  | Cat. 1 | Cat. 2 | Cat. 3 | Cat. 4 | Totals |
| **1st Variable** | Cat. 1 | 28.8 | 16.0 | 22.4 | 12.8 | 80 |
|  | Cat. 2 | 36.0 | 20.0 | 28.0 | 16.0 | 100 |
|  | Cat. 3 | 25.2 | 14.0 | 19.6 | 11.2 | 70 |
|  | Totals | 90 | 50 | 70 | 40 | 250 |

**5.** With $x^2 = 7.566$ and $\alpha = 0.05$, do not reject $H_0$.   **7.** With $x^2 = 28.518$ and $\alpha = 0.01$, reject $H_0$.
**9.** With $x^2 = 15.903$ and $\alpha = 0.01$, reject $H_0$.   **11.** With $x^2 = 26.196$ and $\alpha = 0.01$, reject $H_0$.
**13.** With $x^2 = 21.282$ and $\alpha = 0.05$, reject $H_0$.   **15.** With $x^2 = 23.313$ and $\alpha = 0.05$, reject $H_0$.

### EXERCISES 9-1, SET B   (page 615)

**3.** 0.053, to three places   **5.** 0.060, to three places

### EXERCISES 9-2, SET A   (page 631)

**1. a.** 15   **b.** 85   **c.** 25   **d.** 271   **e.** 151   **f.** 0.99   **3. a.** 32   **b.** 246   **c.** 1   **d.** 33   **e.** 56   **f.** 0.90

**5.**

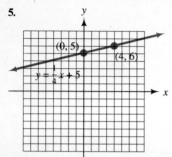

**7.**

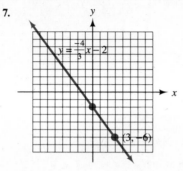

**9.**

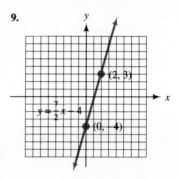

**11. a.** spring constant

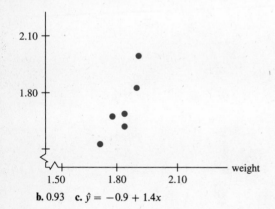

**b.** 0.93   **c.** $\hat{y} = -0.9 + 1.4x$

**13. a.** weld breaking strength

weld
breaking
strength

106 ┼
102 ┼
98 ┼
94 ┼
90 ┼
86 ┼

3.0      4.0      5.0   velocity

**b.** 0.85   **c.** $\hat{y} = 66 + 7x$

**15. a.** $y$ – meters

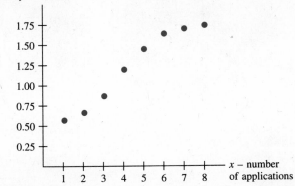

**b.** 0.98  **c.** $\hat{y} = 0.39 + 0.19x$

**17. a.** $y$ – score

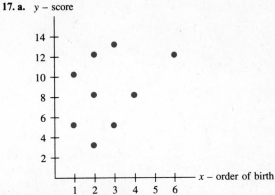

**b.** 0.36  **c.** $\hat{y} = 6.3 + 0.8x$

**19. a.** $y$ – grams

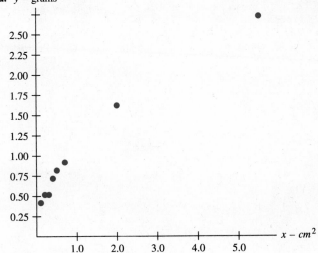

**b.** 0.98  **c.** $\hat{y} = 0.5 + 0.4x$

**21. a.** $y$ – pounds

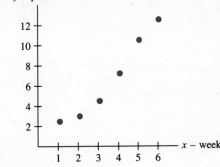

**b.** 0.98  **c.** $\hat{y} = -0.8 + 2.1x$

**EXERCISES 9-2, SET B**  (page 633)

**1. a.**

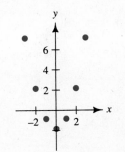

**b.** 0.00  **c.** $y = 9 - 2 = 7$
$y = 4 - 2 = 2$
$y = 1 - 2 = -1$
$y = 0 - 2 = -2$
$y = 1 - 2 = -1$
$y = 4 - 2 = 2$
$y = 9 - 2 = 7$

**3. a.**

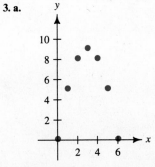

**b.** 0.00  **c.** $y = 6(0) - 0 = 0$
$y = 6(1) - 1 = 5$
$y = 6(2) - 4 = 8$
$y = 6(3) - 9 = 9$
$y = 6(4) - 16 = 8$
$y = 6(5) - 25 = 5$
$y = 6(6) - 36 = 0$

**5. a.**

| x | (A) | (B) |
|---|-----|-----|
| 1 | 1.2 | 1.8 |
| 2 | 2.4 | 3.1 |
| 3 | 3.6 | 4.4 |
| 4 | 4.8 | 5.7 |

**b.**

| (A) | (B) |
|-----|-----|
| 0.2 | 0.8 |
| −0.6 | 0.1 |
| 0.6 | 1.4 |
| −0.2 | 0.7 |

**c.** (A) 0.8; (B) 3.1    **d.** Equation (A)

**7. a.**

| x | (A) | (B) |
|---|-----|-----|
| 1 | 7.0 | 7.4 |
| 2 | 4.5 | 4.8 |
| 3 | 2.0 | 2.2 |
| 4 | −0.5 | −0.4 |

**b.**

| (A) | (B) |
|-----|-----|
| 1.0 | 0.6 |
| 0.5 | 0.2 |
| −2.0 | −2.2 |
| 1.5 | 1.4 |

**c.** (A) 7.5; (B) 7.2    **d.** Equation (B)

## EXERCISES 9-3, SET A    (page 649)

**1. a.** −2.6   **b.** 4.4   **c.** 1.31   **d.** From −2.9 to −2.3   **e.** $\hat{y} = 4.4 - 2.6x$   **f.** −6.0   **g.** From −7.4 to −4.6

**3. a.** 0.5   **b.** −2.3   **c.** 1.18   **d.** From 0.3 to 0.7   **e.** $\hat{y} = -2.3 + 0.5x$   **f.** 0.2   **g.** From −0.9 to 1.3

**5. a.**

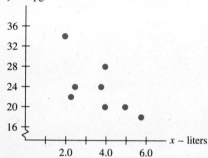

**b.** −0.66   **c.** $\hat{y} = 33.1 - 2.5x$   **d.** −2.5 mpg   **e.** 23.5 mpg
**f.** From 18.7 to 28.5 mpg

**7. a.** 0.0045   **b.** From 1.6 to 2.6   **c.** 2.1   **d.** With $t = 8.468$ and $\alpha = 0.05$, reject $H_0$.
**e.** From 1.9 to 2.3

**9. a.** 8.8   **b.** From 11.1 to 21.1   **c.** 98.8   **d.** With $t = -2.448$, do not reject $H_0$
**e.** From 87.9 to 109.8

**11. a.** With $t = 11.323$, reject $H_0$   **b.** 2.08   **c.** With $t = -8.510$, reject $H_0$

**13. a.** With $t = 1.004$, do not reject $H_0$   **b.** 9   **c.** With $t = -0.222$, do not reject $H_0$

## EXERCISES 9-3, SET B    (page 652)

**1. a.**

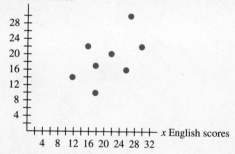

**b.** $r = 0.54$
**c.** A unit increase in an English score $\approx$ a 0.5 increase in mean mathematics score.
**d.** $\hat{y} = 7.6 + 0.5x$
**e.** 19.3
**f.** From 15.5 to 23.2

**3. a.**

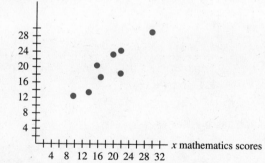

**b.** $r = 0.90$
**c.** A unit increase in a mathematics score $\approx$ a 0.9 increase in mean composite score.
**d.** $\hat{y} = 3.4 + 0.9x$
**e.** 18.8
**f.** From 12.5 to 25.0

## CHAPTER 9 REVIEW EXERCISES (page 658)

**1.** With $x^2 = 13.315$ and $\alpha = 0.05$, reject $H_0$.

**3. a.**    $y$ centimeters            **b.** $\hat{y} = 4.6 + 10.2x$   **c.** 1.00

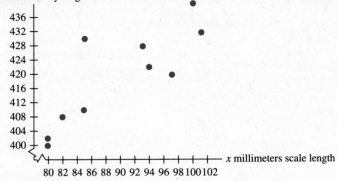

**5. a.** $-1.6$   **b.** 5.0   **c.** 1.9   **d.** From $-2.5$ to $-0.7$   **7. a.** 0.002   **b.** From 0.31 to 0.34   **c.** 0.56   **d.** From 0.52 to 0.61   **e.** 0.95
**f.** from 0.91 to 0.99   **g.** $x^* = 3.0$ is closer to $\bar{x} = 3.2$ than is $x^* = 1.8$.   **h.** From 0.83 to 1.07
**i.** Yes, the estimations would not involve an extrapolation of the observed paired data.

**9. a.**    $y$ millimeters body length

**b.** 897   **c.** 4,190   **d.** 376,660   **e.** 81,069
**f.** 1,757,204   **d.** $\hat{y} = 298 + 1.3x$   **h.** 0.83   **i.** 62.0
**j.** From 0.4 to 2.3   **k.** 423 mm
**l.** From 416 to 431 mm   **m.** 1.3 mm
**n.** From 404 to 443 mm

## CHAPTER 10

### EXERCISES 10-1, SET A (page 686)

**1.** 0.172   **3.** 0.073   **5.** 0.055   **7.** Do not reject $H_0$.   **9.** Reject $H_0$.   **11. a.** 19   **b.** 18 and 30   **13. a.** 33 and 55   **b.** 34
**15.** 96%   **17.** 88.2%   **19.** 3   **21.** 4   **23.** 11   **25.** 23   **27. a.** 3.3 ounces   **b.** 3.0 to 3.6 ounces   **29. a.** $25.75   **b.** $21.95 to $28.50

### EXERCISES 10-1, SET B (page 689)

**3. a.** 6, 6, 5, 2, $-2$, 3, 7, $-1$, 3, 10, 4, 4, 7, 3, 4, 5   **b.** Reject $H_0$.   **5. a.** $-15$, 5, $-4$, $-3$, 2, $-5$, $-18$, $-3$, 3, 1, $-5$, $-19$
**b.** From $-15$ to $+2$ seconds

### EXERCISES 10-2, SET A (page 704)

**1.**

| $W$ | 3 | 4 | 5 | 6 | 7 | 8 | 9 | 10 | 11 |
|-----|---|---|---|---|---|---|---|----|----|
| $P(W)$ | $\frac{1}{15}$ | $\frac{1}{15}$ | $\frac{2}{15}$ | $\frac{2}{15}$ | $\frac{3}{15}$ | $\frac{2}{15}$ | $\frac{2}{15}$ | $\frac{1}{15}$ | $\frac{1}{15}$ |

**3. a.** 0.057  **b.** 0.029  **5. a.** 0.019  **b.** 0.019
**7. a.** 98, ⑩③, ⑩⑤, 108, ⑪⓪, 112, ⑪⑤, 118, 120, 123, 125, ⑫⑧, ⑬③, 135, ⑭②, ⑭⑤
**b.** 73  **c.** Do not reject $H_0$.
**9. a.** ⑨⑧, ⑩②, 105, ⑩⑤, ⑩⑨, 112, ⑪⑧, 121, ⑫①, ⑫⑥, ⑫⑦, ⑬⓪, 131, 138, 141, 142, 148, 150
**b.** 60  **c.** Reject $H_0$.  **11.** 230  **13.** lower $w_c = 536$; upper $w_c = 739$  **15.** 95.2%  **17.** 90.8%  **19. a.** 43 feet  **b.** 94.2%

## EXERCISES 10-3, SET A  (page 716)

**1. a.** 0.375  **b.** 0.391  **c.** 0.392  **d.** 0.034  **3. a.** 0.098  **b.** 0.048  **c.** 0.050  **5.** 0.330  **7.** 0.280

**9.**

| Student | A | B | C | D | E | F | G | H | I | J | K | L |
|---|---|---|---|---|---|---|---|---|---|---|---|---|
| **a.** X | 6 | 11 | 5 | 12 | 2 | 9 | 7 | 10 | 1 | 3 | 4 | 8 |
| **b.** Y | 5 | 10 | 7 | 12 | 3 | 9 | 6 | 8 | 1 | 11 | 4 | 2 |

**c.** 0.608  **d.** Reject $H_0$.

**11. a.**

| Knife | A | B | C | D | E | F | G | H | I | J |
|---|---|---|---|---|---|---|---|---|---|---|
| Quality | 8 | 7 | 10 | 1 | 3 | 5 | 9 | 6 | 4 | 2 |
| Price | 10 | 6 | 8 | 3 | 4 | 1 | 9 | 5 | 7 | 2 |

**b.** 0.758  **c.** Reject $H_0$.

**13.**

| Car | A | B | C | D | E | F | G | H |
|---|---|---|---|---|---|---|---|---|
| **a.** X | 3 | 6 | 8 | 4 | 2 | 1 | 7 | 5 |
| **b.** Y | 4 | 3 | 2 | 5 | 8 | 7 | 1 | 6 |

**c.** −0.857  **d.** Reject $H_0$.
**15.** Do not reject $H_0$.  **17.** Reject $H_0$.

## CHAPTER 10 REVIEW EXERCISES  (page 722)

**1.** 0.090  **3.** Do not reject $H_0$.  **5.** Do not reject $H_0$.  **7.** $j = 5$  **9.** Between 2157 and 2324 seconds  **11.** Reject $H_0$.  **13. a.** 92.6%  **b.** $j = 9$
**15. a.** 0.329  **b.** 0.214  **c.** 0.083  **d.** 0.048

**17. a. and b.**

| Student | A | B | C | D | E | F | G |
|---|---|---|---|---|---|---|---|
| X | 7 | 5 | 1 | 3 | 6 | 2 | 4 |
| Y | 3 | 6 | 5 | 4 | 2 | 1 | 7 |

**c.** −0.071  **d.** Do not reject $H_0$.

# Index

**We would appreciate it if you would take a few minutes to answer these questions. Then cut the page out, fold it, seal it, and mail it. No postage is required. Thank you!**

Which chapters were taught?

How much algebra did you have before this course?        How long ago?

Years in high school (circle)    0    1/2    1    1-1/2    2 or more        _____ last 2 years

Courses in college        0    1    2    3        _____ last 3-5 years

_____ 5 years or more

How would you characterize the quality of the explanations? If there were any that were unclear, please give us the page numbers.

What did you think of the examples (were there enough, right amount of detail, etc.) and exercises? If there is room for improvement, please let us know where.

How helpful were the chapter openers, Statistics in Action, newspaper clippings, chapter summaries, and Career Profiles in the text? Do you have any creative ideas on what else might help you study still more effectively?

What is your major or career goal? _____

What additional math courses, if any, do you plan on taking?_____

What kind of calculator do you own (model and type): _____ Did you use it in class?_____

What did you like most about the book?

FOLD HERE

What did you like the least about the book?

Name_____

College _____ State _____

FOLD HERE

# BUSINESS REPLY MAIL

FIRST CLASS        PERMIT NO. 1537        NEW YORK, NY

Postage will be paid by

HarperCollins Publishers
College Division       Attn: MATH GROUP
1900 East Lake Avenue
Glenview, Illinois   60025

NO POSTAGE
NECESSARY
IF MAILED
IN THE
UNITED STATES

**We would appreciate it if you would take a few minutes to answer these questions. Then cut the page out, fold it, seal it, and mail it. No postage required. Thank you!**

How would you assess the organization and coverage of this book?

How would you characterize the quality of the exposition? Where, if at all, can improvements be made?

How would you assess the quality and quantity of the examples and exercises? Where, if at all, can improvements be made?

How would you assess the pedagogical quality of the chapters?
Do you have any suggestions for improvement?

Your current book (author/title) is _____.

How does this book compare?

Do you have any comments on the supplements? Do you have any ideas for additions?

If you were to adopt this book, what chapters would you cover? Are you willing to be a possible reviewer?
Do you have (or are you aware of someone who has) any plans to write a book?

What did you like most about the book?

What did you like the least about the book?

Name _____

College _____   State _____

## BUSINESS REPLY MAIL

FIRST CLASS          PERMIT NO. 1537          NEW YORK, NY

Postage will be paid by

HarperCollins Publishers
College Division          Attn: MATH GROUP
1900 East Lake Avenue
Glenview, Illinois   60025

**1.** For a population of quantitative data:

$$z = \frac{x - \mu}{\sigma}$$

x is the value to be converted to a z score
$\mu$ is the population mean
$\sigma$ is the population standard deviation

**2.** For a normal random variable $X$:

$$z = \frac{x - \mu_X}{\sigma_X}$$

x is the value of the random variable
$\mu_X$ is the mean of the probability distribution
$\sigma_X$ is the standard deviation of the probability distribution

**3.** For a binomial random variable $X$:

$$z = \frac{by - nP}{\sqrt{nP(1 - P)}}$$

bv is a boundary value
$nP$ is the mean of the probability distribution
$\sqrt{nP(1 - P)}$ is the standard deviation of $X$

**4.** For the random variable $\bar{X}$:

$$z = \frac{\bar{x} - \mu_{\bar{x}}}{\sigma_{\bar{x}}}$$

$\bar{x}$ is a value of the random variable
$\mu_{\bar{x}} = \mu$ (the mean of the sampled population)

$$\sigma_{\bar{x}} = \frac{\sigma}{\sqrt{n}}$$

(the population standard deviation)
(the square root of the sample size)

**5.** For the random variable $\hat{P}$:

$$z = \frac{\hat{p} - \mu_{\hat{p}}}{\sigma_{\hat{p}}}$$

$\hat{p}$ is a value of the random variable
$\mu_{\hat{p}} = P$ (the proportion of the sampled population with the characteristic of interest)

$$\sigma_{\hat{p}} = \sqrt{\frac{P(1 - P)}{n}}$$

($n$ is the sample size)

**6.** For the random variable $\bar{X} - \hat{Y}$:

$$z = \frac{(\bar{x} - \bar{y}) - (\mu_{\bar{x}} - \mu_{\bar{y}})}{\sigma_{\bar{x} - \bar{y}}}$$

$\bar{x} - \bar{y}$ is a value of the random variable
$\mu_{\bar{x}} - \mu_{\bar{y}} = \mu_1 - \mu_2$ (the difference in the means of the sampled populations)

$$\sigma_{\bar{x} - \bar{y}} = \sqrt{\frac{\sigma_1^2}{n_1} + \frac{\sigma_2^2}{n_2}}$$

($\sigma_1^2$ and $\sigma_2^2$ are the variances of the sampled populations)

**7.** For the random variable $\bar{D}$:

$$z = \frac{\bar{d} - \mu_{\bar{D}}}{\sigma_{\bar{D}}}$$

$\bar{d}$ is a value of the random variable
$\mu_{\bar{D}} = \mu_d$ (the mean of the distribution of differences)

$$\sigma_{\bar{D}} = \frac{\sigma_d}{\sqrt{n}}$$

(the standard deviation of the population of differences)
(the square root of the sample size)

**8.** For the random variable $\hat{P}_1 - \hat{P}_2$:

$$z = \frac{(\hat{p}_1 - \hat{p}_2) - \mu_{\hat{p}_1 - \bar{p}_2}}{\sigma_{\bar{p}_1 - \bar{p}_2}}$$

$\hat{p}_1 - \hat{p}_2$ is a value of the random variable
$\mu_{\hat{p}_1 - \hat{p}_2} = P_1 - P_2$ (the difference in the proportions of the sampled populations)

$$\sigma_{\hat{p}_1 - \hat{p}_2} = \sqrt{\frac{P_1(1 - P_1)}{n_1} + \frac{P_2(1 - P_2)}{n_2}}$$

◀ *Equations for z scores are given on the previous page.*

**TABLE 4. The Standard Normal Distribution**

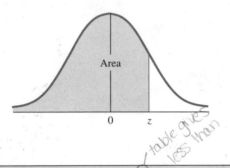

Area

0    z

*table gives less than*

| z | 0.00 | 0.01 | 0.02 | 0.03 | 0.04 | 0.05 | 0.06 | 0.07 | 0.08 | 0.09 |
|---|------|------|------|------|------|------|------|------|------|------|
| | | | | | $P(Z < z)$ | | | | | |
| −3.4 | 0.0003 | 0.0003 | 0.0003 | 0.0003 | 0.0003 | 0.0003 | 0.0003 | 0.0003 | 0.0003 | 0.0002 |
| −3.3 | 0.0005 | 0.0005 | 0.0005 | 0.0004 | 0.0004 | 0.0004 | 0.0004 | 0.0004 | 0.0004 | 0.0003 |
| −3.2 | 0.0007 | 0.0007 | 0.0006 | 0.0006 | 0.0006 | 0.0006 | 0.0006 | 0.0005 | 0.0005 | 0.0005 |
| −3.1 | 0.0010 | 0.0009 | 0.0009 | 0.0009 | 0.0008 | 0.0008 | 0.0008 | 0.0008 | 0.0007 | 0.0007 |
| −3.0 | 0.0013 | 0.0013 | 0.0013 | 0.0012 | 0.0012 | 0.0011 | 0.0011 | 0.0011 | 0.0010 | 0.0010 |
| −2.9 | 0.0019 | 0.0018 | 0.0017 | 0.0017 | 0.0016 | 0.0016 | 0.0015 | 0.0015 | 0.0014 | 0.0014 |
| −2.8 | 0.0026 | 0.0025 | 0.0024 | 0.0023 | 0.0023 | 0.0022 | 0.0021 | 0.0021 | 0.0020 | 0.0019 |
| −2.7 | 0.0035 | 0.0034 | 0.0033 | 0.0032 | 0.0031 | 0.0030 | 0.0029 | 0.0028 | 0.0027 | 0.0026 |
| −2.6 | 0.0047 | 0.0045 | 0.0044 | 0.0043 | 0.0041 | 0.0040 | 0.0039 | 0.0038 | 0.0037 | 0.0036 |
| −2.5 | 0.0062 | 0.0060 | 0.0059 | 0.0057 | 0.0055 | 0.0054 | 0.0052 | 0.0051 | 0.0049 | 0.0048 |
| −2.4 | 0.0082 | 0.0080 | 0.0078 | 0.0075 | 0.0073 | 0.0071 | 0.0069 | 0.0068 | 0.0066 | 0.0064 |
| −2.3 | 0.0107 | 0.0104 | 0.0102 | 0.0099 | 0.0096 | 0.0094 | 0.0091 | 0.0089 | 0.0087 | 0.0084 |
| −2.2 | 0.0139 | 0.0136 | 0.0132 | 0.0129 | 0.0125 | 0.0122 | 0.0119 | 0.0116 | 0.0113 | 0.0110 |
| −2.1 | 0.0179 | 0.0174 | 0.0170 | 0.0166 | 0.0162 | 0.0158 | 0.0154 | 0.0150 | 0.0146 | 0.0143 |
| −2.0 | 0.0228 | 0.0222 | 0.0217 | 0.0212 | 0.0207 | 0.0202 | 0.0197 | 0.0192 | 0.0188 | 0.0183 |
| −1.9 | 0.0287 | 0.0281 | 0.0274 | 0.0268 | 0.0262 | 0.0256 | 0.0250 | 0.0244 | 0.0239 | 0.0233 |
| −1.8 | 0.0359 | 0.0352 | 0.0344 | 0.0336 | 0.0329 | 0.0322 | 0.0314 | 0.0307 | 0.0301 | 0.0294 |
| −1.7 | 0.0446 | 0.0436 | 0.0427 | 0.0418 | 0.0409 | 0.0401 | 0.0392 | 0.0384 | 0.0375 | 0.0367 |
| −1.6 | 0.0548 | 0.0537 | 0.0526 | 0.0516 | 0.0505 | 0.0495 | 0.0485 | 0.0475 | 0.0465 | 0.0455 |
| −1.5 | 0.0668 | 0.0655 | 0.0643 | 0.0630 | 0.0618 | 0.0606 | 0.0594 | 0.0582 | 0.0571 | 0.0559 |
| −1.4 | 0.0808 | 0.0793 | 0.0778 | 0.0764 | 0.0749 | 0.0735 | 0.0722 | 0.0708 | 0.0694 | 0.0681 |
| −1.3 | 0.0968 | 0.0951 | 0.0934 | 0.0918 | 0.0901 | 0.0885 | 0.0869 | 0.0853 | 0.0838 | 0.0823 |
| −1.2 | 0.1151 | 0.1131 | 0.1112 | 0.1093 | 0.1075 | 0.1056 | 0.1038 | 0.1020 | 0.1003 | 0.0985 |
| −1.1 | 0.1357 | 0.1335 | 0.1314 | 0.1292 | 0.1271 | 0.1251 | 0.1230 | 0.1210 | 0.1190 | 0.1170 |
| −1.0 | 0.1587 | 0.1562 | 0.1539 | 0.1515 | 0.1492 | 0.1469 | 0.1446 | 0.1423 | 0.1401 | 0.1379 |
| −0.9 | 0.1841 | 0.1814 | 0.1788 | 0.1762 | 0.1736 | 0.1711 | 0.1685 | 0.1660 | 0.1635 | 0.1611 |
| −0.8 | 0.2119 | 0.2090 | 0.2061 | 0.2033 | 0.2005 | 0.1977 | 0.1949 | 0.1922 | 0.1894 | 0.1867 |
| −0.7 | 0.2420 | 0.2389 | 0.2358 | 0.2327 | 0.2296 | 0.2266 | 0.2236 | 0.2206 | 0.2177 | 0.2148 |
| −0.6 | 0.2743 | 0.2709 | 0.2676 | 0.2643 | 0.2611 | 0.2578 | 0.2546 | 0.2514 | 0.2483 | 0.2451 |
| −0.5 | 0.3085 | 0.3050 | 0.3015 | 0.2981 | 0.2946 | 0.2912 | 0.2877 | 0.2843 | 0.2810 | 0.2776 |